AF517868

ENZYME ASSAYS

Abhilasha Singh

2016
Regency Publications
A Division of
Astral International Pvt. Ltd.
New Delhi - 110 002

First Published, 2003
Reprinted, 2016

ISBN: 978-81-89233-49-5 (Hardbound)

Published by : **Regency Publications**
A Division of
Astral International Pvt. Ltd.
– ISO 9001:2008 Certified Company –
4760-61/23, Ansari Road, Darya Ganj
New Delhi-110 002
Ph. 011-43549197, 23278134
E-mail: info@astralint.com
Website: www.astralint.com

Printed at : **Radiant Printers**

PRINTED IN INDIA

Preface

Due to enormous progress in the field of biology, Biotechnology has shown spectacular development and widens its webs all around. Enzyme biotechnology is one of the major areas of interest. The explosion of genetic information over the past decade has led to the new challenge of identifying the functions of the thousands of proteins whose sequences have been revealed. For example, a significant proportion of the proteins identified as being coded on the human genome are predicted to be enzymes of known function with the complexities of alternative splicing and post-translational modification. How many more as yet unrevealed catalytic activities exist is anybody's guess. In this task, the assay of enzyme activity will play an essential role.

Assay of enzyme catalytic activity is among the most frequently performed procedures in biotechnology, as it is involved with identification of enzymes, estimation of the amount of enzyme present, monitoring the purification of an enzyme, and determining its kinetic parameters.

The present first edition of the book "Enzyme Assays" is designed and written to nourish the basic enzyme assay principles and their utilization in biotechnological techniques. The attempt is made to furnish precise, illustrative, to the point information about techniques in enzyme biotechnology. Simple tabular representation, detailed process, theory of reactions, full information of chemicals is the asset of the book. The range of procedures used to measure the rate of an enzyme-catalyzed reaction is limited only by the nature of the chemical changes and the ingenuity of the investigator. The experimental approaches cover photometric, radiometric, isotopic, electrochemical and HPLC methods. The theory underlying each method is outlined, together with a description of the reaction catalyzed. The chief aim of this particular book is to help the students understand the concepts involved in enzyme biotechnology.

This book is intended to provide a compilation of assay protocols for individual enzymes. As several thousand enzyme catalyzed reactions are known, such a compilation would have been impossible in a book of this size. However the chapters contain experimental protocols that have been carefully chosen to represent the various types of enzyme-catalyzed reactions amenable to assay by a particular technique. This generally notwithstanding, reference is made to over 250 individual enzymes or catalytic activities. These have been compiled in an enzyme index, and which I hope will prove useful to all working in the molecular life sciences.

Author hopes that this book will be beneficial to the beginners, bachelor, master level students in the area of applied, industrial biotechnology, this book is similarly important to faculty engaged in teaching biotechnology at various professional disciplines namely pharmaceutical; biotechnology, biochemistry, molecular biology, organic and analytical chemistry. This book will also be beneficial to the scientists engaged in the research on enzyme biotechnology, analytical biotechnology, organic chemistry at various research institutes and industrial level.

Since this is a maiden effort by the author, suggestions and comments from readers are welcome, which will in fact strengthen her knowledge and help her to gear up for the further improvement of the contents of this book.

Contents

Enzyme Activator Assays

Protein Inhibitor Assays

Introduction to Enzyme Assays

Enzymes are complex naturally occurring proteins that catalyze many biological reactions. Most enzymes are quite specific in their action catalyzing a definite type of chemical reaction or act on a particular compound/substrate and so these have been named after the type of substrate on which these are acting e.g., Lipase, Amylase, Phosphatase etc. Concentration of enzymes is very low and could not be directly measured like Glucose, Protein etc. Therefore the enzymes are indirectly measured by their catalytic activity, which is proportional to their concentration.

Enzymes can be estimated in two different ways:

1. End point assay: Here sample is incubated with the buffered substrate for a fixed period of time & at the end reaction is stopped and the amount of product formed or substrate used is estimated.
2. Rate analysis or Kinetic assay where such changes are measured at short intervals or are continuously monitored. This technique has advantages over the end point assays:

Parameters	*Kinetic assay*	*End point assay*
Range of measurable activity	High	Limited
Reaction Monitoring	5 minutes	Longer incubation period
Detection of high enzyme activity	Within 3 minutes	More than 30 minutes
Standard	Not required	Essential

Thus kinetic assay technique has distinct advantages & the principal is: If the concentration of the substrate (S) is sufficiently high in comparison to enzyme (E) then rate of reaction will be proportional to the concentration of enzyme.

Thus the amount of product (P) formed in a given period of time would be proportional to the amount of active enzyme present, with all other factors remaining constant.

$$E + S — ES — E + P$$

The result of enzyme determination is expressed in terms of product (P) formed per unit of time under specified conditions for a given amount of sample. Thus International Unit (IU) has been adopted by International Union of Pure & Applied Chemistry to standardize the Unit system. *One unit of enzyme is that amount which will catalyze the transformation of 1 micromole of substrate per minute or number of micromole of product formed per minute by the enzyme present in 1 liter of serum i.e. IU/L.*

In practice the reaction rate is not constant with time. There is an initial:

(i) Lag phase (with very little change per unit of time), then comes
(ii) Linear phase of constant change per unit of time and then finaly
(iii) Phase of substrate exhaustion with very little change. In order to obtain the best result we should have a long Linear phase and assay should be stopped before Substrate exhaustion takes place.

Many of the enzyme system involve the conversion of Nicotinamide Adinine Dinucleotide (NAD) to its reduced form NADH or vice-versa. The reduced form NADH has much greater abosrption at 340 UV range and consequently reaction can be followed by measuring change in absorption per minute of time. The change in absorbance per minute when multiplied by the Factor, gives the enzyme activity in IU/L. Factor can also be calculated & is similar to that of End point assays and this can be detected with the help of known calibirators/controls, if required:

$$\text{Factor} = \frac{\text{Value of the known Standard or Calibirator}}{\text{Change in Absorbance of enzyme activity in one min.}} \quad \& \quad \text{Enz. U / L} = \text{Absorb./min} \times \text{Factor}$$

Enzymes are active only over a limited range of pH. But this optimum pH may vary some times with the type of substrate

e.g.: Alk. Phosphatase – Substrate B-Glycerophospahte – pH 8.6

Phenyl phosphate 9.8

Colorimetric tests

Colorimetric tests are commonly used in both biology and chemistry. The idea behind using a colorimetric test is that the amount of compound X can be measured if compound X absorbs light. The higher the concentration of compound X in a solution, the higher the absorbance. In many cases, compound X itself is not a chromophore, but it can be reacted with some other compound and the product is a chromophore. Colorimetric tests are based on the proportional relationship between concentration of a chromophore and the amount of light absorbed by that chromophore. This relationship is described by Beer's law which states

$$A = å \times l \times c$$

where å is a constant called the molar absorptivity(in units of liters/mole cm),
l is path length of the spectrophotometer (usually 1 cm) and
c is concentration which is in terms of molarity (moles/1).

Using this relationship, the concentration of a compound can be determined directly from the absorbance provided molar absorptivity and path length are known. This linear relationship between absorbance and concentration is true for the useful absorbance range; this is generally considered from 0.2 to 1.2. At absorbances greater than 1.2, the linear relationship shows a negative deviation and thus any absorbance value above 1.2 will underestimate the concentration. Absorbances below 0.2 are difficult to read with great accuracy because of the faint color. The higher level of error produces values that are less useful. Molar absorptivity is a constant under a given set of conditions; if pH, temperature or other conditions differ, this value will change. It is also possible to relate the absorbance of a compound to its concentration by use of the standard curve.

The idea of a standard curve is that the relationship between absorbance and concentration is linear provided that the absorbance value is in the useful range. The amount of X in an unknown sample can be measured by comparing the absorbance of the unknown with a series of standards. This is done by plotting the absorbance values vs amount of X and solving for the best straight line which is given in terms of $y=mx + b$ where y is absorbance, x is amount of X, b is the y intercept and m is the slope. The best straight line will give values for the slope and y intercept; it is possible to mathematically solve for amount of X in an unknown sample. Since the standard curve is so important for figuring out the unknown concentration, it is imperative that it be accurate. One way of increasing the accuracy is to use a large number of standards; in this course typically six standards, including a blank, will be used to construct the standard curve.

Lowry assay

Introduction: The Lowry assay for protein is a colorimetric assay for the total amount of proteins. Proteins alone are not chromophores but they can be reacted with compounds and produce chromophores. Since this is an assay for all proteins, any, added reagent must react with all proteins. The Lowry assay is actually a combination of two colorimetric assays. One of the colorimetric reagents reacts specifically with tryosine resides and another colormetric reagent reacts with the peptide bond.

Similar to other colorimetric assays, the absorbance of an unknown solution is related to the concentration by using a standard curve. For Lowry assay, any protein could act as a standard but the most commonly used protein is BSA (bovine serum albumin). This is a protein from cow's blood. It is cheap and readily available and for this reason has become a standard.

Some practical tips to get better results:

1. Incubate the reagents dispensed in quantities required for your assay at specific temperature for 1–2 minutes to avoid strain on the analyzer and keep back the kit in the freeze immediately to preserve its potency.
2. After mixing the sample to reagent feed it immediately to the analyzer so as to observe the Initial Absorbance which must fall with in the range as given in the table below. If not, this shows poor quality of the reagent and it should be avoided.
3. Those instruments which display the absorbance and continue to display the changing abosorbance with the chemical reaction (Enzyme + Substrate) are considered superior for kinetic assays. It can monitor the rate of reaction fast or slow depending on the high or low cons of enzyme and also help in detecting substrate exhaustion along with timings.

4. If we can anticipate higher values of the enzyme (Beyond the linearity range) we can predilute the sample 1 in 5 or 1 in 10 then run the test so as to avoid repetition.
5. We can alter the parameters of the kinetic assay with in the reasonable limits: Initial delay time can be reduced if we have preincubated the reagent. Likewise the number of reading and the timings in linear reaction phase can be altered e.g., from 3 mins. to 2 mins., if you fear substrate exhaustion, while still maintaining the accuracy.
6. Most of the enzymes are tested at 37°C as IU of enzyme is defined at this temperature and India, being a tropical country, it is best suited. We should stick to uniform parameters so to make the results comparable for inter laboratory evaluation.

Enzymatic Assay of ACETATE KINASE (EC 2.7.2.1)

Principle

$$\text{Acetate} + \text{ATP Acetyl} \xrightarrow{\text{Acetate Kinase}} \text{Phosphate} + \text{ADP}$$

$$\text{ADP} + \text{PEP} \xrightarrow{\text{Pyruvate Kinase}} \text{ATP} + \text{Pyruvate}$$

$$\text{Pyruvate} + \beta\text{-NADH} \xrightarrow{\text{Lactic Dehydrogenase}} \text{Lactate} + \beta\text{-NAD}$$

Abbreviations Used

ATP = Adenosine 5'-Triphosphate
ADP = Adenosine 5'-Diphosphate
PEP = Phospho(enol) Pyruvate
β-NADH = β-Nicotinamide Adenine Dinucleotide, Reduced Form
β-NAD = β-Nicotinamide Adenine Dinucleotide, Oxidized Form

Conditions: T = 25°C, pH = 7.6, A340nm, Light path = 1 cm

Method: Continuous Spectrophotometric Rate Determination.

Reagents

A. 100 mM Triethanolamine Buffer, pH 7.6 at 25°C. (Prepare 50 ml in deionized water using Triethanolamine Hydrochloride. Adjust to pH 7.6 at 25°C with 1 M NaOH).
B. 1 M Sodium Acetate Solution (NaOAc) (Prepare 10 ml in deionized water using Sodium Acetate Trihydrate (*Prepare Fresh*).
C. 91 mM Adenosine 5'-Triphosphate Solution (ATP) (Prepare 3 ml in deionized water using Adenosine 5'-Triphosphate, Disodium Salt (*Prepare Fresh*).
D. 56 mM Phospho(enol) pyruvate Solution (PEP) (Prepare 1.5 ml in deionized water using Phospho(enol) pyruvate, Mono(cyclohexylammonium) Salt (*Prepare Fresh*).
E. 200 mM Magnesium Chloride Solution ($MgCl_2$) (Prepare 5 ml in deionized water using Magnesium Chloride, Hexahydrate).
F. 6.4 mM β-Nicotinamide Adenine Dinucleotide, Reduced Form Solution (β-NADH) (Dissolve the contents of one 5 mg vial of β-Nicotinamide Adenine Dinucleotide, Reduced Form, Disodium Salt in the appropriate volume of deionized water (*Prepare Fresh*).
G. PK/LDH Enzymes Suspension (PK/LDH Enzymes Suspension).
H. Myokinase Enzyme Solution (MK) (Immediately before use, prepare a solution containing 2000–3000 units/ml in cold deionized water using Myokinase).
I. Acetate Kinase Enzyme Solution (Immediately before use, prepare a solution containing 0.2–0.5 unit/ml of Acetate Kinase in cold Reagent A).

Procedure

Prepare the reaction cocktail by pipetting (in milliliters) the following reagents into a suitable container:

Reagent A (Buffer)	17.80
Reagent B (NaOAc)	6.00
Reagent E ($MgCl_2$)	1.00
Reagent F (β-NADH)	0.50

Mix and adjust to pH 7.6 at 25°C with 0.1 M HCl or 0.1 M NaOH, if necessary. Pipette (in milliliters) the following reagents into suitable cuvettes:

	Test	Blank
Reaction Cocktail	2.53	2.53
Reagent G (PK/LDH)	0.05	0.05
Reagent H (MK)	0.02	0.02

Mix by inversion and equilibrate to 25°C. Monitor the A340nm until constant, using a suitably thermostatted spectrophotometer. Then add:

Reagent C (ATP)	0.02	0.02
Reagent D (PEP)	0.01	0.01
Reagent A (Buffer)	—	0.01
Reagent I (Enzyme Solution)	0.10	—

Immediately mix by inversion and record the decrease in A340nm for approximately 5 minutes. Obtain the A340nm/minute using the maximum linear rate for both the Test and Blank.

Calculations

$$\text{Units / ml enzyme} = \frac{(\text{A340nm / min Test} - \text{A340nm / min Blank})\ (3.0)\ (\text{df})}{(6.22)\ (0.1)}$$

3.0 = Total volume (in milliliters) of assay.
df = Dilution factor.
6.22 = Millimolar extinction coefficient of β-NADH at 340 nm.
0.1 = Volume (in milliliter) of enzyme used

$$\text{Units / mg solid} = \frac{\text{units / ml enzyme}}{\text{mg solid / ml enzyme}}$$

$$\text{Units / mg protein} = \frac{\text{units / ml enzyme}}{\text{mg protein / ml enzyme}}$$

Unit Definition: *One unit will phosphorylate 1.0 μmole of acetate to acetyl phosphate per minute at pH 7.6 at 25°C.*

Final Assay Concentration

In a 3.00 ml reaction mix, the final concentrations are 63 mM triethanolamine, 200 mM sodium acetate, 6.1 mM adenosine 5'-triphosphate, 1.9 mM phospho(enol) pyruvate, 6.7 mM magnesium chloride, 0.11 mM β-nicotinamide adenine dinucleotide, reduced form, 35 units pyruvate kinase, 50 units lactic dehydrogenase, 40–60 units myokinase, and 0.02–0.05 unit acetate kinase.

Notes

1. Assay not to be used for Acetate Kinase from *Bacillus stearothermophilus.*
2. Contains not less than 700 Pyruvate Kinase units and 1000 Lactic Dehydrogenase units per ml.
3. Pyruvate Kinase Unit Definition: One unit will convert 1.0 μmole of phospho(enol) pyruvate to pyruvate per minute at pH 7.6 at 37°C.
4. Lactic Dehydrogenase Unit Definition: One unit will reduce 1.0 μmole of pyruvate to L-lactate per minute at pH 7.5 at 37°C.
5. Myokinase Unit Definition: One unit will convert 2.0 μmoles of ADP to ATP and AMP per minute at pH 7.6 at 37°C.

Enzymatic Assay of ACETATE KINASE (EC 2.7.2.1) (from *Bacillus stearothermophilus*)

Principle

$$\text{Acetate} + \text{ATP} \xrightarrow{\text{Acetate Kinase}} \text{Acetyle phosphate} + \text{ADP}$$

$$\text{ADP} + \text{PEP} \xrightarrow{\text{Pyruvate Kinase}} \text{ATP} + \text{Pyruvate}$$

$$\text{Pyruvate} + \beta\text{-NADH} \xrightarrow{\text{Lactic Dehydrogenase}} \text{Lactate} + \beta\text{-NAD}$$

Abbreviations Used

ATP = Adenosine 5'-Triphosphate
ADP = Adenosine 5'-Diphosphate
PEP = Phospho(enol) Pyruvate
β-NADH = β-Nicotinamide Adenine Dinucleotide, Reduced Form
β-NAD = β-Nicotinamide Adenine Dinucleotide, Oxidized Form

Conditions: T = 30°C, pH = 7.2, A340nm, Light path = 1 cm

Method: Continuous Spectrophotometric Rate Determination.

Reagents

A. 100 mM Imidazole Solution (Prepare 50 ml in deionized water using Imidazole).
B. Acetic Acid, Glacial (HOAc) (Use Acetic Acid, Glacial).
C. 102 mM Phospho(enol) pyruvate Solution (PEP) (Prepare 5 ml in deionized water using Phospho(enol) pyruvate, Monopotassium Salt. *Prepare Fresh).*
D. 300 mM Adenosine 5'-Triphosphate Solution (ATP) (Prepare 3 ml in deionized water using Adenosine 5'-Triphosphate, Disodium Salt. *Prepare Fresh).*
E. 7.8 mM β-Nicotinamide Adenine Dinucleotide, Reduced Form Solution (β-NADH) (Dissolve the contents of one 10 mg vial of β-Nicotinamide Adenine Dinucleotide, Reduced Form, Disodium Salt, in the appropriate volume of deionized water or dissolve β-Nicotinamide Adenine Dinucleotide, Reduced Form, in the appropriate volume of water).
F. 300 mM Magnesium Chloride Solution ($MgCl_2$) (Prepare 10 ml in deionized water using Magnesium Chloride, Sterile filtered Solution).
G. PK/LDH Enzymes Solution (Use PK/LDH Enzymes Solution in 50% Glycerol).
H. Acetate Kinase Enzyme Solution (Immediately before use, prepare a solution containing 0.50–0.75 unit/ml of Acetate Kinase in cold deionized water).

Procedure

Prepare a reaction cocktail by pipetting (in milliliters) the following reagents into a suitable container:

Reagent A (Imidazole)	18.20
Reagent B (HOAc)	0.74
Reagent C (PEP)	1.00
Reagent D (ATP)	1.00
Reagent F ($MgCl_2$)	2.00

Mix by swirling and adjust to pH 7.2 at 30°C with 1 M KOH. Then add:

Reagent E (β-NADH)	1.00

Mix by swirling. Then add deionized water to make a final volume of 30 ml. Mix by stirring.Pipette (in milliliters) the following reagents into suitable cuvettes:

	Test	Blank'
Reaction Cocktail	2.80	2.80
Reagent G (PK/LDH)	0.10	0.10

Mix by inversion and equilibrate to 30°C. Monitor the A340nm until constant, using a suitably thermostatted spectrophotometer. Then add:

Deionized Water	—	0.10
Reagent H (Enzyme Solution)	0.10	—

Immediately mix by inversion and record the decrease in A340nm for approximately 10 minutes. Obtain the ΔA340nm/minute using the maximum linear rate for both the Test and Blank.

Calculations

$$\text{Units / ml enzyme} = \frac{(\Delta A340\text{nm / min Test} - \Delta A340\text{nm / min Blank})\ (3)\ (\text{df})}{(6.22)\ (0.1)}$$

3 = Total volume (in milliliters) of assay
df = Dilution factor
6.22 = Millimolar extinction coefficient of β-NADH at 340 nm
0.1 = Volume (in milliliter) of enzyme used

$$\text{Units / mg solid} = \frac{\text{units / ml enzyme}}{\text{mg solid / ml enzyme}}$$

$$\text{Units / mg protein} = \frac{\text{units / ml enzyme}}{\text{mg protein / ml enzyme}}$$

Unit Definition: *One unit will phosphorylate 1.0 µmole of acetate to acetyl phosphate per minute at pH 7.2 at 30°C.*

Final Assay Concentration

In a 3.00 ml reaction mix, the final concentrations are 57 mM imidazole, 400 mM acetic acid, 3.2 mM phospho(enol) pyruvate, 9.3 mM adenosine 5'-triphosphate, 19 mM magnesium chloride, 0.24 mM β-nicotinamide adenine dinucleotide, 70 units pyruvate kinase, 100 units lactic dehydrogenase, and 0.05–0.075 unit acetate kinase.

Notes

1. Contains approximately than 700 units/ml of Pyruvate Kinase and 1000 units/ml of Lactic Dehydrogenase.
2. Pyruvate Kinase Unit Definition: One unit will convert 1.0 μmole of phospho(enol) pyruvate to pyruvate per minute at pH 7.6 at 37°C.
3. Lactic Dehydrogenase Unit Definition: One unit will reduce 1.0 μmole of pyruvate to L-lactate per minute at pH 7.5 at 37°C.

Enzymatic Assay of α-ACETOLACTATE DECARBOXYLASE (EC 4.1.1.5)

Principle

$$\text{2-Acetolactate} \xrightarrow{\alpha\text{-Acetolactate Decarboxylase}} \text{Acetoin} + CO_2$$

Conditions: T = 30°C, pH = 6.0, A522nm, Light path = 1 cm

Method: Stopped Spectrophotometric Rate Determination.

Reagents

A. 50 mM MES Buffer, pH 6.0 at 30°C (Prepare 100 ml in deionized water using MES Free Acid, Monohydrate. Adjust to pH 6.0 at 30°C with 1 M NaOH).

B. 15% (w/v) Brij 35 Solution (Brij 35) (Prepare 2 ml in deionized water using Brij 35 Solution, 30% w/v solution. Heat to no more than 60°C to dissolve).

C. 50 mM MES Buffer, pH 6.0 at 30°C with 0.005% (w/v) Brij 35 and 600 mM Sodium Chloride (Buffer C) (Prepare 500 ml by dissolving 4.9 g of MES, Free Acid, Monohydrate and 17.5 g of Sodium Chloride in approximately 450 ml of deionized water. Add 1.7 ml of Reagent B and adjust to pH 6.0 at 30°C with 1 M NaOH).

D. 10 mM α-Acetolactate Substrate Solution (α-AL) (Prepare by diluting 0.05 ml of Ethyl 2-acetoxy-2-methylacetoacetate with 1.5 ml of deionized water and 1.5 ml of 1 N NaOH. Stir 20 minutes, Q.S. to 20 ml with Reagent A. Adjust to pH 6.0 with 0.5 M HCl and then Q.S. to 25 ml with Reagent A.

E. Color Reagent Solution (Color Rgt) (Prepare 250 ml by dissolving 2.5 g of α-Naphthol and 0.25 g of Creatine, Hydrate, in 1 N NaOH (use a minimum amount). Q.S. to 250 ml with 1 N NaOH).

F. 2.27 mM Acetoin Stock Solution (Prepare 100 ml in deionized water using 3-Hydroxy-2-Butanone).

G. α-Acetolactate Decarboxylase Enzyme Solution (Immediately before use, prepare a solution containing approximately 0.03 unit/ml of α- Acetolactate Decarboxylase in Reagent C).

Procedure

Part I Enzyme: Pipette (in milliliters) the following reagents into suitable containers:

	Test	Blank
Reagent D (α-AL)	0.20	0.20
Reagent H (Enz Sol)	0.20	—
Reagent C (Buffer C)	—	0.20

Incubate for 20 minutes at 30°C in a water bath. Then add:

Reagent E (Color Rgt)	4.50	4.50
Reagent H (Enz Sol)	—	0.20

Incubate for 40 minutes at room temperature. Transfer the solutions to suitable cuvettes and record the A522nm for the Tests and Blanks using a suitable spectrophotometer.

Part II Standard Curve: Pipette (in milliliters) the following reagents into suitable 10 ml plastic tubes:

	Std 1	Std 2	Std 3	Std 4	Std 5	Std Blank
Acetoin	0.020	0.040	0.080	0.120	0.160	—
Milli-Q	0.380	0.360	0.320	0.280	0.240	0.400
Color Solution	4.50	4.50	4.50	4.50	4.50	4.50

Color develops in 40 minutes at room temperature. Transfer the solutions to suitable cuvettes and record the A522nm for the Standards and Standard Blank using a suitable spectrophotometer.

Calculations

Standard Curve: ΔA522nm Standard = A522nm Standard–A522nm Standard Blank
Prepare a standard curve by plotting the ΔA522nm of the Standard versus μmoles of acetoin.

Sample Determination: ΔA522nm Test = A522nm Test–A522nm Test Blank
Determine the micromoles of acetoin produced using the standard curve.

$$\text{Units / ml enzyme} = \frac{(\mu\text{moles of acetoin produced})\ (4.9)\ (\text{df})}{(20)\ (0.2)}$$

4.9 = Total volume (in milliliters) of assay
df = Dilution factor
20 = Time (in minutes) of assay
0.2 = Volume (in milliliters) of enzyme used

$$\text{Units / mg solid} = \frac{\text{units / ml enzyme}}{\text{mg solid / ml enzyme}}$$

$$\text{Units / mg protein} = \frac{\text{units / ml enzyme}}{\text{mg protein / ml enzyme}}$$

Unit Definition: *One unit will form one μmole of acetoin from acetolactate per minute at pH 6.0 at 30°C.*

Final Assay Concentration

In a 0.40 ml reaction mix, the final concentrations are 5 mM DL-acetolactate, 48 mM MES (approximately) and 0.006 unit α-acetolactate decarboxylase.

Enzymatic Assay of S-ACETYL COENZYME A SYNTHETASE (EC 6.2.1.1)

Principle

$$\text{Acetate} + \text{CoA} + \text{ATP} \xrightarrow{\text{S-Acetyl CoAS}} \text{Acetyl-CoA} + \text{AMP} + \text{PPi}$$

$$\text{Acetyl-CoA} + NH_2OH \longrightarrow \text{Acetyl-NHOH} + \text{CoA}$$

$$\text{Acetyl-NHOH} + FeCl_3 \longrightarrow \text{Brown-Red Colored Product}$$

Abbreviations Used

CoA = Coenzyme A
ATP = Adenosine 5'-Triphosphate
S-Acetyl CoAS = S-Acetyl Coenzyme A Synthetase
AMP = Adenosine 5'-Monophosphate
PPi = Pyrophosphate
Acetyl-CoA = Acetyl Coenzyme A
Acetyl-NHOH = Acethydroxamic Acid

Conditions: T = 37°C, pH = 7.5, A546nm, Light path = 1 cm

Methods: Colorimetric

Reagents

A. 500 mM Monobasic Potassium Phosphate (KH_2PO_4) (Prepare 100 ml in deionized water using Potassium Phosphate, Monobasic, Anhydrous).
B. 500 mM Dibasic Potassium Phosphate (K_2HPO_4) (Prepare 100 ml in deionized water using PotassiumPhosphate, Dibasic, Trihydrate).
C. 500 mM Potassium Phosphate Buffer, pH 7.5 at 37°C (Prepare 100 ml using Reagent B ($K2HPO_4$). Adjust to pH 7.5 at 37°C with Reagent A (KH_2PO_4)).
D. 100 mM Magnesium Chloride Solution ($MgCl_2$) (Prepare 100 ml in deionized water using Magnesium Chloride, Hexahydrate).
E. 100 mM Adenosine Triphosphate Solution (ATP) (Prepare 10 ml in deionized water using Adenosine 5'-Triphosphate, Disodium Salt).
F. 500 mM Potassium Fluoride Solution (KF) (Prepare 10 ml in deionized water using Potassium Fluoride, Dihydrate).
G. 50 mM Potassium Acetate Solution (KOAc) (Prepare 10 ml in deionized water using Potassium Acetate).
H. 100 mM Reduced Glutathione Solution (GSH) (Prepare 10 ml in deionized water using Glutathione, Reduced Form, Free Acid).
I. 7.8 mM Coenzyme A Solution (CoA) (Prepare 10 ml in deionized water using Coenzyme A, Sodium Salt).
J. 1000 mM Hydroxylamine Solution (HA) (Prepare 20 ml in deionized water using Hydroxylamine, Hydrochloride).

K. 370 mM Iron Chloride and 3.3% (w/v) Trichloroacetic Acid Solution ($FeCl_3$) (Prepare 100 ml by dissolving 10g of Ferric Chloride, Hexahydrate, and 3.3g of Trichloroacetic Acid in 80 ml of deionized water. Add 6.6 ml of hydrochloric Acid, and make up to 100 ml with deionized water).

L. S-Acetyl Coenzyme A Synthetase Enzyme Solution (Immediately before use, prepare a solution containing 0.2–0.4 unit/ml of S-Acetyl Coenzyme A Synthetase in cold Reagent C).

Procedure

Pipette (in milliliters) the following reagents into suitable cuvettes:

	Test	Blank
Reagent C (Buffer)	0.20	0.20
Reagent D ($MgCl_2$)	0.05	0.05
Reagent E (ATP)	0.10	0.10
Reagent F (KF)	0.10	0.10
Reagent G (KOAc)	0.20	0.20
Reagent H (GSH)	0.10	0.10
Reagent I (CoA)	0.05	—
Reagent J (HA)	0.20	0.20
Deionized Water	—	0.05

Mix by inversion and equilibrate to 37°C. Then add:

Reagent L (Enzyme Solution)	0.10	0.10

Immediately mix by inversion and incubate for 20 minutes at 37°C. Then add:

Reagent K ($FeCl_3$)	2.00	2.00

Mix by inversion and transfer to suitable cuvettes. Record the A546nm for both the Test and Blank using a suitable spectrophotometer.

Calculations

$$\text{Units / mg enzyme} = \frac{(\text{A546nm Test} - \text{A546nm Blank})\ (3.1)}{(20)\ (0.98)\ (\text{mg enzyme / RM})}$$

3.1 = Total Volume (in milliliters) of assay
0.98 = Change in absorbance of the formation of one μmole of the brown-red colored product.
20 = Time of assay (in minutes) of acethydroxamic acid under the conditions of the assay.
RM = Reaction Mix

Unit Definition: *One unit will form 1.0 μmole of S-Acetyl Coenzyme A from acetate, ATP and coenzyme A per minute at pH 7.5 at 37°C.*

Final Assay Concentration

In a 1.10 ml reaction mix, the final concentrations are 136 mM potassium phosphate, 4 mM magnesium chloride, 9.1 mM ATP, 45 mM potassium fluoride, 9.1 mM potassium acetate, 9.1 mM reduced glutathione, 0.35 mM coenzyme A, 182 mM hydroxylamine and 0.02–0.04 unit S-acetyl coenzyme A synthetase.

Enzymatic Assay of ACETYLESTERASE (EC 3.1.1.6)

Principle

$$\text{Triacetin} + 3\ H_2O \xrightarrow{\text{Acetylesterase}} \text{Glycerol} + 3\ \text{Acetic Acid}$$

Conditions: T = 30°C, pH 6.5

Method: Titrimetric

Reagents

A. 150 mM Sodium Chloride (Prepare 200 ml in deionized water using SodiumChloride).
B. 20 mM Sodium Hydroxide Solution–Standardized (NaOH) (Prepare 100 ml in deionized water using Sodium Hydroxide, 1.0 N).
C. 5% (v/v) Triacetin Solution, pH 6.5 at 30°C (Prepare 100 ml in Reagent A using Triacetin (C_2:O). Adjust to pH 6.5 with 1 M NaOH).
D. Acetylesterase Enzyme Solution (Immediately before use, prepare a solution containing 20 units/ml of Acetylesterase in cold deionized water).

Procedure

Using a suitable pH meter in conjunction with a magnetic stirrer, pipette (in milliliters) the following reagents into a suitable titration vessel:

	Test	Blank
Reagent C (Triacetin)	20.0	20.0

Equilibrate to 30°C. Then add:

	Test	Blank
Reagent D (Enzyme Solution)	0.05	—
Deionized Water	—	0.05

Run the reaction for 5 minutes. Maintain the pH of the reaction mix at pH 6.5 by the addition of small volumes of Reagent B. Record the time required to add 0.050 ml Reagent B and maintain the pH at 6.5.

Calculation

$$\text{units / ml enzyme} = \frac{(\text{M})\ (\text{ml NaOH for Test} - \text{ml NaOH for Blank})\ (1000)\ (\text{df})}{(\text{Time})\ (0.05)}$$

M = Molarity of NaOH solution.
1000 = Conversion from millimoles to micromoles.
df = Dilution factor.
Time = Time required (in minutes) to maintain pH of 6.5.
0.05 = Volume (in milliliter) of enzyme used.

$$\text{Units / mg protein} = \frac{\text{units / ml enzyme}}{\text{mg protein / ml enzyme}}$$

Unit Definition: *One unit will produce 1.0 μmole of acetic acid from triacetin per minute at pH 6.5 at 30°C.*

Initial Assay Concentration: In a 20.05 ml reaction mix, the initial concentrations are 150 mM sodium chloride, 5% triacetin, and 4 units acetylesterase.

Enzymatic Assay of α-N-ACETYLGALACTOSAMINIDASE (EC 3.2.1.49)

Principle .

$$\text{PNP-NAG} + H_2O \xrightarrow{\alpha\text{-N-Acetylgalactosaminidase}} \text{NAG} + \text{p-Nitrophenol}$$

Abbreviation Used

PNP-NAG = p-Nitrophenyl N-Acetyl- α-D-Galactosaminide.

Conditions: T = 37°, pH = 3.65, A400nm, Light path = 1 cm.

Method: Spectrophotometric Stop Rate Determination.

Reagents

A. 100 mM Citrate and 100 mM Sodium Phosphate Buffer, pH 3.65 at 37°C (Prepare 100 ml in deionized water using Citric Acid, Free Acid, Monohydrate, and Sodium Phosphate, Monobasic, Monohydrate. Adjust to pH 3.65 at 37°C with either 1 M NaOH or 1 M HCl).

B. 100 mM Citrate and 100 mM Sodium Phosphate Buffer with 0.1% (w/v) Bovine Serum Albumin, pH 3.65 at 37°C (Enzyme Diluent) (Prepare 25 ml in Reagent A using Albumin, Bovine).

C. 5.0 mM p-Nitrophenyl N-Acetyl- α-D-Galactosaminide (PNP-NAG) (Prepare 2 ml in Reagent A using p-Nitrophenyl N-Acetyl- α-D-Galactosaminide).

D. 200 mM Borate Buffer, pH 9.8 at 25°C (Borate) (Prepare 25 ml in deionized water using Boric Acid. Adjust to pH 9.8 at 25°C with 1 M NaOH).

E. α-N-Acetylgalactosaminidase Enzyme Solution(Immediately before use, prepare a solution containing 0.02–0.05 unit/ml of α-N-Acetylgalactosaminidase in cold Reagent B).

Procedure

Pipette (in milliliters) the following reagents into suitable containers:

	Test	Blank
Reagent C (PNP-NAG)	0.10	0.10

Equilibrate to 37°C. Then add:

Reagent E (Enzyme Solution)	0.05	—

Immediately mix by inversion and incubate at 37°C for exactly 10 minutes. Then add:

	Test	Blank
Reagent D (Borate)	1.00	1.00
Reagent E (Enzyme Solution)	—	0.05

Mix by inversion and transfer to suitable cuvettes. Record the A400nm for both the Test and Blank using a suitable spectrophotometer.

Calculations

$$\text{Units / ml enzyme} = \frac{(\text{A400nm Test} - \text{A400nm Blank})\ (1.15)\ (\text{df})}{(18)\ (10)\ (0.05)}$$

1.15 = Total volume (in milliliters) of assay
df = Dilution factor
18 = Millimolar extinction coefficient of p-Nitrophenol at 400nm
10 = Time (in minutes) of assay per the Unit Definition
0.05 = Volume (in milliliter) of enzyme used

$$\text{Units / mg solid} = \frac{\text{units / ml enzyme}}{\text{mg solid / ml enzyme}}$$

$$\text{Units / mg protein} = \frac{\text{units / ml enzyme}}{\text{mg protein / ml enzyme}}$$

Unit Definition: *One unit will release 1.0 µmole of p-nitrophenol from p-nitrophenyl-N-acetyl-α-D-galactosaminide per minute at pH 3.65 at 37°C.*

Final Assay Concentration

In a 0.15 ml reaction mix, the final concentrations are 100 mM citric acid, 100 mM sodium phosphate, 0.03% (w/v) bovine serum albumin, 3.3 mM p-nitrophenyl N-acetyl-α-D galactosaminide, and 0.001–0.0025 unit α-Nacetylgalactosaminidase.

Enzymatic Assay of β-N-ACETYLGLUCOSAMINIDASE (EC 3.2.1.30)

Principle

$$\text{PNP-NAG} \xrightarrow{\beta\text{-N-Acetylglucosaminidase}} \text{p-Nitrophenol} + \text{NAG}$$

Abbreviations Used

PNP-NAG = p-Nitrophenyl-N-Acetyl-β-D-Glucosaminide
NAG = N-Acetyl-β-D-Glucosamine

Conditions: T = 25°C, pH = 4.25, A405nm, Light path = 1 cm.

Method: Spectrophotometric Stop Rate Determination

Reagents

A. 100 mM Citrate Buffer with 200 mM Sodium Chloride and 0.02% Albumin, pH 4.25 at 25°C (Prepare 100 ml in deionized water using Citric Acid, Monohydrate, Sodium Chloride, and Albumin, Bovine Serum. Adjust to pH 4.25 at 25°C with 1 M NaOH).

B. 10 mM p-Nitrophenyl-N-Acetyl-β-D-Glucosaminide Solution (PNP-NAG) (Prepare 5 ml in deionized water using p-Nitrophenyl- N-Acetyl-β-D-Glucosaminide).
C. 200 mM Borate Buffer, pH 9.8 at 25°C (Prepare 100 ml in deionized water using Boric Acid. Adjust to pH 9.8 at 25°C with 1 M NaOH).
D. β-N-Acetylglucosaminidase Enzyme Solution (Immediately before use, prepare a solution containing 0.06–0.12 units/ml of β-N-Acetylglucosaminidase in cold deionized water).

Procedure

Pipette (in milliliters) the following reagents into suitable containers:

	Test	Blank
Deionized Water	—	0.10
Reagent A (Buffer)	0.40	0.40
Reagent B (PNP-NAG)	0.50	0.50

Mix and equilibrate to 25°C. Then add:

Reagent D (Enzyme Solution)	0.10	—

Immediately mix and incubate at 25°C for exactly 10 minutes. Then add:

Reagent C (Borate)	3.00	3.00

Mix and transfer to suitable cuvettes. Record the A400nm for both the Test and Blank using a suitable spectrophotometer.

Calculations

$$\text{Units / mg protein} = \frac{(\Delta A405nm\ \text{Test} - \Delta A405nm\ \text{Blank})\ (4)}{(10)\ (18)\ (\text{mg enzyme / RM})}$$

4 = Total volume (in milliliters) of Solution
10 = Time of assay (Unit definition)
18 = Millimolar extinction coefficient of p-Nitrophenol at 405 nm.
RM = Reaction Mix (total volume = 1 ml)

Unit Definition: *One unit will hydrolyze 1.0 μmole of p-nitrophenyl-Nacetyl- β-D-glucosaminide to p-nitrophenol and N-acetyl-β-D-glucosamine per minute at pH 4.25 at 25°C.*

Final Assay Concentration

In a 1 ml reaction mix, the final concentrations are 45 mM citric acid, 90 mM sodium chloride, 0.01% BSA, 4.5 mM p-nitrophenyl-N-Acetyl-β-D-glucosaminide and 0.006–0.012 units β-N-acetylglucosamidase.

Enzymatic Assay of β-N-ACETYLGLUCOSAMINIDASE (EC 3.2.1.30) from Jack Beans

Principle

$$\text{PNP-NAG} + H_2O \xrightarrow{\beta\text{-N-Acetylglucosaminidase}} \text{p-Nitrophenol} + \text{NAG}$$

Abbreviations Used

PNP-NAG = p-Nitrophenyl N-Acetyl-β-D-Glucosaminide
NAG = N-Acetyl-β-D-Glucosamine

Conditions: T = 25°C, pH = 5.0, A400nm, Light path = 1 cm

Method: Spectrophotometric Stop Rate Determination

Reagents

A. 100 mM Citrate Buffer with 200 mM Sodium Chloride and 0.02% (w/v) Bovine Serum Albumin, pH 5.0 at 25°C(Prepare 100 ml in deionized water using Citric Acid, Free Acid, Monohydrate, Sodium Chloride,and Albumin, Bovine. Adjust to pH 5.0 at 25°C with 1 M NaOH).
B. 10 mM p-Nitrophenyl N-Acetyl-β-D-Glucosaminide Solution (PNP-NAG) (Prepare 5 ml in deionized water using p-Nitrophenyl N-Acetyl-β-D-Glucosaminide).
C. 200 mM Borate Buffer, pH 9.8 at 25°C (Prepare 100 ml in deionized water using Boric Acid. Adjust to pH 9.8 at 25°C with 1 M NaOH).
D. β-N-Acetylglucosaminidase Enzyme Solution (Immediately before use, prepare a solution containing 0.05–0.1 unit/ml of β-N-Acetylglucosaminidase in cold Reagent A).

Procedure

Pipette (in milliliters) the following reagents into suitable containers:

	Test	Blank
Reagent A (Buffer)	0.40	0.40
Reagent B (PNP-NAG)	0.50	0.50

Mix by inversion and equilibrate to 25°C. Then add:

Reagent D (Enzyme Solution)	0.10	—

Mix by inversion and incubate for exactly 10 minutes at 25°C. Then add:

Reagent C (Borate Buffer)	3.00	3.00
Reagent D (Enzyme Solution)	—	0.10

Mix by inversion and transfer to suitable cuvettes. Record the A400nm for both the Test and Blank using a suitable spectrophotometer.

Calculations

$$\text{Units / ml enzyme} = \frac{(\text{A400nm Test} - \text{A400nm Blank})\ (4)\ (\text{df})}{(10)\ (18)\ (0.1)}$$

4 = Total volume (in milliliters) of Assay
df = Dilution factor
10 = Time of assay (in minutes) as per the Unit Definition
18 = Millimolar extinction coefficient of p-Nitrophenol at 400nm
0.1 = Volume (in milliliter) of enzyme used

$$\text{Units / mg solid} = \frac{\text{units / ml enzyme}}{\text{mg solid / ml enzyme}}$$

$$\text{Units / mg protein} = \frac{\text{units / ml enzyme}}{\text{mg protein / ml enzyme}}$$

Unit Definition: *One unit will hydrolyze 1.0 µmole of p-nitrophenyl N-acetyl-β-D-glucosaminide to p-nitrophenol and N-acetyl β- D-glucosamine per minute at pH 5.0 at 25°C.*

Final Assay Concentration

In a 1.00 ml reaction mix, the final concentrations are 50 mM citric acid, 100 mM sodium chloride, 0.01% (w/v) bovine serum albumin, 5 mM p-nitrophenyl N-acetyl-β-D-glucosaminide and 0.005–0.01 unit β-Nacetylglucosaminidase.

Enzymatic Assay of β-N-ACETYLGLUCOSAMINIDASE (EC 3.2.1.52)

Principle

$$\text{PNP - NAG } H_2O \xrightarrow{\beta\text{-N-Acetylglucosaminidase}} \text{NAG} + \text{p - Nitrophenol}$$

Abbreviations Used

PNP-NAG = p-Nitrophenyl N-Acetyl-β-D-Glucosaminide
NAG = N-Acetyl-D-Glucosamine

Conditions: T = 25°C, pH = 4.0, A400nm, Light path = 1 cm.

Method: Spectrophotometric Stop Rate Determination.

Reagents

A. 100 mM Citrate Buffer with 0.02% (w/v) Bovine Serum Albumin and 200 mM Sodium Chloride, pH 4.0 at 25°C (Prepare 100 ml in deionized water using Citric Acid, Free Acid, Monohydrate, Albumin, Bovine, and Sodium Chloride. Adjust to pH 4.0 at 25°C with 1 M NaOH).
B. 10 mM p-Nitrophenyl N-Acetyl-β-D-Glucosaminide Solution (PNP-NAG) (Prepare 5 ml in deionized water using p-Nitrophenyl N-Acetyl-β-D-Glucosaminide).
C. 200 mM Borate Buffer, pH 9.8 at 25°C (Borate) (Prepare 100 ml in deionized water using Boric Acid. Adjust to pH 9.8 at 25°C with 1 M NaOH).
D. β-N-Acetylglucosaminidase Enzyme Solution(Immediately before use, prepare a solution containing 0.1–0.2 unit/ml of β-N-Acetylglucosaminidase in cold Reagent A).

Procedure

Pipette (in milliliters) the following reagents into suitable containers:

	Test	Blank
Reagent A (Buffer)	0.40	0.40
Reagent B (PNP-NAG)	0.50	0.50

Mix by swirling and equilibrate to 25°C. Then add:

Reagent D (Enzyme Solution)	0.10	—

Immediately mix by swirling and incubate at 25°C for exactly 10 minutes. Then add:

Reagent C (Borate)	3.00	3.00
Reagent D (Enzyme Solution)	—	0.10

Mix by swirling and transfer to suitable cuvettes. Record the A400nm for both the Test and Blank using a suitable spectrophotometer.

Calculations

$$\text{Units / ml enzyme} = \frac{(\text{A400nm Test} - \text{A400nm Blank})\ (4)\ (\text{df})}{(10)\ (18)\ (0.1)}$$

4 = Total volume (in milliliters) of assay
df = Dilution factor
10 = Time (in minutes) of assay as per the Unit Definition
18 = Millimolar extinction coefficient of p-Nitrophenol at 400 nm
0.1 = Volume (in milliliter) of enzyme used

$$\text{Units / mg protein} = \frac{\text{units / ml enzyme}}{\text{mg protein / ml enzyme}}$$

Unit Definition: *One unit will hydrolyze 1.0 µmole of p-nitrophenyl Nacetyl-β-D-glucosaminide to p-nitrophenol and N-acetyl-Dglucosamine per minute at pH 4.0 at 25°C.*

Final Assay Concentration

In a 1.00 ml reaction mix, the final concentrations are 40 mM citric acid, 80 mM sodium chloride, 0.008% (w/v) Bovine Serum Albumin, 5 mM p-nitrophenyl N-acetyl-β-D-glucosaminide and 0.01–0.02 unit β-Nacetylglucosaminidase.

Enzymatic Assay of β-N-ACETYLGLUCOSAMINIDASE (EC 3.2.1.30)

Principle

$$\text{PNP-NAG} + H_2O \xrightarrow{\beta\text{-N-Acetylglucosaminidase}} \text{p-Nitrophenol} + \text{NAG}$$

Abbreviations Used

PNP-NAG = p-Nitrophenyl N-Acetyl-β-D-Glucosaminide
NAG = N-Acetyl-β-D-Glucosamin.

Conditions: T = 37°C, pH = 4.8, A405nm, Light path = 1 cm.

Method: Spectrophotometric Stop Rate Determination

Reagents

A. 100 mM Citrate Buffer (Prepare 100 ml in deionized water using Citric Acid, Free Acid, Monohydrate).
B. 200 mM Sodium Phosphate Buffer (Prepare 100 ml in deionized water using Sodium Phosphate, Dibasic, Anhydrous).
C. Citrate Phosphate Buffer, pH 4.8 at 37°C (Cit-Phos) (Prepare by adjusting the pH of Reagent A with Reagent B).
D. 2.0 mM p-Nitrophenyl N-Acetyl-β-D-Glucosaminide Solution with 0.1% (w/v) Bovine Serum Albumin (PNP-NAG) (Prepare 10 ml in Reagent C using p-Nitrophenyl NAcetyl- β-D-Glucosaminide and Albumin, Bovine).
E. 1 M Sodium Carbonate Solution (Na_2CO_3) (Prepare 100 ml in deionized water using Sodium Carbonate, Anhydrous).
F. β-N-Acetylglucosaminidase Enzyme Solution (Immediately before use, prepare a solution containing 0.02–0.04 unit/ml of β-N-Acetylglucosaminidase in cold Reagent C).

Procedure

Pipette (in milliliters) the following reagents into suitable containers:

	Test	Blank
Reagent C (Cit-Phos)	0.50	0.50
Reagent D (PNP-NAG)	0.50	0.50

Mix by swirling and equilibrate to 37°C. Then add:

Reagent F (Enzyme Solution)	0.10	—

Mix by swirling and incubate for exactly 5 minutes at 37°C. Then add:

Reagent E (Na_2CO_3)	2.00	2.00
Reagent F (Enzyme Solution)	—	0.10

Mix by swirling and transfer to suitable cuvettes. Record the A405nm for both the Test and Blank using a suitable spectrophotometer.

Calculations

$$\text{Units / ml enzyme} = \frac{(\text{A405nm Test} - \text{A405nm Blank})\,(3.1)\,(\text{df})}{(5)\,(18.5)\,(0.1)}$$

3.1 = Total volume (in milliliters) of assay
df = Dilution factor
5 = Time of assay (in minutes)
18.5 = Millimolar extinction coefficient of p-Nitrophenol at 405nm
0.1 = Volume (in milliliter) of enzyme used

$$\text{Units / mg protein} = \frac{\text{units / ml enzyme}}{\text{mg protein / ml enzyme}}$$

Unit Definition: *One unit will hydrolyze 1.0 μmole of p-nitrophenyl N-acetyl-β-D-glucosaminide to p-nitrophenol and N-acetyl-β-D-glucosamine per minute at pH 4.8 at 37°C.*

Final Assay Concentration

In a 1.10 ml reaction mix, the final concentrations are 0.91 mM p-nitrophenyl N-acetyl-β-D-glucosaminide, 0.05% (w/v) Bovine Serum Albumin and 0.002–0.004 unit of β-Nacetylglucosaminidase. The final concentrations of citric acid and sodium phosphate varies.

Enzymatic Assay of N-ACETYLNEURAMINIC ACID ALDOLASE (EC 4.1.3.3)

Principle

$$\text{NANA} \xrightarrow{\text{NANA-Aldolase}} \text{N-Acetyl-D-Mannosamine} + \text{Pyruvate}$$

$$\text{Pyruvate} + \beta\text{-NADH} \xrightarrow{\text{L-Lactic Dehydrogenase}} \text{L-Lactate} + \beta\text{-NAD}$$

Abbreviations Used

NANA = N-Acetylneuraminic Acid
NANA-Aldolase = N-Acetylneuraminic Acid Aldolase
β-NADH = β-Nicotinamide Adenine Dinucleotide, Reduced Form
β-NAD = β-Nicotinamide Adenine Dinucleotide, Oxidized Form

Conditions: T = 37°C, pH = 7.2, A340nm, Light path = 1 cm

Method: Continuous Spectrophotometric Rate Determination

Reagents

A. 20 mM Potassium Phosphate Buffer, pH 7.2 at 37°C (Prepare 200 ml in deionized water using Potassium Phosphate, Monobasic, Anhydrous. Adjust to pH 7.2 at 37°C with 1 M KOH).

B. 65 mM N-Acetylneuraminic Acid Solution (NANA) (Prepare 2 ml in Reagent A using N-Acetylneuraminic Acid).

C. 1.28 mM β-Nicotinamide Adenine Dinucleotide, Reduced Form Solution (β-NADH) (Dissolve the contents of one 5 mg vial of β-Nicotinamide Adenine Dinucleotide, Reduced Form,Disodium Salt, in the appropriate volume of Reagent A. *Prepare Fresh*).

D. L-Lactic Dehydrogenase Solution (LDH) (Immediately before use, prepare a soiution containing 500 units/ml in cold Reagent A using L-Lactic Dehydrogenase).

E. N-Acetylneuraminic Acid, Aldolase Enzyme Solution(Immediately before use, prepare a solution containing 0.3–0.4 unit/ml of N-Acetylneuraminic Acid Aldolase in cold Reagent A).

Procedure

Pipette (in milliliters) the following reagents into suitable cuvettes:

	Test	Blank
Reagent A (Buffer)	2.50	2.55
Reagent B (NANA)	0.20	0.20
Reagent C (β-NADH)	0.30	0.30
Reagent D (LDH)	0.02	0.02

Mix by inversion and equilibrate to 37°C. Monitor the A340nm until constant, using a suitably thermostatted spectrophotometer. Then add:

Reagent E (Enzyme Solution)	0.05	—

Immediately mix by inversion and record the decrease in A340nm for approximately 5 minutes. Obtain the ΔA340nm/minute using the maximum linear rate for both the Test and Blank.

Calculations

$$\text{Units / ml enzyme} = \frac{(\Delta A340\text{nm / min Test} - \Delta A340\text{nm / min Blank})}{(6.22)\ (\text{mg enzyme / ml RM})}$$

6.22 = Millimolar extinction coefficient of β-NADH at 340 nm
RM = Reaction Mix

Unit Definition: *One unit will release 1.0 μmole of pyruvate from NANA per minute at 37°C at pH 7.2.*

Final Assay Concentration

In a 3.07 ml reaction mix, the final concentrations are 20 mM potassium phosphate, 4.2 mM N-acetylneuraminic acid, 0.12 mM β-NADH, 10 units L-lactic dehydrogenase and 0.015–0.02 unit N-acetylneuraminic acid aldolase.

Note

1. Unit Definition for Lactic Dehydrogenase: One unit will reduce 1.0 μmole of pyruvate to L-lactate per minute at pH 7.5 at 37°C.

Enzymatic Assay of N-ACETYLNEURAMINIC ACID ALDOLASE (EC 4.1.3.3)

Principle

$$\text{NANA} \xrightarrow{\text{NANA-Aldolase}} \text{N-Acetyl-D-Mannosamine} + \text{Pyruvate}$$

$$\text{Pyruvate} + \beta\text{-NADH} \xrightarrow{\text{L-Lactic Dehydrogenase}} \text{L-Lactate} + \beta\text{-NAD}$$

Abbreviations Used

NANA = N-Acetylneuraminic Acid
NANA-Aldolase = N-Acetylneuraminic Acid Aldolase
β-NADH = β-Nicotinamide Adenine Dinucleotide, Reduced Form
β-NAD = β-Nicotinamide Adenine Dinucleotide, Oxidized Form

Conditions: T = 37°C, pH = 7.7, A340nm, Light path = 1 cm

Method: Continuous Spectrophotometric Rate Determination.

Reagents

A. 100 mM Potassium Phosphate Buffer, pH 7.7 at 37°C (Prepare 100 ml in deionized water using Potassium Phosphate, Monobasic, Anhydrous. Adjust to pH 7.7 at 37°C with 1 M KOH).
B. 20 mM N-Acetylneuraminic Acid Solution (NANA) (Prepare 25 ml in Reagent A using N-Acetylneuraminic Acid).
C. 14 mM β-Nicotinamide Adenine Dinucleotide, Reduced Form Solution (β-NADH) (Dissolve the contents of one 10 mg vial of β-Nicotinamide Adenine Dinucleotide, Reduced Form, Disodium Salt, in the appropriate volume of deionized water.
D. L-Lactic Dehydrogenase Solution (LDH) (Immediately before use, prepare a solution containing 2750 units/ml in cold Reagent A using L-Lactic Dehydrogenase).
E. N-Acetylneuraminic Acid, Aldolase Enzyme Solution (Immediately before use, prepare a solution containing 0.2–0.4 unit/ml of N-Acetylneuraminic Acid Aldolase in cold Reagent A).

Procedure

Pipette (in milliliters) the following reagents into suitable cuvettes:

	Test	Blank
Reagent B (NANA)	2.80	2.80
Reagent C (β-NADH)	0.10	0.10
Reagent D (LDH)	0.01	0.01

Mix by inversion and equilibrate to 37°C. Monitor the A340nm until constant, using a suitably thermostatted spectrophotometer. Then add:

Reagent A (Buffer)	—	0.10
Reagent E (Enzyme Solution)	0.10	—

Immediately mix by inversion and record the decrease in A340nm for approximately 5 minutes. Obtain the ΔA340nm/minute using the maximum linear rate for both the Test and Blank.

Calculations

$$\text{Units / ml enzyme} = \frac{(\Delta A340\text{nm / min Test} - \Delta A340\text{nm / min Blank})}{(6.22)\ (\text{mg enzyme / ml RM})}$$

6.22 = Millimolar extinction coefficient of β-NADH at 340nm
RM = Reaction Mix

Unit Definition: *One unit will release 1.0 μmole of pyruvate from NANA per minute at 37°C at pH 7.7.*

Final Assay Concentration

In a 3.01 ml reaction mix, the final concentrations are 97 mM potassium phosphate, 19 mM N-acetylneuraminic acid, 0.47 mM β-NADH, 27.5 units lactic dehydrogenase and 0.02–0.04 unit N-acetylneuraminic acid aldolase.

Note

1. Unit Definition for L-Lactic Dehydrogenase: One unit will reduce 1.0 μmole of pyruvate to L-lactate per minute at pH 7.5 at 37°C.

Enzymatic Assay of ACHROMOPEPTIDASE (EC 3.4.21.50)

Principle

$$\text{Micrococcus lysodeikticus Cells (intact)} \xrightarrow{\text{Achromopeptidase}} \text{Micrococcus lysodeikticus Cells (lysed).}$$

Conditions: T = 37°C, pH 8.0, A600nm, Light path = 1 cm.

Method: Turbidimetric.

Reagents

A. 10 mM Tris HCl Buffer with 10 mM Sodium Chloride, pH 8.0 at 37°C (Prepare 100 ml in deionized water using Trizma Base, and Sodium Chloride. Adjust to pH 8.0 at 37°C with 1 M HCl).

B. *Micrococcus lysodeikticus* Cell Suspension (Substrate) (Prepare 30 ml in Reagent A by resuspending *Micrococcus lysodeikticus,* to a cell density which produces an absorbance of 0.60 at A600nm using a suitable spectrophotometer).

C. Achromopeptidase Enzyme Solution (Immediately before use, prepare a solution containing 350–700 units/ml of Achromopeptidase in cold Reagent A).

Procedure

Pipette (in milliliters) the following reagents into suitable cuvettes:

	Test	Blank
Reagent A (Buffer)	—	0.10
Reagent B (Substrate)	2.90	2.90

Mix by inversion and equilibrate to 37°C. Then add:

Reagent C (Enzyme Solution)	0.10	—

Immediately mix test and the blank by inversion and record the decrease in A600nm for approximately 5 minutes. Obtain the ΔA600nm/minute using the maximum linear rate for both the Test and Blank.

Calculations

$$\text{Units / ml enzyme} = \frac{(\Delta A600nm / min\ Test - \Delta A600nm / min\ Blank)\ (3.0)}{(0.001)\ (0.1)}$$

3.0 = Volume (in milliliters) of assay
0.001 = Change in absorbance at 600nm (Unit Definition)
0.1 = Volume (in milliliter) of enzyme used

$$\text{Units / mg solid} = \frac{\text{units / ml enzyme}}{\text{mg solid / ml enzyme}}$$

Unit Definition: *One unit will produce A600nm of 0.001 per minute per ml at pH 8.0 at 37°C using a suspension of Micrococcus lysodeikticus (M-3770) as substrate.*

Final Assay Concentration

In a 3.00 ml reaction mix, the final concentrations are 10 mM Tris, 10 mM sodium chloride, suspended cells of *Micrococcus lysodeikticus* (producing an absorbance of 0.600 at A600nm) and 35–70 units achromopeptidase.

Enzymatic Assay of ACONITASE (EC 4.2.1.3)

Principle

$$\text{Citrate} \xrightarrow[\text{Fe}^{++},\ \text{Cysteine}]{\text{Aconitase}} \text{Isocitrate}$$

$$\text{Isocitrate} + \beta\text{-NADP} \xrightarrow{\text{ICD}} \alpha\text{-Ketoglutarate} + CO_2 + \beta\text{-NADPH}$$

Abbreviations Used

β-NADP = β-Nicotinamide Adenine Dinucleotide Phosphate, Oxidized Form
β-NADPH = β-Nicotinamide Adenine Dinucleotide Phosphate, Reduced Form
ICD = Isocitric Dehydrogenase

Conditions: T = 25°C, pH = 7.4, A340nm, Light path = 1 cm.

Method: Continuous Spectrophotometric Rate Determination

Reagents

A. 100 mM Tris Buffer, pH 7.4 at 25°C (Prepare 100 ml in deionized water using Trizma Base. Adjust to pH 7.4 at 25°C with 1 M HCl).
B. 2 mM Citric Acid Solution (Cit) (Prepare 20 ml in deionized water using Citric Acid Free Acid, Monohydrate, Adjust to pH 7.4 with 100 mM NaOH).
C. 5.4 mM β-Nicotinamide Adenine Dinucleotide Phosphate Solution (β-NADP) (Dissolve the contents of a 5 mg vial of β-Nicotinamide Adenine Dinucleotide Phosphate, Sodium Salt, in the appropriate volume of deionized water or prepare 1 ml in deionized water using β-Nicotinamide Adenine Dinucleotide Phosphate.(*Prepare Fresh*)).
D. 1 mM Ferrous Ammonium Sulfate Solution ($Fe(NH_4)_2(SO_4)_2$) (Prepare 1 ml in deionized water using Ferrous Ammonium Sulfate, Hexahydrate (*Prepare Fresh*)).

E. 20 mM Manganese Sulfate ($MnSO_4$) (Prepare 2 ml in deionized water using Manganese Sulfate).
F. 50 mM L-Cysteine Solution, pH 7.4 at 25°C (Cys) (Prepare 5 ml in deionized using L-Cysteine Hydrochloride. Adjust to pH 7.4 at 25°C with 1 M NaOH (*Prepare Fresh*)).
G. Activation Buffer (Prepare by combining 4 ml of Reagent A, 0.10 ml of Reagent D, and 0.2 ml of Reagent F. Store at 0°C).
H. Isocitric Dehydrogenase Enzyme Solution (IsoDH) (Immediately before use, prepare a solution containing 14 units/ml in deionized water using Isocitric Dehydrogenase).
I. Aconitase Enzyme (Use 15 mg).

Procedure

Prepare activated Aconitase enzyme solution by combining the following reagents in a suitable vial:

Reagent I (Enzyme)	15 mg
Reagent G (Activation Buffer)	2.15 ml

Mix and incubate at 0°C for 1 hour. Pipette (in milliliters) the following reagents into suitable cuvettes:

	Test	Blank
Deionized Water	1.45	1.45
Reagent A (Buffer)	1.00	1.00
Reagent B (Cit)	0.10	0.10
Reagent C (β-NADP)	0.10	0.10
Reagent E ($MnSO_4$)	0.20	0.20
Reagent H (IsoDH)	0.05	0.05

Mix by inversion and equilibrate at 25°C. Monitor the A340nm until constant, using a suitably thermostatted spectrophotometer. Then add:

Activated Aconitase Enzyme Solution	0.10	—
Reagent G (Activation Buffer)	—	0.10

Immediately mix by inversion and record the increase in A340nm for approximately 5 minutes. Obtain A340nm/minute using the maximum linear rate for both the Test and Blank.

Calculations

$$\text{Units / ml enzyme} = \frac{(\text{A340nm / min Test} - \text{A340nm / min Blank})\ (3)\ (\text{df})}{(6.22)\ (0.1)}$$

3 = Total volume (in milliliters) of assay
df = Dilution factor

6.22 = Millimolar extinction coefficient of β-NADPH at 340 nm
0.1 = Volume (in milliliter) of enzyme used.

$$\text{Units / g solid} = \frac{\text{units / ml enzyme (1000)}}{\text{mg / ml enzyme}}$$

$$\text{Units / mg protein} = \frac{\text{units / ml enzyme}}{\text{mg protein / ml enzyme}}$$

Unit Definition: *One unit will convert 1.0 μmole of citrate (via cis Aaconitate) to isocitrate per minute at pH 7.4 at 25°C.*

Final Assay Concentrations

In a 3.00 ml reaction mix, the final concentrations are 36 mM Tris, 0.07 mM citric acid, 0.18 mM β-nicotinamide adenine dinucleotide phosphate, 1.3 mM manganese sulfate, 0.0008 mM ferrous ammonium sulfate, 0.08 mM L-cysteine, 0.7 unit isocitric dehydrogenase, and 0.70 mg aconitase.

Enzymatic Assay of ACTOMYOSIN (ATPase Assay)

Principle

$$\text{ATP} + H_2O \text{ ATPase} \xrightarrow{\text{Actomyosin}} \text{ADP} + \text{Pi}$$

Abbreviations Used

ATP = Adenosine 5'-Triphosphate
ATPase = Adenosine 5'-Triphosphatase
ADP = Adenosine 5'-Diphosphate
Pi = Inorganic phosphate.

Conditions: T = 25°C, pH = 9.0, A660nm, Light Path = 1 cm.

Method: Spectrophotometric Stop Rate Determination

Reagents

A. 200 mM Glycine Buffer, pH 9.0 at 25°C (Prepare 100 ml in deionized water using Glycine, Free Base. Adjust to pH 9.0 at 25°C with 1 NaOH).
B. 100 mM Calcium Chloride Solution ($CaCl_2$) (Prepare 10 ml in deionized water using Calcium Chloride, Dihydrate).
C. 50 mM Adenosine 5'-Triphosphate Solution (ATP) (Prepare 5 ml in deionized water using Adenosine 5'-Triphosphate, Disodium Salt).
D. 10% (w/v) Ammonium Molybdate Solution (Prepare 15 ml in 10 NH_2SO_4 using Ammonium Molybdate).
E. Taussky-Shorr Reagent (Taussky-Shorr) (Prepare by adding 10 ml of Reagent D to 70 ml of deionized water, then add 5g Ferrous Sulfate, Heptahydrate. Bring the volume to 100 ml with deionized water. Store in an amber container).
F. Phosphorus Standard (P Std) (Use Phosphorus Standard Solution. The concentration is 20 μg/ml, 0.645 μmoles/ml).

G. Actomyosin Solution (Actomyosin) (Immediately before use, prepare a solution containing 2–3 mg/ml of Actomyosin in cold deionized water).

Procedure

Step 1: Pipette (in milliliters) the following reagents into suitable containers:

	Test	Blank
Reagent A (Buffer)	1.00	1.00
Reagent B ($CaCl_2$)	0.20	0.20
Reagent C (ATP)	0.30	0.30
Deionized Water	0.0	0.40

Mix by swirling and equilibrate to 25°C. Then add:

Reagent G (Actomyosin)	0.1	—

Mix by swirling and incubate at 25°C for exactly 5 minutes. Then add:

Reagent E (Taussky-Shorr)	4.00	4.00
Deionized Water	2.00	2.00
Reagent G (Actomyosin)	—	0.10

Mix by swirling and incubate at room temperature for approximately 5 minutes. Transfer the solutions to suitable cuvettes and record the A660nm for both the Test and Blank.

Step 2: Prepare a standard curve by pipetting (in milliliters) the follow reagents into suitable containers:

	Std 1	Std 2	Std 3	Blank
Reagent E (Taussky-Shorr)	4.00	4.00	4.00	4.00
Reagent F (P Std)	0.25	0.50	1.00	—
Deionized Water	3.75	3.50	3.00	4.00

Mix by swirling and incubate at room temperature for approximately 5 minutes. Transfer the solutions to suitable cuvettes and record the A660nm for both the Standards and Standard Blank.

Calculations

Standard Curve: ΔA660nm Standard = A660nm Standard–A660nm Stand Blank) Prepare a standard curve by plotting the ΔA660nm Standard vs µmoles of phosphate).

Sample Determination: (ΔA660nm Test = A660nm Test–A660nm Test Blank) Determine the micromoles of phosphate liberated using the Standard Curve.

$$\text{Units / ml enzyme} = \frac{(\mu\text{moles of phosphate released})\ (8)\ (\text{df})}{(0.1)\ (5)}$$

8 = Total volume (in milliliters) of assay
df = Dilution factor

0.1 = Volume (in milliliter) of Actomyosin used
5 = Time (in minutes) of assay as per the Unit Definition

$$\text{Units / mg protein} = \frac{\text{units / ml enzyme}}{\text{mg protein / ml enzyme}}$$

Unit Definition: *One unit will liberate 1.0 micromole of inorganic phosphorus from adenosine 5'-triphosphate per minute at pH 9.0 at 25°C in the presence of calcium.*

Final Assay Concentration

In a 2.00 ml reaction mixture, the final concentrations are 100 mM glycine, 10 mM calcium chloride, 7.5 mM adenosine 5'-triphosphate, and 0.2–0.3 mg actomyosin.

Enzymatic Assay of ACYL-ACYL CARRIER PROTEIN SYNTHETASE (EC 6.2.1.20)

Principle

$$[I^{14}C]\text{- Palmitic Acid} + \text{ATP} \xrightarrow{\text{ACP}} [I^{14}C]\text{Palmitoyl - ACP} + \text{AMP} + \text{PPi}$$

Abbreviations Used

ACP = Acyl Carrier Protein
ATP = Adenosine 5'-Triphosphate
AMP = Adenosine Monophosphate
PPi = Pyrophosphate

Conditions: T = 37°C, pH = 8.0

Method: Radioactive

Reagents

A. 1000 mM Tris HCl buffer, pH 8.0 at 37°C (Prepare 10 ml in deionized water using Trizma Base. Adjust to pH 8.0 at 37°C with 1 M HCl).
B. 4000 mM Lithium Chloride (LiCl) (Prepare 5 ml in deionized water using Lithium Chloride).
C. 100 mM Adenosine 5'-Triphosphate (ATP) (Prepare 5 ml in deionized water using Adenosine 5'-Triphosphate, Disodium Salt).
D. 200 mM Magnesium Chloride ($MgCl_2$) (Prepare 3 ml in deionized water using Magnesium Chloride, Hexahydrate).
E. 40 mM Dithiothreitol (DTT) (Prepare 3 ml in deionized water using DLDithiothreitol. *Prepare Fresh*).
F. Cocktail Buffer (CB) (Prepare by mixing 1 ml of Reagent A, 1 ml of Reagent B, 0.5 ml of Reagent C, 0.5 ml of Reagent D, and 0.5 ml of Reagent E. Adjust to pH 8.0 at 37°C with Reagent A).
G. 0.05% Acyl Carrier Protein (ACP) (Prepare 2 ml in deionized water using Acyl Carrier Protein).
H. 10% Triton X-100 and 0.6 mM Sodium Hydroxide (Prepare 10 ml in deionized water using Triton X-100, and Sodium Hydroxide).

I. ($I^{14}C$) Palmtic Acid (Dispense 0.05 ml of ($I^{14}C$) Palmitic Acid (100 microCuries/ml, 56 microCuries/micromole into a microcentrifuge tube and dry with nitrogen. Reconstitute in 0.07 ml of Reagent H).
J. 2% Triton X-100 (Prepare 10 ml in deionized water using Triton X-100).
K. Acyl-ACP Synthetase Enzyme Solution (Immediately before use, prepare a solution containing 0.1–0.5 nanomolar units per ml in Reagent J).
L. Chloroform: Methanol: Acetic Acid (3:6:1) (Prepare 400 ml using Methanol, Spectrophotometric Grade, Acetic Acid, Glacial, and Reagent Grade Chloroform).

Procedure

Prepare a palmitic working solution (WS) by pipetting (in microliters) the following reagents into a suitable container:

Reagent I	70

Incubate for 1 hour at 60°C. Then add:

Reagent F (CB)	140

Mix. This is equivalent to 23.8 microCuries/ml of palmitic acid. Pipette (in microliters) the following reagents into suitable microcentrifuge tubes:

	Test	Blank	Background	Total Counts
Deionized Water	—	20	30	30
Palmitic WS	20	20	20	20
Reagent G (ACP)	10	10	—	—
Reagent K (Enzyme)	20	—	—	—

Mix by vortexing. Cap microcentrifuge tubes and incubate at 37°C for 20 minutes. Transfer the Test, Blank and Background liquids to a 1 cm diameter disk of 3 mM Whatman filter paper suspended on a 2" × 22 Gauge needle. Air dry at room temperature. Transfer the entire vial contents of the Total Counts to a scintillation vial. Rinse each filter paper disk twice with 12 ml each time of Reagent L Chloroform:Methanol:Acetic Acid) for 5 minutes. Thoroughly dry the filters and place in scintillation vials containing 7.5 ml of Aquasol scintillation fluid and count with a suitable scintillation counter.

Calculations

$$\text{Nanomolar units / ml / min} = \frac{(\text{CPM Test - CPM Blank})}{(20)\ (0.02)} \times \frac{1}{\text{Total Counts}} \times \frac{8.5 \text{ nanomoles } (I^{14}C)\text{Palmitic Acid}}{\text{RM}}$$

RM = Reaction Mix
0.02 = Volume of Enzyme used
20 = Time of reaction in minutes

Unit Definition: *One unit will produce one nanomole of Palmitoyl-Acyl Carrier protein from Palmitic Acid and Acyl Carrier Protein per minute at pH 8.0 at 37°C.*

Final Assay Concentrations

In a 50 ml microliter reaction mix, the final concentration are 3.81 mM ATP, 76 mM tris, 304 mM lithium chloride, 7.6 mM magnesium chloride, 1.5 mM dithiothreitol, 0.01% acyl carrier protein, 8.5 nanomoles of 114C palmitic acid, 0.002–0.01 nanomolar unit of acyl-ACP synthetase.

Enzymatic Assay of ACYL COENZYME A OXIDASE (EC 1.3.3.6)

Principle

$$\text{Palmitoyl-CoA} + O_2 \xrightarrow{\text{Acyl Coenzyme A Oxidase}} \text{2-Hexadecenoyl-CoA} + H_2O_2$$

$$2H_2O_2 + \text{4-Aminoantipyrine} + \text{Phenol} \xrightarrow{\text{POD}} \text{Quinoneimine dye} + 4H_2O$$

Abbreviation Used

POD = Peroxidase

Conditions: T = 30°C, pH = 8.0, A500nm, Light path = 1 cm

Method: Continuous Spectrophotometric Rate Determination

Reagents

A. 50 mM MES Buffer, pH 8.0 at 30°C (Prepare 200 ml in deionized water using MES Free Acid. Adjust to pH 8.0 at 30°C With 1 M NaOH).

B. 0.5% (w/v) Palmitoyl-CoA Solution (Pal-CoA) (Prepare 10 ml in deionized water using Palmitoyl Coenzyme A, Free Acid).

C. 1.6 mM 4-Aminoantipyrine with 22 mM Phenol Solution (4-AAP) (Prepare 100 ml in Reagent A using 4-Aminoantipyrine Free Base, and Phenol).

D. 1 mM Flavin Adenine Dinucleotide Solution (FAD) (Prepare 5 ml in Reagent A using Flavin Adenine Dinucleotide, Disodium Salt. *Prepare Fresh*).

E. Peroxidase Enzyme Solution (POD) (Immediately before use, prepare a solution containing 100 purpurogallin units/ml in Reagent A using Peroxidase).

F. 10% (v/v) Triton X-100 Solution (X-100) (Prepare 10 ml in deionized water using Triton X-100).

G. Acyl Coenzyme A Oxidase Enzyme Solution (Immediately before use, prepare a solution containing 0.15–0.30 unit/ml of Acyl Coenzyme A Oxidase in cold Reagent A.

Procedure

Prepare a reaction cocktail by pipetting (in milliliters) the following reagents into a suitable container:

Reagent A (Buffer)	13.35
Reagent C (4-AAP)	15.00
Reagent D (FAD)	0.15
Reagent E (POD)	1.50

Mix by swirling. Adjust to pH 8.0 at 30°C with 100 mM HCl or 100 mM NaOH, if necessary. Pipette (in milliliters) the following reagents into suitable cuvettes:

	Test	Blank
Reaction Cocktail	3.00	3.00
Reagent B (Pal-CoA)	0.30	0.30
Reagent F (X-100)	0.03	0.03

Mix by inversion and equilibrate to 30°C. Monitor the A500nm until constant, using a suitably thermostatted spectrophotometer. Then add:

Reagent A (Buffer)	—	0.10
Reagent G (Enzyme Solution)	0.10	—

Immediately mix by inversion and record the increase in A500nm for approximately 5 minutes. Obtain the ΔA500nm/minute using the maximum linear rate for both the Test and Blank.

Calculations

$$\text{Units / ml enzyme} = \frac{(\Delta A500nm / \text{min Test} - \Delta A500nm / \text{min Blank})\,(2)\,(3.43)\,(df)}{(12.78)\,(0.1)}$$

2 = 2 moles H_2O_2 used per mole of dye
3.43 = Total volume (in milliliters) of assay
df = Dilution factor
12.78 = Millimolar extinction coefficient of Quinoneimine Dye at 500nm
0.1 = Volume (in milliliters) of enzyme used

$$\text{Units / mg solid} = \frac{\text{units / ml enzyme}}{\text{mg solid / ml enzyme}}$$

$$\text{Units / mg protein} = \frac{\text{units / ml enzyme}}{\text{mg protein / ml enzyme}}$$

Unit Definition: *One unit will form 1.0 µmole of H_2O_2 and hexadecenoyl-CoA from palmitoyl-CoA per minute at pH 8.0 at 30°C in a peroxidase coupled system.*

Final Assay Concentration

In a 3.43 ml reaction mix, the final concentrations are 45 mM MES, 0.04% (w/v) palmitoyl-CoA, 0.70 mM 4-aminoantipyrine, 0.004 mM flavin adenine dinucleotide, 9.6 mM phenol, 0.09% (v/v) Triton X-100, 15 purpurogallin units peroxidase, and 0.015–0.030 unit acyl coenzyme A oxidase.

Notes

1. Triton is a registered trademark of Union Carbide Chemicals and Plastics Co., Inc.
2. Peroxidase Unit Definition: One unit will form 1.0 mg purpurogallin from pyrogallol in 20 sec. at pH 6.0 at 20°C.

Enzymatic Assay of ACYL COENZYME A SYNTHETASE (EC 6.2.1.3)

Principle

$$\text{CoA} + \text{Oleate} + \text{ATP} \xrightarrow{\text{Acyl Coenzyme A Synthetase}} \text{Oleoyl-CoA} + \text{AMP} + \text{PPi}$$

$$\text{ATP} + \text{AMP} \xrightarrow{\text{MK}} 2\ \text{ADP}$$

$$2\ \text{ADP} + 2\ \text{Phospho(enol) pyruvate} \xrightarrow{\text{Pyruvate Kinase}} 2\ \text{ATP} + 2\ \text{Pyruvate}$$

$$2\ \text{Pyruvate} + 2\beta\text{-NADH} \xrightarrow{\text{LDH}} 2\text{L-Lactate} + 2\beta\text{-NAD}$$

Abbreviations Used

ATP = Adenosine 5'-Triphosphate
AMP = Adenosine 5'-Monophosphate
ADP = Adenosine 5'-Diphosphate
Oleoyl-CoA = Oleoyl Coenzyme A
PPi = Inorganic Pyrophosphate
β-NADH = β-Nicotinamide Adenine Dinucleotide, Reduced Form
β-NAD = β-Nicotinamide Adenine Dinucleotide, Oxidized Form
CoA = Coenzyme A
MK = Myokinase
LDH = L-Lactic Dehydrogenase

Conditions: T = 25°C, pH = 8.1, A340nm, Light path = 1 cm

Method: Continuous Spectrophotometric Rate Determination

Reagents

A. 200 mM Tris Buffer with 20 mM Magnesium Chloride, 2 mM Ethylenediaminetetraacetic Acid (EDTA) and 0.25% (w/v) Triton X-100, pH 8.1 at 25°C (Prepare 100 ml in deionized water using Trizma Base, Magnesium Chloride, Hexahydrate, Ethylenediaminetetraacetic 3Acid, Disodium Dihydrate, and Triton X-100. Adjust to pH 8.1 at 25°C with 1 M HCl).

B. 100 mM Tris Solution, pH 7.5 at 25°C (Prepare 100 ml in deionized water using Trizma Base. Adjust to pH 7.5 at 25°C with 1 M HCl).

C. 14.5 mM Adenosine 5'-Triphosphate Solution (ATP) (Prepare 1 ml in Reagent B using Adenosine 5'-Triphosphate, Disodium Salt. *Prepare Fresh*).

D. 42.7 mM Phospho(enol) pyruvate Solution (PEP) (Prepare 1 ml in Reagent B using Phospho(enol) Pyruvate, Trisodium Salt, Hydrate. *Prepare Fresh*).

E. Myokinase Enzyme Solution (MK) (Immediately before use, prepare a solution containing 72 units/ml in cold Reagent B using Myokinase).

F. PK/LDH Mixed Enzymes (PK/LDH) (Immediately before use, prepare a solution containing 120 units/ml of Pyruvate Kinase in Reagent B using PK/LDH Enzymes Suspension).

G. 49 mM Coenzyme A (CoA) (Prepare 1 ml in Reagent B using Coenzyme A, Sodium Salt. *Prepare Fresh*).

H. 5.3 mM β-Nicotinamide Adenine Dinucleotide, Reduced Form Solution (β-NADH) (Dissolve the contents of one 10 mg vial of β-Nicotinamide Adenine Dinucleotide,

Reduced Form, Disodium Salt, in the appropriate volume of deionized water or prepare 5 ml in deionized water using β-Nicotinamide Adenine Dinucleotide, Reduced Form, Disodium Salt. *Prepare Fresh*).

I. 0.25% (v/v) Triton X-100 (Prepare 20 ml in deionized water using Triton X-100).

J. 0.98 mM Sodium Oleate Solution (Oleate) (Prepare 10 ml in Reagent I using Oleic Acid, Sodium Salt. *Prepare Fresh*).

K. 50 mM Tris Solution, pH 7.5 at 25°C (Enzyme Diluent) (Prepare 100 ml in deionized water using Trizma Base. Adjust to pH 7.5 at 25°C with 1 M HCl).

L. Acyl Coenzyme A Synthetase Enzyme Solution (Immediately before use, prepare a solution containing 0.05–0.25 unit/ml of Acyl Coenzyme A Synthetase in cold Reagent K).

Procedure

Pipette (in milliliters) the following reagents into suitable cuvettes:

	Test	Blank
Reagent A (Buffer)	2.00	2.00
Reagent C (ATP)	0.05	0.05
Reagent D (PEP)	0.05	0.05
Reagent E (MK)	0.05	0.05
Reagent F (PK/LDH)	0.05	0.05
Reagent G (CoA)	0.10	0.10
Reagent H (β-NADH)	0.10	0.10

Mix by inversion and equilibrate to 25°C. Monitor the A340nm until constant, using a suitably thermostatted spectrophotometer. Then add:

Reagent L (Enzyme Solution)	0.20	0.20

Mix by inversion and monitor the A340nm until constant, using a suitably thermostatted spectrophotometer. Then add:

Reagent K (Enzyme Diluent)	—	0.20
Reagent J (Oleate)	0.20	—

Immediately mix by inversion and record the decrease in A340nm for approximately 15 minutes. Obtain the ΔA340nm/minute using the maximum linear rate for both the Test and Blank.

Calculations

$$\text{Units / ml enzyme} = \frac{(\Delta A340nm / min\ Test - \Delta A340nm / min\ Blank)\ (2.8)\ (df)}{(2)\ (6.22)\ (0.2)}$$

2.8 = Total volume (in milliliters) of assay
df = Dilution factor
2 = 2 moles of β-NAD produced per mole of oleate used
6.22 = Millimolar extinction coefficient of β-NADH at 340 nm
0.2 = Volume (in milliliter) of enzyme used

$$\text{Units / mg solid} = \frac{\text{units / ml enzyme}}{\text{mg solid / ml enzyme}}$$

$$\text{Units / mg protein} = \frac{\text{units / ml enzyme}}{\text{mg protein / ml enzyme}}$$

Unit Definition: *One unit will form 1.0 µmole of AMP and oleoyl-CoA from ATP and oleate per minute at pH 8.1 at 25°C, in the presence of CoA.*

Final Assay Concentrations

In a 2.80 ml reaction mix, the final concentrations are 157 mM Tris, 14 mM magnesium chloride, 1.4 mM Ethylenediaminetetraacetic Acid, 0.18% (v/v) Triton X-100, 0.26 mM adenosine 5'- triphosphate, 0.76 mM phospho(enol) pyruvate, 4 units myokinase, 6 units pyruvate kinase, 9 units lactic dehydrogenase, 1.8 mM coenzyme A, 0.19 mM β-nicotinamide adenine dinucleotide, reduced form, 0.07 mM oleate and 0.01–0.05 unit acyl coenzyme A synthetase.

Notes

1. Triton is a registered trademark of Union Carbide Chemicals and Plastics Co., Inc.
2. Myokinase Unit Definition: One unit will convert 2.0 µmoles of ADP to ATP and AMP per minute at pH 7.6 at 37°C.
3. Contains not less than 700 units/ml of Pyruvate Kinase and 1000 units/ml of Lactic Dehydrogenase.
4. Pyruvate Kinase Unit Definition: One unit will convert 1.0 µmole of phospho(enol) pyruvate to pyruvate per minute at pH 7.6 at 37°C.
5. L-Lactic Dehydrogenase Unit Definition: One unit will reduce 1.0 µmole of pyruvate to L-lactate per minute at pH 7.5 at 37°C.

Enzymatic Assay of ACYLASE I (EC 3.5.1.14)

Principle

$$\text{N - Acetyl - L - Methionine} + H_2O \xrightarrow{\text{Acylase I}} \text{L - Methionine} + \text{Acetic Acid}$$

Conditions: T = 37°C, pH = 8.0, A570nm, Light path = 1 cm.

Method: Colorimetric

Reagents

A. 100 mM N-Acetyl-L-Methionine Solution, pH 8.0 at 37°C (NAMET), (Prepare 25 ml by dissolving 478 mg of N-Acetyl-L-Methionine in 10 ml of deionized water and 2 ml of 1 M NaOH. Adjust to pH 8.0 at 25°C with 10 M NaOH and then bring up to a volume of 25 ml with deionized water).

B. 200 mM Citrate Buffer, pH 5.0 at 37°C (Prepare 200 ml in deionized water using Citric Acid, Free Acid, Anhydrous. Adjust to pH 5.0 at 37°C with 1 M NaOH).

C. 1.6% (w/v) Stannous Chloride Solution ($SnCl_2$) (Prepare 12 ml in Reagent B using Stannous Chloride, Anhydrous).

D. Ethylene Glycol Monomethyl Ether (Use Ethylene Glycol Monomethyl Ether).

E. 2% (w/v) Ninhydrin Color Reagent (NCR) (Prepare 50 ml by dissolving 1 g of Ninhydrin, in 25 ml of Reagent D. Then add 25 ml of Reagent B. Store in an amber colored bottle).
F. 50% (v/v) 1-Propanol Solution (Prepare 100 ml in deionized water using 1-Propanol).
G. 0.5 mM Cobalt Chloride Solution ($CoCl_2$) (Prepare 20 ml in deionized water using Cobalt Chloride, Hexahydrate).
H. 100 mM Barbital Buffer, pH 8.0 at 37°C (Prepare 25 ml in deionized water using Barbital, Sodium Salt. Adjust to pH 8.0 at 37°C with 1 M HCl).
I. 0.8 mM L-Methionine Standard Solution (Std. Soln) (Prepare 20 ml in deionized water using L-Methionine).
J. Acylase I Enzyme Solution (Immediately before use, prepare a solution containing 0.1 unit/ml of Acylase I in cold deionized water).

Procedure

Step 1: Pipette (in milliliters) the following reagents into a suitable container:

	Test
Reagent J (Enzyme Solution)	1.00
Reagent H (Barbital Buffer)	2.00
Reagent G ($CoCl_2$)	1.00

Mix by swirling and equilibrate to 37°C. Then add:

Reagent A (NAMET)	1.00

Immediately mix by swirling and incubate at 37°C for exactly 30 minutes. Remove 1 ml from the Test and place into a glass-stoppered test tube. Heat in a boiling water bath for 3 minutes.

Step 2: Pipette (in milliliters) the following reagents into a suitable container.

	Blank
Reagent J (Enzyme Solution)	1.00
Reagent H (Barbital Buffer)	2.00
Reagent G ($CoCl_2$)	1.00

Mix by swirling and equilibrate to 37°C. Then add:

Reagent A (NAMET)	1.00

Mix by swirling and immediately remove 1 ml from the Blank and place into a glass-stoppered test tube. Heat in a boiling water bath for 3 minutes. Cool the Test and Blank solutions which have been boiled by using running water. Then add:

	Test	Test Blank	Std. 1	Std. 2	Std. 3	Std. 4	Std. 5	Std. Blank
Test Solution	—	1.00	—	—	—	—	—	—
Blank Solution	—	—	1.00	—	—	—	—	—
Reagent I (Std Soln)	—	—	0.10	0.20	0.40	0.60	1.00	—
Deionized Water	—	—	0.90	0.80	0.60	0.40	—	1.00
Reagent E (NCR)	2.00	2.00	2.00	2.00	2.00	2.00	2.00	2.00
Reagent C ($SnCl_2$)	0.10	0.10	0.10	0.10	0.10	0.10	0.10	0.10

Mix by swirling and place the vials in a boiling water bath for 20 minutes. Remove the vials and allow to cool at room temperature. Add 10 ml of Reagent F (1-Propanol) to each vial. Mix well and transfer the vial contents to suitable cuvettes. Determine the absorbance at 570 nm for each of the vials using a suitable spectrophotometer.

Standard Curve: (A570nm Standard = A570nm Standard–A570nm Standard Blank) Prepare a standard curve by plotting the A570nm of the L-Methionine Standard Solution versus micromoles of L-Methionine.

Sample Determination: (ΔA570nm Sample = A570nm Test–A570nm Sample Blank) Determine the μmoles of L-Methionine liberated using the Standard curve.

$$\text{Units / ml enzyme} = \frac{(\mu\text{moles of L - Methionine liberated}) (5) (df)}{(1) (1)}$$

5 = Total volume (in milliliters) of assay
df = Dilution factor
1 = Volume (in milliliter) of enzyme used
1 = Volume (in milliliter) of sample used in Colorimetric Determination

$$\text{Units / mg solid} = \frac{\text{units / ml enzyme}}{\text{mg solid / ml enzyme}}$$

$$\text{Units / mg protein} = \frac{\text{units / ml enzyme}}{\text{mg protein / ml enzyme}}$$

Unit Definition: *One unit will hydrolyze 1.0 μmole of N-acetyl-L-methionine per hour at pH 8.0 at 37°C.*

Final Assay Concentration

In a 5.00 ml reaction mix, the final concentrations are 40 mM barbital, 20 mM N-acetyl-L-methionine, 0.1 mM cobalt chloride, and 0.1 unit acylase I.

Enzymatic Assay of ACYLASE I (EC 3.5.1.14) from Porcine

Principle

$$\text{N - Acetyl - L - Methionine} + H_2O \xrightarrow{\text{Acylase I}} \text{L - Methionine} + \text{Acetic Acid}$$

Conditions: T = 25°C, pH = 7.0, A238nm, Light path = 1 cm

Method: Continuous Spectrophotometric Rate Determination

Reagents

A. 100 mM Potassium Phosphate Buffer, pH 7.0 at 25°C (Prepare 100 ml in deionized water using Potassium Phosphate, Monobasic, Anhydrous. Adjust to pH 7.0 at 25°C with 1 M KOH).
B. 15 mM N-Acetyl-L-Methionine Solution (NAMet) (Prepare 20 ml in Reagent A using N-Acetyl-L-Methionine).
C. Acylase I Enzyme Solution (Immediately before use, prepare a solution containing 1500–2500 units/ml of Acylase I in Reagent A. Let stand for 5–10 minutes at room temperature prior to assaying)

Procedure

Pipette (in milliliters) the following reagents into suitable quartz cuvettes:

	Test	Blank
Reagent B (NAMet)	2.90	2.90

Equilibrate to 25°C. Monitor the A238nm until constant, using a suitably thermostatted spectrophotometer. Then add:

Reagent A (Buffer)	—	0.10
Reagent C (Enzyme Solution)	0.10	—

Immediately mix by inversion and record the decrease in A238nm for approximately 5 minutes. Obtain the ΔA238nm/minute using the maximum linear rate for both the Test and Blank.

Calculations

$$\text{Units / ml enzyme} = \frac{(\Delta A238\text{nm / min Test} - \Delta A238\text{nm / min Blank})\,(60)\,(3)\,(\text{df})}{(0.019)\,(0.1)}$$

60 = Conversion factor from minutes to one hour as per the Unit Definition
3 = Total Volume (in milliliters) of assay
df = Dilution factor
0.019 = Millimolar extinction coefficient3 of N-Acetyl-L-Methionine at 238nm
0.1 = Volume (in milliliter) of enzyme used

$$\text{Units / mg solid} = \frac{\text{units / ml enzyme}}{\text{mg solid / ml enzyme}}$$

$$\text{Units / mg protein} = \frac{\text{units / ml enzyme}}{\text{mg protein / ml enzyme}}$$

Unit Definition: *One unit will hydrolyze 1.0 µmole of N-acetyl-L-methionine per hour at pH 7.0 at 25°C.*

Final Assay Concentration

In a 3.00 ml reaction mix, the final concentrations are 100 mM potassium phosphate, 14 mM N-acetyl-L-methionine and 150–250 units acylase I.

Note

1. This assay is not to be used to assay Acylase I, from *Aspergillus melleus.*

Enzymatic Assay of ADENOSINE 5'-TRIPHOSPHATASE (EC 3.6.1.3) From Dog and Rabbit Kidney

Principle

$$ATP + H_2O \xrightarrow{ATPase} ADP + Pi$$

Abbreviations Used

ATPase = Adenosine 5'-Triphosphatase
ATP = Adenosine 5'-Triphosphate
ADP = Adenosine 5'-Diphosphate
Pi = Inorganic Phosphate

Conditions: T = 37°C, pH 7.4, A660nm, Light path = 1 cm

Method: Colorimetric

Reagents

A. 50 mM Tris HCl Buffer, pH 7.4 at 37°C (Prepare 100 ml in deionized water using Trizma Base. Adjust to pH 7.4 at 37°C with 1 M HCl).
B. 15 mM Ouabain Solution (Prepare 10 ml in deionized water using Ouabain, Octahydrate).
C. 600 mM Potassium Chloride Solution (KCl) (Prepare 10 ml in deionized water using Potassium Chloride).
D. 2.5 M Sodium Chloride Solution (NaCl) (Prepare 10 ml in deionized water using Sodium Chloride).
E. 90 mM Magnesium Chloride Solution ($MgCl_2$) Prepare 10 ml in deionized water using Magnesium Chloride Hexahydrate).
F. 80 mM Adenosine 5'-Triphosphate (ATP) (Prepare 10 ml in deionized water using Adenosine 5'-Triphosphate, Tris. Adjust to pH 7.4 at 37°C with 1 M Tris).
G. 10% (w/v) Ammonium Molybdate Solution (Prepare 25 ml in 10 N H_2SO_4 using Molybdic Acid, Ammonia Tetrahydrate Salt).
H. Taussky-Shorr Color Reagent (TSCR) (Prepare by adding 10 ml of Reagent G to 70 ml of deionized water. Add 5 g of Ferrous Sulfate, Heptahydrate. Bring the volume to 100 ml with deionized water).
I. 20% (w/v) Trichloroacetic Acid (TCA) (Prepare 100 ml in deionized water using Trichloroacetic Acid, 6.1 N, approximately 100% (w/v)).
J. Phosphorus Standard (P Std) (Use Phosphorus Standard Solution. The concentration is 20 μg/ml, 0.645 μmoles/ml).
K. Adenosine 5'-Triphosphatase Enzyme Solution (Immediately before use, prepare a solution in cold deionized water containing 4–5 units/ml).

Procedure

Step 1: Pipette (in milliliters) the following reagents into suitable containers:

	Test 1	Blank 1	Test 2	Blank 2
Reagent A (Buffer)	0.90	0.90	0.90	0.90
Reagent B (Ouabain)	—	—	0.10	0.10
Reagent C (KCl)	0.50	0.50	0.50	0.50
Reagent D (NaCl)	0.10	0.10	0.10	0.10
Reagent E ($MgCl_2$)	0.05	0.05	0.05	0.05
Reagent K (Enzyme)	0.10	—	0.10	—
Deionized Water	0.20	0.20	0.10	0.10

Mix and equilibrate for 5 minutes at 37°C. Then add:

Reagent F (ATP)	0.10	0.10	0.10	0.10

Mix and incubate at 37°C for exactly 5 minutes. Then add:

Reagent I (TCA)	1.50	1.50	1.50	1.50

Immediately mix by inversion. Then add:

Reagent K (Enzyme Solution)	—	0.10	—	0.10

Mix by inversion and then centrifuge in a clinical centrifuge for 3 minutes to clarify.

Step 2: Pipette (in milliliters) the following into suitable tubes:

Reagent H (TSCR)	2.00	2.00	2.00	2.00
Test Supernatant	1.00	—	1.00	—
Blank Supernatant	—	1.00	—	1.00
Deionized Water	1.00	1.00	1.00	1.00

Mix and incubate at 25°C for 10 minutes. Read the A660nm for both Tests and Blanks.

Standard Curve: Prepare a standard curve by pipetting the following into suitable tubes (milliliters).

	Std 1	Std 2	Std 3	Blank
Reagent H (TSCR)	2.00	2.00	2.00	2.00
Reagent J (P Std)	0.25	0.50	1.00	—
Reagent I (TCA)	0.50	0.50	0.50	0.50
Deionized Water	1.25	1.00	0.50	1.50

Mix and incubate at 25°C for 10 minutes. Read the A660nm for each standard using a suitable spectrophotometer.

Calculations

Standard Curve: ΔA660nm Standard = A660nm Standard–A660nm Standard Blank. Prepare a standard curve by plotting ΔA660nm Standard vs Phosphate concentration.

Sample Determination: ΔA660nm Test 1 = A660nm Test 1–A660nm Blank 1

ΔA660nm Test 2 = A660nm Test 2–A660nm Blank 2

Determine the micromoles of Phosphate liberated for each Test using the standard curve.

$$\text{Units / ml enzyme} = \frac{(\mu\text{moles of Phosphate released})\,(3.0)\,(\text{df})}{(5)\,(0.1)\,(1.0)}$$

1.0 = Aliquot (in milliliter) of Test Supernatant used in Step 2
3.0 = Total volume (in milliliters) of assay (Step 1)
df = Dilution factor
5 = Time of assay (in minutes) as per the Unit Definition
0.1 = Volume (in milliliter) of enzyme used in Step 1

$$\text{Units / mg solid} = \frac{\text{units / ml enzyme}}{\text{mg solid / ml enzyme}}$$

$$\text{Units / mg protein} = \frac{\text{units / ml enzyme}}{\text{mg protein / ml enzyme}}$$

Test 1 = ATPase, activated (Na, K, Mg)
Test 2 = ATPase, not Ouabain sensitive

Ouabain sensitive = Test 1–Test 2

Unit Definition: *One unit will liberate 1.0 μmole of inorganic phosphorus from ATP per minute at pH 7.4 at 37°C in the presence of Na^+, K^+, and Mg^{++}.*

Final Assay Concentration

In a 1.50 ml reaction mix, the final concentrations are 30 mM Tris, 3 mM magnesium chloride, 5.3 mM adenosine 5'-triphosphate, 167 mM sodium chloride, 20 mM potassium chloride, 1 mM Ouabain (when present), and 0.1–0.2 unit ATPase.

Enzymatic Assay of ADENOSINE 5'-TRIPHOSPHATASE (EC 3.6.1.3)

Principle

$$\text{ATP} + H_2O \xrightarrow{\text{ATPase}} \text{ADP} + \text{Pi}$$

Abbreviations Used

ATPase = Adenosine 5'-Triphosphatase
ATP = Adenosine 5'-Triphosphate
ADP = Adenosine 5'-Diphosphate
Pi = Inorganic Phosphate

Conditions: T = 37°C, pH 7.8, A660nm, Light path = 1 cm

Method: Colorimetric

Reagents

A. 24 mM Tris HCl Buffer with 0.68 mM Ethylenediaminetetraacetic Acid and 6.0 mM Magnesium Chloride, pH 7.8 at 37°C (Prepare 100 ml in deionized water using Trizma Base, Ethylenediaminetetraacetic Acid, Free Acid, and Magnesium Chloride, Hexahydrate. Adjust to pH 7.8 at 37°C with 1 M HCl).
B. 15 mM Ouabain Solution (Ouabain) (Prepare 10 ml in Reagent A using Ouabain Octahydrate).
C. 2 M Sodium Chloride Solution (NaCl) (Prepare 10 ml in deionized water using Sodium Chloride).
D. 45 mM Potassium Chloride and 2 M Sodium Chloride Solution (KCl/NaCl) (Prepare 10 ml in Reagent C, using Potassium Chloride).
E. 80 mM Adenosine 5'-Triphosphate Solution (ATP) (Prepare 10 ml in deionized water using Adenosine 5'-Triphosphate, Tris Salt. Adjust to pH 7.8 at 37°C with 1 M Tris).
F. Taussky-Shorr Reagent (Prepare by adding 10 ml of 10% Ammonium Molybdate, Tetrahydrate, in 10 NH_2SO_4, to 70 ml deionized water; then add a 5 g vial of Ferrous Sulfate, Heptahydrate. Bring the volume to 100 ml with deionized water. Store in an amber container).
G. 20% (w/v) Trichloroacetic Acid (TCA) (Prepare 100 ml in deionized water using Trichloroacetic Acid Solution, approximately 100% (w/v) 6.1 N).
H. Phosphorus Standard (P Std) (Use Phosphorus Standard Solution. The concentration is 20 µg/ml, 0.645 µmole/ml).
I. Adenosine 5'-Triphosphatase Enzyme Solution (Immediately before use, prepare a solution in cold deionized water containing 0.3–0.6 unit/ml).

Procedure

Step 1: Pipette (in milliliters) the following reagents into suitable containers:

	Test 1	Blank 1	Test 2	Blank 2
Reagent A (Buffer)	1.25	1.25	1.15	1.15
Reagent B (Ouabain)	—	—	0.10	0.10
Reagent C (NaCl)	—	—	—	—
Reagent D (NaCl/KCI)	0.10	0.10	0.10	0.10
Reagent I (Enzyme Solution)	0.10	—	0.10	—

Mix and equilibrate for several minutes at 37°C. Then add:

Reagent E (ATP)	0.05	0.05	0.05	0.05

Mix and incubate at 37°C for exactly 15 minutes. Then add:

Reagent G (TCA)	1.50	1.50	1.50	1.50

Immediately mix by inversion. Then add:

Reagent I (Enzyme Solution)	—	0.10	—	0.10

Mix by inversion and then centrifuge for 3 minutes to clarify.

Step 2: Pipette (in milliliters) the following reagents into suitable tubes:

	Std 1	Std 2	Std 3	Blank
Reagent F (Taussky-Shorr)	2.00	2.00	2.00	2.00
Test Supernatant	1.00	—	1.00	—
Blank Supernatant	—	1.00	—	1.00
Deionized Water	1.00	1.00	1.00	1.00

Mix and incubate at 25°C for 5 minutes. Read the A660nm for both Tests and Blanks.

Standard Curve: Prepare a standard curve by pipetting (in milliliters) the following reagents into suitable tubes:

Reagent F (Taussky-Shorr)	2.00	2.00	2.00	2.00
Reagent H (P Std)	0.25	0.50	1.00	—
Reagent G (TCA)	0.50	0.50	0.50	0.50
Deionized Water	1.25	1.00	0.50	1.50

Mix and incubate at 25°C for 10 minutes. Read the A660nm for the Standard and Standard Blank using a suitable spectrophotometer.

Calculations

Standard Curve: ΔA660nm Standard = A660nm Standard–A660nm Standard Blank

Prepare a standard curve by plotting the ΔA660nm of the Standard vs μmoles of Phosphate.

Sample Determination: ΔA660nm Test = A660nm Test–A660nm Test Blank
Determine the micromoles of Phosphate liberated using the standard curve.

$$\text{Units / ml Protein} = \frac{(\mu\text{moles of Phosphate released})\ (3.0)\ (\text{df})}{(15)\ (0.1)\ (1.0)}$$

3.0 = Total volume (in milliliters) of Step 1
df = Dilution factor
15 = Time (in minutes) of assay as per the Unit Definition
1.0 = Aliquot of Test Supernatant used in Step 2
0.1 = Volume (in milliliter) of enzyme used

Test 1 = ATPase, activated (Na, K, Mg)
Test 2 = ATPase, not Ouabain sensitive

Ouabain sensitive = Test 1–Test 2

$$\text{Units / mg solid} = \frac{\text{units / ml enzyme}}{\text{mg solid / ml enzyme}}$$

$$\text{Units / mg protein} = \frac{\text{units / ml enzyme}}{\text{mg protein / ml enzyme}}$$

Unit Definition: *One unit will liberate 1.0 μmole of inorganic phosphorus from ATP per minute at pH 7.8 at 37°C in the presence of* Na^{+}, K^{+}, *and* Mg^{++}.

Final Assay Concentration

In a 1.50 ml reaction mix, the final concentrations are 20 mM Tris, 0.57 mM Ethylenediaminetetraacetic Acid, 5 mM magnesium chloride, 3 mM adenosine 5'-triphosphate, 133 mM sodium chloride and 3 mM potassium chloride, 0.03–0.06 unit adenosine 5'-triphosphatase and 1 mM ouabain (when present).

Enzymatic Assay of ADENOSINE 5'-TRIPHOSPHATE SULFURYLASE (EC 2.7.7.4)

Principle

$$\text{APS} + \text{PPi} \xrightarrow{\text{ATPS}} \text{ATP} + \text{SO}_4$$

$$\text{ATP} + \text{D-Glucose} \xrightarrow{\text{HK}} \text{D-Glucose 6-Phosphate} + \text{ADP}$$

$$\text{D-Glucose 6-Phosphate} + \beta\text{-NADP} \xrightarrow{\text{G-6-PDH}} \text{6-PG} + \beta\text{-NADPH}$$

Abbreviations Used

APS = Adenosine 5'-Phosphosulfate
PPi = Inorganic Pyrophosphate
ATPS = Adenosine 5'-Triphosphate Sulfurylase
ATP = Adenosine 5'-Triphosphate
HK = Hexokinase
ADP = Adenosine 5'-Diphosphate
β-NADP = β-Nicotinamide Adenine Dinucleotide Phosphate, Oxidized Form
G-6-PDH = Glucose 6-Phosphate Dehydrogenase
6-PG = 6-Phospho-D-Gluconate
β-NADPH = β-Nicotinamide Adenine Dinucleotide Phosphate, Reduced Form

Conditions: T = 30°C, pH 8.0, A340nm, Light path = 1 cm.

Method: Continuous Spectrophotometric Rate Determination

Reagents

A. 400 mM Tris Buffer, pH 8.0 at 30°C (Prepare 50 ml in deionized water using Trizma Base. Adjust to pH 8.0 at 30°C with 1 M Acetic Acid).
B. 200 mM β-Nicotinamide Adenine Dinucleotide Phosphate, Oxidized Form, Solution (β-NADP) (Prepare 1 ml in deionized water using β-Nicotinamide Adenine Dinucleotide Phosphate, Sodium Salt).
C. 1 M Magnesium Acetate Solution (MgOAc) (Prepare 1 ml in deionized water using Magnesium Acetate, Tetrahydrate).
D. 1 M D-Glucose Solution (Glucose) (Prepare 2 ml in deionized water using D-(+)Glucose, Anhydrous).
E. 20 mM Adenosine 5'-Phosphosulfate (APS) (Immediately before use, prepare 1 ml in deionized water using Adenosine 5'-Phosphosulfate, Sodium Salt).
F. 100 mM Pyrophosphate Solution, pH 8.0 at 30°C (PPi) (Prepare 5 ml in deionized water using Tetrasodium Pyrophosphate, Decahydrate. Adjust to pH 8.0 at 30°C with 1 M HCl).

G. Hexokinase and Glucose 6-Phosphate Dehydrogenase Enzyme Solution (HK/G-6-PDH) (Immediately before use, prepare a solution containing approximately 20 units/ml of Glucose 6-Phosphate Dehydrogenase using Hexokinase and Glucose 6-Phosphate Dehydrogenase in cold deionized water).

H. Adenosine 5'-Triphosphate Sulfurylase Enzyme Solution (ATPS) (Immediately before use, prepare a solution containing 0.2–0.6 unit/ml of Adenosine 5'-Triphosphate Sulfurylase in cold deionized water).

Procedure

Prepare a reaction cocktail by pipetting (in milliliters) the following reagents into a suitable container.

Reagent A (Buffer)	34.00
Reagent B (β-NADP)	0.34
Reagent C (MgOAc)	0.28
Deionized Water	65.38

Mix by swirling. Adjust to pH 8.0 at 30°C if necessary with either 1 M HCl or 1 M NaOH. Pipette (in milliliters) the following reagents into suitable cuvettes:

	Test	Blank
Reaction Cocktail	2.50	2.50
Reagent D (Glucose)	0.10	0.10
Reagent G (HK/G-6-PDH)	0.10	0.10
Reagent E (APS)	0.05	0.05
Reagent H (ATPS)	0.10	—
Deionized Water	—	0.10

Mix by inversion and equilibrate to 30°C. Monitor the A340nm until constant, using a suitably thermostatted spectrophotometer. Then add:

Reagent F (PPi)	0.10	0.10

Immediately mix by inversion and record the increase in A340nm for approximately 5 minutes. Obtain the ΔA340nm/minute using the maximum linear rate for both the Test and Blank.

Calculations

$$\text{Units / ml enzyme} = \frac{(\Delta\text{A340nm / min Test} - \Delta\text{A340nm / min Blank})\ (2.95)\ (\text{df})}{(6.22)\ (0.1)}$$

2.95 = Total volume (in milliliters) of assay
df = Dilution factor
6.22 = Millimolar extinction coefficient of β-NADPH at 340 nm
0.1 = Volume (in milliliter) of enzyme used

$$\text{Units / mg solid} = \frac{\text{units / ml enzyme}}{\text{mg solid / ml enzyme}}$$

$$\text{Units / mg protein} = \frac{\text{units / ml enzyme}}{\text{mg protein / ml enzyme}}.$$

Unit Definition: *One unit will produce 1.0 µmole of ATP from APS and inorganic pyrophosphate per minute at pH 8.0 at 30°C.*

Final Assay Concentration

In a 2.95 ml reaction mix, the final concentrations are 115 mM Tris, 0.58 mM β-nicotinamide adenine dinucleotide phosphate, 2.4 mM magnesium acetate, 34 mM D-glucose, 4 units hexokinase, 2 units glucose 6-phosphate dehydrogenase, 0.3 mM adenosine 5'-phosphosulfate, 0.02–0.06 unit adenosine 5'-triphosphate sulfurylase, and 3.4 mM pyrophosphate.

Notes

1. Hexokinase Unit Definition: One unit will phosphorylate 1.0 µmole of D-glucose per minute at pH 7.6 at 25°C.
2. Glucose 6-Phosphate Dehydrogenase Unit Definition: One unit will oxidize 1.0 µmole of D-glucose 6-phosphate to 6-phospho-D-gluconate per minute in the presence of NADP at pH 7.4 at 25°C.

Enzymatic Assay of ADENOSINE DEAMINASE (EC 3.5.4.4)

Principle

$$\text{Adenosine} + H_2O \xrightarrow{\text{Adenosine Deaminase}} \text{Inosine} + NH_3$$

Conditions: T = 25°C, pH = 7.5, A265nm, Light path = 1 cm.

Method: Continuous Spectrophotometric Rate Determination.

Reagents

A. 100 mM Potassium Phosphate Buffer, pH 7.5 at 25°C (Prepare 150 ml in deionized water using Potassium Phosphate, Monobasic, Anhydrous. Adjust to pH 7.5 at 25°C with 1 M KOH).

B. 1.35 mM Adenosine Solution, pH 7.0 at 25°C. (Prepare 10 ml in deionized water using Adenosine. Adjust to pH 7.0 at 25°C with 10 mM NaOH).

C. 0.1% (w/v) Bovine Serum Albumin Solution (BSA) (Prepare 50 ml in Reagent A using Albumin, Bovine).

D. Adenosine Deaminase Enzyme Solution (Immediately before use, prepare a solution containing 0.20–0.40 unit/ml of Adenosine Deaminase in cold Reagent C).

Procedure

Pipette (in milliliters) the following reagents into suitable quartz cuvettes:

	Test	Blank
Deionized Water	1.30	1.30
Reagent A (Buffer)	1.50	1.50
Reagent B (Adenosine)	0.10	0.10

Mix by inversion and equilibrate to 25°C. Monitor the A265nm until constant, using a suitably thermostatted spectrophotometer. Then add:

	Test	Blank
Reagent C (BSA)	—	0.10
Reagent D (Enzyme Solution)	0.10	—

Immediately mix by inversion and record the decrease in the A265nm for approximately 5 minutes. Obtain the A265nm/minute using the maximum linear rate for both the Test and Blank.

Calculations

$$\text{Units / ml enzyme} = \frac{(\text{A265nm / min Test} - \text{A265nm / min Blank})\ (3)\ (\text{df})}{(8.1)\ (0.1)}$$

3 = Total volume (in milliliters) of assay
df = Dilution Factor
8.1 = Millimolar extinction coefficient of adenosine at 265nm
0.1 = Volume (in milliliters) of enzyme used

$$\text{Units / mg solid} = \frac{\text{units / ml enzyme}}{\text{mg solid / ml enzyme}}$$

$$\text{Units / mg protein} = \frac{\text{units / ml enzyme}}{\text{mg protein / ml enzyme}}$$

Unit Definition: *One unit will deaminate 1.0 µmole of adenosine to inosine per minute at pH 7.5 at 25°C.*

Final Assay Concentration

In a 3.00 ml reaction mix, the final concentrations are 53.3 mM potassium phosphate, 0.045 mM adenosine, 0.003% (w/v) Bovine Serum Albumin and 0.02–0.04 unit of adenosine deaminase.

Note

1. Substrate concentration should not be significantly increased above that listed in the procedure. Higher concentrations result in a deviation from Beers Law leading to errors in the rate determinations.

Enzymatic Assay of S-ADENOSYL-L-HOMOCYSTEINE HYDROLASE (EC 3.3.1.1)

Principle

$$\text{S-Adenosyl-L-Homocysteine} + H_2O \xrightarrow{\text{SAHase}} \text{Adenosine} + \text{L-Homocysteine}$$

$$\text{Adenosine} + H_2O \xrightarrow{\text{Adenosine Deaminase}} \text{Inosine} + NH_3$$

$$\text{Inosine} + \text{Orthophosphate} \xrightarrow{\text{Nucleoside Phosphorylase}} \text{Hypoxanthine} + \text{Ribose-1-Phosphate}$$

$$\text{Hypoxanthine} \xrightarrow{\text{Xanthine Oxidase}} \text{Xanthine}$$

$$\text{Xanthine} + H_2O + O_2 \xrightarrow{\text{Xanthine Oxidase}} \text{Uric Acid} + H_2O_2$$

Abbreviation Used

SAHase = S-Adenosyl-L-Homocysteine Hydrolase

Conditions: T = 37°C, pH = 7.2, A292nm, Light path = 1 cm

Method: Continuous Spectrophotometric Rate Determination

Reagents

A. 100 mM Potassium Phosphate Buffer, pH 7.2 at 37°C (Prepare 100 ml in deionized water using Potassium Phosphate, Monobasic, Anhydrous. Adjust to pH 7.2 at 37°C with 1 M KOH).

B. 5.2 mM S-Adenosyl-L-Homocysteine Solution (SAH) (Prepare 10 ml in deionized water using S-Adenosyl-L-Homocysteine).

C. Adenosine Deaminase Enzyme Solution (ADA) (Use Adenosine Deaminase, Type VI from Calf Intestinal Mucosa, undiluted, approximately 1500 u/ml).

D. Xanthine Oxidase Enzyme Solution (XO) (Use Xanthine Oxidase, Grade I from Buttermilk, undiluted, approximately 20 u/ml).

E. Nucleoside Phosphorylase Enzyme Solution (NP) (Use Nucleoside Phosphorylase, from Calf Spleen, undiluted, approximately 120 u/ml).

F. S-Adenosyl-L-Homocysteine Hydrolase Enzyme Solution (SAHase) (Use undiluted, containing 7.0–20 units/ml)

Procedure

Pipette (in milliliters) the following reagents into suitable cuvettes:

	Test	Blank
Reagent A (Buffer)	3.00	3.00
Reagent B (SAH)	0.03	0.03
Reagent C (ADA)	0.03	0.03
Reagent D (XO)	0.03	0.03
Reagent E (NP)	0.03	0.03

Mix by inversion and equilibrate to 37°C. Monitor the A292nm until constant, using a suitably thermostatted spectrophotometer. Then add:

	Test	Blank
Reagent F (SAHase)	0.30	—
Deionized Water	—	0.30

Immediately mix by inversion and record the increase in A292nm for approximately 10 minutes. Obtain the ΔA292nm/minute using the maximum linear rate for both the Test and Blank.

Calculations

$$\text{Units / ml enzyme} = \frac{(\Delta A292\text{nm / min Test} - \Delta A292\text{nm / min Blank})\,(3.42)\,(\text{df})}{(0.0121)\,(0.1)}$$

3.42 = Volume (in milliliter) of assay
0.0121 = Micromolar extinction coefficient of Uric Acid
0.1 = Volume (in milliliter) of enzyme used

$$\text{Units / mg protein} = \frac{\text{units / ml enzyme}}{\text{mg protein / ml enzyme}}$$

Unit Definition: *One unit will hydrolyze 1.0 nanomole of S-adenosyl-L-homocysteine to adenosine and L-homocysteine per minute at pH 7.2 at 37°C.*

Final Assay Concentration

In a 3.42 ml reaction mix, the final concentrations are 93 mM potassium phosphate, 48.4 μM S-adenosyl-L-homocysteine, 45 units adenosine deaminase, 0.6 unit xanthine oxidase, 3.6 units nucleoside phosphorylase, and 0.7–2.0 units S-adenosyl-L-homocysteine lyase.

Enzymatic Assay of ADENYLATE CYCLASE (EC 4.6.1.1)

Principle

$$\text{ATP} \xrightarrow{\text{Adenylate Cyclase}} 3',\ 5'\text{-cyclic AMP} + \text{Pyrophosphate}$$

Abbreviations Used

ATP = Adenosine 5'-Triphosphate
3',5'-cyclic AMP = 3',5'-cyclic Adenosine Monophosphate

Condition: T = 37°C, pH 8.5

Method: HPLC Analysis of Products

Reagents

A. 100 mM Tris HCl Buffer, pH 8.5 at 37°C (Prepare 100 ml in deionized water using Trizma. Adjust to pH 8.5 at 37°C with 1 M HCl).
B. 20 mM Magnesium Chloride Solution ($MgCl_2$) (Prepare 10 ml in deionized water using Magnesium Chloride, Anhydrous).
C. 5.0 mM Sodium Fluoride Solution (NaF) (Prepare 10 ml in deionized water using Sodium Fluoride).
D. 15 mM Adenosine 5'-Triphosphate Solution, pH 8.5 at 37°C (ATP) (Prepare 10 ml in deionized water using Adenosine 5'-Triphosphate, Disodium. Adjust to pH 8.5 at 37°C with 0.1 M NaOH).
E. 5.0 mM Phosphocreatine Solution (PCR) (Prepare 1 ml in deionized water using Phosphocreatine, Disodium, Hydrate).

F. 10 mM Potassium Phosphate and 5 mM Tetrabutylammonium Dihydrogen Phosphate Buffer, pH 5.5 at 25°C (Buffer A) (Immediately before use, prepare 1 L in deionized water using Potassium Phosphate, Monobasic, Anhydrous, and Tetrabutylammonium Dihydrogen Phosphate. Adjust to pH 5.5 at 25°C with 1 M KOH. Degas with nitrogen for 10 minutes).

G. 100% (v/v) Acetonitrile (Buffer B) (Use Acetonitrile. Immediately before use, degas with nitrogen for 10 minutes).

H. 1.8 mM Adenosine 5'-Triphosphate Standard Solution (ATP Std) (Prepare 1 ml in deionized water using Adenosine 5'-Triphosphate, Disodium).

I. Adenosine 3':5'-Cyclic Monophosphate Standard Solution (3':5'c-AMP) (Immediately before use, prepare 1 ml each of standards in deionized water containing 0.3 mM, 0.6 mM, 0.9 mM, 1.5 mM, Adenosine 3':5'-Cyclic Monophosphate, Free Acid).

J. Adenylate Cyclase Enzyme Solution (Immediately before use, prepare a solution containing 50–100 units/ml in cold Reagent A).

Procedure

Step 1: Pipette (in milliliters) the following reagents into suitable containers:

	Test	Blank
Reagent A (Buffer)	0.50	0.50
Reagent B ($MgCl_2$)	0.10	0.10
Reagent C (NaF)	0.10	0.10
Reagent D (ATP)	0.10	0.10
Reagent E (PCR)	0.10	0.10

Mix by swirling and incubate in a water bath at 37°C for 10 minutes. Then add:

Reagent J (Enzyme Solution)	0.02	—
Deionized Water	0.08	0.10

Mix by inversion and incubate at 37°C for exactly 60 minutes. Then place the Test, Blank, and Standard solutions into a boiling water bath for one minute. Allow to cool, then filter through a 0.22 μm filter.

Step 2: HPLC Analysis

Column: Vydac C18-TBAP, 25 cm × 4.6 mm.
Injection Volume: 20 μl.
Standard: Reagent I (Adenosine 3':5'-Cyclic Monophosphate)
Flow Rate: 1.5 ml/minute
Wavelength: 260 nm
Buffer A: 10 mM Potassium Phosphate and 5 mM Tetrabutylammonium Dihydrogen Phosphate Buffer, pH 5.5 at 25°C.
Buffer B: 100% (v/v) Acetonitrile Equilibrate the column with 95% Buffer A and 5% Buffer B for 30 minutes.

HPLC Program

	Time (Minutes)					
	0	25	26	35	36	45
Buffer A	95%	70%	0%	0%	95%	95%
Buffer B	5%	30%	100%	100%	5%	5%

Inject 20 µl each of Test, Blank, and Standard. Use at least three different concentrations of the Test. Prepare a standard curve by plotting the peak areas of the standards versus µmoles of 3':5'c-AMP. Determine the micromoles of 3':5'c-AMP liberated using the standard curve.

Calculation

$$\text{Units / ml enzyme} = \frac{(\mu\text{moles of 3' : 5' c-AMP})\,(1)\,(\text{df})}{(60)\,(0.001)\,(0.02)}$$

1 = Volume (in milliliter) of assay
df = Dilution factor
60 = Time (in minutes) of assay as per the Unit Definition
0.001 = Conversion factor from µmoles to nanomoles as per the Unit Definition
0.02 = Volume (in milliliter) of enzyme used

$$\text{Units / mg protein} = \frac{\text{units / ml enzyme}}{\text{mg protein / ml enzyme}}$$

Unit Definition: *One unit will form 1.0 nanomole of cyclic AMP from ATP per minute at pH 8.5 at 37°C.*

Final Assay Concentration

In a 1.00 ml reaction mix, the final concentrations are 52 mM Tris, 2 mM magnesium chloride, 0.5 mM sodium fluoride, 0.5 mM phosphocreatine, 1.5 mM adenosine 5'-triphosphate, and 1–2 units adenylate cyclase.

Notes

1. Sodium Fluoride is included in the assay to inhibit adenosine 5'-triphosphatase.
2. Phosphocreatine is included in the assay to decrease the inhibition from ADP, via phosphocreatine kinase, which is thought to be present in A-0951.

Enzymatic Assay of 5'-ADENYLIC ACID DEAMINASE (EC 3.5.4.6)

Principle

$$\text{5'-AMP} + H_2O \xrightarrow{\text{5'-Adenylic Acid Deaminase}} \text{5'-IMP} + NH_3$$

Abbreviations Used

5'-AMP = Adenosine 5'-Monophosphate
5'-IMP = Inosine 5'-Monophosphate

Conditions: T = 25°C, pH = 6.5, A265nm, Light path = 1 cm.

Method: Continuous Spectrophotometric Rate Determination.

Reagents

A. 10 mM Sodium Citrate Buffer, pH 6.5 at 25°C. (Prepare 1 liter in deionized water using Sodium Citrate, Trisodium Salt, Dihydrate. Adjust to pH 6.5 at 25°C with 1 M HCl).

B. 0.04 mM Adenosine 5'-Monophosphate Solution (5'-AMP) (Prepare 50 ml in Reagent A using Adenosine 5'-Monophosphate, Sodium Salt. *Prepare Fresh*).

C. 1 M Potassium Chloride Solution (KCl) (Prepare 50 ml in deionized water using Potassium Chloride. *Prepare Fresh*).

D. 5'-Adenylic Acid Deaminase Enzyme Solution (Immediately before use, prepare a solution containing 0.2–0.4 unit/ml of 5'-Adenylic Acid Deaminase in cold Reagent C).

Procedure

Pipette (in milliliters) the following reagents into suitable quartz cuvettes:

	Test	Blank
Reagent B (5'-AMP)	2.90	2.90

Equilibrate to 25°C. Monitor the A265nm until constant, using a suitably thermostatted spectrophotometer. Then add:

Reagent C (KCl)	—	0.10
Reagent D (Enzyme Solution)	0.10	—

Immediately mix by inversion and record the decrease in A265nm for approximately 5 minutes. Obtain the ΔA265nm/minute using the maximum linear rate for both the Test and Blank.

Calculations

$$\text{Units / ml enzyme} = \frac{(\text{A265nm / min} - \text{A265nm / min Blank})\ (3)\ (\text{df})}{(8.1)\ (0.1)}$$

3 = Total volume (in milliliters) of assay
df = Dilution factor
8.1 = Millimolar extinction coefficient of 5'-AMP at 265nm
0.1 = Volume (in milliliters) of enzyme used

$$\text{Units / mg solid} = \frac{\text{units / ml enzyme}}{\text{mg solid / ml enzyme}}$$

$$\text{Units / mg protein} = \frac{\text{units / ml enzyme}}{\text{mg protein / ml enzyme}}$$

Unit Definition: *One unit will deaminate 1.0 µmole of 5'-AMP to 5'-IMP per minute at pH 6.5 at 25°C.*

Final Assay Concentration

In a 3.00 ml reaction mix, the final concentrations are 9.7 mM sodium citrate, 0.04 mM adenosine 5'- monophosphate, 33 mM potassium chloride and 0.02–0.04 unit 5'-adenylic acid deaminase.

Note

1. The substrate concentration should not be significantly increased above that listed in the procedure. Higher concentrations result in a deviation from Beers Law leading to errors in the rate determination.

Enzymatic Assay of ADENYLOSUCCINATE LYASE (EC 4.3.2.2)

Principle

$$\text{Adenylosuccinic Acid} \xrightarrow{\text{Adenylosuccinate Lyase}} \text{Fumaric Acid} + \text{5'-AMP}$$

Abbreviation Used

5'-AMP = Adenosine 5'-Monophosphate

Conditions: T = 25°C, pH = 7.0, A280nm, Light path = 1 cm

Method: Continuous Spectrophotometric Rate Determination

Reagents

A. 50 mM Potassium Phosphate Buffer with 1 mM Ethylenediaminetetraacetic Acid (EDTA), pH 7.0 at 25°C (Prepare 100 ml in deionized water using Potassium Phosphate, Monobasic Anhydrous and Ethylenediaminetetraacetic Acid, Disodium Salt, Dihydrate. Adjust to pH 7.0 at 25°C with 1 M KOH or 1 M NaOH).

B. 1.72 mM Adenylosuccinic Acid (ASA) (Prepare 10 ml in Reagent A using Adenylosuccinic Acid, Free Acid. *Prepare Fresh*).

C. Adenylosuccinate Lyase Enzyme Solution (Immediately before use, prepare a solution containing 0.2–0.4 unit/ml of Adenylosuccinate Lyase in cold Reagent A).

Procedure

Pipette (in milliliters) the following reagents into suitable cuvettes:

	Test	Blank
Reagent A (Buffer)	2.80	2.80
Reagent B (ASA)	0.10	0.10

Mix by inversion and equilibrate to 25°C. Monitor the A280nm until constant, using a suitably thermostatted spectrophotometer. Then add:

	Test	Blank
Reagent C (Enzyme Solution)	0.10	—
Reagent A (Buffer)	—	0.10

Immediately mix by inversion and record the decrease in A280nm for approximately 5 minutes. Obtain the ΔA280nm/minute using the maximum linear rate for both the Test and Blank.

Calculations

$$\text{Units/ml enzyme} = \frac{(\Delta A280\text{nm/min} - \Delta A280\text{nm/min Blank})\,(3)\,(\text{df})}{(10.7)\,(0.1)}$$

3 = Total volume (in milliliters) of assay
df = Dilution factor
10.7 = Change in absorbance at 280nm of one micromole per ml of Adenylosuccinic Acid to Fumaric Acid and Adenosine 5'-Monophosphate
0.1 = Volume (in milliliter) of enzyme used

$$\text{Units / mg solid} = \frac{\text{units / ml enzyme}}{\text{mg solid / ml enzyme}}$$

$$\text{Units / mg protein} = \frac{\text{units / ml enzyme}}{\text{mg protein / ml enzyme}}$$

Unit Definition: *One unit will convert 1.0 μmole of adenylosuccinic acid to fumaric acid and 5'-AMP per minute at pH 7.0 at 25°C.*

Final Assay Concentration

In a 3.00 ml reaction mix, the final concentrations are 50 mM potassium phosphate, 1.0 mM Ethylenediaminetetraacetic acid, 0.057 mM adenylosuccinic acid and 0.02–0.04 unit of adenylosuccinate lyase.

Enzymatic Assay of ADP-RIBOSYLCYCLASE

Principle

$$\beta\text{-NAD} \xrightarrow{\text{ADPR}} \text{cyclic ADP-Ribose}$$

Abbreviations Used

β-NAD = β-Nicotinamide Adenine Dinucleotide, Oxidized Form
ADPR = ADP-Ribosylcyclase
cyclic ADP = cyclic Adenosine Diphosphate

Conditions: T = 25°C, pH = 7.0

Method: HPLC Analysis of Products

Reagents

A. 20 mM Potassium Phosphate Buffer, pH 7.0 at 25°C (Prepare 100 ml in deionized water using Potassium Phosphate Monobasic, Anhydrous. Adjust to pH 7.0 at 25°C with 1 M KOH).
B. 6.67 mM β-Nicotinamide Adenine Dinucleotide Solution (β-NAD) (Prepare 1 ml in Reagent A using β-Nicotinamide Adenine Dinucleotide. *Prepare Fresh*).
C. 1 M Acetic Acid Solution (HOAc) (Prepare 10 ml in deionized water using Acetic Acid, Glacial).
D. 10 mM Potassium Phosphate and 500 mM Acetic Acid Solution (Std Dil) (Prepare 5 ml in deionized water using Potassium Phosphate, Monobasic, Anhydrous, and Acetic acid, Glacial).
E. 250 μg/ml Cyclic Adenosine Diphosphate-Ribose Standard Solution (cADP Ribose) (Prepare 1 ml by dissolving a 250 μg vial of Cyclic Adenosine Diphosphate-Ribose, in Reagent D. To confirm the concentration, add 0.1 ml of this solution to 1 ml of Reagent A and determine the concentration using the extinction coefficient.

Centrifuge the remaining solution of C-7323 in an Eppendorf tube at high speed for 10 minutes in a microcentrifuge. This solution is injected into the HPLC instrument as a standard and the concentration of the standard solution should be about 250 μg/ml.).

F. 100 mM Ammonium Phosphate Solution (Prepare 100 ml in deionized water using Ammonium Phosphate, Monobasic).

G. 25% (v/v) Acetonitrile Solution (Prepare 100 ml in deionized water using Acetonitrile).

H. ADP-Ribosylcyclase Enzyme Solution (Immediately before use, prepare a solution containing 0.5–1.0 unit/ml of ADP-Ribosylcyclase in cold Reagent A).

Procedure

Step 1: Pipette (in milliliters) the following reagents into Eppendorf tubes:

	Test	Blank
Reagent B (β-NAD)	0.15	0.15

Equilibrate to 25°C. Then add:

Reagent H (Enzyme Solution)	0.05	—

Mix by swirling and incubate at 25°C for exactly 5 minutes. Then add:

Reagent C (HOAc)	0.20	0.20
Reagent H (Enzyme Solution)	—	0.05

Mix by swirling and place on ice. Centrifuge the Test and Blank at high speed in a microcentrifuge for 10 minutes. Assay the supernatants for cyclic ADP-ribose by the HPLC assay described in Step 2.

Step 2: HPLC Analysis

Column: Supelcosil LC-18, Supelco Catalog No. 5–8298, 25 mm × 4.6 mm.

Injection Volume: 20 μL.

Standard: cADP-Ribose (Reagent E).

Flow Rate: 1.5 ml/minute.

Wavelength: 254 nm.

Attenuation: 8.

Buffer A: Ammonium Phosphate (Reagent F).

Buffer B: Acetonitrile (Reagent G).

HPLC Program: Equilibrate the column with 100% Reagent A.

Time (Minutes)	Function %	Buffer B
0.01	Inject Sample	0
10	Buffer	B 5
20	Buffer	B 15
21	Buffer	B 0
31	Stop	

Calculation

$$\text{Units / ml} = \frac{(\text{SPA} \times 10\text{-}6 - \text{BPA} \times 10\text{-}6)\ Z\ (0.4)\ (\text{df})\ (\text{A})}{(0.05)\ (0.02)}$$

SPA = Sample Peak Area
BPA = Blank Peak Area
0.4 = Volume (in milliliter) of assay
df = Dilution factor
0.05 = Volume (in milliliter) of enzyme used
0.02 = Injection volume (in milliliter) for HPLC

$$\text{A} = \frac{(\text{Concentraion of standard } \mu\text{g / ml divided by } 540\ \mu\text{g} / \mu\text{mole})\ (0.02)}{\text{Standard Peak Area} \times 10\text{-}6}$$

Unit Definition: *One unit of ADP ribosylcyclase will produce 1 µmole of cyclic ADP ribose from β-NAD⁺ in 5 minutes at 25°C and pH 7.0.*

Final Assay Concentration

In a 0.20 ml reaction mix, the final concentrations are 20 mM potassium phosphate, 5 mM β-nicotinamide adenine dinucleotide, and 0.025–0.05 unit ADP-ribosylcyclase.

Notes

1. Calculate the standard concentrations as follows: µg/ml = (A254nm/14.3) (Dilution Factor) (540 µg/µmole). 14.3 is the EmM for cyclic ADP-Ribose.
2. A separate Blank must be run for each Test. Store Blanks on ice before adding enzyme.
3. The factor of injection volume can be left out if the sample and standard volumes are the same.

Enzymatic Assay of AGARASE (EC 3.2.1.81)

Principle

$$\text{Agar} + H_2O \xrightarrow{\text{Agarase}} \text{Reducing Sugar (measured as galactose)}$$

Conditions: T = 43°C, pH = 6.0, A540nm, Light path = 1 cm.

Method: Colorimetric.

Reagents

A. 50 mM Potassium Phosphate Buffer, pH 6.0 at 43°C (Prepare 100 ml in deionized water using Potassium Phosphate, Monobasic, Anhydrous. Adjust to pH 6.0 at 43°C with 1 M KOH).

B. 1.5% (w/v) Agar Substrate Solution (Agar) (Prepare 50 ml in Reagent A using Agar).

C. Agarase Enzyme Solution (Immediately before use, prepare a solution containing 20–40 units/ml of Agarase in cold Reagent A).

D. 16 mM Copper Sulfate, 1.3 M Sodium Sulfate, 226 mM Sodium Carbonate, 190 mM Sodium Bicarbonate, and 43 mM Sodium Potassium Tartrate Solution (Copper Soln) (Prepare 1 liter in deionized water using Cupric Sulfate Pentahydrate, Sodium Sulfate, Anhydrous, Sodium Carbonate, Anhydrous, Sodium Bicarbonate, and Sodium Potassium Tartrate Tetrahydrate).

E. 40 mM Molybdic Acid, 19 mM Arsenic Acid and 756 mM Sulfuric Acid Solution (Ars-Mol) (Prepare 1 liter in deionized water using Molybdic Acid, Ammonium Salt, Tetrahydrate, Arsenic Acid, Sodium Salt, and Sulfuric Acid).

F. 0.5 mg/ml Galactose Standard Solution (Gal Std) (Prepare 10 ml in Reagent A using D(+)Galactose).

Procedure

Pipette (in milliliters) the following reagents into suitable 4 dram vials:

	Test	Blank
Reagent B (Agar)	4.00	4.00

Equilibrate to 43°C. Then add:

Reagent C (Enzyme Solution)	2.00	—
Reagent A (Buffer)	—	2.00

Mix by swirling and incubate at 43°C for 30 minutes. At times 15, 20, 25, and 30 minutes, remove a 1.00 ml aliquot.

Colorimetric Determination: Pipette (in milliliters) the following into suitable tubes (equilibrated to 43°C):

		T_15 Test	T_20 Test	T_25 Test	T_30 Test	Test Blank
T15	Test Solution	1.00	—	—	—	—
T20	Test Solution	—	1.00	—	—	—
T25	Test Solution	—	—	1.00	—	—
T30	Test Solution	—	—	—	1.00	—
Test	Blank	—	—	—	—	1.00
Reagent D (Copper Soln)		1.00	1.00	1.00	1.00	1.00

Immediately mix by swirling. Place a marble over the top of the tubes and transfer the tubes to a boiling water bath. Incubate for 10 minutes. Remove from the boiling water bath and allow to cool to room temperature. Then add:

Reagent E (Ars-Mol)	1.00	1.00	1.00	1.00	1.00

Shake or vortex until the foaming stops and any orange colored precipitate present is dissolved. Then add (in milliliters):

	T_15 Test	T_20 Test	T_25 Test	T_30 Test	Test Blank
Deionized Water	10.00	10.00	10.00	10.00	10.00

Mix by swirling, and transfer to suitable cuvettes. Obtain the A540nm for both Test and Blank, using a suitable spectrophotometer.

Standards: Pipette (in milliliters) the following reagents into suitable tubes (Mix after each addition):

	Std 1	Std 2	Std 3	Std 4	Std 5	Blank
Reagent A (Buffer	0.95	0.80	0.60	0.40	—	1.00
Reagent (Gal std)	0.05	0.20	0.40	0.60	1.00	—
Reagent D (Copper Soln)	1.00	1.00	1.00	1.00	1.00	1.00

Immediately mix by swirling. Place a marble over the top of the tubes and transfer the tubes to a boiling water bath. Incubate for 10 minutes. Remove from the boiling water bath and allow to cool to room temperature. Then add:

	Std 1	Std 2	Std 3	Blank
Reagent E (Ars-Mol)	1.00	1.00	1.00	1.00

Shake or vortex until the foaming stops and any orange colored precipitate present is dissolved. Then add:

Deionized Water	10.00	10.00	10.00	10.00

Mix by swirling, and transfer to suitable cuvettes. Obtain the A540nm for the Standards and Standard Blank, using a suitable spectrophotometer.

Calculations

A540nm Standard = A540nm Standard–A540nm Standard Blank

Prepare a standard curve by plotting the A540nm of the Standard versus micrograms of D-Galactose.

Sample Determination: A540nm Sample = A540nm Test–A540nm Blank

Determine the micrograms of D-Galactose liberated using the standard curve.

$$\text{Units / ml enzyme} = \frac{\text{(Micrograms of D-Galactose liberated) (6.0) (df)}}{\text{(2.0) (T)}}$$

6.0 = total volume (in milliliters) of assay
df = Dilution factor
2.0 = Volume (in milliliters) of enzyme used
T = Time (in minutes) at which Reagent D was added

$$\text{Unit / mg solid} = \frac{\text{unit / ml enzyme}}{\text{mg solid / ml enzyme}}$$

$$\text{Unit / mg protein} = \frac{\text{unit / ml enzyme}}{\text{mg protein / ml enzyme}}$$

Unit Definition: *One unit will produce 1.0 μg of reducing sugar (measured as D-galactose) from agar per minute at pH 6.0 at 43°C.*

Final Assay Concentrations

In a 6.00 ml reaction mix, the final concentrations are 50 mM potassium phosphate, 1.0% (w/v) agar and 40–80 units agarase.

Notes

1. Heat to boiling with stirring to dissolve. Cool to 43°C for the assay. Reagent will solidify below 43°C.
2. Preheat enzyme solution to 43°C immediately before adding to Reagent B (agar) or it will cause the agar to solidify.
3. Sodium Sulfate, Sodium Carbonate, and Sodium Potassium Tartrate are dissolved in approximately 500 ml of deionized water. Cupric Sulfate is dissolved in approximately 100 ml of deionized water and slowly added to the above solution to avoid precipitation. Sodium Bicarbonate is dissolved first in deionized water and then added to the above solution. Dilute the solution to 1 liter. Store in an amber bottle at room temperature and avoid exposure to direct light. If a precipitate forms the solution may be used after the precipitate is allowed to settle. The solution may also be used without allowing the precipitate to settle, but the reaction mix will need to be filtered prior to reading the absorbance of the spectrophotometer.
4. Molybdic Acid is dissolved in approximately 300 ml of deionized water. Add Sulfuric Acid slowly. (Caution: this is an exothermic reaction). Arsenic Acid is dissolved in approximately 300 ml of deionized water and is added to the above solution. The solution is diluted to a total volume of 1 liter and incubated at 37°C for 48–72 hours. Store in an amber bottle at room temperature and avoid exposure to direct light. The solution expires six months after preparation. If a precipitate forms the solution may be used after the precipitate is allowed to settle. The solution may also be used without allowing the precipitate to settle, but the reaction mix will need to be filtered prior to reading the absorbance on the spectrophotometer.
5. Removing the pipet tip from the pipettor and leaving the tip in the reaction mix throughout the incubation allows the tip to equilibrate to the appropriate temperature. Use this same tip to remove 1 ml aliquots at all the time pcints reduces the chance for the agar to solidify in the tip.
6. Do not place in an ice bath to cool to room temperature, it will cause the Agar to solidify.
7. If precipitate is present it can be removed by filtering through an 0.8 μm syringe filter.

Enzymatic Assay of ALANINE AMINOPEPTIDASE (EC 3.4.11.14)

Principle

$$\text{L-Alanine p-Nitroanilide} + H_2O \xrightarrow[Co^{++}]{AAP} \text{L-Alanine} + \text{p-Nitroaniline}$$

Abbreviation Used

AAP = Alanine Aminopeptidase

Conditions: T = 37°C, pH = 7.2, A405nm, Light path = 1 cm

Method: Continuous Spectrophotometric Rate Determination

Reagents

A. 60 mM Potassium Phosphate Buffer, pH 7.2 at 37°C (Prepare 100 ml in deionized water using Potassium Phosphate, Monobasic, Anhydrous. Adjust to pH 7.2 at 37°C with 1 M KOH).

B. 1.66 mM L-Alanine p-Nitroanilide Solution (L-Ala-NA) (Prepare 30 ml in Reagent A using L-Alanine p-Nitroanilide, Hydrochloride. *Prepare Fresh*).

C. 10 mM Tris HCl Buffer with 1 mM Cobalt Chloride, pH 8.0 at 37°C (Activation Buffer) (Prepare 20 ml in deionized water using Trizma Base. Adjust to pH 8.0 at 37°C with 1 M HCl and then add Cobalt Chloride, Hexahydrate).

D. Alanine Aminopeptidase Enzyme Solution, Non-Activated (Enz-Non Act) (Immediately before use, prepare a solution containing 0.3 unit/ml of Alanine Aminopeptidase in cold deionized water).

E. Alanine Aminopeptidase Enzyme Solution, Activated (Enz-Act) (Immediately before use, prepare a solution containing 0.1 unit/ml of Alanine Aminopeptidase in cold Reagent C. Incubate at 37°C for 15 minutes to activate).

Procedure

Pipette (in milliliters) the following reagents into suitable cuvettes:

	Nonactivated		Activated	
	Test	Blank	Test	Blank
Reagent B (L-Ala-NA)	3.00	3.00	3.00	3.00

Equilibrate to 37°C. Monitor the A405nm until constant, using a suitably thermostatted spectrophotometer. Then add:

Deionized Water	—	0.10	—	—
Reagent C (Activation Buffer)	—	—	—	0.10
Reagent D (Enz Non Act)	0.10	—	—	—
Reagent E (Enz-Act)	—	—	0.10	—

Immediately mix by inversion and record the increase in A405nm for approximately 5 minutes. Obtain the ΔA405nm/minute using the maximum linear rate for both the Test and Blank.

Calculations

$$\text{Units / ml enzyme} = \frac{(\Delta A405\text{nm / min} - \Delta A405\text{nm / min Blank})\ (3.1)\ (\text{df})}{(9.62)\ (0.1)}$$

3.1 = Total volume (in milliliters) of assay
df = Dilution factor
9.62 = Millimolar extinction coefficient of p-Nitroaniline at 405nm
0.1 = Volume (in milliliter) of enzyme used

$$\text{Units / mg solid} = \frac{\text{units / ml enzyme}}{\text{mg solid / ml enzyme}}$$

$$\text{Units / mg protein} = \frac{\text{units / ml enzyme}}{\text{mg protein / ml enzyme}}$$

Unit Definition: *One unit will hydrolyze 1.0 µmole of L-alanine p-nitroanilide to L-alanine and p-nitroaniline per minute at pH 7.2 at 37°C.*

Final Assay Concentration

Unactivated: In a 3.10 ml reaction mix, the final concentrations are 58 mM potassium phosphate, 1.61 mM L-alanine p-nitroanilide and 0.03 unit alanine aminopeptidase.

Activated: In a 3.10 ml reaction mix, the final concentrations are 58 mM potassium phosphate, 1.61 mM L-alanine p-nitroanilide, 0.32 mM Tris HCl, 0.03 mM cobalt chloride and 0.01 unit alanine aminopeptidase.

Note

1. This solution should be clear and faint pink in color. If this solution is hazy or green/brown in color, it is not suitable for use.

Enzymatic Assay of L-ALANINE DEHYDROGENASE (EC 1.4.1.1)

Principle

$$\text{L-Alanine} + \beta\text{-NAD} + H_2O \xrightarrow{\text{L-Alanine Dehydrogenase}} \text{Pyruvate} + \beta\text{-NADH} + NH_3$$

Abbreviations Used

β-NAD = β-Nicotinamide Adenine Dinucleotide, Oxidized Form.
β-NADH = β-Nicotinamide Adenine Dinucleotide, Reduced Form.

Conditions: T = 25°C, pH = 10.0, A340nm, Light Path = 1 cm.

Method: Continuous Spectrophotometric Rate Determination

Reagents

A. 50 mM Sodium Bicarbonate Solution ($NaHCO_3$) (Prepare 100 ml in deionized water using Sodium Bicarbonate).
B. 50 mM Sodium Carbonate Solution (Na_2CO_3) (Prepare 150 ml in deionized water using Sodium Carbonate, Anhydrous).
C. 50 mM Sodium Carbonate Buffer, pH 10.0 at 25°C (Prepare 200 ml by adding 100 ml of Reagent B to 100 ml of Reagent A. Adjust to pH 10.0 at 25°C with Reagent B).
D. 500 mM L-Alanine Solution (L-ALA) (Prepare 1.0 ml in deionized water using L-Alanine).
E. 30 mM β-Nicotinamide Adenine Dinucleotide, Oxidized Form Solution (β-NAD) (Dissolve the contents of one 20 mg vial of β-Nicotinamide Adenine Dinucleotide, in the appropriate volume of deionized water. *Prepare Fresh*).
F. L-Alanine Dehydrogenase Enzyme Solution (Immediately before use, prepare a solution containing 0.2–0.5 unit/ml of L-Alanine Dehydrogenase in cold Reagent C).

Procedure

Pipette (in milliliters) the following reagents into suitable cuvettes:

	Test	Blank
Reagent C (Buffer)	2.70	2.70
Reagent D (L-ALA)	0.10	0.10
Reagent E (β-NAD)	0.10	0.10

Mix by inversion and equilibrate to 25°C. Monitor the A340nm until constant, using a suitably thermostatted spectrophotometer. Then add:

	Test	Blank
Reagent C (Buffer)	—	0.10
Reagent F (Enzyme Solution)	0.10	—

Immediately mix by inversion and record the increase in A340nm for approximately 5 minutes. Obtain the ΔA340nm /minute using the maximum linear rate for both the Test and Blank.

Calculations

$$\text{Units / ml enzyme} = \frac{(\Delta A340\text{nm / min Test} - \Delta A340\text{nm / min Blank})\,(3)\,(\text{df})}{(6.22)\,(0.1)}$$

3 = Volume (in milliliters) of assay
df = Dilution factor
6.22 = Millimolar extinction coefficient of β-NADH at 340 nm
0.1 = Volume (in milliliter) of enzyme used

$$\text{Units / mg solid} = \frac{\text{units / ml enzyme}}{\text{mg solid / ml enzyme}}$$

$$\text{Units / mg protein} = \frac{\text{units / ml enzyme}}{\text{mg protein / ml enzyme}}$$

Unit Definition: *One unit will convert 1.0 µmole of L-alanine to pyruvate and NH_3 per minute at pH 10.0 at 25°C.*

Final Assay Concentration

In a 3.00 ml reaction mix, the final concentrations are 47 mM sodium bicarbonate, 17 mM L-alanine, 1.0 mM β-nicotinamide adenine dinucleotide and 0.02–0.05 unit of L-alanine dehydrogenase.

Enzymatic Assay of ALCOHOL DEHYDROGENASE (EC 1.1.1.1)

Principle

$$\text{Ethanol} + \beta\text{-NAD} \xrightarrow{\text{Alcohol Dehydrogenase}} \text{Acetaldehyde} + \beta\text{-NADH}$$

Abbreviations Used

β-NAD = β-Nicotinamide Adenine Dinucleotide, Oxidized Form.
β-NADH = β-Nicotinamide Adenine Dinucleotide, Reduced Form.

Conditions: T = 25°C, pH = 8.8, A340nm, Light path = 1 cm.

Method: Continuous Spectrophotometric Rate Determination.

Reagents

A. 50 mM Sodium Pyrophosphate Buffer, pH 8.8 at 25°C (Prepare 100 ml in deionized water using Sodium Pyrophosphate, Decahydrate. Adjust to pH 8.8 at 25°C with 8% (v/v) Phosphoric Acid).
B. 95% (v/v) Ethanol (Prepare 10 ml in deionized water using 200 Proof USP Ethyl Alcohol).
C. 15 mM β-Nicotinamide Adenine Dinucleotide Solution (β-NAD) (Prepare 15 ml in deionized water using β-Nicotinamide Adenine Dinucleotide . *Prepare Fresh*).
D. 10 mM Sodium Phosphate Solution (Prepare 100 ml in deionized water using Sodium Phosphate, Monobasic, Anhydrous).

E. 10 mM Sodium Phosphate Buffer, pH 7.5 at 25°C (Prepare 100 ml in deionized water using Sodium Phosphate, Dibasic, Anhydrous. Adjust to pH 7.5 at 25°C with Reagent D).

F. 10 mM Sodium Phosphate Buffer with 0.1% (w/v) Bovine Serum Albumin, pH 7.5 at 25°C (Enzyme Diluent) (Prepare 25 ml in Reagent E with Albumin, Bovine).

G. Alcohol Dehydrogenase Enzyme Solution (ADH) (Immediately before use, prepare a solution containing 1 mg/ml of Alcohol Dehydrogenase in cold Reagent E. Further dilute to 0.75 unit/ml with cold Reagent F).

Procedure

Pipette (in milliliters) the following reagents into suitable cuvettes:

	Test	Blank
Reagent A (Buffer)	1.30	1.30
Reagent B (Ethanol)	0.10	0.10
Reagent C (β-NAD)	1.50	1.50

Mix by inversion and equilibrate to 25°C. Monitor the A340nm until constant, using a suitably thermostatted spectrophotometer. Then add:

	Test	Blank
Reagent G (Enzyme Solution)	0.10	—
Reagent F (Enzyme Diluent)	—	0.10

Immediately mix by inversion and record the increase in A340nm for approximately 6 minutes. Obtain the A340nm/minute using the one to six minute range for both the Test and Blank.

Calculations

$$\text{Units / ml enzyme} = \frac{(\text{A340nm / min Test} - \text{A340nm / min Blank})\ (3)\ (\text{df})}{(6.22)\ (0.1)}$$

3 = Total volume (in milliliters) of assay
df = Dilution factor
6.22 = Millimolar extinction coefficient of β-NADH at 340 nm
0.1 = Volume (in milliliter) of enzyme used

$$\text{Units / mg solid} = \frac{\text{units / ml enzyme}}{\text{mg solid / ml enzyme}}$$

$$\text{Units / mg protein} = \frac{\text{units / ml enzyme}}{\text{mg protein / ml enzyme}}$$

Unit Definition: *One unit will convert 1.0 μmole of ethanol to acetaldehyde per minute at pH 8.8 at 25°C.*

Final Assay Concentration

In a 3.00 ml reaction mix, the final concentrations are 22 mM sodium pyrophosphate, 3.2% (v/v) ethanol, 7.5 mM β-nicotinamide adenine dinucleotide, 0.3 mM sodium phosphate, 0.003% (w/v) bovine serum albumin, and 0.075 unit alcohol dehydrogenase.

Notes

1. This assay procedure is not to be used to assay Alcohol Dehydrogenase, Insoluble enzyme attached to beaded agarose.
2. Alcohol Dehydrogenase is very unstable when diluted, and should be assayed immediately following preparations of solutions.

CV Value Calculation

Immediately before use, prepare three independent solutions containing 1 mg solid/ml of Alcohol Dehydrogenase (ADH) in cold Reagent E (10 mM Sodium Phosphate buffer with 0.1% Bovine Serum Albumin, pH 7.5 at 25°C). Further dilute 50 µl of these solutions to either 25 ml or 50 ml with cold Reagent F (Enzyme Diluent). Then assay 100 µl of each dilution and one blank. To calculate the CV Value:

1. Determine the rate (A/min) from one minute to six minutes.
2. Multiply each sample and the blank by 2412A or 4823B, depending on the dilution used. The resulting value will be units/mg solid.
3. Sample (units/mg solid)–Blank (units/mg solid) = Corrected sample (units/mg solid).
4. Determine the mean of the corrected samples.
5. CV Value = [S (Mean corrected sample–corrected sample)](1/2) × 100 Mean corrected sample Dilution factors are as follows:

$$A\ 2412 = \frac{(3.0)\ (500)}{(0.1)\ (6.22)}$$

3.0 = Total volume (in milliliters) in cuvette
0.1 = Volume (in milliliter) of enzyme used
6.22 = Millimolar extinction coefficient of β-NADH at 340nm
500 = Dilution factor (0.05 ml ADH to 25 ml Reagent F)

$$B\ 2423 = \frac{(3.0)\ (1000)}{(0.1)\ (6.22)}$$

3.0 = Total volume in cuvette
0.1 = Volume (in milliliter) of enzyme used
6.22 = Millir olar extinction coefficient of β-NADH at 340nm
1000 = Dilution factor (0.05 ml ADH to 50 ml Reagent F)

Enzymatic Assay of ALCOHOL DEHYDROGENASE, NADP⁺ DEPENDENT (EC 1.1.1.2)

Principle

$$\text{2-Propanol} + \beta\text{-NADP} \xrightarrow{\text{Alcohol Dehydrogenase}} \text{Acetone} + \beta\text{-NADPH}$$

Abbreviations Used

β-NADP = β-Nicotinamide Adenine Dinucleotide Phosphate, Oxidized Form.
β-NADPH = β-Nicotinamide Adenine Dinucleotide Phosphate, Reduced Form

Conditions: T = 40°C, pH = 7.8, A340nm, Light path = 1 cm

Method: Continuous Spectrophotometric Rate Determination

Reagents

A. 100 mM Tris HCl Buffer, pH 7.8 at 40°C (Prepare 100 ml in deionized water using Trizma Hydrochloride. Adjust to pH 7.8 at 40°C with 1 M NaOH).

B. 15 mM β-Nicotinamide Adenine Dinucleotide Phosphate Solution (β-NADP) (Prepare 2.5 ml in Reagent A using β-Nicotinamide Adenine Dinucleotide Phosphate, Sodium Salt or dissolve the contents of one 10 mg vial of β-Nicotinamide Adenine Dinucleotide Phosphate, Sodium Salt, in an appropriate volume of Reagent A. *Prepare Fresh*).

C. 1.5 M 2-Propanol Solution (2-Prop) (Prepare 10 ml in Reagent A using Isopropanol, Anhydrous).

D. Alcohol Dehydrogenase, $NADP^+$ Dependent Enzyme Solution (Immediately before use, prepare a solution containing 0.10–0.40 unit/ml of Alcohol Dehydrogenase, $NADP^+$ Dependent in cold Reagent A).

Procedure

Pipette (in milliliters) the following reagents into suitable cuvettes:

	Test	Blank
Reagent A (Buffer)	2.40	2.60
Reagent B (β-NADP)	0.10	0.10
Reagent C (2-Prop)	0.30	0.30

Mix by inversion and equilibrate to 40°C. Monitor the A340nm until constant, using a suitably thermostatted spectrophotometer. Then add:

Reagent D (Enzyme Solution)	0.20	—

Immediately mix by inversion and record the increase in A340nm for approximately 5 minutes. Obtain the ΔA340nm/minute using the maximum linear rate for both the Test and Blank.

Calculations

$$\text{Units / ml enzyme} = \frac{(\Delta A340\text{nm / min Test} - \Delta A340\text{nm / min Blank})\ (3)\ (\text{df})}{(6.22)\ (0.2)}$$

3 = Total volume (in milliliters) of assay
df = Dilution factor
6.22 = Millimolar extinction coefficient of β-NADPH at 340 nm
0.2 = Volume (in milliliter) of enzyme used

$$\text{Units / mg solid} = \frac{\text{units / ml enzyme}}{\text{mg solid / ml enzyme}}$$

$$\text{Units / mg protein} = \frac{\text{units / ml enzyme}}{\text{mg protein / ml enzyme}}$$

Unit Definition: *One unit will oxidize 1.0 µmole of 2-propanol to acetone per minute at pH 7.8 at 40°C in the presence of β-$NADP^+$.*

Final Assay Concentration

In a 3.00 ml reaction mix, the final concentrations are 100 mM Tris, 0.5 mM β-nicotinamide adenine dinucleotide phosphate, 150 mM 2-propanol and 0.02–0.08 unit alcohol dehydrogenase, β-$NADP^+$ dependent.

Enzymatic Assay of ALCOHOL OXIDASE (EC 1.1.3.13)

Principle

$$\text{Methanol} + O_2 \xrightarrow{\text{Alcohol Oxidase}} \text{Formaldehyde} + H_2O_2$$

$$H_2O_2 + \text{ABTS POD} \rightarrow 2H_2O + \text{Oxidized ABTS}$$

Abbreviations Used

ABTS = 2,2'-Azino-bis(3-Ethylbenzthiazoline-6-Sulfonic Acid).
POD = Peroxidase.

Conditions: T = 25ºC, pH = 7.5, A405nm, Light path = 1 cm.

Method: Continuous Spectrophotometric Rate Determination.

Reagents

A. 100 mM Potassium Phosphate Buffer, pH 7.5 at 25ºC (Prepare 100 ml in deionized water using Potassium Phosphate, Monobasic, Anhydrous. Adjust to pH 7.5 at 25ºC with 1 M KOH).

B. 2 mM 2,2'-Azino-bis(3-Ethylbenzthiazoline-6-Sulfonic Acid) Solution (ABTS) (Prepare 25 ml in Reagent A using 2,2'-Azino-bis(3-Ethylbenzthiazoline-6-Sulfonic Acid), Diammonium Salt. (*Prepare Fresh*). Immediately before use bubble O_2 gas through the solution for approximately 5 minutes).

C. 0.003% (w/w) Hydrogen Peroxide Solution (Prepare 100 ml in deionized water using Hydrogen Peroxide, 30% (w/w) Solution. *Prepare Fresh*).

D. 1.0% (v/v) Methanol Solution (Prepare 10 ml in deionized water using Methanol, Absolute. *Prepare Fresh*).

E. Peroxidase Enzyme Solution (POD) (Immediately before use, prepare a solution containing 250 units/ml of Peroxidase, in cold Reagent A).

F. Alcohol Oxidase Enzyme Solution (Immediately before use, prepare a solution containing 0.1 unit/ml of Alcohol Oxidase in cold Reagent A).

Procedure

Pipette (in milliliters) the following reagents into suitable cuvettes:

	Test	Blank
Reagent B (ABTS)	2.80	2.80
Reagent E (POD)	0.01	0.01

Mix by inversion and equilibrate to 25ºC. Monitor the A405nm until constant, using a suitably thermostatted spectrophotometer. Then add:

	Test	Blank
Reagent A (Buffer)	—	0.10
Reagent F (Alcohol Oxidase)	0.01	—
Reagent C (Hydrogen Peroxide)	0.01	0.01

Mix by inversion and monitor the A405nm until constant. Then add:

	Test	Blank
Reagent D (Methanol)	0.10	0.10

Immediately mix by inversion and record the increase in A405nm for approximately 5 minutes. Obtain the A405nm/minute using the maximum linear rate for both the Test and Blank.

Calculations

$$\text{Units / ml enzyme} = \frac{(\text{A405nm / min Test} - \text{A405nm / min Blank})\ (3.02)\ (\text{df})}{(36.8)\ (0.1)}$$

3.02 = Total volume (in milliliters) of assay
df = Dilution factor
36.8 = Millimolar extinction coefficient of 2,2'-Azino-bis(3-Ethylbenzthiazoline-6-Sulfonic Acid) at 405nm
0.1 = Volume (in milliliters) of enzyme used

$$\text{Units / mg solid} = \frac{\text{units / ml enzyme}}{\text{mg solid / ml enzyme}}$$

$$\text{Units / mg protein} = \frac{\text{units / ml enzyme}}{\text{mg protein / ml enzyme}}$$

Unit Definition: *One unit will oxidize 1.0 mole of methanol to formaldehyde per minute at pH 7.5 at 25°C.*

Final Assay Concentrations

In a 3.02 ml reaction mix, the final concentrations are 96 mM potassium phosphate, 2 mM 2,2'- azino-bis(3-ethylbenzthiazoline-6-sulfonic acid), 0.00001% (w/w) hydrogen peroxide, 0.033% (v/v) methanol, 2.5 units peroxidase and 0.01 unit alcohol oxidase.

Note

1. Peroxidase Unit Definition: One unit will form 1.0 mg purpurogallin from pyrogallol in 20 seconds at pH 6.0 at 20°C. In general, Peroxidase will use 2,2'-Azino-bis(3-ethylbenzthiazoline-6-sulfonic acid) as an electron donor.

Enzymatic Assay of ALDEHYDE DEHYDROGENASE (EC 1.2.1.5)

Principle

$$\text{Acetaldehyde} + \beta\text{-NAD} \xrightarrow{\text{Aldehyde Dehydrogenase}} \text{Acetic Acid} + \beta\text{-NADH}$$

Abbreviations Used

β-NAD = β-Nicotinamide Adenine Dinucleotide, Oxidized Form
β-NADH = β-Nicotinamide Adenine Dinucleotide, Reduced Form

Conditions: T = 25°C, pH = 8.0, A340nm, Light path = 1 cm.

Method: Continuous Spectrophotometric Rate Determination.

Reagents

A. 1 M Tris HCl Buffer, pH 8.0 at 25°C (Prepare 50 ml in deionized water using Trizma Base. Adjust to pH 8.0 at 25°C with 1 M HCl).

B. 20 mM β-Nicotinamide Adenine Dinucleotide, Oxidized Form, Solution (β-NAD) (Prepare 1 ml in deionized water using β-Nicotinamide Adenine Dinucleotide, or dissolve the contents of one 20 mg vial of β-Nicotinamide Adenine Dinucleotide, in the appropriate volume of deionized water. *Prepare Fresh*).

C. 100 mM Acetaldehyde Solution (Acetald) (Prepare 10 ml in deionized water from a 2 M stock solution of Acetaldehyde. *Prepare Fresh*).

D. 3 M Potassium Chloride Solution (KCl) (Prepare 1 ml in deionized water using Potassium Chloride).

E. 1 M 2-Mercaptoethanol Solution (2-ME) (Prepare 1 ml in deionized water using 2-Mercaptoethanol. *Prepare Fresh*).

F. 100 mM Tris HCl Buffer with 0.02% (w/v) Bovine Serum Albumin, pH 8.0 at 25°C (Enz Dil) (Prepare 25 ml in deionized water using Trizma Base, and Albumin, Bovine. Adjust the pH to 8.0 at 25°C with 1 M HCl).

G. Aldehyde Dehydrogenase Enzyme Solution (Immediately before use, prepare a solution containing 0.25–0.5 unit/ml of Aldehyde Dehydrogenase in cold Reagent F).

Procedure

Pipette (in milliliters) the following reagents into suitable cuvettes:

	Test	Blank
Deionized Water	2.32	2.32
Reagent A (Buffer)	0.30	0.30
Reagent B (β-NAD)	0.10	0.10
Reagent D (KCl)	0.10	0.10
Reagent E (2-ME)	0.10	0.10
Reagent C (Acetald)	0.05	0.05

Mix by inversion and equilibrate to 25°C. Monitor the A340nm until constant, using a suitably thermostatted spectrophotometer. Then add:

	Test	Blank
Reagent F (Enz Dil)	—	0.10
Reagent G (Enzyme Solution)	0.10	—

Immediately mix by inversion and record the increase in A340nm for approximately 5 minutes. Obtain the ΔA340nm/minute using the maximum linear rate for both the Test and Blank.

Calculations

$$\text{Units / ml enzyme} = \frac{(\Delta\text{A340nm / min Test} - \Delta\text{A340nm / min Blank})\ (3)\ (\text{df})}{(6.22)\ (0.1)}$$

3 = Total volume (in milliliters) of assay
df = Dilution factor
6.22 = Millimolar extinction coefficient of β-NADH at 340 nm
0.1 = Volume (in milliliter) of enzyme used

$$\text{Units / mg solid} = \frac{\text{units / ml enzyme}}{\text{mg solid / ml enzyme}}$$

$$\text{Units / mg protein} = \frac{\text{units / ml enzyme}}{\text{mg protein / ml enzyme}}$$

Unit Definition: *One unit will oxidize 1.0 µmole of acetaldehyde to acetic acid per minute at pH 8.0 at 25°C in the presence of β-NAD$^+$, potassium and thiols.*

Final Assay Concentration

In a 3.00 ml reaction mix, the final concentrations are 103 mM Tris, 0.67 mM β-nicotinamide adenine dinucleotide, 100 mM potassium chloride, 10 mM 2-mercaptoethanol, mM acetaldehyde, 0.0007% (w/v) Bovine Serum Albumin and 0.025–0.05 unit aldehyde dehydrogenase.

Notes

1. Prepare fresh from a 2M stock solution of acetaldehyde. The 2M stock solution of acetaldehyde is prepared by adding 1.0 ml of 17.9 M acetaldehyde to 7.9 ml of deionized water. In case of any problems with the assay, redistillation of the 17.9 M Acetaldehyde, (from which the 2M stock is made) should be done to alleviate problems with the substrate.
2. Because of the volatility of these reagents, they should be added to the cuvette immediately before running the assay.
3. The enzyme is inhibited by acetaldehyde. Higher rates may be obtained by using lower concentrations of Reagent C (Acetald).

Enzymatic Assay of ALDOLASE (EC 4.1.2.13)

Principle

$$\text{Fructose 1, 6-Diphosphate} + H_2O \xrightarrow{\text{Aldolase}} \text{G3-P} + \text{DHAP}$$

$$\text{G3-P TPI} \longrightarrow \text{DHAP}$$

$$2\ \text{DHAP} + 2\beta\text{-NADH}\ \alpha\text{-GDH} \longrightarrow 2\alpha\text{-Glycerophosphate} + 2\beta\text{-NAD}$$

Abbreviations Used

G 3-P = Glyceraldehyde 3-Phosphate
DHAP = Dihydroxyacetone Phosphate
TPI = Triosephosphate Isomerase
α-GDH = α-Glycerophosphate Dehydrogenase
β-NADH = β-Nicotinamide Adenine Dinucleotide, Reduced Form
β-NAD = β-Nicotinamide Adenine Dinucleotide, Oxidized Form

Conditions: T = 25°C, pH = 7.4, A340nm, Light path = 1 cm.

Method: Continuous Spectrophotometric Rate Determination.

Reagents

A. 100 mM Tris HCl Buffer, pH 7.4 at 25°C (Prepare 250 ml in deionized water using Trizma Base. Adjust to pH 7.4 at 25°C with 1 M HCl).

B. 58 mM Fructose 1,6-Diphosphate Solution (F 1,6-DP) (Prepare 1 ml in deionized water using D-Fructose 1,6-Diphosphate, Tetra(cyclohexylammonium) Salt).

C. 4.0 mM β-Nicotinamide Adenine Dinucleotide, Reduced Form Solution (β-NADH) (Prepare 2 ml in cold deionized water using β-Nicotinamide Adenine Dinucleotide, Reduced Form, Disodium Salt or dissolve the contents of one 5 mg vial of β-Nicotinamide Adenine Dinucleotide, Reduced Form, Disodium Salt, in the appropriate volume of deionized water).

D. α-Glycerophosphate Dehydrogenase/Triosephosphate Isomerase Enzyme Solution (α-GDH/TPI) (Immediately before use, prepare a solution containing 50 α-GDH units/ml of α-Glycerophosphate Dehydrogenase/Triosephosphate Isomerase, in cold deionized water).

E. Aldolase Enzyme Solution (Immediately before use, prepare a solution containing 0.25–0.50 unit/ml of Aldolase in cold Reagent A).

Procedure

Pipette (in milliliters) the following reagents into suitable cuvettes:

	Test	Blank
Reagent A (Buffer)	2.60	2.60
Reagent B (F 1,6-DP)	0.10	0.10
Reagent C (β-NADH)	0.10	0.10
Reagent D (α-GDH/TPI)	0.10	0.10

Mix by inversion and equilibrate to 25°C. Monitor the A340nm until constant, using a suitably thermostatted spectrophotometer. Then add:

Reagent A (Buffer)	—	0.10
Reagent E (Enzyme Solution)	0.10	—

Immediately mix by inversion and record the decrease in A340nm for approximately 5 minutes. Obtain the A340nm/min using the maximum linear rate for both the Test and Blank.

Calculations

$$\text{Units / ml enzyme} = \frac{(\text{A340nm / min Test} - \text{A340nm / min Blank})\ (3)\ (\text{df})}{(2)\ (6.22)\ (0.1)}$$

3 = Total volume (in milliliters) of assay
df = Dilution factor
2 = 2 moles of β-NADH converted to 2 moles of β-NAD per mole of Fructose 1,6-Diphosphate
6.22 = Millimolar extinction coefficient of β-NADH at 340nm
0.1 = Volume (in milliliter) of enzyme used

$$\text{Units / mg solid} = \frac{\text{units / ml enzyme}}{\text{mg solid / ml enzyme}}$$

$$\text{Units / mg protein} = \frac{\text{units / ml enzyme}}{\text{mg protein / ml enzyme}}$$

Unit Definition: *One unit will convert 1.0 µmole of fructose 1,6-diphosphate to dihydroxyacetone phosphate and glyceraldehyde 3-phosphate per minute at pH 7.4 at 25°C.*

Final Assay Concentration

In a 3.00 ml reaction mix, the final concentrations are 90 mM Tris, 1.9 mM fructose 1,6-diphosphate, 0.13 mM β-nicotinamide adenine dinucleotide, 5 units α-glycerophosphate dehydrogenase/triosephosphate isomerase (based on α-glycerophosphate dehydrogenase units) and 0.025–0.050 unit aldolase.

Notes

1. This enzyme assay is not to be used to assay Aldolase, from Staphylococcus aureus, Aldolase, insoluble enzyme attached to polyacrylamide from Rabbit Muscle, and Aldolase from Baker's Yeast.
2. α-Glycerophosphate Dehydrogenase Unit Definition: One unit will convert 1.0 µmole of dihydroxyacetone phosphate to α-glycerophosphate per minute at pH 7.4 at 25°C.
3. Triosephosphate Isomerase Unit Definition: One unit will convert 1.0 µmole of Dglyceraldehyde. 3-phosphate to dihydroxyacetone phosphate per minute at pH 7.6 at 25°C.

Enzymatic Assay of ALLANTOINASE (EC 3.5.2.5)

Principle

$$\text{Allantoin} + H_2O \xrightarrow{\text{Allantoinase}} \text{Allantoate}$$

$$\text{Allantoate} + K_3Fe(CN)_6 + \text{Phenylhydrazine HCl} \xrightarrow[\text{Colored Product}]{} \text{Glyoxylate}$$

Conditions: T = 25°C, pH = 7.0, A540nm, Light path = 1 cm

Method: Colorimetric

Reagents

A. 100 mM Tris HCl Buffer, pH 7.0 at 25°C (Prepare 100 ml in deionized water using Trizma Base. Adjust to pH 7.0 at 25°C with 1 M HCl).
B. 33 mM Allantoin Solution, pH 7.0 at 25°C (Allantoin) (Prepare 50 ml in Reagent A using Allantoin. Adjust to pH 7.0 at 25°C with either 1 M HCl or 1 M NaOH).
C. 69 mM Phenylhydrazine HCl Solution (PH) (Prepare 10 ml in deionized water using Phenylhydrazine Hydrochloride).
D. 25% (v/v) Hydrochloric Acid Solution(Prepare 25 ml in deionized water using Hydrochloric Acid (HCl)).
E. 152 mM Potassium Ferricyanide Solution ($K_3Fe(CN)_6$) (Prepare 10 ml in deionized water using Potassium Ferricyanide).

F. 0.1 mM Glyoxylic Acid Standard Solution (Std) (Prepare 5 ml in deionized water using Glyoxylic Acid, Sodium Salt, Monohydrate).

G. Allantoinase Enzyme Solution (Immediately before use, prepare a solution containing 0.05–0.20 unit/ml of Allantoinase in cold Reagent A).

Procedure

Step 1: Pipette (in milliliters) the following reagents into suitable containers:

	Test	Blank
Reagent B (Allantoin)	4.00	4.00

Equilibrate to 25°C. Then add:

	Test	Blank
Reagent G (Enzyme Solution)	1.00	
Reagent A (Buffer)	—	1.00

Immediately mix by swirling and incubate at 25°C for exactly 5 minutes. Remove aliquots from the Test and Blank at 5, 10, and 15 minutes. Pipette (in milliliters) the following reagents into suitable containers.

	Test 1	Test Blank	Std 1	Std 2	Std 3	Std 4	Std Blank
Reagent D (HCl)	0.02	0.02	0.02	0.02	0.02	0.02	0.02
Reagent C (PH)	0.05	0.05	0.05	0.05	0.05	0.05	0.05
Test Mixture	1.00	—	—	—	—	—	—
Test Blank Mixture	—	1.00	—	—	—	—	—
Reagent F (Std)	—	—	0.20	0.40	0.60	0.80	—
Dieonzed Water	—	—	0.80	0.60	0.40	0.20	1.00

Mix by swirling and boil for 2 minutes. Place on ice. Then add:

Reagent D (HCl)	0.50	0.50	0.50	0.50	0.50	0.50	0.50
Reagent E ($K_3Fe(CN)_6$)	0.05	0.05	0.05	0.05	0.05	0.05	0.05

Allow to incubate on ice for 20 minutes. Then add:

Deionzied Water	2.00	2.00	2.00	2.00	2.00	2.00	2.00

Mix by swirling and transfer the solutions to suitable cuvettes. Record the A540nm for the Tests, Test Blank, Standards, and Standard Blank using a suitable spectrophotometer.

Calculations

Standard Curve: ΔA540nm Standard = A540nm Standard–A540nm Standard Blank
Prepare a standard curve by plotting ΔA540nm Standard vs µmoles of glyoxylic acid.
Sample Determination: ΔA540nm Test = A540nm Test–A540nm Test Blank
Determine the µmoles of glyoxylic acid liberated using the standard curve.

$$\text{Units / ml enzyme} = \frac{(\mu\text{moles of glyoxylic acid released})\ (5)\ (df)}{(1)\ (T)}$$

5 = Total volume (in milliliters) of reaction mix.
df = Dilution factor.
1 = Volume (in milliliter) of enzyme used.
T = Time (in minutes) of assay as per the Unit Definition.

$$\text{Units / mg solid} = \frac{\text{units / ml enzyme}}{\text{mg solid / ml enzyme}}$$

$$\text{Units / mg protein} = \frac{\text{units / ml enzyme}}{\text{mg protein / ml enzyme}}$$

Unit/g protein = unit/mg protein × 1000

Unit Definition: *One unit will hydrolyze 1.0 μmole of allantoin to allantoate (measured as glyoxylate) per minute at pH 7.0 at 25°C.*

Final Assay Concentrations

In a 5.00 ml reaction mix, the final concentrations are 27 mM allantoin, 100 mM Tris, and 0.05–0.20 unit allantoinase.

Note

1. Aliquots from the Test and Blank are to be removed at 5, 10 and 15 minutes time points. A standard curve of glyoxylic acid must be prepared for each time point, as the color intensity may vary with time.

Enzymatic Assay of AMIDASE (EC 3.5.1.4)

Principle

$$\text{Acetamide} + \text{Hydroxylamine} \xrightarrow{\text{Amidase}} \text{Acethydroxamate} + NH_3$$

Conditions: T = 37°C, pH = 7.2, A500nm, Light path = 1 cm

Method: Colorimetric

Reagents

A. 100 mM Sodium Phosphate Buffer, pH 7.2 at 37°C (Prepare 100 ml in deionized water using Sodium Phosphate, Monobasic, Anhydrous. Adjust to pH 7.2 at 37°C with 1 M NaOH).
B. 400 mM Acetamide Solution (Acet) (Prepare 10 ml in deionized water using Acetamide. The solution is stable on ice for 2–3 hours. *Prepare Fresh*).
C. 2 M Hydroxylamine Solution (Hydrox) (Prepare 10 ml in deionized water using Hydroxylamine Hydrochloride. Adjust to pH 7.2 at 37°C with 10 M NaOH).
D. Hydrochloric Acid (HCl) (Use Hydrochloric Acid).
E. 60% (w/v) Ferric Chloride Solution ($FeCl_3$) (Prepare 15 ml in deionized water using Ferric Chloride, Hexahydrate).
F. Color Reagent Solution (CRS) (Prepare 100 ml by adding 5.7 ml of Reagent D to 10 ml of Reagent E. Dilute to 100 ml with deionized water).
G. 7 mM Dithiothreitol Solution (Enz Dil) (Prepare 10 ml in Reagent A using DL-Dithiothreitol).

H. 100 mM Acetamide Solution (Prepare 5 ml in deionized water using Acetamide. *Prepare Fresh*).
I. 5 M Hydroxylamine Solution (Prepare 10 ml in 4.5 M NaOH using Hydroxylamine Hydrochloride).
J. 5 mM Acethydroxamate Standard Solution (Std Soln) (Prepare 100 ml by adding 5 ml of Reagent H to 10 ml of Reagent I. Heat in a boiling water bath for 15 minutes. Cool the solution and adjust to pH 7.2 at 37°C with 5 M HCl, if necessary. Dilute up to 100 ml with deionized water).
K. Amidase Enzyme Solution (Immediately before use, prepare a solution containing 3–6 units/ml of Amidase in cold Reagent G).

Procedure

Prepare a reaction cocktail by pipetting (in milliliters) the following reagents into a suitable container:

Reagent A (Buffer)	16.00
Reagent B (Acet)	8.00
Reagent C (Hydrox)	8.00

Mix by swirling and store on ice. Pipette (in milliliters) the following reagents into suitable containers:

	Test 1	Test Blank	Std 1	Std 2	Std 3	Std 4	Std 5	Std Blank
Reaction Cocktail	1.90	1.90	—	—	—	—	—	—
Reagent J (Std Soln)	—	—	0.20	0.50	1.00	1.50	2.00	—
Deionized Water	—	—	1.80	1.50	1.00	0.50	—	2.00
Equilibrate to 37°C. Then add:								
Reagent K (Enz Soln)	0.10	—	—	—	—	—	—	—
Reagent G (Enz Dil)	—	0.10	—	—	—	—	—	—
Immediately mix by swirling and incubate at 37°C for exactly 5 minutes. Then add:								
Reagent F (CRS)	4.00	4.00	4.00	4.00	4.00	4.00	4.00	4.00

Mix by swirling and transfer the solutions to suitable cuvettes. Tap the cuvettes to dislodge any gas bubbles from the container walls. Record the A500nm for the Test, Test Blank, Standards and Standard Blank using a suitable spectrophotometer.

Calculations

Standard Curve: (ΔA500nm Standard = A500nm Std–A500nm Std Blank). Plot the ΔA500nm of the Standard vs μmoles of Acethydroxamate.

Sample Determination: ΔA500nm Sample = A500nm Test–A500nm Test Blank. Determine the μmoles of Acethydroxamate produced using the Standard Curve.

$$\text{Units / ml enzyme} = \frac{(\mu\text{moles of acethydroxamate produced})\ (\text{df})}{(5)\ (0.1)}$$

df = Dilution factor
5 = Time (in minutes) of assay as per the Unit Definition
0.1 = Volume (in milliliter) of enzyme used

$$\text{Units / mg solid} = \frac{\text{units / ml enzyme}}{\text{mg solid / ml enzyme}}$$

$$\text{Units / mg protein} = \frac{\text{units / ml enzyme}}{\text{mg protein / ml enzyme}}$$

Unit Definition: *One unit will convert 1.0 μmole of acetamide and hydroxylamine to acethydroxamate and ammonia per minute at pH 7.2 at 37ºC.*

Final Assay Concentration

In a 2.00 ml reaction mix, the final concentrations are 53 mM sodium phosphate, 95 mM acetamide, 475 mM hydroxylamine and 0.3–0.6 unit amidase.

Enzymatic Assay of L-AMINO ACID OXIDASE (EC 1.4.3.2)

Principle

$$\text{L-Phenylalanine} + H_2O \xrightarrow{\text{L-Amino Acid Oxidase}} \text{Phenylpyruvate Catalase}$$

Conditions: T = 37°C, pH = 6.5, A308nm, Light path = 1 cm.

Method: Continuous Spectrophotometric Rate Determination.

Reagents

A. 200 mM Sodium Phosphate Buffer, pH 6.5 at 37°C (Prepare 100 ml in deionized water using Sodium Phosphate, Monobasic, Anhydrous. Adjust to pH 6.5 at 37°C with 1 M NaOH).

B. 10 mM L-Phenylalanine Solution (L-Phe) (Prepare 10 ml in deionized water using L-Phenylalanine).

C. 2000 mM Sodium Arsenate Solution (Arsenate) (Prepare 20 ml in Reagent A using Arsenic Acid, Sodium Salt).

D. 2000 mM Boric Acid Solution, pH 6.5 at 37°C (Boric Acid) (Prepare 20 ml in Reagent C using Boric Acid. Adjust to pH 6.5 at 37°C with 5 M HCl).

E. Catalase Enzyme Solution (Catalase) (Immediately before use, prepare a solution containing 60,000 units/ml in cold deionized water using Catalase).

F. L-Amino Acid Oxidase Enzyme Solution (Immediately before use, prepare a solution containing 0.5–1.0 unit/ml of L-Amino Acid Oxidase in cold deionized water).

Procedure

Prepare a reaction cocktail by pipetting (in milliliters) the following reagents into a suitable container:

Reagent A (Buffer)	11.70
Reagent B (L-Phe)	3.00
Reagent D (Boric Acid)	14.00

Mix by stirring and adjust to pH 6.5 at 37°C with 1 M HCl or 1 M NaOH, if necessary. Pipette (in milliliters) the following reagents into suitable cuvettes:

	Test	Blank
Reaction Cocktail	2.87	2.87
Reagent E (Catalase)	0.03	0.03

Mix by inversion and equilibrate to 37°C. Monitor the A308nm until constant, using a suitably thermostatted spectrophotometer. Then add:

Reagent F (Enzyme Solution)	0.10	—
Deionized Water	—	0.10

Immediately mix by inversion and record the increase in A308nm for approximately 10 minutes. Obtain the ΔA308nm /minute using the maximum linear rate for both the Test and Blank.

Calculations

$$\text{Units / ml enzyme} = \frac{(\Delta A308\text{nm / min Test} - \Delta A308\text{nm / min Blank})\,(3)\,(\text{df})}{(5.00)\,(0.1)}$$

3 = Total volume (in milliliters) of assay
df = Dilution factor
5.00 = Millimolar extinction coefficient of the phenylpyruvate keto borate complex at 308nm
0.1 = Volume (in milliliter) of enzyme

$$\text{Units / mg solid} = \frac{\text{units / ml enzyme}}{\text{mg solid / ml enzyme}}$$

$$\text{Units / mg protein} = \frac{\text{units / ml enzyme}}{\text{mg protein / ml enzyme}}$$

Unit Definition: *One unit will oxidatively deaminate 1.0 μmole of L-phenylalanine per minute at pH 6.5 at 37°C.*

Final Assay Concentration

In a 3.00 ml reaction mix, the final concentrations are 171 mM sodium phosphate, 1.0 mM phenylalanine, 933 mM sodium arsenate, 933 mM boric acid, 1800 units catalase and 0.05–0.1 unit L-amino acid oxidase.

Note

1. Catalase Unit Definition: One unit will decompose 1.0 μmole of H_2O_2 per minute at pH 7.0 at 25°C, while the H_2O_2 concentration falls from 10.3 to 9.2 mM.

Enzymatic Assay of D-AMINO ACID OXIDASE (EC 1.4.3.3)

Principle

$$\text{AD-Alanine} + O_2 \xrightarrow{\text{D-Amino Acid Oxidase}} \text{Pyruvate} + NH_3 \text{ Catalase}$$

$$\text{Pyruvate} + \beta\text{-NADH} \xrightarrow{\text{Lactic Acid Dehydrogenase}} \text{Lactate} + \beta\text{-NAD}$$

Abbreviations Used

β-NADH = β-Nicotinamide Adenine Dinucleotide, Reduced Form
β-NAD = β-Nicotinamide Adenine Dinucleotide, Oxidized Form

Conditions: T = 25°C, pH = 8.3, A340nm, Light path = 1 cm

Method: Continuous Spectrophotometric Rate Determination

Reagents

A. 200 mM Tris HCl Buffer, pH 8.3 at 25°C. (Prepare 100 ml in deionized water using Trizma Base. Adjust to pH 8.3 at 25°C with 1 M HCl. Bubble oxygen through the buffer for 5 minutes immediately before use).

B. 224 mM D-Alanine Solution (Prepare 5 ml in deionized water using D-Alanine).

C. 6.4 mM β-Nicotinamide Adenine Dinucleotide, Reduced Form Solution (β-NADH) (Dissolve the contents of one 5 mg vial of β- Nicotinamide Adenine Dinucleotide, Reduced Form, Disodium Salt, in the appropriate volume of Reagent A. *Prepare Fresh*).

D. Catalase Enzyme Solution (Immediately before use, prepare a solution containing 600 units/ml in cold deionized water using Catalase).

E. Lactic Dehydrogenase Enzyme Solution (LDH) (Immediately before use, prepare a solution containing 400 units/ml in cold deionized water using L-Lactic Dehydrogenase).

F. 3.6 M Ammonium Sulfate Solution, pH 6.5 at 25°C (Enz Dil) (Prepare 25 ml in deionized water using Ammonium Sulfate. Adjust to pH 6.5 at 25°C with 5 M NH_4OH).

G. D-Amino Acid Oxidase Enzyme Solution (D-AAO) (Immediately before use, prepare a solution containing 0.4–0.8 unit/ml of D-Amino Acid Oxidase in cold Reagent F).

Procedure

Pipette (in milliliters) the following reagents into suitable cuvettes:

Test	Blank
Reagent A (Buffer)	2.25
Reagent B (D-Alanine Solution)	0.50
Reagent C (β-NADH)	0.050
Reagent D (Catalase)	0.050
Reagent E (LDH)	0.050

Mix by inversion and equilibrate to 25°C. Monitor the A340nm until constant, using a suitably thermostatted spectrophotometer. Then add:

Deionized Water	—	0.10
Reagent G (D-AAO)	0.10	—

Immediately mix by inversion and record the decrease in A340nm for approximately 5 minutes. Obtain the ΔA340nm/minute using the maximum linear rate for both the Test and Blank.

Calculations

$$\text{Units / mg enzyme} = \frac{\Delta A340\text{nm / min Test} - \Delta A340\text{nm / min Blank}}{(6.22)\ (\text{mg enzyme / ml RM})}$$

6.22 = Millimolar extinction coefficient of β-NADH at 340 nm
RM = Reaction Mix

Unit Definition: *One unit will oxidatively deaminate 1.0 μmole of D-alanine to pyruvate per minute at pH 8.3 at 25°C, in the presence of catalase. (This is equivalent to an O_2 uptake of approximately 335 μl in 30 minutes).*

Final Assay Concentration

In a 3.00 ml reaction mix, the final concentrations are 153 mM Tris, 37 mM D-alanine, 0.11 mM β-NADH, 30 units catalase, 20 units lactic acid dehydrogenase, 120 mM ammonium sulfate, and 0.04–0.08 unit D-amino acid oxidase.

Notes

1. Catalase Unit Definition: One unit will decompose 1.0 μmole of H_2O_2 per minute at pH 7.0 at 25°C, while the H_2O_2 concentration falls from 10.3 to 9.2 mM.
2. Lactic Dehydrogenase Unit Definition: One unit will reduce 1.0 μmole of pyruvate to L-lactate per minute at pH 7.5 at 37°C.

Enzymatic Assay of D-AMINO ACID OXIDASE APOENZYME

Principle

$$\text{D-Alanine} + O_2 + H_2O \xrightarrow{\text{D-Amino Acid Oxidase Apoenzyme}} \text{Pyruvate} + NH_3 + H_2O_2$$

$$2H_2O_2 \xrightarrow{\text{Catalase}} 2H_2O + O_2$$

$$\text{Pyruvate} + \beta-\text{NADH} \xrightarrow{\text{LDH}} \text{L-Lactic Acid} + \beta\text{-NAD}$$

Abbreviations Used

β-NADH = β-Nicotinamide Adenine Dinucleotide, Reduced Form
β-NAD = β-Nicotinamide Adenine, Dinucleotide, Oxidized Form

Conditions: T = 25°C, pH = 8.3, A340nm, Light path = 1 cm

Method: Continuous Spectrophotometric Rate Determination

Reagents

A. 200 mM Tris HCl Buffer, pH 8.3 at 25°C(Prepare 100 ml in deionized water using Trizma Base. Adjust to pH 8.3 at 25°C with 1 M HCl).

B. 224 mM D-Alanine Solution (D-Ala) (Prepare 5 ml in deionized water using D-Alanine).
C. 6.4 mM β-Nicotinamide Adenine Dinucleotide, Reduced Form (β-NADH) (Dissolve the contents of one 5 mg vial of β- Nicotinamide Adenine Dinucleotide, Reduced Form, Disodium Salt, in the appropriate volume of Reagent A. *Prepare Fresh*).
D. Catalase Enzyme Solution (Catalase) (Immediately before use, prepare a solution containing 600 units/ml of Catalase, in cold deionized water.
E. L-Lactic Dehydrogenase Enzyme Solution (LDH) (Immediately before use, prepare a solution containing 400 units/ml of L-Lactic Dehydrogenase, in cold deionized water.
F. D-Amino Acid Oxidase Apoenzyme (D-AAO) (Immediately before use, prepare a solution containing 1 mg/ml of D-Amino Acid Oxidase Apoenzyme in cold deionized water).

Procedure

Saturate Reagent A (Buffer) with O_2 by bubbling oxygen gas through Reagent A (Buffer) for 5 minutes immediately before use. Pipette (in milliliters) the following reagents into suitable cuvettes:

	Test	Blank
Reagent A (O_2 Saturated Buffer)	2.25	2.25
Reagent B (D-Ala)	0.50	0.50
Reagent C (β-NADH)	0.05	0.05
Reagent D (Catalase)	0.05	0.05
Reagent E (LDH)	0.05	0.05

Mix by inversion and equilibrate to 25°C. Monitor the A340nm until constant, using a suitably thermostatted spectrophotometer. Then add:

Reagent F (D-AAO)	0.10	—
Deionized Water	—	0.10

Immediately mix by inversion and record the decrease in A340nm for approximately 5 minutes. Obtain the ΔA340nm/minute using the maximum linear rate for both the Test and Blank.

Calculations

$$\text{Units / ml enzyme} = \frac{(\Delta A340\text{nm / min Test} - \Delta A340\text{nm / min Blank})\ (3)\ (df)}{(6.22)\ (0.1)}$$

3 = Total volume (in milliliters) of the assay
df = Dilution factor
6.22 = Millimolar extinction coefficient of β-NADH at 340 nm
0.1 = Volume (in milliliter) of enzyme used in assay

$$\text{Units / mg solid} = \frac{\text{units / ml enzyme}}{\text{mg solid / ml enzyme}}$$

$$\text{Units / mg protein} = \frac{\text{units / ml enzyme}}{\text{mg protein / ml enzyme}}$$

Unit Definition: *One unit will oxidatively deaminate 1.0 µmole of D-alanine to pyruvate per minute at pH 8.3 at 25°C, in the presence of catalase.*

Final Assay Concentrations

In a 3.00 ml reaction mix, the final concentrations are 153 mM Tris, 37 mM D-alanine, 0.11 mM β-nicotinamide adenine dinucleotide, reduced form, 30 units catalase, 20 units L-lactic dehydrogenase, and 0.1 mg D-amino acid oxidase apoenzyme.

Notes

1. L-Lactic Dehydrogenase Unit Definition: One unit will reduce 1.0 µmole of pyruvate to L-lactate per minute at pH 7.5 at 37°C.
2. Catalase Unit Definition: One unit will decompose 1.0 µmole of H_2O_2 per minute at pH 7.0 at 25°C while the H_2O_2 concentration falls from 10.3 to 9.2 mM. The rate of disappearance of H_2O_2 is followed by observing the rate of decrease in absorbance at 240 nm.

Enzymatic Assay of D-AMINO ACID OXIDASE APOENZYME (Reactivation Assay)

Principle

$$\text{D-Alanine} + O_2 + H_2O + \text{D-Amino Acid} \xrightarrow{\text{Oxidase Apoenzyme}} \text{Pyruvate} + NH_3 + H_2O_2$$

$$2H_2O_2 \xrightarrow{\text{Catalase}} 2H_2O + O_2$$

$$\text{Pyruvate} + \beta\text{-NADH} \xrightarrow{\text{LDH}} \text{L-Lactic Acid} + \beta\text{-NAD}$$

Abbreviations Used

FAD = Flavin Adenine Dinucleotide

β-NADH = β-Nicotinamide Adenine Dinucleotide, Reduced Form

β-NAD = β-Nicotinamide Adenine, Dinucleotide, Oxidized Form

Conditions: T = 25°C, pH = 8.3, A340nm, Light path = 1 cm.

Method: Continuous Spectrophotometric Rate Determination

Reagents

A. 200 mM Tris HCl Buffer, pH 8.3 at 25°C (Prepare 100 ml in deionized water using Trizma Base. Adjust to pH 8.3 at 25°C with 1 M HCl).
B. 224 mM D-Alanine Solution (D-Ala) (Prepare 5 ml in deionized water using D-Alanine).
C. 6.4 mM β-Nicotinamide Adenine Dinucleotide, Reduced Form (β-NADH) (Dissolve the contents of one 5 mg vial of β- Nicotinamide Adenine Dinucleotide, Reduced Form, Disodium Salt, in the appropriate volume of Reagent A. *Prepare Fresh*).
D. Catalase Enzyme Solution (Catalase) (Immediately before use, prepare a solution containing 600 units/ml of Catalase, in cold deionized water.

E. L-Lactic Dehydrogenase Enzyme Solution (LDH) (Immediately before use, prepare a solution containing 400 units/ml of L-Lactic Dehydrogenase, in cold deionized water).

F. D-Amino Acid Oxidase Apoenzyme (D-AAO) (Immediately before use, prepare a solution containing 1 mg/ml of D-Amino Acid Oxidase Apoenzyme in cold deionized water).

G. 50 mM Pyrophosphate Buffer, pH 8.5 at 25°C (PPi) (Prepare 5 ml in deionized water using Pyrophosphate, Tetrasodium, Decahydrate. Adjust to pH 8.5 at 25°C with 1 M HCl).

H. 60 mM Flavin Adenine Dinucleotide Solution (FAD) (Prepare 1 ml in Reagent G using Flavin Adenine Dinucleotide, Disodium Salt).

Procedure

Reactivation of D-Amino Acid Oxidase Apoenzyme: Combine 1 ml of Reagent F (D-AAO) with 0.1 ml of Reagent H (FAD). Incubate at 25°C for 30–45 minutes. Saturate Reagent A (Buffer) with O_2 by bubbling oxygen gas through Reagent A (Buffer) for 5 minutes immediately before use. Pipette (in milliliters) the following reagents into suitable cuvettes:

	Test	Blank
Reagent A (O_2 Saturated Buffer)	2.25	2.25
Reagent B (D-Ala)	0.50	0.50
Reagent C (β-NADH)	0.05	0.05
Reagent D (Catalase)	0.05	0.05
Reagent E (LDH)	0.05	0.05

Mix by inversion and equilibrate to 25°C. Monitor the A340nm until constant, using a suitably thermostatted spectrophotometer. Then add:

Reactivated D-Amino Oxidase Apoenzyme	0.10	—
Deionized Water	—	0.10

Immediately mix by inversion and record the decrease in A340nm for approximately 5 minutes. Obtain the ΔA340nm/minute using the maximum linear rate for both the Test and Blank.

Calculations

$$\text{Units / ml enzyme} = \frac{(\Delta A340\text{nm / min Test} - \Delta A340\text{nm / min Blank})\ (3)\ (\text{df})}{(6.22)\ (0.1)}$$

3 = Total volume (in milliliters) of the assay
df = Dilution factor
6.22 = Millimolar extinction coefficient of β-NADH at 340 nm
0.1 = Volume (in milliliter) of enzyme used in assay

$$\text{Units / mg solid} = \frac{\text{units / ml enzyme}}{\text{mg solid / ml enzyme}}$$

$$\text{Units / mg protein} = \frac{\text{units / ml enzyme}}{\text{mg protein / ml enzyme}}$$

Unit Definition: *One unit will oxidatively deaminate 1.0 μmole of D-alanine to pyruvate per minute at pH 8.3 at 25°C, in the presence of catalase.*

Final Assay Concentrations

In a 3.00 ml reaction mix, the final concentrations are 153 mM Tris, 37 mM D-alanine, 0.11 mM β-nicotinamide adenine dinucleotide, reduced form, 30 units catalase, 20 units L-lactic dehydrogenase, 0.2 mM flavin adenine dinucleotide, 0.2 mM pyrophosphate, and 0.1 mg D-amino acid oxidase apoenzyme.

Notes

1. This assay is used to measure the D-Amino Acid Oxidase activity of D-Amino Acid Oxidase Apoenzyme, after it has been reactivated by incubating it with flavin adenine dinucleotide.
2. Catalase Unit Definition: One unit will decompose 1.0 μmole of H_2O_2 per minute at pH 7.0 at 25°C, while the H_2O_2 concentration falls from 10.3 to 9.2 mM. The rate of disappearance of H_2O_2 is followed by observing the rate of decrease in absorbance at 240 nm.
3. L-Lactic Dehydrogenase Unit Definition: One unit will reduce 1.0 μmole of pyruvate to L-lactate per minute at pH 7.5 at 37°C.

Enzymatic Assay of AMINOACYL-tRNA SYNTHETASE (EC 6.1.1.7)

Principle

tRNA + $_{14}$C Arginine + ATP/tRNA–$_{14}$C Arginine + AMP + Pyrophosphate.

Conditions: T = 37°C, pH = 7.6.

Method: Radiolabelled Stop Reaction

Reagents

A. 1 M Tris HCl Buffer with 50 mM Magnesium Chloride, 500 mM Potassium Chloride, 5 mM Ethylenediaminetetraacetic Acid, and 25 mM Adenosine 5'-Triphosphate, pH 7.6 at 37°C (Reaction Mix Buffer) (Prepare 25 ml in deionized water using Trizma Base, Magnesium Chloride, Hexahydrate, Potassium Chloride, Ethylenediaminetetraacetic Acid, Disodium Salt, Dihydrate, and Adenosine 5'-Triphosphate, Disodium Salt. Add all reagents except Adenosine 5'-Triphosphate, adjust the pH to 7.6 and place on ice. Then add the appropriate amount of Adenosine 5'-Triphosphate).

B. $_{14}$C Arginine Solution ($_{14}$C Arg) (Use $_{14}$C L-Arginine, 300 mCi/mmol, 50 μCi/ml).

C. Transfer RNA Solution (tRNA) (Prepare 1 ml in deionized water containing 160 A260 units/ml).

D. 10 mM Tris HCl Buffer, with 50% (v/v) Glycerol, 10 mM Magnesium Chloride, 10 mM Potassium Chloride, and 30 mM 2-Mercaptoethanol, pH 7.2 at 25°C (Enz Dil) (Prepare 25 ml in deionized water using Trizma Base, Glycerol, Potassium Chloride, 2-Mercaptoethanol, and Magnesium Chloride, Hexahydrate).

E. Aminoacyl-tRNA Synthetase Enzyme Solution (Immediately before use, prepare a solution containing 1–4 mg/ml (based on protein) of Aminoacyl-tRNA Synthetase in ice cold Reagent D).

F. 10% (w/v) Trichloroacetic Acid Solution (TCA) (Prepare 10 ml in deionized water using Trichloroacetic Acid, 6.1 N Solution, approximately 100% (w/v). Store on ice).

G. 5% (w/v) Trichloroacetic Acid Solution (Wash Solution) (Prepare 5 ml in deionized water using Reagent F).

H. Methylethyl Cellosolve (Prepare by adding equal volumes of Ethylene Glycol Monoethyl Ether, and Ethylene Glycol Monomethyl Ether).

I. Scintillation Cocktail.

Procedure

Pipette (in milliliters) the following reagents into suitable one dram vials:

	Test	Blank
Deionized Water	0.370	0.370
Reagent A (Reaction Mix Buffer)	0.050	0.050
Reagent C (tRNA)	0.050	0.050
Reagent B ($_{14}$C Arg)	0.020	0.020

Mix by swirling and equilibrate to 37°C. Then add:

Reagent E (Enzyme Solution)	0.01	—
Reagent D (Enz Dil)	—	0.01

Immediately mix by swirling and incubate at 37°C for exactly 10 minutes. Then remove three 0.050 ml aliquots from both the Test and Blank and place into suitable containers containing 0.2 ml of ice cold Reagent F (TCA). Allow the samples to stand for 5–10 minutes on ice. Filter the solutions through 0.45 μm Millipore HA Type filters. Wash the filters 3 times with 0.100 ml each of Reagent G (Wash Solution). Allow the filters to air dry or dry under a heat lamp and transfer them to 2 dram scintillation vials. Dissolve the filters in 2 ml of Reagent H (Methylethyl Cellosolve), then add 5 ml of Reagent I (Scintillation Cocktail). Only use filters that are soluble in Reagent H (Methylethyl Cellosolve). Count the radioactivity in a suitable scintillation counter. Potential DPM (disintegrations per minute) are prepared by pipetting 0.02 ml of Reagent B ($_{14}$C Arg) into a scintillation vial with 2 ml of Reagent H (Methylethyl Cellosolve) and 5 ml of Reagent I (Scintillation Cocktail).

Calculations

$$\text{Potential DPM / pmole} = \frac{\text{DPM of potential}}{\text{Total pmoles of L-Arginine}}$$

$$\text{Units / ml enzyme} = \frac{(\text{DPM Test} - \text{DPM Blank})\,(\text{df})\,(0.5)}{(0.01)\,(\text{Potential DPM / pmole})\,(0.05)}$$

DPM = Disintegrations per minute

df = Dilution factor

0.5 = Volume (in milliliter) of assay

0.01 = Volume (in milliliter) of enzyme used

0.05 = Volume (in milliliter) of reaction mixture which is added to the scintillation cocktail

$$\text{Units / mg protein} = \frac{\text{units / ml enzyme}}{\text{mg protein / ml enzyme}}$$

Unit Definition: *One unit will activate and attach 1.0 picomole (10–12mole) of labeled amino acid to tRNA in 10 minutes at pH 7.6 at 37°C (amino acid used: L-arginine).*

Final Assay Concentration

In a 0.50 ml reaction mix, the final concentrations are 100 mM Tris, 5 mM magnesium chloride, 50 mM potassium chloride, 0.5 mM Ethylenediaminetetraacetic Acid, 2.5 mM adenosine 5'-triphosphate, 8 units t-RNA, 6.67 μM arginine, 1.0% (w/v) glycerol, 0.6 mM 2-mercaptoethanol, and 0.01–0.04 mg aminoacyl tRNA synthetase.

Notes

1. The type of t-RNA used in the assay depends on the source of the t-RNA Synthetase. For Aminoacyl-tRNA Synthetase, Crude from Baker's Yeast, use Ribonucleic Acid, Transfer, Type X from Baker's Yeast, for Aminoacyl-tRNA Synthetase, Crude from Bovine Liver, AminoacyltRNA Synthetase from Rabbit Liver, and for Aminoacyl-tRNA Synthetase, Crude from E. Coli, use Ribonucleic acid, Transfer, Type XXI from *Escherichia coli*).
2. This enzyme is extremely unstable and should not be exposed to room temperature for more than a few minutes. It should also not be exposed to repeated freeze-thaw cycles.

Enzymatic Assay of ν-AMINOLEVULINATE DEHYDRATASE (EC 4.2.1.24)

Principle

$$\text{2d Aminolevulinic Acid} \xrightarrow{\text{AVD}} \text{Porphobilinogen} + 2H_2O$$

Abbreviation Used

AVD = d-Aminolevulinate Dehydratase

Conditions: T = 37°C, pH = 6.5, A555nm, Light path = 1 cm

Method: Colorimetric

Reagents

A. 100 mM Potassium Phosphate Buffer with 20 mM Dithiothreitol, pH 6.5 at 37°C (Prepare 100 ml in deionized water using Potassium Phosphate, Monobasic, Anhydrous, and DL-Dithiothreitol. Adjust to pH 6.5 at 37°C with 1 M KOH).
B. 50 mM d-Aminolevulinic Acid Solution (AV Acid) (Prepare 5 ml in Reagent A using d-Aminolevulinic Acid, Hydrochloride).
C. 0.16 mM Porphobilinogen Solution (Porph) (Prepare 10 ml in Reagent A using Porphobilinogen).
D. 10% (v/v) Trichloroacetic Acid with 100 mM Mercuric Chloride Solution (TCA) (Prepare 15 ml in deionized water using Trichloroacetic Acid Solution, 6.1 N, approximately 100% (w/v), and Mercuric Chloride).
E. Ehrlich's Color Reagent (ECR) (Prepare by adding 1 g of p-Dimethylaminobenzaldehyde, to 42 ml of Acetic Acid, Glacial. Then add 8 ml of Perchloric Acid).
F. d-Aminolevulinate Dehydratase (Enzyme Soln) (Immediately before use, prepare a solution containing 0.15 unit/ml in cold Reagent A).

Procedure

Pipette (in milliliters) the following reagents into suitable containers:

	Test 1	Test 2	Test 3	Blank	Std 1	Std 2	Std 3	Std 4	Std 5	Std Blank
Reagent A (Buffer)	0.70	0.65	0.60	0.90	0.80	0.75	0.70	0.65	0.60	1.00
Reagent F (Enzyme Soln)	0.20	0.25	0.30	0.10	—	—	—	—	—	—

Mix by vortexing and equilibrate at 37°C for 10 minutes (time required for enzyme activation). Then add:

	Test 1	Test 2	Test 3	Blank	Std 1	Std 2	Std 3	Std 4	Std 5	Std Blank
Reagent C (Porph)	—	—	—	—	0.20	0.25	0.30	0.35	0.40	—
Reagent B (AV Acid)	0.10	0.10	0.10	—	—	—	—	—	—	—

Mix by vortexing and incubate at 37°C for 60 minutes. Then add:

	Test 1	Test 2	Test 3	Blank	Std 1	Std 2	Std 3	Std 4	Std 5	Std Blank
Reagent D (TCA)	1.00	1.00	1.00	1.00	1.00	1.00	1.00	1.00	1.00	1.00

Mix by vortexing and centrifuge to clarify. Pipette (in milliliters) the following into suitable cuvettes:

	Test 1	Test 2	Test 3	Blank	Std 1	Std 2	Std 3	Std 4	Std 5	Std Blank
Test Supernatant	0.50	0.50	0.50	—	—	—	—	—	—	—
Blank Supernatant	—	—	0.50	—	—	—	—	—	—	—
Std Supernatant	—	—	—	—	0.50	0.50	0.50	0.50	0.50	—
Std Blk Supernatant	—	—	—	—	—	—	—	—	—	0.50
Reagent E (EÇR)	0.50	0.50	0.50	0.50	0.50	0.50	0.50	0.50	0.50	0.50

Mix by inversion and let the precipitate settle. Read the A555nm for the Test, Blank, Standards, and Standard Blank after 10 minutes.

Calculations

Standard Curve: ΔA555nm Standard = A555nm Standard–A555nm Standard Blank

Prepare a Standard Curve by plotting the ΔA555nm Standard vs the μmoles of porphobilinogen.

Sample Determination: ΔA555nm Test = A555nm Test–A555nm Blank

Determine the μmoles of porphobilinogen produced using the Standard curve.

$$\text{Units/ml enzyme} = \frac{(\mu\text{mole of porphobilinogen produced})}{(0.25)}$$

0.25 = Volume (in milliliter) of enzyme used (may also be 0.20 or 0.30 depending upon the enzyme volume)

$$\text{Units/mg solid} = \frac{\text{units/ml enzyme}}{\text{mg solid/ml enzyme}}$$

$$\text{Units/mg protein} = \frac{\text{units/ml enzyme}}{\text{mg protein/ml enzyme}}$$

Unit Definition: *One unit will produce 1.0 μmole of porphobilinogen from d-aminolevulinic acid in 60 minutes at pH 6.5 at 37°C.*

Final Assay Concentrations

In a 1.00 ml reaction mix, the final concentrations are 100 mM potassium phosphate, 20 mM dithiothreitol, 5 mM d-aminolevulinic acid, and 0.03–0.045 unit d-aminolevulinate dehydratase.

Note

1. This precipitate does not interfere with the assay and the solution is not centrifuged after formation of the precipitate. This eliminates contaminating additional laboratory glassware withmercuric chloride.

Enzymatic Assay of AMINOPEPTIDASE (EC 3.4.11.10)

Principle

$$\text{L-Leucine p-Nitroanilide} \xrightarrow{AP} \text{L-Leucine} + \text{Nitroaniline}$$

Abbreviation Used

AP = Aminopeptidase

Conditions: T = 25°C, pH = 8.0, A405nm, Light path = 1 cm.

Method: Continuous Spectrophotometric Rate Determination

Reagents

A. 1 mM Tricine Solution (Prepare 100 ml in deionized water using Tricine).
B. Methanol (Use Methanol, Absolute).
C. 50 mM L-Leucine p-Nitroanilide Stock Solution (Prepare 1 ml in Reagent B using L-Leucine p-Nitroanilide, Free Base).
D. 1.0 mM L-Leucine p-Nitroanilide Solution (Leu-NA) (Prepare 5 ml by adding 0.1 ml of Reagent C to 4.9 ml of Reagent A).
E. 200 mM Tricine Buffer, pH 8.0 at 25°C (Buffer) (Prepare 100 ml in deionized water using Tricine. Adjust to pH 8.0 at 25°C with 1 M NaOH).
F. 20 mM Tricine Buffer with 0.05% (w/v) Bovine Serum Albumin, pH 8.0 at 25°C (Enz Dil) (Prepare 25 ml in deionized water using Reagent E and Albumin, Bovine. Adjust to pH 8.0 at 25°C with either 1 M HCl or 1 M NaOH if necessary).
G. Aminopeptidase Enzyme Solution (Immediately before use, prepare a solution containing 0.02–0.04 unit/ml of Aminopeptidase in cold Reagent F).

Procedure

Prepare a reaction cocktail by pipetting (in milliliters) the following reagents into a suitable container:

Reagent D (Leu-NA)	2.00
Reagent E (Buffer)	1.00
Deionized Water	7.00

Mix by swirling. Pipette (in milliliters) the following reagents into suitable cuvettes:

	Test	Blank
Reaction Cocktail	0.90	0.90

Equilibrate to 25°C. Then add:

Reagent G (Enzyme Solution)	0.10	—
Reagent F (Enz Dil)	—	0.10

Immediately mix by inversion and record the increase in A405nm for approximately 5 minutes. Obtain the ΔA405nm/minute using the maximum linear rate for both the Test and Blank.

Calculations

$$\text{Units / ml enzyme} = \frac{(\Delta A405nm / min\ Test - \Delta A405nm / min\ Blank)\ (1)\ (df)}{(10.8)\ (0.1)}$$

1 = Total volume (in milliliter) of assay
df = Dilution factor
10.8 = Millimolar extinction coefficient of p-Nitroaniline at A405nm
0.1 = Volume (in milliliter) of enzyme used

$$\text{Units / mg solid} = \frac{\text{units / ml enzyme}}{\text{mg solid / ml enzyme}}$$

$$\text{Units / mg protein} = \frac{\text{units / ml enzyme}}{\text{mg protein / ml enzyme}}$$

Unit Definition: *One unit will hydrolyze 1.0 μmole of L-leucine p-nitroanilide to L-leucine and p-nitroaniline per minute at pH 8.0 at 25°C.*

Final Assay Concentrations

In a 1.00 ml reaction mix, the final assay concentrations are 20 mM Tricine, 0.4% (v/v) methanol, 0.18 mM L-leucine p-nitroanilide, 0.005% (w/v) bovine serum albumin, and 0.002–0.004 unit aminopeptidase.

Enzymatic Assay of AMINOPEPTIDASE I

Principle

$$\text{L-Leucine p-Nitroanilide} + H_2O \xrightarrow{\text{AP I}} \text{L-Leucine} + \text{p-Nitroaniline}$$

Abbreviation Used

AP I = Aminopeptidase I

Conditions: T = 25°C, pH = 8.0, A405nm, Light path = 1 cm

Method: Continuous Spectrophotometric Rate Determination

Reagents

A. 50 mM Tris HCl Buffer with 100 mM Sodium Chloride and 1 mM Calcium Chloride, pH 8.0 at 25°C (Prepare 100 ml in deionized water using Trizma Hydrochloride,

Sodium Chloride and Calcium Chloride, Dihydrate. Adjust to pH 8.0 at 25°C with 1 M NaOH).

B. Methanol (Use Methanol, Absolute).

C. 150 mM L-Leucine p-Nitroanilide Stock Solution (Prepare 1 ml in Reagent B using L-Leucine p-Nitroanilide, Free Base).

D. 3 mM L-Leucine p-Nitroanilide Solution (Leu-NA) (Prepare 30 ml in Reagent A using Reagent C).

E. Aminopeptidase I Enzyme Solution (Immediately before use, prepare a solution containing 0.1–0.5 unit/ml of Aminopeptidase I in cold Reagent A).

Procedure

Prepare (in milliliters) the following reagents into suitable cuvettes:

	Test	Blank
Reagent D (Leu-NA)	2.95	2.95

Equilibrate to 25°C. Then add:

Reagent E (Enzyme Solution)	0.05	—
Reagent A (Buffer)	—	0.05

Immediately mix by inversion and record the increase in A405nm for approximately 5 minutes. Obtain the ΔA405nm/min using the maximum linear rate for both the Test and Blank.

Calculations

$$\text{Units / ml enzyme} = \frac{(\Delta A405nm / \min \text{ Test} - \Delta A405nm / \min \text{ Blank})\ (3)\ (df)}{(10.6)\ (0.05)}$$

3 = Total volume (in milliliters) of assay
df = Dilution factor
10.6 = Millimolar extinction coefficient of p-Nitroaniline at A405nm
0.05 = Volume (in milliliter) of enzyme used

$$\text{Units / mg solid} = \frac{\text{units / ml enzyme}}{\text{mg solid / ml enzyme}}$$

$$\text{Units / mg protein} = \frac{\text{units / ml enzyme}}{\text{mg protein / ml enzyme}}$$

Unit Definition: *One unit will hydrolyze 1.0 μmole of L-leucine p-nitroanilide to L-leucine and p-nitroaniline per minute at pH 8.0, at 25°C, and 3.0 mM substrate concentration.*

Final Assay Concentrations

In a 3.00 ml reaction mix, the final assay concentrations are 49 mM Tris, 98 mM sodium chloride, 1 mM calcium chloride, 2% (v/v) methanol, 3 mM L-leucine p-nitroanilide and 0.005–0.025 unit aminopeptidase.

Note

1. The substrate is close to its solubility limit in this buffer, so care must be taken to assure it is completely dissolved. Upon diluting Reagent C (Stock) with Reagent A (Buffer), mix vigorously using a vortex mixer.

Enzymatic Assay of α-AMYLASE (EC 3.2.1.1)

Principle

$$\text{Starch} + H_2O \xrightarrow{\alpha\text{-Amylase}} \text{Reducing Groups (Maltose)}$$

Conditions: T = 20°C, pH = 6.9, A540nm, Light path = 1 cm.

Method: Colorimetric

Reagents

A. 20 mM Sodium Phosphate Buffer with 6.7 mM Sodium Chloride, pH 6.9 at 20°C (Prepare 100 ml in deionized water using Sodium Phosphate, Monobasic, Anhydrous and Sodium Chloride. Adjust to pH 6.9 at 20°C with 1 M NaOH).

B. 1.0% (w/v) Soluble Starch Solution (Starch) (Prepare 25 ml in Reagent A using Starch Potato Soluble. Facilitate solubilization by heating the starch solution in a glass beaker directly on a heating/stir plate using constant stirring. Bring to boil and maintain the solution at this temperature for 15 minutes. Allow the starch solution to cool to room temperature with stirring. Return the starch solution to its original volume (25 ml) by the addition of water and dispense samples for assay while stirring).

C. Sodium Potassium Tartrate Solution (Dissolve 12.0 grams of Sodium Potassium Tartrate, Tetrahydrate, in 8.0 ml of 2 M NaOH. Heat directly on a heating/stir plate using constant stirring to dissolve. *Do not boil*).

D. 96 mM 3,5-Dinitrosalicylic Acid Solution (Prepare 20 ml in deionized water using 3,5-Dinitrosalicylic Acid. Heat directly on a heating/stir plate using constant stirring to dissolve. *Do not boil*).

E. Color Reagent Solution (Clr Rgt Soln) (With stirring, slowly add Reagent C to Reagent D. Dilute to 40 ml with deionized water. If not completely dissolved, the reagents should dissolve when mixed. The solution should be stored in an amber bottle at room temperature. The Color Reagent Solution is stable for 6 months).

F. 0.2% (w/v) Maltose Standard Solution (Prepare 10 ml in deionized water using Maltose, Monohydrate).

G. α-Amylase Solution (Immediately before use, prepare a solution containing 1 unit/ml of α-Amylase in cold deionized water).

Procedure

Pipette (in milliliters) the following reagents into suitable containers:

	Test	Blank
Reagent B (Starch)	1.00	1.00

Mix by swirling and equilibrate to 20°C. Then add:

	Test	Blank
Reagent G (Enzyme Solution)	1.00	—

Mix by swirling and incubate for exactly 3.0 minutes at 20°C. Then add:

Reagent E (Clr Rgt Soln)	1.00	1.00
Reagent G (Enzyme Solution)	—	1.00

Cap and place in a boiling water bath for exactly 15 minutes, then cool on ice to room temperature and add:

Deionized water	9.00	9.00

Mix by inversion and record the A540nm for both the Test and Blank using a suitable spectrophotometer.

Standard Curve: A standard curve is made by pipetting (in milliliters) the following reagents into suitable containers:

	Std 1	Std 2	Std 3	Std 4	Std 5	Std Blank
Reagent F (Std Soln)	0.20	0.40	0.60	0.80	1.00	—
Deionized Water	1.80	1.60	1.40	1.20	1.00	2.00
Reagent E (Clr Rgt Soln)	1.00	1.00	1.00	1.00	1.00	1.00

Cap and place in a boiling water bath for exactly 15 minutes, then cool on ice to room temperature and add:

Deionized Water	9.00	9.00	9.00	9.00	9.00	9.00

Mix by inversion and record the A540nm for the Standards and Standard Blank using a suitable spectrophotometer.

Calculations

Standard Curve: ΔA540nm Standard = A540nm Std–A540nm Std Blank

Plot the ΔA540nm of the Standards vs milligrams of Maltose.

Sample Determination: ΔA540nm Sample = A540nm Test–A540nm Blank

Determine the milligrams of Maltose liberated using the Standard Curve.

$$\text{Units / ml enzyme} = \frac{(\text{mg of Maltose released})(\text{df})}{(1)}$$

df = dilution factor

1 = Volume (in milliliter) of enzyme used

$$\text{Units / mg solid} = \frac{\text{units / ml enzyme}}{\text{mg solid / ml enzyme}}$$

$$\text{Units / mg protein} = \frac{\text{units / ml enzyme}}{\text{mg protein / ml enzyme}}$$

Unit Definition: *One unit will liberate 1.0 mg of maltose from starch in 3 minutes at pH 6.9 at 20°C.*

Final Assay Concentrations

In a 2.00 ml reaction mix, the final concentrations are 10 mM sodium phosphate, 0.50% (w/v) starch, 3.4 mM sodium chloride and 1 unit α-amylase.

Enzymatic Assay of β-AMYLASE (EC 3.2.1.2)

Principle

$$\text{Starch} + H_2O \xrightarrow{\beta\text{-Amylase}} \text{Reducing Groups (Maltose)}$$

Conditions: T = 20°C, pH = 4.8, A540nm, Light path = 1 cm.

Method: Colorimetric

Reagents

A. 16 mM Sodium Acetate Buffer, pH 4.8 at 20°C (Prepare 100 ml in deionized water using Sodium Acetate, Trihydrate. Adjust to pH 4.8 at 20°C with 1 M HCl).

B. 1.0% (w/v) Soluble Starch Solution (Starch) (Prepare 25 ml in Reagent A using Starch Potato, Soluble. Facilitate solubilization by heating the starch solution in a glass beaker, directly on a heating/stir plate using constant stirring. Bring to a boil and maintain the solution at this temperature for 15 minutes. Allow the starch solution to cool to room temperature with stirring. Return the starch solution to its original volume (25 ml) by the addition of deionized water and dispense samples for assay while stirring).

C. Sodium Potassium Tartrate Solution (Dissolve 12.0 grams of Sodium Potassium Tartrate, Tetrahydrate, in 8.0 ml of 2 M NaOH. Heat directly on a heating/stir plate using constant stirring to dissolve. *Do not boil*).

D. 96 mM 3,5-Dinitrosalicylic Acid Solution (Prepare 20 ml in deionized water using 3,5-Dinitrosalicylic Acid. Heat directly on a heating/stir plate using constant stirring to dissolve. *Do not boil*).

E. Color Reagent Solution (Clr Rgt Soln) (With stirring, slowly add Reagent C to Reagent D. Dilute to 40 ml with deionized water. If not completely dissolved, the reagents should dissolve when mixed. The solution should be stored in an amber bottle at room temperature. The Color Reagent Solution is stable for 6 months).

F. 0.2% (w/v) Maltose Standard Solution (Prepare 10 ml in deionized water using Maltose, Monohydrate).

G. β-Amylase Enzyme Solution (Immediately before use, prepare a solution containing 1 unit/ml of β-Amylase in cold deionized water).

Procedure

Pipette (in milliliters) the following reagents into suitable containers:

	Test	Blank
Reagent B (Starch)	1.00	1.00

Mix by swirling and equilibrate to 20°C. Then add:

Reagent G (Enzyme Solution)	1.00	—

Mix by swirling and incubate for exactly 3.0 minutes at 20°C. Then add:

Reagent E (Clr Rgt Soln)	1.00	1.00
Reagent G (Enzyme Solution)	—	1.00

Cap and place in a boiling water bath for exactly 15 minutes, then cool on ice to room temperature and add:

Reagent E (Clr Rgt Soln)	1.00	1.00
Deionized water	9.00	9.00

Mix by inversion and record the A540nm for both the Test and Blank using a suitable spectrophotometer.

Standard Curve: A standard curve is made by pipetting (in milliliters) the following reagents into suitable containers:

	Std 1	Std 2	Std 3	Std 4	Std 5	Std Blank
Reagent F (Std Soln)	0.20	0.40	0.60	0.80	1.00	—
Deionized Water	1.80	1.60	1.40	1.20	1.00	2.00
Reagent E (Clr Rgt Soln)	1.00	1.00	1.00	1.00	1.00	1.00

Place in a boiling water bath for exactly 15 minutes, then cool on ice to room temperature and add:

Deionized Water	9.00	9.00	9.00	9.00	9.00	9.00

Mix by inversion and record the A540nm for the Standards and Standard Blank using a suitable spectrophotometer.

Calculations

Standard Curve: ΔA540nm Standard = A540nm Std–A540nm Std Blank

Plot the ΔA540nm of the Standards vs milligrams of Maltose.

Sample Determination: ΔA540nm Sample = A540nm Test–A540nm Test Blank

Determine the milligrams of Maltose liberated using the Standard Curve.

$$\text{Units / ml enzyme} = \frac{(\text{mg of Maltose released})\ (\text{df})}{(1)}$$

df = Dilution Factor

1 = Volume (in milliliter) of enzyme used

$$\text{Units / mg solid} = \frac{\text{units / ml enzyme}}{\text{mg solid / ml enzyme}}$$

$$\text{Units / mg protein} = \frac{\text{units / ml enzyme}}{\text{mg protein / ml enzyme}}$$

Unit Definition: *One unit will liberate 1.0 mg of maltose from starch in 3 minutes at pH 4.8 at 20°C.*

Final Assay Concentrations

In a 2.00 ml reaction mix, the final concentrations are 8 mM sodium acetate, 0.50% (w/v) starch and 1 unit β-amylase.

Suitability Assay for Starch Azure as a Substrate for α-AMYLASE

Principle

$$\text{Insoluble Starch Azure} + H_2O \xrightarrow{\alpha\text{-Amylase}} \text{Soluble Colored Reaction Product}$$

Conditions: T = 37°C, pH = 7.0, A595nm, Light path = 1 cm

Method: Colorimetric

Reagents

A. 20 mM Sodium Phosphate Buffer with 50 mM Sodium Chloride, pH 7.0 at 37°C (Prepare 100 ml in deionized water using Sodium Phosphate, Monobasic, Anhydrous, and Sodium Chloride. Adjust to pH 7.0 at 37°C with 1 M NaOH).

B. 2.0% (w/v) Starch Azure Suspension (Starch Azure) (Prepare 4.5 ml in Reagent A using Starch Azure).

C. 2.75 M Acetic Acid Solution (Acetic Acid) (Prepare 100 ml in deionized water using Acetic Acid, Glacial).

D. α-Amylase Enzyme Solution (Immediately before use, prepare a solution containing 0.25 mg/ml of α-Amylase, in cold deionized water).

Procedure

Pipette (in milliliters) the following reagents into suitable containers:

	Test	Blank
Reagent B (Starch Azure)	4.50	4.50

Equilibrate to 37°C. Then add:

	Test	Blank
Reagent D (Enzyme Solution)	0.50	—

Mix by swirling and incubate at 37°C for exactly 15 minutes in a water bath where constant shaking can be maintained. Then add:

	Test	Blank
Reagent C (Acetic Acid)	2.00	2.00
Reagent D (Enzyme Solution)	—	0.50

Mix by swirling. Filter the suspension through Whatman 54 filter paper. Then transfer the filtered solutions to suitable cuvettes and record the A595nm for both the Test and Blank using a suitable spectrophotometer.

Calculations

$$\Delta A595nm = A595nm\ Test - A595nm\ Blank$$

$$\Delta A595nm / 15\ min / RM = \frac{\Delta A595nm}{(15)\ (RM)}$$

15 = Time (in minutes) of assay as per the Unit Definition
RM = Reaction Mixture

Specification: Compare the value of the ΔA595nm/15 min/RM to that of a control sample. The values should be similar.

Final Concentrations

In a 5.00 ml reaction mix, the final concentrations are 18 mM sodium phosphate, 45 mM sodium chloride, 1.8% (w/v) starch azure, and 0.125 mg α-amylase.

Enzymatic Assay of AMYLOGLUCOSIDASE (EC 3.2.1.3)

Principle

Step 1

$$\text{Starch} + H_2O \xrightarrow{\text{Amyloglucosidase}} \text{D-Glucose}$$

Step 2

$$\text{D-Glucose} + \text{ATP} \xrightarrow{\text{Hexokinase}} \text{Glucose 6-Phosphate} + \text{ADP}$$

$$\text{Glucose 6-Phosphate} + \beta\text{-NADP} \xrightarrow{\text{G-6-PDH}} \text{6-PG} + \beta\text{-NADPH}$$

Abbreviations Used

ATP = Adenosine 5'-Triphosphate
ADP = Adenosine 5'-Diphosphate
G-6-PDH = Glucose-6-Phosphate Dehydrogenase
β-NADP = β-Nicotinamide Adenine Dinucleotide Phosphate, Oxidized Form
β-NADPH = β-Nicotinamide Adenine Dinucleotide Phosphate, Reduced Form
6-PG = 6-Phospho-D-Gluconate

Conditions: T = 55°C, pH = 4.5, A340nm, Light path = 1 cm

Method: Spectrophotometric Stop Rate Determination

Reagents

A. 50 mM Sodium Acetate Buffer, pH 4.5 at 55°C (Prepare 50 ml in deionized water using Sodium Acetate, Trihydrate. Adjust to pH 4.5 at 55°C with 1 M HCl).

B. 1% (w/v) Starch Solution (Starch) (Prepare 10 ml in Reagent A using Starch, Potato, Soluble. Facilitate solubilization by heating. *Do not boil*).

C. Amyloglucosidase Enzyme Solution (Amylogluc) (Immediately before use, prepare a solution containing 0.3–0.6 unit/ml of Amyloglucosidase in cold deionized water).
D. 50% (w/v) Trichloroacetic Acid Solution (TCA) (Prepare 5 ml in deionized water using Trichloroacetic Acid, 6.1 N Solution, approximately 100% (w/v)).
E. Glucose (HK) Determination Vial (16–10) (Immediately before use, dissolve the contents of one vial of Glucose (HK) 50 in 50 ml of deionized water).

Procedure

Step 1: Pipette (in milliliters) the following reagents into suitable tubes:

	Test	Blank
Reagent B (Starch Solution)	1.00	1.00

Equilibrate to 55°C. Then add:

Reagent C (Amylogluc)	1.00	—

Immediately mix by swirling and incubate at 55°C for exactly 3 minutes. Then add:

Reagent D (TCA)	0.30	0.30
Reagent C (Amylogluc)	—	1.00

Mix by swirling and adjust to pH 7.0 with solid Sodium Bicarbonate. Centrifuge the solutions to clarify and use the supernatant in Step 2.

Step 2: Pipette (in milliliters) the following reagents into suitable cuvettes:

	Test	Blank
Reagent E (16–10)	2.80	2.80

Equilibrate to 25°C. Monitor the A340nm until constant, using a suitably thermostatted spectrophotometer. Record this value as the initial A340nm. Then add:

Test Supernatant (Step 1)	0.20	—
Blank Supernatant (Step 2)	—	0.20

Immediately mix by inversion. Monitor the A340nm until the A340nm is constant (approximately 5–10 minutes at room temperature). Obtain the final A340nm for both the Test and Blank.

Calculations

$$\Delta A340nm = A340nm\ Final - A340nm\ Initial$$

$$\text{Units / ml enzyme} = \frac{(\Delta A340nm\ Test - \Delta A308nm\ Blank)\ (180)\ (2.3)\ (3.0)\ (df)}{(6.22)\ (1000)\ (1)\ (0.2)}$$

180 = Micrograms of glucose per micromole of glucose
2.3 = Total volume (in milliliters) of Step 1
3.0 = Total volume (in milliliters) of Step 2
df = Dilution factor
6.22 = Millimolar extinction coefficient of β-NADPH at 340nm

1000 = Conversion factor from micrograms to milligrams
1 = Volume (in milliliter) of enzyme used in Step 1
0.2 = Volume (in milliliter) from Step 1 used in Step 2

$$\text{Units / mg solid} = \frac{\text{units / ml enzyme}}{\text{mg solid / ml enzyme}}$$

$$\text{Units / g solid} = \frac{\text{units / ml enzyme}}{\text{g solid / ml enzyme}}$$

$$\text{Units / mg protein} = \frac{\text{units / ml enzyme}}{\text{mg protein / ml enzyme}}$$

Unit Definition: *One unit will liberate 1.0 milligram of glucose from starch in three minutes at pH 4.5 at 55°C.*

Final Assay Concentration

In a 2.00 ml reaction mix, the final concentrations are 25 mM sodium acetate, 0.5% (w/v) starch and 0.3–0.6 unit amyloglucosidase.

Enzymatic Assay of ANCROD (EC 3.4.21.28)

Principle

$$\text{Fibrinogen} \xrightarrow{\text{Ancrod}} \text{Fibrin}$$

Conditions: T = 37°C, pH 7.35

Method: Fibrometer

Reagents

A. 143 mM 5,5-Diethylbarbiturate and 143 mM Sodium Acetate Solution (Soln A) (Prepare 275 ml in deionized water using Sodium Acetate, Trihydrate, and Barbital, Sodium Salt).
B. 4.25% (w/v) Sodium Chloride Solution (Prepare 300 ml in deionized water using Sodium Chloride).
C. 36 mM 5,5-Diethylbarbiturate, 36 mM Sodium Acetate, 0.85% (w/v) Sodium Chloride Solution, pH 7.35 at 25°C (Soln B) (Prepare 1 liter by combining 250 ml of Reagent A, 200 ml of Reagent B, 217 ml of 0.1 M HCl, and 333 ml of deionized water. Adjust the pH to 7.35 at 37°C with either 1 M HCl or 1 M NaOH).
D. 25.7 mM Sodium Citrate Solution (Prepare 100 ml in deionized water using Citric Acid Trisodium Salt, Dihydrate).
E. 0.85% (w/v) Sodium Chloride Solution (NaCl) (Prepare 300 ml in deionized water using Reagent B).
F. 7.2 mM 5,5-Diethylbarbiturate, 7.2 mM Sodium Acetate, 5.1 mM Sodium Citrate, 0.68% (w/v) Sodium Chloride, 1.0% (w/v) Bovine Serum Albumin, and 0.5% (w/v) Polyethylene Glycol (Enz Dil) (Prepare 500 ml by combining 100 ml of Reagent C, 100 ml of Reagent D, 300 ml of Reagent E, Albumin Bovine, and Polyethylene Glycol).

G. Plasma Solution (Plasma) (Immediately before use, reconstitute 2.0 ml of Human Plasma, lyophilized with 2 ml of deionized water. Then add 2 ml of Reagent E. Keep at room temperature).

H. Ancrod Enzyme Solution (Ancrod) (Immediately before use, reconstitute the Ancrod vial with 1 ml of deionized water. Further dilute to 2.7–5.1 units/ml of Ancrod in Reagent F, in glass tubes. Serial dilutions should be made in Reagent F so that clotting times are between 15–25 seconds. If the clotting time recorded is too long, increase the Ancrod concentration. If the time recorded is too short, decrease the Ancrod concentration. Subsequentdilutions are also to be made in glass tubes).

I. NIH Thrombin Standard Solution (Std) (Use Thrombin Reference Standard (Lot J) which has been diluted in Reagent F.

Note: The current Standard Curve has been made with N.I.H. Thrombin Standard diluted in glass test tubes.

Procedure

Step 1: Pipette (in milliliters) the following reagents into fibrometer cups, which have been pre-warmed to 37°C.

	Test
Reagent G (Plasma)	0.20

Allow to stand at 37°C for thirty seconds. Then add:

Reagent H (Ancrod)	0.10

Record the clotting time using a suitably thermostatted fibrometer. A 0.3 ml fibrometer probe should be used. Compare the clotting times of the Tests against a NIH Standard Curve for Thrombin clotting times versus units of Thrombin/ml. Determine the units of Ancrod of the Test from the Standard Curve.

Calculations

Units/ml enzyme = (Units of Ancrod from Std Curve) × (df)
df = Dilution factor

$$\text{Units / mg solid} = \frac{\text{units / ml enzyme}}{\text{mg solid / ml enzyme}}$$

$$\text{Units / mg protein} = \frac{\text{units / ml enzyme}}{\text{mg protein / ml enzyme}}$$

Unit Definition: *Activity is expressed in NIH units obtained by direct comparison to a NIH Thrombin.*

Final Assay Concentrations

In a 0.30 ml reaction mix, the final concentrations are 33% (v/v) plasma, and 0.27–0.51 unit ancrod (components of the enzyme diluent are not included).

Note

1. Standards should be used immediately and kept on ice after they are dissolved.

Enzymatic Assay of ANGIOTENSIN CONVERTING ENZYME (EC 3.4.15.1)

Principle

$$\text{Hip-L-His-L-Leu} + H_2O \xrightarrow{\text{ACE}} \text{Hippuric Acid} + \text{L-His-L-Leu}$$

Abbreviations Used

Hip-L-His-L-Leu = Hippuryl-L-Histidyl-L-Leucine
ACE = Angiotensin Converting Enzyme
L-His-L-Leu = L-Histidyl-L-Leucine

Conditions: T = 37°C, pH = 8.3, A214nm, Light path = 1 cm

Method: HPLC Stopped Analysis of Products

Reagents

A. 100 mM Tris HCl Buffer with 300 mM Sodium Chloride, and 10 μM Zinc Chloride, pH 8.3 at 37°C (Prepare 100 ml in deionized water using Trizma Base,Sodium Chloride, and Zinc Chloride. Adjust to pH 8.3 at 37°C with 1 M HCl).
B. 50 mM Hippuryl-L-Histidyl-L-Leucine Solution (HHL) (Prepare 1 ml in Reagent A using Hippuryl-His-Leu, Free Base. *Prepare Fresh*).
C. 0.02% (w/v) Hippuric Acid Standard Solution (Hipp Std) (Prepare 1 ml in Reagent A using Hippuric Acid, Free Acid. Further dilute with Reagent A to standards containing the following concentrations (mg/ml): 0.05, 0.075, 0.10, and 0.15).
D. Angiotensin Converting Enzyme Solution (Immediately before use, prepare a solution containing 0.02–0.10 unit/ml of Angiotensin Converting Enzyme in cold Reagent A).

Procedure

Step 1: Pipette (in milliliters) the following reagent into suitable microcentrifuge tubes:

	Test	Blank
Reagent A (Buffer)	0.08	0.09
Reagent B (Enz Sol)	0.01	—

Equilibrate to 37°C. Then add:

Reagent B (HHL)	0.01	0.01

Incubate for exactly 30 minutes at 37°C. Terminate the reaction by heating at 100°C for 4 minutes and then microcentrifuge at 15,000 g for 10 minutes. Transfer 0.08 ml of both the Test and Blank to HPLC vials.

Step 2: Analysis of Products. The substrate and products are separated by reverse phase HPLC chromatography as follows:

1. **Column:** Supelcosil LC-18, Supeclo Prod. No. 5–8298, Particle size: 5 μm, 25 cm × 4.6 mm.
2. **Mobile Phase**

Time (min)	0	15	25
Reagent	4.5	30	30
Reagent	95.5	70	70

3. Pressure: 2150 PSI, Flow rate 1.5 ml/min, Detection: 214nm, Sample Injection volume: 20 μl.
4. Injection blank and then standards of Reagent C (Hipp Std) and samples. A comparison can then be made between the standard curve of hippuric acid and hippuric acid generated from the sample reaction. Convert mg of hippuric acid to μmoles.

Calculations

$$\text{Units / ml enzyme} = \frac{(\mu\text{moles of hippuric acid})\,(\text{df})\,(0.1)}{(0.01)\,(30)}$$

0.1 = Total volume (in milliliter) of enyzme assay
0.01 = Volume (in milliliter) of enzyme used
30 = Time (in minutes of assay)

$$\text{Units / mg solid} = \frac{\text{units / ml enzyme}}{\text{mg solid / ml enzyme}}$$

$$\text{Units / mg protein} = \frac{\text{units / ml enzyme}}{\text{mg protein / ml enzyme}}$$

Unit Definition: *One unit will produce 1.0 μmole of hippuric acid from hippuryl-His-Leu per min in 100 mM Tris HCl, 300 mM NaCl, and 10 μM $ZnCl_2$ at pH 8.3 at 37°C.*

Final Assay Concentrations

In a 0.10 ml reaction mix, the final concentrations are 100 mM Tris, 300 mM Sodium Chloride, 10 μM zinc chloride, 5 mM hippuryl-L-histidy-L-leucine and 0.001–0.0002 unit angiotensin converting enzyme.

Enzymatic Assay of ANGIOTENSIN CONVERTING ENZYME (EC 3.4.15.1)

Principle

$$\text{Hip-L-His-L-Leu} + H_2O \xrightarrow{\text{ACE}} \text{Hippuric acid} + \text{L-His-L-Leu}$$

Abbreviations Used

Hip-L-His-L-Leu = Hippuryl-L-Histidyl-L-Leucine
ACE = Angiotensin Converting Enzyme
L-His-L-Leu = L-Histidyl-L-Leucine

Conditions: T = 37°C, pH = 8.3, A228nm, Light path = 1 cm

Method: Spectrophotometric Stop Rate Determination.

Reagents

A. 50 mM HEPES HCl Buffer with 300 mM Sodium Chloride, pH 8.3 at 37°C (Prepare 100 ml in deionized water using HEPES Sodium Salt, and Sodium Chloride. Adjust to pH 8.3 at 37°C with 1 M HCl).
B. 0.3% (w/v) Hippuryl-L-Histidyl-L-Leucine Solution (HHL) (Prepare 2 ml in Reagent A using Hippuryl-His-Leu, Free Base. *Prepare Fresh*).
C. 1 M Hydrochloric Acid Solution (HCl) (Prepare 100 ml in deionized water using Hydrochloric Acid).
D. Ethyl Acetate.
E. Angiotensin Converting Enzyme Solution (Immediately before use, prepare a solution containing 0.33 unit/ml of Angiotensin Converting Enzyme in cold deionized water).

Procedure

Pipette (in milliliters) the following reagents into suitable containers:

	Test	Blank
Reagent B (HHL)	0.20	0.20
Reagent C (HCl)	—	0.25

Mix by swirling and equilibrate to 37°C. Then add:

Reagent E (Enzyme Solution)	0.05	0.05

Immediately mix by swirling and incubate for 15 minutes at 37°C. Then add:

Reagent C (HCl)	0.25	—
Reagent D (Ethyl Acetate)	2.00	2.00

Shake vigorously for 60 seconds. Centrifuge for 2 minutes. Pipette 1.0 ml of the clear upper layer of each vial into corresponding 4 dram vials. Place vials in a boiling water bath (approximately 1 inch depth) for 15 minutes in a hood. After the ethyl acetate has evaporated, then add:

Deionized Water	3.00	3.00

Mix by inversion, but do not shake, until the residue is dissolved and transfer to suitable quartz cuvettes. Record the A228nm for both the Test and Blank, using a suitable spectrophotometer.

Calculations

$$\text{Units / ml enzyme} = \frac{(\text{A228nm Test} - \text{A228nm Blank})\,(2)\,(3)}{(9.8)\,(15)\,(0.91)\,(0.05)}$$

2 = Conversion factor since the hippuric acid detected is ½ of the total amount produced in the assay (2 ml of ethyl acetate is added and 1 ml of the organic layer containing the product, hippuric acid, is removed).

3 = Total volume of hippuric acid solution.
9.8 = Millimolar extinction coefficient of hippuric acid at 228 nm.
15 = Time (in minutes) of the assay as per the Unit Definition.
0.91 = Extraction efficiency of Ethyl Acetate.
0.05 = Volume (in milliliter) of enzyme used.

$$\text{Units / mg solid} = \frac{\text{units / ml enzyme}}{\text{mg solid / ml enzyme}}$$

$$\text{Units / mg protein} = \frac{\text{units / ml enzyme}}{\text{mg protein / ml enzyme}}$$

Unit Definition: *One unit will produce 1.0 μmole of hippuric acid from hippuryl-his-leu per minute in 50 mM HEPES and 300 mM NaCl at pH 8.3 at 37°C.*

Final Assay Concentration

In a 0.25 ml reaction mix, the final concentrations are 40 mM HEPES, 240 mM sodium chloride, 0.2% (w/v) hippuryl-L-histidyl-L-leucine and 0.016 unit angiotensin converting enzyme.

Note

1. The A228nm (for 20 minutes) must be in the range of 0.4–1.0 for the Test and in the range of 0.1–0.15 for the Blank.

Enzymatic Assay of APOTRYPTOPHANASE

Principle

$$\text{PRP} + \text{L-Tryptophan} + H_2O \xrightarrow{\text{Apotryptophanase}} \text{Indole} + \text{Pyruvate} + NH_3$$

Abbreviation Used

PRP = Pyridoxal 5-Phosphate

Conditions: T = 37°C, pH = 8.3, A540nm, Light path = 1 cm

Method: Colorimetric

Reagents

A. 1 M Potassium Phosphate, Dibasic, Solution (Prepare 100 ml in deionized water using Potassium Phosphate, Dibasic, Trihydrate).
B. 1 M Potassium Phosphate, Monobasic, Solution (Prepare 100 ml in deionized water using Potassium Phosphate, Monobasic, Anhydrous).
C. 1 M Potassium Phosphate Buffer, pH 8.3 at 37°C (Prepare by equilibrating 100 ml of Reagent A to 37°C and adjusting the pH to 8.3 at 37°C with Reagent B).
D. 0.81 mM Pyridoxal 5-Phosphate (PRP) (Prepare 5 ml in deionized water using Pyridoxal 5-Phosphate. *Prepare Fresh*).
E. 50 mM L-Tryptophan Solution, pH 10.2 at 37°C (Prepare 25 ml in deionized water using L-Tryptophan. Adjust to pH 10.2 at 37°C with 1 M NaOH).
F. 100% (w/v) Trichloroacetic Acid (TCA) (Use Trichloroacetic Acid, 6.1 N Solution, approximately 100%).

G. Toluene (Use Toluene).

H. 95% (v/v) Ethanol (Prepare 125 ml in deionized water using, 200 Proof USP Ethyl Alcohol).

I. 5% (w/v) p-Dimethylaminobenzaldehyde Solution (DMAB) (Prepare 20 ml in Reagent H using p-Dimethylaminobenzaldehyde. *Prepare Fresh* and protect from light).

J. Hydrochloric Acid (Use Hydrochloric Acid).

K. 859 mM Hydrochloric Acid-Alcohol Reagent (Acid- Alcohol) (Prepare by adding 8 ml of Reagent J to 100 ml of Reagent H).

L. 0.43 mM Indole Standard Solution (Indole Std) (Prepare 100 ml in deionized water using Indole).

M. Apotryptophanase Enzyme Solution (Immediately before use, prepare a solution containing 2–4 mg/ml in Reagent C).

Procedure

Pipette (in milliliters) the following reagents into suitable containers:

	Test	Test Blank	Substrate Blank
Reagent C (Buffer)	0.20	0.20	0.40
Reagent D (PRP)	0.10	0.10	0.10
Reagent M (Enzyme Solution)	0.20	0.20	—
Deionized Water	1.30	1.50	1.30

Mix by inversion and equilibrate to 37°C. Then add:

Reagent E (L-Tryptophan Soln)	0.20	—	0.20

Immediately mix by inversion and incubate at 37°C for exactly 10 minutes. Then add:

Reagent F (TCA)	0.20	0.20	0.20

Mix by inversion. Add 2.0 ml of Reagent G (Toluene) to the Test, Test Blank, and Substrate Blank. Mix vigorously to phase extract the liberated indole. Allow the phases to separate. The toluene layer will be the upper layer. Pipette (in milliliters) the following reagents into suitable containers:

	Test	Test Blank	Substrate Blank
Aliquot from Toluene Layer	0.20	0.20	0.20
Reagent I (DMAB)	1.00	1.00	1.00
Reagent K (Acid-Alcohol)	8.80	8.80	8.80

Mix by inversion and allow to equilibrate at 25°C for 10 minutes. Transfer the solutions to suitable cuvettes and record the A540nm for the Test, Test Blank, and Substrate Blank using a suitably thermostatted spectrophotometer.

Colorimetric Assay: Prepare a standard curve by pipetting (in milliliters) the following reagents into suitable containers:

	Std 1	Std 2	Std 3	Std Blank
Reagent C (Buffer)	0.20	0.20	0.20	0.20
Reagent L (Indole Std)	0.10	0.20	0.30	—
Deionized Water	1.70	1.60	1.50	1.80

Mix by inversion and then add:

Reagent F (TCA)	0.20	0.20	0.20	0.20

Mix by inversion and add 2.0 ml of Reagent G (Toluene) to each container. Mix vigorously to phase extract the liberated indole. The toluene layer will be the upper layer. Allow the phases to separate and then pipette (in milliliters) the following reagents into suitable containers.

Aliquot from Toluene Layer	0.20	0.20	0.20	0.20
Reagent I (DMAB)	1.00	1.00	1.00	1.00
Reagent K (Acid-Alcohol)	8.80	8.80	8.80	8.80

Mix by inversion and allow to equilibrate at 25°C for 10 minutes. Transfer the solutions to suitable cuvettes and record the A540nm for the Standards and Standard Blank, using a suitably thermostatted spectrophotometer.

Calculations

Standard Curve: ΔA540nm Standard = A540nm Standard–A540nm Standard Blank

Prepare a standard curve by plotting the ΔA540nm of the Standard vs micrograms of indole.

Sample Determination: ΔA540nm =A540nm Test–(A540nm Test Blank + A540nm Substrate Blank)

Determine the total micrograms of indole liberated using the Standard curve.

$$\text{Units / ml enzyme} = \frac{(\mu\text{g indole released})\ (\text{df})}{(0.2)}$$

df = Dilution factor

0.2 = Volume (in milliliter) of enzyme used

$$\text{Units / mg solid} = \frac{\text{units / ml enzyme}}{\text{mg solid / ml enzyme}}$$

Activity: One mg will release 75–150 μg of indole from L-tryptophan in 10 minutes at pH 8.3 at 37°C in the presence of 4 × 10–2 mM pyridoxal 5'-phosphate.

Final Assay Concentration

In a 2.00 ml reaction mix, the final concentrations are 200 mM potassium phosphate, 0.041 mM pyridoxal 5- phosphate, 5 mM tryptophan, and 0.4–0.8 mg apotryptophanase.

Enzymatic Assay of APYRASE (EC 3.6.1.5) ATP as Substrate

Principle

$$ATP + H_2O \xrightarrow{\text{Apyrase}} ADP + Pi$$

Abbreviations Used

ATP = Adenosine 5'-Triphosphate.
ADP = Adenosine 5'-Diphosphate.
Pi = Inorganic Phosphate.

Conditions: T = 30°C, pH 6.5, A660nm, Light path = 1 cm.

Method: Colorimetric

Reagents

A. 40 mM Succinate Buffer with 4 mM Calcium Chloride, pH 6.6 at 30°C (Prepare 100 ml in deionized water using Succinic Acid, Free Acid, and Calcium Chloride, Dihydrate. Adjust to pH 6.6 at 30°C with 5 M KOH).
B. 2.0 mM Adenosine 5'-Triphosphate Solution (ATP) (Prepare 15 ml in Reagent A using Adenosine 5'-Triphosphate, Disodium Salt. Adjust to pH 6.5 at 30°C using 1 M NaOH).
C. Apyrase Enzyme Solution (Immediately before use, prepare a solution containing 0.50–1.5 ATPase units/ml of Apyrase in cold deionized water).
D. Phosphorus Standard Solution (Use Phosphorus Standard Solution. The concentration is 20 µg/ml, 0.645 µmoles/ml).
E. 10% (w/v) Ammonium Molybdate Solution (Amm. Moly). (Prepare 25 ml in 10 N H_2SO_4 using Molybdic Acid, Ammonium Tetrahydrate Salt).
F. Taussky-Shorr Reagent (TSCR) (Prepare by adding 10 ml of Reagent E to 70 ml of deionized water. Then add a 5 g vial of Ferrous Sulfate Heptahydrate, and mix until dissolved. Add enough deionized water for a final volume of 100 ml).

Procedure

Pipette (in milliliters) the following reagents into suitable containers:

	Test	Blank
Reagent B (ATP)	1.90	1.90

Equilibrate to 30°C. Then add:

	Test	Blank
Deionized Water	—	0.10
Reagent C (Enzyme Solution)	0.10	—

Immediately mix by swirling and incubate at 30°C for exactly 10 minutes. Then add:

	Test	Blank
Reagent F (TSCR)	5.00	5.00
Deionized Water	3.00	3.00

Mix by swirling and immediately (within 1–2 minutes) transfer to suitable cuvettes and record the A660nm for both the Test and Blank using a suitable spectrophotometer.

Colorimetric Assay

Standard Curve: Prepare a standard curve by pipetting (in milliliters) the following reagents into suitable containers:

	Std 1	Std 2	Std 3	Std 4	Std 5	Std Blank
Deionized Water	4.50	4.00	3.50	3.00	2.50	5.00
Reagent D (Phosphorus Std)	0.50	1.00	1.50	2.00	2.50	0.00
Reagent F (TSCR)	5.00	5.00	5.00	5.00	5.00	5.00

Mix by swirling and immediately (within 1–2 minutes) transfer to suitable cuvettes and record the A660nm for the Standards and Blank.

Calculations

Standard Curve: ΔA660nm Standard = A660nm Standard–A660nm Standard Blank

Prepare a standard curve by plotting ΔA660nm Standard versus μmoles of Phosphate.

Sample Determination: ΔA660nm Test = A660nm Test–A660nm Blank

Determine the micromoles of phosphate liberated using the Standard curve.

$$\text{Units / ml enzyme} = \frac{(\mu\text{moles of Phosphate released})\ (\text{df})}{(10)\ (0.1)}$$

df = Dilution factor.
10 = Time (in minutes) of assay as per the Unit Definition.
0.1 = Volume (in milliliter) of enzyme used.

$$\text{Units / mg solid} = \frac{\text{units / ml enzyme}}{\text{mg solid / ml enzyme}}$$

$$\text{Units / mg protein} = \frac{\text{units / ml enzyme}}{\text{mg protein / ml enzyme}}$$

Unit Definition: *One unit will liberate 1.0 μmole of inorganic phosphate from adenosine 5'-triphosphate per minute at pH 6.5 at 30°C.*

Final Assay Concentration

In a 2.00 ml reaction mix, the final concentrations are 38 mM sodium succinate, 3.8 mM calcium chloride, 1.9 mM adenosine 5'-triphosphate, and 0.05–0.15 unit apyrase.

Enzymatic Assay of APYRASE (EC 3.6.1.5) ADP as Substrate

Principle

$$\text{ADP} + H_2O \xrightarrow{\text{Apyrase}} \text{AMP} + \text{Pi}$$

Abbreviations Used

ADP = Adenosine 5'-Diphosphate.
AMP = Adenosine 5'-Monophosphate.
Pi = Inorganic Phosphate.

Conditions: T = 30°C, pH 6.5, A660nm, Light path = 1 cm

Method: Colorimetric

Reagents

A. 40 mM Succinate Buffer with 4 mM Calcium Chloride, pH 6.6 at 30°C (Prepare 100 ml in deionized water using Succinic Acid, Free Acid, and Calcium Chloride, Dihydrate. Adjust to pH 6.6 at 30°C with 5 M NaOH).

B. 2.0 mM Adenosine 5'-Diphosphate Solution (ADP) (Prepare 15 ml in Reagent A using Adenosine 5'-Diphosphate, Di(Monocyclohexylammonium) Salt. Adjust to pH 6.5 at 30°C using 1 M NaOH).

C. Apyrase Enzyme Solution (Immediately before use, prepare a solution containing 0.5–1.5 ADPase units/ml of Apyrase in cold deionized water).

D. Phosphorus Standard Solution (Use Phosphorus Standard Solution, The concentration is 20 µg/ml, 0.645 µmoles/ml).

E. 10% (w/v) Ammonium Molybdate Solution (Amm. Moly). (Prepare 25 ml in 10 N H_2SO_4 using Molybdic Acid, Ammonium Tetrahydrate Salt).

F. Taussky-Shorr Reagent (TSCR) (Prepare by adding 10 ml of Reagent E to 70 ml of deionized water. Then add a 5 g vial of Ferrous Sulfate, Heptahydrate, and mix until dissolved. Add enough deionized water to a final volume of 100 ml).

Procedure

Pipette (in milliliters) the following reagents into suitable containers:

	Test	Blank
Reagent B (ADP)	1.90	1.90

Equilibrate to 30°C. Then add:

	Test	Blank
Deionized Water	—	0.10
Reagent C (Enzyme Solution)	0.10	—

Immediately mix by swirling and incubate at 30°C for exactly 10 minutes. Then add:

	Test	Blank
Reagent F (TSCR)	5.00	5.00
Deionized Water	3.00	3.00

Mix by swirling and immediately (within 1–2 minutes) transfer to suitable cuvettes and record the A660nm for both Test and Blank using a suitable spectrophotometer.

Colorimetric Assay

Standard Curve: Prepare a standard curve by pipetting (in milliliters) the following reagents into suitable containers:

	Std 1	Std 2	Std 3	Std 4	Std 5	Std Blank
Deionized Water	4.50	4.00	3.50	3.00	2.50	5.00
Reagent D (Phosphorus Std)	0.50	1.00	1.50	2.00	2.50	0.00
Reagent F (TSCR)	5.00	5.00	5.00	5.00	5.00	5.00

Mix by swirling and immediately (within 1–2 minutes) transfer to suitable cuvettes and record the A660nm for Standards and Blank.

Calculations

Standard Curve: ΔA660nm Standard = A660nm Standard–A660nm Standard Blank
Prepare a standard curve by plotting ΔA660nm Standard versus µmoles of Phosphate.

Sample Determination: ΔA660nm Test = A660nm Test–A660nm Blank
Determine the micromoles of phosphate liberated using the Standard curve.

$$\text{Units / ml enzyme} = \frac{(\mu\text{moles of Phosphate released})\ (\text{df})}{(10)\ (0.1)}$$

df = Dilution factor
10 = Time (in minutes) of assay as per the Unit Definition
0.1 = Volume (in milliliter) of enzyme used

$$\text{Units / mg solid} = \frac{\text{units / ml enzyme}}{\text{mg solid / ml enzyme}}$$

$$\text{Units / mg protein} = \frac{\text{units / ml enzyme}}{\text{mg protein / ml enzyme}}$$

Unit Definition: *One unit will liberate 1.0 µmole of inorganic phosphate from adenosine 5'-diphosphate per minute at pH 6.5 at 30°C.*

Final Assay Concentration

In a 2.00 ml reaction mix, the final concentrations are 38 mM succinate, 3.8 mM calcium chloride, 1.9 mM adenosine 5'-diphosphate, and 0.05–0.15 unit apyrase.

Enzymatic Assay of APYRASE (EC 3.6.1.5) AMP as Substrate

Principle

$$\text{AMP} + H_2O \xrightarrow{\text{Apyrase}} \text{Adenosine} + \text{Pi}$$

Abbreviations Used

AMP = Adenosine 5'-Monophosphate
Pi = Inorganic Phosphate

Conditions: T = 30°C, pH 6.5, A660nm, Light path = 1 cm

Method: Colorimetric

Reagents

A. 40 mM Succinate Buffer with 4 mM Calcium Chloride, pH 6.5 at 30°C (Prepare 100 ml in deionized water using Succinic Acid, Free Acid, and Calcium Chloride, Dihydrate. Adjust to pH 6.5 at 30°C with 1 M KOH).

B. 2 mM Adenosine 5'-Monophosphate Solution (AMP) (Prepare 15 ml in Reagent A using Adenosine 5'-Monophosphate, Sodium Salt. Adjust to pH 6.5 at 25°C using 10 mM NaOH).

C. 0.1% Albumin Solution (BSA) (Prepare 50 ml in deionized water using Albumin Bovine Serum).

D. Apyrase Enzyme Solution (Immediately before use, prepare a solution containing 5.0–10 AMPase units/ml of Apyrase in cold Reagent C).

E. Phosphorus Standard Solution (Use Phosphorus Standard Solution. The concentration is 20 µg/ml, 0.645 µmoles/ml).

F. 10% (w/v) Ammonium Molybdate Solution (Amm. Moly). (Prepare 25 ml in 5 M Sulfuric Acid using Molybdic Acid, Ammonium Tetrahydrate Salt).

G. Taussky-Shorr Reagent (TSCR) (Prepare by adding 10 ml Reagent F to 70 ml of deionized water. Then add 5 g Ferrous Sulfate, Heptahydrate, and mix until dissolved. Add enough deionized water to a final volume of 100 ml).

Procedure

Pipette (in milliliters) the following reagents into suitable containers:

	Test	Blank
Reagent B (AMP)	1.90	1.90

Equilibrate to 30°C. Then add:

Reagent C (BSA)	—	0.10
Reagent D (Enzyme Solution)	0.10	—

Immediately mix and incubate at 30°C for exactly 10 minutes. Then add:

Deionized Water	3.00	3.00
Reagent G (TSCR)	5.00	5.00

Mix and incubate at 25°C for 5 minutes. Transfer to suitable cuvettes and record the A660nm for both Test and Blank using a suitable spectrophotometer.

Colorimetric Assay

Standard Curve: A standard curve is made by pipetting (in milliliters) the following reagents into suitable containers:

	Std 1	Std 2	Std 3	Std 4	Std 5	Blank
Deionized Water	4.50	4.00	3.50	3.00	2.50	5.00
Reagent E (Phosphorus Std)	0.20	0.40	0.60	0.80	1.00	0.00
Reagent G (TSCR)	5.00	5.00	5.00	5.00	5.00	5.00

Transfer to suitable cuvettes and record the A660nm for Standards and Blank.

Calculations

Find the slope from the plot of the A660nm of the Standards vs Phosphorus concentration. Use the slope (M) to determine the concentration of the test mixture.

$$\text{Units / ml enzyme} = \frac{\text{A660nm Test} - \text{A660nm Blank}}{(10)\ (M)\ (\text{mg enzyme / RM})}$$

10 = Time of Assay (Unit Definition)
M = Slope
RM = Reaction Mix (volume = 2.00 ml)

Unit Definition: *One unit will liberate 1.0 μmole of inorganic phosphate from adenosine 5'-monophosphate per minute at pH 6.5 at 30°C.*

Final Assay Concentration

In a 2 ml reaction mix, the final concentrations are 38 mM sodium succinate, 3.8 mM calcium chloride, 1.9 mM AMP, 0.005% BSA, and 0.5–1.0 units apyrase.

Enzymatic Assay of ARGINASE (EC 3.5.3.1)

Principle

$$\text{L-Arginine} + H_2O \xrightarrow{\text{Arginase}} \text{L-Ornithine} + \text{Urea}$$

Conditions: T = 37°C, pH = 9.5, A535nm, Light path = 1 cm.

Method: Colorimetric

Reagents

A. 50 mM Manganese Maleate Activation Buffer, pH 7.0 at 37°C. The following components are made separately:
   1. 100 mM Manganese Sulfate Solution (Prepare 60 ml in deionized water using Manganese Sulfate. *Prepare Fresh*).
   2. 125 mM Maleic Acid Solution, pH 8.0 at 37°C (Prepare 50 ml in deionized water using Maleic Acid. Adjust the pH to 8.0 at 37°C using 2 M NaOH). Combine 50 ml of Component 1 and 40 ml of Component 2 and mix. Equilibrate to 37°C and adjust to pH 7.0 using 0.1 M HCl. Dilute to a final volume of 100 ml with deionized water.

B. 713 mM L-Arginine Substrate Solution (Prepare 50 ml in deionized water using L-Arginine, Free Base. Equilibrate to 37°C and adjust to pH 9.5 using 5 M HCl. The L-Arginine will dissolve upon the addition of the HCl. *Prepare Fresh*).

C. Arginase Enzyme Solution (Prepare a solution containing 40–60 units/ml in Reagent A. Activate the enzyme by incubating for 4 hours at 37°C. Dilute 0.1 ml of activated enzyme to 50 ml using deionized water, immediately prior to use).

D. 20 mM Urea Stock Solution, pH 9.5. (Prepare 50 ml in deionized water using Urea. Adjust to pH 9.5 using 0.1 N NaOH).

E. 4 mM Urea Standard Solution, pH 9.5 (Prepare 10 ml by diluting 2 ml of Reagent D to 10 ml with deionized water).

F. BUN Acid-Color Reagent (BUN) (Immediately before use, add 60 ml of BUN Acid Reagent to 40 ml of BUN Color Reagent, and mix).

Procedure

Pipette (in milliliters) the following reagents into suitable vials

	Test	Test Blank	Std 1	Std 2	Std 3	Std 4	Std 5	Std Blank
Reagent C (Enz Soln)	0.30	—	—	—	—	—	—	—
Reagent E (Urea Standard Soln)	—	—	0.03	0.05	0.10	0.20	0.30	—
Deionized Water	0.30	0.30	0.57	0.55	0.50	0.40	0.30	0.60

Mix by inversion and equilibrate to 37°C. Then add:

Reagent B (Substrate Soln)	0.40	0.40	0.40	0.40	0.40	0.40	0.40	0.40

Mix by inversion and incubate at 37°C for exactly 30 minutes. Then add:

Reagent F (BUN)	8.00	8.00	8.00	8.00	8.00	8.00	8.00	8.00
Reagent C (Enz Soln)	—	0.30	—	—	—	—	—	—

Mix by inversion and place all the vials in a boiling water bath for 12 minutes. Remove and place the vials in an ice bath for 3 minutes. Transfer the solutions to suitable cuvettes and read the absorbance at 535nm for each of the vials using deionized water as a reference.

Calculations

Standard Curve: ΔA535nm Standard = A535nm Standard–A535nm Standard blank.
Plot the ΔA535nm of the Standards vs μmoles of urea
Sample Determination: ΔA535nm Sample = A535nm Test–A535nm Test Blank

Determine the μmoles of urea liberated using the Standard Curve.

$$\text{Units / ml enzyme} = \frac{(\mu\text{moles of urea liberated})\ (\text{df})}{(30)\ (0.3)}$$

df = Dilution factor
30 = Time of assay in minutes
0.3 = Volume (in milliliter) of enzyme used

$$\text{Units / mg solid} = \frac{\text{units / ml enzyme}}{\text{mg solid / ml enzyme}}$$

$$\text{Units / mg protein} = \frac{\text{units / ml enzyme}}{\text{mg protein / ml enzyme}}$$

Unit Definition: *One unit will convert 1.0 micromole of L-arginine to ornithine and urea per minute at pH 9.5 at 37°C.*

Final Assay Concentration

In a 1.00 ml reaction mix, the final concentrations are 0.03 mM manganese malate, 285 mM L-arginine and 0.024–0.036 unit arginase.

Enzymatic Assay of L-ARGININE DECARBOXYLASE (EC 4.1.1.19)

Principle

$$\text{L-Arg} + \text{Pyridoxyl 5-Phosphate} \xrightarrow{\text{L-Arginine Decarboxylase}} \text{Agmatine} + CO_2$$

Abbreviation Used

L-Arg = L-Arginine

Conditions: T = 37°C, pH 5.2

Method: Manometric Assay using Warburg Flasks

Reagents

A. 500 mM Sodium Acetate Buffer, pH 5.2 at 37°C (Prepare 100 ml in deionized water using Sodium Acetate, Trihydrate. Adjust pH to 5.2 with 1 M NaOH).
B. 100 mM L-Arginine Hydrochloride Solution (Prepare 50 ml in Reagent A using L-Arginine, Hydrochloride).
C. 10 mM Pyridoxal 5-Phosphate Solution (PRP) (Prepare 25 ml in Reagent A using Pyridoxal 5-Phosphate).
D. L-Arginine Decarboxylase Solution (Immediately before use, prepare a solution containing 3–5 units/ml of L-Arginine Decarboxylase in Reagent A).

Procedure

Pipette (in milliliters) the following reagents into Warburg flasks.

Main Chamber	Thermo Barometer Substrate Flask	Enzyme Blank	Test	Blank
Reagent A (Buffer)	2.8	0.30	0.30	2.80
Reagent B (Ormithine HCl) Side Arm	—	2.50	2.50	—
Reagent D (Enzyme Solution)	—	—	0.10	0.10
Reagent C (PRP)	—	0.10	0.10	0.10
Reagent A (Buffer)	0.20	0.10	—	—

Be sure to confirm the stability of the pressure with the flask sealed off, before proceeding with the assay. This is to insure temperature equilibrium and the absence of leaks in the flask. The enzyme activity is determined by calculation of the rate of production of CO_2 at 37°C. The reaction rate should be linear for about 20 minutes.

Calculations

$$\frac{\text{Units}}{\text{ml Arginine Decarboxylase}} = 1000\left(22.4\frac{(C)\,L^{(k)}}{\text{mole}}\cdot 1\right)^{(\text{Dilution Factor})} \quad (\text{ml Arginine Decarboxylase})$$

C = mm of CO_2 gas evolved/minute.
k = Warburg flask constant in μL/mm.
22.4 L = Volume gas occupies under STP conditions.
1000 = Conversion factor for L to ml

Unit Definitions: *One unit will release 1.0 μmole of CO_2 from L-arginine per minute at pH 5.2 at 37°C.*

Final Assay Concentrations

In a 3.0 ml reaction mix, the final concentrations are 500 mM sodium acetate, 83 mM L-arginine, 0.33 mM pyridoxal 5-phosphate, and 0.3–0.5 units L-arginine decarboxylase.

Notes

1. The tests are done at least in triplicate since it is common for the flasks to have small leaks.
2. The mm of CO_2 gas evolved (C) is corrected for any temperature and barometric changes (ThermoBarometer) during the experiment and also for the Substrate Blank and Enzyme Blank mm CO_2 corrected = mm CO_2 measured Test–mm CO_2 measured of [Thermobarometer + Substrate Blank + Enzyme Blank] Values of the corrected mm CO_2 produced are plotted versus time. The best straight line is drawn, not necessarily through the origin. The slope, C = mm CO_2/ time, is obtained.
3. The Flask constant, k, is calculated according to the equation:

$$l = \frac{\left[(Vg)\left(\frac{273}{T}2\right) + V_1\, a\right]}{P}$$

Where, P = Standard pressure as mm of manometer fluid.
Vg = Gas volume in flask and manometer.
V_1 = Volume of liquid in flask.
T = Absolute temperature.
a = Solubility of gas; for CO_2 at 37°C, a = 0.57.

Enzymatic Assay of ARGININE KINASE (EC 2.7.3.3)

Principle

$$\text{ATP} + \text{L-Arginine} \xrightarrow{\text{AK}} \text{ADP} + \text{N-Phospho-L-Arginine}$$

$$\text{ADP} + \text{PEP} \xrightarrow{\text{PK}} \text{Pyruvate} + \text{ATP}$$

$$\text{Pyruvate} + \beta\text{-NADH} \xrightarrow{\text{LDH}} \text{Lactate} + \beta\text{-NAD}$$

Abbreviations Used

ATP = Adenosine 5'-Triphosphate
AK = Arginine Kinase
ADP = Adenosine 5'-Diphosphate
PEP = Phospho(enol) pyruvate
PK = Pyruvate Kinase
β-NADH = β-Nicotinamide Adenine Dinucleotide, Reduced Form
LDH = L-Lactic Dehydrogenase
β-NAD = β-Nicotinamide Adenine Dinucleotide, Oxidized Form

Conditions: T = 30°C, pH = 8.6, A340nm, Light path = 1 cm

Method: Continuous Spectrophotometric Rate Determination

Reagents

A. 250 mM Glycine Buffer, pH 8.6 at 30°C (Prepare 100 ml in deionized water using Glycine, Free Base. Adjust to pH 8.6 at 30°C with 1 M NaOH).

B. 100 mM Glycine buffer, with 10 mM 2-Mercaptoethanol (Enz Dil) (Prepare 50 ml in deionized water using Reagent A and 2-Mercaptoethanol).

C. 200 mM Magnesium Sulfate Solution ($MgSO_4$) (Prepare 10 ml in deionized water using Magnesium Sulfate, Heptahydrate).

D. 2 M Potassium Chloride Solution (KCl) (Prepare 10 ml in deionized water using Potassium Chloride).

E. 300 mM Phospho(enol) pyruvate Solution (PEP) (Prepare 3 ml in deionized water using Phospho(enol) pyruvate, Trisodium Salt, Hydrate).

F. 200 mM Adenosine 5'-Triphosphate Solution (ATP) (Prepare 2 ml in deionized water using Adenosine 5'-Triphosphate, Disodium Salt).

G. 500 mM L-Arginine Solution (ARG) (Prepare 2 ml in deionized water using L-Arginine, Free Base).

H. 7.5 mM β-Nicotinamide Adenine Dinucleotide, Reduced Form, Solution (β-NADH) (Dissolve the contents of one 5 mg vial of β-Nicotinamide Adenine Dinucleotide, Reduced Form, Disodium Salt, in the appropriate volume of Reagent A. *Prepare Fresh*).

I. PK/LDH Enzyme Solution (PK/LDH) (Immediately before use, prepare a solution containing 20 units/ml of Pyruvate Kinase in cold Reagent A using PK/LDH Enzymes Suspension).

J. Arginine Kinase Enzyme Solution (AK) (Prepare a solution containing 0.2–0.4 unit/ml of Arginine Kinase in Reagent B. Let stand at room temperature for 15 minutes before use to obtain full activation).

Procedure

Prepare a reaction cocktail by pipetting (in milliliters) the following reagents into a suitable container:

Reagent A (Buffer)	19.50
Reagent C ($MgSO_4$)	2.00
Reagent D (KCl)	2.00
Reagent E (PEP)	2.00
Reagent F (ATP)	1.00
Reagent H (β-NADH)	0.50

Mix by swirling and adjust to pH 8.6 at 30°C with 1 M HCl or 1 M NaOH. Pipette (in milliliters) the following reagents into suitable cuvettes.

	Test	Blank
Reaction Cocktail	2.70	2.70
Reagent I (PK/LDH)	0.10	0.10
Reagent G (ARG)	0.10	0.10

Mix by inversion and equilibrate to 30°C. Monitor the A340nm until constant, using a suitably thermostatted spectrophotometer. Then add:

	Test	Blank
Reagent J (AK)	0.10	—
Reagent B (Enz Dil)	—	0.10

Immediately mix by inversion and record the decrease in the A340nm for approximately 5 minutes. Obtain the ΔA340nm/minute using the maximum linear rate for both the Test and Blank.

Calculations

$$\text{Units / ml enzyme} = \frac{(\Delta\text{A340nm / min Test} - \Delta\text{A340nm / min Blank})\ (3)\ (\text{df})}{(6.22)\ (0.1)}$$

3 = Total volume (in milliliters) of assay
df = Dilution factor
6.22 = Millimolar extinction coefficient of β-NADH at 340nm
0.1 = Volume (in milliliter) of enzyme used

$$\text{Units / mg solid} = \frac{\text{units / ml enzyme}}{\text{mg solid / ml enzyme}}$$

$$\text{Units / mg protein} = \frac{\text{units / ml enzyme}}{\text{mg protein / ml enzyme}}$$

Unit Definition: *One unit will convert 1.0 μmole of L-arginine and ATP to N-phospho-L-arginine and ADP per minute at pH 8.6 at 30°C.*

Final Assay Concentrations

In a 3.00 ml reaction mix, the final concentrations are 178 mM glycine, 0.33 mM 2-mercaptoethanol, 13 mM magnesium sulfate, 133 mM potassium chloride, 20 mM phospho(enol) pyruvate, 6.7 mM adenosine 5'-triphosphate, 0.13 mM β-nicotinamide adenine dinucleotide, reduced form, 2 units pyruvate kinase, 3 units lactic dehydrogenase, 17 mM L-arginine, and 0.02–0.04 unit arginine kinase.

Notes

1. Contains not less than 700 Pyruvate Kinase units and 1000 L-Lactic Dehydrogenase units per ml.
2. Pyruvate Kinase Unit Definition: One unit will convert 1.0 μmole of phospho(enol) pyruvate to pyruvate per minute at pH 7.6 at 37°C.
3. L-Lactic Dehydrogenase Unit Definition: One unit will reduce 1.0 μmole of pyruvate to L-lactate per minute of pH 7.5 at 37°C.

Enzymatic Assay of ARGININOSUCCINATE LYASE (EC 4.3.2.1)

Principle

$$\text{L-Argininosuccinate} \xrightarrow{\text{AL}} \text{L-Arginine} + \text{Fumarate}$$

Abbreviation Used

AL = Argininosuccinate Lyase

Conditions: T = 37°C, pH 7.5, A240nm, Light path = 1 cm

Method: Continuous Spectrophotometric Rate Determination.

Reagents

A. 100 mM Potassium Phosphate Buffer, pH 7.5 at 37°C (Prepare 100 ml in deionized water using Potassium Phosphate, Monobasic, Anhydrous. Adjust to pH 7.5 at 37°C with 2 M KOH).
B. 1 M Potassium Phosphate Solution (KH_2PO_4) (Prepare 1 ml in deionized water using Potassium Phosphate, Monobasic, Anhydrous).
C. 11.7 mM Argininosuccinic Acid Substrate Solution (ASA) (Prepare 10 ml by dissolving Argininosuccinic Acid, Sodium Salt, in 9.85 ml of deionized water. Add 0.15 ml of Reagent B (KH_2PO_4), mix by swirling).
D. Argininosuccinate Lyase Enzyme Solution (Immediately before use, prepare a solution containing 0.5–1.5 units/ml of Argininosuccinate Lyase in cold deionized water).

Procedure

Pipette (in milliliters) the following reagents into suitable quartz cuvettes:

	Test	Blank
Reagent A (Buffer)	2.00	2.00
Deionized Water	0.65	0.75
Reagent D (Enzyme Solution)	0.10	—

Mix by inversion and equilibrate to 37°C. Monitor the A240nm until constant, using a suitably thermostatted spectrophotometer. Then add:

	Test	Blank
Reagent C (ASA)	0.25	0.25

Immediately mix by inversion and record the increase in A240nm for approximately 5 minutes. Obtain the ΔA240nm/minute using the maximum linear rate for both the Test and Blank.

Calculations

$$\text{Units / ml enzyme} = \frac{(\Delta A240\text{nm / min Test} - \Delta A240\text{nm / min Blank})\,(3.0)\,(\text{df})}{(2.44)\,(0.1)}$$

3.0 = Total volume (in milliliters) of assay
df = Dilution factor
2.44 = Millimolar extinction coefficient of fumarate at 240 nm
0.1 = Volume (in milliliter) of enzyme used

$$\text{Units / mg solid} = \frac{\text{units / ml enzyme}}{\text{mg solid / ml enzyme}}$$

$$\text{Units / mg protein} = \frac{\text{units / ml enzyme}}{\text{mg protein / ml enzyme}}$$

Unit Definition: *One unit will form 1.0 µmole each of L-arginine and fumarate from L-argininosuccinate per minute at pH 7.5 and 37°C.*

Final Assay Concentration

In a 3.00 ml reaction mix, the final concentrations are 68 mM potassium phosphate, 0.98 mM argininosuccinic acid, and 0.05–0.15 unit argininosuccinate lyase.

Enzymatic Assay of (R)-AROMATIC ALCOHOL DEHYDROGENASE, NADP⁺ Dependent (EC 1.1.1.2)

Principle

$$\text{2-Hexanol} + \beta\text{-NADP} \xrightarrow{\text{RAADH}} \text{2-Hexanone} + \beta\text{-NADPH} + Mg_{2+}$$

Abbreviations Used

β-NADP = β-Nicotinamide Adenine Dinucleotide Phosphate, Oxidized Form
RAADH = (R)-Aromatic Alcohol Dehydrogenase
β-NADPH = β-Nicotinamide Adenine Dinucleotide Phosphate, Reduced Form

Conditions: T = 50°C, pH = 7.8, A340nm, Light path = 1 cm

Method: Continuous Spectrophotometric Rate Determination

Reagents

A. 50 mM HEPES Buffer with 2.0 mM Magnesium Chloride and 2.0 mM Dithiothreitol, pH 7.8 at 50°C (Prepare 100 ml in deionized water using HEPES, Free Acid, Magnesium Chloride, Hexahydrate, and DLDithiothreitol. Adjust to pH 7.8 at 50°C with 1 M NaOH. *Prepare Fresh*).
B. 26 mM β-Nicotinamide Adenine Dinucleotide Phosphate, Oxidized Form (β-NADP) (Prepare 1 ml in deionized water using β-Nicotinamide Adenine Dinucleotide Phosphate, Sodium Salt).
C. 100 mM 2-Hexanol Solution (2-Hex) (Prepare 5 ml in deionized water using 2-Hexanol).
D. (R)-Aromatic Alcohol Dehydrogenase Enzyme Solution (Immediately before use, prepare a solution containing 1.66–3.32 unit/ml of (R)-Aromatic Alcohol Dehydrogenase, NADP⁺ Dependent in cold deionized water).

Procedure

Equilibrate Reagent A (Buffer) and the sample cell of the spectrophotometer to 50°C. Pipette (in milliliters) the following reagents into suitable cuvettes by adding the reagents as a bead of liquid adhering to the side of the cuvette:

	Test	Blank
Reagent D (Enzyme Solution)	0.01	—
Deionized Water	—	0.01

Then add:

	Test	Blank
Reagent A (Buffer)	1.00	1.00

Direct the flow from the pipette onto the bead of Reagent D to ensure effective mixing. Then add:

Reagent B (β-NADP)	0.10	0.10

Mix by inversion and equilibrate to 50°C. Monitor the A340nm until constant, using a suitably thermostatted spectrophotometer. Then add:

Reagent C (2-Hex)	0.10	0.10

Immediately mix by inversion and record the increase in A340nm for approximately 5 minutes. Obtain the ΔA340nm/minute using the maximum linear rate for both the Test and Blank.

Calculations

$$\text{Units / ml enzyme} = \frac{(\Delta A340\text{nm / min Test} - \Delta A340\text{nm / min Blank})\,(1.21)\,(\text{df})}{(6.22)\,(0.01)}$$

1.21 = Volume (in milliliters) of assay
df = Dilution factor
6.22 = Millimolar extinction coefficient of β-NADPH at 340nm
0.01 = Volume (in milliliter) of enzyme used

$$\text{Units / mg solid} = \frac{\text{units / ml enzyme}}{\text{mg solid / ml enzyme}}$$

$$\text{Units / mg protein} = \frac{\text{units / ml enzyme}}{\text{mg protein / ml enzyme}}$$

Unit Definition: *One unit will oxidize 1.0 μmole of 2-hexanol to 2-hexanone per minute at pH 7.8 at 50°C.*

Final Assay Concentration

In a 1.21 ml reaction mix, the final concentration are 41 mM HEPES, 1.6 mM magnesium chloride, 1.6 mM DL-dithiothreitol, 2.1 mM β-nicotinamide adenine dinucleotide phosphate, 8.3 mM 2-hexanol, and 0.0166–0.0332 unit (R)-aromatic alcohol dehydrogenase.

Enzymatic Assay of (S)-AROMATIC ALCOHOL DEHYDROGENASE, NADP⁺ Dependent(EC 1.1.1.2)

Principle

$$\text{2-Propanol} + \beta\text{-NADP} \xrightarrow{\text{SAADH}} \text{Acetone} + \beta\text{-NADPH}$$

Abbreviations Used

β-NADP = β-Nicotinamide Adenine Dinucleotide Phosphate, Oxidized Form
SAADH = (S)-Aromatic Alcohol Dehydrogenase, $NADP^+$ dependent
β-NADPH = β-Nicotinamide Adenine Dinucleotide Phosphate, Reduced Form

Conditions: T = 50°C, pH 9.0, A340nm, Light path = 1 cm
Method: Continuous Spectrophotometric Rate Determination

Reagents

A. 50 mM Glycine Buffer with 2.0 mM DL-Dithiothreitol, pH 9.0 at 50°C (Prepare 100 ml in deionized water using Glycine, Free Base and DL-Dithiothreitol. Adjust to pH 9.0 at 50°C with 1 M NaOH).
B. 26 mM β-Nicotinamide Adenine Dinucleotide Phosphate, Oxidized Form Solution (β-NADP) (Dissolve the contents of one 10 mg vial of β-Nicotinamide Adenine Dinucleotide Phosphate, Sodium Salt, Preweighed Vial, in the appropriate volume of deionized water).
C. 650 mM 2-Propanol Solution (2-Prop) (Prepare 1 ml in deionized water using Isopropanol, Anhydrous).
D. (S)-Aromatic Alcohol Dehydrogenase, $NADP^+$ Dependent Enzyme Solution (Immediately before use, prepare a solution containing 0.5–1.0 unit/ml of S-Aromatic Alcohol Dehydrogenase, $NADP^+$ Dependent in cold deionized water).

Procedure

Pipette (in milliliters) the following reagents into suitable cuvettes:

	Test	Blank
Reagent D (Enzyme Solution)	0.02	—
Deionized Water	—	0.02
Reagent A (Buffer)	1.00	1.00
Reagent C (2-Prop)	0.10	0.10

Mix by inversion and equilibrate to 50°C. Monitor the A340nm until constant, using a suitably thermostatted spectrophotometer. Then add:

Reagent B (β-NADP)	0.10	0.10

Immediately mix by inversion and record the increase in A340nm for approximately 5 minutes. Obtain the ΔA340nm/minute using the maximum linear rate for both the Test and Blank.

Calculations

$$\text{Units / ml enzyme} = \frac{(\Delta A340\text{nm / min Test} - \Delta A340\text{nm / min Blank})\,(1.22)\,(\text{df})}{(6.22)\,(0.02)}$$

1.22 = Total volume (in milliliters) of assay
df = Dilution factor

6.22 = Millimolar extinction coefficient of β-NADPH at 340nm
0.02 = Volume (in milliliter) of enzyme used

$$\text{Units / mg solid} = \frac{\text{units / ml enzyme}}{\text{mg solid / ml enzyme}}$$

$$\text{Units / mg protein} = \frac{\text{units / ml enzyme}}{\text{mg protein / ml enzyme}}$$

Unit Definition: *One unit will oxidize 1.0 µmole of 2-propanol to acetone per minute at pH 9.0 at 50°C.*

Final Assay Concentrations

In a 1.22 ml reaction mix, the final concentrations are 41 mM glycine, 1.6 mM DL-dithiothreitol, 2.1 mM β-nicotinamide adenine dinucleotide phosphate, 53 mM 2-propanol, and 0.01–0.02 unit (S)-aromatic alcohol dehydrogenase, $NADP^+$ dependent.

Enzymatic Assay of ARYL ACYLAMIDASE (EC 3.5.1.13)

Principle

$$\text{N-Acetyl-p-Aminophenol} + H_2O \xrightarrow{\text{Aryl Acylamidase}} \text{p-Aminophenol} + \text{Acetate}$$

Conditions: T = 37°C, pH = 9.0, A615nm, Light path = 1 cm

Method: Colorimetric

Reagents

A. 100 mM Tris HCl Buffer, pH 9.0 at 37°C (Prepare 100 ml in deionized water using Trizma Base. Adjust to pH 9.0 at 37°C with 1 M HCl).

B. 100 mM N-Acetyl-p-Aminophenol Solution (NAAP) (Prepare 50 ml in Reagent A using 4-Acetamidophenol).

C. 5.0 mM p-Aminophenol Standard Solution (Std) (Prepare by dissolving 54.56 mg of p-Aminophenol, in 80 ml of deionized water. Adjust the pH to 11.0 at 25°C with 0.1 M NaOH. The solution should be clear and purple at this point. Adjust the pH back to 9.0 using 0.1 M HCl. The color of the solution should turn brown. Quantitatively transfer to a 100 ml volumetric flask and dilute to 100 ml with deionized water. Protect the solution from light and use immediately after preparation).

D. 270 mM Ammonium Hydroxide Solution (NH_4OH) (Prepare 100 ml in deionized water using Ammonium Hydroxide).

E. 1.28 mM Cupric Sulfate Solution ($CuSO_4$) (Prepare 100 ml in Reagent D using Cupric Sulfate, Pentahydrate).

F. 0.41% (v/v) O-Cresol Solution (Prepare 48.2 ml by adding 0.2 ml of O-Cresol to 48 ml of deionized water. *Prepare Fresh*).

G. Color Reagent Solution (CRS) (Prepare 54.2 ml by adding 48.2 ml of Reagent F to 6 ml of Reagent E. Mix well. Prepare fresh and use within 30 minutes).

H. Aryl Acylamidase Enzyme Solution (Immediately before use, prepare a solution containing approximately 0.13–0.25 unit/ml of Aryl Acylamidase in cold Reagent A).

Procedure

Pipette (in milliliters) the following reagents into suitable containers:

	Test	Blank
Reagent A (Buffer)	—	0.10
Reagent B (NAAP)	2.90	2.90

Mix by swirling and equilibrate to 37°C. Then add:

Reagent H (Enzyme Solution)	0.10	—

Immediately mix by inversion and incubate at 37°C for exactly 10 minutes. Pipette (in milliliters) the following reagents into suitable containers:

Reagent G (CRS)	2.50	2.50
Test Mixture	1.00	—
Blank Mixture	—	1.00

Mix by swirling and incubate at room temperature for exactly 5 minutes. Transfer the solutions to suitable cuvettes and record the A615nm for the Test and Blank using a suitable spectrophotometer.

Standard Curve: Prepare a standard curve by pipetting (in milliliters) the following reagents into suitable containers:

	Std 1	Std 2	Std 3	Std 4	Std 5	Std Blank
Reagent C (Std)	0.02	0.04	0.06	0. 08	0.10	—
Deionized Water	0.08	0.06	0.04	0.02	—	0.10
Reagent B (NAAP)	2.90	2.90	2.90	2.90	2.90	2.90

Mix by swirling and incubate at 37°C for exactly 10 minutes. Pipette (in milliliters) the following reagents into suitable containers:

	Std 1	Std 2	Std 3	Std 4	Std 5	Std Blank
Reagent G (CRS)	2.50	2.50	2.50	2.50	2.50	2.50
Std 1	1.00	—	—	—	—	—
Std 2	—	1.00	—	—	—	—
Std 3	—	—	1.00	—	—	—
Std 4	—	—	—	1.00	—	—
Std 5	—	—	—	—	1.00	—
Blank	—	—	—	—	—	1.00

Mix by swirling and incubate at room temperature for exactly 5 minutes. Transfer the solutions to suitable cuvettes and record the A615nm for the Standards and Standard Blank using a suitable spectrophotometer.

Calculations

Standard Curve: ΔA615nm Standard = A615nm Standard–A615nm Standard Blank
Plot the ΔA615nm Standard vs μmoles of p-Aminophenol.

Sample Determination: ΔA615nm Sample = A615nm Test–A615nm Test Blank. Determine the μmoles of p-Aminophenol liberated using the Standard curve.

$$\text{Units / ml enzyme} = \frac{(\mu\text{moles of p - Aminophenol released})\,(\text{df})}{(10)\,(0.1)}$$

df = Dilution factor
10 = Time (in minutes) of assay as per the Unit Definition
0.1 = Volume (in milliliter) of enzyme used in the assay

$$\text{Units / mg solid} = \frac{\text{units / ml enzyme}}{\text{mg solid / ml enzyme}}$$

$$\text{Units / mg protein} = \frac{\text{units / ml enzyme}}{\text{mg protein / ml enzyme}}$$

Unit Definition: *One unit will convert 1.0 μmole of N-acetyl-p-aminophenol (acetaminophen) to p-aminophenol per minute at pH 9.0 at 37°C.*

Final Assay Concentrations

In a 3.00 ml reaction mix, the final concentrations are 100 mM Tris, 97 mM N-acetyl-p-aminophenol, and 0.0125–0.025 unit aryl acylamidase.

Enzymatic Assay of ARYLAMIDASE (EC 3.4.11.2)

Principle

$$\text{L - Alanine } \beta\text{ - Naphthylamide} \xrightarrow{\text{Arylamidase}} \beta\text{ - Naphthylamine} + \text{L - Alanine}$$

$$\beta\text{ - Naphthylamine} + \text{p - Dimethylaminocinnamaldehyde} \xrightarrow[\text{(Red Colored)}]{} \text{Schiff's Base}$$

Conditions: T = 37°C, pH = 7.0, A540nm, Light path = 1 cm.

Method: Spectrophotometric Stop Rate Determination

Reagents

A. 100 mM Sodium Phosphate Buffer, pH 7.0 at 37°C (Prepare 100 ml in deionized water using Sodium Phosphate, Monobasic, Anhydrous. Adjust to pH 7.0 at 37°C with 1 M NaOH).

B. 0.50 mM L-Alanine-β-Naphthylamide Solution (L-Ala-Naphth) (Prepare 10 ml in Reagent A using L-Alanine β-Naphthylamide, Free Base).

C. 95% (v/v) Ethanol (EtOH) (Prepare 25 ml in deionized water using 200 Proof USP Ethyl Alcohol).

D. 260 mM Hydrochloric Acid/Ethanol Solution (HCl) (Prepare 25 ml in Reagent C using Hydrochloric Acid).
E. 0.06% (w/v) p-Dimethylaminocinnamaldehyde Solution (DMAC) (Prepare 20 ml in Reagent C using p-Dimethylaminocinnamaldehyde).
F. Arylamidase Enzyme Solution (Immediately before use, prepare a solution containing 0.05–0.10 unit/ml of Arylamidase in cold Reagent A).

Procedure

Pipette (in milliliters) the following reagents into suitable cuvettes:

	Test	Blank
Reagent B (L-Ala-Naphth)	1.00	1.00

Equilibrate to 37°C. Then add:

Reagent F (Enzyme Solution)	0.10	—

Mix by inversion and incubate at 37°C for exactly 5 minutes. Then add:

Reagent D (HCl)	1.00	1.00

Cool in an ice bath. Then add:

Reagent F (Enzyme Solution)	—	0.10
Reagent E (DMAC)	1.00	1.00

Mix by inversion and let the color develop for approximately 10 minutes in an ice bath. Record the A540nm for both the Test and Blank using a suitable spectrophotometer.

Calculations

$$\text{Units / ml enzyme} = \frac{(\text{A540nm Test} - \text{A540nm Blank})\ (3.1)\ (\text{df})}{(42.3)\ (5)\ (0.1)}$$

3.1 = Total volume (in milliliters) of assay
df = Dilution factor
42.3 = Millimolar extinction coefficient of the azo product (Schiff's Base) at 540 nm
5 = Time (in minutes) of assay
0.1 = Volume (in milliliter) of enzyme used in assay

$$\text{Units / mg protein} = \frac{\text{units / ml enzyme}}{\text{mg protein / ml enzyme}}$$

Unit Definition: *One unit will hydrolyze 1.0 μmole of L-alanine-β-naphthylamide to L-alanine and β-naphthylamine per minute at pH 7.0 at 37°C.*

Final Assay Concentration

In a 1.10 ml reaction mix, the final concentrations are 100 mM sodium phosphate, 0.45 mM L-alanine-βnaphthylamide, and 0.005–0.01 unit arylamidase.

Enzyme Assay of ARYLAMINE ACETYLTRANSFERASE (EC 2.3.1.5)

Principle

$$\text{Acetyl-CoA} + \text{Dye} \xrightarrow{\text{AAT}} \text{CoASH} + \text{Acetylated Dye}$$

Abbreviations Used

Acetyl-CoA = Acetyl Coenzyme A
Dye = p-Nitroaniline
AAT = Arylamine Acetyltransferase
CoASH = Coenzyme A

Conditions: T = 25°C, pH = 8.0, A400nm, Light path = 1 cm

Methods: Continuous Spectrophotometric Rate Determination

Reagents

A. 100 mM Potassium Phosphate Buffer, pH 8.0 at 25°C (Prepare 100 ml in deionized water using Potassium Phosphate, Monobasic, Anhydrous. Adjust to pH 8.0 at 25°C with 1 M KOH).
B. 100 mM β-Mercaptoethanol Solution (β-ME) (Prepare 5 ml in deionized water using 2- Mercaptoethanol).
C. 30 mM Ethylenediaminetetraacetic Acid Solution (EDTA) (Prepare 5 ml in deionized water using Ethylenediaminetetraacetic Acid, Disodium Salt, Dihydrate).
D. 1.5 mM p-Nitroaniline Solution (Dye) (Prepare 5 ml in deionized water using p-Nitroaniline).
E. 6.0 mM Acetyl Coenzyme A Solution (Acetyl CoA) (Prepare 1 ml in deionized water using Acetyl Coenzyme A (C:0), Sodium Salt. *Prepare Fresh*).
F. Arylamine Acetyltransferase Enzyme Solution (Immediately before use, prepare a solution containing 100 mg/ml of Arylamine Acetyltransferase in cold Reagent A).

Procedure

Pipette (in milliliters) the following reagents into suitable containers:

	Test	Blank
Reagent A (Buffer)	1.50	1.50
Reagent B (β-ME)	0.15	0.15
Reagent C (EDTA)	0.10	0.10
Reagent D (Dye)	0.20	0.20
Reagent F (Enzyme Solution)	1.00	1.00

Mix by inversion and equilibrate to 25°C. Monitor the A400nm until constant (approx. 4 minutes) using a suitably thermostatted spectrophotometer. Then add:

Reagent E (Acetyl CoA)	0.05	—
Deionized Water	—	0.05

Immediately mix by inversion and record the increase in A400nm for approximately 5 minutes. Obtain the ΔA400nm/minute using the maximum linear rate for both the Test and Blank.

Calculations

$$\text{Units / ml enzyme} = \frac{(\Delta A400nm\ Test - \Delta A400nm\ Blank)\ (1000)\ (3)\ (df)}{(11.19)\ (1)}$$

1000 = Conversion from µmoles to nanomoles
3 = Volume (in milliliters) of assay
df = Dilution factor
11.19 = Millimolar extinction coefficient of acetylated dye
1 = Volume (in milliliters) of enzyme used

$$\text{Units / mg solid} = \frac{\text{units / ml enzyme}}{\text{mg solid / ml enzyme}}$$

$$\text{Units / mg protein} = \frac{\text{units / ml enzyme}}{\text{mg protein / ml enzyme}}$$

Unit Definition: *One unit will acetylate 1.0 nanomole of p-nitroaniline per minute at pH 8 at 25°C.*

Final Concentration

In a 3.00 ml reaction mix, the final concentrations are 83 mM potassium phosphate, 5 mM β-mercaptoethanol, 1 mM Ethylenediaminetetraacetic Acid, 0.1 mM p-nitroaniline, 0.1 mM acetyl coenzyme A and 100 mg of arylamine acetyltransferase.

Enzymatic Assay of ASCORBATE OXIDASE (EC 1.10.3.3)

Principle

$$\text{2L - Ascorbic Acid} + O_2 \xrightarrow{\text{Ascorbate Oxidase}} \text{2L - Dehydroascorbic Acid} + 2H_2O$$

Conditions: T = 25°C, pH = 5.6, A245nm, Light path = 1 cm

Method: Spectrophotometric Stop Rate Determination

Reagents

A. 100 mM Potassium Phosphate, and 4 mM Sodium Phosphate Buffer with 0.5 mM Ethylenediaminetetraacetic (EDTA), pH 5.6 at 25°C (Prepare 100 ml in deionized water using Potassium Phosphate, Monobasic; Sodium Phosphate, Dibasic,and Ethylenediaminetetraacetic Acid, Disodium Salt, Dihydrate. Adjust to pH 5.6 with 1 M HCl or 1 M KOH at 25°C if necessary).

B. 200 mM Hydrochloric Acid (HCl) (Prepare 100 ml in deionized water using Hydrochloric Acid).

C. 0.5 mM L-Ascorbic Acid Solution (Ascorbate Substrate) (Prepare 50 ml in Reagent A using L-Ascorbic Acid, Free Acid, *Prepare Fresh.* To standardize, mix 1 ml of 0.5 mM L-Ascorbic Acid with 3 ml of Reagent B immediately before measurement and read the absorbance at 245 nm against deionized water. The A245nm should be 1.25 " 0.05. If the absorbance reading is not in this range, dilute the solution with Reagent A or with 0.5 mM L-Ascorbic Acid until the A245nm equals 1.25 " 0.05).

D. 4 mM Sodium Phosphate Buffer with 0.05% (w/v) Bovine Serum Albumin, pH 5.6 at 25°C (Enzyme Diluent) (Prepare 100 ml in deionized water using Sodium Phosphate, Dibasic and Albumin, Bovine. Adjust the pH to 5.6 at 25°C with 1 M HCl).

E. Ascorbate Oxidase Enzyme Solution (Immediately before use, prepare a solution containing 0.12–0.24 unit/ml of Ascorbate Oxidase in cold Reagent D).

Procedure

Pipette (in milliliters) the following reagents into suitable vials:

	Test	Blank
Reagent C (Ascorbate Substrate)	1.00	1.00

Equilibrate for several minutes at 25°C. Then add:

Reagent E (Enzyme Solution)	0.10	—

Mix by swirling and incubate at 25°C for exactly 5 minutes. Then add:

Reagent B (HCl)	3.00	3.00

Mix by swirling and then add:

Reagent E (Enzyme Solution)	—	0.10

Mix by swirling. Transfer the solutions to suitable cuvettes and record the A245nm of the Test and Blank using a suitable spectrophotometer.

Calculations

$$\text{Units / ml enzyme} = \frac{(\text{A245nm Blank} - \text{A245nm Test})\ (4.1)\ (\text{df})}{(10)\ (5)\ (0.1)}$$

5 = Time (in minutes) of assay as per the Unit Definition
10 = Millimolar extinction coefficient of L-Ascorbic Acid at 245nm under the assay conditions
0.1 = Volume (in milliliter) of enzyme used
4.1 = Total volume (in milliliters) of assay
df = Dilution factor

$$\text{Units / mg solid} = \frac{\text{units / ml enzyme}}{\text{mg solid / ml enzyme}}$$

$$\text{Units / mg protein} = \frac{\text{units / ml enzyme}}{\text{mg protein / ml enzyme}}$$

Unit Definition: *One unit will oxidize 1.0 µmole of L-ascorbate to dehydroascorbate per minute at pH 5.6 at 25°C.*

Final Assay Concentration

In a 1.10 ml reaction mix, the final concentrations are 91 mM potassium phosphate, 0.45 mM Lascorbic acid, 0.5 mM Ethylenediaminetetraacetic Acid, 4 mM sodium phosphate, 0.005% (w/v) bovine serum albumin, and 0.012–0.024 unit of ascorbate oxidase.

Enzymatic Assay of ASPARAGINASE (EC 3.5.1.1)

Principle

$$\text{L-Asparagine} + H_2O \xrightarrow{\text{Asparaginase}} \text{L-Aspartate} + NH_3$$

Conditions: T = 37°C, pH = 8.6, A436nm, Light path = 1 cm

Method: Spectrophotometric Stop Rate Determination

Reagents

A. 50 mM Tris Buffer, pH 8.6 at 37°C (Prepare 100 ml in deionized water using Trizma Base. Adjust to pH 8.6 at 37°C with 1 M HCl).
B. 189 mM L-Asparagine Solution (Prepare 10 ml in deionized water using L-Asparagine, Anhydrous).
C. 6 mM Ammonium Sulfate Standard Solution ($(NH_4)_2SO_4$ Std) (Prepare 100 ml deionized water using Ammonium Sulfate, Grade I).
D. 1.5 M Trichloroacetic Acid (TCA) (Prepare 10 ml in deionized water using Trichloroacetic Acid, 6.1 N Solution).
E. Ammonia Color Reagent (Use Nessler's Reagent).
F. Asparaginase Enzyme Solution (Immediately before use, prepare a solution containing 2.0–4.0 units/ml of Asparaginase in cold deionized water).

Procedures

Step 1: Pipette (in milliliters) the following reagents into suitable tubes:

	Test	Std Blank	Std 1	Std 2	Std 3	Blank
Reagent A (Buffer)	1.00	1.00	1.00	1.00	1.00	1.00
Reagent B (L-Asparagine Soln)	0.10	0.10	—	—	—	—
Reagent C ($(NH_4)_2SO_4$ Std)	—	—	0.25	0.50	1.00	—
Deionized Water	0.90	0.90	085	0.60	0.10	1.10

Equilibrate to 37°C. Then add:

	Test	Std Blank	Std 1	Std 2	Std 3	Blank
Reagent F (Enzyme Solution)	0.10	—	—	—	—	—

Immediately mix by inversion and incubate at 37°C for 30 minutes. Then add:

	Test	Std Blank	Std 1	Std 2	Std 3	Blank
Deionized Water	0.90	0.90	085	0.60	0.10	1.10
Reagent D (TCA)	0.10	0.10	0.10	0.10	0.10	0.10
Reagent F (Enzyme Solution)	—	0.10	—	—	—	—

Mix by inversion. Centrifuge for 2 minutes to clarify.

Step 2: Pipette (in milliliters) the following reagents into suitable containers:

Deionized Water	4.30	4.30	4.30	4.30	4.30	4.30
Supernatant (from Step 1)	0.20	0.20	0.20	0.20	0.20	0.20
Reagent E (Ammonia Color Reagent)	0.50	0.50	0.50	0.50	0.50	0.50

Immediately mix by inversion and after 1 minute record the A436nm for Standards, Tests, and Blanks.

Calculations

Standard Curve: ΔA436nm Standard = A436nm Standard–A436nm Standard Blank
Prepare a standard curve by plotting the ΔA436nm of the Standard versus Ammonia (NH_3) concentration. Note that 1 mole of Ammonium Sulfate corresponds to 2 moles of Ammonia, therefore a 6 mM Ammonium Sulfate standard is equivalent to a 12 mM ammonium standard.

Sample Determination: ΔA436nm Test = A436nm Test–A436nm Test Blank
Determine the μmoles of NH_3 liberated using the standard curve.

$$\text{Units / ml enzyme} = \frac{(\mu\text{moles of } NH_3 \text{ liberated})\ (2.20)}{(0.2)\ (30)\ (0.1)}$$

2.20 = Volume of Step 1
0.2 = Volume of Step 1 used in Step 2
30 = Time of assay in minutes
0.1 = Volume (in milliliters) of enzyme used

$$\text{Units / mg solid} = \frac{\text{units / ml enzyme}}{\text{mg solid / ml enzyme}}$$

$$\text{Units / mg protein} = \frac{\text{units / ml enzyme}}{\text{mg protein / ml enzyme}}$$

Unit Definition: *One unit will liberate 1.0 μmole of ammonia from L-asparagine per minute at pH 8.6 at 37°C.*

Final Assay Concentration

In a 2.20 ml reaction mix, the final concentrations are, 23 mM Tris, 8.6 mM L-asparagine and 0.2–0.4 units of asparaginase.

Enzymatic Assay of L-ASPARTASE (EC 4.3.1.1)

Principle

$$\text{L-Aspartate} \xrightarrow{\text{L-Aspartase}} \text{Fumarate} + NH_3$$

Conditions: T = 30°C, pH 8.5, A240nm, Light path = 1 cm

Method: Continuous Spectrophotometric Rate Determination

Reagents

A. 150 mM Tris HCl Buffer, pH 8.5 at 30°C (Prepare 100 ml in deionized water using Trizma Base. Adjust to pH 8.5 at 30°C with 1 M HCl).

B. 60 mM Magnesium Sulfate Solution ($MgSO_4$) (Prepare 1 ml in deionized water using Magnesium Sulfate, Heptahydrate).

C. 3.0 mM Ethylenediaminetetraacetic Acid Solution (EDTA) (Prepare 1 ml in Reagent A using Ethylenediaminetetraacetic Acid, Disodium Salt, Dihydrate).

D. 500 mM L-Aspartate Solution (L-Asp) (Prepare 10 ml in deionized water using L-Aspartic Acid, Monopotassium).

E. 5 mM Potassium Phosphate buffer, pH 7.0 at 25°C (Enz Dil) (Prepare 25 ml in deionized water using Potassium phosphate, Monobasic, Anhydrous. Adjust to pH 7.0 at 25°C with 1 M KOH).

F. L-Aspartase Enzyme Solution (Immediately before use, prepare a solution containing 2 units/ml of L-Aspartase in ice cold Reagent E. Place on ice).

Procedure

Pipette (in milliliters) the following reagents into suitable cuvettes:

	Test	Blank
Reagent A (Buffer)	1.00	1.00
Reagent B ($MgSO_4$)	0.10	0.10
Reagent C (EDTA)	0.10	0.10
Reagent D (L-Asp)	0.30	0.30
Deionized Water	1.40	1.40

Mix by inversion and equilibrate to 30°C. Then add:

Reagent F (Enzyme Solution)	0.10	—
Reagent E (Enz Dil)	—	0.10

Immediately mix by inversion and record the increase in A240nm for approximately 5 minutes. Obtain the ΔA240nm/minute using the maximum linear rate for both the Test and Blank.

Calculations

$$\text{Units / ml enzyme} = \frac{(\Delta A240\text{nm / min Test} - \Delta A240\text{nm / min Blank})\ (3)\ (\text{df})}{(2.53)\ (0.1)}$$

3 = Total volume (in milliliters) of assay
df = Dilution factor
2.53 = Millimolar extinction coefficient of potassium fumarate at 240nm
0.1 = Volume (in milliliter) of enzyme used

$$\text{Units / mg solid} = \frac{\text{units / ml enzyme}}{\text{mg solid / ml enzyme}}$$

$$\text{Units / mg protein} = \frac{\text{units / ml enzyme}}{\text{mg protein / ml enzyme}}$$

Unit Definition: *One unit will convert 1.0 μmole of L-aspartate to fumarate per minute at pH 8.5 at 30°C.*

Final Assay Concentrations

In a 3.00 ml reaction mix, the final concentrations are 55 mM Tris, 2 mM magnesium sulfate, 0.1 mM Ethylenediaminetetraacetic Acid, 50 mM L-aspartate, 0.2 mM potassium phosphate, and 0.2 unit L-aspartase.

Enzymatic Assay of ASPARTATE TRANSCARBAMYLASE (EC 2.1.3.2)

Principle

$$\text{Carbamyl Phosphate} + \text{L-Aspartate} \xrightarrow{AT} \text{N-Carbamyl-Aspartate} + \text{Orthophosphate}$$

Abbreviation Used

AT= Aspartate Transcarbamylase

Conditions: T = 37°C, pH = 8.5, A466nm, Light path = 1 cm

Method: Colorimetric

Reagents

A. 1 M Tris HCl Buffer, pH 8.5 at 37°C (Prepare 100 ml in deionized water using Trizma Base. Adjust to pH 8.5 at 37°C with 1 M HCl).

B. 100 mM Carbamyl Phosphate Solution (Carbam Phos) (Prepare 1 ml in cold (0–5°C) deionized water using Carbamyl Phosphate, Dilithium Salt. Store at 0–5°C. *Prepare Fresh*).

C. 200 mM L-Aspartic Acid Solution (L-Asp) (Prepare 5 ml in deionized water using L-Aspartic Acid, Free Acid. Adjust to pH 8.0–8.5 at 37°C with 1 M KOH. *Prepare Fresh*).

D. 5% (w/v) Trichloroacetic Acid Solution (TCA) (Prepare 25 ml in deionized water using Trichloroacetic Acid, 6.1 N Solution, approximately 100% (w/v)).

E. 18 N Sulfuric Acid Solution (H_2SO_4) (Prepare 25 ml in deionized water using Sulfuric Acid. Store at 0–5°C).

F. 0.5% (w/v) Antipyrine Solution (AP) (Prepare 25 ml in Reagent E using Antipyrine. Store at 0–5°C).

G. 5% (v/v) Glacial Acetic Acid Solution (Acetic Acid) (Prepare 50 ml in deionized water using Acetic Acid, Glacial. Store at 0–5°C).

H. 15.8 mM 2,3-Butanedione Monoxime Solution (Prepare 15 ml in Reagent G using 2,3-Butanedione Monoxime. Store at 0–5°C).

I. Color Reagent solution (CRS) (Immediately before use, prepare 30 ml by cautiously mixing 20 ml of Reagent F with 10 ml of Reagent H).

J. 10 mM N-Carbamyl-DL-Aspartate Standard Solution (Std Soln) (Prepare 10 ml in Reagent A using N-Carbamyl- DL-Aspartic Acid. Adjust to pH 7.0–9.0 with 10 N KOH. *Prepare Fresh*).

K. Aspartate Transcarbamylase Enzyme Solution (Immediately before use, prepare a solution containing 0.1–0.2 unit/ml of Aspartate Transcarbamylase in cold deionized water).

Procedure

Pipette (in milliliters) the following reagents into suitable containers:

	Test	Blank
Reagent A (Buffer)	0.10	0.10
Reagent B (Carbam Phos)	0.10	0.10
Reagent C (L-Asp)	0.50	0.50
Deionized Water	0.20	0.20

Mix by swirling and equilibrate to 37°C. Then add:

Reagent K (Enzyme Solution)	0.10	—

Mix by swirling and incubate at 37°C for exactly 30 minutes. Then add:

Reagent D (TCA)	2.00	2.00
Reagent K (Enzyme Solution)	—	0.10

Mix by swirling.

Color Development

Standard Curve: Prepare a standard curve by pipetting (in milliliters) the following reagents into suitable containers:

	Std 1	Std 2	Std 3	Std 4	Std Blank
Reagent J (Std Soln)	0.25	0.50	0.75	1.00	—
Deionized Water	0.75	0.50	0.25	—	1.00
Reagent I (CRS)	2.00	2.00	2.00	2.00	2.00

Mix by swirling and heat at 60°C in a water bath for 120 minutes. Remove from the water bath and cool to 0–4°C in an ice bath.

Sample: Pipette (in milliliters) the following reagents into suitable containers:

	Test	Blank
Test Aliquot	1.00	—
Blank Aliquot	—	1.00
Reagent I (CRS)	2.00	2.00

Mix by swirling and heat at 60°C in a water bath for 120 minutes. Remove from the water bath and cool to 0–4°C in an ice bath. Transfer the solutions to suitable cuvettes and record the A466nm for the Test, Blank, Standards, and Standard Blank using a suitable spectrophotometer.

Calculations

Standard Curve: ΔA466nm Standard = A466nm Standard–A466nm Standard Blank

Plot the ΔA466nm Standard versus µmoles of N-Carbamyl Aspartate.

Sample Determination: ΔA466nm Sample = A466nm Test–A466nm Test Blank

Determine the µmoles of N-Carbamyl-Aspartate produced using the Standard Curve.

$$\text{Units / ml enzyme} = \frac{(\mu\text{moles of Carbamyl - Aspartate produced})\ (3)\ (\text{df})}{(30)\ (0.1)}$$

3 = Total volume (in milliliters) of stopped reaction
df = Dilution factor
30 = Time (in minutes) of assay as per the Unit Definition
0.1 = Volume (in milliliter) of enzyme used

$$\text{Units / mg solid} = \frac{\text{units / ml enzyme}}{\text{mg solid / ml enzyme}}$$

$$\text{Units / mg protein} = \frac{\text{units / ml enzyme}}{\text{mg protein / ml enzyme}}$$

Unit Definition: *One unit will catalyze the formation of 1.0 µmole of N-carbamyl-L-aspartate per minute at pH 8.5 at 37°C.*

Final Assay Concentrations

In a 1.00 ml reaction mix, the final concentrations are 100 mM Tris, 10 mM carbamyl phosphate, 100 mM L-aspartic acid, and 0.01–0.02 unit aspartate transcarbamylase.

Note

1. Carbamyl Phosphate is very unstable in solution and therefore should be made fresh before each assay is performed. Material should be weighed in a dry atmosphere.

Enzymatic Assay of BILIRUBIN OXIDASE (EC 1.3.3.5)

Principle

$$\text{Bilirubin} + O_2 \xrightarrow{\text{Bilirubin Oxidase}} \text{Biliverdin} + H_2O$$

Conditions: T = 37°C, pH = 8.4, A440nm, Light path 1 cm.

Method: Continuous Spectrophotometric Rate Determination

Reagents

A. 200 mM Tris HCl Buffer, pH 8.4 at 37°C (Prepare 100 ml in deionized water using Trizma Hydrochloride. Adjust to pH 8.4 at 37°C with 1 M NaOH).
B. 0.002% (w/v) Bilirubin Solution (Bilirubin) (Prepare 50 ml by initially suspending 1 mg of Bilirubin, Mixed Isomers, in 1 ml of deionized water. Dissolve it by adding one drop of 2 N NaOH and then bring to a final volume of 50 ml with Reagent A).
C. Bilirubin Oxidase Enzyme Solution (Immediately before use, prepare a solution containing 0.025–0.050 unit/ml of Bilirubin Oxidase in cold deionized water).

Procedure

Pipette (in milliliters) the following reagents into suitable cuvettes:

	Test	Blank
Reagent B (Bilirubin)	2.90	2.90

Equilibrate to 37°C. Then add:

Reagent C (Enzyme Solution)	0.10	—
Deionized Water	—	0.10

Immediately mix by inversion and record the decrease in A440nm for approximately 5 minutes. Obtain the ΔA440nm/minute using the maximum linear rate for both the Test and Blank.

Calculations

$$\text{Units / ml enzyme} = \frac{(\Delta A440\text{nm / min Test} - \Delta A440\text{nm / min Blank})\ (3)\ (\text{df})}{(56.3)\ (0.10)}$$

3 = Total volume (in milliliters) of assay
df = Dilution factor
56.3 = Millimolar extinction coefficient of Bilirubin at 440nm
0.10 = Volume (in milliliter) of enzyme used

$$\text{Units / mg solid} = \frac{\text{units / ml enzyme}}{\text{mg solid / ml enzyme}}$$

$$\text{Units / mg protein} = \frac{\text{units / ml enzyme}}{\text{mg protein / ml enzyme}}$$

Unit Definition: *One unit will oxidize 1.0 μmole of bilirubin per minute at pH 8.4 at 37°C.*

Final Assay Concentrations

In a 3.00 ml reaction mix, the final concentrations are 193 mM Tris, 0.002% (w/v) bilirubin, and 0.0025–0.0050 unit bilirubin oxidase.

Enzymatic Assay of BROMELAIN (EC 3.4.22.4)

Principle

$$\text{N}\alpha\text{ - CBZ - L - Lysine p - Nitrophenyl Ester} \xrightarrow{\text{Bromelain}} \text{p - Nitrophenol} + \text{N}\alpha\text{ - CBZ - L - Lysine}$$

Abbreviation Used

CBZ = Carbobenzoxy

Conditions: T = 25°C, pH = 4.6, A340nm, Light path = 1 cm

Method: Continuous Spectrophotometric Rate Determination

Reagents

A. 30 mM Sodium Acetate Buffer with 100 mM Potassium Chloride and 1.0 mM L-Cysteine, pH 4.6 at 25°C (Prepare 100 ml in deionized water using Sodium Acetate, Trihydrate, L-Cysteine, Hydrochloride, Monohydrate, and Potassium Chloride. Adjust to pH 4.6 with 1 M HCl).

B. 50 mM Nα-CBZ-L-Lysine p-Nitrophenyl Ester (LNPE) (Prepare in 0.2 ml of Acetonitrile, using Nα-CBZ-L-Lysine P-Nitrophenyl Ester, Hydrochloride. Dilute to 1.0 ml with deionized water. *Prepare Fresh*).

C. Bromelain Enzyme Solution (Prepare a solution containing 0.2–0.4 units/ml of Bromelain in Reagent A).

Procedure

Pipette (in milliliters) the following reagents into suitable cuvettes:

	Test	Blank
Reagent A (Buffer)	2.60	2.70
Reagent C (Enzyme Solution)	0.10	—

Mix by inversion and equilibrate to 25°C. Then add:

Reagent B (LNPE)	0.10	0.10

Immediately mix by inversion and record the increase in A340nm for approximately 5 minutes. Obtain the ΔA340nm/minute using the maximum linear rate for both the Tests and Blank.

Calculations

$$\text{Units / ml enzyme} = \frac{(\Delta A340\text{nm / min Test} - \Delta A340\text{nm / min Blank})\ (2.8)\ (\text{df})}{(6.32)\ (0.1)}$$

2.8 = Volume (in milliliters) of assay
df = Dilution factor
6.32 = Millimolar extinction coefficient of p-Nitrophenol at 340nm
0.1 = Volume (in milliliter) of enzyme used

$$\text{Units / mg solid} = \frac{\text{units / ml enzyme}}{\text{mg solid / ml enzyme}}$$

$$\text{Units / mg protein} = \frac{\text{units / ml enzyme}}{\text{mg protein / ml enzyme}}$$

Unit Definition: *One unit will release 1.0 mmole of p-nitrophenol from Nα-CBZ-L-lysine p-nitrophenyl ester per minute at pH 4.6 at 25°C.*

Final Assay Concentration

In a 2.80 ml reaction mix, the final concentrations are 29 mM sodium acetate, 96 mM potassium chloride, 0.96 mM α-cysteine, 1.8 mM Nα-CBZ-L-lysine p-nitrophenyl ester, and 0.02–0.04 units bromelain.

Notes

1. Solutions should be used as soon as possible since they hydrolyze on standing.
2. Allow enzyme solution to set on ice for approximately 2 hours.

Enzymatic Assay of BROMELAIN (EC 3.4.22.32)

Principle

$$\text{Gelatin} + H_2O \xrightarrow{\text{Bromelain}} \text{Amino Acids} + \text{Oligopeptides}$$

Conditions: T = 45°C, pH = 4.5

Method: Titrimetric

Reagents

A. 100 mM Sodium Acetate Buffer with 2.6 M Sodium Chloride, pH 4.5 at 45°C (Prepare 100 ml in deionized water using Acetic Acid, Glacial and Sodium Chloride. Adjust to pH 4.5 at 45°C with 1 M NaOH).
B. 5.0% (w/v) Gelatin Solution, pH 4.5 at 45°C (Prepare 100 ml in deionized water using Gelatin, Type A from Porcine Skin, approximately 300 bloom. Dissolve gelatin in approximately 40 ml boiling water. Stir until the gelatin is dissolved. Add approximately 60 ml water and place in an 80°C water bath for about 20 minutes. Cool to 45°C).
C. 3% (v/v) Hydrogen Peroxide Solution (H_2O_2) (Prepare 20 ml in deionized water using Hydrogen Peroxide, 30% Solution).
D. 0.05 N Sodium Hydroxide Standardized Solution (NaOH) (Prepare 100 ml in deionized water using Sodium Hydroxide, Volumetric Standard, 0.5 N).
E. 37% Formaldehyde Solution (Use Formaldehyde, 37% Solution).
F. Bromelain Enzyme Solution (Immediately before use, prepare a solution containing 2.0–4.0 units/ml Bromelain in Reagent A).

Procedure

Using a suitable pH meter in conjunction with a magnetic stirrer, pipette (in milliliters) the following reagents into a suitably thermostatted titration vessel:

	Test	Blank
Reagent B (Gelatin Solution)	25.0	25.0

Equilibrate to 45°C. Then add:

Reagent F (Enzyme Solution)	1.0	—

Incubate at 45°C for 20 minutes. Then add:

Reagent C (H_2O_2)	0.10	0.10
Reagent F (Enzyme Solution)	—	1.0

Adjust both the Test and Blank to pH 6.9 with Reagent D. Then add:

Reagent E (Formaldehyde Solution)	10.0	10.0

Titrate both the Test and Blank to pH 7.8 with Reagent D. Record the volume of Reagent D required to titrate both the Test and Blank form pH 6.9 to pH 7.8.

Calculation

$$\text{Units / g enzyme} = \frac{\text{(Volume for Test – Volume for Blank) (N) (14) (1000)}}{\text{mg enzyme / RM}}$$

N = Normality of Reagent D
14 = mg Nitrogen per millimole Nitrogen
1000 = milligrams/gram
RM = Reaction Mix

Unit Definition: *One unit will hydrolyze 1.0 mg of amino nitrogen from gelatin in 20 minutes at pH 4.5 at 45°C.*

Final Assay Concentrations

In a 26.0 ml reaction mix, the final concentrations are 3.8 mM sodium acetate, 5.8 mM sodium chloride, 4.8% gelatin and 2.0–4.0 units bromelain.

Note

1. Allow enzyme solution to stand on ice (0–4°C) for 1 hour prior to assay.

Enzymatic Assay of BROMOPEROXIDASE

Principle

$$H_2O_2 + Br^- \xrightarrow[V^{+5}]{\text{Bromoperoxidase}} \text{Intermediates} \xrightarrow{\text{MCD}} \text{MCD Br} + H_2O$$

Abbreviations Used

MCD = Monochlorodimedon
MCD = Brominated Monochlorodimedon

Conditions: T = 25°C, pH = 6.4, A290nm, Light path = 1 cm

Method: Continuous Spectrophotometric Rate Determination

Reagents

A. 50 mM MES Buffer, pH 6.4 at 25°C (Prepare 250 ml in deionized water using MES, Free Acid, Hydrate. Adjust to pH 6.4 at 25°C with 1 M NaOH).
B. 100% (v/v) Ethanol (Use Ethyl Alcohol, Anhydrous).
C. 0.1 mM Monochlorodimedon with 200 mM Potassium Bromide (MCD) (Prepare by dissolving 1.2 gram of Potassium Bromide in 50 ml of Reagent A. Then dissolve 21.75 mg of Monochlorodimedon in 0.5 ml of Reagent B. Add 20 µl of the MCD solution to the KBR solution).

D. 100 mM Hydrogen Peroxidase (H_2O_2) (Prepare by diluting 0.284 ml of Hydrogen Peroxidase, 30% (w/w) solution to 25 ml with Reagent C).

E. Bromoperoxidase Enzyme Solution (Prepare a solution containing 0.2 mg/ml of Bromoperoxidase in Reagent A. Incubate at 25°C for 30 minutes prior to use.

Pipette (in milliliters) the following reagents into 1 ml quartz cuvettes:

	Test	Blank
Reagent (MCD)	0.89	0.89

Incubate for 3 minutes at 25°C. Then add:

Reagent E (Enzyme Solution)	0.01	—
Reagent A (Buffer)	—	0.01

Mix by inversion and equilibrate to 25°C. Monitor the A290nm until constant, using a suitably thermostatted spectrophotometer. Then add:

Reagent D (H_2O_2)	0.10	0.10

Mix by inversion and record the decrease in A290nm for several minutes. Obtain the A290nm/minute using the maximum linear rate for both the Test and Blank.

Calculations

$$\text{Units / ml enzyme} = \frac{(\text{A290nm / min Test} - \text{A290nm / min Blank})\,(1)\,(\text{df})}{(0.01)\,(19.9)}$$

1 = Total volume (in milliliter) of assay
df = Dilution factor
0.01 = Volume (in milliliter) of enzyme used
19.9 = Millimolar excitation coefficient of monochlorodimedon

$$\text{Units / mg solid} = \frac{\text{units / ml enzyme}}{\text{mg solid / ml enzyme}}$$

$$\text{Units / mg protein} = \frac{\text{units / ml enzyme}}{\text{mg protein / ml enzyme}}$$

Unit Definition: *One unit will catalyze the conversion of 1.0 micromole of monochlorodimedon to monobromochlorodimedon per min at pH 6.4 at 25°C.*

Final Assay Concentration

In a 1.00 ml reaction mix, the final concentrations are 45 mM MES, 0.09 mM monochlorodimedon, 178 mM potassium bromide, 10 mM hydrogen peroxide, and 2 μg bromoperoxidease.

Note

1. A linear decrease of 0.3–0.7 absorbance units per minute should be observed.

Enzymatic Assay of CALCINEURIN (EC 3.1.3.16)

Principle

The assay for Calcineurin is based on its ability to bind to Calmodulin. Calmodulin which is bound to Calcineurin is no longer available to activate Phosphodiesterase 3':5'-Cyclic Nucleotide. This loss of activation is a measure of the amount of Calcineurin present in the assay.

$$\text{PDE} + \text{CaM} \xrightarrow{\text{Calcineurin}} \text{CaM.Calcineurin}$$

$$3' : 5'\text{-cAMP} + H_2O \xrightarrow{\text{CaM.PDE}} 5'\text{-AMP}$$

Abbreviations Used

CaM = Calmodulin (Phosphodiesterase 3':5'-Cyclic Nucleotide Activator)
CaM.Calcineurin = Calmodulin-Calcineurin Complex
PDE = Phosphodiesterase 3':5'-Cyclic Nucleotide
3':5'-cAMP = Adenosine 3':5'-Cyclic Monophosphate
CaM.PDE = Activated form of Phosphodiesterase 3':5'-Cyclic Nucleotide
5'-AMP = Adenosine 5'-Monophosphate
ATP = Adenosine 5'-Triphosphate
MK = Myokinase
ADP = Adenosine 5'-Diphosphate
PEP = Phospho(enol) pyruvate
PK = Pyruvate Kinase
β-NADH = β-Nicotinamide Adenine Dinucleotide, Reduced Form
LDH = L-Lactic Dehydrogenase
β-NAD = β-Nicotinamide Adenine Dinucleotide, Oxidized Form

Conditions: T = 30°C, pH = 7.5, A340nm, Light path = 1 cm

Method: Continuous Spectrophotometric Rate Determination

Discussion

General: Before assaying for Calcineurin, the activity of the PDE and CaM working solutions must be determined. This has to be done every time and a new solution should be made and at the start of each day. The working range for PDE is 0.012–0.016 units per assay and for CaM, 2.0 to 2.2 units per assay. An activated control, either fully activated PDE (when assaying for CaM) or PDE activated with 2 units of CaM (when assaying for Calcineurin), must be run with each set of 4 Tests. The assay is very sensitive to the amount of PDE added. This is the reason, a control is run with each set of Tests. The PDE activity generally changes slowly during the day but doesn't usually cause a problem as long as the controls are run with each set of cuvettes.

Assay for PDE: The preparation of the Working PDE Solution (Reagent O) is carried out in two steps. The first step is the estimation of the activity of the PDE Stock

Solution (Reagent N). Based on this estimation, the second step is the dilution of a portion of the PDE Stock Solution (Reagent N) to 0.24–0.32 unit/ml with Reagent J (BSA). The Working PDE Solution (Reagent O) is assayed and adjusted to 0.24–0.32 unit/ml, if necessary. Once the proper Working PDE Solution (Reagent O) is prepared, proceed to the assay for CaM (Procedure Part II).

Assay for CaM: Once Reagent O (Working PDE Solution) is prepared, the final solution required before assaying for Calcineurin is the Reagent Q (Working CaM Solution). The activity of Reagent Q (Working CaM Solution) has to be determined so that the number of CaM units added to the Test during the assay for Calcineurin is known. The activity of CaM present in the Reagent Q (Working CaM Solution) is determined graphically. The ratio of Corrected ΔA340/min (CaM unknown) and Corrected ΔA340/min (fully activated) is plotted against volumes used of the CaM unknown. For the results to be valid, three different volumes with ratios between 0.35–0.65 are required with at least one ratio above 0.5 and another one below 0.5. With every set of cuvettes, one of the four MUST be a fully activated PDE assay.

Assay for Calcineurin: The order of addition of CaM, Calcineurin, and PDE will affect the activity. CaM and Calcineurin are mixed together first and incubated at 30°C for 5 minutes before adding PDE. The amount of Calcineurin (Reagent R) present in the sample is also determined graphically. The ratio of Corrected ΔA340/min (Calcineurin unknown) Corrected ΔA340/min (PDE Activated by 2 units of CaM) is plotted against the volumes of the sample used. Three different volumes with ratios between 0.35–0.65 are required with at least one of the 3 valid ratios above 0.5 and another one below 0.5. With every set of cuvettes, one of the four must be a PDE Activated by 2 units CaM assay.

Reagents

A. 200 mM Tris HCl Buffer, pH 7.5 at 30°C (Prepare 100 ml in deionized water, using Trizma Base. Adjust to pH 7.5 at 30°C with 1 M HCl).
B. 1 M Potassium Chloride Solution (KCl) (Prepare 100 ml in deionized water using Potassium Chloride).
C. 60.0 mM Magnesium Sulfate Solution ($MgSO_4$) (Prepare 100 ml in deionized water using Magnesium Sulfate, Anhydrous).
D. 78 mM Tris HCl with 65 mM Potassium Chloride and 7.8 mM Magnesium Sulfate (Assay Buffer) (Prepare 57.3 ml by combining 22.5 ml of Reagent A, 3.75 ml of Reagent B, 7.5 ml of Reagent C, and 20 ml of deionized water. Equilibrate mixture to 30°C and adjust to pH 7.5 with 2 M HCl. Bring to a final volume of 57.3 ml with deionized water).
E. 83.5 mM Phospho(enol) pyruvate Solution (PEP) (Prepare 1 ml in deionized water using Phospho(enol) pyruvate, Trisodium Salt, Hydrate).
F. 42.0 mM Adenosine 5'-Triphosphate Solution (ATP) (Prepare 5 ml in deionized water using Adenosine 5'-Triphosphate, Disodium Salt. Adjust to pH 7.5 at 30°C with solid Sodium Bicarbonate).
G. 76.8 mM Adenosine 3':5'-Cyclic Monophosphate Solution (3':5'-cAMP) (Prepare 2 ml in deionized water using Adenosine 3':5'-Cyclic Monophosphate, Sodium Salt).

H. 14 mM Calcium Chloride Solution ($CaCl_2$) (Prepare 300 ml in deionized water using Calcium Chloride, Dihydrate).

I. 18.3 mM β-Nicotinamide Adenine Dinucleotide, Reduced Form Solution (β-NADH) (Dissolve the contents of one 5 mg vial of β-Nicotinamide Adenine Dinucleotide, Reduced Form, Disodium Salt, in the appropriate volume of deionized water).

J. 0.3% (w/v) Bovine Serum Albumin Solution (BSA) (Prepare 10 ml in Reagent D using Albumin, Bovine).

K. PK/LDH Enzyme Suspension (Use PK/LDH Enzymes Suspension).

L. 3.2 M Ammonium Sulfate Solution (NH_4SO_4) (Prepare 10 ml in deionized water using Ammonium Sulfate).

M. Myokinase Enzyme Solution (Immediatley before use, prepare a solution containing 1667 units/ml in cold Reagent L using Myokinase).

N. Phosphodiesterase 3':5'-Cyclic Nucleotide Stock Solution (PDE Stock) (Immediately before use, prepare 0.6 ml of a solution containing 0.83 unit/ml by dissolving a 0.5 unit vial of Phosphodiesterase 3':5'-Cyclic Nucleotide, in 0.6 ml of Reagent D).

O. Phosphodiesterase 3':5'-Cyclic Nucleotide Working Solution (PDE Work) (Immediately before use prepare a solution containing 0.24–0.32 unit/ml of Phosphodiesterase 3':5'-Cyclic Nucleotide by diluting Reagent N in cold Reagent J).

P. Calmodulin Stock Solution (CaM Stock) (Prepare a solution containing 1000 units/ml Phosphodiesterase 3':5'-Cyclic Nucleotide Activator, in cold deionized water).

Q. Calmodulin Working Solution (CaM Work) (Prepare a solution of Phosphodiesterase 3':5'-Cyclic Nucleotide Activator by diluting Reagent P such that under the conditions of this assay system, the solution contains 40–44 units/ml. A starting point for dilution is 0.08 ml of Reagent P with 1.0 ml of Reagent J. Follow the assay procedure for Calmodulin and determine the units/ml. Based on this information, adjust the solution to 40–44 units/ml and repeat the determination for the Calmodulin activity).

R. Calcineurin Solution (Calcin).

Procedure

Due to the length of the assay (approximately 30 minutes) the reaction cocktail is prepared as needed, 4 cuvette volumes at a time. Prepare a reaction cocktail by pipetting (in milliliters) the following reagents into a suitable container:

Reagent D (Assay Buffer)	11.45
Reagent E (PEP)	0.10
Reagent F (ATP)	0.10
Reagent H ($CaCl_2$)	0.10
Reagent I (β-NADH)	0.15
Reagent K (PK/LDH)	0.07
Reagent M (Myokinase)	0.02

Part I: Assay for PDE

A. Estimation of Stock PDE Solution (Reagent N) activity. The Stock PDE solution (Reagent N) is assayed with 25 units of (CaM Stock) Reagent P (these are Label Units). Pipette (in milliliters) the following reagents into a suitable cuvette:

	Activated Test
Reaction Cocktail	2.80
Deionized Water	0.315
Reagent P (CaM Stock)	0.025
Reagent N (PDE Stock)	0.01

Mix by inversion and equilibrate to 30°C for 5 minutes. Monitor the A340nm until constant, using a suitably thermostatted spectrophotometer. Then add:

Reagent G (3':5'-cAMP)	0.05

Immediately mix by inversion and record the decrease in A340nm for approximately 15 minutes. Obtain the ΔA340nm/minute using the maximum linear rate.

Calculation

$$\text{Units / ml PDE (Fully activated)} = \frac{(\Delta A340 / \text{min Test})\ (3.2)}{(6.22)\ (2)\ (0.01)}$$

3.2 = Total volume (in milliliters) of assay
6.22 = Millimolar extinction coefficient of β-NADH at 340nm
2 = 2 μmoles of β-NAD produced per mole of 3':5'-cAMP hydrolyzed
0.01 = volume (in milliliter) of enzyme used

B. Determination of the Working PDE Solution Activity (Reagent O) (Based on the information obtained above)dilute Reagent N (PDE Stock) with Reagent J (BSA) to 0.24–0.32 unit/ml. Assay Reagent O (PDE Work) to verify its concentration by pipetting (in milliliters) the following reagents into suitable cuvettes:

	Non-Activated Test	Activated Test
Reaction Cocktail	2.80	2.80
Deionized Water	0.30	0.275
Reagent P (CaM Stock)	—	0.025
Reagent O (PDE Work)	0.05	0.05

Mix by inversion and equilibrate to 30°C for 5 minutes. Monitor the A340nm until constant, using a suitably thermostatted spectrophotometer. Then add:

Reagent G (3':5'-cAMP)	0.05	0.05

Immediately mix by inversion and record the decrease in A340nm for approximately 15 minutes. Obtain the ΔA340nm/minute using the maximum linear rate for both the Non-Activated Test and Activated Test.

Calculation

For Activated and Non-activated:

$$\text{Units / ml PDE} = \frac{(\Delta A340 / \text{min Test})\ (3.2)}{(6.22)\ (2)\ (0.05)}$$

3.2 = Total volume (in milliliters) of assay
6.22 = Millimolar extinction coefficient of β-NADH at 340nm

2 = 2 µmoles of β-NAD produced per mole of 3':5'-cAMP hydrolyzed
0.05 = Volume (in milliliter) of enzyme used

Determine the number of units/ml for both the Activated and Non-Activated PDE. There should be 0.24–0.32 unit/ml of activated PDE, if not, adjust to 0.24–0.32 unit/ml and repeat Procedure Part I, Section B. Activated Test, If the ratio of is less than 5, a new PDENon- Activated Test working solution (Reagent O) must be prepared.

Part II, Assay for Calmodulin (CaM)

Pipette (in milliliters) the following reagents into suitable cuvettes:

	Test 1	Test 2	Test 3	Fully Activated Test 4
Reaction Cocktail	2.80	2.80	2.80	2.80
Deionized Water	0.28	0.275	0.27	0.275
Reagent P (CaM Stock)	—	—	—	0.025
Reagent Q (CaM Work)	0.02	0.025	0.03	—
Reagent O (PDE Work)	0.05	0.05	0.05	0.05

Mix by inversion and equilibrate to 30°C for 5 minutes. Monitor the A340nm until constant, using a suitably thermostatted spectrophotometer. Then add:

Reagent G (3':5'-cAMP)	0.05	0.05	0.05	0.05

Immediately mix by inversion and record the decrease in A340nm for approximately 15 minutes. Obtain the ΔA340nm/minute using the maximum linear rate for each Test.

Calculations

1. Determine the corrected ΔA340/min for each cuvette by
   [ΔA340/min]–[ΔA340/min] nonactivated PDE = Corrected ΔA340/min
2. Determine the ratio of Corrected
   ΔA340/min (CaM unknown) = Ratio Corrected ΔA340/min (Fully Activated)
   Tests 1, 2, and 3 are the CaM unknowns and Tests 4 is the fully activated. Only ratio values between 0.35–0.65 can be used. Repeat the assay with different volumes of Reagent Q (CaM Work) until at least three ratios are found that meet the 0.35–0.65 requirements.
3. Plot the ratio values versus the volumes (ml) of Working CaM Solution used. Draw the best straight line through these points.
4. Find the volume which corresponds exactly to the ratio value of 0.50. This volume is V0.5.
5. Determine the U/ml of the Working CaM Solution by

$$\frac{1.0 \text{ CaM unit}}{V0.5} \text{U / ml of Working CaM Solution}$$

V0.5 = Value (in milliliter) of CaM giving activation up to 50% of the maximum activity as per the Unit Definition If the working CaM solution is not between 40 and 44 units/ml, adjust the concentration of Reagent Q (CaM Work) and reassay accordingly.

Part III: Assay for Calcineurin

Note: The order of addition, CaM and Calcineurin, then PDE is very important and must not be changed. The amount of Calcineurin activity added should be between 0.8 and 1.2 units per assay. Pipette (in milliliters) the following reagents into suitable cuvettes:

	Test 1	Test 2	Test 3	Fully Activated Test 4
Reaction Cocktail	2.80	2.80	2.80	2.80
Deionized Water*	—	—	—	0.25
Reagent R (Calcin) up to	0.25	0.25	0.25	—
Reagent P (CaM Stock)	—	—	—	0.025
Reagent Q (CaM Work)	0.05	0.05	0.05	0.05

* The volume of deionized water and Reagent R added should have a combined volume of 0.25 ml.

Mix by inversion and equilibrate at 30°C for 5 minutes. Then add:

Reagent O (PDE Work)	0.05	0.05	0.05	0.05

Mix by inversion and equilibrate to 30°C for 5 minutes. Monitor the A340nm until constant, using a suitably thermostatted spectrophotometer. Then add:

Reagent G (3':5'-cAMP)	0.05	0.05	0.05	0.05

Immediately mix by inversion and record the decrease in A340nm for approximately 15 minutes. Obtain the ΔA340nm/minute using the maximum linear rate for each Test.

Calculations

1. Determine the corrected ΔA340/min for each of the four Tests by:
   Corrected ΔA340/min = [ΔA340nm/min]–[ΔA340nm/min]nonactivated PDE
2. Determine the ratio of Corrected ΔA340/min (Calcineurin unknown) Corrected ΔA340/min (Fully Activated by 2 units CaM) Only ratio values between 0.35–0.65 can be used. Repeat the assay with different volumes of the Calcineurin unknown until at least three ratios are found that meet the 0.35–0.65 requirement.
3. Plot the ratio values versus the volumes (ml) of Calcineurin unknown used. Draw the best straight line though these points.
4. Find the volume which corresponds exactly to the ratio value of 0.50. The volume is V0.5.
5. Determine the units/ml of Calcineurin by:

$$\text{Units / ml} = \frac{\text{(Working CaM Concentration) (0.05) (df)}}{2\ (\text{V0.5})}$$

Working CaM Concentration = units/ml of Reagent Q (CaM Work)
0.05 = Volume (in milliliter) of Reagent Q (CaM Work)
df = Dilution factor
2 = Units of activator (as per the Unit Definition)
V0.5 = The volume (in milliliter) of Calcineurin which will cause 50% inhibition (as per the Unit Definition)

$$\text{Units / mg solid} = \frac{\text{units / ml enzyme}}{\text{mg solid / ml enzyme}}$$

$$\text{Units / mg protein} = \frac{\text{units / ml enzyme}}{\text{mg protein / ml enzyme}}$$

Unit Definitions (Calcineurin): *One unit will cause a 50% inhibition of the activated phosphodiesterase, 3':5'-cyclic nucleotide activity when assayed with two units of activator and 0.1 mM* Ca^{++} *in an enzyme coupled system at pH 7.5 at 30°C.*

Final Assay Concentrations (Calcineurin): In a 3.20 ml reaction mix the final concentrations are 65 mM Tris, 54 mM potassium chloride, 6.5 mM magnesium sulfate, 0.61 mM phospho(enol) pyruvate, 0.31 mM adenosine 5'- triphosphate, 0.10 mM calcium chloride, 0.13 mM β-nicotinamide adenine dinucleotide, reduced form, 11.4 units pyruvate kinase, 16.3 units L-lactic dehydrogenase, 7.8 units myokinase, 2.0–2.2 units calmodulin, 0.012–0.016 unit phosphodiesterase 3':5'-cyclic nucleotide, 1.2 mM adenosine 3':5'-cyclic monophosphate, and varying amounts of units of calcineurin.

Notes

1. Contains not less than 700 units/ml of Pyruvate Kinase units and 1000 units/ml of L-Lactic Dehydrogenase units/ml.
2. Pyruvate Kinase Unit Definition: One unit will convert 1.0 μmole of phospho(enol) pyruvate to pyruvate per minute at pH 7.6 at 37°C.
3. L-Lactic Dehydrogenase Unit Definition: One unit will reduce 1.0 μmole of pyruvate to L-lactate per minute at pH 7.5 at 37°C.
4. Myokinase Unit Definition: One unit will convert 2.0 μmoles of ADP to ATP and AMP per minute at pH 7.6 at 37°C.
5. Phosphodiesterase 3':5'-Cyclic Nucleotide Unit Definition: One unit will hydrolyze 1.0 μmole of 3':5'-cyclic-AMP to 5'-AMP per minute at pH 7.5 at 30°C.
6. Phosphodiesterase 3':5'-Cyclic Nucleotide Activator (Calmodulin) Unit Definition: One unit will stimulate 0.016 activated unit of phosphodiesterase, 3':5'-cyclic nucleotide, P-0520, in a 3 ml reaction volume at pH 7.5 and 30°C, to 50% of the maximum activity of the enzyme when saturated with activator, in the presence of 0.01 mM Ca^{++}.

Enzymatic Assay of CARBAMATE KINASE (EC 2.7.2.2)

Principle

$$\text{ADP} + \text{Carbamyl Phosphate} \xrightarrow{\text{CK}} \text{ATP} + NH_3 + CO_2$$

$$\text{D-Glucose} + \text{ATP} \xrightarrow{\text{Hexokinase}} \text{D-Glucose 6-Phosphate} + \text{ADP} + Mg^{++}$$

$$\text{D-Glucose 6-Phosphate} + \beta\text{-NADP} \xrightarrow{\text{G-6-PDH}} \text{6-PG-}\beta\text{-NADPH}$$

Abbreviations Used

ADP = Adenosine 5'-Phosphate
CK = Carbamate Kinase
ATP = Adenosine 5'-Triphosphate
ADP = Adenosine 5'-Diphosphate
β-NADP = β-Nicotinamide Adenine Dinucleotide Phosphate, Oxidized Form
G-6-PDH = Glucose 6-Phosphate Dehydrogenase
6-PG = 6-Phospho-D-Gluconate
β-NADPH = β-Nicotinamide Adenine Dinucleotide Phosphate, Reduced Form

Conditions: T = 37°C, pH = 8.3, A340nm, Light path = 1 cm

Method: Continuous Spectrophotometric Rate Determination

Reagents

A. 20 mM Tris HCl Buffer, pH 8.3 at 37°C (Prepare 100 ml in deionized water using Trizma Base. Adjust to pH 8.3 at 37°C with 1 M HCl).

B. 10 mM β-D(+)Glucose, 34 mM Magnesium Chloride, 1.0 mM β-Nicotinamide Adenine Dinucleotide Phosphate, Oxidized Form, and 5.0 mM Adenosine 5'-Diphosphate Solution, pH 8.3 at 37°C (Reagent Soln) (Prepare 10 ml in Reagent A using β-D(+)Glucose, Magnesium Chloride, Hexahydrate, β-Nicotinamide Adenine Dinucleotide Phosphate, and Adenosine 5'-Diphosphate, Sodium Salt. Adjust to pH 8.3 at 37°C with either 1 M HCl or 1 M NaOH if necessary).

C. 30 mM Carbamyl Phosphate Solution (Carbamyl Phosphate) (Immediately before use prepare 1 ml in deionized water using Carbamyl Phosphate, Dilithium Salt. This solution is unstable. *Prepare Fresh*).

D. 0.05% (w/v) Bovine Serum Albumin Solution (BSA) (Prepare 10 ml in Reagent A using Albumin, Bovine).

E. Glucose-6-P-Dehydrogenase and Hexokinase Enzyme Solution (G-6-PDH/HK) (Immediately before use, prepare 1 ml of a solution containing 100 units/ml each of Glucose-6-Phosphate Dehydrogenase, and Hexokinase, in cold Reagent A).

F. Carbamate Kinase Enzyme Solution (CK) (Immediately before use, prepare a solution containing 50 units/ml of Carbamate Kinase in cold deionized water. Then dilute to 0.3–0.4 unit/ml with Reagent D).

Procedure

Pipette (in milliliters) the following reagents into suitable containers:

	Test	Blank
Reagent C (Carbamyl Phosphate)	0.10	0.10
Reagent B (Reagent Soln)	2.70	2.70
Reagent E (G-6-PDH/HK)	0.10	0.10

Mix by inversion and equilibrate to 37°C. Monitor the A340nm until constant using a suitably thermostatted spectrophotometer. Then add:

	Test	Blank
Reagent F (CK)	0.10	—
Reagent D (BSA)	—	0.10

Immediately mix by inversion and record the increase in 340nm for approximately 5 minutes. Obtain the ΔA340nm/minute using the maximum linear rate for both the Test and Blank.

Calculations

$$\text{Units / ml enzyme} = \frac{(\Delta A340\text{nm / min Test} - \Delta A340\text{nm / min Blank})\,(3)\,(\text{df})}{(6.22)\,(0.1)}$$

3 = Total volume (in milliliters) of assay
df = Dilution factor
6.22 = Millimolar extinction coefficient of β-NADPH at 340nm
0.1 = Volume (in milliliter) of carbamate kinase used

$$\text{Units / mg solid} = \frac{\text{units / ml enzyme}}{\text{mg solid / ml enzyme}}$$

$$\text{Units / mg protein} = \frac{\text{units / ml enzyme}}{\text{mg protein / ml enzyme}}$$

Unit Definition: *One unit will form 1.0 μmole of ATP from ADP and carbamyl phosphate per minute at pH 8.3 at 37°C.*

Final Assay Concentrations

In a 3.00 ml reaction mix, the final concentrations are 19 mM Tris, 9 mM β-D(+)glucose, 31 mM magnesium chloride, 0.9 mM β-nicotinamide adenine dinucleotide phosphate, 4.5 mM adenosine 5'-diphosphate, 1 mM carbamyl phosphate, 0.002% bovine serum albumin, 10 units glucose-6-phosphate dehydrogenase, 10 units hexokinase, and 0.03–0.04 unit carbamate kinase.

Notes

1. Hexokinase Unit Definition: One unit will phosphorylate 1.0 μmole of D-glucose per minute at pH 7.6 at 25°C.
2. Glucose 6-Phosphate Dehydrogenase Unit Definition: One unit will oxidize 1.0 μmole of glucose 6-phosphate to 6-phospho-D-gluconate per minute in the presence of NAD at pH 7.8 at 30°C.

Enzymatic Assay of CARBONIC ANHYDRASE (EC 4.2.1.1) Tris-Sulfate Buffer

Principle

$$\text{p-Nitrophenyl Acetate} + H_2O \xrightarrow{\text{Carbonic Anhydrase}} \text{p-Nitrophenol} + \text{Acetate}$$

Conditions: T = 0°C, pH = 7.6, A348nm, Light path = 1 cm

Method: Continuous Spectrophotometric Rate Determination

Reagents

A. 15 mM Tris Sulfate Buffer, pH 7.6 at 0°C (Prepare 100 ml in deionized water using Trizma Sulfate. Adjust to pH 7.6 at 0°C with 1 M NaOH).

B. 3 mM p-Nitrophenyl Acetate Solution (PNPA) (Prepare 25 ml in deionized water using p-Nitrophenyl Acetate. Facilitate by first dissolving 13.6 mg of p-Nitrophenyl Acetate in 1 ml of Acetone. Then raise the volume to 25 ml with deionized water. *Prepare Fresh*).

C. Carbonic Anhydrase Enzyme Solution (Immediately before use, prepare a solution containing 100–200 units/ml of Carbonic Anhydrase in Reagent A).

Procedure

Pipette (in milliliters) the following reagents into suitable cuvettes:

	Test	Blank
Reagent A (Buffer)	1.90	2.00
Reagent B (PNPA)	1.00	1.00

Mix by inversion and equilibrate to 0°C. Then add:

Reagent C (Enzyme Solution)	0.10	—

Immediately mix by inversion and record the increase in A348nm for approximately 5 minutes. Obtain the ΔA348nm/minute using the maximum linear rate for both the Test and Blank.

Calculations

$$\text{Units / mg enzyme} = \frac{(\Delta A348nm / min\ Test - \Delta A348nm / min\ Blank)\ (1000)}{(5.00)\ (mg\ enzyme / ml\ RM)}$$

1000 = Conversion to micromoles
5.00 = Millimolar extinction coefficient of p-Nitrophenol at pH 7.6 at 0°C
RM = Reaction Mix

Unit Definition: *Multiply Units/mg by 1.5 to obtain Units. One unit will cause the pH of a 0.012 M Veronal buffer to drop from 8.3 to 6.3 per minute at 0°C.*

Final Assay Concentration

In a 3 ml reaction mix, the final concentrations are 10 mM Tris sulfate, 1 mM p-nitrophenyl acetate and 10–20 units carbonic anhydrase.

Enzymatic Assay of CARBONIC ANHYDRASE (EC 4.2.1.1)

Principle

$$CO_2 + H_2O \xrightarrow{\text{Carbonic Anhydrase}} H_2CO_3$$

Conditions: T = 0°C, pH 8.3

Methods: Titrimetric

Reagents

A. 20 mM Tris Sulfate Buffer, pH 8.3 at 25°C (Prepare 100 ml in deionized water using Trizma Base. Adjust to pH 8.3 at 25°C with 2 N H_2SO_4. Store buffer on ice).
B. CO_2 Saturated H_2O Solution (Substrate Solution) (Prepare 200 ml in deionized water by bubbling CO_2 (dry ice in Erlenmeyer flask) through the H_2O for 30 minutes. Perform this step on ice. Wait 20 minutes before using).
C. Carbonic Anhydrase Enzyme Solution (Prepare a solution containing 1 mg/ml of solid Carbonic Anhydrase in cold deionized water. Immediately prior to use, dilute the enzyme to 10–60 units/ml with cold deionized water).

Procedure

Immediately prior to performing the assay on the blank, pipette (in milliliters) the following reagents into a 4 dram vial which is on ice.

	Blank
Reagent A (Buffer)	3.00
Deionized Water	0.05
Reagent B (Substrate)	2.00

Immediately after the addition of Reagent B (Substrate), insert the probe of a standardized pH meter into the Blank vial. Record the time (T) required for the pH to change from 8.3 to 6.3, or if the pH drops below 8.3 before the pH probe is placed in reaction mixture. Begin recording the time upon addition of water. Continue repeating the Blank measurements until at least three values of T are within a 15 second time window and the average of these three values is T Blank,avg between 70–100 seconds. Immediately prior to performing the assay on the Test, pipette (in milliliters) the following reagents into a 4 dram vial which is on ice.

	Test
Reagent A (Buffer)	3.00
Reagent C (Enzyme)	0.05
Reagent B (Substrate)	2.00

Immediately after the addition of Reagent B (Substrate), insert the probe of a standardized pH meter into the test vials. Record the time (T) required for the pH to change from 8.3 to 6.3. Perform in triplicate. If the pH drops below 8.3 before the pH probe is placed in the reaction mixture, begin recording time immediately upon addition of enzyme. Repeat the Blank measurement after running the test vials. If the Blank is more than 20% from (T Blank, avg), Reagent B (substrate) may have deteriorated and the assay must be repeated.

Calculations

$$\text{Units / ml enzyme} = \frac{(\text{T Blank, avg} - \text{T Test, avg})\ (\text{df})}{(\text{Test, avg})\ (0.05)}$$

df = Dilution factor
T = Time (in seconds) required for the pH to change from 8.3 to 6.3 as per the Unit Definition
0.05 = Volume (in milliliter) of enzyme used

$$\text{Units / mg solid} = \frac{\text{units / ml enzyme}}{\text{mg solid / ml enzyme}}$$

$$\text{Units / mg protein} = \frac{\text{units / ml enzyme}}{\text{mg protein / ml enzyme}}$$

Unit Definition: *One Wilbur-Anderson (W-A) unit will cause the pH of a 0.02 M Trizma buffer to drop from 8.3 to 6.3 per minute at 0°C. (One W-A unit is essentially equivalent to one Roughton-Booth unit).*

Final Assay Concentration

In a 5.05 reaction mix, the final concentrations are 12 mM Tris and 0.5–3 units carbonic anhydrase.

Enzymatic Assay of CARBOXYPEPTIDASE A (EC 3.4.17.1)

Principle

$$\text{Hippuryl-L-Phe} + H_2O \xrightarrow{\text{Carboxypeptidase A}} \text{Hippuric acid} + \text{L-Phenylalanine}$$

Abbreviation Used

Hippuryl-L-Phe = Hippuryl-L-Phenylalanine

Conditions: T = 25°C, pH = 7.5, A254nm, Light path = 1 cm

Method: Continuous Spectrophotometric Rate Determination

Reagents

A. 25 mM Tris HCl Buffer with 500 mM Sodium Chloride, pH 7.5 at 25°C, and Sodium Chloride. Adjust to pH 7.5 at 25°C with 1 M NaOH).
B. 1.0 mM Hippuryl-L-Phenylalanine Solution (Hippuryl-LPhe) (Prepare 100 ml in Reagent A using Hippuryl-Phe. Approximatley 30 minutes of stirring is required for this product to dissolve completely. *Prepare Fresh*).
C. 1.0 M Sodium Chloride Solution (NaCl) (Prepare 10 ml in deionized water using Sodium Chloride).
D. Carboxypeptidase A Enzyme Solution (Immediately before use, prepare a solution containing 6–12 units/ml of Carboxypeptidase A in cold Reagent C).

Procedure

Pipette (in milliliters) the following reagents into suitable quartz cuvettes:

	Test	Blank
Reagent B (Hippuryl-L-Phe)	2.90	2.90

Equilibrate to 25°C. Monitor the A254nm until constant, using a suitably thermostatted spectrophotometer. Then add:

Reagent C (NaCl)	—	0.10
Reagent D (Enzyme Solution)	0.10	—

Immediately mix by inversion and record the increase in A254nm for approximately 5 minutes. Obtain the ΔA254nm/minute using the maximum linear rate for both the Test and Blank.

Calculations

$$\text{Units / ml enzyme} = \frac{(\Delta A254nm / \text{min Test} - \Delta A254nm / \text{min Blank})\,(3)\,(df)}{(0.36)\,(0.10)}$$

3 = Total volume (in milliliters) of assay
df = Dilution factor
0.36 = Millimolar extinction coefficient of hippuric acid at 254nm
0.10 = Volume (in milliliters) of enzyme used

$$\text{Units / mg solid} = \frac{\text{units / ml enzyme}}{\text{mg solid / ml enzyme}}$$

$$\text{Units / mg protein} = \frac{\text{units / ml enzyme}}{\text{mg protein / ml enzyme}}$$

Unit Definition: *One unit will hydrolyze 1.0 µmole of hippuryl-Lphenylalanine per minute at pH 7.5 at 25°C.*

Final Assay Concentration

In a 3.00 ml reaction mix, the final concentrations are 24 mM Tris, 0.97 mM hippuryl-L-phenylalanine, 520 mM sodium chloride, and 0.6–1.2 units carboxypeptidase A.

Note

1. Hippuryl-L-Phe solution is unstable and must be used within 3 hours.

Enzymatic Assay of CARBOXYPEPTIDASE B (EC 3.4.17.2)

Principle

$$\text{Hippuryl - L - Arg} + H_2O \xrightarrow{\text{Carboxypeptidase B}} \text{Hippuric acid} + \text{L - Arginine}$$

Abbreviation Used

Hippuryl-L-Arg = Hippuryl-L-Arginine

Conditions: T = 25°C, pH = 7.65, A254nm, Light path = 1 cm

Method: Continuous Spectrophotometric Rate Determination

Reagents

A. 25 mM Tris HCl Buffer with 100 mM Sodium Chloride, pH 7.65 at 25°C (Prepare 100 ml in deionized water using Trizma Hydrochloride, and Sodium Chloride. Adjust to pH 7.65 at 25°C with 1 M NaOH).

B. 1.0 mM Hippuryl-L-Arginine Solution (Hippuryl-L-Arg) (Prepare 50 ml in Reagent A using Hippuryl-L-Arginine. *Prepare Fresh*).

C. Carboxypeptidase B Enzyme Solution (Immediately before use, prepare a solution containing 4–8 units/ml of Carboxypeptidase B in cold deionized water).

Procedure

Pipette (in milliliters) the following reagents into suitable quartz cuvettes:

	Test	Blank
Reagent B (Hippuryl-L-Arg)	2.90	2.90

Equilibrate to 25°C. Monitor the A254nm until constant, using a suitably thermostatted spectrophotometer and Reagent B. Then add:

	Test	Blank
Deionized Water	—	0.10
Reagent C (Enzyme Solution)	0.10	—

Immediately mix by inversion and record the increase in A254nm for approximately 5 minutes. Obtain the ΔA254nm/minute using the maximum linear rate for both the Test and Blank.

Calculations

$$\text{Units / ml enzyme} = \frac{(\Delta A254nm / min\ Test - \Delta A254nm / min\ Blank)\ (3)\ (df)}{(0.36)\ (0.1)}$$

3 = Total volume (in milliliters) of assay
df = Dilution factor
0.36 = Millimolar extinction coefficient of hippuric acid at 254 nm
0.1 = Volume (in milliliters) of enzyme used

$$\text{Units / mg solid} = \frac{\text{units / ml enzyme}}{\text{mg solid / ml enzyme}}$$

$$\text{Units / mg protein} = \frac{\text{units / ml enzyme}}{\text{mg protein / ml enzyme}}$$

Unit Definition: *One unit will hydrolyze 1.0 μmole of hippuryl-L-arginine per minute at pH 7.65 at 25°C.*

Final Assay Concentration

In a 3.00 ml reaction mix, the final concentrations are 24 mM Tris, 0.97 mM hippuryl-L-arginine, 97 mM sodium chloride, and 0.4–0.8 unit carboxypeptidase B.

Note

1. The substrate solution has a high initial A254nm which requires the use of Reagent B rather than air as the reference.

Enzymatic Assay of CARBOXYPEPTIDASE G (EC 3.4.19.9)

Principle

$$\text{(+)Amethopterin} + H_2O \xrightarrow{\text{Carboxypeptidase G}} \text{L - Glutamic Acid} + \text{4 - Amino - } N_{10} \text{ - Methylpteroic Acid}$$

Conditions: T = 30°C, pH 7.3, A320nm, Light path = 1 cm

Method: Continuous Spectrophotometric Rate Determination

Reagents

A. 50 mM Tris HCl Buffer, with 0.1 mM Zinc Chloride, pH 7.3 at 30°C (Prepare 100 ml in deionized water using Trizma Base and Zinc Chloride. Adjust to pH 7.3 at 30°C with 1 M HCl).

B. 1.8 mM (+)Amethopterin Substrate Solution (Prepare 10 ml by dissolving (+)Amethopterin, Hydrate in 0.1 ml of 1 M NaOH. Dilute to 10 ml with Reagent A).
C. Carboxypeptidase G Enzyme Solution (Immediately before use, prepare a solution containing 0.3–0.6 unit/ml of Carboxypeptidase G in cold deionized water).

Procedure

Pipette (in milliliters) the following reagents into suitable cuvettes:

	Test	Blank
Reagent A (Buffer)	2.80	2.80
Reagent B (Substrate Soln)	0.10	0.10

Mix by inversion and equilibrate to 30°C. Monitor the A320nm until constant, using a suitably thermostatted spectrophotometer. Then add:

Reagent C (Enzyme Solution)	0.10	—
Deionized Water	—	0.10

Immediately mix by inversion and record the decrease in ΔA320nm/minute using the maximum linear rate for both the Test and the Blank.

Calculation

$$\text{Units / ml enzyme} = \frac{(\Delta A320\text{nm / min Test} - \Delta A320\text{nm / min Blank})\ (3)\ (\text{df})}{(8.3)\ (0.1)}$$

3 = Volume (in milliliters) of assay
df = Dilution factor
8.3 = The difference in the millimolar extinction coefficients between the substrate and product at 320 nm.
0.1 = Volume (in milliliters) of enzyme used

$$\text{Units / mg solid} = \frac{\text{units / ml enzyme}}{\text{mg solid / ml enzyme}}$$

$$\text{Units / mg protein} = \frac{\text{units / ml enzyme}}{\text{mg protein / ml enzyme}}$$

Unit Definition: *One unit will hydrolyze 1.0 µmole of L-glutamic acid from (+)amethopterin per minute at pH 7.3 at 30°C.*

Final Assay Concentrations

In a 3.00 reaction mix, the final concentrations are 48 mM Tris, 0.1 mM zinc chloride, 0.06 mM (+)amethopterin, and 0.03–0.06 unit carboxypeptidase G.

Enzymatic Assay of CARBOXYPEPTIDASE P (EC 3.4.17.16)

Principle

$$\text{N-CBZ-Glu-Tyrosine} + H_2O \xrightarrow{\text{Carboxypeptidase P}} \text{N-CBZ-L-Glutamic Acid} + \text{L-Tyrosine}$$

Abbreviation Used

N-CBZ = N-Carbobenzoxy

Conditions: T = 30°C, pH = 3.7, A570nm, Light path = 1 cm

Method: Colorimetric

Reagents

A. 50 mM Sodium Acetate Buffer with 0.02% (v/v) Triton X-100, pH 3.7 at 30°C (Prepare 100 ml in deionized water using Sodium Acetate, Trihydrate and Triton X-100. Adjust the pH to 3.7 at 30°C with 1 M HCl).

B. 300 mM Sodium Hydroxide Solution (NaOH) (Prepare 100 ml in deionized water using Sodium Hydroxide Solution, 1.0 Normal).

C. 2.5% (v/v) Acetic Acid Solution (HOAc) (Prepare 100 ml in deionized water using Acetic Acid, Glacial).

D. 500 mM Sodium Citrate Solution, pH 5.0 at 30°C (Sod Cit) (Prepare 100 ml in deionized water using Citric Acid, Free Acid, Anhydrous. Adjust the pH to 5.0 at 30°C using 1 M NaOH).

E. Ninhydrin Color Reagent (NCR) (Prepare 60 ml by adding 0.5 g Ninhydrin, to 59 ml of Ethylene Glycol Monomethyl Ether. Then add 1 ml of Reagent F).

F. 10 mM Potassium Cyanide Solution (KCN) (Prepare 2 ml in deionized water using Potassium Cyanide).

G. 65% (v/v) Ethanol (EtOH) (Prepare 25 ml in deionized water using Ethyl Alcohol HPLC Grade).

H. 0.1 mM Tyrosine Standard Solution (Std Soln) (Prepare 10 ml in deionized water using L-Tyrosine, Free Base. Heat gently to dissolve).

I. 1 mM N-CBZ-Glu-Tyrosine Substrate Solution (Prepare 10 ml in Reagent A using N-CBZ-Glu-Tyr. Heat gently to dissolve).

K. Carboxypeptidase P Enzyme Solution (Immediately before use, prepare a solution containing 0.03–0.06 unit/ml Carboxypeptidase P in Reagent A).

Procedure

Pipette (in milliliters) the following reagents into suitable containers:

	Test	Blank
Reagent I (Substrate Soln)	0.50	0.50
Reagent A (Buffer)	0.30	0.30

Mix by swirling and equilibrate to 30°C. Then add:

Reagent K (Enzyme Soln)	0.20	—

Mix by swirling and incubate at 30°C for exactly 20 minutes. Then add:

Reagent B (NaOH)	0.20	0.20
Reagent K (Enzyme Soln)	—	0.20

Mix by swirling and incubate at 30°C for 30 minutes.

Color Development: Pipette (in milliliters) the following reagents into suitable containers:

	Test	Blank	Std 1	Std 2	Std 3	Std 4	Std 5	Std Blank
Test Solution	1.20	—	—	—	—	—	—	—
Blank Solution	—	1.20	—	—	—	—	—	—
Reagent H (Std Soln)	—	—	0.20	0.40	0.80	1.00	1.20	—
Deionized Water	—	—	1.00	0.80	0.40	0.20	—	1.20
Reagent C (HOAC)	0.20	0.20	0.20	0.20	0.20	0.20	0.20	0.20
Reagent D (Sod Cit)	2.00	2.00	2.00	2.00	2.00	2.00	2.00	2.00
Reagent E (NCR)	1.00	1.00	1.00	1.00	1.00	1.00	1.00	1.00

Mix by swirling and place vented caps on each container. Place the containers in a boiling water bath for 15 minutes. Cool on ice, and then add:

Reagent G (EtOH)	0.60	0.60	0.60	0.60	0.60	0.60	0.60	0.60

Mix by swirling and transfer the contents of the containers to suitable cuvettes. Read the absorbance at 570nm for each of the cuvettes using a suitable spectrophotometer.

Calculation

Standard Curve: ΔA570nm Standard = A570nm Standard–A570nm Standard Blank
Plot the ΔA570nm Standard vs μmoles Tyrosine.

Sample Determination: ΔA570nm Sample = A570nm Test–A570nm Test Blank
Determine the μmoles of Tyrosine liberated using the standard curve.

$$\text{Units / ml enzyme} = \frac{(\mu\text{mole Tyrosine liberated})\ (\text{df})}{(20)\ (0.2)}$$

df = Dilution factor
20 = Time of assay (in minutes) as per the Unit Definition
0.2 = Volume (in milliliters) of enzyme used

$$\text{Units / mg solid} = \frac{\text{units / ml enzyme}}{\text{mg solid / ml enzyme}}$$

$$\text{Units / mg protein} = \frac{\text{units / ml enzyme}}{\text{mg protein / ml enzyme}}$$

Unit Definition: *One unit will hydrolyze 1.0 μmole of N-CBZ-Glu-Tyr to NCBZ-L-glutamic acid and L-tyrosine per minute at pH 3.7 at 30°C.*

Final Assay Concentration

In a 1.00 ml reaction mix, the final concentrations are 50 mM sodium acetate, 0.02% (v/v) Triton X-100, 0.5 mM N-CBZ-Glu-Tyrosine, and 0.006–0.012 unit carboxypeptidase P.

Enzymatic Assay of CARBOXYPEPTIDASE W (EC 3.4.16.1)

Principle

$$\text{N-CBZ-Phe-Ala} + H_2O \xrightarrow{\text{Carboxypeptidase W}} \text{N-CBZ-Phe} + \text{Ala}$$

Abbreviation Used

N-CBZ = N-Carbobenzoxy

Conditions: T = 30°C, pH = 4.0, A570nm, Light path = 1 cm

Method: Colorimetric

Reagents

A. 50 mM Sodium Acetate Buffer, pH 4.0 at 30°C (Enz Dil) (Prepare 100 ml in deionized water using Sodium Acetate, Trihydrate. Adjust to pH 4.0 at 30°C with 5 M HCl).

B. 500 mM Citrate Buffer, pH 5.0 at 30°C (Cit Buffer) (Prepare 100 ml in deionized water using Citric Acid, Free Acid, Anhydrous. Adjust to pH 5.0 at 30°C with 5 M NaOH).

C. 500 mM Sodium Acetate Buffer, pH 6.5 at 30°C. (Prepare 100 ml in deionized water using Sodium Acetate, Trihydrate. Adjust to pH 6.5 at 30°C with 1 M HCl).

D. 2 mM N-CBZ-Phe-Ala Solution (Substrate) (Prepare 10 ml in Reagent A using N-CBZ-Phe- la. Dissolve by heating in a boiling water bath and mix by swirling).

E. 1 mM L-Tyrosine Standard Solution (Std Soln) (Prepare 10 ml in deionized water using L-Tyrosine, Free Base. Heat gently (do not boil) until the tyrosine dissolves and cool to room temperature).

F. Ninhydrin Color Reagent (NCR) (Prepare 100 ml by dissolving 2 g of Ninhydrin and 300 mg of Hydrindantin, in 75 ml of Ethylene Glycol Monomethyl Ether. (Note: Bubble N_2 gas through the Ethylene Glycol Monomethyl Ether before adding the reagents. Mix by swirling and bubble N_2 gas through the solution (the solution should be bright yellow). Add 25 ml of Reagent C and mix by swirling. The solution should turn red. Place into another amber bottle and purge with N_2 gas. Seal the bottle tightly. The ninhydrin color reagent must be in the reduced form (red in color) for use in this assay. The solution is stable for 2–3 days).

G. Carboxypeptidase W Enzyme Solution (Immediately before use, prepare a solution containing 0.075–0.15 unit/ml of Carboxypeptidase W in cold Reagent A).

Procedure

Pipette (in milliliters) the following reagents into suitable containers:

	Test	Blank
Reagent D (Substrate)	0.50	0.50
Reagent A (Enz Dil)	0.40	0.40

Mix by swirling and equilibrate to 30°C. Then add:

Reagent G (Enz Soln)	0.10	—

Mix by swirling and incubate for exactly 20 minutes at 30°C.

Color Development: Pipette (in milliliters) the following reagents into suitable containers:

	Test	Blank	Std 1	Std 2	Std 3	Std 4	Std 5	Std Blank
Reagent F (NCR)	1.00	1.00	1.00	1.00	1.00	1.00	1.00	1.00
Reagent B (Cit Buffer)	3.00	3.00	3.00	3.00	3.00	3.00	3.00	3.00
Test Solution	1.00	—	—	—	—	—	—	—
Blank Solution	—	0.90	—	—	—	—	—	—
Reagent G (Enz Soln)	—	0.10	—	—	—	—	—	—
Reagent E (Std Soln)	—	—	0.05	0.10	0.15	0.25	0.30	—
Deionized Water	—	—	0.95	0.90	0.85	0.75	0.70	1.00

Mix by swirling and place vented caps on each container. Place these containers in a boiling water bath for 15 minutes. Cool on ice for 10 minutes. Transfer the contents of the containers to suitable cuvettes. Record the absorbance at A570 nm for each of the cuvettes using a suitable spectrophotometer.

Calculation

Standard Curve: ΔA570nm Standard = A570nm Standard–A570nm Standard Blank

Plot the ΔA570nm Standard vs µmoles Tyrosine

Sample Determination: ΔA570nm Sample = A570nm Test–A570nm Sample Blank

Determine the µmoles of Tyrosine equivalents using the standard curve.

$$\text{Units / ml enzyme} = \frac{(\mu\text{mole of Tyrosine equivalents})\ (\text{df})}{(20)\ (0.1)}$$

df = Dilution factor

20 = Time (in minutes) of assay as per the Unit Definition

0.1 = Volume (in milliliter) of enzyme used

$$\text{Units / mg solid} = \frac{\text{units / ml enzyme}}{\text{mg solid / ml enzyme}}$$

$$\text{Units / mg protein} = \frac{\text{units / ml enzyme}}{\text{mg protein / ml enzyme}}$$

Unit Definition: *One unit will hydrolyze 1.0 µmole of N-CBZ-Phe-Ala to N-CBZ-L-phenylalanine and L-alanine (equivalent in ninhydrin color to 1.0 µmole of tyrosine) per minute at pH 4.0 at 30°C.*

Final Concentration

In a 1.00 ml reaction mix, the final concentrations are 50 mM sodium acetate, 1 mM N-CBZ-phe-ala and 0.0075–0.015 unit carboxypeptidase W.

Enzymatic Assay of CARBOXYPEPTIDASE Y

Principle

$$\text{N-CBZ-Phe-Ala} + H_2O \xrightarrow{\text{Carboxypeptidase Y}} \text{N-CBZ-L-Phe} + \text{L-Alanine}$$

Abbreviations Used

N-CBZ-Phe-Ala = N-Carbobenzoxy-L-Phenylalanine-L-Alanine
N-CBZ-L-Phe = N-Carbobenzoxy-L-Phenylalanine

Conditions: T = 25°C, pH = 6.75, A230nm, Light path = 1 cm

Method: Continuous Spectrophotometric Rate Determination

Reagents

A. 50 mM 2-[N-Morpholino]Ethanesulfonic Acid (MES) Buffer, pH 6.75 at 25°C (Prepare 200 ml in deionized water using MES, Free Acid. Adjust to pH 6.75 at 25°C with 1 M NaOH).
B. 2.0 mM N-CBZ-L-Phenylalanine-L-Alanine Solution (CBZ-Phe-Ala) (Prepare 100 ml in Reagent A using N-CBZ-Phe-Ala. Facilitate solubilization by first dissolving in 2 ml of Methanol. Adjust to pH 6.75 at 25°C with 1 M HCl or 1 M NaOH, if necessary).
C. Carboxypeptidase Y Enzyme Solution (Immediately before use prepare a solution containing 36–72 units/ml of Carboxypeptidase Y in cold Reagent A).

Procedure

Pipette (in milliliters) the following reagents into suitable quartz cuvettes:

	Test	Blank
Reagent A (Buffer)	—	0.01
Reagent B (CBZ-Phe-Ala)	3.00	3.00

Mix by inversion and equilibrate to 25°C. Monitor the A230nm until constant, using a suitably thermostatted spectrophotometer. Then add:

	Test	Blank
Reagent C (Enzyme Solution)	0.01	—

Immediately mix by inversion and record the decrease in A230nm for approximately 5 minutes. Obtain the ΔA230nm/minute using the maximum linear rate for both the Test and Blank.

Calculations

$$\text{Units / mg enzyme} = \frac{\Delta A230\text{nm / min Test} - \Delta A230\text{nm / min Blank}}{(0.1915)\ (\text{mg enzyme / ml RM})}$$

0.1915 = Millimolar extinction coefficient of N-CBZ-Phe- Ala at 230 nm
RM = Reaction Mix

Unit Definition: *One unit will hydrolyze 1.0 μmole of N-CBZ-Phe-Ala to N-CBZ-L-phenylalanine and L-alanine per minute at pH 6.75 at 25°C, based on EM 2301 = 191.5.*

Final Assay Concentration

In a 3.01 ml reaction mix, the final concentrations are 50 mM MES, 2.0 mM N-CBZ-Phe-Ala, 2% (w/v) methanol, and 0.36–0.72 unit carboxypeptidase Y.

Enzymatic Assay of CARNITINE ACETYLTRANSFERASE (EC 2.3.1.7)

Principle

$$\text{ATPAcetyl-L-Carnitine} + \text{CoA} \xrightarrow{\text{Carnitine Acetyltransferase}} \text{Acetyl-CoA} + \text{L-Carnitine}$$

Abbreviations Used

CoA = Coenzyme A
Acetyl-CoA = Acetyl Coenzyme A

Conditions: T = 25°C, pH = 8.0, A233nm, Light path = 1 cm

Method: Continuous Spectrophotometric Rate Determination

Reagents

A. 100 mM Tris Buffer, pH 8.0 at 25°C (Prepare 100 ml in deionized water using Trizma Base. Adjust to pH 8.0 at 25°C with 1 M HCl).
B. 11 mM Coenzyme A (Prepare 2 ml in deionized water using Coenzyme A, Sodium Salt).
C. 83.4 mM Acetyl-DL-Carnitine (Prepare 10 ml in deionized water using Acetyl-DLCarnitine Hydrochloride).
D. Carnitine Acetyltransferase Enzyme Solution (Immediately before use, prepare a solution containing 0.3–0.6 unit/ml of Carnitine Acetyltransferase in cold Reagent A).

Procedure

Pipette (in milliliters) the following reagents into suitable cuvettes:

	Test	Blank
Reagent A (Buffer)	2.65	2.75
Reagent B (Coenzyme A)	0.05	0.05
Reagent C (Acetyl-DL-Carnitine)	0.20	0.20

Mix by inversion and equilibrate to 25°C. Monitor the A233nm until constant, using a suitably thermostatted spectrophotometer. Then add:

Reagent D (Enzyme Solution)	0.10	—

Immediately mix by inversion and record the increase in A233nm for approximately 5 minutes. Obtain the ΔA233nm/minute using the maximum linear rate for both the Test and Blank.

Calculations

$$\text{Units / ml enzyme} = \frac{(\Delta A233nm\text{ / min Test} - \Delta A233nm\text{ / min Blank})\ (3)\ (df)}{(4.5)\ (0.1)}$$

3 = Total volume (in milliliters) of assay
df = Dilution factor
4.5 = Millimolar extinction coefficient of Acetyl-CoA at 233nm
0.1 = Volume (in milliliter) of enzyme used

$$\text{Units / mg solid} = \frac{\text{units / ml enzyme}}{\text{mg solid / ml enzyme}}$$

$$\text{Units / mg protein} = \frac{\text{units / ml enzyme}}{\text{mg protein / ml enzyme}}$$

Unit Definition: *One unit will convert 1.0 μmole of acetyl-L-carnitine and CoA to L-carnitine and acetyl-CoA per minute at pH 8.0 at 25°C.*

Final Assay Concentration

In a 3.00 ml reaction mix, the final concentrations are 92 mM Tris, 0.18 mM coenzyme A, 5.6 mM acetyl-DL-carnitine and 0.03–0.06 unit of carnitine acetyltransferase.

Enzymatic Assay of CASEIN KINASE II

Principle

$$\text{Casein} + {}_{32}\text{P - ATP} \xrightarrow{\text{Casein Kinase II}} {}_{32}\text{P - Phosphorylated Casein} + \text{ADP}$$

Conditions: T = 37°C, pH 7.5

Method: Radiolabelled Stop Reaction

Reagents

A. 100 mM HEPES Buffer, pH 7.5 at 37°C (Enzyme Diluent) (Prepare 25 ml in deionized water using HEPES, Free Acid. Adjust to pH 7.5 at 37°C with 1 M NaOH).

B. 200 mM HEPES Buffer with 50 mM Magnesium Chloride, 0.5 mM Adenosine 5'-Triphosphate, 650 mM Potassium Chloride, 25 mM Dithiothreitol, and 100 μCi [ν-$_{32}$P] Adenosine 5-Triphosphate/ml, pH 7.5 at 37°C (5X Reaction Buffer) (Prepare 0.25 ml in deionized water using HEPES, Free Acid, Magnesium Chloride, 4.9 M Solution, Dithiothreitol, Potassium Chloride, Adenosine 5'- Triphosphate, Disodium Salt, and ν-$_{32}$P-Adenosine 5'-Triphosphate, 3000 Ci/mmole. Adjust to pH 7.5 at 37°C with 1 M NaOH).

C. 1% (w/v) Casein Solution (Casein) (Prepare 0.25 ml in deionized water using Casein, 5% Solution).

D. 50 μg/ml Heparin Solution (Heparin) (Prepare 0.25 ml in deionized water using Heparin, Sodium).

E. 10% (w/v) Trichloroacetic Acid with 10 mM Sodium Pyrophosphate (TCA) (Prepare 250 ml in deionized water using Trichloroacetic Acid, 6.1 N Solution, 100% (w/v), and Sodium Pyrophosphate, Decahydrate).

F. Ethanol (EtOH) (Use Ethyl Alcohol, Denatured).

G. Acetone.

H. Chromatography Paper (Whatman 3mm paper) (Use Whatman 3mm Chromatography Paper).

I. Casein Kinase II Enzyme Solution (Immediately before use, prepare a solution containing 0.3 unit/ml of Casein Kinase II in cold Reagent A).

Procedure

Pipette (in milliliters) the following reagents into an Eppendorf tube.

	Test	Blank
Reagent C (Casein)	0.01	0.01
Deionized Water	0.01	0.01
Reagent I (Enzyme Solution)	0.02	—
Reagent A (Enzyme Diluent)	—	0.02
Reagent B (5x Reaction Buffer)	0.01	0.01

Immediately mix by gently vortexing for a few seconds and then incubate for exactly 10 minutes at 37°C. Pipette (in milliliter) a 0.04 ml aliquot of the Test and Blank onto a separate piece of Reagent H (Whatman 3 mm paper). Soak the pieces in Reagent E (TCA) for 2–5 minutes at room temperature. Wash the pieces of paper 4 times with 10 ml of Reagent E (TCA). Agitate gently throughout each wash for 15 minutes. Repeat each wash 4 times with Reagent F (EtOH) and Reagent G (Acetone). Dry the paper pieces at room temperature or under a heat lamp. Count the radioactivity incorporated into the precipitated casein using the Cerenkov method (i.e. count the β-emission without scintillation fluid using the 3H Channel).

Calculations

1. Count the radioactivity (R) of 0.01 ml of Reagent B (5X Reaction Buffer), in order to obtain the total radioactivity in cpm (counts per minute) per assay.
2. Divide the above value (R) by the amount of Adenosine 5'-Triphosphate present in the assay (5 nmole), inorder to obtain the specific radioactivity (SR). SR=R/5
3. Subtract the Blank value from the Test and multiply the result by the factor 5/4 in order to obtain the total counts per tube, C.
   C = CTest–CBlank X 5/4
   5/4 = Correction factor since only 0.04 out of 0.05 ml reaction mix is counted.

$$\text{Units / ml enzyme} = \frac{C}{(SR)\,(10)\,(0.02)}$$

C = Radioactivity of Test in cpm
SR = Specific Radioactivity
10 = Time (in minutes) of assay
0.02 = Volume (in milliliter) of enzyme used

Unit Definition: *One unit of Casein Kinase II will transfer 1 nanomole of phosphate from ATP to casein per minute at 37°C at pH 7.5.*

Final Assay Concentration

In a 0.05 ml reaction mix, the final concentrations are 80 mM HEPES, 0.2% (w/v) casein, 10 mM magnesium chloride, 5 mM dithiothreitol, 0.1 mM adenosine 5'-triphosphate, 130 mM potassium chloride, and 0.006 unit casein kinase II.

Note

1. Run a second Test sample to be used as a control. Replace the 0.01 ml of deionized water with 0.01 ml of Reagent D (Heparin). A 10 µg/ml concentration of Heparin should inhibit the casein kinase II by more than 95%.

Enzymatic Assay of CATALASE (EC 1.11.1.6)

Principle

$$2H_2O_2 \xrightarrow{\text{Catalase}} 2H_2O + O_2$$

Conditions: T = 25°C, pH = 7.0, A240nm, Light Path = 1 cm.

Method: Continuous Spectrophotometric Rate Determination

Reagents

A. 50 mM Potassium Phosphate Buffer pH 7.0 at 25°C(Prepare 200 ml in deionized water using Potassium Phosphate, Monobasic, Anhydrous. Adjust to pH 7.0 at 25°C using 1 M KOH).

B. 0.036% (w/w) Hydrogen Peroxide Solution (H_2O_2) (Substrate Solution) (Prepare in Reagent A using Hydrogen Peroxide, 30% (w/w). Determine the A240nm of this solution using Reagent A as a blank. The A240nm should be between 0.550 and 0.520 absorbance units. Add hydrogen peroxide to increase the absorbance and Reagent A to decrease the absorbance).

C. Catalase Solution (Immediately before use prepare a solution containing 50–100 units per ml in cold Reagent A).

Procedure

Pipette (in milliliters) the following reagents into suitable cuvettes.

	Test	Blank
Reagent A	—	3.00

Equilibrate to 25°C. Monitor the A240nm until constant and record the A240nm using a suitably thermostatted spectrophotometer. Then add:

Reagent C (Catalase Solution)	0.1	—

Immediately mix by inversion and record the time required for the A240nm to decrease from 0.45 to 0.40 absorbance units.

Calculation

$$\text{Units / ml enzyme} = \frac{(3.45)\ (\text{df})}{(\text{min})\ (0.1)}$$

3.45 = Corresponds to the decomposition of 3.45 micromoles of hydrogen peroxide in a 3.0 ml reaction mixture producing a decrease in the A240nm from 0.45 to 0.40 absorbance units.

df = Dilution factor.
min = Time in minutes required for the A240nm to decrease from 0.45 to 0.40 absorbance units
0.1 = Volume (in milliliter) of enzyme used.

$$\text{Units / mg solid} = \frac{\text{units / ml enzyme}}{\text{mg solid / ml enzyme}}$$

$$\text{Units / mg protein} = \frac{\text{units / ml enzyme}}{\text{mg protein / ml enzyme}}$$

Unit Definition: *One unit will decompose 1.0 µmole of H_2O_2 per minute at pH 7.0 at 25°C, while the H_2O_2 concentration falls from 10.3 mM to 9.2 mM. The rate of disappearance of H_2O_2 is followed by observing the rate of decrease in the absorbance at 240 nm.*

Final Assay Concentration

In a 3.00 ml reaction mix, the final concentrations are 50 mM potassium phosphate, 0.035% (w/w) hydrogen peroxide, and 5–10 units catalase.

Notes

1. To obtain the proper activity, catalase should be completely dissolved before assaying. Catalase products supplied as a crystalline suspension may require incubation at 37°C upon dilution to obtain complete dissolution.
2. The initial A240nm will exceed 0.450 absorbance units and start to decrease. Start timing the reaction when the A240nm reaches 0.450 absorbance units.

Enzymatic Assay of CATECHOL OXIDASE (EC 1.14.18.1)

Principle

$$\text{Catechol} + O_2 \xrightarrow{\text{Catechol Oxidase}} \text{O-Benzoquinone} + H_2O$$

$$\text{O-Benzoquinone} + \text{Ascorbic Acid} \rightarrow \text{Catechol} + \text{Dehydro-Ascorbic Acid}$$

Conditions: T = 25°C, pH = 6.5, A265nm, Light path = 1 cm

Method: Continuous Spectrophotometric Rate Determination

Reagents

A. 50 mM Potassium Phosphate Buffer, pH 6.5 at 25°C (Prepare 100 ml in deionized water using Potassium Phosphate, Monobasic. Adjust to pH 6.5 at 25°C with 1 M NaOH).
B. 5 mM Catechol Solution (Prepare 10 ml in Reagent A using Catechol. *Prepare Fresh*).
C. 2.1 mM L-Ascorbic Acid Solution (Prepare 10 ml in Reagent A using L-Ascorbic Acid, Sodium Salt. *Prepare Fresh*).
D. 0.065 mM Ethylenediaminetetraacetic Acid Solution (EDTA) (Prepare 10 ml in Reagent A using Ethylenediaminetetraacetic Acid, Disodium Salt).
E. Catechol Oxidase Enzyme Solution (Immediately before use, prepare a solution containing 500–1000 units/ml of Catechol Oxidase in cold Reagent A).

Procedure

Pipette (in milliliters) the following reagents into suitable quartz cuvettes:

	Test	Blank
Reagent A (Buffer)	2.60	2.80
Reagent B (Catechol)	0.10	0.10
Reagent C (Ascorbic Acid)	0.10	—
Reagent D (EDTA)	0.10	0.10

Mix by inversion and equilibrate to 25°C. Monitor the A265nm until constant, using a suitably thermostatted spectrophotometer. Then add:

Reagent E (Enzyme Solution)	0.10	—

Immediately mix by inversion and record the decrease in A265nm for approximately 5 minutes. Obtain the ΔA265nm/minute using the maximum linear rate for both the Test and Blank.

Calculations

$$\text{Units / mg enzyme} = \frac{(\Delta A265\text{nm / min Test} - \Delta A265\text{nm / min Blank})}{(0.001)\ (\text{mg enzyme / RM})}$$

0.001 = The change in A265nm/minute per unit of Catechol Oxidase at pH 6.5 at 25°C in a 3 ml reaction mix

RM = Reaction Mix (final volume = 3 ml)

Unit Definition: *One unit will cause the change in A265nm of 0.001 per minute at pH 6.5 and 25°C in a 3 ml reaction mix containing catechol and L-ascorbic acid.*

Final Assay Concentration

In a 3 ml reaction mix, the final concentrations are 50 mM potassium phosphate, 0.17 mM catechol, 0.07 mM L-ascorbic acid and 50–100 units of catechol oxidase.

Note

1. Ascorbic acid is not included in the blank to produce a true blank. Possible interaction with catechol and ascorbic acid might give false readings.

Enzymatic Assay of CATECHOL-O-METHYL TRANSFERASE (EC 2.1.1.6)

Principle

$$\text{SAM} + \text{DHBA} \xrightarrow{\text{Catechol-O-Methyl Transferase}} \text{SAH} + \text{3-O-MHBA}$$

Abbreviations Used

SAM = S-Adenosyl-Methionine
SAH = S-Adenosyl-Homocysteine
DHBA = 3,4-Dihydroxybenzoic Acid
3-O-MHBA = 3-O-Methyl,4-hydroxybenzoic Acid

Conditions: T = 37°C, pH = 7.9

Method: Radioactive.

Reagents

A. 100 mM Tris buffer, pH 7.9 at 37°C (Prepare 50 ml in deionized water using Trizma Base. Adjust to pH 7.9 at 37°C with 6 M HCl).
B. 100 mM Magnesium Chloride ($MgCl_2$) (Prepare 10 ml in deionized water using Magnesium Chloride).
C. 100 mM Dithiothreitol (DTT) (Prepare 10 ml in deionized water using Dithiothreitol).
D. 10 mM Ethylenediaminetetraacetic Acid (EDTA) (Prepare 10 ml in deionized water using Ethylenediaminetetraacetic Acid, Tetrasodium Salt).
E. 100 mM 3,4-Dihydroxybenzoic Acid (DHBA) (Prepare 20 ml in O_2 free water using 3,4-Dihydroxybenzoic Acid. *Prepare Fresh and Protected from Light*).
F. S-Adenosyl-L-(methyl-$_{14}$C)-Methionine (2 μM/ml; 1 μCi/ml)
G. 0.1% Albumin Buffer (BSA) (Prepare 25 ml in deionized water using Albumin, Bovine Serum).
H. Catechol-O-Methyl Transferase Enzyme Solution (Immediately before use prepare a solution containing 500 units/ml of Catechol-O-Methyl Transferase in cold Reagent G.
I. Universal LSC Cocktail for Aqueous Samples).

Procedure

Prepare a reaction cocktail by pipetting (in milliliters) the following reagents into a suitable container:

Deionized Water	3.25
Reagent A (Buffer)	5.00
Reagent B ($MgCl_2$)	0.10
Reagent C (DTT)	0.05
Reagent D (EDTA)	0.10

Purge material for 5 minutes using N_2. Then add:

Reagent E (DHBA)	1.00
Reagent F ($_{14}$C-methyl)-SAM)	0.50

Mix by swirling. Equilibrate to 37°C. Inject into serum vials (flushed with Nitrogen and sealed with serum caps) the following (in milliliters):

	Test	Blank
Reaction Mix	0.50	0.50
Reagent G (Buffer)	—	0.10
Reagent H (Enzyme Solution)	0.10	—

Mix by inversion. Incubate at 37°C for 15 minutes. Withdraw 0.1 ml from each vial at time zero, 5 minutes, and 15 minutes, using a syringe and place in suitable dram vials containing 1.0 ml of 0.1 M HCl. Add (in milliliters) to each dram vial:

Ethyl Acetate	5.00

Shake well and centrifuge to separate the two phases. Withdraw 4.0 ml of the ethyl acetate layer (upper layer) from each vial and place it in suitable scintillation counting vials. To a separate scintillation vial add 0.1 ml of the reaction mix. This is the total possible counts/reaction (TCR). To each scintillation vial add (in milliliters):

Reagent I (LSC Cocktail)	15

Shake and count in a suitable scintillation counter for.

Calculations

CPM = Counts per minute of Sample – 0 time counts per minute.

$$\text{CPM / nMole} = \frac{\text{TCR}}{(0.1)(83.3)}$$

$$\text{Units / mg Protein} = \frac{\text{CPM} \times 60}{(\text{CPM / nMole})(\text{mg Protein / reaction})(0.56)}$$

CPM = Net counts per minute of sample
CPM/nMole = Total potential counts of S-Adenosyl-L-(methyl-${}_{14}$C)-Methionine.
TCR = Total possible counts/reaction
0.1 = volume of reaction mix used
83.3 = nMoles of SAM per ml in reaction mix
60 = conversion to hour time frame from minute
0.56 = Efficiency of recovery using ethyl acetate

Unit Definition: *One unit will catalyze the methylation of 1.0 nanomole of protocatechuic acid per hour at pH 7.9 at 37°C using S-adenosyl-L-[methyl${}_{14}$C]-methionine as the methyl donor.*

Final Assay Concentrations

In a 0.6 ml reaction mix, the final concentrations are 42 mM Tris, 0.83 mM magnesium chloride, 0.42 mM dithiothreitol, 0.083 mM EDTA, 8.3 mM 3,4-dihydroxybenzoic acid, 0.025 μCi S-adenosyl-L-(methyl-${}_{14}$C)-methionine and 50 units catechol-O-methyl transferase.

Notes

1. The TCR determines the number of counts in the reaction mix before enzyme reaction.
2. The recovery of ${}_{14}$C material using the Ethyl Acetate extraction and correcting for the removal of only 4 ml of the layer is only 56% efficient, so a correction must be made for the counts.

Enzymatic Assay of CATECHOL-O-METHYL TRANSFERASE (EC 2.1.1.6)

Principle

$${}_{14}\text{C(CH}_3\text{)-SAM} + \text{Protocatechuic Acid} \xrightarrow{\text{CMT}} \text{SAH} + {}_{14}\text{C-Methyl Protocatechuic Acid}$$

Abbreviations Used

$_{14}C(CH_3)$-SAM = S-Adenosyl-L(Methyl-$_{14}$C)-Methionine
CMT = Catechol-O-Methyl Transferase
SAH = S-Adenosyl-Homocysteine

Conditions: T = 37°C, pH 7.9

Method: Radiolabelled Stop Reaction

Reagents: Prepare all reagents in deionized water which has been boiled for 10 minutes. Purge all reagents with nitrogen gas.

A. 500 mM Tris HCl Buffer, pH 7.9 at 37°C (Prepare 100 ml in deionized water using Trizma Base. Adjust to pH 7.9 at 37°C with 1 M HCl).
B. 100 mM Magnesium Chloride Solution ($MgCl_2$) (Prepare 10 ml in deionized water using Magnesium Chloride, Hexahydrate).
C. 100 mM DL-Dithiothreitol Solution (DTT) (Prepare 10 ml in deionized water using DL-Dithiothreitol).
D. 10 mM Ethylenediaminetetraacetic Acid Solution (EDTA) (Prepare 10 ml in deionized water using Ethylenediaminetetraacetic Acid, Tetrasodium Salt, Hydrate).
E. 10 mM Protocatechuic Acid Solution (PCA) (Prepare 20 ml in deionized water using Protocatechuic Acid).
F. S-Adenosyl-L-(Methyl-$_{14}$C)-Methionine Solution (Hot SAM) (Use 60 mCi/mmol, 25 μCi/ml) G. 0.2% (w/v) S-Adenosyl-L-Methionine (Cold SAM) (Immediatley, prior to use, prepare 5 ml in Reagent A using S-Adenosyl-L-Methionine, p-Toluenesulfonate Salt, *Prepare Fresh*).
H. 0.1% (w/v) Bovine Serum Albumin Solution with 11 mM DL-Dithiothreitol (Enz Dil) (Prepare 10 ml in deionized water using Albumin Bovine, and DL-Dithiothreitol).
I. Catechol-O-Methyl Transferase Enzyme Solution (Immediately before use, prepare a solution containing at least 1300–2000 units/ml of Catechol-O-Methyl Transferase in cold Reagent H).
J. Scintillation Cocktail (Scint).
K. Ethyl Acetate Solution (EtOAc).

Procedure

Prepare a reaction cocktail by pipetting (in milliliters) the following reagents into a suitable container:

Reagent A (Buffer)	1.00
Reagent F (Hot SAM)	0.120
Reagent G (Cold SAM)	0.80
Reagent E (PCA)	1.00
Reagent B ($MgCl_2$)	0.10
Reagent C (DTT)	0.05
Reagent D (EDTA)	0.10
Deionized Water	6.83

Mix by swirling and equilibrate 100 µl of the reaction cocktail to 37°C. Keep in a N_2 environment. Inject (in milliliters) the following reagents into serum vials (which have been flushed with nitrogen and sealed with serum caps):

	Test	Blank
Reaction Cocktail	0.10	0.10
Deionized Water	—	0.01
Reagent I (Enzyme Solution)	0.01	—

Mix by swirling and purge with nitrogen gas. Incubate at 37°C for 15 minutes. Then add:

1 M HCl	1.00	1.00
Reagent K (EtOAc)	5.00	5.00

Shake well for several minutes and centrifuge to separate the two phases. Withdraw 2 ml of the ethyl acetate layer (upper layer) from each vial and place into a suitable scintillation vial. To a separate scintillation vial, add 0.05 ml of the reaction cocktail and 2 ml of Reagent K (EtOAc). This is the total potential counts. Then to each vial add:

Reagent J (Scint)	5.00	5.00

Mix thoroughly and count in a suitable scintillation counter.

Calculations

$$\text{Units / ml enzyme} = \frac{(\text{dpm Test} - \text{dpm Blank})\ (5)\ (4)\ (\text{df})}{(\text{SA of SAM})\ (2)\ (0.70)\ (0.01)}$$

5 = Volume (in milliliters) of ethyl acetate added
4 = Conversion factor from minutes to hours (takes reaction time into account)
df = Dilution factor
SA = Specific activity in dpm/nmole from total potential count vials.
2 = Volume (in milliliters) of organic phase which was counted
0.70 = Efficiency of the extraction of the methylated product
0.01 = Volume (in milliliter) of enzyme used

Unit Definition: *One unit will catalyze the methylation of 1.0 nanomole of protocatechuic acid per hour at pH 7.9 at 37°C using S-adenosyl-L(methyl $_{14}$C)-methionine as the methyl donor.*

Final Assay Concentration

In a 0.11 ml reaction mix, the final concentrations are 82 mM Tris, 0.9 mM magnesium chloride, 1.5 mM DL-dithiothreitol, 0.1 mM Ethylenediaminetetraacetic Acid, 0.9 mM protocatechuic acid, 0.4 mM S-adenosyl-L-methionine, 0.01% (w/v) bovine serium albumin, and 13–20 units catechol-omethyl transferase.

Note

1. The concentration of 0.2% (w/v) is based upon the pure S-adenosyl-L-methionine. This does not take into account salts, solvent, water, etc. that may be present.

Enzymatic Assay of CATHEPSIN B (EC 3.4.22.1)

Principle

$$\text{Na-CBZ-L-Lysine p-Nitrophenyl Ester} + H_2O \xrightarrow{\text{Cathepsin B}} \text{Na-CBZ-L-Lysine} + \text{p-Nitrophenol}$$

Abbreviation Used

CBZ = N-Carbobenzoxy

Conditions: T = 25°C, pH = 5.0, A326nm, Light path = 1 cm

Method: Continuous Spectrophotometric Rate Determination

Reagents

A. 20 mM Sodium Acetate Buffer with 1.0 mM Ethylenediaminetetraacetic Acid and 5.0 mM L-Cysteine (Prepare 100 ml in deionized water using Sodium Acetate, Trihydrate, LCysteine, Hydrochloride, Monohydrate and Ethylenediaminetetraacetic Acid, Disodium Salt, Dihydrate. Adjust to pH 5.0 at 25°C with 1 M NaOH).

B. Dimethyl Sulfoxide Solution (DMSO) (Use Dimethyl Sulfoxide).

C. 5.2 mM Na-CBZ-L-Lysine p-Nitrophenyl Ester Solution (Substrate) (Prepare 2 ml in Reagent B using Na-CBZ-L-Lysine p-Nitrophenyl Ester, Hydrochloride).

D. Cathepsin B Enzyme Solution (Immediately before use, prepare a solution containing 2.5–5.0 units/ml of Cathepsin B in Reagent A).

Procedure

Pipette (in milliliters) the following reagents into suitable cuvettes:

	Test	Blank
Reagent A (Buffer)	3.00	3.00
Reagent C (Substrate)	0.05	0.05

Mix by inversion and equilibrate to 25°C. Monitor the rate of increase in the absorbance at 326 nm for at least two minutes but no more than three minutes using a suitably thermostatted spectrophotometer. This rate should be approximately 0.03 absorbance units per minute. Then add:

	Test	Blank
Reagent D (Enzyme Solution)	0.01	—
Reagent A (Buffer)	—	0.01

Immediately mix by inversion and record the increase in A326nm for approximately 3 minutes. Obtain the ΔA326nm/minute using the maximum linear rate for both the Test and the Blank.

Calculations

$$\text{Units / ml enzyme} = \frac{(\Delta A326nm / min\ Test - \Delta A326nm / min\ Blank)\ (3.06)\ (df)}{(7.58)\ (0.01)}$$

3.06 = Total volume (in milliliters) of assay
df = Dilution factor
7.58 = Millimolar extinction coefficient of p-nitrophenol at 326nm
0.01 = Volume (in milliliter) of enzyme used

$$\text{Units / mg solid} = \frac{\text{units / ml enzyme}}{\text{mg solid / ml enzyme}}$$

$$\text{Units / mg protein} = \frac{\text{units / ml enzyme}}{\text{mg protein / ml enzyme}}$$

Unit Definition: *One unit will hydrolyze 1 μmole of Na-CBZ-lysine p-nitrophenyl ester per minute at pH 5.0 at 25°C.*

Final Assay Concentration

In a 3.06 ml reaction mix, the final concentrations are 20 mM sodium acetate, 0.98 mM Ethylenediaminetetraacetic Acid, 4.9 mM L-cysteine, 0.08 mM Na-CBZ-L-lysine p-nitrophenyl ester, 2% (v/v) dimethyl sulfoxide, and 0.025–0.050 unit cathepsin B.

Enzymatic Assay of CATHEPSIN B (EC 3.4.22.1)

Principle

$$\text{Na-CBZ-Arg-Arg-7-Amido-4-Methylcoumarin} + H_2O \xrightarrow{\text{Cathepsin B}} \text{Arg-Arg} + \text{7-AMC}$$

Abbreviations Used

CBZ = Carbobenzoxy
Arg-Arg = Arginylarginine
7-AMC = 7-Amino-4-Methylcoumarin

Conditions: T = 40°C, pH = 6.0, Excitation = 348nm, Emission = 440nm, Light path = 1 cm

Method: Fluorometric Rate Determination

Reagents

A. 352 mM Potassium Phosphate Buffer with 48 mM Sodium Phosphate and 4.0 mM Ethylenediaminetetraacetic Acid, pH 6.0 at 40°C (Prepare 100 ml in deionized water using Potassium Phosphate, Monobasic, Anhydrous, Sodium Phosphate, Dibasic, Anhydrous, and Ethylenediaminetetraacetic Acid, Disodium salt, Dihydrate. Adjust to pH 6.0 at 40°C with 1 M HCl).

B. 8.0 mM L-Cysteine HCl Solution, pH 6.0 at 40°C (L-Cys) (Prepare 50 ml in Reagent A using L-Cysteine Hydrochloride, Monohydrate. Adjust to pH 6.0 at 40°C with 1 M NaOH. *Prepare Fresh*).

C. 0.1% (v/v) Brij 35 Solution (Brij 35) (Prepare 100 ml in deionized water using Brij 35 Solution, 30% (w/v) solution).

D. 0.02 mM Na-CBZ-Arg-Arg-7-Amido-4-Methylcoumarin (Arg-Arg-7-AMC) (Prepare by dissolving 1 mg of Na-CBZ-Arg-Arg-7-Amido- 4-Methylcoumarin, Hydrochloride in 0.14 ml of Dimethyl Sulfoxide. Dilute to 70 ml with Reagent C).

E. 5.0 μM 7-Amino-4-Methylcoumarin (7-AMC) (Prepare by dissolving 1 mg of 7-Amino-4-Methylcoumarin in 1.0 ml of Dimethyl Sulfoxide. Dilute to 5.0 μM (approximately 1125 fold dilution) with Reagent C).
F. Cathepsin B Enzyme Solution (Immediately before use, prepare a solution containing 5–10 units Cathepsin B in cold Reagent C).

Procedure

Pipette (in milliliters) the following reagents into suitable fluorometric cuvettes:

	Test	Blank
Reagent B (L-Cys)	0.75	0.75
Reagent C (Brij 35)	0.90	1.00
Reagent F (Enzyme Solution)	0.10	—

Mix by inversion and equilibrate to 40°C. Monitor the fluorescence intensity at the excitation wavelength of 348 nm and the emission wavelength of 440 nm until constant using a suitably thermostatted fluorometer. Then add:

Reagent D (Arg-Arg-7-AMC)	0.75	0.75

Immediately mix by inversion and record the increase in fluorescence intensity at the excitation wavelength of 348 nm and the emission wavelength of 440 nm for 5 minutes. Obtain the Δ intensity/5 minutes by using the maximum linear rate for both the Test and Blank.

Standard Curve: Pipette (in milliliters) the following reagents into suitable quartz fluorometric cuvettes:

	Std 1	Std 2	Std 3	Std 4	Std 5	Std Blank
Reagent B (L-Cys)	0.75	0.75	0.75	0.75	0.75	0.75
Reagent C (Brij 35)	1.55	1.35	1.15	0.95	0.75	1.75
Reagent E (7-AMC)	0.20	0.40	0.60	0.80	1.00	—

Mix by inversion and equilibrate to 40°C. Obtain the fluorescence intensity for each Standard and the Standard Blank at the excitation wavelength of A348nm and the emission wavelength of A440nm.

Calculations

Standard Curve: Δ Intensity Standard = Intensity Std–Intensity Std Blk

Construct a Standard curve by plotting the ΔIntensity of the Standards versus nanomoles of 7-Amino-4- Methylcoumarin.

Sample Determination: Determine the nanomoles of 7-Amino-4-Methylcoumarin liberated using the standard curve.

$$\text{nanomoles liberated} = \frac{\Delta\text{Intensity Sample/5 min} - \Delta\text{ Intensity Blank/5 min}}{\text{Intensity/nanomoles of 7-AMC}}$$

$$\text{Units / ml enzyme} = \frac{(\text{nanomoles of 7 - Amino - 4 - Methylcoumarin liberated}) (\text{df})}{(5)(1)}$$

df = Dilution factor
5 = Reaction time
1 = Volume (in milliliter) of enzyme used

$$\text{Units / mg protein} = \frac{\text{units / ml enzyme}}{\text{mg protein / ml enzyme}}$$

Unit Definition: *One unit will liberate 1.0 nanomole of 7-Amino-4-Mmethylcoumarin from Na-CBZ-L-arginyl-L-arginine 7-Amido-4-Methylcoumarin per minute at pH 6.0 at 40°C.*

Final Assay Concentration

In a 2.50 ml reaction mix, the final concentrations are 106 mM potassium phosphate, 14 mM sodium phosphate, 1.2 mM Ethylenediaminetetraacetic Acid, 2.4 mM L-cysteine, 0.07% (v/v) Brij 35, 0.006 mM Na-CBZ-Arg-Arg-7-Amino-4-Methylcoumarin, 0.06% (v/v) dimethyl sulfoxide, and 0.05–0.1 mg cathepsin B.

Note

1. This solution must be made fresh and used within 3 hours of preparation.

Enzymatic Assay of CATHEPSIN C (EC 3.4.14.1)

Principle

$$\text{GPAA} + \text{Hydroxylamine} \xrightarrow{\text{Cathepsin C}} \text{Glycyl - phenylalanylhydroxamic acid} + NH_3$$

Abbreviation Used

GPAA = Glycyl-L-phenylalaninamide

Conditions: T = 37°C, pH 6.8, A = 510 nm, Light path = 1 cm

Method: Colorimetric

Reagents

A. 2000 mM Hydroxylamine HCl Buffer, pH 6.8 at 37°C (Prepare 100 ml by first making a 4000 mM stock solution of Hydroxylamine HCl in deionized water using Hydroxylamine Hydrochloride. Adjust to pH 6.8 at 37°C with concentrated NaOH and dilute to 2000 mM just prior to use. *Prepare Fresh*).

B. 250 mM Glycyl-L-Phenylalaninamide Acetate Solution, pH 6.8 at 37EC (GPAA) (Prepare 10 ml in deionized water using GLY-PHE Amide, Acetate Salt. Adjust to pH 6.8 at 37°C with 0.1 M NaOH. *Prepare Fresh*).

C. 125 mM β-Mercaptoethylamine Solution, pH 6.8 at 37°C (2-MEA) (Prepare 10 ml in deionized water using 2-Mercaptoethylamine Hydrochloride. Adjust to pH 6.8 at 37°C with 0.1 M NaOH. *Prepare Fresh*).

D. 20% (v/v) Trichloracetic Acid Solution (TCA) (Prepare 10 ml in deionized water using Trichloroacetic Acid, 6.1 N Solution).

E. 5% (w/v) Ferric Chloride Solution ($FeCl_3A_6H_2O$) (Prepare 25 ml in 0.1 M HCl using Ferric Chloride, Hexahydrate).

F. Cathepsin C Enzyme Solution (Immediately before use, prepare a solution containing 1.5–3.0 units/ml of Cathepsin C in cold deionized water).

Procedure

Pipette (in milliliters) the following reagents into suitable containers:

	Test	Blank
Reagent A (Buffer)	0.10	0.10
Reagent B (GPAA)	0.10	0.10
Reagent C (2-MEA)	0.10	0.10
Deionized Water	0.10	0.10

Mix by inversion and equilibrate to 37°C, for no more than 5 minutes. Then add:

Reagent F (Enzyme Solution)	0.10	—

Mix by inversion and incubate at 37°C for exactly 10 minutes. Then add:

Reagent D (TCA)	0.50	0.50
Reagent E ($FeCl_3A_6H_2O$)	0.50	0.50
Reagent F (Enzyme)	—	0.10
Deionized Water	1.50	1.50

Mix by inversion. If turbidity occurs upon the addition of the TCA, samples should be centrifuged. Transfer the solution to suitable cuvettes and record the A510nm within 10 minutes for both the Test and Blank using a suitable spectrophotometer.

Calculations

$$\text{Units / ml enzyme} = \frac{(\text{A510nm Test} - \text{A510nm Blank})\,(3.0)\,(\text{df})}{(0.37)\,(10)\,(0.1)}$$

3.0 = Volume of assay (ml)
df = Dilution factor
0.37 = Millimolar extinction coefficient of phenylalanine hydroxamate at 510nm.
10 = Time of assay (in minutes) as per "Unit Definition"
0.1 = Volume (in milliliter) of enzyme used

$$\text{Units / mg solid} = \frac{\text{units / ml enzyme}}{\text{mg solid / ml enzyme}}$$

$$\text{Units / mg protein} = \frac{\text{units / ml enzyme}}{\text{mg protein / ml enzyme}}$$

Unit Definition: *One unit will produce 1.0 μmole of hydroxamic acid from glycyl-L-phenylalaninamide and hydroxylamine per minute at pH 6.8 at 37°C using DL-phenylalanine hydroxamate as the standard. In addition to its hydrolytic properties cathepsin C catalyzes the polymerization of dipeptide amides.*

Final Assay Concentration

In a 0.50 ml reaction mix, the final concentrations are 400 mM hydroxylamine, 50 mM glycyl-Lphenylalaninamide acetate, 25 mM 2-mercaptoethylamine and 0.15–0.30 unit cathepsin C.

Enzymatic Assay of CATHEPSIN D (EC 3.4.23.5)

Principle

$$\text{Hemoglobin} + H_2O \xrightarrow{\text{Cathepsin D}} \text{TCA - Soluble Peptides}$$

Abbreviation Used

TCA = Trichloroacetic Acid

Conditions: T = 37°C, pH = 3.0, A280nm, Light path = 1 cm

Method: Spectrophotometric Stop Rate Determination

Reagents

A. 400 mM Citrate Buffer, pH 2.8 at 37°C (Prepare 50 ml in deionized water using Citric Acid, Trisodium Salt, Dihydrate. Adjust to pH 2.8 at 37°C with 1 M HCl).
B. 2.5% (w/v) Hemoglobin Solution (Hb) (Prepare 20 ml in deionized water using Hemoglobin, Bovine. Filter the solution through glass wool to remove particulates).
C. 5% (v/v) Trichloroacetic Acid Solution (TCA) (Prepare 100 ml in deionized water using Trichloroacetic Acid, 6.1 N Solution, approximately 100% (w/v)).
D. Cathepsin D Enzyme Solution (Immediately before use, prepare a solution containing 5–10 units/ml of Cathepsin D in cold deionized water).

Procedure

Prepare a reaction cocktail by pipetting the following reagents into a suitable container:

Deionized Water	10.00
Reagent A (Buffer)	2.50
Reagent B (Hb)	10.00

Mix by swirling and equilibrate at 37°C. Adjust to pH 3.0 with either 100 mM HCl or 100 mM NaOH. Pipette (in milliliters) the following reagents into suitable containers:

	Test	Blank
Reaction Cocktail	2.25	2.25

Equilibrate at 37°C. Then add:

	Test	Blank
Reagent D (Enzyme Solution)	0.01	—

Mix by inversion and incubate for exactly 10 minutes at 37°C. Then add:

Reagent C (TCA)	2.25	2.25
Reagent D (Enzyme Solution)	—	0.01

Mix by inversion and filter the solutions through a 0.45 μm syringe filter. Transfer solutions to suitable quartz cuvettes and record the A280nm for both the Test and Blank using a suitable spectrophotometer.

Calculation

$$\text{Units / ml enzyme} = \frac{(\text{A280nm Test} - \text{A280nm Blank})\ (4.51)\ (\text{df})}{(10)\ (0.01)\ (1)}$$

4.51 = Total volume (in milliliters) of assay
df = Dilution factor
10 = Time of assay (in minutes)
0.01 = Volume (in milliliter) of enzyme used
1 = Increase in A280nm per minute as per the Unit Definition

$$\text{Units / mg solid} = \frac{\text{units / ml enzyme}}{\text{mg solid / ml enzyme}}$$

$$\text{Units / mg protein} = \frac{\text{units / ml enzyme}}{\text{mg protein / ml enzyme}}$$

Unit Definition: *One unit will produce an increase in A280nm of 1.0 per minute per ml at pH 3.0 at 37°C measured as TCA-soluble products using hemoglobin as substrate.*

Final Assay Concentration

In a 2.26 ml reaction mix, the final concentrations are 44 mM citrate, 1.1% (w/v) hemoglobin and 0.05–0.10 unit cathepsin D.

Note

1. Glass wool is used to filter the hemoglobin solution since filter paper and syringe filters may become easily clogged.

Enzymatic Assay of CATHEPSIN D (EC 3.4.23.5)

Principle

$$\text{Acid Denatured Hemoglobin} \xrightarrow{\text{Cathepsin D}} \text{TCA - Soluble Peptides}$$

Conditions: T = 37°C, pH = 3.3, A280nm, Light path = 1 cm

Method: Spectrophotometric Stop Rate Determination.

Reagents

A. 100 mM Formic Acid Buffer, pH 3.3 at 37°C (Prepare 100 ml in deionized water using Formic Acid, Sodium Salt. Adjust to pH 3.3 at 37°C with 1 M HCl or 1 M NaOH).
B. 2.5% (w/v) Hemoglobin Solution (Hb) (Prepare 10 ml in Reagent A using Hemoglobin).
C. 10% (v/v) Trichloroacetic Acid Solution (TCA) (Prepare 100 ml in deionized water using Trichloroacetic Acid, 6.1 N solution, approximately 100% (2/v)).
D. Cathepsin D Enzyme Solution (Immediately before use, prepare a solution containing 5–15 units/ml if Cathepsin D in cold Reagent A).

Procedure

Pipette (in milliliters) the following reagents into suitable containers:

	Test	Blank
Reagent A (Buffer)	0.73	0.73
Reagent B (Hb)	0.05	0.05

Equilibrate to 37°C. Then add:

Regent D (Enzyme Solution)	0.02	—

Mix by inversion and incubate for exactly 30 minutes at 37°C. Then add:

Reagent (TCA)	0.50	0.50
Reagent D (Enzyme Solution)	—	0.02

Mix by inversion and incubate for 10 minutes at 37°C. Filter the solutions through a 0.2 μM syringe filter. Transfer the solution to sutiable cuvettes and record the A280nm for both the Test and Blank using a sutiable spectrophotometric.

Calculations

$$\text{Units / vial enzyme} = \frac{(\text{A280nm Test} - \text{A280nm Blank})\ (1.3)\ (\text{df})}{(1)\ (0.02)}$$

1.3 = Total volume (in milliliters) of assay
df = Dilution factor of vial
1 = Increase in A280nm for 30 minutes as per the Unit Definition

$$\text{Units / mg protein} = \frac{\text{units / ml enzyme}}{\text{mg protein / ml enzyme}}$$

Unit Definition: *One unit will produce an increase in A280nm of 1.0 in 30 minutes at pH 3.3 at 37°C measured as TCA soluble products using acid denatured hemoglobin as substrate (1 cm light path).*

Final Assay Concentration

In a 0.8 ml reaction mix, the final concentrations are 100 mM formic acid, 0.16% (w/v) hemoglobin and 0.1–0.3 units cathepsin B.

Enzymatic Assay of CATHEPSIN G (EC 3.4.21.20)

Principle

$$\text{N-Succinyl-Ala-Ala-Pro-Phe p-NA} + H_2O \xrightarrow{\text{Cathepsin G}} \text{N-Succinyl-Ala-Ala-Pro-Phe} + \text{p-nitroaniline}$$

Abbreviation Used

p-NA = p-nitroanilide

Conditions: T = 37°C, pH = 7.5, A410nm, Light path = 1 cm

Method: Continuous Spectrophotometric Rate Determination

Reagents

A. 100 mM HEPES NaOH Buffer, pH 7.5 at 37EC (Prepare 100 ml in deionized water using HEPES, Free Acid. Adjust to pH 7.5 at 37°C with 1 M NaOH).
B. Dimethyl Sulfoxide (DMSO).
C. 20 mM N-Succinyl-Ala-Ala-Pro-Phe p-Nitroanilide Solution (Substrate) (Prepare 2 ml in Reagent B using N-Succinyl-Ala-Ala-Pro-Phe p-Nitroanilide).
D. Cathepsin G Enzyme Solution (Immediately before use, prepare a solution containing 1.25–2.50 units/ml of Cathepsin G in cold deionized water).

Procedure

Pipette (in milliliters) the following reagents into suitable containers:

	Test	Blank
Reagent A (Buffer)	1.60	1.60
Reagent C (Substrate)	0.20	0.20

Mix by inversion and equilibrate to 37°C. Then add:

Reagent D (Enzyme Solution)	0.025	—
Deionized Water	—	0.025

Immediately mix by inversion and record the increase in A410nm for approximately 5 minutes. Obtain the ΔA410nm/minute using the maximum linear rate for both the Test and Blank.

Calculations

$$\text{Units / ml enzyme} = \frac{(\Delta A410nm / \text{min Test} - \Delta A410nm / \text{min Blank})\,(1000)\,(1.825)\,(df)}{(8.8)\,(0.025)\,(60)}$$

1000 = Factor for converting μmoles to nanomoles as per the Unit Definition
1.825 = Total volume (in milliliters) of assay
df = Dilution factor
8.8 = Millimolar extinction coefficient of p-nitroaniline at 410nm
0.025 = Volume (in milliliters) of enzyme used
60 = Factor for converting minutes to seconds as per the Unit Definition

$$\text{Units / mg protein} = \frac{\text{units / ml enzyme}}{\text{mg protein / ml enzyme}}$$

Unit Definition: *One unit will release 1.0 nanomole of p-nitroaniline per second from N-Succinyl-Ala-Ala-Pro-Phe-pnitroanilide at pH 7.5 at 37°C.*

Final Assay Concentration

In a 1.825 ml reaction mix, the final concentrations are 88 mM HEPES, 2.2 mM N-Succinyl-Ala-Ala-Pro-Phe-p-nitroanilide, 11% (v/v) dimethyl sulfoxide and 0.03–0.06 unit Cathepsin G.

Enzymatic Assay of CELLULASE (EC 3.2.1.4)

Principle

$$\text{Cellulose} + H_2O \xrightarrow{\text{Cellulase}} \text{D-Glucose}$$

$$\text{D-Glucose} + \text{ATP} \xrightarrow{\text{Hexokinase}} \text{D-Glucose 6-Phosphate} + \text{ADP}$$

$$\text{D-Glucose 6-Phosphate} + \beta\text{-NAD G-6PDH} \rightarrow \text{6-PG} + \beta\text{-NADH}$$

Abbreviations Used

ATP = Adenosine 5'-Triphosphate

ADP = Adenosine 5'-Diphosphate

β-NAD = β-Nicotinamide Adenine Dinucleotide, Oxidized Form

β-NADH = β-Nicotinamide Adenine Dinucleotide, Reduced Form

G-6PDH = Glucose 6-Phosphate Dehydrogenase

G-PG = 6-Phospho-D-Gluconate

Conditions: T = 37°C, pH = 5.0, A340nm, Light path = 1 cm.

Method: Spectrophotometric Stop Rate Determination

Reagents

A. 50 mM Sodium Acetate Buffer, pH 5.0 at 37°C (Prepare 200 ml in deionized water using Sodium Acetate, Trihydrate. Adjust to pH 5.0 at 37°C with 1 M HCl).

B. 5% (w/v) Cellulose Solution (Prepare 100 ml in Reagent A using Cellulose. Mix and heat gently to make a uniform suspension).

C. Cellulase Enzyme Solution (Cellulase) (Immediately before use, prepare a solution containing 2–6 units/ml of Cellulase in cold deionized water).

D. Glucose (HK) Determination Vial (16–10) (Use Glucose (HK). Dissolve the contents in 10 ml of deionized water).

Procedure

Step 1: Pipette (in milliliters) the following reagents into suitable tubes:

	Test	Blank
Reagent B	4.00	4.00

Equilibrate to 37°C. Then add:

	Test	Blank
Reagent C (Cellulase)	1.00	—
Deionized Water	—	1.00

Immediately mix by swirling and incubate at 37°C for exactly 120 minutes with moderate shaking. Immediately transfer into an ice bath. Allow to stand until the suspension is settled. Centrifuge for 2 minutes to clarify. Use the supernatant in Step 2.

Step 2: Pipette (in milliliters) the following reagents into suitable cuvettes:

Reagent D (16–10)	3.00	3.00

Equilibrate to 25°C. Monitor the A340nm until constant, using a suitably thermostatted spectrophotometer. Record the initial A340nm for both the Test and Blank. Then add:

Test Supernatant (Step 1)	0.10	—
Blank Supernatant (Step 1)	—	0.10

Immediately mix by inversion and record the increase in A340nm until complete (for approximately 5 minutes). Obtain the final A340nm for both the Test and Blank.

Calculations

$$\Delta\text{A340nm Test} = \text{A340nm Test Final} - \text{A340nm Test Initial}$$

$$\Delta\text{A340nm Blank} = \text{A340nm Blank Final} - \text{A340nm Blank Initial}$$

$$\text{Units / ml enzyme} = \frac{(\Delta\text{A340nm Test} - \Delta\text{A340nm Blank})\,(3.1)\,(\text{df})}{(6.22)\,(2)\,(1)\,(0.1)\,(5)}$$

3.1 = Final volume (in milliliters) of Step 2
5 = Total volume (in milliliters) of reaction mix (Step 1)
df = Dilution factor
6.22 = Millimolar extinction coefficient of β-NADH at 340nm
2 = Conversion factor from 2 hours to 1 hour as per the Unit Definition
1 = Volume (in milliliter) of cellulase used in Step 1
0.1 = Volume (in milliliter) from Step 1 used in Step 2

$$\text{Units / mg solid} = \frac{\text{units / ml enzyme}}{\text{mg solid / ml enzyme}}$$

Unit Definition: *One unit will liberate 1.0 μmole of glucose from cellulose in one hour at pH 5.0 at 37°C (2 hour incubation time).*

Final Assay Concentration

In a 5.00 ml reaction mix, the final concentrations are 40 mM sodium acetate and 2–6 units of cellulase.

Enzymatic Assay of CERAMIDE GLYCANASE (EC 3.2.1.123)

Principle

$$\text{Monosialoganglioside} \xrightarrow{\text{Ceramide Glycanase}} \text{Monosaccharides}$$

$$\text{Monosaccharides} \xrightarrow{\text{Park-Johnson Reaction}} \text{Green Colored Dye Complex}$$

Conditions: T = 37°C, pH 5.0, A690nm, Light path = 1 cm

Method: Colorimetric

Reagents

A. 50 mM Sodium Acetate Buffer, pH 6.0 at 37°C (Buffer) (Prepare 100 ml in deionized water using Sodium Acetate, Trihydrate. Adjust to pH 6.0 at 37°C with 2 M Acetic Acid).

B. 50 mM Sodium Acetate Buffer with 2.3 mM Sodium Cholate, pH 5.0 at 37°C (Prepare 25 ml in deionized water using Sodium Acetate, Trihydrate, and Cholic Acid, Sodium Salt, Hydrate. Adjust to pH 5.0 at 37°C with 2 M Acetic Acid).

C. 50% (v/v) Chloroform and 50% (v/v) Methanol Solution (Prepare 5 ml using Chloroform and Methanol)

D. 0.047% (w/v) Monosialoganglioside Solution (GM) (Prepare by dissolving 1 mg of Monosialoganglioside (GM) from Bovine Brain, Boehringer Mannheim, in 2.15 ml of Reagent C. Evaporate the solution under nitrogen gas. This should be done under a vacuum hood. Dissolve the residue with 2.15 ml of Reagent B and sonicate (4 × 30 sec) with pauses to adjust the temperature of the solution to 4°C. Store at 4°C).

E. 50 mM Sodium Carbonate and 10 mM Potassium Cyanide Solution (Prepare 100 ml in deionized water using Sodium Carbonate, Anhydrous, and Potassium Cyanide. Caution: Potassium Cyanide is toxic).

F. 1.5 mM Potassium Ferricyanide Solution (Pot Ferr) (Prepare 100 ml in deionized water using Potassium Ferricyanide).

G. 50 mM Sulfuric Acid Solution (Prepare 200 ml in deionized water using Sulfuric Acid).

H. 3.1 mM Ferric Ammonium Sulfate and 3.5 mM Sodium Dodecyl Sulfate Solution (Prepare 200 ml in Reagent G using Ferric Ammonium Sulfate, Dodecahydrate and Lauryl Sulfate, Sodium Salt).

I. 1.11 mM Glucose Standard Solution (Std) (Prepare 1 ml in deionized water using Glucose Standard Solution, 1000 mg/l).

J. Ceramide Glycanase Enzyme Solution (Immediately before use, prepare a solution containing 0.007–0.014 unit/ml of Ceramide Glycanase in cold deionized water).

Procedure

Pipette (in milliliters) the following reagents into suitable containers:

	Test	Std	Blank
Reagent D (GM)	0.05	—	0.05
Reagent I (Std)	—	0.01	—
Deionized Water	0.03	0.09	0.05

Mix by swirling and equilibrate to 37°C. Then add:

Reagent J (Enzyme Soln)	0.02	—	—

Immediately mix by swirling and incubate at 37°C for exactly 30 minutes. Then add:

Reagent A (Buffer)	0.20	0.20	0.20
Reagent E	0.125	0.125	0.125
Reagent F (Pot Ferr)	0.125	0.125	0.125

Mix by swirling and incubate at 100°C for 15 minutes in a boiling water bath. Remove from the boiling water bath and equilibrate to room temperature. Then add:

Reagent H	1.00	1.00	1.00

Mix by swirling and incubate at 25°C for 15 minutes. If necessary, clarify the solutions by filtering through a 0.45 μM Millipore filter. Transfer to suitable cuvettes and record the A690nm for the Test, Standard, and Blank.

Calculations

Standard: ΔA690nm Standard = A690nm Standard–A690nm Blank
Sample Determination: ΔA690nm Test = A690nm Test–A690nm Blank
Determine the μmoles of monosialoganglioside hydrolyzed by comparing it to the standard.

$$\text{Units / ml enzyme} = \frac{(\mu\text{moles of monosialoganglioside hydrolyzed})\ (\text{df})}{(30)\ (0.02)}$$

df = Dilution factor
0.02 = Volume (in milliliter) of enzyme used
30 = Time (in minutes) of assay as per the Unit Definition

$$\text{Units / mg protein} = \frac{\text{units / ml enzyme}}{\text{mg protein / ml enzyme}}$$

Unit Definition: *One unit will hydrolyze 1.0 micromole of monosialoganglioside GM per minute at pH 5.0 at 37°C.*

Final Assay Concentration

In a 0.10 ml reaction mix, the final concentrations are 0.024% (w/v) monosialoganglioside, 25 mM sodium acetate, 1.2 mM sodium cholate, and 0.00014–0.00028 unit ceramide glycanase.

Enzymatic Assay of CERULOPLASMIN

Principle

$$\text{N,N - Dimethyl - p - phenylenediamine} + H_2O \xrightarrow{\text{Ceruloplasmin}} \text{Oxidized product}$$

Conditions: T = 37°C, pH = 5.5, A550nm, Light path = 1 cm

Method: Continuous Spectrophotometric Rate Determination

Reagents

A. 200 mM Sodium Acetate Buffer, pH 5.5 at 37°C (Prepare 100 ml in deionized water using Sodium Acetate, Trihydrate. Adjust to pH 5.5 at 37°C with 1 M Acetic Acid).

B. 153 mM N,N-Dimethyl-p-phenylenediamine Solution (DPD) (Prepare 10 ml in deionized water using N,N-Dimethyl- p-phenylenediamine, Monohydrochloride. *Prepare Fresh. Store on Ice. Keep away from Light*).

C. 100 mM Sodium Chloride Solution (Enzyme Diluent) (Prepare 100 ml in deionized water using Sodium Chloride).

D. Ceruloplasn..in Solution (Ceruloplasmin) (Immediately before use, prepare a solution containing approximately 30 units/ml of Ceruloplasmin in cold Reagent C).

Procedure

Pipette (in milliliters) the following reagents into suitable cuvettes:

	Test	Blank
Reagent A (Buffer)	2.00	2.00
Deionized Water	0.80	0.80
Reagent C (Enzyme Diluent)	—	0.10
Reagent D (Ceruloplasmin)	0.10	—

Mix by inversion and equilibrate to 37°C. Monitor the A550nm until constant, using a suitably thermostatted spectrophotometer. Then add:

Reagent B (DPD)	0.10	0.10

Immediately mix by inversion and record the increase in A550nm for approximately 5 minutes. Obtain the ΔA550nm/minute using the maximum linear rate for both the Test and Blank.

Calculations

$$\text{Units / ml enzyme} = \frac{(\Delta A550nm / \text{min Test} - \Delta A550nm / \text{min Blank})\ (3)\ (df)}{(0.01)\ (7)\ (0.1)}$$

3 = Total volume (in milliliters) of assay

df = Dilution factor

0.01 = Change in Absorbance at 550nm (Unit Definition)

7 = Conversion Factor to published Unit Definition of a 7 ml reaction volume
0.1 = Volume (in milliliters) of enzyme used

$$\text{Units / mg solid} = \frac{\text{units / ml enzyme}}{\text{mg solid / ml enzyme}}$$

$$\text{Units / mg protein} = \frac{\text{units / ml enzyme}}{\text{mg protein / ml enzyme}}$$

Unit Definition: *One unit is arbitrarily defined as that amount of "Oxidase" which will cause a A550nm of 0.01 per minute using N,N-dimethyl-p-phenylenediamine as substrate at pH 5.5 and 37°C, in a 7 ml reaction volume.*

Final Assay Concentration

In a 3.00 ml reaction mix, the final concentrations are 133 mM sodium acetate, 5.1 mM N,N-dimethyl-pphenylenediamine hydrochloride, 3.3 mM sodium chloride and 3.0 units ceruloplasmin.

Notes

1. Do not use HCl for adjusting the pH. The chloride ion is an inhibitor of human ceruloplasmin and the concentration must be kept constant. The NaCl is necessary for diluting the enzyme solution.
2. The solution is stable for approximately 2–3 hours. A fresh solution should be prepared if an increase in absorbance is seen in the Blank.

Enzymatic Assay of CERULOPLASMIN

Principle

$$\text{N, N - Dimethyl - p - phenylenediamine} + H_2O \xrightarrow{\text{Ceruloplasmin}} \text{Oxidized product}$$

Conditions: T = 37°C, pH = 6.4, A550nm, Light path = 1 cm

Method: Continuous Spectrophotometric Rate Determination

Reagents

A. 200 mM Sodium Acetate Buffer, pH 6.4 at 37°C (Prepare 100 ml in deionized water using Sodium Acetate, Trihydrate. Adjust to pH 6.4 at 37°C with 1 M Acetic Acid).
B. 153 mM N,N-Dimethyl-p-phenylenediamine Solution (DPD) (Prepare 10 ml in deionized water using N,N-Dimethyl-p-phenylenediamine, Monohydrochloride. *Prepare Fresh. Store on Ice. Keep away from Light*).
C. 100 mM Sodium Chloride Solution (Enzyme Diluent) (Prepare 100 ml in deionized water using Sodium Chloride).
D. Ceruloplasmin Solution (Ceruloplasmin) (Immediately before use, prepare a solution containing approximately 30 units/ml of Ceruloplasmin in cold Reagent C).

Procedure

Pipette (in milliliters) the following reagents into suitable cuvettes:

	Test	Blank
Reagent A (Buffer)	2.00	2.00
Deionized Water	0.80	0.80
Reagent C (Enzyme Diluent)	—	0.10
Reagent D (Ceruloplasmin)	0.10	—

Mix by inversion and equilibrate to 37°C. Monitor the A550nm until constant, using a suitably thermostatted spectrophotometer. Then add:

Reagent B (DPD)	0.10	0.10

Immediately mix by inversion and record the increase in A550nm for approximately 5 minutes. Obtain the ΔA550nm/minute using the maximum linear rate for both the Test and Blank.

Calculations

$$\text{Units / ml enzyme} = \frac{(\Delta\text{A550nm / min Test} - \Delta\text{A550nm / min Blank})\ (3)\ (\text{df})}{(0.01)\ (0.1)\ (7)}$$

3 = Total volume (in milliliters) of assay
df = Dilution factor
0.01 = Change in Absorbance at 550nm (Unit Definition)
0.1 = Volume (in milliliters) of enzyme used
7 = Conversion Factor to published Unit Definition of a 7 ml reaction volume

$$\text{Units / mg solid} = \frac{\text{units / ml enzyme}}{\text{mg solid / ml enzyme}}$$

$$\text{Units / mg protein} = \frac{\text{units / ml enzyme}}{\text{mg protein / ml enzyme}}$$

Unit Definition: *One unit is arbitrarily defined as that amount of "Oxidase" which will cause a A550nm of 0.01 per minute using N,N-dimethyl-p-phenylenediamine as substrate at pH 6.4 and 37°C, in a 7 ml reaction volume.*

Final Assay Concentration

In a 3.00 ml reaction mix, the final concentrations are 133 mM sodium acetate, 5.1 mM N,N-dimethyl-pphenylenediamine hydrochloride, 3.3 mM sodium chloride and 3 units ceruloplasmin.

Notes

1. Do not use HCl for adjusting the pH. The chloride ion is an inhibitor of human ceruloplasmin and the concentration must be kept constant. The NaCl is necessary for diluting the enzyme solution.
2. This solution is stable for approximately 2–3 hours. A fresh solution should be prepared if an increase in absorbance is seen in the blank.

Enzymatic Assay of CHITINASE (EC 3.2.1.14)

Principle

$$\text{Chitin} + H_2O \xrightarrow{\text{Chitinase}} \text{Chitobiose}$$

$$\text{Chitobiose} + H_2O \xrightarrow{\beta\text{-N-Acetylglucosaminidase}} \text{N-Acetyl-D-Glucosamine}$$

Conditions: T = 25°C, pH = 6.0, A540nm, Light path = 1 cm

Method: Colorimetric

Reagents

A. 200 mM Potassium Phosphate Buffer, pH 6.0 at 25°C with 2 mM Calcium Chloride. (Prepare 100 ml in deionized water using Potassium Phosphate, Monobasic, Anhydrous and Calcium Chloride, Dihydrate. Adjust to pH 6.0 at 25°C with 1 M KOH).

B. 1.25% (w/v) Chitin Suspension (Chitin) (Prepare 25 ml in Reagent A using Chitin. Stir on a magnetic stirrer for approximately 15 minutes to ensure that large particles are broken apart).

C. 5.3 M Sodium Potassium Tartrate Solution (Dissolve 12.0 grams of Sodium Potassium Tartrate, Tetrahydrate, in 8.0 ml of 2 M NaOH. Heat in a boiling water bath to dissolve, but *Do not boil*).

D. 96 mM 3,5-Dinitrosalicylic Acid Solution (Dissolve 438 mg of 3,5-Dinitrosalicylic Acid, in 20 ml of deionized water. Heat in a boiling water bath to dissolve, but *Do not boil*).

E. Color Reagent Solution (Clr Rgt Soln) (With stirring, slowly add Reagent C solution to Reagent D. Dilute to 40 ml with deionized water. If not completely dissolved, the reagents should dissolve when mixed. The solution should be stored in an amber bottle at room temperature. The Color Reagent Solution is stable for approximately 6 months).

F. Chitinase Enzyme Solution (Immediately before use, prepare a solution containing 1 unit/ml of Chitinase in cold Reagent A).

G. β-N-Acetylglucosaminidase (NAGase) (Immediately before use, prepare a solution containing 35 units/ml in cold Reagent A using β-N-Acetylglucosaminidase).

H. 0.1% (w/v) N-Acetyl-D-Glucosamine Standard Solution (NAG) (Prepare 10 ml in deionized water using N-Acetyl-D-Glucosamine).

Procedure

Pipette (in milliliters) the following reagents into 4 dram vials:

	Test	Blank
Reagent B (Chitin)	2.00	2.00
Deionized Water	—	0.50
Reagent F (Enzyme Solution)	0.50	—

Close each container securely with a cap. The vials should be placed on their sides on a rotary platform at a speed sufficient to keep the chitin in suspension. Incubate the Test and Blank for two hours at 25°C. After two hours, place the vials into a boiling water bath for 5 minutes. Cool to room temperature by placing the vials in a cold water bath. Add 1 unit of Reagent G (NAGase) to each vial. Incubate the vials for 30 minutes at 25°C with mixing as before. Centrifuge the suspensions and retain the supernatant liquid.

Colorimetric Assay

Standard Curve: A standard curve is prepared by pipetting (in milliliters) the following reagents into suitable containers:

	Std 1	Std 2	Std 3	Std 4	Std 5	Std Blank
Reagent H (NAG Std Soln)	0.1	0.2	0.3	0.4	0.5	0.0
Deionized Water	2.9	2.8	2.7	2.6	2.5	3.0
Reagent E (Clr Rgt Soln)	1.5	1.5	1.5	1.5	1.5	1.5

Sample: Pipette the following reagents into suitable vials:

	Test	Blank
Test Supernatant	1.00	—
Blank Supernatant	—	2.00
Deionized water	2.00	1.00
Reagent E (Clr Rgt Soln)	1.50	1.50

Place all the containers in a boiling water bath for 5 minutes. Remove and allow the containers to cool to room temperature. Transfer the solutions to suitable cuvettes and record the A540nm for each of the containers.

Calculations

Standard Curve: ΔA540nm Standard = A540nm Std–A540nm Std Blank

Plot the ΔA540nm of the standards versus milligrams of NAG.

Sample Determination: ΔA540nm Sample = A540nm Test–A540nm Test Blank

Determine the milligrams of NAG liberated using the Standard Curve.

$$\text{Units / ml enzyme} = \frac{(\text{mg NAG released})\ (2.5 + \text{Volume of NAGase})}{(2)\ (1)\ (0.5)}$$

2.5 = Initial reaction volume of assay

2 = Conversion factor for converting 2 hours to 1 hour as per the Unit Definition

1 = Volume (in milliliter) of supernatant used in colorimetric determination

0.5 = Volume (in milliliter) of enzyme used

$$\text{Units / g solid} = \frac{\text{units / ml enzyme}}{\text{g solid / ml enzyme}}$$

Unit Definition: *One unit will liberate 1.0 mg of N-acetyl-D-glucosamine from chitin per hour at pH 6.0 at 25°C in a two step reaction with β-N-acetylglucosaminidase from Aspergillus niger (2 hour assay).*

Final Assay Concentrations

In a 2.50 ml reaction mix, the final concentrations are 160 mM potassium phosphate, 2 mM calcium chloride, 1% (w/v) chitin, 1 unit of β-N-acetylglucosaminidase and 0.5 unit chitinase.

Notes

1. The color reaction is inhibited by ammonium sulfate. The enzyme is an ammonium sulfate suspension. Before being used in this assay β-N-Acetylglucosaminidase, must be centrifuged and the supernatant liquid removed.The pellet must be resuspended in cold Reagent A immediately prior to use.
2. Centrifuge the 4-dram glass vials for 10 minutes at 5400 rpm. Remove 1.5 ml of supernatant and add it to the microcentrifuge tubes. Centrifuge for 10 minutes at 10,000 rpm. Remove 1.0 ml of supernatant for color development.
3. The reaction volume needs to be adjusted according to the amount of Reagent E (NAGase used) in the assay.
4. β-N-Acetylglucosaminidase Unit Definition: One unit will hydrolyze 1.0 μmole of p-nitrophenyl N-acetyl-β-D-glucosaminide to p-nitrophenol and N-acetyl-D-glucosamine per minute at pH 4.0 at 25°C.

Enzymatic Assay of CHITOSANASE (EC 3.2.1.132)

Principle

$$\text{Chitosan} + H_2O \xrightarrow{\text{Chitosanase}} \text{N-acetyl-D-glucosamine} + \text{D-glucosamine}$$

Conditions: T = 37°C, pH = 5.0, Excitation = 395nm, Emission = 493 nm, Light path = 1 cm

Method: Fluorimetric Determination

Reagents

A. 700 mM Sodium Phosphate Buffer, pH 8.4 at 37°C (Phos. Buffer) (Prepare 100 ml in deionized water using Sodium Phosphate, Monobasic, Anhydrous. Adjust to pH 8.4 at 37°C with 1 M NaOH).

B. 100 mM Sodium Acetate Buffer, pH 5.0 at 37°C (Acet. Buffer) (Prepare 100 ml in deionized water using Sodium Acetate, Trihydrate. Adjust to pH 5.0 at 37°C with 1 M HCl).

C. 1 M Potassium Hydroxide Solution (KOH) (Prepare 10 ml in deionized water using Potassium Hydroxide Solution, 8.0 N).

D. 50 mM Acetic Acid Solution (HOAc) (Prepare 100 ml in deionized water using Acetic Acid, Glacial).

E. 0.4% (w/v) Chitosan Substrate Solution (Chitosan) (Dissolve 400 mg of Chitosan, that has been passed through a sieve (50–100) mesh, in 100 ml of Reagent D. Adjust to pH 5.0 at 37°C with Reagent C. Store the viscous solutions at 4°C).

F. 1 mM Glucosamine Standard Solution (Std) (Prepare 10 ml in deionized water using Glucosamine, Hydrochloride).

G. Acetone.

H. 0.3% (w/v) Fluorescamine Solution (Prepare 10 ml in Reagent G using Fluorescamine).

I. 0.1% (w/v) Bovine Serum Albumin Solution (Enzyme Diluent) (Prepare 10 ml in deionized water using Albumin, Bovine).

J. Chitosanase Enzyme Solution (Immediately before use, prepare a solution containing 0.01–0.07 unit/ml Chitosanase in cold Reagent I).

Procedure

Pipette (in milliliters) the following reagents into Eppendorf tubes:

	Test	Blank
Reagent B (Acet. Buffer)	0.10	0.10
Reagent E (Chitosan)	0.10	—

Mix by swirling and equilibrate to 37°C. Then add:

Reagent J (Enzyme Solution)	0.10	0.10

Mix by swirling and incubate at 37°C for exactly 30 minutes. Then add:

Reagent C (KOH)	0.05	0.05
Reagent E (Chitosan)	—	0.10

Place in an ice bath for 30 minutes then centrifuge for 10 minutes. Remove and save the supernatants. Pipette (in milliliters) the following reagents into fluorometer cuvettes:

Test Supernatant	0.10	—
Blank Supernatant	—	0.10
Reagent A (Phos. Buffer)	1.90	1.90
Reagent H (Fluorescamine)	0.05	0.05

Mix by inversion. Record the increase in fluorescence intensity at the excitation wavelength of 395nm and the emission wavelength of 493 nm.

Standard Curve: Pipette (in milliliters) the following reagents into suitable fluorimeter cuvettes:

	Std 1	Std 2	Std 3	Std 4	Std 5	Std Blank
Reagent F (Glucosamine)	0.02	0.04	0.06	0.08	0.10	—
Deionized Water	0.08	0.06	0.04	0.02	—	0.10
Reagent A (Phos Buffer)	1.90	1.90	1.90	1.90	1.90	1.90
Reagent H (Fluorescamine)	0.05	0.05	0.05	0.05	0.05	0.05

Mix by inversion. Record the increase in fluorescence intensity at the excitation wavelength of 395nm and the emission wavelength of 493 nm.

Calculations

Standard Curve: ΔIntensityStandard = IntensityStandard–IntensityStandard Blank

Prepare a standard curve by plotting the ΔIntensityStandard versus micromoles of glucosamine.

Sample Determination: ΔIntensitySample = IntensityTest–IntensityBlank

Determine the micromoles of glucosamine equivalents liberated from the Standard curve.

$$\text{Units / ml enzyme} = \frac{\text{(nanomoles of glucosamine equivalent liberated) (df) (0.35)}}{(0.1)\,(0.1)}$$

df = Dilution factor
0.35 = Volume (in milliliter) of stopped reaction
0.1 = Volume (in milliliter) of reaction mix used in fluorimetric determination
0.1 = Volume (in milliliter) of enzyme used

Unit Definition: *One unit will release 1.0 micromole of glucosamine from chitosan per minute at pH 5.0 at 37°C.*

Final Assay Concentration

In a 0.30 reaction mix, the final concentrations are 33 mM sodium acetate, 0.13% (w/v) chitosan, 16.7 mM acetic acid, 0.03% (w/v) bovine serum albumin and 0.001–0.007 unit chitosanase.

Notes

1. The chitosan should be ground to a fine powder (50–100 mesh) before preparing a solution.
2. The same cuvettes should be used for the standards and the samples.

Enzymatic Assay of CHLORAMPHENICOL ACETYLTRANSFERASE (EC 2.3.1.28)

Principle

$$\text{Acetyl CoA} + \text{Chloramphenicol} \xrightarrow{\text{CAT}} \text{CoA} + \text{Chloramphenicol 3 - Acetate}$$

$$\text{CoA} + \text{DTNB} \rightarrow \text{TNB} + \text{CoA derivative}$$

Abbreviations Used

Acetyl CoA = Acetyl Coenzyme A
CAT = Chloramphenicol Acetyltransferase
CoA = Coenzyme A
DTNB = 5,5'-Dithio-bis(2-Nitrobenzoic Acid)
TNB = 5-Thio-2-Nitrobenzoic Acid

Conditions: T = 25°C, pH = 7.8, A412nm, Light path = 1 cm

Method: Continuous Spectrophotometric Rate Determination

Reagents

A. 100 mM Tris Buffer, pH 7.8 at 25°C (Prepare 100 ml in deionized water using Trizma Base. Adjust to pH 7.8 at 25°C with 1 M HCl).
B. 2.5 mM 5,5'-Dithio-bis(2-Nitrobenzoic Acid) Solution, pH 7.8 at 25°C (DTNB) (Prepare 10 ml in Reagent A using 5,5'-Dithio-bis(2-Nitrobenzoic Acid. Facilitate

dissolution by adding 1 M NaOH until dissolved. Adjust to pH 7.8 at 25°C with 1 M HCl).

C. 5.0 mM Acetyl Coenzyme A Solution (Acetyl CoA) (Prepare 10 ml in deionized water using Acetyl Coenzyme A, Sodium Salt. *Prepare Fresh*).

D. 0.3% (w/v) Chloramphenicol Solution (Prepare 10 ml in deionized water using Chloramphenicol. Facilitate dissolution by first dissolving in 2 drops of methanol).

E. 10 mM Tris Buffer, pH 7.8 at 25°C (Enzyme Diluent) (Prepare 100 ml in deionized water using Trizma Base. Adjust to pH 7.8 at 25°C with 1 M HCl).

F. Chloramphenicol Acetyltransferase Enzyme Solution (Immediately before use, prepare a solution containing 500 units/ml of Chloramphenicol Acetyltransferase in cold Reagent E).

Procedure

Pipette (in milliliters) the following reagents into suitable cuvettes:

	Test	Blank
Reagent A (Buffer)	2.70	2.70
Reagent B (DTNB)	0.10	0.10
Reagent C (Acetyl CoA)	0.10	0.10
Reagent E (Enzyme Diluent)	—	0.05
Reagent F (Enzyme Solution)	0.05	—

Mix by inversion and equilibrate to 25°C. Monitor the A412nm until constant, using a suitably thermostatted spectrophotometer. Then add:

Reagent D (Chloramphenicol)	0.05	0.05

Immediately mix by inversion and record the increase in A412nm for approximately 5 minutes. Obtain the ΔA412nm/minute using the maximum linear rate for both the Test and Blank.

Calculations

$$\text{Units / ml enzyme} = \frac{(\Delta A412\text{nm / min Test} - \Delta A412\text{nm / min Blank})\ (3)\ (\text{df})}{(0.0136)\ (0.05)}$$

3 = Total volume (in milliliters) of assay
df = Dilution Factor
0.0136 = Micromolar extinction coefficient for TNB at 412nm
0.05 = Volume (in milliliters) of enzyme used

$$\text{Units / mg solid} = \frac{\text{units / ml enzyme}}{\text{mg solid / ml enzyme}}$$

$$\text{Units / mg protein} = \frac{\text{units / ml enzyme}}{\text{mg protein / ml enzyme}}$$

Unit Definition: *One unit will convert 1.0 nanomole of chloramphenicol and acetyl coenzyme A to chloramphenicol 3-acetate and coenzyme A per minute at pH 7.8 at 25°C.*

Final Assay Concentration

In a 3.00 ml reaction mix, the final concentrations are 94 mM Tris, 0.083 mM 5,5'-dithio-bis(2-nitrobenzoic acid), 0.17 mM acetyl coenzyme A, 0.005% (w/v) chloramphenicol and 25 units chloramphenicol acetyltransferase.

Enzymatic Assay of CHLOROPEROXIDASE (EC 1.11.1.10)

Principle

$$\text{Monochlorodimedon} + H_2O + Cl^- + H^+ \xrightarrow{\text{Chloroperoxidase}} \text{Dichlorodimedon} + H_2O$$

Conditions: T = 25°C, pH = 2.75, A278nm, Light path = 1 cm

Method: Continuous Spectrophotometric Rate Determination

Reagents

A. 100 mM Citric Acid Buffer with 100 mM Potassium Phosphate, pH 2.75 at 25°C (Prepare 200 ml in deionized water using Citric Acid, Free Acid, Anhydrous and Potassium Phosphate, Dibasic. Adjust to pH 2.75 at 25°C with 1 M HCl).
B. 0.1 mM Monochlorodimedon with 20 mM Potassium Chloride Solution, pH 2.75 at 25°C (Monochlorodimedon) (Prepare 100 ml in Reagent A using Monochlorodimedon, and Potassium Chloride. If necessary, adjust to pH 2.75 at 25°C with 1 M HCl or 1 M NaOH).
C. 0.3% (v/v) Hydrogen Peroxide Solution (Prepare 10 ml in deionized water using Hydrogen Peroxide, 30% (w/w). *Prepare Fresh*).
D. Chloroperoxidase Enzyme Solution (Immediately before use, prepare a solution containing 0.2–1.0 unit/ml of Chloroperoxidase in cold Reagent A).

Procedure

Pipette (in milliliters) the following reagents into suitable quartz cuvettes:

	Test	Blank
Reagent B (Monochlorodimedon)	2.90	2.90
Reagent C (H_2O_2)	0.06	0.06

Mix by inversion and equilibrate to 25°C. Monitor the A278nm until constant, using a suitably thermostatted spectrophotometer. Then add:

Reagent A (Buffer)	—	0.05
Reagent D (Enzyme Solution)	0.05	—

Immediately mix by inversion and record the decrease in A278nm for approximately 5 minutes. Obtain the ΔA278nm/minute using the maximum linear rate for both the Test and Blank.

Calculations

$$\text{Units / mg enzyme} = \frac{(\Delta A278\text{nm / min Test} - \Delta A278\text{nm / min Blank})}{(12.2)\ (\text{mg enzyme / ml RM})}$$

12.2 = Millimolar extinction coefficient of Monochlorodimedon at 278 nm
RM = Reaction Mix

Unit Definition: *One unit will catalyze the conversion of 1.0 μmole of monochlorodimedon to dichlorodimedon per minute at pH 2.75 at 25°C in the presence of potassium chloride and* H_2O_2.

Final Assay Concentration

In a 3.01 ml reaction mix, the final concentrations are 98 mM citric acid, 98 mM potassium phosphate, 0.096 mM monochlorodimedon, 19 mM potassium chloride, 0.006% (v/v) hydrogen peroxide and 0.01–0.05 unit of chloroperoxidase.

Enzymatic Assay of CHOLESTEROL ESTERASE (EC 3.1.1.13)

Principle

$$\text{Cholesterol Oleate} \xrightarrow{\text{Cholesterol Esterase}} \text{Cholesterol} + \text{Oleic Acid}$$

$$\text{Cholesterol} + O_2 + H_2O \xrightarrow{\text{Cholesterol Oxidase}} H_2O_2 + \text{Cholestenone}$$

$$2H_2O_2 + \text{4-AAP} + \xrightarrow{\text{Phenol Peroxidase}} 4H_2O + \text{Quinoneimine Dye}$$

Abbreviation Used

4-AAP = 4-Aminoantipyrine

Conditions: T = 37°C, pH = 7.0, A500nm, Light path = 1 cm

Method: Continuous Spectrophotometric Rate Determination

Reagents

A. 400 mM Potassium Phosphate Buffer, pH 7.0 at 37°C (Prepare 100 ml in deionized water using Potassium Phosphate, Monobasic, Anhydrous and Potassium Phosphate Dibasic, Trihydrate. Adjust to pH 7.0 at 37°C with 1 M NaOH).
B. 0.9% (w/v) Sodium Chloride Solution (Prepare 25 ml in deionized water using SodiumChloride).
C. 8.6 mM Cholesteryl Oleate Solution (Chol-Oleate) (Prepare by first dissolving 56.0 mg of Cholesteryl Oleate, in 1 ml of Polyoxyethylene Ether, 9-Lauryl Ether. While stirring, add 9 ml of hot Reagent B. Store at room temperature).
D. 15% (w/v) Taurocholic Acid Solution (Tauro) (Prepare 10 ml in deionized water using Taurocholic Acid, Sodium Salt).
E. 1.75% (w/v) 4-Aminoantipyrine Solution (4-AAP) (Prepare 1 ml in deionized water using 4- Aminoantipyrine, Free Base).
F. 6% (w/v) Phenol Solution (Prepare 10 ml in deionized water using Phenol).
G. Cholesterol Oxidase Enzyme Solution (Chol Oxid) (Immediately before use, prepare a solution containing 20–30 units/ml of Cholesterol Oxidase, in cold Reagent A).
H. Peroxidase Enzyme Solution (POD) (Immediately before use, prepare a solution containing 5 mg/ml of Peroxidase Type II from Horseradish,in cold deionized water).

I. Cholesterol Esterase Enzyme Solution (Immediately before use, repare a solution containing 0.25–0.50 unit/ml of Cholesterol sterase in cold Reagent A).

Procedure

Prepare a reaction cocktail by pipetting (in milliliters) the following reagents into a suitable container:

	Test
Reagent A (Buffer)	31.80
Reagent C (Chol-Oleate)	7.50
Reagent D (Tauro)	1.50
Reagent E (4-AAP)	0.75
Reagent F (Phenol)	1.50
Reagent G (Chol Oxid)	0.30
Reagent H (POD)	0.15

Mix by inversion and equilibrate to 37° C. Pipette (in milliliters) the following reagents into a suitable container:

	Test	Blank
Reaction Cocktail	2.90	2.90
Reagent I (Cholesterol Esterase)	0.10	—
Reagent A (Buffer)	—	0.10

Immediately mix by inversion and record the increase in A500nm for approximately 5 minutes. Obtain the ΔA500nm/minute using the maximum linear rate for both the Test and Blank.

Calculations

$$\text{Units / mg enzyme} = \frac{(\Delta A500\text{nm / min Test} - \Delta A500\text{nm / min Blank})}{(0.5)\,(13.78)\,(\text{mg enzyme / ml RM})}$$

13.78 = Millimolar extinction coefficient of Quinoneimine Dye at 500 nm under the assay conditions
RM = Reaction Mix
0.5 = Conversion factor based on one mole of H_2O_2 produces half a mole of Quinoneimine Dye.

Unit Definition: *One unit will hydrolyze 1.0 μmole of cholesteryl oleate to cholesterol and oleic acid per minute at pH 7.0 at 37°C in the presence of taurocholate.*

Final Assay Concentration

In a 3.00 ml reaction mix, the final concentrations are 299 mM potassium phosphate, 0.50% (w/v) taurocholic acid, 0.05 mg peroxidase, 1.4 mM cholesteryl oleate, 1.7% (w/v) polyoxyethylene 9-Lauryl ether, 0.14% (w/v) sodium chloride, 0.2% (w/v) phenol, 0.4–0.6 unit cholesterol oxidase and 0.025–0.05 unit cholesterol esterase.

Enzymatic Assay of CHOLESTEROL OXIDASE (EC 1.1.3.6)

Principle

$$\text{Cholesterol} + O_2 \xrightarrow{\text{Cholesterol Oxidase}} \text{4-Cholesten-3-one} + H_2O_2$$

Conditions: T = 37°C, pH = 5.0, A243nm, Light path = 1 cm

Method: Stopped Spectrophotometric Rate Determination

Reagents

A. 0.5% (w/v) Cholesterol Substrate Solution (Prepare 10 ml in Reagent B using Cholesterol).
B. Ethyl Alcohol (EtOH).
C. 500 mM Sodium Acetate Buffer, pH 5.0 at 37°C (Prepare 100 ml in deionized water using Sodium Acetate, Trihydrate. Adjust to pH 5.0 at 37°C with 1 M HCl).
D. 10 mM Sodium Phosphate Solution, pH 7.3 at 37°C (Enz Dil) (Prepare 50 ml in deionized water using Sodium Phosphate, Monobasic, Anhydrous. Adjust to pH 7.3 at 37°C with 1 M NaOH).
E. Cholesterol Oxidase Enzyme Solution (Immediately before use, prepare a solution 0.003–0.009 unit/ml of Cholesterol Oxidase in Reagent C).

Procedure

Pipette (in milliliters) the following reagents into suitable containers.

	Test	Blank
Reagent A (Substrate Solution)	0.10	0.10
Reagent C (Buffer)	0.40	0.40

Mix by inversion and equilibrate to 37°C. Then add:

Reagent E (Enzyme Solution)	0.50	—

Immediately mix by inversion and incubate at 37°C for exactly 30 minutes. Then add:

Reagent B (EtOH)	3.00	3.00
Reagent E (Enzyme Solution)	—	0.50

Mix by inversion and transfer to suitable cuvettes. Record the A243nm for both the Test and Blank in a suitable spectrophotometer.

Calculations

$$\text{Units / ml enzyme} = \frac{(\text{A243nm Test} - \text{A243nm Blank})\ (4)\ (\text{df})}{(30)\ (18)\ (0.5)}$$

4 = Volume (in milliliters) of the stopped reaction
df = Dilution factor
30 = Time (in minutes) of the assay as per the Unit Definition

18 = Millimolar extinction coefficient of 4-cholesten-3-one
0.5 = Volume (in milliliter) of enzyme used

$$\text{Units / mg solid} = \frac{\text{units / ml enzyme}}{\text{mg solid / ml enzyme}}$$

$$\text{Units / mg protein} = \frac{\text{units / ml enzyme}}{\text{mg protein / ml enzyme}}$$

Unit Definition: *One unit will convert 1.0 μmole of cholesterol to 4- cholesten-3-one per minute at pH 5.0 at 37°C.*

Final Assay Concentration

In a 1.0 ml reaction mix, the final concentrations are 0.05% (w/v) cholesterol, 10% (v/v) ethyl alcohol, 450 Mm sodium acetate, and 0.0015–0.0045 unit cholesterol oxidase.

Enzymatic Assay of CHOLESTEROL OXIDASE (EC 1.1.3.6)

Principle

$$\text{Cholesterol} + O_2 \xrightarrow{\text{Cholesterol Oxidase}} H_2O_2 + \text{4-Cholesten-3-One}$$

$$2H_2O_2 + \text{4-AAP} + \text{Phenol} \xrightarrow{\text{Peroxidase}} 4H_2O + \text{Quinoneimine Dye}$$

Abbreviation Used

4-AAP = 4-Aminoantipyrine

Conditions: T = 37°C, pH 7.0, A500nm, Light path = 1 cm

Method: Continuous Spectrophotometric Rate Determination

Reagents

A. 100 mM Potassium Phosphate Solution (Prepare 100 ml in deionized water using Potassium Phosphate, Monobasic, Anhydrous).
B. 100 mM Potassium Phosphate Solution (Prepare 100 ml in deionized water using Potassium Phosphate, Dibasic, Anhydrous).
C. 100 mM Potassium Phosphate Buffer, pH 7.0 at 37°C (Prepare using Reagent B. Adjust to pH 7.0 at 37°C with Reagent A).
D. 13.6 mM Cholesterol Solution (Add 5 ml of Triton X-100, to a 250 ml wide mouth Erlenmeyer flask and stir constantly with a stir bar while on a hot plate using the low heat settings. Add 500 mg of Cholesterol, and stir constantly until all the Cholesterol is dissolved. Do not allow the Triton X-100 to yellow or boil. Once the Cholesterol is dissolved, continue stirring and gradually add 90 ml of deionized water by pipetting the water along the flask walls slowly. Cover the flask with aluminum foil and increase the heat to the high setting. Stir and allow to boil for 1 minute. Do not allow the Cholesterol solution to boil over the top. The solution will be cloudy. Immediately cool under cold running water with gentle agitation to achieve a clear solution. Then add 2.6 g of Taurocholic Acid, Sodium Salt. Continue to cool and mix well.

E. Color Reaction Mix (CRM) (Prepare by adding 200 mg of Phenol, to 100 ml of Reagent C. Mix thoroughly to dissolve by stirring. Then add 32 mg of 4-Aminoantipyrine, Free Base. Continue to mix by stirring. Transfer the solution into an amber bottle to protect from light).

F. Cholesterol Oxidase Enzyme Solution (Cholesterol Oxidase) (Prepare a solution containing 1 mg/ml of Cholesterol Oxidase in Reagent C. Immediately before use, dilute to 0.5–0.75 unit/ml with Reagent C).

G. Peroxidase Enzyme Solution (POD) (Immediately before use, prepare a solution containing 1900 units/ml in Reagent C using Peroxidase from Horseradish).

Procedure

Pipette (in milliliters) the following reagents into suitable cuvettes.

	Test	Blank
Reagent E (CRM)	2.80	2.80
Reagent D (Cholesterol Solution)	0.20	0.20
Reagent G (POD)	0.01	0.01

Mix by inversion and equilibrate to 37°C. Monitor the baseline at 500 nm. There should be no blank rate. Then add:

	Test	Blank
Reagent F (Cholesterol Oxidase)	0.02	—
Reagent C (Buffer)	—	0.02

Immediately mix by inversion and record the increase in A500nm for approximately 5 minutes. Obtain the ΔA500nm/minute using maximum linear rate for both the Test and Blank.

Calculations

$$\text{Units / ml enzyme} = \frac{(\Delta A500\text{nm / min Test} - \Delta A500\text{nm / min Blank})\ (3.03)\ (\text{df})}{(0.5)\ (13.78)\ (0.02)}$$

3.03 = Total volume (in milliliters) of assay
df = Dilution factor
13.78 = Millimolar extinction coefficient of Quinoneimine Dye at 500 nm under the assay conditions
0.5 = Conversion factor based on one mole of H_2O_2 produces half a mole of Quinoneimine Dye
0.02 = Volume (in milliliter) of enzyme used

$$\text{Units / mg solid} = \frac{\text{units / ml enzyme}}{\text{mg solid / ml enzyme}}$$

$$\text{Units / mg protein} = \frac{\text{units / ml enzyme}}{\text{mg protein / ml enzyme}}$$

Unit Definition: *One unit will convert 1.0 μmole of cholesterol to 4-cholesten-3-one per minute at pH 7.0 at 37°C.*

Final Assay Concentration

In a 3.03 ml reaction mix, the final concentrations are 94 mM potassium phosphate, 0.35% Triton X-100, 3.4 mM taurocholic acid, 0.9 mM cholesterol, 19.8 mM phenol, 1.5 mM aminoantipyrine, 19 units peroxidase and 0.01–0.015 unit cholesterol oxidase.

Note

1. Peroxidase Unit Definition: One unit will form 1.0 mg purpurogallin from pyrogallol in 20 sec at pH 6.0 at 20°C.

Enzymatic Assay of CHOLESTEROL OXIDASE (EC 1.1.3.6)

Principle

$$\text{Cholesterol} + O_2 \xrightarrow{\text{Cholesterol Oxidase}} H_2O_2 + \text{4-Cholesten-3-One}$$

$$2H_2O_2 + \text{O-Dianisidine (reduced)} \xrightarrow{\text{Peroxidase}} 2H_2O + \text{O-Dianisidine (oxidized)}$$

Conditions: T = 25°C, pH = 7.5, A500nm, Light path = 1 cm

Method: Continuous Spectrophotometric Rate Determination

Reagents

A. 50 mM Potassium Phosphate Buffer, pH 7.5 at 25°C (Prepare 100 ml in deionized water using Potassium Phosphate, Monobasic, Anhydrous. Adjust to pH 7.5 at 25°C with 1 M KOH).

B. 1% (w/v) O-Dianisidine Solution (ODA) (Dissolve the contents of a 50 mg vial of O-Dianisidine Dihydrochloride, in 5 ml of deionized water or prepare 5 ml in deionized water using O-Dianisidine Dihydrochloride. *Prepare Fresh*).

C. 0.5% (w/v) Cholesterol with 10% (v/v) Triton X-100 Solution (Chol) (Prepare by initially dissolving the Cholesterol, in 10 ml of Triton X-100. Heat until the solution clarifies. Then add 90 ml of deionized water. Stir well. Store the solution at 4°C).

D. Peroxidase Enzyme Solution (POD) (Immediately before use, prepare a solution containing 100 pyrogallol units/ml in deionized water using Peroxidase from Horseradish).

E. Cholesterol Oxidase Enzyme Solution (Immediately before use, prepare a solution containing 0.1–0.2 unit/ml of Cholesterol Oxidase in cold Reagent A).

Procedure

Prepare a reaction cocktail by pipetting (in milliliters) the following reagents into a suitable container:

Reagent A (Buffer)	40.0
Reagent B (ODA)	0.50

Mix thoroughly by swirling and adjust to pH 7.5 at 25°C with 100 mM HCl or 100 mM KOH, if necessary. Then add Reagent A to a final volume of 50 ml. Mix by swirling thoroughly and oxygenate for approximately 10 minutes immediately before use. Pipette (in milliliters) the following reagents into suitable cuvettes:

	Test	Blank
Reaction Cocktail	2.70	2.70
Reagent C (Chol)	0.10	0.10
Reagent D (POD)	0.10	0.10

Mix by inversion and equilibrate to 25°C. Monitor the A500nm until constant, using a suitably thermostatted spectrophotometer. Then add:

Reagent A (Buffer)	—	0.10
Reagent E (Enzyme Solution)	0.10	—

Immediately mix by inversion and record the increase in A500nm for approximately 5 minutes. Obtain the ΔA500nm/minute using the maximum linear rate for both the Test and Blank.

Calculations

$$\text{Units / ml enzyme} = \frac{(\Delta A500nm / min\ Test - \Delta A500nm / min\ Blank)\ (3)\ (df)}{(7.5)\ (0.1)}$$

3 = Volume (in milliliters) of assay
df = Dilution factor
7.5 = Millimolar extinction coefficient of O-Dianisidine (Oxidized) at 500nm
0.1 = Volume (in milliliters) of enzyme used

$$\text{Units / mg solid} = \frac{\text{units / ml enzyme}}{\text{mg solid / ml enzyme}}$$

$$\text{Units / mg protein} = \frac{\text{units / ml enzyme}}{\text{mg protein / ml enzyme}}$$

Unit Definition: *One unit will convert 1.0 μmole of cholesterol to 4-cholesten-3-one per minute at pH 7.5 at 25°C.*

Final Assay Concentration

In a 3.00 ml reaction mix, the final concentrations are 38 mM potassium phosphate, 0.009% (w/v) O-Dianisidine, 0.017% (w/v) cholesterol, 0.33% (v/v) Triton X-100, 10 units peroxidase and 0.01–0.02 unit cholesterol oxidase.

Notes

1. This assay is not to be used to assay Cholesterol Oxidase from Schizophyllum commune and Cholesterol Oxidase from Brevibacterium sp.
2. The solution can still be used if it becomes cloudy upon addition of the water.
3. Peroxidase Unit Definition: One unit will form 1.0 mg purpurogallin from pyrogallol in 20 seconds at pH 6.0 at 20°C.

Enzymatic Assay of CHOLINE ACETYLTRANSFERASE (EC 2.3.1.6)

Principle

$$\text{Choline} + {}_{14}\text{C - Acetyl CoA} \xrightarrow{\text{Choline Acetyltransferase}} \text{CoA} + {}_{14}\text{C - Acetyl Choline}$$

Abbreviations Used

CoA = Coenzyme A

${}_{14}$C-Acetyl CoA = ${}_{14}$C-Acetyl Coenzyme A

Conditions: T = 37°C, pH 7.5

Method: Radiolabelled Stop Reaction

Reagents

A. 75 mM Sodium Phosphate Buffer with 0.133 mM Eserine Salicylate, pH 7.5 at 37°C (Prepare 1 ml in deionized water using Sodium Phosphate Dibasic, Anhydrous and Eserine, Salicylate Salt. Adjust to pH 7.5 with 1 M H_3PO_4 at 37°C).

B. 10 mM Choline Chloride Solution (Chol Chlor) (Prepare 1 ml in Reagent A (Buffer) using Choline, Chloride Salt).

C. 167 nmole/ml ${}_{14}$C-Acetyl Coenzyme A Solution (${}_{14}$C-Acetyl CoA) (Prepare 0.300 ml in deionized water using 2–10 mCi/mmol ${}_{14}$C-Acetyl Coenzyme A).

D. 88 mM Tetraphenylboron Solution (TPB) (Prepare 1 ml in 3-Butenenitrile, using Tetraphenylboron, Sodium Salt).

E. Choline Acetyltransferase Enzyme Solution (Immediately before use, prepare a solution containing 15–30 units/ml of Choline Acetyltransferase in cold deionized water).

F. Scintillation Cocktail.

Procedure

Pipette (in milliliters) the following reagents into an Eppendorf tube:

	Test	Blank
Reagent C (${}_{14}$C-Acetyl CoA)	0.030	0.030
Reagent B (Chol Chlor)	0.030	0.030

Mix by swirling and equilibrate for 5 minutes at 37°C. Then add:

	Test	Blank
Reagent E (Enzyme Solution)	0.005	—
Deionized Water	—	0.005

Immediately mix by swirling and incubate for exactly 10 minutes at 37°C. Then add:

	Test	Blank
Reagent D (TPB)	0.050	0.050

Centrifuge both the Test and Blank solutions. Remove 0.030 ml of the upper phase from the Test and Blank and place in 7.0 ml glass scintillation vials with 5 ml of Reagent F. Mix by swirling and count for 2 minutes with a suitable scintillation counter. To determine the total potential decays per minute (dpm), pipette 0.03 ml of Reagent C (${}_{14}$C-Acetyl CoA), 0.030 ml of Reagent B (Chol Chlor) and 0.005 ml of Reagent E (Enzyme Solution) into a 7.0 ml glass scintillation vial with 5 ml of Reagent F. Mix by swirling and count for 2 minutes with a suitable scintillation counter.

Calculations

$$\text{nmoles of }_{14}\text{C - Acetyl Choline} = \frac{(\mu\text{Ci / ml})\,(1000)\,(0.030)}{(\text{mCi / nmole})\,(5.0)}$$

1000 = Conversion factor from µmoles to nmoles
0.030 = Aliquot volume for $_{14}$C-Acetyl CoA
5 = Volume of scintillation fluid

$$\text{Potential dpm / nmole} = \frac{\text{dpm of reaction mix}}{\text{total nmoles of }_{14}\text{C - Acetyl CoA}}$$

$$\text{Units / mg enzyme} = \frac{(\text{dpm Test} - \text{dpm Blank})\,(0.065)\,(0.115)}{(0.030)\,(\text{potential dpm / nmole})\,(\text{mg enzyme / RM})\,(10)}$$

0.065 = Total volume of reaction mixture
0.115 = Volume of assay counted
10 = Time of assay (in minutes)
0.030 = Volume of organic layer counted
RM = Reaction Mixture

Unit Definition: *One unit will catalyze the transfer of 1.0 nanomole of acetate from $_{14}$C-acetyl CoA to choline per minute at pH 7.5 at 37°C.*

Final Assay Concentrations

In a 0.065 ml reaction mix, the final concentrations are 5.01 nmoles of $_{14}$C-acetyl coenzyme A, 35 mM sodium phosphate, 0.061 mM eserine salicylate, 4.6 mM choline chloride, and 0.075–0.15 unit choline acetyltransferase.

Enzymatic Assay of CHOLINE KINASE (EC 2.7.1.32)

Principle

$$\text{Choline} + \text{ATP} \xrightarrow{\text{CK}} \text{O - Phosphocholine} + \text{ADP}$$

$$\text{ADP} + \text{PEP} \xrightarrow{\text{PK}} \text{ATP} + \text{Pyruvate}$$

$$\text{Pyruvate} + \beta\text{ - NADH} \xrightarrow{\text{LDH}} \text{Lactate} + \beta\text{ - NAD}$$

Abbreviations Used

ATP = Adenosine 5'-Triphosphate
CK = Choline Kinase
ADP = Adenosine 5'-Diphosphate
PEP = Phospho(enol) pyruvate
PK = Pyruvate Kinase
β-NADH = β-Nicotinamide Adenine Dinucleotide, Reduced Form
LDH = Lactic Dehydrogenase
β-NAD = β-Nicotinamide Adenine Dinucleotide, Oxidized Form

Conditions: T = 25°C, pH = 8.5, A340nm, Light path = 1 cm

Method: Continuous Spectrophotometric Rate Determination

Reagents

A. 100 mM Glycylglycine Buffer with 60 mM Potassium Chloride, 10 mM Magnesium Chloride, 4 mM Ethylenediaminetetraacetic Acid, and 0.2 mM β-Nicotinamide Adenine Dinucleotide, Reduced Form, pH 8.5 at 25°C (React Cocktail) (Prepare 30 ml in deionized water using Gly-Gly, Free Base, Potassium Chloride, Magnesium Chloride, Hexahydrate, Ethylenediaminetetraacetic Acid, Tetrasodium Salt, Hydrate, and β-Nicotinamide Adenine Dinucleotide, Reduced Form, Disodium Salt. Adjust to pH 8.5 at 25°C with either 1 M HCl or 1 M NaOH).

B. 49 mM Phospho(enol) pyruvate Solution (PEP) (Prepare 1 ml in deionized water using Phospho(enol) pyruvate, Monopotassium Salt).

C. 30 mM Adenosine 5'-Triphosphate Solution (ATP) (Prepare 1 ml in deionized water using Adenosine 5'-Triphosphate, Disodium Salt).

D. 100 mM Choline Chloride Solution (CC) (Prepare 1 ml in deionized water using Choline Chloride Salt).

E. PK/LDH Enzymes Solution (PK/LDH).

F. 10 mM Tris Acetate Buffer with 13 mM β- Mercaptoethanol, 1 mM Ethylenediaminetetraacetic Acid, and 15 mM Magnesium Chloride, pH 7.2 at 25°C (Enz Dil) (Prepare 25 ml in deionized water using TrizmaAcetate, 2-Mercaptoethanol, Ethylenediaminetetraacetic Acid, Tetrasodium Salt, Hydrate, and Magnesium Chloride, Hexahydrate. Adjust to pH 7.2 at 25°C with either 1 M HCl or 1 M NaOH).

G. Choline Kinase Enzyme Solution (CK) (Immediately before use, prepare a solution containing 0.25–0.50 unit/ml of Choline Kinase in cold Reagent F).

Procedure

Pipette (in milliliters) the following reagents into suitable cuvettes:

	Test	Blank
Reagent A (React Cocktail)	2.80	2.80
Reagent E (PK/LDH)	0.05	0.05
Reagent C (ATP)	0.10	0.10
Reagent B (PEP)	0.10	0.10
Reagent G (CK)	0.10	—
Reagent F (Enz Dil)	—	0.10

Mix by inversion and equilibrate to 25°C. Monitor the ΔA340nm until constant, using a suitably thermostatted spectrophotometer. Then add:

Reagent D (CC)	0.10	0.10

Immediately mix by inversion and record the decrease in A340nm for approximately 5 minutes. Obtain the ΔA340nm/minute using the maximum linear rate for both the Test and Blank.

Calculations

$$\text{Units / ml enzyme} = \frac{(\Delta A340\text{nm / min Test} - \Delta A340\text{nm / min Blank})\ (3.25)\ (\text{df})}{(6.22)\ (0.1)}$$

3.25 = Total volume (in milliliters) of assay
df = Dilution factor
6.22 = Millimolar extinction coefficient of β-NADH at 340nm
0.1 = Volume (in milliliter) of enzyme used

$$\text{Units / mg solid} = \frac{\text{units / ml enzyme}}{\text{mg solid / ml enzyme}}$$

$$\text{Units / mg protein} = \frac{\text{units / ml enzyme}}{\text{mg protein / ml enzyme}}$$

Unit Definition: *One unit will catalyze the phosphorylation of 1.0 μmole of choline to choline phosphate by ATP per minute at pH 8.5 at 25°C.*

Final Assay Concentrations

In a 3.25 ml reaction mix, the final concentrations are 86 mM glycylglycine, 52 mM potassium chloride, 9.1 mM magnesium chloride, 3 mM Ethylenediaminetetraacetic Acid, 0.2 mM β-nicotinamide adenine dinucleotide, reduced form, 1.5 mM phospho(enol) pyruvate, 0.9 mM adenosine 5'-triphosphate, 3.1 mM choline chloride, 35 units pyruvate kinase, 50 units lactic dehydrogenase, 0.3 mM Tris acetate, 0.4 mM 2-mercaptoethanol, and 0.025–0.05 unit choline kinase.

Notes

. Contains not less than 700 Pyruvate Kinase units and 1000 Lactic Dehydrogenase units per ml.
2. These reagents must be pipetted fresh for each set of samples, since the phospho(enol) pyruvate may break down in the reaction mixture and give a secondary reaction.
3. Pyruvate Kinase Unit Definition: One unit will convert 1.0 μmole of phospho(enol) pyruvate to pyruvate per minute at 7.6 at 37°C.
4. L-Lactic Dehydrogenase Unit Definition: One unit will reduce 1.0 μmole of pyruvate to L-lactate per minute at pH 7.5 at 37°C.

Enzymatic Assay of CHOLINE OXIDASE (EC 1.1.3.17)

Principle

$$\text{Choline} + O_2 \xrightarrow{\text{Choline Oxidase}} \text{Betaine Aldehyde} + H_2O_2$$

$$\text{Betaine Aldehyde} + O_2 + H_2 \xrightarrow{\text{Choline Oxidase}} \text{Betaine} + H_2O_2$$

$$2H_2O_2 + 4\text{-Aminoantipyrine} + \text{Phenol} \xrightarrow{\text{POD}} \text{Quinoneimine dye} + 4H_2O$$

Abbreviation Used

POD = Peroxidase

Conditions: T = 37°C, pH = 8.0, A500nm, Light path = 1 cm

Method: Continuous Spectrophotometric Rate Determination

Reagents

A. 100 mM Tris HCl Buffer, pH 8.0 at 37°C (Prepare 100 ml in deionized water using Trizma Base. Adjust to pH 8.0 at 37°C with 1 M HCl).

B. 2.1% (w/v) Choline Chloride Solution (Choline) (Prepare 100 ml in Reagent A using Choline, Chloride Salt).
C. 1% (w/v) 4-Aminoantipyrine Solution (4-AAP) (Prepare 2 ml in deionized water using 4-Aminoantipyrine, Free Base).
D. 1% (w/v) Phenol Solution (Phenol) (Prepare 5 ml in deionized water using Phenol).
E. 10 mM Tris HCl with 2.0 mM Ethylenediaminetetraacetic Acid and 134 mM Potassium Chloride Solution (Enz Dil) (Prepare 100 ml in deionized water using Trizma Base, Ethylenediaminetetraacetic Acid, Disodium Salt, Dihydrate, and Potassium Chloride. Adjust to pH 8.0 at 37°C with 1 M HCl).
F. Peroxidase Enzyme (POD).
G. Choline Oxidase Enzyme Solution (Choline Oxidase) (Immediately before use, prepare a solution containing 0.1–0.5 unit/ml of Choline Oxidase in cold Reagent E).

Procedure

Prepare a reaction cocktail by pipetting (in milliliters) the following reagents into a suitable amber container:

Reagent B (Choline)	97.00
Reagent C (4-AAP)	1.00
Reagent D (Phenol)	2.00
Reagent F (POD, Purpurogallin units)	500

Mix by swirling. Pipette (in milliliters) the following reagents into suitable cuvettes:

	Test	Blank
Reaction Cocktail	3.00	3.00

Equilibrate to 37°C. Monitor the A500nm until constant, using a suitably thermostatted spectrophotometer. Then add:

Reagent E (Enz Dil)	—	0.05
Reagent G (Choline Oxidase)	0.05	—

Immediately mix by inversion and record the increase in A500nm for approximately 5 minutes. Obtain the ΔA500nm/minute using the maximum linear rate for both the Test and Blank.

Calculations

$$\text{Units / ml enzyme} = \frac{(\Delta A500nm / \min \text{ Test} - \Delta A500nm / \min \text{ Blank})\ (3.05)\ (df)}{(12)\ (0.5)\ (0.05)}$$

3.05 = Volume (in milliliters) of assay
df = Dilution factor
12 = Millimolar extinction coefficient of Quinoneimine Dye at 500nm under the conditions of the assay
0.5 = μmole of Quinoneimine Dye formed per μmole of H_2O_2
0.05 = Volume (in milliliter) of choline oxidase used in assay

$$\text{Units / mg solid} = \frac{\text{units / ml enzyme}}{\text{mg solid / ml enzyme}}$$

$$\text{Units / mg protein} = \frac{\text{units / ml enzyme}}{\text{mg protein / ml enzyme}}$$

Unit Definition: *One unit will form 1.0 μmole of H_2O_2 from the oxidation of 1 μmole of choline to betaine aldehyde per minute at pH 8.0 at 37°C.*

Final Assay Concentration

In a 3.05 ml reaction mix, the final concentrations are 96 mM Tris, 2.0% (w/v) choline, 0.01% (w/v) 4-aminoantipyrine, 0.02% (w/v) phenol, 15 units peroxidase, 0.03 mM Ethylenediaminetetraacetic Acid, 2 mM potassium chloride and 0.005–0.025 unit choline oxidase.

Note

1. Peroxidase Unit Definition: One unit will form 1.0 mg purpurogallin from pyrogallol in 20 seconds at pH 6.0 at 20°C.

Enzymatic Assay of CHOLINESTERASE, ACETYL (EC 3.1.1.7)

Principle

$$\text{Acetylcholine} + H_2O \xrightarrow{\text{Cholinesterase, Acetyl}} \text{Choline} + \text{Acetic Acid}$$

Conditions: T = 37°C, pH = 8.0

Method: Titrimetric

Reagents

A. 1600 mM Magnesium Chloride Solution (Prepare 15 ml in deionized water using Magnesium Chloride, Hexahydrate).
B. 1000 mM Sodium Chloride Solution (Prepare 50 ml in deionized water using Sodium Chloride).
C. 4 mM Acetylcholine Chloride Solution (Acetylcholine) (Prepare 400 ml in 350 ml deionized water, 10 ml Reagent A, and 40 ml Reagent B using Acetylcholine Chloride).
D. 20 mM Sodium Hydroxide Solution-Standardized (NaOH) (Prepare 50 ml in cold deionized water using Sodium Hydroxide. Standardize according to the ACS Reagent Procedure).
E. Cholinesterase, Acetyl Enzyme Solution (Immediately before use, prepare a solution containing 3–5 units/ml of Cholinestase, Acetyl in cold deionized water).

Procedure

Using a suitable pH meter in conjunction with a magnetic stirrer, pipette (in milliliters) the following reagents into a suitably thermostatted titration vessel:

	Test
Reagent C (Acetylcholine)	50.00

Equilibrate to 37°C. Adjust to pH 8.3 with Reagent D, using a burette. Then add:

Reagent E (Enzyme Solution)	0.4

Run the reaction for 1–5 minutes. Record the time when the pH reaches 8.0. Maintain the pH of the reaction mix at pH 8.0 by the addition of small volumes (0.05 ml) of Reagent D. Record the volume of Reagent D used to maintain the pH at 8.0 and the time required.

Calculation

$$\text{Units / ml enzyme} = \frac{\text{(Molarity of NaOH) (NaOH) (1000) (df)}}{\text{(T) (0.4)}}$$

NaOH = Volume (in milliliters) of Reagent D used in the assay
1000 = Conversion from millimoles to micromoles (Unit Definition)
df = Dilution factor
T = Time of assay (Unit Definition)
0.4 = Volume (in milliliter) of enzyme used

$$\text{Units / mg solid} = \frac{\text{units / ml enzyme}}{\text{mg solid / ml enzyme}}$$

$$\text{Units / mg protein} = \frac{\text{units / ml enzyme}}{\text{mg protein / ml enzyme}}$$

Unit Definition: *One unit will hydrolyze 1.0 µmole of acetylcholine to choline and acetate per minute at pH 8.0 at 37°C.*

Initial Assay Concentrations

In a 50.4 ml reaction mix, the initial concentrations are 40 mM magnesium chloride, 100 mM sodium chloride, 4 mM acetylcholine chloride and 1.2–2 units cholinesterase acetyl.

Enzymatic Assay of CHOLINESTERASE, BUTYRYL (EC 3.1.1.8)

Principle

$$\text{Butyrylcholine} + H_2O \xrightarrow{\text{Cholinesterase, Butyryl}} \text{Butyrate} + \text{Choline}$$

Conditions: T = 37°C, pH = 8.0

Method: Titrimetric

Reagents

A. 1600 mM Magnesium Chloride Solution (Prepare 15 ml in deionized water using Magnesium Chloride, Hexahydrate).
B. 1000 mM Sodium Chloride Solution (Prepare 50 ml in deionized water using Sodium Chloride).

C. 4 mM Butyrylcholine Chloride Solution (BuCholine) (Prepare by dissolving 336 mg of Butyrylcholine Chloride, in 350 ml deionized water. Then add 10 ml Reagent A, and 40 ml Reagent B for a total volume of 400 ml).
D. 20 mM Sodium Hydroxide Solution-Standardized (NaOH) (Prepare 50 ml in cold deionized water using Sodium Hydroxide).
E. Cholinesterase, Butyryl Enzyme Solution (Immediately before use, prepare a solution containing 30–60 units/ml of Cholinesterase, Butyryl in cold deionized water).

Procedure

Using a suitable pH meter in conjunction with a magnetic stirrer, pipette (in milliliters) the following reagents into a suitably thermostatted titration vessel:

	Test
Reagent C (BuCholine)	50.00

Equilibrate to 37°C. Adjust to pH 8.5 with Reagent D (NaOH), using a burette. Then add:

Reagent E (Enzyme Solution)	0.40

Run the reaction for 1–5 minutes. Record the time when the pH reaches 8.0. Maintain the pH of the reaction mix at pH 8.0 by the addition of small volumes (0.05 ml) of Reagent D. Record the volume of Reagent D used to maintain the pH at 8.0 and the time required.

Calculation

$$\text{Units / mg enzyme} = \frac{(\text{Molarity of NaOH})\ (\text{NaOH})\ (1000)}{(\text{T})\ (\text{mg enzyme / RM})}$$

NaOH = Volume (in milliliters) of Reagent D used in the assay
1000 = Conversion from millimoles to micromoles (Unit definition)
T = Time of assay
RM = Reaction Mix

Unit Definition: *One unit will hydrolyze 1.0 μmole of butyrylcholine to choline and butyrate per minute at pH 8.0 at 37°C.*

Initial Assay Concentrations

In a 50.4 ml reaction mix, the initial concentrations are 40 mM magnesium chloride, 99 mM sodium chloride, 4 mM butyrylcholine chloride and 12–24 units cholinesterase, butyryl.

Enzymatic Assay of CHOLOYLGLYCINE HYDROLASE (EC 3.5.1.24)

Principle

$$\text{Glycocholic Acid} + H_2O \xrightarrow{\text{CH}} \text{Glycine} + \text{Cholic Acid}$$

Abbreviation Used

CH = Choloylglycine Hydrolase

Conditions: T = 37°C, pH = 5.6, A600nm, Light path = 1 cm

Method: Colorimetric

Reagents

A. 25 mM Sodium Acetate Buffer, pH 5.6 at 37°C (Buffer) (Prepare 100 ml in deionized water using Sodium Acetate, Trihydrate. Adjust to pH 5.6 at 37°C with 1 M HCl).
B. 144 mM Sodium Glycocholate Solution (Glycoch) (Prepare 1 ml in Reagent A using Glycocholic Acid, Sodium Salt. *Prepare Fresh*).
C. 107 mM β-Mercaptoethanol Solution (2-ME) (Prepare 20 ml in deionized water using 2-Mercaptoethanol. *Prepare Fresh*).
D. 55 mM Ethylenediaminetetraacetic Acid Solution (EDTA) (Prepare 5 ml in deionized water using Ethylenediaminetetraacetic Acid, Disodium Salt, Dihydrate).
E. Ethylene Glycol Monomethyl Ether (Use Ethylene Glycol Monomethyl Ether).
F. 4% (w/v) Ninhydrin Solution (Prepare 25 ml in Reagent E using Ninhydrin).
G. 200 mM Sodium Citrate Buffer, pH 5.0 at 37°C (Prepare 50 ml in deionized water using Citric Acid, Free Acid, Anhydrous. Adjust to pH 5.0 at 37°C with 1 M NaOH).
H. 0.16% (w/v) Stannous Chloride Solution (Prepare 25 ml in Reagent G using Stannous Chloride, Anhydrous).
I. 50% (v/v) n-Propanol Solution (Prop) (Prepare 25 ml in deionized water using 1-Propanol).
J. Color Reagent Solution (CRS) (Prepare 60 ml by combining 20 ml each of Reagent F, Reagent H, and Reagent I).
K. 2.66 mM Glycine Standard Solution (Std) (Prepare 5 ml in deionized water using Glycine, Free Base).
L. 20% (w/v) Trichloroacetic Acid Solution (TCA) (Prepare 20 ml in deionized water using Trichloroacetic Acid, 6.1 N Solution, approximately 100% (w/v)).
M. 5 mM Sodium Phosphate Buffer, pH 7.0 at 37°C (Enz Dil) (Prepare 25 ml in deionized water using Sodium Phosphate, Monobasic, Anhydrous. Adjust to pH 7.0 at 37°C with 1 M NaOH).
N. Choloylglycine Hydrolase Enzyme Solution (Immediately before use, prepare a solution containing 15–30 units/ml of Choloylglycine Hydrolase in cold Reagent M).

Procedure

Pipette (in milliliters) the following reagents into suitable containers:

	Test	Blank
Reagent A (Buffer)	0.20	0.20
Reagent B (Glycoch)	0.20	0.20
Reagent C (2-ME)	0.20	0.20
Reagent D (EDTA)	0.20	0.20
Deionized Water	0.10	0.10

Mix by swirling and equilibrate to 37°C. Then add:

	Test	Blank
Reagent N (Enzyme Solution)	0.10	—

Mix by swirling and incubate at 37°C for exactly 5 minutes. Then add:

Reagent L (TCA)	1.00	1.00
Reagent N (Enzyme Solution)	—	0.10

Mix by swirling and centrifuge for 10 minutes. Use the supernatant in the color development.

Color Development

Standard Curve: Prepare a standard curve by pipetting (in milliliters) the following reagents into suitable vials:

	Std 1	Std 2	Std 3	Std 4	Std 5	Std Blank
Reagent K (Std Soln)	0.03	0.05	0.10	0.15	0.20	—
Deionized Water	0.17	0.15	0.10	0.05	—	0.20
Reagent J (CRS)	2.00	2.00	2.00	2.00	2.00	2.00

Mix by swirling and heat in a boiling water bath for 20 minutes. Cool to room temperature and add 10 ml of Reagent I (Prop). Mix by swirling.

Sample: Pipette (in milliliters) the following reagents into a suitable container:

	Test	Blank
Test Supernatant	0.20	—
Blank Supernatant	—	0.20
Reagent J (CRS)	2.00	2.00

Mix by swirling and heat in a boiling water bath for 20 minutes. Cool to room temperature and add 10 ml of Reagent I (Prop). Mix by swirling. Transfer to suitable cuvettes and determine the A600nm for the Test, Blank, Standards, and Standard Blank.

Calculations

Standard Curve: ΔA600nm Standard = A600nm Standard–A600nm Standard Blank

Plot the ΔA600nm Standard vs μmoles Glycine.

Sample Determination: ΔA600nm Test = A600nm Test–A600nm Blank

Determine the μmoles of Glycine liberated using the Standard curve.

$$\text{Units / ml enzyme} = \frac{(\mu\text{moles of glycine})\ (2)\ (\text{df})}{(0.1)\ (0.2)}$$

2 = Volume (in milliliters) of stopped reaction

df = Dilution factor

0.1 = Volume (in milliliter) of enzyme used

0.2 = Volume (in milliliter) of the Test used in color development.

$$\text{Units / mg solid} = \frac{\text{units / ml enzyme}}{\text{mg solid / ml enzyme}}$$

$$\text{Units / mg protein} = \frac{\text{units / ml enzyme}}{\text{mg protein / ml enzyme}}$$

Unit Definition: *One unit will hydrolyze 1.0 μmole of glycocholic acid to glycine and cholic acid in 5 minutes at pH 5.6 at 37°C.*

Final Assay Concentrations

In a 1.00 ml reaction mix, the final concentrations are 10 mM sodium acetate, 29 mM sodium glycocholate, 21 mM β-mercaptoethanol, 11 mM Ethylenediaminetetraacetic Acid, 0.5 mM sodium phosphate, and 1.5–3.0 units choloylglycine hydrolase.

Enzymatic Assay of CHONDRO-4-SULFATASE (EC 3.1.6.9)

Principle

$$\text{Di-4S} + H_2O \xrightarrow{CS} \text{Di-OS} + SO_4$$

Abbreviations Used

Di-4S = (4-Deoxy-β-D-Gluc-4-enuronosyl-[1 >3]-N-acetyl-D-galactosamine-4-sulfate)
Di-OS = Unsaturated Disaccharide
CS = Chondro-4-Sulfatase

Conditions: T = 37°C, pH = 7.5, A360nm, Light path = 1 cm

Method: Turbidimetric

Reagents

A. 200 mM Tris and 250 mM Sodium Acetate Buffer with 250 mM Sodium Chloride and 0.05% (w/v) Bovine Serum Albumin, pH 7.5 at 37°C (Prepare 100 ml in deionized water using Trizma Base, Sodium Acetate, Trihydrate, Sodium Chloride and Albumin, Bovine or equivalent. Adjust to pH 7.5 with either 1 M HCl or 1 M NaOH).
B. 300 mM Hydrochloric Acid Solution (HCl) (Prepare 100 ml in deionized water using Hydrochloric Acid).
C. 1.47 mM Cetylpyridinium Chloride Solution (Cetyl Pyr) (Prepare 100 ml in Reagent B (HCl) using Cetylpyridinium Chloride. Clarify by warming before use. *Prepare Fresh*).
D. 10.2 N Hydrochloric Acid Solution (HCl) (Prepare 5 ml in deionized water using Hydrochloric Acid).
E. 20 mM Barium Chloride and 4.8% (w/v) Gelatin Solution ($BaCl_2$/Gelatin) (Prepare by dissolving 0.5 g of Gelatin in 100 ml of deionized water with stirring at 60°C. Let stand overnight at 4°C. Then add 1.45 ml of Reagent D (HCl) and 0.5 g of Barium Chloride, Dihydrate).

F. 0.01% (w/v) Bovine Serum Albumin Solution (BSA) (Prepare 10 ml in deionized water using Albumin, Bovine or equivalent).

G. 1% (w/v) Di-4S Substrate Solution (Di-4S) (Prepare 1 ml in deionized water using Di-4S, ICN Immunobiologicals).

H. 5 mM Potassium Sulfate Solution (K_2SO_4 Std) (Prepare 10 ml in deionized water using Potassium Sulfate).

I. Chondro-4-Sulfatase Enzyme Solution (Immediately before use, prepare a solution containing 0.1–0.3 unit/ml of Chondro-4-Sulfatase in cold Reagent F).

Procedure

Step 1: Pipette (in milliliters) the following reagents into suitable containers:

	Test	Blank
Reagent A (Buffer)	0.05	0.05
Reagent I (Enzyme Solution)	0.10	—
Reagent F (BSA)	—	0.10

Mix by swirling and equilibrate to 37°C. Then add:

Reagent G (Di-4S)	0.05	0.05

Mix by swirling and incubate at 37° for exactly 5 minutes. Then add:

Reagent C (Cetyl Pyr)	0.20	0.20

Mix by swirling and let stand for 10 minutes at 37°C.

Step 2: Pipette (in milliliters) the following reagents into a suitable container:

Test Mixture	0.30	—
Blank Mixture	—	0.30
Reagent E ($BaCl_2$/Gelatin)	0.70	0.70

Mix by swirling and let stand for 10 minutes at 25°C. Transfer the solution to suitable cuvettes and record the A360nm for the Test and Blank.

Colorimetric Assay

	Std 1	Std 2	Std 3	Std 4	Std 5	Std Blank
Reagent H (K_2SO_4 Std)	0.02	0.03	0.05	0.07	0.10	—
Reagent E ($BaCl_2$/Gelatin)	0.70	0.70	0.70	0.70	0.70	0.70
Deionized Water	0.28	0.27	0.25	0.23	0.20	0.30

Mix by swirling and let stand for 10 minutes at 25°C. Transfer the Standards and Standard Blank to suitable cuvettes and record the A360nm for Standards and Standard Blank using a suitable spectrophotometer.

Calculations

Standard Curve: ΔA360nm Standard = A360nm Standard–A360nm Standard Blank

Prepare a standard curve by plotting the ΔA360nm Standard vs μmoles of potassium sulfate.

Sample Determination: ΔA360nm Sample = A360nm Test–A360nm Blank

Calculations

Determine the concentration of inorganic sulfate liberated using the Standard curve.

$$\text{Units / mg enzyme} = \frac{(\mu\text{moles of inorganic sulfate liberated})\,(0.4)\,(\text{df})}{(0.1)}$$

0.4 = Total volume (in milliliters) of Step 1
df = Dilution factor
0.1 = Volume (in milliliter) of enzyme used in Step 1

$$\text{Units / ml solid} = \frac{\text{units / ml enzyme}}{\text{mg solid / ml enzyme}}$$

$$\text{Units / ml protein} = \frac{\text{units / ml enzyme}}{\text{mg protein / ml enzyme}}$$

Unit Definition: *One unit will liberate 1.0 μmole of inorganic sulfate from 4-deoxy-β-D-gluc-4-enuronosyl-[1>3]-N-acetyl-Dgalactosamine 4-sulfate per minute at pH 7.5 at 37°C.*

Final Assay Concentration

In a 0.20 ml reaction mix, the final concentrations are 50 mM Tris, 62 mM sodium acetate, 62 mM sodium chloride, 0.02% (w/v) bovine serum albumin, 0.2% (w/v) 4-deoxy-β-Dgluc-4-enuronosyl-[1 >3]-N-acetyl-D-galactosamine 4-sulfate and 0.01–0.03 unit chondro-4-sulfatase.

Enzymatic Assay of CHONDRO-6-SULFATASE (EC 3.1.6.10)

Principle

$$\text{Di-6S} + H_2O \xrightarrow{\text{CS}} \text{Di-OS} + SO_4$$

Abbreviations Used

Di-6S = 4-Deoxy-β-D-Gluc-4-enuronosyl-[1 >3]-N-acetyl-D-galactosamine-6-sulfate
Di-OS = Unsaturated Disaccharide
CS = Chondro-6-Sulfatase

Conditions: T = 37°C, pH = 7.5, A360nm, Light path = 1 cm

Method: Turbidimetric

Reagents

A. 200 mM Tris and 250 mM Sodium Acetate Buffer with 250 mM Sodium Chloride and 0.05% (w/v) Bovine Serum Albumin, pH 7.5 at 37°C (Prepare 100 ml in deionized water using Trizma Base, Sodium Acetate, Trihydrate, Sodium Chloride and Albumin, Bovine. Adjust to pH 7.5 with either 1 M HCl or 1 M NaOH).

B. 300 mM Hydrochloric Acid Solution (HCl) (Prepare 100 ml in deionized water using Hydrochloric Acid).

C. 14.7 mM Cetylpyridinium Chloride Solution (Cetyl Pyr) (Prepare 100 ml in Reagent B using Cetylpyridinium Chloride. Clarify by warming before use. *Prepare Fresh*).

D. 10.2 N Hydrochloric Acid Solution (HCl) (Prepare 5 ml in deionized water using Hydrochloric Acid).

E. 20 mM Barium Chloride and 0.49% (w/v) Gelatin Solution ($BaCl_2$/Gelatin) (Prepare by dissolving 0.5 g of Gelatin, in 100 ml of deionized water with stirring at 60°C. Let stand overnight at 4°C. Then add 1.45 ml of Reagent D (HCl) and 0.5 g of Barium Chloride, Dihydrate).

F. 0.01% (w/v) Bovine Serum Albumin Solution (BSA) (Prepare 10 ml in deionized water using Albumin, Bovine).

G. 1% (w/v) Di-6S Substrate Solution (Di-6S) (Prepare 1 ml in deionized water using Di-6S, ICN Immunobiologicals).

H. 5 mM Potassium Sulfate Solution (K_2SO_4 Std) (Prepare 10 ml in deionized water using Potassium Sulfate).

I. Chondro-6-Sulfatase Enzyme Solution (Immediately before use, prepare a solution containing 0.75–1.5 unit/ml of Chondro-4-Sulfatase in cold Reagent F).

Procedure

Step 1: Pipette (in milliliters) the following reagents into suitable containers:

	Test	Blank
Reagent A (Buffer)	0.05	0.05
Reagent I (Enzyme Solution)	0.05	—
Reagent F (BSA)	0.05	0.10

Mix by swirling and equilibrate to 37°C. Then add:

	Test	Blank
Reagent G (Di-6S)	0.05	0.05

Mix by swirling and incubate at 37° for exactly 5 minutes. Then add:

	Test	Blank
Reagent C (Cetyl Pyr)	0.20	0.20

Mix by swirling and let stand for 10 minutes at 37°C.

Step 2: Pipette (in milliliters) the following reagents into a suitable container:

	Test	Blank
Test Mixture (From Step 1)	0.30	—
Blank Mixture (From Step 1)	—	0.30
Reagent E ($BaCl_2$/Gelatin)	0.70	0.70

Mix by swirling and let stand for 10 minutes at 25°C. Transfer the solutions to suitable cuvettes and record the A360nm for the Test and Blank.

Colorimetric Assay

	Std 1	Std 2	Std 3	Std 4	Std 5	Std Blank
Reagent H (K_2SO_4 Std)	0.02	0.03	0.05	0.07	0.10	—
Reagent E ($BaCl_2$/Gelatin)	0.70	0.70	0.70	0.70	0.70	0.70
Deionized Water	0.28	0.27	0.25	0.23	0.20	0.30

Calculations

Standard Curve: ΔA360nm Standard = A360nm Standard–A360nm Standard Blank

Prepare a standard curve by plotting the ΔA360nm Standard vs μmoles of potassium sulfate.

Sample Determination: ΔA360nm Sample = A360nm Test–A360nm Blank

Calculations

Determine the concentration of inorganic sulfate liberated using the Standard curve.

$$\text{Units / ml enzyme} = \frac{(\mu\text{moles of inorganic sulfate liberated})\ (0.4)\ (\text{df})}{(0.05)}$$

0.4 = Total volume (in milliliter) of Step 1
df = Dilution factor
0.05 = Volume (in milliliter) of enzyme used in Step 1

$$\text{Units / mg solid} = \frac{\text{units / ml enzyme}}{\text{mg solid / ml enzyme}}$$

$$\text{Units / mg protein} = \frac{\text{units / ml enzyme}}{\text{mg protein / ml enzyme}}$$

Unit Definition: *One unit will liberate 1.0 μmole of inorganic sulfate from 4-deoxy-β-D-gluc-4-enuronosyl-[1 >3]-N-acetyl-Dgalactosamine 6-sulfate per minute at pH 7.5 at 37°C.*

Final Assay Concentration

In a 0.20 ml reaction mix, the final concentrations are 50 mM Tris, 62 mM sodium acetate, 62 mM sodium chloride, 0.02% (w/v) bovine serum albumin, 0.2% (w/v) 4-deoxy-β-Dgluc-4-enuronosyl-[1>3]-N-acetyl-D-galactosamine 6-sulfate and 0.04–0.08 unit chondro-6-sulfatase.

Enzymatic Assay of CHONDROITINASE ABC (EC 4.2.2.4)

Principle

$$\text{Chondroitin Sulfate} + H_2O \xrightarrow{\text{Chondroitinase ABC}} \text{Unsaturated Disaccharides}$$

Conditions: T = 37°C, pH = 8.0, A232nm, Light path = 1 cm

Method: Spectrophotometric Stop Rate Determination

Reagents

A. 250 mM Tris HCl and 300 mM Sodium Acetate Buffer with 0.05% (w/v) Bovine Serum Albumin, pH 8.0 at 37°C (Prepare 200 ml in deionized water using Trizma Base, Sodium Acetate, Trihydrate, and Albumin, Bovine. Adjust to pH 8.0 at 37°C with 1 M HCl or 1 M NaOH).
B. 0.5% (w/v) Chondroitin Sulfate A Solution (Chon A) (Prepare 5 ml in Reagent A using Chondroitin Sulfate A, Sodium Salt).
C. 0.5% (w/v) Chondroitin Sulfate B Solution (Chon B) (Prepare 5 ml in Reagent A using Chondroitin Sulfate B, Sodium Salt).
D. 0.5% (w/v) Chondroitin Sulfate C Solution (Chon C) (Prepare 5 ml in Reagent A using Chondroitin Sulfate C, Sodium Salt).
E. 0.01% (w/v) Bovine Serum Albumin Solution (BSA) (Prepare 20 ml in deionized water using Albumin Bovine Serum).
F. 50 mM Potassium Chloride Solution, pH 1.8 at 25°C (KCl) (Prepare 50 ml in deionized water using Potassium Chloride. Adjust to pH 1.8 at 25°C with 1 M HCl).
G. Chondroitinase ABC Enzyme Solution (Immediately before use, prepare a solution containing 0.06–0.10 unit/ml of Chondroitinase ABC in cold Reagent E).

Procedure

Pipette (in milliliters) the following reagents into suitable tubes:

	Test 1	Test 2	Test 3
Reagent G (Enzy Soln)	0.80	0.80	0.80

Equilibrate to 37°C for 2 minutes. Then add:

Reagent B (Chon A)	0.20	—	—
Reagent C (Chon B)	—	0.20	—
Reagent D (Chon C)	—	—	0.20

Mix by inversion. Incubate at 37°C for 21 minutes. At 0, 3, 6, 9, 12, 15, 18, and 21 minutes (T_T), transfer 0.10 ml from each Test to separate tubes containing 0.90 ml of Reagent F (KCl). The Blank for this assay is T_0. Incubate each tube for an additional 10 minutes at 37°C. Centrifuge for 10 minutes and transfer the Test and Blank supernatants to suitable quartz cuvettes. Record the absorbance at 232nm for each tube.

Calculations

T_T = Time intervals (minutes) for enzyme assay incubation
ΔA232nm Test (T_T) = A232nm Test (T_T)–A232nm Blank (T_0)
Plot the A232nm Test (T_T) vs. Time. Determine the A232nm/min Test from a linear portion of the graph.

$$\text{Units/ml} = \frac{(\text{A232nm / min Test})\ (1.0)\ (\text{df})}{(\text{EmM})\ (0.10)\ (0.8)}$$

df = Dilution factor
EmM = Millimolar extinction coefficient of unsaturated disaccharides: 5.1 for products from Chondroitin A and Chondroitin B, and 5.5 for products from Chondroitin C.
0.10 = Volume (in milliliter) of reaction mix used
T = Time (in minutes) of the assay as per the Unit Definition
1.0 = Total volume (in milliliter) of assay
0.8 = Volume (in milliliter) of enzyme used

$$\text{Units / mg solid} = \frac{\text{units / ml enzyme}}{\text{mg solid / ml enzyme}}$$

Unit Definition: *One unit will liberate 1.0 μmole of 2-acetamido-2-deoxy-3-O-(β-D-gluc-4-ene-pyranosyluronic acid)-4-O-sulfo-Dgalactose from chondroitin sulfate A or 1.0 μmole of 2-acetamido-2-deoxy-3-O-(β-D-gluc-4-ene-pyranosyluronic acid)-6-O-sulfo-D-galactose from chondroitin sulfate C per minute at pH 8.0 at 37°C.*

Final Assay Concentration

In a 1.00 ml reaction mix, the final concentrations are 50 mM Tris, 60 mM sodium acetate, 0.02% (w/v) bovine serum albumin, 0.1% (w/v) chondroitin sulfate A, B or C, and 0.048–0.08 unit chondroitinase ABC.

Enzymatic Assay of CHONDROITINASE AC (EC 4.2.2.5)

Principle

$$\text{Chondroitin Sulfate} + H_2O \xrightarrow{\text{Chondroitinase AC}} \text{Unsaturated Disaccharides}$$

Conditions: T = 37°C, pH = 7.3, A232nm, Light path = 1 cm

Method: Continuous Spectrophotometric Rate Determination

Reagents

A. 250 mM Tris HCl and 75 mM Sodium Acetate Buffer, pH 7.3 at 37°C (Prepare 200 ml in deionized water using Trizma Base and Sodium Acetate, Trihydrate. Adjust to pH 7.3 at 37°C with 1 M HCl).

B. 0.06% (w/v) Bovine Serum Albumin Solution (BSA) (Prepare 50 ml in deionized water using Albumin, Bovine or equivalent).

C. 0.5% (w/v) Chondroitin Sulfate A Solution (Chon A) (Prepare 5 ml in Reagent B using Chondroitin Sulfate A, Sodium Salt).

D. 0.5% (w/v) Chondroitin Sulfate B Solution (Chon B) (Prepare 5 ml in Reagent B using Chondroitin Sulfate B, Sodium Salt).

E. 0.5% (w/v) Chondroitin Sulfate C Solution (Chon C) (Prepare 5 ml in Reagent B using Chondroitin Sulfate C, Sodium Salt).

F. Chondroitinase AC Enzyme Solution (Enz) (Immediately before use, prepare a solution containing 0.5–1.0 unit/ml of Chondroitinase AC in cold Reagent B).

Procedure

Pipette (in milliliters) the following reagents into suitable cuvettes:

	Test 1	Blank 1	Test 2	Blank 2	Test 3	Blank 3
Reagent A (Buffer)	2.40	2.40	2.40	2.40	2.40	2.40
Reagent B (BSA)	0.10	0.20	0.10	0.20	0.10	0.20
Reagent C (Chon A)	0.50	0.50	—	—	—	—
Reagent D (Chon B)	—	—	0.50	0.50	—	—
Reagent E (Chon C)	—	—	—	—	0.50	0.50

Mix by inversion and equilibrate to 37°C. Monitor the A232nm until constant, using a suitably thermostatted spectrophotometer. Then add:

Reagent E (Enz)	0.10	—	0.10	—	0.10	—

Immediately mix by inversion and record the increase in A232nm for approximately 10 minutes. Obtain the ΔA232nm/minute using the maximum linear rate for Tests and Blanks.

Calculations

$$\text{Units / mg enzyme} = \frac{(\Delta\text{A232nm / min Test} - \Delta\text{A232nm / min Blank})}{(1.0)\ (\text{mg enzyme / RM})}$$

1.0 = Absorbance change per Unit Definition
RM = Reaction Mix

Unit Definition: *One unit will cause a Δ A232nm of 1.0 per minute due to the release of unsaturated disaccharides, from chondroitin sulfate A at pH 7.3 at 37°C.*

Final Assay Concentration

In a 3.10 reaction mix, the final concentrations are 194 mM Tris, 58 mM sodium acetate, 0.08% (w/v) chondroitin sulfate A, B or C, 0.01% (w/v) BSA and 0.01–0.10 unit chondroitinase AC.

Enzymatic Assay of CHONDROITINASE B

Principle

$$\text{Chondroitin Sulfate B} + H_2O \xrightarrow{\text{Chondroitinase B}} \text{Unsaturated Uronic Acid}$$

Conditions: T = 25°C, pH 7.5, A232nm, Light path = 1 cm

Method: Spectrophotometric Stop Rate Determination

Reagents

A. 20 mM Tris HCl Buffer with 50 mM Sodium Chloride, 4 mM Calcium Chloride, and 0.01% (w/v) Bovine Serum Albumin, pH 7.5 at 25°C (Prepare 100 ml in deionized water using Trizma Base, Sodium Chloride, Calcium Chloride, Dihydrate, and Albumin Bovine. Adjust to pH 7.5 at 25°C with 1 M HCl).

B. 2.0% (w/v) Chondroitin Sulfate B Solution (Chon B) (Prepare 5 ml in deionized water using Chondroitin Sulfate B, Sodium Salt).

C. 20 mM Hydrochloric Acid Solution (HCl) (Prepare 100 ml in deionized water using Hydrochloric Acid).

D. Chondroitinase B Enzyme Solution (Immediately before use, prepare a solution containing 25–50 units/ml of Chondroitinase B in cold Reagent A).

Procedure

Pipette (in milliliters) the following reagents into suitable cuvettes:

	Test	Blank
Reagent A (Buffer)	0.40	0.40
Reagent B (Chon B)	0.05	0.05

Mix by inversion and equilibrate to 25°C. Then add:

Reagent D (Enzyme Solution)	0.05	—

Immediately mix by inversion and incubate at 25°C for exactly 20 minutes. Then add:

Reagent C (HCl)	2.50	2.50
Reagent D (Enzyme Solution)	—	0.05

Mix by inversion. Centrifuge the solutions then transfer the Test and Blank to suitable cuvettes. Record the absorbance at A232nm.

Calculations

$$\text{Units / ml enzyme} = \frac{(\text{A232nm Test} - \text{A232nm Blank})\,(3)\,(3)\,(\text{df})}{(0.55)\,(0.05)}$$

3 = Total volume (in milliliters) of assay
3 = Time factor to convert to 60 minutes as per the Unit Definition
df = Dilution factor
0.55 = Millimolar extinction coefficient of the reaction product multiplied by 0.11
0.05 = Volume (in milliliters) of enzyme used

$$\text{Units / mg solid} = \frac{\text{units / ml enzyme}}{\text{mg solid / ml enzyme}}$$

$$\text{Units / mg protein} = \frac{\text{units / ml enzyme}}{\text{mg protein / ml enzyme}}$$

Unit Definition: *One unit will form 0.1 μmole of unsaturated uronic acid per hour at pH 7.5 at 25°C using chondroitin sulfate B as substrate.*

Final Assay Concentrations

In a 0.50 ml reaction mix, the final concentrations are 18 mM Tris, 45 mM sodium chloride, 4 mM calcium chloride, 0.009% (w/v) bovine serum albumin, 0.2% (w/v) chondroitin sulfate B and 1.25–2.5 units chondroitinase B.

Enzymatic Assay of CHONDROITINASE C (EC 4.2.2)

Principle

$$\text{Chondroitin Sulfate C} + H_2O \xrightarrow{\text{Chondroitinase C}} \text{Unsaturated Uronic Acid}$$

Conditions: T = 25°C, pH = 8.0, A232nm, Light path = 1 cm

Method: Spectrophotometric Stop Rate Determination

Reagents

A. 50 mM Tris HCl Buffer, pH 8.0 at 25°C (Prepare 100 ml in deionized water using Trizma Base. Adjust to pH 8.0 at 25°C with 1 M HCl).
B. 2.0% (w/v) Chondroitin Sulfate C Solution (CS) (Immediately before use, prepare 1.0 ml in deionized water using Chondroitin Sulfate C, Sodium Salt).
C. 50 mM Hydrochloric Acid Solution (HCl) (Prepare 50 ml in deionized water using Hydrochloric Acid).
D. Chondroitinase C Enzyme Solution (Immediately before use, prepare a solution containing 80–120 units/ml of Chondroitinase C in cold Reagent A).

Procedure

Pipette (in milliliters) the following reagent into suitable containers:

	Test	Blank
Reagent A (Buffer)	0.40	0.40
Reagent D (Enz Sol)	0.05	—

Mix by swirling and equilibrate to 25°C. Then add:

Reagent B (CS)	0.05	0.05

Immediately mix by swirling and incubate at 25°C for exactly 10 minutes. Then add:

Reagent C (HCl)	2.50	2.50
Reagent D (Enz Sol)	—	0.05

Mix by swirling and centrifuge for 10 minutes at 13,000 rpm in a microcentrifuge. Transfer the supernatant to suitable quartz cuvettes and record the A232nm for both the Test and Blank using a suitable spectrophotometer.

Calculations

$$\text{Units / ml enzyme} = \frac{(\text{A232nm Test} - \text{A232nm Blank})\ (3)\ (10)\ (6)\ (\text{df})}{(5.50)\ (0.05)}$$

3 = Total volume (in milliliters) of stopped assay
10 = Conversion factor from 1 μmole to 0.1 μmole as per the Unit Definition
6 = Conversion factor from 10 minutes to 1 hour
df = Dilution factor
5.50 = Millimolar extinction coefficient of the unsaturated uronic acid product at A232nm.
0.05 = Volume (in milliliter) of enzyme used

$$\text{Units / mg solid} = \frac{\text{units / ml enzyme}}{\text{mg solid / ml enzyme}}$$

Unit Definition: *One unit will form 0.1 μmole of unsaturated uronic acid per hour at pH 8.0 at 25°C.*

Final Assay Concentration

In a 0.50 ml reaction mix, the final concentrations are 45 mM Tris, 0.2% (w/v) chondroitin sulfate C and 4–6 units chondroitinase.

Enzymatic Assay of CHYMOPAPAIN (EC 3.4.22.6)

Principle

$$\text{BAEE} + H_2O \xrightarrow{\text{Chymopapain}} \text{Na - Benzoyl - L - Arginine} + \text{Ethanol}$$

Abbreviation Used

BAEE = Na-Benzoyl-L-Arginine Ethyl Ester

Conditions: T = 25°C, pH = 6.2

Method: Titrimetric

Reagents

A. 80 mM Na-Benzoyl-L-Arginine Ethyl Ester (BAEE) (Prepare 100 ml in deionized water using Na-Benzoyl-L-Arginine Ethyl Ester, Hydrochloride).
B. 20 mM Ethylenediaminetetraacetic Acid (EDTA) (Prepare 100 ml in deionized water using Ethylenediaminetetraacetic Acid, Disodium Salt, Dihydrate).
C. 50 mM L-Cysteine Solution, pH 6.2 at 25°C (Cys) (Prepare 50 ml in Reagent B using L-Cysteine, Hydrochloride, Monohydrate. Adjust to pH 6.2 with 1 M NaOH).
D. 3.0 M Sodium Chloride Solution (NaCl) (Prepare 50 ml in deionized water using Sodium Chloride).
E. 100 mM Sodium Hydroxide Solution-Standardized (NaOH) (Prepare 50 ml in cold deionized water using Sodium Hydroxide, Anhydrous. Standardize according to the ACS Reagent Procedure).
F. Chymopapain Enzyme Solution (Immediately before use, prepare a solution containing 0.75–1.5 units/ml of Chymopapain in deionized water).

Procedure

Pipette (in milliliters) the following reagents into a suitable container:

	Test
Reagent A (BAEE)	7.00
Reagent C (Cys)	1.00
Reagent D (NaCl)	1.00

Mix by swirling and equilibrate to 25°C. Adjust the pH to 6.3 at 25°C with 0.1 M NaOH. Then add:

Reagent F (Enzyme Solution)	1.00

When the pH drops to 6.2, begin timing the reaction. Run the reaction for 1–5 minutes. Maintain the pH of the reaction mix at pH 6.2 by the addition of small volumes (0.025–0.050 ml) of Reagent E. Record the volume of Reagent E used to maintain the pH at 6.2 and the time required.

Calculation

$$\text{Units / ml enzyme} = \frac{\text{(Molarity of NaOH) (NaOH) (1000) (df)}}{\text{(T) (1)}}$$

NaOH = Volume (in milliliters) of Reagent E used to maintain the pH at 6.2
1000 = Conversion factor to micromoles from millimoles as Per the Unit Definition
df = Dilution factor
T = Time (in minutes) required to maintain the pH at 6.2
1 = Volume (in milliliters) of enzyme used

$$\text{Units / mg solid} = \frac{\text{units / ml enzyme}}{\text{mg solid / ml enzyme}}$$

$$\text{Units / mg protein} = \frac{\text{units / ml enzyme}}{\text{mg protein / ml enzyme}}$$

Unit Definition: *One unit will hydrolyze 1.0 μmole of BAEE to Na-benzoyl-Larginine per minute at pH 6.2 at 25°C.*

Initial Assay Concentrations

In a 10.00 ml reaction mix, the initial concentrations are 56 mM Na-benzoyl-L-arginine ethyl ester, 5 mM L-cysteine, 2 mM Ethylenediaminetetraacetic Acid, 300 mM sodium chloride and 0.75–1.5 units chymopapain.

Enzymatic Assay of CHYMOTRYPSIN (EC 3.4.21.1)

Principle

$$\text{BTEE} + H_2O \xrightarrow{\text{Chymotrypsin}} \text{N-Benzoyl-L-Tyrosine} + \text{Ethanol}$$

Abbreviation Used

BTEE = N-Benzoyl-L-Tyrosine Ethyl Ester

Conditions: T = 25°C, pH = 7.8, A256nm, Light path = 1 cm

Method: Continuous Spectrophotometric Rate Determination

Reagents

A. 80 mM Tris HCl Buffer, pH 7.8 at 25°C (Prepare 100 ml in deionized water using Trizma Base. Adjust to pH 7.8 at 25°C with 1 M HCl).
B. 1.18 mM N-Benzoyl-L-Tyrosine Ethyl Ester Solution (BTEE) (Prepare 50 ml using N-Benzoyl-L-Tyrosine Ethyl Ester, by initially dissolving in 31.7 ml of Methanol, Absolute. Add enough deionized water to make the final volume 50 ml).

C. 2 M Calcium Chloride Solution ($CaCl_2$) (Prepare 5 ml in deionized water using Calcium Chloride, Dihydrate).
D. 1 mM Hydrochloric Acid Solution (HCl) (Prepare 50 ml in deionized water using Hydrochloric Acid).
E. Chymotrypsin Enzyme Solution (Immediately before use, prepare a solution containing 2–5 units/ml of Chymotrypsin in cold Reagent D).

Procedure

Pipette (in milliliters) the following reagents into suitable quartz cuvettes:

	Test	Blank
Reagent A (Buffer)	1.42	1.42
Reagent B (BTEE)	1.40	1.40
Reagent C ($CaCl_2$)	0.08	0.08

Mix by inversion and equilibrate to 25°C. Monitor the A256nm until constant, using a suitably thermostatted spectrophotometer. Then add:

	Test	Blank
Reagent D (HCl)	—	0.10
Reagent E (Enzyme Solution)	0.10	—

Immediately mix by inversion and record the increase in A256nm for approximately 5 minutes. Obtain the ΔA256nm/minute using the maximum linear rate for both the Test and Blank.

Calculations

$$\text{Units / ml enzyme} = \frac{(\Delta A256\text{nm / min Test} - \Delta A256\text{nm / min Blank})\ (3)\ (\text{df})}{(0.965)\ (0.1)}$$

3 = Volume (in milliliters) of assay
df = Dilution factor
0.965 = Millimolar extinction coefficient of N-Benzoyl-L-Tyrosine at 256nm.
0.1 = Volume (in milliliters) of enzyme used

$$\text{Units / mg solid} = \frac{\text{units / ml enzyme}}{\text{mg solid / ml enzyme}}$$

$$\text{Units / mg protein} = \frac{\text{units / ml enzyme}}{\text{mg protein / ml enzyme}}$$

Unit Definition: *One unit will hydrolyze 1.0 μmole of BTEE per minute at pH 7.8 at 25°C.*

Final Assay Concentration

In a 3.00 ml reaction mix, the final concentrations are 38 mM Tris, 0.55 mM N-benzoyl-L-tyrosine ethyl ester, 30% (v/v) methanol, 53 mM calcium chloride, 0.03 mM hydrochloric acid, and 0.2–0.5 unit chymotrypsin.

Enzymatic Assay of α-CHYMOTRYPSIN (Insoluble) (EC 3.4.21.1)

Principle

$$\text{ATEE} + H_2O \xrightarrow{\alpha\text{-Chymotrypsin}} \text{N-Acetyl-L-Tyrosine} + \text{Ethanol}$$

Abbreviation Used

ATEE = N-Acetyl-L-Tyrosine Ethyl Ester

Conditions: T = 30°C, pH 8.0

Method: Titrimetric

Reagents

A. 50% (w/w) Methanol Solution (MeOH) (Prepare 25 ml in deionized water using Methanol, Absolute).
B. 50 mM N-Acetyl-L-Tyrosine Ethyl Ester (ATEE) (Prepare 20 ml in Reagent A using N-Acetyl-L-Tyrosine Ethyl Ester).
C. 500 mM Calcium Chloride Solution ($CaCl_2$) (Prepare 25 ml in deionized water using Calcium Chloride, Dihydrate).
D. 100 mM Sodium Hydroxide Solution Standardized (NaOH) (Prepare 100 ml in deionized water using Sodium Hydroxide, Anhydrous. Standardize according to the ACS Reagent Procedure).
E. α-Chymotrypsin Enzyme, Insoluble (Enzyme) (Use α-Chymotrypsin, Insoluble, 5–10 mg).

Procedure

Weigh (in milligrams) the following reagents into a suitable container:

	Test
Reagent E (Enzyme)	5.00–10.00

Then pipette (in milliliters) the following reagents:

Deionized Water	6.00
Reagent C ($CaCl_2$)	2.00

Mix by swirling and adjust the pH to 8.4 with Reagent D. Equilibrate to 30°C and then add:

Reagent B (ATEE)	2.00

Mix by swirling and incubate at 30°C. When the pH reaches 8.0 begin timing the reaction. Run the reaction for 1–5 minutes. Maintain the pH of the reaction mix at 8.0 by the addition of small volumes (0.10 ml) of Reagent D. Record the volume of Reagent D used to maintain the pH at 8.0 and the time required.

Calculations

$$\text{Units / mg solid} = \frac{(\text{Molarity of NaOH})\,(\text{NaOH})\,(1000)}{(\text{mg solid})\,(\text{T})}$$

NaOH = Volume (in milliliters) of Reagent D used in the assay
1000 = Conversion factor to µmoles from millimoles
mg solid = mg solid of α-chymotrypsin, insoluble, used in the assay
T = Time of assay (in minutes) required to maintain thepH at 8.0

Unit Definition: *One unit will hydrolyze 1.0 µmole of ATEE per minute at pH 8.0 at 30°C.*

Final Assay Concentrations

In a 10.00 ml reaction mix, the final concentrations are 100 mM calcium chloride, 10 mM N-acetyl-L-tyrosine ethyl ester, and 5–10 mg of α-chymotrypsin, insoluble.

Enzymatic Assay of α-CHYMOTRYPSINOGEN A

Principle

$$\text{Chymotrypsinogen} \xrightarrow{\text{Trypsin}} \text{Chymotrypsin} + \text{Peptide Fragments}$$

$$\text{BTEE} + H_2O \xrightarrow{\text{Chymotrypsin}} \text{Na - Benzoyl - L - Tyrosine} + \text{Ethanol}$$

Abbreviation Used

BTEE = Na-Benzoyl-L-Tyrosine Ethyl Ester

Conditions: T = 25°C, pH = 7.8, A256nm, Light path = 1 cm

Method: Continuous Spectrophotometric Rate Determination

Reagents

A. 50 mM Potassium Phosphate Buffer, pH 8.0 at 25°C (Prepare 100 ml in deionized water using Potassium Phosphate Monobasic, Anhydrous. Adjust to pH 8.0 at 25°C with 1 M KOH).

B. 1 mM Hydrochloric Acid Solution (HCl) (Prepare 50 ml in deionized water using Hydrochloric Acid).

C. 80 mM Tris HCl Buffer, pH 7.8 at 25°C (Tris Buffer) (Prepare 100 ml in deionized water using Trizma Base. Adjust to pH 7.8 at 25°C with 1 M HCl).

D. 1.18 mM N-Benzoyl-L-Tyrosine Ethyl Ester (BTEE) (Prepare 50 ml using N-Benzoyl-L-Tyrosine Ethyl Ester, by initially dissolving in 31.7 ml of Methanol, Absolute. Add enough deionized water to make a final volume of 50 ml).

E. 2 M Calcium Chloride Solution ($CaCl_2$) (Prepare 5 ml in deionized water using Calcium Chloride, Dihydrate).

F. 0.1% (w/v) Trypsin Enzyme Solution (Trypsin) (Immediately before use, prepare 10 ml in Reagent A using Trypsin, Type III).

G. α-Chymotrypsinogen A Solution (Chymo A) (Immediately before use, prepare a solution as described below in Step 1 and Step 2).

Procedure

Step 1: Free α-Chymotrypsin Activity Immediately before use, dissolve 10 mg of α-Chymotrypsinogen A in 10 ml of Reagent B. Immediately proceed with the chymotrypsin assay.
Step 2: Activated α-Chymotrypsin Activity Immediately before use, dissolve 10 mg of α-Chymotrypsinogen A in 10 ml of Reagent A. To 9 ml of this solution, add 1 ml of Reagent F (Trypsin) and incubate at 25°C for 3 hours. Then dilute 0.5 ml of this solution to 10.5 ml With Reagent B. Proceed with the chymotrypsin assay.
Step 3: Chymotrypsin Assay: Pipette (in milliliters) the following reagents into suitable cuvettes:

	Test	Blank
Reagent C (Tris Buffer)	1.42	1.42
Reagent D (BTEE)	1.40	1.40
Reagent E ($CaCl_2$)	0.08	0.08

Mix by inversion and equilibrate to 25°C. Monitor the A256nm until constant, using a suitably thermostatted spectrophotometer. Then add:

Reagent B (HCl)	—	0.10
Enzyme Solution (from Step 1 or Step 2)	0.10	—

Immediately mix by inversion and record the increase in A256nm for approximately 5 minutes. Obtain the A256nm/minute using the maximum linear rate for both the Test and Blank.

Calculations

The calculations shown below can be used to determine both the free α-Chymotrypsin activity and also the α-Chymotrypsin activity after activation by trypsin.

$$\text{Units / ml enzyme} = \frac{(\text{A256nm / min Test} - \text{A256nm / min Blank})\ (3)\ (\text{df})}{(0.964)\ (0.1)}$$

3 = Volume (in milliliters) of assay
df = Dilution factor
0.964 = Millimolar extinction coefficient of N-Benzoyl-L-Tyrosine at 256nm
0.1 = Volume (in milliliter) of enzyme used in assay

$$\text{Units / mg solid} = \frac{\text{units / ml enzyme}}{\text{mg solid / ml enzyme}}$$

$$\text{Units / mg protein} = \frac{\text{units / ml enzyme}}{\text{mg protein / ml enzyme}}$$

Unit Definition: *One unit will hydrolyze 1.0 μmole of BTEE per minute at pH 7.8 at 25°C.*

Final Assay Concentration

In a 3.00 ml reaction mix, the final concentrations are 38 mM Tris, 0.55 mM N-benzoyl-L-tyrosine ethyl ester, 30% (v/v) methanol, 53 mM calcium chloride, 0.03 mM hydrochloric acid, 0.48 μg trypsin, and 4.3 μg of α-chymotrypsinogen A.

Enzymatic Assay of CITRATE LYASE (EC 4.1.3.6)

Principle

$$\text{Citrate} + H_2O \xrightarrow{\text{Citrate Lyase}} \text{Oxalacetate} + \text{Acetate}$$

$$\text{Oxalacetate} + \beta\text{-NADH} \xrightarrow{\text{Malic Dehydrogenase}} \text{L-Malate} + \beta\text{-NAD}$$

$$(\text{Oxalacetate} \xrightarrow{\text{Oxalacetate Decarboxylase}} \text{Pyruvate} + CO_2)$$

$$\text{Pyruvate} + \beta\text{-NADH} \xrightarrow{\text{L-Lactic Dehydrogenase}} \text{L-Lactate} + \beta\text{-NAD}$$

Abbreviations Used

β-NADH = β-Nicotinamide Adenine Dinucleotide, Reduced Form
β-NAD = β-Nicotinamide Adenine Dinucleotide, Oxidized Form

Conditions: T = 25°C, pH = 7.6, A340nm, Light path = 1 cm

Method: Continuous Spectrophotometric Rate Determination

Reagents

A. 100 mM Triethanolamine Buffer, pH 7.6 at 25°C (Prepare 100 ml in deionized water using Triethanolamine Hydrochloride. Adjust to pH 7.6 at 25°C with 1 M NaOH).
B. 29 mM Zinc Chloride Solution ($ZnCl_2$) (Prepare 5 ml in deionized water using Zinc Chloride).
C. 14 mM β-Nicotinamide Adenine Dinucleotide, Reduced Form Solution (β-NADH) (Prepare 1 ml in deionized water using β-Nicotinamide Adenine Dinucleotide, Reduced Form, Disodium Salt or dissolve the contents of a 10 mg vial of β-Nicotinamide Adenine Dinucleotide, Reduced Form, Disodium Salt in the appropriate volume of deionized water. *Prepare Fresh*).
D. 100 mM Sodium Citrate Solution (Cit) (Prepare 10 ml in deionized water using Citric Acid, Trisodium Salt, Dihydrate).
E. L-Lactic Dehydrogenase Enzyme Solution (LDH) (Use L-Lactic Dehydrogenase).
F. Malic Dehydrogenase Enzyme Solution (MDH) .
G. 10 mM Triethanolamine with 0.3 mM Zinc Chloride and 454 mM Ammonium Sulfate Solution (Enz Dil) (Prepare 100 ml in deionized water using Triethanolamine Hydrochloride, Zinc Chloride, and Ammonium Sulfate. Adjust to pH 7.6 at 25°C with 1 M NaOH).
H. Citrate Lyase Enzyme Solution (Immediately before use, prepare a solution containing 0.1–0.3 unit/ml of Citrate Lyase in cold Reagent G).

Procedure

Prepare a reaction cocktail by pipetting (in milliliters) the following reagents into a suitable container:

Reagent A (Buffer)	28.50
Reagent B ($ZnCl_2$)	0.50
Reagent C (β-NADH)	0.50
Reagent D (Cit)	0.20

Mix by inversion and adjust to pH 7.6 at 25°C with 1 M HCl or 1 M NaOH, if necessary. Pipette (in milliliters) the following reagents into suitable cuvettes:

	Test	Blank
Reaction Cocktail	2.88	2.88
Reagent E (LDH)	0.01	0.01
Reagent F (MDH)	0.01	0.01

Mix by inversion and equilibrate to 25°C. Monitor the A340nm until constant, using a suitably thermostatted spectrophotometer. Then add:

Reagent G (Enz Dil)	—	0.10
Reagent H (Enzyme Solution)	0.10	—

Immediately mix by inversion and record the decrease in A340nm for approximately 5 minutes. Obtain the ΔA340nm/minute using the maximum linear rate for both the Test and Blank.

Calculations

$$\text{Units / ml enzyme} = \frac{(\Delta\text{A340nm / min Test} - \Delta\text{A340nm / min Blank})\ (3)\ (\text{df})}{(6.22)\ (0.1)}$$

3 = Total volume (in milliliters) of assay
df = Dilution factor
6.22 = Millimolar extinction coefficient of β-NADH at 340nm
0.1 = Volume (in milliliter) of enzyme used

$$\text{Units / mg solid} = \frac{\text{units / ml enzyme}}{\text{mg solid / ml enzyme}}$$

Unit Definition: *One unit will convert 1.0 μmole of citrate to oxalacetate per minute at pH 7.6 at 25°C.*

Final Assay Concentration

In a 3.00 ml reaction mix, the final concentrations are 96 mM triethanolamine, 0.5 mM zinc chloride, 0.23 mM β-NADH, 0.67 mM sodium citrate, 100 units L-lactic dehydrogenase, 50 units malic dehydrogenase, 15 mM ammonium sulfate and 0.01–0.03 unit citrate lyase.

Notes

1. The addition of L-Lactic Dehydrogenase ensures that any oxalacetate that is converted to pyruvate is measured. This conversion to pyruvate occurs by either oxalacetate decarboxylase which may be present in citrate lyase or by a nonenzymatic reaction.
2. Contains not less than 10,000 L-Lactic Dehydrogenase units per ml.
3. Contains not less than 5,000 Malic Dehydrogenase units per ml.
4. L-Lactic Dehydrogenase Unit Definition: One unit will reduce 1.0 μmole of pyruvate to L-lactate per minute at pH 7.5 at 37°C.
5. Malic Dehydrogenase Unit Definition: One unit will convert 1.0 μmole of oxalacetate and β-NADH to L malate and β-NAD per minute at pH 7.5 at 25°C.

Enzymatic Assay of CITRATE SYNTHASE (EC 4.1.3.7)

Principle

$$\text{L-Malate} + \beta\text{-NAD} \xrightarrow{\text{MDH}} \text{Oxalacetate} + \beta\text{-NADH}$$

$$\text{Acetyl CoA} + \text{Oxalacetate} \xrightarrow{\text{CS}} \text{Citrate} + \text{CoA-CS}$$

Abbreviations Used

MDH = Malic Dehydrogenase
β-NAD = β-Nicotinamide Adenine Dinucleotide, Oxidized Form
β-NADH = β-Nicotinamide Adenine Dinucleotide, Reduced Forr
CoA = Coenzyme A
CS = Citrate Synthase

Conditions: T = 37°C, pH = 8.0, A340nm, Light path = 1 cm

Method: Continuous Spectrophotometric Rate Determination

Reagents

A. 100 mM Tris Buffer, pH 8.0 at 37°C (Prepare 100 ml in deionized water using Trizma Base. Adjust to pH 8.0 at 37°C with 1 M HCl).
B. 100 mM L-Malic Acid (Prepare 10 ml in Reagent A using L(-)Malic Acid, Monosodium Salt).
C. 50 mM β-Nicotinamide Adenine Dinucleotide, Oxidized Form Solution (β-NAD) (Prepare 2 ml in deionized water using β-Nicotinamide Adenine Dinucleotide).
D. 2.0 mM Acetyl Coenzyme A Solution (Acetyl CoA) (Prepare 2.0 ml in Reagent A using Acetyl Coenzyme A, Sodium Salt).
E. Malic Dehydrogenase Enzyme Solution (MDH) (Prepare 1.0 ml in Reagent A containing approximately 560 units/ml Malic Dehydrogenase).
F. Citrate Synthase Enzyme Solution (CS) (Prepare a solution containing 0.2–0.4 units of Citrate Synthase in cold Reagent A).

Procedure

Prepare a reaction cocktail by pipetting (in milliliters) the following reagents into a suitable container:

Reagent A (Buffer)	7.50
Reagent B (L-Malic Acid)	2.00
Reagent C (β-NAD)	1.00
Reagent D (Acetyl CoA)	2.00
Reagent E (MDH)	1.00
Deionized Water	15.50

Mix and adjust to pH 8.0 at 37°C with 1 M HCl or 1 M NaOH, if necessary. Pipette (in milliliters) the following reagents into suitable cuvettes:

	Test	Blank
Reaction Cocktail	2.90	2.90

Mix by inversion and equilibrate to 37°C. Monitor the A340nm until constant using a suitably thermostatted spectrophotometer. Then add:

Reagent F (CS)	0.10	—
Reagent A (Buffer)	—	0.10

Immediately mix by inversion and record the increase in A340nm for approximately 5 minutes. Obtain the ΔA340nm/minute using the maximum linear rate for both the Test and Blank.

Calculations

$$\text{Units / ml enzyme} = \frac{(\Delta A340nm / min\ Test - \Delta A340nm / min\ Blank)\ (3)\ (df)}{(6.22)\ (0.1)}$$

3 = Volume (in milliliters) of the assay
df = Dilution factor
6.22 = Millimolar extinction coefficient of β-NADH at 340nm
0.1 = Volume (in milliliter) of enzyme used

$$\text{Units / mg solid} = \frac{\text{units / ml enzyme}}{\text{mg solid / ml enzyme}}$$

$$\text{Units / mg protein} = \frac{\text{units / ml enzyme}}{\text{mg protein / ml enzyme}}$$

Unit Definition: *One unit will form 1.0 μmole of citrate from oxalacetate and acetyl CoA per minute at pH 8.0 at 37°C.*

Final Assay Concentration

In a 3.0 ml reaction mix, the final concentrations are 46 mM Tris, 6.7 mM L-malic acid, 1.7 mM β-nicotinamide adenine dinucleotide, 0.13 mM acetyl CoA, 56 units malic dehydrogenase, and 0.02–0.04 unit citrate synthase.

Enzymatic Assay of CLOSTRIPAIN (EC 3.4.22.8)

Principle

$$BAEE + H_2O \xrightarrow{\text{Clostripain}} N\text{-}\alpha\text{-Benzoyl-L-Arginine} + \text{Ethanol}$$

Abbreviation Used

BAEE = N-α-Benzoyl-L-Arginine Ethyl Ester

Conditions: T = 25°C, pH = 7.6, A253nm, Light path = 1 cm

Method: Continuous Spectrophotometric Rate Determination

Reagents

A. 10 mM MOPS HCl Buffer with 2.5 mM Dithiothreitol and 1.0 mM Calcium Chloride, pH 7.4 at 25°C (Activation Soln) (Prepare 100 ml in deionized water using MOPS, Sodium Salt, DL-Dithiothreitol and Calcium Chloride, Dihydrate. Adjust the pH to 7.4 at 25°C with 1 M HCl).

B. 25 mM Sodium Phosphate Buffer with 2.5 mM Dithiothreitol and 0.25 mM N-α-Benzoyl-LArginine Ethyl Ester, pH 7.6 at 25°C (Assay Cocktail) (Prepare 30 ml in deionized water using Sodium Phosphate, Monobasic, Anhydrous, DL-Dithiothreitol, and N-α-Benzoyl-L-Arginine Ethyl Ester, Hydrochloride. Adjust the pH to 7.6 at 25°C with 1 M NaOH).

C. Clostripain Enzyme Solution (Approximately 2–3 hours before use, prepare a solution containing 0.5–1.5 unit/ml of Clostripain in Reagent A. Store at room temperature. The enzyme should be activated after this time period).

Procedure

Pipette (in milliliters) the following reagents into suitable quartz cuvettes:

	Test	Blank
Reagent B (Assay Cocktail)	3.00	3.00

Equilibrate to 25°C. Monitor the A253nm until constant, using a suitably thermostatted spectrophotometer. Then add:

Reagent C (Enzyme Soln)	0.05	—
Reagent A (Activation Soln)	—	0.05

Immediately mix by inversion and record the increase in A253nm for approximately 5 minutes. Obtain the ΔA253nm/minute using the maximum linear rate for both the Test and Blank.

Calculations

$$\text{Units / ml enzyme} = \frac{(\Delta A253\text{nm / min Test} - \Delta A253\text{nm / min Blank})\ (3.05)\ (\text{df})}{(1.15)\ (0.05)}$$

3.05 = Total Volume (in milliliters) of assay
df = Dilution factor
1.15 = Millimolar extinction coefficient of N-α-Benzoyl-L-Arginine at 253nm
0.05 = Volume (in milliliters) of enzyme used.

$$\text{Units / mg solid} = \frac{\text{units / ml enzyme}}{\text{mg solid / ml enzyme}}$$

$$\text{Units / mg protein} = \frac{\text{units / ml enzyme}}{\text{mg protein / ml enzyme}}$$

Unit Definition: *One unit will hydrolyze 1.0 µmole of BAEE per minute at pH 7.6 at 25°C in the presence of 2.5 mM DTT.*

Final Assay Concentrations

In a 3.05 ml reaction mix, the final concentrations are 25 mM sodium phosphate, 0.25 mM N-abenzoyl-L-arginine ethyl ester, 0.02 mM calcium chloride, 0.16 mM MOPS, 2.5 mM dithiothreitol, and 0.025–0.075 unit clostripain.

Note

1. The clostripain enzyme solution can also be activated by storing a 0.50–1.5 µm/ml stock solution in Reagent A overnight at 5°C.

Suitability Assay for COLLAGEN

Principle

$$\text{Collagen} + H_2O \xrightarrow{\text{Collagenase}} \text{Peptides}$$

Conditions: T = 37°C, pH = 7.4, A570nm, Light path = 1 cm

Method: Colorimetric

Reagents

A. 50 mM TES Buffer with 0.36 mM Calcium Chloride, pH 7.4 at 37°C (Prepare 1000 ml in deionized water using TES Free Acid, and Calcium Chloride, Dihydrate. Adjust the pH to 7.4 at 37°C with 1 M NaOH).
B. Collagen (Different types of Collagen will produce varying amounts of enzyme activity when used as a substrate for collagenase).
C. Collagenase Enzyme Solution (Immediately before use, prepare a solution containing 0.1 mg/ml Collagenase in Buffer A).
D. Ethylene Glycol Monoethyl Ether.
E. 4% (w/v) Ninhydrin Solution (Prepare 100 ml in Reagent D, using Ninhydrin).
F. 200 mM Citrate Buffer with 0.16% (w/v) Stannous Chloride, pH 5.0 at 25°C (Prepare 100 ml in deionized water using Citric Acid, Free Acid, Anhydrous. Adjust to pH 5.0 at 25°C with 1 M NaOH. Then add the Stannous Chloride, Anhydrous).
G. 50% (v/v) 1-Propanol Solution (Prepare 100 ml in deionized water using 1-Propanol).
H. Ninhydrin Color Reagent (NCR) (Immediately before use, combine equal volumes of Reagent E and Reagent F).
I. 10 mM Hydrochloric Acid Solution (Prepare 50 ml in deionized water using Hydrochloric Acid).
K. 4.0 mM L-Leucine Standard Solution (Std Soln) (Prepare 20 ml in Reagent I using L-Leucine. *Prepare Fresh*).

Procedure

Weigh the following reagent into suitable containers:

	Test	Blank
Reagent B (Collagen)	25.00mg	25.00mg

Then add (in milliliters) the following reagent:

Reagent A (Buffer)	5.00	5.00

Incubate the vials at 37°C until equilibrated. Then add:

Reagent A (Buffer)	—	0.10
Reagent C (Enzyme Solution)	0.10	—

Mix well and incubate at 37°C. Swirl the containers for 10–15 seconds at 1.5 and 3.5 hours. After 5 hours, filter the contents of the containers through a Whatman #54 filter paper or a 0.8 μm syringe filter into clean containers. Use the filtrates for color development.

Color Development

Standard Curve: Prepare a standard curve by pipetting the following reagents (in milliliters) into suitable vials.

	Std 1	Std 2	Std 3	Std 4	Std Blank
Reagent K (Std Soln)	0.05	0.10	0.15	0.20	0.00
Deionized Water	0.15	0.10	0.05	0.00	0.20
Reagent H (NCR)	2.00	2.00	2.00	2.00	2.00

Sample: Pipette (in milliliters) the following reagents into suitable vials:

	Test	Blank
Test Filtrate	0.20	—
Blank Filtrate	—	0.20
Reagent H (NCR)	2.00	2.00

Mix well and place vented caps on each vial. Place the vials in a boiling water bath for 20 minutes. Remove the vials and allow to cool to room temperature. Add 10 ml of Reagent G (50% 1-Propanol) to each vial. Mix well and transfer the vial contents to suitable cuvettes. Determine the absorbance at 570 nm for each of the vials using a suitable spectrophotometer.

Calculations

Standard Curve: ΔA570nm Standard = A570nm Standard–A570nm Standard blank

Prepare a standard curve by plotting the ΔA570nm of the L-Leucine Standard Solution versus micromoles of L-Leucine.

Sample Determination: ΔA570nm Sample = A570nm Test–A570nm Sample blank

Determine the μmoles of L-Leucine equivalents liberated using the Standard curve.

$$\text{Units / ml enzyme} = \frac{(\mu\text{moles of L - Leucine equivalents liberated})\ (5.1)\ (\text{df})}{(0.2)\ (0.1)}$$

df = Dilution factor
5.1 = Total volume (in milliliters) of Assay
0.2 = Total volume (in milliliter) of sample used in Colorimetric Determination
0.1 = Volume (in milliliter) of enzyme used

$$\text{Units / mg solid} = \frac{\text{units / ml enzyme}}{\text{mg solid / ml enzyme}}$$

Unit Definition: *One unit liberates peptides from collagen equivalent in ninhydrin color to 1.0 μmole of leucine in 5 hours at pH 7.4 at 37°C in the presence of calcium ions.*

Specification: Suitable for use as a substrate for Collagenase.

Final Assay Concentration

In a 5.10 ml reaction mix, the final concentrations are 50 mM TES, 0.36 mM calcium chloride, 25 mg collagen and 0.01 mg collagenase.

Enzymatic Assay of COLLAGENASE (EC 3.4.24.3)
Collagen Digestion Assay

Principle

$$\text{Collagen} + H_2O \xrightarrow{\text{Collagenase}} \text{Peptides}$$

Conditions: T = 37°C, pH = 7.4, A570nm, Light path = 1 cm.

Method: Colorimetric

Reagents

A. 50 mM TES Buffer with 0.36 mM Calcium Chloride, pH 7.4 at 37°C (Prepare 1000 ml in deionized water using TES Free Acid, and Calcium Chloride, Dihydrate. Adjust the pH to 7.4 at 37°C with 1 M NaOH).
B. Collagen Type I (Use Collagen, Type I. Different lots of collagen will produce varying amounts of enzyme activity when used as a substrate for collagenase).
C. Collagenase Enzyme Solution (Immediately before use, prepare a solution containing 0.05–0.1 mg/ml Collagenase in Buffer A).
D. Ethylene Glycol Monoethyl Ether (Use Ethylene Glycol Monoethyl Ether).
E. 4% (w/v) Ninhydrin Solution (Prepare 100 ml in Reagent D, using Ninhydrin).
F. 200 mM Citrate Buffer with 0.16% (w/v) Stannous Chloride, pH 5.0 at 25°C (Prepare 100 ml in deionized water using Citric Acid, Free Acid, Anhydrous. Adjust to pH 5.0 at 25°C with 5 M NaOH; then add the Stannous Chloride, Anhydrous).
G. 50% (v/v) Isopropanol Solution (Isopropanol) (Prepare 100 ml in deionized water using Isopropanol, Anhydrous).
H. Ninhydrin Color Reagent (NCR) (Immediately before use, combine equal volumes of Reagent E and Reagent F).

I. 10 mM Hydrochloric Acid Solution (Prepare 50 ml in deionized water using Hydrochloric Acid).
J. 4.0 mM L-Leucine Standard Solution (Std Soln) (Prepare 20 ml in Reagent I using L-Leucine. *Prepare Fresh*).

Procedure

Weigh the following reagent into suitable containers:

	Test	Blank
Reagent B (Collagen)	25.00 mg	25.00 mg

Then add (in milliliters) the following reagent:

Reagent A (Buffer)	5.00	5.00

Incubate the vials at 37°C until equilibrated. Then add:

Reagent A (Buffer)	—	0.10
Reagent C (Enzyme Solution)	0.10	—

Mix well and incubate at 37°C. Swirl the containers for 10–15 seconds at 1.5 and 3.5 hours. After 5 hours, filter the contents of the containers through a Whatman #54 filter paper or a 0.8 in syringe J filter into clean containers. Use the filtrates for color development.

Color Development

Standard Curve: Prepare a standard curve by pipetting the following reagents (in milliliters) into suitable containers.

	Std 1	Std 2	Std 3	Std 4	Std Blank
Reagent K (Std Soln)	0.05	0.10	0.15	0.20	0.00
Deionized Water	0.15	0.10	0.05	0.00	0.20
Reagent H (NCR)	2.00	2.00	2.00	2.00	2.00

Sample: Pipette (in milliliters) the following reagents into suitable containers:

	Test	Blank
Test Filtrate	0.20	—
Blank Filtrate	—	0.20
Reagent H (NCR)	2.00	2.00

Mix well and place vented caps on each container. Place the containers in a boiling water bath for 30 minutes. Remove the containers and allow to cool to room temperature. Add 10 ml of Reagent G (Isopropanol) to each container. Mix well and transfer the container contents to suitable cuvettes. Determine the absorbance at 570 nm for each of the containers using a suitable spectrophotometer.

Calculations

Standard Curve: ΔA570nm Standard = A570nm Standard–A570nm Standard Blank
Prepare a standard curve by plotting the ΔA570nm of the L-Leucine Standard Solution versus micromoles of L-Leucine.

Sample Determination: ΔA570nm Sample = A570nm Test–A570nm Test Blank
Determine the ìmoles of L-Leucine equivalents liberated using the Standard curve.

$$\text{Units / ml enzyme} = \frac{(\text{imoles of L - Leucine equivalents liberated})\ (5.1)\ (\text{df})}{(0.2)\ (0.1)}$$

df = Dilution factor
5.1 = Total volume (in milliliters) of Assay
0.2 = Total volume (in milliliter) of sample used in Colorimetric Determination
0.1 = Volume (in milliliter) of enzyme used

$$\text{Units / mg solid} = \frac{\text{units / ml enzyme}}{\text{mg solid / ml enzyme}}$$

Unit Definition: *One unit liberates peptides from collagen (C-9879) equivalent in ninhydrin color to 1.0 μmole of leucine in 5 hours at pH 7.4 at 37°C in the presence of calcium ions.*

Final Assay Concentration

In a 5.10 ml reaction mix, the final concentrations are 50 mM TES, 0.36 mM calcium chloride, 25 mg collagen and 0.005–0.01 mg collagenase.

Enzymatic Assay of COLLAGENASE using N-(3-(2-FURYL)ACRYLOYL)-LEU-GLY-PRO-ALA (FALGPA) as the Substrate (EC 3.4.24.3)

Principle

$$\text{FALGPA} \xrightarrow{\text{Collagenase}} \text{FAL} + \text{Gly - Pro - Ala}$$

Abbreviations Used

FALGPA = N-(3-[2-Furyl]Acryloyl)-Leu-Gly-Pro-Ala
FAL = N-(3-[2-Furyl]Acryloyl)-Leu

Conditions: T = 25°C, pH 7.5, A345nm 1, Light path = 1 cm

Method: Continuous Spectrophotometric Rate Determination

Reagents

A. 50 mM Tricine Buffer with 10 mM Calcium Chloride and 400 mM Sodium Chloride, pH 7.5 at 25°C (Prepare 500 ml in deionized water using Tricine, Calcium Chloride, Dihydrate, and Sodium Chloride. Adjust to pH 7.5 at 25°C with 1 M NaOH).
B. 1.0 mM N-(3-[2-Furyl]Acryloyl)-Leu-Gly-Pro-Ala Solution (FALGPA) (Prepare 50 ml in Reagent A using N-(3-[2-Furyl]Acryloyl)-Leu-Gly-Pro-Ala. Approximately 30 minutes of stirring is required for this product to dissolve completely. Adjust to pH 7.5 at 25°C with either 1 M NaOH or 1 M HCl).
C. Collagenase Enzyme Solution (Immediately before use, prepare a solution containing 2 units/ml of Collagenase in cold deionized water).

Procedure

Pipette (in milliliters) the following reagents into suitable cuvettes:

	Test	Blank
Reagent B (FALGPA)	2.90	2.90
Deionized Water	—	0.10

Mix by inversion and equilibrate to 25°C. Monitor the A345nm until constant, using a suitably thermostatted spectrophotometer. Then add:

Reagent C (Enzyme Solution)	0.10	—

Immediately mix by inversion and record the decrease in A345nm for approximately 5 minutes. Obtain the ΔA345nm/minute by using the maximum linear rate for both the Test and Blank.

Calculation

$$\text{Units / ml enzyme} = \frac{(\Delta A345nm / \text{min Test} - \Delta A345nm / \text{min Blank})\,(3)\,(df)}{(0.53)\,(0.1)}$$

3 = Total volume (in milliliters) of assay
df = Dilution factor
0.53 = Millimolar extinction coefficient of FALGPA at 345nm
0.1 = Volume (in milliliters) of enzyme used

$$\text{Units / mg solid} = \frac{\text{units / ml enzyme}}{\text{mg solid / ml enzyme}}$$

$$\text{Units / mg protein} = \frac{\text{units / ml enzyme}}{\text{mg protein / ml enzyme}}$$

Unit Definition: *One unit hydrolyzes 1.0 µmole of furylacryloyl-Leu-Gly-Pro-Ala (FALGPA, F-5135) per minute at 25°C at pH 7.5 in the presence of calcium ions.*

Final Assay Concentration

In a 3.00 ml reaction mix, the final concentrations are 48 mM tricine, 9.7 mM calcium chloride,387 mM sodium chloride, 0.97 mM N-(3-[2-furyl]acryloyl)-Leu-Gly-Pro-Ala, and 0.20 unit collagenase.

Note

1. It is critical that the wavelength is set at exactly 345nm.

Enzymatic Assay of COLLAGENASE Pz-Pro-Leu-Gly-Pro-D-Arg as the Substrate (EC 3.4.24.3)

Principle

$$\text{Pz-Pro-Leu-Gly-Pro-D-Arg} + H_2O \xrightarrow{\text{Collagenase}} \text{Pz-Pro-Leu} + \text{Gly-Pro-D-Arg}$$

Abbreviation Used

Pz = 4-Phenylazobenzyloxycarbonyl

Conditions: T = 37°C, pH = 7.1, A320nm, Light path = 1 cm

Method: Spectrophotometric Stop Rate Determination

Reagents

A. 100 mM Tris HCl Buffer, pH 7.1 at 37°C (Prepare 100 ml in deionized water using Trizma Base. Adjust to pH 7.1 at 37°C with 1 M HCl).
B. Methanol (Use Methanol, Absolute).
C. 1.3 mM Pz-Pro-Leu-Gly-Pro-D-Arg Substrate Solution (Prepare by dissolving 10 mg of Pz-Pro-Leu-Gly-Pro-DArg, in 0.2 ml of Reagent B. Bring to a total volume of 10 ml with Reagent A).
D. 100 mM Calcium Chloride Solution ($CaCl_2$) (Prepare 100 ml in deionized water using Calcium Chloride, Dihydrate).
E. 25 mM Citric Acid Solution (Citric Acid) (Prepare 100 ml in deionized water using Citric Acid, Free Acid, Monohydrate).
F. Ethyl Acetate (EtOAc) (Use Ethyl Acetate).
G. Sodium Sulfate (Na_2SO_4) (Use Sodium Sulfate, Anhydrous).
H. 20 mM Calcium Chloride Solution (Enz Dil) (Prepare 100 ml in deionized water using Reagent D).
I. Collagenase Enzyme Solution (Immediately before use, prepare a solution containing 200–500 units/ml of Collagenase in cold Reagent H).

Procedure

Step 1: Pipette (in milliliters) the following reagents into suitable containers:

	Test	Blank
Reagent C (Substrate Solution)	2.00	2.00
Reagent D ($CaCl_2$)	0.40	0.40

Mix by swirling and equilibrate to 37°C. Then add:

Reagent I (Enzyme Solution)	0.10	—
Reagent H (Enz Dil)	—	0.10

Immediately mix by swirling and incubate for exactly 15 minutes at 37°C.

Step 2: Pipette (in milliliters) the following reagents into suitable containers:

	Test	Blank
Test Mixture (from Step 1)	0.50	—
Blank Mixture (from Step 1)	—	0.50
Reagent E (Citric Acid)	1.00	1.00
Reagent F (EtOAc)	5.00	5.00

Mix by shaking for 15 seconds. Remove the ethyl acetate phase (upper layer) from both the Test and Blank and place into separate tubes, each containing 300 mg of Reagent G

(Na_2SO_4). Mix by shaking and filter through Whatman #54 filter paper. Transfer the solutions to suitable cuvettes and record the ΔA320nm for both the Test and Blank using a suitable spectrophotometer.

Calculations

$$\text{Units / ml enzyme} = \frac{(\Delta A320nm\ Test - \Delta A320nm\ Blank)\ (2.5)\ (df)}{(0.042)\ (0.5)\ (0.1)}$$

2.5 = Total volume (in milliliters) of assay in Step 1
df = Dilution factor
0.042 = Extinction coefficient of 0.01 micromole of Pz-Pro-Leu in 5 ml of ethyl acetate
0.5 = Volume (in milliliter) of reaction mixture in Step 1 used in Step 2
0.1 = Volume (in milliliter) of enzyme used in Step 1

$$\text{Units / mg solid} = \frac{\text{units / ml enzyme}}{\text{mg solid / ml enzyme}}$$

Units Definition: *One unit will liberate 0.01 µmole of Pz-Pro-Leu from Pz- Pro-Leu-Gly-Pro-D-Arg (P-8383) in 15 minutes at pH 7.1 at 37°C.*

Final Assay Concentrations

In a 2.50 ml reaction mix, the final concentrations are 1 mM Pz-Pro-Leu-Gly-Pro-D-Arg, 1.6% (v/v) methanol, 78 mM Tris, 17 mM calcium chloride, and 20–50 units collagenase.

Enzymatic Assay of CREATINASE (EC 3.5.3.3)

Principle

$$\text{Creatine} + H_2O \xrightarrow{\text{Creatinase}} \text{Sarcosine} + \text{Urea}$$

$$\text{Urea} + \xrightarrow{\text{DAB}} \text{Yellow Dye Product}$$

Abbreviation Used

DAB = p-Dimethylaminobenzaldehdye

Conditions: T = 37°C, pH = 7.5, A435nm, Light path = 1 cm

Method: Spectrophotometric Stopped Rate Determination

Reagents

A. 50 mM Potassium Phosphate Buffer, pH 7.5 at 37°C (Prepare 100 ml in deionized water using Potassium Phosphate, Monobasic, Anhydrous. Adjust to pH 7.5 at 37°C with 5 M NaOH).
B. 100 mM Creatine Solution (Creatine) (Prepare 10 ml in Reagent A using Creatine, Hydrate. Adjust to pH 7.5 at 37°C, if necessary, with either 1 M HCl or 1 M NaOH).
C. Dimethyl Sulfoxide (Use Dimethyl Sulfoxide).
D. Hydrochloric Acid (Use Hydrochloric Acid).
E. 117 mM p-Dimethylaminobenzaldehyde Solution (DMAB) (Prepare by dissolving 1 g of p-Dimethylaminobenzaldehyde, in 50 ml of Reagent C. Then carefully add 7.5 ml of Reagent D and mix by swirling. Protect from light. *Prepare Fresh*).

F. Creatinase Enzyme Solution (Immediately before use, prepare a solution containing 2–4 units/ml of Creatinase in cold Reagent A).

Procedure

Pipette (in milliliters) the following reagents into suitable cuvettes:

	Test	Blank
Reagent B (Creatine)	0.90	0.90

Equilibrate to 37°C using a suitably thermostatted spectrophotometer. Then add:

Reagent F (Enzyme Solution)	0.10	—

Immediately mix by inversion and incubate at 37°C for exactly 10 minutes. Then add:

Reagent E (DMAB)	2.00	2.00
Reagent F (Enzyme Solution)	—	0.10

Mix by swirling and let stand for 20 minutes at 25°C. Record the A435nm for both the Test and Blank.

Calculations

$$\text{Units / ml enzyme} = \frac{(\text{A435nm Test} - \text{A435nm Blank})\ (3)\ (\text{df})}{(0.321)\ (10)\ (0.1)}$$

3 = Total volume (in milliliters) of assay
df = Dilution factor
0.321 = Millimolar extinction coefficient of the yellow dye
10 = Time (in minutes) of assay as per the Unit Definition
0.1 = Volume (in milliliter) of enzyme used

$$\text{Units / mg solid} = \frac{\text{units / ml enzyme}}{\text{mg solid / ml enzyme}}$$

$$\text{Units / mg protein} = \frac{\text{units / ml enzyme}}{\text{mg protein / ml enzyme}}$$

Unit Definition: *One unit will hydrolyze 1.0 μmole of creatine to urea and sarcosine per minute at pH 7.5 at 37°C.*

Final Assay Concentrations

In a 1.00 ml reaction mix, the final concentrations are 50 mM potassium phosphate, 90 mM creatine, and 0.2–0.4 unit creatinase.

Enzymatic Assay of CREATINE PHOSPHOKINASE (EC 2.7.3.2)

Principle

$$\text{Phosphocreatine} + \text{ADP} \xrightarrow{\text{Creatine Phosphokinase}} \text{Creatine} + \text{ATP}$$

$$\text{D-Glucose} + \text{ATP} \xrightarrow{\text{Hexokinase}} \text{Glucose-6-Phosphate} + \text{ADP}$$

$$\text{Glucose 6-Phosphate} + \beta\text{-NADP} \xrightarrow{\text{G-6-PDH}} \text{6-PG} + \beta\text{-NADPH}$$

Abbreviations Used

ADP = Adenosine 5'-Diphosphate
ATP = Adenosine 5'-Triphosphate
G-6-PDH = Glucose-6-Phosphate Dehydrogenase
β-NADP = β-Nicotinamide Adenine Dinucleotide Phosphate, Oxidized Form
β-NADPH = β-Nicotinamide Adenine Dinucleotide Phosphate, Reduced Form
6-PG = 6-Phospho-D-Gluconate

Conditions: T = 30°C, pH = 7.4, A340nm, Light path = 1 cm

Method: Continuous Spectrophotometric Rate Determination

Reagents

A. 250 mM Glycylglycine Buffer with 0.10% (w/v) Bovine Serum Albumin, pH 7.4 at 30°C. (Prepare 100 ml in deionized water using Glycylglycine, Free Base, and Albumin Bovine Serum. Adjust to pH 7.4 at 30°C with 1 M NaOH).
B. 400 mM Phosphocreatine Solution (Prepare 10 ml in deionized water using Phosphocreatine, Disodium Salt Hydrate).
C. 40 mM Adenosine 5'-Diphosphate Solution (ADP) (Prepare 10 ml in deionized water using Adenosine 5'-Diphosphate, Sodium Salt).
D. 1000 mM D-Glucose Solution (Prepare 10 ml in deionized water using β-D(+)Glucose).
E. 20 mM β-Nicotinamide Adenine Dinucleotide Phosphate Solution (β-NADP) (Dissolve the contents of one 30 mg vial of β-Nicotinamide Adenine Dinucleotide Phosphate, Sodium Salt, in the appropriate volume of deionized water).
F. 300 mM Magnesium Acetate (Prepare 5 ml in deionized water using Magnesium Acetate, Tetrahydrate).
G. Hexokinase Solution (Immediately before use, prepare a solution containing 300 units/ml of Hexokinase, in cold deionized water).
H. Glucose-6-Phosphate Dehydrogenase Enzyme Solution (G-6-PDH) (Immediately before use, prepare a solution containing 10 units/ml of Glucose-6-Phosphate Dehydrogenase, in cold deionized water).
I. Creatine Phosphokinase Enzyme Solution (Creat PPK) (Immediately before use, prepare a solution containing 0.3 unit/ml of Creatine Phosphokinase in cold Reagent A).

Procedure

Prepare a reaction cocktail by pipetting (in milliliters) the following reagents into a suitable container:

Deionized Water	36.00
Reagent A (Buffer)	10.00
Reagent B (Phosphocreatine)	2.00
Reagent C (ADP)	2.00
Reagent D (D-Glucose)	2.00
Reagent E (β-NADP)	1.20
Reagent F (Magnesium Acetate)	0.80

Mix and adjust to pH 7.4 at 30°C with 1 M NaOH. Pipette (in milliliters) the following reagents into suitable cuvettes:

	Test	Blank
Reaction Cocktail	2.70	2.70
Reagent G (Hexokinase)	0.10	0.10
Reagent H (G-6-PDH)	0.10	0.10

Mix by inversion and equilibrate to 30°C. Monitor the A340nm until constant, using a suitably thermostatted spectrophotometer. Then add:

Reagent A (Buffer)	—	0.10
Reagent I (Creat PPK)	0.10	—

Immediately mix by inversion and record the increase in A340nm for approximately 5 minutes. Obtain the ΔA340nm/minute using the maximum linear rate for both the Test and Blank.

Calculations

$$\text{Units / ml enzyme} = \frac{(\Delta A340\text{nm / min Test} - \Delta A340\text{nm / min Blank})\ (3)\ (\text{df})}{(6.22)\ (0.1)}$$

3.0 = Total volume (in milliliters) of assay
df = Dilution factor
6.22 = Millimolar extinction coefficient of β-NADPH at 340nm
0.1 = Volume (in milliliters) of enzyme used

$$\text{Units / mg solid} = \frac{\text{units / ml enzyme}}{\text{mg solid / ml enzyme}}$$

$$\text{Units / mg protein} = \frac{\text{units / ml enzyme}}{\text{mg protein / ml enzyme}}$$

Unit Definition: *One unit will transfer 1.0 μmole of phosphate from phosphocreatine to ADP per minute at pH 7.4 at 30°C.*

Final Assay Concentration

In a 3.00 ml reaction mix, the final concentrations are 50 mM glycylglycine, 0.020% (w/v) Bovine Serum Albumin, 13 mM phosphocreatine, 1.3 mM adenosine 5'-diphosphate, 33 mM D-glucose, 0.40 mM β-nicotine adenine dinucleotide phosphate, oxidized form, 4.0 mM magnesium acetate, 30 units hexokinase, 1.0 unit glucose-6-phosphate dehydrogenase and 0.03 unit creatine phosphokinase.

Note

1. Ammonium sulfate and chloride are strong inhibitors of Creatine Phosphokinase.

Enzymatic Assay of CREATININASE (EC 3.5.2.10)

Principle

Conditions: T = 37°C, pH = 6.5, A525nm, Light path = 1 cm.

Method: Colorimetric

Reagents

A. 300 mM Potassium Phosphate Buffer, pH 6.5 at 37°C (Prepare 100 ml in deionized water using Potassium Phosphate, Monobasic, Anhydrous. Adjust to pH 6.5 at 37°C with Reagent B).
B. 300 mM Potassium Phosphate Solution (Prepare 100 ml in deionized water using Potassium Phosphate, Dibasic, Trihydrate).
C. 10 mM Potassium Phosphate Buffer, pH 8.0 at 37°C (Enzyme Diluent) (Prepare 25 ml in deionized water using Reagent A. Adjust to pH 8.0 at 37°C with Reagent B).
D. 0.1 mM Mercury Chloride and 189 mM Sodium Carbonate Solution (Stop Rgt) (Prepare 100 ml in deionized water using Mercuric Chloride and Sodium Carbonate, Anhydrous).
E. 139 mM α-Naphthol Solution (α-Naphthol) (Prepare 100 ml in Reagent F using α-Naphthol).
F. 95% (v/v) Ethanol (EtOH) (Prepare 100 ml in deionized water using 200 Proof USP Ethyl Alcohol).
G. 300 mM Sodium Hydroxide and 300 mM Sodium Carbonate Solution (Alk Soln) (Prepare 100 ml in deionized water using Sodium Hydroxide, Anhydrous and Sodium Carbonate, Anhydrous).
H. 0.05% (v/v) Diacetyl Solution (Diacet) (Prepare 50 ml in deionized water using Diacetyl).
I. 100 mM Creatinine Substrate Solution (Creatinine) (Prepare 25 ml in deionized water using Creatinine, Free Base, Anhydrous).
J. 4 mM Creatine Standard Solution (Creat Std) (Prepare 10 ml in Reagent C using Creatine, Hydrate).
K. Creatininase Enzyme Solution (Immediately before use, prepare a solution containing 5–10 units/ml of Creatininase in cold Reagent C).

Procedure

Step 1: Pipette (in milliliters) the following reagents into suitable containers:

	Test	Blank
Reagent A (Buffer)	0.10	0.10
Reagent I (Creatinine)	0.80	0.80

Mix by swirling and equilibrate to 37°C. Then add:

Reagent K (Enzyme Soln)	0.10	—

Immediately mix by swirling and incubate at 37°C for exactly 10 minutes. Then add:

Reagent D (Stop Rgt)	2.00	2.00
Reagent K (Enzyme Soln)	—	0.10

Procedure

Step 2: Pipette (in milliliters) the following reagents into suitable containers:

	Test	Blank
Deionized Water	0.90	0.90
Reagent E (α-Naphthol)	0.50	0.50
Reagent G (Alk Soln)	0.50	0.50
Reagent H (Diacet)	0.50	0.50

Mix by swirling and then add:

	Test	Blank
Test Solution	0.10	—
Blank Solution	—	0.10

Mix by swirling and incubate at 25°C for 1 hour. Then add:

	Test	Blank
Deionized Water	2.50	2.50

Mix by swirling. Transfer the Test, Blank, Standards and Standard Blank to suitable cuvettes and read the absorbance at 525nm.

Colour Development

Standard Curve: Prepare a standard curve by pipetting (in milliliters) the following reagents into suitable containers:

	Std 1	Std 2	Std 3	Std 4	Std 5	Std Blank
Deionized Water	0.98	0.96	0.94	0.92	0.90	1.00
Reagent E (α-Naphthol)	0.50	0.50	0.50	0.50	0.50	0.50
Reagent G (Alk Soln)	0.50	0.50	0.50	0.50	0.50	0.50
Reagent H (Diacet)	0.50	0.50	0.50	0.50	0.50	0.50

Mix by swirling and then add:

	Std 1	Std 2	Std 3	Std 4	Std 5	Std Blank
Reagent J (Creat Std)	0.02	0.04	0.06	0.08	0.10	—

Mix by swirling and incubate at 25°C for one hour. Then add:

	Std 1	Std 2	Std 3	Std 4	Std 5	Std Blank
Deionized Water	2.50	2.50	2.50	2.50	2.50	2.50

Mix by swirling. Transfer the Standards and Standard Blank to suitable cuvettes and read the absorbance at 525nm.

Calculations

Standard Curve: ΔA525nm Standard = A525nm Standard–A525nm Standard Blank

Plot the ΔA525nm Standard vs μmoles creatine.

Sample Determination: ΔA525nm Sample = A525nm Test–A525nm Sample Blank

Determine the μmoles of creatinine hydrolyzed using the Standard Curve.

$$\text{Units / ml enzyme} = \frac{(\mu\text{moles Creatinine hydrolyzed})\ (3)\ (\text{df})}{(10)\ (0.1)\ (0.1)}$$

3 = Volume (in milliliters) of stopped reaction
df = Dilution factor
10 = Time of assay (in minutes) as per the Unit Definition
0.1 = Volume (in milliliters) of stopped reaction used in Colorimeteric Determination
0.1 = Volume (in milliliters) of enzyme used

$$\text{Units / mg solid} = \frac{\text{units / ml enzyme}}{\text{mg solid / ml enzyme}}$$

$$\text{Units / mg protein} = \frac{\text{units / ml enzyme}}{\text{mg protein / ml enzyme}}$$

Unit Definition: *One unit will hydrolyze 1.0 μmole of creatinine to creatine per minute at pH 6.5 at 37°C.*

Final Assay Concentration

In a 1.00 ml reaction mix, the final concentrations are 31 mM potassium phosphate, 80 mM creatinine, and 0.5–1.0 unit creatininase.

Enzymatic Assay of CREATININASE (EC 3.5.2.10) From *Pseudomonas species*

Principle

$$\text{Creatinine} + H_2O \xrightarrow{\text{Creatininase}} \text{Creatine}$$

$$\text{Creatine} + \text{ATP} \xrightarrow{\text{CPK}} \text{Creatine-P} + \text{ADP}$$

$$\text{ADP} + \text{PEP} \xrightarrow{\text{PK}} \text{ATP} + \text{Pyruvate}$$

$$\text{Pyruvate} + \beta\text{-NADH} \xrightarrow{\text{LDH}} \text{L-Lactate} + \beta\text{-NAD}$$

Abbreviations Used

ATP = Adenosine 5'-Triphosphate
CPK = Creatine Phosphokinase
Creatine-P = Phosphocreatine
ADP = Adenosine 5'-Diphosphate
PEP = Phospho(enol) pyruvate
PK = Pyruvate Kinase
β-NADH = β-Nicotinamide Adenine Dinucleotide, Reduced Form
LDH = Lactic Dehydrogenase
β-NAD = β-Nicotinamide Adenine Dinucleotide, Oxidized Form

Conditions: T = 25°C, pH = 8.0, A340nm, Light path = 1 cm

Method: Continuous Spectrophotometric Rate Determination

Reagents

A. 100 mM Glycylglycine Buffer, pH 8.0 at 25°C (Prepare 100 ml in deionized water using Gly-Gly, Free Base. Adjust to pH 8.0 at 25°C with 1 M NaOH).

B. 12 mM β-Nicotinamide Adenine Dinucleotide, Reduced Form, Solution (β-NADH) (Dissolve the contents of one 10 mg vial of β-Nicotinamide Adenine Dinucleotide, Reduced Form, Disodium Salt in the appropriate volume of Reagent A. *Prepare Fresh*).

C. 16 mM Adenosine 5'-Triphosphate Solution (ATP) (Prepare 5 ml in Reagent A using Adenosine 5'-Triphosphate, Disodium Salt).

D. 20 mM Phospho(enol) pyruvate Solution (PEP) (Prepare 5 ml in Reagent A using Phospho(enol) pyruvate, Tri (cyclohexylammonium) Salt).

E. 440 mM Creatinine Solution (Substrate) (Prepare 5 ml in deionized water using Creatinine, Free Base, Anhydrous).

F. 100 mM Magnesium Chloride Solution ($MgCl_2$) (Prepare 5 ml in deionized water using Magnesium Chloride, Hexahydrate).

G. PK/LDH Mixed Enzymes (PK/LDH) (Use PK/LDH Enzymes Suspension).

H. 100 mM Glycylglycine Buffer with 200 mM Potassium Chloride, pH 8.0 at 25°C (CPK Enz Diluent) (Prepare 25 ml in deionized water using Gly-Gly, Free Base and Potassium Chloride. Adjust to pH 8.0 at 25°C with 1 M NaOH).

I. Creatine Phosphokinase Enzyme Solution (CPK) Immediately before use, prepare a solution containing 900 units/ml of Creatine Phosphokinase in Reagent H).

J. Creatininase Enzyme Solution (Creatininase) (Immediately before use, prepare a solution containing 0.30–0.50 unit/ml of Creatininase in cold Reagent A).

Procedure

Pipette (in milliliters) the following reagents into suitable cuvettes:

	Test	Blank
Reagent A (Buffer)	2.30	2.30
Reagent B (β-NADH)	0.10	0.10
Reagent C (ATP)	0.20	0.20
Reagent D (PEP)	0.10	0.10
Reagent E (Substrate)	0.20	0.20
Reagent F ($MgCl_2$)	0.10	0.10
Reagent G (PK/LDH)	0.04	0.04
Reagent I (CPK)	0.10	0.10

Mix by inversion and equilibrate to 25°C. Monitor the A340nm until constant, using a suitably thermostatted spectrophotometer. Then add:

Reagent J (Creatininase)	0.05	—
Reagent A (Buffer)	—	0.05

Immediately mix by inversion and record the decrease in the A340nm for approximately 5 minutes. Obtain the ΔA340nm/minute using the maximum linear rate for both the Test and Blank.

Calculation

$$\text{Units / ml enzyme} = \frac{(\Delta A340nm / min\ Test - \Delta A340nm / min\ Blank)\ (3.19)\ (df)}{(6.22)\ (0.05)}$$

3.19 = Total Volume (in milliliters) of assay
df = Dilution factor
6.22 = Millimolar extinction coefficient of β-NADH at 340nm
0.05 = Volume (in milliliter) of enzyme used

$$\text{Units / mg solid} = \frac{\text{units / ml enzyme}}{\text{mg solid / ml enzyme}}$$

$$\text{Units / mg protein} = \frac{\text{units / ml enzyme}}{\text{mg protein / ml enzyme}}$$

Unit Definition: *One unit will hydrolyze 1.0 µmole of creatinine to creatine per minute at pH 8.0 at 25°C.*

Final Assay Concentration

In a 3.19 ml reaction mix, the final concentrations are 88 mM glycylglycine, 0.38 mM β-nicotinamide adeninedinucleotide, 1.0 mM adenosine 5'-triphosphate, 0.63 mM phospho(enol) pyruvate, 28 mM creatinine, 3.1 mM magnesium chloride, 28 units pyruvate kinase, 40 units lactic dehydrogenase, 90 units creatine phosphokinase, 6.1 mM potassium chloride, and 0.015–0.025 unit creatininase.

Notes

1. Contains not less than 700 Pyruvate Kinase units and 1000 Lactic Dehydrogenase units per ml.
2. Pyruvate Kinase Unit Definition: One unit will convert 1.0 µmole of phospho(enol) pyruvate to pyruvate per minute at pH 7.6 at 37°C.
3. L-Lactic Dehydrogenase Unit Definition: One unit will reduce 1.0 µmole of pyruvate to L-lactate per minute at pH 7.5 at 37°C.
4. Creatine Phosphokinase Unit Definition: One unit will transfer 1.0 µmole of phosphate from phosphocreatine to ADP per minute at pH 7.4 at 30°C.

Enzymatic Assay of CREATININE DEIMINASE (EC 3.5.4.21)

Principle

$$\text{Creatinine} + H_2O \xrightarrow{\text{Creatinine Deiminase}} \text{N - Methylhydantoin} + NH_3$$

$$NH_4^+ + \alpha\text{ - Ketoglutarate} + \beta\text{ - NADPH GDH} \rightarrow \text{Glutamate} + H_2O + \beta\text{ - NADP}$$

Abbreviations Used

β-NADPH = β-Nicotinamide Adenine Dinucleotide Phosphate, Reduced Form
GDH = L-Glutamic Dehydrogenase (NADP)
β-NADP = β-Nicotinamide Adenine Dinucleotide Phosphate, Oxidized Form

Conditions: T = 37°C, pH = 7.5, A340nm, Light path = 1 cm

Method: Continuous Spectrophotometric Rate Determination

Reagents

A. 50 mM Potassium Phosphate Buffer, pH 7.5 at 37°C (Prepare 100 ml in deionized water using Potassium Phosphate, Monobasic, Anhydrous. Adjust to pH 7.5 at 37°C with 1 M KOH).

B. 50 mM Creatinine Solution (Creatinine) (Prepare 25 ml in Reagent A using Creatinine, Free Base, Anhydrous).

C. 3.0 mM β-Nicotinamide Adenine Dinucleotide Phosphate, Reduced Form Solution (β-NADPH) (Dissolve the contents of one 5 mg vial of β-Nicotinamide Adenine Dinucleotide Phosphate, Reduced Form, Tetrasodium Salt, in the appropriate volume of Reagent A. *Prepare Fresh*).

D. 10.0 mM α-Ketoglutarate Solution (α-KG) (Prepare 3 ml in Reagent A using α-Ketoglutaric Acid, Free Acid. *Prepare Fresh*).

E. L-Glutamic Dehydrogenase (NADP) Enzyme Solution (GDH) (Immediately before use, prepare a solution containing 1000 units/ml of L-Glutamic Dehydrogenase (NADP), in cold deionized water).

F. Creatinine Deiminase Enzyme Solution (Creat Deiminase) (Immediately before use, prepare a solution containing 0.06–0.12 unit/ml of Creatinine Deiminase in cold Reagent A).

Procedure

Pipette (in milliliters) the following reagents into suitable cuvettes:

	Test	Blank
Reagent B (Creatinine)	2.40	2.40
Reagent C (β-NADPH)	0.30	0.30
Reagent D (α-KG)	0.30	0.30
Reagent E (GDH)	0.05	0.05

Mix by inversion and equilibrate to 37°C. Monitor the A340nm until constant, using a suitably thermostatted spectrophotometer. Then add:

Reagent F (Creat Deiminase)	0.10	—
Reagent A (Buffer)	—	0.10

Immediately mix by inversion and record the decrease in A340nm for approximately 5 minutes. Obtain the ΔA340nm/minute using the maximum linear rate for both the Test and Blank.

Calculations

$$\text{Units / ml enzyme} = \frac{(\Delta A340\text{nm / min Test} - \Delta A340\text{nm / min Blank})\ (3.15)\ (\text{df})}{(6.22)\ (0.1)}$$

3.15 = Total volume (in milliliters) of assay
df = Dilution factor
6.22 = Millimolar extinction coefficient of β-NADPH at 340nm
0.1 = Volume (in milliliter) of enzyme used

$$\text{Units / mg solid} = \frac{\text{units / ml enzyme}}{\text{mg solid / ml enzyme}}$$

$$\text{Units / mg protein} = \frac{\text{units / ml enzyme}}{\text{mg protein / ml enzyme}}$$

Unit Definition: *One unit will hydrolyze 1.0 µmole of creatinine to N-methylhydantoin and NH_3 per minute at pH 7.5 at 37°C in a coupled system with L-glutamic dehydrogenase.*

Final Assay Concentrations

In a 3.15 ml reaction mix, the final concentrations are 49 mM potassium phosphate, 38 mM creatinine, 0.29 mM β-nicotinamide adenine dinucleotide phosphate, reduced form, 0.95 mM α-ketoglutaric acid, 50 units Lglutamic dehydrogenase (NADP), and 0.006–0.012 unit creatinine deiminase.

Note

1. L-Glutamic Acid Dehydrogenase (NADP) Unit Definition: One unit will reduce 1.0 µmole of α-ketoglutarate to L-glutamate per minute at pH 8.3 at 30°C in the presence of ammonium ions and NADPH.

Enzymatic Assay of CROTONASE (EC 4.2.1.17)

Principle

$$\text{Trans-Crotonoyl CoA} + H_2O \xrightarrow{\text{Crotonase}} \text{Hydroxybutyryl CoA}$$

Abbreviation Used

CoA = Coenzyme A

Conditions: T = 25°C, pH = 7.5, A263nm, Light path = 1 cm

Method: Continuous Spectrophotometric Rate Determination

Reagents

A. 1000 mM Tris HCl, pH 7.5 at 25°C (Prepare 100 ml in deionized water using Trizma Base. Adjust to pH 7.5 with 1.0 N HCl).
B. 0.1% w/v Albumin, Chicken Egg (Prepare 50 ml in deionized water using Albumin, Chicken Egg).
C. 3.2 mM Crotonoyl CoA, Lithium Salt (Prepare 10 ml in deionized water using Crotonoyl CoA, Lithium).
D. 100 mM Ethylenediaminetetraacetic Acid, Dipotassium (Prepare 50 mL in deionized water).
E. Assay Buffer (Prepare 3 ml using 1 ml of Reagent A, 1 ml of Reagent B, 0.15 ml of Reagent D, and 0.85 ml of deionized water).
F. Crotonase Enzyme Solution (Immediately before use, prepare a solution containin 0.33 units per ml in ice cold deionized water).

Procedure

Pipette (in milliliters) the following reagents into suitable cuvettes:

	Test	Blank
Reagent E (Assay Buffer)	0.4	0.4
Deionized Water	2.4	2.4
Reagent C (Crotonoyl CoA, Lithium)	0.1	0.1

Mix by inversion and incubate to 25°C using a suitably thermostatted spectrophotometer (The absorbance should be between 0.5 and 0.7). Monitor the A263 until constant then add:

Reagent F (Enzyme Solution)	0.1	—
Deionized water	—	0.1

Immediately mix by inversion and record the decrease in A263. The decrease in A263 should not exceed 0.06 A263/min. Obtain the ΔA263/minute using the maximum linear rate for both the Test and Blank.

Calculations

$$\text{Units / mg enzyme} = \frac{(\Delta A263\text{nm / min Test} - \Delta A263\text{nm / min Blank})}{(6.7)\text{ (mg enzyme / ml RM)}}$$

6.7 = Millimolar extinction coefficient of Crotonoyl CoA at 263nm
RM = Reaction Mix

Unit Definition: *One unit will hydrate 1.0 μmole of crotonoyl coenzyme A to hydroxybutyryl coenzyme A per minute at 7.5 at 25°C.*

Final Assay Concentration

In a 3 ml reaction mix, the final concentrations are 44 mM Tris HCl, 0.0044% Albumin, Chicken egg, 0.67 mM EDTA, dipotassium, and 0.107 mM Crotonoyl CoA, Lithium.

Enzymatic Assay of CUCUMISIN

Principle

$$\text{Casein} + H_2O \xrightarrow{\text{Protease}} \text{Amino Acids}$$

Conditions: T = 37°C, pH = 10.0, A275nm, Light path = 1 cm

Method: Spectrophotometric Stop Rate Determination

Reagents

A. 100 mM Sodium Tetraborate Solution (Prepare 10 ml in deionized water using Borax).
B. 0.6% (w/v) Casein with 10.0 mM Sodium Tetraborate, pH 11.0 at 37°C (Casein). (Prepare by dissolving 600 mg of Casein in 4 ml of 0.1 M NaOH in a hot water bath

(60°C). Cool to room temperature and then add 10 ml of Reagent A and 80 ml of deionized water. Adjust to pH 10.0 at 37°C with 1 M NaOH and bring to a volume of 100 ml with deionized water).

C. 110 mM Trichloroacetic Acid with 220 mM Sodium Acetate and 330 mM Acetic Acid (TCA) (Prepare 40 ml in deionized water using Trichloroacetic Acid, 6.1 N, approximately 100% (w/v), Sodium Acetate Trihydrate, and Acetic Acid, Glacial).

D. 10.0 mM Sodium Tetraborate Buffer, pH 10.0 at 37°C (Enz Dil I) Prepare 10 ml in deionized water using Borax. Ajust to pH 10.0 at 37°C with 1 M NaOH. Adjust to pH 10.0 at 37°C with 1 M NaOH).

E. 2.0 mM Calcium Acetate Solution (Enz Dil II) (Prepare 10 ml in deionized water using Calcium Acetate).

F. Cucumisin Enzyme Solution (Prepare by dissolving 1 mg of Cucumisin in 2 ml of ice-cold Reagent D. Immediately before use, dilute to 0.20–0.40 unit/ml of Cucumisin with Reagent E).

Procedure

Pipette (in milliliters) the following reagents into suitable containers:

	Test	Blank
Reagent B (Casein)	3.00	3.00

Equilibrate to 37°C. Then add:

Reagent F (Enzyme Solution)	0.50	—

Immediately mix by swirling and incubate at 37°C for exactly 10 minutes. Then add:

Reagent C (TCA)	3.20	3.20
Reagent F (Enzyme Solution)	—	0.50

Mix by swirling and incubate at 37°C for 20 minutes. Filter the solutions through Whatman #50 filter paper and transfer the solutions to suitable cuvettes and record the A275nm for both the Test and Blank using a suitable spectrophotometer.

Calculations

$$\text{Units / ml enzyme} = \frac{(\text{A275nm Test} - \text{A275nm Blank})\ (6.7)\ (\text{df})}{(10)\ (1.34)\ (0.5)}$$

6.7 = Total volume (in milliliters) of assay
df = Dilution factor
10 = Time (in minutes) of the assay per the Unit Definition
1.34 = Millimolar extinction coefficient of tyrosine at 275nm under the conditions of this assay
0.5 = Volume (in milliliter) of enzyme used

$$\text{Units / mg solid} = \frac{\text{units / ml enzyme}}{\text{mg solid / ml enzyme}}$$

$$\text{Units / mg protein} = \frac{\text{units / ml enzyme}}{\text{mg protein / ml enzyme}}$$

Unit Definition: *One unit will hydrolyze casein to produce peptides equivalent to 1.0 μmole of tyrosine per minute at pH 10.0 at 37°C.*

Final Assay Concentrations

In a 3.50 ml reaction mix, the final concentrations are 0.5% (w/v) casein, 9 mM sodium tetraborate, 0.3 mM calcium acetate, and 0.10–0.2 unit cucumisin.

Enzymatic Assay of CYTOCHROME-c REDUCTASE (EC 1.6.99.3)

Principle

$$\underset{\text{(oxidized)}}{\text{Cytochrome-c}} + \beta\text{-NADH} \xrightarrow{\text{Cytochrome-c}} \underset{\text{(reduced)}}{\text{Cytochrome-c}} + \beta\text{-NAD Reductase}$$

Abbreviations Used

β-NADH = β-Nicotinamide Adenine Dinucleotide, Reduced Form
β-NAD = β-Nicotinamide Adenine Dinucleotide, Oxidized Form

Conditions: T = 25°C, pH.= 8.5, A550nm, Light path = 1 cm

Method: Continuous Spectrophotometric Rate Determination

Reagents

A. 300 mM Glycylglycine Buffer, pH 8.5 at 25°C (Gly Gly) (Prepare 100 ml in deionized water using Gly-Gly, Hydrochloride. Adjust pH to 8.5 at 25°C with 1 M NH_4OH).
B. 1% (w/v) Cytochrome-c Solution (Prepare 10 ml in deionized water using Cytochrome-c from Horse Heart).
C. 7.05 mM β-Nicotinamide Adenine Dinucleotide, Reduced Form (β-NADH) (Dissolve the contents of one vial of β-Nicotinamide Adenine Dinucleotide, Reduced Form, Disodium Salt, in 1 ml of Reagent A).
D. 20 mM Potassium Bicarbonate Buffer, pH 8.5 at 25°C ($KHCO_3$) (Prepare 100 ml in deionized water using Potassium Bicarbonate. Adjust pH 8.5 at 25°C with 1 M HCl).
E. Cytochrome c Reductase Enzyme Solution (CCR) (Immediately before use, prepare a solution containing 0.05–0.25 units/ml of Cytochrome-c Reductase in cold Reagent D).

Procedure

Pipette (in milliliters) the following reagents into suitable cuvettes:

	Test	Blank
Deionized Water	2.50	2.50
Reagent A (Gly Gly)	0.20	0.20
Reagent B (Cytochrome-c)	0.10	0.10
Reagent C (β-NADH)	0.10	0.10

Mix by inversion and equilibrate to 25°C. Monitor the A550nm until constant, using a suitably thermostatted spectrophotometer. Then add:

	Test	Blank
Reagent E (Enzyme Solution)	0.10	—
Reagent D ($KHCO_3$)	—	0.10

Immediately mix by inversion and record the increase in A550nm for approximately 5 minutes. Obtain the ΔA550nm / minute using the maximum linear rate for both the Test and Blank.

Calculations

$$\text{Units / mg enzyme} = \frac{(\Delta A550nm / \text{min Test} - \Delta A550nm / \text{min Blank})}{(21.0)\ (\text{mg enzyme / ml RM})}$$

21.0 = Δ millimolar extinction coefficient between oxidized and reduced cytochrome-c at pH 8.5
RM = Reaction Mix

Unit Definition: *One unit will reduce 1.0 µmole of oxidized cytochrome-c per minute at pH 8.5 at 25°C.*

Final Assay Concentration

In a 3 ml reaction mix, the final concentrations are 30 mM glycylglycine, 0.033% cytochrome-c, 0.24 mM β-NADH and 0.005–0.025 units of cytochrome c reductase.

Enzymatic Assay of CYTOCHROME OXIDASE (EC 1.9.3.1)

Principle

$$\text{Cytochrome-c (reduced)} + 2H^+ + \tfrac{1}{2}O_2 \xrightarrow{\text{Cytochrome Oxidase}} \text{Cytochrome-c (oxidized)} + H_2O$$

Conditions: T = 37°C, pH = 7.0, A550nm, Light path = 1 cm

Method: Continuous Spectrophotometric Rate Determination

Reagents

A. 100 mM Potassium Phosphate Buffer, pH 7.0 at 37°C (Prepare 200 ml in deionized water using Potassium Phosphate, Monobasic, Anhydrous. Adjust to pH 7.0 at 37°C with 1 M KOH).

B. 10 mM Potassium Phosphate Buffer, pH 7.0 at 37°C (Prepare 2 liters in deionized water using Reagent A. Adjust to pH 7.0 at 37°C with 1 M KOH or 1 M HCl).

C. 1.0% (w/v) Cytochrome-c, Reduced, Solution (Prepare 10 ml by dissolving 100 mg of Cytochrome-c, in 8 ml of Reagent B. Reduce the Cytochrome-c by adding 3–5 mg of L-Ascorbic Acid, Sodium Salt. The excess ascorbate is removed by dialyzing against Reagent B for 18–24 hours at 0–4°C with three changes of buffer. Remove from dialysis and bring up to a final volume of 10 ml with Reagent B. This Cytochrome-c remains reduced for several months if stored at 0–4°C).

D. 100 mM Potassium Ferricyanide Solution (Prepare 10 ml in Reagent A using Potassium Ferricyanide. *Handle with Extreme Caution*).

E. 250 mM Sucrose Solution with 1% (v/v) Tween 80 (Prepare 50 ml in Reagent A using Sucrose, and Polyoxyethylenesorbitan Monooleate, Tween 80).

F. Cytochrome Oxidase Enzyme Solution (Immediately before use, prepare a solution containing 0.1 mg solid/ml (0.02 mg protein/ml) of Cytochrome Oxidase in cold Reagent E).

Procedure

Pipette (in milliliters) the following reagents into suitable cuvettes:

	Test	Blank
Reagent B (Buffer)	2.70	2.70
Reagent C (Cytochrome-c, Reduced)	0.20	0.20
Reagent D (Potassium Ferricyanide)	—	0.10

Mix by inversion and equilibrate to 37°C. Then add:

Reagent F (Enzyme Solution)	0.10	—

Immediately mix by inversion and monitor the decrease in A550nm for 10 minutes, recording the A550nm values at one minute intervals.

Calculations

$$E = Et–Ef$$

Et = the A550nm value at the time points 0 through 10 minutes for the test
Ef = the A550nm value at the 10 minute time point for the blank

Plot in(E) *vs* Time. Determine the slope (M). The activity is calculated using the following equation:

$$\text{Units / mg enzyme} = \frac{(M)}{(\text{mg solid / RM})}$$

M = Slope of ln (DE) versus time
RM = Reaction Mix

Unit Definition: *One unit will oxidize 1.0 µmole of reduced cytochrome-c to oxidized cytochrome-c (using 0.5 µmole of oxygen) per minute at pH 7.0 at 37°C.*

Final Assay Concentrations

In a 3.00 ml reaction mix, the final concentrations are 13 mM potassium phosphate, 0.07% (w/v) cytochrome-c, 8.3 mM sucrose, 0.03% (v/v) Tween 80, and 0.01 mg solid cytochrome oxidase.

Note

1. Potassium Ferricyanide is included in the blank only in order to completely oxidize the reduced cytochrome-c.

Enzymatic Assay of CYTOCHROME P450 REDUCTASE (EC 1.6.2.4)

Principle

$$\text{NADPH} + 2\ \text{Ferricytochrome} \xrightarrow{\text{Cytochrome P450 Reductase}} \text{NADP} + 2\ \text{Ferrocytochrome}$$

Conditions: T = 30°C, pH = 7.7, A550nm, Light path = 1 cm

Method: Continuous Spectrophotometric Rate Determination

Reagents

A. 1 M Potassium Phosphate Buffer, pH 7.7 at 30°C (Prepare 10 ml in deionized water using Potassium Phosphate, Monobasic, Anhydrous. Adjust to pH 7.7 at 30°C with 1 M KOH).
B. 100 mM Ethylenediaminetetraacetic Acid (Prepare 10 ml in deionized water using Ethylenediaminetetraacetic Acid, Disodium Salt, Dihydrate).
C. 300 mM Potassium Phosphate Buffer with 0.1 mM Ethylenediaminetetraacetic Acid, pH 7.7 at 30°C (Working Buffer) (Prepare 20 ml by combining 6 ml of Reagent A and 0.02 ml of Reagent B. Dilute to 20 ml with deionized water).
D. 0.036 mM Cytochrome-c (Cyt-c) (Prepare 10 ml in Reagent C using Cytochrome-c).
E. 1 mM Nicotinamide Adenine Dinucleotide Phosphate, Reduced Form (NADPH) (Prepare 5 ml in deionized water using Nicotinamide Adenine Dinucleotide Phosphate, Reduced Form, Disodium Salt).
F. 0.05% (w/v) Bovine Serum Albumin (Enzyme Diluent) (Prepared 5 ml in Reagent C using Albumin, Bovine).
G. Cytochrome P450 Reductase Enzyme Solution (Immediately before use, prepare a solution containing 0.1–0.25 unit/ml of Cytochrome P450 Reductase in cold Reagent F).

Procedure

Pipette (in milliliters) the following reagents into suitable cuvettes:

	Test	Blank
Reagent D (Cyt-c)	0.92	0.92
Reagent G (Enz Soln)	0.08	—
Reagent F (Enz Dil)	—	0.08

Equilibrate to 30°C. Then add:

Reagent E (NADPH)	0.10	0.10

Immediately mix by inversion and record the increase in A550nm for 2 minutes. Obtain the ΔA550nm/minute using the maximum linear rate for both the Test and Blank.

Calculations

$$\text{Units / ml enzyme} = \frac{(\Delta A550nm / min\ Test - \Delta A550nm / min\ Blank)\ (1.1)\ (df)}{(21)\ (0.08)}$$

1.1 = Total volume (in milliliters) of assay
df = Dilution factor
21 = Difference in the millimolar extinction coefficient between oxidized and reduced cytochrome-c.
0.08 = Volume (in milliliter) of enzyme used

$$\text{Units / mg protein} = \frac{\text{units / ml enzyme}}{\text{mg protein / ml enzyme}}$$

Unit Definition: *One unit will catalyze the reduction of 1.0 μmole of cytochrome-c by NADPH per minute at pH 7.7 at 30°C.*

Final Assay Concentration

In a 1.10 ml reaction mix, the final concentrations are 273 mM potassium phosphate, 0.09 mM Ethylenediaminetetraacetic Acid, 0.03 mM cytochrome c, 0.09 mM nicotinamide adenine dinucleotide phosphate, reduced form, 0.004% (w/v) bovine serum albumin, and 0.008–0.02 unit cytochrome P450 reductase.

Enzymatic Assay of CYTOHELICASE (Laminarinase Activity)

Principle

$$\text{Laminarin} + H_2O \xrightarrow{\text{Cytohelicase}} \text{Reducing Sugar (measured as glucose)}$$

Conditions: T = 50°C, pH = 4.6, A540nm, Light path = 1 cm

Method: Colorimetric

Reagents

A. 50 mM Citrate Buffer, pH 4.6 at 50°C (Prepare 50 ml in deionized water using Citric Acid, Free Acid, Anhydrous. Adjust to pH 4.6 at 50°C with 1 M NaOH).

B. 0.5% (w/v) Laminarin Substrate Solution (Laminarin) (Prepare 20 ml in Reagent A using Laminarin).

C. Cytohelicase Enzyme Solution (Immediately before use, prepare a solution containing 0.10–0.20 mg solid/ml of Cytohelicase in cold deionized water).

D. 15 mM Copper Sulfate, 1.3 M Sodium Sulfate, 226 mM Sodium Carbonate, 190 mM Sodium Bicarbonate and 43 mM Potassium Tartrate Solution (Copper Soln) (Prepare 1 liter in deionized water using Cupric Sulfate Pentahydrate, Sodium Bicarbonate, Sodium Sulfate, Anhydrous, Sodium Carbonate, Anhydrous, and Sodium Potassium Tartrate Tetrahydrate).

E. 40 mM Molybdic Acid, 19 mM Arsenic Acid and 756 mM Sulfuric Acid Solution (Ars-Mol) (Prepare 1 liter in deionized water using Molybdic Acid, Ammonium Salt Tetrahydrate, Arsenic Acid, Sodium Salt, and Sulfuric Acid).

F. 1 mg/ml Sigma Glucose Standard Solution (Glucose Std).

Procedure

Pipette (in milliliters) the following reagents into suitable test tubes:

	Test	Blank
Deionized Water	—	0.50
Reagent B (Laminarin)	0.50	0.50

Mix by swirling and equilibrate to 50°C. Then add:

Reagent C (Enzyme Solution)	0.50	—

Mix by swirling and incubate at 50°C for exactly 60 minutes. Then add:

	Test	Test Blank	Std 1	Std 2	Std 3	Std 4	Std Blank
Deionized Water	—	—	0.90	0.80	0.70	0.60	1.00
Reagent F (Glucose Std)	—	—	0.10	0.20	0.30	0.40	—
Reagent D (Copper Soln)	1.00	1.00	1.00	1.00	1.00	1.00	1.00

Immediately mix by swirling. Place a marble over the top of the tubes and transfer the tubes to a boiling water bath. Incubate for 10 minutes. Remove from the boiling water bath and allow to cool to room temperature. Then add:

Reagent E (Ars-Mol)	1.00	1.00	1.00	1.00	1.00	1.00	1.00

Shake or vortex the tubes until the foaming stops and any precipitate present is dissolved. Then add:

Deionized Water	1.00	1.00	1.00	1.00	1.00	1.00	1.00

Mix by swirling and transfer the solutions to suitable cuvettes. Obtain the A540nm for Test, Blank and Standards, using a suitable spectrophotometer.

Calculations

Standard Curve: ΔA540nm Std = A540nm Std–A540nm Std Blank Prepare a standard curve by plotting the ΔA540nm of the Standard vs the milligrams of glucose.

Sample Determination: ΔA540nm Sample = A540nm Test–A540nm Test Blank

Determine the milligrams of glucose liberated using the Standard Curve.

$$\text{Units / ml enzyme} = \frac{\text{(milligrams of glucose liberated) (df)}}{(0.5)\,(0.5)}$$

df = Dilution factor
0.5 = Milligram of glucose liberated per unit as per the Unit Definition
0.5 = Volume (in milliliter) of enzyme used

$$\text{Units / mg solid} = \frac{\text{units / ml enzyme}}{\text{mg solid / ml enzyme}}$$

$$\text{Units / mg protein} = \frac{\text{units / ml enzyme}}{\text{mg protein / ml enzyme}}$$

Unit Definition: *One unit will liberate 0.5 milligram of glucose per hour at pH 4.6 at 50°C.*

Final Assay Concentrations

In a 1.00 ml reaction mix, the final concentrations are 25 mM citric acid, 0.3% (w/v) laminarin and 0.05–0.1 mg solid cytohelicase.

Notes

1. Sodium Sulfate, Sodium Carbonate, and Sodium Potassium Tartrate are dissolved in approximately 50 ml of deionized water. Cupric Sulfate is dissolved in approximately 100 ml of deionized water and is slowly added to the above solution to avoid precipitation. Sodium Bicarbonate is dissolved

first in deionized water and then added to the above solution. Dilute the solution to 1 liter. If a precipitate forms, it should be removed by filtration prior to use. Store in an amber bottle and avoid exposure to direct sunlight. Store at room temperature.

2. Molydic Acid is dissolved in approximately 300 ml of deionized water. Add Sulfuric Acid slowly. Caution, this is an exothermic reaction! Arsenic Acid is dissolved in approximately 300 ml of deionized water and is added to the above solution. The solution is diluted to a total volume of 1 liter and incubated at 37°C for 48–72 hours. If a precipitate forms, it should be removed by filtration prior to use. Store in an amber bottle and avoid exposure to direct sunlight. The solution expires six months after preparation. Store at room temperature in an exhaust hood.

Enzymatic Assay of DEOXYRIBONUCLEASE I (EC 3.1.21.1)

Principle

$$DNA + H_2O \xrightarrow{\text{Deoxyribonuclease I}} \text{5'-Oligodeoxyribonucleotides}$$

Abbreviation Used

DNA = Deoxyribonucleic Acid

Conditions: T = 25°C, pH = 5.0, A260nm, Light path = 1 cm

Method: Continuous Spectrophotometric Rate Determination

Reagents

A. 1 M Sodium Acetate Buffer, pH 5.0 at 25°C (Prepare 100 ml in deionized water using Sodium Acetate, Trihydrate. Adjust to pH 5.0 at 25°C with 5 M HCl).

B. 100 mM Magnesium Sulfate Solution ($MgSO_4$) (Prepare 5 ml in deionized water using Magnesium Sulfate, Heptahydrate).

C. 0.033% (w/v) Deoxyribonucleic Acid Solution (DNA) (Dissolve a 1 mg vial of Deoxyribonucleic Acid, Sodium, in 3.1 ml of deionized water).

D. 150 mM Sodium Chloride (NaCl) (Prepare 50 ml in deionized water using Sodium Chloride).

E. Deoxyribonuclease I Enzyme Solution (DNase I) (Immediately before use, prepare a solution containing 400–500 Kunitz units/ml in cold Reagent D).

F. Standard DNase Solution (Std Soln) (Reconstitute a vial of Deoxyribonuclease I, Standardized vial, containing 2000 Kunitz units, with 1 ml of cold Reagent D. Immediately before use, dilute to 400–500 Kunitz units/ml with cold Reagent D).

Procedure

Prepare a reaction cocktail by pipetting (in milliliters) the following reagent into a suitable container:

Reagent A (Buffer)	2.50
Reagent B ($MgSO_4$)	1.25
Reagent C (DNA)	3.00
Deionized Water	18.25

Mix by gentle stirring and adjust to pH 5.0 at 25°C with 1 M HCl or 1 M NaOH, if necessary. Then pipette (in milliliters) the following reagents into suitable quartz cuvettes:

	Test	Std	Blank
Reagent D (NaCl)	—	—	0.50
Reaction Cocktail	2.50	2.50	2.50

Mix by inversion and equilibrate to 25°C. Monitor the A260nm until constant, using a suitably thermostatted spectrophotometer. Then add:

Reagent E (DNase I)	0.50	—	—
Reagent F (Std Soln)	—	0.50	—

Immediately mix by inversion and record the increase in A260nm for approximately 10 minutes. Obtain the ΔA260nm/minute using the maximum linear rate for the Test, Standard, and Blank.

Calculations

$$\text{Units / ml Test Soln} = \frac{(\Delta A260\text{nm / min Test} - \Delta A260\text{nm / min Blank})\ (3)\ (\text{df})}{(0.001)\ (0.5)}$$

$$\text{Units / ml Std Soln} = \frac{(\Delta A260\text{nm / min Std} - \Delta A260\text{nm / min Blank})\ (3)\ (\text{df})}{(0.001)\ (0.5)}$$

3 = Volume (in milliliters) of assay
df = Dilution factor (includes dilution of the standard vial)
0.001 = A260nm as per the Unit Definition
0.5 = Volume (in milliliter) of enzyme used

$$\text{CF} = \frac{(2000)}{\text{Experimental units / Std vial}}$$

CF = Correction Factor
2000 = Theoretical units/Std vial
Corrected Units/ml enzyme = (CF) (Units/ml Test Solution)

$$\text{Units / mg solid} = \frac{\text{units / ml enzyme}}{\text{mg solid / ml enzyme}}$$

$$\text{Units / mg protein} = \frac{\text{units / ml enzyme}}{\text{mg protein / ml enzyme}}$$

Unit Definition: *One Kunitz unit will produce a ΔA260 of 0.001 per minute per ml at pH 5.0 at 25°C, using DNA, Types I or III as substrate.* $[Mg^{++}]$ *= 4.2 mM.*

Final Assay Concentration

In a 3.00 ml reaction mix, the final concentrations are 83 mM sodium acetate, 4.2 mM magnesium sulfate, 25 mM sodium chloride, 0.003% (w/v) deoxyribonucleic acid, and 200–250 units deoxyribonuclease I.

Note

1. There is no absolute standard for the assay of Dnase.

Enzymatic Assay of DEOXYRIBONUCLEASE II (EC 3.1.22.1)

Principle

$$DNA + H_2O \xrightarrow{\text{Deoxyribonuclease II}} \text{3'-Oligodeoxyribonucleotides}$$

Abbreviation Used

DNA = Deoxyribonucleic Acid

Conditions: T = 25°C, pH 4.6, A260nm, Light path = 1 cm

Method: Continuous Spectrophotometric Rate Determination

Reagents

A. 1 M Sodium Acetate Buffer (Prepare 100 ml in deionized water using Sodium Acetate, Trihydrate. Adjust to pH 4.6 at 25°C with 6 N HCl).
B. 20 mM Magnesium Sulfate Solution ($MgSO_4$) (Prepare 25 ml in deionized water using Magnesium Sulfate, Heptahydrate).
C. 0.033% (w/v) Deoxyribonucleic Acid Solution (DNA) (Prepare 30 ml in deionized water using Deoxyribonucleic Acid, Sodium Salt. Gently mix by stirring at 4°C for 15–24 hours).
D. 150 mM Sodium Chloride Solution (NaCl) (Prepare 25 ml in deionized water using Sodium Chloride).
E. Standard DNase Solution (Std Soln) (Immediately before use, reconstitute a vial of Deoxyribonuclease II, Standardized vial, containing approximately 1500 Kunitz units, with 2 ml of cold deionized water).
F. Deoxyribonuclease II Enzyme Solution (DNase II) (Immediately before use, prepare a solution containing 500–1000 Kunitz units/ml in cold deionized water).

Procedure

Prepare a reaction cocktail by pipetting (in milliliters) the following reagents into a suitable container:

Reagent A (Buffer)	7.50
Reagent B ($MgSO_4$)	3.75
Reagent C (DNA)	9.00
Deionized water	54.75

Mix by swirling and equilibrate to 25°C. Pipette (in milliliters) the following reagents into suitable cuvettes:

	Test	Std	Blank
Reaction Cocktail	2.50	2.50	2.50
Reagent D (NaCl)	0.40	0.40	0.40

Mix by inversion and equilibrate to 25°C. Monitor the A260nm until constant, using a suitably thermostatted spectrophotometer. Then add:

	Test	Std	Blank
Reagent E (Std Soln)	—	0.10	—
Reagent F (DNase II)	0.10	—	—
Deionized Water	—	—	0.10

Immediately mix by inversion and record the increase in A260nm for approximately 10 minutes. Obtain the ΔA260nm/minute using the maximum linear rate for the Test, Standard, and Blank.

Calculations

$$\text{Units / ml Test Solution} = \frac{(\Delta A260\text{nm / min Test} - \Delta A260\text{nm / min Blank})\ (3)\ (\text{df})}{(0.001)\ (0.1)}$$

$$\text{Units / ml Std Solution} = \frac{(\Delta A260\text{nm / min Std} - \Delta A260\text{nm / min Blank})\ (3)\ (\text{df})}{(0.001)\ (0.1)}$$

$$\text{CF} = \frac{750}{\text{Experimental Units / ml Std Soln}}$$

3 = Volume (in milliliters) of assay
df = Dilution factor
0.001 = A260nm per unit as per the Unit Definition
0.1 = Volume (in milliliter) of enzyme used
CF = Correction factor
750 = Theoretical Units/ml Std Soln

Corrected Units/ml enzyme = (CF) (Units/ml Test Solution)

$$\text{Units / mg solid} = \frac{\text{units / ml enzyme}}{\text{mg solid / ml enzyme}}$$

$$\text{Units / mg protein} = \frac{\text{units / ml enzyme}}{\text{mg protein / ml enzyme}}$$

Unit Definition: *One Kunitz unit will produce a A260 of 0.001 per minute per ml at pH 4.6 at 25°C. [Mg^{++}] = 0.83 mM.*

Final Assay Concentrations

In a 3.00 ml reaction mix, the final concentrations are 83 mM sodium acetate, 0.83 mM magnesium sulfate, 0.0033% (w/v) deoxyribonucleic acid, 20 mM sodium chloride, and 50–100 units deoxyribonuclease II.

Note

1. There is no absolute standard for the assay of DNase.

Enzymatic Assay of DEXTRAN SUCRASE (EC 2.4.1.5)

Principle

$$\text{Sucrose} + (1,6\text{-}\alpha\text{-D-glucosyl})_n \xrightarrow{\text{n Dextran Sucrase}} \text{Fructose} + (1,6\text{-}\alpha\text{-D-glucosyl})\ n+1$$

Conditions: T = 30°C, pH = 5.2, A546nm, Light path = 1 cm

Method: Colorimetric

Reagents

A. 50 mM Sodium Acetate Buffer with 1 mM Calcium Chloride, pH 5.2 at 30°C (Prepare 100 ml in deionized water using Sodium Acetate, Trihydrate and Calcium Chloride, Dihydrate. Adjust to pH 5.2 at 30°C using 1 M HCl).
B. 12.5% (w/v) Sucrose Substrate Solution (Sucrose) (Prepare 10 ml in Reagent A using Sucrose).
C. Dextran Sucrase Enzyme Solution (Enz Soln) (Immediately before use, prepare a solution containing 0.1–0.2 unit/ml of Dextran Sucrase in cold Reagent A).
D. 26 mM Copper Sulfate, 1.3 M Sodium Sulfate, 225 mM Sodium Carbonate, 190 mM Sodium Bicarbonate and 43 mM Sodium Potassium Tartrate Solution (Copper Soln) (Prepare 1 liter in deionized water using Cupric Sulfate Pentahydrate, Sodium Bicarbonate, Sodium Sulfate, Anhydrous, Sodium Carbonate, Anhydrous, and Sodium Potassium Tartrate Tetrahydrate).
E. 40 mM Molybdic Acid, 19 mM Arsenic acid and 756 mM Sulfuric Acid Solution (Ars-Mol) (Prepare 1 liter in deionized water using Molybdic Acid, Ammonium Salt Tetrahydrate, Arsenic Acid, Sodium Salt and Sulfuric Acid).
F. 1.39 mM Fructose Standard Solution (Std Soln) (Prepare 10 ml in deionized water using Fructose).

Procedure

Pipette (in milliliters) the following reagents into suitable test tubes:

	Test	Test Blank	Std 1	Std 2	Std 3	Std 4	Std 5	Std Blank
Reagent B (Sucrose)	0.40	0.40	—	—	—	—	—	—
Reagent F (Std Soln)	—	—	0.10	0.20	0.30	0.40	0.50	—
Deionized Water	—	—	0.40	0.30	0.20	0.10	—	0.50

Mix by swirling and equilibrate to 30°C. Then add:

Reagent C (Enz Soln)	0.10	—	—	—	—	—	—	—

Immediately mix by swirling and incubate at 30°C for exactly 15 minutes. Then add:

Reagent D (Copper Soln)	0.50	0.50	0.50	0.50	0.50	0.50	0.50	0.50
Reagent C (Enz Soln)	—	0.10	—	—	—	—	—	—

Mix by swirling. Place a marble over the top of each tube and transfer the tubes to a boiling water bath. Incubate for 15 minutes. Remove from the boiling water bath and allow to cool to room temperature. Then add:

Reagent E (Ars-Mol)	0.50	0.50	0.50	0.50	0.50	0.50	0.50	0.50

Shake or vortex until the foaming stops and any precipitate present is dissolved. Then add:

Deionized Water	10.00	10.00	10.00	10.00	10.00	10.00	10.00	10.00

Mix by swirling and transfer the solutions to cuvettes. Record the A546nm for the Test, Blank and Standards using a suitable spectrophotometer.

Calculations

Standard Curve: ΔA546nm Std = A546nm Std–A546nm Std Blank.
Prepare a standard curve by plotting the ΔA546nm Standard versus the μmoles of fructose.

Sample Determination: ΔA546nm Sample = A546nm Test–A546nm Test Blank
Determine the μmoles of fructose using the Standard Curve.

$$\text{Units / ml enzyme} = \frac{(\mu\text{moles of fructose liberated})\ (\text{df})}{(15)\ (0.1)}$$

df = Dilution factor
15 = Time (in minutes) of the assay as per the Unit Definition
0.1 = Volume (in milliliter) of enzyme used

$$\text{Units / mg solid} = \frac{\text{units / ml enzyme}}{\text{mg solid / ml enzyme}}$$

$$\text{Units / mg protein} = \frac{\text{units / ml enzyme}}{\text{mg protein / ml enzyme}}$$

Unit Definition: *One unit will liberate 1.0 μmole of fructose from sucrose per minute at pH 5.2 at 30°C.*

Final Assay Concentration

In a 0.50 ml reaction mix, the final concentrations are 50 mM sodium acetate, 1 mM calcium chloride, 10% (w/v) sucrose and 0.01–0.02 unit dextran sucrase.

Notes

1. Sodium Sulfate, Sodium Carbonate, and Sodium Potassium Tartrate are dissolved in approximately 500 ml of deionized water. Cupric Sulfate is dissolved in approximately 10 ml of deionized water and is slowly added to the above solution to avoid precipitation. Sodium Bicarbonate is dissolved first in deionized water and then added to the above solution. Dilute the solution to 1 liter. If a precipitate forms, it should be removed by filtration prior to use. Store in an amber bottle and avoid exposure to direct sunlight. Store at room temperature.
2. Molybdic Acid is dissolved in approximately 300 ml of deionized water. Add Sulfuric acid slowly. Caution, this is an exothermic reaction. Arsenic Acid is dissolved in approximately 300 ml of deionized water and is added to the above solution. The solution is diluted to a total volume of 1 liter and incubated at 37°C for 48–72 hours. If a precipitate forms, it should be removed by filtration prior to use. Store in an amber bottle and avoid exposure to direct sunlight. The solution expires six months after preparation. Store at room temperature in an exhaust hood.

Enzymatic Assay of DEXTRANASE (EC 3.2.1.11)

Principle

$$\text{Dextran} \xrightarrow{\text{Dextranase}} \text{Iso - Maltose}$$

Conditions: T = 37°C, pH = 6.0, A540nm, Light path = 1 cm

Method: Colorimetric

Reagents

A. 100 mM Potassium Phosphate Buffer, pH 6.0 at 37°C (Prepare 100 ml in deionized water using Potassium Phosphate, Monobasic, Anhydrous. Adjust to pH 6.0 at 37°C with 1 M NaOH).

B. 2.0% (w/v) Dextran with 0.01% Thimerosal Solution (Prepare 25 ml in Reagent A using Dextran, Industrial Grade, average molecular weight 200,000–300,000 and Thimerosal. Adjust to pH 6.0 at 37°C with 100 mM HCl or 100 mM KOH, if necessary).

C. Color Reagent Solution containing 30% (w/v) Sodium Potassium Tartrate, and 1.1% (w/v) 3,5- Dinitrosalicylic Acid (Clr Rgt Soln) The following components are made separately:

   1. Sodium Potassium Tartrate Solution (Dissolve 12.0 grams of Sodium Potassium Tartrate, Tetrahydrate, in 8.0 ml of 2 M NaOH. Heat in boiling water bath to dissolve, but *Do not boil*).
   2. 3,5-Dinitrosalicylic Acid Solution (Dissolve 438 mg of 3,5-Dinitrosalicylic Acid in 20 ml of deionized water. Heat in boiling water bath to dissolve, but *Do not boil*).

   With stirring, slowly add the Sodium Potassium Tartrate solution to the 3,5-Dinitrosalicylic Acid solution. Dilute to 40 ml with deionized water. If not completely dissolved, the reagents should dissolve when mixed. The solution should be stored in an amber bottle at room temperature. The Color Reagent Solution should be stable for at least 6 months.

D. 3 mM Maltose Solution (Std Soln) (Prepare 10 ml in deionized water using Maltose, Monohydrate).

E. Dextranase Enzyme Solution (Immediately before use, prepare a solution containing 1–2 units/ml of Dextranase in cold deionized water).

Procedure

Pipette (in milliliters) the following reagents into suitable containers:

	Test	Blank
Reagent B (Dextran)	0.95	0.95

Equilibrate to 37°C. Then add:

	Test	Blank
Deionized Water	—	0.05
Reagent E (Enzyme Solution)	0.05	—

Incubate for exactly 30 minutes at 37°C. Then add:

	Test	Blank
Reagent C (Clr Rgt Soln)	1.00	1.00

Mix by inversion and place in a boiling water bath for 15 minutes, then cool and add:

Deionized Water	10.00	10.00

Mix by inversion and record the A540nm for both the Test and Blank using a suitable spectrophotometer.

Standard Curve: A standard curve is made by pipetting (in milliliters) the following reagents into suitable containers:

	Std 1	Std 2	Std 3	Std 4	Std 5	Std Blank
Reagent D (Std Soln)	0.10	0.30	0.50	0.70	1.00	0.00
Deionized Water	0.90	0.70	0.50	0.30	—	1.00
Reagent C (Clr Rgt Soln)	1.00	1.00	1.00	1.00	1.00	1.00

Place in a boiling water bath for 15 minutes. Remove the containers, cool in ice bath and then add:

Deionized Water	10.00	10.00	10.00	10.00	10.00	10.00

Mix by inversion and record the A540nm for the Standards and Standard Blank using a suitable spectrophotometer.

Calculations

Standard Curve: ΔA540nm Standard = A540nm Std–A540nm Std Blank. Find the slope (M) by plotting the ΔA540nm Standards vs Maltose concentration.

Sample Determination: Determine the concentration of Maltose liberated using the following equation:

$$\text{mg Maltose released} = \frac{\Delta\text{A540nm Test} - \text{A540nm Blank}}{(\text{RM})}$$

$$\text{Units / mg enzyme} = \frac{(\text{mg Maltose released})}{(30)\ (\text{mg enzyme / RM})}$$

30 = Time of assay, 30 minutes, while unit definition is per minute
RM = Reaction Mix, Step 1 (volume = 1.0 ml)

Unit Definition: *One unit will liberate 1.0 µmole of isomaltose (measured as maltose) per minute at pH 6.0 at 37°C, using dextran as substrate.*

Final Assay Concentrations

In a 1 ml reaction mix, the final concentrations are 95 mM potassium phosphate, 2.0% dextran, 0.01% thimerosal and 0.05–0.1 units dextranase.

Enzymatic Assay of DIACETYL REDUCTASE (EC 1.1.1.5)

Principle

$$\text{Diacetyl} + \beta\text{-NADH} \xrightarrow{\text{Diacetyl Reductase}} \beta\text{-NAD} + \text{Acetoin}$$

Abbreviations Used

β-NADH = β-Nicotinamide Adenine Dinucleotide, Reduced Form
β-NAD = β-Nicotinamide Adenine Dinucleotide, Oxidized Form

Conditions: T = 25°, pH 6.1, A340nm, Light path = 1 cm

Method: Continuous Spectrophotometric Rate Determination

Reagents

A. 100 mM Potassium Phosphate Buffer, pH 6.1 at 25°C (Prepare 100 ml in deionized water using Potassium Phosphate, Dibasic, Trihydrate. Adjust to pH 6.1 at 25°C with 1 M HCl).
B. 0.25 mM β-Nicotinamide Adenine Dinucleotide, Reduced Form Solution (β-NADH) (Dissolve the contents of one 1 mg vial of β-Nicotinamide Adenine Dinucleotide, Reduced Form, Disodium Salt, Preweighed Vial, in the appropriate volume of Reagent A).
C. 5% (w/v) Diacetyl Solution (Diacetyl) (Prepare 1 ml in deionized water using Diacetyl).
D. Diacetyl Reductase Enzyme Solution (Immediately before use, prepare a solution containing 0.3–0.6 unit/ml of Diacetyl Reductase in cold Reagent A).

Procedure

Pipette (in milliliters) the following reagents into suitable cuvettes:

	Test	Blank
Reagent B (b-NADH)	0.98	0.98

Incubate at 25°C for 5 minutes. Then add:

Reagent D (Enzyme Solution)	0.01	—
Reagent A (Buffer)	—	0.01

Mix by inversion and monitor the A340nm until constant using a suitably thermostatted spectrophotometer. Then add:

Reagent C (Diacetyl)	0.01	0.01

Immediately mix by inversion and record the decrease in A340nm for approximately 5 minutes. Obtain the ΔA340nm/minute using the maximum linear rate for both the Test and Blank.

Calculations

$$\text{Units/ml enzyme} = \frac{(\Delta A340\text{nm/min Test} - \Delta A340\text{nm/min Blank})\ (1)\ (\text{df})}{(6.22)\ (0.01)}$$

1 = Total volume (in milliliter) of assay
df = Dilution factor
6.22 = Millimolar extinction coefficient of β-NADH at 340nm
0.01 = Volume (in milliliter) of enzyme used

$$\text{Units / mg solid} = \frac{\text{units / ml enzyme}}{\text{mg solid / ml enzyme}}$$

$$\text{Units / mg protein} = \frac{\text{units / ml enzyme}}{\text{mg protein / ml enzyme}}$$

Unit Definition: *One unit will oxidize 1 μmole of NADH per minute at pH 6.1 at 25°C with diacetyl as substrate.*

Final Assay Concentrations

In a 1.00 ml reaction mix, the final concentrations are 99 mM potassium phosphate, 0.25 mM β-nicotinamide adenine dinucleotide, reduced form, 0.05% (w/v) diacetyl, and 0.003–0.006 unit diacetyl reductase.

Enzymatic Assay of (S)-DIACETYL REDUCTASE

Principle

$$\text{Diacetyl} + \beta\text{-NADPH} \xrightarrow{\text{(S)-Diacetyl Reductase}} \beta\text{-NADP} + \text{(S)} = \text{Acetoin}$$

Abbreviations Used

β-NADPH = β-Nicotinamide Adenine Dinucleotide Phosphate, Reduced Form
β-NADP = β-Nicotinamide Adenine Dinucleotide Phosphate, Oxidized Form

Conditions: T = 25°C, pH = 6.9, A340nm, Light path = 1 cm

Method: Continuous Spectrophotometric Rate Determination

Reagents

A. 100 mM Potassium Phosphate Buffer, pH 6.9 at 25°C (Prepare 100 ml in deionized water using Potassium Phosphate, Monobasic, Anhydrous. Adjust to pH 6.9 at 25°C with 1 M KOH).
B. 1 mM DL-Dithiothreitol Solution (DTT) (Prepare 25 ml in Reagent A using DL-Dithiothreitol).
C. 114 mM Diacetyl Solution (Diacetyl) (Prepare 10 ml in deionized water using Diacetyl).
D. 1.10 mM β-Nicotinamide Adenine Dinucleotide Phosphate, Reduced Form Solution (β-NADPH) (Dissolve the contents of one 1 mg vial of β-Nicotinamide Adenine Dinucleotide Phosphate, Tetrasodium Salt, Preweighed Vial in the appropriate volume of deionized water).
E. (S)-Diacetyl Reductase Enzyme Solution (Immediately before use, prepare a solution containing 0.05–0.10 unit/ml of (S)-Diacetyl Reductase in cold Reagent A).

Procedure

Pipette (in milliliters) the following reagents into suitable cuvettes:

	Test	Blank
Reagent B (DTT)	0.75	0.75
Reagent D (β-NADPH)	0.10	0.10
Reagent E (Enzyme Solution)	0.10	0.10

Mix by inversion and equilibrate to 25°C. Monitor the A340nm until constant using a suitably thermostatted spectrophotometer. Then add:

Reagent C (Diacetyl)	0.10	—
Deionized Water	—	0.10

Immediately mix by inversion and record the decrease in A340nm for approximately 5 minutes. Obtain the ΔA340nm/min using the maximum linear rate for both the Test and Blank.

Calculations

$$\text{Units / ml enzyme} = \frac{(\Delta A340nm / \min \text{ Test} - \Delta A340nm / \min \text{ Blank})\,(1.05)\,(df)}{(6.22)\,(0.1)}$$

1.05 = Total volume (in milliliters) of assay
df = Dilution factor
6.22 Millimolar extinction coefficient of β-NADPH at 340nm
0.1 = Volume (in milliliter) of enzyme used in assay

$$\text{Units / mg solid} = \frac{\text{units / ml enzyme}}{\text{mg solid / ml enzyme}}$$

$$\text{Units / mg protein} = \frac{\text{units / ml enzyme}}{\text{mg protein / ml enzyme}}$$

Unit Definition: *One unit will convert 1 μmole of diacetyl and β-NADPH to (S)-acetoin and β-NADP per minute at pH 6.9 at 25°C.*

Final Assay Concentrations

In a 1.05 ml reaction mix, the final concentrations are 81 mM potassium phosphate, 0.7 mM DL-dithiothreitol, 1.1 mM β-nicotinamide adenine dinucleotide phosphate, reduced form, 11 mM diacetyl, and 0.005–0.01 unit (S)-diacetyl reductase.

Enzymatic Assay of DIAMINE OXIDASE (EC 1.4.3.6)

Principle

$$\text{Putrescine} + H_2O + O_2 \xrightarrow{\text{Diamine Oxidase}} H_2O_2 + \text{4-Aminobutanal} + NH_3$$

$$H_2O_2 + \text{O-Dianisidine} \xrightarrow[\text{(oxidized)}]{\text{Peroxidase}} 2H_2O + \text{O-Dianisidine (reduced)}$$

Conditions: T = 37°C, pH = 7.2, A440nm, Light path = 1 cm

Method: Continuous Spectrophotometric Rate Determination

Reagents

A. 100 mM Sodium Phosphate Buffer, pH 7.2 at 37°C (Prepare 50 ml in deionized water using Sodium Phosphate, Monobasic. Adjust to pH 7.2 at 37°C with 1 M NaOH).
B. 75 mM Putrescine Solution (Put) (Prepare 1 ml in Reagent A using Putrescine Dihydrochloride).
C. 16 mM O-Dianisidine Solution (ODA) (Dissolve the contents of a 50 mg vial of Odianisidine Dihydrochloride, in 10 ml of deionized water or prepare 5 ml in deionized water using O-Dianisidine Dihydrochloride. *Prepare Fresh*).
D. Peroxidase Enzyme Solution (POD) (Immediately before use, prepare a solution containing 2000 pyrogallol units/ml in cold Reagent A using Peroxidase Type II from Horseradish).
E. Diamine Oxidase Enzyme Solution (Immediately before use, prepare a solution containing 3–6 units/ml of Diamine Oxidase in cold Reagent A).

Procedure

Pipette (in milliliters) the following reagents into suitable cuvettes:

	Test	Blank
Reagent A (Buffer)	2.50	2.60
Reagent B (Put)	0.20	0.20
Reagent C (ODA)	0.10	0.10
Reagent D (POD)	0.01	0.01

Mix by inversion and equilibrate to 37°C. Monitor the A440nm until constant, using a suitably thermostatted spectrophotometer. Then add:

Reagent E (Enzyme Solution)	0.20	—

Immediately mix by inversion and record the increase in A440nm for approximately 5 minutes. Obtain the ΔA440nm/minute using the maximum linear rate for both the Test and Blank.

Calculations

$$\text{Units / mg enzyme} = \frac{(\Delta A440\text{nm / min Test} - \Delta A440\text{nm / min Blank})\ (60)}{(11.3)\ (\text{mg enzyme / ml RM})}$$

60 = conversion from units/min to units/hour (Unit Definition)
11.3 = Millimolar extinction coefficient of O-Dianisidine (Oxidized) at 440 nm
RM = Reaction Mix

Unit Definition: *One unit will oxidize 1.0 μmole of putrescine per hour at pH 7.2 at 37°C.*

Final Assay Concentration

In a 3.01 ml reaction mix, the final concentrations are 97 mM sodium phosphate, 5.0 mM putrescine, 0.53 mM O-Dianisidine, 20 units peroxidase, and 0.6–1.2 units diamine oxidase.

Enzymatic Assay of DIAPHORASE (EC 1.8.1.4)

Principle

$$\text{DPIP} + \beta\text{-NADH} \xrightarrow{\text{Diaphorase}} \beta\text{-NAD} + \text{reduced DPIP}$$

Abbreviations Used

DPIP = 2,6-Dichlorophenol-Indophenol
β-NADH = β-Nicotinamide Adenine Dinucleotide, Reduced Form
β-NAD = β-Nicotinamide Adenine Dinucleotide, Oxidized Form

Conditions: T = 25°C, pH = 7.5, A600nm, Light path = 1 cm

Method: Continuous Spectrophotometric Rate Determination

Reagents

A. 20 mM Tris HCl Buffer, pH 7.5 at 25°C (Prepare 100 ml in deionized water using Trizma Base. Adjust to pH 7.5 at 25°C with 1 M HCl).
B. 0.23 mM β-Nicotinamide Adenine Dinucleotide, Reduced Form Solution (β-NADH) (Dissolve the contents of a 1 mg vial of β-Nicotinamide Adenine Dinucleotide, Reduced Form, Disodium Salt, in 60 ml of Reagent A or prepare 6 ml in Reagent A using β-Nicotinamide Adenine Dinucleotide, Reduced Form, Disodium Salt. *Prepare Fresh*).
C. 1.17 mM 2,6-Dichlorophenol-Indophenol Solution (DPIP) (Prepare 100 ml in deionized water using 2,6-Dichlorophenol-Indophenol, Sodium Salt).
D. 200 mM Tris HCl Buffer with 294 mM Potassium Chloride, 0.54 mM Flavin Mononucleotide and 0.025% (w/v) Bovine Serum Albumin, pH 7.5 at 25°C (Enzyme Diluent) (Prepare 100 ml in deionized water using Trizma Base, Potassium Chloride, Flavin Mononucleotide, Sodium Salt, and Albumin, Bovine. Adjust to pH 7.5 at 25°C with 1 M HCl).
E. Diaphorase Enzyme Solution (Immediately before use, prepare a solution containing 0.05–0.1 unit/ml of Diaphorase in cold Reagent D).

Procedure

Pipette (in milliliters) the following reagents into suitable cuvettes:

	Test	Blank
Reagent B (β-NADH)	2.80	2.80
Reagent C (DPIP)	0.10	0.10

Mix by inversion and equilibrate to 25°C. Monitor the A600nm until constant, using a suitably thermostatted spectrophotometer. Then add:

Reagent D (Enzyme Diluent)	—	0.10
Reagent E (Enzyme Solution)	0.10	—

Immediately mix by inversion and record the decrease in the A600nm for 5–10 minutes. Obtain the ΔA600nm/minute using the maximum linear rate for both the Test and Blank.

Calculations

$$\text{Units / ml enzyme} = \frac{(\Delta A600\text{nm / min Test} - \Delta A600\text{nm / min Blank})(3)(\text{df})}{(21)(0.1)}$$

3 = Total volume (in milliliters) of assay
df = Dilution factor
21 = Millimolar extinction coefficient of DPIP at 600nm
0.1 = Volume (in milliliters) of enzyme used

$$\text{Units / mg solid} = \frac{\text{units / ml enzyme}}{\text{mg solid / ml enzyme}}$$

$$\text{Units / mg protein} = \frac{\text{units / ml enzyme}}{\text{mg protein / ml enzyme}}$$

Unit Definition: *One unit of either "diaphorase" or "lipoyl" dehydrogenase will oxidize 1.0 μmole of β-NADH per minute at pH 7.5 at 25°C, with the corresponding reduction of the appropriate electron acceptor.*

Final Assay Concentration

In a 3.00 ml reaction mix, the final concentrations are 25 mM Tris, 0.22 mM β-nicotinamide adenine dinucleotide, reduced form, 0.039 mM 2,6-dichlorophenol-indophenol, 9.8 mM potassium chloride, 0.018 mM flavin mononucleotide, 0.00083% (w/v) bovine serum albumin, and 0.005–0.01 unit diaphorase.

Note

1. This assay is not to be used for Diaphorase, from Porcine Heart.

Enzymatic Assay of DIHYDROFOLATE REDUCTASE (EC 1.5.1.3)

Principle

$$\text{DHFA} + \beta\text{-NADPH} \xrightarrow{\text{Dihydrofolate Reductase}} \text{THFA} + \beta\text{-NADP}$$

Abbreviations Used

DHFA = 7,8-Dihydrofolic acid
THFA = 5,6,7,8-Tetrahydrofolic Acid
β-NADPH = β-Nicotinamide Adenine Dinucleotide Phosphate, Reduced Form
β-NADP = β-Nicotinamide Adenine Dinucleotide Phosphate, Oxidized Form

Conditions: T = 25°C, pH = 6.5, A340nm, Light Path = 1 cm

Method: Continuous Spectrophotometric Rate Determination

Reagents

A. 50 mM Potassium Phosphate Buffer, pH 6.5 at 25°C (Prepare 100 ml in deionized water using Potassium Phosphate, Monobasic, Anhydrous. Adjust to pH 6.5 at 25°C with 1 M KOH).
B. 0.11 mM β-Nicotinamide Adenine Dinucleotide Phosphate, Reduced Form Solution (β-NADPH) (Dissolve the contents of one 5 mg vial of β-Nicotinamide Adenine

Dinucleotide Phosphate, Reduced Form, Tetrasodium Salt, in the appropriate volume of Reagent A. *Prepare Fresh*).

C. 2.3 mM Dihydrofolic Acid Solution (DHFA) (Immediately before use, prepare 1 ml in Reagent A using Dihydrofolic Acid. Facilitate dissolution with 1 M KOH (0.020 ml of 1 M KOH per mg of Dihydrofolic Acid. *Prepare Fresh*. This solution is stable for only 10 minutes).

D. 0.1% (w/v) Bovine Serum Albumin Solution (BSA) (Prepare 100 ml in Reagent A using Albumin, Bovine).

E. Dihydrofolate Reductase Enzyme Solution (Immediately before use, prepare a solution containing 0.12–0.25 unit/ml of Dihydrofolate Reductase in cold Reagent D).

Procedure

Pipette (in milliliters) the following reagents into suitable cuvettes:

	Test	Blank
Reagent B (β-NADPH)	3.00	3.00
Reagent C (DHFA)	0.10	0.10

Mix by inversion and equilibrate to 25°C. Monitor the A340nm until constant, using a suitably thermostatted spectrophotometer. Then add:

Reagent E (Enzyme Solution)	0.10	—
Reagent D (BSA)	—	0.10

Immediately mix by inversion and record the decrease in A340nm for approximately 5 minutes. Obtain the ΔA340nm /minute using the maximum linear rate for both the Test and Blank.

Calculations

$$\text{Units / ml enzyme} = \frac{(\Delta A340\text{nm / min Test} - \Delta A340\text{nm / min Blank})\ (3.2)\ (\text{df})}{(12.3)\ (0.1)}$$

3.2 = Total volume (in milliliters) of assay
df = Dilution factor
12.3 = Difference in the millimolar extinction coefficients of the substrates and the products in the dihydrofolate reductase reaction
0.1 = Volume (in milliliters) of assay

$$\text{Units / mg solid} = \frac{\text{units / ml enzyme}}{\text{mg solid / ml enzyme}}$$

$$\text{Units / mg protein} = \frac{\text{units / ml enzyme}}{\text{mg protein / ml enzyme}}$$

Unit Definition: *One unit will convert 1.0 μmole of 7,8-dihydrofolate and β-NADPH to 5,6,7,8-tetrahydrofolate and β-NADP per minute at pH 6.5 at 25°C.*

Final Assay Concentration

In a 3.20 ml reaction mix, the final concentrations are 50 mM potassium phosphate, 0.072 mM dihydrofolic acid, 0.10 mM β-nicotinamide adenine dinucleotide phosphate, reduced form, 0.003% (w/v) bovine serum albumin and 0.012–0.025 unit dihydrofolate reductase.

Enzymatic Assay of DIHYDROPTERIDINE REDUCTASE (EC 1.6.99.7)

Principle

$$6,7\text{-DMTHP} + H_2O_2 \xrightarrow{\text{Peroxidase}} 6,7\text{-DMDHP (Quinonoid Isomer)}$$

$$6,7\text{-DMDHP (Quinonoid Isomer)} + \beta\text{-NADH} \xrightarrow{\text{DHPR}} 6,7\text{-DMTHP} + \beta\text{-NAD}$$

Abbreviations Used

DHPR = Dihydropteridine Reductase
6,7-DMTHP = 6,7-Dimethyltetrahydropterine
6,7-DMDHP = 6,7-Dimethyldihydropterine
β-NADH = β-Nicotinamide Adenine Dinucleotide, Reduced Form
β-NAD = β-Nicotinamide Adenine Dinucleotide, Oxidized Form

Conditions: T = 25°C, pH = 7.2, A340nm, Light Path = 1 cm

Method: Continuous Spectrophotometric Rate Determination

Reagents

A. 100 mM Tris HCl Buffer, pH 7.2 at 25°C (Prepare 100 ml in deionized water using Trizma Base. Adjust to pH 7.2 at 25°C with 1 M HCl).
B. 0.43 mM β-Nicotinamide Adenine Dinucleotide, Reduced Form Solution (β-NADH) (Dissolve the contents of one 1 mg vial of β-Nicotinamide Adenine Dinucleotide, Reduced Form, Disodium Salt, in the appropriate volume of deionized water. *Prepare Fresh*).
C. 0.3% (v/v) Hydrogen Peroxide Solution (H_2O_2) (Prepare 10 ml in deionized water using Hydrogen Peroxide, 30% (w/w)).
D. 2.15 mM 6,7-Dimethyltetrahydropterine Solution (6,7-DMTHP) (Immediately before use prepare 2 ml in cold Reagent A using 6,7-Dimethyl-5,6,7,8-Tetrahydropterine, Monohydrochloride).
E. Peroxidase Enzyme Solution (Immediately before use prepare a solution containing 100 purpurogallin units/ml of Peroxidase in cold deionized water).
F. Dihydropteridine Reductase Enzyme Solution (DHPR) (Immediately before use prepare a solution containing 0.08–0.32 unit/ml using cold Reagent A).

Procedure

Pipette (in milliliters) the following reagents into suitable cuvettes:

	Test	Blank
Reagent A (Buffer)	1.50	1.50
Reagent C (H_2O_2)	0.03	0.03
Reagent D (6,7-DMTHP)	0.05	0.05
Reagent B (β-NADH)	1.00	1.00
Reagent E (Peroxidase Enzyme Solution)	0.05	0.05
Deionized Water	0.27	0.27

Mix by inversion and equilibrate to 25°C. Monitor the A340nm until constant, using a suitably thermostatted spectrophotometer. Then add:

	Test	Blank
Reagent F (DHPR)	0.10	—
Reagent A (Buffer)	—	0.10

Immediately mix by inversion and record the decrease in A340nm for approximately 5 minutes. Obtain the A340nm/minute using the maximum linear rate for both the Test and Blank.

Calculations

$$\text{Units / mg enzyme} = \frac{(\text{A340nm / min Test} - \text{A340nm / min Blank})}{(6.22)\ (\text{mg enzyme / ml RM})}$$

6.22 = Millimolar extinction coefficient of β-NADH at 340nml
RM = Reaction Mix

Unit Definition: *One unit will oxidize 1.0 µmole of β-NADH to β-NAD with 6,7-Dimethyldihydropterine (quinonoid isomer) as the non-nucleotide substrate per minute at pH 7.2 at 25°C.*

Final Assay Concentration

In a 3.00 ml reaction mix, the final concentrations are 55 mM Tris, 0.14 mM β-NADH, 0.003% (w/w) hydrogen peroxide, 0.04 mM 6,7-dimethyl-5,6,7,8-tetrahydropterine, 5 units peroxidase, and 0.008–0.032 unit dihydropteridine reductase.

Enzymatic Assay of DIPEPTIDYL PEPTIDASE (EC 3.4.14.5)

Principle

$$\text{Gly-Pro-4-Nitroanilide} + H_2O \xrightarrow{\text{Dipeptidyl Peptidase}} \text{Gly-Pro} + \text{p-Nitroaniline}$$

Conditions: T = 37°C, pH = 8.0, A405nm, Inc. = 15 minutes

Method: Stopped Rate Spectrophotometric Determination using a Microplate Reader

Reagents

A. 0.1 M Tris, pH 8.0 at 37°C (Prepare 100 ml in deionized water using Trizma Base. Adjust to pH 8.0 at 37°C using 1 M HCl).
B. 1 mM Gly-Pro-pNA Solution (Prepare 10 ml in Reagent A, using Gly-Pro-p-nitroaniline, Hydrochloride. *Prepare Fresh*).
C. 1 mM p-Nitroaniline Solution (pNA) (Prepare 10 ml in Reagent A, using p-Nitroaniline. *Prepare Fresh*).
D. Dipeptidyl Peptidase Enzyme Solution (Immediately before use, prepare the following in cold reagent A. Reconstitute 0.75 unit vial in 1 ml. Dilute to obtain 0.04–0.08 units/ml).

Procedure

1. Set up microtitre 96-well plate with the following standards:
S1 = 20 µl 1 mM pNA (20 nmols)
S2 = 40 µl 1 mM pNA (40 nmols)
S3 = 60 µl 1 mM pNA (60 nmols)
S4 = 80 µl 1 mM pNA (80 nmols)
S5 = 100 µl 1 mM pNA (100 nmols)
Make all wells up to 0.1 ml with Reagent A.

2. Pipette in enzyme samples as follows:
T1 = 10 µl
T2 = 20 µl
T3 = 30 µl
T4 = 40 µl
T5 = 50 µl
Make all wells up to 0.1 ml with Reagent A.

3. Add 0.1 ml of Reagent B to each well (including standards) to start the reaction.

4. Incubate at 37°C for 15 minutes.

5. Read absorbance at 405 nm in a microplate reader.

Calculations

Plot a standard curve of A405nm versus nmoles of pNA. Calculate nmoles/minute and hence from this determine µmoles/min/ml of enzyme.

Unit Definition: *One unit produces 1.0 µmole of 4-Nitroaniline from Gly-Pro-4-nitroaniline per minute in 0.1 M Tris/HCl at pH 8.0 at 37°C.*

Final Assay Concentration

In a 0.2 ml reaction mix, the final concentrations are 100 mM Tris, 0.5 mM Gly-Pro-pNA, and varying amounts of enzyme.

Enzymatic Assay of DOPAMINE β-HYDROXYLASE (EC 1.14.17.1)

Principle

$$\text{Tyramine} + \text{Ascorbate} + O_2 \xrightarrow{\text{Dopamine } \beta\text{-Hydroxylase}} \text{Octopamine}$$

$$\text{Octopamine} + \text{Periodate} \rightarrow \text{p-Hydroxybenzaldehyde}$$

Conditions: T = 37°C, pH = 5.0, A330nm, Light path = 1 cm

Method: Spectrophotometric Stopped Rate Determination

Reagents

A. 1 M Sodium Acetate Buffer, pH 5.0 at 37°C (Prepare 100 ml in deionized water using Sodium Acetate, Trihydrate. Adjust to pH 5.0 at 37°C with 1 M HCl).

B. 400 mM Tyramine Hydrochloride Solution (Prepare 5 ml in deionized water using Tyramine Hydrochloride).

C. 200 mM Sodium Ascorbate Solution (Asc) (Prepare 10 ml in deionized water using L-Ascorbic Acid, Sodium Salt. *Prepare Fresh*).
D. 200 mM Sodium Fumarate Solution (Fum) (Prepare 10 ml in deionized water using Fumaric Acid, Disodium Salt. *Prepare Fresh*).
E. 200 mM N-Ethylmaleimide Solution (NEM) (Prepare 10 ml in deionized water using N-Ethylmaleimide. *Prepare Fresh*).
F. 94 mM Sodium m-Periodate Solution (Per) (Prepare 10 ml in deionized water using Sodium m-Periodate. *Prepare Fresh*).
G. 530 mM Sodium Bisulfite Solution (Bisul) (Prepare 10 ml in deionized water using Sodium Bisulfite. *Prepare Fresh*).
H. 0.5 mM DL-Octopamine Standard Solution (Octo) (Prepare 100 ml in deionized water using DL-Octopamine Hydrochloride).
I. Catalase Enzyme Solution (Cat) (Immediately before use, prepare a solution containing 35,000 units/ml in deionized water using Catalase).
J. Dopamine β-Hydroxylase Enzyme Solution (Immediately before use, prepare a solution containing 0.15–0.30 unit/ml of Dopamine β-Hydroxylase in cold deionized water).
K. Dowex 50W Ion Exchange Resin (Prepare 5 ml using Dowex 50W Hydrogen Resin, by washing with deionized water until the eluate is clear and colorless).
L. 3 M Trichloroacetic Acid Solution (TCA) (Prepare 5 ml in deionized water using Trichloroacetic Acid, 6.1 N Solution, approximately 100% (w/v).
M. 4 M Ammonium Hydroxide Solution (NH_4OH) (Prepare 50 ml in deionized water using Ammonium Hydroxide).

Procedure

Prepare 0.5 ml Dowex 50W columns for each Test, Standard, and Blank. Prepare a reaction cocktail by pipetting (in milliliters) the following reagents into a suitable container:

Deionized Water	3.50
Reagent A (Buffer)	2.00
Reagent B (Tyramine)	0.50
Reagent C (Asc)	0.50
Reagent D (Fum)	0.50
Reagent E (NEM)	1.50
Reagent I (Cat)	0.50

Mix by swirling and adjust to pH 5.0 at 37°C with either 1 M HCl or 1 M NaOH, if necessary. Pipette (in milliliters) the following reagents into suitable containers:

	Test	Std 1	Std 2	Blank
Reaction Cocktail	0.90	0.90	0.90	0.90

Equilibrate to 37°C with agitation (using a metabolic shaker) while exposing to air in order to aerate the solution. Then add:

	Test	Std 1	Std 2	Blank
Deionized Water	—	—	0.05	0.10
Reagent J (Enzyme Soln.)	0.10	—	—	—
Reagent H (Octo)	—	0.10	0.05	—

Mix by swirling and incubate with agitation for exactly 5 minutes. Then add:

Reagent L (TCA)	0.20	0.20	0.20	0.20

Mix by inversion. Pour the contents of the Test, Standards, and Blank into Dowex 50W columns appropriately labeled. Wash each column 10 times using 0.5 ml of water for each washing and discard the washings. Elute each column with 2.5 ml of Reagent M using 0.5 ml volumes and save the eluates for Test, Standards, and Blanks. Pipette (in milliliters) the following reagents into suitable containers:

Test Eluate	2.50	—	—	—
Standard 1 Eluate	—	2.50	—	—
Standard 2 Eluate	—	—	2.50	—
Blank Eluate	—	—	—	2.50
Reagent F (Per)	0.25	0.25	0.25	0.25

Mix by swirling and incubate at 25°C for exactly 10 minutes. Then add:

Reagent G (Bisul)	0.25	0.25	0.25	0.25

Mix by swirling. Transfer to suitable cuvettes and record the absorbance at A330nm for the Test, Standards, and Blank, using a suitable spectrophotometer.

Calculations

Standard Determination

$$\Delta A330nm = A330nm = Standard - A330nm\ Blank$$

Sample Determination

$$\text{Units / ml enzyme} = \frac{(\Delta A330nm\ Test - \Delta A330nm\ Blank)\ (df)}{(5)\ (\Delta A330nm / \mu moles / Std)\ (0.1)}$$

df = Dilution factor
5 = Time (in minutes) of assay as per the Unit Definition
0.1 = Volume (in milliliters) of enzyme used

$$\text{Units / mg solid} = \frac{\text{units / ml enzyme}}{\text{mg solid / ml enzyme}}$$

$$\text{Units / mg protein} = \frac{\text{units / ml enzyme}}{\text{mg protein / ml enzyme}}$$

Unit Definition: *One unit will convert 1.0 μmole of tyramine to octopamine per minute at pH 5.0 at 37°C.*

Final Assay Concentrations

In a 1.00 ml reaction mix, the final concentrations are 200 mM sodium acetate, 20 mM tyramine, 10 mM sodium ascorbate, 10 mM sodium fumarate, 30 mM N-ethylmaleimide, 1750 units catalase and 0.015–0.030 unit dopamine β-hydroxylase.

Note

1. A simple column can be constructed by putting a porous polyethylene plug in a 1 ml graduated disposable syringe. The Dowex 50W resin is made into a slurry using deionized water and poured into the column until the resin slurry settles to the 0.5 ml mark of the syringe.

Enzymatic Assay of XYLANASE Activity in DRISELASE (EC 3.2.1.8)

Principle

$$\text{Xylan} + H_2O \xrightarrow{\text{Xylanase}} \text{Reducing Sugar (measured as glucose)}$$

Conditions: T = 37°C, pH = 4.5, A410nm, Light path = 1 cm

Method: Colorimetric

Reagents

A. 100 mM Sodium Acetate Buffer, pH 4.5 at 37°C (Prepare 50 ml in deionized water using Sodium Acetate, Trihydrate. Adjust to pH 4.5 at 37°C with 5 M HCl).
B. 2.5% (w/v) Xylan Substrate solution (Xylan) (Prepare 15 ml in Reagent A using Xylan).
C. Driselase (Xylanase) Enzyme Solution (Immediately before use, prepare a solution containing 0.075 – 0.15 units/ml of Xylanase in cold deionized water).
D. 500 mM Sodium Hydroxide Solution (Prepare 200 ml using deionized water and Sodium Hydroxide (1.0 N)).
E. 0.5% (w/v) p-Hydroxybenzoic Acid Hydrazide Solution (PAHBAH) (Prepare 200 ml in Reagent D using p-Hydroxybenzoic Acid Hydrazide. *Prepare Fresh before use*).
F. 0.2 mg/ml Glucose Standard Solution (Glucose Std) (Prepare by diluting Glucose Standard Solution, with deionized water).

Procedure

Pipette (in milliliters) the following reagents into suitable containers:

	Test	Blank
Deionized Water	—	1.00
Reagent A (Buffer)	3.00	3.00
Reagent B (Xylan)	1.00	1.00

Mix by swirling and equilibrate to 37°C. Then add:

	Test	Blank
Reagent C (Enzyme Solution)	1.00	—

Mix by swirling and incubate at 37°C for exactly 60 minutes. Pipette (in milliliters) the following reagents into suitable test tubes:

	Test	Blank	Std 1	Std 2	Std 3	Std 4	Std 5	Std Blank
Deionized water	0.97	0.97	0.95	0.90	0.80	0.60	0.40	1.00
Test Solution	0.03	—	—	—	—	—	—	—
Blank Solution	—	0.03	—	—	—	—	—	—
Reagent F (Std)	—	—	0.05	0.10	0.20	0.40	0.60	—
Reagent E (PAHBAH)	9.00	9.00	9.00	9.00	9.00	9.00	9.00	9.00

Immediately mix by swirling and transfer the tubes to a boiling water bath. Incubate for 5 minutes. Remove the tubes from the boiling water bath and allow to cool to room temperature. Mix by inversion and transfer the solutions to suitable cuvettes. Obtain the A410nm for the Test, Blank, and Standards using a suitable spectrophotometer.

Calculations

Standard Curve: ΔA410nm Std = A410nm Std–A410nm Std Blank

Prepare a standard curve by plotting the ΔA410nm Standard vs the milligrams of Glucose.

Sample Determination: ΔA410nm Sample = A410nm Test–A410nm Blank

Determine the milligrams of glucose using the standard curve.

$$\text{Units / ml enzyme} = \frac{\text{(milligrams of glucose liberated) (5) (df)}}{(60)\,(1)\,(0.03)}$$

5 = Volume (in milliliters) of assay
df = Dilution factor
60 = Time of assay (in minutes) as per Unit Definition
1 = Volume (in milliliters) of enzyme used
0.03 = Volume (in milliliter) used for reducing sugar determination

$$\text{Units / mg solid} = \frac{\text{units / ml enzyme}}{\text{mg solid / ml enzyme}}$$

$$\text{Units / mg protein} = \frac{\text{units / ml enzyme}}{\text{mg protein / ml enzyme}}$$

Unit Definition: *One unit will liberate 1.0 milligram of reducing sugar from xylan (measured as glucose) per minute at pH 4.5 at 37°C.*

Final Assay Concentrations

In a 5.00 ml reaction mix, the final concentrations are 60 mM sodium acetate, 0.50% (w/v) xylan and 0.075–0.15 unit xylanase.

Enzymatic Assay of ELASTASE (EC 3.4.21.36)

Principle

$$\text{Insoluble Elastin - Orcein} + H_2O \xrightarrow{\text{Elastase}} \text{Soluble Hydrolysis Products}$$

Conditions: T = 37°C, pH = 8.8, A590nm, Light path = 1 cm

Method: Spectrophotometric Stop Rate Determination

Reagents

A. 200 mM Tris HCl Buffer, pH 8.8 at 37°C (Prepare 100 ml in deionized water using Trizma Base. Adjust to pH 8.8 at 37°C with 1 M HCl).
B. Elastin-Orcein Substrate (El-Or).
C. Elastase Enzyme Solution (Immediately before use, prepare a solution containing approximately 1500–1800 units/ml of Elastase in cold Reagent A).

Procedure

Step 1: Standard Curve: Weigh (in milligrams) the following reagents:

	Std 1	Blank 1	Std 2	Blank 2	Std 3	Blank 3	Std 4	Blank 4
Reagent B (EL-OR)	5.00	5.00	10.00	10.00	15.00	15.00	20.00	20.00

Then add (in milliliters):

Reagent A (Buffer)	6.00	6.01	6.00	6.01	6.00	6.01	6.00	6.01

Mix and equilibrate to 37°C. Then add:

Reagent C (Enzyme)	0.01	—	0.01	—	0.01	—	0.01	—

Mix by swirling and place all the containers in a suitably thermostatted metabolic shaker. Incubate the Standards and Standard Blanks at 37°C for 12–16 hours. Filter the solutions through 0.8 μm syringe filter. Record the A590nm for the Standards and Standard Blanks using a suitable spectrophotometer.

Step 2: Samples: Weigh (in milligrams) the following reagents:

	Test 1	Test 2	Test 3	Test Blank
Reagent B (El-Or)	50.00	50.00	50.00	50.00

Then add (in milliliters):

Reagent A (Buffer)	6.00	5.995	5.99	6.01

Mix and equilibrate to 37°C. Then add:

Reagent C (Enzyme)	0.01	0.015	0.02	—

Mix by swirling and place all the containers in a suitably thermostatted metabolic shaker. Incubate the Tests and Test Blank at 37°C for 20 minutes. Filter the solution throughs 0.8 μm syringe filters. Record the A590nm for the Tests and Test Blank using a suitable spectrophotometer.

Calculations

Standard Curve: ΔA590nm Standard = A590nm Std–A590nm Std Blank

Prepare a standard curve by plotting the ΔA590nm of the Standards vs milligrams of solubilized elastinorcein.

Sample Determination: ΔA590nm Sample = A590nm Test–A590nm Blank

Determine the mg of solubilized elastin-orcein using the Standard Curve.

$$\text{Units / ml enzyme} = \frac{\text{(mg of elastin solubilized) (df)}}{0.01}$$

df = Dilution Factor

0.01 = Volume (in milliliters) of enzyme used

$$\text{Units / mg solid} = \frac{\text{units / ml enzyme}}{\text{mg solid / ml enzyme}}$$

$$\text{Units / mg protein} = \frac{\text{units / ml enzyme}}{\text{mg protein / ml enzyme}}$$

Unit Definition: *One unit will solubilize 1 mg of elastin in 20 minutes at pH 8.8 at 37°C.*

Final Assay Concentration

In a 6.01 ml reaction mix, the final concentrations are 200 mM Tris, 50 mg Elastin-Orcein and 15–36 units elastase.

Notes

1. This assay procedure is not to be used to assay Elastase, Leukocyte.
2. The Metabolic shaker should be adjusted to approximately 60 shake cycles/minute.
3. If the filtrate is hazy, it can be centrifuged to remove the haziness.
4. Due to the variance of Elastin-Orcein lots, the standard curve may not be consistent. The amount of enzyme utilized in the assay is not affected by this variance.

Enzymatic Assay of ENDONUCLEASE from *Neurospora crassa*

Principle

$$\text{DNA} + H_2O \xrightarrow{\text{Endonuclease}} \text{Acid - Soluble Oligonucleotides}$$

Abbreviation Used

DNA = Deoxyribonucleic Acid

Conditions: T = 37°C, pH 8.0, A260nm, Light path = 1 cm

Method: Spectrophotometric Stop Rate Determination

Reagents

A. 200 mM Tris HCl Buffer with 20 mM Magnesium Chloride and 300 mM Sodium Chloride, pH 8.0 at 37°C (Prepare 100 ml in deionized water using Trizma Base,

Magnesium Chloride, Hexahydrate, and Sodium Chloride. Adjust to pH 8.0 at 37°C with 1 M HCl).

B. 0.08% (w/v) Deoxyribonucleic Acid .(DNA) (Prepare by dissolving 15 mg of Deoxyribonucleic Acid, Sodium Salt, in 5 ml of deionized water. Mix by stirring for 4–5 hours at room temperature, then heat in a boiling water bath for 15 minutes. Cool to room temperature and then remove 2.7 ml of the DNA solution and combine with 5 ml of Reagent A and 2.3 ml of deionized water).

C. 1 M Perchloric Acid Solution ($HClO_4$) (Prepare 10 ml in deionized water using Perchloric Acid).

D. Endonuclease Enzyme Solution (Immediately before use, prepare a solution containing 16–48 units/ml of Endonuclease in cold deionized water).

Procedure

Pipette (in milliliters) the following reagents into suitable containers:

	Test	Blank
Reagent B (DNA)	1.00	1.00
Deionized Water	0.50	0.50

Mix by swirling and equilibrate to 37°C. Then add:

Reagent D (Enzyme Solution)	0.10	—
Deionized Water	—	0.10

Mix by inversion and incubate for exactly 30 minutes at 37°C. Remove 0.30 ml from both the Test and Blank and transfer into Eppendorf tubes each containing 0.50 ml of Reagent C ($HClO_4$) and 0.20 ml of deionized water. Mix by swirling and centrifuge for 10 minutes. Remove 0.10 ml of the supernatant from both the Test and Blank and add 2.90 ml of deionized water to each tube. Record the A260nm for both the Test and Blank in a suitable spectrophotometer.

Calculation

$$\text{Units / ml enzyme} = \frac{(\text{A260nm Test} - \text{A260nm Blank})\,(3)\,(1.60)\,(\text{df})}{(0.1)\,(0.3)\,(0.1)}$$

3 = Total volume (in milliliters) used to measure the A260nm
1.60 = Volume (in milliliters) of reaction mix
0.1 = Volume (in milliliter) of enzyme used
0.3 = Volume (in milliliter) removed from the reaction mix
0.1 = Volume (in milliliter) removed from the stopped reaction
df = Dilution Factor

$$\text{Units / mg protein} = \frac{\text{units / ml enzyme}}{\text{mg protein / ml enzyme}}$$

Unit Definition: *One unit will produce acid-soluble oligonucleotides equivalent to a A260nm of 1.0 in 30 minutes at pH 8.0 at 37°C in a reaction volume of 1.60 ml.*

Final Assay Concentrations

In a 1.60 ml reaction mix, the final concentrations are 63 mM Tris, 6 mM magnesium chloride, 94 mM sodium chloride, 0.05% (w/v) deoxyribonucleic acid, and 1.6–4.8 units endonuclease.

Enzymatic Assay of ENDONUCLEASE

Principle

$$DNA + H_2O \xrightarrow{\text{Endonuclease}} \text{Acid - Soluble Oligonucleotides}$$

Abbreviation Used

DNA = Deoxyribonucleic Acid

Conditions: T = 37°C, pH = 8.0, A260nm, Light path = 1 cm

Method: Spectrophotometric Stop Rate Determination

Reagents

A. 50 mM Tris HCl Buffer with 1 mM Magnesium Chloride and 0.1% (w/v) Bovine Serum Albumin, pH 8.0 at 37°C (Prepare 100 ml in deionized water using Trizma Base, Magnesium Chloride, Hexahydrate, and Albumin, Bovine or equivalent. Adjust to pH 8.0 at 37°C with 1 M HCl).
B. 0.1% (w/v) Deoxyribonucleic Acid (DNA) (Prepare 25 ml in Reagent A, using Deoxyribonucleic Acid).
C. 4% (v/v) Perchloric Acid Solution ($HClO_4$) (Prepare 10 ml in deionized water using Perchloric Acid).
D. Endonuclease Enzyme Solution Immediately before use, prepare a solution containing 40 units/ml of Endonuclease in cold Reagent A).

Procedure

Pipette (in milliliters) the following reagents into suitable tubes:

	Test	Blank
Reagent B (DNA)	2.50	2.50

Equilibrate to 37°C. Then add:

Reagent D (Enzyme Solution)	0.125	—
Reagent A (Buffer)	—	0.125

Mix by inversion and incubate for exactly 30 minutes at 37°C. Remove 0.50 ml from both the Test and Blank and transfer into Eppendorf tubes containing 0.50 ml of ice cold Reagent C ($HClO_4$). Cool on ice for 60 minutes. Centrifuge for 5 minutes at 4°C. Obtain the A260nm of the supernatant for both the Test and Blank in a suitable spectrophotometer using air.

Calculation

$$\text{Units / ml enzyme} = \frac{(\text{A260nm Test} - \text{A260nm Blank})\ (2.625)\ (\text{df})}{(1)\ (0.5)\ (0.125)}$$

2.625 = Total volume (in milliliters) of assay
df = Dilution factor
1 = A260nm of one unit as per the Unit Definition
0.5 = Volume (in milliliter) of supernatant used in assay
0.125 = Volume (in milliliter) of enzyme used

Unit Definition: *One unit will produce acid-soluble oligonucleotides equivalent to a A260nm of 1.0 in 30 minutes at pH 8.0 at 37°C (reaction volume 2.625 ml).*

Final Assay Concentration

In a 2.625 ml reaction mix, the final concentrations are 50 mM Tris, 1 mM magnesium chloride, 0.1% (w/v) bovine serum albumin, 0.095% (w/v) deoxyribonucleic acid and 5 units endonuclease.

Enzymatic Assay of ENOLASE (EC 4.2.1.11)

Principle

$$\text{2 - Phosphoglycerate} + H_2O \xrightarrow{\text{Enolase}} \text{PEP}$$

$$\text{PEP} + \text{ADP} \xrightarrow{\text{Pyruvate Kinase}} \text{Pyruvate} + \text{ATP}$$

$$\text{Pyruvate} + \beta\text{ - NADH} \xrightarrow{\text{L-Lactic Dehydrogenase}} \text{L - Lactate} + \beta - \text{NAD}$$

Abbreviations Used

PEP = Phospho(enol) Pyruvate
ADP = Adenosine 5'-Diphosphate
ATP = Adenosine 5'-Triphosphate
β-NADH = β-Nicotinamide Adenine Dinucleotide, Reduced Form
β-NAD = β-Nicotinamide Adenine Dinucleotide, Oxidized

Conditions: T = 25°C, pH = 7.4, A340nm, Light path = 1 cm

Method: Continuous Spectrophotometric Rate Determination

Reagents

A. 100 mM Triethanolamine Buffer, pH 7.4 at 25°C (Prepare 50 ml in deionized water using Triethanolamine Hydrochloride. Adjust to pH 7.4 at 25°C with 1 M NaOH).
B. 56 mM 2-Phosphoglycerate Solution (DPG) (Prepare 2 ml in deionized water using D(+)2-Phosphoglyceric Acid, Sodium Salt, Hydrate).
C. 7 mM β-Nicotinamide Adenine Dinucleotide, Reduced Form Solution (β-NADH) (Dissolve the contents of one 5 mg vial of β-Nicotinamide Adenine Dinucleotide, Reduced Form, Disodium Salt, in 1 ml of Reagent A or prepare 1 ml in deionized water using β-Nicotinamide Adenine Dinucleotide, Reduced Form, Disodium salt. *Prepare Fresh*).

D. 500 mM Magnesium Sulfate with 2 M Potassium Chloride Solution ($MgSO_4$/KCl) (Prepare 5 ml in deionized water using Magnesium Sulfate, Anhydrous, and Potassium Chloride).
E. 20 mM Adenosine 5'-Diphosphate Solution (ADP) (Prepare 1 ml in deionized water using Adenosine 5'-Diphosphate, Sodium Salt. *Prepare Fresh*).
F. PK/LDH Mixed Enzymes.
G. 15 mM Tris HCl with 0.02% (w/v) Bovine Serum Albumin (Enzyme Diluent) (Prepare 100 ml in deionized water using Trizma Base, and Albumin, Bovine A-4503, or equivalent. Adjust to pH 7.4 at 25°C with 1 M HCl).
H. Enolase Enzyme Solution (Immediately before use, prepare a solution containing 0.25–0.5 unit/ml of Enolase in cold Reagent G).

Procedure

Pipette (in milliliters) the following reagents into suitable cuvettes:

	Test	Blank
Reagent A (Buffer)	2.39	2.39
Reagent B (DPG)	0.10	0.10
Reagent C (β-NADH)	0.05	0.05
Reagent D ($MgSO_4$/KCl)	0.15	0.15
Reagent E (ADP)	0.20	0.20
Reagent F (PK/LDH)	0.01	0.01

Mix by inversion and equilibrate to 25°C. Monitor the A340nm until constant, using a suitably thermostatted spectrophotometer. Then add:

Reagent G (Enzyme Diluent)	—	0.10
Reagent H (Enzyme Solution)	0.10	—

Immediately mix by inversion and record the decrease in A340nm for approximately 5 minutes. Obtain the ΔA340nm/minute using the maximum linear rate for both the Test and Blank.

Calculations

$$\text{Units / ml enzyme} = \frac{(\Delta\text{A340nm / min Test} - \Delta\text{A340nm / min Blank})\ (3)\ (\text{df})}{(6.22)\ (0.1)}$$

3 = Volume (in milliliters) of assay
df = Dilution factor
6.22 = Millimolar extinction coefficient of β-NADH at 340nm
0.1 = Volume (in milliliter) of enzyme used

$$\text{Units / mg solid} = \frac{\text{units / ml enzyme}}{\text{mg solid / ml enzyme}}$$

$$\text{Units / mg protein} = \frac{\text{units / ml enzyme}}{\text{mg protein / ml enzyme}}$$

Unit Definition: *One unit will convert 1.0 μmole of 2-phosphoglycerate to phospho(enol) pyruvate per minute at pH 7.4 at 25°C.*

Final Assay Concentration

In a 3.00 ml reaction mix, the final concentrations are 81 mM triethanolamine, 1.9 mM 2-phosphoglycerate, 0.12 mM β-nicotinamide adenine dinucleotide, reduced form, 25 mM magnesium sulfate, 100 mM potassium chloride, 1.3 mM adenosine 5'-diphosphate, 7 units pyruvate kinase, 10 units L-lactic dehydrogenase and 0.025–0.05 unit Enolase.

Notes

1. Contains approximately 700 units/ml of Pyruvate Kinase and 1000 units/ml of Lactic Dehydrogenase.
2. Pyruvate Kinase unit definition: One unit will convert 1.0 µmole of phospho(enol) pyruvate to pyruvate per minute at pH 7.6 at 37°C.
3. Lactic Dehydrogenase unit definition: One unit will reduce 1.0 µmole of pyruvate to L-lactate per minute at pH 7.5 at 37°C.

Enzymatic Assay of ENTEROKINASE (EC 3.4.21.9)

Principle

Step 1

$$\text{Trypsinogen} + H_2O \xrightarrow{\text{Enterokinase}} \text{Trypsin} + \text{Val-Asp-Asp-Asp-Asp-Lys}$$

Step 2

$$\text{BAEE} + H_2O \xrightarrow{\text{Trypsin}} \text{Na-Benzoyl-L-Arginine} + \text{EtOH}$$

Abbreviations Used

BAEE = Na-Benzoyl-L-Arginine Ethyl Ester
EtOH = Ethanol

Conditions: T = 25°C, pH = 5.6, A253nm, Light path = 1 cm

Method: Continuous Spectrophotometric Rate Determination

Reagents

A. 40 mM Succinate Buffer, pH 5.6 at 25°C (Prepare 100 ml in deionized water using Succinic Acid, Free Acid. Adjust to pH 5.6 at 25°C with 1 M NaOH).
B. 1 mM Hydrochloric Acid with 5 mM Calcium Chloride Solution (Prepare 100 ml in deionized water using Hydrochloric Acid, 1.0 N, and Calcium Chloride, Dihydrate).
C. 0.1% (w/v) Trypsinogen Solution (Trypsinogen) (Immediately before use, prepare 25 ml in cold Reagent B using Trypsinogen).
D. Enterokinase Enzyme Solution (Immediately before use, prepare a solution containing 2–5 units/ml of Enterokinase in cold deionized water).
E. 67 mM Sodium Phosphate Buffer, pH 7.6 at 25°C (Prepare 1 liter in deionized water using Sodium Phosphate, Monobasic, Anhydrous. Adjust to pH 7.6 at 25°C with 1 M NaOH).
F. 0.248 mM Na-Benzoyl-L-Arginine Ethyl Ester Solution (BAEE) (Prepare 100 ml in Reagent E using Na-Benzoyl-Larginine Ethyl Ester, Hydrochloride. *Prepare Fresh*).
G. 40 mM Hydrochloric Acid with 5 mM Calcium Chloride Solution (HCl-$CaCl_2$) (Prepare 1 liter in deionized water using Hydrochloric Acid, 1.0 N, and Calcium Chloride, Dihydrate).

Procedure

Step 1: Pipette (in milliliters) the following reagents into suitable tubes:

	Test Mix	Blank Mix
Reagent A (Buffer)	1.80	1.80
Reagent C (Trypsinogen)	0.50	0.50

Mix by inversion and equilibrate to 25°C. Then add:

Reagent D (Enterokinase)	0.10	—
Deionized Water	—	0.10

Immediately mix by inversion and incubate at 25°C for exactly 15 minutes. Then add:

Reagent G (HCl-$CaCl_2$)	3.00	3.00

Step 2: Pipette (in milliliters) the following reagents into suitable quartz cuvettes:

	Test	Blank
Reagent F (BAEE)	3.00	3.00

Equilibrate to 25°C. Monitor the A253nm until constant, using a suitably thermostatted spectrophotometer. Then add:

Test Mix (Step 1)	0.20	—
Blank Mix (Step 1)	—	0.20

Immediately mix by inversion and record the increase in A253nm for approximately 5 minutes. Obtain the ΔA253nm/minute using the maximum linear rate for both the Test and Blank.

Calculations

$$\text{Units / ml enzyme} = \frac{(\Delta\text{A253nm / min Test} - \Delta\text{A253nm / min Blank})\ (5.4)\ (\text{df})}{(0.001)\ (0.20\ \text{ml})\ (\text{P.A.})\ (0.024)\ (15)\ (0.1)}$$

5.4 = Volume (in milliliters) of Step 1
df = Dilution factor
0.001 = The change in A253nm/minute per unit of Trypsin as per the Unit Definition
0.20 = Volume (in milliliter) from Step 1 used in Step 2
P.A. = Potential activity of Trypsinogen
0.024 = mg trypsin/nanomole trypsin
15 = Time (in minutes) for Step 1 as per the Unit Definition
0.1 = Volume (in milliliter) of enterokinase used

$$\text{Units / mg solid} = \frac{\text{units / ml enzyme}}{\text{mg solid / ml enzyme}}$$

$$\text{Units / mg protein} = \frac{\text{units / ml enzyme}}{\text{mg protein / ml enzyme}}$$

Unit Definition: *One unit will produce 1.0 nanomole of trypsin from trypsinogen per minute at pH 5.6 at 25°C.*

Final Assay Concentration

In a 2.40 ml reaction mix, the final concentrations are 30 mM succinate, 1 mM calcium chloride, 0.2 mM hydrochloric acid, 0.5 mg trypsinogen and 0.2–0.5 unit enterokinase.

Notes

1. The potential activity is a reported value found on the product label of Trypsinogen.
2. This unit corresponds to approximately 2.7 units of the assay at 5°C. One unit would activate 0.065 mg of trypsinogen per hour at pH 5.8 at 5°C.

Enzymatic Assay of ESTERASE (EC 3.1.1.1)

Principle

$$\text{o-Nitrophenyl Butyrate} + H_2O \xrightarrow{\text{Esterase}} \text{o-Nitrophenol} + \text{Butyric Acid}$$

Conditions: T = 25°C, pH = 7.5, A420nm, Light path = 1 cm

Method: Continuous Spectrophotometric Rate Determination

Reagents

A. 50 mM Potassium Phosphate Buffer, pH 7.5 at 25°C (Prepare 100 ml in deionized water using Potassium Phosphate, Monobasic, Anhydrous. Adjust to pH 7.5 with 1 M KOH).
B. 100 mM o-Nitrophenyl Butyrate Solution (ONPB) (Prepare 10 ml in Dimethylsulfoxide using o-Nitrophenyl Butyrate, and Dimethyl Sulfoxide).
C. Esterase Enzyme Solution (Immediately before use, prepare a solution containing 0.5–1.0 unit/ml of Esterase in cold Reagent A).

Procedure

Pipette (in milliliters) the following reagents into suitable cuvettes:

	Test	Blank
Reagent A (Buffer)	2.87	2.87
Reagent B (ONPB)	0.03	0.03

Mix by inversion and equilibrate to 25°C. Monitor the A420nm until constant, using a suitably thermostatted spectrophotometer. Then add:

	Test	Blank
Reagent C (Enzyme Solution)	0.10	—
Reagent A (Buffer)	—	0.10

Immediately mix by inversion and record the increase in A420nm for approximately 5 minutes. Obtain the ΔA420nm/minute using the maximum linear rate for both the Test and Blank.

Calculations

$$\text{Units / ml enzyme} = \frac{(\Delta A420nm / min\ Test - \Delta A420nm / min\ Blank)\ (3)\ (df)}{(5.00)\ (0.1)}$$

3 = Total volume (in milliliters) of assay
df = Dilution factor
5.00 = Millimolar extinction coefficient of o-Nitrophenol at 420 nm at pH 7.5
0.1 = Volume (in milliliter) of enzyme used

$$\text{Units / mg solid} = \frac{\text{units / ml enzyme}}{\text{mg solid / ml enzyme}}$$

$$\text{Units / mg protein} = \frac{\text{units / ml enzyme}}{\text{mg protein / ml enzyme}}$$

Unit Definition: *One unit will hydrolyze 1.0 μmole of o-nitrophenyl butyrate to butyric acid and o-nitrophenol per minute at pH 7.5 at 25°C.*

Final Assay Concentration

In a 3.00 ml reaction mix, the final concentrations are 50 mM potassium phosphate, 1 mM 0-nitrophenyl butyrate and 0.05–0.10 unit of esterase.

Enzymatic Assay of ESTERASE (EC 3.1.1.1)

Principle

$$\text{Ethyl Butyrate} + H_2O \xrightarrow{\text{Esterase}} \text{Butyric Acid} + \text{Ethanol}$$

Conditions: T = 25°C, pH = 8.0

Method: Titrimetric

Reagents

A. 10 mM Borate Buffer, pH 8.0 at 25°C (Prepare 500 ml in deionized water using Boric Acid. Adjust to pH 8.0 with 1 M NaOH).
B. 0.1% (v/v) Ethyl Butyrate Solution (Prepare 200 ml in Reagent A using Butyric Acid Ethyl Ester).
C. Ethyl Butyrate (Use Butyric Acid Ethyl Ester).
D. 10 mM Sodium Hydroxide Solution-Standardized (NaOH) (Prepare 50 ml in cold deionized water using Sodium Hydroxide, Pellets. Standardize according to the ACS Reagent Procedure).
E. Esterase Enzyme Solution (Immediately before use, prepare a solution containing 50 units/ml of Esterase in cold Reagent A).

Procedure

Using a suitable pH meter in conjunction with a magnetic stirrer, pipette (in milliliters) the following reagents into a suitable titration vessel:

	Test
Reagent B (Ethyl Butyrate Soln)	25.0

Equilibrate to 25°C. Then add:

	Test
Reagent C (Ethyl Butyrate)	0.025

Adjust the pH to 8.1 using Reagent D (NaOH). Then add:

Reagent E (Enzyme Solution)	0.10

Start a timer when the pH reaches 8.0. Run the reaction for 1–5 minutes. Maintain the pH of the reaction mix at pH 8.0 by the addition of small volumes (0.050 ml) of Reagent D. Record the volume of Reagent D used to maintain the pH at 8.0 and the time required to consume the added Reagent D.

Calculations

$$\text{Units / ml enzyme} = \frac{(\text{Molarity of NaOH})\ (\text{NaOH})\ (1000)\ (\text{df})}{(\text{T})\ (0.10)}$$

NaOH = Volume (in milliliters) of Reagent D used in the assay
1000 = Conversion factor from millimoles to micromoles (Unit definition)
df = Dilution factor
T = Time required (in minutes) to consume the added Reagent D (NaOH) while maintaining the pH at 8.0
0.1 = Volume (in milliliter) of enzyme used

$$\text{Units / mg protein} = \frac{\text{units / ml enzyme}}{\text{mg protein / ml enzyme}}$$

Unit Definition: *One unit will hydrolyze 1.0 μmole of ethyl butyrate to butyric acid and ethanol per minute at pH 8.0 at 25°C.*

Initial Assay Concentrations

In a 25.125 ml reaction mix, the initial concentrations are 10 mM borate, 0.2% (v/v) ethyl butyrate and 5 units esterase.

Enzymatic Assay of FERREDOXIN-NADP+ REDUCTASE (EC 1.18.1.2)

Principle

$$\text{Ferredoxin Oxidized} + \text{DPIred} \xrightarrow[\text{FD: NADP red}]{\text{Light}} \text{Ferreodoxin Reduced} + \text{DPIPox}$$

$$\text{Reduced Ferredoxin Reduced Ferredoxin} + \text{NADP} + \xrightarrow{\text{Chloroplasts}} \text{Oxidized Ferredoxin} + \text{NADPH}_2$$

Abbreviations Used

DPIPred = 2,6 Dichlorophenolindophenol, reduced
DPIPox = 2,6 Dichlorophenolindophenol, oxidized
Fd:NADP red = Ferredoxin-NADP+ Reductase

Conditions: T = 25°C, pH 7.6, A340nm, Light path

Method: Spectrophotometric Rate Determination

Reagents

A. 100 mM Tris HCl Buffer, pH 7.8 at 25°C (Prepare 200 ml in deionized water using Trizma Base. Adjust to pH 7.8 at 25°C using 1 M HCl).

B. 100 mM L-Ascorbic Acid Solution (Ascorbic) (Prepare 10 ml in Reagent A using Ascorbic Acid Sodium Salt).

C. 2 mM 2,6-Dichlorophenolindophenol Solution (DPIP) (Prepare 5 ml in Reagent A using 2,6-Dichloroindophenol, Sodium Salt).

D. 1 mM β-Nicotinamide Adenine Dinucleotide Phosphate Solution (NADP) (Prepare by dissolving β-Nicotinamide Adenine Dinucleotide Phosphate, Sodium Salt in the appropriate volume of Reagent A).

E. 0.5% (w/v) Ferredoxin-$NADP^+$ Reductase Enzyme Solution (Immediately before use, prepare a 2 ml solution containing approximatley 5 mg solid/ml in deionized water).

F. Fresh Chloroplasts (Chloroplast Susp). (Prepare fresh "treated" chloroplasts as described in the notes).

G. 0.2% (w/v) Ferredoxin Solution (FD) (Prepare 10 ml in Reagent A using Ferredoxin).

H. 350 mM Sodium Chloride Solution (Prepare 100 ml in deionized water using Sodium Chloride).

I. 200 mM Tris HCl Buffer, pH 8.0 at 25°C (Prepare 100 ml in deionized water using Trizma Base. Adjust to pH 8.0 at 25°C with 1 M HCl).

J. 4 mM Tris HCl with 35 mM Sodium Chloride, pH 8.0 at 25°C (Prepare 100 ml in deionized water using Trizma Base and Sodium Chloride. Adjust to pH 8.0 at 25°C at 1 M HCl).

K. 50 mM Tris HCl Buffer, pH 7.6 at 25°C (Prepare 100 ml in deionized water using Trizma Base. Adjust to pH 7.6 at 25°C with 1 M HCl).

L. NaCl–Tris Solution (Prepare 100 ml by adding 10 ml of Reagent H and 2 ml of Reagent I to 88 ml of deionized water).

Procedure

Set up the light box, then pipette (in milliliters) the following reagents into suitable cuvettes:

	Test	Blank
Reagent A (Buffer)	1.10	1.10
Deionized Water	1.00	1.00
Reagent B (Ascorbate)	0.10	0.10
Reagent C (DPIP)	0.10	0.10
Reagent F (Chloroplast Susp).	0.10	0.10

Mix by inversion and equilibrate to 25°C. Insure the Test and Blank have a positive reading at A340nm, using a suitably thermostatted spectrophotometer. Then add:

Reagent A (Buffer)	—	0.50
Reagent D (NADP)	0.50	—
Reagent G (FD)	0.10	0.10

Mix by inversion and equilibrate to 25°C. Keep the cuvettes in the dark. Read the record A340 at time zero and at 2 minute intervals for 6 minutes. Next, place the cuvettes in front of the light box. Read and record A340 at time zero and at 2 minute intervals for 6 minutes. Then add:

Reagent E	2 μl	—

Mix by inversion and equilibrate to 25°C. Keep the cuvettes in the dark. Read the record A340 at time zero and at 2 minute intervals for 6 minutes. Next, place the cuvettes in front of the light box. Read and record A340 at time zero and at 2 minute intervals for 6 minutes.

Assay Conditions	All reagents, no reductase; dark Col. 1		All reagents, no reductase; light Col. 2		All reagents, no reductase; dark Col. 3		All reagents, no reductase; light Col. 4	
Minutes	A_{340}	A_{340}	A_{340}	A_{340}	A_{340}	A_{340}	A_{340}	A_{340}
0								
2								
4								
6								

Column 1, 2 and 3 should show no A340.

Unit Definition: *One unit will reduce 1.0 μmole of NADP per minute at pH 7.6 at 25°C in the presence of treated chloroplasts, spinach ferredoxin, 2,6 dichloropheno-lindophenol, L-ascorbic acid and NAD$^+$.*

Final Assay Concentrations

In a 3.00 ml reaction mix, the final concentrations are 36.6 mM Tris, 3.33 mM ascorbic acid, 0.067 mM 2,6-dichlorophenolindophenol, 0.167 mM β-nicotinamide adenine dinucleotide phosphate, 33.3 μg/ml chlorophyll, and 0.0033 mg/ml unit ferredoxin-NADP$^+$ reductase.

Notes

1. The preparation of chloroplasts is as follows:
   A. Weigh approximately 35 g of fresh spinach leaves from a grocery store, that have been depetiolated.
   B. Homogenize the spinach in a Waring blender using 100 ml of Reagent H and 10 ml of Reagent I.
   C. Filter the homogenate through glass wool. Then centrifuge at 5500 rpm (5000 g) for 5 minutes using a Sorvall GSA rotor at 0°C.
   D. Discard the supernatant and resuspend the pellet with 100 ml of Reagent H and 10 ml of Reagent I, using a Q-tip.
   E. Centrifuge the suspension again at 0°C at 5500 rpm (5000 g) for 5 minutes.
   F. Discard the supernatant (light green) and resuspend the pellet in 8 ml of Reagent L.
   G. Filter the suspension through glass wool. Store the filtrate at 0–5°C in the dark (untreated chloroplasts).
   H. Prepare treated chloroplasts by diluting 4–5 ml of the untreated chloroplasts 10 × with Reagent K. Store in the dark at room temperature overnight (15 hours).
   I. Centrifuge in a Sorvall GSA rotor at 0°C at 10,500 rpm (18,000 g) for 10 minutes.

J. Discard the supernatant (pale yellow) and resuspend the precipitate in 50 ml of Reagent Repeat step I and step J.

K. Resuspend the pellet in 4–5 ml of Reagent K. Filter through glass wool. Keep in the dark and store the green filtrate at 0–5°C.

2. Determination of Chlorophyll concentration in Chloroplast

A. Based on equation for 80% acetone extracts.
Total Chl (µg/ml) = 6.45 (A665) + 17.72 (A649)

B. In a quartz cuvette, combine 2.4 ml acetone (Z6375) with 0.6 ml DI water. Keep covered to avoid evaporation.

C. In a 25 ml graduated cylinder, add 4.5 ml DI water and 0.5 ml chloroplast suspension. Then adjust the volume to 25 ml with 100% acetone. Filter immediately through Whatman #54 filter paper. Read the absorbance at 665 nm and 649 nm against 80% aqueous acetone in a quartz cuvette.
Chl (µg/ml) = [6.45 (A665) + 17.72 (A649)] × 50

D. In a 25 ml graduated cylinder, add 4.75 ml DI water and 0.25 ml chloroplast suspension. Then adjust the volume to 25 ml with 100% acetone. Filter immediately through Whatman #54 filter paper. Read absorbance at 665 nm and 649 nm against 80% aqueous acetone reference in a quartz cuvette.
Chl (Fg/ml) = [6.45 (A665) + 17.72 (A649)] × 100

E. Concentration of chlorophyll in chlroplast suspension is the average from the results of Step C and D. The chlorophyll concentration should be approximatley 1000 µg/ml. Dilute with Reagent K if needed.

Enzymatic Assay of FICIN (EC 3.4.22.3)

Principle

$$\text{Casein} + H_2O \xrightarrow{\text{Ficin}} \text{Amino Acids}$$

Conditions: T = 37°C, pH = 7.0, A280nm, light path = 1 cm

Method: Spectrophotometric Stop Rate Determination

Reagents

A. 100 mM Potassium Phosphate Buffer, pH 7.6 at 37°C (Prepare 100 ml in deionized water using Potassium Phosphate, Monobasic, Anhydrous. Adjust to pH 7.6 at 37°C with 5 M KOH).

B. 2.0% (w/v) Casein Solution (Casein) Prepare 50 ml in Reagent A using Casein. Heat to dissolve, but *Do not boil*. Cool to 37°C).

C. 250 mM L-Cysteine Solution, pH 7.0 at 37°C, (L-Cys) (Prepare 5 ml in deionized water using L-Cysteine, Hydrochloride, Monohydrate. Adjust to pH 7.0 at 37°C with solid Sodium Bicarbonate).

D. 250 mM Ethylenediaminetetraacetic Acid Solution (EDTA) (Prepare 10 ml in deionized water using Ethylenediaminetetraacetic Acid, Tetrasodium Salt, Hydrate).

E. 250 mM Sodium Hydroxide Solution (NaOH) (Prepare 10 ml in deionized water using Sodium Hydroxide Solution, 1.0 N).

F. 1 M Potassium Phosphate Buffer, pH 7.0 at 37°C (Enzyme Diluent) (Prepare 100 ml in deionized water using Potassium Phosphate, Monobasic, Anhydrous. Adjust to pH 7.0 at 37°C with 5 M KOH).

G. 5.0% (w/v) Trichloroacetic Acid Solution (TCA) (Prepare 100 ml in deionized water using Trichloroacetic Acid, 6.1 N Solution, approximately 100% (w/v)).

H. Ficin Enzyme Solution (Immediately before use, prepare a solution containing 0.04–0.08 unit/ml of Ficin in cold Reagent F).

Procedure

Pipette (in milliliters) the following reagents into suitable containers:

	Test	Blank
Reagent B (Casein)	2.00	2.00
Reagent C (L-Cys)	0.20	0.20
Reagent D (EDTA)	0.20	0.20
Reagent E (NaOH)	0.20	0.20
Reagent F (Enzyme Diluent)	0.60	1.00
Deionized Water	0.40	0.40

Mix by swirling and equilibrate to 37°C. Then add:

Reagent H (Enzyme Solution)	0.40	—

Immediately mix by swirling and incubate for exactly 20 minutes at 37°C. Then add:

Reagent G (TCA)	6.00	6.00

Mix by swirling and incubate at 37°C for 60 minutes. Filter through Whatman #50 filter paper. Transfer the solutions to suitable quartz cuvettes and record the A280nm for both the Test and Blank.

Calculations

$$\text{Units / ml enzyme} = \frac{(\text{A280nm Test} - \text{A280nm Blank})\,(\text{df})}{(20)\,(0.4)\,(1.0)}$$

df = Dilution factor
20 = Time of assay (in minutes)
0.4 = Volume (in milliliter) of enzyme used
1.0 = Change in A280nm per unit as per the Unit Definition

$$\text{Units / mg solid} = \frac{\text{units / ml enzyme}}{\text{mg solid / ml enzyme}}$$

$$\text{Units / mg protein} = \frac{\text{units / ml enzyme}}{\text{mg protein / ml enzyme}}$$

Unit Definition: *One unit will produce a A280 of 1.0 per minute at pH 7.0 at 37°C when measuring TCA soluble products from casein in a final volume of 10 ml.*

Final Assay Concentration

In a 4.00 ml reaction volume, the final concentrations are 1.0% (w/v) casein, 300 mM potassium phosphate, 12 mM L-cysteine, 12 mM Ethylenediaminetetraacetic Acid, 12 mM sodium hydroxide, and 0.02–0.03 unit ficin.

Enzymatic Assay of FK-BINDING PROTEIN Peptidyl-Prolyl Isomerase Activity

Principle

$$\text{Suc} - \text{Ala} - \text{Leu} - \text{Cis} - \text{Pro} - \text{Phe} - \text{pNA} \xrightarrow{\text{FKBP}} \text{Suc - Ala - Leu - trans - Pro - PhepNA}$$

$$\text{Suc} - \text{Ala} - \text{Leu} - \text{trans} - \text{Pro} - \text{Phe} - \text{pNA} + H_2O \xrightarrow{\text{Chymotrypsin}} \text{Suc - Ala - Leu - trans - Pro - Phe} + \text{p - Nitroaniline}$$

Abbreviations Used

pNA = p-Nitroanilide
FKBP = FK Binding Protein

Conditions: T = 1°C, pH 8.0, A405nm, Light path = 1 cm

Method: Continuous Spectrophotometric Rate Determination

Reagents

A. 50 mM HEPES Buffer with 100 mM Sodium Chloride, pH 8.0 at 1°C (Prepare 100 ml in deionized water using HEPES, Free Acid and Sodium Chloride. Adjust to pH 8.0 at 1°C with 1 M NaOH).

B. 2, 2, 2-Trifluoroethanol (TFE) (Use 2,2,2-Trifluoroethanol. Before use, dry overnight with Calcium Chloride, Anhydrous and Molecular Sieves, 3Å. Filter through a Whatman #5 filter paper in a dry room).

C. 470 mM Lithium Chloride Solution (LiCl/TFE) (Prepare 10 ml in Reagent B using Lithium Chloride, Anhydrous. Dry the Lithium Chloride before use in a vacuum oven at 150°C for 24 hours. The Lithium Chloride Solution is at or near saturation at 460 mM in Reagent B. If the solution is turbid or not all of the Lithium Chloride goes into solution with stirring (1 hour) then filter through a Whatman #1 filter paper before use. *Prepare Fresh*).

D. 6.0 mM Suc-Ala-Leu-Pro-Phe-pNA Solution (Substrate) (Immediately before use, prepare 0.5 ml in Reagent C using Suc-Ala-Leu-Pro-Phe-pNA. Keep on ice).

E. 1 mM Hydrochloric Acid Solution (HCl) (Prepare 10 ml in deionized water using Hydrochloric Acid).

F. Chymotrypsin Enzyme Solution (Chym) (Immediately before use, prepare a solution containing 60 mg/ml of α-Chymotrypsin in cold Reagent E).

G. FK Binding Protein Solution (FKBP) (Immediately before use, prepare a solution containing 1 unit/ml of FK Binding Protein in cold Reagent A).

Procedure

Pipette (in milliliters) the following reagents into suitable cuvettes:

	Test	Blank
Reagent A (Buffer)	0.865	0.875
Reagent G (FKBP)	0.01	—
Reagent F (Chym)	0.10	0.10

Mix by inversion and equilibrate to 1°C. Monitor the A405nm until constant, using a suitably thermostatted spectrophotometer. Then add:

	Test	Blank
Reagent D (Substrate)	0.025	0.025

Immediately mix by inversion and record the increase in A405 for approximately 1 minute. Obtain the ΔA405nm/minute using the maximum linear rate for both the Test and Blank.

Calculations

$$\text{Units / ml enzyme} = \frac{(\Delta A405\text{nm / min Test} - \Delta A405\text{nm / min Blank})\ (1)\ (\text{df})}{(0.01)\ (9.3)}$$

1 = Total volume (in milliliter) of assay
df = Dilution factor
0.01 = Volume (in milliliter) of FKBP used
9.3 = Millimolar extinction coefficient 4 of p-Nitroaniline at 405nm under the conditions of the assay

$$\text{Units / mg solid} = \frac{\text{units / ml enzyme}}{\text{mg solid / ml enzyme}}$$

$$\text{Units / mg protein} = \frac{\text{units / ml enzyme}}{\text{mg protein / ml enzyme}}$$

Unit Definition: *One unit will catalyze the isomerization of 1.0 micromole of Suc-Ala-Leu-cis-Pro-Phe-p-Nitroanilide to Suc-Ala-Leutrans- Pro-Phe-p-Nitroanilide per minute at pH 8.0 at 1°C.*

Final Assay Concentrations

In a 1.00 ml reaction mix, the final concentrations are 44 mM HEPES, 88 mM sodium chloride, 0.15 mM Suc-Ala-Leu- Pro-Phe-p-nitroanilide, 12 mM lithium chloride, 2.5% (v/v) 2,2,2-trifluoroethanol, 0.1 mM hydrochloric acid, 6 mg chymotrypsin, and 0.01 unit FK-binding protein.

Notes

1. In order to dry Reagent B (TFE), add the calcium chloride and molecular sieves each, at a concentration of 10% (w/v).
2. The presence of Reagent C (LiCl/TFE) insures that the majority of the substrate is converted to the cis isomer.
3. A sharp increase in absorbance ("burst phase") will occur almost instantly after addition of the substrate as the chymotrypsin hydrolyzes the excess trans isomer. The cis to trans conversion catalyzed by the FK Binding Protein is mostly linear within the first minute of the reaction after the "burst phase."

Enzymatic Assay of FORMALDEHYDE DEHYDROGENASE (EC 1.2.1.1)

Principle

$$\text{Formaldehyde} + \beta\text{-NAD} \xrightarrow{\text{FDH}} \text{Formate} + \beta\text{-NADH}$$

Abbreviations Used

β-NAD = β-Nicotinamide Adenine Dinucleotide, Oxidized Form
β-NADH = β-Nicotinamide Adenine Dinucleotide, Reduced Form
FDH = Formaldehyde Dehydrogenase

Conditions: T = 37°C, pH = 7.5, A340nm, Light path = 1 cm

Method: Continuous Spectrophotometric Rate Determination

Reagents

A. 50 mM Potassium Phosphate Buffer, pH 7.5 at 37°C (Prepare 50 ml in deionized water using Potassium Phosphate, Monobasic, Anhydrous. Adjust to pH 7.5 at 37°C with 1 M KOH).
B. 6.0 mM β-Nicotinamide Adenine Dinucleotide Solution (β-NAD) (Dissolve the contents of one 20 mg vial of β-Nicotinamide Adenine Dinucleotide, in the appropriate volume of deionized water. *Prepare Fresh*).
C. 0.08% (v/v) Formaldehyde Solution (Formaldehyde) (Prepare by adding 0.1 ml of Formaldehyde, 37% Solution, to 45 ml of deionized water).
D. 60 mM Reduced Glutathione Solution (Glutathione) (Prepare 2 ml in deionized water using Glutathione, Reduced Form, Free Acid).
E. Formaldehyde Dehydrogenase Enzyme Solution (Immediately before use, prepare a solution containing 0.25–0.50 unit/ml of Formaldehyde Dehydrogenase in cold Reagent A).

Procedure

Pipette (in milliliters) the following reagents into suitable cuvettes:

	Test	Blank
Deionized Water	0.20	0.20
Reagent A (Buffer)	2.00	2.10
Reagent B (β-NAD)	0.50	0.50
Reagent C (Formaldehyde)	0.10	0.10
Reagent D (Glutathione)	0.10	0.10

Mix by inversion and equilibrate to 37°C. Monitor the A340nm until constant, using a suitably thermostatted spectrophotometer. Then add:

Reagent E (Enzyme Solution)	0.10	—

Immediately mix by inversion and record the increase in A340nm for approximately 5 minutes. Obtain the ΔA340nm/minute using the maximum linear rate for both the Test and Blank.

Calculations

$$\text{Units / mg solid} = \frac{(\Delta A340\text{nm / min Test} - \Delta A340\text{nm / min Blank})}{(6.22)\ (\text{mg solid / ml RM})}$$

6.22 = Millimolar extinction coefficient of β-NADH at 340nm
RM = Reaction Mix

Unit Definition: *One unit will oxidize 1.0 μmole of formaldehyde to formic acid per minute at pH 7.5 at 37°C in the presence of reduced glutathione.*

Final Assay Concentration

In a 3.00 ml reaction mix, the final concentrations are 35 mM potassium phosphate, 1.0 mM β-NAD, 2.0 mM glutathione, 0.0027% (v/v) formaldehyde and 0.025–0.050 unit formaldehyde dehydrogenase.

Enzymatic Assay of FORMALDEHYDE DEHYDROGENASE (EC 1.2.1.46)

Principle

$$\text{Formaldehyde} + \beta\text{-NAD} \xrightarrow{\text{FDH}} \text{Formate} + \beta\text{-NADH}$$

Abbreviations Used

β-NAD = β-Nicotinamide Adenine Dinucleotide, Oxidized Form
β-NADH = β-Nicotinamide Adenine Dinucleotide, Reduced Form
FDH = Formaldehyde Dehydrogenase

Conditions: T = 37°C, pH = 7.5, A340nm, Light path = 1 cm

Method: Continuous Spectrophotometric Rate Determination

Reagents

A. 50 mM Potassium Phosphate Buffer, pH 7.5 at 37°C (Prepare 50 ml in deionized water using Potassium Phosphate, Monobasic, Anhydrous. Adjust to pH 7.5 at 37°C with 1 M KOH).
B. 5.7 mM β-Nicotinamide Adenine Dinucleotide Solution (β-NAD) (Prepare 5 ml in deionized water using β-Nicotinamide Adenine Dinucleotide or dissolve the contents of one 20 mg vial of β-Nicotinamide Adenine Dinucleotide, in the appropriate volume of deionized water. *Prepare Fresh*).
C. 0.08% (v/v) Formaldehyde Solution (Formaldehyde) (Prepare by adding 0.1 ml of Formaldehyde, 37% (w/w) Solution (Formalin, to 45 ml of deionized water).
D. Formaldehyde Dehydrogenase Enzyme Solution (Immediately before use, prepare a solution containing 0.5–1.0 unit/ml of Formaldehyde Dehydrogenase in cold Reagent A).

Procedure

Pipette (in milliliters) the following reagents into suitable cuvettes:

	Test	Blank
Deionized Water	0.30	0.30
Reagent A (Buffer)	2.00	2.05
Reagent B (β-NAD)	0.50	0.50
Reagent C (Formaldehyde)	0.10	0.10

Mix by inversion and equilibrate to 37°C. Monitor the A340nm until constant using a suitably thermostatted spectrophotometer. Then add:

Reagent D (Enzyme Solution)	0.05	—

Immediately mix by inversion and record the increase in A340nm for approximately 5 minutes. Obtain the ΔA340nm/minute using the maximum linear rate for both the Test and Blank.

Calculations

$$\text{Units / ml enzyme} = \frac{(\Delta A340nm / min\ Test - \Delta A340nm / min\ Blank)\ (2.95)\ (df)}{(6.22)\ (0.05)}$$

2.95 = Total volume (in milliliters) of assay
df = Dilution factor
6.22 = Millimolar extinction coefficient of β-NADH at 340 nm
0.05 = Volume (in milliliter) of enzyme used

$$\text{Units / mg solid} = \frac{\text{units / ml enzyme}}{\text{mg solid / ml enzyme}}$$

$$\text{Units / mg protein} = \frac{\text{units / ml enzyme}}{\text{mg protein / ml enzyme}}$$

Unit Definition: *One unit will oxidize 1.0 μmole of formaldehyde to formic acid per minute at pH 7.5 at 37°C.*

Final Assay Concentration

In a 2.95 ml reaction mix, the final concentrations are 35 mM potassium phosphate, 1.0 mM β-nicotinamide adenine dinucleotide, 0.003% (v/v) formaldehyde, and 0.03–0.05 unit formaldehyde dehydrogenase.

Enzymatic Assay of FORMATE DEHYDROGENASE (EC 1.2.1.2) from *Candida boidinii*

Principle

$$\text{Formate} + \beta\text{-NAD} \xrightarrow{\text{Formate Dehydrogenase}} CO_2 + \beta\text{-NADH}$$

Abbreviations Used

β-NAD = β-Nicotinamide Adenine Dinucleotide, Oxidized Form
β-NADH = β-Nicotinamide Adenine Dinucleotide, Reduced Form

Conditions: T = 37°C, pH = 7.5, A340nm, Light path = 1 cm

Method: Continuous Spectrophotometric Rate Determination

Reagents

A. 200 mM Sodium Phosphate Buffer, pH 7.5 at 37°C (Prepare 100 ml in deionized water using Sodium Phosphate, Monobasic, Anhydrous. Adjust the pH to 7.5 with 1 M NaOH).

B. 200 mM Sodium Formate Solution (Form) (Prepare 10 ml in deionized water using Formic Acid, Sodium Salt. *Prepare Fresh*).
C. 10.6 mM β-Nicotinamide Adenine Dinucleotide Solution (β-NAD) (Prepare 3 ml in deionized water using β-Nicotinamide Adenine Dinucleotide or dissolve the contents of one 20 mg vial of β-Nicotinamide Adenine Dinucleotide, in the appropriate volume of deionized water. *Prepare Fresh*).
D. 1.5 mM β-Nicotinamide Adenine Dinucleotide Solution (Enzyme Diluent) (Prepare 50 ml in Reagent A using β-Nicotinamide Adenine Dinucleotide. *Prepare Fresh*).
E. Formate Dehydrogenase Enzyme Solution (Immediately before use, prepare a solution containing 0.25–0.50 unit/ml of Formate Dehydrogenase in Cold Reagent D).

Procedure

Pipette (in milliliters) the following reagents into suitable cuvettes:

	Test	Blank
Deionized Water	1.10	1.10
Reagent A (Buffer)	0.75	0.75
Reagent B (Form)	0.75	0.75

Mix by inversion and equilibrate to 37°C. Monitor the A340nm until constant, using a suitably thermostatted spectrophotometer. Then add:

Reagent C (β-NAD)	0.30	0.30
Reagent D (Enzyme Diluent)	—	0.10
Reagent E (Enzyme Solution)	0.10	—

Immediately mix by inversion and record the increase in A340nm for approximately 5 minutes. Obtain the ΔA340nm/minute using the maximum linear rate for both the Test and Blank.

Calculations

$$\text{Units / ml enzyme} = \frac{(\Delta A340\text{nm / min Test} - \Delta A340\text{nm / min Blank})\ (3)\ (\text{df})}{(6.22)\ (0.1)}$$

3 = Total volume (in milliliters) of assay
df = Dilution factor
6.22 = Millimolar extinction coefficient of β-NADH at 340nm
0.1 = Volume (in milliliter) of enzyme used

$$\text{Units / mg solid} = \frac{\text{units / ml enzyme}}{\text{mg solid / ml enzyme}}$$

$$\text{Units / mg protein} = \frac{\text{units / ml enzyme}}{\text{mg protein / ml enzyme}}$$

Unit Definition: *One unit will oxidize 1.0 μmole of formate to CO_2 per minute in the presence of β-NAD, at pH 7.5 at 37°C.*

Final Assay Concentrations

In a 3.00 ml reaction mix, the final concentrations are 57 mM sodium phosphate, 50 mM formate, 1.1 mM β-nicotinamide adenine dinucleotide and 0.025–0.050 unit formate dehydrogenase.

Note

1. The linearity of the assay may be increased by increasing the β-NAD concentration 10 fold.

Enzymatic Assay of FORMATE DEHYDROGENASE (EC 1.2.1.2)

Principle

$$\text{Formate} + \beta\text{-NAD} \xrightarrow{\text{Formate Dehydrogenase}} CO_2 + \beta\text{-NADH}$$

Abbreviations Used

β-NAD = β-Nicotinamide Adenine Dinucleotide, Oxidized Form
β-NADH = β-Nicotinamide Adenine Dinucleotide, Reduced Form

Conditions: T = 37°C, pH = 7.0, A340nm, Light path = 1 cm

Method: Continuous Spectrophotometric Rate Determination

Reagents

A. 200 mM Sodium Phosphate Buffer, pH 7.0 at 37°C (Prepare 100 ml in deionized water using Sodium Phosphate, Monobasic, Anhydrous. Adjust the pH to 7.0 with 1 M NaOH).
B. 200 mM Sodium Formate Solution (Form) (Prepare 10 ml in deionized water using Formic Acid, Sodium Salt. *Prepare Fresh*).
C. 10.5 mM β-Nicotinamide Adenine Dinucleotide Solution (β-NAD) (Prepare 3 ml in deionized water using β-Nicotinamide Adenine Dinucleotide or dissolve the contents of one 20 mg vial of β-Nicotinamide Adenine Dinucleotide, in the appropriate volume of deionized water. *Prepare Fresh*).
D. 1.5 mM β-Nicotinamide Adenine Dinucleotide Solution (Enzyme Diluent) (Prepare 50 ml in Reagent A using β-Nicotinamide Adenine Dinucleotide. *Prepare Fresh*).
E. Formate Dehydrogenase Enzyme Solution (Immediately before use, prepare a solution containing 0.25–0.50 unit/ml of Formate Dehydrogenase in Cold Reagent D).

Procedure

Pipette (in milliliters) the following reagents into suitable cuvettes:

	Test	Blank
Deionized Water	1.10	1.10
Reagent A (Buffer)	0.75	0.75
Reagent B (Form)	0.75	0.75

Mix by inversion and equilibrate to 37°C. Monitor the A340nm until constant, using a suitably thermostatted spectrophotometer. Then add:

	Test	Blank
Reagent C (β-NAD)	0.30	0.30
Reagent D (Enzyme Diluent)	—	0.10
Reagent E (Enzyme Solution)	0.10	—

Immediately mix by inversion and record the increase in A340nm for approximately 5 minutes. Obtain the ΔA340nm/minute using the maximum linear rate for both the Test and Blank.

Calculations

$$\text{Units / ml enzyme} = \frac{(\Delta\text{A340nm / min Test} - \Delta\text{A340nm / min Blank})\,(3)\,(\text{df})}{(6.22)\,(0.1)}$$

3 = Total volume (in milliliters) of assay
df = Dilution factor
6.22 = Millimolar extinction coefficient of β-NADH at 340nm
0.1 = Volume (in milliliter) of enzyme used

$$\text{Units / mg solid} = \frac{\text{units / ml enzyme}}{\text{mg solid / ml enzyme}}$$

$$\text{Units / mg protein} = \frac{\text{units / ml enzyme}}{\text{mg protein / ml enzyme}}$$

Unit Definition: *One unit will oxidize 1.0 μmole of formate to CO_2 per minute in the presence of β-NAD, at pH 7.0 at 37°C.*

Final Assay Concentration

In a 3.00 ml reaction mix, the final concentrations are 57 mM sodium phosphate, 50 mM formate, 1.1 mM β-Nicotinamide adenine dinucleotide and 0.025–0.050 unit formate dehydrogenase.

Note

1. The linearity of the assay may be increased by increasing the β-NAD concentration 10 fold.

Enzymatic Assay of FORMATE DEHYDROGENASE (EC 1.2.1.2)

Principle

$$\text{Formate} + \beta\text{-NAD} \xrightarrow{\text{Formate Dehydrogenase}} CO_2 + \beta\text{-NADH}$$

Abbreviations Used

β-NAD = β-Nicotinamide Adenine Dinucleotide, Oxidized Form
β-NADH = β-Nicotinamide Adenine Dinucleotide, Reduced Form

Conditions: T = 37°C, pH = 7.6, A340nm, Light path = 1 cm

Method: Continuous Spectrophotometric Rate Determination

Reagents

A. 100 mM Potassium Phosphate Monobasic Solution (KH_2PO_2) (Prepare 100 ml in deionized water using Potassium Phosphate, Monobasic, Anhydrous).

B. 100 mM Potassium Phosphate Dibasic Solution (K_2HPO_4) (Prepare 100 ml in deionized water using Potassium Phosphate, Dibasic, Trihydrate).

C. 100 mM Potassium Phosphate Buffer, pH 7.6 at 37°C (Prepare 100 ml by adjusting the pH of Reagent B to 7.6 at 37°C with Reagent A).

D. 1.0 M Sodium Formate Solution (Form) (Prepare 10 ml in deionized water using Formic Acid, Sodium Salt. *Prepare Fresh*).

E. 60 mM β-Nicotinamide Adenine Dinucleotide Solution (β-NAD) (Prepare 3 ml in deionized water using β-Nicotinamide Adenine Dinucleotide Lithium Salt. *Prepare Fresh*).

F. Formate Dehydrogenase Enzyme Solution (Immediately before use, prepare a solution containing 0.45–0.90 unit/ml of Formate Dehydrogenase in cold deionized water).

Procedure

Pipette (in milliliters) the following reagents into suitable cuvettes:

	Test	Blank
Reagent C (Buffer)	2.50	2.50
Reagent D (Form)	0.50	0.50
Reagent E (β-NAD)	0.10	0.10

Mix by inversion and equilibrate to 37°C. Monitor the A340nm until constant, using a suitably thermostatted spectrophotometer. Then add:

Reagent F (Enzyme Solution)	0.10	—
Deionized Water	—	0.10

Immediately mix by inversion and record the increase in A340nm for approximately 5 minutes. Obtain the ΔA340nm/minute using the maximum linear rate for both the Test and Blank.

Calculations

$$\text{Units / ml enzyme} = \frac{(\Delta A340\text{nm / min Test} - \Delta A340\text{nm / min Blank})\ (3.2)\ (\text{df})}{(6.22)\ (0.1)}$$

3.2 = Total Volume (in milliliters) of assay

df = Dilution factor

6.22 = Millimolar extinction coefficient of β-NADH at 340nm

0.1 = Volume (in milliliter) of enzyme used

$$\text{Units / mg solid} = \frac{\text{units / ml enzyme}}{\text{mg solid / ml enzyme}}$$

$$\text{Units / mg protein} = \frac{\text{units / ml enzyme}}{\text{mg protein / ml enzyme}}$$

Unit Definition: *One unit will oxidize 1.0 µmole of formate to CO_2 per minute in the presence of β-NAD, at pH 7.6 at 37°C.*

Final Assay Concentration

In a 3.20 ml reaction mix, the final concentrations are 78 mM potassium phosphate, 156 mM formate, 1.9 Mm β-nicotinamide adenine dinucleotide and 0.045–0.090 unit formate dehydrogenase.

Enzymatic Assay of FORMIMINO-L-GLUTAMIC ACID TRANSFERASE (EC 2.1.2.5)

Principle

$$\text{THF} + \text{FIGLU} \xrightarrow{\text{FIGLU Transferase}} \text{L-Glutamate} + \text{5-Formimino-THF}$$

Abbreviations Used

THF = Tetrahydrofolic Acid
FIGLU = Formimino-L-Glutamic Acid
5-Formimino-THF = 5-Formimino-Tetrahydrofolic Acid

Conditions: T = 25°C, pH = 7.2, A365nm, Light path = 1 cm

Method: Colorimetric

Reagents

A. 500 mM Potassium Phosphate Buffer with 500 mM 2-Mercaptoethanol, pH 7.4 at 25°C (Prepare 100 ml in deionized water using Potassium Phosphate, Monobasic, Anhydrous and 2-Mercaptoethanol. Adjust to pH 7.4 at 25°C with 1 M KOH).
B. 5.6 mM Tetrahydrofolic Acid Solution (THF) (Prepare 10 ml in Reagent A using Tetrahydrofolic Acid. *Prepare Immediately Before Use*).
C. 36 mM Formimino-L-Glutamic Acid Solution (FIGLU) (Prepare 10 ml in deionized water using Formimino-L-Glutamic Acid, Hemibarium Salt. Add small amounts of Sodium Sulfate, Anhydrous to precipitate barium sulfate. Centrifuge after each sodium sulfate addition. Add sodium sulfate until no precipitate forms. Save the supernatant).
D. 10% (v/v) Perchloric Acid Solution (Per Acid) (Prepare 25 ml in deionized water using Perchloric Acid).
E. Formimino-L-Glutamic Acid Transferase Enzyme Solution (Immediately before use, prepare a solution containing 1.0–2.0 units/ml of Formimino-L-Glutamic Acid Transferase in cold deionized water).

Procedure

Pipette (in milliliters) the following reagents into suitable containers:

	Test	Blank
Reagent B (THF)	0.60	0.60
Reagent C (FIGLU)	0.30	—
Deionized water	2.00	2.30

Mix by inversion and equilibrate to 25°C. Then add:

	Test	Blank
Reagent E (Enzyme Solution)	0.010	0.010

Mix by inversion and incubate at 25°C for exactly 5 minutes. Then add:

Reagent D (Per Acid)	1.0	1.0

Mix by swirling and place in a boiling water bath for 1 minute. Then cool in an ice bath and centrifuge to clarify. Transfer the solutions to suitable cuvettes and record the A365nm for both the Test and Blank using a suitable spectrophotometer.

Calculation

$$\text{Units / ml enzyme} = \frac{(\text{A365nm Test} - \text{A365nm Blank})\,(3.91)}{(5)\,(22.1)\,(\text{mg enzyme / RM})}$$

5 = Time of assay (in minutes) as per unit definition
22.1 = Millimolar extinction coefficient of 5,10-Methenyltetrahydrofolic Acid at 365 nm (5-Formimino-THF is converted to 5,10-Methenyltetrahydrofolic Acid under acidic conditions)
3.91 = Volume of colorimetric assay
RM = Reaction Mix

Unit Definition: *One unit will convert 1.0 μmole of FIGLU and THF to L-glutamic acid and 5-formimino-THF per minute at pH 7.2 at 25°C (measured as 5,10-methenyl-THF after perchloric acid treatment).*

Final Assay Concentrations

In a 2.91 ml reaction mix, the final concentrations are 103 mM potassium phosphate, 103 mM 2-mercaptoethanol, 1.2 mM THF, 3.7 mM FIGLU, and 0.010–0.020 unit formimino-L-glutamic acid transferase.

Enzymatic Assay of FRUCTOSE-6-PHOSPHATE KINASE, PYROPHOSPHATE DEPENDENT (EC 2.7.1.90) from Mung Bean

Principle

$$\text{PPi} + \text{F-6-P} \xrightarrow[\text{F-2, 6-DP1}]{\text{PPi-PFK}} \text{F-1, 6-DP} + \text{Pi}$$

$$\text{F-1, 6-DP} \xrightarrow{\text{Aldolase}} \text{GAP} + \text{DHAP}$$

$$\text{GAP} \xrightarrow{\text{TPI}} \text{DHAP}$$

$$\text{2DHAP} + 2\beta\text{-NADH} \xrightarrow{\text{GDH}} \text{2 Glycerol-3-Phosphate} + 2\beta\text{-NAD}$$

Abbreviations Used

PPi = Pyrophosphate
F-6-P = D-Fructose-6-Phosphate
F-2, 6-DP = Fructose 2,6-Diphosphate
PPi-PFK = Fructose-6-Phosphate Kinase, Pyrophosphate Dependent
F-1, 6-DP = D-Fructose-1, 6-Diphosphate

Pi = Inorganic Phosphate
GAP = D-Glyceraldehyde-3-Phosphate
TPI = Triosephosphate Isomerase
DHAP = Dihydroxyacetone Phosphate
β-NADH = β-Nicotinamide Adenine Dinucleotide, Reduced Form
GDH = α-Glycerophosphate Dehydrogenase
β-NAD = β-Nicotinamide Adenine Dinucleotide, Oxidized Form

Conditions: T = 30°C, pH 7.6, A340nm, Light path = 1 cm

Method: Continuous Spectrophotometric Rate Determination

Reagents

A. 100 mM Imidazole HCl Buffer with 1 mM Magnesium Chloride and 0.2 mM Ethylenediaminetetraacetic Acid, pH 7.6 at 30°C (Prepare 100 ml in deionized water using Imidazole, Magnesium Chloride, Hexahydrate and Ethylenediaminetetraacetic Acid, Disodium Salt, Dihydrate. Adjust to pH 7.6 at 30°C with 1 M HCl).

B. 100 mM D-Fructose-6-Phosphate Solution (F-6-P) (Prepare 10 ml in Reagent A using D-Fructose 6-Phosphate, Dipotassium Salt).

C. 5.0 mM β-Nicotinamide Adenine Dinucleotide, Reduced Form Solution (β-NADH) (Dissolve the contents of one 5 mg vial of β-Nicotinamide Adenine Dinucleotide, Reduced Form, Disodium Salt, in the appropriate volume of Reagent A. *Prepare Fresh*).

D. Aldolase Enzyme Solution (Immediately before use, prepare a solution containing 50 units/ml of Aldolase in cold Reagent A).

E. α-Glycerophosphate Dehydrogenase/Triosephosphate Isomerase Enzyme Solution (α-GDH/TPI) (Immediately before use, prepare a solution containing 50 α-GDH units/ml of α-Glycerophosphate Dehydrogenase-Triosephosphate Isomerase Type X from Rabbit Muscle in cold Reagent A).

F. 30 mM Pyrophosphate Solution (PPi) (Prepare 10 ml in Reagent A using Pyrophosphate, Disodium).

G. 30 μM Fructose 2,6-Diphosphate Solution (F-2,6-DP) (Prepare 1 ml in Reagent A using D-Fructose 2,6-Diphosphate, Sodium Salt).

H. Fructose-6-Phosphate Kinase, Pyrophosphate Dependent Enzyme Solution (PPi-PFK) (Immediately before use, prepare a solution containing 0.02–0.07 unit/ml in deionized water).

Procedure

Pipette (in milliliters) the following reagents into suitable cuvettes:

	Test	Blank
Reagent A (Buffer)	1.45	1.45
Reagent B (F-6-P)	1.00	1.00
Reagent C (β-NADH)	0.10	0.10
Reagent D (Aldolase)	0.10	0.10
Reagent E (α-GDH/TPI)	0.10	0.10
Reagent G (F-2,6-DP)	0.10	0.10
Reagent H (PPi-PFK)	0.10	—
Deionized Water	—	0.10

Mix by inversion and equilibrate to 30°C. Monitor the A340nm until constant, using a suitably thermostatted spectrophotometer. Then add:

	Test	Blank
Reagent F (PPi)	0.10	0.10

Immediately mix by inversion and record the decrease in A340nm for approximately 5 minutes. Obtain the ΔA340nm using the maximum linear rate for both the Test and Blank.

Calculations

$$\text{Units / ml enzyme} = \frac{(\Delta A340nm / \min \text{Test} - \Delta A340nm / \min \text{Blank})\ (3.05)\ (df)}{(6.22)\ (2)\ (0.1)}$$

3.05 = Total volume (in milliliters) of assay
df = Dilution factor
6.22 = Millimolar extinction coefficient of β-NADH at 340nm
2 = Factor accounting for 2 moles of β-NADH oxidized per mole of pyrophosphate converted
0.1 = Volume (in milliliter) of enzyme used

$$\text{Units / mg solid} = \frac{\text{units / ml enzyme}}{\text{mg solid / ml enzyme}}$$

$$\text{Units / mg protein} = \frac{\text{units / ml enzyme}}{\text{mg protein / ml enzyme}}$$

Unit Definition: *One unit will convert 1.0 μmole of pyrophosphate and fructose 6-phosphate to fructose 1,6-diphosphate and inorganic phosphate per minute at pH 7.6 at 30°C in a coupled system with aldolase, α-glycerophosphate dehydrogenase, triosephosphate isomerase, and 1 μmole fructose 2,6-diphosphate.*

Final Assay Concentration

In a 3.05 ml reaction mix, the final concentrations are 97 mM imidazole buffer, 1 mM magnesium chloride, 0.2 mM Ethylenediaminetetraacetic Acid, 33 mM D-fructose- 6-phosphate, 0.16 mM β-nicotinamide adenine dinucleotide, reduced form, 0.98 mM pyrophosphate, 5 units aldolase, 5 units α-glycerophosphate dehydrogenase, 40 units of triosephosphate isomerase, 0.002–0.007 unit fructose-6-phosphate kinase, pyrophosphate dependent, and 0.98 μM fructose 2,6-diphosphate.

Notes

1. Fructose 2,6-Diphosphate is an activator of Fructose- 6-Phosphate Kinase, Pyrophosphate Dependent from Mung Bean.
2. The Triosephosphate Isomerase activity is approximately 8 fold that of the α-Glycerophosphate Dehydrogenase.
3. Aldolase Unit Definition: One unit will convert 1.0 μmole of fructose 1,6-diphosphate to dihydroxyacetone phosphate and glyceraldehyde 3- phosphate per minute at pH 7.4 at 25°C.
4. α-Glycerophosphate Dehydrogenase Unit Definition: One unit will convert 1.0 μmole of dihydroxyaceton phosphate to α-glycerophosphate per minute at pH 7.4 at 25°C.
5. Triosephosphate Isomerase Unit Definition: One unit will convert 1.0 μmole of D-glyceraldehyde 3-phosphate to dihydroxyacetone phosphate per minute at pH 7.6 at 25°C.

Enzymatic Assay of FRUCTOSE-6-PHOSPHATE KINASE, PYROPHOSPHATE DEPENDENT (EC 2.7.1.90)

Principle

$$PPi + F\text{-}6\text{-}P \xrightarrow{PPi\text{-}PFK} F\text{-}1, 6\text{-}DP + Pi$$

$$F\text{-}1, 6\text{-}DP \xrightarrow{Aldolase} GAP + DHAP$$

$$GAP \xrightarrow{TPI} DHAP$$

$$2DHAP + 2\,\beta\text{-}NADH \longrightarrow 2Glycerol\text{-}3\text{-}Phosphate + 2\,\beta\text{-}NAD$$

Abbreviations Used

PPi = Pyrophosphate
F-6-P = D-Fructose-6-Phosphate
PPi-PFK = Fructose-6-Phosphate Kinase, Pyrophosphate Dependent
F-1, 6-DP = D-Fructose-1, 6-Diphosphate
Pi = Inorganic Phosphate
GAP = D-Glyceraldehyde-3-Phosphate
TPI = Triosephosphate Isomerase
DHAP = Dihydroxyacetone Phosphate
GDH = α-Glycerophosphate Dehydrogenase
β-NADH = β-Nicotinamide Adenine Dinucleotide (reduced form)
β-NAD = β-Nicotinamide Adenine Dinucleotide (oxidized form)

Conditions: T = 30°C, pH 7.4, A340nm, Light path = 1 cm

Method: Continuous Spectrophotometric Rate Determination

Reagents

A. 150 mM Imidazole Buffer with 1 mM Magnesium Chloride, pH 7.4 at 30°C (Prepare 100 ml in deionized water using Imidazole and Magnesium Chloride, Hexahydrate. Adjust to pH 7.4 at 30°C with 1 M HCl).

B. 90 mM D-Fructose-6-Phosphate Solution (F-6-P) (Prepare 10 ml in Reagent A using D-Fructose 6-Phosphate, Dipotassium Salt).

C. 6.4 mM β-Nicotinamide Adenine Dinucleotide, Reduced Form Solution (β-NADH) (Dissolve the contents of one 5 mg vial of β-Nicotinamide Adenine Dinucleotide, Reduced Form, Disodium Salt in 1 ml of Reagent A. *Prepare Fresh*).

D. Aldolase Enzyme Solution (Immediately before use, prepare a solution containing 50 units/ml of Aldolase, in cold Reagent A).

E. α-Glycerophosphate Dehydrogenase/Triosephosphate Isomerase Enzyme Solution (α-GDH/TPI) (Immediately before use, prepare a solution containing 25 α-GDH units/ml of α-Glycerophophate Dehydrogenase Triosephosphate Isomerase Type X from Rabbit Muscle in cold Reagent A).

F. 100 mM Pyrophosphate Solution, pH 7.4 at 30°C (PPi) (Prepare 10 ml in Reagent A using Pyrophosphate, Disodium. Adjust to pH 7.4 at 30°C with either 1 M HCl or 1 M NaOH).

G. Fructose-6-Phosphate Kinase, Pyrophosphate Dependent Enzyme Solution (PPi-PFK) (Immediately before use, prepare a solution containing 0.1–0.5 units/ml in deionized water).

Procedure

Pipette (in milliliters) the following reagents into suitable cuvettes:

	Test	Blank
Reagent A (Buffer)	2.50	2.50
Reagent B (F-6-P)	0.10	0.10
Reagent C (β-NADH)	0.10	0.10
Reagent D (Aldolase)	0.10	0.10
Reagent E (α-GDH/TPI)	0.10	0.10
Reagent F (PPi)	0.05	0.05

Mix by inversion and equilibrate to 30°C. Monitor the A340nm until constant, using a suitably thermostatted spectrophotometer. Then add:

Reagent G (PPi-PFK)	0.050	—
Deionized Water	—	0.050

Immediately mix by inversion and record the decrease in A340nm for approximately 5 minutes. Obtain the ΔA340nm using the maximum linear rate for both the Test and Blank.

Calculations

$$\text{Units / ml enzyme} = \frac{(\Delta A340\text{nm / min Test} - \Delta A340\text{nm / min Blank})\ (3)\ (\text{df})}{(6.22)\ (2)\ (0.05)}$$

3 = Total volume (in milliliters) of assay
df = Dilution factor
6.22 = Millimolar extinction coefficient of β-NADH at 340nm
2 = Factor accounting for 2 moles of β-NADH oxidized per mole of pyrophosphate converted
0.05 = Volume (in milliliters) of enzyme used

$$\text{Units / mg solid} = \frac{\text{units / ml enzyme}}{\text{mg solid / ml enzyme}}$$

$$\text{Units / mg protein} = \frac{\text{units / ml enzyme}}{\text{mg protein / ml enzyme}}$$

Unit Definition: *One unit will convert 1.0 μmole of pyrophosphate and fructose 6-phosphate to fructose 1,6-diphosphate and inorganic phosphate per minute at pH 7.4 at 30°C in a coupled system with aldolase, glycerophosphate dehydrogenase and triosephosphate isomerase.*

Final Assay Concentration

In a 3.0 ml reaction mix, the final concentrations are 47.5 mM imidazole buffer, 0.98 mM magnesium chloride, 3.0 mM D-fructose-6-phosphate, 0.21 mM β-NADH, 1.67 mM pyrophosphate, 5.0 units aldolase, 2.5 units α-GDH/TPI, and .005–0.025 units PPi-PFK.

Enzymatic Assay of FRUCTOSE-6-PHOSPHATE KINASE, PYROPHOSPHATE DEPENDENT (EC 2.7.1.90) from Potato Tubers

Principle

$$PPi + F\text{-}6\text{-}P \xrightarrow[F\text{-}2,\,6\text{-}DP]{PPi\text{-}PFK} F\text{-}1,\,6\text{-}DP + Pi$$

$$F\text{-}1,\,6\text{-}DP \xrightarrow{Aldolase} GAP + DHAP$$

$$GAP \xrightarrow{TPI} DHAP$$

$$2DHAP + 2\,\beta\text{-}NADH \xrightarrow{GDH} 2\,\alpha\text{-}Glycerophosphate + 2\,\beta\text{-}NAD$$

Abbreviations Used

PPi = Pyrophosphate
F-6-P = D-Fructose 6-Phosphate
F-2,6-DP = Fructose 2,6-Diphosphate
PPi-PFK = Fructose-6-Phosphate Kinase, Pyrophosphate Dependent
F-1,6-DP = D-Fructose 1,6-Diphosphate
Pi = Inorganic Phosphate
GAP = D-Glyceraldehyde 3-Phosphate
TPI = Triosephosphate Isomerase
DHAP = Dihydroxyacetone Phosphate
β-NADH = β-Nicotinamide Adenine Dinucleotide, Reduced Form
GDH = α-Glycerophosphate Dehydrogenase
β-NAD = β-Nicotinamide Adenine Dinucleotide, Oxidized Form

Conditions: T = 30°C, pH 8.0, A340nm, Light path = 1 cm

Method: Continuous Spectrophotometric Rate Determination

Reagents

A. 75 mM Tris HCl Buffer with 7.5 mM Magnesium Chloride, pH 8.0 at 30°C (Prepare 100 ml in deionized water using Trizma and Magnesium Chloride, Hexahydrate. Adjust to pH 8.0 at 30°C with 1 M HCl).

B. 75 mM D-Fructose 6-Phosphate Solution (F-6-P) (Prepare 10 ml in deionized water using D-Fructose 6-Phosphate, Dipotassium Salt).

C. 4.3 mM β-Nicotinamide Adenine Dinucleotide, Reduced Form Solution (β-NADH) (Dissolve the contents of one 10 mg vial of β-Nicotinamide Adenine Dinucleotide, Reduced Form, Disodium Salt in the appropriate volume of Reagent A. *Prepare Fresh*).

D. Aldolase Enzyme Solution (Immediately before use, prepare a solution containing 10 units/ml of Aldolase in cold deionized water).

E. α-Glycerophosphate Dehydrogenase/Triosephosphate Isomerase Enzyme Solution2 (α-GDH/TPI) (Immediately before use, prepare a solution containing 10 α-GDH units/ml of α-Glycerophosphate Dehydrogenase- Triosephosphate Isomerase Type X from Rabbit Muscle in cold deionized water).

F. 75 mM Pyrophosphate Buffer, pH 8.0 at 30°C (PPi) (Prepare 10 ml in deionized water using Sodium Pyrophosphate, Decahydrate. Adjust to pH 8.0 at 30°C with 1 M HCl).
G. 30 μM Fructose 2,6-Diphosphate Solution (F-2, 6-DP) (Prepare 1 ml in Reagent A using D-Fructose 2,6-Diphosphate, Sodium Salt).
H. 175 mM Glucose 6-Phosphate Solution (G-6-P) (Prepare 5 ml in deionized water using D-Glucose 6-Phosphate, Monosodium Salt).
I. Fructose-6-Phosphate Kinase, Pyrophosphate Dependent Enzyme Solution (PPi-PFK) (Immediately before use, prepare a solution containing 0.2–0.5 unit/ml of Fructose-6-Phosphate Kinase, Pyrophosphate Dependent, in cold deionized water).

Procedure

Pipette (in milliliters) the following reagents into suitable cuvettes:

	Test	Blank
Reagent A (Buffer)	2.00	2.00
Reagent B (F-6-P)	0.10	0.10
Reagent C (β-NADH)	0.10	0.10
Reagent D (Aldolase)	0.10	0.10
Reagent E (α-GDH/TPI)	0.10	0.10
Reagent G (F-2,6-DP)	0.10	0.10
Reagent H (G-6-P)	0.30	0.30
Reagent I (PPi-PFK))	0.03	—
Deionized Water	0.07	0.10

Mix by inversion, and equilibrate to 30°C for 5 minutes. Monitor the A340nm until constant, using a suitably thermostatted spectrophotometer. Then add:

Reagent F (PPi)	0.10	0.10

Immediately mix by inversion and record the decrease in A340nm for approximately 5 minutes. Obtain the ΔA340nm/minute using the maximum linear rate for both the Test and Blank.

Calculations

$$\text{Units / ml enzyme} = \frac{(\Delta A340nm / min\ Test - \Delta A340nm / min\ Blank)\ (3)\ (df)}{(6.22)\ (2)\ (0.1)}$$

3 = Total volume (in milliliters) of assay
df = Dilution factor
6.22 = Millimolar extinction coefficient of β-NADH at 340nm
2 = Factor accounting for 2 moles of β-NADH oxidized per mole of pyrophosphate converted
0.1 = Volume (in milliliter) of enzyme used

$$\text{Units / mg solid} = \frac{\text{units / ml enzyme}}{\text{mg solid / ml enzyme}}$$

$$\text{Units / mg protein} = \frac{\text{units / ml enzyme}}{\text{mg protein / ml enzyme}}$$

Unit Definition: *One unit will convert 1.0 μmole of pyrophosphate and fructose 6-phosphate to fructose 1,6-diphosphate and inorganic phosphate per minute at pH 8.0 at 30°C in the presence of 1 μM fructose 2,6-diphosphate and 17 mM glucose 6-phosphate in a coupled assay system using aldolase, α-glycerophosphate dehydrogenase and triosephosphate isomerase.*

Final Assay Concentration

In a 3.00 ml reaction mix, the final concentrations are 55 mM Tris, 5.5 mM magnesium chloride, 17 mM D-glucose-6- phosphate, 2.5 mM D-fructose-6-phosphate, 0.14 mM β-nicotinamide adenine dinucleotide, reduced form, 2.5 mM pyrophosphate, 1 unit aldolase, 1 unit α-glycerophosphate dehydrogenase, 8 units of triosephosphate isomerase, 0.006–0.015 unit fructose-6-phosphate kinase, pyrophosphate dependent, and 1 μM fructose 2,6-diphosphate.

Notes

1. Fructose 2,6-Diphosphate is an activator of Fructose-6-Phosphate Kinase, Pyrophosphate Dependent from Potato Tubers.
2. The Triosephosphate Isomerase activity is approximately 8 fold that of the α-Glycerophosphate Dehydrogenase activity.
3. Aldolase Unit Definition: One unit will convert 1.0 μmole of fructose 1,6-diphosphate to dihydroxyacetone phosphate and glyceraldehyde 3-phosphate per minute at pH 7.4 at 25°C.
4. α-Glycerophosphate Dehydrogenase Unit Definition: One unit will convert 1.0 μmole of dihydroxyacetone phosphate to α-glycerophosphate per minute at pH 7.4 at 25°C.
5. Triosephosphate Isomerase Unit Definition: One unit will convert 1.0 μmole of D-glyceraldehyde 3-phosphate to dihydroxyacetone phosphate per minute at pH 7.6 at 25°C.

Enzymatic Assay of D-FRUCTOSE-1, 6-DIPHOSPHATASE (EC 3.1.3.11)

Principle

$$\text{F-1, 6-P2} + H_2O \xrightarrow{\text{FDP}} \text{D-Fructose 6-Phosphate} + \text{Pi}$$

$$\text{D-Fructose 6-Phosphate} \xrightarrow{\text{PGI}} \text{D-Glucose 6-Phosphate}$$

$$\text{D-Glucose 6-Phosphate} + \beta\text{-NADP} \xrightarrow{\text{G-6-PDH}} \text{Gluconate 6-Phosphate} + \beta\text{-NADPH}$$

Abbreviations Used

F-1, 6-P2 = D-Fructose 1,6-Diphosphate
Pi = Inorganic Phosphate
FDP = D-Fructose-1,6-Diphosphatase
PGI = Phosphoglucose Isomerase
β-NADP = β-Nicotinamide Adenine Dinucleotide, Oxidized Form
G-6-PDH = Glucose-6-Phosphate Dehydrogenase
β-NADPH = β-Nicotinamide Adenine Dinucleotide, Reduced Form

Conditions: T = 25°C, pH 9.5, A340nm, Light path = 1 cm

Method: Continuous Spectrophotometric Rate Determination

Reagents

A. 100 mM Glycine Buffer, pH 9.5 at 25°C (Prepare 100 ml in deionized water using Glycine, Free Base . Adjust to pH 9.5 at 25°C with 1 M NaOH).

B. 41 mM D-Fructose 1,6-Diphosphate Solution (F-1,6-P2) Prepare 1 ml in deionized water using D-Fructose 1,6- Diphosphate, Tetra(cyclohexylammonium) Salt).

C. 51 mM Manganese Chloride Solution ($MnCl_2$) (Prepare 1 ml in deionized water using Manganese Chloride, Tetrahydrate).

D. Phosphoglucose Isomerase Enzyme Solution (PGI) (Immediately before use, prepare a solution containing 700 units/ml of Phosphoglucose Isomerase in cold deionized water).

E. Glucose-6-Phosphate Dehydrogenase Enzyme Solution (G-6-PDH) (Immediately before use, prepare a solution containing 300 units/ml of Glucose-6-Phosphate Dehydrogenase in cold deionized water).

F. 12 mM β-Nicotinamide Adenine Dinucleotide Phosphate, Oxidized Form, Solution (β-NADP) (Prepare 2 ml in deionized water using β-Nicotinamide Adenine Dinucleotide Phosphate, Sodium Salt).

G. D-Fructose-1,6-Diphosphatase Enzyme Solution (FDP) (Immediately before use, prepare a solution containing 1.5–3.0 units/ml of D-Fructose-1,6-Diphosphatase in cold Reagent A).

Procedure

Pipette (in milliliters) the following reagents into suitable cuvettes:

	Test	Blank
Reagent A (Buffer)	2.74	2.74
Reagent B (F-1,6-P2)	0.10	0.10
Reagent C ($MnCl_2$)	0.02	0.02
Reagent D (PGI)	0.02	0.02
Reagent E (G-6-PDH)	0.01	0.01

Mix by inversion and equilibrate to 25°C. Monitor the A340nm until constant, using a suitably thermostatted spectrophotometer. Then add:

Reagent F (β-NADP)	0.10	0.10
Reagent G (FDP)	0.02	—
Reagent A (Buffer)	—	0.02

Immediately mix by inversion and record the increase inA340nm for approximately 5 minutes. Obtain the ΔA340nm/minute using the maximum linear rate for both the Test and Blank.

Calculations

$$\text{Units / ml enzyme} = \frac{(\Delta A340\text{nm / min Test} - \Delta A340\text{nm / min Blank})\,(3.01)\,(\text{df})}{(6.22)\,(0.02)}$$

3.01 = Total volume (in milliliters) of assay
df = Dilution factor
6.22 = Millimolar extinction coefficient of β-NADPH at 340nm
0.02 = Volume (in milliliter) of enzyme used

$$\text{Units / mg solid} = \frac{\text{units / ml enzyme}}{\text{mg solid / ml enzyme}}$$

$$\text{Units / mg protein} = \frac{\text{units / ml enzyme}}{\text{mg protein / ml enzyme}}$$

Unit Definition: *One unit will convert 1.0 μmole of fructose 1,6-diphosphate to fructose 6-phosphate and inorganic phosphate per minute at pH 9.5 at 25°C.*

Final Assay Concentration

In a 3.01 ml reaction mix, the final concentrations are 92 mM glycine, 1.4 mM fructose 1,6-diphosphate, 0.34 mM manganese chloride, 0.4 mM β-nicotinamide adenine dinucleotide phosphate, 14 units phosphoglucose isomerase, 3 units glucose-6-phosphate dehydrogenase and 0.03–0.06 unit fructose-1,6-diphosphatase.

Notes

1. Phosphoglucose Isomerase Unit Definition: One unit will convert 1.0 μmole of fructose 6-phosphate to glucose 6-phosphate per minute at pH 7.4 at 25°C.
2. Glucose-6-Phosphate Dehydrogenase Unit Definition: One unit will oxidize 1.0 μmole of D-glucose 6-phosphate to 6-phospho-D-gluconate per minute in the presence of β-NADP at pH 7.4 at 25°C.

Enzymatic Assay of FRUCTOSE-6-PHOSPHATE KINASE (EC 2.7.1.11)

Principle

$$\text{F-6-P} + \text{ATP} \xrightarrow{\text{F-6-PK}} \text{F-1, 6-DP} + \text{ADP}$$

$$\text{ADP} + \text{PEP} \xrightarrow{\text{Pyruvate Kinase}} \text{ATP} + \text{Pyruvate}$$

$$\text{Pyruvate} + \beta\text{-NADH Lactic Dehydrogenase} \rightarrow \text{Lactate} + \beta\text{-NAD}$$

Abbreviations Used

F-6-P = Fructose 6-Phosphate
ATP = Adenosine 5'-Triphosphate
F-6-PK = Fructose-6-Phosphate Kinase
F-1,6-DP = Fructose 1,6-Diphosphate
ADP = Adenosine 5'-Diphosphate
PEP = Phospho(enol) Pyruvate
β-NADH = β-Nicotinamide Adenine Dinucleotide, Reduced Form
β-NAD = β-Nicotinamide Adenine Dinucleotide, Oxidized Form

Conditions: T = 37°C, pH = 8.0, A340nm, Light path = 1 cm

Method: Continuous Spectrophotometric Rate Determination

Reagents

A. 100 mM Tris Buffer, pH 8.0 at 37°C. (Prepare 100 ml in deionized water using Trizma Base. Adjust to pH 8.0 at 37°C with 1 M HCl).
B. 40 mM Magnesium Sulfate with 140 mM Potassium Chloride Solution ($MgSO_4$.KCl) (Prepare 5 ml in deionized water using Magnesium Sulfate, Heptahydrate and Potassium Chloride).

C. 30 mM Phospho(enol) pyruvate Solution (PEP) (Prepare 5 ml in deionized water using Phospho(enol) pyruvate, Monopotassium Salt. *Prepare Fresh*).
D. 20 mM Fructose 1,6-Diphosphate Solution (F-1,6-DP) (Prepare 2 ml in deionized water using Fructose 1,6-Diphosphate, Tetrasodium Salt, Amorphous. *Prepare Fresh*).
E. 30 mM Fructose 6-Phosphate Solution (F-6-P) (Prepare 5 ml in deionized water using D-Fructose 6-Phosphate, Disodium Salt. *Prepare Fresh*).
F. 20 mM Adenosine 5'-Triphosphate Solution (ATP) (Prepare 5 ml in deionized water using Adenosine 5'-Triphosphate, Disodium Salt. *Prepare Fresh*).
G. 7.0 mM β-Nicotinamide Adenine Dinucleotide, Reduced Form Solution (β-NADH) (Prepare 1 ml in deionized water using β-Nicotinamide Adenine Dinucleotide, Reduced Form, Disodium Salt or dissolve the contents of one 10 mg vial of β-Nicotinamide Adenine Dinucleotide, Reduced Form, Disodium Salt in the appropriate volume of deionized water. *Prepare Fresh*).
H. PK/LDH Enzyme Suspension (PK/LDH).
I. Fructose-6-Phosphate Kinase Enzyme Solution (Immediately before use, prepare a solution containing 0.3–0.6 unit/ml of Fructose-6-Phosphate Kinase in cold Reagent A).

Procedure

Prepare a reaction cocktail by pipetting (in milliliters) the following reagents into a suitable container:

Reagent A (Buffer)	22.00
Reagent B ($MgSO_4$.KCl)	1.00
Reagent C (PEP)	1.00
Reagent D (F-1,6-DP)	1.00
Reagent E (F-6-P)	2.00
Reagent F (ATP)	2.00
Reagent G (β-NADH)	0.60

Mix and adjust to pH 8.0 at 37°C with 100 mM HCl or 100 mM NaOH, if necessary. Pipette (in milliliters) the following reagents into suitable cuvettes:

	Test	Blank
Reaction Cocktail	2.96	2.96
Reagent H (PK/LDH)	0.04	0.04

Mix by inversion and equilibrate to 37°C. Monitor the ΔA340nm until constant, using a suitably thermostatted spectrophotometer. Then add:

Reagent A (Buffer)	—	0.10
Reagent I (Enzyme Solution)	0.10	—

Immediately mix by inversion and record the decrease in A340nm for approximately 10 minutes. Obtain the ΔA340nm/minute using the maximum linear rate for both the Test and Blank.

Calculations

$$\text{Units / ml enzyme} = \frac{(\Delta A340\text{nm / min Test} - \Delta A340\text{nm / min Blank})\ (3.1)\ (\text{df})}{(6.22)\ (0.10)}$$

3.1 = Total volume (in milliliters) of assay
df = Dilution factor
6.22 = Millimolar extinction coefficient of β-NADH at 340nm
0.10 = Volume (in milliliter) of enzyme used

$$\text{Units / mg protein} = \frac{\text{units / ml enzyme}}{\text{mg protein / ml enzyme}}$$

Unit Definition: *One unit will convert 1.0 µmole of fructose 6-phosphate and ATP to fructose 1,6-diphosphate and ADP per minute at pH 8.0 at 37°C.*

Final Assay Concentrations

In a 3.10 ml reaction mix, the final concentrations are 74 mM Tris, 1.3 mM magnesium sulfate, 4.5 mM potassium chloride, 0.97 mM phospho(enol) pyruvate, 0.65 mM fructose 1,6-diphosphate, 1.9 mM fructose 6-phosphate, 1.3 mM adenosine 5'-triphosphate, 0.14 mM β-nicotinamide adenine dinucleotide, 28 units pyruvate kinase, 40 units lactic dehydrogenase and 0.03–0.06 unit fructose-6-phosphate kinase.

Notes

1. Pyruvate Kinase Unit Definition: One unit will convert 1.0 µmole of phos(enol) pyruvate to pyruvate per minute at pH 7.6 at 37°C.
2. L-Lactic Dehydrogenase Unit Definition: One unit will reduce 1.0 µmole of pyruvate to L-lactate per minute at pH 7.5 at 37°C.

Enzymatic Assay of FRUCTOSE-6-PHOSPHATE KINASE (EC 2.7.1.11) from *Bacillus stearothermophilus*

Principle

$$\text{F-6-P} + \text{ATP} \xrightarrow{\text{F-6-PK}} \text{F-1, 6-DP} + \text{ADP}$$

$$\text{ADP} + \text{PEP} \xrightarrow{\text{Pyruvate Kinase}} \text{ATP} + \text{Pyruvate}$$

$$\text{Pyruvate} + \beta\text{-NADH} \xrightarrow{\text{Lactic Dehydrogenase}} \text{Lactate} + \beta\text{-NAD}$$

Abbreviations Used

F-6-P = Fructose 6-Phosphate
ATP = Adenosine 5'-Triphosphate
F-6-PK = Fructose-6-Phosphate Kinase
F-1,6-DP = Fructose 1,6-Diphosphate
ADP = Adenosine 5'-Diphosphate
PEP = Phospho(enol) Pyruvate
β-NADH = β-Nicotinamide Adenine Dinucleotide, Reduced Form
β-NAD = β-Nicotinamide Adenine Dinucleotide, Oxidized Form

Conditions: T = 30°C, pH = 9.0, A340nm, Light path = 1 cm

Method: Continuous Spectrophotometric Rate Determination

Reagents

A. 100 mM Tris Buffer, pH 9.0 at 30°C. (Prepare 100 ml in deionized water using Trizma Base. Adjust to pH 9.0 at 30°C with 1 M HCl).
B. 100 mM Magnesium Sulfate Solution ($MgSO_4$) (Prepare 10 ml in deionized water using Magnesium Sulfate, Heptahydrate).
C. 56 mM Phospho(enol) pyruvate Solution (PEP) (Prepare 5 ml in deionized water using Phospho(enol) pyruvate, Trisodium Salt, Hydrate. *Prepare Fresh*).
D. 2.5 M Potassium Chloride Solution (KCl) (Prepare 5 ml in deionized water using Potassium Chloride).
E. 500 mM Fructose 6-Phosphate Solution (F-6-P) (Prepare 10 ml in deionized water using D-Fructose 6-Phosphate, Disodium Salt. *Prepare Fresh*).
F. 100 mM Adenosine 5'-Triphosphate Solution (ATP) (Prepare 8.2 ml in deionized water using Adenosine 5'-Triphosphate, Disodium Salt and then add 1.8 ml of 1 N NaOH. *Prepare Fresh*).
G. 13.1 mM β-Nicotinamide Adenine Dinucleotide, Reduced Form Solution (β-NADH) (Prepare 1 ml in deionized water using β-Nicotinamide Adenine Dinucleotide, Reduced Form, Disodium Salt or dissolve the contents of one 10 mg vial of β-Nicotinamide Adenine Dinucleotide, Reduced Form, Disodium Salt in the appropriate volume of deionized water. *Prepare Fresh*).
H. PK/LDH Enzyme Suspension.
I. 50 mM Tris HCl Buffer, pH 8.5 at 30°C (Enz Dil) (Prepare 50 ml in deionized water using Trizma Base, Adjust to pH 8.5 at 30°C with 1 M HCl).
J. Fructose-6-Phosphate Kinase Enzyme Solution (Immediately before use, prepare a solution containing 0.2–0.4 unit/ml of Fructose-6-Phosphate Kinase in cold Reagent I).

Procedure

Prepare a reaction cocktail by pipetting (in milliliters) the following reagents into a suitable container:

Reagent A (Buffer)	27.36
Reagent F (ATP)	0.30
Reagent C (PEP)	0.39
Reagent G (β-NADH)	0.60
Reagent E (F-6-P)	0.60
Reagent D (KCl)	0.06
Reagent B ($MgSO_4$)	0.60
Reagent H (PK/LDH)	0.10

Mix by swirling and equilibrate to 30°C. Pipette (in milliliters) the following reagents into suitable cuvettes:

	Test	Blank
Reaction Cocktail	3.00	3.00

Mix by inversion and equilibrate to 30°C. Monitor the ΔA340nm until constant, using a suitably thermostatted spectrophotometer. Then add:

Reagent I (Enz Dil)	—	0.10
Reagent J (Enzyme Solution)	0.10	—

Immediately mix by inversion and record the decrease in A340nm for approximately 10 minutes. Obtain the ΔA340nm/minute using the maximum linear rate for both the Test and Blank.

Calculations

$$\text{Units / ml enzyme} = \frac{(\Delta A340\text{nm / min Test} - \Delta A340\text{nm / min Blank})\ (3.1)\ (\text{df})}{(6.22)\ (0.1)}$$

3.1 = Volume (in milliliters) of assay
df = Dilution factor
6.22 = Millimolar extinction coefficient of β-NADH at 340nm
0.1 = Volume (in milliliter) of enzyme used

Unit Definition: *One unit will convert 1.0 μmole of fructose 6-phosphate and ATP to fructose 1,6-diphosphate and ADP per minute at pH 9.0 at 30°C.*

Final Assay Concentrations

In a 3.10 ml reaction mix, the final concentrations are 90 mM Tris, 1.9 mM magnesium sulfate, 4.8 mM potassium chloride, 0.70 mM phospho(enol) pyruvate, 9.7 mM fructose 6-phosphate, 0.97 mM adenosine 5'-triphosphate, 0.25 mM β-nicotinamide adenine dinucleotide, 7 units pyruvate kinase, 10 units lactic dehydrogenase, and 0.02–0.04 unit fructose-6-phosphate kinase.

Notes

1. The NaOH is added in order to increase the pH of the adenosine 5'-triphosphate solution.
2. Pyruvate Kinase Unit Definition: One unit will convert 1.0 μmole of phospho(enol) pyruvate to pyruvate per minute at pH 7.6 at 37°C.
3. L-Lactic Dehydrogenase Unit Definition: One unit will reduce 1.0 μmole of pyruvate to L-lactate per minute at pH 7.5 at 37°C.

Enzymatic Assay of FRUCTOSE-6-PHOSPHATE KINASE (EC 2.7.1.11) from Rabbit Liver

Principle

$$\text{ATP} + \text{D-Fructose 6-Phosphate} \xrightarrow[\text{Mg}_2^+]{\text{PFK}} \text{F-1, 6-DP} + \text{ADP}$$

$$\text{F-1, 6-DP} \xrightarrow{\text{Aldolase}} \text{GAP} + \text{DHAP}$$

$$GAP \xrightarrow{TPI} DHAP$$

$$2DHAP + 2\beta\text{-}NADH \xrightarrow{GDH} 2\ \text{Glycerol-3-Phosphate} + 2\beta\text{-}NAD$$

Abbreviations Used

ATP = Adenosine 5'-Triphosphate
PFK = Fructose-6-Phosphate Kinase
F-1,6-DP = D-Fructose 1,6-Diphosphate
ADP = Adenosine 5'-Diphosphate
GAP = D-Glyceraldehyde 3-Phosphate
DHAP = Dihydroxyacetone Phosphate
TPI = Triosephosphate Isomerase
β-NADH = β-Nicotinamide Adenine Dinucleotide, Reduced Form
GDH = α-Glycerophosphate Dehydrogenase
β-NAD = β-Nicotinamide Adenine Dinucleotide, Oxidized Form

Conditions: T = 30°C, pH = 8.0, A340nm, Light path = 1 cm

Method: Continuous Spectrophotometric Rate Determination

Reagents

A. 60 mM Tris HCl Buffer, pH 8.0 at 30°C (Prepare 100 ml in deionized water using Trizma Base. Adjust to pH 8.0 at 30°C with 1 M HCl).
B. 30 mM DL-Dithiothreitol Solution (DTT) (Prepare 10 ml in deionized water using DL-Dithiothreitol).
C. 60 mM Adenosine 5'-Triphosphate Solution (ATP) (Prepare 10 ml in deionized water using Adenosine 5'-Triphosphate, Disodium Salt).
D. 6 mM β-Nicotinamide Adenine Dinucleotide, Reduced Form Solution (β-NADH) (Dissolve the contents of one 5 mg vial of β- Nicotinamide Adenine Dinucleotide, Reduced Form, Disodium Salt in the appropriate volume of deionized water).
E. 750 mM Magnesium Chloride Solution ($MgCl_2$) (Prepare 10 ml in deionized water using Magnesium Chloride, Hexahydrate).
F. Aldolase Enzyme Solution (Aldolase) (Immediately before use, prepare a solution containing 100 units/ml of Aldolase, in cold deionized water).
G. α-Glycerophosphate Dehydrogenase/Triosephosphate Isomerase Enzyme Solution (GDH/TPI) (Immediately before use, prepare a solution containing 100 GDH units/ml using α-Glycerophosphate Dehydrogenase-Triosephosphate Isomerase, in cold deionized water).
H. 120 mM D-Fructose 6-Phosphate Solution (F-6P) (Prepare 10 ml in deionized water using D-Fructose 6-Phosphate, Disodium Salt).
I. Fructose-6-Phosphate Kinase Enzyme Solution (PKF) (Immediately before use, prepare a solution containing 0.5–1.0 unit/ml in cold deionized water).

Procedure

Pipette (in milliliters) the following reagents into suitable cuvettes:

	Test	Blank
Reagent A (Buffer)	2.50	2.50
Reagent B (DTT)	0.10	0.10
Reagent C (ATP)	0.10	0.10
Reagent D (β-NADH)	0.10	0.10
Reagent E ($MgCl_2$)	0.02	0.02
Reagent F (Aldolase)	0.02	0.02
Reagent G (α-GDH/TPI)	0.01	0.01
Reagent I (PFK)	0.05	—
Deionized Water	—	0.05

Mix by inversion and equilibrate to 30°C. Monitor the A340nm until constant, using a suitably thermostatted spectrophotometer. Then add:

Reagent H (F-6P)	0.10	0.10

Immediately mix by inversion and record the decrease in A340nm for approximately 5 minutes. Obtain the ΔA340nm/min using the maximum linear rate for both the Test and Blank.

Calculations

$$\text{Units / ml enzyme} = \frac{(\Delta\text{A340nm / min Test} - \Delta\text{A340nm / min Blank})\ (3)\ (\text{df})}{(6.22)\ (0.05)\ (2)}$$

3 = Total volume (in milliliters) of assay
df = Dilution factor
6.22 = Millimolar extinction coefficient of β-NADH at 340nm
0.05 = Volume (in milliliter) of enzyme used
2 = Conversion factor accounting for 2 moles of β-NADH oxidized per mole of D-fructose-1,6-diphosphate produced

$$\text{Units / mg solid} = \frac{\text{units / ml enzyme}}{\text{mg solid / ml enzyme}}$$

$$\text{Units / mg protein} = \frac{\text{units / ml enzyme}}{\text{mg protein / ml enzyme}}$$

Unit Definition: *One unit will convert 1.0 μmole of fructose 6-phosphate and ATP to fructose 1,6-diphosphate and ADP per minute at pH 8.0 at 30°C in a coupled system with aldolase, glycerophosphate dehydrogenase and triosephosphate isomerase.*

Final Assay Concentrations

In a 3.00 ml reaction mix, the final concentrations are 50 mM Tris, 1 mM DL-dithiothreitol, 2 mM adenosine 5'-triphosphate, 0.2 mM β-nicotinamide adenine dinucleotide, reduced Form, 5 mM magnesium chloride, 2 units aldolase, 1 unit α-glycerophosphate dehydrogenase, 8 units triosephosphate isomerase, 4 mM fructose 6-phosphate, and 0.025–0.05 unit fructose-6-phosphate kinase.

Notes

1. The Triosephosphate Isomerase activity is approximately 8 fold that of the α-Glycerophosphate Dehydrogenase activity.
2. Aldolase Unit Definition: One unit will convert 1.0 μmole of fructose 1,6-diphosphate to dihydroxyacetone phosphate and glyceraldehyde 3-phosphate per minute at pH 7.4 at 25°C.
3. α-Glycerophosphate Dehydrogenase Unit Definition: One unit will convert 1.0 μmole of dihydroxyacetone phosphate to α-glycerophosphate per minute at pH 7.4 at 25°C.
4. Triosephosphate Isomerase Unit Definition: One unit will convert 1.0 μmole of D-glyceraldehyde 3-phosphate to dihydroxyacetone phosphate per minute at pH 7.6 at 25°C.

Enzymatic Assay of D-FRUCTOSE DEHYDROGENASE (EC 1.1.99.11)

Principle

$$\text{D-Fuc} + 2\,K_3Fe(CN) \xrightarrow{\text{D-Fructose Dehydrogenase}} \text{5-K-D-Fruc} + 2\,K_4Fe(CN)_6$$

$$2\,K_4Fe(CN)_6 + \text{Ferric Sulfate} \rightarrow 2\ \text{Prussian Blue} + 3K_2SO_4$$

Abbreviations Used

D-Fruc = D(-)Fructose
5-K-D-Fruc = 5-Keto-D-Fructose

Conditions: T = 37°C, pH = 4.5, A660nm, Light path = 1 cm

Method: Spectrophotometric Stop Rate Determination

Reagents

A. 200 mM Sodium Phosphate with 0.23% (v/v) Triton X-100 Solution (Prepare 100 ml in deionized water using Sodium Phosphate, Dibasic, and Triton X-100).
B. 100 mM Citric Acid Solution (Prepare 100 ml in deionized water using Citric Acid, Free Acid, Anhydrous).
C. McIlvaine Buffer with 0.1% Triton X-100, pH 4.5 at 37°C (Prepare by combining 90 ml of Reagent A with 110 ml of Reagent B. Adjust the pH to 4.5 at 37°C with either Reagent A or Reagent B as needed).
D. 1 M D-Fructose Solution (D-Fructose) (Prepare 10 ml in Reagent C using D(-)Fructose).
E. 100 mM Potassium Ferricyanide Solution ($K_3Fe(CN)_6$) (Prepare 10 ml in Reagent C using Potassium Ferricyanide).
F. 0.5% (w/v) Ferric Sulfate with 0.3% (w/v) Sodium Dodecyl Sulfate and 8.1% (v/v) Phosphoric Acid Solution (Stop Soln) (Prepare 50 ml in deionized water using Ferric Sulfate, Lauryl Sulfate, Sodium Salt and Phosphoric Acid).
G. D-Fructose Dehydrogenase Enzyme Solution (Immediately before use, prepare a solution containing 1.0–2.0 units/ml of D-Fructose Dehydrogenase in cold Reagent C).

Procedure

Pipette (in milliliters) the following reagents into suitable tubes:

	Test	Blank
Reagent C (Buffer)	0.70	0.80
Reagent D (D-Fructose)	0.10	—
Reagent G (Enzyme Solution)	0.10	0.10

Mix by swirling and equilibrate to 37°C. Then add:

Reagent E ($K_3Fe(CN)_6$)	0.10	0.10

Immediately mix by swirling and incubate for exactly 5 minutes at 37°C. Then add:

Reagent F (Stop Soln)	0.50	0.50

Immediately mix by swirling and incubate for 20 minutes. Then add:

Deionized Water	3.50	3.50

Mix by swirling and transfer the solutions to suitable cuvettes. Record the A660nm for both the Test and Blank using a suitable spectrophotometer.

Calculations

$$\text{Units / ml enzyme} = \frac{(\text{A660nm Test} - \text{A660nm Blank})\ (5)\ (\text{df})}{(2)\ (2)\ (5)\ (0.1)}$$

5 = Total volume (in milliliters) of assay
df = Dilution factor
2 = Millimolar extinction coefficient of Prussian Blue at 660 nm
2 = Two moles of Prussian Blue produced per mole of D-Fructose oxidized
5 = Time (in minutes) of assay as per the Unit Definition
0.1 = Volume (in milliliters) of enzyme used

$$\text{Units / mg solid} = \frac{\text{units / ml enzyme}}{\text{mg solid / ml enzyme}}$$

$$\text{Units / mg protein} = \frac{\text{units / ml enzyme}}{\text{mg protein / ml enzyme}}$$

Unit Definition: *One unit will convert 1.0 μmole of D-fructose to 5-keto-D-fructose per minute at pH 4.5 at 37°C.*

Final Assay Concentrations

In a 1.00 ml reaction mix, the final concentrations are 90 mM sodium phosphate, 55 mM citric acid, 0.1% (v/v) Triton X-100, 100 mM D(-)fructose, 10 mM potassium ferricyanide and 0.1–0.2 unit D-fructose dehydrogenase.

Enzymatic Assay of L-FUCOSE DEHYDROGENASE (EC 1.1.1.122) from Porcine Liver

Principle

$$\text{L-Fucose} + \beta\text{-NAD} \xrightarrow{\text{L-Fucose Dehydrogenase}} \text{L-Fucono-1, 5-lactone} + \beta\text{-NADH}$$

Abbreviations Used

β-NAD = β-Nicotinamide Adenine Nucleotide, Oxidized Form
β-NADH = β-Nicotinamide Adenine Nucleotide, Reduced Form

Conditions: T = 37°C, pH = 8.7, A340nm, Light path = 1 cm

Method: Continuous Spectrophotometric Rate Determination

Reagents

A. 55 mM Tris HCl Buffer, pH 8.7 at 37°C (Prepare 50 ml in deionized water using Trizma Base. Adjust to pH 8.7 at 37°C with 1 M HCl).
B. 5.5 mM α-L-Fucose Solution (Fucose) (Prepare 25 ml in Reagent A using α-L(-) Fucose. *Prepare Fresh*).
C. 30 mM β-Nicotinamide Adenine Dinucleotide Solution (β-NAD) (Prepare 1 ml in deionized water using β-Nicotinamide Adenine Dinucleotide. *Prepare Fresh*).
D. L-Fucose Dehydrogenase Enzyme Solution (Immediately before use, prepare a solution containing 0.2–0.5 unit/ml of L-Fucose Dehydrogenase in cold Reagent A).

Procedure

Pipette (in milliliters) the following reagents into suitable cuvettes:

	Test	Blank
Reagent B (Fucose)	2.80	2.80
Reagent C (b-NAD)	0.10	0.10

Mix by inversion and equilibrate to 37°C. Monitor the A340nm until constant, using a suitably thermostatted spectrophotometer. Then add:

Reagent A (Buffer)	—	0.10
Reagent D (Enzyme Solution)	0.10	—

Immediately mix by inversion and record the increase in A340nm for approximately 5 minutes. Obtain the ΔA340nm/minute using the maximum linear rate for both the Test and Blank.

Calculations

$$\text{Units / mg enzyme} = \frac{\Delta A340\text{nm / min Test} - \Delta A340\text{nm / min Blank}}{(6.22)\ (\text{mg enzyme / ml RM})}$$

6.22 = Millimolar extinction coefficient of β-NADH at 340nm
RM = Reaction Mix

Unit Definition: *One unit will oxidize 1.0 μmole of L-fucose to L-fucono-1,5-lactone per minute at pH 8.7 at 37°C.*

Final Assay Concentrations

In a 3.00 ml reaction mix, the final concentrations are 53 mM Tris, 5.1 mM L-fucose, 1.0 mM β-NAD and 0.02–0.05 unit L-fucose dehydrogenase.

Enzymatic Assay of L-FUCOSE DEHYDROGENASE From *Pseudomonas sp.*

Principle

$$\text{L-Fucose} + \beta\text{-NADP} \xrightarrow{\text{L-Fucose Dehydrogenase}} \text{L-Fucono-1, 5-lactone} + \beta\text{-NADPH}$$

Abbreviations Used

β-NADP = β-Nicotinamide Adenine Dinucleotide Phosphate, Oxidized Form
β-NADPH = β-Nicotinamide Adenine Dinucleotide Phosphate, Reduced Form

Conditions: T = 37°C, pH = 9.5, A340nm, Light path = 1 cm

Method: Continuous Spectrophotometric Rate Determination

Reagents

A. 120 mM Tris, 120 mM Imidazole and 100 mM Acetate HCl Buffer, pH 9.5 at 37°C (Prepare 100 ml in deionized water using Trizma Base, Imidazole and Sodium Acetate, Trihydrate. Adjust to pH 9.5 at 37°C with 5 M HCl).
B. 150 mM L-Fucose Solution (Fucose) (Prepare 25 ml in Reagent A using L(-)Fucose. *Prepare Fresh*).
C. 15 mM β-Nicotinamide Adenine Dinucleotide Phosphate Solution (β-NADP) (Prepare 5 ml in cold deionized water using β-Nicotinamide Adenine Dinucleotide Phosphate, Sodium Salt).
D. L-Fucose Dehydrogenase Enzyme Solution (Immediately before use, prepare a solution containing 0.15–0.25 unit/ml of L-Fucose Dehydrogenase in cold deionized water).

Procedure

Pipette (in milliliters) the following reagents into suitable cuvettes:

	Test	Blank
Reagent A (Buffer)	2.50	2.50
Reagent B (Fucose)	0.20	—
Reagent C (β-NADP)	0.20	0.20
Deionized Water	—	0.20

Mix by inversion and equilibrate to 37°C. Monitor the A340nm until constant, using a suitably thermostatted spectrophotometer. Then add:

Reagent D (Enzyme Solution)	0.10	0.10

Immediately mix by inversion and record the increase in A340nm for approximately 10 minutes. Obtain the ΔA340nm/minute using the maximum linear rate for both the Test and Blank.

Calculations

$$\text{Units/ml enzyme} = \frac{(\Delta A340\text{nm/min Test} - \Delta A340\text{nm/min Blank})\ (3)\ (\text{df})}{(6.22)\ (0.1)}$$

3 = Volume (in milliliters) of assay
df = Dilution factor
6.22 = Millimolar extinction coefficient of β-NADPH at 340 nm
0.1 = Volume (in milliliter) of enzyme used

$$\text{Units / mg solid} = \frac{\text{units / ml enzyme}}{\text{mg solid / ml enzyme}}$$

Unit Definition: *One unit will oxidize 1.0 μmole of L-fucose to L-fucono-1,5-lactone per minute at pH 9.5 at 37°C in the presence of NADP.*

Final Assay Concentrations

In a 3.00 ml reaction mix, the final concentrations are 108 mM Tris, 108 mM imidazole, 90 mM sodium acetate, 10 mM L-fucose, 1.0 mM β-nicotinamide adenine dinucleotide phosphate and 0.015–0.025 unit L-fucose dehydrogenase.

Note

1. This solution should be used within 1 hour. Store on ice.

Enzymatic Assay of α-L-FUCOSIDASE (EC 3.2.1.51)

Principle

$$\text{PNP } \alpha\text{-L-Fucoside} + H_2O \xrightarrow{\alpha\text{-L-Fucosidase}} \text{L-Fucose} + \text{p-Nitrophenol}$$

Abbreviation Used

PNP α-L-Fucoside = p-Nitrophenyl α-L-Fucoside

Conditions: T = 37°C, pH = 4.0, A400nm, Light path = 1 cm

Method: Spectrophotometric Stop Rate Determination

Reagents

A. 50 mM Sodium Citrate Buffer, pH 4.5 at 37°C (Prepare 100 ml in deionized water using Citric Acid, Trisodium Salt. Adjust to pH 4.5 at 37°C with HCl. *Prepare Fresh*).
B. 2.0 mM p-Nitrophenyl α-L-Fucopyranoside Solution (PNP-FUC) (Prepare 5 ml in Reagent A using p-Nitrophenyl α-L-Fucopyranoside).
C. 200 mM Borate Solution, pH 9.8 at 25°C (Borate) (Prepare 20 ml in deionized water using Boric Acid. Adjust to pH 9.8 at 25°C with 1 M NaOH).
D. α-L-Fucosidase Enzyme Solution (Immediately before use, prepare a solution containing 0.50–1.0 unit/ml of α-L-Fucosidase in cold Reagent A).

Procedure

Pipette (in milliliters) the following reagents into suitable cuvettes:

	Test	Blank
Reagent B (PNP-FUC)	0.25	0.25

Mix by inversion and equilibrate to 37°C. Then add:

	Test	Blank
Reagent D (Enzyme Solution)	0.10	—

Immediately mix by inversion and incubate for exactly 15 minutes at 37°C. Then add:

Reagent C (Borate)	2.20	2.20
Reagent D (Enzyme Solution)	—	0.10

Mix by inversion and transfer to suitable cuvettes. Record the A400nm for both the Test and Blank in a suitable spectrophotometer.

Calculations

$$\text{Units / mg enzyme} = \frac{(\text{A400nm Test} - \text{A400nm Blank})\ (2.55)}{(17.7)\ (15)\ (\text{mg enzyme / RM})}$$

17.7 = Millimolar extinction coefficient of p-Nitrophenol at 400nm
2.55 = Total volume of enzyme assay
15 = Time of assay (in minutes) as per the Unit Definition
RM = Reaction mix

Unit Definition: *One unit will hydrolyze 1.0 µmole of p-nitrophenyl α-L-fucoside to p-nitrophenol and L-fucose per minute at pH 4.5 at 37°C.*

Final Assay Concentrations

In a 0.35 ml reaction mix, the final concentrations are 50 mM citrate, 1.4 mM p-nitrophenyl α-L-fucoside and 0.05–0.10 unit α-L-fucosidase.

Enzymatic Assay of α-L-FUCOSIDASE (EC 3.2.1.51)

Principle

$$\text{PNP } \alpha\text{-L-Fucoside} + H_2O \xrightarrow{\alpha\text{-L-Fucosidase}} \text{L-Fucose} + \text{p-Nitrophenol}$$

Abbreviation Used

PNP α-L-Fucoside = p-Nitrophenyl α-L-Fucoside

Conditions: T = 25°C, pH = 5.0, A400nm, Light path = 1 cm

Method: Spectrophotometric Stop Rate Determination

Reagents

A. 100 mM Sodium Citrate Buffer, pH 5.5 at 25°C (Prepare 100 ml in deionized water using Citric Acid, Trisodium Salt. Adjust to pH 5.5 at 25°C with 1 M HCl. *Prepare Fresh*).

B. 10 mM p-Nitrophenyl α-L-Fucopyranoside Solution (PNP-FUC) (Prepare 5 ml in deionized water using p-Nitrophenyl α-L-Fucopyranoside).

C. 200 mM Borate Solution, pH 9.8 at 25°C (Borate) (Prepare 20 ml in deionized water using Boric Acid. Adjust to pH 9.8 at 25°C with 1 M NaOH).

D. α-L-Fucosidase Enzyme Solution (Immediately before use, prepare a solution containing 0.02–0.04 unit/ml of α-L-Fucosidase in cold Reagent A).

Procedure

Pipette (in milliliters) the following reagents into suitable cuvettes:

	Test	Blank
Reagent A (Buffer)	0.40	0.40
Reagent B (PNP-FUC)	0.50	0.50

Mix by inversion and equilibrate to 25°C. Then add:

Reagent D (Enzyme Solution)	0.10	—

Immediately mix by inversion and incubate for exactly 10 minutes at 25°C. Then add:

Reagent C (Borate)	3.00	3.00
Reagent D (Enzyme Solution)	—	0.10

Mix by inversion and transfer to suitable cuvettes. Record the A400nm for both the Test and Blank in a suitable spectrophotometer.

Calculations

$$\text{Units / mg enzyme} = \frac{(\text{A400nm Test} - \text{A400nm Blank})\,(4)}{(18)\,(10)\,(\text{mg enzyme / RM})}$$

18 = Millimolar extinction coefficient of p-Nitrophenol at 400nm
10 = Time of assay (in minutes) as per the Unit Definition
4 = Total volume of enzyme assay
RM = Reaction mix

Unit Definition: *One unit will hydrolyze 1.0 μmole of p-nitrophenyl α-L-fucoside to p-nitrophenol and L-fucose per minute at pH 5.5 at 25°C.*

Final Assay Concentrations

In a 1.00 ml reaction mix, the final concentrations are 50 mM citrate, 5.0 mM p-nitrophenyl α-L-fucoside and 0.002–0.004 unit α-L-fucosidase.

Enzymatic Assay of α-L-FUCOSIDASE (EC 3.2.1.51)from Human Placenta

Principle

$$\text{PNP } \alpha\text{-L-Fucoside} + H_2O \xrightarrow{\alpha\text{-L-Fucosidase}} \text{L-Fucose} + \text{p-Nitrophenol}$$

Abbreviation Used

PNP α-L-Fucoside = p-Nitrophenyl α-L-Fucoside

Conditions: T = 37°C, pH = 5.6, A400nm, Light path = 1 cm

Method: Spectrophotometric Stop Rate Determination

Reagents

A. 100 mM MES Buffer, pH 5.6 at 37°C (Prepare 100 ml in deionized water using MES, Free Acid. Adjust to pH 5.6 at 37°C with 1 M NaOH).

B. 2.0 mM p-Nitrophenyl α-L-Fucoside Solution (PNP-FUC) (Prepare 10 ml in Reagent A, using p-Nitrophenyl α-L-Fucopyranoside).
C. 1 M Sodium Bicarbonate Solution ($NaHCO_3$) (Prepare 20 ml in deionized water using Sodium Bicarbonate).
D. α-L-Fucosidase Enzyme Solution (Immediately before use, prepare a solution containing 0.05–0.10 unit/ml of α-L-Fucosidase in cold Reagent A).

Procedure

Pipette (in milliliters) the following reagents into suitable cuvettes:

	Test	Blank
Reagent B (PNP-FUC)	1.00	1.00

Equilibrate to 37°C. Then add:

Reagent D (Enzyme Solution)	0.10	—

Immediately mix by inversion and incubate for exactly 10 minutes at 37°C. Then add:

Reagent C ($NaHCO_3$)	1.00	1.00
Reagent D (Enzyme Solution)	—	0.10

Mix by inversion and transfer to suitable cuvettes. Record the A400nm for both the Test and Blank in a suitable spectrophotometer.

Calculations

$$\text{Units / mg enzyme} = \frac{(\text{A400nm Test} - \text{A400nm Blank})\,(2.1)\,(\text{df})}{(18.3)\,(10)\,(0.1)}$$

2.1 = Total volume (in milliliters) of stopped reaction
df = Dilution factor
18.3 = Millimolar extinction coefficient of p-Nitrophenol at 400nm
10 = Time (in minutes) of assay as per the Unit Definition
0.1 = Volume (in milliliters) of enzyme used

$$\text{Units / mg solid} = \frac{\text{units / ml enzyme}}{\text{mg solid / ml enzyme}}$$

$$\text{Units / mg protein} = \frac{\text{units / ml enzyme}}{\text{mg protein / ml enzyme}}$$

Unit Definition: *One unit will hydrolyze 1.0 μmole of p-nitrophenyl α-L-fucoside to p-nitrophenol and L-fucose per minute at pH 5.6 at 37°C.*

Final Assay Concentration

In a 1.10 ml reaction mix, the final concentrations are 100 mM MES, 1.8 mM p-nitrophenyl α-L-fucopyranoside, and 0.005–0.01 unit α-L-fucosidase.

Enzymatic Assay of α-L-FUCOSIDASE (EC 3.2.1.51)

Principle

$$\text{PNP } \alpha\text{-L-Fucoside} + H_2O \xrightarrow{\alpha\text{-L-Fucosidase}} \text{L-Fucose} + \text{p-Nitrophenol}$$

Abbreviation Used

PNP α-L-Fucoside = p-Nitrophenyl α-L-Fucoside

Conditions: T = 25°C, pH = 6.0 A400nm, Light path = 1 cm

Method: Spectrophotometric Stop Rate Determination

Reagents

A. 100 mM Sodium Citrate Buffer, pH 6.5 at 25°C (Prepare 100 ml in deionized water using Citric Acid, Trisodium Salt, Dihydrate. Adjust to pH 6.5 at 25°C with 1 M HCl. *Prepare Fresh*).

B. 10 mM p-Nitrophenyl α-L-Fucopyranoside Solution (PNP-FUC) (Prepare 5 ml in deionized water using p-Nitrophenyl α-L-Fucopyranoside. Heat gently to dissolve).

C. 200 mM Borate Solution, pH 9.8 at 25°C (Borate) (Prepare 50 ml in deionized water using Boric Acid. Adjust to pH 9.8 at 25°C with 1 M NaOH).

D. α-L-Fucosidase Enzyme Solution (Immediately before use, prepare a solution containing 0.05–0.10 unit/ml of α-L-Fucosidase in cold Reagent A).

Procedure

Pipette (in milliliters) the following reagents into suitable cuvettes:

	Test	Blank
Reagent A (Buffer)	0.40	0.40
Reagent B (PNP-FUC)	0.50	0.50

Mix by swirling and equilibrate to 25°C. Then add:

Reagent D (Enzyme Solution)	0.10	—

Immediately mix by swirling and incubate for exactly 10 minutes at 25°C. Then add:

Reagent C (Borate)	3.00	3.00
Reagent D (Enzyme Solution)	—	0.10

Mix by swirling and transfer to suitable cuvettes. Record the A400nm for both the Test and Blank in a suitable spectrophotometer.

Calculations

$$\text{Units / ml enzyme} = \frac{(\text{A400nm Test} - \text{A400nm Blank})\,(4)\,(\text{df})}{(18)\,(10)\,(0.1)}$$

4 = Total volume (in milliliters) of enzyme assay
df = Dilution factor

18 = Millimolar extinction coefficient of p-Nitrophenol at 400nm
10 = Time of assay (in minutes) as per the Unit Definition
0.1 = Volume (in milliliter) of enzyme used

$$\text{Units / mg protein} = \frac{\text{units / ml enzyme}}{\text{mg protein / ml enzyme}}$$

Unit Definition: *One unit will hydrolyze 1.0 μmole of p-nitrophenyl α-L-fucoside to p-nitrophenol and L-fucose per minute at pH 6.5 at 25°C.*

Final Assay Concentrations

In a 1.00 ml reaction mix, the final concentrations are 50 mM citrate, 5.0 mM p-nitrophenyl α-L-fucoside and 0.005–0.010 unit α-L-fucosidase.

Enzymatic Assay of FUMARASE (EC 4.2.1.2)

Principle

$$\text{L-Malate} \xrightarrow{\text{Fumarase}} \text{Fumarate} + H_2O$$

Conditions: T = 25°C, pH = 7.6, A240nm, Light path = 1 cm.

Method: Continuous Spectrophotometric Rate Determination.

Reagents

A. 100 mM Potassium Phosphate Buffer, pH 7.6 at 25°C (Prepare 100 ml in deionized water using Potassium Phosphate, Monobasic. Adjust to pH 7.6 with 1 M KOH).
B. 50 mM L-Malic Acid, pH 7.6 at 25°C (Prepare 100 ml in Reagent A using L-Malic Acid, Monosodium Salt, If necessary, adjust to pH 7.6 at 25°C using 1 M KOH).
C. 0.1% Bovine Serum Albumin (Prepare 10 ml in deionized water using Albumin Bovine).
D. Fumarase Enzyme Solution (Immediately before use, prepare a solution containing 1 unit/ml of Fumarase in cold Reagent C. The enzyme solution must be clear and colorless before addition to the reaction mix. Any cloudiness indicates the enzyme is not completely solubilized).

Procedure

Pipette (in milliliters) the following reagents into suitable quartz cuvettes:

	Test	Blank
Reagent B (L-Malic Acid)	2.9	2.9

Mix by inversion and equilibrate to 25°C. Monitor the A240nm until constant, using a suitable thermostatted spectrophotometer. Then add:

Reagent C (BSA)	—	0.10
Reagent D (Enzyme Solution)	0.10	—

Immediately mix by inversion and record the increase in A240nm/minutes for 5–10 minutes. Obtain the A240nm/minutes using the maximum linear rate for both the Test and Blank.

Calculations

$$\text{Units / ml enzyme} = \frac{(\text{A240nm / min Test} - \text{A240nm / min Blank})\,(3)\,(\text{df})}{(2.44)\,(0.1)}$$

3 = Volume (in milliliters) of assay
df = Dilution factor
2.44 = Millimolar extinction coefficient of Fumarate at 240nm
0.1 = Volume (in milliliter) of enzyme used.

$$\text{Units / mg protein} = \frac{\text{units / ml enzyme}}{\text{mg protein / ml enzyme}}$$

Unit Definition: *One unit will convert 1.0 μ-mole of L-Malate to Fumarate per minute at pH 7.6 at 25°C.*

Final Assay Concentrations

In a 3 ml reaction mix, the final concentrations are 96.7 mM potassium phosphate, 48.3 mM L-malic acid and 0.1 unit fumarase.

Enzymatic Assay of GABASE

Principle

$$\gamma\text{ - Amino - n - Butyrate} + \alpha\text{ - Ketoglutarate} \xrightarrow{\text{GABGT}} \text{Succinic Semialdehyde} + \text{Glutamate}$$

$$\text{Succinic Semialdehyde} + \beta\text{ - NADP} \xrightarrow{\text{SSDH}} \text{Succinate } \beta\text{ - NADPH}$$

Abbreviations Used

GABGT = γ-Aminobutyric Glutamic Transaminase
β-NADP = β-Nicotinamide Adenine Dinucleotide Phosphate, Oxidized Form
SSDH = Succinic Semialdehyde Dehydrogenase
β-NADPH = β-Nicotinamide Adenine Dinucleotide Phosphate, Reduced Form

Conditions: T = 25°C, pH = 8.6, A340nm, Light path = 1 cm

Method: Continuous Spectrophotometric Rate Determination

Reagents

A. 100 mM Potassium Pyrophosphate Buffer, pH 8.6 at 25°C (Prepare 100 ml in deionized water using Pyrophosphate, Tetrapotassium, Anhydrous. Adjust to pH 8.6 at 25°C with 1 M HCl).
B. 100 mM 2-Mercaptoethanol Solution (2-ME) (Prepare 10 ml in deionized water using 2-Mercaptoethanol).
C. 25 mM β-Nicotinamide Adenine Dinucleotide Phosphate Solution (β-NADP) (Dissolve the contents of one 30 mg vial of β-Nicotinamide Adenine Dinucleotide Phosphate, Sodium Salt, Preweighed Vial in the appropriate volume of deionized water. Neutralize to pH 7.0 with solid Sodium Bicarbonate).
D. 60 mM γ-Amino-n-Butyric Acid Solution (GABA) (Prepare 10 ml in Reagent A using γ-Amino-n-Butyric Acid).

E. 100 mM α-Ketoglutarate Solution (α-Ketoglut) (Prepare 2 ml in deionized water using α-Ketoglutaric Acid, Monosodium Salt).

F. 75 mM Potassium Phosphate Buffer with 25% (v/v)Glycerol, pH 7.2 at 25°C (Enz Dil) (Prepare 25 ml in deionized water using Potassium Phosphate, Monobasic, Anhydrous and Glycerol. Adjust to pH 7.2 at 25°C with 1 M KOH).

G. Gabase Enzyme Solution (Immediately before use, prepare a solution containing 1–2 units/ml of Gabase in cold Reagent F).

Procedure

Pipette (in milliliters) the following reagents into suitable cuvettes:

	Test	Blank
Reagent A (Buffer)	2.30	2.30
Reagent B (2-ME)	0.10	0.10
Reagent E (α-Ketoglut)	0.15	0.15
Reagent C (β-NADP)	0.15	0.15
Reagent D (GABA)	0.30	0.30

Mix by inversion and equilibrate to 25°C. Monitor the A340nm until constant, using a suitably thermostatted spectrophotometer. Then add:

Reagent G (Enzyme Solution)	0.02	—
Reagent F (Enz Dil)	—	0.02

Immediately mix by inversion and record the increase in A340nm for approximately 5 minutes. Obtain the ΔA340nm/minute using the maximum linear rate for both the Test and Blank.

Calculations

$$\text{Units / ml enzyme} = \frac{(\Delta\text{A340nm / min Test} - \Delta\text{A340nm / min Blank})\ (3.02)\ (\text{df})}{(6.22)\ (0.02)}$$

3.02 = Total volume (in milliliters) of assay
df = Dilution factor
6.22 = Millimolar extinction coefficient of β-NADPH at 340nm
0.02 = Volume (in milliliter) of enzyme used

$$\text{Units / mg solid} = \frac{\text{units / ml enzyme}}{\text{mg solid / ml enzyme}}$$

$$\text{Units / mg protein} = \frac{\text{units / ml enzyme}}{\text{mg protein / ml enzyme}}$$

Unit Definition: *One unit will convert 1.0 μmole of α-aminobutyric acid (GABA) to succinic semialdehyde and then to succinate per minute with a stoichiometric reduction of 1.0 μmole of NADP⁺ at pH 8.6 at 25°C.*

Final Assay Concentrations

In a 3.02 ml reaction mix, the final concentrations are 86 mM potassium pyrophosphate, 3.3 mM 2-mercaptoethanol, 1.2 mM β-nicotinamide adenine dinucleotide phosphate,

5 mM α-ketoglutarate, 6.0 mM γ-amino-n-butyric acid, 0.50 mM potassium phosphate, 0.17% (v/v) glycerol, and 0.02–0.04 unit gabase.

Enzymatic Assay of GALACTOKINASE (EC 2.7.1.6)

Principle

$$\text{D(+)-Galactose} + \text{ATP} \xrightarrow{\text{GK}} \alpha\text{-D-Galactose-1-Phosphate} + \text{ADP}$$

$$\text{ADP} + \text{PEP} \xrightarrow[\text{Mg}^{++},\, \text{K}^{+}]{\text{PK}} \text{Pyruvate} + \text{ATP}$$

$$\text{Pyruvate} + \beta\text{-NADH} \xrightarrow{\text{LDH}} \text{Lactate} + \beta\text{-NAD}$$

Abbreviations Used

ATP = Adenosine 5'-Triphosphate
ADP = Adenosine 5'-Diphosphate
PEP = Phospho(enol) pyruvate
PK = Pyruvate Kinase
β-NADH = β-Nicotinamide Adenine Dinucleotide, Reduced Form
β-NAD = β-Nicotinamide Adenine Dinucleotide, Oxidized Form
LDH = Lactic Dehydrogenase

Conditions: T = 25°C, pH = 7.0, A340nm, Light path = 1 cm.

Method: Continuous Spectrophotometric Rate Determination.

Reagents

A. 160 mM Potassium Phosphate Buffer, pH 7.0 at 25°C (Prepare 100 ml in deionized water using Potassium Phosphate, Monobasic. Adjust to pH 7.0 at 25°C with 1 M NaOH).

B. 100 mM D-Galactose Substrate Solution (Prepare 1.0 ml in deionized water using (+)- Galactose).

C. 5.9 mM Adenosine 5'-Triphosphate Solution (ATP) (Prepare 5.0 ml in deionized water using Adenosine 5'-Triphosphate, Disodium Salt).

D. 16.2 mM Phospho(enol) pyruvate (PEP) (Prepare 5 ml in deionized water using Phospho(enol) pyruvate, Trisodium Salt).

E. 800 mM Potassium Chloride Solution (KCl) (Prepare 5 ml in deionized water using Potassium Chloride).

F. 100 mM Magnesium Chloride ($MgCl_2$) (Prepare 5 ml in deionized water using Magnesium Chloride, Hexahydrate).

G. 20 mM Ethylenediaminetetraacetic Acid (EDTA) (Prepare 5 ml in deionized water using Ethylenediaminetetraacetic Acid, Tetrasodium Salt).

H. 3.76 mM β-Nicotinamide Adenine Dinucleotide, Reduced Form, (β-NADH) (Dissolve the contents of one 5 mg vial of β-Nicotinamide Adenine Dinucleotide, Reduced Form, Disodium Salt, in 1.70 ml of Reagent A).

I. PK/LDH Mixed Enzyme Solution (PK/LDH) (Use PK/LDH Enzyme Suspension).

J. Galactokinase Enzyme Solution (GK) (Immediately before use, prepare a solution containing 0.25–0.75 units/ml of Galactokinase in cold, deionized water).

Procedure

Prepare a reaction cocktail by pipetting (in milliliters) the following reagents into a suitable container:

Reagent A (Buffer)	11.5
Reagent C (ATP)	3.0
Reagent D (PEP)	3.0
Reagent E (KCL)	3.0
Reagent F ($MgCl_2$)	3.0
Reagent G (EDTA)	3.0
Reagent H (β-NADH)	1.0

Mix and adjust to pH 7.0 at 25°C with 1 M HCl or 1 M NaOH, if necessary. Pipette (in milliliters) the following reagents into suitable cuvettes:

	Test	Blank
Reaction Mix Cocktail	2.75	2.75
Reagent I (PK/LDH)	0.05	0.05

Mix by inversion and equilibrate to 25°C. Monitor the A340nm until constant, using a suitably thermostatted spectrophotometer. Then add:

Reagent J (GK)	0.10	—
Deionized Water	—	0.10

Mix by inversion and monitor the A340nm until constant. Then add:

Reagent B (D-Galactose)	0.10	0.10

Immediately mix by inversion and record the decrease in A340nm for approximately 5 minutes. Obtain the A340nm/minute using the maximum linear rate for both the Test and Blank.

Calculations

$$\text{Units / mg enzyme} = \frac{(\text{A340nm / min Test} - \text{A340nm / min Blank})}{(6.22)\ (\text{mg enzyme / ml RM})}$$

6.22 = Millimolar extinction coefficient of β-NADH at 340nm
RM = Reaction Mix

Unit Definition: *One unit will convert 1.0 μmole of D-galactose to galactose 1-phosphate per minute at pH 7.0 at 25°C.*

Final Assay Concentrations

In a 3 ml reaction mix, the final concentrations are 67 mM potassium phosphate, 0.6 mM ATP, 1.6 mM PEP, 80 mM KCL, 10 mM $MgCl_2$, 2 mM EDTA, 0.13 mM β-NADH, 3.3 mM D(+)-galactose, 35 units pyruvate kinase, 50 units lactic dehydrogenase and 0.025–0.075 units galactokinase.

Enzymatic Assay of β-GALACTOSE DEHYDROGENASE (EC 1.1.1.48)

Principle

$$\beta\text{-D-Galactose} + \beta\text{-NAD} \xrightarrow{\beta\text{-Galactose Dehydrogenase}} \text{D-Galactonate} + \beta\text{-NADH}$$

Abbreviations Used

β-NAD = β-Nicotinamide Adenine Dinucleotide, Oxidized Form
β-NADH = β-Nicotinamide Adenine Dinucleotide, Reduced Form

Conditions: T = 25°C, pH = 8.6, A340nm, Light path = 1 cm

Method: Continuous Spectrophotometric Rate Determination

Reagents

A. 100 mM Tris HCl Buffer, pH 8.6 at 25°C (Prepare 50 ml in deionized water using Trizma Base. Adjust to pH 8.6 at 25°C with 1 M HCl).
B. 72 mM β-Nicotinamide Adenine Dinucleotide Solution (β-NAD) (Dissolve the contents of one 50 mg vial of β-Nicotinamide Adenine Dinucleotide in the appropriate volume of deionized water. *Prepare Fresh*).
C. 10% (w/v) D(+)Galactose Solution (Gal) (Prepare 1 ml in deionized water using D(+)Galactose).
D. 1 M Ammonium Sulfate (Enzyme Diluent) (Prepare 10 ml in deionized water using Ammonium Sulfate).
E. β-Galactose Dehydrogenase Enzyme Solution (Immediately before use, prepare a solution containing 0.15–0.30 unit/ml of β-Galactose Dehydrogenase in cold Reagent D).

Procedure

Pipette (in milliliters) the following reagents into suitable cuvettes:

	Test	Blank
Reagent A (Buffer)	3.00	3.00
Reagent B (β-NAD)	0.10	0.10
Reagent C (Gal)	0.10	0.10

Mix by inversion and equilibrate to 25°C. Monitor the A340nm until constant, using a suitably thermostatted spectrophotometer. Then add:

Reagent D (Enzyme Diluent)	—	0.10
Reagent E (Enzyme Solution)	0.10	—

Immediately mix by inversion and record the increase in A340nm for approximately 5 minutes. Obtain the ΔA340nm/minute using the maximum linear rate for both the Test and Blank.

Calculations

$$\text{Units / ml enzyme} = \frac{(\Delta A340nm / \min \text{ Test} - \Delta A340nm / \min \text{ Blank})\ (3.3)\ (df)}{(6.22)\ (0.1)}$$

3.3 = Volume (in milliliters) of assay
df = Dilution factor
0.1 = Volume (in milliliter) of enzyme used
6.22 = Millimolar extinction coefficient of β-NADH at 340nm

$$\text{Units / mg solid} = \frac{\text{units / ml enzyme}}{\text{mg solid / ml enzyme}}$$

$$\text{Units / mg protein} = \frac{\text{units / ml enzyme}}{\text{mg protein / ml enzyme}}$$

Unit Definition: *One unit will convert 1.0 μmole of D-galactose to D-galactonate per minute at pH 8.6 at 25°C.*

Final Assay Concentrations

In a 3.30 ml reaction mix, the final concentrations are 91 mM Tris, 2.2 mM β-nicotinamide adenine dinucleotide, 0.30% (w/v) D-galactose, 30 mM ammonium sulfate and 0.015–0.030 unit β-galactose dehydrogenase.

Note

1. Mutorotation of galactose does not affect the activity of this enzyme assay.

Enzymatic Assay of GALACTOSE OXIDASE (EC 1.1.3.9)

Principle

$$\text{D(+)Galactose} + O_2 \xrightarrow{\text{Galactose Oxidase}} \text{D-Galacto-hexodialdose} + H_2O_2$$

$$H_2O_2 + \text{O-Tolidine (reduced) O} \xrightarrow{\text{Peroxidase}} \text{O-Tolidine (oxidized)} + H_2O$$

Conditions: T = 25°C, pH = 6.0, A425nm, Light path = 1 cm

Method: Continuous Spectrophotometric Rate Determination

Reagents

A. 100 mM Potassium Phosphate Buffer, pH 6.0 at 25°C (Prepare 100 ml in deionized water using Potassium Phosphate, Monobasic, Anhydrous. Adjust to pH 6.0 at 25°C with 1 M KOH).
B. 0.5% (w/v) O-Tolidine Solution (Prepare 1 ml in Methanol using O-Tolidine).
C. Dye Buffer Solution (DB) (Prepare by adding 0.2 ml of Reagent B to 24.0 ml of Reagent A. Mix thoroughly).
D. 555 mM D(+)Galactose Solution (D-Gal) (Prepare 15 ml in deionized water using D(+)Galactose).
E. Peroxidase Solution (POD) (Immediately before use, prepare a solution containing 5 purpurogallin units/ml in cold deionized water using Peroxidase, Horseradish).

F. Galactose Oxidase Enzyme Solution (Immediately before use, prepare a solution containing 0.50 unit/ml of Galactose Oxidase in cold Reagent A).

Procedure

Pipette (in milliliters) the following reagents into suitable cuvettes:

	Test	Blank
Reagent C (DB)	1.70	1.70
Reagent D (D-Gal)	1.50	1.50
Reagent E (POD)	0.10	0.10

Mix by inversion and equilibrate to 25°C. Monitor the A425nm until constant, using a suitably thermostatted spectrophotometer. Then add:

Deionized Water	—	0.10
Reagent F (Enzyme Solution)	0.10	—

Immediately mix by inversion and record the increase in A425nm for approximately 5 minutes. Obtain the ΔA425nm/minute using the maximum linear rate for both the Test and Blank.

Calculations

$$\text{Units / ml enzyme} = \frac{(\Delta A425\text{nm / min Test} - \Delta A425\text{nm / min Blank})\ (3.4)\ (\text{df})}{(1.0)\ (0.1)}$$

3.4 = Volume (in milliliters)
df = Dilution factor
1.0 = The change in A425nm/minute per unit Galactose Oxidase at pH 6.0 at 25°C in a 3.4 ml reaction mix
0.1 = volume (in milliliters) of assay

$$\text{Units / mg solid} = \frac{\text{units / ml enzyme}}{\text{mg solid / ml enzyme}}$$

$$\text{Units / mg protein} = \frac{\text{units / ml enzyme}}{\text{mg protein / ml enzyme}}$$

Unit Definition: *One unit will produce an ΔA425nm of 1.0 per minute at pH 6.0 at 25°C, in a peroxidase and O-tolidine system.*

Final Assay Concentrations

In a 3.40 ml reaction mix, the final concentrations are 50 mM potassium phosphate, 0.002% (w/v) O-tolidine, 0.4% (v/v) methanol, 245 mM D(+)galactose, 0.5 unit peroxidase and 0.050 unit galactose oxidase.

Notes

1. O-Tolidine is a possible carcinogen. Take Appropriate Precautions.
2. Allow to stand 2 hours at 25°C to allow for mutarotation.

Enzymatic Assay of GALACTOSE-1-PHOSPHATE URIDYL TRANSFERASE (EC 2.7.7.12)

Principle

$$\alpha\text{-D-Galactose 1-Phosphate} + \text{UDPG} \xrightarrow{\text{GalPUT}} \alpha\text{-D-Glucose 1-Phosphate} + \text{UDPGal}$$

$$\alpha\text{-D-Glucose 1-Phosphate} \xleftrightarrow{\text{Phosphoglucomutase}} \alpha\text{-D-Glucose 6-Phosphate}$$

$$\alpha\text{-D-Glucose 6-Phosphate} + \beta\text{-NADP} \xrightarrow{\text{G-6-PDH}} \text{6-Phosphogluconate} + \beta\text{-NADPH}$$

Abbreviations Used

UDPG = Uridine 5'-Diphosphoglucose
GalPUT = Galactose-1-Phosphate Uridyl Transferase
UDPGal = Uridine 5'-Diphosphogalactose
β-NADP = β-Nicotinamide Adenine Dinucleotide Phosphate, Oxidized Form
G-6-PDH = Glucose-6-Phosphate Dehydrogenase
β-NADPH = β-Nicotinamide Adenine Dinucleotide Phosphate, Reduced Form

Conditions: T = 25°C, pH = 8.7, A340nm, Light path = 1 cm

Method: Continuous Spectrophotometric Rate Determination

Reagents

A. 1000 mM Glycine Buffer, pH 8.7 at 25°C (Prepare 25 ml in deionized water using Glycine Free Base. Adjust to pH 8.7 at 25°C with 1 M NaOH).
B. 40 mM Galactose 1-Phosphate Solution (Gal 1-P) (Prepare 2 ml in deionized water using α-D-Galactose 1-Phosphate, Dipotassium Salt).
C. 10 mM Uridine 5'-Diphosphoglucose Solution (UDPG) (Prepare 3 ml in deionized water using Uridine 5'-Diphosphoglucose, Disodium Salt).
D. 0.2 mM Glucose 1,6-Diphosphate Solution (G 1,6-DiP) (Prepare 1 ml in deionized water using α-D-Glucose 1,6-Diphosphate, Cyclohexylammonium Salt, Hydrate).
E. 20 mM β-Nicotinamide Adenine Dinucleotide Phosphate Solution (β-NADP) (Prepare 2 ml in deionized water using β-Nicotinamide Adenine Dinucleotide Phosphate, Sodium Salt. *Prepare Fresh*).
F. 300 mM Magnesium Chloride Solution ($MgCl_2$) (Prepare 5 ml in deionized water using Magnesium Chloride, Hexahydrate).
G. 200 mM L-Cysteine Hydrochloride Solution (Cys) (Prepare 2 ml in deionized water using L-Cysteine Hydrochloride, Monohydrate. Neutralize by adding solid Sodium Bicarbonate).
H. Phosphoglucomutase Solution (PGLUM) (Immediately before use, prepare a solution containing 15 units/ml in cold deionized water using Phosphoglucomutase).
I. Glucose-6-Phosphate Dehydrogenase Solution (G-6-PDH) (Immediately before use, prepare a solution containing 15 units/ml in cold deionized water using Glucose-6-Phosphate Dehydrogenase).

J. 100 mM Citrate Solution, pH 7.5 at 25°C (Enzyme Diluent) (Prepare 100 ml in deionized water using Citric Acid, Trisodium Salt, Dihydrate. Adjust to pH 7.5 at 25°C with 1 M HCl).

K. Galactose-1-Phosphate Uridyl Transferase Enzyme Solution (GalPUT) (Immediately before use, prepare a solution containing 0.2–0.8 unit/ml of Galactose-1-Phosphate Uridyl Transferase in cold Reagent J).

Procedure

Prepare a reaction cocktail by pipetting (in milliliters) the following reagents into a suitable container:

Deionized water	17.50
Reagent A (Buffer)	3.00
Reagent B (Gal 1-P)	1.00
Reagent C (UDPG)	2.00
Reagent D (G 1,6-DP)	1.00
Reagent E (β-NADP)	1.00
Reagent F ($MgCl_2$)	1.00
Reagent G (Cys)	1.50

Mix and adjust to pH 8.7 at 25°C with 100 mM HCl or 100 mM NaOH, if necessary. Pipette (in milliliters) the following reagents into suitable cuvettes:

	Test	Blank
Reaction Cocktail	2.80	2.80
Reagent H (PGLUM)	0.05	0.05
Reagent I (G-6-PDH)	0.05	0.05

Mix by inversion and equilibrate to 25°C. Monitor the A340nm until constant, using a suitably thermostatted spectrophotometer. Then add:

Reagent J (Enzyme Diluent)	—	0.10
Reagent K (GalPUT)	0.10	—

Immediately mix by inversion and record the increase in A340nm for approximately 5 minutes. Obtain the ΔA340nm/minute using the maximum linear rate for both the Test and Blank.

Calculations

$$\text{Units / mg enzyme} = \frac{(\Delta A340nm / min\ Test - \Delta A340nm / min\ Blank)}{(6.22)\ (mg\ enzyme / ml\ RM)}$$

6.22 = Millimolar extinction coefficient of β-NADPH at 340nm
RM = Reaction Mix

Unit Definition: *One unit will form 1.0 μmole of glucose 1-phosphate from UDP-glucose, galactose 1-phosphate and NADP⁺ per minute at pH 8.7 at 25°C as detected by a coupled system using phosphoglucomutase and β-NADP.*

Final Assay Concentrations

In a 3.00 ml reaction mix, the final concentrations are 100 mM glycine, 1.3 mM galactose 1-phosphate, 0.67 mM uridine 5'-diphosphoglucose, 0.0067 mM glucose-1, 6- diphosphate, 0.67 mM b-NADP, 10 mM $MgCl_2$, 10 mM L-cysteine hydrochloride, 3.3 mM citrate, 0.75 unit phosphoglucomutase, 0.75 unit glucose-6-phosphate dehydrogenase and 0.02–0.08 unit galactose-1-phosphate uridyl transferase.

Notes

1. Glucose-6-Phosphate Dehydrogenase Unit Definition: One unit will oxidize 1.0 μmole of D-glucose 6-phosphate to 6-phospho-D-gluconate per minute in the presence of β-NADP at pH 7.4 at 25°C.
2. Phosphoglucomutase Unit Definition: One unit will convert 1.0 μmole of α-D-glucose 1-phosphate to α-D-glucose 6-phosphate per minute at pH 7.4 at 30°C.

Enzymatic Assay of β-GALACTOSIDASE (EC 3.2.1.23) from *E. coli* β-Lactose as Substrate

Principle

$$\beta\text{-Lactose} + H_2O \xrightarrow{\beta\text{-Galactosidase}} \text{D-Glucose} + \text{D-Galactose}$$

$$\text{D-Glucose} + O_2 + H_2O \xrightarrow{\text{Glucose Oxidase}} \text{D-Gluconic Acid} + H_2O_2$$

$$H_2O_2 + \text{O-Dianisidine(Reduced)} \xrightarrow{\text{Peroxidase}} H_2O + \text{O-Dianisidine(Oxidized)}$$

Conditions: T = 37°C, pH = 7.3, A510nm, Light path = 1 cm

Method: Colorimetric

Reagents

A. 100 mM Potassium Phosphate Monobasic Solution (Prepare 100 ml in deionized water using Potassium Phosphate, Anhydrous, Monobasic).

B. 100 mM Potassium Phosphate Buffer, pH 7.3 at 37°C (Prepare 100 ml in deionized water using Potassium Phosphate, Dibasic, Trihydrate. Adjust to pH 7.3 at 37°C with Reagent A).

C. 1.0% (w/v) β-Lactose Solution (Lac) (Prepare 40 ml in Reagent A using β-Lactose).

D. 100 mM Magnesium Sulfate Solution ($MgSO_4$) (Prepare 5 ml in deionized water using Magnesium Sulfate, Heptahydrate).

E. 4.2% (v/v) Perchloric Acid Solution (PCA) (Prepare 10 ml in deionized water using Perchloric Acid).

F. 6.3 mM O-Dianisidine Solution (O-Dianisidine) (Prepare 1 ml in deionized water using O-Dianisidine, Dihydrochloride).

G. PGO Enzymes Solution (PGO) (Immediately before use, dissolve 1 capsule of PGO Enzymes in 100 ml of deionized water).

H. 5.6 mM Glucose Standard (Gluc Std) (Use Glucose Standard Solutions).

I. β-Galactosidase Enzyme Solution (Immediately before use, prepare a solution containing 3–5 units/ml of β-Galactosidase in cold Reagent B).

Procedure

Colorimetric Assay

Step 1: Pipette (in milliliters) the following reagents into suitable containers:

	Test	Blank
Reagent B (Buffer)	0.70	0.70
Reagent C (Lac)	4.00	4.00
Reagent D ($MgSO_4$)	0.20	0.20

Mix by swirling and equilibrate to 37°C. Then add:

Reagent I (Enzyme Solution)	0.10	—

Mix by swirling and incubate at 37°C for exactly 30 minutes. Then add:

Reagent E (PCA)	1.00	1.00
Reagent I (Enzyme Solution)	—	0.10

Mix by swirling and centrifuge to clarify, if the solution is turbid.

Step 2: Pipette (in milliliters) the following reagents into suitable vials:

	Test	Blank
Test Supernatant	0.20	—
Blank Supernatant	—	0.20
Reagent F (O-Dianisidine)	0.10	0.10
Reagent G (PGO)	6.00	6.00

Mix by swirling and incubate at 37°C for 30 minutes. Transfer the Test and Blank solutions to suitable cuvettes and record the A510nm for each using a suitable spectrophotometer.

Standard Curve: A standard curve is made by pipetting (in milliliters) the following reagents into suitable containers:

	Std 1	Std 2	Std 3	Std 4	Std 5	Std Blank
Reagent H (Gluc Std)	0.005	0.01	0.05	0.10	0.20	—
Deionized Water	0.195	0.19	0.15	0.10	—	0.20
Reagent F (O-Dianisidine)	0.100	0.10	0.10	0.10	0.10	0.10
Reagent G (PGO)	6.00	6.00	6.00	6.00	6.00	6.00

Mix by swirling and incubate at 37°C for 30 minutes. Transfer the Standards and Standard Blank solutions to suitable cuvettes and record the A510nm for the Standards and Standard Blank.

Calculations

Standard Curve: ΔA510nm Standard = A510nm Std–A510nm Std Blank

Determine the μmoles of Glucose liberated using the standard curve.

Sample Determination: ΔA510nm Sample = A510nm Test–A510nm Test Blank
Determine the µmoles of Glucose liberated using the standard curve.

$$\text{Units / ml enzyme} = \frac{(\mu\text{moles glucose released})\ (6)}{(0.2)\ (30)\ (0.1)}$$

6 = Total volume (in milliliters) of Step 1
0.2 = Volume (in milliliter) used in Step 2
30 = Time of assay (in minutes) as per the Unit Definition
0.1 = Volume (in milliliter) of enzyme used

$$\text{Units / mg solid} = \frac{\text{units / ml enzyme}}{\text{mg solid / ml enzyme}}$$

$$\text{Units / mg protein} = \frac{\text{units / ml enzyme}}{\text{mg protein / ml enzyme}}$$

Unit Definition: *One unit will hydrolyze 1.0 µmole of β-lactose to glucose and galactose per minute at pH 7.3 at 37°C.*

Final Assay Concentrations

In a 5.00 ml reaction mix, the final concentrations are 96 mM potassium phosphate, 0.8% (w/v) lactose, 4 mM magnesium sulfate and 0.3–0.5 unit β-galactosidase.

Enzymatic Assay of β-GALACTOSIDASE (EC 3.2.1.23) β-Lactose as Substrate

Principle

$$\beta\text{-Lactose} + H_2O \xrightarrow{\beta\text{-Galactosidase}} \text{D-Glucose} + \text{D-Galactose}$$

$$\text{D-Glucose} + O_2 + H_2O \xrightarrow{\text{Glucose Oxidase}} \text{D-Gluconic Acid} + H_2O_2$$

$$H_2O_2 + \text{O-Dianisidine(Reduced)} \xrightarrow{\text{Peroxidase}} H_2O + \text{O-Dianisidine(Oxidized)}$$

Conditions: T= 30°C, pH = 4.5, A510nm, Light path = 1 cm

Method: Colorimetric

Reagents

A. 10 mM Citric Acid (Prepare 100 ml in deionized water using Citric Acid, Free Acid, Anhydrous).
B. 20 mM Sodium Phosphate Solution (Prepare 100 ml in deionized water using Sodium Phosphate, Dibasic, Anhydrous).
C. Phosphate Citrate Buffer, pH 4.5 at 30°C (Phos-Cit Buffer) (Prepare 100 ml using Reagent B. Adjust to pH 4.5 at 30°C with Reagent A).
D. 5.0% (w/v) β-Lactose Solution (Lac) (Prepare 10 ml in Reagent C using β-Lactose).
E. 4.2% (v/v) Perchloric Acid Solution (PCA) (Prepare 10 ml in deionized water using Perchloric Acid).
F. 6.3 mM O-Dianisidine Solution (Dian) (Prepare by dissolving one 50 mg pre-weighed vial of O-Dianisidine, Hydrochloride in the appropriate volume of deionized water).

G. PGO Enzymes Solution (PGO) (Immediately before use, dissolve 1 capsule of PGO Enzymes in 100 ml of deionized water).

H. 2.0 mM Glucose Solution (Gluc Std) (Prepare 10 ml in deionized water using β-D(+)-Glucose).

I. β-Galactosidase Enzyme Solution (Immediately before use, prepare a solution containing 5.0–10 units/ml β-Galactosidase in cold deionized water).

Procedure

Colorimetric Assay

Step 1: Pipette (in milliliters) the following reagents into suitable containers:

	Test	Blank
Reagent C (Phos-Cit Buffer)	0.20	0.20
Reagent D (Lac)	4.70	4.70

Mix by swirling and equilibrate to 30°C . Then add:

Reagent I (Enzyme Solution)	0.10	—

Mix by swirling and incubate at 30°C for exactly 10 minutes. Then add:

Reagent E (PCA)	1.00	1.00
Reagent I (Enzyme solution)	—	0.10

(If solution is hazy, centrifuge to clarify).

Step 2: Pipette (in milliliters) the following reagents into suitable containers:

Blank Supernatant	—	0.20
Test Supernatant	0.20	—
Reagent F (Dian)	0.10	0.10
Reagent G (PGO)	6.00	6.00

Mix by swirling and incubate at 30°C for 30 minutes. Transfer the Test and Blank solutions to suitable cuvettes and record the A510nm for the Test and Blank using a suitable spectrophotometer.

Standard Curve: A standard curve is made by pipetting (in milliliters) the following reagents into suitable containers:

	Std 1	Std 2	Std 3	Std 4	Std 5	Std Blank
Reagent H (Gluc Std)	0.005	0.01	0.05	0.10	0.20	—
Deionized Water	0.195	0.19	0.15	0.10	—	0.20
Reagent F (O-Dianisidine)	0.100	0.10	0.10	0.10	0.10	0.10
Reagent G (PGO)	6.00	6.00	6.00	6.00	6.00	6.00

Mix by swirling and incubate at 30°C for 30 minutes. Transfer the solutions to suitable cuvettes and record the A510nm for the Standards and Standard Blank using a suitable spectrophotometer.

Calculations

Standard Curve: ΔA510nm Standard = A510nm Std–A510nm Std Blank

Prepare a standard curve by plotting the ΔA510nm of the Standards vs μmoles of Glucose.

Sample Determination: ΔA510nm Sample = A510nm Test–A510nm Test Blank

Determine the μmoles of Glucose liberated using the standard curve.

$$\text{Units / ml enzyme} = \frac{(\mu\text{moles glucose released})\,(6)\,(\text{df})}{(0.2)\,(10)\,(0.1)}$$

6 = Total volume (in milliliters) of assay in Step 1
df = Dilution factor
0.2 = Volume (in milliliter) used in Colorimetric Determination
10 = Time of assay (in minutes) as per the Unit Definition
0.1 = Volume (in milliliter) of enzyme used

$$\text{Units / mg solid} = \frac{\text{units / ml enzyme}}{\text{mg solid / ml enzyme}}$$

$$\text{Units / mg protein} = \frac{\text{units / ml enzyme}}{\text{mg protein / ml enzyme}}$$

Unit Definition: *One unit will hydrolyze 1.0 μmole of β-lactose to glucose and galactose per minute at pH 4.5 at 30°C.*

Final Assay Concentrations

In a 5.00 ml reaction mix, the final concentrations are 4.7% (w/v) lactose, and 0.5–1.0 unit β-galactosidase. The concentration of citric acid and sodium phosphate are not exactly known.

Enzymatic Assay of β-GALACTOSIDASE (EC 3.2.1.23)

Principle

$$\text{ONP-}\beta\text{-D-Galactopyranoside} \xrightarrow{\beta\text{-Galactosidase}} \text{o-Nitrophenol} + \beta\text{-D-Galactose}$$

Abbreviation Used

ONP-β-D-Galactopyranoside =o-Nitrophenyl-β-Galactopyranoside

Conditions: T = 25°C, pH 4.0, A410nm, Light path = 1 cm

Method: Spectrophotometric Stop Rate Determination

Reagents

A. 400 mM Citrate Buffer, pH 4.0 at 25°C (Prepare 100 ml in deionized water using Citric Acid, Free Acid, Monohydrate. Adjust to pH 4.0 at 25°C with 1 M NaOH).
B. 10 mM o-Nitrophenyl-β-D-Galactoside Substrate Solution (ONP-Gal) (Prepare 5 ml in deionized water using o-Nitrophenyl-β-D-Galactopyranoside).
C. 200 mM Borate Buffer, pH 9.8 at 25°C (Prepare 100 ml in deionized water using Boric Acid. Adjust to pH 9.8 at 25°C with 1 M NaOH).

D. β-Galactosidase Enzyme Solution (Immediately before use, prepare a solution containing 0.05–0.10 units/ml of β-Galactosidase in cold deionized water).

Procedure

Pipette (in milliliters) the following reagents into suitable cuvettes:

	Test	Blank
Reagent A (Citrate Buffer)	0.40	0.40
Reagent B (ONP-Gal)	0.50	0.50

Mix by inversion and equilibrate to 25°C. Monitor the A410nm until constant, using a suitably thermostatted spectrophotometer. Then add:

	Test	Blank
Reagent D (Enzyme Solution)	0.10	—
Deionized Water	—	0.10

Immediately mix by inversion and incubate for exactly 10 minutes. Then add:

	Test	Blank
Reagent C (Borate Buffer)	3.00	3.00

Mix and record the ΔA410nm for both the Test and Blank.

Calculations

$$\text{Units / mg enzyme} = \frac{\Delta A410\text{nm / min Test} - \Delta A410\text{nm / min Blank } (4.0)}{(10)\ (4.6)\ (\text{mg enzyme / RM})}$$

4.0 = Total volume of assay
10 = Conversion factor for 10 minutes to 1 minute
4.6 = Millimolar extinction coefficient of o-Nitrophenol at pH 9.8
RM = Reaction Mix

Unit Definition: *One unit will hydrolyze 1.0 μmole of o-nitrophenyl β-D-galactoside to o-nitrophenol and D-galactose per minute at pH 4.0 at 25°C.*

Final Assay Concentration

In a 1.0 ml reaction mix, the final concentrations are 160 mM citrate buffer, 5.0 mM ONP-Gal and 0.005–0.010 units β-Galactosidase.

Enzymatic Assay of β-GALACTOSIDASE (EC 3.2.1.23) o-Nitrophenyl β-D-Galactopyranoside as Substrate

Principle

$$\text{ONP } \beta\text{-D-Galactopyranoside} + H_2O \xrightarrow{\beta\text{-Galactosidase}} \text{o-Nitrophenol} + \beta\text{-D-Galactose}$$

Abbreviation Used

ONP β-D-Galactopyranoside = o-Nitrophenyl β-D-Galactopyranoside

Conditions: T = 30°C, pH = 4.5, A410nm, Light path = 1 cm

Method: Spectrophotometric Stop Rate Determination

Reagents

A. 10 mM Citric Acid Solution (Prepare 100 ml in deionized water using Citric Acid, Free Acid, Anhydrous).
B. 20 mM Sodium Phosphate Solution (Prepare 100 ml in deionized water using Sodium Phosphate, Dibasic, Anhydrous).
C. 20 mM Phosphate-Citrate Buffer, pH 4.5 at 30°C (Phos-Cit Buffer) (Prepare 100 ml using Reagent B. Adjust to pH 4.5 at 30°C with Reagent A).
D. 10 mM o-Nitrophenyl β-D-Galactoside Solution (ONP-Gal) (Prepare 10 ml in Reagent C using o-Nitrophenyl β-D-Galactopyranoside).
E. 200 mM Borate Buffer, pH 9.8 at 30°C (Prepare 100 ml in deionized water using Boric Acid. Adjust to pH 9.8 at 30°C with 1 M NaOH).
F. β-Galactosidase Enzyme Solution (Immediately before use, prepare a solution containing 0.2–0.4 unit/ml of β-Galactosidase in cold deionized water).

Procedure

Pipette (in milliliters) the following reagents into suitable cuvettes:

	Test	Blank
Reagent C (Phos-Cit Buffer)	0.40	0.40
Reagent D (ONP-Gal)	0.50	0.50

Mix by inversion and equilibrate to 30°C, using a suitable spectrophotometer. Then add:

Reagent F (Enzyme Solution)	0.10	—

Immediately mix by inversion and incubate at 30°C for exactly 10 minutes. Then add:

Reagent E (Borate Buffer)	3.00	3.00
Reagent F (Enzyme Solution)	—	0.10

Mix by inversion and record the A410nm for both the Test and Blank.

Calculations

$$\text{Units / ml enzyme} = \frac{(\text{A410nm / min Test} - \text{A410nm / min Blank})\ (4)\ (\text{df})}{(10)\ (4.6)\ (0.1)}$$

4 = Total volume (in milliliters) of the assay
df = Dilution factor
10 = Time of assay (in minutes) as per the Unit Definition
4.6 = Millimolar extinction coefficient of o-Nitrophenol at 410 nm
0.1 = Volume (in milliliters) of enzyme used

Unit Definition: *One unit will hydrolyze 1.0 μmole of o-nitrophenyl β-D-galactoside to o-nitrophenol and β-D-galactose per minute at pH 4.5 at 30°C.*

Final Assay Concentrations

In a 1.00 ml reaction mix, the final concentrations are 18 mM sodium phosphate, 5 mM o-nitrophenyl β-D-galactopyranoside, and 0.02–0.04 unit β-galactosidase.

Enzymatic Assay of α-GALACTOSIDASE (EC 3.2.1.22)

Principle

$$\text{ONP } \alpha\text{-D-Galactopyranoside} + H_2O \xrightarrow{\alpha\text{-Galactosidase}} \text{o-Nitrophenol} + \text{D-Galactose}$$

Abbreviation Used

ONP α-D-Galactopyranoside =o-Nitrophenyl α-Galactopyranoside

Conditions: T = 25°C, pH 4.0, A410nm, Light path = 1 cm

Method: Spectrophotometric Stop Rate Determination

Reagents

A. 100 mM Citrate Buffer, pH 4.0 at 25°C (Prepare 100 ml in deionized water using Citric Acid, Trisodium salt, Dihydrate. Adjust to pH 4.0 at 25°C with 1 M HCl).
B. 10 mM o-Nitrophenyl α-D-Galactopyranoside Solution (ONP-Gal) (Prepare 5 ml in deionized water using o-Nitrophenyl α-D-Galactopyranoside).
C. 200 mM Borate Solution, pH 9.8 at 25°C (Borate) (Prepare 50 ml in deionized water using Boric Acid. Adjust to pH 9.8 at 25°C with 1 M NaOH).
D. α-Galactosidase Enzyme Solution (Immediately before use, prepare a solution containing 0.15–0.30 unit/ml of α-Galactosidase in cold deionized water).

Procedure

Pipette (in milliliters) the following reagents into suitable cuvettes:

	Test	Blank
Reagent A (Citrate Buffer)	0.40	0.40
Reagent B (ONP-Gal)	0.50	0.50

Mix by inversion and equilibrate to 25°C. Then add:

Reagent D (Enzyme Solution)	0.10	—

Immediately mix by inversion and incubate at 25°C for exactly 10 minutes. Then add:

Reagent C (Borate Buffer)	3.00	3.00
Reagent D (Enz Sol)	—	0.10

Mix by inversion and transfer to suitable cuvettes. Record the A410nm for both the Test and Blank using a suitable spectrophotometer.

Calculations

$$\text{Units / ml enzyme} = \frac{(\text{A410nm Test} - \text{A410nm Blank})\ (4)\ (\text{df})}{(4.6)\ (10)\ (0.1)}$$

4 = Total volume of assay
df = Dilution factor
4.6 = millimolar extinction coefficient of o-Nitrophenol at 410nm

10 = Time of assay (in minutes) as per Unit Definition
0.1 = Volume (in milliliter) of enzyme used

$$\text{Units / mg solid} = \frac{\text{units / ml enzyme}}{\text{mg solid / ml enzyme}}$$

$$\text{Units / mg protein} = \frac{\text{units / ml enzyme}}{\text{mg protein / ml enzyme}}$$

Unit Definition: *One unit will hydrolyze 1.0 μmole of o-nitrophenyl α-D-galactoside to o-nitrophenol and D-galactose per minute at pH 4.0 at 25°C.*

Final Assay Concentrations

In a 1.00 ml reaction mix, the final concentrations are 43 mM citrate, 5.3 mM o-nitrophenyl α-D-galactoside and 0.015–0.03 unit α-galactosidase.

Enzymatic Assay of β-GALACTOSIDASE (EC 3.2.1.23)

Principle

$$\text{ONP } \beta\text{-D-Galactopyranoside} + H_2O \xrightarrow{\beta\text{-Galactosidase}} \text{o-Nitrophenol} + \beta\text{-D-Galactose}$$

Abbreviation Used

ONP β-D-Galactopyranoside = o-Nitrophenyl β-DGalactopyranoside

Conditions: T = 37°C, pH 6.0, A405nm, Light path = 1 cm

Method: Spectrophotometric Stop Rate Determination

Reagents

A. 100 mM Sodium Acetate Buffer, pH 6.0 at 37°C (Prepare 100 ml in deionized water using Sodium Acetate, Trihydrate. Adjust to pH 6.0 at 37°C with 1 M Acetic Acid).
B. 2.0 mM o-Nitrophenyl β-D-Galactoside with 0.01% (w/v) Bovine Serum Albumin, pH 6.0 (Substrate Solution) (Prepare 100 ml in Reagent A using o-Nitrophenyl β-D-Galactopyranoside and Albumin, Bovine. Adjust to pH 6.0 at 30°C, if necessary, with 1 M Acetic Acid or 1 M NaOH).
C. 1000 mM Sodium Carbonate Solution (Na_2CO_3) (Prepare 100 ml in deionized water using Sodium Carbonate, Anhydrous).
D. β-Galactosidase Enzyme Solution (Immediately before use, prepare a solution containing 0.02–0.04 unit/ml of β-Galactosidase in cold deionized water).

Procedure

Pipette (in milliliters) the following reagents into suitable cuvettes:

	Test	Blank
Reagent B (Substrate solution)	0.50	0.50
Deionized Water	0.30	0.30

Mix by inversion and equilibrate to 37°C. Then add:

	Test	Blank
Reagent D (Enzyme Solution)	0.20	—

Immediately mix by inversion and incubate for exactly 10 minutes. Then add:

Reagent C (Na_2CO_3)	4.00	4.00
Reagent D (Enzyme Solution)	—	0.20

Mix by inversion and record the A405nm for both the Test and Blank.

Calculations

$$\text{Units / ml enzyme} = \frac{(\text{A405nm Test} - \text{A405nm Blank})\,(5)\,(\text{df})}{(10)\,(4.6)\,(0.2)}$$

5 = Total volume (in milliliters) of assay
df = Dilution factor
10 = Time of assay (in minutes) as per the Unit Definition
4.6 = Millimolar extinction coefficient of o-Nitrophenol at 405nm
0.2 = Volume (in milliliters) of enzyme used

$$\text{Units / mg solid} = \frac{\text{units / ml enzyme}}{\text{mg solid / ml enzyme}}$$

$$\text{Units / mg protein} = \frac{\text{units / ml enzyme}}{\text{mg protein / ml enzyme}}$$

Unit Definition: *One unit will hydrolyze 1.0 µmole of o-nitrophenyl β-D-galactoside to o-nitrophenol and D-galactose per minute at pH 6.0 at 37°C.*

Final Assay Concentrations

In a 1.00 ml reaction mix, the final concentrations are 50 mM sodium acetate, 0.005% (w/v) Bovine Serum Albumin, 1.0 mM o-nitrophenyl β-D-galactopyranoside and 0.004–0.008 unit β-galactosidase.

Enzymatic Assay of β-GALACTOSIDASE (EC 3.2.1.23)

Principle

$$\text{ONP } \beta\text{-D-Galactopyranoside} + H_2O \xrightarrow{\beta\text{-Galactosidase}} \text{o-Nitrophenol} + \beta\text{-D-Galactose}$$

Abbreviation Used

ONP β-D-Galactopyranoside = o-Nitrophenyl β-D-Galactopyranoside

Conditions: T = 37°C, pH = 7.2, A410nm, Light path = 1 cm

Method: Continuous Spectrophotometric Rate Determination

Reagents

A. 100 mM Potassium Phosphate Buffer, pH 7.2 at 37°C (Prepare 100 ml in deionized water using Potassium Phosphate, Dibasic, Trihydrate. Adjust to pH 7.2 at 37°C with 100 mM Potassium Phosphate, Monobasic, Anhydrous).

B. 68 mM o-Nitrophenyl β-D-Galactopyranoside Solution (ONP-Gal) (Prepare 5 ml in Reagent A using o-Nitrophenyl β-D-Galactopyranoside. Warm gently to solubilize, keep at 37°C).

C. 30 mM Magnesium Chloride Solution ($MgCl_2$) (Prepare 10 ml in deionized water using Magnesium Chloride, Hexahydrate).

D. 3.36 M 2-Mercaptoethanol Solution (2-ME) (Prepare 2 ml in deionized water using 2-Mercaptoethanol).

E. β-Galactosidase Enzyme Solution (Immediately before use, prepare a solution containing 2.0–4.0 units/ml of β-Galactosidase in cold Reagent A).

Procedure

Pipette (in milliliters) the following reagents into suitable cuvettes:

	Test	Blank
Reagent A (Buffer)	2.60	2.70
Reagent C ($MgCl_2$)	0.10	0.10
Reagent D (2-ME)	0.10	0.10
Reagent E (Enzyme Solution)	0.10	—

Mix by inversion and equilibrate to 37°C. Monitor the A410nm until constant, using a suitably thermostatted spectrophotometer. Then add:

Reagent B (ONP-Gal)	0.10	0.10

Immediately mix by inversion and record the increase in A410nm for approximately 5 minutes. Obtain the ΔA410nm/minute using the maximum linear rate for both the Test and Blank.

Calculations

$$\text{Units / mg enzyme} = \frac{(\Delta A410\text{nm Test} - \Delta A410\text{nm / min Blank})\ (3)\ (\text{df})}{(3.5)\ (0.1)}$$

3 = Total volume (in milliliters) of assay
df = Dilution factor
3.5 = Millimolar extinction coefficient of o-Nitrophenol at 410nm
0.1 = Volume (in milliliter) of enzyme used

$$\text{Units / mg solid} = \frac{\text{units / ml enzyme}}{\text{mg solid / ml enzyme}}$$

$$\text{Units / mg protein} = \frac{\text{units / ml enzyme}}{\text{mg protein / ml enzyme}}$$

Unit Definition: *One unit will hydrolyze 1.0 μmole of o-nitrophenyl β-D-galactoside to o-nitrophenol and D-galactose per minute at pH 7.2 at 37°C.*

Final Assay Concentrations

In a 3.00 ml reaction mix, the final concentrations are 93 mM potassium phosphate, 2.3 mM onitrophenyl β-D-galactopyranoside, 1.0 mM magnesium chloride, 112 mM 2-mercaptoethanol and 0.20–0.40 unit β-galactosidase.

Enzymatic Assay of β-GALACTOSIDASE (EC 3.2.1.23)

Principle

$$\text{ONP } \beta\text{-D-Galactopyranoside} \xrightarrow{\beta\text{-Galactosidase}} \text{o-Nitrophenol} + \beta\text{-D-Galactose}$$

Abbreviation Used

ONP β-D-Galactopyranoside = o-Nitrophenyl β-D-Galactopyranoside

Conditions: T = 37°C, pH = 7.3, A410nm, Light path = 1 cm

Method: Continuous Spectrophotometric Rate Determination

Reagents

A. 100 mM Sodium Phosphate Solution (Prepare 100 ml in deionized water using Sodium Phosphate, Monobasic, Anhydrous).
B. 100 mM Sodium Phosphate Buffer, pH 7.3 at 37°C (Prepare 100 ml in deionized water using Sodium Phosphate, Dibasic, Anhydrous. Adjust to pH 7.3 at 37°C with Reagent A).
C. 68 mM o-Nitrophenyl β-D-Galactoside Solution (ONP-Gal) (Prepare 5 ml in Reagent B using o-Nitrophenyl β-D-Galactopyranoside).
D. 30 mM Magnesium Chloride Solution ($MgCl_2$) (Prepare 10 ml in deionized water using Magnesium Chloride, Hexahydrate).
E. 3.36 M 2-Mercaptoethanol Solution (2-ME) (Prepare 2 ml in deionized water using 2-Mercaptoethanol).
F. β-Galactosidase Enzyme Solution (Immediately before use, prepare a solution containing 0.2–1.0 unit/ml of β-Galactosidase in cold Reagent B).

Procedure

Pipette (in milliliters) the following reagents into suitable cuvettes:

	Test	Blank
Reagent B (Buffer)	2.60	2.70
Reagent D ($MgCl_2$)	0.10	0.10
Reagent E (2-ME)	0.10	0.10
Reagent F (Enzyme Solution)	0.10	—

Mix by inversion and equilibrate to 37°C. Monitor the A410nm until constant, using a suitably thermostatted spectrophotometer. Then add:

Reagent C (ONP-Gal)	0.10	0.10

Immediately mix by inversion and record the increase in A410nm for approximately 5 minutes. Obtain the ΔA410nm/minute using the maximum linear rate for both the Test and Blank.

Calculations

$$\text{Units / mg enzyme} = \frac{(\Delta\text{A410nm Test} - \Delta\text{A410nm / min Blank})\ (3)\ (\text{df})}{(3.5)\ (0.1)}$$

3 = Total volume (in milliliters) of assay
df = Dilution factor
3.5 = Millimolar extinction coefficient of o-Nitrophenol at 410nm
0.1 = Volume (in milliliter) of enzyme used

$$\text{Units / mg solid} = \frac{\text{units / ml enzyme}}{\text{mg solid / ml enzyme}}$$

$$\text{Units / mg protein} = \frac{\text{units / ml enzyme}}{\text{mg protein / ml enzyme}}$$

Unit Definition: *One unit will hydrolyze 1.0 μmole of o-nitrophenyl β-D-galactoside per minute at pH 7.3 at 37°C.*

Final Assay Concentrations

In a 3.00 ml reaction mix, the final concentrations are 93 mM sodium phosphate, 2.3 mM o-nitrophenyl β-D-galactoside, 1.0 mM magnesium chloride, 112 mM 2-mercaptoethanol and 0.02–0.1 unit β-galactosidase.

Note

1. The equilibration time must not exceed 3 minutes.

Enzymatic Assay of β-GALACTOSIDASE (EC 3.2.1.23)

Principle

$$\text{PNP - β - D - Galactopyranoside} \xrightarrow{\text{β-Galactosidase}} \text{p - Nitrophenol} + \text{D - Galactose}$$

Abbreviation Used

PNP-β-D-Galactopyranoside = p-Nitrophenyl-β-Galactopyranoside

Conditions: T = 25°C, pH = 4.4, A400nm, Light path = 1 cm

Method: Spectrophotometric Stop Rate Determination

Reagents

A. 200 mM Sodium Acetate Buffer, with 0.25% (w/v) Bovine Serum Albumin and 100 mM Sodium Chloride, pH 4.4 at 25°C. (Prepare 100 ml in deionized water using Sodium Acetate Trihydrate, Albumin Bovine and Sodium Chloride. Adjust to pH 4.4 at 25°C with 1 M Acetic Acid).
B. 10 mM p-Nitrophenyl-β-D-Galactoside Solution (PNP-Gal) Prepare 5 ml in deionized water using p-Nitrophenyl β-D-Galactopyranoside).
C. 200 mM Borate Buffer, pH 9.8 at 25°C (Prepare 100 ml in deionized water using Boric Acid. Adjust to pH 9.8 at 25°C with 1 M NaOH).
D. β-Galactosidase Enzyme Solution (Immediately before use, prepare a solution containing approximately 1.0 unit/ml of β-Galactosidase in cold deionized water).

Procedure

Pipette (in milliliters) the following reagents into suitable cuvettes:

	Test	Blank
Reagent A (Sodium Acetate Buffer)	0.40	0.40
Reagent D (Enzyme Solution)	0.02	—

Mix by inversion and equilibrate to 25°C. Monitor the A400nm until constant, using a suitably thermostatted spectrophotometer. Then add:

	Test	Blank
Reagent B (PNP-Gal)	0.50	0.50

Immediately mix by inversion and incubate at 25° C for exactly 10 minutes. Then add:

	Test	Blank
Reagent C (Borate Buffer)	3.00	3.00
Reagent D (Enzyme Solution)	—	0.02

Mix and record the A400nm for both the Test and Blank.

Calculations

$$\text{Units / mg enzyme} = \frac{\text{A400nm Test} - \text{A400nm Blank (3.92)}}{\text{(10) (18.0) (mg enzyme / RM)}}$$

3.92 = Total volume of assay
10 = Conversion factor for 10 minutes to 1 minute
18.0 = Millimolar extinction coefficient of p-Nitrophenol at 400nm
RM = Reaction Mix

Unit Definition: *One unit will hydrolyze 1.0 μmole of p-nitrophenyl β-D-galactopyrano-side to p-nitrophenol and D-galactose per minute at pH 4.4 at 25°C.*

Final Assay Concentrations

In a 0.92 ml reaction mix, the final concentrations are 87 mM sodium acetate, 5.4 mM p-nitrophenyl β-galactopyranoside, 0.11% Bovine Serum Albumin, 43.5 mM sodium chloride, and 0.02 units β-galactosidase.

Enzymatic Assay of α-GALACTOSIDASE (EC 3.2.1.22)

Principle

$$\text{PNP } \alpha\text{-D-Galactopyranoside} + H_2O \xrightarrow{\alpha\text{-Galactosidase}} \text{p-Nitrophenol} + \text{D-Galactose}$$

Abbreviation Used

PNP α-D-Galactopyranoside =p-Nitrophenyl α-Galactopyranoside

Conditions: T = 25°C, pH = 6.5, A405nm, Light path = 1 cm

Method: Spectrophotometric Stop Rate Determination

Reagents

A. 100 mM Potassium Phosphate Monobasic Solution (Prepare 100 ml in deionized water using Potassium Phosphate, Monobasic, Anhydrous).

B. 100 mM Potassium Phosphate Dibasic Solution (Prepare 100 ml in deionized water using Potassium Phosphate, Dibasic, Trihydrate).

C. 100 mM Potassium Phosphate Buffer, pH 6.5 at 25°C (Prepare 100 ml by adjusting 50 ml of Reagent A to pH 6.5 at 25°C by adding Reagent B).

D. 9.9 mM p-Nitrophenyl α-D-Galactopyranoside Solution (PNP-Gal) (Prepare 4 ml in deionized water using p-Nitrophenyl α-D-Galactopyranoside).

E. 200 mM Borate Buffer, pH 9.8 at 25°C (Prepare 100 ml in deionized water using Boric Acid. Adjust to pH 9.8 at 25°C with 1 M NaOH).

F. α-Galactosidase Enzyme Solution (Immediately before use, prepare a solution containing 0.05–0.10 units/ml of α-Galactosidase in cold Reagent C).

Procedure

Pipette (in milliliters) the following reagents into suitable cuvettes:

	Test	Blank
Reagent C (Potassium Phosphate Buffer)	0.70	0.70
Reagent D (PNP-Gal)	0.20	0.20

Mix by swirling and equilibrate to 25°C. Then add:

Reagent F (Enzyme Solution)	0.10	—

Immediately mix by swirling and incubate at 25°C for exactly 5 minutes. Then add:

Reagent E (Borate Buffer)	2.00	2.00
Reagent F (Enzyme Solution)	—	0.10

Mix by swirling and record the A405nm for both the Test and Blank, using a suitably thermostatted spectrophotometer.

Calculations

$$\text{Units / mg enzyme} = \frac{(\text{A405nm Test} - \text{A405nm Blank})\,(3.0)\,(\text{df})^{\bullet}}{(18.5)\,(5)\,(0.1)}$$

3.0 = Total volume of assay
df = Dilution factor
5 = Conversion factor for 5 minutes to 1 minute
18.5 = Millimolar extinction coefficient of p-Nitrophenol at 405nm
0.1 = Volume (in milliliter) of enzyme used

$$\text{Units / mg solid} = \frac{\text{units / ml enzyme}}{\text{mg solid / ml enzyme}}$$

$$\text{Units / mg protein} = \frac{\text{units / ml enzyme}}{\text{mg protein / ml enzyme}}$$

Unit Definition: *One unit will hydrolyze 1.0 μmole of p-nitrophenyl α-D-galactoside to p-nitrophenol and D-galactose per minute at pH 6.5 at 25°C.*

Final Assay Concentration

In a 1.00 ml reaction mix, the final concentrations are 80 mM potassium phosphate, 2.0 mM p-nitrophenyl α-D-galctopyranoside, and 0.005–0.01 units agalactosidase.

Enzymatic Assay of β-GALACTOSIDASE (EC 3.2.1.23)

Principle

$$\text{ONP } \beta\text{-D-Galactopyranoside} \xrightarrow{\beta\text{-Galactosidase}} \text{o-Nitrophenol} + \beta\text{-D-Galactose}$$

Abbreviation Used

ONP β-D-Galactopyranoside =o-Nitrophenyl β-Galactopyranoside

Conditions: T = 25°C, pH 3.5, A410nm, Light path = 1 cm

Method: Spectrophotometric Stop Rate Determination

Reagents

A. 400 mM Citrate Buffer, pH 3.5 at 25°C (Prepare 100 ml in deionized water using Citric Acid, Free Acid, Monohydrate. Adjust to pH 3.5 at 25°C with 1 M NaOH).
B. 10 mM o-Nitrophenyl β-D-Galactoside Substrate Solution (ONP-Gal) (Prepare 5 ml in deionized water using o-Nitrophenyl β-D-Galactopyranoside).
C. 200 mM Borate Buffer, pH 9.8 at 25°C (Prepare 100 ml in deionized water using Boric Acid. Adjust to pH 9.8 at 25°C with 1 M NaOH).
D. β-Galactosidase Enzyme Solution (Immediately before use, prepare a solution containing 0.05–0.10 unit/ml of β-Galactosidase in cold deionized water).

Procedure

Pipette (in milliliters) the following reagents into suitable cuvettes:

	Test	Blank
Reagent A (Citrate Buffer)	0.40	0.40
Reagent B (ONP-Gal)	0.50	0.50

Mix by inversion and equilibrate to 25°C. Monitor the A410nm until constant, using a suitably thermostatted spectrophotometer. Then add:

Reagent D (Enzyme Solution)	0.10	—
Deionized Water	—	0.10

Immediately mix by inversion and incubate for exactly 10 minutes. Then add:

Reagent C (Borate Buffer)	3.00	3.00

Mix and record the ΔA410nm for both the Test and Blank.

Calculations

$$\text{Units / ml enzyme} = \frac{(\Delta A410\text{nm Test} - \Delta A410\text{nm Blank})\ (4)\ (\text{df})}{(10)\ (4.6)\ (0.1)}$$

4 = Total volume (in milliliters) of assay
df = Dilution factor
10 = Time of assay (in minutes) as per the Unit Definition
4.6 = Millimolar extinction coefficient of o-Nitrophenol at pH 9.8
0.1 = Volume (in milliliters) of enzyme used

$$\text{Units / mg solid} = \frac{\text{units / ml enzyme}}{\text{mg solid / ml enzyme}}$$

$$\text{Units / mg protein} = \frac{\text{units / ml enzyme}}{\text{mg protein / ml enzyme}}$$

Unit Definition: *One unit will hydrolyze 1.0 μmole of o-nitrophenyl β-D-galactoside to o-nitrophenol and D-galactose per minute at pH 3.5 at 25°C.*

Final Assay Concentration

In a 1.00 ml reaction mix, the final concentrations are 160 mM citrate buffer, 5.0 mM o-nitrophenyl β-Dgalactoside and 0.005–0.010 unit β-galactosidase.

Enzymatic Assay of GALACTOSYLTRANSFERASE (EC 2.4.1.22)

Principle

$$\text{UDP - Galactose} + \text{D - Glucose} \xrightarrow{\text{Galactosyltransferase}} \text{UDP} + \text{Lactose}$$

$$\text{UDP} + \text{PEP} \xrightarrow{\text{PK}} \text{Pyruvate} + \text{UTP}$$

$$\text{Pyruvate} + \beta\text{- NADH} \xrightarrow{\text{LDH}} \text{Lactate} + \beta\text{- NAD}$$

Abbreviations Used

UDP-Galactose = Uridine 5'-Diphosphogalactose
UDP = Uridine 5'-Diphosphate
PEP = Phospho(enol) pyruvate
PK = Pyruvate Kinase
β-NADH = β-Nicotinamide Adenine Dinucleotide, Reduced Form
LDH = Lactic Dehydrogenase
β-NAD = β-Nicotinamide Adenine Dinucleotide, Oxidized Form

Conditions: T = 30°C, pH = 8.4, A340nm, Light path = 1 cm

Method: Continuous Spectrophotometric Rate Determination

Reagents

A. 250 mM Glycylglycine Buffer, pH 8.6 at 30°C. (Prepare 50 ml in deionized water using Gly-Gly, Hydrochloride. Adjust to pH 8.6 at 30°C with 1 M NaOH).
B. 0.70 mM β-Nicotinamide Adenine Dinucleotide, Reduced Form (β-NADH) (Dissolve the contents of one 5 mg vial of β-Nicotinamide Adenine Dinucleotide, Reduced Form, Disodium Salt in the appropriate volume of deionized water. *Prepare Fresh*).
C. 6.4 mM Phospho(enol) pyruvate Solution (PEP) (Prepare 10 ml in deionized water using Phospho(enol) Pyruvate Tri(Cyclohexylammonium) Salt).

D. 100 mM Manganese Chloride Tetrahydrate with 1 M Potassium Chloride Solution ($MnCl_2$/KCl) (Prepare 5.0 ml in deionized water using Manganese Chloride Tetrahydrate and Potassium Chloride).
E. 5.65 mM Uridine 5'-Diphosphogalactose Solution (UDPGalactose) (Prepare 10 ml in deionized water using Uridine 5'-Diphosphogalactose, Sodium Salt).
F. 50 mM Glycylglycine Buffer, pH 8.0 at 30°C. (Prepare 10 ml in deionized water using Gly-Gly, Hydrochloride. Adjust to pH 8.0 at 30°C with 1 M NaOH).
G. 0.6% (w/v) α-Lactalbumin Solution (Prepare 2 ml in Reagent F using α-Lactalbumin).
H. 286 mM D-Glucose Solution (Prepare 10 ml in deionized water using β-D(+)Glucose).
I. PK/LDH Enzymes Suspension.
J. 20 mM Tris HCl Buffer with 2 mM Ethylenediaminetetraacetic Acid and 2 mM 2-Mercaptoethanol, pH 7.5 at 30°C (Enz Dil) (Prepare 50 ml in deionized water using Trizma Hydrochloride, Ethylenediaminetetraacetic Acid, Tetrasodium Salt and 2-Mercaptoethanol. Adjust to pH 7.5 at 30°C with 1 M NaOH).
K. Galactosyltransferase Enzyme Solution (Immediately before use, prepare a solution containing 0.1–0.2 unit/ml of Galactosyltransferase in cold Reagent J).

Procedure

Prepare a reaction cocktail by pipetting (in milliliters) the following reagents into a suitable container:

Reagent A (Buffer)	5.00
Reagent B (β-NADH)	5.00
Reagent C (PEP)	5.00
Reagent D ($MnCl_2$/KCl)	1.25

Mix and adjust to pH 8.4 at 30°C with 1 M HCl or 1 M NaOH, if necessary. Pipette (in milliliters) the following reagents into suitable cuvettes:

	Test	Blank
Deionized Water	0.50	0.50
Reaction Cocktail	2.00	2.00
Reagent E (UDP-Galactose)	0.20	0.20
Reagent G (α-Lactalbumin)	0.10	0.10
Reagent I (PK/LDH Suspension)	0.025	0.025
Reagent J (Enz Dil)	—	0.04
Reagent K (Enzyme Solution)	0.04	—

Mix by inversion and equilibrate to 30°C. Monitor the A340nm until constant, using a suitably thermostatted spectrophotometer. Then add:

	Test	Blank
Reagent H (D-Glucose)	0.20	0.20

Immediately mix by inversion and record the decrease in A340nm for approximately 10 minutes. Obtain the ΔA340nm/minute using the maximum linear rate for both the Test and Blank.

Calculations

$$\text{Units / ml enzyme} = \frac{(\Delta A340nm / min\ Test - \Delta A340nm / min\ Blank)\ (3.065)\ (df)}{(6.22)\ (0.04)}$$

3.065 = Total volume (in milliliters) of assay
df = Dilution factor
6.22 = Millimolar extinction coefficient of β-NADH at 340nm
0.04 = Volume (in milliliter) of enzyme used

$$\text{Units / mg solid} = \frac{\text{units / ml enzyme}}{\text{mg solid / ml enzyme}}$$

$$\text{Units / mg protein} = \frac{\text{units / ml enzyme}}{\text{mg protein / ml enzyme}}$$

Unit Definition: *One unit will transfer 1.0 μmole of galactose from UDP-galactose to D-glucose per minute at pH 8.4 at 30°C in the presence of 0.2 mg of α-lactalbumin per ml of reaction mixture.*

Final Assay Concentrations

In a 3.065 ml reaction mix, the final concentrations are 52 mM glycylglycine, 0.14 mM β-nicotinamide adenine dinucleotide, 1.3 mM phospho(enol) pyruvate, 5.0 mM manganese chloride, 50 mM potassium chloride, 0.37 mM uridine 5'-diphosphogalactose, 0.02% (w/v) α-lactalbumin, 19 mM glucose, 0.26 mM Tris, 0.03 mM Ethylenediaminetetraacetic Acid, 0.03 mM 2-mercaptoethanol, 17.5 units pyruvate kinase, 25 units lactic dehydrogenase and 0.004–0.008 unit galactosyltransferase.

Notes

1. Lactic Dehydrogenase Unit Definition: One unit will reduce 1.0 μmole of pyruvate to L-lactate per minute at pH 7.5 at 37°C.
2. Pyruvate Kinase Unit Definition: One unit will convert 1.0 μmole of phospho(enol) pyruvate to pyruvate per minute at pH 7.6 at 37°C.
3. α-Lactalbumin is included in the assay since, according to Fitzgerald *et al.*, it lowers the apparent Km of glucose.

Enzymatic Assay of GLUCOKINASE (EC 2.7.1.2)

Principle

$$\beta\text{-D(+)Glucose} + \text{ATP} \xrightarrow{\text{Glucokinase}} \text{D-Glucose 6-Phosphate} + \text{ADP}$$

$$\text{D-Glucose 6-Phosphate} + \beta\text{-NADP} \xrightarrow{\text{G-6PDH}} \text{6-Phospho-D-gluconate} + \beta\text{-NADPH}$$

Abbreviations Used

ATP = Adenosine 5'-Triphosphate
ADP = Adenosine 5'-Diphosphate
β-NADP = β-Nicotinamide Adenine Dinucleotide Phosphate, Oxidized Form
β-NADPH = β-Nicotinamide Adenine Dinucleotide Phosphate, Reduced Form
G-6PDH = Glucose-6-Phosphate Dehydrogenase

Conditions: T = 30°C, pH = 9.0, A340nm, Light path = 1 cm.

Method: Continuous Spectrophotometric Rate Determination.

Reagents

A. 75 mM Tris HCl Buffer, pH 9.0 at 30°C (Prepare 100 ml in deionized water using Trizma Base. Adjust to pH 9.0 at 30°C with 1 M HCl).

B. 600 mM Magnesium Chloride Solution ($MgCl_2$) (Prepare 10 ml in deionized water using Magnesium Chloride, Hexahydrate).

C. 120 mM Adenosine Triphosphate Solution (ATP) (Prepare 10 ml in deionized water using Adenosine 5'-Triphosphate, Disodium Salt).

D. 360 mM β-D(+)Glucose Solution (Glucose) (Prepare 10 ml in deionized water using β-D(+)Glucose).

E. 27 mM β-Nicotinamide Adenine Dinucleotide Phosphate, Oxidized Form Solution (β-NADP) (Dissolve the contents of one 30 mg vial of β-Nicotinamide Adenine Dinucleotide Phosphate, Sodium Salt in the appropriate volume of deionized water).

F. Glucose-6-Phosphate Dehydrogenase Enzyme Solution (G-6PDH) (Immediately before use, prepare a solution containing 100 units/ml of Glucose-6-Phosphate Dehydrogenase, in cold deionized water).

G. 50 mM Tris HCl Buffer, pH 8.5 at 30°C (Enzyme Diluent) (Prepare 50 ml in deionized water using Trizma Base. Adjust to pH 8.5 at 30°C with 1 M HCl).

H. Glucokinase Enzyme Solution (GLCK) Immediately before use, prepare a solution containing 0.25–0.50 unit/ml of Glucokinase in cold Reagent G).

Procedure

Prepare a reaction cocktail by pipetting (in milliliters) the following reagents into a suitable container:

Reagent A (Buffer)	24.00
Reagent B ($MgCl_2$)	1.00
Reagent C (ATP)	1.00
Reagent D (Glucose)	1.00
Reagent E (β-NADP)	1.00

Mix by swirling and adjust to pH 9.0 at 30°C with 1 M HCl or 1 M NaOH, if necessary. Pipette (in milliliters) the following reagents into suitable cuvettes:

	Test	Blank
Reaction Cocktail	2.80	2.80
Reagent F (6-GPDH)	0.10	0.10

Mix by inversion and equilibrate to 30°C. Monitor the A340nm until constant using a suitably thermostatted spectrophotometer. Then add:

Reagent H (GLCK)	0.10	—
Reagent G (Enzyme Diluent)	—	0.10

Immediately mix by inversion and record the increase in A340nm for approximately 5 minutes. Obtain the A340nm/minute using the maximum linear rate for both the Test and Blank.

Calculations

$$\text{Units / ml enzyme} = \frac{(\text{A340nm / min Test} - \text{A340nm min Blank})\ (3)}{(6.22)\ (0.1)}$$

3 = Volume (in milliliters) of assay
6.22 = Millimolar extinction coefficient of β-NADPH at 340nm
0.1 = Volume (in milliliters) of enzyme used

$$\text{Units / mg solid} = \frac{\text{units / ml enzyme}}{\text{mg solid / ml enzyme}}$$

$$\text{Units / mg protein} = \frac{\text{units / ml enzyme}}{\text{mg protein / ml enzyme}}$$

Unit Definition: *One unit will phosphorylate 1.0 μmole of D-glucose to D-glucose 6-phosphate per minute at pH 9.0 at 30°C.*

Final Assay Concentrations

In a 3.00 ml reaction mix, the final concentrations are 60 mM Tris, 20 mM magnesium chloride, 4.0 mM adenosine 5'-triphosphate, 12.0 mM glucose, 0.9 mM β-nicotinamide adenine dinucleotide phosphate, 10 units glucose 6-phosphate dehydrogenase and 0.025–0.050 unit glucokinase.

Note

1. Glucose-6-Phosphate Dehydrogenase Unit Definition: One unit will oxidize 1.0 μmole of D-glucose 6-phosphate to 6-phospho-D-gluconate per minute in the presence of NADP at pH 7.4 at 25°C.

Enzymatic Assay of GLUCONATE DEHYDROGENASE, NAD(P) Independent (EC 1.1.99.3)

Principle

$$\text{D-Gluconate} + \text{DCIP} \xrightarrow{\text{GDH}} \text{2-Dehydro-D-Gluconate} + \text{Reduced DCIP}$$

Abbreviations Used

DCIP = 2,6-Dichlorophenol-Indophenol
GDH = Gluconate Dehydrogenase, NAD(P) Independent

Conditions: T = 25°C, pH 6.0, A600nm, Light path = 1 cm

Method: Continuous Spectrophotometric Rate Determination

Reagents

A. 135 mM Potassium Phosphate Buffer, pH 6.0 at 25°C (Prepare 50 ml in deionized water using Potassium Phosphate, Monobasic, Anhydrous. Adjust to pH 6.0 at 25°C with 1 M KOH).

B. 13 mM Phenazine Methosulfate Solution (PMS) (Prepare 1 ml in deionized water using Phenazine Methosulfate).

C. 2.2 mM 2,6-Dichlorophenol-Indophenol Solution (DPIP) (Prepare 1 ml in deionized water using 2,6-Dichlorophenol-Indophenol, Sodium Salt).

D. 165 mM Sodium Gluconate Solution (Gluconate) (Prepare 2 ml in Reagent A using D-Gluconic Acid, Sodium Salt).

E. 135 mM Potassium Phosphate Buffer with 0.05% (w/v) Bovine Serum Albumin, pH 6.0 at 25°C (Enz Dil) (Prepare 10 ml in Reagent A using Albumin, Bovine. Adjust to pH 6.0 at 25°C with either 1 M NaOH or 1 M HCl).

F. Gluconate Dehydrogenase, NAD(P) Independent Enzyme Solution (Immediately before use, prepare a solution containing 0.08–0.25 unit/ml of Gluconate Dehydrogenase, NAD(P) Independent in cold Reagent E).

Procedure

Pipette (in milliliter) the following reagents into suitable cuvettes:

	Test	Blank
Reagent A (Buffer)	0.50	0.50
Reagent D (Gluconate)	0.20	0.20
Reagent C (DPIP)	0.10	0.10
Reagent B (PMS)	0.10	0.10

Mix by inversion and equilibrate to 25°C. Monitor the A600nm until constant, using a suitably thermostatted spectrophotometer. Then add:

Reagent F (Enzyme Solution)	0.10	—
Reagent E (Enz Dil)	—	0.10

Immediately mix by inversion and record the decrease in A600nm for approximately 5 minutes. Obtain the ΔA600nm/minute using the maximum linear rate for both the Test and Blank.

Calculations

$$\text{Units / ml enzyme} = \frac{(\Delta A600\text{nm Test} - \Delta A600\text{nm Blank})\ (1)\ (\text{df})}{(10)\ (0.1)}$$

1 = Total volume (in milliliter) of assay
df = Dilution factor
10 = Millimolar extinction coefficient of 2,6-dichlorophenol-indophenol at 600nm
0.1 = Volume (in milliliter) of enzyme used

$$\text{Units / mg protein} = \frac{\text{units / ml enzyme}}{\text{mg protein / ml enzyme}}$$

Unit Definition: *One unit will reduce 1.0 μmole of 2,6-dichlorophenolindophenol per minute at 25°C at pH 6.0 in the presence of sodium gluconate.*

Final Assay Concentrations

In a 1.00 ml reaction mix, the final concentrations are 108 mM potassium phosphate, 33 mM sodium gluconate, 0.22 mM 2,6-dichlorophenol-indophenol, 1.3 mM phenazine methosulfate, 0.005% (w/v) Bovine Serum Albumin, and 0.008–0.025 unit gluconate dehydrogenase, NAD(P) independent.

Notes

1. Phenazine Methosulfate is used as an electron acceptor in the assay.
2. This enzyme exhibits a lag phase before the initial rate.

Enzymatic Assay of GLUCONATE KINASE (EC 2.7.1.12)

Principle

$$\text{D-Gluconate} + \text{ATP} \xrightarrow{\text{Gluconate Kinase}} \text{6-Phospho-D-Gluconate} + \text{ADP}$$

$$\text{6-Phospho-D-gluconate} + \beta\text{-NADP} \xrightarrow[\text{Mg}_2^+]{\text{G-PGDH}} \text{D-Ribulose-5'-P} + \beta\text{-NADPH} + \text{CO}_2$$

Abbreviations Used

ATP = Adenosine 5'-Triphosphate
ADP = Adenosine 5'-Diphosphate
β-NADP = β-Nicotinamide Adenine Dinucleotide Phosphate, Oxidized Form
G-PGDH = 6-Phosphogluconic Dehydrogenase
D-Ribulose-5'-P = D-Ribulose 5'-Phosphate
β-NADPH = β-Nicotinamide Adenine Dinucleotide Phosphate, Reduced Form

Conditions: T = 30°C, pH = 8.0, A340nm, Light path = 1 cm

Method: Continuous Spectrophotometric Rate Determination

Reagents

A. 100 mM Glycylglycine Buffer, pH 8.0 at 30°C (Prepare 100 ml in deionized water using Gly-Gly, Free Base. Adjust to pH 8.0 at 30°C with 1 M NaOH).
B. 100 mM Magnesium Chloride Solution ($MgCl_2$) (Prepare 10 ml in deionized water using Magnesium Chloride, Hexahydrate).
C. 110 mM Adenosine 5'-Triphosphate Solution (ATP) (Prepare 10 ml in deionized water using Adenosine 5'-Triphosphate, Disodium Salt. Neutralize with Sodium Bicarbonate until the effervescence ceases).
D. 13 mM β-Nicotinamide Adenine Dinucleotide Phosphate, Oxidized Form, Solution (β-NADP) (Prepare 5 ml in deionized water using β-Nicotinamide Adenine Dinucleotide Phosphate, Sodium Salt).
E. 92 mM D-Gluconic Acid Solution (Gluconic Acid) (Prepare 5 ml in deionized water using D-Gluconic Acid, Sodium Salt).
F. 6-Phosphogluconic Dehydrogenase Enzyme Solution (6- PGDH) (Immediately before use, prepare a solution containing 50 units/ml of 6-Phosphogluconic Dehydrogenase, in cold deionized water).
G. Gluconate Kinase Enzyme Solution (GK) (Immediately before use, prepare a solution containing 0.2–0.4 unit/ml of Gluconate Kinase in cold deionized water).

Procedure

Prepare a reaction cocktail by pipetting (in milliliters) the following reagents into a suitable container:

Reagent A (Buffer)	24.00
Reagent B ($MgCl_2$)	1.00
Reagent C (ATP)	1.00
Reagent D (β-NADP)	1.00
Reagent E (Gluconic Acid)	1.00

Mix by swirling and equilibrate to 30°C. Adjust to pH 8.0 with either 1 M NaOH or 1 M HCl. Pipette (in milliliters) the following reagents into suitable cuvettes:

	Test	Blank
Reaction Cocktail	2.80	2.80
Reagent F (6-PGDH)	0.10	0.10

Mix by inversion and equilibrate to 30°C. Monitor the A340nm until constant, using a suitably thermostatted spectrophotometer. Then add:

Reagent G (GK)	0.10	—
Deionized Water	—	0.10

Immediately mix by inversion and record the increase in A340nm for approximately 15 minutes. Obtain the ΔA340nm/minute using the maximum linear rate for both the Test and Blank.

Calculations

$$\text{Units / ml enzyme} = \frac{(\Delta A340nm / min\ Test - \Delta A340nm / min\ Blank)\ (3)\ (df)}{(6.22)\ (0.1)}$$

3 = Total volume (in milliliters) of assay
df = Dilution factor
6.22 = Millimolar extinction coefficient of β-NADPH at 340nm
0.1 = Volume (in milliliter) of enzyme used

$$\text{Units / mg solid} = \frac{\text{units / ml enzyme}}{\text{mg solid / ml enzyme}}$$

$$\text{Units / mg protein} = \frac{\text{units / ml enzyme}}{\text{mg protein / ml enzyme}}$$

Unit Definition: *One unit will convert 1.0 μmole of D-gluconate to 6-phospho-D-gluconate per minute at pH 8.0 at 30°C.*

Final Assay Concentrations

In a 3.00 ml reaction mixture, the final concentrations are 80 mM glycylglycine, 3.3 mM magnesium chloride, 3.7 mM adenosine 5'-triphosphate, 0.43 mM β-nicotinamide adenine dinucleotide phosphate, 3.1 mM D-gluconic acid, 5 units 6-phosphogluconic dehydrogenase, and 0.02–0.04 unit gluconate kinase.

Note

1. 6-Phosphogluconic Dehydrogenase Unit Definition: One unit will oxidize 1.0 µmole of 6-phospho-D-gluconate to D-ribulose 5'-phosphate and CO_2 per minute at pH 7.4 at 37°C in the presence of NADP.

Enzymatic Assay of GLUCOSE DEHYDROGENASE (EC 1.1.1.47)

Principle

$$\beta\text{-D(+)Glucose} + \beta\text{-NAD} \xrightarrow{\text{Glucose Dehydrogenase}} \text{D-Glucono-}\delta\text{-Lactone} + \beta\text{-NADH}$$

Abbreviations Used

β-NAD = β-Nicotinamide Adenine Dinucleotide, Oxidized Form
β-NADH = β-Nicotinamide Adenine Dinucleotide, Reduced Form

Conditions: T = 25°C, pH = 7.6, A340nm, Light path = 1 cm

Method: Continuous Spectrophotometric Rate Determination

Reagents

A. 100 mM Potassium Phosphate Buffer, pH 7.6 at 25°C (Prepare 100 ml in deionized water using Potassium Phosphate, Monobasic, Anhydrous. Adjust to pH 7.6 at 25°C with 5 M KOH).
B. 1000 mM β-D(+)Glucose Solution (Glucose) (Prepare 10 ml in deionized water using β-D(+) Glucose).
C. 20 mM β-Nicotinamide Adenine Dinucleotide Solution (β-NAD) (Prepare 1 ml in deionized water using β-Nicotinamide Adenine Dinucleotide or dissolve the contents of one 20 mg vial of β-Nicotinamide Adenine Dinucleotide in the appropriate volume of deionized water. *Prepare Fresh*).
D. Glucose Dehydrogenase Enzyme Solution (Immediately before use, prepare a solution containing 0.3–0.6 unit/ml of Glucose Dehydrogenase in cold deionized water).

Procedure

Pipette (in milliliters) the following reagents into suitable cuvettes:

	Test	Blank
Reagent A (Buffer)	1.80	1.80
Reagent B (Glucose)	1.00	1.00
Reagent C (β-NAD)	0.10	0.10

Mix by inversion and equilibrate to 25°C. Monitor the A340nm until constant, using a suitably thermostatted spectrophotometer. Then add:

Reagent D (Enzyme Solution)	0.10	—
Deionized Water	—	0.10

Immediately mix by inversion and record the increase in A340nm for approximately 10 minutes. Obtain the ΔA340nm/minute using the maximum linear rate for both the Test and Blank.

Calculations

$$\text{Units / mg enzyme} = \frac{\Delta A340\text{nm / min Test} - \Delta A340\text{nm / min Blank}}{(6.22)\ (\text{mg enzyme / ml RM})}$$

6.22 = Millimolar extinction coefficient of β-NADH at 340 nm
RM = Reaction Mix

Unit Definition: *One unit will oxidize 1.0 μmole of β-D-glucose to D-glucono-δ-lactone per minute at pH 7.6 at 25°C.*

Final Assay Concentration

In a 3.00 ml reaction mix, the final concentrations are 60 mM potassium phosphate, 333 mM glucose, 0.67 mM β-NAD and 0.03–0.06 unit glucose dehydrogenase.

Note

1. This enzyme is only to be assayed using β-D(+)glucose. It is highly specific for the β-form.

Enzymatic Assay of GLUCOSE DEHYDROGENASE (EC 1.1.1.47)

Principle

$$\beta\text{-D(+) Glucose} + \beta\text{-NAD} \xrightarrow{\text{Glucose Dehydrogenase}} \text{D-Glucono-}\delta\text{-Lactone} + \beta\text{-NADH}$$

Abbreviations Used

β-NAD = β-Nicotinamide Adenine Dinucleotide, Oxidized Form
β-NADH = β-Nicotinamide Adenine Dinucleotide, Reduced Form

Conditions: T = 25°C, pH = 7.6, A340nm, Light path = 1 cm

Method: Continuous Spectrophotometric Rate Determination

Reagents

A. 120 mM Potassium Phosphate Buffer with 150 mM Sodium Chloride, pH 7.6 at 25°C (Prepare 100 ml in deionized water using Potassium Phosphate, Monobasic, Anhydrous and Sodium Chloride. Adjust to pH 7.6 at 25°C with 5 M NaOH).
B. 556 mM β-D(+) Glucose Solution (Glucose) (Prepare 5 ml in Reagent A using β-D(+) Glucose).
C. 120 mM β-Nicotinamide Adenine Dinucleotide Solution (β-NAD) (Prepare 1 ml in deionized water using β-Nicotinamide Adenine Dinucleotide or dissolve the contents of one 50 mg vial of β- Nicotinamide Adenine Dinucleotide in the appropriate volume of deionized water. *Prepare Fresh*).
D. Glucose Dehydrogenase Enzyme Solution (Immediately before use, prepare a solution containing 0.3–0.6 unit/ml of Glucose Dehydrogenase in cold Reagent A).

Procedure

Pipette (in milliliters) the following reagents into suitable cuvettes:

	Test	Blank
Reagent A (Buffer)	2.00	2.10
Reagent B (Glucose)	1.00	1.00
Reagent D (Enzyme Solution)	0.10	—

Mix by inversion and equilibrate to 25°C. Monitor the A340nm until constant, using a suitably thermostatted spectrophotometer. Then add:

Reagent C (β-NAD)	0.05	0.05

Immediately mix by inversion and record the increase in A340nm for approximately 20 minutes. Obtain the ΔA340nm/minute using the maximum linear rate for both the Test and Blank.

Calculations

$$\text{Units / ml enzyme} = \frac{(\Delta A340\text{nm / min Test} - \Delta A340\text{nm / min Blank})\,(3.15)\,(\text{df})}{(6.22)\,(0.1)}$$

3.15 = Total volume (in milliliter) of enzyme assay
df = Dilution factor
6.22 = Millimolar extinction coefficient of β-NADH at 340nm
0.1 = Volume (in milliliter) of enzyme used

$$\text{Units / mg solid} = \frac{\text{units / ml enzyme}}{\text{mg solid / ml enzyme}}$$

$$\text{Units / mg protein} = \frac{\text{units / ml enzyme}}{\text{mg protein / ml enzyme}}$$

Unit Definition: *One unit will oxidize 1.0 μmole of β-D-glucose to D-glucono-δ-lactone per minute at pH 7.6 at 25°C.*

Final Assay Concentration

In a 3.15 ml reaction mix, the final concentrations are 118 mM potassium phosphate, 148 mM sodium chloride, 177 mM glucose, 2 mM β-nicotinamide adenine dinculeotide, and 0.03–0.06 unit glucose dehydrogenase.

Enzymatic Assay of GLUCOSE DEHYDROGENASE (EC 1.1.1.47)

Principle

$$\beta\text{-D-(+)-Glucose} + \beta\text{-NADP} \xrightarrow{\text{Glucose Dehydrogenase}} \text{D-Glucono-1, 4-Lactone} + \beta\text{-NADPH}$$

Abbreviations Used

β-NADP = β-Nicotinamide Adenine Dinucleotide Phosphate, Oxidized Form
β-NADPH = β-Nicotinamide Adenine Dinucleotide Phosphate, Reduced Form

Conditions: T = 37°C, pH 7.0, A340nm, Light path = 1 cm

Method: Continuous Spectrophotometric Rate Determination

Reagents

A. 50 mM Sodium Phosphate Buffer, pH 7.0 at 37°C (Prepare 100 ml in deionized water using Sodium Phosphate, Monobasic, Monohydrate. Adjust to pH 7.0 at 37°C with 1 M NaOH).
B. 500 mM β-D(+)Glucose Solution (Glucose) (Prepare 10 ml in deionized water using β-D(+)Glucose. *Prepare Fresh*).
C. 4.0 mM β-Nicotinamide Adenine Dinucleotide, Phosphate, Oxidized Form Solution (β-NADP) (Prepare by dissolving the contents of one 10 mg vial of β-Nicotinamide Adenine Dinucleotide Phosphate in the appropriate volume of deionized water. *Prepare Fresh*).
D. Glucose Dehydrogenase Enzyme Solution (Immediately before use, prepare a solution containing 0.3–0.6 unit/ml of Glucose Dehydrogenase in cold Reagent A).

Procedure

Pipette (in milliliters) the following reagents into suitable cuvettes:

	Test	Blank
Reagent A (Buffer)	0.70	0.80
Reagent B (Glucose)	0.10	0.10
Reagent C (β-NADP)	0.10	0.10

Mix by inversion and equilibrate to 37°C. Monitor the A340nm until constant, using a suitably thermostatted spectrophotometer. Then add:

Reagent D (Enzyme Solution)	0.10	—

Immediately mix by inversion and record the increase in A340nm for approximately 5 minutes. Obtain the ΔA340nm/min using the maximum linear rate for both the Test and Blank.

Calculations

$$\text{Units / ml enzyme} = \frac{(\Delta A340\text{nm / min Test} - \Delta A340\text{nm / min Blank})\,(1)\,(\text{df})}{(6.22)\,(0.1)}$$

1 = Volume (in milliliter) of assay
df = Dilution factor
6.22 = Millimolar extinction coefficient of β-NADPH at 340nm
0.1 = Volume (in milliliter) of enzyme used

$$\text{Units / mg protein} = \frac{\text{units / ml enzyme}}{\text{mg protein / ml enzyme}}$$

Unit Definition: *One unit will oxidize 1.0 µmole per minute of β-D-glucose to D-glucono-1, 4-lactone or D-galactose to D-galactonate at pH 7.0 at 37°C in the presence of NADP⁺.*

Final Assay Concentrations

In a 1.00 ml reaction mix, the final concentrations are 40 mM sodium phosphate, 50 mM β-D(+)glucose, 0.4 mM β-nicotinamide adenine dinucleotide phosphate, and 0.03–0.06 unit glucose dehydrogenase.

Enzymatic Assay of GLUCOSE DEHYDROGENASE, NADP Dependent (EC 1.1.1.119)

Principle

$$\text{D-Glucose} + \beta\text{-NADP} \xrightarrow{\text{Glucose Dehydrogenase}} \text{D-Glucono-1, 4-Lactone} + \beta\text{-NADPH}$$

Abbreviations Used

β-NADP = β-Nicotinamide Adenine Dinucleotide, Phosphate Oxidized Form
β-NADPH = β-Nicotinamide Adenine Dinucleotide, Phosphate Reduced Form

Conditions: T = 37°C, pH 8.0, A340nm, Light path = 1 cm

Method: Continuous Spectrophotometric Rate Determination

Reagents

A. 200 mM Tris HCl Buffer, pH 8.0 at 37°C (Prepare 100 ml in deionized water using Trizma Base. Adjust to pH 8.0 at 37°C with 1 M HCl).
B. 10 mM β-Nicotinamide Adenine Dinucleotide, Phosphate, Oxidized Form, Solution (NADP) (Prepare by dissolving the contents of one 10 mg vial of β-Nicotinamide Adenine Dinucleotide, Phosphate, Disodium in the appropriate volume of deionized water. *Prepare Fresh*).
C. 10% (w/v) Bovine Serum Albumin Solution (BSA) (Prepare 1 ml in deionized water using Albumin, Bovine).
D. 10% (v/v) Triton X-100 Solution (X-100) (Prepare 10 ml in deionized water using Triton X-100).
E. 250 mM β-D-(+)Glucose Solution (Glucose) (Prepare 10 ml in deionized water using β-D- (+)Glucose).
F. Glucose Dehydrogenase, NADP Dependent Enzyme Solution (Immediately before use, prepare a solution containing 0.25–0.50 unit/ml of Glucose Dehydrogenase, NADP dependent in cold Reagent A).

Procedure

Pipette (in milliliters) the following reagents into suitable cuvettes:

	Test	Blank
Reagent A (Buffer)	0.60	0.70
Reagent B (NADP)	0.15	0.15
Reagent C (BSA)	0.03	0.03
Reagent D (X-100)	0.15	0.15
Reagent E (Glucose)	0.60	0.60
Deionized Water	1.37	1.37

Mix by inversion and equilibrate to 37°C. Monitor the A340nm until constant, using a suitably thermostatted spectrophotometer. Then add:

	Test	Blank
Reagent F (Enzyme Solution)	0.10	—

Immediately mix by inversion and record the increase in A340nm for approximately 15 minutes. Obtain the ΔA340nm/minute using the maximum linear rate for both the Test and Blank.

$$\text{Units / ml enzyme} = \frac{(\Delta A340\text{nm / min Test} - \Delta A340\text{nm / min Blank})\ (3)\ (\text{df})}{(6.22)\ (0.1)}$$

3 = Volume (in milliliters) of assay
df = Dilution factor
6.22 = Millimolar extinction coefficient for β-NADPH at 340nm
0.1 = Volume (in milliliter) of enzyme used

$$\text{Units / mg solid} = \frac{\text{units / ml enzyme}}{\text{mg solid / ml enzyme}}$$

Unit Definition: *One unit will oxidize 1.0 µmole of β-D-glucose to D-glucono-1, 4-lactone per minute at pH 8.0 at 37°C in the presence of NADP.*

Final Assay Concentrations

In a 3.00 ml reaction mix, the final concentrations are 47 mM Tris, 0.5 mM β-nicotinamide adenine dinucleotide phosphate, 0.1% (w/v) Bovine Serum Albumin, 0.5% (v/v) Triton X-100, 50 mM β-D(+)glucose, and 0.025–0.05 unit glucose dehydrogenase, NADP dependent.

Enzymatic Assay of GLUCOSE-6-PHOSPHATASE (EC 3.1.3.9)

Principle

$$\text{Glucose 6 - Phosphate} + H_2O \xrightarrow{\text{Glucose-6-Phosphatase}} \text{Glucose} + \text{Pi}$$

Abbreviation Used

Pi = Inorganic Phosphate

Conditions: T = 37°C, pH 6.5, A660nm, Light path = 1 cm

Method: Colorimetric

Reagents

A. 100 mM Cacodylate Buffer, pH 6.5 at 37°C (Prepare 50 ml in deionized water using Cacodylic Acid, Sodium Salt. Adjust to pH 6.5 at 37°C with 1 M HCl).
B. Glucose-6-Phosphatase Enzyme Solution (Immediately before use, prepare a solution containing 1–2 units/ml of Glucose-6-Phosphatase in cold deionized water).
C. 200 mM Glucose 6-Phosphate Solution (G-6-P) (Prepare 5 ml in deionized water using D-Glucose 6-Phosphate, Monosodium Salt).

D. 20% (w/v) Trichloroacetic Acid Solution (TCA) (Prepare 10 ml in deionized water using Trichloroacetic Acid, 6.1 N Solution).

E. Phosphorus Standard (Use Phosphorus Standard Solution. The concentration is 20 µg/ml).

F. 10% (w/v) Ammonium Molybdate Solution (Amm. Moly). (Prepare 25 ml in Reagent G (H_2SO_4) using Molybdic Acid, Ammonium Salt, Tetrahydrate).

G. 5000 mM Sulfuric Acid (H_2SO_4) (Prepare 50 ml in deionized water using Sulfuric Acid).

H. Taussky-Shorr Color Reagent (TSCR) (Prepare by adding 10 ml Reagent F (Amm Moly) to 70 ml of deionized water. Then add 5 g Ferrous Sulfate, Heptahydrate and mix until dissolved. Add enough deionized water for a final volume of 100 ml. *Prepare Fresh*).

Procedure

In suitable containers marked T0, T5, T10 and T15, pipette (in milliliters) 0.9 ml of Reagent D (TCA). Pipette (in milliliters) the following reagents into suitable containers:

	Test	Blank
Deionized Water	—	0.30
Reagent A (Buffer)	12.00	12.00
Reagent B (Enzyme Solution)	0.30	—

Mix by swirling and equilibrate to 37°C. Then add:

	Test	Blank
Reagent C (G-6-P)	3.00	3.00

Immediately mix by swirling and incubate at 37°C for 15 minutes. Pipette (in milliliters) 3 ml aliquots at times 0, 5, 10, and 15 minutes to appropriately labeled containers (T0, T5, T10, T15). Mix by swirling. Incubate at 25°C for 5 minutes. Centrifuge all containers for 5 minutes.

Colorimetric Assay

Standard Curve: A standard curve is made by pipetting (in milliliters) the following reagents into suitable containers:

	Std 1	Std 2	Std 3	Std 4	Std 5	Std Blank
Deionized Water	1.80	1.60	1.40	1.20	1.00	2.00
Reagent E (Phosphorus Std)	0.20	0.40	0.60	0.80	1. 00	0.00
Reagent H (TSCR)	2.00	2.00	2.00	2.00	2.00	2.00

Sample: Pipette (in milliliters) the following into suitable containers:

	Times T0		Time T5		Time T10		Time T15	
	Test	Blank	Test	Blank	Test	Blank	Test	Blank
T0 Test	2.00	—	—	—	—	—	—	—
T0 Blank	—	2.00	—	—	—	—	—	—
T5 Test	—	—	2.00	—	—	—	—	—
T5 Blank	—	—	—	2.00	—	—	—	—
T10 Test	—	—	—	—	2.00	—	—	—
T10 Blank	—	—	—	—	—	2.00	—	—
T15 Test	—	—	—	—	—	—	2.00	—
T15 Blank	—	—	—	—	—	—	—	2.00
Reagent H	2.00	2.00	2.00	2.00	2.00	2.00	2.00	2.00

Mix by swirling and incubate at 25°C for 5 minutes. Transfer to suitable cuvettes and record the A660nm for Standards, Tests and Blanks.

Calculations

Standard Curve: ΔA660nm Standard = A66nm Standard–A660nm Standard Blank. Find the slope by plotting the ΔA660nm Standards vs µmoles of Phosphorus. Use the slope (M) to determine the µmoles of Phosphorus liberated.

Sample Determination: ΔA660nm Sample = A660nm Test T–A660nm Blank T

$$\text{Units / mg enzyme} = \frac{(\Delta A660\text{nm T})\,(15.3)\,(3.9)}{(T)\,(3.0)\,(2.0)\,(M)\,(\text{mg enzyme / RM})}$$

15.3 = Total volume of Reaction Mix
3.9 = Volume of Stopped Reaction
T = Time (in minutes) at which assay was stopped
3.0 = Volume from Reaction Mix used in Colorimetric Determination
2.0 = Volume used for Phosphorus Determination
RM = Reaction Mix
M = Slope

Unit Definition: *One unit will release 1.0 µmole of inorganic phosphate from glucose 6-phosphate per minute at pH 6.5 at 37°C.*

Final Assay Concentration

In a 15.30 ml reaction mix, the final concentrations are 78.4 mM sodium cacodylate, 39.2 mM glucose 6-phosphate and 0.3–0.6 unit glucose-6-phosphatase.

Note

1. Glucose-6-Phosphatase also catalyzes the following two reactions:
   a. PPi or Nucleoside di- or Triphosphate + Glucose → Glucose 6-Phosphate + Pi or Nucleoside Monophosphate or Diphosphate.
   b. PPi + H_2O → 2 Pi.

Enzymatic Assay of GLUCOSE-6-PHOSPHATE DEHYDROGENASE (EC 1.1.1.49)

Principle

$$\text{D-Glucose 6-Phosphate} + \beta\text{-NADP} \xrightarrow[\text{Mg}_2^+]{\text{G-6-PDH}} \text{6-Phospho-D-gluconate} + \beta + \text{NADPH}$$

Abbreviations Used

G-6-PDH = Glucose-6-Phosphate Dehydrogenase
β-NADP = β-Nicotinamide Adenine Dinucleotide Phosphate, Oxidized Form
β-NADPH = β-Nicotinamide Adenine Dinucleotide Phosphate, Reduced Form

Conditions: T = 25°C, pH = 7.4, A340nm, Light path = 1 cm

Method: Continuous Spectrophotometric Rate Determination

Reagents

A. 250 mM Glycylglycine Buffer, pH 7.4 at 25°C (Prepare 100 ml in deionized water using Glycylglycine, Free Base. Adjust to pH 7.4 at 25°C with 1 M NaOH).
B. 60 mM D-Glucose 6-Phosphate Solution (G 6-P) (Prepare 2 ml in deionized water using D-Glucose 6-Phosphate, Monosodium Salt).
C. 20 mM β-Nicotinamide Adenine Dinucleotide Phosphate Solution (β-NADP) (Prepare 2 ml in deionized water using β-Nicotinamide Adenine Dinucleotide Phosphate, Sodium Salt).
D. 300 mM Magnesium Chloride Solution ($MgCl_2$) (Prepare 2 ml in deionized water using Magnesium Chloride Hexahydrate).
E. Glucose-6-Phosphate Dehydrogenase Enzyme Solution (Immediately before use, prepare a solution containing 0.3–0.6 unit/ml of Glucose-6-Phosphate Dehydrogenase in cold Reagent A).

Procedure

Prepare a reaction cocktail by pipetting (in milliliters) the following reagents into a suitable container:

Deionized Water	21.0
Reagent A (Buffer)	5.0
Reagent B (G-6-P)	1.0
Reagent C (β-NADP)	1.0
Reagent D ($MgCl_2$)	1.0

Mix and equilibrate to 25°C. Adjust the pH of the reaction cocktail to 7.4 with 1 M NaOH. Pipette (in milliliters) the following reagents into suitable cuvettes:

	Test	Blank
Reaction Cocktail	2.90	2.90

Equilibrate to 25°C. Monitor the A340nm until constant, using a suitably thermostatted spectrophotometer. Then add:

Reagent A (Buffer)	—	0.10
Reagent E (Enzyme Solution)	0.10	—

Immediately mix by inversion and record the increase in A340nm for approximately 5 minutes. Obtain the ΔA340nm/minute using the maximum linear rate for both the Test and Blank.

Calculations

$$\text{Units / ml enzyme} = \frac{(\Delta A340nm / \min \text{ Test} - \Delta A340nm / \min \text{ Blank})\ (3)\ (df)}{(6.22)\ (0.1)}$$

3 = Total volume (in milliliters) of assay
df = Dilution factor
6.22 =Millimolar extinction coefficient of β-NADPH at 340nm
0.1 = Volume (in milliliters) of enzyme used

$$\text{Units / mg solid} = \frac{\text{units / ml enzyme}}{\text{mg solid / ml enzyme}}$$

$$\text{Units / mg protein} = \frac{\text{units / ml enzyme}}{\text{mg protein / ml enzyme}}$$

Unit Definition: *One unit will oxidize 1.0 μmole of D-glucose 6-phosphate To 6-phospho-D-gluconate per minute in the presence of β-NADP at pH 7.4 at 25°C.*

Final Assay Concentration

In a 3.00 ml reaction mix, the final concentrations are 50 mM glycylglycine, 2 mM D-glucose 6-phosphate, 0.67 mM β-nicotinamide adenine dinucleotide phosphate, 10 mM magnesium chloride, and 0.03–0.06 unit glucose-6-phosphate dehydrogenase.

Notes

1. Enzyme assay not to be used for Glucose-6-Phosphate Dehydrogenase from Leuconostoc mesenteroides or from Bacillus stearothermophilus.
2. It is important to maintain the pH at 7.4 to ensure reproducibility.
3. Glucose-6-Phosphate Dehydrogenase is subject to sulfate ion inhibition.

Enzymatic Assay of GLUCOSE-6-PHOSPHATE DEHYDROGENASE (EC 1.1.1.49)

Principle

$$\text{D-Glucose 6-Phosphate} + \beta\text{-NADP} \xrightarrow[Mg_2^+]{\text{G-6-PDH}} \text{6-Phospho-D-gluconate} + \beta\text{-NADPH}$$

Abbreviations Used

β-NADP = β-Nicotinamide Adenine Dinucleotide Phosphate, Oxidized Form
G-6-PDH = Glucose-6-Phosphate Dehydrogenase
β-NADPH = β-Nicotinamide Adenine Dinucleotide Phosphate, Reduced Form

Conditions: T = 30°C, pH = 9.0, A340nm, Light path = 1 cm

Method: Continuous Spectrophotometric Rate Determination

Reagents

A. 100 mM Tris HCl Buffer, pH 9.0 at 30°C (Prepare 100 ml in deionized water using Trizma Hydrochloride. Adjust to pH 9.0 at 30°C with 1 M NaOH).
B. 100 mM D-Glucose 6-Phosphate Solution (G -6-P) (Prepare 2 ml in deionized water using D-Glucose 6-Phosphate, Monopotassium Salt).

C. 20 mM β-Nicotinamide Adenine Dinucleotide Phosphate Solution (β-NADP) (Dissolve the contents of one 10 mg vial of β-Nicotinamide Adenine Dinucleotide Phosphate, Sodium Salt in the appropriate volume of deionized water).
D. 600 mM Magnesium Chloride Solution ($MgCl_2$) (Prepare 2 ml in deionized water using Magnesium Chloride Hexahydrate).
E. Glucose-6-Phosphate Dehydrogenase Enzyme Solution (Immediately before use, prepare a solution containing 0.3–0.6 unit/ml of Glucose-6-Phosphate Dehydrogenase in cold deionized water).

Procedure

Pipette (in milliliters) the following reagents into suitable cuvettes:

	Test	Blank
Reagent A (Buffer)	2.52	2.52
Reagent B (G-6-P)	0.08	0.08
Reagent C (β-NADP)	0.10	0.10
Reagent D ($MgCl_2$)	0.20	0.20

Mix by inversion and equilibrate to 30°C. Monitor the A340nm until constant, using a suitably thermostatted spectrophotometer. Then add:

Deionized Water	—	0.10
Reagent E (Enzyme Solution)	0.10	—

Immediately mix by inversion and record the increase in A340nm for approximately 5 minutes. Obtain the ΔA340nm/minute using the maximum linear rate for both the Test and Blank.

Calculation

$$\text{Units/ml enzyme} = \frac{(\Delta A340\text{nm/min Test} - \Delta A340\text{nm/min Blank})\,(3)\,(\text{df})}{(6.22)\,(0.1)}$$

3 = Total volume (in milliliters) of assay
df = Dilution factor
6.22 = Millimolar extinction coefficient of β-NADPH at 340nm
0.1 = Volume (in milliliters) of enzyme used

$$\text{Units/mg solid} = \frac{\text{units/ml enzyme}}{\text{mg solid/ml enzyme}}$$

$$\text{Units/mg protein} = \frac{\text{units/ml enzyme}}{\text{mg protein/ml enzyme}}$$

Unit Definition: *One unit will oxidize 1.0 μmole of D-glucose 6-phosphate To 6-phospho-D-gluconate per minute in the presence of β-NADP at pH 9.0 at 30°C.*

Final Assay Concentration

In a 3.00 ml reaction mix, the final concentrations are 84 mM Tris, 2.7 mM D-glucose 6-phosphate, 0.67 mM β-nicotinamide adenine dinucleotide phosphate, 40 mM magnesium chloride, and 0.03–0.06 unit glucose-6-phosphate dehydrogenase.

Enzymatic Assay of α-GLUCOSIDASE (EC 3.2.1.20)

Principle

$$\text{Maltose} + H_2O \xrightarrow{\alpha\text{-Glucosidase}} 2\ \text{D-Glucose}$$

$$\text{D-Glucose} + O_2 + H_2O \xrightarrow{\text{Glucose Oxidase}} \text{D-Gluconic Acid} + H_2O_2$$

$$H_2O_2 + \text{o-Dianisidine (Reduced)} \xrightarrow{\text{Peroxidase}} H_2O + \text{o-Dianisidine (Oxidized)}$$

Conditions: T = 37°C, pH 4.0, A500nm, Light path = 1 cm

Method: Colorimetric

Reagents

A. 50 mM Sodium Acetate Buffer, pH 4.0 at 37°C (Prepare 100 ml in deionized water using Acetic Acid, Glacial. Adjust to pH 4.0 at 37°C with 1 M NaOH).

B. 4.0% (w/v) Maltose Substrate Solution (Maltose) (Prepare 50 ml in Reagent A using Maltose).

C. o-Dianisidine Color Reagent (Dian) (Prepare 25 ml in deionized water using o-Dianisidine, Dihydrochloride).

D. PGO Enzymes Solution (PGO) (Immediately before use, dissolve 1 capsule of PGO Enzymes in 100 ml of deionized water. Store in amber bottle, on ice).

E. 4.2% (v/v) Perchloric Acid Solution (PCA) (Prepare 20 ml in deionized water using Perchloric Acid).

F. α-Glucosidase Enzyme Solution (α-Glucosidase) (Immediately before use, prepare a solution containing 1.0–2.0 units/ml of α-Glucosidase in cold deionized water).

Procedure

Step 1: Pipette (in milliliters) the following reagents into suitable containers:

	Test	Blank
Reagent B (Maltose)	5.00	5.00

Equilibrate to 37°C. Then add:

Reagent F (α-Glucosidase)	0.10	—

Immediately mix by swirling and incubate at 37°C for 30 minutes. Then add:

Reagent E (PCA)	1.00	1.00
Reagent F (α-Glucosidase)	—	0.10

Mix by swirling. If the solution is turbid, centrifuge to clarify.

Step 2: Pipette (in milliliters) the following reagents into suitable containers:

	Test	Blank
Blank Supernatant	—	0.20
Test Supernatant	0.20	—
Reagent D (PGO)	6.00	6.00
Reagent C (Dian)	0.10	0.10

Mix by swirling and incubate at 37°C for 30 minutes. Transfer the solutions to suitable cuvettes and record the A500nm for Test and Blank using a suitable spectrophotometer.

Calculations

$$\text{Units / ml enzyme} = \frac{(\text{A500nm Test} - \text{A500nm Blank})\,(6.1)\,(6.3)\,(\text{df})}{(30)\,(0.1)\,(0.2)\,(7.5)\,(2)}$$

6.1 = Volume (in milliliters) of the stopped reaction (Step1)
6.3 = Volume (in milliliters) of the Colorimetric Determination (Step 2)
df = Dilution factor
30 = Time (in minutes) of the assay as per the Unit Definition
0.1 = Volume (in milliliters) of enzyme used
0.2 = Volume (in milliliters) of stopped reaction used in Colorimetric Determination
7.5 = Millimolar extinction coefficient of o-dianisidine (oxidized)
2 = Moles of glucose produced per mole of maltose

$$\text{Units / mg solid} = \frac{\text{units / ml enzyme}}{\text{mg solid / ml enzyme}}$$

$$\text{Units / mg protein} = \frac{\text{units / ml enzyme}}{\text{mg protein / ml enzyme}}$$

Unit Definition: *One unit will convert 1.0 μmole of maltose to 2.0 μmoles of D-glucose per minute at pH 4.0 at 37°C.*

Final Assay Concentration

In a 5.10 ml reaction mix, the final concentrations are 49 mM sodium acetate, 3.9% (w/v) maltose, and 0.1–0.2 unit of α-glucosidase.

Notes

1. Peroxidase Unit Definition: One unit will form 1.0 mg purpurogallin from pyrogallol in 20 seconds at pH 6.0 at 20°C.
2. Glucose Oxidase Unit Definition: One unit will oxidize 1.0 μmole of β-D-glucose to D-gluconic acid and H_2O_2 per minute at pH 5.1 at 35°C, equivalent to an O_2 uptake of 22.4 μl per minute.

Enzymatic Assay of β-GLUCOSIDASE (EC 3.2.1.21)

Principle

$$\beta\text{-D-Glucoside} + H_2O \xrightarrow{\beta\text{-Glucosidase}} \text{D-Glucose} + \text{an Alcohol}$$

Method: Stopped Spectrophotometric Endpoint Determination

Conditions: T = 37°C, pH = 5.0, A340nm, Light path = 1 cm

Reagents

A. 100 mM Sodium Acetate Buffer, pH 5.0 at 37°C (Prepare 100 ml in deionized water using Sodium Acetate, Trihydrate. Adjust to pH 5.0 at 37°C with 1 M HCl).
B. 1.0% (w/v) Salicin Substrate Solution (Salicin) (Prepare 50 ml in Reagent A using Salicin).
C. β-Glucosidase Enzyme Solution (Immediately before use, prepare a solution containing 0.5 unit/ml of β-glucosidase in cold deionized water).
D. Glucose (HK) Reagent Solution (HK) (Prepare 50 ml in deionized water using Glucose (HK) 50 Reagent Vials).
E. 500 mM Carbonate-Bicarbonate Buffer Solution (Carb/Biocarb) (Prepare 100 ml by dissolving the contents of 10 Carbonate-Bicarbonate Buffer Capsules in 100 ml of deionized water or by dissolving 2.95 g of Sodium Carbonate, Anhydrous and 1.90 g of Sodium Bicarbonate into 100 ml of deionized water).

Procedure

Step 1: Pipette (in milliliters) the following reagents into 4 dram vials:

	Test	Blank
Deionized Water	—	1.00
Reagent B (Salicin)	4.00	4.00

Incubate at 37°C for 5 minutes. Then add:

Reagent C (Enz Sol)	1.00	—

Mix by inversion and incubate at 37°C for exactly 10 minutes. Then add:

Reagent E (Carb/Bicarb)	5.00	5.00

This stopped reaction mixture is then used in Step 2.

Step 2: Pipette (in milliliters) the following reagents into suitable cuvettes:

Reagent D (HK)	1.80	1.80

Obtain the initial A340nm using a suitable spectrophotometer. Then add:

Blank (Stopped reaction from Step 1)	—	0.20
Test (Stopped reaction from Step 2)	0.20	—

Mix by inversion and monitor the reaction at A340nm. Once it has reached the endpoint (approximately 5–10 minutes at room temperature) record the final A340nm using a suitable spectrophotometer.

Calculations

$$\Delta A340nm = A340nm\ Final - A340nm\ Initial$$

$$\text{Units / ml enzyme} = \frac{(\Delta A340nm\ \text{Test} - \Delta A340nm\ \text{Blank})\ (2)\ (10)\ (df)}{(6.22)\ (10)\ (0.2)\ (1)}$$

2 = Final assay volume (in milliliters) of Step 2
10 = Initial assay volume (in milliliters) of Step 1
df = Dilution factor
6.22 = Millimolar extinction coefficient of β-NADH at A340nm
10 = Time of assay in minutes (as per the Unit Definition)
0.2 = Volume (in milliliters) of Test used in Step 2
1 = Volume (in milliliter) of enzyme used in Step 1

Unit Definition: *One unit will liberate 1.0 µmole of glucose from salicin per minute at pH 5.0 at 37°C.*

Final Assay Concentration

In a 5.00 ml reaction mix, the final concentrations are 80 mM sodium acetate, 0.8% (w/v) salicin, and 0.5 unit β-glucosidase.

Enzymatic Assay of β-GLUCURONIDASE (EC 3.2.1.31)

Principle

$$\text{PheP-Gluc} + H_2O \xrightarrow{\beta\text{-Glucuronidase}} \text{D-Glucuronate} + \text{Phenolphthalein}$$

Abbreviation Used

PheP-Gluc = Phenolphthalein Glucuronide

Conditions: T = 37°C, pH = 3.8, A540nm, Light path = 1 cm

Method: Spectrophotometric Stop Rate Determination

Reagents

A. 100 mM Sodium Acetate Buffer, pH 3.8 at 37°C (Prepare 50 ml in deionized water using Sodium Acetate, Trihydrate. Adjust to pH 3.8 at 37°C with 1 M HCl).
B. 1.2 mM Phenolphthalein Glucuronide Substrate Solution (PheP-Gluc) (Prepare 10 ml in deionized water using Phenolphthalein Glucuronic Acid, Sodium Salt).
C. 200 mM Glycine Buffer, pH 10.4. (Use Glycine Buffer Solution or prepare 100 ml in deionized water using Glycine Free Base. Adjust to pH 10.4 at 37°C with 1 M NaOH).
D. 0.2% (w/v) Sodium Chloride with 0.1% (w/v) Bovine Serum Albumin (Prepare 20 ml in deionized water using Sodium Chloride and Albumin, Bovine Serum).
E. β-Glucuronidase Enzyme Solution (Immediately before use, prepare a solution containing 300 units/ml of β-Glucuronidase in cold Reagent D).
F. 95% (v/v) Ethanol (Prepare 20 ml in deionized water).
G. 0.05% (w/v) Phenolphthalein Standard Solution (Std Soln) (Prepare 5 ml by dissolving 2.5 mg of Phenolphthalein in 5 ml of Reagent F).

Procedure

Pipette (in milliliters) the following reagents into suitable containers:

	Test	Blank
Reagent A (Buffer)	0.70	0.70
Reagent B (PheP-Gluc)	0.70	0.70
Reagent C (Glycine Buffer)	—	0.10

Mix by inversion and equilibrate to 37°C. Then add:

Reagent E (Enzyme Solution)	0.10	—

Mix by inversion and incubate at 37°C for exactly 30 minutes. Then add:

Reagent C (Glycine Buffer)	5.00	5.00
Reagent E (Enzyme Solution)	—	0.10

Immediately mix by inversion. Transfer the solutions to suitable cuvettes and determine the A540nm for both the Test and Blank using a suitably thermostatted spectrophotometer.

Colorimetric Assay

Standard Curve: Prepare a standard curve by pipetting (in milliliters) the following reagents into suitable containers:

	Std 1	Std 2	Std 3	Std 4	Blank
Reagent A (Buffer)	0.70	0.70	0.70	0.70	0.70
Reagent B (PheP-Gluc)	0.70	0.70	0.70	0.70	0.70
Reagent F (Ethanol)	0.07	0.05	0.03	—	0.10
Reagent G (Std Soln)	0.03	0.05	0.07	0.10	—
Reagent C (Glycine Buffer)	5.00	5.00	5.00	5.00	5.00

Mix by inversion and transfer the standards to suitable cuvettes. Determine the A540nm for each standard using a suitable spectrophotometer.

Calculations

Standard Curve: ΔA540 Standard = A540 Standard–A540 Standard blank
Prepare a standard curve by plotting ΔA540 Standard vs micrograms of Phenolphthalein.
Sample Determination: ΔA540 Sample = A540 Sample–A540 Sample blank
Determine the total micrograms of phenolphthalein liberated using the Standard curve.

$$\text{Units / ml enzyme} = \frac{(\mu\text{g phenolphthalein liberated})\ (2)\ (\text{df})}{0.1}$$

2 = Time correction of assay (Unit Definition = 1 hour)
df = Dilution factor
0.1 = Volume (in milliliters) of enzyme used

$$\text{Units / g solid} = \frac{\text{units / ml enzyme}}{\text{g solid / ml enzyme}}$$

$$\text{Units / g protein} = \frac{\text{units / ml enzyme}}{\text{g protein / ml enzyme}}$$

Unit Definition: *One unit will liberate 1.0 microgram of phenolphthalein from phenolphthalein glucuronide per hour at pH 3.8 at 37°C.*

Final Assay Concentration

In a 1.50 ml reaction mix, the final concentrations are 47 mM sodium acetate, 0.56 mM phenolphthalein glucuronic acid, 30 units β-glucuronidase, 0.01% (w/v) sodium chloride, and 0.007% (w/v) Bovine Serum Albumin.

Enzymatic Assay of β-GLUCURONIDASE (EC 3.2.1.31) from *E. coli*

Principle

$$\text{PheP-Gluc} + H_2O \xrightarrow{\beta\text{-Glucuronidase}} \text{D-Glucuronate} + \text{Phenolphthalein}$$

Abbreviation Used

PheP-Gluc = Phenolphthalein Glucuronide

Conditions: T = 37°C, pH = 6.8, A540nm, Light path = 1 cm

Method: Spectrophotometric Stop Rate Determination

Reagents

A. 75 mM Potassium Phosphate Buffer, with 1.0% (w/v) Bovine Serum Albumin, pH 6.8 at 37°C (Prepare 100 ml in deionized water using Potassium Phosphate, Monobasic, Anhydrous and Albumin, Bovine. Adjust to pH 6.8 at 37°C with 1 M KOH).

B. 3.0 mM Phenolphthalein Glucuronide Substrate Solution (PheP-Gluc) (Prepare 10 ml in deionized water using Phenolphthalein Glucuronic Acid, Free Acid).

C. 200 mM Glycine Buffer Solution, pH 10.4. (Use Glycine Buffer Solution or prepare 100 ml in deionized water using Glycine Free Base. Adjust to pH 10.4 at 37°C with 1 M NaOH).

D. β-Glucuronidase Enzyme Solution (Immediately before use, prepare a solution containing 400–800 units/ml of β-Glucuronidase in cold Reagent A).

E. 95% (v/v) Ethanol (Prepare 20 ml in deionized water).

F. 0.05% (w/v) Phenolphthalein Standard Solution (Std Soln) (Prepare 5 ml by dissolving 2.5 mg of Phenolphthalein in 5 ml of Reagent E or use Phenolphthalein Standard Solution).

Procedure

Pipette (in milliliters) the following reagents into suitable containers:

	Test	Blank
Deionized Water	0.65	0.65
Reagent A (Buffer)	0.50	0.50
Reagent B (PheP-Gluc)	0.25	0.25

Mix by inversion and equilibrate to 37°C. Then add:

	Test	Blank
Reagent D (Enzyme Solution)	0.10	—

Mix by inversion and incubate at 37°C for exactly 30 minutes. Then add:

Reagent C (Glycine Buffer)	5.00	5.00
Reagent D (Enzyme Solution)	—	0.10

Immediately mix by inversion. Transfer the solutions to suitable cuvettes and record the A540nm for both the Test and Blank using a suitable spectrophotometer.

Colorimetric Assay

Standard Curve: Prepare a standard curve by pipetting (in milliliters) the following reagents into suitable containers:

	Std 1	Std 2	Std 3	Std 4	Std 5	Std Blank
Deionized Water	0.65	0.65	0.65	0.65	0.65	0.65
Reagent A (Buffer)	0.50	0.50	0.50	0.50	0.50	0.50
Reagent B (PheP-Gluc)	0.25	0.25	0.25	0.25	0.25	0.25
Reagent F (Std Soln)	0.02	0.03	0.05	0.07	0.10	—
Reagent E (Ethanol)	0.08	0.07	0.05	0.03	—	0.10
Reagent C (Glycine Buffer)	5.00	5.00	5.00	5.00	5.00	5.00

Mix by inversion and transfer the standards to suitable cuvettes. Record the A540nm for each standard using a suitable spectrophotometer.

Calculations

Standard Curve: ΔA540 Standard = A540 Standard–A540 Standard blank

Prepare a standard curve by plotting the ΔA540 for the Standard vs micrograms of Phenolphthalein.

Sample Determination: ΔA540 Sample = A540 Sample–A540 Sample blank

Determine the total micrograms of phenolphthalein liberated using the Standard curve.

$$\text{Units / ml enzyme} = \frac{(\mu\text{g phenolphthalein released})\ (2)\ (\text{df})}{0.1}$$

2 = Time correction of assay (Unit Definition = 1 hour)
df = Dilution factor
0.1 = Volume (in milliliter) of enzyme used

$$\text{Units / g solid} = \frac{\text{units / ml enzyme}}{\text{g solid / ml enzyme}}$$

$$\text{Units / g protein} = \frac{\text{units / ml enzyme}}{\text{g protein / ml enzyme}}$$

Unit Definition: *One unit will liberate 1.0 μg of phenolphthalein from phenolphthalein glucuronide per hour at pH 6.8 at 37°C.*

Final Assay Concentration

In a 1.50 ml reaction mix, the final concentrations are 30 mM potassium phosphate, 0.50 mM phenolphthalein glucuronic acid, and 40–80 units β-glucuronidase.

Enzymatic Assay of β-GLUCURONIDASE (EC 3.2.1.31) From Helix pomatia and Bovine Liver

Principle

$$\text{PheP-Gluc} + H_2O \xrightarrow{\beta\text{-Glucuronidase}} \text{D-Glucuronate} + \text{Phenolphthalein}$$

Abbreviation Used

PheP-Gluc = Phenolphthalein Glucuronide

Conditions: T = 37°C, pH = 5.0, A540nm, Light path = 1 cm

Method: Spectrophotometric Stop Rate Determination

Reagents

A. 100 mM Sodium Acetate Buffer, pH 5.0 at 37°C (Prepare 50 ml in deionized water using Sodium Acetate, Trihydrate. Adjust to pH 5.0 at 37°C with 1 M HCl).

B. 1.2 mM Phenolphthalein Glucuronide Substrate Solution (PheP-Gluc) (Prepare 10 ml in deionized water using of Phenolphthalein Glucuronic Acid, Sodium Salt).

C. 200 mM Glycine Buffer Solution, pH 10.4 (Use Glycine Buffer Solution or prepare 100 ml in deionized water using Glycine Free Base. Adjust to pH 10.4 at 37°C with 1 M NaOH).

D. 0.2% (w/v) Sodium Chloride Solution (NaCl) (Prepare 20 ml in deionized water using Sodium Chloride).

E. β-Glucuronidase Enzyme Solution (Immediately before use, prepare a solution containing 250–500 units/ml of β-Glucuronidase in cold Reagent D).

F. 95% (v/v) Ethanol (Prepare 20 ml in deionized water).

G. 0.05% (w/v) Phenolphthalein Standard Solution (Std Soln) (Prepare 2 ml by dissolving 1.0 mg of Phenolphthalein in Reagent F).

Procedure

Pipette (in milliliters) the following reagents into suitable containers:

	Test	Blank
Reagent A (Buffer)	0.70	0.70
Reagent B (PheP-Gluc)	0.70	0.70

Mix by inversion and equilibrate to 37°C. Then add:

Reagent E (Enzyme Solution)	0.10	—

Mix by inversion and incubate at 37°C for exactly 30 minutes. Then add:

Reagent C (Glycine Buffer)	5.00	5.00
Reagent E (Enzyme Solution)	—	0.10

Immediately mix by inversion. Transfer the solutions to suitable cuvettes and record the A540nm for both the Test and Blank using a suitable spectrophotometer.

Colorimetric Assay

Standard Curve: Prepare a standard curve by pipetting (in milliliters) the following reagents into suitable containers:

	Std 1	Std 2	Std 3	Std 4	Std Blank
Reagent A (Buffer)	0.70	0.70	0.70	0.70	0.70
Reagent B (PheP-Gluc)	0.70	0.70	0.70	0.70	0.70
Reagent G (Std Soln)	0.03	0.05	0.07	0.10	—
Reagent F (Ethanol)	0.07	0.05	0.03	—	0.10
Reagent C (Glycine Buffer)	5.00	5.00	5.00	5.00	5.00

Mix by inversion and transfer the standards to suitable cuvettes. Record the A540nm for each standard using a suitable spectrophotometer.

Calculations

Standard Curve: ΔA540 Standard = A540 Standard–A540 Standard blank. Prepare a standard curve by plotting the ΔA540 for the Standard vs micrograms of Phenolphthalein.

Sample Determination: ΔA540 Sample = A540 Sample–A540 Sample blank. Determine the total micrograms of phenolphthalein liberated using the Standard curve.

$$\text{Units / ml enzyme} = \frac{(\mu\text{g phenolphthalein released})\ (2)\ (\text{df})}{0.1}$$

2 = Conversion factor from 30 minutes to 1 hour as per the Unit Definition
df = Dilution factor
0.1 = Volume (in milliliter) of enzyme used

$$\text{Units / g solid} = \frac{\text{units / ml enzyme}}{\text{g solid / ml enzyme}}$$

$$\text{Units / g protein} = \frac{\text{units / ml enzyme}}{\text{g protein / ml enzyme}}$$

Unit Definition: *One unit will liberate 1.0 microgram of phenolphthalein from phenolphthalein glucuronide per hour at pH 5.0 at 37°C.*

Final Assay Concentration

In a 1.50 ml reaction mix, the final concentrations are 47 mM sodium acetate, 0.56 Mm phenolphthalein glucuronic acid, and 25–50 units β-glucuronidase.

Enzymatic Assay of L-GLUTAMATE OXIDASE (EC 1.4.3.11)

Principle

$$\text{L-Glutamate} + O_2 + H_2O \xrightarrow{\text{L-Glutamate Oxidase}} \alpha\text{-Ketoglutaric Acid} + NH_3 + H_2O_2$$

$$2H_2O_2 + H_2O \xrightarrow{\text{Catalase}} 2H_2O + O_2$$

Conditions: T = 30°C, pH = 7.4, A316nm, Light path = 1 cm

Method: Colorimetric

Reagents

A. 100 mM Potassium Phosphate Buffer, pH 7.4 at 30°C (Prepare 100 ml in deionized water using Potassium Phosphate, Monobasic, Anhydrous. Adjust to pH 7.4 at 30°C with 1 M KOH).
B. 100 mM L-Glutamate Solution (L-Glu) (Prepare 2 ml in deionized water using L-Glutamic Acid, Monosodium Salt).
C. 25% (w/v) Trichloroacetic Acid Solution (TCA) (Prepare 5 ml in deionized water using Trichloroacetic Acid, 6.1 N Solution, approximately 100% (w/v)).
D. 1 M Acetate Buffer, pH 5.0 at 37°C (Acetate) (Prepare 25 ml in deionized water using Sodium Acetate, Trihydrate. Adjust to pH 5.0 at 37°C with 1 M HCl).
E. 0.10% (w/v) 3-Methyl-2-Benzothiazolinone Hydrazone Solution (MBTH) (Prepare 10 ml in deionized water using 3-Methyl-2-Benzothiazolinone Hydrazone, Hydrochloride, Hydrate).
F. 2.6 mM α-Ketoglutaric Acid Solution (KG) (Prepare 1 ml in Reagent A using α-Ketoglutaric Acid, Disodium Salt).
G. Catalase Enzyme Solution (Catalase) (Immediately before use, prepare a solution containing 300 units/ml of Catalase in cold deionized water).
H. L-Glutamate Oxidase Enzyme Solution (L-Glu Ox) (Immediately before use, prepare a solution containing 0.05–0.10 unit/ml of L-Glutamate Oxidase in cold Reagent A).

Procedure

Pipette (in milliliters) the following reagents into suitable containers:

	Test	Blank
Reagent A (Buffer)	0.70	0.80
Reagent G (Catalase)	0.10	0.10
Reagent B (L-Glu)	0.10	0.10

Mix by swirling and equilibrate to 30°C. Then add:

Reagent H (L-Glu Ox)	0.10	—

Immediately mix by swirling and incubate at 30°C for exactly 20 minutes with shaking. Then add:

Reagent C (TCA)	0.10	0.10
Reagent D (Acetate)	1.90	1.90
Reagent E (MBTH)	0.80	0.80

Mix by swirling and incubate at 50°C for 30 minutes. Then allow to stand at room temperature for 20 minutes. Transfer the solutions to suitable cuvettes and record the A316nm for both the Test and Blank using a suitable spectrophotometer.

Colorimetric Assay

Standard Curve: Prepare a standard curve by pipetting (in milliliters) the following reagents into suitable containers:

	Std 1	Std 2	Std 3	Std 4	Std 5	Std Blank
Reagent F (KG)	0.02	0.04	0.06	0.08	0.10	—
Reagent A (Buffer)	0.98	0.96	0.94	0.92	0.90	1.00

Mix by swirling and incubate at 30°C for exactly 20 minutes with shaking. Then add:

	Std 1	Std 2	Std 3	Std 4	Std 5	Std Blank
Reagent C (TCA)	0.10	0.10	0.10	0.10	0.10	0.10
Reagent D (Acetate)	1.90	1.90	1.90	1.90	1.90	1.90
Reagent E (MBTH)	0.80	0.80	0.80	0.80	0.80	0.80

Mix by swirling and incubate at 50°C for 30 minutes. Then allow to stand at room temperature for 20 minutes. transfer the solutions to suitable cuvettes and record the 316nm for the Standards and Standard Blank using a suitable spectrophotometer.

Calculations

Standard Curve: ΔA316nm Standard = A316nm Standard–A316nm Standard Blank. Prepare a standard curve by plotting the ΔA316nm for the Standard versus μmoles of α-Ketoglutarate.

Sample Determination: ΔA316nm Sample = A316nm Test–A316nm Blank. Determine the μmoles of α-Ketoglutarate produced using the Standard curve.

$$\text{Units / ml enzyme} = \frac{(\mu\text{moles of } \alpha\text{ - Ketoglutarate produced) (df)}}{(20)\,(0.1)}$$

df = Dilution factor
20 = Time (in minutes) of assay as per the Unit Definition
0.1 = Volume (in milliliter) of enzyme used in assay

$$\text{Units / mg solid} = \frac{\text{units / ml enzyme}}{\text{mg solid / ml enzyme}}$$

$$\text{Units / mg protein} = \frac{\text{units / ml enzyme}}{\text{mg protein / ml enzyme}}$$

Unit Definition: *One unit will form 1.0 μmole of α-ketoglutaric acid from L-glutamic acid per minute at pH 7.4 at 30°C.*

Final Assay Concentration

In a 1.00 ml reaction mix, the final concentrations are 80 mM potassium phosphate, 10 mM L-glutamic acid, 30 units catalase, and 0.005–0.01 unit L-glutamate oxidase.

Note

1. Catalase Unit Definition: One unit will decompose 1.0 μmole of H_2O_2 per minute at pH 7.0 at 25°C, while the H_2O_2 concentration falls from 10.3 to 9.2 mM.

Enzymatic Assay of L-GLUTAMIC DECARBOXYLASE (EC 4.1.1.15)

Principle

$$\text{L-Glutamic Acid} \xrightarrow[\text{PRP}]{\text{L-Glutamic Decarboxylase}} \text{4-Aminobutanoate} + CO_2$$

Abbreviation Used

PRP = Pyridoxal 5-Phosphate

Conditions: T = 37°C, pH = 5.0

Method: Manometric Assay using Warburg Flasks

Reagents

A. 100 mM Sodium Acetate Buffer, pH 5.0 at 37°C (Prepare 100 ml in deionized water using Sodium Acetate, Trihydrate. Adjust to pH 5.0 at 37°C with 1 M HCl).

B. 100 mM L-Glutamic Acid Solution (L-Glu) (Prepare 50 ml in Reagent A using L-Glutamic Acid, Monosodium Salt. Adjust to pH 5.0 at 37°C with either 1 M NaOH or 1 M HCl).

C. 10 mM Pyridoxal 5-Phosphate Solution (PRP) (Prepare 25 ml in Reagent A using Pyridoxal 5- Phosphate).

D. L-Glutamic Decarboxylase Enzyme Solution (Immediately before use, prepare a solution containing 3–5 units/ml of L-Glutamic Decarboxylase in cold Reagent A).

Procedure

Pipette (in milliliters) the following reagents into Warburg Flasks:

Main Chamber

	Thermo-Barometer Substrate Flask	Enzyme Blank	Test	Blank
Reagent A (Buffer)	2.80	2.80	0.30	0.30
Reagent B (L-Glu)	—	—	2.50	2.50

Side Arm

	Thermo-Barometer Substrate Flask	Enzyme Blank	Test	Blank
Reagent A (Buffer)	0.20	—	—	0.10
Reagent C (PRP)	—	0.10	0.10	0.10
Reagent D (Enz Solution)	—	0.10	0.10	—

Be sure to confirm the stability of the pressure with the flask sealed off and before proceeding with the assay. This is to ensure temperature equilibrium and the absence of leaks in the flask. The enzyme activity is determined by calculation of the rate of production of CO_2 at 37°C. The reaction rate should be linear for about 20 minutes.

Calculations

$$\frac{\text{Units}}{\text{L-Glutamic Decarboxylase}} = \frac{(C)(K)(df)}{(22.4\ l/\text{mole})(\text{ml L-Glutamic Decarboxylase})}$$

C = mm of CO_2 gas evolved/minute
K = Warburg flask constant in µl/mm³
df = Dilution factor
22.4 l = Volume gas occupies under STP conditions

Unit Definition: *One unit will release 1.0 µmole of CO_2 from L-glutamic acid per minute at pH 5.0 at 37°C.*

Final Assay Concentrations

In a 3.00 ml reaction mix, the final concentrations are 100 mM sodium acetate, 83 mM L-glutamic acid, 0.33 mM pyridoxal 5-phosphate, and 0.3–0.5 unit L-glutamic acid decarboxylase.

Notes

1. The tests are done, at least in triplicate, since it is common for the flasks to have small leaks.
2. The mm of CO_2 gas evolved (C) is corrected for any temperature and barometric changes (Thermobarometer) during the experiment and also for the Substrate and Enzyme Blank:mm CO_2 corrected = mm CO_2 measured Test–mm CO_2 measured of [Thermobarometer + Substrate Blank + Enzyme Blank] Values of the corrected mm CO_2 produced are plotted versus time. The best straight line is drawn not necessarily through the origin. The slope, c = mm CO_2/time, is obtained.
3. The flask constant, K, is calculated according to the equation:

$$K = \frac{\left[(V_g)\left(\frac{223}{T}\right) + V_f\, a\right]}{P_o}$$

Po = Standard pressure as mm of manometer fluid
V_g = Gas volume (in milliliters) in flask and manometer
V_f = Volume of Liquid in flask
T = Absolute temperature
a = Solubility of gas; (for CO_2 at 37°C, a = 0.57)
223 = Constant

Enzymatic Assay of L-GLUTAMIC DEHYDROGENASE (EC 1.4.1.3)

Principle

$$\alpha\text{-KG} + \beta\text{-NADH} + NH_4 + \xrightarrow{\text{L-Glutamic Dehydrogenase}} \text{L-Glu} + \beta\text{-NAD} + H_2O$$

Abbreviations Used

α-KG = α-Ketoglutarate
L-Glu = L-Glutamate
β-NADH = β-Nicotinamide Adenine Dinucleotide, Reduced Form
β-NAD = β-Nicotinamide Adenine Dinucleotide, Oxidized Form

Conditions: T = 25°C, pH = 7.3, A340nm, Light path = 1 cm

Method: Continuous Spectrophotometric Rate Determination

Reagents

A. 100 mM Triethanolamine Buffer, pH 7.3 at 25°C (Prepare 100 ml in deionized water using Triethanolamine Hydrochloride. Adjust to pH 7.3 at 25°C with 1 M NaOH).
B. 200 mM α-Ketoglutarate Solution (α-KG) (Prepare 10 ml in deionized water using α-Ketoglutaric Acid. Adjust to pH 6.5–7.5 using solid Sodium Bicarbonate).
C. 3.2 M Ammonium Acetate Solution (NH_4OAc) (Prepare 5 ml in deionized water using Ammonium Acetate).
D. 10 mM β-Nicotinamide Adenine Dinucleotide, Reduced Form Solution (β-NADH) (Prepare 1 ml in deionized water using β-Nicotinamide Adenine Dinucleotide, Reduced Form, Disodium Salt. *Prepare Fresh*).
E. 25 mM Ethylenediaminetetraacetic Acid Solution (EDTA) (Prepare 1 ml in deionized water using Ethylenediaminetetraacetic Acid, Tetrasodium Salt, Hydrate).
F. L-Glutamic Dehydrogenase Enzyme Solution (Immediately before use, prepare a solution containing 0.3–0.6 unit/ml of L-Glutamic Dehydrogenase in cold Reagent A).

Procedure

Prepare a reaction cocktail by pipetting (in milliliters) the following reagents into a suitable containers.

Reagent A (Buffer)	26.0
Reagent B (α-KG)	2.0
Reagent C (NH_4OAc)	0.5
Reagent D (β-NADH)	0.3
Reagent E (EDTA)	0.3

Mix by stirring and equilibrate to 25°C. Adjust to pH 7.3 at 25°C with 1 M NaOH or 1 M HCl. Then add:

	Test	Blank
Reaction Cocktail	2.90	2.90

Equilibrate to 25°C. Then add:

	Test	Blank
Reagent F (Enzyme Solution)	0.10	—
Reagent A (Buffer)	—	0.10

Immediately mix by inversion and record the decrease in A340nm for approximately 5–10 minutes. Obtain the ΔA340nm/minute using the maximum linear rate for both the Test and Blank.

Calculations

$$\text{Units/ml enzyme} = \frac{(\Delta A340\text{nm/min Test} - \Delta A340\text{nm/min Blank})\ (3)\ (\text{df})}{(6.22)\ (0.1)}$$

3 = Volume (in milliliters) of assay
df = Dilution factor
6.22 = Millimolar extinction coefficient of β-NADH at 340nm
0.1 = Volume (in milliliter) of enzyme used

$$\text{Units / mg solid} = \frac{\text{units / ml enzyme}}{\text{mg solid / ml enzyme}}$$

$$\text{Units / mg protein} = \frac{\text{units / ml enzyme}}{\text{mg protein / ml enzyme}}$$

Unit Definition: *One unit will reduce 1.0 µmole of α-ketoglutarate to L-glutamate per minute at pH 7.3 at 25°C, in the presence of ammonium ions.*

Final Assay Concentration

In a 3.00 ml reaction mix, the final concentrations are 90 mM triethanolamine hydrochloride, 13 mM a α-ketoglutarate, 53 mM ammonium acetate, 0.06 mM β-NADH, 0.25 mM EDTA and 0.03–0.06 unit L-glutamic dehydrogenase.

Enzymatic Assay of L-GLUTAMIC DEHYDROGENASE (β-NADP) (EC 1.4.1.4)

Principle

$$\alpha\text{-Ketoglutarate} + NH_3 + \beta\text{-NADPH} \xrightarrow{\text{L-Glut DH}} \text{L-Glutamate} + \beta\text{-NADP} + H_2O$$

Abbreviations Used

β-NADPH = β-Nicotinamide Adenine Dinucleotide Phosphate, Reduced Form
β-NADP = β-Nicotinamide Adenine Dinucleotide Phosphate
L-GlutDH = L-Glutamic Dehydrogenase

Conditions: T = 30°C, pH = 8.3, A340nm, Light path = 1 cm

Method: Continuous Spectrophotometric Rate Determination

Reagents

A. 100 mM Tris HCl Buffer, pH 8.3 at 30°C (Prepare 100 ml in deionized water using Trizma Hydrochloride. Adjust to pH 8.3 at 30°C with 1 M NaOH).
B. 3.3 M Ammonium Chloride Solution (NH_4Cl) (Prepare 2 ml in deionized water using Ammonium Chloride).
C. 225 mM α-Ketoglutarate Solution (α-KG) (Prepare 5 ml in deionized water using α-Ketoglutaric Acid, Free Acid. Adjust to pH 8 with 1 M NaOH).
D. 7.5 mM β-Nicotinamide Adenine Dinucleotide Phosphate, Reduced Form Solution (β-NADPH) (Dissolve the contents of one 10 mg vial of β-Nicotinamide Adenine Dinucleotide Phosphate, Reduced Form, Tetrasodium Salt in 1.6 ml of deionized water *or* prepare 1 ml in deionized water using β- Nicotinamide Adenine Dinucleotide Phosphate, Reduced Form, Tetrasodium Salt. *Prepare Fresh*).
E. 50 mM Potassium Phosphate with 50 mM Ethylenediaminetetraacetic Acid Solution, pH 6.6 at 30°C (Prepare 20 ml in deionized water using Potassium Phosphate, Monobasic and Ethylenediaminetetraacetic Acid, Tetrasodium Salt. Adjust to pH 6.6 at 30°C using 1 M NaOH).

F. L-Glutamic Dehydrogenase Enzyme Solution (Immediately before use, prepare a solution containing 0.3–1.0 units/ml of L-Glutamic Dehydrogenase in cold Reagent E).

Procedure

Pipette (in milliliters) the following reagents into suitable cuvettes:

	Test	Blank
Reagent A (Buffer)	2.50	2.50
Reagent B (NH_4Cl)	0.20	0.20
Reagent C (α-KG)	0.10	0.10
Reagent D (β-NADPH)	0.10	0.10

Mix by inversion and equilibrate to 30°C. Monitor the A340nm until constant, using a suitably thermostatted spectrophotometer. Then add:

Reagent E (Diluent)	—	0.05
Reagent F (Enzyme Solution)	0.05	—

Immediately mix by inversion and record the decrease in A340nm for approximately 5 minutes. Obtain the ΔA340nm/minute using the maximum linear rate for both the Test and Blank.

Calculations

$$\text{Units / mg enzyme} = \frac{\Delta A340\text{nm / min Test} - \Delta A340\text{nm / min Blank}}{(6.22)\ (\text{mg enzyme / ml RM})}$$

6.22 = Millimolar extinction coefficient of β-NADPH at 340 nm
RM = Reaction Mix

Unit Definition: *One unit will reduce 1.0 μmole of α-ketoglutarate to L-glutamate per minute at pH 8.3 at 30°C in the presence of amminium ions and β-NADPH.*

Final Assay Concentration

In a 2.95 ml reaction mix, the final concentrations are 83 mM Tris HCl, 112 mM ammonium chloride, 7.7 mM ketoglutarate, 0.25 mM β-NADPH, 0.83 mM potassium phosphate, 0.83 mM EDTA and 0.015–0.05 units L-glutamic dehydrogenase.

Enzymatic Assay of L-GLUTAMIC DEHYDROGENASE (NADP) (EC 1.4.1.4)

Principle

$$\alpha\text{-KG} + \beta\text{-NADPH} + NH_4^+ \xrightarrow{\text{L-Glutamic Dehydrogenase}} \text{L-Glu} + \beta\text{-NADP} + H_2O$$

Abbreviations Used

α-KG = α-Ketoglutarate
L-Glu = L-Glutamate
β-NADPH = β-Nicotinamide Adenine Dinucleotide Phosphate, Reduced Form
β-NADP = β-Nicotinamide Adenine Dinucleotide Phosphate, Oxidized Form

Conditions: T = 30°C, pH = 8.3, A340nm, Light path = 1 cm

Method: Continuous Spectrophotometric Rate Determination

Reagents

A. 100 mM Tris HCl Buffer, pH 8.3 at 30°C (Prepare 100 ml in deionized water using Trizma, Hydrochloride. Adjust to pH 8.3 at 30°C with 1 M NaOH).

B. 225 mM α-Ketoglutarate Solution (α-KG) (Prepare 10 ml in Reagent A using α-Ketoglutaric Acid or α-Ketoglutaric Acid. Adjust to pH 7.0–9.0 at 30°C with 1 M NaOH).

C. 3300 mM Ammonium Chloride Solution (NH_4Cl) (Prepare 5 ml in deionized water using Ammonium Chloride).

D. 7.5 mM β-Nicotinamide Adenine Dinucleotide Phosphate, Reduced Form Solution (β-NADPH) (Prepare 1 ml in deionized water using β-Nicotinamide Adenine Dinucleotide Phosphate, Reduced Form, Tetrasodium Salt or equivalent. *Prepare Fresh*).

E. 50 mM Potassium Phosphate with 50 mM Ethylenediaminetetraacetic Acid Solution, pH 6.6 at 30°C (Enzyme Diluent) (Prepare 100 ml in deionized water using Potassium Phosphate, Dibasic, Trihydrate and Ethylenediaminetetraacetic Acid, Disodium Salt, Dihydrate. Adjust to pH 6.6 at 30°C with 1 M KOH).

F. L-Glutamic Dehydrogenase Enzyme Solution (Immediately before use, prepare a solution containing 0.3–0.6 unit/ml of L-Glutamic Dehydrogenase in cold Reagent E).

Procedure

Pipette (in milliliters) the following reagents into suitable cuvettes:

	Test	Blank
Reagent A (Buffer)	2.50	2.50
Reagent B (α-KG)	0.10	0.10
Reagent C (NH_4Cl)	0.20	0.20
Reagent D (β-NADPH)	0.10	0.10

Mix by inversion and equilibrate to 30°C. Monitor the A340nm until constant, using a suitably thermostatted spectrophotometer. Then add:

Reagent F (Enzyme Solution)	0.10	—
Reagent E (Enzyme Diluent)	—	0.10

Immediately mix by inversion and record the decrease in A340nm for approximately 5 minutes. Obtain the ΔA340nm/minute using the maximum linear rate for both the Test and Blank.

Calculations

$$\text{Units / ml enzyme} = \frac{(\Delta A340\text{nm / min Test} - \Delta A340\text{nm / min Blank})\ (3)\ (df)}{(6.22)\ (0.1)}$$

3 = Total volume (in milliliters) of assay
df = Dilution factor
6.22 = Millimolar extinction coefficient of β-NADPH at 340nm
0.1 = Volume (in milliliters) of enzyme used

$$\text{Units / mg solid} = \frac{\text{units / ml enzyme}}{\text{mg solid / ml enzyme}}$$

$$\text{Units / mg protein} = \frac{\text{units / ml enzyme}}{\text{mg protein / ml enzyme}}$$

Unit Definition: *One unit will reduce 1.0 μmole of α-ketoglutarate to L-glutamate per minute at pH 8.3 at 30°C, in the presence of ammonium ions and NADPH.*

Final Assay Concentration

In a 3.00 ml reaction mix, the final concentrations are 83 mM Tris, 7.5 mM α-ketoglutarate, 220 mM ammonium chloride, 0.25 mM β-nicotinamide adenine dinucleotide phosphate, reduced form, 1.7 mM potassium phosphate, 1.7 mM Ethylenediaminetetraacetic Acid, and 0.03–0.06 unit L-glutamic dehydrogenase.

Enzymatic Assay of GLUTAMIC-PYRUVIC TRANSAMINASE (EC 2.6.1.2)

Principle

$$\text{L-Alanine} + \alpha\text{-Ketoglutaric Acid} \xrightarrow{\text{GPT}} \text{Pyruvate} + \text{L-Glutamate}$$

$$\text{Pyruvate} + \beta\text{-NADH} \xrightarrow{\text{Lactic Acid Dehydrogenase}} \text{Lactate} + \beta\text{-NAD}$$

Abbreviations Used

GPT = Glutamic-Pyruvic Transaminase
β-NADH = β-Nicotinamide Adenine Dinucleotide, Reduced Form
β-NAD = β-Nicotinamide Adenine Dinucleotide

Conditions: T = 37°C, pH = 7.4, A340nm, Light path = 1 cm

Method: Continuous Spectrophotometric Rate Determination

Reagents

A. 100 mM Tris Buffer, pH 7.4 at 37°C. (Prepare 100 ml in deionized water using Trizma Base. Adjust to pH 7.4 at 37°C with 1 M HCl).
B. 100 mM α-Ketoglutaric Acid Solution (α-KGA) (Prepare 10 ml in Reagent A using α-Ketoglutaric Acid, Monosodium Salt).
C. 1 mM L-Alanine Solution (Prepare 10 ml in Reagent A using L-Alanine).
D. 6.4 mM β-Nicotinamide Adenine Dinucleotide, Reduced Form Solution (β-NADH) (Dissolve the contents of one 10 mg vial of β-Nicotinamide Adenine Dinucleotide, Reduced Form, Disodium Salt, Preweighed Vial in the appropriate volume of Reagent A).
E. Lactic Dehydrogenase Enzyme Solution (LDH) (Immediately before use, prepare a solution containing 400–600 units/ml in cold deionized water using Lactic Dehydrogenase).

F. Glutamic-Pyruvic Transaminase Enzyme Solution (Immediately before use, prepare a solution containing 0.3–0.6 units/ml of Glutamic-Pyruvic Transaminase in cold deionized water).

Procedure

Prepare a reaction cocktail by pipetting (in milliliters) the following reagents into a suitable container:

Reagent A (Buffer)	18.5
Reagent B (α-KGA)	3.0
Reagent C (L-Alanine)	6.0
Reagent D (β-NADH)	0.5
Reagent E (LDH)	1.0

Mix and adjust to pH 7.4 at 37°C with 1 M NaOH or 1 M HCl, if necessary. Pipette (in milliliters) the following reagents into suitable cuvettes:

	Test	Blank
Reaction Cocktail	2.90	2.90

Equilibrate to 37°C. Monitor the A340nm until constant, using a suitably thermostatted spectrophotometer. Then add:

Deionized Water	—	0.10
Reagent F (Enzyme Solution)	0.10	—

Immediately mix by inversion and record the decrease in A340nm for approximately 5 minutes. Obtain the ΔA340nm/minute using the maximum linear rate for both the Test and Blank.

Calculations

$$\text{Units / mg enzyme} = \frac{(\Delta A340nm / min\ Test - \Delta A340nm / min\ Blank)\ (3)\ (df)}{(6.22)\ (0.1)}$$

3 = Total volume (in milliliters) of assay
df = Dilution factor
6.22 = Millimolar extinction coefficient of NADH at 340nm

$$\text{Units / mg solid} = \frac{\text{units / ml enzyme}}{\text{mg solid / ml enzyme}}$$

$$\text{Units / mg protein} = \frac{\text{units / ml enzyme}}{\text{mg protein / ml enzyme}}$$

Unit Definition: *One unit will convert 1.0 μmole of α-ketoglutarate to L-glutamate per minute at pH 7.4 at 37°C, in the presence of L-alanine.*

Final Assay Concentration

In a 3.00 ml reaction mix, the final concentrations are 93 mM Tris, 10 mM α-ketoglutarate, 200 mM L-alanine, 0.11 mM β-NADH, 13–20 units lactic dehydrogenase and 0.03–0.06 units glutamic pyruvic transaminase.

Note

1. Lactic Dehydrogenase Unit Definition: One unit will reduce 1.0 μmole of pyruvate to L-lactate per minute at pH 7.5 at 37°C.

Enzymatic Assay of GLUTAMINASE (EC 3.5.1.2) (From *E. coli*)

Principle

$$\text{L-Glutamine} + H_2O \xrightarrow{\text{Glutaminase}} \text{Glutamate} + NH_3$$

Conditions: T = 37°C, pH = 4.9, A340nm, Light path = 1 cm

Method: Spectrophotometric Stop Rate Determination

Reagents

A. 100 mM Sodium Acetate Buffer, pH 4.9 at 25°C (Prepare 100 ml in deionized water using Sodium Acetate, Trihydrate. Adjust to pH 4.9 at 25°C with 1 M HCl).
B. 5 mM Sodium Acetate Buffer, pH 6.0 at 25°C (Prepare 10 ml in deionized water using Sodium Acetate, Trihydrate. Adjust to pH 6.0 at 25°C with 1 M HCl).
C. 80 mM L-Glutamine Solution (Prepare 10 ml in Reagent A using L-Glutamine. *Prepare Fresh*).
D. Glutaminase Solution (Immediately before use, prepare a solution containing 5 units/ml of Glutaminase in cold Reagent B).
E. Ammonia Reagent.
F. L-Glutamate Dehydrogenase.

Procedure

Step 1: Pipette (in milliliters) the following reagents into suitable tubes:

	Test	Blank
Reagent A (Sodium Acetate Buffer)	0.4	0.4
Reagent C (L-Glutamine)	0.5	0.5

Equilibrate to 37°C. Then add:

Reagent D (Glutaminase)	0.1	—
Reagent B (Sodium Acetate Buffer)	—	0.1

Immediately mix by swirling and incubate at 37°C for 15 minutes. Then add:

Cold Deionized Water	9.0	9.0

Immediately use as Test and Blank solutions in Step 2.

Step 2: Pipette (in milliliters) the following reagents into suitable cuvettes:

Reagent E (Ammonia Reagent)	1.0	1.0

Equilibrate to 25°C. Then add:

	Test	Blank
Test Solution (10 ml)	0.1	—
Blank Solution (10 ml)	—	0.1

Monitor the A340nm until constant, using a suitably thermostatted spectrophotometer. Record the initial A340nm for both Test and Blank. Then add:

Reagent F (L-Glutamate Dehydrogenase)	0.01	0.01

Immediately mix by inversion and record the increase in A340nm until complete (approximately 5 minutes). Obtain the final A340nm for both the Test and Blank.

Calculations

ΔA340nm Test = A340nm Test Final–A340nm Test Initial
ΔA340nm Blank = A340nm Blank Final–A340nm Blank Initial

$$\text{Units / mg enzyme} = \frac{\Delta\text{A340nm Test} - \Delta\text{A340nm Blank (10 ml) (1.11)}}{\text{(6.22) (15) (mg enzyme / RM) (0.1 ml)}}$$

6.22 = Millimolar extinction coefficient of β-NADH at 340nm
15 = Reaction time of Step 1
RM = Reaction Mixture of Step 1
10 = Final volume of Step 1
0.1 = Volume from Step 1 used in Step 2
1.11 = Step 2 Reaction Mix volume

Unit Definition: *One unit will deaminate 1.0 μmole of L-glutamine per minute at 37°C at pH 4.9.*

Final Assay Concentration

In a 1.0 ml reaction mix, the final concentrations are 90 mM sodium acetate, 40 mM L-glutamine and 0.5 units glutaminase.

Note

1. This assay is not for the use with glutaminase from porcine kidney.

Enzymatic Assay of GLUTAMINASE (EC 3.5.1.2)

Principle

$$\text{Glutamine} + H_2O \xrightarrow{\text{Glutaminase}} \text{Glutamic Acid} + NH_3$$

$$\text{Glutamic Acid} + \beta\text{-NAD} \xrightarrow{\text{GDH}} 2\text{-Ketoglutarate} + \beta\text{-NADH} + NH_3$$

Abbreviations Used

β-NAD = β-Nicotinamide Adenine Dinucleotide, Oxidized Form
GDH = Glutamic Dehydrogenase
β-NADH = β-Nicotinamide Adenine Dinucleotide, Reduced Form

Conditions: T = 37°C, pH = 8.6, A340nm, Light path = 1 cm

Method: Spectrophotometric Stop Rate Determination

Reagents

A. 206 mM Potassium Phosphate and 63 mM Tris HCl Buffer with 0.25 mM Ethylenediaminetetraacetic Acid, pH 8.6 at 37°C (Phos/Tris) (Prepare 100 ml in deionized water using Potassium Phosphate, Dibasic, Trihydrate, Trizma Base and Ethylenediaminetetraacetic Acid. Adjust to pH 8.6 at 37°C with 1 M HCl).

B. 100 mM L-Glutamine Solution (Glutamine) (Prepare 10 ml in deionized water using L-Glutamine. Store on ice. *Prepare Fresh*).

C. 3000 mM Hydrochloric Acid Solution (HCl) (Prepare 30 ml in deionized water using Hydrochloric Acid).

D. 400 mM Potassium Phosphate and 10 mM Boric Acid Buffer, pH 8.0 at 37°C (Phos/Boric Acid) (Prepare 100 ml in deionized water using Potassium Phosphate, Dibasic, Trihydrate and Boric Acid. Adjust to pH 8.0 at 37°C with 1 M H_3PO_4).

E. 130 mM Tris HC1 Buffer, pH 9.4 at 37°C (Tris/HCl) (Prepare 100 ml in deionized water using Trizma Base. Adjust to pH 9.4 at 37°C with 1 M HCl).

F. 20 mM β-Nicotinamide Adenine Dinucleotide Solution (β-NAD) (Prepare 10 ml in deionized water using β-Nicotinamide Adenine Dinucleotide).

G. 20 mM Adenosine 5'-Diphosphate Solution (ADP) (Prepare 10 ml in deionized water using Adenosine 5'-Diphosphate, Sodium Salt).

H. 3.0% (v/v) Hydrogen Peroxide Solution (H_2O_2) (Prepare 10 ml in deionized water using Hydrogen Peroxide, 30% (w/w) Solution).

I. 5.0 mM Glutamic Acid Standard Solution (Glu Std) (Prepare 1 ml in deionized water using L-Glutamic Acid, Hydrochloride. Store in an ice bath. *Prepare Fresh*).

J. Glutaminase Enzyme Solution (Glutaminase) (Immediately before use, prepare a solution containing 0.5–1.0 unit/ml of Glutaminase in Reagent D).

K. L-Glutamic Dehydrogenase Enzyme Solution (GDH) (Immediately before use, prepare a solution containing 70 units/ml of L-Glutamic Dehydrogenase in cold deionized water).

Procedure

Step 1: Pipette (in milliliters) the following reagents into suitable containers:

	Test	Blank
Reagent A (Phos/Tris)	0.70	0.70
Reagent B (Glutamine)	0.20	0.20

Mix by inversion and equilibrate to 37°C. Then add:

Reagent J (Glutaminase)	0.10	—
Reagent D (Phos/Boric Acid)	—	0.10

Immediately mix by inversion and incubate for exactly 30 minutes at 37°C. Then add:

Reagent C (HCl)	0.10	0.10

Mix by vortexing and cool to 0–5°C in an ice bath for 5 minutes.

Step 2: Pipette (in milliliters) the following reagents into suitable containers:

	Test	Blank	Std	Std Blank
Reagent E (Tris/HCl)	2.25	2.25	2.25	2.25
Reagent F (β-NAD)	0.30	0.30	0.30	0.30
Reagent G (ADP)	0.06	0.06	0.06	0.06
Reagent H (H_2O_2)	0.03	0.03	0.03	0.03
Reagent K (GDH)	0.30	0.30	0.30	0.30
Reagent I (Glu Std)	—	—	0.06	—
Deionized Water	—	—	—	0.06

Mix by inversion and equilibrate to 37°C. Monitor the A340nm until constant using a suitable spectrophotometer and record the initial A340nm for the Test, Blank, Standard, and Standard Blank. Then add:

Test Solution (Step 1)	0.06	—	—	—
Blank Solution (Step 1)	—	0.06	—	—

Mix by inversion and record the increase in A340nm until complete (approximately 30 minutes). Record the final A340nm for the Test, Blank, Standard and Standard Blank.

Calculations

$$\Delta A340nm\ Test = A340nm\ Test\ Final - A340nm\ Test\ Initial$$

$$\Delta A340nm\ Blank = A340nm\ Blank\ Final - A340nm\ Blank\ Initial$$

$$\text{Units / mg enzyme} = \frac{(A340nm\ Test - A340nm\ Blank)\ (1.1)\ (3.0)}{(6.22)\ (mg\ enzyme / RM)\ (0.06)\ (30)}$$

1.1 = Volume of assay (ml) in Step 1
3.0 = Volume of assay (ml) in Step 2
6.22 = Millimolar extinction coefficient of β-NADH at 340nm.
RM = Reaction Mix
0.06 = Volume of Step 1 used in Step 2
30 = Time of assay (in minutes) as per the Unit Definition

$$\Delta A340nm\ Standard = A340nm\ Standard - A340nm\ Standard\ Blank$$

The theoretical ΔA340nm for the Standard is 0.62. It should be between 0.560 and 0.680.

Unit Definition: *One unit will deaminate 1.0 μmole of L-glutamine per min at pH 8.6 at 37°C.*

Final Assay Concentration

In a 1.00 ml reaction mix, the final concentrations are 184 mM potassium phosphate, 44 mM Tris, 0.18 mM Ethylenediaminetetraacetic Acid, 1.0 mM boric acid, 20 mM glutamine, and 0.05–0.10 unit glutaminase.

Enzymatic Assay of GLUTAMINE SYNTHETASE (EC 6.3.1.2)

Principle

$$\text{Glutamate} + NH_4^+ + \text{ATP} \xrightarrow{\text{Glutamine Synthetase}} \text{L-Glutamine} + \text{ADP} + \text{Pi}$$

$$\text{ADP} + \text{Phospho(enol) pyruvate} \xrightarrow{\text{Pyruvate Kinase}} \text{ATP} + \text{Pyruvate}$$

$$\text{Pyruvate} + \beta\text{-NADH} \xrightarrow{\text{L-Lactic Dehydrogenase}} \text{L-Lactate} + \beta\text{-NAD}$$

Abbreviations Used

ATP = Adenosine 5'-Triphosphate
ADP = Adenosine 5'-Diphosphate
Pi = Inorganic Phosphate
β-NADH = β-Nicotinamide Adenine Dinucleotide, Reduced Form
β-NAD = β-Nicotinamide Adenine Dinucleotide, Oxidized Form

Conditions: T = 37°C, pH = 7.1, A340nm, Light path = 1 cm

Method: Continuous Spectrophotometric Rate Determination

Reagents

A. 100 mM Imidazole HCl Buffer, pH 7.1 at 37°C (Prepare 200 ml in deionized water using Imidazole. Adjust to pH 7.1 at 37°C with 1 M HCl).
B. 3 M Sodium Glutamate Solution (Glu) (Prepare 10 ml in deionized water using L-Glutamic Acid, Monosodium Salt).
C. 250 mM Adenosine 5'-Triphosphate Solution (ATP) (Prepare 5 ml in deionized water using Adenosine 5'-Triphosphate, Disodium Salt. *Prepare Fresh*).
D. 33 mM Phospho(enol) pyruvate Solution (PEP) (Prepare 10 ml in deionized water using Phospho(enol) pyruvate, Trisodium Salt, Hydrate. *Prepare Fresh*).
E. 900 mM Magnesium Chloride Solution ($MgCl_2$) (Prepare 10 ml in deionized water using Magnesium Chloride, Hexahydrate).
F. 1 M Potassium Chloride Solution (KCl) (Prepare 5 ml in deionized water using Potassium Chloride).
G. 1.2 M Ammonium Chloride Solution (NH_4Cl) (Prepare 5 ml in deionized water using Ammonium Chloride).
H. 12.8 mM β-Nicotinamide Adenine Dinucleotide Solution, Reduced Form (β-NADH) (Dissolve the contents of one 10 mg vial of β-Nicotinamide Adenine Dinucleotide, Reduced Form, Disodium Salt in the appropriate volume of Reagent A. *Prepare Fresh*).
I. PK/LDH Enzymes Solution (PK/LDH) (Use PK/LDH Enzymes Solution in 50% Glycerol).
J. Glutamine Synthetase Enzyme Solution (Immediately before use, prepare a solution containing 4–8 units/ml of Glutamine Synthetase in cold deionized water).

Procedure

Prepare a Reaction Cocktail by pipetting (in milliliters) the following reagents into a suitable container:

Deionized Water	20.60
Reagent A (Buffer)	17.20
Reagent B (Glu)	1.80
Reagent C (ATP)	1.80
Reagent E ($MgCl_2$)	3.55
Reagent F (KCl)	0.90
Reagent G (NH_4Cl)	1.80

Mix by stirring and adjust to pH 7.1 at 37°C with 0.1 N HCl or 0.1 N NaOH, if necessary. Pipette (in milliliters) the following reagents into a suitable cuvettes:

	Test	Blank
Reaction Cocktail	2.70	2.70
Reagent D (PEP)	0.10	0.10
Reagent H (β-NADH)	0.06	0.06

Mix by inversion and equilibrate to 37°C. Monitor the A340nm until constant, using a suitably thermostatted spectrophotometer. Then add:

Reagent I (PK/LDH)	0.04	0.04

Mix by inversion and equilbrate to 37°C. Monitor the A340nm until constant, using a suitably thermostatted spectrophotometer. Then add:

Deionized water	—	0.10
Reagent J (Enzyme Solution)	0.10	—

Immediately mix by inversion and record the decrease in A340nm for approximately 10 minutes. Obtain the ΔA340nm/min using the maximum linear rate for both the Test and Blank.

Calculations

$$\text{Units / ml enzyme} = \frac{(\Delta A340nm / min\ Test - \Delta A340nm / min\ Blank)\ (3)\ (15)}{(6.22)\ (0.1)}$$

3 = Total volume (in milliliters) of assay
15 = Conversion factor to 15 minutes (Unit Definition)
6.22 = Millimolar extinction coefficient of β-NADH at 340nm
0.1 = Volume (in milliliter) of enzyme used

$$\text{Units / mg solid} = \frac{\text{units / ml enzyme}}{\text{mg solid / ml enzyme}}$$

$$\text{Units / mg protein} = \frac{\text{units / ml enzyme}}{\text{mg protein / ml enzyme}}$$

Unit Definition: *One unit will convert 1.0 μmole of L-glutamate to L-glutamine in 15 minutes at pH 7.1 at 37°C.*

Final Assay Concentrations

In a 3.00 ml reaction mix, the final concentrations are 34.1 mM imidazole, 102 mM sodium glutamate, 8.5 mM adenosine 5'-triphosphate, 1.1 mM phosphoenolpyruvate, 60 mM magnesium chloride, 18.9 mM potassium chloride, 45 mM ammonium chloride, 0.25 mM β-nicotinamide adenine dinucleotide, 28 units pyruvate kinase, 40 units L-lactic dehydrogenase and 0.4–0.8 unit glutamine synthetase.

Notes

1. Contains approximately 700 units/ml pyruvate kinase and 1,000 units/ml lactic dehydrogenase.
2. L-Lactic Dehydrogenase Unit Definition: One unit will reduce 1.0 μmole of pyruvate to L-lactate per minute at pH 7.5 at 37°C.
3. Pyruvate Kinase Unit Definition: One unit will convert 1.0 μmole of phospho(enol) pyruvate to pyruvate per minute at pH 7.6 at 37°C.

Enzymatic Assay of GLUTATHIONE PEROXIDASE (EC 1.11.1.9)

Principle

$$2GSH + H_2O_2 \xrightarrow{\text{Glutathione Peroxidase}} GSSG + 2H_2O$$

$$GSSG + \beta\text{-NADPH} \xrightarrow{\text{Glutathione Reductase}} \beta\text{-NADP} + 2GSH$$

Abbreviations Used

GSH = Glutathione, Reduced Form
GSSG = Glutathione, Oxidized Form
β-NADPH = β-Nicotinamide Adenine Dinucleotide Phosphate, Reduced Form
β-NADP = β-Nicotinamide Adenine Dinucleotide Phosphate, Oxidized Form

Conditions: T = 25°C, pH = 7.01, A340nm, Light path = 1 cm

Method: Continuous Spectrophotometric Rate Determination

Reagents

A. 50 mM Sodium Phosphate Buffer with 0.40 mM Ethylenediaminetetra acetic Acid (EDTA), pH 7.0 at 25°C (Prepare 100 ml in deionized water using Sodium Phosphate, Monobasic, Anhydrous and Ethylenediaminetetraacetic Acid, Tetrasodium Salt. Adjust to pH 7.0 at 25°C with 1 M NaOH).
B. 1.0 mM Sodium Azide Solution (Buffer w/Azide) (Prepare 50 ml in Reagent A using Sodium Azide).
C. β-Nicotinamide Adenine Dinucleotide Phosphate, Reduced Form (β-NADPH) (Use 1.0 mg vial of β-Nicotinamide Adenine Dinucleotide Phosphate, Reduced Form, Tetrasodium Salt).
D. Glutathione Reductase Enzyme Solution (GR) (Immediately before use, prepare a solution containing 100 units/ml of Glutathione Reductase, in cold deionized water).
E. 200 mM Glutathione, Reduced (GSH) (Prepare 5 ml in deionized water using Glutathione, Free Acid, Reduced Form).
F. 10.0 mM Sodium Phosphate Buffer with 1.0 mM Dithiothreitol, pH 7.0 at 25°C (Buffer w/DTT) (Prepare 100 ml in deionized water using Sodium Phosphate,

Monobasic, Anhydrous and DL-Dithiothreitol. Adjust to pH 7.0 at 25°C with 1 M NaOH).

G. Glutathione Peroxidase Enzyme Solution (Immediately before use, prepare a solution containing 1.5–3.0 units/ml of Glutathione Peroxidase in cold Reagent F).

H. 0.042% (w/w) Hydrogen Peroxide (H_2O_2) (Prepare 5 ml in deionized water using Hydrogen Peroxide, 30% (w/w) Solution. *Prepare Fresh*).

Procedure

Prepare a reaction cocktail by pipetting (in milliliters) the following reagents into Reagent C (β-NADPH vial):

Reagent B (Buffer w/Azide)	9.20
Reagent D (GR)	0.10
Reagent E (GSH)	0.05

Mix by inversion and adjust to pH 7.0 at 25°C with 1 M HCl or 1 M NaOH, if necessary. Pipette (in milliliters) the following reagents into suitable cuvettes:

	Test	Blank
Reaction Cocktail	3.00	3.00
Reagent F (Buffer w/DTT)	—	0.05
Reagent G (Glutathione Peroxidase)	0.05	—

Mix by inversion and equilibrate to 25°C. Monitor the A340nm until constant, using a suitably thermostatted spectrophotometer. Then add:

Reagent H (H_2O_2)	0.05	0.05

Immediately mix by inversion and record the decrease in A340nm for approximately 5 minutes. Obtain the ΔA340nm/minute using the maximum linear rate for both the Test and Blank.

Calculations

$$\text{Units / ml enzyme} = \frac{(\Delta\text{A340nm / min Test} - \Delta\text{A340nm / min Blank})\,(2)\,(3.1)\,(\text{df})}{(6.22)\,(0.05)}$$

2 = 2 μmoles of GSH produced per μmole of β-NADPH oxidized
3.1 = Total volume (in milliliters) of assay
df = Dilution factor
6.22 = Millimolar extinction coefficient of β-NADPH at 340nm
0.05 = Volume (in milliliters) of enzyme used

$$\text{Units / mg solid} = \frac{\text{units / ml enzyme}}{\text{mg solid / ml enzyme}}$$

$$\text{Units / mg protein} = \frac{\text{units / ml enzyme}}{\text{mg protein / ml enzyme}}$$

Unit Definition: *One unit will catalyze the oxidation by H_2O_2 of 1.0 μmole of reduced glutathione to oxidized glutathione per minute at pH 7.0 at 25°C.*

Final Assay Concentration

In a 3.05 ml reaction mix, final concentrations are 48 mM sodium phosphate, 0.38 mM Ethylenediaminetetraacetic Acid, 0.12 mM β-nicotinamide adenine dinucleotide phosphate, reduced form, 0.95 mM sodium azide, 3.2 units of glutathione reductase, 1 mM glutathione, 0.02 mM DL-dithiothreitol, 0.0007% (w/w) hydrogen peroxide and 0.075–0.15 unit of glutathione peroxidase.

Notes

1. The enzyme activity at the pH optimum of 8.8 is almost 10 times that at pH 7.0. However, non-enzymatic oxidation of GSH increases at alkaline pH, complicating the assay considerably.
2. Sodium Azide is included to inhibit the competitive catalase activity of impurities such as haemoglobin. If haemoglobin is not present, then sodium azide is not required.
3. Hydrogen peroxidase, 30% (w/w) solution, has a limited shelf-life. The A340nm of a 0.042% (w/w) hydrogen peroxide solution should be 0.540.
4. Glutathione Reductase Unit Definition: One unit will reduce 1.0 μmole of oxidized glutathione per minute at pH 7.6 at 25°C.

Enzymatic Assay of GLUTATHIONE REDUCTASE (EC 1.6.4.2)

Principle

$$\beta\text{-NADPH} + \text{GSSG} \xrightarrow{\text{Glutathione Reductase}} \beta\text{-NADP} + 2\text{GSH}$$

Abbreviations Used

β-NADPH = β-Nicotinamide Adenine Dinucleotide Phosphate, Reduced Form
β-NADP = β-Nicotinamide Adenine Dinucleotide Phosphate, Oxidized Form
GSSG = Glutathione, Oxidized Form
GSH = Glutathione, Reduced Form

Conditions: T = 25°C, pH = 7.6, A340nm, Light Path = 1 cm

Method: Continuous Spectrophotometric Rate Determination

Reagents

A. 100 mM Potassium Phosphate Buffer with 3.4 mM Ethylenediaminetetraacetic Acid (EDTA), pH 7.6 at 25°C (Prepare 200 ml in deionized water using Potassium Phosphate, Monobasic, Anhydrous and Ethylenediaminetetraacetic Acid, Dipotassium Salt. Adjust to pH 7.6 at 25°C with 1 M KOH).
B. 30 mM Glutathione Substrate Solution (GSSG) (Prepare 5 ml in deionized water using Glutathione, Oxidized Form, Disodium Salt).
C. 0.8 mM β-Nicotinamide Adenine Dinucleotide Phosphate, Reduced Form Solution (β-NADPH) (Prepare 5 ml in cold Reagent A using β-Nicotinamide Adenine Dinucleotide Phosphate, Tetrasodium Salt).

D. 1.0% (w/v) Bovine Serum Albumin (BSA) (Prepare 100 ml in Reagent A using Albumin, Bovine. This solution should be kept cold).

E. Glutathione Reductase Enzyme Solution (Immediately before use, prepare a solution containing 0.30–0.60 unit/ml of Glutathione Reductase in cold Reagent D).

Procedure

Pipette (in milliliters) the following reagents into suitable cuvettes:

	Test	Blank
Deionized Water	0.65	0.65
Reagent A (Buffer)	1.50	1.50
Reagent B (GSSG)	0.10	0.10
Reagent C (β-NADPH)	0.35	0.35
Reagent D (BSA)	0.30	0.40

Mix by inversion and equilibrate to 25°C. Monitor the A340nm until constant, using a suitably thermostatted spectrophotometer. Then add:

Reagent E (Enzyme Solution)	0.10	—

Immediately mix by inversion and record the decrease in the A340nm for approximately 5 minutes. Obtain the ΔA340nm / minute using the maximum linear rate for both the Test and Blank.

Calculations

$$\text{Units / ml enzyme} = \frac{(\Delta A340nm / \text{min Test} - \Delta A340nm / \text{min Blank})\,(3)\,(df)}{(6.22)\,(0.1)}$$

3 = Total volume (in milliliters) of assay
df = Dilution factor
6.22 = Millimolar extinction coefficient of β-NADPH at 340nm
0.1 = Volume (in milliliters) of enzyme used

$$\text{Units / mg solid} = \frac{\text{units / ml enzyme}}{\text{mg solid / ml enzyme}}$$

$$\text{Units / mg protein} = \frac{\text{units / ml enzyme}}{\text{mg protein / ml enzyme}}$$

Unit Definition: *One unit will reduce 1.0 µmole of oxidized glutathione per minute at pH 7.6 at 25°C.*

Final Assay Concentration

In a 3.00 ml reaction mix, the final concentrations are 75 mM potassium phosphate, 2.6 mM Ethylenediaminetetraacetic Acid, 1 mM glutathione, 0.09 mM β-nicotinamide adenine dinucleotide phosphate, reduced form, 0.13% (w/v) Bovine Serum Albumin, and 0.03–0.06 unit of glutathione reductase.

Enzymatic Assay of GLUTATHIONE REDUCTASE (EC 1.6.4.2) Coenzyme A-Glutathione Reductase Activity

Principle

$$\text{CoA-S-S-G} + \text{NADPH} \xrightarrow{\text{Glutathione Reductase}} \text{CoA-SH} + \text{GSH} + \text{NADP}$$

Abbreviations Used

CoA-S-S-G = Coenzyme A Glutathione Disulfide
NADPH = Nicotinamide Adenine Nucleotide Phosphate, Reduced Form
CoA-SH = Coenzyme A, Reduced Form
GSH = Glutathione, Reduced Form
NADP = Nicotinamide Adenine Dinucleotide Phosphate, Oxidized Form

Conditions: T = 25°C, pH = 5.5, A340nm, Light path = 1 cm

Method: Continuous Spectrophotometric Rate Determination

Reagents

A. 75 mM Potassium Phosphate Buffer with 0.15% (w/v) Bovine Serum Albumin, pH 5.5 at 25°C (Prepare 100 ml in deionized water using Potassium Phosphate, Monobasic, Anhydrous and Albumin, Bovine. Adjust to pH 5.5 at 25°C with 2 N NaOH).
B. 6.0 mM Coenzyme A Glutathione Disulfide Solution (CoA-S-S-G) (Prepare 5 ml in deionized water using Coenzyme A Glutathione Disulfide, Sodium Salt).
C. 4.5 mM β-Nicotinamide Adenine Dinucleotide Phosphate, Reduced Form Solution (β-NADPH) (Dissolve the contents of one 5 mg vial of β-Nicotinamide Adenine Dinucleotide Phosphate, Reduced Form, Tetrasodium Salt, in the appropriate volume of deionized water or prepare 1 ml in deionized water using β-Nicotinamide Adenine Dinucleotide Phosphate, Reduced Form, Tetrasodium Salt).
D. Glutathione Reductase Enzyme Solution (Immediately before use, prepare a solution containing 0.25–0.50 unit/ml of Glutathione Reductase in cold deionized water).

Procedure

Pipette (in milliliters) the following reagents into suitable cuvettes:

	Test	Blank
Reagent A (Buffer)	2.00	2.00
Reagent B (CoA-S-S-G)	0.50	0.50
Reagent C (β-NADPH)	0.10	0.10
Deionized Water	0.30	0.40

Mix by inversion and equilibrate to 25°C. Monitor the A340nm until constant, using a suitably thermostatted spectrophotometer. Then add:

Reagent D (Enzyme Solution)	0.10	—

Immediately mix by inversion and record the decrease in the A340nm for approximately 5 minutes. Obtain the ΔA340nm/minute using the maximum linear rate for both the Test and Blank.

Calculations

$$\text{Units / ml enzyme} = \frac{(\Delta A340nm / min\ Test - \Delta A340nm / min\ Blank)\ (3)\ (df)}{(6.22)\ (0.1)}$$

3 = Total volume (in milliliters) of assay
df = Dilution factor
6.22 = Millimolar extinction coefficient of β-NADPH at 340nm
0.1 = Volume (in milliliter) of enzyme used

$$\text{Units / mg solid} = \frac{\text{units / ml enzyme}}{\text{mg solid / ml enzyme}}$$

$$\text{Units / mg protein} = \frac{\text{units / ml enzyme}}{\text{mg protein / ml enzyme}}$$

Unit Definition: *One unit will reduce 1.0 µmole of CoA-S-S-G per minute at pH 5.5 at 25°C.*

Final Assay Concentrations

In a 3.00 ml reaction mix, the final concentrations are 50 mM potassium phosphate, 1.0 mM coenzyme A glutathione disulfide, 0.15 mM β-nicotinamide adenine dinucleotide phosphate, reduced form, 0.01% (w/v) Bovine Serum Albumin, and 0.025–0.050 unit glutathione reductase.

Enzymatic Assay of GLUTATHIONE-S-TRANSFERASE (EC 2.5.1.18) p-Nitrobenzyl Chloride as Substrate

Principle

$$\text{G-SH} + \text{PNBC} \xrightarrow{\text{Glutathione-S-Transferase}} \text{G-PNBC Conjugate}$$

Abbreviations Used

G-SH = Glutathione, Reduced Form
PNBC = p-Nitrobenzyl Chloride
G-PNBC Conjugate = Glutathione p-Nitrobenzyl Chloride Conjugate

Conditions: T = 25°C, pH = 6.5, A310nm, Light path = 1 cm

Method: Continuous Spectrophotometric Rate Determination

Reagents

A. 100 mM Potassium Phosphate Buffer with 1.0 mM Ethylenediaminetetraacetic Acid, pH 6.5 at 25°C (Prepare 100 ml in deionized water using Potassium Phosphate, Monobasic, Anhydrous and Ethylenediaminetetraacetic Acid, Tetrasodium Salt: Hydrate. Adjust to pH 6.5 at 25°C with 1 M KOH).

B. 150 mM Glutathione, Reduced Solution (G-SH) (Prepare 10 ml in Reagent A using Glutathione, Free Acid, Reduced Form. *Prepare Fresh*).
C. 30 mM p-Nitrobenzyl Chloride Solution (PNBC) (Prepare 1 ml in Reagent D using p-Nitrobenzyl Chloride. *Prepare Fresh*).
D. 95% Ethanol (Nondenatured)
E. Glutathione-S-Transferase Enzyme Solution (Immediately before use, prepare a solution containing 0.4–0.8 units/ml of Glutathione-Stransferase in cold Reagent A).

Procedure

Pipette (in milliliters) the following reagents into suitable quartz cuvettes:

	Test	Blank
Reagent A (Buffer)	2.70	2.80
Reagent B (G-SH)	0.10	0.10
Reagent C (PNBC)	0.10	0.10

Equilibrate to 25°C. Monitor the A310nm until constant, using a suitably thermostatted spectrophotometer. Then add:

Reagent E (Enzyme Solution)	0.10	—

Immediately mix by inversion and record the increase in A310nm for approximately 5 minutes. Obtain the ΔA310nm/minute using the maximum linear rate for both the Test and Blank.

Calculations

$$\text{Units / ml enzyme} = \frac{(\Delta A310\text{nm / min Test} - \Delta A310\text{nm / min Blank})\,(3)\,(\text{df})}{(1.9)\,(0.1)}$$

3 = Total volume (in milliliters) of assay
df = Dilution factor
1.9 = Millimolar extinction coefficient of Glutathione-p-Nitrobenzyl Conjugate at 310nm
0.1 = Volume (in milliliter) of enzyme used

$$\text{Units / mg solid} = \frac{\text{units / ml enzyme}}{\text{mg solid / ml enzyme}}$$

$$\text{Units / mg protein} = \frac{\text{units / ml enzyme}}{\text{mg protein / ml enzyme}}$$

Unit Definition: *One unit will conjugate 1.0 μmole of p-nitrobenzyl chloride with reduced glutathione per minute at pH 6.5 at 25°C.*

Final Assay Concentration

In a 3.00 ml reaction mix, the final concentrations are 97 mM potassium phosphate, 0.97 mM EDTA, 5.0 mM glutathione, reduced, 1.0 mM p-nitrobenzyl chloride, 3.2% (v/v) ethanol and 0.04–0.08 unit glutathione-S-transferase.

Enzymatic Assay of GLUTATHIONE-S-TRANSFERASE (EC 2.5.1.18) 1,2-Epoxy-3-(p-Nitrophenoxy) propane as Substrate

Principle

$$\text{EPNPP} + \text{G-SH} \xrightarrow{\text{Glutathione-S-Transferase}} \text{G-EPNPP Conjugate}$$

Abbreviations Used

G-SH = Glutathione, Reduced Form
EPNPP = 1,2-Epoxy-3-(p-Nitrophenoxy)propane
G-EPNPP Conjugate = Glutathione 1,2-Epoxy-3-(p-Nitrophenoxy)propane Conjugate

Conditions: T = 25°C, pH = 6.5, A360nm, Light path = 1 cm

Method: Continuous Spectrophotometric Rate Determination

Reagents

A. 100 mM Potassium Phosphate Buffer with 1.0 mM Ethylenediaminetetraacetic Acid, pH 6.5 at 25°C (Prepare 100 ml in deionized water using Potassium Phosphate, Monobasic, Anhydrous and Ethylenediaminetetraacetic Acid, Tetrasodium Salt: Hydrate. Adjust to pH 6.5 at 25°C with 1 M KOH).
B. 150 mM Glutathione, Reduced Solution (G-SH) (Prepare 10 ml in Reagent A using Glutathione, Free Acid, Reduced Form. *Prepare Fresh*).
C. 15 mM 1,2-Epoxy-3-(p-Nitrophenoxy)propane Solution (EPNPP) (Prepare 1 ml in Reagent D using 1,2-Epoxy-3-(p-Nitrophenoxy)propane. *Prepare Fresh*).
D. 95% Ethanol (Nondenatured).
E. Glutathione-S-Transferase Enzyme Solution (Immediately before use, prepare a solution containing 1.0–2.0 units/ml of Glutathione-Stransferase in cold Reagent A).

Procedure

Pipette (in milliliters) the following reagents into suitable cuvettes:

	Test	Blank
Reagent A (Buffer)	2.70	2.80
Reagent B (G-SH)	0.10	0.10
Reagent C (EPNPP)	0.10	0.10

Equilibrate to 25°C. Monitor the A360nm until constant, using a suitably thermostatted spectrophotometer. Then add:

Reagent E (Enzyme Solution)	0.10	—

Immediately mix by inversion and record the increase in A360nm for approximately 5 minutes. Obtain the ΔA360nm/minute using the maximum linear rate for both the Test and Blank.

Calculations

$$\text{Units / ml enzyme} = \frac{(\Delta A360nm / \text{min Test} - \Delta A360nm / \text{min Blank})(3)(df)}{(0.5)(0.1)}$$

3 = Total volume (in milliliters) of assay
df = Dilution factor
0.5 = Millimolar extinction coefficient of Glutathione-EPNPP conjugate at 360nm
0.1 = Volume (in milliliter) of enzyme used

$$\text{Units / mg solid} = \frac{\text{units / ml enzyme}}{\text{mg solid / ml enzyme}}$$

$$\text{Units / mg protein} = \frac{\text{units / ml enzyme}}{\text{mg protein / ml enzyme}}$$

Unit Definition: *One unit will conjugate 1.0 μmole of 1,2-epoxy-3-(p-nitrophenoxy)propane with reduced glutathione per minute at pH 6.5 at 25°C.*

Final Assay Concentration

In a 3.00 ml reaction mix, the final concentrations are 97 mM potassium phosphate, 0.97 mM EDTA, 5.0 mM glutathione, reduced, 0.50 mM 1,2-epoxy-3-(p-nitrophenoxy)propane, 3.2% (v/v) ethanol and 0.10–0.20 unit glutathione-S-transferase.

Enzymatic Assay of GLYCERALDEHYDE-3-PHOSPHATE DEHYDROGENASE (EC 1.2.1.12) from *Bacillus stearothermophilus*

Principle

$$\text{3-PGA} + \text{ATP} \xrightarrow{\text{3-PGK}} \text{Glycerate-1, 3 Diphosphate} + \text{ADP}$$

$$\text{Glycerate-1, 3 Diphosphate} + \beta\text{-NADH} \xrightarrow{\text{GAPDH}} \text{G-3-P} + \beta\text{-NAD} + \text{Pi}$$

Abbreviations Used

3-PGA = 3-Phosphoglyceric Acid
ATP = Adenosine 5'-Triphosphate
3-PGK = 3-Phosphoglyceric Phosphokinase
ADP = Adenosine 5'-Diphosphate
β-NADH = β-Nicotinamide Adenine Dinucleotide, Reduced Form
GAPDH = Glyceraldehyde-3-Phosphate Dehydrogenase
G-3-P = Glyceraldehyde 3-Phosphate
β-NAD = β-Nicotinamide Adenine Dinucleotide, Oxidized Form
Pi = Inorganic Phosphate

Conditions: T = 30°C, pH = 7.6, A340nm, Light path = 1 cm

Method: Continuous Spectrophotometric Rate Determination

Reagents

A. 100 mM Triethanolamine Buffer, pH 7.6 at 30°C (Prepare 100 ml in deionized water using Triethanolamine Hydrochloride. Adjust to pH 7.6 at 30°C with 1 M NaOH).

B. 100 mM 3-Phosphoglyceric Acid Solution (3-PGA) (Prepare 2 ml in deionized water using D(-)3-Phosphoglyceric Acid, Tri(cyclohexylammonium) Salt).

C. 200 mM L-Cysteine HCl Solution (Cys) (Prepare 1 ml in deionized water using L-Cysteine Hydrochloride, Monohydrate. Neutralize the solution by adding solid Sodium Bicarbonate. *Prepare Fresh*).

D. 100 mM Magnesium Sulfate Solution ($MgSO_4$) (Prepare 10 ml in deionized water using Magnesium Sulfate, Heptahydrate).

E. 7.0 mM β-Nicotinamide Adenine Dinucleotide, Reduced Form Solution (β-NADH) (Prepare 1 ml in deionized water using β-Nicotinamide Adenine Dinucleotide, Reduced Form, Disodium Salt or dissolve the contents of one 5 mg vial of β-Nicotinamide Adenine Dinucleotide, Reduced Form, Disodium Salt in the appropriate volume of deionized water. *Prepare Fresh*).

F. 34 mM Adenosine 5'-Triphosphate Solution (ATP) (Prepare 1 ml in deionized water using Adenosine 5'-Triphosphate, Disodium Salt. *Prepare Fresh*).

G. 3-Phosphoglyceric Phosphokinase Enzyme Solution (3- PGK) Immediately before use, prepare a solution containing 100 units/ml in cold deionized water using 3-Phosphoglyceric Phosphokinase).

H. Glyceraldehyde-3-Phosphate Dehydrogenase Enzyme Solution (GAPDH) (Immediately before use, prepare a solution containing 0.2–0.4 unit/ml of Glyceraldehyde-3-Phosphate Dehydrogenase in cold Reagent A).

Procedure

Pipette (in milliliters) the following reagents into suitable cuvettes:

	Test	Blank
Reagent A (Buffer)	2.40	2.50
Reagent B (3-PGA)	0.20	0.20
Reagent F (ATP)	0.10	0.10
Reagent D ($MgSO_4$)	0.05	0.05
Reagent E (β-NADH)	0.05	0.05
Reagent C (Cys)	0.05	0.05

Mix by inversion and equilibrate to 30°C. Then add:

Reagent G (3-PGK)	0.05	0.05

Mix by inversion and equilibrate to 30°C. Monitor the A340nm until constant, using a suitably thermostatted spectrophotometer. Then add:

Reaction H (GAPDH)	0.10	—

Immediately mix by inversion and record the decrease in A340nm for approximately 5 minutes. Obtain the ΔA340nm/minute using the maximum linear rate for both the Test and Blank.

Calculations

$$\text{Units / ml enzyme} = \frac{(\Delta A340\text{nm / min Test} - \Delta A340\text{nm / min Blank})\ (3)\ (\text{df})}{(6.22)\ (0.1)}$$

3 = Total volume (in milliliters) of assay
df = Dilution factor
6.22 = Millimolar extinction coefficient of β-NADH at 340nm
0.1 = Volume (in milliliter) of assay

$$\text{Units / mg solid} = \frac{\text{units / ml enzyme}}{\text{mg solid / ml enzyme}}$$

$$\text{Units / mg protein} = \frac{\text{units / ml enzyme}}{\text{mg protein / ml enzyme}}$$

Unit Definition: *One unit will reduce 1.0 µmole of 3-phosphoglycerate to D-glyceraldehyde 3-phosphate per minute in a coupled system with 3-phosphoglyceric phosphokinase at pH 7.6 at 30°C.*

Final Assay Concentration

In a 3.00 ml reaction mix, the final concentrations are 83 mM triethanolamine, 6.7 mM 3-phosphoglyceric acid, 3 mM L-cysteine, 2 mM magnesium sulfate, 0.1 mM β-nicotinamide adenine dinucleotide, reduced form, 1.1 mM adenosine 5'-triphosphate, 5 units 3-phosphoglyceric phosphokinase and 0.02–0.04 unit glyceraldehyde-3-phosphate dehydrogenase.

Notes

1. Adjust the concentration of the enzyme so that the A340nm/minute is less than 0.09.
2. Do not change the order in which the reagents are added to the cuvettes.
3. 3-Phosphoglyceric Phosphokinase Unit Definition: One unit will convert 1.0 µmole of 1,3-diphosphoglycerate to 3-phosphoglycerate per minute at pH 6.9 at 25°C.

Enzymatic Assay of GLYCERALDEHYDE-3-PHOSPHATE DEHYDROGENASE (EC 1.2.1.12)

Principle

$$\text{3-PGA} + \text{ATP} \xrightarrow{\text{3-PGK}} \text{Glycerate-1, 3 Diphosphate} + \text{ADP}$$

$$\text{Glycerate-1, 3 Diphosphate} + \beta\text{-NADH} \xrightarrow{\text{GAPDH}} \text{G-3-P} + \beta\text{NAD} + \text{Pi}$$

Abbreviations Used

3-PGA = 3-Phosphoglyceric Acid
ATP = Adenosine 5'-Triphosphate
3-PGK = 3-Phosphoglyceric Phosphokinase
ADP = Adenosine 5'-Diphosphate
β-NADH = β-Nicotinamide Adenine Dinucleotide, Reduced Form
GAPDH = Glyceraldehyde-3-Phosphate Dehydrogenase
G-3-P = Glyceraldehyde 3-Phosphate

β-NAD = β-Nicotinamide Adenine Dinucleotide, Oxidized Form
Pi = Inorganic Phosphate

Conditions: T = 25°C, pH = 7.6, A340nm, Light path = 1 cm

Method: Continuous Spectrophotometric Rate Determination

Reagents

A. 100 mM Triethanolamine Buffer, pH 7.6 at 25°C (Prepare 100 ml in deionized water using Triethanolamine Hydrochloride. Adjust to pH 7.6 at 25°C with 1 M NaOH. *Prepare fresh*).
B. 100 mM 3-Phosphoglyceric Acid Solution (3-PGA) (Prepare 2 ml in deionized water using D(-)3-Phosphoglyceric Acid, Tri(cyclohexylammonium) Salt).
C. 200 mM L-Cysteine HCl Solution (Cys) (Prepare 1 ml in deionized water using L-Cysteine Hydrochloride, Monohydrate. Neutralize the solution by adding solid Sodium Bicarbonate. *Prepare Fresh*).
D. 100 mM Magnesium Sulfate Solution ($MgSO_4$) (Prepare 10 ml in deionized water using Magnesium Sulfate, Heptahydrate).
E. 7.0 mM β-Nicotinamide Adenine Dinucleotide, Reduced Form Solution (β-NADH) (Prepare 1 ml in deionized water using β-Nicotinamide Adenine Dinucleotide, Reduced Form, Disodium Salt or dissolve the contents of one 5 mg vial of β-Nicotinamide Adenine Dinucleotide, Reduced Form, Disodium Salt, in the appropriate volume of deionized water. *Prepare Fresh*).
F. 34 mM Adenosine 5'-Triphosphate Solution (ATP) (Prepare 1 ml in deionized water using Adenosine 5'-Triphosphate, Disodium Salt. *Prepare Fresh*).
G. 3-Phosphoglyceric Phosphokinase Enzyme Solution (3-PGK) (Immediately before use, prepare a solution containing 200 units/ml in cold deionized water using 3-Phosphoglyceric Phosphokinase).
H. Glyceraldehyde-3-Phosphate Dehydrogenase Enzyme Solution (GAPDH) (Immediately before use, prepare a solution containing 0.3–0.6 unit/ml of Glyceraldehyde-3-Phosphate Dehydrogenase in cold Reagent A).

Procedure

Pipette (in milliliters) the following reagents into suitable cuvettes:

	Test	Blank
Reagent A (Buffer)	2.40	2.50
Reagent B (3-PGA)	0.20	0.20
Reagent C (Cys)	0.05	0.05
Reagent D ($MgSO_4$)	0.05	0.05
Reagent E (β-NADH)	0.05	0.05
Reagent F (ATP)	0.10	0.10
Reagent G (3-PGK)	0.05	0.05

Mix by inversion and equilibrate to 25°C. Monitor the A340nm until constant, using a suitably thermostatted spectrophotometer. Then add:

	Test	Blank
Reaction H (Enzyme Solution)	0.10	—

Immediately mix by inversion and record the decrease in A340nm for approximately 5 minutes. Obtain the ΔA340nm/minute using the maximum linear rate for both the Test and Blank.

Calculations

$$\text{Units / ml enzyme} = \frac{(\Delta\text{A340nm / min Test} - \Delta\text{A340nm / min Blank})\ (3)\ (\text{df})}{(0.1)\ (6.22)}$$

3 = Volume (in milliliters) of assay
df = Dilution factor
0.1 = Volume (in milliliter) of enzyme used
6.22 = Millimolar extinction coefficient of β-NADH at 340nm

$$\text{Units / mg solid} = \frac{\text{units / ml enzyme}}{\text{mg solid / ml enzyme}}$$

$$\text{Units / mg protein} = \frac{\text{units / ml enzyme}}{\text{mg protein / ml enzyme}}$$

Unit Definition: *One unit will reduce 1.0 μmole of 3-phosphoglycerate to D-glyceraldehyde-3-phosphate per minute in a coupled system with 3-phosphoglyceric phosphokinase at pH 7.6 at 25°C.*

Final Assay Concentration

In a 3.00 ml reaction mix, the final concentrations are 83 mM triethanolamine, 6.7 mM 3-phosphoglyceric acid, 3 mM L-cysteine, 2 mM magnesium sulfate, 0.1 mM β-NADH, 1.1 mM ATP, 10 units 3-phosphoglyceric phosphokinase and 0.03–0.06 unit glyceraldehyde-3-phosphate dehydrogenase.

Notes

1. Not to be used to assay activity of Glyceraldehyde-3-Phosphate Dehydrogenase, from Bacillus Stearothermophilus.
2. 3-Phosphoglyceric Phosphokinase unit definition: One unit will convert 1.0 μmole of 1,3-diphosphoglycerate to 3-phosphoglycerate per minute at pH 6.9 at 25°C.

Enzymatic Assay of GLYCEROL DEHYDROGENASE (EC 1.1.1.6)

Principle

$$\text{Glycerol} + \beta\text{-NAD} \xrightarrow{\text{Glycerol Dehydrogenase}} \text{Dihydroxyacetone} + \beta\text{-NADH}$$

Abbreviations Used

β-NAD = β-Nicotinamide Adenine Dinucleotide, Oxidized Form
β-NADH = β-Nicotinamide Adenine Dinucleotide, Reduced Form

Conditions: T = 25°C, pH 10.0, A340nm, Light path = 1 cm

Method: Continuous Spectrophotometric Rate Determination

Reagents

A. 500 mM Potassium Bicarbonate Solution ($KHCO_3$) (Prepare 100 ml in deionized water using Potassium Bicarbonate).

B. 500 mM Potassium Carbonate Buffer, pH 10.0 at 25°C (Prepare 100 ml in deionized water using Potassium Carbonate. Adjust to pH 10.0 at 25°C with Reagent A ($KHCO_3$)).

C. 1 M Ammonium Sulfate Solution ($(NH_4)_2SO_4$) (Prepare 10 ml in deionized water using Ammonium Sulfate).

D. 10 mM β-Nicotinamide Adenine Dinucleotide, Oxidized Form, Solution (β-NAD) (Prepare 2 ml in deionized water using β-Nicotinamide Adenine Dinucleotide, or dissolve the contents of one 50 mg vial of β-Nicotinamide Adenine Dinucleotide in the appropriate volume of deionized water. *Prepare Fresh)*.

E. 1 M Glycerol Solution (Glycerol) (Prepare 100 ml in deionized water using Glycerol).

F. 50 mM Potassium Phosphate with 0.1% (w/v) Bovine Serum Albumin and 0.05 mM Manganese Chloride Solution, pH 7.5 at 25°C (Enzyme Diluent) (Prepare 100 ml in deionized water using Potassium Phosphate, Monobasic, Anhydrous, Albumin, Bovine, or equivalent and Manganese Chloride, Tetrahydrate. Adjust to pH 7.5 at 25°C with 1 M KOH).

G. Glycerol Dehydrogenase Enzyme Solution (Immediately before use, prepare a solution containing 0.2–0.6 unit/ml of Glycerol Dehydrogenase in cold Reagent F).

Procedure

Pipette (in milliliters) the following reagents into suitable cuvettes:

	Test	Blank
Deionized Water	1.80	1.80
Reagent B (Buffer)	0.60	0.60
Reagent C ($(NH_4)_2SO_4$)	0.10	0.10
Reagent D (β-NAD)	0.10	0.10
Reagent E (Glycerol)	0.30	0.30

Mix by inversion and equilibrate to 25°C. Monitor the A340nm until constant, using a suitably thermostatted spectrophotometer. Then add:

Reagent F (Enzyme Diluent)	—	0.10
Reagent G (Enzyme Solution)	0.10	—

Immediately mix by inversion and record the increase in A340nm for approximately 5 minutes. Obtain the ΔA340nm/minute using the maximum linear rate for both the Test and Blank.

Calculations

$$\text{Units / ml enzyme} = \frac{(\Delta A340\text{nm / min Test} - \Delta A340\text{nm / min Blank})\ (3)\ (df)}{(6.22)\ (0.1)}$$

3 = Total volume (in milliliters) of assay
df = Dilution factor

6.22 = Millimolar extinction coefficient of β-NADH at 340nm
0.1 = Volume (in milliliter) of enzyme used

$$\text{Units / mg solid} = \frac{\text{units / ml enzyme}}{\text{mg solid / ml enzyme}}$$

$$\text{Units / mg protein} = \frac{\text{units / ml enzyme}}{\text{mg protein / ml enzyme}}$$

Unit Definition: *One unit will oxidize 1.0 μmole of glycerol to dihydroxyacetone per minute at pH 10.0 at 25°C.*

Final Assay Concentrations

In a 3.00 ml reaction mix, the final concentrations are 100 mM potassium carbonate, 33 mM ammonium sulfate, 0.33 mM β-nicotinamide adenine dinucleotide, 100 mM glycerol, 1.7 mM potassium phosphate, 0.003% (w/v) Bovine Serum Albumin, 0.002 mM manganese chloride and 0.02–0.06 unit glycerol dehydrogenase.

Note

1. Do not use if solution becomes light brown or hazy light brown.

Enzymatic Assay of GLYCEROL DEHYDROGENASE (EC 1.1.1.6)

Principle

$$\text{Glycerol} + \beta\text{-NAD} \xrightarrow{\text{Glycerol Dehydrogenase}} \text{Dihydroxyacetone} + \beta\text{-NADH}$$

Abbreviations Used

β-NAD = β-Nicotinamide Adenine Dinucleotide, Oxidized Form
β-NADH = β-Nicotinamide Adenine Dinucleotide, Reduced Form

Conditions: T = 25°C, pH 8.0, A340nm, Light path = 1 cm

Method: Continuous Spectrophotometric Rate Determination

Reagents

A. 100 mM Potassium Phosphate Buffer with 0.1% (w/v) Bovine Serum Albumin, pH 8.8 at 25°C (Prepare 100 ml in deionized water using Potassium Phosphate, Monobasic, Anhydrous and Albumin Bovine or equivalent. Adjust to pH 8.8 at 25°C with 1 M KOH).

B. 2565 mM Glycerol with 100 mM Glycine, 100 mM Potassium Chloride Solution, pH 8.8 at 25°C (Glycerol) (Prepare 100 ml in deionized water using Glycerol, Glycine Free Base, and Potassium Chloride. Adjust to pH 8.8 at 25°C with 1 M NaOH).

C. 52 mM β-Nicotinamide Adenine Dinucleotide Solution (β-NAD) (Prepare 2 ml in deionized water using β-Nicotinamide Adenine Dinucleotide or dissolve the contents of one 50 mg vial of β-Nicotinamide Adenine Dinucleotide in the appropriate volume of deionized water. *Prepare Fresh).*

D. Glycerol Dehydrogenase Enzyme Solution (Immediately before use, prepare a solution containing 0.35–0.70 unit/ml of Glycerol Dehydrogenase in cold Reagent A).

Procedure

Pipette (in milliliters) the following reagents into suitable cuvettes:

	Test	Blank
Reagent B (Glycerol)	2.50	2.50
Reagent C (β-NAD)	0.15	0.15

Mix by inversion and equilibrate to 25°C. Monitor the A340nm until constant, using a suitably thermostatted spectrophotometer. Then add:

Reagent A (Buffer)	—	0.05
Reagent D (Enzyme Solution)	0.05	—

Immediately mix by inversion and record the increase in A340nm for approximately 5 minutes. Obtain the ΔA340nm/minute using the maximum linear rate for both the Test and Blank.

Calculations

$$\text{Units / mg enzyme} = \frac{\Delta A340\text{nm / min Test} - \Delta A340\text{nm / min Blank}}{(6.22)\ (\text{mg enzyme / ml RM})}$$

6.22 = Millimolar extinction coefficient of β-NADH at 340nm
RM = Reaction Mix

Unit Definition: *One unit will oxidize 1.0 μmole of glycerol to dihydroxyacetone per minute at pH 8.8 at 25°C.*

Final Assay Concentrations

In a 2.70 ml reaction mix, the final concentrations are 93 mM glycine, 93 mM potassium chloride, 2375 mM glycerol, 2.9 mM β-NAD, 2 mM potassium phosphate, 0.002% (w/v) BSA and 0.018–0.035 unit glycerol dehydrogenase.

Enzymatic Assay of GLYCEROL-3-PHOSPHATE OXIDASE (EC 1.1.3.21) from *Streptococcus thermophilius*

Principle

$$\text{Dl-}\alpha\text{-Glycerophosphate} + O_2 \xrightarrow{\text{GPO}} \text{Dihydroxyacetone Phosphate} + H_2O_2$$

$$2H_2O_2 + 4\text{-AAP} + \text{Phenol} \xrightarrow{\text{Peroxidase}} \text{Quinoneimine Dye} + H_2O$$

Abbreviations Used

GPO = Glycerol-3-Phosphate Oxidase
4-AAP = 4-Aminoantipyrine

Conditions: T = 37°C, pH 7.0, A505nm, Light path = 1 cm

Method: Continuous Spectrophotometric Rate Determination

Reagents

A. 500 mM Potassium Phosphate Buffer, pH 7.0 at 37°C (Prepare 50 ml in deionized water using Potassium Phosphate, Monobasic, Anhydrous. Adjust to pH 7.0 at 37°C with 1 M NaOH).

B. 0.3% (w/v) 4-Aminoantipyrene Solution (4-AAP) (Prepare 10 ml in deionized water using 4-Aminoantipyrine, Free Base).

C. 0.2% (w/v) Phenol Solution (Phenol) (Prepare 10 ml in deionized water using Phenol).

D. 2 M DL- α-Glycerophosphate Solution (Substrate) (Prepare 50 ml in deionized water using DL- α-Glycerophosphate, Disodium Salt, Hexahydrate).

E. Peroxidase Enzyme Solution (POD) (Immediately before use, prepare a solution containing 50 units/ml of Peroxidase in cold deionized water).

F. Glycerol-3-Phosphate Oxidase Enzyme Solution (GPO) (Immediately before use prepare a solution containing 2–3 units/ml of Glycerol-3-Phosphate Oxidase in cold deionized water).

Procedure

Prepare a reaction cocktail by pipetting (in milliliters) the following reagents into a suitable container:

Reagent A (Buffer)	20.00
Reagent B (4-AAP)	10.00
Reagent C (Phenol)	10.00
Reagent D (Substrate)	50.00

Mix by swirling and adjust to pH 7.0 at 37°C with either 1 M NaOH or 1 M HCl. Pipette (in milliliters) the following reagents into suitable cuvettes:

	Test	Blank
Reaction Cocktail	0.90	0.90

Equilibrate to 37°C. Then add:

Reagent E (POD)	0.10	0.10

Mix by swirling and equilibrate to 37°C. Monitor the A505nm until constant, using a suitably thermostatted spectrophotometer. Then add:

Reagent F (GPO)	0.01	—
Deionized Water	—	0.01

Immediately mix by inversion and record the increase in A505nm for approximately 5 minutes. Obtain the ΔA505nm/minute using the maximum linear rate for both the Test and Blank.

Calculations

$$\text{Units / ml enzyme} = \frac{(\Delta A505\text{nm / min Test} - \Delta A505\text{nm / min Blank})\,(1.01)\,(\text{df})}{(13.3)\,(0.01)\,(0.5)}$$

1.01 = Total volume (in milliliters) of assay
df = Dilution factor
13.3 = Millimolar extinction coefficient of Quinoneimine Dye at 505 nm under the assay conditions
0.01 = Volume (in milliliter) of enzyme used
0.5 = Conversion factor based on one mole of H_2O_2 produces half a mole of Quinoneimine Dye

$$\text{Units / mg solid} = \frac{\text{units / ml enzyme}}{\text{mg solid / ml enzyme}}$$

$$\text{Units / mg protein} = \frac{\text{units / ml enzyme}}{\text{mg protein / ml enzyme}}$$

Unit Definition: *One unit will oxidize 1.0 μmole of L-glycerol 3-phosphate to dihydroxyacetone phosphate with the formation of H_2O_2 per minute at 37°C at pH 7.0.*

Final Assay Concentrations

In a 1.01 ml reaction mix, the final concentrations are 99 mM potassium phosphate, 0.03% (w/v) 4-aminoantipyrene, 0.02% (w/v) phenol, 990 mM DL-α-glycerophosphate, 5 units peroxidase, and 0.02–0.03 unit glycerol-3-phosphate oxidase.

Note

1. Peroxidase Unit Definition: One unit will form 1.0 mg purpurogallin from pyrogallol in 20 seconds at pH 6.0 at 20°C.

Enzymatic Assay of GLYCEROL-3-PHOSPHATE OXIDASE (EC 1.1.3.21)

Principle

$$\text{Dl-}\alpha\text{-Glycerophosphate} + O_2 \xrightarrow{\text{GPO}} \text{Dihydroxyacetone phosphate} + H_2O_2$$

$$2H_2O_2 + \text{4-AAP} + \text{Phenol} \xrightarrow{\text{POD}} \text{Quinoneimine Dye} + 4H_2O$$

Abbreviations Used

GPO = Glycerol-3-Phosphate Oxidase
4-AAP = 4-Aminoantipyrine
POD = Peroxidase

Conditions: T = 37°C, pH = 8.1, A500nm, Light path = 1 cm

Method: Spectrophotometric Stop Rate Determination

Reagents

A. 125 mM Tris HCl Buffer with 0.125% (v/v) Triton X-100, pH 8.1 at 37°C (Prepare 100 ml in deionized water using Trizma Base and Triton X-100. Adjust to pH 8.1 at 37°C with 1 M HCl).
B. 200 mM DL-α-Glycerophosphate Solution (Substrate) (Prepare 60 ml in Reagent A using DL-α-Glycerophosphate, Disodium Salt, Hexahydrate. Readjust pH to 8.1 at 37°C with 1 M HCl or 1 N NaOH, if necessary).
C. 0.1% (w/v) 4-Aminoantipyrine Solution (4-AAP) (Prepare 15 ml in deionized water using 4-Aminoantipyrine, Free Base).

D. 0.1% (w/v) Phenol Solution (Phenol).
E. Peroxidase Enzyme Solution (POD) (Immediately before use, prepare a solution containing 25 units/ml of Peroxidase from Horseradish, in cold deionized water).
F. 0.25% (w/v) Sodium Dodecyl Sulfate Solution (SDS) (Prepare 25 ml in deionized water using Lauryl Sulfate, Sodium Salt).
G. 20 mM Tris HCl Buffer with 0.2% (w/v) Bovine Serum Albumin, pH 7.5 at 37°C (Enzyme Diluent) (Prepare 50 ml in deionized water using Trizma Base and Albumin, Bovine. Adjust to pH 7.5 at 37°C with 1 M HCl).
H. Glycerol-3-Phosphate Oxidase Enzyme Solution (GPO) (Immediately before use, prepare a solution containing 0.20–0.45 unit/ml of Glycerol-3-Phosphate Oxidase in cold Reagent G).

Procedure

Prepare a reaction cocktail by pipetting (in milliliters) the following reagents into a suitable container:

Reagent B (Substrate)	25.00
Reagent C (4-AAP)	5.00
Reagent D (Phenol)	10.00
Reagent E (POD)	10.00

Mix by swirling and equilibrate to 37°C. Adjust the pH to 8.1 with either 1 M HCl or 1 M NaOH if necessary. Pipette (in milliliters) the following reagents into suitable cuvettes:

	Test	Blank
Reaction Cocktail	1.00	1.00

Equilibrate to 37°C. Then add:

Reagent H (GPO)	0.02	—
Reagent G (Enzyme Diluent)	—	0.02

Immediately mix by inversion and incubate at 37°C for exactly 10 minutes. Then add:

Reagent F (SDS)	2.00	2.00

Mix by swirling. Transfer the solutions to suitable cuvettes and record the A500nm for both the Test and Blank.

Calculations

$$\text{Units / ml enzyme} = \frac{(\text{A500nm Test} - \text{A500nm Blank})\ (3.02)\ (\text{df})}{(10)\ (13.3)\ (0.5)\ (0.02)}$$

3.02 = Total volume (in milliliters) of stopped reaction
df = Dilution factor
10 = Time (in minutes) of assay as per the Unit Definition
13.3 = Millimolar extinction coefficient of Quinoneimine Dye at 500 nm under the assay conditions
0.5 = Conversion factor based on one mole of H_2O_2 produces half a mole of Quinoneimine Dye
0.02 = Volume (in milliliters) of enzyme used

$$\text{Units / mg solid} = \frac{\text{units / ml enzyme}}{\text{mg solid / ml enzyme}}$$

$$\text{Units / mg protein} = \frac{\text{units / ml enzyme}}{\text{mg protein / ml enzyme}}$$

Unit Definition: *One unit will oxidize 1.0 µmole of L-glycerol 3-phosphate to dihydroxyacetone phosphate with the formation of H_2O_2 per minute at 37°C at pH 8.1.*

Final Assay Concentration

In a 1.05 ml reaction mix, the final concentrations are 60 mM Tris, 0.06% (v/v) Triton, 0.01% (w/v) 4-aminoantipyrine, 0.02% (w/v) phenol, 5 units peroxidase, 95 mM DL-α-glycerophosphate, 0.01% (w/v) Bovine Serum Albumin, and 0.004–0.009 unit glycerol-3-phosphate oxidase.

Notes

1. This assay is not to be used to assay Glycerol-3-Phosphate Oxidase, from Streptococcus Thermophilus.
2. Initial dilution of enzyme should be to 1 mg/ml or higher.
3. Peroxidase Unit Definition: One unit will form 1.0 mg purpurogallin from pyrogallol in 20 seconds at pH 6.0 at 20°C.

Enzymatic Assay of α-GLYCEROPHOSPHATE DEHYDROGENASE (EC 1.1.1.8)

Principle

$$\text{DHAP} + \beta\text{-NADH} \xrightarrow{\alpha\text{-Glycerophosphate Dehydrogenase}} \alpha\text{-Glycerophosphate} + \beta\text{-NAD}$$

Abbreviations Used

DHAP = Dihydroxyacetone Phosphate
β-NADH = β-Nicotinamide Adenine Dinucleotide, Reduced Form
β-NAD = β-Nicotinamide Adenine Dinucleotide, Oxidized Form

Conditions: T = 25°C, pH = 7.4, A340nm, Light path = 1 cm

Method: Continuous Spectrophotometric Rate Determination

Reagents

A. 300 mM Triethanolamine HCl Buffer, pH 7.4 at 25°C (Prepare 100 ml in deionized water using Triethanolamine Hydrochloride. Adjust to pH 7.4 at 25°C with 1 M NaOH).
B. 76 mM Dihydroxyacetone Phosphate Solution (DHAP) (Prepare 1 ml in deionized water using Dihydroxyacetone Phosphate, Lithium Salt).
C. 4 mM β-Nicotinamide Adenine Dinucleotide, Reduced Form Solution (β-NADH) (Prepare 1 ml in Reagent A using β-Nicotinamide Adenine Dinucleotide, Reduced Form, Disodium Salt).
D. α-Glycerophosphate Dehydrogenase Enzyme Solution (Immediately before use, prepare a solution containing 0.15–0.30 unit/ml of α-Glycerophosphate Dehydrogenase in cold Reagent A).

Procedure

Pipette (in milliliters) the following reagents into suitable cuvettes:

	Test	Blank
Reagent A (Buffer)	2.70	2.70
Reagent B (DHAP)	0.10	0.10
Reagent C (β-NADH)	0.10	0.10

Mix by inversion and equilibrate to 25°C. Monitor the A340nm until constant, using a suitably thermostatted spectrophotometer. Then add:

Reagent D (Enzyme Solution)	0.10	—
Reagent A (Buffer)	—	0.10

Immediately mix by inversion and record the decrease in A340nm for approximately 5 minutes. Obtain the ΔA340nm/minute using the maximum linear rate for both the Test and Blank.

Calculations

$$\text{Units / ml enzyme} = \frac{(\Delta A340\text{nm / min Test} - \Delta A340\text{nm / min Blank})\ (3)\ (\text{df})}{(6.22)\ (0.10)}$$

3 = Volume (in milliliter) of assay
df = Dilution factor
6.22 = Millimolar extinction coefficient of β-NADH at 340nm
0.10 = Volume (in milliliter) of enzyme used

$$\text{Units / mg solid} = \frac{\text{units / ml enzyme}}{\text{mg solid / ml enzyme}}$$

$$\text{Units / mg protein} = \frac{\text{units / ml enzyme}}{\text{mg protein / ml enzyme}}$$

Unit Definition: *One unit will convert 1.0 μmole of dihydroxyacetone phosphate to α-glycerophosphate per minute at pH 7.4 at 25°C.*

Final Assay Concentration

In a 3.00 ml reaction mix, the final concentrations are 290 mM triethanolamine, 2.5 mM dihydroxyacetone phosphate, 0.13 mM β-nicotinamide adenine dinucleotide reduced form and 0.015–0.03 unit α-glycerophosphate dehydrogenase.

Enzymatic Assay of GLYCEROPHOSPHORYLCHOLINE PHOSPHODIESTERASE (EC 3.1.4.2)

Principle

$$\text{L-}\alpha\text{-Glycerophosphorylcholine} + H_2O \xrightarrow{\text{GPCP}} \text{Glycerophosphate} + \text{Choline}$$

$$\text{Choline} + O_2 \xrightarrow{\text{Choline Oxidase}} \text{Betaine Aldehyde} + H_2O_2$$

$$\text{Betaine Aldehyde} \xrightarrow{\text{Choline Oxidase}} \text{Betaine} + H_2O_2$$

$$2H_2O_2 + \text{4-AAP} + \text{Phenol} \xrightarrow{\text{POD}} \text{Quinoneimine Dye} + 2H_2O$$

Abbreviations Used

GPCP = Glycerophosphorylcholine Phosphodiesterase
4-AAP = 4-Aminoantipyrine
POD = Peroxidase

Conditions: T = 37°C, pH = 8.0, A500nm, Light path = 1 cm

Method: Colorimetric

Reagents

A. 200 mM Tris HCl Buffer, pH 8.0 at 37°C (Prepare 100 ml in deionized water using Trizma Base. Adjust to pH 8.0 at 37°C with 1 M HCl).
B. 10 mM L-α-Glycerophosphorylcholine Solution (GPC) (Prepare 5 ml by pipetting 2.57 ml of L-α-Glycerophosphorylcholine, into a 25 ml beaker. Evaporate the methanol under nitrogen gas and then reconstitute with 5 ml of Reagent A).
C. 10 mM Calcium Chloride Solution ($CaCl_2$) (Prepare 10 ml in deionized water using Calcium Chloride, Dihydrate).
D. 100 mM Ethylenediaminetetraacetic Acid Solution (EDTA) (Prepare 10 ml in deionized water using Ethylenediaminetetraacetic Acid, Tetrasodium Salt, Hydrate).
E. 0.3% (w/v) 4-Aminoantipyrine Solution (4-AAP) (Prepare 2 ml in deionized water using 4-Aminoantipyrine).
F. 0.2% (w/v) Phenol Solution (Phenol) (Prepare 2 ml in deionized water using Phenol).
G. Choline Oxidase Enzyme Solution (Choline Ox) (Immediately before use, prepare a solution containing 60 units/ml of Choline Oxidase in cold deionized water).
H. Peroxidase Enzyme Solution (POD) (Immediately before use, prepare a solution containing 50 units/ml of Peroxidase in cold deionized water).
I. 10 mM Tris HCl Buffer, pH 8.0 at 37°C (Enz Dil) (Prepare 10 ml in deionized water using Trizma Base. Adjust to pH 8.0 at 37°C with 1 M HCl).
J. Glycerophosphate Phosphodiesterase Enzyme Solution (GPCP) (Immediately before use, prepare a solution containing approximately 0.30 unit/ml of Glycerophosphate Phosphodiesterase in cold Reagent I).

Procedure

Prepare a reaction cocktail by pipetting (in milliliters) the following reagents into a suitable container:

Reagent A (Buffer)	1.00
Reagent D (EDTA)	2.00
Reagent E (4-AAP)	1.00
Reagent F (Phenol)	1.00
Reagent G (Choline Ox)	1.00
Reagent H (POD)	1.00
Deionized Water	3.00

Pipette (in milliliters) the following reagents into suitable cuvettes:

	Test	Blank
Reagent A (Buffer)	0.10	0.10
Reagent B (GPC)	0.05	0.05
Reagent C ($CaCl_2$)	0.05	0.05
Deionized Water	0.25	0.25

Equilibrate to 37°C. Then add:

Reagent J (GPCP)	0.05	—
Reagent I (Enz Dil)	—	0.05

Mix by inversion and incubate at 37°C for exactly 10 minutes. Then add:

Reaction Cocktail	1.00	1.00

Mix by inversion and incubate at 37°C for 20 minutes. Then add:

Deionized Water	1.50	1.50

Mix by inversion and record the A500nm for both the Test and Blank using a suitable spectrophotometer.

Calculations

$$\text{Units / ml enzyme} = \frac{(\text{A500nm Test} - \text{A500nm Blank})\ (3)\ (\text{df})}{(10)\ (12)\ (0.5)\ (2)\ (0.05)}$$

3 = Total volume (in milliliters) of assay
df = Dilution factor
10 = Time (in minutes) of the assay as per the Unit Definition
12 = Millimolar extinction coefficient of quinoneimine dye at 500 nm
0.5 = Conversion factor derived from the fact that 2 μmoles of H_2O_2 produce 1 μmole of quinoneimine dye at 500nm
2 = Conversion factor derived from the fact that 1 μmole of glycerophosphorylcholine produces 2 μmoles of H_2O_2
0.05 = Volume (in milliliter) of enzyme used

$$\text{Units / mg solid} = \frac{\text{units / ml enzyme}}{\text{mg solid / ml enzyme}}$$

$$\text{Units / mg protein} = \frac{\text{units / ml enzyme}}{\text{mg protein / ml enzyme}}$$

Unit Definition: *One unit will produce 1.0 μmole of choline from L-α-glycerophosphorylcholine, G-4007, per minute at pH 8.0 at 37°C.*

Final Assay Concentrations

In a 0.50 ml reaction mixture, the final concentrations are 25 mM Tris, 1 mM glycerophosphorylcholine, 10 mM calcium chloride, and 0.015 unit glycerophosphorylcholine phosphodiesterase.

Notes

1. Choline Oxidase Unit Definition: One unit will form 1.0 μmole of H_2O_2 from choline and H_2O per minute at pH8.0 at 37°C.
2. Peroxidase Unit Definition: One unit will form 1.0 mg purpurogallin from pyrogallol in 20 seconds at pH 6.0 at 20°C.

Enzymatic Assay of GLYCOGEN SYNTHASE (EC 2.4.1.11)

Principle

$$\text{UDPG} + (\text{Glycogen})_n \xrightarrow{\text{Glycogen Synthase}} \text{UDP} + (\text{Glycogen})_{n+1}$$

$$\text{UDP} + \text{PEP} \xrightarrow{\text{PK}} \text{UTP} + \text{Pyruvate}$$

$$\text{Pyruvate} + \beta\text{-NADH} \xrightarrow{\text{LDH}} \text{Lactate} + \beta\text{-NAD}$$

Abbreviations Used

UDPG = Uridine 5'-Diphosphoglucose
UDP = Uridine 5'-Diphosphate
PEP = Phospho(enol) pyruvate
UTP = Uridine 5'-Triphosphate
PK = Pyruvate Kinase
LDH–Lactic Dehydrogenase
β-NADH = β-Nicotinamide Adenine Dinucleotide, Oxidized Form
β-NAD = Nicotinamide Adenine Dinucleotide, Reduced Form

Conditions: T = 30°C, pH 8.2, A340nm, Light path = 1 cm

Method: Spectrophotometric Stop Rate Determination

Reagents

A. 500 mM Tris HCl Buffer, pH 8.2 at 30°C (Step 1 Buffer) (Prepare 100 ml in deionized water using Trizma Base. Adjust to pH 8.2 at 30°C with 1 M HCl).
B. 200 mM Tris HCl Buffer, pH 7.5 at 30°C (Step 2 Buffer) (Prepare 100 ml in deionized water using Trizma Base. Adjust to pH 7.5 at 30°C with 1 M HCl).
C. 300 mM Magnesium Chloride Solution ($MgCl_2$) (Prepare 25 ml in deionized water using Magnesium Chloride, Hexahydrate).
D. 100 mM Ethylenediaminetetraacetic Acid Solution (EDTA) (Prepare 25 ml in deionized water using Ethylenediaminetetraacetic Acid, Tetrasodium Salt, Hydrate).
E. 100 mM β-Mercaptoethanol Solution (β-ME) (Prepare 10 ml in deionized water using 2-Mercaptoethanol).
F. 1000 mM Potassium Chloride Solution (KCl) (Prepare 10 ml in deionized water using Potassium Chloride).
G. 60 mM Magnesium Sulfate Solution ($MgSO_4$) (Prepare 10 ml in deionized water using Magnesium Sulfate, Anhydrous).
H. 1.0% (w/v) Glycogen Solution (Glycogen) (Prepare 200 ml in Reaction Cocktail (Step 1) using Glycogen).
I. 10 mM D-Glucose 6-Phosphate Solution (G-6P) (Prepare 200 ml in Reaction Cocktail (Step 1) using D-Glucose 6-Phosphate, Disodium salt, Hydrate).

J. 40 mM phospho(enol) pyruvate Solution (PEP) (Prepare 5 ml in deionized water using Phospho(enol) pyruvate, Monopotassium Salt).
K. 3.8 mM Uridine 5'-Diphosphoglucose Solution (UDPG) (Prepare 200 ml in Reaction Cocktail (Step 1) using Uridine 5'-Diphosphoglucose, Disodium Salt).
L. 7.1 mM β-Nicotinamide Adenine Dinucleotide Reduced Form Solution (β-NADH) (Prepare 1 ml in deionized water using β-Nicotinamide Adenine Dinucleotide, Reduced Form).
M. PK/LDH Enzyme Suspension (PK/LDH) (Use PK/LDH Enzyme Suspension).
N. Glycogen Synthase Enzyme Solution (Glycogen Synth) (Immediately before use, prepare a solution containing 2.0–4.0 units/ml of Glycogen Synthase in cold deionized water).

Procedure

Step 1: Prepare a reaction cocktail by pipetting (in milliliters) the following reagents into a suitable container:

Reagent A (Step 1 Buffer)	20.00
Reagent C ($MgCl_2$)	8.50
Reagent D (EDTA)	2.00
Reagent E (β-ME)	5.00
Deionized water	164.50

Mix by swirling and then add (in milligrams) the following reagents to the aforementioned solution in order to make the reaction cocktail.

Reagent K (UDPG)	511.50
Reagent H (Glycogen)	2000.00
Reagent I (G-6P)	564.00

Pipette (in milliliters) the following reagents into suitable cuvettes:

	Test	Blank
Reaction Cocktail (Step 1)	3.00	3.00

Equilibrate to 30°C. Then add:

Reagent N (Glycogen Synth)	0.10	—
Deionized water	—	0.10

Immediately mix by inversion and incubate for exactly 5 minutes at 30°C. Stop the reaction by heating both the Test and Blank for 5 minutes at 100°C. Cool with running tap water then transfer both the Blank and Test solutions to Eppendorf tubes and centrifuge.

Step 2: Prepare a reaction cocktail by pipetting (in milliliters) the following reagents into a suitable container:

Reagent B (Step 2 Buffer)	17.00
Reagent F (KCl)	2.50
Reagent G ($MgSO_4$)	5.00
Reagent J (PEP)	0.75
Reagent D (EDTA)	0.25
Reagent L (β-NADH)	1.00
Deionized water	22.50

Mix by swirling. Pipette (in milliliters) the following reagents into suitable cuvettes:

	Test	Blank
Reaction Cocktail (Step 2)	2.80	2.80
Reagent M (PK/LDH)	0.01	0.01

Mix by inversion and equilibrate to 30°C. Monitor the A340nm until constant, using a suitable spectrophotometer. Then add:

Test Supernatant (Step 1)	0.10	—
Blank Supernatant (Step 2)	—	0.10

Immediately mix by inversion and record the decrease in A340nm for approximately 5 minutes. Obtain the final A340nm for both the Test and Blank Supernatant.

Calculations

ΔA340nm Test = A340nm Test Initial–A340nm Test Final
ΔA340nm Blank = A340nm Blank Initial–A340nm Blank Final

$$\text{Units / mg enzyme} = \frac{\text{A340nm Test} - \text{A340nm Blank (2.91)}}{(5)\ (6.22)\ (0.1)\ (\text{mg enzyme / ml RM})}$$

2.91 = Final volume of Step 2
0.1 = Volume form Step 1 used in Step 2
6.22 = Millimolar extinction coefficient of β-NADH at 340 nm
5 = Time of Reaction (in minutes) of Step 1
RM = Reaction Mix

Unit Definition: *One unit will catalyze the incorporation of 1.0 μmole of glucose from UDP-glucose into glycogen per minute at pH 8.2 at 30°C, yielding 1.0 μmole of UDP which is measured in a PK/LDH/NADH system.*

Final Assay Concentration

In a 3.10 ml reaction mix, the final concentrations are 48 mM Tris, 12.4 $MgCl_2$, 1.00 mM EDTA, 2.4 mM 2-mercaptoethanol, 3.63 mM UDPG, 9.7 mM glucose 6-phosphate, 0.2–0.4 unit glycogen synthase.

Enzymatic Assay of GLYCOGEN SYNTHASE KINASE-3B

Principle

$$\text{PI-2-}\gamma\text{-}_{32}\text{P-ATP} \xrightarrow{\text{GSK}} [_{32}\text{P}]\text{-Phosphorylated PI-2} + \text{ADP}$$

Abbreviations Ùsed

PI-2 = Phosphatase Inhibitor-2
γ-$_{32}$P-ATP = Adenosine 5-Triphosphate γ-$_{32}$P label
GSK = Glycogen Synthase Kinase-3B
ADP = Adenosine 5'-Triphosphate

Conditions: T = 30°C, pH = 7.5

Method: Radioactive

Reagents

A. 20 mM Tris HCL Buffer, pH 7.5 at Room Temperature (Enz Dil) (Prepare 10 ml in deionized water using Trizma Base. Adjust to pH 7.5 at room temperature using 1 M HCl).
B. γ-$_{32}$P-Adenosine 5-Triphosphate Solution (γ-$_{32}$-P-ATP) (Use product with a specific activity of 3000 curies/mmol).
C. 80 mM Tris HCl Buffer, with 40 mM Magnesium Chloride, 20 mM Dithiothreitol and 0.8 mM Adenosine 5'-Triphosphate, pH 7.5 at room temperature (4X Reaction Buffer) (Prepare 25 ml in deionized water using Trizma Base, Magnesium Chloride, Hexahydrate, DL-Dithiothreitol and Adenosine 5-Triphosphate, Disodium Salt. Adjust to pH 7.5 at room temperature with 1 M HCl. Before use, add Reagent B (γ-$_{32}$P-ATP) to a final concentration of 200 µCi per ml of the 4X Reaction Buffer).
D. 10 mM Sodium Pyrophosphate Solution (Prepare 200 ml in deionized water using Sodium Pyrophosphate, Disodium Salt, Anhydrous).
E. 10% (w/v) Trichloracetic Acid Solution (TCA Wash Solution) (Prepare 200 ml in Reagent D using Trichloroacetic Acid, 6.1 N Solution, approximately 100% (w/v)).
F. Chromatography Paper (Using Whatman 3 mm chromatography paper, cut into 1 × 2 cm squares).
G. Phosphatase Inhibitor-2 Solution (PI-2) (Immediately before use, reconstitute a 100 µg vial of Phosphastase Inhibitor-2, with 200 µl of deionized water. Determine protein concentration by the Bradford method. Dilute to 0.5 mg/ml).
H. Glycogen Synthase Kinase 3B Enzyme Solution (Immediately before use, reconstitute a vial with 100 µl of deionized water. Further dilute 5 fold in Reagent A a final concentration of 400 units/ml).

Procedure

Pipette (in milliliters) the following reagents with suitable containers:

	Test	Blank
Reagent H (Enzyme Solution)	0.005	—
Reagent A (Enz Dil)	—	0.005
Reagent G (PI-2)	0.01	0.01
Reagent C (4x Reaction Buffer)	0.005	0.005

Vortex gently for a few seconds and incubate for 10 minutes at 30°C. Remove a 0.015 ml aliquot from both the Test and Blank reaction mixture and place on Reagent F (1 × 2 cm chrom. paper). Soak the paper rectangles in Reagent E (TCA Wash Solution) at room temperature for 15 minutes. Wash the paper rectangles 4 times with Reagent E. Each wash should consist of 10 ml of Reagent E per paper rectangle. Agitate gently throughout each wash for 15 minutes. This is then followed by a single wash with ethanol and another wash with acetone. Dry the paper pieces at room temperature or under a lamp. Then count the radioactivity that has been incorporated into precipitated phosphatase inhibitor-2 using the Cerenkov mode (i.e. Count the β-emission without scintillation fluid using the 3H channel).

Calculations

1. Count **R**, the radioactivity of 5 µl of the 4X Reaction Buffer, in order to obtain the total radioactivity in cpm per assay tube (perform in duplicate).
2. Divide the above value (**R**) by the amount of ATP present in the assay tube (4000 pmol), in order to obtain the specific radioactivity, **SR**. **SR** = **R**/4000 cpm/pmol.
3. Subtract the blank value from the count of the sample and multiply the result by a factor of 4/3 (to adjust for aliquots taken from the reaction) in order to obtain the total counts per reaction **C**.

$$\text{Units / ml} = \frac{(C)\,(df)}{(SR)\,(10)\,(0.005)}$$

df = Dilution factor
10 = Conversion factor to convert to a one minute rate.
0.005 = Volume (in milliliters) of enzyme used
C = Count per reaction
SR = Specific Radioactivity

$$\text{Units / mg solid} = \frac{\text{units / ml enzyme}}{\text{mg solid / ml enzyme}}$$

Unit Definition: *One unit will transfer one pmol of phosphate from ATP to phosphate inhibtor 2 per minute at pH 7.5 at 30°C.*

Final Assay Concentration

In a 0.020 ml reaction mix, the final concentrations are 25 mM Tris, 10 mM magnesium chloride, 5 mM DL-dithiothreitol, 0.2 mM adenosine 5-triphosphate, 5 µg phosphatase inhibitor P-2 and 2 units glycogen synthase kinase 3B.

Enzymatic Assay of GLYCOLATE OXIDASE (EC 1.1.3.15)

Principle

$$\text{Glycolate} + O_2 \xrightarrow{\text{Glycolate Oxidase}} \text{Glyoxylate}$$

$$\text{Glyoxylate} + \text{Phenylhydrazline} \rightarrow \text{Glyoxylate Phenylhydrazone}$$

Conditions: T = 25°C, pH = 7.8, A324nm, Light path = 1 cm

Method: Continuous Spectrophotometric Rate Determination

Reagents

A. 100 mM Phosphate Buffer, pH 7.8 at 25°C (Prepare 100 ml in deionized water using Potassium Phosphate, Monobasic, Anhydrous. Adjust to pH 7.8 at 25°C with 1 M KOH).

B. 40 mM Glycolate Solution (Prepare 25 ml in Reagent A using Glycolic Acid. Adjust to pH 8.0 at 25°C with 1 M KOH).
C. 100 mM L-Cysteine HCl Solution (Prepare 25 ml in Reagent A using L-Cysteine Hydrochloride, Monohydrate. Adjust to pH 6.0 at 25°C with 1 M KOH. *Prepare Fresh*).
D. 100 mM Phenylhydrazine HCl Solution (Prepare 25 ml in Reagent A using Phenylhydrazine Hydrochloride. Adjust to pH 6.0 with 1 M KOH. *Prepare Fresh and Keep away from Light*).
E. 1 mM Flavin Mononucleotide Solution (FMN) (Prepare 10 ml in deionized water using Flavin Mononucleotide, Sodium Salt. *Prepare Fresh and Keep away from Light*).
F. Glycolate Oxidase Enzyme Solution (Immediately before use, prepare a solution containing 0.2 unit/ml of Glycolate Oxidase in cold Reagent A).

Procedure

Prepare a reaction cocktail by pipetting (in milliliters) the following reagents into a suitable amber bottle (*Prepare Fresh. Solution is Stable for Only is a Few Hours*):

Reagent A (Buffer)	22.00
Reagent B (Glycolate)	5.00
Reagent C (L-Cysteine)	1.00
Reagent D (Phenylhydrazine)	1.00

Mix and adjust to pH 7.8 at 25°C with either 1 M HCl or 1 M KOH, if necessary. Oxygenate by bubbling 99.9% pure O_2 through the solution for 5 to 7 minutes (*this must be done prior to each assay*). Immediately pipette (in milliliters) into suitable quartz cuvettes:

	Test	Blank
Oxygenated Reaction Cocktail	2.90	2.90
Reagent A (Buffer)	—	0.10
Reagent E (FMN)	0.10	0.10

Mix by inversion and equilibrate to 25°C. Monitor the A324nm until constant, using a suitably thermostatted spectrophotometer. Then add:

Reagent F (Enzyme Solution)	0.10	—

Immediately mix by inversion and record the increase in A324nm for approximately 15 minutes. Obtain the ΔA324nm/minute using the maximum linear rate for both the Test and Blank.

Calculations

$$\text{Units / ml enzyme} = \frac{(\Delta A324nm / min\ Test - \Delta A324nm / min\ Blank)\ (3.1)\ (df)}{(17)\ (0.1)}$$

3.1 = Total volume (in milliliters) of assay
df = dilution factor

17 = Millimolar extinction coefficient of Glyoxylate Phenylhydrazone at 324nm
0.1 = Volume (in milliliters) of enzyme used

$$\text{Units / mg solid} = \frac{\text{units / ml enzyme}}{\text{mg solid / ml enzyme}}$$

$$\text{Units / mg protein} = \frac{\text{units / ml enzyme}}{\text{mg protein / ml enzyme}}$$

Unit Definition: *One unit will produce 1.0 µmole of glyoxylate from glycolate per minute at pH 7.8 at 25°C, in the presence of phenylhydrazine.*

Final Assay Concentration

In a 3.10 ml reaction mix, the final concentrations are 97 mM potassium phosphate, 6.5 mM glycolic acid, 3.22 mM cysteine, 3.22 mM phenylhydrazine, 0.03 mM flavin mononucleotide and 0.02 unit glycolate oxidase.

Note

1. The maximum linear rate is generally found between 10–15 minutes.

Enzymatic Assay of GLYCOLATE OXIDASE (EC 1.1.3.15)

Principle

$$\text{Glycolate} + O_2 \xrightarrow{\text{Glycolate Oxidase}} \text{Glyoxylate}$$

$$\text{Glyoxylate} + \text{Phenylhydrazine} \rightarrow \text{Glyoxylate Phenylhydrazone}$$

Conditions: T = 25°C, pH = 8.3, A324nm, Light path = 1 cm

Method: Continuous Spectrophotometric Rate Determination

Reagents

A. 100 mM Phosphate Buffer, pH 8.3 at 25°C (Prepare 100 ml in deionized water using Potassium Phosphate, Monobasic, Anhydrous. Adjust to pH 8.3 at 25°C with 1 M KOH).

B. 40 mM Glycolic Acid Solution (Glycolic Acid) (Prepare 25 ml in Reagent A using Glycolic Acid. Adjust to pH 8.0 at 25°C with 1 M KOH).

C. 100 mM L-Cysteine HCl Solution (L-Cysteine) (Prepare 25 ml in Reagent A using L-Cysteine Hydrochloride, Monohydrate. Adjust to pH 6.0 at 25°C with 1 M KOH. *Prepare Fresh*).

D. 100 mM Phenylhydrazine HCl Solution (Phenylhydrazine) (Prepare 25 ml in Reagent A using Phenylhydrazine Hydrochloride. Adjust to pH 6.0 at 25°C with 1 M KOH. *Prepare Fresh and Keep away from Light*).

E. 1 mM Flavin Mononucleotide Solution (FMN) (Prepare 5 ml in deionized water using Flavin Mononucleotide, Sodium Salt. *Prepare Fresh*).

F. 5 mM Flavin Mononucleotide with 0.1% (w/v) Bovine Serum Albumin (Enzyme Dil) (Prepare 10 ml in deionized water using Flavin Mononucleotide, Sodium Salt, and Albumin, Bovine. *Prepare Fresh and Keep away from Light*).

G. Glycolate Oxidase Enzyme Solution (Immediately before use, prepare a solution containing 0.1–0.2 unit/ml of Glycolate Oxidase in cold Reagent F).

Procedure

To prepare a reaction cocktail, pipette (in milliliters) the following reagents into a suitable amber bottle (*Prepare Fresh-Solution is Stable for Only is a Few Hours*):

Reagent A (Buffer)	22.00
Reagent B (Glycolic Acid)	5.00
Reagent C (L-Cysteine)	1.00
Reagent D (Phenylhydrazine)	1.00

Mix by swirling and equilibrate to 25°C. If necessary, adjust to pH 8.3 with either 1 M KOH or 1 M HCl. Oxygenate by bubbling pure O_2 through the solution for 5 to 7 minutes (*this must be done prior to each assay*). Immediately pipette (in milliliters) into suitable quartz cuvettes:

	Test	Blank
Oxygenated Reaction Cocktail	2.9	2.9
Reagent E (FMN)	0.1	0.1
Reagent F (Enzyme Dil)	—	0.1

Immediately mix by inversion. Monitor the A324nm until constant, using a suitably thermostatted spectrophotometer. Then add:

Reagent G (Enzyme Solution)	0.1	—

Immediately mix by inversion and record the increase in A324nm for approximately 15 minutes. Obtain the ΔA324nm/min using the maximum linear rate for both the Test and Blank.

Calculations

$$\text{Units / mg enzyme} = \frac{(\Delta A324\text{nm / min Test} - \Delta A324\text{nm / min Blank})\,(3.1)\,(\text{df})}{(17)\,(0.1)}$$

3.1 = Total volume (in milliliters) of assay
df = Dilution factor
17 = Millimolar extinction coefficient of Glyoxylate Phenylhydrazone at 324nm
0.1 = Volume (in milliliter) of enzyme used

$$\text{Units / mg solid} = \frac{\text{units / ml enzyme}}{\text{mg solid / ml enzyme}}$$

$$\text{Units / mg protein} = \frac{\text{units / ml enzyme}}{\text{mg protein / ml enzyme}}$$

Unit Definition: *One unit will produce 1.0 µmole of glyoxylate from glycolate per minute at pH 8.3 at 25°C, in the presence of phenylhydrazine.*

Final Assay Concentration

In a 3.10 ml reaction mix, the final concentrations are 94 mM potassium phosphate, 6.5 mM glycolic acid, 3.2 mM cysteine, 3.2 mM phenylhydrazine, 0.2 mM flavin mononucleotide, 0.003% (w/v) Bovine Serum Albumin and 0.01–0.02 unit glycolate oxidase.

Note

1. The maximum linear rate is generally found between 10–15 minutes.

Enzymatic Assay of GLYOXALASE I (EC 4.4.1.5)

Principle

$$\text{Methylglyoxal} + \text{GSH} \xrightarrow{\text{Glyoxalase I}} \text{S} - \text{Lactoyl Glutathione}$$

Abbreviation Used

GSH = Reduced Glutathione

Conditions: T = 25°C, pH 6.6, A240nm, Light path = 1 cm

Method: Continuous Spectrophotometric Rate Determination

Reagents

A. 100 mM Potassium Phosphate Buffer, pH 6.6 at 25°C (Prepare 100 ml in deionized water using Potassium Phosphate, Monobasic, Anhydrous. Adjust to pH 6.6 with 1 M KOH).
B. 0.2% (v/v) Methylglyoxal Solution (Meth-Glyox) (Prepare 50 ml in deionized water using Methylglyoxal.)
C. 2.0% (w/v) Reduced Glutathione Solution, pH 6.6 at 25°C (GSH) (Prepare 10 ml in deionized water using Glutathione, Reduced Form, Free Acid. Adjust to pH 6.6 at 25°C with solid Sodium Bicarbonate. *Prepare Fresh)*.
D. 10 mM Potassium Phosphate Buffer with 0.1% (w/v) Bovine Serum Albumin, pH 7.4 at 25°C (Enzyme Diluent) (Prepare 10 ml in deionized water using Potassium Phosphate, Monobasic, Anhydrous, and Albumin, Bovine. Adjust to pH 7.4 at 25°C with 1 M KOH).
E. Glyoxalase I Enzyme Solution (Glyoxalase I) (Immediately before use, prepare a solution containing 0.25–0.50 unit/ml of Glyoxalase I in Reagent D).

Procedure

Pipette (in milliliters) the following reagents into suitable cuvettes:

	Test	Blank
Reagent A (Buffer)	2.35	2.35
Reagent B (Meth-Glyox)	0.50	0.50
Reagent C (GSH)	0.05	0.05

Mix by inversion and equilibrate to 25°C. Monitor the A240nm until constant, using a suitably thermostatted spectrophotometer. Then add:

	Test	Blank
Reagent E (Glyoxalase I)	0.10	—
Reagent D (Enzyme Diluent)	—	0.10

Mix by inversion and record the increase in A240nm for approximately 5 minutes. Obtain the ΔA240nm/minute using the maximum linear rate for both the Test and Blank.

Calculations

$$\text{Units / ml enzyme} = \frac{(\Delta A240nm / min\ Test - \Delta A240nm / min\ Blank)\ (3)\ (df)}{(3.37)\ (0.1)}$$

3 = Total volume (in milliliters) of assay
df = Dilution factor
3.37 = Millimolar extinction coefficient of S-Lactoylglutathione at 240nm
0.1 = Volume (in milliliter) of enzyme used

$$\text{Units / mg solid} = \frac{\text{units / ml enzyme}}{\text{mg solid / ml enzyme}}$$

$$\text{Units / mg protein} = \frac{\text{units / ml enzyme}}{\text{mg protein / ml enzyme}}$$

Unit Definition: *One unit will form 1.0 μmole of S-lactoylglutathione from methylglyoxal and reduced glutathione per minute at pH 6.6 at 25°C.*

Final Assay Concentration

In a 3.00 ml reaction mix, the final concentrations are 79 mM potassium phosphate, 0.033% (w/v) reduced glutathione, 0.003% (w/v) Bovine Serum Albumin, and 0.025–0.05 unit of glyoxalase I.

Note

1. Wear a respirator when handling methylglyoxal directly from the bottle.

Enzymatic Assay of GLYOXALASE II (EC 3.1.2.6)

Principle

$$\text{S-Lactoylglutathione} + H_2O \xrightarrow{\text{Glyoxalase II}} \text{GSH} + \text{DL-Lactic Acid}$$

Abbreviation Used

GSH = Reduced Glutathione

Conditions: T = 25°C, pH = 7.4, A240nm, Light path = 1 cm

Method: Continuous Spectrophotometric Rate Determination

Reagents

A. 50 mM Tris HCl Buffer, pH 7.4 at 25°C (Prepare 100 ml in deionized water using Trizma Base. Adjust to pH 7.4 at 25°C using 1 M HCl).

B. 0.76% (w/v) S-Lactoylglutathione Solution (SLG) (Prepare 10 ml in deionized water using S-Lactoylglutathione, Free Acid).

C. Glyoxalase II Enzyme Solution (Glyox II) (Immediately before use, prepare a solution containing 0.2–0.4 unit/ml of Glyoxalase II in cold deionized water).

Procedure

Pipette (in milliliters) the following reagents into suitable cuvettes:

	Test	Blank
Reagent A (Buffer)	2.90	2.90
Reagent B (SLG)	0.07	0.07

Procedure

Mix by inversion and equilibrate to 25°C. The A240nm should be 0.6–0.8. If not, add more Reagent B so that the A240nm is within this range. Then add:

	Test	Blank
Reagent C (Enzyme Solution)	0.10	0.10
Deionized Water	—	0.10

Immediately mix by inversion and record the decrease in A240nm for approximately 3 minutes. Obtain the ΔA240nm/minute using the maximum linear rate for both the Test and Blank.

Calculations

$$\text{Units / ml enzyme} = \frac{(\Delta A240nm / \min \text{ Test} - \Delta A240nm / \min \text{ Blank})\ (3.07)\ (df)}{(3.37)\ (0.1)}$$

3.07 = Total volume (in milliliters) of assay
df = Dilution factor
3.37 = Millimolar extinction coefficient of S-Lactoylglutathione at 240nm
0.1 = Volume (in milliliter) of enzyme used

$$\text{Units / mg solid} = \frac{\text{units / ml enzyme}}{\text{mg solid / ml enzyme}}$$

$$\text{Units / mg protein} = \frac{\text{units / ml enzyme}}{\text{mg protein / ml enzyme}}$$

Unit Definition: *One unit will hydrolyze 1.0 µmole of S-lactoylglutathione per minute at pH 7.4 at 25°C.*

Final Assay Concentrations

In a 3.07 ml reaction mix, the final concentrations are 48 mM Tris, 0.0176% (w/v) D-lactoylgutathione and 0.02–0.04 unit glyoxalase II.

Enzymatic Assay of GLYOXYLATE REDUCTASE (EC 1.1.1.26)

Principle

$$\text{Glyoxylate} + \beta\text{-NADH} \xrightarrow{\text{Glyoxylate Reductase}} \text{Glycolate} + \beta\text{-NAD}$$

Abbreviations Used

β-NADH = β-Nicotinamide Adenine Dinucleotide, Reduced Form
β-NAD = β-Nicotinamide Adenine Dinucleotide, Oxidized Form

Conditions: T = 25°C, pH = 6.4, A340nm, Light path = 1 cm

Method: Continuous Spectrophotometric Rate Determination

Reagents

A. 50 mM Potassium Phosphate Buffer, pH 6.4 at 25°C (Prepare 200 ml in deionized water using Potassium Phosphate, Monobasic, Anhydrous. Adjust to pH 6.4 at 25°C with 1 M KOH).
B. 702 mM Glyoxylic Acid solution (Glyox) (Prepare 1 ml in Reagent A using Glyoxylic Acid, Monohydrate, Sodium Salt).
C. 12.8 mM β-Nicotinamide Adenine Dinucleotide, Reduced Form Solution (β-NADH) (Dissolve the contents of one 5 mg vial of β- Nicotinamide Adenine Dinucleotide, Reduced Form, Disodium Salt in the appropriate volume of cold Reagent A *or* prepare 1 ml in cold Reagent A using β-Nicotinamide Adenine Dinucleotide, Reduced Form, Disodium Salt. *Prepare Fresh*).
D. Glyoxylate Reductase Enzyme Solution (Immediately before use, prepare a solution containing 1–2 units/ml of Glyoxylate Reductase in cold deionized water).

Procedure

Pipette (in milliliters) the following reagents into suitable cuvettes:

	Test	Blank
Reagent A (Buffer)	2.44	2.44
Reagent B (Glyox)	0.50	0.50
Reagent C (β-NADH)	0.05	0.05

Mix by inversion and equilibrate to 25°C. Monitor the A340nm until constant, using a suitably thermostatted spectrophotometer. Then add:

Deionized Water	—	0.10
Reagent D (Enzyme Solution)	0.10	—

Immediately mix by inversion and record the decrease in A340nm for approximately 5 minutes. Obtain the ΔA340nm/minute using the maximum linear rate for both the Test and Blank.

Calculations

$$\text{Units / ml enzyme} = \frac{(\Delta A340nm / \min \text{Test} - \Delta A340nm / \min \text{Blank}) (3.07) (df)}{(6.22) (0.1)}$$

3.07 = Total volume (in milliliters) of assay
df = Dilution Factor
6.22 = Millimolar extinction coefficient of β-NADH at 340nm
0.1 = Volume (in milliliters) of enzyme used

$$\text{Units / mg solid} = \frac{\text{units / ml enzyme}}{\text{mg solid / ml enzyme}}$$

$$\text{Units / mg protein} = \frac{\text{units / ml enzyme}}{\text{mg protein / ml enzyme}}$$

Unit Definition: *One unit will convert 1.0 µmole of glyoxylate to glycolate per minute at pH 6.4 at 25°C.*

Final Assay Concentration

In a 3.09 ml reaction mix, the final concentrations are 48 mM sodium phosphate, 114 mM glyoxylate, 0.2 mM β nicotinamide adenine dinucleotide, reduced form, and 0.1–0.2 unit glyoxylate reductase.

Enzymatic Assay of GLUTATHIONE-S-TRANSFERASE (EC 2.5.1.18) 1-Chloro-2,4-Dinitrobenzene as Substrate

Principle

$$\text{G-SH} + \text{CDNB} \xrightarrow{\text{Glutathione-S-Transferase}} \text{G-SDNB Conjugate} + \text{HCl}$$

Abbreviations Used

G-SH = Glutathione, Reduced Form
CDNB = 1-Chloro-2,4-Dinitrobenzene
G-SDNB Conjugate = Glutathione-2,4-Dinitrobenzene

Conditions: T = 25°C, pH = 6.5, A340nm, Light path = 1 cm

Method: Continuous Spectrophotometric Rate Determination

Reagents

A. 100 mM Potassium Phosphate Buffer with 1.0 mM Ethylenediaminetetraacetic Acid, pH 6.5 at 25°C (Prepare 100 ml in deionized water using Potassium Phosphate, Monobasic, Anhydrous, and Ethylenediaminetetraacetic Acid, Tetrasodium Salt: Hydrate. Adjust to pH 6.5 at 25°C with 1 M KOH).
B. 75 mM Glutathione, Reduced Solution (G-SH) (Prepare 10 ml in Reagent A using Glutathione, Free Acid, Reduced Form. *Prepare Fresh*).

C. 30 mM 1-Chloro-2,4-Dinitrobenzene Solution (CDNB) (Prepare 1 ml in Reagent D using 1-Chloro-2,4- Dinitrobenzene. *Prepare Fresh*).
D. 95% Ethanol (Nondenatured)
E. Glutathione-S-Transferase Enzyme Solution (Immediately before use, prepare a solution containing 0.075–0.15 unit/ml of Glutathione-S-Transferase in cold Reagent A).

Procedure

Pipette (in milliliters) the following reagents into suitable cuvettes:

	Test	Blank
Reagent A (Buffer)	2.70	2.80
Reagent B (G-SH)	0.10	0.10
Reagent C (CDNB)	0.10	0.10

Equilibrate to 25°C. Monitor the A340nm until constant, using a suitably thermostatted spectrophotometer. Then add:

Reagent E (Enzyme Solution)	0.10	—

Immediately mix by inversion and record the increase in A340nm for approximately 5 minutes. Obtain the ΔA340nm/minute using the maximum linear rate for both the Test and Blank.

Calculations

$$\text{Units / ml enzyme} = \frac{(\Delta A340\text{nm / min Test} - \Delta A340\text{nm / min Blank})\ (3.0)\ (\text{df})}{(9.6)\ (0.10)}$$

3.0 = Total volume (in milliliters) of assay
df = Dilution factor
9.6 = Millimolar extinction coefficient of Glutathione-1-Chloro-2,4-Dinitrobenzene conjugate at 340nm.
0.10 = Volume (in milliliter) of enzyme used.

$$\text{Units / mg solid} = \frac{\text{units / ml enzyme}}{\text{mg solid / ml enzyme}}$$

$$\text{Units / mg protein} = \frac{\text{units / ml enzyme}}{\text{mg protein / ml enzyme}}$$

Unit Definition: *One unit will conjugate 1.0 μmole of 1-chloro-2,4-dinitrobenzene with reduced glutathione per minute at pH 6.5 at 25°C.*

Final Assay Concentration

In a 3.00 ml reaction mix, the final concentrations are 97 mM potassium phosphate, 0.97 mM Ethylenediaminetetraacetic Acid, 2.5 mM glutathione, reduced, 1.0 mM 1-chloro-2,4-dinitrobenzene, 3.2% (v/v) ethanol and 0.0075–0.015 unit glutathione-S-transferase.

Enzymatic Assay of GUANASE (EC 3.5.4.3)

Principle

$$\text{Guanine} + H_2O \xrightarrow{\text{Guanase}} \text{Xanthine} + NH_3$$

$$\text{Xanthine} + H_2O + O_2 \xrightarrow{\text{XOD}} \text{Uric Acid} + H_2O_2$$

Conditions: T = 25°C, pH = 8.0, A290nm, Light path = 1 cm

Method: Continuous Spectrophotometric Rate Determination

Reagents

A. 100 mM Tris HCl Buffer, pH 8.0 at 25°C (Prepare 100 ml in deionized water using Trizma Base. Adjust to pH 8.0 at 25°C with 1 M HCl).

B. 0.001% (w/v) Guanine Solution (Guanine) (Prepare by dissolving 10 mg of Guanine, in 10 ml of 1 M NaOH. Then add 1 ml of the Guanine Solution to 100 ml of Reagent A. Adjust to pH 8.0 at 25°C with either 1 M HCl or 1 M NaOH, if necessary).

C. Xanthine Oxidase Enzyme Solution (XOD) (Prepare a solution containing 0.3 unit/ml of Xanthine Oxidase, in cold deionized water).

D. Guanase Enzyme Solution (Guanase) (Immediately before use, prepare a solution containing 0.10–0.15 unit/ml of Guanase in cold deionized water).

Procedure

Pipette (in milliliters) the following reagents into suitable cuvettes:

	Test	Blank
Reagent B (Guanine)	2.80	2.80
Reagent C (XOD)	0.10	0.10

Mix by inversion and equilibrate to 25°C. Monitor the A290nm until constant, using a suitably thermostatted spectrophotometer. Then add:

	Test	Blank
Reagent D (Guanase)	0.10	—
Deionized Water	—	0.10

Immediately mix by inversion and record the increase in A290nm for approximately 10 minutes. Obtain the ΔA290nm/minute using the maximum linear rate for both the Test and Blank.

Calculations

$$\text{Units / ml enzyme} = \frac{(\Delta A290\text{nm / min Test} - \Delta A290\text{nm / min Blank})\,(3)\,(\text{df})}{(12.2)\,(0.1)}$$

3 = Total volume (in milliliters) of assay
df = Dilution factor
12.2 = Millimolar extinction coefficient1 of Uric Acid at 290nm
0.1 = Volume (in milliliter) of enzyme used

$$\text{Units / mg protein} = \frac{\text{units / ml enzyme}}{\text{mg protein / ml enzyme}}$$

Unit Definition: *One unit will deaminate 1.0 µmole of guanine to xanthine per minute at pH 8.0 at 25°C.*

Final Assay Concentrations

In a 3.00 ml reaction mix, the final concentrations are 93 mM Tris, 0.0009% (w/v) guanine, 62 mM sodium hydroxide, 0.03 unit xanthine oxidase, and 0.01–0.015 unit guanase.

Note

1. Xanthine Oxidase Unit Definition: One unit will convert 1.0 µmole of xanthine to uric acid per minute at pH 7.5 at 25°C.

Enzymatic Assay of GUANYLATE KINASE (EC 2.7.4.8)

Principle

$$\text{GMP} + \text{ATP} \xrightarrow{\text{Guanylate Kinase}} \text{GDP} + \text{ADP}$$

$$\text{ADP} + \text{PEP} \xrightarrow{\text{Pyruvate Kinase}} \text{ATP} + \text{Pyruvate}$$

$$\text{GDP} + \text{PEP} \xrightarrow{\text{Pyruvate Kinase}} \text{GTP} + \text{Pyruvate}$$

$$2\ \text{Pyruvate} + 2\ \beta\text{-NADH} \xrightarrow{\text{Lactic Dehydrogenase}} 2\ \text{Lactate} + 2\ \beta\text{-NAD}$$

Abbreviations Used

GMP = Guanosine 5'-Monophosphate
ATP = Adenosine 5'-Triphosphate
GDP = Guanosine 5'-Diphosphate
ADP = Adenosine 5'-Diphosphate
PEP = Phospho(enol) phosphate
β-NADH = β-Nicotinamide Adenine Dinucleotide, Reduced Form
β-NAD = β-Nicotinamide Adenine Dinucleotide, Oxidized Form

Conditions: T = 30°C, pH = 7.5, A340nm, Light path = 1 cm

Method: Continuous Spectrophotometric Rate Determination

Reagents

A. 200 mM Tris HCl Buffer, pH 7.5 at 30°C (Prepare 50 ml in deionized water using Trizma Base. Adjust to pH 7.5 at 30°C with 1 M HCl).
B. 1 M Potassium Chloride Solution (KCl) (Prepare 10 ml in deionized water using Potassium Chloride).
C. 60 mM Magnesium Sulfate Solution ($MgSO_4$) (Prepare 20 ml in deionized water using Magnesium Sulfate, Heptahydrate).
D. 40 mM Phospho(enol) pyruvate Solution (PEP) (Prepare 50 ml in deionized water using Phospho(enol) Pyruvate, Trisodium Salt, Hydrate. *Prepare Fresh*).

E. 100 mM Ethylenediaminetetraacetic Acid Solution (EDTA) (Prepare 10 ml in deionized water using Ethylenediaminetetraacetic Acid, Tetrasodium Salt, Hydrate).
F. 3.8 mM β-Nicotinamide Adenine Dinucleotide, Reduced Form (β-NADH) (Prepare 2 ml in deionized water using β-Nicotinamide Adenine Dinucleotide, Reduced Form, Dipotassium Salt. *Prepare Fresh*).
G. 30 mM Adenosine 5'-Triphosphate Solution (ATP) (Prepare 2 ml in deionized water using Adenosine 5'-Triphosphate, Disodium Salt. *Prepare Fresh*).
H. PK/LDH Enzymes Suspension (PK/LDH).
I. 100 mM Guanosine 5'-Monophosphate Solution (GMP) (Prepare 5 ml in deionized water using Guanosine 5'-Monophosphate, Disodium Salt. *Prepare Fresh*).
J. Guanylate Kinase Enzyme Solution (Immediately before use, prepare a solution containing 1.5–3.0 units/ml of Guanylate Kinase in cold deionized water).

Procedure

Prepare a reaction cocktail by pipetting (in milliliters) the following reagents into a suitable container:

Deionized Water	12.50
Reagent A (Buffer)	10.20
Reagent B (KCl)	1.50
Reagent C ($MgSO_4$)	3.00
Reagent D (PEP)	0.45
Reagent E (EDTA)	0.15
Reagent F (β-NADH)	1.00

Mix by swirling and adjust to pH 7.5 at 30°C with 100 mM HCl or 100 mM NaOH, if necessary. Pipette (in milliliters) the following reagents into suitable cuvettes:

	Test	Blank
Deionized Water	—	0.008
Reaction Cocktail	2.70	2.70
Reagent G (ATP)	0.10	0.10
Reagent H (PK/LDH)	0.05	0.05
Reagent J (Enzyme Solution)	0.008	—

Mix by inversion and equilibrate to 30°C. Monitor the A340nm until constant, using a suitably thermostatted spectrophotometer. Then add:

Reagent I (GMP)	0.10	0.10

Immediately mix by inversion and record the decrease in A340nm for approximately 10 minutes. Obtain the ΔA340nm/minute using the maximum linear rate for both the Test and Blank.

Calculations

$$\text{Units/ml enzyme} = \frac{(\Delta A340nm/min\ Test - \Delta A340nm/min\ Blank)\ (2.958)\ (df)}{(2)\ (6.22)\ (0.008)}$$

2.958 = Total volume (in milliliters) of enzyme assay
df = Dilution factor
2 = 2 moles of β-NAD produced per mole of GMP utilized
6.22 = Millimolar extinction coefficient of β-NADH at 340nm
0.008 = Volume (in milliliters) of enzyme used

$$\text{Units / mg solid} = \frac{\text{units / ml enzyme}}{\text{mg solid / ml enzyme}}$$

$$\text{Units / mg protein} = \frac{\text{units / ml enzyme}}{\text{mg protein / ml enzyme}}$$

Unit Definition: *One unit will convert 1.0 µmole each of GMP and ATP to GDP and ADP per minute at pH 7.5 at 30°C.*

Final Assay Concentrations

In a 2.958 ml reaction mix, the final concentrations are 64.7 mM Tris, 47.5 mM potassium chloride, 5.7 mM magnesium sulfate, 0.57 mM phospho(enol) pyruvate, 0.48 mM Ethylenediaminetetraacetic Acid, 0.12 mM β-nicotinamide adenine dinucleotide, reduced form, 1.0 mM adenosine 5'-triphosphate, 35 units pyruvate kinase, 50 units lactic dehydrogenase, 3.4 mM guanosine 5'-monophosphate and 0.012–0.024 unit guanylate kinase.

Notes

1. Contains not less than 700 Pyruvate Kinase units and 1000 L-Lactic Dehydrogenase units per ml.
2. L-Lactic Dehydrogenase Unit Definition: One unit will reduce 1.0 µmole of pyruvate to L-lactate per minute at pH 7.5 at 37°C.
3. Pyruvate Kinase Unit Definition: One unit will convert 1.0 µmole of phospho(enol) pyruvate to pyruvate per minute at pH 7.6 at 37°C.

Enzymatic Assay of HEMICELLULASE

Principle

$$\text{Hemicellulose} + H_2O \xrightarrow{\text{Hemicellulase}} \text{D-Galactose}$$

$$\text{D-Galactose} + \beta\text{-NAD} \xrightarrow{\beta\text{-Galactose Dehydrogenase}} \text{Galactonate} + \beta\text{-NADH}$$

Abbreviations Used

β-NAD = β-Nicotinamide Adenine Dinucleotide, Oxidized Form
β-NADH = β-Nicotinamide Adenine Dinucleotide, Reduced Form

Conditions: T = 37°C, pH = 5.5, A340nm, Light path = 1 cm

Method: Spectrophotometric Stop Rate Determination

Reagents

A. 50 mM Sodium Acetate Buffer, pH 5.5 at 37°C (Prepare 50 ml in deionized water using Sodium Acetate, Trihydrate. Adjust to pH 5.5 at 37°C with 1 M HCl).

B. 1% (w/v) Locust Bean Gum Suspension (Hemicel) (Prepare 100 ml in Reagent A using Gum, Locust Bean. Mix and heat gently to facilitate dissolving the Locust Bean Gum. *Do not Boil*).

C. Hemicellulase Enzyme Solution (Immediately before use, prepare a solution containing 0.2–0.6 unit/ml of Hemicellulase in cold deionized water).

D. 19 mM β-Nicotinamide Adenine Dinucleotide Solution (β-NAD) (Dissolve the contents of one 20 mg vial of β-Nicotinamide Adenine Dinucleotide, in 1.58 ml of deionized water *or* prepare 1.5 ml in deionized water using β-Nicotinamide Adenine Dinucleotide. *Prepare Fresh*).

E. 100 mM Tris HCl Buffer, pH 8.6 at 25°C (Tris) (Prepare 100 ml in deionized water using Trizma Base. Adjust to pH 8.6 at 25°C with 1 M HCl).

F. β-Galactose Dehydrogenase (β-GalDH) (Use β-Galactose Dehydrogenase).

Procedure

Step 1: Pipette (in milliliters) the following reagents into suitable tubes:

	Test	Blank
Reagent B (Hemicel)	4.00	4.00

Equilibrate to 37°C. Then add:

Reagent C (Enzyme Solution)	1.00	—

Immediately mix by swirling and incubate at 37°C for exactly 120 minutes with moderate shaking. Then add:

Reagent C (Enzyme Solution)	—	1.00

Immediately transfer the tubes into an ice bath. Allow to stand until the suspension is settled. Centrifuge for 2 minutes to clarify.

Step 2: Pipette (in milliliters) the following reagents into suitable cuvettes:

	Test	Blank
Test Supernatant	0.50	—
Blank Supernatant	—	0.50
Reagent E (Tris)	2.50	2.50
Reagent D (β-NAD)	0.10	0.10

Mix by swirling and equilibrate to 25°C. Monitor the A340nm until constant, using a suitably thermostatted spectrophotometer. Record the initial A340nm for both the Test and Blank. Then add:

Reagent F (β-GalDH)	0.05	0.05

Immediately mix by inversion and record the increase in A340nm until complete (for approximately 30 minutes). Obtain the A340nm for both the Test and Blank.

Calculations

$$\Delta A340nm\ Test = A340nm\ Test\ Final - A340nm\ Test\ Initial$$

$$\Delta A340nm\ Blank = A340nm\ Blank\ Final - A340nm\ Blank\ Initial$$

$$Units/mg\ enzyme = \frac{(\Delta A340nm\ Test - \Delta A340nm\ Blank)\ (3.15)}{(6.22)\ (2)\ (mg\ enzyme/ml\ Step\ 1\ RM)\ (0.5)}$$

6.22 = Millimolar extinction coefficient of β-NADH at 340nm
2 = Conversion factor from 2 hours to 1 hour (Unit Definition)
3.15 = Final volume of Step 2
0.5 = Volume from Step 1 used in Step 2
RM = Reaction Mixture

Unit Definition: *One unit will liberate 1.0 µmole of D-galactose from hemicellulose in one hour at pH 5.5 at 37°C.*

Final Assay Concentration

In a 5.00 ml reaction mix, the final concentrations are 40 mM sodium acetate, 0.8% (w/v) locust bean gum and 0.2–0.6 unit of hemicellulase.

Enzymatic Assay of HEPARINASE I (EC 4.2.2.7)

Principle

$$Heparin + H_2O \xrightarrow{Heparinase\ I} Unsaturated\ Uronic\ Acid$$

Conditions: T = 25°C, pH = 7.5, A235nm, Light path = 1 cm

Method: Spectrophotometric Stop Rate Determination

Reagents

A. 20 mM Tris HCl Buffer with 50 mM Sodium Chloride, 4 mM Calcium Chloride and 0.01% (w/v) Bovine Serum Albumin, pH 7.5 at 25°C (Prepare 50 ml in deionized water using Trizma Base, Sodium Chloride, Calcium Chloride, Dihydrate, and Albumin, Bovine. Adjust to pH 7.5 at 25°C with 1 M HCl).

B. 2% (w/v) Heparin Solution (Prepare 2 ml in deionized water using Heparin, Sodium Salt).

C. 50 mM Hydrochloric Acid Solution (Prepare 50 ml in deionized water using Hydrochloric Acid).

D. Heparinase I Enzyme Solution (Immediately before use, prepare a solution containing 100–300 units/ml of Heparinase I in cold Reagent A).

Procedure

Pipette (in milliliters) the following reagents into suitable containers:

	Test	Blank
Reagent A (Buffer)	0.375	0.375
Reagent B (Heparin)	0.05	0.05

Mix by inversion and equilibrate to 25°C. Then add:

	Test	Blank
Reagent D (Enzyme Solution)	0.075	—

Immediately mix by inversion and incubate at 25°C for exactly 10 minutes. Then add:

Reagent C (HCl)	2.50	2.50
Reagent D (Enzyme Solution)	—	0.075

Immediately mix by inversion and transfer to suitable quartz cuvettes. Record the A235nm for both the Test and Blank.

Calculations

$$\text{Units / ml enzyme} = \frac{(\text{A235nm Test} - \text{A235nm Blank})\ (6)\ (3)\ (10)}{(5.50)\ (0.075)}$$

6 = Time correction factor from 10 minutes to 1 hour as per the Unit Definition
3 = Volume (in milliliters) of assay
10 = Conversion factor from 1 μmole to 0.1 μmole as per the Unit Definition
5.50 = Millimolar extinction coefficient1 of the Unsaturated Uronic Acid products at 235nm
0.075 = Volume (in milliliter) of enzyme used

$$\text{Units / mg solid} = \frac{\text{units / ml enzyme}}{\text{mg solid / ml enzyme}}$$

$$\text{Units / mg protein} = \frac{\text{units / ml enzyme}}{\text{mg protein / ml enzyme}}$$

Unit Definition: *One unit will form 0.1 μmole of unsaturated uronic acid per hour at pH 7.5 at 25°C.*

Final Assay Concentration

In a 0.50 ml reaction mix, the final concentrations are 18 mM Tris, 45 mM sodium chloride, 4 mM calcium chloride, 0.01% (w/v) Bovine Serum Albumin, 0.2% (w/v) heparin and 7.5–23 units heparinase I.

Enzymatic Assay of HEPARINASE II

Principle

$$\text{Heparin} + H_2O \xrightarrow{\text{Heparinase II}} \text{Unsaturated Uronic Acid}$$

Conditions: T = 25°C, pH = 7.0, A235nm, Light path = 1 cm

Method: Spectrophotometric Stop Rate Determination

Reagents

A. 100 mM Sodium Acetate Buffer, with 0.01% (w/v) Bovine Serum Albumin pH 7.0 at 25°C (Prepare 50 ml in deionized water using Sodium Acetate, Trihydrate and Albumin Bovine. Adjust to pH 7.0 at 25°C with 0.1 M HCl).

B. 5.0% (w/v) Heparin Solution (Heparin) (Prepare 4 ml in deionized water using Heparin, Sodium Salt).

C. 10 mM Calcium Acetate Solution ($Ca(OAc)_2$) (Prepare 10 ml in deionized water using Calcium Acetate).

D. 30 mM Hydrochloric Acid Solution (Prepare 50 ml using Hydrochloric Acid).

E. Heparinase II Enzyme Solution (Immediately before use, prepare a solution containing 30–40 units/ml of Heparinase II in cold Reagent A).

Procedure

Pipette (in milliliters) the following reagents into suitable containers:

	Test
Reagent A (Buffer)	0.13
Reagent B (Heparin)	0.04
Reagent C ($Ca(OAc)_2$)	0.03

Mix by swirling and equilibrate to 25°C. Then add:

Reagent E (Enzyme Solution)	0.10

Immediately mix by swirling and transfer 0.05 ml of the Test solution to 3.00 ml of Reagent D (T_i). Incubate at 25°C for exactly 60 minutes. Then transfer another aliquot (0.05 ml) of the Test solution to 3.00 ml of Reagent D (T_f).Record the A235nm for both the T_i and T_f solutions.

Calculations

$$\text{Units / ml enzyme} = \frac{(\text{A235nm } T_f - \text{A235nm } T_i)\,(0.3)\,(3.05)\,(10)\,(\text{df})}{(5.50)\,(0.1)\,(0.05)}$$

0.05 = Aliquot from reaction mix used in the final volume
0.3 = Volume of reaction mix in assay
10 = 1 μmole to 0.1 μmole conversion according to unit definition
df = Dilution factor
5.50 = Millimolar extinction coefficient of the Unsaturated Uronic Acid products at 235nm
3.05 = Final volume of assay
0.1 = Volume (in milliliter) of enzyme used

$$\text{Units / mg solid} = \frac{\text{units / ml enzyme}}{\text{mg solid / ml enzyme}}$$

Unit Definition: *One unit will form 0.1 μmole of unsaturated uronic acid per hour at pH 7.0 at 25°C.*

Final Assay Concentration

In a 0.30 ml reaction mix, the final concentrations are 77 mM sodium acetate, 0.7% (w/v) heparin, 1 mM calcium acetate, 0.008% (w/v) BSA and 3.0–4.0 units heparinase II.

Enzymatic Assay of HEPARINASE III (EC 4.2.2.8)

Principle

$$\text{Heparan Sulfate} \xrightarrow{\text{Heparinase III}} \text{Unsaturated Uronic Acid}$$

Conditions: T = 25°C, pH = 7.5, A235nm, Light path = 1 cm

Method: Spectrophotometric Stop Rate Determination

Reagents

A. 20 mM Tris HCl with 50 mM Sodium Chloride, 4 mM Calcium Chloride and 0.01% (w/v) Bovine Serum Albumin, pH 7.5 at 25°C (Prepare 50 ml in deionized water using Trizma Base, Sodium Chloride, Calcium Chloride, Dihydrate, and Albumin, Bovine. Adjust to pH 7.5 at 25°C with 1 M HCl).
B. 1.0% (w/v) Heparan Sulfate Solution (HS) (Immediately before use, prepare 0.5 ml in deionized water using Heparan Sulfate, Sodium Salt).
C. 50 mM Hydrochloric Acid Solution (HCl) (Prepare 50 ml in deionized water using Hydrochloric Acid).
D. Heparinase III Enzyme Solution (Immediately before use, prepare a solution containing 75–100 units/ml of Heparinase III in cold Reagent A).

Procedure

Pipette (in milliliters) the following reagents into suitable containers:

	Test	Blank
Reagent A (Buffer)	0.23	0.23
Reagent B (HS)	0.05	0.05

Mix by swirling and equilibrate to 25°C. Then add:

Reagent D (Enzyme Solution)	0.02	—

Immediately mix by swirling and incubate at 25°C for exactly 30 minutes. Then add:

Reagent C (HCl)	2.70	2.70
Reagent D (Enzyme Solution)	—	0.02

Mix by swirling and centrifuge for 3 minutes. Transfer the supernatants to suitable quartz cuvettes. Record the A235nm for both the Test and Blank using a suitable spectrophotometer.

Calculations

$$\text{Units / ml enzyme} = \frac{(\text{A235nm Test} - \text{A235nm Blank})\,(2)\,(10)\,(3)\,(\text{df})}{(5.50)\,(0.02)}$$

2 = Conversion factor from 30 minutes to 1 hour as per the Unit Definition
10 = 1 μmole to 0.1 μmole conversion factor as per the Unit Definition
3 = Total volume (in milliliters) of assay
df = Dilution factor
5.50 = Millimolar extinction coefficient of the Unsaturated Uronic Acid products at 235nm
0.02 = Volume (in milliliter) of enzyme used

$$\text{Units / mg solid} = \frac{\text{units / ml enzyme}}{\text{mg solid / ml enzyme}}$$

$$\text{Units / mg protein} = \frac{\text{units / ml enzyme}}{\text{mg protein / ml enzyme}}$$

Unit Definition: *One unit will form 0.1 μmole of unsaturated uronic acid per hour at pH 7.5 at 25°C.*

Final Assay Concentration

In a 0.30 ml reaction mix, the final concentrations are 17 mM Tris, 42 mM sodium chloride, 3 mM calcium chloride, 0.008% (w/v) Bovine Serum Albumin, 0.2% (w/v) heparan sulfate and 1.5–2 units heparinase III.

Enzymatic Assay of HESPERIDINASE (EC 3.2.1.40)

Principle

$$\text{Hesperidin} + H_2O \xrightarrow{\text{Hesperidinase}} \text{Reducing Sugar (measured as glucose)}$$

Conditions: T = 40°C, pH = 3.8, A540nm, Light path = 1 cm

Method: Colorimetric

Reagents

A. 200 mM Sodium Phosphate Solution (Prepare 100 ml in deionized water using Sodium Phosphate, Monobasic, Anhydrous).

B. 100 mM Citric Acid Solution (Prepare 100 ml in deionized water using Citric Acid, Free Acid, Monohydrate).

C. McIlvaine Buffer, pH 3.8 at 40°C (Prepare 100 ml using Reagent A. Adjust to pH 3.8 at 40°C with Reagent B).

D. 666 mM Sodium Hydroxide Solution (NaOH) (Prepare 100 ml in deionized water using Sodium Hydroxide, Anhydrous).

E. 2.0 mM Hesperidin Substrate Solution (Hesperidin) (Prepare by dissolving 120 mg of Hesperidin, in 30 ml of Reagent D. Add 40 ml of Reagent C and adjust to pH 3.8 with 1 M HCl).

F. Hesperidinase Enzyme Solution (Immediately before use, prepare a solution containing 0.03–0.06 unit/ml of Hesperidinase in cold deionized water).

G. 16 mM Copper Sulfate, 1.26 M Sodium Sulfate, 226 mM Sodium Carbonate, 190 mM Sodium Bicarbonate, and 43 mM Sodium Potassium Tartrate Solution (Copper Soln) (Prepare 1 liter in deionized water using Cupric Sulfate Pentahydrate, Sodium Bicarbonate, Sodium Sulfate, Anhydrous, Sodium Carbonate, Anhydrous, and Sodium Potassium Tartrate Tetrahydrate).

H. 40 mM Molybdic Acid, 19 mM Arsenic Acid and 756 mM Sulfuric Acid Solution (Ars-Mol) (Prepare 1 liter in deionized water using Molybdic Acid, Ammonium Salt Tetrahydrate, Arsenic Acid, Sodium Salt, and Sulfuric Acid).

I. 5.55 mM Glucose Standard Solution (Std).

Procedure

Pipette (in milliliters) the following reagents into suitable test tubes:

	Test	Blank
Reagent E (Hesperidin)	4.00	4.00

Equilibrate to 40°C. Then add:

Deionized Water	—	1.00
Reagent F (Enzyme)	1.00	—

Mix by swirling and incubate at 40°C for exactly 30 minutes. Then add:

	Test	Test Blank	Std 1	Std 2	Std 3	Std 4	Std Blank
Test Solution	1.00	—	—	—	—	—	—
Test Blank Solution	—	1.00	—	—	—	—	—
Deionized Water	—	—	0.95	0.90	0.85	0.80	1.00
Reagent I (Std)	—	—	0.05	0.10	0.15	0.20	—
Reagent G (Copper Solution)	1.00	1.00	1.00	1.00	1.00	1.00	1.00

Immediately mix by swirling. Place a marble over the top of the tubes and transfer the tubes to a boiling water bath. Incubate for 10 minutes. Remove from the boiling water bath and allow to cool to room temperature. Then add:

Reagent H (Ars-Mol)	1.00	1.00	1.00	1.00	1.00	1.00	1.00

Shake or vortex the tubes until the foaming stops and any precipitate present is dissolved. Then add:

Deionized Water	10.00	10.00	10.00	10.00	10.00	10.00	10.00

Mix by swirling and transfer the solutions to suitable cuvettes. Record the A540nm for Test, Blank and Standards, using a suitable spectrophotometer.

Calculations

Standard Curve: ΔA540nm Std = A540nm Std–A540nm Std Blank
Prepare a standard curve by plotting the ΔA540nm of the Standard vs the μmoles of glucose.
Sample Determination: ΔA540nm Sample = A540nm Test–A540nm Test Blank
Determine the μmoles of glucose liberated using the Standard Curve.

$$\text{Units / ml enzyme} = \frac{[\mu\text{moles of reducing sugar (as glucose) liberated}]\,(\text{df})\,(5)}{(30)\,(1)\,(1)}$$

df = Dilution factor
5 = Volume (in milliliters) of reaction mix
30 = Time (in minutes) of assay per the Unit Definition
1 = Volume (in milliliter) of enzyme used
1 = Volume (in milliliter) of reaction mixture used in the colorimetric determination of the reducing sugar

$$\text{Units / mg solid} = \frac{\text{units / ml enzyme}}{\text{mg solid / ml enzyme}}$$

$$\text{Units / mg protein} = \frac{\text{units / ml enzyme}}{\text{mg protein / ml enzyme}}$$

Unit Definition: *One unit will liberate 1.0 μmole of reducing sugar (as glucose) from hesperidin per minute at pH 3.8 at 40°C.*

Final Assay Concentrations

In a 5.00 ml reaction mix, the final concentrations are 1.6 mM hesperidin and 0.03–0.06 unit hesperidinase. (Note: Concentration of the buffers are not exact due to the method of preparation of the McIlvaine's Buffer).

Notes

1. Hesperidin substrate solution only remains soluble at pH 3.8 (Assay Conditions) for 5–20 minutes. Maintain the pH of the hesperidin solution to pH 3.8, this solution should be dispensed into reaction vessels (4.0 ml each).
2. Sodium Sulfate, Sodium Carbonate, and Sodium Potassium Tartrate are dissolved in approximately 500 ml of deionized water. Cupric Sulfate is dissolved in approximately 100 ml of deionized water and is slowly added to the above solution to avoid precipitation. Sodium Bicarbonate is dissolved first in deionized water and then added to the above solution. Dilute the solution to 1 liter. If a precipitate forms, it should be removed by filtration prior to use. Store in an amber bottle at room temperature and avoid exposure to direct sunlight.
3. Molybdic Acid is dissolved in approximately 300 ml of deionized water. Add Sulfuric Acid slowly. Caution, this is an exothermic reaction! Arsenic Acid is dissolved in approximately 300 ml of deionized water and is added to the above solution. The solution is diluted to a total volume of 1 liter and incubated at 37°C for 48–72 hours. Stir vigorously for 1 hour prior to each use. Store in an amber bottle at room temperature and avoid exposure to direct sunlight. The solution expires six months after preparation.
4. Since the hesperidin substrate will precipitate out of solution, the reaction vials should be swirled often during the reaction time to keep the substrate in suspension.
5. Copper Solution (Reagent G) should be dispensed into vials ahead of time to facilitate stopping the reaction.
6. This solution should be centrifuged prior to reading in spectrophotometer.

Enzymatic Assay of HEXOKINASE (EC 2.7.1.1)

Principle

$$\text{Glucose} + \text{ATP} \xrightarrow{\text{Hexokinase}} \text{Glucose 6-Phosphate} + \text{ADP} + \text{H}^+$$

$$\text{Cresol Red} + \text{H}^+ \rightarrow \text{Reduced Cresol Red}$$

Abbreviations Used

ATP = Adenosine 5'-Triphosphate

ADP = Adenosine 5'-Diphosphate

Conditions: T = 25°C, pH 8.5, A560nm, Light path = 1 cm

Method: Continuous Spectrophotometric Rate Determination

Reagents

A. 100 mM Glycylglycine Buffer, pH 8.5 at 25°C. (Prepare 100 ml in deionized water using Glycylglycine, Free Base. Adjust to pH 8.5 at 25°C with 1 M HCl).

B. 200 mM Adenosine 5'-Triphosphate Solution (ATP) (Prepare 10 ml in deionized water using Adenosine 5'-Triphosphate, Disodium Salt. *Prepare Fresh*).
C. 200 mM Glucose Solution (Gluc) (Prepare 10 ml in deionized water using β-(+)-Glucose).
D. 0.01% Cresol Red with 128 mM Magnesium Chloride Solution (Cresol Red) (Prepare 200 ml in deionized water using Cresol Red, Sodium Salt and Magnesium Chloride, Hexahydrate. Facilitate by first dissolving Cresol Red into 6.6 ml of 95% ethanol. Transfer this solution to a 200 ml graduated cylinder and add 52 g of Magnesium Chloride, Hexahydrate. Dilute to 200 ml with deionized water).
E. 100 mM Hydrochloric Acid Standardized Solution (HCl) (Prepare 1 liter in deionized water using Concentrated Hydrochloric Acid. Standardize against Tris Base. Color change is from orange to pink).
F. 0.5% Glucose Solution .
G. Hexokinase Enzyme Solution (Immediately before use, prepare a solution containing 10 units/ml of Hexokinase in Reagent F).

Procedure

Prepare a reaction cocktail by pipetting (in milliliters) the following reagents into a suitable container:

Reagent B (ATP)	5.00
Reagent D (Cresol Red)	6.60

Mix and slowly add 0.1 M NaOH until the solution just turns from red to purple (pH about 8.2). Then add:

Deionized Water	33.40
Reagent A (Buffer)	5.00

Mix. Adjust to pH 8.5 at 25°C with 100 mM HCl or 100 mM NaOH, if necessary.

Titer Determination: Determine titer of reaction cocktail by pipetting (in milliliters) the following reagents into a suitable cuvette:

Reaction Cocktail	2.50
Reagent C (Glucose)	0.40

Mix by inversion and equilibrate to 25°C. Monitor the A560nm until constant, using a suitably thermostatted spectrophotometer. Record the initial A560nm. Then add:

Reagent D (HCl)	0.10

Mix and immediately record final A560nm.

$$\text{Titer} = \frac{(\text{A560nm initial} - \text{A560nm final})\ (2.9)}{(1000)\ (0.1)\ (\text{Molarity of Reagent D})}$$

2.9 = Volume of Titer reaction Mix
1000 = Coversion from millimolar to micromolar
0.1 = Volume of Reagent D used

Sample: Pipette (in milliliter) the following reagents into suitable cuvettes:

	Test	Blank
Reaction Cocktail	2.50	2.50
Reagent C (Gluc)	0.40	0.40

Mix by inversion and equilibrate to 25°C. Monitor the A560nm until constant, using a suitably thermostatted spectrophotometer. Then add:

Reagent F (Diluent)	—	0.10
Reagent G (Enzyme Solution)	0.10	—

Immediately mix by inversion and record the decrease in A560nm for approximately 5 minutes. Obtain the ΔA560nm/minute using the maximum linear rate for both the Test and Blank.

Calculations

$$\text{Units / mg enzyme} = \frac{\Delta A560\text{nm / min Test} - \Delta A560\text{nm / min Blank}}{\text{(titer) (mg enzyme / ml RM)}}$$

RM = Reaction Mix

Unit Definition: *One unit will phosphorylate 1.0 μmole of glucose per minute at pH 8.5 (±0.5) at 25°C.*

Final Assay Concentrations

In a 3 ml reaction mix, the final concentrations are 8.3 mM glycylglycine, 17 mM ATP, 0.0011% cresol red, 14 mM magnesium chloride, 27 mM glucose and 1 unit hexokinase.

Enzymatic Assay of HEXOKINASE INSOLUBLE (EC 2.7.1.1)

Principle

$$\text{D-Glucose} + \text{ATP} \xrightarrow{\text{Hexokinase}} \text{D-Glucose 6-Phosphate} + \text{ADP}$$

$$\text{D-Glucose 6-Phosphate} + \beta\text{-NADP} \xrightarrow{\text{G-6-PDH}} \text{6-PG} + \beta\text{-NADPH}$$

Abbreviations Used

ATP = Adenosine 5'-Triphosphate
ADP = Adenosine 5'-Diphosphate
G-6-PDH = Glucose-6-Phosphate Dehydrogenase
β-NADP = β-Nicotinamide Adenine Dinucleotide Phosphate, Oxidized Form
β-NADPH = β-Nicotinamide Adenine Dinucleotide Phosphate, Reduced Form
6-PG = 6-Phospho-D-Gluconate

Conditions: T = 30°C, pH = 7.4, A340nm, Light path = 1 cm

Method: Spectrophotometric Rate Determination

Reagents

A. 250 mM Glycylglycine Buffer, pH 7.4 at 30°C (Prepare 100 ml in deionized water using Gly-Gly, Hydrochloride. Adjust to pH 7.4 at 30°C with 1 M NaOH).

B. 1 M Glucose Solution (Glucose) (Prepare 25 ml in deionized water using β-D(+)Glucose).

C. 20 mM β-Nicotinamide Adenine Dinucleotide Phosphate, Oxidized Form, Solution (β-NADP) (Prepare 10 ml in deionized water using β-Nicotinamide Adenine Dinucleotide Phosphate, Sodium Salt. *Prepare Fresh*).

D. 300 mM Magnesium Chloride Solution ($MgCl_2$) (Prepare 10 ml in deionized water using Magnesium Chloride, Hexahydrate).

E. 30 mM Adenosine 5'-Triphosphate Solution (ATP) (Prepare 10 ml in deionized water using Adenosine 5'-Triphosphate, Disodium Salt).

F. Glucose-6-Phosphate Dehydrogenase Enzyme Solution (G-6-PDH) (Immediately before use, prepare a solution containing 50 units/ml of Glucose-6-Phosphate Dehydrogenase in cold Reagent A).

G. Hexokinase Insoluble Enzyme Suspension (Hex-Insol) (Immediately before use, prepare a suspension containing 0.5–1.0 unit/ml of Hexokinase, Insoluble Enzyme Attached to Beaded Agarose in cold deionized water).

Procedure

Prepare a reaction cocktail by pipetting (in milliliters) the following reagents into a suitable container:

Deionized Water	41.40
Reagent A (Buffer)	10.00
Reagent E (ATP)	2.00
Reagent B (Glucose)	4.00
Reagent C (β-NADP)	2.00
Reagent D ($MgCl_2$)	2.00
Reagent F (G-6-PDH)	2.00

Mix by swirling and adjust to pH 7.4 at 30°C with either 1 M HCl or 1 M NaOH if necessary. Incubate at 30°C for 5 minutes and then pipette (in milliliters) the following reagents into 50 ml beakers:

	Test	Blank
Reaction Cocktail	20.00	20.00

Equilibrate to 30°C. Then add:

Reagent G (Hex-Insol)	0.15	—
Deionized Water	—	0.15

Mix continuously by stirring and incubate at 30°C for exactly 5 minutes. Immediately filter the samples through Whatman #54 filter paper and transfer the filtrates from both the Test and Blank to suitable cuvettes and record the A340nm for approximately

3 minutes using a suitable spectrophotometer. The final A340 after the 5 minute reaction time (prior to filtration) is the result of the total Hexokinase activity (which is the sum of the soluble and insoluble Hexokinase activities). The soluble activity is indicated by the increasing linear change in A340 after filtration. The insoluble activity is the difference between the total activity and the soluble activity.

Calculations

Use the initial spectrophotometric readings of the filtered Test and Blank to calculate the total activity.

$$\text{Total Activity (Units / ml enzyme)} = \frac{(\text{A340nm Test} - \text{A340nm Blank})\,(20.15)\,(\text{df})}{(5)\,(6.22)\,(0.15)}$$

Use the ΔA340/min determined for the filtered Test and Blank to calculate the soluble activity.

$$\text{Soluble Activity (Units / ml enzyme)} = \frac{(\Delta\text{A340nm / min Test} - \Delta\text{A340nm / min Blank})\,(20.15)\,(\text{df})}{(6.22)\,(0.15)}$$

20.15 = Total volume (in milliliters) of assay
df = Dilution factor
5 = Time (in minutes) of assay
6.22 = Millimolar extinction coefficient of β-NADH at 340 nm
0.15 = Volume (in milliliter) of enzyme used

Insoluble Activity (Units/ml enzyme) = Total Activity – Soluble Activity

$$\text{Insoluble Activity (Units / g)} = \frac{\text{units / ml enzyme (1000)}}{\text{mg dry agarose / ml enzyme solution}}$$

Unit Definition: *One unit will phosphorylate 1.0 μmole of glucose per minute at pH 7.4 at 30°C.*

Final Assay Concentration

In a 20.15 ml reaction mix, the final concentrations are 39 mM glycylglycine, 63 mM glucose, 0.63 mM β-nicotinamide adenine dinucleotide phosphate, 9.4 mM magnesium chloride, 0.94 mM adenosine 5'-triphosphate, 32 units glucose-6-phosphate dehydrogenase, and 0.075–0.15 unit hexokinase, insoluble enzyme attached to beaded agarose.

Note

1. Glucose-6-Phosphate Dehydrogenase Unit Definition: One unit will oxidize 1.0 μmole of D-glucose 6-phosphate to 6-phospho-D-gluconate per minute in the presence of β-NADP at pH 7.4 at 25°C.

Enzymatic Assay of HEXOKINASE (EC 2.7.1.1)

Principle

$$\text{D-Glucose} + \text{ATP} \xrightarrow{\text{Hexokinase}} \text{D-Glucose 6-Phosphate} + \text{ADP}$$

$$\text{D-Glucose 6-Phosphate} + \beta\text{-NADP} \xrightarrow{\text{G-6-PDH}} \text{6-PG} + \beta\text{-NADPH}$$

Abbreviations Used

ATP = Adenosine 5'-Triphosphate
ADP = Adenosine 5'-Diphosphate
G-6-PDH = Glucose-6-Phosphate Dehydrogenase
β-NADP = β-Nicotinamide Adenine Dinucleotide Phosphate, Oxidized Form
β-NADPH = β-Nicotinamide Adenine Dinucleotide Phosphate, Reduced Form
6-PG = 6-Phospho-D-Gluconate

Conditions: T = 25°C, pH = 7.6, A340nm, Light path = 1 cm

Method: Continuous Spectrophotometric Rate Determination

Reagents

A. 50 mM Triethanolamine Buffer, pH 7.6 at 25°C (Prepare 100 ml in deionized water using Triethanolamine Hydrochloride. Adjust to pH 7.6 at 25°C with 1 M NaOH).
B. 555 mM D-Glucose Solution (Prepare 10 ml in Reagent A using D-(+)-Glucose, Anhydrous).
C. 19 mM Adenosine 5'-Triphosphate Solution (ATP) (Prepare 10 ml in deionized water using Adenosine 5'-Triphosphate, Disodium Salt. *Prepare Fresh*).
D. 100 mM Magnesium Chloride Solution ($MgCl_2$) (Prepare 5 ml in deionized water using Magnesium Chloride, Hexahydrate).
E. 14 mM β-Nicotinamide Adenine Dinucleotide Phosphate, Oxidized Form, Solution (β-NADP) (Dissolve the contents of two 10 mg vials of β-Nicotinamide Adenine Dinucleotide Phosphate, Sodium Salt, in the appropriate volume of deionized water *or* prepare 10 ml in deionized water using β-Nicotinamide Adenine Dinucleotide Phosphate,Sodium Salt. *Prepare Fresh*).
F. Glucose-6-Phosphate Dehydrogenase Enzyme Solution (G-6-PDH) (Immediately before use, prepare a solution containing 125 units/ml of Glucose-6-Phosphate Dehydrogenase, in cold Reagent A).
G. Hexokinase Enzyme Solution (Immediately before use, prepare a solution containing 0.5–1.0 unit/ml of Hexokinase in cold deionized water).

Procedure

Pipette (in milliliters) the following reagents into suitable cuvettes:

	Test	Blank
Reagent A (Buffer)	1.00	1.00
Reagent B (D-Glucose)	1.00	1.00
Reagent C (ATP)	0.10	0.10
Reagent D ($MgCl_2$)	0.20	0.20
Reagent E (β-NADP)	0.20	0.20
Reagent F (G-6-PDH)	0.02	0.02

Mix by inversion and equilibrate to 25°C. Monitor the A340nm until constant, using a suitably thermostatted spectrophotometer. Then add:

Deionized Water	—	0.05
Reagent G (Enzyme Solution)	0.05	—

Immediately mix by inversion and record the increase in A340nm for approximately 5 minutes. Obtain the ΔA340nm/minute using the maximum linear rate for both the Test and Blank.

Calculations

$$\text{Units / ml enzyme} = \frac{(\Delta A340\text{nm / min Test} - \Delta A340\text{nm / min Blank})\ (2.57)\ (\text{df})}{(6.22)\ (0.05)}$$

2.57 = Total volume (in milliliters) of assay
df = Dilution factor
6.22 = Millimolar extinction coefficient of β-NADPH at 340 nm
0.05 = Volume (in milliliter) of enzyme used

$$\text{Units / mg solid} = \frac{\text{units / ml enzyme}}{\text{mg solid / ml enzyme}}$$

$$\text{Units / mg protein} = \frac{\text{units / ml enzyme}}{\text{mg protein / ml enzyme}}$$

Unit Definition: *One unit will phosphorylate 1.0 μmole of D-glucose per minute at pH 7.6 at 25°C.*

Final Assay Concentration

In a 2.57 ml reaction mix, the final concentrations are 39 mM triethanolamine, 216 mM D-glucose, 0.74 mM adenosine 5'-triphosphate, 7.8 mM magnesium chloride, 1.1 mM β-nicotinamide adenine dinucleotide phosphate, 2.5 units glucose-6-phosphate dehydrogenase, and 0.025–0.05 unit of hexokinase.

Notes

1. Glucose-6-Phosphate Dehydrogenase unit definition: One unit will oxidize 1.0 μmole of D-glucose 6-phosphate to 6-phospho-D-gluconate per minute in the presence of β-NADP at pH 7.4 at 25°C.
2. Other types of glucose-6-phosphate dehydrogenase may contain varying amounts of hexokinase as an impurity.

Enzymatic Assay of HEXOKINASE (EC 2.7.1.1)

Principle

$$\text{D-Glucose} + \text{ATP} \xrightarrow{\text{Hexokinase}} \text{D-Glucose 6-Phosphate} + \text{ADP}$$

$$\text{D-Glucose 6-Phosphate} + \beta\text{-NADP} \xrightarrow{\text{G-6-PDH}} \text{6-PG} + \beta\text{-NADPH}$$

Abbreviations Used

ATP = Adenosine 5'-Triphosphate
ADP = Adenosine 5'-Diphosphate
G-6-PDH = Glucose-6-Phosphate Dehydrogenase
β-NADP = β-Nicotinamide Adenine Dinucleotide Phosphate, Oxidized Form
β-NADPH = β-Nicotinamide Adenine Dinucleotide Phosphate, Reduced Form
6-PG = 6-Phospho-D-Gluconate

Conditions: T = 30°C, pH = 7.6, A340nm, Light path = 1 cm

Method: Continuous Spectrophotometric Rate Determination

Reagents

A. 50 mM Tris HCl Buffer, pH 7.6 at 30°C (Prepare 500 ml in deionized water using Trizma Hydrochloride. Adjust to pH 7.6 at 30°C with 1 M NaOH).
B. 50 mM D-Glucose Solution (D-Glucose) (Prepare 10 ml in Reagent A using D-(+)-Glucose, Anhydrous, Mixed Anomers).
C. 30 mM Adenosine 5'-Triphosphate Solution (ATP) (Prepare 10 ml in Reagent A using Adenosine 5'-Triphosphate, Disodium Salt. *Prepare Fresh*).
D. 200 mM Magnesium Chloride Solution ($MgCl_2$) (Prepare 5 ml in Reagent A using Magnesium Chloride, Hexahydrate).
E. 1 mM β-Nicotinamide Adenine Dinucleotide Phosphate Solution (β-NADP) (Dissolve the contents of one 10 mg vial of β-Nicotinamide Adenine Dinucleotide Phosphate, Sodium Salt, in the appropriate volume of Reagent A. *Prepare Fresh*).
F. Glucose-6-Phosphate Dehydrogenase Enzyme Solution (G-6-PDH) (Immediately before use, prepare a solution containing 500 units/ml in cold Reagent A using Glucose-6-Phosphate Dehydrogenase).
G. 100 mM D-Glucose Solution (Enzyme Diluent) (Prepare 50 ml in Reagent A using D-(+)-Glucose, Anhydrous, Mixed Anomers).
H. Hexokinase Enzyme Solution (HK) (Immediately before use, prepare a solution containing 0.25 unit/ml of Hexokinase in cold Reagent G).

Procedure

Prepare a reaction cocktail by pipetting (in milliliters) the following reagents into a suitable container:

Reagent A (Buffer)	11.00
Reagent B (D-Glucose)	3.00
Reagent C (ATP)	10.00
Reagent D ($MgCl_2$)	3.00

Mix and adjust to pH 7.6 at 30°C with 0.1 M HCl or 0.1 M NaOH, if necessary. Pipette (in milliliters) the following reagents into suitable cuvettes:

	Test	Blank
Reaction Cocktail	2.70	2.70
Reagent E (β-NADP)	0.30	0.30
Reagent F (G-6-PDH)	0.01	0.01

Mix by inversion and equilibrate to 30°C. Monitor the A340nm until constant, using a suitably thermostatted spectrophotometer. Then add:

	Test	Blank
Reagent G (Enzyme Diluent)	—	0.10
Reagent H (Enzyme Solution)	0.10	—

Immediately mix by inversion and record the increase in A340nm for approximately 5 minutes. Obtain the ΔA340nm/minute using the maximum linear rate for both the Test and Blank.

Calculations

$$\text{Units / ml enzyme} = \frac{(\Delta A340\text{nm / min Test} - \Delta A340\text{nm / min Blank})\ (3.11)\ (\text{df})}{(6.22)\ (0.1)}$$

3.11 = Total volume (in milliliters) of assay
df = Dilution factor
6.22 = Millimolar extinction coefficient of β-NADPH at 340nm
0.1 = Volume (in milliliters) of enzyme used

$$\text{Units / mg solid} = \frac{\text{units / ml enzyme}}{\text{mg solid / ml enzyme}}$$

$$\text{Units / mg protein} = \frac{\text{units / ml enzyme}}{\text{mg protein / ml enzyme}}$$

Unit Definition: *One unit will phosphorylate 1.0 μmole of D-glucose per minute at pH 7.6 at 30°C.*

Final Assay Concentration

In a 3.11 ml reaction mix, the final concentrations are 50 mM Tris, 8.0 mM D-glucose, 9.6 mM adenosine 5'-triphosphate, 19 mM magnesium chloride, 0.1 mM β-nicotinamide adenine dinucleotide phosphate, 5.0 units glucose-6-phosphate dehydrogenase and 0.025 unit hexokinase.

Note

1. Glucose-6-Phosphate Dehydrogenase unit definition: One unit will oxidize 1.0 μmole of D-glucose 6-phosphate to 6-phospho-D-gluconate per minute in the presence of β-NADP at pH 7.8 at 30°C.

Enzymatic Assay of HISTIDASE (EC 4.3.1.3)

Principle

$$\text{L-Histidine} \xrightarrow{\text{Histidase}} \text{Urocanic Acid} + NH_3$$

Conditions: T = 25°C, pH 9.0, A277nm, Light path = 1 cm

Method: Continuous Spectrophotometric Rate Determination

Reagents

A. 200 mM Tris HCl Buffer, pH 9.0 at 25°C (Prepare 100 ml in deionized water using Trizma Base. Adjust to pH 9.0 at 25°C with 1 M HCl).
B. 10 mM Magnesium Chloride Solution, with 10 mM Tris HCl, pH 9.0 at 25°C ($MgCl_2$) (Prepare 10 ml in deionized water using Magnesium Chloride, Hexahydrate and Reagent A. Adjust to pH 9.0 at 25° with 1 M HCl or 1 M NaOH, if necessary).
C. 100 mM Reduced Glutathione Solution, pH 9.0 at 25°C (GSH) (Prepare 10 ml in Reagent A using Glutathione, Reduced Form, Free Acid. Adjust to pH 9.0 at 25°C with 100 mM NaOH).

D. 100 mM Tris HCl Buffer, pH 9.0 at 25°C (Enzyme Diluent) (Prepare 50 ml in deionized water using Reagent A. Adjust to pH 9.0 at 25°C with 1 M HCl or 1 M NaOH, if necessary).

E. 200 mM L-Histidine Substrate Solution, pH 9.0 at 25°C (Hist) (Prepare 10 ml in Reagent A using L-Histidine, Monohydrochloride, Monohydrate. Adjust to pH 9.0 at 25°C with 100 mM NaOH or 100 mM HCl. *Prepare Fresh*).

F. Histidase Enzyme Solution (Immediately before use, prepare a solution containing 200–500 units/ml of Histidase in cold Reagent D).

Procedure

Pipette (in milliliters) the following reagents into suitable cuvettes:

	Test	Blank
Reagent A (Buffer)	0.50	0.50
Reagent B ($MgCl_2$)	1.00	1.00
Reagent C (GSH)	0.10	0.10
Reagent F (Enzyme Solution)	0.05	—
Reagent D (Enzyme Diluent)	—	0.05

Mix by inversion and incubate at 25°C for 30 minutes. This amount of time is required to activate the enzyme. Monitor the A277nm until constant, using a suitably thermostatted spectrophotometer. Then add:

Reagent E (Hist)	0.40	0.40

Immediately mix by inversion and record the increase in A277nm for approximately 5 minutes. Obtain the ΔA277nm/minute using the maximum linear rate for both the Test and Blank.

Calculations

$$\text{Units / ml enzyme} = \frac{(\Delta A277nm / \text{min Test} - \Delta A277nm / \text{min Blank})\ (3)\ (df)}{(0.0188)\ (0.05)}$$

3 = Volume (in milliliters) of assay
df = Dilution factor
0.0188 = μmolar extinction coefficient of Urocanic Acid at 277nm
0.05 = Volume (in milliliters) of enzyme used

$$\text{Units / mg solid} = \frac{\text{units / ml enzyme}}{\text{mg solid / ml enzyme}}$$

$$\text{Units / mg protein} = \frac{\text{units / ml enzyme}}{\text{mg protein / ml enzyme}}$$

Unit Definition: *One unit will deaminate 1.0 nanomole (10–9 mole) of L-histidine to urocanic acid per minute at pH 9.0 at 25°C.*

Final Assay Concentration

In a 3.00 ml reaction mix, the final concentrations are 71 mM Tris, 3.3 mM magnesium chloride, 3.3 mM reduced glutathione, 27 mM L-histidine, and 10–25 units histidase.

Enzymatic Assay of L-HISTIDINE DECARBOXYLASE (EC 4.1.1.22)

Principle

$$\text{L-His} \xrightarrow{\text{Histidine Decarboxylase}} \text{Histamine} + CO_2 + \text{PRP}$$

Abbreviations Used

L-His = L-Histidine
PRP = Pyridoxal 5-Phosphate

Conditions: T = 37°C, pH 4.5

Method: Manometric Assay using Warburg Flasks

Reagents

A. 80 mM Sodium Acetate Buffer, pH 4.5 at 37°C (Prepare 100 ml in deionized water using Sodium Acetate Trihydrate. Adjust to pH 4.5 at 37°C with 1 M HCl).
B. 100 mM L-Histidine Solution (L-His) (Prepare 50 ml in Reagent A using L-Histidine, Free Base. Adjust the pH to 4.5 with 1 M HCl, if necessary).
C. 61 mM Pyridoxal 5-Phosphate Solution (PRP) (Prepare 25 ml in Reagent A using Pyridoxal 5-Phosphate. Adjust the pH to 4.5 at 37°C with either 1 M HCl or 1 M NaOH).
D. Histidine Decarboxylase Enzyme Solution (Immediately before use, prepare a solution containing 4–10 units/ml of L-Histidine Decarboxylase in cold deionized water).

Procedure

Pipette (in milliliters) the following reagents into Warburg flasks:

Main Chamber	Thermo-barometer Flask	Enzyme Blank	Test	Substrate Blank
Reagent A (Buffer)	2.80	2.80	0.30	0.30
Reagent B (L-His)	—	—	2.50	2.50
Reagent D (Enzyme Soluion)	—	—	0.10	0.10
Reagent C (PRP)	—	0.10	0.10	0.10
Reagent A (Buffer)	0.20	—	—	0.10

Be sure to confirm the stability of the pressure with the flask sealed off before proceeding with the assay. This is to ensure temperature equilibrium and the absence of leaks in the flask. The enzyme activity is determined by calculation of the rate of production of CO_2 at 37°C. The reaction rate should be linear for about 20 minutes.

Calculations

$$\frac{\text{Units}}{\text{ml L-Histidine Decarboxylase}} = \frac{(C)(K)\text{ Dilution Factor}}{\left(22.4\frac{1}{\text{mole}}1\right)(\text{ml L-Histidine Decarboxylase})}$$

C = mm of CO_2 gas evolved/minute
K = Warburg flask constant in μl/mm
22.4 l = Volume gas occupies under STP conditions

Unit Definitions: *One unit will release 1.0 µmole of CO_2 from L-histidine per minute at pH 4.5 at 37°C.*

Final Assay Concentrations

In a 3.00 ml reaction mix, the final concentrations are 77 mM sodium acetate, 83 mM L-histidine, 2 mM pyridoxal 5-phosphate, and 0.4–1.0 unit L-histidine decarboxylase.

Notes

1. The tests are done in triplicate, since it is common for the flasks to have small leaks.
2. The mm of CO_2 gas evolved (C) is corrected for any temperature and barometric changes (Thermobarometer) during the experiment and also for the Substrate Blank and Enzyme Blank: mm CO_2 corrected = mm CO_2 measured Test–mm CO_2 measured for [Thermobarometer + Substrate Blank + Enzyme Blank] Values of the corrected mm CO_2 produced are plotted versus time. The best straight line is drawn not necessarily through the origin. The slope, C = mm CO_2/time, is obtained.
3. The flask constant, K, is calculated according to the equation:

$$K = \frac{\left[(V_g)\left(\frac{273}{T}2\right) + V_f\, a\right]}{P_o}$$

where
P_o = Standard pressure as mm of manometer fluid
V_g = Volume (in milliliters) of gas in flask and manometer
V_f = Folume (in milliliters) of liquid in flask
T = Absolute temperature
a = Solubility of gas; (for CO_2 at 37°C, a = 0.57)

Enzymatic Suitability Assay of HIV PROTEASE

Principle

$$\text{HIV Protease Substrate} \xrightarrow{\text{HIV Protease}} \text{Product A} + \text{Product B}$$

Abbreviations Used

HIV Protease Substrate =His-Lys-Ala-Arg-Val-Leu-p-nitro-Phe-Glu-Ala-le-Ser-NH_2
Product A = His-Lys-Ala-Arg-Val-Leu
Product B = p-nitro-Phe-Glu-Ala-Nle-Ser-NH_2

Conditions: T = 37°C, pH = 5.5, A215nm

Method: HPLC Analysis of Proteolytic Cleavage Products

Reagents

A. 50 mM Sodium Acetate Buffer, pH 5.5 at 37°C (Prepare 10 mL in deionized water using Sodium Acetate, Trihydrate, Adjust the pH to 5.5 at 37°C with 1 M HCl).
B. 50 mM Sodium Acetate Buffer, pH 5.5 at 37°C with 1.085 M Glycerol, 1 mM Dithiothreitol, 1.0 M Urea, and 100 mM Ethylenediaminetetraacetic Acid Disodium (Refolding Buffer) (Prepare 100 mL in deionized water using Sodium Acetate, Trihydrate, Glycerol, DL-Dithiothreitol, Urea, and Ethylenediaminetetraacetic Acid, Disodium Dihydrate. Adjust the pH to 5.5 at 37°C with 1 M HCl).

C. 1.5 mM HIV Protease Substrate in 50 mM Sodium Acetate, pH 5.5 at 37°C (Substrate Solution) (Prepare 2 ml in Reagent A using HIV Protease Substrate).
D. HIV Protease Solution (Immediately before use, prepare a solution containing 0.05 mg/ml HIV Protease in Refolding Buffer (Reagent B)).
E. 0.1% Trifluoroacetic Acid in Acetonitrile/Water (3:1) (Solvent A) (Prepare 500 ml in deionized water using Trifluoroacetic Acid, Acetonitrile, and deionized water).
F. 0.1% Trifluoroacetic Acid in water (Solvent B) (Prepare 500 ml in deionized water using Trifluoroacetic Acid).

Procedure

Step 1: Pipette (in milliliters) the following reagents into a suitable tube:

Reagent D (Enzyme Solution)	0.09
Reagent C (Substrate Solution)	0.02

Mix gently and incubate for 120 minutes at 37°C. Remove tube and put in an ice bath to stop the reaction. Refrigerate until the time of HPLC analysis.

Step 2: HPLC Conditions: Column: Vydac Reverse Phase C18, Particle size: 5 μm, 25 cm × 4.6 mm.

Sample: 10 μL

Solvents: A. 0.1% Trifluoroacetic Acid in Acetonitrile/Water (3:1)

B. 0.1% Trifluoroacetic Acid in water

Gradient: 10% to 85% over 20 minutes

Flow Rate: 1.5 ml/min

Detection: 215 nm

Suitable for cleaving HIV substrate III. Yields 2 peaks which elute before the substrate peak when analyzed by reversed phase HPLC.

Final Assay Concentration

In a 0.11 ml reaction mix, the final concentrations are 0.27 mM HIV Protease substrate III, 50 mM sodium acetate, 888 mM glycerol, 0.82 mM dithiothreitol, 818 mM urea, 82 mM Ethylenediaminetetraacetic Acid, 4.5 μg HIV protease.

Enzymatic Assay of HYALURONATE LYASE (EC 4.2.2.1)

Principle

$$\text{Hyaluronic Acid} \xrightarrow{\text{Hyaluronate Lyase}} \text{Depolymerization Products} + \text{Hyaluronic Acid}$$

Conditions: T = 37°C, pH = 5.35, %T600nm, Light Path = 1 cm

Method: Turbidimetric

Reagents

A. 300 mM Sodium Phosphate pH 5.35 at 37°C. (Prepare 50 ml in deionized water using Sodium Phosphate, Monobasic. Adjust pH to 5.35 at 37°C using 1 M NaOH).
B. 0.3 mg/ml Hyaluronic Acid (Substrate Solution) (Prepare 20 ml of a 0.3 mg/ml solution using Hyaluronic Acid in Reagent A. Heat gently and shake to dissolve).

C. 20 mM Sodium Phosphate pH 7.0 at 37°C with 77 mM Sodium Chloride and 0.01% (w/v) Bovine Serum Albumin (Enzyme Diluent) (Prepare 250 ml in deionized water using Sodium Phosphate, Monobasic, Sodium Chloride and Bovine Serum Albumin. Adjust to pH 7.0 using 1 M NaOH).
D. 24 mM Sodium Acetate with 79 mM Acetic Acid 0.1% (w/v) Bovine Serum Albumin pH 3.75 at 37°C containing 1:1 HCl:H_2O (Acid Albumin Solution) (Prepare 1 L in deionized water using Sodium Acetate, Trihydrate, Glacial Acetic Acid, Bovine Serum Albumin. Adjust to pH 3.75 at 37°C using 5 N HCl).
E. Hyaluronate Lyase Solution (Immediately before use prepare a solution containing 5–10 units per ml in cold Reagent C (Enzyme Diluent)).
F. NF Standard Hyaluronidase Solution (Immediately before use prepare a solution containing 15 units NF Standard Hyaluronidase per ml in cold Reagent C (Enzyme Diluent)).

Procedure

Pipette (in milliliters) the following reagents into suitable tubes.

	Test 1	Test 2	Test 3	Blank	Std 1	Std 2	Std 3	Std 4	Std 5
Reagent B	1.00	1.00	1.00	1.00	1.00	1.00	1.00	1.00	1.00

Incubate at 37°C for 10 minutes. Then add at time zero:

	Test 1	Test 2	Test 3	Blank	Std 1	Std 2	Std 3	Std 4	Std 5
Reagent E	1.00	0.75	0.50	—	—	—	—	—	—
Reagent F	—	—	—	—	1.00	0.80	0.60	0.40	0.20
Reagent C	—	0.25	0.50	1.00	—	0.20	0.40	0.60	0.80

Mix by inversion and incubate at 37°C for 45 minutes. Then add:

	Test 1	Test 2	Test 3	Blank	Std 1	Std 2	Std 3	Std 4	Std 5
Reagent D	10.00	10.00	10.00	10.00	10.00	10.00	10.00	10.00	10.00

Immediately mix tubes by inversion and let stand at room temperature or 10 minutes, then transfer to suitable cuvettes. Read the percent transmittance (%T) versus deionized water in a suitably thermostatted spectrophotometer at 600 nm.

Calculation

$$\text{Units / mg enzyme} = \frac{\Delta\%\ \text{Ttest}}{(\Delta\%\ \text{Tstandard / NF units / RM})\ (\text{mg enzyme / RM})}$$

Δ% Ttest = % Ttest–% Tblank
Δ% Tstandard = % Tstandard–% Tblank
RM = Reaction Mix

Unit Definition: *One unit is equivalent to one NF unit.*

Final Assay Concentrations

In a 2 ml reaction mix, the final concentrations are 160 mM sodium phosphate, 39 mM sodium chloride, 0.005% (w/v) Bovine Serum Albumin, 0.15 mg/ml hyaluronic acid, and 2.5–10 units hyauronate lyase.

Enzymatic Assay of HYALURONIDASE (EC 3.2.1.35)

Principle

$$\text{Hyaluronic Acid} \xrightarrow{\text{Hyaluronidase}} \text{Depolymerization Products} + \text{Hyaluronic Acid}$$

Conditions: T = 37°C, pH = 5.90, A600nm, Light Path = 1 cm

Method: Turbidimetric

Reagents

A. 300 mM Sodium Phosphate pH 5.35 at 37°C. (Prepare 100 ml in deionized water using Sodium Phosphate, Monobasic, Anhydrous. Adjust the pH to 5.35 at 37°C using 1 M NaOH).

B. 0.30% (w/v) Hyaluronic Acid Solution (Prepare 5–10 ml of a 3 mg/ml solution using Hyaluronic Acid, in Reagent A. Heat until just before boiling and stir for approximately 20 minutes to dissolve).

C. 0.03% (w/v) Hyaluronic Acid Solution (Substrate Solution) (Prepare 10 ml in Reagent A using Reagent B. Prewarm to 37°C immediately before use).

D. 20 mM Sodium Phosphate pH 7.0 at 37°C with 77 mM Sodium Chloride and 0.01% (w/v) Bovine Serum Albumin (Enz Dil) (Prepare 150 ml in deionized water using Sodium Phosphate, Monobasic, Anhydrous, Sodium Chloride, and Albumin, Bovine. Adjust to pH 7.0 using 1 M NaOH).

E. 24 mM Sodium Acetate with 79 mM Acetic Acid and 0.1% (w/v) Bovine Serum Albumin pH 3.75 at 37°C containing 1:1 HCl:H_2O (Acid Albumin Soln) (Prepare 100 ml in deionized water using Sodium Acetate, Trihydrate, Glacial Acetic Acid, and Albumin, Bovine. Adjust to pH 3.75 at 37°C using 5 N HCl).

F. Hyaluronidase Enzyme Solution (Immediately before use prepare a solution containing 3–5 units per ml in cold Reagent D).

G. National Formulary (NF) Standard Hyaluronidase Solution (NF Std) (Immediately before use prepare a solution containing 10 units NF Standard Hyaluronidase per ml in cold Reagent D).

Procedure

Pipette (in milliliters) the following reagents into suitable tubes.

	Test 1	Test 2	Test 3	Blank	Std 1	Std 2	Std 3	Std 4	Std 5	Std 6
Reagent F (Enzyme)	1.00	0.75	0.50	—	—	—	—	—	—	—
Reagent G (NF Std)	—	—	—	—	0.10	0.20	0.30	0.40	0.50	0.60
Reagent D (Enz Dil)	—	0.25	0.50	1.00	0.90	0.80	0.70	0.60	0.50	0.40

Indubate at 37°C for 10 minutes. Then add at time zerox:

Reagent G (Substrate)	1.00	1.00	1.00	1.00	1.00	1.00	1.00	1.00	1.00	1.00

Mix by inversion and incubate at 37°C for exactly 45 minutes. Pipette 0.50 ml of the above reaction mixture into a cuvette containing 2.5 ml of Reagent E (Acid Albumin Solution). Immediately mix by inversion. Allow the cuvettes to stand at room temperature for exactly 10 minutes. Read the absorbance (A600nm) of the standard curve versus the blank on the spectrophotometer and analyze it using quadratic regression. Read the absorbance (A600nm) of the tests versus the same blank. Set up the spectrophotometer to determine the Hyaluronidase units for the tests versus the standard curve.

Calculations

$$\Delta A600nm \text{ Sample} = A600nm \text{ Test} - A600nm \text{ Blank}$$

$$\Delta A600nm \text{ Standard} = A600nm \text{ Standard} - A600nm \text{ Blank}$$

$$\text{Units / ml enzyme} = \frac{(\text{Hyaluronidase units from Std Curve})(df)}{(\text{Enzyme aliquot})}$$

df = Dilution factor
Enzyme aliquot = Volume (in milliliters) of enzyme used in assay

$$\text{Units / mg solid} = \frac{\text{units / ml enzyme}}{\text{mg solid / ml enzyme}}$$

Unit Definition: *One unit is equivalent to one NF unit*.

Final Assay Concentrations

In a 2.00 ml reaction mix, the final concentrations are 160 mM sodium phosphate, 39 mM sodium chloride, 0.005% (w/v) Bovine Serum Albumin, 0.015% (w/v) hyaluronic acid, and 1.5–5 units hyaluronidase.

Enzymatic assay of D-HYDANTOINASE (EC 3.5.22)

Principle

$$\text{Hydantoin} + H_2O \xrightarrow{\text{D-Hydantoinase}} \text{N-carbamylglycine}$$

Conditions: T = 40°C, pH = 9.0, A450nm, Light path = 1 cm

Method: Spectrophotometric Stopped Rate Determination

Reagents

A. 100 mM Sodium Borate Buffer with 50 mM Sodium Chloride, pH 9.0 at 40°C (Prepare 100 ml in deionized water using Boric Acid and Sodium Chloride. Adjust to pH 9.0 at 40°C with 5 M NaOH).

B. 100 mM Hydantoin Solution (Hydantoin) (Prepare 5 ml in Reagent A using Hydantoin).

C. 12% (w/v) Trichloroacetic Acid Solution (TCA) (Prepare 25 ml in deionized water using Trichloroacetic Acid, 6.1 N Solution, approx. 100%(w/v)).

D. 10% (w/v) p-Dimethylaminobenzaldehyde Solution (DMABA) (Prepare 5 ml in 6 M HCl using p-Dimethylaminobenzaldehyde).

E. 250 mM Hydantoic Acid Solution (Hydant. Std) (Prepare 5 ml in deionized water using Hydantoic Acid).

F. Hydantoinase Enzyme Solution (Immediately before use, prepare a solution containing 10 mg/ml of Hydantoinase in cold deionized water).

Procedure

Equilibrate Reagent B (Hydantoin) to 40°C in a water bath. Pipette (in milliliters) the following reagents into suitable tubes:

	Test	Blank
Reagent F (Enzyme Solution)	0.10	0.10

Equilibrate to 40°C. Then add:

Reagent B (Hydantoin)	0.30	—

Cap the tubes and incubate at 40°C for exactly 10 minutes. Then add:

Reagent C (TCA)	0.70	0.70
Reagent B (Hydantoin)	—	0.30

Then add:

Reagent D (DMBA)	0.15	0.15

Cap the tubes, mix by inversion and then centrifuge for 2 minutes in a microcentrifuge. Transfer the supernatant to suitable cuvettes and record the A450nm for the Test and Blank using a suitable spectrophotometer.

Standard Curve: A standard curve is prepared by pipetting (in milliliters) the following reagents into suitable tubes:

	Std 1	Std 2	Std 3	Std 4	Std Blank
Reagent E (Hydant. Std)	0.02	0.04	0.08	0.10	0.00
Deionized Water	0.38	0.36	0.32	0.30	0.40
Reagent C (TCA)	0.70	0.70	0.70	0.70	0.70
Reagent D (DMABA)	0.15	0.15	0.15	0.15	0.15

Cap the tubes, mix by inversion and then centrifuge for 2 minutes in a microcentrifuge. Transfer the supernatant to suitable cuvettes and record the A450nm for the Standards and Standard Blank using a suitable spectrophotometer.

Calculations

Standard Curve: ΔA450nm Standard = A450nm Standard–A450nm Standard Blank

Plot the ΔA450nm of the standards versus μmoles of Hydantoic acid.

Sample Determination: ΔA450nm Test = A450nm Test–A450nm Test Blank

Determine the μmoles of hydantoic acid produced using the Standard Curve.

$$\text{Units / ml enzyme} = \frac{(\mu\text{moles of hydantoic acid produced})}{(10)(0.1)}$$

10 = Time (in minutes) of assay
0.1 = Volume (in milliliters) of enzyme used

$$\text{Units / mg solid} = \frac{\text{units / ml enzyme}}{\text{mg solid / ml enzyme}}$$

$$\text{Units/mg solid} = 1000 \times \text{units/mg solid}$$

Unit Definition: *One unit will catalyze the formation of 1 μmole of N-carbamylglycine from hydantoin per minute at pH 9.0 at 40°C.*

Final Assay Concentration

In a 0.4 ml reaction mix, the final concentrations are 75 mM boric acid, 38 mM sodium chloride, 75 mM hydantion, and 1 mg hydantoinase.

Note

1. Reagent B (Hydantoin) must be equilibrated to 40°C immediately prior to addition.

Enzymatic Assay of β-HYDROXYACYL-CoA DEHYDROGENASE (EC 1.1.1.35)

Principle

$$\text{S-Acetoacetyl-CoA} + \beta\text{-NADH} \xrightarrow{\text{HOADH}} \beta\text{-Hydroxybutyryl-CoA} + \beta\text{-NAD}$$

Abbreviations Used

β-NADH = β-Nicotinamide Adenine Dinucleotide, Reduced Form
HOADH = β-Hydroxyacyl-CoA Dehydrogenase
β-NAD = β-Nicotinamide Adenine Dinucleotide, Oxidized Form
CoA = Coenzyme A

Conditions: T = 37°C, pH = 7.3, A340nm, Light path = 1 cm

Method: Continuous Spectrophotometric Rate Determination

Reagents

A. 100 mM Potassium Phosphate Buffer, pH 7.3 at 37°C (Prepare 100 ml in deionized water using Potassium Phosphate, Monobasic, Anhydrous,Adjust to pH 7.3 at 37°C with 1 M KOH).
B. 5.4 mM S-Acetoacetyl Coenzyme A Solution (SAAC) (Prepare 1 ml in Reagent A using S-Acetoacetyl Coenzyme A, Sodium Salt. Store on ice).
C. 6.4 mM β-Nicotinamide Adenine Dinucleotide, Reduced Form (β-NADH) (Prepare 1 ml in cold Reagent A using β-Nicotinamide Adenine Dinucleotide, Reduced Form, Disodium Salt. *Prepare Fresh*).
D. β-Hydroxyacyl-CoA Dehydrogenase Enzyme Solution (Immediately before use, prepare a solution containing 0.2–0.7 unit/ml of β-Hydroxyacyl-CoA Dehydrogenase in cold Reagent A).

Procedure

Pipette (in milliliters) the following reagents into suitable cuvettes:

	Test	Blank
Reagent A (Buffer)	2.80	2.80
Reagent B (SAAC)	0.05	0.05
Reagent C (β-NADH)	0.05	0.05

Mix by inversion and equilibrate to 37°C. Monitor the A340nm until constant, using a suitably thermostatted spectrophotometer. Then add:

Reagent D (Enzyme Solution)	0.10	—
Deionized Water	—	0.10

Immediately mix by inversion and record the decrease in A340nm for approximately 5 minutes. Obtain the ΔA340nm/minute using the maximum linear rate for both the Test and Blank.

Calculations

$$\text{Units / ml enzyme} = \frac{(\Delta A340nm / min\ Test - \Delta A340nm / min\ Blank)\ (3)\ (df)}{(6.22)\ (0.1)}$$

3 = Total volume (in milliliters) of assay
df = Dilution factor
6.22 = Millimolar extinction coefficient of β-NADH at 340nm
0.1 = Volume (in milliliters) of enzyme used

$$\text{Units / mg solid} = \frac{\text{units / ml enzyme}}{\text{mg solid / ml enzyme}}$$

$$\text{Units / mg protein} = \frac{\text{units / ml enzyme}}{\text{mg protein / ml enzyme}}$$

Unit Definition: *One unit will convert 1.0 μmole of acetoacetyl-CoA to β-hydroxybutyryl-CoA per minute at pH 7.3 at 37°C in the presence of β-NADH.*

Final Assay Concentration

In a 3.00 ml reaction mix, the final concentrations are 97 mM potassium phosphate, 0.09 mM S-acetoacetyl-coenzyme A, 0.1 mM β-nicotinamide adenine dinucleotide, reduced form and 0.02–0.07 unit β-hydroxyacyl-CoA dehydrogenase.

Enzymatic Assay of p-HYDROXYBENZOATE HYDROXYLASE (EC 1.14.13.2)

Principle

$$\text{p-Hydroxybenzoate} + \beta\text{-NADH} + O_2 \xrightarrow{\text{PHBH}} \text{Protocatechuate} + \beta\text{-NADP} + H_2O + \text{FAD}$$

Abbreviations Used

β-NADPH = β-Nicotinamide Adenine Dinucleotide Phosphate, Reduced Form
PHBH = p-Hydroxybenzoate Hydroxylase

FAD = Flavin Adenine Dinucleotide
β-NADP = β-Nicotinamide Adenine Dinucleotide Phosphate, Oxidized Form

Method: T = 37°C, pH = 8.2, A340nm, Light path = 1 cm

Conditions: Continuous Spectrophotometric Rate Determination

Reagents

A. 50 mM Tris Maleate Buffer, pH 8.2 at 37°C (Prepare 100 ml in deionized water using Trizma Maleate. Adjust to pH 8.2 at 37°C with 1 M NaOH).
B. 5.0 mM p-Hydroxybenzoate Solution (p-Hydroxybenz) (Prepare 10 ml in Reagent A using p-Hydroxybenzoic Acid. *Prepare Fresh*).
C. 0.20 mM Flavin Adenine Dinucleotide Solution (FAD) (Prepare 10 ml in Reagent A using Flavin Adenine Dinucleotide, Disodium Salt. *Prepare Fresh*).
D. 3.0 mM β-Nicotinamide Adenine Dinucleotide Phosphate Reduced Form Solution (β-NADPH) (Dissolve the contents of one 5 mg vial of β-Nicotinamide Adenine Dinucleotide Phosphate, Reduced Form, Tetrasodium Salt, in the appropriate volume of Reagent A or dissolve the appropriate amount of β-Nicotinamide Adenine Dinucleotide Phosphate, Reduced Form, Tetrasodium Salt in the appropriate volume of Reagent A. *Prepare Fresh*).
E. 50 mM Potassium Phosphate Buffer, pH 6.0 at 37°C (Enz Dil) (Prepare 50 ml in deionized water using Potassium Phosphate, Monobasic, Anhydrous. Adjust to pH 6.0 at 37°C with 1 M KOH).
F. p-Hydroxybenzoate Hydroxylase Enzyme Solution (Immediately before use, prepare a solution containing 0.20–0.50 unit/ml of p-Hydroxybenzoate Hydroxylase in cold Reagent E).

Procedure

Prepare a reaction cocktail in an amber bottle by pipetting (in milliliters) the following reagents into a suitable container (*Prepare Fresh*):

Reagent A (Buffer)	21.00
Reagent B (p-Hydroxybenz)	3.00
Reagent C (FAD)	3.00
Reagent D (β-NADPH)	3.00

Mix by stirring and adjust to pH 8.2 at 37°C with either 1 M HCl or 1 M NaOH. Pipette (in milliliters) the following reagents into suitable cuvettes:

	Test	Blank
Reaction Cocktail	3.00	3.00

Equilibrate to 37°C. Monitor the A340nm until constant, using a suitably thermostatted spectrophotometer. Then add:

Reagent F (Enzyme Solution)	0.05	—
Reagent E (Enz Dil)	—	0.05

Immediately mix by inversion and record the decrease in A340nm for approximately 5–10 minutes. Obtain the ΔA340nm/minute using the maximum linear rate for both the Test and Blank.

Calculations

$$\text{Units / ml enzyme} = \frac{(\Delta\text{A340nm / min Test} - \Delta\text{A340nm / min Blank})\ (3.05)\ (\text{df})}{(6.22)\ (0.5)}$$

3.05 = Total volume (in milliliters) of assay
df = Dilution factor
6.22 = Millimolar extinction coefficient of β-NADPH at 340nm
0.05 = Volume (in milliliter) of enzyme used in assay

$$\text{Units / mg solid} = \frac{\text{units / ml enzyme}}{\text{mg solid / ml enzyme}}$$

$$\text{Units / mg protein} = \frac{\text{units / ml enzyme}}{\text{mg protein / ml enzyme}}$$

Unit Definition: *One unit will hydroxylate 1.0 μmole of p-hydroxybenzoate to protocatechuate per minute at pH 8.2 at 37°C in the presence of β-NADPH.*

Final Assay Concentration

In a 3.05 ml reaction mix, the final concentrations are 49 mM Tris maleate, 0.49 mM p-hydroxybenzoate, 0.02 mM flavin adenine dinucleotide, 0.3 mM β-nicotinamide adenine dinucleotide phosphate reduced form, 0.8 mM potassium phosphate, and 0.01–0.025 unit p-hydroxybenzoate hydroxylase.

Enzymatic Assay of β-HYDROXYBUTYRATE DEHYDROGENASE (EC 1.1.1.30)

Principle

$$\text{D-}\beta\text{-Hydroxybutyrate} + \beta\text{-NAD} \xrightarrow{\beta\text{-HBDB}} \text{Acetoacetate} + \beta\text{-NADH}$$

Abbreviations Used

β-NAD = β-Nicotinamide Adenine Dinucleotide
β-NADH = β-Nicotinamide Adenine Dinucleotide, Reduced Form
β-HBDB = β-Hydroxybutyrate Dehydrogenase

Conditions: T = 37°C, pH = 7.8, A340nm, Light path = 1 cm

Method: Continuous Spectrophotometric Rate Determination

Reagents

A. 100 mM Tris HCl Buffer, pH 7.8 at 37°C (Prepare 50 ml in deionized water using Trizma Base. Adjust to pH 7.8 at 37°C with 1 M HCl).
B. 160 mM β-Hydroxybutyric Acid Solution (β-HB) (Prepare 5 ml in Reagent A using DL-β-Hydroxybutyric Acid).

C. 30 mM β-Nicotinamide Adenine Dinucleotide Solution (β-NAD) (Dissolve the contents of one 50 mg vial of β-Nicotinamide Adenine Dinucleotide, in 2.5 ml of deionized water *or* prepare 2 ml in deionized water using β-Nicotinamide Adenine Dinucleotide. *Prepare Fresh*).

D. β-Hydroxybutyrate Dehydrogenase Enzyme Solution (Immediately before use, prepare a solution containing 0.3–0.6 units/ml of β-Hydroxybutyrate Dehydrogenase in cold Reagent A).

Procedure

Pipette (in milliliters) the following reagents into suitable cuvettes:

	Test	Blank
Reagent A (Buffer)	2.20	2.30
Reagent B (β-HB)	0.50	0.50
Reagent C (β-NAD)	0.20	0.20

Mix by inversion and equilibrate to 37°C. Monitor the A340nm until constant, using a suitably thermostatted spectrophotometer. Then add:

Reagent D (Enzyme Solution)	0.10	—

Immediately mix by inversion and record the increase in A340nm for approximately 5 minutes. Obtain the ΔA340nm/minute using the maximum linear rate for both the Test and Blank.

Calculations

$$\text{Units / mg enzyme} = \frac{\Delta\text{A340nm / min Test} - \Delta\text{A340nm / min Blank}}{(6.22)\ (\text{mg enzyme / ml RM})}$$

6.22 = Millimolar extinction coefficient of β-NADH at 340nm
RM = Reaction Mix

Unit Definition: *One unit will oxidize 1.0 μmole of D-β-hydroxybutyrate to acetoacetate per minute at pH 7.8 at 37° C.*

Final Assay Concentration

In a 3 ml reaction mix, the final concentrations are 93 mM Tris, 27 mM D-β-hydroxybutyrate, 2 mM β-NAD and 0.03–0.06 units β-hydroxybutyrate dehydrogenase.

Enzymatic Assay of β-HYDROXYBUTYRATE DEHYDROGENASE (EC 1.1.1.30) from *Pseudomonas lemoignei*

Principle

$$\text{D-}\beta\text{-Hydroxybutyrate} + \beta\text{-NAD} \xrightarrow{\beta\text{-HBDB}} \text{Acetoacetate} + \beta\text{-NADH}$$

Abbreviations Used

β-NAD = β-Nicotinamide Adenine Dinucleotide, Oxidized Form
β-NADH = β-Nicotinamide Adenine Dinucleotide, Reduced Form
β-HBDB = β-Hydroxybutyrate Dehydrogenase

Conditions: T = 37°C, pH = 7.8, A340nm, Light path = 1 cm

Method: Continuous Spectrophotometric Rate Determination

Reagents

A. 100 mM Tris HCl Buffer, pH 7.8 at 37°C (Prepare 50 ml in deionized water using Trizma Base. Adjust to pH 7.8 at 37°C with 1 M HCl).
B. 100 mM β-Hydroxybutyric Acid Solution (β-HB) (Prepare 5 ml in deionized water using DL-β-Hydroxybutyric Acid, Sodium Salt).
C. 50 mM β-Nicotinamide Adenine Dinucleotide Solution (β-NAD) (Dissolve the contents of one 50 mg vial of β-Nicotinamide Adenine Dinucleotide, in 1.5 ml of deionized water *or* prepare 2 ml in deionized water using β-Nicotinamide Adenine Dinucleotide. *Prepare Fresh*).
D. β-Hydroxybutyrate Dehydrogenase Enzyme Solution (Immediately before use, prepare a solution containing 0.3–0.6 units/ml of β-Hydroxybutyrate Dehydrogenase in cold Reagent A).

Procedure

Pipette (in milliliters) the following reagents into suitable cuvettes:

	Test	Blank
Reagent A (Buffer)	2.60	2.70
Reagent B (β-HB)	0.10	0.10
Reagent C (β-NAD)	0.20	0.20

Mix by inversion and equilibrate to 37°C. Monitor the A340nm until constant, using a suitably thermostatted spectrophotometer. Then add:

Reagent D (Enzyme Solution)	0.10	—

Immediately mix by inversion and record the increase in A340nm for approximately 5 minutes. Obtain the ΔA340nm/minute using the maximum linear rate for both the Test and Blank.

Calculations

$$\text{Units / mg enzyme} = \frac{\Delta A340\text{nm / min Test} - \Delta A340\text{nm / min Blank}}{(6.22)\ (\text{mg enzyme / ml RM})}$$

6.22 = Millimolar extinction coefficient of β-NADH at 340nm
RM = Reaction Mix

Unit Definition: *One unit will oxidize 1.0 μmole of D-β-hydroxybutyrate to acetoacetate per minute at pH 7.8 at 37° C.*

Final Assay Concentration

In a 3 ml reaction mix, the final concentrations are 90 mM Tris, 3.3 mM D-β-hydroxybutyrate, 3.3 mM β-NAD and 0.03–0.06 units β-hydroxybutyrate dehydrogenase.

Enzymatic Assay of HYDROXYSTEROID DEHYDROGENASE
α-Hydroxysteroid Dehydrogenase Activity

Principle

$$\text{Androsterone} + \beta\text{-NAD} \xrightarrow{\alpha\text{-HSDH}} \beta\text{-NADH} + 5\alpha\text{-Androstane-3, 17-dione}$$

Abbreviations Used

β-NAD = β-Nicotinamide Adenine Dinucleotide, Oxidized Form
α-HSDH = α-Hydroxysteroid Dehydrogenase
β-NADH = β-Nicotinamide Adenine Dinucleotide, Reduced Form

Conditions: T = 25°C, pH = 8.9, A340nm, Light path = 1 cm

Method: Continuous Spectrophotometric Rate Determination

Reagents

A. 100 mM Pyrophosphate Buffer, pH 8.9 at 25°C (Prepare 100 ml in deionized water using Pyrophosphate, Tetrasodium, Decahydrate. Adjust to pH 8.9 at 25°C with 1 M HCl).
B. Methanol (Use Absolute Methanol).
C. 0.015% (w/v) Androsterone Solution (Androsterone) (Prepare 2 ml in Reagent B using Androsterone).
D. 6 mM β-Nicotinamide Adenine Dinucleotide, Oxidized Form Solution (β-NAD) (Prepare 2 ml in deionized water using β-Nicotinamide Adenine Dinucleotide, or dissolve the contents of one 20 mg vial of β- Nicotinamide Adenine Dinucleotide, in the appropriate volume of deionized water. Adjust to pH 7.0 at 25°C with Sodium Bicarbonate).
E. 10 mM Potassium Phosphate Buffer with 0.1% (w/v) Bovine Serum Albumin, pH 7.2 at 25°C (Enzyme Diluent) (Prepare 50 ml in deionized water using Potassium Phosphate, Monobasic, Anhydrous and Albumin, Bovine).
F. α-Hydroxysteroid Dehydrogenase Enzyme Solution (Immediately before use, prepare a solution containing 0.2–0.4 unit/ml of Hydroxysteroid Dehydrogenase in cold Reagent E).

Procedure

Pipette (in milliliters) the following reagents into suitable cuvettes:

	Test	Blank
Deionized Water	2.00	2.00
Reagent A (Buffer)	0.60	0.60
Reagent D (β-NAD)	0.20	0.20
Reagent F (Enzyme Solution)	0.10	—
Reagent E (Enz Dil)	—	0.10

Mix by inversion and equilibrate to 25°C. Monitor the A340nm until constant, using a suitably thermostatted spectrophotometer. Then add:

Reagent C (Androsterone)	0.10	0.10

Immediately mix by inversion and record the increase in A340nm for approximately 5 minutes. Obtain the ΔA340nm/minute using the maximum linear rate for both the Test and Blank.

Calculations

$$\text{Units / ml enzyme} = \frac{(\Delta A340nm / \min \text{ Test} - \Delta A340nm / \min \text{ Blank})(df)}{(6.22)(0.1)}$$

6.22 = Millimolar extinction coefficient of β-NADH at 340nm
0.1 = Volume (in milliliter) of enzyme used

$$\text{Units / mg solid} = \frac{\text{units / ml enzyme}}{\text{mg solid / ml enzyme}}$$

$$\text{Units / mg protein} = \frac{\text{units / ml enzyme}}{\text{mg protein / ml enzyme}}$$

Unit Definition: *One unit will oxidize 1.0 μmole of androsterone per minute in the presence of β-NAD at pH 8.9 at 25°C.*

Final Assay Concentrations

In a 3.00 ml reaction mix, the final concentrations are 20 mM pyrophosphate, 0.0005% (w/v) androsterone, 3.3% (v/v) methanol, 0.4 mM β-nicotinamide adenine dinucleotide, 0.33 mM potassium phosphate, 0.003% (w/v) Bovine Serum Albumin, and 0.02–0.04 unit hydroxysteroid dehydrogenase.

Enzymatic Assay of β-HYDROXYSTEROID DEHYDROGENASE (EC 1.1.1.51)

Principle

$$\text{Testosterone} + \beta\text{-NAD} \xrightarrow[\text{Dehydrogenase}]{\beta\text{-Hydroxysteroid}} 5\text{-}\alpha\text{-Androsterone-3, 17-dione} + \beta\text{-NADH}$$

Abbreviations Used

β-NAD = Nicotinamide Adenine Dinucleotide, Oxidized Form
β-NADH = Nicotinamide Adenine Dinucleotide, Reduced Form

Conditions: T = 25°C, pH = 8.9, A340nm, Light path = 1 cm

Method: Continuous Spectrophotometric Rate Determination

Reagents

A. 100 mM Sodium Pyrophosphate Buffer, pH 8.9 at 25°C (Prepare 100 ml in deionized water using Tetrasodium Pyrophosphate Decahydrate. Adjust to pH 8.9 at 25°C with 1 M HCL).
B. 6 mM β-Nicotinamide Adenine Dinucleotide, Oxidized Form, Solution (β-NAD) (Prepare 5 ml in deionized water using β-Nicotinamide Adenine Dinucleotide, Oxidized Form).
C. 0.015% (w/v) Testosterone Substrate Solution (Prepare 5 ml in Reagent F using Testosterone).
D. 10 mM Potassium Phosphate Buffer, pH 7.2 at 25°C (Prepare 50 ml in deionized water using Potassium Phosphate, Monobasic. Adjust to pH 7.2 with 1 M KOH).
E. β-Hydroxysteroid Dehydrogenase Enzyme Solution (Immediately before use, prepare a solution containing 0.3 units/ml in cold Reagent D).
F. Absolute Methanol (MeOH) (Use Absolute Methanol).

Procedure

Pipette (in milliliters) the following reagents into a suitable cuvettes:

	Test	Blank
Reagent A (Buffer)	0.60	0.60
Reagent B (β-NAD)	0.20	0.20
Reagent E (Enzyme Solution)	0.10	0.10
Deionized Water	2.00	2.00

Mix by inversion and equilibrate to 25°C. Monitor the A340nm until constant, using a suitably thermostatted spectrophotometer. Then add:

Reaction C (Testosterone)	0.10	—
Reagent F (MeOH)	—	0.10

Immediately mix by inversion and record the increase in A340nm for approximately 5 minutes. Obtain the ΔA340nm/minute using the maximum linear rate for both the Test and Blank.

Calculations

$$\text{Units / mg enzyme} = \frac{(\Delta A340\text{nm / min Test} - \Delta A340\text{nm / min Blank})}{(6.22)\ (\text{mg enzyme / ml RM})}$$

6.22 = Millimolar extinction coefficient of β-NADH at 340nm
RM = Reaction Mix

Unit Definition: *One unit will oxidize 1.0 μmole of testosterone per minute at pH 8.9 at 25°C, in the presence of β-NAD$^+$.*

Final Assay Concentration

In a 3.0 ml reaction mix, the final concentrations are 20 mM sodium pyrophosphate, 0.4 mM β-NAD, 0.3 mM sodium phosphate, 0.0005% testosterone, and 0.03 units β-hydroxysteroid dehydrogenase.

Enzymatic Assay of 3α,20β-HYDROXYSTEROID DEHYDROGENASE (EC 1.1.1.53)

Principle

$$\text{Cortisone} + \beta\text{-NADH} \xrightarrow{3\alpha,\ 20\beta\text{-HSDH}} \text{20-Dihydrocortisone} + \beta\text{-NAD}$$

Abbreviations Used

β-NADH = β-Nicotinamide Adenine Dinucleotide, Reduced Form
3α,20β-HSDH = 3α,20β-Hydroxysteroid Dehydrogenase
β-NAD = β-Nicotinamide Adenine Dinucleotide, Oxidized Form

Conditions: T = 25°C, pH 7.6, A340nm, Light path = 1 cm

Method: Continuous Spectrophotometric Rate Determination

Reagents

A. 50 mM Triethanolamine HCl Buffer, pH 7.6 at 25°C (Prepare 100 ml in deionized water using Triethanolamine Hydrochloride. Adjust to pH 7.6 at 25°C with 1 M NaOH).
B. Methanol (Use Absolute Methanol).
C. 28 mM Cortisone Solution (Cortisone) (Prepare 5 ml in Reagent B using Cortisone).
D. 6.4 mM β-Nicotinamide Adenine Dinucleotide, Reduced Form Solution (β-NADH) (Dissolve the contents of one 10 mg vial of β-Nicotinamide Adenine Dinucleotide, Reduced Form, in the appropriate volume of deionized water. *Prepare Fresh*).
E. 3α,20β-Hydroxysteroid Dehydrogenase Enzyme Solution (3α,20β-HSDH) (Immediately before use, prepare a solution containing 0.17–0.35 unit/ml of 3α, 20β-Hydroxysteroid Dehydrogenase in cold deionized water).

Procedure

Pipette (in milliliters) the following reagents into suitable cuvettes:

	Test	Blank
Reagent A (Buffer)	2.75	2.75
Reagent C (Cortisone)	0.10	0.10
Reagent D (β-NADH)	0.05	0.05

Mix by inversion and equilibrate to 25°C. Monitor the A340nm until constant, using a suitably thermostatted spectrophotometer. Then add:

Reagent E (3α,20β-HSDH)	0.10	—
Deionized Water	—	0.10

Immediately mix by inversion and record the decrease in A340nm for approximately 5 minutes. Obtain the ΔA340nm/minute using the maximum linear rate for both the Test and Blank.

Calculations

$$\text{Units / ml enzyme} = \frac{(\Delta\text{A340nm / min Test} - \Delta\text{A340nm / min Blank})\ (3)\ (\text{df})}{(6.22)\ (0.1)}$$

3 = Total volume (in milliliters) of assay
df = Dilution factor
6.22 = Millimolar extinction coefficient of β-NADH at 340nm
0.1 = Volume (in milliliter) of enzyme used

$$\text{Units / mg solid} = \frac{\text{units / ml enzyme}}{\text{mg solid / ml enzyme}}$$

$$\text{Units / mg protein} = \frac{\text{units / ml enzyme}}{\text{mg protein / ml enzyme}}$$

Unit Definition: *One unit will reduce 1.0 μmole of cortisone to 20-dihydrocortisone per minute at pH 7.6 at 25°C, in the presence of β-NADH.*

Final Assay Concentration

In a 3.00 ml reaction mix, the final concentrations are 46 mM triethanolamine, 0.93 mM cortisone, 0.1 mM β-nicotinamide adenine dinucleotide, reduced form, 0.017–0.035 unit 3α, 20β-hydroxysteroid dehydrogenase, and 3.3% (v/v) methanol.

Enzymatic Assay of HYPOXANTHINE-GUANINE PHOSPHORIBOSYL TRANSFERASE (EC 2.4.2.8)

Principle

$$\text{PRPP} + \text{Guanine} \xrightarrow{\text{HGPRT}} \text{GMP} + \text{Pyrophosphate} + \text{Mg}^{++}$$

$$\text{GMP} + \text{ATP} \xrightarrow{\text{GK}} \text{GDP} + \text{ADP}$$

$$\text{ADP} + \text{PEP} \xrightarrow{\text{PK}} \text{ATP} + \text{Pyruvate}$$

$$\text{GDP} + \text{PEP} \xrightarrow{\text{PK}} \text{GTP} + \text{Pyruvate}$$

$$\text{2Pyruvate} + 2\beta\text{-NADH} \xrightarrow{\text{LDH}} \text{2 Lactate} + 2\ \beta\text{-NAD}$$

Abbreviations Used

PRPP = 5-Phosphorylribose 1-Pyrophosphate
HGPRT = Hypoxanthine-Guanine Phosphoribosyl Transferase
GMP = Guanosine 5'-Monophosphate
ATP = Adenosine 5'-Triphosphate
GK = Guanylate Kinase
GDP = Guanosine 5'-Diphosphate
ADP = Adenosine 5'-Diphosphate

PEP = Phospho(enol) pyruvate
PK = Pyruvate Kinase
GTP = Guanosine 5'-Triphosphate
β-NADH = β-Nicotinamide Adenine Dinucleotide, Reduced Form
LDH = Lactic Dehydrogenase
β-NAD = β-Nicotinamide Adenine Dinucleotide, Oxidized Form

Conditions: T = 37°C, pH = 7.5, A340nm, Light path = 1 cm

Method: Continuous Spectrophotometric Rate Determination

Reagents

A. 71 mM Tris HCl Buffer, pH 7.5 at 37°C (Prepare 50 ml in deionized water using Trizma Hydrochloride. Adjust to pH 7.5 at 37°C with 1 M NaOH).
B. 1.2 M Magnesium Sulfate Solution ($MgSO_4$) (Prepare 5 ml in deionized water using Magnesium Sulfate, Heptahydrate).
C. 3.9 M Potassium Chloride Solution (KCl) (Prepare 5 ml in deionized water using Potassium Chloride).
D. 42 mM 5-Phosphorylribose 1-Pyrophosphate Solution (PRPP) (Prepare 2 ml in deionized water using 5-Phosphorylribose 1-Pyrophosphate, Sodium Salt).
E. 12 mM Guanine Solution (Guanine) (Prepare 2 ml in deionized water using Guanine).
F. 18 mM Adenosine 5'-Triphosphate Solution (ATP) (Prepare 2 ml in deionized water using Adenosine 5'-Triphosphate, Disodium Salt).
G. 30 mM Phospho(enol) pyruvate Solution (PEP) (Prepare 2 ml in deionized water using Phospho(enol) pyruvate, Trisodium Salt, Hydrate).
H. 4.8 mM β-Nicotinamide Adenine Dinucleotide, Reduced Form Solution (β-NADH) (Prepare 2 ml in deionized water using β-Nicotinamide Adenine Dinucleotide, Reduced Form, Disodium Salt).
I. Guanylate Kinase Enzyme Solution (GK) (Use Guanylate Kinase).
J. PK/LDH Enzyme Suspension (PK/LDH).
K. Hypoxanthine-Guanine Phosphoribosyl Transferase Enzyme Solution (HGPRT) (Immediately before use, prepare a solution containing 200–250 unit/ml of Hypoxanthine-Guanine Phosphoribosyl Transferase in cold deionized water).

Procedure

Prepare a reaction cocktail by pipetting (in milliliters) the following reagents into a suitable container:

Reagent A (Buffer)	21.00
Reagent B ($MgSO_4$)	1.00
Reagent C (KCl)	1.00
Reagent E (Guanine)	0.50
Reagent F (ATP)	1.00
Reagent G (PEP)	1.00
Reagent H (β-NADH)	1.00

Mix by swirling and adjust to pH 7.5 at 37°C with 100 mMHCl or 100 mM NaOH, if necessary. Pipette (in milliliters) the following reagents into suitable cuvettes:

	Test	Blank
Reaction Cocktail	2.65	2.65
Reagent I (GK)	0.10	0.10
Reagent J (PK/LDH)	0.05	0.05
Reagent K (HGPRT)	0.10	0.10

Mix by inversion and equilibrate to 37°C. Monitor the A340nm until constant, using a suitably thermostatted spectrophotometer. Then add:

Reagent D (PRPP)	0.10	—
Deionized Water	—	0.10

Immediately mix by inversion and record the decrease in A340nm for approximately 5 minutes. Obtain the ΔA340nm/minute using the maximum linear rate for both the Test and Blank.

Calculations

$$\text{Units / ml enzyme} = \frac{(\text{A340nm / min Test} - \text{A340nm / min Blank})\ (3)\ (\text{df})}{(2)\ (0.00622)\ (0.1)}$$

3 = Total volume (in milliliters) of assay
df = Dilution factor
2 = 2 moles of β-NAD produced per mole of GMP produced
0.00622 = μmolar extinction coefficient of β-NADH at 340nm
0.1 = Volume (in milliliter) of enzyme used

$$\text{Units / mg solid} = \frac{\text{units / ml enzyme}}{\text{mg solid / ml enzyme}}$$

$$\text{Units / mg protein} = \frac{\text{units / ml enzyme}}{\text{mg protein / ml enzyme}}$$

Unit Definition: *One unit will catalyze the formation of 1 nmole of guanosine 5'-monophosphate (GMP) per minute from guanine and phosphoribosyl pyrophosphate at pH 7.5 at 37°C.*

Final Assay Concentration

In a 3.00 ml reaction mix, the final concentrations are 49.7 mM Tris, 40 mM magnesium sulfate, 130 mM potassium chloride, 0.2 mM guanine, 0.6 mM adenosine 5'- triphosphate, 1 mM phospho(enol) pyruvate, 0.16 mM β-nicotinamide adenine dinucleotide, reduced form, 1.4 mM phosphoribosyl pyrophosphate, 35 units pyruvate kinase, 50 units L-lactic ehydrogenase, 1–2 units guanylate kinase, and 20–25 units hypoxanthine-guanine phosphoribosyl transferase.

Notes

1. Guanylate Kinase unit Definition: One unit will convert 1.0 µmole each of GMP and ATP to GDP and ADP per minute at pH 7.5 at 30°C.
2. L-Lactic Dehydrogenase Unit Definition: One unit will reduce 1.0 µmole of pyruvate to L-lactate per minute at pH 7.5 at 37°C.
3. Pyruvate Kinase Unit Definition: One unit will convert 1.0 µmole of phospho(enol) pyruvate to pyruvate per minute at pH 7.6 at 37°C.

Enzymatic Assay of INORGANIC PYROPHOSPHATASE (EC 3.6.1.1)

Principle

$$\text{Pyrophosphate} + H_2O \xrightarrow{\text{Inorganic Pyrophosphatase}} 2\ \text{Orthophosphate}$$

Conditions: T = 25°C, pH 9.0, A660nm, Light path = 1 cm

Method: Colorimetric

Reagents

A. 50 mM Tris HCl Buffer with 2 mM Magnesium Chloride and 2 mM Sodium Pyrophosphate, pH 9.0 at 25°C (Substrate Solution) (Prepare 50 ml in deionized water using Trizma Base, Magnesium Chloride and Pyrophosphate, Tetrasodium, Decahydrate. Adjust to pH 9.0 at 25°C with 1 M HCl).

B. 100 mM Tris HCl Buffer with 2 mM Magnesium Chloride, pH 7.5 at 25°C (Enz Dil) (Prepare 100 ml in deionized water using Trizma Base and Magnesium Chloride. Adjust to pH 7.5 at 25°C with 1 M HCl).

C. 10% (w/v) Ammonium Molybdate Solution (Prepare 25 ml in 10 N H_2SO_4 using Molybdic Acid, Ammonium, Tetrahydrate Salt).

D. Taussky-Shorr Color Reagent (TCSR) (Prepare by adding 10 ml of Reagent C to 70 ml of deionized water. Add 5 g of Ferrous Sulfate, Heptahydrate. Bring the volume to 100 ml with deionized water).

E. Phosphorus Standard (Use Phosphorus Standard Solution. The phosphorus concentration is 20 µg/ml, 0.645 µmoles/ml).

F. Inorganic Pyrophosphatase Enzyme Solution (Prepare a solution containing 100–200 µg/ml of Inorganic Pyrophosphatase in cold Reagent B and incubate for 2 hours at 0–5°C. Immediately before use, dilute to 4–5 units/ml in cold Reagent B).

Procedure

Pipette (in milliliters) the following reagents into suitable containers:

	Test	Blank
Reagent A (Substrate Solution)	5.00	5.00

Equilibrate to 25°C. Then add:

Reagent B (Enz Dil)	—	0.10
Reagent F (Enzyme Solution)	0.10	—

Immediately mix by swirling and incubate at 25°C for exactly 4 minutes. Pipette (in milliliters) the following reagents into suitable containers:

	Test	Blank	Std 1	Std 2	Std 3	Std 4	Std 5	Std Blank
Reagent D (TSCR)	5.00	5.00	5.00	5.00	5.00	5.00	5.00	5.00
Test Mixture	1.00	—	—	—	—	—	—	—
Blank Mixture	—	1.00	—	—	—	—	—	—
Reagent E (Standard)	—	—	0.10	0.30	0.50	0.70	1.00	—
Deionized Water	4.00	4.00	4.90	4.70	4.50	4.30	4.00	5.00

Mix by swirling and incubate at 25°C for exactly 5 minutes. Transfer to suitable cuvettes and record the A660nm for Test, Test Blank, Standards, and Standard Blank.

Calculations

Standard Curve: ΔA660nm Standard = A660nm Standard–A660nm Standard Blank

Prepare a standard curve by plotting ΔA660nm Standard vs micromoles of Phosphate.

Sample Determination: ΔA660nm Test = A660nm Test–A660nm Test Blank

Determine the micromoles of Phosphate liberated using the standard curve.

$$\text{Units / ml enzyme} = \frac{(\mu\text{moles of Phosphate released})\,(5.1)\,(\text{df})}{(1)\,(4)\,(0.1)}$$

5.1 = Total volume (in milliliters) of assay

df = Dilution factor

1 = Volume (in milliliter) of Test Mixture used in colorimetric determination

4 = Time of assay (in minutes) as per the Unit Definition

0.1 = Volume (in milliliter) of enzyme used

$$\text{Units / mg solid} = \frac{\text{units / ml enzyme}}{\text{mg solid / ml enzyme}}$$

$$\text{Units / mg protein} = \frac{\text{units / ml enzyme}}{\text{mg protein / ml enzyme}}$$

Unit Definition: *One unit will liberate 1.0 μmole of inorganic orthophosphate per minute at pH 9.0 at 25°C.*

Final Assay Concentration

In a 5.10 ml reaction mix, the final concentrations are 51 mM Tris, 2 mM sodium pyrophosphate, 2 mM magnesium chloride and 0.4–0.5 unit inorganic pyrophosphatase.

Enzymatic Assay of INORGANIC PYROPHOSPHATASE (EC 3.6.1.1) from Baker's Yeast

Principle

$$\text{Pyrophosphate} + H_2O \xrightarrow{\text{Inorganic Pyrophosphatase}} 2\ \text{Orthophosphate}$$

Conditions: T = 25°C, pH = 7.2, A660nm, Light path = 1 cm

Method: Colorimetric

Reagents

A. 50 mM Tris HCl Buffer, pH 7.2 at 25°C (Prepare 100 ml in deionized water using Trizma Base. Adjust to pH 7.2 at 25°C with 1 M HCl).

B. 20 mM Tris HCl Solution, pH 7.2 at 25°C (Enzyme Diluent) (Prepare 100 ml in deionized water using Trizma Base. Adjust to pH 7.2 at 25°C with 1 M HCl).

C. 10 mM Sodium Pyrophosphate Solution (Prepare 50 ml in deionized water using Pyrophosphate, Tetrasodium, Decahydrate).

D. 10 mM Magnesium Chloride Solution ($MgCl_2$) (Prepare 50 ml in deionized water using Magnesium Chloride, Hexahydrate).

E. 10% (w/v) Ammonium Molybdate Solution (Prepare 25 ml in 10 N H_2SO_4 using Molybdic Acid, Ammonium Tetrahydrate Salt).

F. Taussky-Shorr Color Reagent (TSCR) (Prepare by adding 10 ml of Reagent E to 70 ml of deionized water. Add 5 g of Ferrous Sulfate, Heptahydrate. Bring the volume to 100 ml with deionized water).

G. Phosphorus Standard (Use Phosphorus Standard Solution. The phosphorus concentration is 20 μg/ml, 0.645 μmoles/ml).

H. Inorganic Pyrophosphatase Enzyme Solution (Immediately before use, prepare a solution containing 5–10 units/ml of Inorganic Pyrophosphatase in cold Reagent B).

Procedure

Pipette (in milliliters) the following reagents into suitable containers:

	Test	Test Blank
Reagent A (Buffer)	4.00	4.00
Reagent B (Enz Dil)	—	0.10
Reagent C (Pyrophosphate)	1.00	1.00
Reagent D ($MgCl_2$)	1.00	1.00

Mix by swirling and equilibrate to 25°C. Then add:

Reagent H (Enzyme Solution)	0.10	—

Immediately mix by swirling and incubate at 25°C for exactly 10 minutes. Pipette (in milliliters) the following reagents into suitable containers:

	Test	Test Blank	Std 1	Std 2	Std 3	Std Blank
Reagent F (TSCR)	5.00	5.00	5.00	5.00	5.00	5.00
Test Mixture	1.00	—	—	—	—	—
Test Blank Mixture	—	1.00	—	—	—	—
Reagent G (Standard)	—	—	0.50	1.00	1.50	—
Deionized Water	4.00	4.00	4.50	4.00	3.50	5.00

Mix by swirling and incubate at 25°C for exactly 5 minutes. Transfer to suitable cuvettes and record the A660nm for Test, Test Blank, Standards, and Standard Blank.

Calculations

Standard Curve: ΔA660nm Standard = A660nm Standard–A660nm Standard Blank

Prepare a standard curve by plotting ΔA660nm Standard vs μmoles of Phosphate.

Sample Determination: ΔA660nm Test = A660nm Test–A660nm Test Blank

Determine the μmoles of Phosphate liberated using the standard curve.

$$\text{Units / ml enzyme} = \frac{(\mu\text{moles of Phosphate released})\ (6.1)\ (\text{df})}{(1)\ (10)\ (0.1)}$$

6.1 = Total volume (in milliliters) of assay
df = Dilution factor
10 = Time of assay (in minutes) as per the Unit Definition
1 = Volume (in milliliter) of Test Mixture used in Colorimetric Determination
0.1 = Volume (in milliliter) of enzyme used

$$\text{Units / mg solid} = \frac{\text{units / ml enzyme}}{\text{mg solid / ml enzyme}}$$

$$\text{Units / mg protein} = \frac{\text{units / ml enzyme}}{\text{mg protein / ml enzyme}}$$

Unit Definition: *One unit will liberate 1.0 μmole of inorganic orthophosphate per minute at pH 7.2 at 25°C.*

Final Assay Concentration

In a 6.10 ml reaction mix, the final concentrations are 33 mM Tris, 1.6 mM sodium pyrophosphate, 1.6 mM magnesium chloride and 0.5–1.0 unit inorganic pyrophosphatase.

Enzymatic Assay of INORGANIC PYROPHOSPHATASE (EC 3.6.1.1) From *Bacillus stearothermophilus*

Principle

$$\text{Pyrophosphate} + H_2O \xrightarrow{\text{Inorganic Pyrophosphatase}} 2\ \text{Orthophosphate}$$

Conditions: T = 50°C, pH 9.0, A660nm, Light path = 1 cm

Method: Colorimetric

Reagents

A. 50 mM Glycine Buffer, pH 9.0 at 50°C (Prepare 100 ml in deionized water using Glycine, Free Base. Adjust to pH 9.0 at 50°C with 1 M NaOH).

B. 50 mM Tris HCl, pH 8.0 at 25°C (Enzyme Diluent) (Prepare 50 ml in deionized water using Trizma, Hydrochloride. Adjust to pH 8.0 at 25°C with 1 M KOH).

C. 10 mM Sodium Pyrophosphate Solution (Pyrophosphate) (Prepare 25 ml in deionized water using Pyrophosphate, Tetrasodium, Decahydrate).

D. 10 mM Magnesium Chloride Solution ($MgCl_2$) (Prepare 50 ml in deionized water using Magnesium Chloride, Hexahydrate).

E. 10% (w/v) Ammonium Molybdate Solution (Prepare 25 ml in 10 N H_2SO_4 using Molybdic Acid, Ammonium, Tetrahydrate Salt).

F. Taussky-Shorr Color Reagent (TSCR) (Prepare by adding 10 ml of Reagent E to 70 ml of deionized water. Add 5 g of Ferrous Sulfate, Heptahydrate. Bring the volume to 100 ml with deionized water).

G. Phosphorus Standard (Use Phosphorus Standard Solution. The phosphorus concentration is 20 μg/ml, 0.645 μmoles/ml).

H. Inorganic Pyrophosphatase Enzyme Solution (Immediately before use, prepare a solution containing 6–12 units/ml of Inorganic Pyrophosphatase in cold Reagent B).

Procedure

Pipette (in milliliters) the following reagents into suitable containers:

	Test	Blank
Reagent A (Buffer)	4.00	4.00
Reagent B (Enzyme Diluent)	—	0.10
Reagent D ($MgCl_2$)	1.00	1.00
Reagent H (Enzyme Solution)	0.10	—

Mix by swirling and equilibrate to 50°C. Then add:

	Test	Blank
Reagent C (Pyrophosphate)	1.00	—
Deionized Water	—	1.00

Immediately mix by swirling and incubate at 50°C for exactly 10 minutes. Pipette (in milliliters) the following reagents into suitable containers:

	Test	Test Blank	Std 1	Std 2	Std 3	Std 4	Std 5	Std Blank
Reagent F (TSCR)	5.00	5.00	5.00	5.00	5.00	5.00	5.00	5.00
Test Mixture	1.00	—	—	—	—	—	—	—
Test Blank Mixture	—	1.00	—	—	—	—	—	—
Reagent G (Standard)	—	—	0.50	0.70	1.00	1.20	1.50	—
Deionized Water	4.00	4.00	4.50	4.30	4.00	3.80	3.50	5.00

Mix by swirling and incubate at 25°C for 10 minutes. Transfer to suitable cuvettes and record the A660nm for Test, Test Blank, Standards, and Standard Blank.

Calculations

Standard curve: ΔA660nm Standard = A660nm Standard–A660nm Standard Blank

Prepare a standard curve by plotting the ΔA660nm Standard vs micromoles of Phosphate.

Sample Determination: ΔA660nm Test = A660nm Test–A660nm Test Blank

Determine the micromoles of Phosphate liberated using the standard curve.

$$\text{Units / ml enzyme} = \frac{(\mu\text{moles of Phosphate released})\ (6.1)\ (\text{df})}{(1)\ (10)\ (0.1)}$$

6.1 = Total volume (in milliliters) of assay
df = Dilution factor
10 = Time of assay (in minutes) as per the Unit Definition
1 = Volume (in milliliter) of Test Mixture used in Colorimetric Determination
0.1 = Volume (in milliliter) of enzyme used

$$\text{Units / mg solid} = \frac{\text{units / ml enzyme}}{\text{mg solid / ml enzyme}}$$

$$\text{Units / mg protein} = \frac{\text{units / ml enzyme}}{\text{mg protein / ml enzyme}}$$

Unit Definition: *One unit will liberate 1.0 μmole of inorganic orthophosphate per minute at pH 9.0 at 50°C.*

Final Assay Concentration

In a 6.10 ml reaction mix, the final concentrations are 33 mM glycine, 1.6 mM sodium pyrophosphate, 1.6 mM magnesium chloride, 0.82 mM Tris and 0.6–1.2 unit inorganic pyrophosphatase.

Enzymatic Assay of MYO-INOSITOL DEHYDROGENASE (EC 1.1.1.18)

Principle

$$\text{myo-Inositol} + \beta\text{-NAD} \xrightarrow{\text{myo-Inositol Dehydrogenase}} \text{scyllo-Inosose} + \beta\text{-NADH}$$

Abbreviations Used

β-NAD = β-Nicotinamide Adenine Dinucleotide
β-NADH = β-Nicotinamide Adenine Dinucleotide, Reduced Form

Conditions: T = 25°C, pH = 9.0, A340nm, Light path = 1 cm

Method: Continuous Spectrophotometric Rate Determination

Reagents

A. 100 mM Sodium Pyrophosphate Buffer, pH 9.0 at 25°C (Prepare 100 ml in deionized water using Tetrasodium Pyrophosphate, Anhydrous. Adjust to pH 9.0 at 25°C with 1 M HCl).
B. 5 mM β-Nicotinamide Adenine Dinucleotide Solution (β- NAD) (Prepare 5 ml in deionized water using β-Nicotinamide Adenine Nucleotide. Adjust to pH 7.0 at 25°C with solid Sodium Bicarbonate).
C. 250 mM myo-Inositol Solution (Substrate) (Prepare 5 ml in deionized water using myo-Inositol).
D. 20 mM Potassium Phosphate Buffer, pH 7.0 at 25°C (Prepare 100 ml in deionized water using Potassium Phosphate, Monobasic. Adjust to pH 7.0 at 25°C with 1 M NaOH).
E. myo-Inositol Dehydrogenase Solution (Immediately before use, prepare a solution containing 4 units/ml of myo-Inositol Dehydrogenase in cold Reagent D).

Procedure

Pipette (in milliliters) the following reagents into suitable cuvettes:

	Test	Blank
Reagent A (Buffer)	0.30	0.30
Reagent B (β-NAD)	0.30	0.30
Reagent C (Substrate)	0.30	0.30
Deionized Water	2.10	2.10

Mix by inversion and equilibrate to 25°C. Monitor the A340nm until constant, using a suitably thermostatted spectrophotometer. Then add:

Reagent A (Buffer)	—	0.01
Reagent E (Enzyme Solution)	0.01	—

Immediately mix by inversion and record the increase in A340nm for approximately 5 minutes. Obtain the ΔA340nm/minute using the maximum linear rate for both the Test and Blank.

Calculations

$$\text{Units / ml enzyme} = \frac{(\Delta A340\text{nm / min Test} - \Delta A340\text{nm / min Blank})\,(3.01)\,(\text{df})}{(6.22)\,(0.01)}$$

3.01 = Total volume (in milliliters) of assay
df = Dilution factor
6.22 = Millimolar extinction coefficient of β-NADH at 340nm
0.01 = Volume (in milliliter) of enzyme used

$$\text{Units / mg solid} = \frac{\text{units / ml enzyme}}{\text{mg solid / ml enzyme}}$$

$$\text{Units / mg protein} = \frac{\text{units / ml enzyme}}{\text{mg protein / ml enzyme}}$$

Unit Definition: *One unit will convert 1.0 µmole of myo-inositol and β-NAD to scyllo-inosose and β-NADH per minute at pH 9.0 at 25°C.*

Final Assay Concentration

In a 3.01 ml reaction mix, the final concentrations are 10 mM sodium pyrophosphate, 0.5 mM β-nicotinamide adenine dinucleotide, 25 mM myo-inositol and 0.04 unit myoinositol dehydrogenase.

Enzymatic Assay of INVERTASE (EC 3.2.1.26)

Principle

$$\text{Sucrose} + H_2O \xrightarrow{\text{Invertase}} \text{Glucose} + \text{Fructose}$$

Conditions: T = 25°C, pH = 4.6, A340nm, Light path = 1 cm

Method: Colorimetric

Reagents

A. 100 mM Sodium Acetate Buffer, pH 4.6 at 25°C (Prepare 200 ml in deionized water using Sodium Acetate, Trihydrate. Adjust to pH 4.6 at 25°C with 1 M HCl).
B. 10.0% (w/v) Sucrose Substrate Solution (Sucrose) (Prepare 100 ml in deionized water using Sucrose. *Prepare Fresh*).
C. 300 mM Trizma Base (Prepare 100 ml in deionized water using Trizma Base, Reagent Grade).
D. Invertase Enzyme Solution (Immediately before use, prepare a solution containing 5–7 units/ml of Invertase in cold deionized water).
E. Glucose Assay Reagent (HK Kit) (Prepare 50 ml in deionized water according to packaged insert instructions).

Procedure

Step 1: Pipette (in milliliters) the following reagents into suitable test tubes:

	Test	Blank
Reagent A	1.00	1.00
Reagent B (Sucrose)	0.50	0.50

Mix by swirling and equilibrate to 25°C. Then add:

Reagent D (Enzyme)	0.10	—

Mix by swirling and incubate at 25°C for exactly 5 minutes. Then add:

Reagent C (Trizma)	0.40	0.40
Reagent D (Enzyme)	—	0.10

Immediately mix by swirling and place on ice.

Step 2: Add 3.00 ml of Reagent E (HK Reagent) to separate suitable 3 ml cuvettes (one for each blank and one for each test used in Step 1 above). Measure and record the absorbance at 340 nm for each of the cuvettes with HK Reagent using a suitable spectrophotometer. (This is the Initial absorbance). Transfer 0.030 ml from each of the tubes in Step 1 to the corresponding cuvette in Step 2. Monitor the absorbance at 340 nm until the absorbance is constant (approximately 10–15 minutes). (This is the endpoint of the HK reaction). Measure and record the absorbance at 340 nm for each of the cuvettes with HK Reagent using a suitable spectrophotometer. (This is the Final absorbance).

Calculations

ΔA340nm Test = A340nm Test Final–A340nm Test Initial
ΔA340nm Blank = A340nm Blank Final–A340nm Blank Initial
Corrected ΔA340nm Test = ΔA340nm Test–ΔA340nm Blank

$$\text{Units / ml enzyme} = \frac{(\text{Corrected } \Delta A340\text{nm Test})\ (2.0)\ (3.03)\ (\text{df})}{(6.22)\ (5)\ (0.10)\ (0.03)}$$

2.0 = Reaction mixture volume (in milliliters) of Step 1
3.03 = Total volume (in milliliters) of reaction mixture Step 2
df = Dilution factor
6.22 = Millimolar extinction coefficient of NADP to NADPH at 340nm
5 = Time (in minutes) of assay
0.10 = Volume (in milliliter) of enzyme used
0.03 = Volume (in milliliter) of reaction mixture used in Step 2

$$\text{Units / mg solid} = \frac{\text{units / ml enzyme}}{\text{mg solid / ml enzyme}}$$

$$\text{Units / mg protein} = \frac{\text{units / ml enzyme}}{\text{mg protein / ml enzyme}}$$

Unit Definition: *One unit will hydrolyze 1.0 μmole of sucrose to glucose and fructose per minute at pH 4.6 at 25°C.*

Final Assay Concentrations

In a 1.60 ml reaction mix, the final concentrations are 63 mM sodium acetate, 3.1% (w/v) sucrose and 0.5–0.70 unit invertase.

Enzymatic Assay of INVERTASE (EC 3.2.1.26)

Principle

$$\text{Sucrose} + H_2O \xrightarrow{\text{Invertase}} \text{Glucose} + \text{Fructose}$$

Conditions: T = 55°C, pH = 4.5, A410nm, Light path = 1 cm

Method: Colorimetric

Reagents

A. 100 mM Sodium Acetate Buffer, pH 4.5 at 55°C (Prepare 200 ml in deionized water using Sodium Acetate, Trihydrate. Adjust to pH 4.5 at 55°C with 1 M HCl).
B. 1.0% (w/v) Sucrose Substrate Solution (Sucrose) (Prepare 100 ml in Reagent A using Sucrose. *Prepare Fresh*).
C. Invertase Enzyme Solution (Immediately before use, prepare a solution containing 0.30–0.40 unit/ml of Invertase in cold deionized water).
D. 0.5 M NaOH Solution (Prepare a 100 ml solution by the addition of 50 ml of 1.0 N Sodium Hydroxide to 50 ml deionized water).
E. 0.5% (w/v) p-Hydroxybenzoic Acid Hydrazide Solution (PAHBAH) (Prepare 50 ml in Reagent D using p-Hydroxybenzoic Acid Hydrazide. *Prepare Fresh*).
F. 5.56 mM Glucose Standard Solution (Std Soln) (Prepare 25 ml in deionized water using Glucose Standard Solution).

Procedure

Pipette (in milliliters) the following reagents into suitable test tubes:

	Test	Test Blank	Std 1	Std 2	Std 3	Std 4	Std 5	Std 6	Std Blank
Deionized water	—	0.10	—	—	—	—	—	—	—
Reagent B (Sucrose)	0.90	0.90	—	—	—	—	—	—	—
Reagent F (Std Soln)	—	—	0.05	0.10	0.20	0.30	0.40	0.50	—
Deionized water	—	—	0.95	0.90	0.80	0.70	0.60	0.50	1.00

Mix by swirling and equilibrate to 55°C. Then add:

Reagent C (Enzyme)	0.10	—	—	—	—	—	—	—	—

Mix by swirling and incubate at 55°C for exactly 20 minutes. Swirl and invert tubes after the 20 minutes incubation. Remove 0.10 ml of the reaction mixtures and place in labeled separate tubes containing:

Reagent E (PAHBAH)	2.90	2.90	2.90	2.90	2.90	2.90	2.90	2.90	2.90

Immediately mix by swirling. Place a marble over the top of the tube and transfer the tubes to a boiling water bath. Incubate the tubes for 5 minutes. Tubes can be boiled together within 15 minutes of stopping the reaction. Remove from the boiling water bath and allow to cool to room temperature. Mix by swirling and transfer the solutions to suitable cuvettes. Obtain the A410nm for Test, Blank, and Standards, using a suitable spectrophotometer.

Calculations

Standard Curve: ΔA410nm Std = A410nm Std–A410nm Std Blank

Prepare a standard curve by plotting the ΔA410nm Standard vs the μmoles of glucose.

Test Determination: ΔA410nm Test = A410nm Test – A410nm Blank

Determine the μmoles of glucose using the Standard Curve.

Calculations

$$\text{Units / ml enzyme} = \frac{(\mu\text{moles of glucose liberated})\ (\text{df})}{(20)\ (0.1)\ (2)}$$

df = Dilution factor

20 = Time (in minutes) of assay

0.1 = Volume (in milliliter) of enzyme used

2 = Conversion factor for 1 μmole of sucrose being hydrolyzed to glucose and fructose

$$\text{Units / mg solid} = \frac{\text{units / ml enzyme}}{\text{mg solid / ml enzyme}}$$

$$\text{Units / mg protein} = \frac{\text{units / ml enzyme}}{\text{mg protein / ml enzyme}}$$

Unit Definition: *One unit will hydrolyze 1.0 μmole of sucrose to invert sugar per minute at pH 4.5 at 55°C.*

Final Assay Concentrations

In a 1.00 ml reaction mix, the final concentrations are 90 mM sodium acetate, 0.90% (w/v) sucrose and 0.030–0.040 unit invertase.

Notes

1. When setting up for assay, set up dram vials (or equivalent) in duplicate. One set for the sucrose and standard incubation step and the second for the PAHBAH solution step.
2. Both glucose and fructose are detected as reducing sugars.

Enzymatic Assay of ISOAMYLASE (EC 3.2.1.68)

Principle

$$\text{Starch} + H_2O \xrightarrow{\text{Isoamylase}} \text{Maltotriose} + \text{Higher MW Maltosaccharides}$$

Abbreviation Used

MW = Molecular Weight

Conditions: T = 40°C, pH = 3.5, A610nm, Light path = 1 cm

Method: Colorimetric

Reagents

A. 500 mM Sodium Acetate Buffer, pH 3.5 at 40°C (Prepare 100 ml in deionized water using Sodium Acetate, Trihydrate. Adjust to pH 3.5 at 40°C with 5 M HCl).
B. 1% (w/v) Starch Solution (Prepare 50 ml in deionized water using Starch, Rice. The Starch Solution must be heated at 80°C for 30 minutes).
C. 10 mM Iodine/Potassium Iodide Solution (I_2-KI) (Prepare 50 ml in deionized water using Iodine and Potassium Iodide. Stir overnight and store in the dark at room temperature).
D. 50 mM Sodium Acetate Buffer, with 0.1% (w/v) Bovine Serum Albumin, pH 3.5 at 40°C (Enzyme Diluent) (Prepare 100 ml in deionized water using Sodium Acetate, Trihydrate and Albumin, Bovine. Adjust to pH 3.5 at 40°C with 5 M HCl).
E. Isoamylase Enzyme Solution (Prepare a solution containing 0.0005 mg protein/ml of Isoamylase in cold Reagent D).
F. Inactivated Isoamylase Solution (Prepare a solution containing 0.0005–0.001 mg protein/ml of Isoamylase in cold Reagent D. Boil the solution in a boiling water bath for 10 minutes to inactiviate the enzyme).

Procedure

Pipette (in milliliters) the following reagents into suitable tubes:

	Test	Blank
Reagent A (Buffer)	0.075	0.075
Reagent B (Starch)	0.350	0.350

Mix and equilibrate to 40°C. Then add:

	Test	Blank
Reagent E (Enzyme Solution)	0.075	—
Reagent F (Inactivated Isoamylase)	—	0.075

Mix and incubate for exactly 15 minutes at 40°C. Then add:

Reagent C(I_2-KI)	0.50	0.50
Deionized Water	11.50	11.50

Allow vials to stand at room temperature for 15 minutes then centrifuge for approximately 10 minutes. Filter the supernatant from the Test and Blank through a 0.8 μm syringe filter. Record the A610nm for both the Test and Blank using a suitable spectrophotometer.

Calculations

$$\text{Units / ml enzyme} = \frac{(\text{A610nm Test} - \text{A610nm Blank})\,(4)\,(\text{df})\,(12.5)}{(0.1)\,(0.075)}$$

12.5 = Total volume (in milliliters) of assay
4 = Time (in minutes) conversion factor from 15 minutes to 60 minutes
df = Dilution factor
0.1 = Increase in A610nm per hour
0.075 = Volume (in milliliter) of enzyme used

$$\text{Units / mg protein} = \frac{\text{units / ml enzyme}}{\text{mg protein / ml enzyme}}$$

Unit Definition: *One unit causes an increase in A610nm of 0.1 in 1 hour using rice starch as a substrate.*

Final Assay Concentrations

In a 0.5 ml reaction mix, the final concentrations are 82 mM sodium acetate, 0.7% (w/v) starch and 0.038 μg of isoamylase.

Enzymatic Assay of ISOCITRATE LYASE (EC 4.1.3.1)

Principle

$$\text{Isocitrate} \xrightarrow{\text{Isocitrate Lyase}} \text{Succinate} + \text{Glyoxylate}$$

$$\text{Glyoxylate} + \text{Phenylhydrazine} \rightarrow \text{Phenylhydrazine Glyoxylate}$$

Conditions: T = 30°C, pH = 6.8, A324nm, Light Path = 1 cm

Method: Continuous Spectrophotometric Rate Determination

Reagents

A. 50 mM Imidazole Buffer, pH 6.8 at 30°C (Prepare 100 ml in deionized water using Imidazole. Adjust to pH 6.8 at 30°C with 1 M HCl).
B. 50 mM Magnesium Chloride Solution ($MgCl_2$) (Prepare 10 ml in deionized water using Magnesium Chloride, Hexahydrate).

C. 10 mM Ethylenediaminetetraacetic Acid Solution (EDTA) (Prepare 10 ml in deionized water using Ethylenediaminetetraacetic Acid, Disodium Salt, Dihydrate).
D. 40 mM Phenylhydrazine HCl Solution (Phenylhydrazine) (Prepare 10 ml in deionized water using Phenylhydrazine Hydrochloride).
E. 10 mM DL-Isocitric Acid Solution (Isocitrate) (Prepare 10 ml in deionized water using DL-Isocitric Acid, Trisodium Salt).
F. Isocitrate Lyase Enzyme Solution (Immediately before use, prepare a solution containing 0.05–0.07 unit/ml of Isocitrate Lyase in cold Reagent A).

Procedure

Pipette (in milliliters) the following reagents into suitable cuvettes:

	Test	Blank
Reagent A (Buffer)	0.50	0.50
Reagent B ($MgCl_2$)	0.10	0.10
Reagent C (EDTA)	0.10	0.10
Reagent D (Phenylhydrazine)	0.10	0.10
Reagent E (Isocitrate)	0.10	0.10

Mix by inversion and equilibrate to 30°C. Monitor the A324nm until constant using a suitably thermostatted spectrophotometer. Then add:

Reagent F (Enzyme Solution)	0.10	—
Reagent A (Buffer)	—	0.10

Immediately mix by inversion and record the increase in A324nm for approximately 5 minutes. Obtain the ΔA324nm/minute using the maximum linear rate for both the Test and Blank.

Conditions

$$\text{Units / ml enzyme} = \frac{(\Delta\text{A324nm / min Test} - \Delta\text{A324nm / min Blank})\ (1)\ (\text{df})}{(16.8)\ (0.1)}$$

1 = Volume (in milliliter) of assay
df = Dilution factor
16.8 = Millimolar extinction coefficient of Phenylhydrazine Glyoxylate at 324nm
0.1 = Volume (in milliliter) of enzyme used

$$\text{Units / mg protein} = \frac{\text{units / ml enzyme}}{\text{mg protein / ml enzyme}}$$

Unit Definition: *One unit catalyzes the formation of 1 μmole of glyoxylate per minute at pH 6.8 at 30°C.*

Final Assay Concentrations

In a 1.00 ml reaction mix, the final concentrations are 30 mM imidazole, 5 mM magnesium chloride, 1 mM Ethylenediaminetetraacetic Acid, 4 mM phenylhydrazine, 1 mM DL-isocitric acid, and 0.005–0.007 unit isocitrate lyase.

Enzymatic Assay of ISOMALTASE (EC 3.2.1.10)

Principle

$$\text{Isomaltose} + H_2O \xrightarrow{\text{Isomaltase}} 2\ \text{Glucose}$$

$$\text{Glucose} + \text{ATP} \xrightarrow{\text{Hexokinase}} \text{Glucose 6-Phosphate}$$

$$\text{Glucose 6-Phosphate} + \beta\text{-NADP} \xrightarrow{\text{G-6-P-DH}} \text{Gluconate} + \beta\text{-NADPH}$$

Abbreviations Used

ATP = Adenosine 5'-Triphosphate
G-6-P-DH = Glucose 6-Phosphate Dehydrogenase
β-NADP = β-Nicotinamide Adenine Dinucleotide Phosphate, Oxidized Form
β-NADPH = β-Nicotinamide Adenine Dinucleotide Phosphate, Reduced Form

Conditions: T = 25°C, pH 6.8, A340nm, Light path = 1 cm

Method: Spectrophotometric Stop Rate Determination

Reagents

A. 50 mM Potassium Phosphate Buffer, pH 6.8 at 25°C (Phosphate Buffer) (Prepare 100 ml in deionized water using Potassium Phosphate, Monobasic, Anhydrous. Adjust to pH 6.8 at 25°C with 1 M KOH).
B. 100 mM Isomaltose Solution (Isomaltose) (Prepare 10 ml in deionized water using Isomaltose).
C. 300 mM Triethanolamine HCl Buffer, pH 7.6 at 25°C (TEA Buffer) (Prepare 100 ml in deionized water using Triethanolamine, Hydrochloride. Adjust to pH 7.6 at 25°C with 1 M NaOH).
D. 100 mM Magnesium Chloride Solution ($MgCl_2$) (Prepare 10 ml in deionized water using Magnesium Chloride, 4.9 M Solution).
E. 16 mM Adenosine 5'-Triphosphate Solution (ATP) (Prepare 10 ml in deionized water using Adenosine 5'-Triphosphate, Disodium Salt).
F. 12 mM β-Nicotinamide Adenine Dinucleotide Phosphate, Oxidized Form, Solution (β-NADP) (Prepare 10 ml in deionized water using β-Nicotinamide Adenine Dinucleotide Phosphate, Sodium Salt).
G. Hexokinase and Glucose-6-Phosphate Dehydrogenase Enzyme Solution (Hex/G-6-PDH) (Immediately before use, prepare a solution containing 200 Hexokinase units/ml of Hexokinase and Glucose-6-Phosphate Dehydrogenase in cold Reagent C).
H. Isomaltase Enzyme Solution (Isomaltase) (Immediately before use, prepare a solution containing 0.25–0.50 unit/ml of Isomaltase in cold deionized water).

Procedure

Step 1: Pipette (in milliliters) the following reagents into suitable containers:

	Test	Blank
Reagent A (Phosphate Buffer)	0.40	0.40
Reagent B (Isomaltose)	0.50	0.50

Mix by swirling and equilibrate to 25°C. Then add:

	Test	Blank
Reagent H (Isomaltase)	0.10	—
Deionized Water	—	0.10

Immediately mix by swirling and incubate at 25°C for exactly 30 minutes. Then place in a boiling water bath for 3 minutes. Cool to room temperature. Then add:

Deionized Water	9.00	9.00

Mix by swirling.

Step 2: Prepare a reaction cocktail by combining the following reagents (in milliliters):

Reagent C (TEA Buffer)	25.00
Reagent D ($MgCl_2$)	1.00
Reagent E (ATP)	1.00
Reagent F (β-NADP)	1.00
Reagent G (Hex/G-6-PDH)	0.20

Pipette (in milliliters) the following reagents into suitable cuvettes:

	Test	Blank
Reaction Cocktail	2.50	2.50

Equilibrate to 25°C. Then add:

Diluted Test from Step 1	0.50	—
Diluted Blank from Step 1	—	0.50

Mix by inversion and record the increase in A340nm for approximately 5–10 minutes until constant using a suitably thermostatted spectrophotometer. Obtain the final ΔA340nm for both the Test and Blank solutions.

Calculations

$$\text{Units / ml enzyme} = \frac{(\Delta\text{A340nm Test} - \Delta\text{A340nm Blank})\ (10)\ (3)\ (\text{df})}{(2)\ (6.22)\ (0.1)\ (0.5)\ (30)}$$

10 = Total volume (in milliliters) of assay in Step 1
3 = Total volume (in milliliters) of assay in Step 2
df = Dilution factor
2 = Moles of glucose release per mole of isomaltose
6.22 = Millimolar extinction coefficient of β-NADPH at 340nm
0.1 = Volume (in milliliters) of enzyme used in Step 1
0.5 = Volume (in milliliters) of diluted test used in Step 2
30 = Reaction time (in minutes) of Step 1

$$\text{Units / mg solid} = \frac{\text{units / ml enzyme}}{\text{mg solid / ml enzyme}}$$

$$\text{Units / mg protein} = \frac{\text{units / ml enzyme}}{\text{mg protein / ml enzyme}}$$

Unit Definition: *One unit will convert 1.0 μmole of isomaltose to 2.0 μmoles of D-glucose per minute at pH 6.8 at 25°C (liberated glucose determined at pH 7.6).*

Final Assay Concentrations

In a 1.00 ml reaction mix, the final concentrations are 20 mM potassium phosphate, 50 mM isomaltose and 0.025–0.050 unit isomaltase.

Notes

1. Hexokinase Unit Definition: One unit will phosphorylate 1.0 μmole of D-glucose per minute at pH 7.6 at 25°C.
2. Glucose-6-Phosphate Dehydrogenase Unit Definition: One unit will oxidize 1.0 μmole of D-glucose 6-phosphate to 6-phospho-D-gluconate per minute in the presence of NADP at pH 7.4 at 25°C.

Enzymatic Assay of KALLIKREIN (EC 3.4.21.34) from Human Plasma

Principle

$$\text{BAEE} \xrightarrow{\text{Kallikrein}} \text{Na - Benzoyl - L - arginine} + \text{Ethanol}$$

$$\text{Ethanol} + \beta\text{-NAD} \xrightarrow{\text{ADH}} \text{Acetaldehyde} + \beta\text{-NADH}$$

Abbreviations Used

BAEE = Na-Benzoyl-L-Arginine Ethyl Ester
β-NAD = β-Nicotinamide Adenine Dinucleotide, Oxidized Form
ADH = Alcohol Dehydrogenase
β-NADH = β-Nicotinamide Adenine Dinucleotide, Reduced Form

Conditions: T = 25°C, pH = 8.7, A340nm, Light path = 1 cm

Method: Continuous Spectrophotometric Rate Determination

Reagents

A. 335 mM Pyrophosphate, 197 mM Semicarbazide and 53 mM Glycine Buffer, pH 8.7 at 25°C (Prepare 100 ml in deionized water using Pyrophosphate, Tetrasodium, Anhydrous, Semicarbazide Hydrochloride and Glycine, Free Base. Adjust to pH 8.7 at 25°C with 2 M NaOH).
B. 30 mM β-Nicotinamide Adenine Dinucleotide, Oxidized Form, Solution (β-NAD) (Dissolve the contents of one 50 mg vial of β-Nicotinamide Adenine Dinucleotide, in the appropriate volume of deionized water. *Prepare Fresh*).
C. 6 mM Na-Benzoyl-L-Arginine Ethyl Ester Solution (BAEE) (Prepare 10 ml in deionized water using Na-Benzoyl-L-Arginine Ethyl Ester, Hydrochloride).
D. 2.4 mM Ammonium Sulfate Solution $(NH_4)_2SO_4$ (Prepare 25 ml in deionized water using Ammonium Sulfate).

E. Alcohol Dehydrogenase Enzyme Solution (ADH) (Immediately before use, prepare a solution containing 10,000 units/ml of Alcohol Dehydrogenase in cold Reagent D).

F. Kallikrein Solution (Immediately before use, prepare a solution containing 0.15–0.30 unit/ml of Kallikrein in cold deionized water).

Procedure

Pipette (in milliliters) the following reagents into suitable cuvettes:

	Test	Blank
Reagent A (Buffer)	2.30	2.30
Reagent B (β-NAD)	0.10	0.10
Reagent C (BAEE)	0.50	0.50
Reagent E (ADH)	0.02	0.02

Mix by inversion and equilibrate to 25°C. Monitor the A340nm until constant, using a suitably thermostatted spectrophotometer. Then add:

Reagent F (Kallikrein)	0.10	—
Deionized Water	—	0.10

Immediately mix by inversion and record the increase in A340nm for approximately 5 minutes. Obtain the ΔA340nm/minute using the maximum linear rate for both the Test and Blank.

Calculations

$$\text{Units / ml enzyme} = \frac{(\Delta\text{A340nm / min Test} - \Delta\text{A340nm / min Blank})\,(3.02)\,(\text{df})}{(6.22)\,(0.1)}$$

3.02 = Total volume (in milliliters) of assay
df = Dilution factor
6.22 = Millimolar extinction coefficient of β-NADH at 340 nm
0.1 = Volume (in milliliter) of enzyme used

$$\text{Units / mg solid} = \frac{\text{units / ml enzyme}}{\text{mg solid / ml enzyme}}$$

$$\text{Units / mg protein} = \frac{\text{units / ml enzyme}}{\text{mg protein / ml enzyme}}$$

Unit Definition: *One unit will hydrolyze 1.0 μmole of BAEE to Na-benzoyl-Larginine and ethanol per minute at pH 8.7 at 25°C.*

Final Assay Concentrations

In a 3.02 ml reaction mix, the final concentrations are 255 mM pyrophosphate, 150 mM semicarbazide, 40 mM glycine, 1 mM Na-benzoyl-L-arginine ethyl ester, 1 mM β-nicotinamide adenine dinucleotide, 16 mM ammonium sulfate, 200 units alcohol dehydrogenase and 0.015–0.030 unit kallikrein.

Note

1. Alcohol Dehydrogenase Unit Definition: One unit will convert 1.0 μmole of ethanol to acetaldehyde per minute at pH 8.8 at 25°C.

Enzymatic Assay of KANAMYCIN 6'-ACETYLTRANSFERASE (EC 2.3.1.55)

Principle

$$\text{Acetyl-CoA} + \xrightarrow{\text{Kanamycin KAT}} \text{CoA} + \text{N-6'-Acetylkanamycin}$$

$$\text{CoA} + \text{DTNB} \rightarrow \text{TNB} + \text{CoA Derivative}$$

Abbreviations Used

Acetyl-CoA = Acetyl Coenzyme A
KAT = Kanamycin 6'-Acetyltransferase
CoA = Coenzyme A
DTNB = 5,5'-Dithio-bis-(2-Nitrobenzoic Acid)
TNB = 5-Thio-2-Nitrobenzoic Acid

Conditions: T = 37°C, pH = 5.7, A412nm, Light path = 1 cm

Method: Continuous Spectrophotometric Rate Determination

Reagents

A. 100 mM MES Buffer, pH 5.7 at 37°C (Prepare 100 ml in deionized water using MES, Free Acid. Adjust to pH 5.7 at 37°C With 1 M NaOH).
B. 17 mM 5,5'-Dithio-bis(2-Nitrobenzoic Acid) (DTNB) (Immediately before use, prepare 3 ml in deionized water using 5,5'-Dithio-bis (2-Nitrobenzoic Acid). Solubilize by the addition of a minimal volume of 0.1 M NaOH. The color of this solution must be a very pale yellow).
C. 5 mM Acetyl Coenzyme A Solution (Acet CoA) (Prepare 10 ml in deionized water using Acetyl Coenzyme A, Sodium Salt).
D. 0.3% (w/v) Kanamycin Sulfate Solution (KS) (Prepare 1 ml in deionized water using Kanamycin, Monosulfate).
E. 10 mM Tris HCl, pH 7.8 at 37°C (Enz Dil) (Prepare 50 ml in deionized water using Trizma Base. Adjust to pH 7.8 at 37°C with 1 M HCl).
F. Kanamycin 6'-Acetyltransferase Enzyme Solution (Immediately before use, prepare a solution containing 50 units/ml of Kanamycin 6'-Acetyltransferase in cold Reagent E).

Procedure

Prepare a reaction cocktail by pipetting (in milliliters) the following reagents into a suitable container:

Reagent B (DTNB)	3.00
Reagent C (Acet CoA)	1.00
Reagent A (Buffer)	46.00

Mix by swirling. Pipette (in milliliters) the following reagents into suitable cuvettes:

	Test	Blank
Reaction Cocktail	2.90	2.90
Reagent F (Enzyme Solution)	0.05	—
Reagent E (Enz Dil)	—	0.05

Mix by inversion and equilibrate to 37°C. Monitor the A412nm until constant, using a suitably thermostatted spectrophotometer. Then add:

Reagent D (KS)	0.05	0.05

Immediately mix by inversion and record the increase in A412nm for approximately 5 minutes. Obtain the ΔA412nm/minute using the maximum linear rate for both the Test and Blank.

Calculations

$$\text{Units / ml enzyme} = \frac{(\Delta A412\text{nm / min Test} - \Delta A412\text{nm / min Blank})\ (3)\ (\text{df})}{(0.0136)\ (0.05)}$$

3 = Total volume (in milliliters) of assay
df = Dilution factor
0.0136 = μmolar extinction coefficient for 5-Thio- 2-Nitrobenzoic Acid at 412nm
0.05 = Volume (in milliliter) of enzyme used

$$\text{Units / mg solid} = \frac{\text{units / ml enzyme}}{\text{mg solid / ml enzyme}}$$

$$\text{Units / mg protein} = \frac{\text{units / ml enzyme}}{\text{mg protein / ml enzyme}}$$

Unit Definition: *One unit will convert 1.0 nanomole of kanamycin and acetyl-CoA to kanamycin 6'-acetate and CoA per minute at pH 5.7 at 37°C.*

Final Assay Concentrations

In a 3.00 ml reaction mix, the final concentrations are 89 mM MES, 1 mM 5,5'-dithio-bis-(2-nitrobenzoic acid), 0.1 mM acetyl coenzyme A, 0.005% (w/v) kanamycin sulfate, 0.2 mM Tris, and 2.5 units kanamycin 6'-acetyltransferase.

Enzymatic Assay of KERATANASE (EC 3.2.1.103)

Principle

$$\text{Keratan} + H_2O \xrightarrow{\text{Keratanase}} \text{D-Galactose}$$

Conditions: T = 37°C, pH = 7.4, A690nm, Light path = 1 cm.

Method: Colorimetric

Reagents

A. 50 mM Tris HCl Buffer, pH 7.4 at 37°C (Prepare 100 ml in deionized water using Trizma Base. Adjust to pH 7.4 at 37°C with 1 M HCl).

B. 0.06% (w/v) Keratan Sulfate Solution (Ker) (Prepare 5 ml in cold Reagent A using Keratan Sulfate).
C. 0.05% (w/v) Ferricyanide Solution ($K_3Fe(CN)_6$) (Prepare 10 ml in deionized water using Potassium Ferricyanide. *Store in an Amber Bottle — Light Sensitive*).
D. 50 mM Sodium Carbonate with 10 mM Potassium Cyanide Solution (Na_2CO_3.KCN) (Prepare 1 liter in deionized water using Sodium Carbonate, Anhydrous, and Potassium Cyanide).
E. 3.1 mM Ferric Ammonium Sulfate with 0.1% (w/v) SDS and 25 mM Sulfuric Acid Solution (Fe(III)) (Prepare 1 liter in 25 mM Sulfuric Acid using Ferric Ammonium Sulfate, Lauryl Sulfate, Sodium Salt (SDS, and Sulfuric Acid).
F. 0.1 mM (w/v) Galactose Standard Solution (Gal) (Prepare 10 ml in deionized water using D(+)Galactose).
G. Keratanase Enzyme Solution (Immediately before use, prepare a solution containing 5 units/ml of Keratanase in cold Reagent A).

Procedure

Pipette (in milliliters) the following reagents into suitable vials:

	Test	Blank
Reagent B (Ker)	0.18	0.20

Equilibrate to 37°C. Then add:

Reagent G (Enzyme Solution)	0.02	—

Mix by swirling and incubate at 37°C for exactly 15 minutes. After 15 minutes, place the Test and Blank vials into a boiling water bath for 2 minutes. Remove the vials and allow to cool to room temperature. Then add:

Deionized Water	0.80	0.80

Mix by inversion. Then add:

Reagent C ($K_3Fe(CN)_6$)	1.00	1.00
Reagent D (Na_2CO_3.KCN)	1.00	1.00

Mix by swirling and place the vials into a boiling water bath for 15 minutes. Allow to cool to room temperature. Then add:

Reagent E (Fe (III))	5.00	5.00

Mix by swirling and let stand at 25°C for 15 minutes. Transfer to suitable cuvettes and record the A690nm for both the Test and Blank using a suitable spectrophotometer.

Standard Curve: A standard curve is made by pipetting (in milliliters) the following reagents into suitable containers:

	Std 1	Std 2	Std 3	Std 4	Std 5	Blank
Reagent F (Gal)	0.10	0.30	0.50	0.70	1.00	—
Deionized Water	0.90	0.70	0.50	0.30	—	1.00
Reagent C ($K_3Fe(CN)_6$)	1.00	1.00	1.00	1.00	1.00	1.00
Reagent D (Na_2CO_3-KCN)	1.00	1.00	1.00	1.00	1.00	1.00

Place the containers into a boiling water bath for 15 minutes. Allow to cool to room temperature. Then add:

Reagent E (Fe (III)) ($K_3Fe(CN)_6$)	5.00	5.00	5.00	5.00	5.00	5.00

Mix by swirling and let stand at 25°C for 15 minutes. Transfer to suitable cuvettes and record the A690nm for the Standards and Blank.

Calculations

Standard Curve: A690nm Standard = A690nm Std–A690nm Std Blank
Plot the A690nm Standards vs µmoles of reducing sugar.
Sample Determination: ΔA690nm Sample = A690nm Test–A690nm Test Blank
Determine the µmoles of reducing sugar liberated using the Standard Curve.

$$\text{Units / vial enzyme} = \frac{(\mu\text{moles of reducing Sugar liberated})\ (4)\ (\text{df})}{(0.02)}$$

4 = Conversion factor for minutes to hours as per the Unit Definition
0.02 = Volume (in milliliter) of enzyme used in assay
df = Dilution factor

Unit Definition: *One unit will liberate 1.0 µmole of reducing sugar (measured as galactose) from keratan sulfate per hour at pH 7.4 at 37°C.*

Final Assay Concentrations

In a 0.20 ml reaction mix, the final concentrations are 50 mM Tris, 0.05% (w/v) keratan and 0.1 unit keratanase.

Enzymatic Assay of α-KETOGLUTARATE DEHYDROGENASE

Principle

$$\alpha\text{-KGA} + \text{Coenzyme A} + \beta\text{-NAD} \xrightarrow{\alpha\text{-KGDH}} \text{Succinyl-CoA} + \beta\text{-NADH} + CO_2$$

Abbreviations Used

α-KGDH = α-Ketoglutarate Dehydrogenase
β-NAD = β-Nicotinamide Adenine Dinucleotide, Oxidized Form
β-NADH = β-Nicotinamide Adenine Dinucleotide, Reduced Form
α-KGA = α-Ketoglutaric Acid
CoA = Coenzyme A

Conditions: T = 30°C, pH 7.4, A340nm, Light path = 1 cm

Method: Continuous Spectrophotometric Rate Determination

Reagents

A. 150 mM MOPS HCl Buffer, pH 7.4 at 30°C (Prepare 100 ml in deionized water using MOPS, Sodium Salt. Adjust to pH 7.4 at 30°C with 1 M HCl).

B. 12 mM Magnesium Chloride Solution ($MgCl_2$) (Prepare 10 ml in deionized water using Magnesium Chloride, Hexahydrate).

C. 0.6 mM Calcium Chloride Solution ($CaCl_2$) (Prepare 100 ml in deionized water using Calcium Chloride).

D. 18 mM Cocarboxylase Solution (Cocarboxylase) (Prepare 10 ml in deionized water using Cocarboxylase).

E. 0.72 mM Coenzyme A Solution (CoA) (Prepare 10 ml in deionized water using Coenzyme A, Sodium Salt).

F. 20 mM β-Nicotinamide Adenine Dinucleotide, Oxidized Form (β-NAD) (Prepare 10 ml in deionized water using β-Nicotinamide Adenine Dinucleotide).

G. 15.6 mM L-Cysteine Solution (Cys-HCl) (Prepare 10 ml in deionized water using L-Cysteine, Hydrochloride, Monohydrate. Adjust to pH 7.4 at 30°C with 1 M NaOH).

H. 75 mM α-Ketoglutaric Acid Solution (α-KGA) (Prepare 10 ml in deionized water using α-Ketoglutaric Acid, Monosodium Salt).

I. 50 mM MOPS HCl Buffer, pH 7.4 at 30°C (Enz Dil) (Prepare 100 ml in deionized water using MOPS, Sodium Salt. Adjust to pH 7.4 at 30°C with 1 M HCl).

J. α-Ketoglutarate Dehydrogenase Enzyme Solution (α-KGDH) (Immediately before use, prepare a solution containing 0.5–1.0 units/ml of α-Ketoglutarate Dehydrogenase in cold Reagent I).

Procedure

Pipette (in milliliters) the following reagents into suitable cuvettes:

	Test	Blank
Reagent A (Buffer)	1.00	1.00
Reagent B ($MgCl_2$)	0.05	0.05
Reagent C ($CaCl_2$)	0.05	0.05
Reagent D (Cocarboxylase)	0.05	0.05
Reagent E (Coenzyme A)	0.50	0.50
Reagent F (β-NAD)	0.30	0.30
Reagent G (Cys-HCl)	0.50	0.50
Reagent I (α-KGDH)	0.05	0.05
Deionized Water	0.30	0.30

Mix by inversion and equilibrate to 30°C. Monitor the A340nm until constant, using a suitably thermostatted spectrophotometer. Then add:

Reagent H (α-KGA)	0.20	—
Deionized Water	—	0.20

Immediately mix by inversion and record the increase in A340nm for approximately 5 minutes. Obtain the ΔA340nm/minute using the maximum linear rate for both the Test and Blank.

Calculations

$$\text{Units / ml enzyme} = \frac{(\Delta A340\text{nm / min Test} - \Delta A340\text{nm / min Blank})\,(3)\,(\text{df})}{(6.22)\,(0.05)}$$

$$\text{Units / mg protein} = \frac{\text{units / ml enzyme}}{\text{mg protein / ml enzyme}}$$

3 = Total volume (in milliliters) of assay
df = Dilution factor
6.22 = Millimolar extinction coefficient of β-NADH at 340nm
0.05 = Volume (in milliliter) of enzyme used

Unit Definition: *One unit will convert 1.0 μmole of β-NAD to β-NADH per minute at pH 7.4 at 30°C in the presence of saturating levels of Coenzyme A.*

Final Assay Concentration

In a 3.00 ml reaction mix, the final concentrations are 50.8 mM MOPS, 0.2 mM $MgCl_2$, 0.01 mM $CaCl_2$, 0.3 mM cocarboxylase, 0.12 mM coenzyme A, 2.0 mM β-nicotinamide adenine dinucleotide, 2.6 mM L-cysteine, 5.0 mM α-Ketoglutaric acid, and 0.025–0.05 units aketoglutarate dehydrogenase.

Note

1. Cocarboxylase is also known as thiamine pyrophosphate.

Enzymatic Assay of LACCASE (EC 1.10.3.2)

Principle

$$\text{Syringaldzine} + O_2 + \xrightarrow{\text{Laccase}} \text{Oxidized Syringaldazine} + 2H_2O$$

Conditions: T = 30°C, pH = 6.5, A530nm, Light path = 1 cm.

Method: Continuous Spectrophotometric Rate Determination.

Reagents

A. 100 mM Potassium Phosphate Buffer, pH 6.5 at 30°C (Prepare 100 ml in deionized water using Potassium Phosphate, Monobasic, Anhydrous. Adjust to pH 6.5 at 30°C with 1 M KOH).

B. 0.216 mM Syringaldazine Solution (Prepare 3 ml in absolute methanol using Syringaldazine).

C. Laccase Enzyme Solution (Immediately before use, prepare a solution containing 25–50 units/ml of Laccase in cold deionized water).

Procedure

Pipette (in milliliters) the following reagents into suitable cuvettes:

	Test	Blank
Deionized Water	—	0.50
Reagent A (Buffer)	2.20	2.20
Reagent C (Enzyme Solution)	0.50	—

Equilibrate to 30°C. Monitor the A530nm until constant, using a suitably thermostatted spectrophotometer. Then add:

Reagent B (Syringaldazine)	0.30	0.30

Immediately mix by inversion and record the increase in A530nm for approximately 10 minutes. Obtain the A530nm/minute using the maximum linear rate for both the Test and Blank.

Calculations

$$\text{Units / ml enzyme} = \frac{(\text{A530nm / min Test} - \text{A530nm / min Blank})\ (\text{df})}{(0.001)\ (0.5)}$$

df = Dilution factor
0.001 = The change in A530nm/minute per unit of Laccase at pH 6.5 at 30°C in a 3 ml reaction mix.
0.5 = Volume (in milliliter) of enzyme used.

$$\text{Units / mg solid} = \frac{\text{units / ml enzyme}}{\text{mg solid / ml enzyme}}$$

Unit Definition: *One unit will produce a A530nm of 0.001 per minute at pH 6.5 at 30°C in a 3 ml reaction volume using syringaldazine as substrate.*

Final Assay Concentration

In a 3 ml reaction mix, the final concentrations are 73 mM potassium phosphate, 0.02 mM syringaldazine, 10% methanol, and 12.5–25.0 units laccase.

Suitability Assay for α-LACTALBUMIN

Principle

$$\text{UDP - Galactose} + \text{D - Glucose} \xrightarrow{\text{Galactosyltransferase}} \text{UDP} + \text{Lactose}$$

$$\text{UDP} + \text{PEP PK} \xrightarrow{\alpha-\text{Lactalbumin}} \text{Pyruvate} + \text{UTP}$$

$$\text{Pyruvate} + \beta\text{ - NADH} \xrightarrow{\text{LDH}} \text{Lactate} + \beta\text{ - NAD}$$

Abbreviations Used

UDP-Galactose = Uridine 5'-Diphosphogalactose
UDP = Uridine 5'-Diphosphate

PEP = Phospho(enol) pyruvate
UTP = Uridine 5'-Triphosphate
PK = Pyruvate Kinase
β-NADH = β-Nicotinamide Adenine Dinucleotide, Reduced Form
LDH = Lactic Dehydrogenase
β-NAD = β-Nicotinamide Adenine Dinucleotide, Oxidized Form

Conditions: T = 30°C, pH = 8.4, A340nm, Light path = 1 cm

Method: Continuous Spectrophotometric Rate Determination

Reagents

A. 250 mM Glycylglycine Buffer, pH 8.6 at 30°C. (Prepare 50 ml in deionized water using Gly-Gly, Hydrochloride. Adjust to pH 8.6 at 30°C with 1 M NaOH).
B. 0.32 mM β-Nicotinamide Adenine Dinucleotide, Reduced Form (β-NADH) (Dissolve the contents of one 5 mg vial of β-Nicotinamide Adenine Dinucleotide, Reduced Form, Disodium Salt, in the appropriate volume of deionized water. *Prepare Fresh*).
C. Phospho(enol) pyruvate (PEP) (Use Phospho(enol) Pyruvate Tri(cyclohexylammonium) Salt).
D. 100 mM Manganese Chloride Tetrahydrate with 1 M Potassium Chloride Solution ($MnCl_2$/KCl) (Prepare 5.0 ml in deionized water using Manganese Chloride Tetrahydrate and Potassium Chloride).
E. 4.8 mM Uridine 5'-Diphosphogalactose Solution (UDPGalactose) (Prepare 10 ml in deionized water using Uridine 5'-Diphosphogalactose, Sodium Salt).
F. 50 mM Glycylglycine Buffer, pH 8.0 at 30°C. (Prepare 10 ml in deionized water using Gly-Gly, Hydrochloride. Adjust to pH 8.0 at 30°C with 1 M NaOH).
G. 0.3% (w/v) α-Lactalbumin Solution (Prepare 2 ml in Reagent F using α-Lactalbumin).
H. 1.0 M D-Glucose Solution (Prepare 2 ml in deionized water using β-D(+)Glucose).
I. PK/LDH Enzymes Suspension.
J. 20 mM Tris HCl Buffer with 2 mM Ethylenediaminetetraacetic Acid and 2 mM 2-Mercaptoethanol, pH 7.5 at 30°C (Enz Dil) (Prepare 50 ml in deionized water using Trizma Hydrochloride, Ethylenediaminetetraacetic Acid, Tetrasodium Salt and 2-Mercaptoethanol. Adjust to pH 7.5 at 30°C with 1 M NaOH).
K. Galactosyltransferase Enzyme Solution (Immediately before use, prepare a solution containing 0.05–0.15 unit/ml of Galactosyltransferase, in cold Reagent J).

Procedure

Prepare a reaction cocktail by pipetting (in milliliters) the following reagents into a suitable container:

Reagent A (Buffer)	10.00
Reagent B (β-NADH)	20.00
Reagent D ($MnCl_2$/KCl)	2.50
Reagent C (PEP)	30 mg

Mix and adjust to pH 8.4 at 30°C with 1 M HCl or 1 M NaOH, if necessary. Pipette (in milliliters) the following reagents into suitable cuvettes:

	Test	Blank
Deionized Water	0.25	0.25
Reaction Cocktail	2.00	2.00
Reagent E (UDP-Galactose)	0.20	0.20
Reagent G (α-Lactalbumin)	0.20	0.20
Reagent I (PK/LDH Suspension)	0.05	0.05
Reagent H (D-Glucose)	0.20	0.20

Mix by inversion and equilibrate to 30°C. Monitor the A340nm until constant, using a suitably thermostatted spectrophotometer. Then add:

Reagent K (Galactosyltransferase)	0.10	—
Reagent J (Enz Dil)	—	0.10

Immediately mix by inversion and record the decrease in A340nm for approximately 10 minutes. Obtain the ΔA340nm/minute using the maximum linear rate for both the Test and Blank.

Calculations

$$\text{Units / ml enzyme} = \frac{(\text{A340nm / min Test} - \text{A340nm / min Blank})\,(3)\,(\text{df})}{(6.22)\,(0.01)}$$

3 = Total volume (in milliliters) of assay
df = Dilution factor
6.22 = Millimolar extinction coefficient of β-NADH at 340nm
0.1 = Volume (in milliliter) of enzyme used

$$\text{Units / mg solid} = \frac{\text{units / ml enzyme}}{\text{mg solid / ml enzyme}}$$

$$\text{Units / mg protein} = \frac{\text{units / ml enzyme}}{\text{mg protein / ml enzyme}}$$

Unit Definition: *One unit will transfer 1.0 µmole of galactose from UDP-galactose to D-glucose per minute at pH 8.4 at 30°C in the presence of 0.2 mg of α-lactalbumin per ml of reaction mixture.*

Final Assay Concentrations

In a 3.00 ml reaction mix, the final concentrations are 55 mM glycylglycine, 0.13 mM β-nicotinamide adenine dinucleotide, reduced form, 1.3 mM phospho(enol) pyruvate, 5.1 mM manganese chloride, 51.2 mM potassium chloride, 0.32 mM uridine 5'-diphosphogalactose, 0.02% (w/v) α-lactalbumin, 67 mM glucose, 0.67 mM Tris, 0.07 mM Ethylenediaminetetraacetic Acid, 0.07 mM 2-mercaptoethanol, 35 units pyruvate kinase, 50 units lactic dehydrogenase, and 0.005–0.015 unit galactosyltransferase.

Notes

1. L-Lactic Dehydrogenase Unit Definition: One unit will reduce 1.0 µmole of pyruvate to L-lactate per minute at pH 7.5 at 37°C.
2. Pyruvate Kinase Unit Definition: One unit will convert 1.0 µmole of phospho(enol) pyruvate to pyruvate per minute at pH 7.6 at 37°C.
3. α-Lactalbumin is included in the assay since it lowers the apparent Km of glucose.

4. Galactosyltransferase Unit Definition: One unit will transfer 1.0 µmole of galactose from UDP-galactose to D-glucose per minute at pH 8.4 at 30°C in the presence of 0.2 mg α-lactalbumin per ml of reaction mixture.

Enzymatic Assay of LACTATE OXIDASE

Principle

$$\text{L-Lactate} + O_2 \xrightarrow{\text{LOX}} \text{Pyruvate} + H_2O_2$$

$$2H_2O_2 + \text{4-AAP} + \text{DMA} \xrightarrow{\text{POD}} \text{Quinonedimine dye} + H_2O$$

Abbreviations Used

LOX = L-Lactate Oxidase
4-AAP = Aminoantipyrine
DMA = N,N-Dimethylaniline
POD = Peroxidase.

Conditions: T = 37°C, pH = 6.5, A565nm, Light path = 1 cm.

Method: Spectrophotometric Stop Rate Determination

Reagents

A. 200 mM 3,3-Dimethylglutaric Acid-NaOH Buffer, pH 6.5 at 37°C (DMGA) (Prepare 5 ml in deionized water using 3,3-Dimethylglutaric Acid. Adjust to pH 6.5 at 37°C with 1 M NaOH).

B. 15 mM 4-Aminoantipyrine Solution (4-AAP) (Prepare 1 ml in deionized water using 4-Aminoantipyrine, Free Base).

C. 500 mM L(+)Lactic Acid Solution, pH 6.5 at 37°C (Lactic Acid) (Prepare 1 ml in deionized water using L(+)Lactic Acid, Free Acid. Adjust to pH 6.5 with 1 M NaOH).

D. Peroxidase Enzyme Solution (POD) (Immediately before use, prepare a solution containing 50 Purpurogallin units/ml of Peroxidase Type II from Horseradish, in cold deionized water).

E. 10 mM Potassium Phosphate Buffer with 0.010 mM Flavin Adenine Dinucleotide (FAD), pH 7.0 at 37°C (Enzyme Diluent) (Prepare 50 ml in deionized water using Potassium Phosphate, Monobasic, Anhydrous, and Flavin Adenine Dinucleotide, Disodium Salt Adjust to pH 7.0 at 37°C with 1 M NaOH. *Prepare Fresh*).

G. 0.2% (v/v) N,N-Dimethylaniline Solution (DMA) (Prepare 10 ml in deionized water using N,N-Dimethylaniline).

H. 0.25% (w/v) Dodecylbenzenesulfonic Acid Solution (DBS) (Prepare 5 ml in deionized water using Dodecylbenzenesulfonic Acid, Sodium Salt).

I. Lactate Oxidase Enzyme Solution (LOX) (Immediately before use, prepare a solution containing 0.1–0.2 units/ml of Lactate Oxidase in cold Reagent E).

Procedure

Prepare a reaction cocktail by pipetting (in milliliters) the following reagents into a suitable container.

Reagent A (DMGA)	2.00
Reagent D (POD)	1.00
Reagent B (4-AAP)	1.00
Reagent C (Lactic Acid)	1.00
Deionized Water	3.00

Mix by inversion and equilibrate to 37°C.

Pipette (in milliliters) the following reagents into a suitable cuvette:

	Test	Blank
Reaction Cocktail	0.80	0.80
Reagent G (DMA)	0.20	0.20

Mix by inversion and equilibrate to 37°C. Then add:

Reagent I (LOX)	0.020	—
Reagent E (Enzyme Diluent)	—	0.020

Immediately mix by inversion and incubate at 37°C for exactly 10 minutes. Then add:

Reagent H (DBS)	2.00	2.00

Mix by inversion and record the A565nm for both the Test and Blank using a suitable spectrophotometer.

Calculations

$$\text{Units / mg enzyme} = \frac{(\text{A565nm Test} - \text{A565nm Blank})\ (3.02)\ (\text{df})}{(35.33)\ (0.5)\ (10)\ (0.02)}$$

3.02 = Total volume of assay.
df = Dilution factor.
35.33 = Millimolar extinction coefficient of Quinonediimine dye at 565nm.
0.5 = Conversion factor based on one mole of H_2O_2 produces half a mole of Quinonediimine dye.
10 = Time of assay (in minutes) as per unit definition
0.02 = Volume (in milliliter) of enzyme used.

$$\text{Units / mg solid} = \frac{\text{units / ml enzyme}}{\text{mg solid / ml enzyme}}$$

Unit Definition: *One unit will oxidize 1.0 mole of L-lactate to pyruvate and H_2O_2 per minute at pH 6.5 at 37°C.*

Final Assay Concentrations

In a 1.02 ml reaction mix, the final concentrations are 39 mM 3,3 dimethylglutaric acid, 5 units peroxidase, 1.5 mM 4-aminoantipyrine, 49 mM L(+)lactic acid, 0.04% (v/v) N,N-dimethylaniline, 0.20 mM potassium phosphate, 0.20 m FAD and 0.002–0.004 unit lactate oxidase.

Note

1. Unit Definition for Peroxidase: One unit will form 1.0 mg purpurogallin from pyrogallol in 20 seconds at pH 6.0 at 20°C. This purpurogallin (20 seconds) unit is equivalent to approximately 18 M units per minute at 25°C.

Enzymatic Assay of D-LACTIC DEHYDROGENASE (EC 1.1.1.28)

Principle

$$\text{Pyruvate} + \beta\text{-NADH} \xrightarrow{\text{D-LacticDehydrogenase}} \text{D-Lactate} + \beta\text{-NAD}$$

Abbreviations Used

β-NADH = β-Nicotinamide Adenine Dinucleotide, Reduced Form
β-NAD = β-Nicotinamide Adenine Dinucleotide, Oxidized Form

Conditions: T = 25°C, pH = 7.0, A340nm, Light path = 1 cm

Method: Continuous Spectrophotometric Rate Determination

Reagents

A. 100 mM Potassium Phosphate Buffer, pH 7.0 at 25°C (Prepare 200 ml in deionized water using Potassium Phosphate, Monobasic, Anhydrous. Adjust to pH 7.0 at 25°C with 1 M KOH).
B. 11 mM β-Nicotinamide Adenine Dinucleotide, Reduced Form Solution (β-NADH) (Prepare 1 ml in cold deionized water using β-Nicotinamide Adenine Dinucleotide, Reduced Form, Disodium Salt. *Prepare Fresh*).
C. 20 mM Sodium Pyruvate Solution (Pyruvate) (Prepare 1.0 ml in cold deionized water using Pyruvic Acid, Sodium Salt).
D. 1.0% (w/v) Bovine Serum Albumin Solution (BSA) (Prepare 50 ml in Reagent A using Albumin, Bovine or equivalent).
E. D-Lactic Dehydrogenase Enzyme Solution (Immediately before use, prepare a solution containing 0.3–0.60 unit/ml of D-Lactic Dehydrogenase in cold Reagent D. *Prepare Fresh*).

Procedure

Pipette (in milliliters) the following reagents into suitable cuvettes:

	Test	Blank
Reagent A (Buffer)	2.50	2.50
Reagent B (β-NADH)	0.05	0.05
Reagent C (Pyruvate)	0.10	0.10

Mix by inversion and equilibrate to 25°C. Monitor the A340nm until constant, using a suitably thermostatted spectrophotometer. Then add:

Reagent D (BSA)	—	0.05
Reagent E (Enzyme Solution)	0.05	—

Immediately mix by inversion and record the decrease in A340nm for approximately 5 minutes. Obtain the ΔA340nm/minute using the maximum linear rate for both the Test and Blank.

Calculations

$$\text{Units / ml enzyme} = \frac{(\Delta A340nm / \text{min Test} - \Delta A340nm / \text{min Blank})\ (2.7)\ (df)}{(6.22)\ (0.05)}$$

2.7 = Total volume (in milliliters) of the assay
df = Dilution factor
6.22 = Millimolar extinction coefficient of β-NADH at 340nm
0.05 = Volume (in milliliters) of assay

$$\text{Units / mg solid} = \frac{\text{units / ml enzyme}}{\text{mg solid / ml enzyme}}$$

$$\text{Units / mg protein} = \frac{\text{units / ml enzyme}}{\text{mg protein / ml enzyme}}$$

Unit Definition: *One unit will reduce 1.0 μmole of pyruvate to D-lactate per minute at pH 7.0 at 25°C.*

Final Assay Concentrations

In a 2.75 ml reaction mix, the final concentrations are 94 mM potassium phosphate, 0.20 mM β-nicotinamide adenine dinucleotide, reduced form, 0.74 mM pyruvate, 0.015–0.03 unit D-lactic dehydrogenase.

Enzymatic Assay of L-LACTIC DEHYDROGENASE (CYTOCHROME) (EC 1.1.2.3)

Principle

$$\text{Pyruvate} + 2\text{Ferricyanide} \xrightarrow{\text{L-Lactic Dehydrogenase}} \text{L-Lactate} + 2\text{Ferrocyanide}$$

Conditions: T = 37°C, pH = 8.4, A420nm, Light path = 1 cm

Method: Continuous Spectrophotometric Rate Determination

Reagents

A. 100 mM Sodium Pyrophosphate Buffer, pH 8.4 at 37°C (Prepare 100 ml in deionized water using Tetrasodium Pyrophosphate, Decahydrate. Adjust to pH 8.4 at 37°C with 1 M HCl).

B. 67 mM Sodium Phosphate with 1 mM Ethylenediaminetetraacetic Acid Solution, pH 7.4 at 37°C (Enzyme Diluent). (Prepare 100 ml in deionized water using Sodium Phosphate, Monobasic, Anhydrous and Ethylenediaminetetraacetic Acid, Disodium Salt, Dihydrate. Adjust to pH 7.4 at 37°C with 1 M NaOH).

C. 500 mM Lactic Acid Solution (Lactic) (Prepare 5.0 ml in deionized water using L(+)-Lactic Acid, Free Acid. *Prepare Fresh*).

D. 8.3 mM Potassium Ferricyanide Solution ($K_3Fe(CN)_6$) (Prepare 5 ml in deionized water using Potassium Ferricyanide).

E. 10 mM Ethylenediaminetetraacetic Acid Solution (EDTA) (Prepare 10 ml in deionized water using Ethylenediaminetetraacetic Acid, Disodium Salt, Dihydrate).

F. L-Lactic Dehydrogenase Enzyme Solution (Immediately before use, prepare a solution containing 1–6 units/ml of L-Lactic Dehydrogenase in cold Reagent B).

Procedure

Prepare a reaction cocktail by pipetting (in milliters) the following reagents into a suitable container:

Reagent A (Buffer)	20.00
Reagent C (Lactic)	5.00
Reagent D ($K_3Fe(CN)_6$)	1.00
Reagent E (EDTA)	3.00

Mix by swirling and adjust to pH 8.4 at 37°C with 1M HCl or 1M NaOH, if necessary. Pipette (in milliliters) the following reagents into suitable cuvettes:

	Test	Blank
Reaction Cocktail	2.90	2.90

Mix by inversion and equilibrate to 37°C. Monitor the A420nm until constant, using a suitably thermostatted spectrophotometer. Then add:

Reagent B (Enzyme Diluent)	—	0.10
Reagent F (Enzyme Solution)	0.10	—

Immediately mix by inversion and record the decrease in A420nm for approximately 5 minutes. Obtain the ΔA420nm/minute using the maximum linear rate for both the Test and Blank.

Calculations

$$\text{Units / ml enzyme} = \frac{(\Delta A420\text{nm / min Test} - \Delta A420\text{nm / min Blank})\ (3)\ (\text{df})}{(1.04)\ (0.1)}$$

3 = Total volume (in milliliters) of assay
df = Dilution factor
1.04 = Millimolar extinction coefficient of Potassium Ferricyanide at 420nm
0.1 = Volume (in milliliters) of enzyme used

$$\text{Units / mg solid} = \frac{\text{units / ml enzyme}}{\text{mg solid / ml enzyme}}$$

$$\text{Units / mg protein} = \frac{\text{units / ml enzyme}}{\text{mg protein / ml enzyme}}$$

Unit Definition: *One unit will reduce 1.0 μmole of ferricyanide and will oxidize 0.5 μmole of L-lactate to pyruvate per minute at pH 8.4 at 37°C.*

Final Assay Concentrations

In a 3.00 ml reaction mix, the final concentrations are 2.2 mM sodium phosphate, 1 mM Ethylenediaminetetraacetic Acid, 83 mM lactic acid, 0.28 mM potassium ferricyanide, 67 mM sodium pyrophosphate, and 0.1–0.6 unit L-lactic dehydrogenase (cytochrome).

Enzymatic Assay of L-LACTIC DEHYDROGENASE (EC 1.1.1.27)

Principle

$$\text{Pyruvate} + \beta\text{-NADH} \xrightarrow{\text{L-Lactic Dehydrogenase}} \text{L-Lactate} + \beta\text{-NAD}$$

Abbreviations Used

β-NADH = β-Nicotinamide Adenine Dinucleotide, Reduced Form
β-NAD = β-Nicotinamide Adenine Dinucleotide, Oxidized form

Conditions: T = 30°C, pH = 6.0, A340nm, Light path = 1 cm

Method: Continuous Spectrophotometric Rate Determination

Reagents

A. 100 mM Sodium Acetate Buffer, pH 6.0 at 30°C (Prepare 100 ml in deionized water using Sodium Acetate, Trihydrate. Adjust to pH 6.0 at 30°C with 1 M Acetic Acid).
B. 250 mM Pyruvate Solution (Pyruvate) (Prepare 10 ml in deionized water using Pyruvic, Acid, Sodium Salt).
C. 13 mM β-Nicotinamide Adenine Dinucleotide, Reduced Form Solution (β-NADH) (Prepare 1 ml in cold deionized water using β-Nicotinamide Adenine Dinucleotide, Reduced Form, Disodium Salt. *Prepare Fresh*).
D. 50 mM Potassium Phosphate Buffer, pH 7.5 at 30°C (Enzyme Diluent) (Prepare 100 ml in deionized water using Potassium Phosphate, Monobasic, Anhydrous. Adjust to pH 7.5 at 30°C with 1 M KOH).
E. L-Lactic Dehydrogenase Enzyme Solution (Immediately before use, prepare a solution containing 0.5–1.0 unit/ml of L-Lactic Dehydrogenase in cold Reagent D).

Procedure

Pipette (in milliliters) the following reagents into suitable cuvettes:

	Test	Blank
Reagent A (Buffer)	2.85	2.85
Reagent B (Pyruvate)	0.12	0.12
Reagent C (β-NADH)	0.06	0.06

Mix by inversion and equilibrate to 30°C. Monitor the A340nm until constant, using a suitably thermostatted spectrophotometer. Then add:

Reagent D (Enzyme Diluent)	—	0.05
Reagent E (Enzyme Solution)	0.05	—

Immediately mix by inversion and record the decrease in A340nm for approximately 5 minutes. Obtain the ΔA340nm/minute using the maximum linear rate for both the Test and Blank.

Calculations

$$\text{Units / ml enzyme} = \frac{(\Delta A340\text{nm / min Test} - \Delta A340\text{nm / min Blank})\ (3.08)\ (\text{df})}{(6.22)\ (0.05)}$$

3.08 = Total volume (in milliliters) of assay
df = Dilution factor
6.22 = Millimolar extinction coefficient of β-NADH at 340nm
0.05 = Volume (in milliliter) of assay

$$\text{Units / mg solid} = \frac{\text{units / ml enzyme}}{\text{mg solid / ml enzyme}}$$

$$\text{Units / mg protein} = \frac{\text{units / ml enzyme}}{\text{mg protein / ml enzyme}}$$

Unit Definition: *One unit will reduce 1.0 μmole of pyruvate to L-lactate per minute at pH 6.0 at 30°C.*

Final Assay Concentrations

In a 3.08 reaction mix, the final concentrations are 93 mM sodium acetate, 9.7 mM pyruvate, 0.25 mM β-nicotinamide adenine dinucleotide, reduced form, 0.81 mM potassium phosphate, and 0.025–0.05 unit L-lactic dehydrogenase.

Enzymatic Assay of L-LACTIC DEHYDROGENASE (EC 1.1.1.27)

Principle

$$\text{Pyruvate} + \beta\text{-NADH} \xrightarrow{\text{L-Lactic Dehydrogenase}} \text{L-Lactate} + \beta\text{-NAD}$$

Abbreviations Used

β-NADH = β-Nicotinamide Adenine Dinucleotide, Reduced Form
β-NAD = β-Nicotinamide Adenine Dinucleotide, Oxidized Form

Conditions: T = 37°C, pH = 7.5, A340nm, Light path = 1 cm

Method: Continuous Spectrophotometric Rate Determination

Reagents

A. 100 mM Sodium Phosphate Buffer, pH 7.5 at 37°C (Prepare 200 ml in deionized water using Sodium Phosphate, Monobasic, Anhydrous. Adjust to pH 7.5 at 37°C with 1 M NaOH).

B. 0.13 mM β-Nicotinamide Adenine Dinucleotide, Reduced Form Solution (β-NADH) (Prepare 10 ml in cold Reagent A using β-Nicotinamide Adenine Dinucleotide, Reduced Form, Disodium Salt or dissolve the contents of 1 mg vial of β-Nicotinamide Adenine Dinucleotide, Reduced Form, Disodium Salt in the appropriate volume of Reagent A. *Prepare Fresh*).

C. 69 mM Sodium Pyruvate Solution (Pyruvate) (Prepare 1.0 ml in cold Reagent A using Pyruvic Acid, Sodium Salt).

D. 1.0% (w/v) Bovine Serum Albumin Solution (BSA) (Prepare 50 ml in Reagent A using Albumin, Bovine. *Prepare Fresh*).

E. L-Lactic Dehydrogenase Enzyme Solution (Immediately before use, prepare a solution containing 0.25–0.75 unit/ml of L-Lactic Dehydrogenase in cold Reagent D).

Procedure

Pipette (in milliliters) the following reagents into suitable cuvettes:

	Test	Blank
Reagent B (β-NADH)	2.80	2.80
Reagent C (Pyruvate)	0.10	0.10

Mix by inversion and equilibrate to 37°C. Monitor the A340nm until constant, using a suitably thermostatted spectrophotometer. Then add:

	Test	Blank
Reagent D (BSA)	—	0.10
Reagent E (Enzyme Solution)	0.10	—

Immediately mix by inversion and record the decrease in A340nm for approximately 5 minutes. Obtain the ΔA340nm/minute using the maximum linear rate for both the Test and Blank.

Calculations

$$\text{Units / ml enzyme} = \frac{(\Delta A340nm / \text{min Test} - \Delta A340nm / \text{min Blank})\ (3)\ (df)}{(6.22)\ (0.01)}$$

3 = Total volume (in milliliters) of assay
df = Dilution factor
6.22 = Millimolar extinction coefficient of β-NADH at 340nm
0.1 = Volume (in milliliter) of enzyme used

$$\text{Units / mg solid} = \frac{\text{units / ml enzyme}}{\text{mg solid / ml enzyme}}$$

$$\text{Units / mg protein} = \frac{\text{units / ml enzyme}}{\text{mg protein / ml enzyme}}$$

Unit Definition: *One unit will reduce 1.0 µmole of pyruvate to L-lactate per minute at pH 7.5 at 37°C.*

Final Assay Concentration

In a 3.00 ml reaction mix, the final concentrations are 100 mM sodium phosphate, 0.12 mM β-nicotinamide adenine dinucleotide, reduced form, 2.3 mM pyruvate, 0.033% (w/v) Bovine Serum Albumin and 0.025–0.075 unit L-lactic dehydrogenase.

Enzymatic Assay of LACTOPEROXIDASE (EC 1.11.1.7)

Principle

$$\underset{\text{(donor)}}{H_2O_2 + \text{Pyrogallol}} \xrightarrow{\text{Lactoperoxidase}} 2H_2O + \underset{\text{(oxidized donor)}}{\text{Purpurogallin}}$$

Conditions: T = 20°C, pH = 6.0, A420nm, Light path = 1 cm

Method: Continuous Spectrophotometric Rate Determination

Reagents

A. 100 mM Potassium Phosphate Buffer, pH 6.0 at 20°C (Prepare 100 ml in deionized water using Potassium Phosphate, Monobasic, Anhydrous. Adjust to pH 6.0 at 20°C with 1.0 M KOH).

B. 0.50% (w/w) Hydrogen Peroxide Solution (H_2O_2) (Prepare 50 ml in deionized water using Hydrogen Peroxide, 30% (w/w) Solution. *Prepare Fresh*).

C. 5% (w/v) Pyrogallol Solution (Prepare 10 ml in deionized water using Pyrogallol. *Prepare Fresh and Keep away from Light*).

D. 0.10% (w/v) Bovine Serum Albumin Solution (Enzyme Diluent) Prepare 50 ml in Reagent A using Albumin, Bovine).

E. Lactoperoxidase Enzyme Solution (Immediately before use, prepare a solution containing 0.4–0.7 unit/ml of Peroxidase in cold Reagent D).

Procedure

Pipette (in milliliters) the following reagents into suitable cuvettes:

	Test	Blank
Deionized Water	2.10	2.10
Reagent A (Buffer)	0.32	0.32
Reagent B (H_2O_2)	0.16	0.16
Reagent C (Pyrogallol)	0.32	0.32

Mix by inversion and equilibrate to 20°C. Monitor the A420nm until constant, using a suitably thermostatted spectrophotometer. Then add:

Reagent D (Enzyme Diluent)	—	0.10
Reagent E (Enzyme Solution)	0.10	—

Immediately mix by inversion and record the increase in A420nm for approximately 5 minutes. Obtain the ΔA420nm/20 seconds using the maximum linear rate for both the Test and Blank.

Calculation

$$\text{Units / ml enzyme} = \frac{(\Delta A420\text{nm} / 20 \text{ sec Test} - \Delta A420\text{nm} / 20 \text{ sec Blank})\ (3)\ (\text{df})}{(12)\ (0.1)}$$

sec = seconds
3 = Volume (in milliliters) of assay
df = Dilution factor
12 = Extinction coefficient of 1 mg/ml of Purpurogallin at 420nm
0.1 = Volume (in milliliter) of enzyme used

$$\text{Units / mg solid} = \frac{\text{units / ml enzyme}}{\text{mg solid / ml enzyme}}$$

$$\text{Units / mg protein} = \frac{\text{units / ml enzyme}}{\text{mg protein / ml enzyme}}$$

Unit Definition: *One unit will form 1.0 milligram of purpurogallin from pyrogallol in 20 seconds at pH 6.0 at 20°C.*

Final Assay Concentrations

In a 3.00 ml reaction mix, the final concentrations are 14 mM potassium phosphate, 0.027% (w/w) hydrogen peroxide, 0.5% (w/v) pyrogallol, 0.003% (w/v) Bovine Serum Albumin, and 0.04–0.07 unit lactoperoxidase.

Notes

1. This assay procedure is not to be used to assay Lactoperoxidase-Biotin labeled.
2. The enzyme concentration may have to be modified in order for the rate, A420nm/20 seconds, to be within the specified range of 0.16–0.28.

Enzymatic Assay of LACTOPEROXIDASE (EC 1.11.1.7) 2,2'-Azino-bis (3-Ethylbenzthiazoline-6-Sulfonic Acid) as a Substrate

Principle

$$H_2O_2 + \text{Reduced ABTS} \xrightarrow{\text{Peroxidase}} 2H_2O + \text{Oxidized ABTS}$$

Abbreviation Used

ABTS = 2,2'-Azino-bis(3-Ethylbenzthiazoline-6-Sulfonic Acid)

Conditions: T = 25°C, pH = 5.5, A436nm, Light path = 1 cm

Method: Continuous Spectrophotometric Rate Determination

Reagents

A. 100 mM Potassium Phosphate, Monobasic Solution (Prepare 100 ml in deionized water using Potassium Phosphate, Monobasic, Anhydrous).
B. 100 mM Potassium Phosphate Dibasic Solution (Prepare 100 ml in deionized water using Potassium Phosphate, Dibasic, Trihydrate).
C. 100 mM Potassium Phosphate buffer, pH 5.5 at 25°C (Prepare 100 ml by adjusting the pH of Reagent A to 5.5 at 25°C with Reagent B).
D. 100 mM 2,2'-Azino-bis(3-Ethylbenzthiazoline-6-Sulfonic Acid) Substrate Solution (ABTS) (Prepare 30 ml in deionized water using 2,2'-Azino-bis(3-Ethylbenzthiazoline-6-Sulfonic Acid), Diammonium Salt. *Prepare Fresh*).

E. 0.025% (w/w) Hydrogen Peroxide Solution (H_2O_2) (Prepare 50 ml in cold deionized water using Hydrogen Peroxide, 30% (w/w) Solution.

F. 200 mM Potassium Phosphate Solution (Enz Dil) (Prepare 100 ml in deionized water using Potassium Phosphate, Dibasic, Trihydrate).

G. Lactoperoxidase Enzyme Solution (Immediately before use, prepare a solution containing 0.15–0.25 unit/ml of Peroxidase in cold Reagent F).

Procedure

Pipette (in milliliters) the following reagents into suitable cuvettes:

	Test	Blank
Reagent C (Buffer)	2.20	2.20
Reagent D (ABTS)	0.70	0.70
Reagent E (H_2O_2)	0.10	0.10

Mix by inversion and equilibrate to 25°C. Monitor the A436nm until constant, using a suitably thermostatted spectrophotometer. Then add:

Reagent G (Enzyme Solution)	0.05	—
Reagent F (Enz Dil)	—	0.05

Immediately mix by inversion and record the increase in A436nm for approximately 2 minutes. Obtain the ΔA436nm/minute using the maximum linear rate for both the Test and Blank.

Calculations

$$\text{Units / ml enzyme} = \frac{(\Delta A436nm / \min \text{Test} - \Delta A436nm / \min \text{Blank})\ (3.05)\ (df)}{(29.3)\ (0.05)}$$

3.05 = Total volume (in milliliters) of assay
df = Dilution factor
29.3 = Millimolar extinction coefficient of oxidized ABTS at 436nm
0.05 = Volume (in milliliter) of enzyme used

$$\text{Units / mg solid} = \frac{\text{units / ml enzyme}}{\text{mg solid / ml enzyme}}$$

$$\text{Units / mg protein} = \frac{\text{units / ml enzyme}}{\text{mg protein / ml enzyme}}$$

Unit Definition: *One unit will oxidize 1.0 µmole of 2,2'-azino-bis (3-ethylbenzthiazoline-6-sulfonic acid) per minute at pH 5.5 at 25°C.*

Final Assay Concentrations

In a 3.05 ml reaction mix, the final concentrations are 75 mM potassium phosphate, 23 mM 2,2'-azino-bis(3-ethylbenzthiazoline-6-sulfonic acid), 0.0008% (w/w) hydrogen peroxide, and 0.0075–0.0125 unit peroxidase.

Notes

1. This assay is to be used to assay the activity of Lactoperoxidase from Bovine Milk. This assay is also used to determine the suitability of 2,2'-Azino-bis(3-ethylbenzthiazoline-6-sulfonic acid), Diammonium Salt as a substrate for lactoperoxidase.
2. ABTS is a registered trademark of Boehringer Mannheim GmbH.
3. The absorbance of this solution should be approximately 0.4 measured at A240nm.
4. The rate should be measured at approximately 120 readings/minute.
5. The maximum linear rate occurs within the first minute of the reaction.

Suitability Assay for LAMINARIN AZURE as a Substrate for LAMINARINASE

Principle

$$\text{Laminarin Azure} + H_2O \xrightarrow{\text{Laminarinase}} \text{Soluble Colored Reaction Products}$$

Conditions: T = 37°C, pH = 5.0, A595nm, Light path = 1 cm

Method: Colorimetric

Reagents

A. 100 mM Sodium Acetate Buffer, pH 5.0 at 37°C (Prepare 100 ml in deionized water using Sodium Acetate, Trihydrate. Adjust to pH 5.0 at 37°C using 1 M HCl).
B. Laminarin Azure (LA) (Weigh 30 mg of Laminarin Azure, each into 4 suitable vials).
C. 95% (v/v) Ethanol (EtOH) (Prepare 100 ml in deionized water using Ethyl Alcohol, Denatured).
D. Laminarinase Enzyme Solution (Immediately before use, prepare a solution containing 10 units/ml of Laminarinase in cold Reagent A).

Procedure

Pipette (in milliliters) the following reagents into suitable vials containing Reagent B (LA):

	Test 1	Test 2	Test 3	Blank
Reagent A (Buffer)	0.90	0.90	0.90	1.00

Incubate for approximately 5 minutes at 37°C. Then add:

Reagent D (Enz Sol)	0.10	0.10	0.10	—

Allow to incubate at 37°C. Then stop the reaction at the following time intervals by adding:

5 minutes	6.00	—	—	—
10 minutes	—	6.00	—	6.00
15 minutes	—	—	6.00	—

Mix by swirling and filter through a 0.45 μM filter. Transfer the filtered solutions to suitable cuvettes and record the A595nm for the Tests and Blank.

Calculations

$$\Delta A595nm / min / unit\ Laminarinase = \frac{A595nm\ Test - A595nm\ Blank}{(T)\ (units\ of\ Laminarinase / RM)}$$

T = Time (in minutes) of assay. Time is 5, 10, or 15 minutes
RM = Reaction Mix

The activity is expressed as the change in absorbance at A595nm per minute per unit Laminarinase. Blank values should be less than 0.10 absorbance units at A595nm.

Enzymatic Assay of LAMINARINASE (EC 3.2.1.6)

Principle

$$Laminarin + H_2O \xrightarrow{Laminarinase} Reducing\ Sugar\ (measured\ as\ glucose)$$

Conditions: T = 37°C, pH = 5.0, A540nm, Light path = 1 cm

Method: Colorimetric

Reagents

A. 100 mM Sodium Acetate Buffer, pH 5.0 at 37°C (Prepare 50 ml in deionized water using Sodium Acetate, Trihydrate. Adjust to pH 5.0 at 37°C with 1 M HCl).
B. 2.5% (w/v) Laminarin Substrate Solution (Laminarin) (Prepare 5 ml in deionized water using Laminarin).
C. Laminarinase Enzyme Solution (Immediately before use, prepare a solution containing 0.075–0.15 unit/ml of Laminarinase in cold deionized water).
D. 16 mM Copper Sulfate, 1.3 M Sodium Sulfate, 226 mM Sodium Carbonate, 190 mM Sodium Bicarbonate and 43 mM Sodium Potassium Tartrate Solution (Copper Soln) (Prepare 1 liter in deionized water using Cupric Sulfate Pentahydrate, Sodium Bicarbonate, Sodium Sulfate, Anhydrous, Sodium Carbonate, Anhydrous and Sodium Potassium Tartrate Tetrahydrate).
E. 40 mM Molybdic Acid, 19 mM Arsenic Acid and 756 mM Sulfuric Acid Solution (Ars-Mol) (Prepare 1 liter in deionized water using Molybdic Acid, Ammonium Salt Tetrahydrate, Arsenic Acid, Sodium Salt and Sulfuric Acid).
F. 1 mg/ml Glucose Standard Solution (Glucose Std).

Procedure

Pipette (in milliliters) the following reagents into suitable containers:

	Test	Blank
Deionized Water	—	1.00
Reagent A (Buffer)	3.00	3.00
Reagent B (Laminarin)	1.00	1.00

Mix by swirling and equilibrate to 37°C. Then add:

	Test	Blank
Reagent C (Enzyme Solution)	1.00	—

Mix by swirling and incubate at 37°C for exactly 60 minutes. Pipette (in milliliters) the following reagents into suitable test tubes:

	Test	Test Blank	Std 1	Std 2	Std 3	Std 4	Std Blank
Deionized water	0.90	0.90	0.90	0.80	0.70	0.60	1.00
Test Solution	0.10	—	—	—	—	—	—
Blank Solution	—	0.10	—	—	—	—	—
Reagent F (Glucose Std)	—	—	0.10	0.20	0.30	0.40	—
Reagent D (Copper Soln)	1.00	1.00	1.00	1.00	1.00	1.00	1.00

Immediately mix by swirling. Place a marble over the top of the tubes and transfer the tubes to a boiling water bath. Incubate for 10 minutes. Remove the tubes from the boiling water bath and allow to cool to room temperature. Then add:

Reagent E (Ars-Mol)	1.00	1.00	1.00	1.00	1.00	1.00	1.00

Shake or vortex the tubes until foaming stops and any precipitate present is dissolved. Then add:

Deionized water	10.00	10.00	10.00	10.00	10.00	10.00	10.00

Mix by inversion and transfer the solutions to suitable cuvettes. Obtain the A540nm for Test, Blank and Standards, using a suitable spectrophotometer.

Calculations

Standard Curve: ΔA540nm Std = A540nm Std–A540nm Std Blank

Prepare a standard curve by plotting the ΔA540nm of the Standard vs the milligrams of Glucose.

Sample Determination: ΔA540nm Test = A540nm Test–A540nm Blank

Determine the mg of reducing sugar released as glucose equivalents using the standard curve.

$$\text{Units / ml enzyme} = \frac{(\text{mg of Glucose released})\,(5)\,(\text{df})}{(60)\,(1)\,(0.1)}$$

5 = Total volume (in milliliters) of assay
df = Dilution factor
60 = Time of assay (in minutes) as per the Unit Definition
1 = Volume (in milliliter) of enzyme used
0.1 = Volume (in milliliter) of Test used in Colorimetric Determination

$$\text{Units / mg solid} = \frac{\text{units / ml enzyme}}{\text{mg solid / ml enzyme}}$$

$$\text{Units / mg protein} = \frac{\text{units / ml enzyme}}{\text{mg protein / ml enzyme}}$$

Unit Definition: *One unit will liberate 1.0 milligram of reducing sugar (measured as glucose) from laminarin per minute at pH 5.0 at 37°C.*

Final Assay Concentrations

In a 5.00 ml reaction mix, the final concentrations are 60 mM sodium acetate, 0.50% (w/v) laminarin and 0.075–0.15 unit laminarinase.

Notes

1. This assay can also be used as a substrate suitability assay for Laminarin.
2. Sodium Sulfate, Sodium Carbonate, and Sodium Potassium Tartrate are dissolved in approximately 500 ml of deionized water. Cupric Sulfate is dissolved in approximately 100 ml of deionized water and is slowly added to the above solution to avoid precipitation. Sodium Bicarbonate is dissolved first in deionized water and then added to the above solution. Dilute the solution to 1 liter. If a precipitate forms, it should be removed by filtration prior to use. Store in an amber bottle and avoid exposure to direct sunlight. Store at room temperature.
3. Molybdic Acid is dissolved in approximately 300 ml of deionized water. Add Sulfuric Acid slowly. Caution, this is an exothermic reaction! Arsenic Acid is dissolved in approximately 300 ml of deionized water and is added to the above solution. The solution is diluted to a total volume of 1 liter and incubated at 37°C for 48–72 hours. If a precipitate forms, it should be removed by filtration prior to use. Store in an amber bottle and avoid exposure to direct sunlight. The solution expires six months after preparation. Store at room temperature in an exhaust hood.
4. Reagent D (Copper Solution) and deionized water should already be present in all of the tubes before the Test, Blank, and Reagent F (Glucose Std) are added.

Enzymatic Assay of LAR-PROTEIN TYROSINE PHOSPHATASE

Principle

$$\text{p-Nitrophenyl Phosphate} + H_2O \xrightarrow{\text{LAR-TP}} \text{p-Nitrophenol} + \text{Pi}$$

Abbreviations Used

LAR-TP = LAR-Protein Tyrosine Phosphatase
Pi = Inorganic phosphate

Conditions: T = 30°C, pH = 7.0, A405nm, Light path = 1 cm

Method: Spectrophotometric Stop Rate Determination

Reagents

A. 50 mM Imidazole HCl Buffer with 100 mM p-Nitrophenyl Phosphate, 5 mM Ethylenediaminetetraacetic Acid, 100 mM Sodium Chloride, 10 mM Dithiothreitol and 0.02% (w/v) Bovine Serum Albumin, pH 7.0 at 30°C (2x Reaction Buffer) (Prepare 5 ml in deionized water using Imidazole Phosphatase Substrate, Sodium Chloride, DL-Dithiothreitol, Ethylenediaminetetraacetic Acid, Disodium Salt, Dihydrate and Albumin, Bovine. Adjust to pH 7.0 at 30°C with 1 M HCl.

B. 10 mM Imidazole HCl Buffer, pH 7.0 at 30°C (Enz Dil) (Prepare 50 ml in deionized water using Imidazole. Adjust to pH 7.0 at 25°C with 1 M HCl).

C. 200 mM Sodium Hydroxide Solution (NaOH) (Prepare 25 ml in deionized water using Sodium Hydroxide, Anhydrous).

D. LAR-Protein Tyrosine Phosphatase Enzyme Solution (Immediately before use, prepare a solution containing 20–40 units/ml of LAR-Protein Tyrosine Phosphatase in cold Reagent B).

Procedure

Pipette (in milliliters) the following reagents into a suitable Eppendorf tube:

	Test	Blank
Reagent B (Enz Dil)	—	0.10
Reagent A (2x Reaction Buffer)	0.10	0.10

Equlibrate to 30°C. Then add:

Enzyme Solution	0.10	—

Immediately mix by swirling and incubate at 30°C for exactly 10 minutes. Then add:

Reagent C (NaOH)	0.80	0.80

Mix by inversion and record the A405nm for both the Test and Blank in a suitable spectrophotometer.

Calculations

$$\text{Units / ml enzyme} = \frac{(\text{A405nm Test} - \text{A405nm Blank})\,(1.0)\,(\text{df})}{(0.018)\,(10)\,(0.1)}$$

1.0 = Total volume (in milliliter) of assay
df = Dilution factor
0.018 = Micromolar extinction coefficient of p-Nitrophenol at 405nm
10 = Time (in minutes) of assay
0.1 = Volume (in milliliter) of enzyme used

$$\text{Units / mg solid} = \frac{\text{units / ml enzyme}}{\text{mg solid / ml enzyme}}$$

$$\text{Units / mg protein} = \frac{\text{units / ml enzyme}}{\text{mg protein / ml enzyme}}$$

Unit Definition: *One unit will hydrolyze 1.0 nmole of p-nitrophenyl phosphate per minute at pH 7.0 at 30°C.*

Final Assay Concentration

In a 0.20 ml reaction mix, the final concentrations are 30 mM imidazole, 2.5 mM Ethylenediaminetetraacetic Acid, 5 mM dithiothreitol, 50 mM sodium chloride, 0.01% (w/v) Bovine Serum Albumin, 50 mM p-nitrophenyl phosphate, and 2–4 unit LAR-protein tyrosine phosphatase.

Enzymatic Assay of LEUCINE AMINOPEPTIDASE, CYTOSOL (EC 3.4.11.1)

Principle

$$\text{L-Leucinamide} + H_2O \xrightarrow{\text{Leucine Aminopeptidase}} \text{L-Leucine} + NH_3$$

Conditions: T = 25°C, pH = 8.5, A238nm, Light path = 1 cm

Method: Continuous Spectrophotometric Rate Determination

Reagents

A. 500 mM Tris HCl (Buffer), pH 8.5 at 25°C (Prepare 50 ml in deionized water using Trizma Base. Adjust to pH 8.5 at 25°C with 1 M HCl).

B. 500 mM Tris HCl Buffer, pH 8.5 at 40°C (Act Soln) (Prepare 50 ml in deionized water using Trizma Base. Adjust to pH 8.5 at 40°C with 1 M HCl).

C. 125 mM L-Leucinamide Solution, pH 8.5 at 25°C (Leuamide) (Prepare 10 ml in deionized water using L-Leucinamide Hydrochloride. Adjust to pH 8.5 at 25°C with 1 M HCl or 1 M KOH).

D. 25 mM Manganese Chloride Solution ($MnCl_2$) (Prepare 10 ml in deionized water using Manganese Chloride, Tetrahydrate).

E. 125 mM Magnesium Chloride Solution ($MgCl_2$) (Prepare 10 ml in deionized water using Magnesium Chloride, Hexahydrate).

F. Leucine Aminopeptidase, Cytosol Enzyme Solution (Immediately before use, prepare a solution containing 500–550 units/ml of Leucine Aminopeptidase, Cytosol in cold deionized water).

Procedure

Prepare the activated enzyme solution (Act Enz Soln) by pipetting (in milliliters) the following reagents into a suitable container:

Reagent B (Act Soln)	0.10
Reagent D ($MnCl_2$)	0.10
Reagent F (Enzyme Solution)	0.20
Deionized Water	1.90

Mix by inversion and incubate at 40°C for 2–3 hours. Pipette (in milliliters) the following reagents into suitable cuvettes:

	Test	Blank
Reagent A (Buffer)	0.10	0.10
Reagent C (Leu-amide)	1.00	1.00
Reagent E ($MgCl_2$)	0.10	0.10
Deionized Water	1.20	1.40

Mix by inversion and equilibrate to 25°C. Monitor the A238nm until constant, using a suitably thermostatted spectrophotometer. Then add:

Act Enz Soln	0.20	—

Immediately mix by inversion and record the decrease in A238nm for approximately 10 minutes. Obtain the ΔA238nm/minute using the maximum linear rate for both the Test and Blank.

Calculations

$$\text{Units / ml enzyme} = \frac{(\Delta A238nm / \text{min Test} - \Delta A238nm / \text{min Blank})(2.6)(df)}{(0.0117)(0.2)}$$

0.0117 = Millimolar extinction coefficient of L-Leucinamide at 238nm
2.6 = Volume (in milliliters) of assay
df = Dilution factor
0.2 = Volume (in milliliters) of enzyme used

$$\text{Units / mg solid} = \frac{\text{units / ml enzyme}}{\text{mg solid / ml enzyme}}$$

$$\text{Units / mg protein} = \frac{\text{units / ml enzyme}}{\text{mg protein / ml enzyme}}$$

Unit Definition: *One unit will hydrolyze 1.0 µmole of L-leucinamide to L-leucine and NH_3 per minute at pH 8.5 at 25°C.*

Final Assay Concentrations

In a 2.60 ml reaction mix, the final concentrations are 21 mM Tris, 48 mM L-leucinamide, 0.084 mM manganese chloride, 4.8 mM magnesium chloride and 8.7–9.6 units leucine aminopeptidase, cytosol.

Enzymatic Assay of LEUCINE AMINOPEPTIDASE, MICROSOMAL (EC 3.4.11.2)

Principle

$$\text{L-Leucine p-Nitroanilide} + H_2O \xrightarrow{\text{LAP}} \text{L-Leucine} + \text{p-Nitroaniline}$$

Abbreviation Used

LAP = Leucine Aminopeptidase, Microsomal

Conditions: T = 37°C, pH = 7.2, A405nm, Light path = 1 cm

Method: Continuous Spectrophotometric Rate Determination

Reagents

A. 50 mM Sodium Phosphate Buffer, pH 7.2 at 37°C (Prepare 100 ml in deionized water using Sodium Phosphate, Dibasic, Anhydrous. Adjust to pH 7.2 at 37°C with 1 M HCl).
B. Methanol.
C. 24 mM L-Leucine p-Nitroanilide Solution (LeuNA) (Prepare 1 ml in Reagent B using L-Leucine p-Nitroanilide, Free Base).
D. Leucine Aminopeptidase, Microsomal Enzyme Solution (Immediately before use, prepare a solution containing 0.10–0.15 unit/ml of Leucine Aminopeptidase, Microsomal in cold deionized water).

Procedure

Pipette (in milliliters) the following reagents into suitable cuvettes:

	Test	Blank
Reagent A (Buffer)	2.80	2.80
Reagent C (LeuNA)	0.10	0.10

Mix by inversion and equilibrate to 37°C. Monitor the A405nm until constant, using a suitably thermostatted spectrophotometer. Then add:

Deionized Water	—	0.10
Reagent D (Enzyme Solution)	0.10	—

Immediately mix by inversion and record the increase in A405nm for approximately 5 minutes. Obtain the ΔA405nm/minute using the maximum linear rate for both the Test and Blank.

Calculations

$$\text{Units / ml enzyme} = \frac{(\Delta A405\text{nm / min Test} - \Delta A405\text{nm / min Blank})\ (3)\ (df)}{(9.9)\ (0.1)}$$

3 = Total volume (in milliliters) of assay
df = Dilution factor
9.9 = Millimolar extinction coefficient1 of p-Nitroaniline at 405nm
0.1 = Volume (in milliliter) of enzyme used

$$\text{Units / mg solid} = \frac{\text{units / ml enzyme}}{\text{mg solid / ml enzyme}}$$

$$\text{Units / mg protein} = \frac{\text{units / ml enzyme}}{\text{mg protein / ml enzyme}}$$

Unit Definition: *One unit will hydrolyze 1.0 μmole of L-leucine pnitroanilide to L-leucine and p-nitroaniline per minute at pH 7.2 at 37°C.*

Final Assay Concentration

In a 3.00 ml reaction mix, the final concentrations are 47 mM sodium phosphate, 0.80 mM L-leucine p-nitroanilide, 3.3% (v/v) methanol and 0.01–0.015 unit leucine aminopeptidase, microsomal.

Enzymatic Assay of LEUCINE DEHYDROGENASE (EC 1.4.1.9)

Principle

$$\text{L-Leucine} + \beta\text{-NAD} \xrightarrow{\text{Leucine Dehydrogenase}} \alpha\text{-Ketoisocaproate} + \beta\text{-NADH}$$

Abbreviations Used

β-NAD = β-Nicotinamide Adenine Dinucleotide, Oxidized Form
β-NADH = β-Nicotinamide Adenine Dinucleotide, Reduced Form

Conditions: T = 37°C, pH = 10.5, A340nm, Light path = 1 cm

Method: Continuous Spectrophotometric Rate Determination

Reagents

A. 200 mM Glycine with 200 mM Potassium Chloride Buffer, pH 10.5 at 37°C (Prepare 100 ml in deionized water using Glycine Free Base and Potassium Chloride. Adjust to pH 10.5 at 25°C with 1 M KOH).

B. 20 mM L-Leucine Solution (Leu) (Prepare 20 ml in Reagent A using L-Leucine).

C. 12.5 mM β-Nicotinamide Adenine Dinucleotide Solution (β-NAD) (Prepare 2 ml in deionized water using β-Nicotinamide Adenine Dinucleotide or dissolve the contents of one 20 mg vial of β-Nicotinamide Adenine Dinucleotide in the appropriate volume of deionized water. *Prepare Fresh*).

D. 25 mM Potassium Phosphate Solution, pH 7.2 at 37°C (Enzyme Diluent) (Prepare 100 ml in deionized water using Potassium Phosphate, Monobasic, Anhydrous. Adjust to pH 7.2 at 37°C with 1 M KOH).

E. Leucine Dehydrogenase Enzyme Solution (Immediately before use, prepare a solution containing 0.1–0.5 unit/ml of Leucine Dehydrogenase in cold Reagent D).

Procedure

Pipette (in milliliters) the following reagents into suitable cuvettes:

	Test	Blank
Reagent B (Leu)	3.00	3.00
Reagent C (β-NAD)	0.30	0.30

Mix by inversion and equilibrate to 37°C. Monitor the A340nm until constant, using a suitably thermostatted spectrophotometer. Then add:

Reagent D (Enzyme Diluent)	—	0.05
Reagent E (Enzyme Solution)	0.05	—

Immediately mix by inversion and record the increase in A340nm for approximately 5 minutes. Obtain the ΔA340nm/minute using the maximum linear rate for both the Test and Blank.

Calculations

$$\text{Units / mg enzyme} = \frac{\Delta A340nm / min\ Test - \Delta A340nm / min\ Blank}{(6.22)\ (mg\ enzyme / ml\ RM)}$$

6.22 = Millimolar extinction coefficient of β-NADH at 340nm
RM = Reaction Mix

Unit Definition: *One unit will convert 1.0 μmole of L-leucine to α-Ketoisocaproate per minute at pH 10.5 at 37°C.*

Final Assay Concentration

In a 3.35 ml reaction mix, the final concentrations are 179 mM glycine, 179 mM potassium chloride, 18 mM L-leucine, 1.1 mM β-NAD, 0.37 mM potassium phosphate and 0.005–0.025 unit leucine dehydrogenase.

Enzymatic Assay of LEUKOTRIENE D_4 HYDROLASE (EC 3.4.13.19)

Principle

$$\text{Gly-D-Phe} + H_2O \xrightarrow{\text{Leukotriene } D_4 \text{ Hydrolase}} \text{D-Phe} + \text{Gly}$$

Abbreviations Used

D-Phe = Phenylalanine
Gly = Glycine

Conditions: T = 37°C, pH = 8.0, A214nm, Light path = 1 cm

Method: Stopped HPLC Analysis of Products

Reagents

A. 100 mM Tris HCl Buffer, pH 8.0 at 37°C (Prepare 100 ml in deionized water using Trizma Base. Adjust to pH 8.0 at 37°C with 1 M HCl).
B. 10 mM Gly-D-Phe Solution (Gly-D-Phe) (Prepare 1 ml in Reagent A using Gly-D-Phe).
C. 2.4 mM D-Phenylalanine Solution (Phe Std) (Prepare 5 ml in Reagent A using D-Phenylalanine. Initially prepare a stock solution containing 0.40 mg/ml. Further dilute to stock concentrations of (mg/ml): 0.30, 0.20, 0.10, and 0.050).
D. Leukotriene D_4 Hydrolase Enzyme Solution (Immediately before use, prepare a solution containing 0.05–0.1 unit/ml in cold Reagent A).
E. 0.08% (w/v) Phosphoric Acid Solution (Buffer A) (Prepare 200 ml in deionized water using Phosphoric Acid. Adjust to pH 2.5 with 1M NaOH).
F. Acetonitrile (Buffer B).

Procedure

Step 1: Pipette (in milliliters) the following reagents into suitable microcentrifuge tubes:

	Test	Blank
Reagent A (Buffer)	0.08	0.09
Reagent D (Enzyme Solution)	0.01	—

Equilibrate to 37°C. Then add:

Reagent B (Gly-D-Phe)	0.01	0.01

Incubate for exactly 30 minutes at 37°C. Then terminate the reaction by heating at 100°C for 4 minutes. Microcentrifuge at 15,000 g for 10 minutes and then transfer 0.08 ml of the supernatant from both the Test and Blank to HPLC vials.

Step 2: HPLC analysis of reaction products

1. Column: Supelcosil LC-18, 4.6 × 150 mm, 5 µm particle size.
2. Mobile Phase Time (min)

	0	15	25
Buffer B (Reagent F)	4.5%	30%	30%
Buffer A (Reagent E)	95.5%	70%	70%

Pressure: 2150 PSI, flow rate 1.5 ml/min, detection: 214nm, Sample volume injected: 20 µl.

3. Inject blank and standards of D-phenylalanine (Reagent C). A comparison can then be made between the standard curve of D-phenylalanine and D-phenylalanine generated from the sample reaction.

Calculations

$$\text{Units / ml enzyme} = \frac{(\mu\text{moles of D-phenylalanine})\ (0.1)}{(0.01)\ (30)}$$

0.1 = Volume (in milliliter) of assay
0.01 = Volume (in milliliter) of enzyme used
30 = Time (in minutes) of assay

$$\text{Units / mg solid} = \frac{\text{units / ml enzyme}}{\text{mg solid / ml enzyme}}$$

$$\text{Units / mg protein} = \frac{\text{units / ml enzyme}}{\text{mg protein / ml enzyme}}$$

Unit Definition: *One unit will produce 1.0 µmole of D-phenylalanine from Gly-D-Phe per min at pH 8.0 at 37°C.*

Final Assay Concentration

In a 0.10 ml reaction mix, the final concentrations are 100 mM Tris, 1 mM gly-D-phe, and 0.0005–0.001 unit of leukotriene D_4 hydrolase.

Enzymatic Assay of LIPASE (EC 3.1.1.3)

Principle

$$\text{Triglyceride} + H_2O \xrightarrow{\text{Lipase}} \text{Glycerol} + 3 \text{ Fatty Acids}$$

$$\text{Glycerol} + \text{ATP} \xrightarrow{\text{Glycerokinase}} \text{Glycerol-3-P} + \text{ADP}$$

$$\text{Glycerol-3-P} + O_2 \xrightarrow{\text{GPO}} \text{Dihydroxyzcetone-P} + H_2O_2$$

$$2H_2O_2 + \text{4-AAP} + \text{NNDT} \xrightarrow{\text{Peroxidase}} \text{Quinoneimine Dye}$$

Abbreviations Used

ATP = Adenosine 5'-Triphosphate
Glycerol-3-P = Glycerol 3-Phosphate
ADP = Adenosine 5'-Diphosphate
GPO = Glycerol-3-Phosphate Oxidase
Dihydroxyacetone-P = Dihydroxyacetone Phosphate
4-AAP = 4-Aminoantipyrine
NNDT = N,N-Diethyl-m-Toluidine.

Conditions: T = 37°C, pH = 7.0, A545nm, Light path = 1 cm.

Method: Continuous Spectrophotometric Rate Determination

Reagents

A. 5.0% (v/v) Triton X-100 Solution (X-100) (Prepare 100 ml in deionized water using Triton X-100).

B. 4.0% (w/v) Bovine Serum Albumin (BSA) (Prepare 100 ml in deionized water using Albumin, Bovine. A-4503 or equivalent).

C. 100 mM Potassium Phosphate Buffer, pH 7.0 at 37°C (Prepare 100 ml in deionized water using Potassium Phosphate, Monobasic, Anhydrous. Adjust to pH 7.0 at 37°C with 1 M KOH).

D. 11% (w/v) Olive Oil Emulsion (Olive Oil) (Prepare by adding 5 g of Olive Oil, to 5 ml of Reagent A. Sonicate the mixture. To the oil emulsion add 25 ml of Reagent B and 15 ml of Reagent C).

E. 50 mM MES NaOH Buffer, 6.5 at 37°C (Prepare 100 ml in deionized water using MES, Free Acid . Adjust to pH 6.5 at 37°C with 1 M NaOH).

F. 200 mM Trichloroacetic Acid Solution (TCA) (Prepare 50 ml in deionized water using Trichloroacetic Acid, 6.1 N Solution).

G. N,N-Diethyl-m-Toluidine (NNDT) (Use N,N-Diethyl-m-Toluidine).

H. 4-Aminoantipyrene (4-AAP) (Use 4-Aminoantipyrene, Free Base).

I. Adenosine 5'-Triphosphate (ATP) (Use Adenosine 5'-Triphosphate, Disodium Salt).

J. Magnesium Chloride ($MgCl_2$) (Use Magnesium Chloride, Hexahydrate).

K. Glycerokinase (GK) (Use Glycerokinase).

L. Peroxidase (POD) (Use Peroxidase).

M. Glycerol-3-Phosphate Oxidase (GPO) (Use Glycerol-3-Phosphate Oxidase).

N. Color Reagent (CR) (Prepare 200 ml in Reagent E (MES) by dissolving the following products in this order: 4.0 ml of Reagent A (Triton), 0.04 ml of Reagent G (NNDT), 4.0 mg of Reagent H (4-AAP), 24.2 mg of Reagent I (ATP), 40.7 mg of Reagent J ($MgCl_2$), 200 units of Reagent K (GK), 500 units of Reagent M (GPO) and 300 units of Reagent L (POD)).

O. 20 mM Potassium Phosphate Buffer with 2 mM Magnesium Chloride and 0.5 mM Ethylenediaminetetraacetic Acid, pH 7.5 at 37°C (Prepare 25 ml in deionized water using Potassium Phosphate, Monobasic, Anhydrous, Magnesium Chloride, Hexahydrate, and Ethylenediaminetetraacetic Acid, Trisodium Salt. Adjust to pH 7.5 at 37°C with 1 M KOH).

P. Lipase Enzyme Solution (Lipase) (Immediately before use, prepare a solution containing 0.7–1.4 units/ml of Lipase in cold Reagent O).

Procedure

Step 1: Pipette (in milliliters) the following reagents into suitable containers:

	Test	Blank
Reagent D (Olive Oil)	2.00	2.00

Equilibrate to 37°C. Then add:

Reagent P (Lipase)	0.20	—

Mix by inversion and incubate for exactly 15 minutes. Then add:

	Test	Blank
Reagent F (TCA)	2.00	2.00
Reagent P (Lipase)	—	0.20

Mix by inversion and filter the solutions through Whatman No. 42 filter paper.
Step 2: Pipette (in milliliters) the following reagents into suitable containers:

Test Filtrate	0.05	—
Blank Filtrate	—	0.05
Reagent N (CR)	3.00	3.00

Mix by inversion and incubate at 37°C for 15 minutes. Transfer to suitable cuvettes and record the A545nm for both the Test and Blank using a suitable spectrophotometer.

Calculation

$$\text{Units / mg enzyme} = \frac{(\text{A545nm Test} - \text{A545nm Blank})\,(3.05)\,(4.20)}{(15)\,(28.2)\,(0.5)\,(0.05)\,(\text{mg enzyme / RM})}$$

4.20 = Total volume (in milliliters) of Step 1
3.05 = Volume (in milliliters) of Colorimetric Assay in Step 2
15 = Time of assay (in minutes) as per the Unit Definition.
28.2 = Millimolar extinction coefficient of Quinoneimine Dye at 545nm under the assay conditions.
0.5 = Conversion factor based on one mole of H_2O_2 produces half a mole of Quinoneimine Dye.
0.05 = Volume from Step 1 used in Step 2.
RM = Reaction Mix

Unit Definition: *One unit will produce 1.0 μmole of glycerol from a triglyceride per minute at pH 7.0 at 37°C in the presence of Bovine Serum Albumin. Assayed in a coupled system with glycerokinase.*

Final Assay Concentration

In a 2.20 ml reaction mix, the final concentrations are 32 mM potassium phosphate, 10% (w/v) olive oil, 0.50% (v/v) Triton X-100, 0.05 mM Ethylenediaminetetraacetic Acid, 2.0% (w/v) Bovine Serum Albumin, 0.14–0.28 unit lipase.

Notes

1. Triton is a registered trademark of Union Carbide.
2. Sonicate at 300 watts, for 3 minutes (do this 5 times).
3. Glycerokinase Unit Definition: One unit will convert 1.0 μmole of glycerol and ATP to L-α-glycerophosphate and ADP per min at pH 9.8 at 25°C is a coupled system with PK/LDH.
4. Glycerol-3-Phosphate Oxidase Unit Definition: One unit will oxidize 1.0 μmole of L-glycerol 3-phosphate to dihydroxyacetone phosphate with the formation of H_2O_2 per minute at 37°C at the appropriate pH.
5. Peroxidase Unit Definition: One unit will form 1.0 mg purpurogallin from pyrogallol in 20 sec at pH 6.0 at 20°C.

Enzymatic Assay of LIPASE (EC 3.1.1.3)

Principle

$$\text{Triglyceride} + H_2O \xrightarrow{\text{Lipase}} \text{Diglyceride} + \text{Fatty Acid}$$

Conditions: T = 37°C, pH = 7.2

Method: Titrimetric

Reagents

A. 200 mM Tris HCl Buffer, pH 7.2 at 37°C (Prepare 100 ml in deionized water using Trizma Base. Adjust to pH 7.2 at 37°C with 1 M HCl).
B. Olive Oil Substrate Solution (Olive Oil).
C. 95% Ethanol (Nondenatured) (Prepare 50 ml in deionized water using 200 Proof USP Ethyl Alcohol).
D. 0.9% (w/v) Thymolphthalein Indicator Solution (TPH Indic) (Use Thymolphthalein Indicator Solution, or prepare 15 ml in Reagent C using Thymolphthalein).
E. 50 mM Sodium Hydroxide Solution-Standardized (NaOH) (Prepare 100 ml in deionized water using Sodium Hydroxide, Anhydrous. Standardize according to the ACS Reagent Procedure).
F. Lipase Enzyme Solution (Immediately before use, prepare a solution containing approximately 500–1000 units/ml of Lipase in cold deionized water).

Procedure

Pipette (in milliliters) the following reagents into suitable containers:

	Test	Blank
Deionized Water	2.50	2.50
Reagent A (Buffer)	1.00	1.00
Reagent B (Olive Oil)	3.00	3.00

Mix by swirling and equilibrate to 37°C. Then add:

	Test	Blank
Reagent F (Enzyme Solution)	1.00	—

Mix by swirling and incubate at 37°C for exactly 30 minutes. Immediately after starting the incubation, pipette (in milliliters) 1.00 ml of Reagent F (Enzyme Solution) into a 50 ml Erlenmeyer flask marked "Blank" and store at 0–4°C. After 30 minutes transfer the Test solution to a 50 ml Erlenmeyer flask and the Blank solution to the 50 ml Erlenmeyer flask labeled "Blank." Then add:

	Test	Blank
Reagent C (95% Ethanol)	3.00	3.00

Mix by swirling and then add 4 drops of Reagent D (TPH Indic) to both the Test and Blank solutions. Titrate each solution with Reagent E (NaOH) to a light blue color. Use a 25 ml burette with 0.1 ml graduations for the titration.

Calculations

$$\text{Units / ml enzyme} = \frac{\text{(NaOH) (Molarity of NaOH) (1000) (2) (df)}}{(1)}$$

NaOH = Volume (in milliliters) of Reagent E used for Test minus volume (in milliliters) of Reagent E used for Blank.
1000 = Conversion factor from milliequivalent to microequivalent
2 = Time conversion factor from 30 minutes to 1 hour (Unit Definition).
df = Dilution factor
1 = Volume (in milliliter) of enzyme used.

$$\text{Units / mg solid} = \frac{\text{units / ml enzyme}}{\text{mg solid / ml enzyme}}$$

$$\text{Units / mg protein} = \frac{\text{units / ml enzyme}}{\text{mg protein / ml enzyme}}$$

Unit Definition: *One unit will hydrolyze 1.0 microequivalent of fatty acid from a triglyceride in one hour at pH 7.2 at 37°C. (This is equivalent to approximately 10 microliters of CO_2 in 30 minutes).*

Final Assay Concentration

In a 7.50 ml reaction mix, the final concentrations are 26.7 mM Tris, 40% (v/v) olive oil, and 500–1,000 units lipase.

Enzymatic Assay of LIPASE (EC 3.1.1.3)

Principle

$$\text{Triglyceride} + H_2O \xrightarrow{\text{Lipase}} \text{Diglyceride} + \text{Fatty Acid}$$

Conditions: T = 37°C, pH = 7.7

Method: Titrimetric

Reagents

A. Olive Oil Substrate (Olive Oil).
B. 3000 mM Sodium Chloride Solution (NaCl) (Prepare 100 ml in deionized water using Sodium Chloride).
C. 1.5% (w/v) Sodium Taurocholate Solution (Tauro) (Prepare 25 ml in deionized water using Taurocholic Acid, Sodium Salt).
D. 75 mM Calcium Chloride Solution ($CaCl_2$) (Prepare 25 ml in deionized water using Calcium Chloride, Dihydrate).
E. 10 mM Sodium Hydroxide Solution-Standardized (NaOH) (Prepare 50 ml in cold deionized water using Sodium Hydroxide, Anhydrous).
F. 5 mM Calcium Chloride Solution (Prepare 25 ml in deionized water using Calcium Chloride, Dihydrate).
G. Lipase Enzyme Solution (Immediately before use, prepare a suspension containing 20,000–30,000 units/ml of Lipase in cold Reagent F).

Procedure

Prepare a reaction cocktail by pipetting (in milliliters) the following reagents into a suitable container:

Deionized Water	50.00
Reagent A (Olive Oil)	50.00
Reagent B (NaCl)	20.00
Reagent C (Tauro)	20.00
Reagent D ($CaCl_2$)	10.00

Mix by swirling and adjust to pH 8.0 at 37°C with Reagent E (NaOH). Pipette (in milliliters) the following reagents into a 50 ml beaker:

	Test	Blank
Reaction Cocktail	15.00	15.00

Equilibrate to 37°C. Then add:

Reagent G (Enzyme Solution)	0.010	—
Deionized water	—	0.010

When the pH reaches 7.7, begin timing the reaction. Run the reaction for 1–5 minutes. Maintain the pH of the reaction mix at pH 7.7 by the addition of small volumes (0.025 ml) of Reagent E (NaOH). Record the volume of Reagent E (NaOH) used to maintain the pH at 7.7 and the time required.

Calculations

$$\text{Units / ml enzyme} = \frac{(\text{Molarity of NaOH})\ (\text{NaOH})\ (60)\ (\text{df})}{(\text{T})\ (0.01)}$$

NaOH = Volume (in microliters) of Reagent E used in the Test.
60 = Time conversion from minutes to hours as per the Unit Definition.
df = Dilution factor.
T = Time (in minutes) required to consume the added Reagent E (NaOH) while maintaining the pH at 7.7
0.01 = Volume (in milliliter) of enzyme used.

$$\text{Units / mg solid} = \frac{\text{units / ml enzyme}}{\text{mg solid / ml enzyme}}$$

$$\text{Units / mg protein} = \frac{\text{units / ml enzyme}}{\text{mg protein / ml enzyme}}$$

Unit Definition: *One unit will hydrolyze 1.0 microequivalent of fatty acid from a triglyceride in one hour at pH 7.7 at 37°C, using olive oil as substrate.*

Final Assay Concentrations

In a 15.01 ml reaction mix, the final concentrations are 33% (v/v) olive oil, 400 mM sodium chloride, 0.2% (w/v) sodium taurocholate, 5 mM calcium chloride and 200–300 units lipase.

Enzymatic Assay of LIPASE (EC 3.1.1.3) from Human Pancreas

Principle

$$\text{Triglyceride} + H_2O \xrightarrow{\text{Lipase}} \text{Diglyceride} + \text{Fatty Acid}$$

Conditions: T = 37ºC, pH = 9.5, A340nm, Light path = 1 cm.

Method: Continuous Spectrophotometric Rate Determination

Reagents

A. 100 mM Glycine Buffer with 19 mM Sodium Deoxycholate, pH 9.5 at 37ºC (Buffer) (Prepare 100 ml in deionized water using Glycine, Free Base, and Deoxycholic Acid, Sodium Salt. Adjust to pH 9.5 at 37ºC with 1 M NaOH).
B. Ethyl Alcohol (EtOH) (Use Ethyl Alcohol, Denatured, HPLC grade).
C. n-Butanol (BuOH) (Use n-Butanol).
D. 0.3 mM Triolein Solution (Triolein) (Prepare 101.5 ml by adding 3.5 ml of Triolein, 5 ml of Reagent C. Bring to a total volume of 100 ml using Reagent B. Then add 1.5 ml of this mixture (dropwise) to 100 ml of Reagent A).
E. 0.0003% (w/v) Colipase Substrate Working Solution (CSWS) (Prepare 101.5 ml in Reagent D using Colipase).
F. Lipase Enzyme Solution (Immediately before use, prepare a solution containing 325–650 units/ml of lipase in cold deionized water).

Procedure

Pipette (in milliliters) the following reagents into suitable cuvettes:

	Test	Blank
Reagent E (CSWS)	2.00	2.00

Equilibrate to 37ºC. Then add:

Reagent F (Enzyme Solution)	0.02	—
Deionized Water	—	0.02

Immediately mix by inversion and record the increase in A340nm for approximately 6 minutes. Obtain the A340nm/minute using the maximum linear rate for both the Test and Blank.

Calculations

$$\text{Units / ml enzyme} = \frac{(\text{A340nm / min Test} - \text{A340nm / min Blank})\ (\text{df})}{(0.001)\ (0.02)}$$

df = Dilution factor
0.001 = Change in absorbance per unit as per the Unit Definition
0.02 = Volume (in milliliter) of enzyme used.

$$\text{Units / mg protein} = \frac{\text{units / ml enzyme}}{\text{mg protein / ml enzyme}}$$

Unit Definition: *One unit will cause a change in A340nm of 0.001 per minute at pH 9.5 at 37°C (reaction volume = 2.02 ml).*

Final Assay Concentration

In a 2.02 ml reaction mix, the final concentrations are 98 mM glycine, 19 mM sodium deoxycholate, 0.3 mM triolein, 0.07% (v/v) n-Butanol, 1.3% (v/v) ethanol, 0.0003% (w/v) colipase and 6.5–13 units lipase.

Notes

1. Reagent A (Buffer) must be 37°C in order for the deoxycholate to dissolve. The solution should be clear and colorless.
2. Allow the triolein to sit for 25 minutes at room temperature before using. After adding the triolein to Reagent C (BuOH), allow the solution to stir for 10–15 minutes before adding Reagent B (EtOH). Mix by swirling for 10–15 minutes and continue to mix by swirling until the solution is colorless. After addition of Reagent A (Buffer) the solution will be hazy, but it should be homogenous.

Enzymatic Assay of LIPASE (EC 3.1.1.3) (Olive Oil as Substrate)

Principle

$$\text{Triglyceride} + H_2O \xrightarrow{\text{Lipase}} \text{Diglyceride} + \text{Fatty Acid}$$

Conditions: T = 37°C, pH = 7.7

Method: Titrimetric

Reagents

A. 200 mM Tris HCl Buffer, pH 7.7 at 37°C (Prepare 100 ml in deionized water using Trizma Base. Adjust to pH 7.7 at 37°C with 1 M HCl).
B. Olive Oil Substrate Solution (Olive Oil)
C. 95% Ethanol (Nondenatured) (Prepare 50 ml in deionized water using 200 Proof USP Ethyl Alcohol).
D. 0.9% (w/v) Thymolphthalein Indicator Solution (TPH Indic) (Use Thymolphthalein Indicator Solution, or prepare 15 ml in Reagent C using Thymolphthalein).
E. 50 mM Sodium Hydroxide Solution-Standardized (NaOH) (Prepare 100 ml in deionized water using Sodium Hydroxide, Anhydrous. Standardize according to the ACS Reagent Procedure).
F. Lipase Enzyme Solution (Immediately before use, prepare a solution containing 500 units/ml of Lipase in cold deionized water).

Procedure

Pipette (in milliliters) the following reagents into suitable containers:

	Test	Blank
Deionized Water	2.50	2.50
Reagent A (Buffer)	1.00	1.00
Reagent B (Olive Oil)	3.00	3.00

Mix by swirling and equilibrate to 37°C. Then add:

	Test	Blank
Reagent F (Enzyme Solution)	1.00	—

Mix by vigorously swirling and incubate at 37°C for exactly 30 minutes. Immediately after starting the incubation, pipette (in milliliters) 1.00 ml of Reagent F (Enzyme Solution) into a 50 ml Erlenmeyer flask marked "Blank" and store at 0–4°C. After 30 minutes transfer the Test solution to a 50 ml Erlenmeyer flask and the Blank solution to the 50 ml Erlenmeyer flask labeled "Blank." Then add:

Reagent C (95% Ethanol)	3.00	3.00

Mix by swirling and then add 4 drops of Reagent D (TPH Indic) to both the Test and Blank solutions. Titrate each solution with Reagent E (NaOH) to a light blue color. Use a 25 ml burette with 0.1 ml graduations for the titration.

Calculations

$$\text{Units / ml enzyme} = \frac{(\text{NaOH})\ (\text{Molarity of NaOH})\ (1000)\ (2)\ (\text{df})}{(1)}$$

NaOH = Volume (in milliliters) of Reagent E used for Test minus volume (in milliliters) of Reagent E used for Blank.
1000 = Conversion factor from milliequivalent to microequivalent
2 = Time conversion factor from 30 minutes to 1 hour (Unit Definition)
df = Dilution factor
1 = Volume (in milliliter) of enzyme used

$$\text{Units / mg solid} = \frac{\text{units / ml enzyme}}{\text{mg solid / ml enzyme}}$$

$$\text{Units / mg protein} = \frac{\text{units / ml enzyme}}{\text{mg protein / ml enzyme}}$$

Unit Definition: *One unit will hydrolyze 1.0 microequivalent of fatty acid from a triglyceride in one hour at pH 7.7 at 37°C. (This is equivalent to approximately 10 microliters of CO_2 in 30 minutes).*

Final Assay Concentration

In a 7.50 ml reaction mix, the final concentrations are 26.7 mM Tris, 40% (v/v) olive oil, and 500 units lipase.

Enzymatic Assay of LIPASE (EC 3.1.1.3) (Triacetin as Substrate)

Principle

$$\text{Triacetin} + H_2O \xrightarrow{\text{Lipase}} \text{Glycerol} + \text{Acetic Acid}$$

Conditions: T = 37°C, pH = 7.4

Method: Titrimetric

Reagents

A. 150 mM Tris HCl Buffer with 330 mM Triacetin, pH 7.4 at 37°C (Buffered Triacetin) (Prepare 100 ml by adding 6.21 ml of Triacetin and 1.82 g of Trizma Base to 75 ml of deionized water. Adjust to pH 7.4 at 37°C with 1 M HCl and dilute to a final volume of 100 ml. Stir vigorously for approximately 30 minutes or until a homogenous dispersion is obtained. Readjust to pH 7.4 at 37°C with either 1 M HCl or 1 M NaOH if necessary).

B. 95% (v/v) Ethanol (Prepare 20 ml in deionized water using 200 Proof USP Ethyl Alcohol).

C. 0.9% (w/v) Thymolphthalein Indicator Solution (TPH Indic) (Use Thymolphthalein Indicator Solution or prepare 15 ml in Reagent B using Thymolphthalein).

D. 100 mM Sodium Hydroxide Solution-Standardized (NaOH) (Prepare 100 ml in deionized water using Sodium Hydroxide, Anhydrous, Preweighed Vial. Standardize according to the ACS Reagent Procedure).

E. Lipase Enzyme Solution (Immediately before use, prepare a solution containing 150–300 units/ml of Lipase in cold deionized water).

Procedure

Pipette (in milliliters) the following reagents into suitable containers:

	Test	Blank
Reagent A (Buffered Triacetin)	3.00	3.00

Equilibrate to 37°C. Then add:

Reagent E (Enzyme Solution)	1.00	—

Mix by swirling and incubate at 37°C for exactly 60 minutes. Then add 4 drops of Reagent C (TPH Indic) to both the Test and Blank. Immediately titrate (with a graduated buret) the Test with Reagent D (NaOH) to a pale blue end point. Record the volume of Reagent D (NaOH) required. Then add:

Reagent E (Enzyme Solution)	—	1.00

Immediately titrate the Blank with Reagent D (NaOH) to a pale blue end point. Record the volume of Reagent D (NaOH) required.

Calculations

$$\text{Units / ml enzyme} = \frac{(\text{NaOH})\ (\text{Molarity of NaOH})\ (1000)\ (\text{df})}{(1)}$$

NaOH = Volume (in milliliters) of Reagent D (NaOH) used for the Test minus the volume (in milliliters) of Reagent D (NaOH) used for the Blank

1000 = Conversion factor from milliequivalent to microequivalent

df = Dilution factor

1 = Volume (in milliliter) of enzyme used in assay

Unit Definition: *One unit will hydrolyze 1.0 microequivalent of fatty acid from a triglyceride in 1 hour at pH 7.4 at 37°C.*
(Incubation time: 60 minutes). (This is equivalent to approximately 10 microliters of CO_2 in 30 minutes).

Final Assay Concentrations

In a 4.00 ml reaction mix, the final concentrations are 113 mM Tris, 248 mM triacetin, and 150–300 units lipase.

Enzymatic Assay of LIPOAMIDE DEHYDROGENASE (EC 1.8.1.4)

Principle

$$\beta\text{-NADH} + \text{DL-Lipoamide} \xrightarrow{\text{Lipoamide Dehydrogenase}} \beta\text{-NAD} + \text{Dehydrolipoamide}$$

Abbreviations Used

β-NADH = β-Nicotinamide Adenine Dinucleotide, Reduced Form
β-NAD = β-Nicotinamide Adenine Dinucleotide, Oxidized Form
DL-Lipoamide = DL-6,8-Thioctic Acid Amide
Dihydrolipoamide = DL-6,8-Dihydrothioctic Acid Amide

Conditions: T = 25°C, pH = 6.5, A340nm, Light Path = 1 cm

Method: Continuous Spectrophotometric Rate Determination

Reagents

A. 50 mM Sodium Phosphate Buffer, pH 6.5 at 25°C (Prepare 200 ml in deionized water using Sodium Phosphate, Monobasic, Anhydrous. Adjust to pH 6.5 at 25°C with 1 M NaOH).
B. 28 mM DL-Thioctic Acid Amide Solution (DL-Thio) (Prepare by dissolving 20 mg of DL-6,8-Thioctic Acid Amide in 2 ml Ethanol (Nondenatured). Dilute this solution with 1.5 ml Reagent A. *Prepare Fresh*).
C. 7.0 mM β-Nicotinamide Adenine Dinucleotide, Reduced Form Solution (β-NADH) (Prepare 2 ml in deionized water using β-Nicotinamide Adenine Dinucleotide, Reduced Form, Disodium Salt or dissolve the contents of one 10 mg vial of β-Nicotinamide Adenine Dinucleotide, Reduced Form, Disodium Salt in the appropriate volume of deionized water. *Prepare Fresh*).
D. 20 mM β-Nicotinamide Adenine Dinucleotide, Oxidized Form Solution (β-NAD) (Dissolve the contents of one 5.0 mg vial of β-Nicotinamide Adenine Dinucleotide in the appropriate volume of deionized water or prepare 2.0 ml in deionized water using β-Nicotinamide Adenine Dinucleotide. *Prepare Fresh*).
E. 300 mM Ethylenediaminetetraacetic Acid with 2.0% (w/v) Albumin Solution, pH 7.0 at 25°C (EDTA) (Prepare 50 ml in deionized water using Ethylenediaminetetraacetic Acid, Tetrasodium Salt, Hydrate and Albumin Bovine. Adjust to pH 7.0 at 25°C with 5 M HCl).
F. Lipoamide Dehydrogenase Enzyme Solution (Immediately before use, prepare a solution containing 0.3–0.6 unit/ml of Lipoamide Dehydrogenase in cold Reagent A).

Procedure

Pipette (in milliliters) the following reagents into suitable cuvettes:

	Test	Blank
Reagent A (Buffer)	2.50	2.60
Reagent B (DL-Thio)	0.10	0.10
Reagent C (β-NADH)	0.10	0.10
Reagent D (β-NAD)	0.10	0.10
Reagent E (EDTA)	0.10	0.10

Mix by inversion and equilibrate to 25°C. Monitor the A340nm until constant, using a suitably thermostatted spectrophotometer. Then add:

Reagent F (Enzyme Solution)	0.10	—

Immediately mix by inversion and record the decrease in A340nm for approximately 5–10 minutes. Obtain the ΔA340nm/minute using the maximum linear rate for both the Test and Blank.

Calculations

$$\text{Units / ml enzyme} = \frac{(\Delta A340\text{nm / min Test} - \Delta A340\text{nm / min Blank})\,(3)\,(\text{df})}{(6.22)\,(0.1)}$$

3 = Volume (in milliliters) of assay
df = Dilution factor
6.22 = Millimolar extinction coefficient of β-NADH at 340nm
0.1 = Volume (in milliliter) of enzyme used

$$\text{Units / mg protein} = \frac{\text{units / ml enzyme}}{\text{mg protein / ml enzyme}}$$

Unit Definition: *One unit will reduce 1.0 μmole of DL-lipoamide to DL-dihydrolipoamide per minute at pH 6.5 at 25°C.*

Final Assay Concentrations

In a 3.00 ml reaction mix, the final concentrations are 44 mM sodium phosphate, 0.93 mM DL-thioctic acid amide, 1.9% (v/v) ethanol, 0.2 mM β-nicotinamide adenine dinucleotide, reduced form, 0.67 mM β-nicotinamide adenine dinucleotide, 10 mM Ethylenediaminetetraacetic Acid, 0.07% (w/v) Bovine Serum Albumin and 0.03–0.06 unit lipoamide dehydrogenase.

Enzymatic Assay of LIPOPROTEIN LIPASE (EC 3.1.1.34)

Principle

$$\text{PNPB} + H_2O \xrightarrow{\text{LPL}} \text{p-Nitrophenol} + \text{Butyric Acid}$$

Abbreviations Used

PNPB = p-Nitrophenyl Butyrate
LPL = Lipoprotein Lipase

Conditions: T = 37°C, pH = 7.2, A400nm, Light path = 1 cm

Method: Continuous Spectrophotometric Rate Determination

Reagents

A. 100 mM Sodium Phosphate Buffer with 150 mM Sodium Chloride and 0.5% (v/v) Triton X-100, pH 7.2 at 37°C. (Prepare 100 ml in deionized water using Sodium Phosphate, Monobasic, Anhydrous, Sodium Chloride and Triton X-100. Adjust to pH 7.2 at 37°C with 1 M NaOH).
B. Acetonitrile.
C. 50 mM p-Nitrophenyl Butyrate (PNPB) (Prepare 1.0 ml in Reagent B using p-Nitrophenyl Butyrate).
D. Lipoprotein Lipase Enzyme Solution (Immediately before use, prepare a solution containing 60–70 units/ml of Lipoprotein Lipase in cold Reagent A).

Procedure

Pipette (in milliliters) the following reagents into suitable cuvettes:

	Test	Blank
Reagent A (Buffer)	0.90	0.90
Reagent D (Enzyme Solution)	0.10	0.10

Mix by inversion and equilibrate to 37°C. Monitor the A400nm until constant, using a suitably thermostatted spectrophotometer. Then add:

Reagent C (PNPB)	0.010	—
Deionized Water	—	0.010

Immediately mix by inversion and record the increase at A400nm for approximately 5 minutes. Obtain the ΔA400nm/minute using the maximum linear rate for both the Test and Blank.

Calculations

$$\text{Units/ml enzyme} = \frac{(\Delta A400nm/\min \text{ Test} - \Delta A400nm/\min \text{ Blank})(1.01)(df)}{(0.0148)(0.1)}$$

1.01 = Volume (in milliliters) of assay
df = Dilution factor
0.0148 = Micromolar extinction coefficient of p-Nitrophenol at 400nm
0.1 = Volume (in milliliter) of enzyme used

$$\text{Units/mg protein} = \frac{\text{units/ml enzyme}}{\text{mg protein/ml enzyme}}$$

Unit Definition: *One unit will release 1.0 nanomole (10–9 mole) of p-nitrophenol per minute at pH 7.2 at 37°C using p-nitrophenyl butyrate as substrate.*

Final Assay Concentrations

In a 1.01 ml reaction mix the final concentrations are 99 mM sodium phosphate, 149 mM sodium chloride, 0.5% (v/v) Triton X-100, 0.50 mM p-nitrophenyl butyrate, 1% (v/v) acetonitrile and 6–7 units lipoprotein lipase.

Enzymatic Assay of LIPOXIDASE (EC 1.13.11.12)

Principle

$$\text{Linoleic Acid} + O_2 \xrightarrow{\text{Lipoxidase}} \text{Peroxide of Linoleic Acid}$$

Abbreviation Used

Peroxide of Linoleic Acid = (9Z,11E)-(13S)-13-Hydroperoxyoctadeca-9,11-dienoate

Conditions: T = 25°C, pH=9.0, A234nm, Light path = 1 cm

Method: Continuous Spectrophotometric Rate Determination

Reagents

A. 200 mM Borate Buffer, pH 9.0 at 25°C (Prepare 100 ml in deionized water using Boric Acid. Adjust to pH 9.0 at 25°C with 1 M NaOH).

B. 95% (v/v) Ethyl Alcohol (EtOH) (Prepare 10 ml in deionized water using 200 Proof USP Ethyl Alcohol).

C. 0.017% (v/v) Linoleic Acid Solution (Linoleic Acid) (Prepare by combining 0.05 ml of Reagent B and 0.05 ml of Linoleic Acid into a suitable container. Vortex to dissolve completely. Bring to a volume of 50 ml by slowly adding Reagent A (foaming may occur). Mix by stirring until the solution is homogeneous. Then combine 5.0 ml of the Linoleic Acid/Ethanol solution with 20 ml of Reagent A and 5.0 ml of deionized water. Mix by stirring).

D. Lipoxidase Enzyme Solution (Immediately before use, prepare a solution containing 5,000–10,000 units/ml of Lipoxidase in Reagent A).

Procedure

Pipette (in milliliters) the following reagents into suitable cuvettes:

	Test	Blank
Reagent A (Buffer)	0.90	1.00
Reagent C (Linoleic Acid)	2.00	2.00

Mix by inversion and equilibrate to 25°C. Monitor the A234nm until constant, using a suitably thermostatted spectrophotometer. Then add:

Reagent D (Enzyme Solution)	0.10	—

Immediately mix by inversion and record the increase in A234nm for approximately 5 minutes. Obtain the ΔA234nm/minute using the maximum linear rate for both the Test and Blank.

Calculations

$$\text{Units / ml enzyme} = \frac{(\Delta A234nm / \min \text{ Test} - \Delta A234nm / \min \text{ Blank})\ (df)}{(0.001)\ (0.1)}$$

df = Dilution factor

0.001 = Change in A234nm per minute at pH 9.0 at 25°C when linoleic acid is the substrate in a 3.00 ml reaction volume (as per the unit definition)

0.1 = Volume (in milliliter) of enzyme used

$$\text{Units / mg solid} = \frac{\text{units / ml enzyme}}{\text{mg solid / ml enzyme}}$$

$$\text{Units / mg protein} = \frac{\text{units / ml enzyme}}{\text{mg protein / ml enzyme}}$$

Final Assay Concentrations

In a 3.00 ml reaction mix, the final concentrations are 178 mM boric acid, 0.011% (v/v) linoleic acid, 0.01% (v/v) ethanol and 500–1000 units lipoxidase.

Unit Definition: *One unit will cause an increase in A234nm of 0.001 per minute at pH 9.0 at 25°C when linoleic acid is the substrate in 3.0 ml volume (1 cm light path). One A234nm unit is equivalent to the oxidation of 0.12 μmole of linoleic acid.*

Enzymatic Assay of LUCIFERASE (EC 1.13.12.7) from Firefly

Principle

$$\text{Luciferin} + O_2 + \text{ATP} \xrightarrow{\text{Luciferase}} \text{Oxyluciferin} + CO_2 + \text{AMP} + \text{Pyrophosphate} + \text{Light} + Mg^{++}$$

Abbreviations Used

ATP = Adenosine 5'-Triphosphate
AMP = Adenosine 5'-Monophosphate

Conditions: T = 25°C, pH = 7.6

Method: Luminometric

Reagents

A. 1 M Glycine-Tris Buffer with 10 mM Ethylenediaminetetraacetic Acid and 100 mM Magnesium Sulfate, pH 7.0 at 25°C (Enz Dil) (Prepare 50 ml in deionized water using Glycine, Free Base, Ethylenediaminetetraacetic Acid, Disodium Salt, Dihydrate and Magnesium Sulfate, Heptahydrate. Adjust to pH 7.0 at 25°C with solid Trizma Base).

B. 0.15 mM Luciferin, 50 mM glycine, 1 mM Tris, 5 mM $MgSO_4$, 0.5 mM Ethylenediaminetetraacetic Acid, 0.1% (w/v) Bovine Serum Albumin, and 0.1% (w/v) Sodium Azide (Prepare by reconstituting 1 vial of Firefly Diluent, Glycine Diluent with added Luciferin with 10 ml of deionized water. Store on ice).

C. 1 mM Adenosine 5'-Triphosphate Solution (Prepare 10 ml in deionized water using Adenosine 5'-Triphosphate, Disodium Salt).

D. 0.01 mM Adenosine 5'-Triphosphate Solution (ATP) (Prepare 10.0 ml by adding 0.1 ml of Reagent C to 9.9 ml of deionized water).

E. 0.15 mM Luciferin, 50 mM Glycine, 1 mM Tris, 5 mM Magnesium Sulfate, 0.5 mM Ethylenediaminetetraacetic Acid, 0.1% (w/v) Bovine Serum Albumin, 0.1% (w/v) Sodium Azide, and 100 nM Adenosine 5'-Triphosphate Solution (Reaction Cocktail)

(Immediately before use, prepare by adding 0.010 ml of Reagent D to 0.99 ml of Reagent B).

F. $_{14}$C 1.0 μCi Standard ($_{14}$C Std) (Use $_{14}$C 1.0 μCi Counting Standard).

G. Luciferase Enzyme Solution (Immediately before use, prepare a solution containing 3,000,000–4,000,000 units/ml of Luciferase in cold Reagent A).

Procedure

Set Reagent F($_{14}$C Std) equal to 50% on the 100 scale of a photometer. Pipette (in milliliters) the following reagents into suitable containers:

	Test
Reagent E (Reaction Cocktail)	0.050

Incubate at 25°C for 10 minutes. Then add:

Reagent G (Enzyme Solution)	0.001

Immediately after adding Reagent G (Enzyme Solution), shake the tube twice and record the peak height of the luminescence.

Calculations

$$\text{Units / mg enzyme} = \frac{\text{(Photometer Reading) (df)}}{(0.001)}$$

Photometer Reading = One light unit will produce one count at the peak height of luminescence
df = Dilution factor
0.001 = Volume (in milliliter) of enzyme used

$$\text{Units / mg solid} = \frac{\text{units / ml enzyme}}{\text{mg solid / ml enzyme}}$$

$$\text{Units / mg protein} = \frac{\text{units / ml enzyme}}{\text{mg protein / ml enzyme}}$$

Unit Definition: *Light units are measured in 50 ml of assay mixture containing 5 picomoles of ATP and 7.5 nanomoles luciferin in glycine Tris buffer, pH 7.6 at 25°C. One light unit produces a biometer peak height equivalent to 0.02 mCi of $_{14}$C in PPO/ POPOP cocktail.*

Final Assay Concentrations

In a 0.051 ml reaction mix, the final concentrations are 0.15 mM luciferin, 49 mM glycine, 0.98 mM Tris, 7 mM magnesium sulfate, 0.7 mM Ethylenediaminetetraacetic Acid, 0.1% (w/v) Bovine Serum Albumin, 100 nM adenosine 5'-triphosphate and 3,000–4,000 units luciferase.

Enzymatic Assay of L-LYSINE DECARBOXYLASE (EC 4.1.1.18)

Principle

$$\text{L-Lysine} \xrightarrow{\text{L-Lysine Decarboxylase}} \text{Cadaverine} + CO_2 + \text{PRP}$$

Abbreviation Used

PRP = Pyridoxal 5-Phosphate

Conditions: T = 37°C, pH 6.0

Method: Manometric Assay using Warburg Flasks

Reagents

A. 500 mM Sodium Acetate Buffer, pH 6.0 at 37°C (Prepare 100 ml in deionized water using Sodium Acetate, Trihydrate. Adjust to pH 6.0 at 37°C with 1 M HCl).
B. 100 mM L-Lysine Solution (L-Lys) (Prepare 50 ml in Reagent A using L-Lysine, Monohydrochloride. Adjust to pH 6.0 at 37°C, if necessary, with either 1 M NaOH or 1 M HCl).
C. 50 mM Pyridoxal 5-Phosphate Solution (PRP) (Prepare 10 ml in Reagent A using Pyridoxal 5-Phosphate).
D. L-Lysine Decarboxylase Enzyme Solution (Immediately before use, prepare a solution containing 2–4 units/ml of L-Lysine Decarboxylase in cold Reagent A).

Procedure

Main Chamber

	Thermo-barometer Flask	Enzyme Blank	Test	Substrate Blank
Reagent A (Buffer)	2.80	2.80	0.30	0.30
Reagent B (L-Lys)	—	—	2.50	2.50
Side Arm				
Reagent A (Buffer)	0.20	—	—	0.10
Reagent C (PRP)	—	0.10	0.10	0.10
Reagent D (Enzyme Solution)	—	0.10	0.10	—

Be sure to confirm the stability of the pressure with the flask sealed off before proceeding with the assay. This is to ensure temperature equilibrium and the absence of leaks in the flask. The enzyme activity is determined by calculation of the rate of production of CO_2 at 37°C. The reaction rate should be linear for about 20 minutes. Obtain the maximum linear rate.

Calculations

$$\frac{\text{Units}}{\text{ml L - Lysine Decarboxylase}} = \frac{(C)(K)(df)}{\left(22.4\frac{1}{\text{mole}}1\right)(\text{ml}\cdot\text{L - Lysine Decarboxylase})}$$

C = mm of CO_2 gas evolved/minute
K = Warburg flask constant in μl/mm
df = Dilution factor
22.4 1 = Volume gas occupies under STP conditions

Unit Definition: *One unit will release 1.0 μmole of CO_2 from L-lysine per minute at pH 6.0 at 37°C.*

Final Assay Concentrations

In a 3.00 ml reaction mix, the final concentrations are 500 mM sodium acetate, 83 mM L-lysine, 1.7 mM pyridoxal 5-phosphate, and 0.2–0.4 unit L-lysine decarboxylase.

Notes

1. The tests are done in triplicate, since it is common for the flasks to have small leaks.
2. The mm of CO_2 gas evolved (C) is corrected for any temperature and barometric changes (Thermobarometer) during the experiment and also for the Substrate Blank and Enzyme Blank: mm CO_2 corrected = mm CO_2 measured Test – mm CO_2 measured for [Thermobarometer + Substrate Blank + Enzyme Blank]
Values of the corrected mm CO_2 produced are plotted versus time. The best straight line is drawn not necessarily through the origin. The slope, C = mm CO_2/time, is obtained.
3. The Flask constant, K, is calculated according to the equation:

$$K = \frac{\left[(V_g)\left(\frac{273}{T}2\right) + V_f\ a\right]}{P_o}$$

Po = Standard pressure as mm of manometer fluid
Vg = Volume (in milliliters) of gas in flask and manometer
Vf = Volume (in milliliters) of liquid in flask
T = Absolute temperature
a = Solubility of gas; (for CO_2 at 37°C, a = 0.57)

Enzymatic Assay of LYSINE OXIDASE (EC 1.4.3.14)

Principle

$$\text{L-Lysine} + O_2 + H_2O \xrightarrow{\text{Lysine Oxidase}} \text{6-Amino-2-Oxohexanoate} + NH_3 + H_2O_2$$

$$H_2O_2 + \text{6-Amino-2-Oxohexanoate} \xrightarrow{\text{Catalase}} \text{1-Piperidine-2-Carboxylate}$$

$$\text{1-Piperidine-2-Carboxylate} \xrightarrow{\text{MBTH}} \text{UV absorbing complex}$$

Abbreviation Used

MBTH = 3-Methyl-2-Benzothiazolinone Hydrazone

Conditions: T = 37°C, pH = 8.0, A317nm, Light path = 1 cm

Method: Spectrophotometric Stop Rate Determination

Reagents

A. 100 mM L-Lysine Solution (L-Lys) (Prepare 5 ml in deionized water using L-Lysine Monohydrochloride).
B. 200 mM Potassium Phosphate Buffer, pH 8.0 at 37°C (Buffer I) (Prepare 100 ml in deionized water using Potassium Phosphate, Monobasic, Anhydrous. Adjust to pH 8.0 at 37°C with 1 M KOH).
C. 0.1% (w/v) 3-Methyl-2-Benzothialinone Hydrazone Solution (MBTH) (Prepare 10 ml in deionized water using 3-Methyl-2-Benzothiazolinone Hydrazone, Hydrochloride, Hydrate. This reagent is Toxic. *Prepare Fresh*).

D. 12.5% (w/v) Trichloroacetic Acid Solution (TCA) (Prepare 2 ml in deionized water using Trichloroacetic Acid, 6.1 N Solution, approximately 100% (w/v)).

E. 10 mM α-Ketobutyric Acid Solution (α-KA) (Prepare 2 ml in deionized water using α-Ketobutyric Acid, Free Acid).

F. 1 M Sodium Acetate Buffer, pH 5.0 at 37°C (Buffer II) (Prepare 50 ml in deionized water using Sodium Acetate, Trihydrate. Adjust to pH 5.0 at 37°C with 1 M HCl).

G. Catalase Enzyme Solution (Catalase) (Immediately before use, prepare a solution containing 3,500 units/ml in cold deionized water using Catalase).

H. Lysine Oxidase Enzyme Solution (Immediately before use, prepare a solution containing 0.03–0.1 unit/ml in cold deionized water).

Procedure

Step 1: Pipette (in milliliters) the following reagents into an Eppendorf tube:

	Test	Blank
Reagent A (L-Lys)	0.10	0.10
Reagent B (Buffer I)	0.35	0.35
Reagent G (Catalase)	0.10	0.10
Deionized Water	0.35	0.45

Equilibrate at 37°C for 2 minutes. Then add:

Reagent H (Lysine Oxidase)	0.10	—

Immediately mix by inversion and incubate at 37°C for exactly 20 minutes. Then add:

Reagent D (TCA)	0.10	0.10

Centrifuge the Test and Blank for 5 minutes.

Step II: Colorimetric Assay: Pipette (in milliliters) the following reagents into suitable containers.

	Test	Blank	Std 1	Std 2	Std 3	Std 4	Std 5	Std Blank
Reagent C (MBTH)	0.40	0.40	0.40	0.40	0.40	0.40	0.40	0.40
Reagent F (Buffer II)	1.00	1.00	1.00	1.00	1.00	1.00	1.00	1.00
Test Supernatant	0.50	—	—	—	—	—	—	—
Blank supernatent	—	0.50	—	—	—	—	—	—
Reagent E (α-KA)	—	—	0.02	0.05	0.10	0.15	0.25	—
Deionized Water	1.10	1.10	1.58	1.55	1.50	1.45	1.35	1.60

Mix by swirling and incubate at 50°C for exactly 30 minutes. Cool to room temperature and record the A317nm for the Test, Blank, Standards, and Standard Blank, using a suitable spectrophotometer.

Calculations

Standard Curve: A317nm Standard = A317nm Standard–A317nm Standard Blank

Prepare a standard curve by plotting the A317nm Standard versus micromoles of α-ketobutyric acid.

Sample Determination: ΔA317nm Test = A317nm Test–A317nm Test Blank

Determine the micromoles of α-keto acid equivalents formed using the standard curve.

$$\text{Units / ml enzyme} = \frac{(\mu\text{moles of } \alpha\text{ - keto acid equivalents formed})\ (1.1)\ (\text{df})}{(0.5)\ (20)\ (0.1)}$$

1.1 = Volume (in milliliter) of stopped assay
df = Dilution factor
0.5 = Volume (in milliliter) of stopped assay used in the colorimetric assay
20 = Time (in minutes) of assay
0.1 = Volume (in milliliter) of L-lysine oxidase used in the assay

$$\text{Units / mg solid} = \frac{\text{units / ml enzyme}}{\text{mg solid / ml enzyme}}$$

$$\text{Units / mg protein} = \frac{\text{units / ml enzyme}}{\text{mg protein / ml enzyme}}$$

Unit Definition: *One unit will catalyze the formation of 1 µmole of 6-amino-2-oxohexanoic acid from L-lysine per minute at 37°C at pH 8.0.*

Final Assay Concentrations

In a 1.00 ml reaction mix, the final concentrations are 10 mM L-lysine, 70 mM potassium phosphate, 350 units catalase, and 0.003–0.01 unit L-lysine oxidase.

Note

1. Catalase Unit Definition: One unit will decompose 1.0 µmole of H_2O_2 per minute at pH 7.0 at 25°C, while the H_2O_2 concentration falls from 10.3 to 9.2 mM. The rate of disappearance of H_2O_2 is followed by observing the rate of decrease in absorbance at 240 nm.

Enzymatic Assay of LYSOSTAPHIN

Principle

$$\text{Staphylococcal Cells (intact)} \xrightarrow{\text{Lysostaphin}} \text{Staphylococcal Cells (lysed)}$$

Conditions: T = 37°C, pH 7.5, A620nm, Light path = 1 cm

Method: Turbidimetric Rate Determination

Reagents

A. 50 mM Tris HCl Buffer with 145 mM Sodium Chloride, pH 7.5 at 37°C (Prepare 1 liter in deionized water using Trizma Base and Sodium Chloride. Adjust to pH 7.5 at 37°C with 1 M HCl).

B. Cell Suspension (Prepare using the following steps:
   1. Prepare 100 ml of a 30% (w/v) Tryptic Soy Broth in Reagent A using Tryptic Soy Broth. In a 250 ml flask, inoculate 50 ml of the Tryptic Soy Broth with 1 ml of a 1 mg/ml culture of *Staphylococcus aureus*, American Type Culture Collection #29737. Incubate, with gentle shaking, at 37°C for 18–24 hours.

2. Centrifuge the culture to pellet the cells. Remove the supernatant liquid and suspend the cells in cold Reagent A. Repeat the centrifugation and resuspend the cells in a minimum volume of cold Reagent A.
3. Adjust the cell density to an A620nm of 1.3–1.5 by diluting with cold Reagent A. Keep the diluted suspension on ice until ready to assay samples. *The Final Diluted Suspension (Step 3) must be Prepared Fresh. Use the Step 3 Cell Suspension within 15–20 Minutes*).

C. Lysostaphin Enzyme Solution (Immediately before use, prepare a solution containing 0.05–0.25 unit/ml of Lysostaphin in cold Reagent A).

Procedure

Pipette (in milliliters) the following reagents into suitable vials:

	Test	Blank
Reagent A (Buffer)	—	5.00
Reagent C (Enzyme Solution)	5.00	—

Equilibrate to 37°C for approximately 5 minutes. Then add:

Reagent B (Cell Suspension)	1.00	1.00

Mix by swirling and immediately determine at zero time, (T0), the A620nm, for both the Test and the Blank using Reagent A as a reference. Incubate the cuvette at 37°C for exactly 10 minutes. Mix by swirling and determine at ten minutes, (T10), the A620nm for both the Test and the Blank.

Calculations

ΔA620nm = A620nm Test–A620nm Blank

ΔA620nm T0–ΔA620nm T10

% Lysis = × 100%ΔA620nm T0

ΔA620nm T0–ΔA620nm T10

$$\text{Units / ml} = \frac{(\text{df})}{(0.125)\ (5)}$$

df = Dilution factor

5 = Volume (in milliliters) of enzyme assay

0.125 = The change in turbidity produced by 1 unit of of Lysostaphin (as per the Unit Definition)

$$\text{Units / mg solid} = \frac{\text{units / ml enzyme}}{\text{mg solid / ml enzyme}}$$

$$\text{Units / mg protein} = \frac{\text{units / ml enzyme}}{\text{mg protein / ml enzyme}}$$

Unit Definition: *One unit will reduce the turbidity (A620nm) of a suspension of S. aureus cells from 0.250 to 0.125 in 10 minutes at pH 7.5 at 37°C in a 6.0 ml reaction mixture.*

Final Assay Concentrations

In a 6.00 ml reaction mix, the final concentrations are 50 mM Tris, 145 mM sodium chloride and 0.25–1.25 units lysostaphin.

Notes

1. Reproducibility of this assay is very dependent on the freshness and quality of the cell suspension.
2. This procedure incorporates the use of a potentially infectious microorganism. Containment and/or disinfection of all such products, solutions and test reagents is imperative. Heat or chemical sterilization of all contaminated reagents and any spills or releases, with subsequent disposal as Biohazardous Waste, must be employed.
3. The A620nm at zero time (T0) should fall between 0.2–0.3 absorbance unit.

Enzymatic Assay of LYSOZYME (EC 3.2.1.17)

Principle

$$\underset{\text{(Intact)}}{\textit{Micrococcus lysodeikticus}\text{ Cells}} \xrightarrow{\text{Lysozyme}} \underset{\text{(Lysed)}}{\textit{Micrococcus lysodeikticus}\text{ Cells}}$$

Conditions: T = 25°C, pH = 5.0, A450nm, Light path = 1 cm.

Method: Turbidimetric

Reagents

A. 100 mM Sodium Phosphate Buffer, pH 5.0 at 25°C (Prepare 100 ml in deionized water using Sodium Phosphate, Monobasic, Anhydrous. Adjust to pH 5.0 at 25°C with 1 M NaOH).

B. 0.01% (w/v) *Micrococcus lysodeikticus* Cell Suspension (Substrate) (Prepare 25 ml in Reagent A using *Micrococcus lysodeikticus,* ATCC 4698 lyophilized cells. The A450nm of this suspension should be between 0.6 and 0.7).

C. Lysozyme Enzyme Solution (Immediately before use, prepare a solution containing 300–1000 units/ml of lysozyme in cold Reagent A).

Procedure

Pipette (in milliliters) the following reagents into suitable cuvettes:

	Test	Blank
Reagent B (Substrate)	0.97	0.97

Equilibrate to 25°C. Monitor the A450nm until constant, using a suitably thermostatted spectrophotometer. Then add:

Reagent C (Enzyme Solution)	0.03	—
Reagent A (Buffer)	—	0.03

Immediately mix by inversion and record the decrease in A450nm for approximately 5 minutes. Obtain the A450nm/minute using the maximum linear rate for both the Test and Blank.

Calculations

$$\text{Units / ml enzyme} = \frac{(\text{A450nm / min Test} - \text{A450nm / min Blank})\ (\text{df})}{(0.001)\ (0.03)}$$

df = Dilution factor
0.001 = Change in absorbance at A450nm as per the Unit Definition
0.03 = Volume (in milliliters) of enzyme used.

$$\text{Units / mg solid} = \frac{\text{units / ml enzyme}}{\text{mg solid / ml enzyme}}$$

$$\text{Units / mg protein} = \frac{\text{units / ml enzyme}}{\text{mg protein / ml enzyme}}$$

Unit Definition: *One unit will produce a A450nm of 0.001 per minute at pH 5.0 at 25°C using a suspension of Micrococcus lysodeikticus as substrate, in a 1.0 ml reaction mixture.*

Final Assay Concentrations

In a 1.00 ml reaction mix, the final concentrations are 100 mM sodium phosphate, 0.01% (w/v) *Micrococcus lysodeikticus* cell suspension and 9–30 units lysozyme.

Enzymatic Assay of LYSOZYME (EC 3.2.1.17)

Principle

Micrococcus lysodeikticus Cells (Lysed) $\xrightarrow{\text{Lysozyme}}$ *Micrococcus lysodeikticus* Cells (Intact)

Conditions: T = 25°C, pH = 6.24, A450nm, Light path = 1 cm.

Method: Turbidimetric Rate Determination.

Reagents

A. 66 mM Potassium Phosphate Buffer, pH 6.24 at 25°C (Prepare 100 ml in deionized water using Potassium Phosphate, Monobasic, Anhydrous. Adjust to pH 6.24 at 25°C with 1 M KOH).
B. 0.015% (w/v) *Micrococcus lysodeikticus* Cell Suspension (Substrate) (Prepare 25 ml in Reagent A using *Micrococcus lysodeikticus,* ATCC 4698 lyophilized cells. The A450nm of this suspension should be between 0.6 and 0.7).
C. Lysozyme Enzyme Solution (Immediately before use, prepare a solution containing 200–400 units/ml of lysozyme in cold Reagent A).

Procedure

Pipette (in milliliters) the following reagents into suitable cuvettes:

	Test	Blank
Reagent B (Substrate)	2.50	2.50

Equilibrate to 25°C. Monitor the A450nm until constant, using a suitably thermostatted spectrophotometer. Then add:

Reagent C (Enzyme Solution)	0.10	—
Reagent A (Buffer)	—	0.10

Immediately mix by inversion and record the decrease in A450nm for approximately 5 minutes. Obtain the A450nm/minute using the maximum linear rate for both the Test and Blank.

Calculations

$$\text{Units / ml enzyme} = \frac{(\text{A450nm / min Test} - \text{A450nm / min Blank})\ (\text{df})}{(0.001)\ (0.1)}$$

df = Dilution factor
0.001 = Change in absorbance at A450nm as per the Unit Definition
0.1 = Volume (in milliliter) of enzyme used

$$\text{Units / mg solid} = \frac{\text{units / ml enzyme}}{\text{mg solid / ml enzyme}}$$

$$\text{Units / mg protein} = \frac{\text{units / ml enzyme}}{\text{mg protein / ml enzyme}}$$

Unit Definition: *One unit will produce A450nm of 0.001 per minute at pH 6.24 at 25°C using a suspension of Micrococcus lysodeikticus as substrate, in a 2.6 ml reaction mixture.*

Final Assay Concentrations

In a 2.60 ml reaction mix, the final concentrations are 66 mM potassium phosphate, 0.014% (w/v) *Micrococcus lysodeikticus* cell suspension and 20–40 units lysozyme.

Note

1. This assay procedure is not to be used to assay Lysozyme, Bovine Recombinant Expressed in Pichia pastoris, Lysozyme, Human, Recombinant Expressed in Pichia pastoris, and Lysozyme Insoluble Enzyme on Agarose,

Enzymatic Assay of LYSOZYME (EC 3.2.1.17)

Principle

$$\textit{Micrococcus lysodeikticus}\ \text{Cells}\ (\text{Intact}) \xrightarrow{\text{Lysozyme}} \textit{Micrococcus lysodeikticus}\ \text{Cells}\ (\text{Lysed})$$

Conditions: T = 30°C, pH = 7.0, A450nm, Light path = 1 cm

Method: Turbidimetric

Reagents

A. 30 mM Potassium Phosphate Buffer with 0.1% (w/v) Bovine Serum Albumin, pH 7.4 at 30°C (Prepare 100 ml in deionized water using Potassium Phosphate, Monobasic, Anhydrous and Albumin, Bovine. Adjust to pH 7.4 at 30°C with 1 M KOH).

B. 0.01% (w/v) *Micrococcus lysodeikticus* Cell Suspension (Substrate) (Prepare 25 ml in Reagent A using *Micrococcus lysodeikticus,* ATCC 4698 lyophilized cells. The A450nm of this suspension should be between 0.6 and 0.7).

C. Lysozyme Enzyme Solution (Immediately before use, prepare a solution containing 5000–10,000 units/ml of lysozyme in cold Reagent A).

Procedure

Pipette (in milliliters) the following reagents into suitable cuvettes:

	Test	Blank
Reagent B (Substrate)	1.00	1.00

Equilibrate to 30°C. Monitor the A450nm until constant, using a suitably thermostatted spectrophotometer. Then add:

Reagent C (Enzyme Solution)	0.01	—
Reagent A (Buffer)	—	0.01

Immediately mix by inversion and record the decrease in A450nm for approximately 5 minutes. Obtain the ΔA450nm/minute using the maximum linear rate for both the Test and Blank.

Calculations

$$\text{Units / ml enzyme} = \frac{(\Delta A450nm / \text{min Test} - \Delta A450nm / \text{min Blank})\,(df)}{(0.001)\,(0.1)}$$

df = Dilution factor
0.001 = Change in absorbance at A450nm as per the Unit Definition
0.01 = Volume (in milliliter) of enzyme used

$$\text{Units / mg solid} = \frac{\text{units / ml enzyme}}{\text{mg solid / ml enzyme}}$$

$$\text{Units / mg protein} = \frac{\text{units / ml enzyme}}{\text{mg protein / ml enzyme}}$$

Unit Definition: *One unit will produce a ΔA450nm of 0.001 per minute at pH 7.4 at 30°C using a suspension of Micrococcus lysodeikticus as substrate, in a 1.0 ml reaction mixture.*

Final Assay Concentrations

In a 1.00 ml reaction mix, the final concentrations are 30 mM potassium phosphate, 0.01% (w/v) *Micrococcus lysodeikticus* cell suspension, 0.1% (w/v) Bovine Serum Albumin, and 50–1000 units lysozyme.

Notes

1. Store the *Micrococcus lysodeikticus* cell suspension until homogenous. This may take up to 15 minutes.
2. The initial absorbance should be between 0.5 and 0.7.
3. Run one test at a time. The reaction is too fast to enable multiple tests to be run simultaneously.

Enzymatic Assay of LYTICASE

Principle

$$\text{Yeast Cells (intact)} \xrightarrow{\text{Lyticase}} \text{Yeast Cells (lysed)}$$

Conditions: T = 25°C, pH 7.5, A800nm, Light path = 1 cm

Method: Turbidimetric Rate Determination

Reagents

A. 67 mM Potassium Phosphate Buffer, pH 7.5 at 25°C (Prepare 100 ml in deionized water using Potassium Phosphate, Monobasic, Anhydrous. Adjust to pH 7.5 at 25°C with 1 M KOH).

B. 0.2% (w/v) Bakers Yeast Substrate Suspension (YSC-1) (Prepare 100 ml in deionized water using Yeast, Saccharomyces cerevisiae, Bakers Yeast. Allow solution to stir for 2 hours before use).

C. Lyticase Enzyme Solution (Immediately before use, prepare a solution containing approximately 500 units/ml of Lyticase in cold deionized water).

Procedure

Pipette (in milliliters) the following reagents into suitable cuvettes:

	Test	Test 1	Test 2	Blank
Deionized Water	0.55	0.54	0.53	0.60
Reagent A (Buffer)	1.50	1.50	1.50	1.50
Reagent B (YSC-1)	0.90	0.90	0.90	0.90
Reagent C (Enzy Soln)	0.05	0.06	0.07	—

Mix the Tests and Blank samples by inversion. Monitor the decrease in A800nm for approximately 10 minutes. Obtain the ΔA800nm/minute using the maximum linear rate for both the Tests and Blank. The ΔA800nm for the blank should cover only the range of the times chosen for each of the Test samples.

Calculations

$$\text{Units / ml enzyme} = \frac{(\Delta A800\text{nm / min Test} - \Delta A800\text{nm / min Blank})\ (\text{df})}{(0.001)\ (0.1)}$$

df = Dilution factor
0.001 = Change in absorbance at 800nm as per the Unit Definition
0.1 = Volume (in milliliter) of enzyme used

$$\text{Units / mg solid} = \frac{\text{units / ml enzyme}}{\text{mg solid / ml enzyme}}$$

$$\text{Units / mg protein} = \frac{\text{units / ml enzyme}}{\text{mg protein / ml enzyme}}$$

Unit Definition: *One unit will produce a A800nm of 0.001 per minute at pH 7.5 at 25°C, using a suspension of yeast as substrate in a 3 ml reaction mix.*

Final Assay Concentrations

In a 3.00 ml reaction mix, the final concentrations are 34 mM potassium phosphate, 0.1% (w/v) baker's yeast, and 25, 30, and 35 units lyticase.

Notes

1. YSC-1 substrate solution was found to be stable for up to 8 hours after preparation.
2. Preparation of 500 units/ml is crucial to obtaining valid activity in this procedure. There is an inverse relationship between the enzyme concentration and the activity.
3. The amount of enzyme added to each of the three cuvettes is critical. Thus, aliquots of 0.05, 0.06, and 0.07 ml should be followed.
4. It is important that both the test samples and the blank are inverted due to the settling of the suspended yeast in the reaction mixture.
5. It is crucial that a 2 minute time window be chosen in order to achieve consistent results. Any smaller time window may result in a falsely high rate due to settling of the substrate or other sources of noise in the assay. Additionally, any larger time window may gradually decrease the true maximum rate of the enzyme. Also, the ΔA800nm/min must fall between 0.025 and 0.035 in order for the assay to be valid.
6. The blank is an important measurement due to the settling of the substrate in the cuvettes over time and its potential to create a false rate. Thus, if the maximized ΔA800nm/min of the various enzyme levels are not in the exact same time window, each level needs to be coupled with separately calculated ΔA800nm/min for the blank.

Enzymatic Assay of MALATE SYNTHASE (EC 4.1.3.2)

Principle

$$\text{Acetyl CoA} + H_2O + \text{Glyoxylate} \xrightarrow{\text{Malate Synthase}} \text{L-Malate} + \text{CoA}$$

$$\text{CoA} + \text{DTNB} \rightarrow \text{TNB} + \text{CoA derivative}$$

Abbreviations Used

Acetyl CoA = Acetyl Coenzyme A
CoA = Coenzyme A
DTNB = 5,5'-Dithio-bis (2-Nitrobenzoic Acid)
TNB = 5-Thio-2-Nitrobenzoic Acid

Conditions: T = 30°C, pH 8.0, A412nm, Light path = 1 cm

Method: Continuous Spectrophotometric Rate Determination

Reagents

A. 50 mM Imidazole Buffer, pH 8.0 at 30°C (Prepare 100 ml in deionized water using Imidazole. Adjust to pH 8.0 at 30°C with 1 M HCl).
B. 100 mM Magnesium Chloride Solution ($MgCl_2$) (Prepare 10 ml in deionized water using Magnesium Chloride, Hexahydrate).
C. 2.5 mM Acetyl CoA Solution (Acetyl CoA) (Prepare 1 ml in deionized water using Acetyl Coenzyme A (C2:0), Sodium Salt).
D. 10 mM Glyoxylic Acid Solution (Glyoxylic Acid) (Prepare 10 ml in deionized water using Glyoxylic Acid, Sodium Salt, Monohydrate).

E. 95% (v/v) Ethanol (EtOH) (Prepare 10 ml in deionized water using 200 Proof USP Ethyl Alcohol).

F. 2 mM 5,5'-Dithio-bis (2-Nitrobenzoic Acid) Solution (DTNB) (Prepare 10 ml in Reagent E using 5,5'-Dithio-bis (2-Nitrobenzoic Acid)).

G. Malate Synthase Enzyme Solution (Immediately before use, prepare a solution containing 0.07–0.09 unit/ml of Malate Synthase in cold Reagent A).

Procedure

Pipette (in milliliters) the following reagents into suitable cuvettes:

	Test	Blank
Reagent A (Buffer)	0.50	0.50
Reagent B ($MgCl_2$)	0.10	0.10
Reagent C (Acetyl CoA)	0.10	0.10
Reagent D (Glyoxylic Acid)	0.10	0.10
Reagent F (DTNB)	0.10	0.10

Mix by inversion and equilibrate to 30°C. Monitor the A412nm until constant, using a suitably thermostatted spectrophotometer. Then add:

Reagent G (Enzyme Solution)	0.10	—
Reagent A (Buffer)	—	0.10

Immediately mix by inversion and record the increase in A412nm for approximately 5 minutes. Obtain the ΔA412nm/minute using the maximum linear rate for both the Test and Blank.

Calculations

$$\text{Units / ml enzyme} = \frac{(\Delta A412\text{nm / min Test} - \Delta A412\text{nm / min Blank})\ (1)\ (\text{df})}{(13.6)\ (0.1)}$$

1 = Volume (in milliliter) of assay
df = Dilution factor
13.6 = Millimolar extinction coefficient for TNB at 412nm
0.1 = Volume (in milliliter) of assay

$$\text{Units / mg protein} = \frac{\text{units / ml enzyme}}{\text{mg protein / ml enzyme}}$$

Unit Definition: *One unit will catalyze the cleavage of 1.0 µmole of acetyl-CoA per minute at pH 8.0 at 30°C, in the presence of glyoxylate.*

Final Assay Concentrations

In a 1.00 ml reaction mix, the final concentrations are 30 mM imidazole, 10 mM magnesium chloride, 0.25 mM acetyl-CoA, 1 mM glyoxylic acid, 0.2 mM 5,5'-dithio-bis (2-nitrobenzoic Acid), and 0.007–0.009 unit malate synthase.

Enzymatic Assay of MALIC DEHYDROGENASE (EC 1.1.1.37)

Principle

$$\text{Oxalacetate} + \beta\text{-NADH} \xrightarrow{\text{Malic Dehydrogenase}} \text{L-Malic} + \beta\text{-NAD}$$

Abbreviations Used

β-NADH = β-Nicotinamide Adenine Dinucleotide, Reduced Form
β-NAD = β-Nicotinamide Adenine Dinucleotide, Oxidized Form

Conditions: T = 25°C, pH = 7.5, A340nm, Light Path = 1 cm

Method: Continuous Spectrophotometric Rate Determination

Reagents

A. 100 mM Potassium Phosphate Buffer, pH 7.5 at 25°C (Prepare 100 ml in deionized water using Potassium Phosphate, Monobasic, Anhydrous. Adjust to pH 7.5 at 25°C with 1 M KOH).

B. 0.14 mM β-Nicotinamide Adenine Dinucleotide, Reduced Form Solution (β-NADH) (Dissolve the contents of one 5 mg vial of β-Nicotinamide Adenine Dinucleotide, Reduced Form, Disodium Salt in 50 ml of Reagent A *or* prepare 20 ml in Reagent A using β-Nicotinamide Adenine Dinucleotide, Reduced Form, Disodium Salt. *Prepare Fresh*).

C. 7.6 mM Oxalacetic Acid Solution (OAA) (Prepare 10 ml in Reagent A using Oxalacetic Acid. Prepare immediately prior to use and keep on ice, as it is very unstable in solution).

D. Malic Dehydrogenase Solution (Immediately before use, prepare a solution containing 0.2–0.5 units/ml Malic Dehydrogenase in cold Reagent A).

Procedure

Pipette (in milliliters) the following reagents into suitable cuvettes:

	Test	Blank
Reagent B (β-NADH)	2.80	2.80

Mix by inversion and equilibrate to 25°C. Monitor the A340nm until constant, using a suitably thermostatted spectrophotometer. Then add:

	Test	Blank
Reagent C (OAA)	0.10	0.10
Reagent A (Buffer)	—	0.10
Reagent D (Enzyme Solution)	0.10	—

Immediately mix by inversion and record the decrease in A340nm for approximately 2 minutes. Obtain the ΔA340nm/minute using the maximum linear rate for both the Test and Blank.

Calculations

$$\text{Units / ml enzyme} = \frac{(\Delta A340nm / \text{min Test} - \Delta A340nm / \text{min Blank})\ (3)\ (df)}{(6.22)\ (0.1)}$$

3 = Volume (in milliliters) of assay
df = Dilution factor
6.22 = Millimolar extinction coefficient of β-NADH at 340nm
0.1 = Volume (in milliliter) of enzyme used

$$\text{Units / mg solid} = \frac{\text{units / ml enzyme}}{\text{mg solid / ml enzyme}}$$

$$\text{Units / mg protein} = \frac{\text{units / ml enzyme}}{\text{mg protein / ml enzyme}}$$

Unit Definition: *One unit will convert 1.0 µmole of oxalacetate and β-NADH to L-malate and β-NAD per minute at pH 7.5 at 25°C.*

Final Assay Concentrations

In a 3.00 ml reaction mix, the final concentrations are 100 mM potassium phosphate, 0.13 mM β-nicotinamide adenine dinucleotide, reduced form, 0.25 mM oxalacetic acid and 0.02–0.03 units malic dehydrogenase.

Enzymatic Assay of MALIC ENZYME (E.C. 1.1.1.40)

Principle

$$\text{L-Malate} + \beta\text{-NADP} \xrightarrow{\text{Malic Enzyme}} \text{Pyruvate} + CO_2 + \beta\text{-NADPH}$$

Abbreviations Used

β-NADP = β-Nicotinamide Adenine Dinucleotide Phosphate, Oxidized Form
β-NADPH = β-Nicotinamide Adenine Dinucleotide Phosphate, Reduced Form

Conditions: T = 25°C, pH = 7.4, A340nm, Light path = 1 cm

Method: Continuous Spectrophotometric Rate Determination

Reagents

A. 100 mM Triethanolamine HCl Buffer, pH 7.4 at 25°C. (Prepare 100 ml in deionized water using Triethanolamine, Hydrochloride. Adjust to pH 7.4 at 25°C with 1 M NaOH).
B. 100 mM L-Malic Acid Solution (Malic Acid) (Prepare 5 ml in deionized water using L(-)Malic Acid, Free Acid).
C. 20 mM β-Nicotinamide Adenine Dinucleotide Phosphate, Oxidized Form, Solution (NADP) (Prepare 2 ml in deionized water using β-Nicotinamide Adenine Dinucleotide Phosphate, Sodium Salt or β-Nicotinamide Adenine Dinucleotide Phosphate, Sodium Salt).
D. 20 mM Manganese Chloride Solution ($MnCl_2$) (Prepare 25 ml in deionized water using Manganese Chloride, Tetrahydrate).
E. Malic Enzyme Solution (Immediately before use, prepare a solution containing 0.25–0.50 unit/ml of Malic Enzyme in cold deionized water).

Procedure

Pipette (in milliliters) the following reagents into suitable cuvettes:

	Test	Blank
Reagent A (Buffer)	2.00	2.00
Reagent B (Malic Acid)	0.10	0.10
Reagent C (NADP)	0.05	0.05
Reagent D ($MnCl_2$)	0.75	0.75

Mix by inversion and equilibrate to 25°C. Monitor the A340nm until constant, using a suitably thermostatted spectrophotometer. Then add:

Reagent E (Enzyme Solution)	0.10	—
Deionized Water	—	0.10

Immediately mix by inversion and monitor the increase in A340nm for approximately 5–10 minutes. Obtain the ΔA340nm/minute using the maximum linear rate for both the Test and Blank.

Calculations

$$\text{Units / ml enzyme} = \frac{(\Delta\text{A340nm / min Test} - \Delta\text{A340nm / min Blank})\ (3)\ (\text{df})}{(6.22)\ (0.1)}$$

3 = Total volume (in milliliters) of assay
df = Dilution factor
6.22 = Millimolar extinction coefficient of β-NADPH at 340nm
0.1 = Volume (in milliliter) of enzyme used

$$\text{Units / mg solid} = \frac{\text{units / ml enzyme}}{\text{mg solid / ml enzyme}}$$

$$\text{Units / mg protein} = \frac{\text{units / ml enzyme}}{\text{mg protein / ml enzyme}}$$

Unit Definition: *One unit will convert 1.0 μmole of L-malate and NADP to pyruvate, CO_2 and NADPH per minute at pH 7.4 at 25°C.*

Final Assay Concentrations

In a 3.00 ml reaction mix, the final concentrations are 67 mM triethanolamine, 3.3 mM L-malic acid, 0.3 mM β-nicotinamide adenine dinucleotide phosphate, 5.0 mM manganese chloride and 0.025–0.050 unit malic enzyme.

Enzymatic Assay of MALTASE (EC 3.2.1.20)

Principle

$$\text{Maltose} + H_2O \xrightarrow{\text{Maltase}} 2\ \text{Glucose}$$

$$\text{Glucose} + \text{ATP} \xrightarrow{\text{Hexokinase}} \text{Glucose 6 - Phosphate}$$

$$\text{Glucose 6 - Phosphate} + \beta\text{- NADP} \xrightarrow{\text{G-6-P-DH}} \text{Gluconate} + \beta\text{- NADPH}$$

Abbreviations Used

ATP = Adenosine 5'-Triphosphate
β-NADP = β-Nicotinamide Adenine Dinucleotide Phosphate, Oxidized Form
G-6-P-DH = Glucose 6-Phosphate Dehydrogenase
β-NADPH = β-Nicotinamide Adenine Dinucleotide Phosphate, Reduced Form

Conditions: T = 25°C, pH 6.0, A340nm, Light path = 1 cm.

Method: Spectrophotometric Stop Rate Determination

Reagents

A. 50 mM Potassium Phosphate Buffer, pH 6.0 at 25°C (Phosphate Buffer) (Prepare 100 ml in deionized water using Potassium Phosphate, Monobasic, Anhydrous. Adjust to pH to 6.0 at 25°C with 1 M KOH).
B. 300 mM Maltose Solution (Maltose) (Prepare 10 ml in deionized water using Maltose, Monohydrate).
C. 300 mM Triethanolamine HCl Buffer, pH 7.6 at 25°C (TEA Buffer) (Prepare 100 ml in deionized water using Triethanolamine Hydrochloride. Adjust to pH 7.6 at 25°C with 1 M NaOH).
D. 100 mM Magnesium Chloride Solution ($MgCl_2$) (Prepare 10 ml in deionized water using Magnesium Chloride, 4.9 M Solution).
E. 16 mM Adenosine 5'-Triphosphate Solution (ATP) (Prepare 10 ml in deionized water using Adenosine 5'-Triphosphate, Disodium Salt).
F. 12 mM β-Nicotinamide Adenine Dinucleotide Phosphate, Oxidized Form, Solution (β-NADP) (Prepare 10 ml in deionized water using β-Nicotinamide Adenine Dinucleotide Phosphate, Sodium Salt).
G. Hexokinase and Glucose-6-Phosphate Dehydrogenase Enzyme Solution (Hex/G-6-PDH) (Immediately before use, prepare a solution containing 200 Hexokinase units/ml of Hexokinase and Glucose-6-Phosphate Dehydrogenase).
H. Maltase Enzyme Solution (Maltase) (Immediately before use, prepare a solution containing 0.25–0.50 unit/ml of Maltase in cold Reagent A).

Procedure

Step 1: Pipette (in milliliters) the following reagents into suitable containers:

	Test	Blank
Reagent A (Phosphate Buffer)	0.40	0.40
Reagent B (Maltose)	0.50	0.50

Mix by swirling and equilibrate to 25°C. Then add:

Reagent H (Maltase)	0.10	—
Reagent A (Phosphate Buffer)	—	0.10

Immediately mix by swirling and incubate at 25°C for exactly 30 minutes. Then place in a boiling water bath for 3 minutes. Cool to room temperature. Then add:

Deionized Water	9.00	9.00

Mix by swirling.

Step 2: Prepare a reaction cocktail by combining the following reagents (in milliliters):

Reagent C (TEA Buffer)	25.00
Reagent D ($MgCl_2$)	1.00
Reagent E (ATP)	1.00
Reagent F (β-NADP)	1.00
Reagent G (Hex/G-6-PDH)	0.20

Pipette (in milliliters) the following reagents into suitable cuvettes:

	Test	Blank
Reaction Cocktail	2.50	2.50

Equilibrate to 25°C. Then add:

Diluted Test from Step 1	0.50	—
Diluted Blank from Step 1	—	0.50

Mix by inversion and record the increase in A340nm for approximately 5–10 minutes until constant, using a suitably thermostatted spectrophotometer. Obtain the final A340nm for both the Test and Blank solutions.

Calculations

$$\text{Units / ml enzyme} = \frac{(\text{A340nm Test} - \text{A340nm Blank})\,(3)\,(\text{df})}{(2)\,(6.22)\,(0.1)\,(0.5)\,(30)}$$

3 = Total volume (in milliliters) of assay in Step 2
df = Dilution factor
2 = Moles of glucose released per mole of maltose
6.22 = Millimolar extinction coefficient of β-NADPH at 340nm
0.1 = Volume (in milliliters) of enzyme used in Step 1
0.5 = Volume (in milliliters) of diluted test used in Step 2
30 = Reaction time (in minutes) of Step 1

$$\text{Units / mg solid} = \frac{\text{units / ml enzyme}}{\text{mg solid / ml enzyme}}$$

$$\text{Units / mg protein} = \frac{\text{units / ml enzyme}}{\text{mg protein / ml enzyme}}$$

Unit Definition: *One unit will convert 1.0 μmole of maltose to 2.0 μmoles of D-glucose per minute at pH 6.0 at 25°C (liberated glucose determined at pH 7.6).*

Final Assay Concentrations

In a 1.00 ml reaction mix, the final concentrations are 25 mM potassium phosphate, 150 mM maltose and 0.025–0.050 unit maltase.

Notes

1. Hexokinase Unit Definition: One unit will phosphorylate 1.0 μmole of D-glucose per minute at pH 7.6 at 25°C.
2. Glucose-6-Phosphate Dehydrogenase: One unit will oxidize 1.0 μmole of D-glucose 6-phosphate to 6-phospho-D-gluconate per minute in the presence of NADP at pH 7.4 at 25°C.

Enzymatic Assay of MANDELONITRILE LYASE (EC 4.1.2.10)

Principle

$$\text{Mandelonitrile} \xrightarrow{\text{ML}} \text{Benzaldehyde} + \text{HCN}$$

Abbreviations Used

ML = Mandelonitrile Lyase
HCN = Hydrogen Cyanide

Conditions: T = 25°C, pH = 5.4, A275nm, Light path = 1 cm

Method: Continuous Spectrophotometric Rate Determination

Reagents

A. 100 mM Sodium Acetate Buffer (Prepare 100 ml in deionized water using Sodium Acetate Trihydrate. Adjust to pH 5.4 at 25°C with 1 M HCl).
B. 95% (v/v) Ethanol (Prepare 10 ml in deionized water using 200 Proof USP Ethyl Alcohol).
C. 168 mM DL-Mandelonitrile Solution (Substrate) (Prepare 10 ml in cold 95% Ethanol using DL-Mandelonitrile. *Prepare Fresh*).
D. 0.1% (w/v) Bovine Serum Albumin Solution (Enzyme Diluent) (Prepare 10 ml in cold deionized water using Albumin, Bovine).
E. Mandelonitrile Lyase Enzyme Solution (Immediately before use, prepare a solution containing 1–2 units/ml of Mandelonitrile Lyase in cold Reagent D).

Procedure

Pipette (in milliliters) the following reagents into suitable cuvettes:

	Test	Blank
Reagent A (Buffer)	2.85	2.85
Reagent C (Substrate)	0.05	0.05

Mix by inversion and equilibrate to 25°C. Monitor the A275nm until constant, using a suitably thermostatted spectrophotometer. Then add:

	Test	Blank
Reagent E (Enzyme Solution)	0.10	—
Reagent D (Enzyme Diluent)	—	0.10

Immediately mix by inversion and record the increase in A275nm for approximately 5 minutes. Obtain the ΔA275/minute using the maximum linear rate for both Test and Blank.

Calculations

$$\text{Units / ml enzyme} = \frac{(\Delta A275\text{nm / min Test} - \Delta A275\text{nm / min Blank})\ (3)\ (\text{df})}{(1.2)\ (0.1)}$$

3 = Volume (in milliliters) of assay
df = Dilution factor
1.2 = Millimolar extinction coefficient of Benzaldehyde at 275nm
0.1 = Volume (in milliliter) of enzyme used

$$\text{Units / mg protein} = \frac{\text{units / ml enzyme}}{\text{mg protein / ml enzyme}}$$

Unit Definition: *One unit will form 1.0 µmole of benzaldehyde and HCN from mandelonitrile per minute at pH 5.4 at 25°C.*

Final Assay Concentrations

In a 3.00 ml reaction mix, the final concentrations are 95 mM sodium acetate, 2.8 mM DL-mandelonitrile, 0.003% (w/v) Bovine Serum Albumin, 1.6% (v/v) ethanol, and 0.1–0.2 unit of mandelonitrile lyase.

Enzymatic Assay of MANNITOL DEHYDROGENASE (EC 1.1.1.67)

Principal

$$\text{D(-) Fructose} + \beta\text{-NADH} \xrightarrow{\text{Mannitol Dehydrogenase}} \text{D-Mannitol} + \beta\text{-NAD}$$

Abbreviations Used

β-NADH = β-Nicotinamide Adenine Dinucleotide, Reduced Form
β-NAD = β-Nicotinamide Adenine Dinucleotide, Oxidized Form

Conditions: T = 30°C, pH = 5.3, A340nm, Light path = 1 cm

Method: Continuous Spectrophotometric Rate Determination

Reagents

A. 50 mM Sodium Acetate Buffer, pH 5.3 at 30°C (Prepare 100 ml in deionized water using Sodium Acetate, Trihydrate. Adjust to pH 5.3 at 30°C with Acetic Acid, Glacial).
B. 1000 mM D(-) Fructose Solution (Fruc) (Prepare 10 ml in deionized water using D(-)Fructose).
C. 10 mM β-Nicotinamide Adenine Dinucleotide, Reduced Form Solution (β-NADH) (Prepare 1 ml in cold deionized water using β-Nicotinamide Adenine Dinucleotide, Reduced Form, Disodium Salt. *Prepare Fresh*).
D. Mannitol Dehydrogenase Enzyme Solution (Immediately before use, prepare a solution containing 0.1–0.5 unit/ml of Mannitol Dehydrogenase in cold Reagent A).

Procedure

Pipette (in milliliters) the following reagents into suitable cuvettes:

	Test	Blank
Deionized Water	0.630	0.630
Reagent A (Buffer)	0.400	0.400
Reagent C (β-NADH)	0.025	0.025

Mix by inversion and equilibrate to 30°C. Monitor the A340nm until constant, using a suitably thermostatted spectrophotometer. Then add:

	Test	Blank
Deionized Water	—	0.120
Reagent B (Fruc)	0.120	—
Reagent D (Enzyme Solution)	0.025	0.025

Immediately mix by inversion and record the decrease in A340nm for approximately 5 minutes. Obtain the ΔA340nm/minute using the maximum linear rate for both the Test and Blank.

Calculations

$$\text{Units / ml enzyme} = \frac{(\Delta A340\text{nm / min Test} - \Delta A340\text{nm / min Blank})\,(1.2)\,(\text{df})}{(6.22)\,(0.025)}$$

1.2 = Total volume (in milliliters) of assay
df = Dilution factor
6.22 = Millimolar extinction coefficient of β-NADH at 340nm
0.025 = Volume (in milliliter) of enzyme used

$$\text{Units / mg solid} = \frac{\text{units / ml enzyme}}{\text{mg solid / ml enzyme}}$$

$$\text{Units / mg protein} = \frac{\text{units / ml enzyme}}{\text{mg protein / ml enzyme}}$$

Unit Definition: *One unit will reduce 1.0 μmole of D(-)fructose per minute in the presence of β-NADH at pH 5.3 at 30°C.*

Final Assay Concentration

In a 1.20 ml reaction mix, the final concentrations are 18 mM sodium acetate, 100 mM D(-)fructose, 0.21 mM β-nicotinamide adenine dinculeotide, 0.0025–0.0125 unit mannitol dehydrogenase.

Enzymatic Assay of MANNITOL DEHYDROGENASE (EC 1.1.1.67) From *Actinobacillus sp.*

Principle

$$\text{D-Mannitol} + \beta\text{-NAD} \xrightarrow{\text{Mannitol Dehydrogenase}} \text{D-Fructose} + \beta\text{-NADH}$$

Abbreviations Used

β-NAD = β-Nicotinamide Adenine Dinucleotide, Oxidized Form
β-NADH = β-Nicotinamide Adenine Dinucleotide, Reduced Form

Conditions: T = 37°C, pH = 7.6, A340nm, Light path = 1 cm

Method: Continuous Spectrophotometric Rate Determination

Reagents

A. 60 mM Potassium Phosphate Buffer, pH 7.6 at 37°C (Prepare 100 ml in deionized water using Potassium Phosphate, Monobasic, Anhydrous. Adjust to pH 7.6 at 37°C with 1 M KOH).
B. 500 mM Mannitol Solution (Mannitol) (Prepare 5 ml in deionized water using D-Mannitol).
C. 5.0 mM β-Nicotinamide Adenine Dinucleotide, Oxidized Form, Solution (β-NAD) (Dissolve the contents of a 50 mg vial of β-Nicotinamide Adenine Dinucleotide in the appropriate volume of deionized water).
D. Mannitol Dehydrogenase Enzyme Solution (Immediately before use, prepare a solution containing 0.10–0.20 unit/ml of Mannitol Dehydrogenase in cold Reagent A).

Procedure

Pipette (in milliliters) the following reagents into suitable cuvettes:

	Test	Blank
Reagent A (Buffer)	1.00	1.00
Reagent B (Mannitol)	0.20	0.20
Reagent C (β-NAD)	1.00	1.00
Deionized Water	0.80	0.80

Mix by inversion and equilibrate to 37°C. Monitor the A340nm until constant, using a suitably thermostatted spectrophotometer. Then add:

Reagent D (Enzyme solution)	0.10	—
Reagent A (Buffer)	—	0.10

Immediately mix by inversion and record the increase in A340nm for approximately 5 minutes. Obtain the ΔA340nm/minute using the maximum linear rate for both the Test and Blank.

Calculations

$$\text{Units / ml enzyme} = \frac{(\Delta A340\text{nm / min Test} - \Delta A340\text{nm / min Blank})\,(3.1)\,(\text{df})}{(6.22)\,(0.1)}$$

3.1 = Total volume (in milliliters) of assay
df = Dilution factor
6.22 = Millimolar extinction coefficient of β-NADH at 340nm
0.1 = Volume (in milliliter) of enzyme used

$$\text{Units / mg solid} = \frac{\text{units / ml enzyme}}{\text{mg solid / ml enzyme}}$$

$$\text{Units / mg protein} = \frac{\text{units / ml enzyme}}{\text{mg protein / ml enzyme}}$$

Unit Definition: *One unit will convert 1.0 μmole of D-mannitol to D-fructose per minute at pH 7.6 at 37°C.*

Final Assay Concentrations

In a 3.10 ml reaction mix, the final concentrations are 21 mM potassium phosphate, 32 mM mannitol, 1.6 mM β-nicotinamide adenine dinucleotide and 0.01–0.02 unit mannitol dehydrogenase.

Enzymatic Assay of α-MANNOSIDASE (EC 3.2.1.24)

Principle

$$\text{p-Nitrophenyl-}\alpha\text{-D-Mannoside} + H_2O \xrightarrow{\alpha\text{-Mannosidase}} \text{D-Mannose} + \text{p-Nitrophenol}$$

Conditions: T = 25°C, pH = 4.5, A405nm, Light path = 1 cm.

Method: Spectrophotometric Stop Rate Determination

Reagents

A. 100 mM Citrate Buffer, pH 4.5 at 25°C (Prepare 100 ml in deionized water using Citric Acid, Free Acid, Monohydrate. Adjust to pH 4.5 at 25°C with 1 M NaOH).
B. 10 mM p-Nitrophenyl α-D-Mannoside Solution (PNP-Man) (Prepare 5 ml in deionized water using p-Nitrophenyl α-D-Mannopyranoside).
C. 200 mM Borate Buffer, pH 9.8 at 25°C (Prepare 100 ml in deionized water using Boric Acid. Adjust to pH 9.8 at 25°C with 1 M NaOH).
D. α-Mannosidase Enzyme Solution (Immediately before use, prepare a solution containing 0.05–0.1 unit/ml of α-Mannosidase in cold deionized water).

Procedure

Pipette (in milliliters) the following reagents into suitable cuvettes:

	Test	Blank
Deionized Water	—	0.10
Reagent A (Buffer)	0.50	0.50
Reagent B (PNP-Man)	0.50	0.50

Mix by inversion and equilibrate to 25°C. Then add:

Reagent D (Enzyme Solution)	0.10	—

Immediately mix by inversion and incubate at 25°C for exactly 5 minutes in a suitable spectrophotometer. Then add:

Reagent C (Borate)	2.00	2.00

Mix by inversion and record the A405nm/min for both the Test and Blank.

Calculations

$$\text{Units / mg enzyme} = \frac{(\text{A405nm / min Test} - \text{A405nm / min Blank})\ (3.1)}{(5)\ (18.5)\ (\text{mg enzyme / RM})}$$

3.1 = Total volume (in milliliters) of Solution
5 = Time of assay (Unit definition)
18.5 = Millimolar extinction coefficient of p-Nitrophenol at 405nm
RM = Reaction Mix

Unit Definition: *One unit will hydrolyze 1.0 μmole of p-nitrophenyl α-D-mannoside to p-nitrophenol and D-mannose per minute at pH 4.5 at 25°C.*

Final Assay Concentration

In a 1.10 ml reaction mix, the final concentrations are 45 mM citric acid, 4.5 mM p-nitrophenyl α-D-mannoside and 0.005–0.01 unit α-mannosidase.

Enzymatic Assay of β-MANNOSIDASE (EC 3.2.1.25)

Principle

$$\text{p-Nitrophenyl } \beta\text{-D-Mannoside} + H_2O \xrightarrow{\beta\text{-Mannosidase}} \text{D-Mannose} + \text{p-Nitrophenol}$$

Conditions: T = 25°C, pH = 4.0, A400nm, Light path = 1 cm.

Method: Spectrophotometric Stop Rate Determination

Reagents

A. 100 mM Citrate Buffer, pH 4.0 at 25°C. (Prepare 100 ml in deionized water using Citric Acid, Free Acid, Monohydrate. Adjust to pH 4.0 at 25°C with 1 M NaOH).
B. 10 mM p-Nitrophenyl β-D-Mannoside Solution (PNP-β-Man) (Prepare 5 ml in deionized water using p-Nitrophenyl β-D-Mannopyranoside).
C. 200 mM Borate Buffer, pH 9.8 at 25°C (Prepare 100 ml in deionized water using Boric Acid. Adjust to pH 9.8 at 25°C with 1 M NaOH).
D. β-Mannosidase Enzyme Solution (Immediately before use, prepare a solution containing 0.05–0.1 unit/ml of β-Mannosidase in cold deionized water).

Procedure

Pipette (in milliliters) the following reagents into suitable containers:

	Test	Blank
Reagent A (Buffer)	0.40	0.40
Reagent D (Enzyme Solution)	0.10	—

Mix by swirling and equilibrate to 25°C. Then add:

Reagent B (PNP-β-Man)	0.50	0.50

Immediately mix by swirling and incubate at 25°C for exactly 10 minutes. Then add:

Reagent C (Borate)	3.00	3.00
Reagent D (Enzyme Solution)	—	0.10

Mix by swirling and transfer to suitable cuvettes. Record the A400nm for both the Test and Blank using a suitable spectrophotometer.

Calculations

$$\text{Units / mg enzyme} = \frac{(\text{A400nm / min Test} - \text{A400nm / min Blank})\ (4)}{(10)\ (18)\ (\text{mg enzyme / RM})}$$

4 = Total volume (in milliliters) of Solution
10 = Time of assay (in minutes) as per the Unit Definition
18 = Millimolar extinction coefficient of p-Nitrophenol at 400nm
RM = Reaction Mix

Unit Definition: *One unit will hydrolyze 1.0 μmole of p-nitrophenyl β-D-mannopyranoside to p-nitrophenol and D-mannopyranoside per minute at pH 4.0 at 25°C.*

Final Assay Concentration

In a 1.00 ml reaction mix, the final concentrations are 40 mM citric acid, 5.0 mM p-nitrophenyl β-D-mannoside and 0.005–0.01 unit β-mannosidase.

Enzymatic Assay of METALLOENDOPEPTIDASE

Principle

$$\text{Casein} + H_2O \xrightarrow{\text{Metalloendopeptidase}} \text{Amino Acids}$$

Conditions: T = 30°C, pH 11.0, A275nm, Light path = 1 cm.

Method: Spectrophotometric Stopped Rate Determination

Reagents

A. 100 mM Sodium Tetraborate Solution (Prepare 10 ml in deionized water using Borax).

B 0.6% (w/v) Casein with 10.0 mM Sodium Tetraborate, pH 11.0 at 30°C (Casein). Prepare by dissolving 600 mg of Casein, in 4 ml of 0.1 M NaOH in a hot water bath (60°C). Cool to room temperature and then add 10 ml of Reagent A and 80 ml of deionized water. Adjust to pH 11.0 at 30°C with 1 M NaOH and bring to a volume of 100 ml with deionized water).

C. 110 mM Trichloroacetic Acid Solution with 220 mM Sodium Acetate and 330 mM Acetic Acid (TCA) (Prepare 40 ml in deionized water using Trichloroacetic Acid, 6.1 N, approximately 100% (w/v), Sodium Acetate, Trihydrate and Acetic Acid, Glacial).

D. 50 mM Potassium Phosphate Buffer, pH 7.0 at 30°C (Enz Dil) (Prepare 25 ml in deionized water using Potassiums Phosphate, Monobasic, Anhydrous, Adjust to pH 7.0 at 30°C with 1 M KOH).

E. Metalloendopeptidase Enzyme Solution (Immediately before use, prepare a solution containing approximately 0.25 unit/ml of Metalloendopeptidase in cold Reagent D).

Procedure

Pipette (in milliliters) the following reagents into suitable containers:

	Test	Blank
Reagent B (Casein)	3.00	3.00

Equilibrate to 30°C. Then add:

Reagent E (Enzyme Solution)	0.10	—

Immediately mix by swirling and incubate at 30°C for exactly 3 minutes. Then add:

Reagent C (TCA)	3.20	3.20
Reagent E (Enzyme Solution)	—	0.10

Mix by swirling and incubate at 37°C for 20 minutes. Filter the solutions through Whatman #50 filter paper. Transfer the solutions to suitable cuvettes and record the A275nm for both the Test and Blank using a suitable spectrophotometer.

Calculations

$$\text{Units / ml enzyme} = \frac{(\text{A275nm Test} - \text{A275nm Blank})\,(6.3)\,(\text{df})}{(3)\,(1.34)\,(0.1)}$$

6.3 = Total volume (in milliliters) of the stopped reaction.
df = Dilution factor.
3 = Time (in minutes) of assay per the Unit Definition.
1.34 = Millimolar extinction coefficient of tyrosine under the conditions of this assay.
0.1 = Volume (in milliliter) of enzyme used.

Unit Definition: *One unit will hydrolyze casein to produce peptide equivalent to 1.0 μmole of tyrosine per minute at pH 11.0 at 30°C.*

Final Assay Concentration

In a 3.10 ml reaction mix, the final concentrations are 0.58% (w/v) casein, 9.7 mM sodium tetraborate, 1.6 mM potassium phosphate, and 0.025 unit metalloendopeptidase.

Note

1. The assay is also run for 2 minutes. The assay time which gives the most linear response is used in the calculation.

Enzymatic Assay of L-METHIONINE GAMMA-LYASE (EC 4.4.1.11)

Principle

$$\text{L-Methionine} \xrightarrow{\text{L-Methionine Gamma-Lyase}} \text{Methanethiol} + \text{2-Ketobutyrate} + NH_3$$

$$\text{2-Ketobutyrate} + \text{MBTH} \rightarrow \text{Azine Derivative}$$

Abbreviation Used

MBTH = 3-Methyl-2-Benzothiazolinone

Conditions: T = 37°C, pH = 8.0, A320nm, Light path = 1 cm

Method: Stopped Spectrophotometric Rate Determination

Reagents

A. 100 mM Potassium Phosphate Buffer with 25 mM L-Methionine and 0.01 mM Pyridoxal 5-Phosphate, pH 8.0 at 37°C (Prepare 100 ml in deionized water using Potassium Phosphate, Monobasic, Anhydrous, L-Methionine and Pyridoxal 5-Phosphate. Adjust to pH 8.0 at 37°C with 5 M KOH).

B. 50% (w/v) Trichloroacetic Acid Solution (TCA) (Prepare 5 ml in deionized water using Trichloroacetic Acid, 6.1 N Solution, approximately 100% (w/v)).

C. 1 M Sodium Acetate Buffer, pH 5.0 at 37°C (NaOAc) (Prepare 100 ml in deionized water using Sodium Acetate, Trihydrate. Adjust to pH 5.0 at 37°C with 5 M HCl).

D. 0.1% (w/v) 3-Methyl-2-Benzothiazolinone Hydrazone (MBTH) (Prepare 10 ml in deionized water using 3-Methyl-2-Benzothiazolinone Hydrazone, Hydrochloride Hydrate).

E. 100 mM Potassium Phosphate Buffer with 1 mM Ethylenediaminetetraacetic Acid, 0.01% (v/v) 2-Mercaptoethanol and 0.02 mM Pyridoxal 5-Phosphate, pH 7.2 at 37°C (Enzyme Diluent) (Prepare 10 ml in deionized water using Potassium Phosphate, Monobasic, Anhydrous, 2-Mercaptoethanol, Ethylenediaminete-traacetic Acid, Tetrasodium Salt, Hydrate, and Pyridoxal 5-Phosphate. Adjust to pH 7.2 at 37°C with 5 M KOH. Store on ice).

F. L-Methionine Gamma-Lyase Enzyme Solution (Immediately before use, prepare a solution containing 0.08–0.4 unit/ml of L-Methionine Gamma-Lyase in ice-cold Reagent E).

Procedure

Step 1: Pipette (in milliliters) the following reagents into suitable cuvettes:

	Test	Blank
Reagent A (Buffer)	2.00	2.00

Equilibrate to 37°C. Then add:

	Test	Blank
Reagent F (Enzyme Solution)	0.02	—
Reagent E (Enzyme Diluent)	—	0.02

Mix by inversion and incubate at 37°C for exactly 10 minutes. Then add:

	Test	Blank
Reagent B (TCA)	0.25	0.25

Mix by inversion.

Step 2: Pipette (in milliliters) the following reagents into suitable container:

	Test	Blank
Reagent C (NaOAC)	2.00	2.00
Test Reaction Mix from Step 1	1.00	—
Blank Reaction Mix from Step 1	—	1.00
Reagent D (MBTH)	0.80	0.80

Mix by inversion and incubate at 50°C for exactly 30 minutes. Then incubate at 25°C for 30 minutes. Transfer the Test and Blank solutions to suitable cuvettes and record the A320nm using a suitable spectrophotometer.

Calculations

$$\text{Units / vial enzyme} = \frac{(\text{A320nm Test} - \text{A320nm Blank})\,(2.27)\,(3.8)\,(\text{df})}{(15.74)\,(10)\,(0.02)\,(1)}$$

2.27 = Volume (in milliliters) of assay in Step 1
3.8 = Volume (in milliliters) of assay in Step 2
df = Dilution factor
15.74 = Millimolar extinction coefficient of the azine derivative
10 = Time (in minutes) of assay as per the Unit Definition
0.02 = Volume (in milliliter) of enzyme used in Step 1
1 = Volume (in milliliter) of Step 1 used in Step 2

Unit Definition: *One unit will release 1.0 μM of α-ketobutyrate from L-methionine per minute at pH 8.0 at 37°C.*

Reference

In a 2.02 ml reaction mix, the final concentrations are 100 mM potassium phosphate, 25 mM L-methionine, 0.01 mM pyridoxal 5-phosphate, 0.0001% (v/v) 2-mercaptoethanol, 0.01 mM Ethylenediaminetetraacetic Acid, and 0.0016–0.008 unit L-methionine gamma-lyase.

Note

1. This reagent may be prepared by first making stock solutions of the separate components and then combining them. This is necessary since extremely small weigh-ups are required.

Enzymatic Assay of 5,10-METHYLENETETRAHYDROFOLATE DEHYDROGENASE (EC 1.5.1.5)

Principle

$$\text{5,10-Methylene-FH}_4 + \beta\text{-NADP} \xrightarrow{\text{5,10-MeFH}_4\text{DH}} \text{5,10-Methenyl-FH}_4 + \beta\text{-NADPH}$$

Abbreviations Used

5,10-Methylene-FH_4 = 5,10-Methylenetetrahydrofolate
5,10-Methenyl-FH_4 = 5,10-Methenyltetrahydrofolate
5,10-MeFH_4DH = 5,10-Methylenetetrahydrofolate Dehydrogenase
β-NADP = β-Nicotinamide Adenine Dinucleotide Phosphate, Oxidized Form
β-NADPH = β-Nicotinamide Adenine Dinucleotide Phosphate, Reduced Form

Conditions: T = 25°C, pH = 7.5, A340nm, Light path = 1 cm

Method: Continuous Spectrophotometric Rate Determination

Reagents

A. 50 mM Potassium Phosphate Buffer with 100 mM Potassium Chloride, pH 7.5 at 25°C (Prepare 200 ml in deionized water using Potassium Phosphate, Monobasic, Anhydrous and Potassium Chloride. Adjust to pH 7.5 at 25°C with 1 M NaOH).
B. 0.002% (w/v) Tetrahydrofolic Acid with 0.002% (v/v) Formaldehyde and 0.1% (v/v) 2-Mercaptoethanol Solution (FH_4) (Immediately before used, prepare 100 ml

in Reagent A using Tetrahydrofolic Acid, Formaldehyde, 37% Solution and 2-Mercaptoethanol).

C. 20 mM β-Nicotinamide Adenine Dinucleotide Phosphate (β-NADP) (Prepare 2 ml in deionized water using β-Nicotinamide Adenine Dinucleotide Phosphate, Sodium Salt. *Prepare Fresh*).

D. 5,10-Methylenetetrahydrofolate Dehydrogenase Enzyme Solution (Immediately before use, prepare a solution containing 0.01–0.03 unit/ml of 5,10-Methylenetetrahydrofolate Dehydrogenase in cold Reagent A).

Procedure

Pipette (in milliliters) the following reagents into suitable cuvettes:

	Test	Blank
Reagent B (FH_4)	2.80	2.80
Reagent D (Enzyme Solution)	0.10	0.10

Mix by inversion and equilibrate to 25°C. Monitor the A340nm until constant using a suitably thermostatted spectrophotometer. Then add:

Reagent C (β-NADP)	0.10	—
Deionized Water	—	0.10

Immediately mix by inversion and record the increase in A340nm for approximately 5 minutes. Obtain the ΔA340nm/minute using the maximum linear rate for both the Test and Blank.

Calculations

$$\text{Units / mg enzyme} = \frac{\Delta A340\text{nm / min Test} - \Delta A340\text{nm / min Blank}}{(7.1)\ (\text{mg enzyme / ml RM})}$$

7.1 = Millimolar extinction coefficient of β-NADPH at 340nm under the conditions of the assay.
RM = Reaction Mix

Unit Definition: *One unit will convert 1.0 μmole 5,10-methylenetetrahydrofolate and NADP to 5,10-methenyltetrahydrofolate and NADPH per minute at pH 7.5 at 25°C.*

Final Assay Concentrations

In a 3.00 ml reaction mix, the final concentrations are 48 mM potassium phosphate, 97 mM potassium chloride, 0.002% (v/v) formaldehyde, 0.09% (v/v) 2-mercaptoethanol, 0.002% (w/v) tetrahydrofolic acid, 0.67 mM β-NADP and 0.001–0.003 unit/ml 5,10-methylenetetrahydrofolate dehydrogenase.

Notes

1. Weigh the Tetrahydrofolate in the dark. Allow the solution to stand at room temperature, protected from light, for 5 minutes to generate 5,10-Methylenetetrahydrofolic acid.
2. The millimolar extinction coefficient of NADP under the conditions of this assay is 7.1.

Determination of the Concentration and Molecular Weight of METHYLGLYOXAL

Principle

$$\text{Methylglyoxal} + \text{GSH} \xrightarrow{\text{Glyoxalase I}} \text{S-Lactoylglutathione}$$

Abbreviation Used

GSH = Reduced Glutathione

Conditions: T = 25°C, pH 6.6, A240nm, Light path = 1 cm

Method: Spectrophotometric

Reagents

A. 1 M Potassium Phosphate Buffer, pH 6.6 at 25°C (Prepare 100 ml in deionized water using Potassium Phosphate, Monobasic, Anhydrous. Adjust to pH 6.6 with 5 M KOH).
B. 0.92 mM Methylglyoxal Solution (Meth-Glyox) (Immediately before use, prepare using deionized water).
C. 2.0% (w/v) Reduced Glutathione Solution, pH 6.6 at 25°C (GSH) (Prepare 10 ml in deionized water using Glutathione, Reduced Form, Free Acid. Adjust to pH 6.6 at 25°C with solid Sodium Bicarbonate).
D. 10 mM Potassium Phosphate Buffer with 1% (w/v) Bovine Serum Albumin, pH 7.4 at 25°C (Enzyme Diluent) (Prepare 10 ml in deionized water using Potassium Phosphate, Monobasic, Anhydrous and Albumin, Bovine or equivalent. Adjust to pH 7.4 at 25°C with 1 M KOH).
E. Glyoxalase I Enzyme Solution (Glyoxalase I) (Immediately before use, prepare a solution containing 2.5 units/ml of Glyoxalase I, in Reagent D).

Procedure

Pipette (in milliliters) the following reagents into suitable cuvettes:

	Test	Blank
Deionized Water	2.25	2.75
Reagent A (Buffer)	0.10	0.10
Reagent B (Meth-Glyox)	0.50	—
Reagent C (GSH)	0.05	0.05

Mix by inversion and equilibrate for 5 minutes at 25°C. Record the initial A240nm for both the Test and Blank using a suitably thermostatted spectrophotometer. Then add:

Reagent E (Glyoxalase I)	0.10	0.10

Mix by inversion and allow the reaction to proceed for 10 minutes. Record the final ΔA240nm for both the Test and Blank.

Calculations

$$\Delta A240nm = Af\ 240nm - Ai\ 240nm \quad Ai = \text{Initial Absorbance}$$

Af = Final Absorbance

$$\text{Micromoles Methylglyoxal / weighed sample} = \frac{(\Delta A)\ (3.00)\ (df)}{(3.37)\ (0.5)}$$

3.00 = Total volume of Reaction Mix
df = Dilution factor
3.37 = Millimolar extinction coefficient of S-Lactoylglutathione at 240nm
0.5 = Volume of Methyglyoxal used in assay

$$\text{Apparent molecular weight} = \frac{\text{mg sample weighed} \times 1000}{\mu\text{moles Methylglyoxal / weighed sample}}$$

$$\text{\% Methylglyoxal in sample} = \frac{\text{theoretical molecular weight}}{\text{apparent molecular weight}}$$

Final Assay Concentrations

In a 3.0 ml reaction mix, the final concentrations are 34 mM potassium phosphate, 0.033% (w/v) reduced glutathione, 0.03% (w/v) Bovine Serum Albumin, and 0.25 unit of glyoxalase I.

Enzymatic Assay of MICROPEROXIDASE Suitability Assay

Principle

$$H_2O_2 + \text{Mesidine (donor)} \xrightarrow{\text{Microperoxidase}} 2H_2O + \text{Mesidine (oxidized donor)}$$

Conditions: T = 10°C, pH = 4.9, A490nm, Light path = 1 cm

Method: Continuous Spectrophotometric Rate Determination

Reagents

A. 100 mM Sodium Acetate Buffer, pH 4.9 at 10°C (Prepare 100 ml in deionized water using Sodium Acetate, Trihydrate. Adjust to pH 4.9 at 10°C with 1 M HCl. Store on ice).

B. 0.34% (w/w) Hydrogen Peroxide Solution (H_2O_2) (Prepare 10 ml in deionized water using Hydrogen Peroxide, 30% (w/w) Solution. *Prepare Fresh.* Store on ice).

C. 200 mM Mesidine HCl Solution (Mesidine) (Prepare 100 ml by dissolving 2.75 g of 2,4,6-Trimethylaniline in 25 ml of 1 M HCl. Dilute to 100 ml with deionized water. Store on ice).

D. Microperoxidase Solution (Immediately before use, prepare a solution containing 0.4 mg/ml of Microperoxidase in cold deionized water. Store on ice).

Procedure

Pipette (in milliliters) the following reagents into 4 dram vials:

	Test	Blank
Reagent A (Buffer)	7.00	7.00
Reagent B (H_2O_2)	0.40	0.40
Reagent C (Mesidine)	0.60	0.60

Mix by inversion and place on ice for several minutes. Place empty cuvettes into a thermostatted spectrophotometer and equilibrate to 10°C. Add 0.20 ml of Reagent D (Microperoxidase) to the Test and 0.20 ml of cold deionized water to the Blank 4 dram vials. Immediately mix by inversion and remove 3 ml from the 4 dram vials and place into the appropriate cuvettes in the spectrophotometer. Record the increase in A490nm for approximately 20 minutes. Obtain the ΔA490nm/minute using the maximum linear rate for both the Test and Blank.

Calculations

$$\Delta A490nm / min / \mu M = \frac{(\Delta A490nm\ Test - \Delta A490nm\ Blank)}{\mu M}$$

$$\mu M = \frac{mg / ml}{Molecular\ Weight} \times 1000$$

mg = Weight of the Microperoxidase sample used in the assay

Final Assay Concentrations

In a 8.20 ml reaction mix, the final concentrations are 85 mM sodium acetate, 0.017% (w/w) hydrogen peroxide, 15 mM mesidine, and 0.08 mg microperoxidase.

Notes

1. Ensure that no condensation forms on the cuvette walls.
2. The molecular weight of the Microperoxidase that is used in the calculation should include the water and salt content.

Enzymatic Assay of MUTANOLYSIN

Principal

$$Streptococcus\ faecalis\ \text{Cell Walls (Insoluble)} \xrightarrow{\text{Mutanolysin}} \text{Soluble Products}$$

Conditions: T = 37°C, pH = 6.0, A600nm, Light Path = 1 cm

Method: Continuous Spectrophotometric Rate Determination

Reagents

A. 50 mM MES Buffer with 1 mM Magnesium Chloride pH 6.0 at 37°C (Assay Buffer) (Prepare 50 ml in deionized water using MES and Magnesium Chloride, Hexahydrate. Adjust to pH 6.0 at 37°C using 1 M NaOH).

B. 50 mM TES with 1 mM Magnesium Chloride pH 7.0 at 37°C. (Enzyme Diluent) (Prepare 50 ml in deionized water using TES and Magnesium Chloride, Hexahydrate. Adjust to pH 7.0 at 37°C using 1 M NaOH).

C. *Streptococcus faecalis* Cell Wall Suspension (Substrate Solution) (A cell wall suspension of log-phase *Streptococcus faecalis* STF-3 (ATCC 12784) cells in Reagent A are diluted to an A600nm of 0.48 to 0.52 using Reagent A).

D. Mutanolysin Enzyme Solution (Immediately before use, prepare a solution containing 100–200 units per ml in cold Reagent B).

Procedure

Pipette (in milliliters) the following reagents into suitable cuvettes:

	Test	Blank
Reagent C (Substrate Solution)	3.00	3.00

Equilibrate at 37°C and monitor the A600nm until constant using a suitably thermostatted spectrophotometer. Then add:

Reagent D (Enzyme Solution)	0.05	—
Reagent B (Enzyme Diluent)	—	0.05

Immediately mix by inversion and record the decrease in A600nm for 20 minutes. Obtain the ΔA600nm/min using the maximum linear rate for both the Test and Blank.

Calculation

$$\text{Units / ml enzyme} = \frac{(\Delta A600\text{nm / min Test} - \Delta A600\text{ nm / min Blank})\ (3.05)\ (\text{df})}{(0.01)\ (0.05)}$$

3.05 = Volume (in milliliters) of reaction mix
df = Dilution factor
0.01 = Decrease in absorbance per minute as defined by the unit definition
0.05 = Volume (in milliliter) of enzyme used

$$\text{Units / mg solid} = \frac{\text{units / ml enzyme}}{\text{mg solid / ml enzyme}}$$

$$\text{Units / mg protein} = \frac{\text{units / ml enzyme}}{\text{mg protein / ml enzyme}}$$

Unit Definition: *One unit per ml will produce a A600nm of 0.01 per minute at pH 6.0 at 37°C using a suspension of Streptococcus faecalis cell walls as substrate.*

Final Assay Concentrations

In a 3.05 ml reaction mix, the final concentrations are 49 mM MES, 1 mM $MgCl_2$, 0.82 mM TES, and 5–10 units mutanolysin.

Enzymatic Assay of MUTAROTASE (EC 5.1.3.3)

Principle

$$\alpha\text{-D(+)-Glucose} + H_2O \xleftrightarrow{\text{Mutarotase}} \beta\text{-D(+)-Glucose}$$

Conditions: T = 25°C, pH = 7.4, 589nm

Method: Continuous Polarimetric

Reagents

A. 5 mM Ethylenediaminetetraacetic Acid Solution, pH 7.4 at 25°C (EDTA) (Prepare 500 ml in deionized water using Ethylenediaminetetraacetic Acid, Tetrasodium Salt, Hydrate. Adjust to pH 7.4 at 25°C with 1 M NaOH).

B. α-D-(+)-Glucose Tube (Glucose Tube) (Weigh 100 mg into suitable tubes using D-(+)-Glucose, Anhydrous. *Note*: This product contains 6–7% of the (β-) anomer).

C. Mutarotase Enzyme Solution (Immediately before use, prepare a solution containing 125–250 units/ml of Mutarotase in cold Reagent A).

Procedure

Spontaneous Optical Rotation Determination Using a suitable polarimeter: Take a 10 cm cell at 589nm (Na Lamp) at 25°C. At zero time add 10.00 ml of Reagent A (EDTA) to Reagent B (Glucose Tube) and dissolve rapidly by agitation. Transfer to the polarimeter cell and begin to record the rotation at 1.0 minute intervals for the first 10 minutes (earlest reading should be taken 3.5 minutes after the addition of Reagent A), at 5 minute intervals for the second 10 minutes and at 15 minute intervals until the rotation is constant.

Test Optical Rotation Determination Using a suitable polarimeter: Take a 10 cm cell at 589 nm (Na Lamp) at 25°C. Add 9.90 ml of Reagent A (EDTA) to Reagent B (Glucose Tube) and dissolve rapidly by agitation. At zero time, add 0.10 ml of Reagent C (Enzyme Solution). Mix by swirling and transfer to the polarimeter cell as rapidly as possible. Begin to record the rotation at 30 second intervals for 10 minutes.

Calculations

Graphs

1. Spontaneous Rotation Plot using the spontaneous rotation data, plot the rotation vs. time in minutes on semi-log paper. Extrapolate to zero time to obtain the initial rotation. The initial rotation corresponds to an α-D(+)-Glucose concentration of 555 µmoles per reaction mixture.
2. Blank Rotation Plot using the spontaneous rotation data, plot the rotation, vs time in 1.0 minute intervals. Determine the change in rotation for 5.0 minutes.
3. Standard Curve using the spontaneous rotation data, plot the rotation, vs µmoles α-D(+)-glucose. This plot consists of two points; the rotation at 555 µmoles (Initial Rotation) and the rotation at 195 µmoles (Final Rotation).
4. Test Plot Using the test rotation data, plot the rotation, vs time in minutes and determine the change in rotation for 5 minutes.

Calculations

Obtain the following values:

A = Initial rotation from Graph #1

B = Δ Rotation per 5.0 minutes from Graph #2

C = Blank rotation per 5.0 minutes, C = A–B

D = Conversion of C to µmoles α-D-(+)-Glucose from graph #3

E = Concentration of α-D(+)-Glucose per reaction mix at Initial Rotation, E = 555 µmole

F = Spontaneous or blank rate in µmoles α-D-(+)-Glucose/min,

F = E–D, 5.0 minutes

G = Test Rotation 5.0 minutes from Graph #4

H = Test Rotation after 5.0 minutes, H = A–G

I = Conversion of H to µmoles α-D-(+)-Glucose from Graph #3

J = Test rate in µmoles α-D-(+)-Glucose/min, J = E–I – F, 5.0 minutes

$$\text{Units / ml enzyme} = \frac{J}{(5)\,(\text{mg enzyme / RM})}$$

RM = Reaction Mix

5 = Time of assay (in minutes) as per unit definition

Unit Definition: *One unit will increase the spontaneous mutarotation of α-D-glucose to β-D-glucose by 1.0 μmole per minute at pH 7.4 at 25°C.*

Final Assay Concentrations

In a 10.00 ml reaction mix, the final concentrations are 5 mM EDTA, 1% (w/v) α-D-(+)-glucose and 12.5–25 units mutarotase.

Note

1. Average time for equilibration is 3.5–4 hours and will be at +53°C.

Enzymatic Assay of MYCODEXTRANASE (EC 3.2.1.61)

Principle

$$\text{Nigeran} \xrightarrow{\text{Mycodextranase}} \text{Nigerose} + 4\text{-}\alpha\text{-D-Nigerosylglucose}$$

Conditions: T = 37°C, pH = 4.5, A540nm, Light path = 1 cm

Method: Colorimetric

Reagents

A. 100 mM Sodium Acetate Buffer, pH 4.5 at 37°C (Prepare 100 ml in deionized water using Sodium Acetate, Trihydrate. Adjust to pH 4.5 at 37°C with 1 M HCl).

B. 1.25% (w/v) Nigeran Solution (Nigeran) (Immediately before use, prepare 5 ml in boiling deionized water using Nigeran. Continue boiling until dissolved. Allow to cool to 37°C before using in the assay).

C. Mycodextranase Enzyme Solution (Immediately before use, prepare a solution containing 0.5–1.0 unit/ml of Mycodextranase in cold deionized water).

D. 16 mM Copper Sulfate, 1300 mM Sodium Sulfate, 226 mM Sodium Carbonate, 190 mM Sodium Bicarbonate, and 43 mM Sodium Potassium Tartrate Solution (Copper Soln) (Prepare 1 liter in deionized water using Cupric Sulfate Pentahydrate, Sodium Bicarbonate, Sodium Sulfate, Anhydrous, Sodium Carbonate, Anhydrous and Sodium Potassium Tartrate Tetrahydrate).

E. 40 mM Molybdic Acid, 19 mM Arsenic Acid, and 756 mM Sulfuric Acid Solution (Ars-Mol Soln) (Prepare 1 liter in deionized water using Molybdic Acid, Ammonium Salt Tetrahydrate, Arsenic Acid, Sodium Salt, and Sulfuric Acid).

F. Glucose Standard Solution (Glucose).

Procedure

Step 1: Pipette (in milliliters) the following reagents into suitable test tubes:

	Test	Blank
Reagent A (Buffer)	1.00	1.00
Deionized Water	0.80	0.90
Reagent C (Enzyme Solution)	0.10	—

Mix by swirling and equilibrate to 37°C. Then add:

	Test	Blank
Reagent B (Nigeran)	0.10	0.10

Immediately mix by swirling and incubate at 37° for exactly 10 minutes.

Step 2: Immediately transfer 1 ml of the reaction mixture into a suitable container containing 1 ml of Reagent D (Copper Soln) as indicated below. Pipette (in milliliters) the following reagents into suitable containers:

	Test	Test Blank	Std 1	Std 2	Std 3	Std 4	Std 5	Std Blank
Test Solution	1.00	—	—	—	—	—	—	—
Test Blank Solution	—	1.00	—	—	—	—	—	—
Deionized Water	—	—	0.97	0.95	0.93	0.90	0.80	1.00
Reagent F (Glucose)	—	—	0.03	0.05	0.07	0.10	0.20	—
Reagent D (Copper Soln)	1.00	1.00	1.00	1.00	1.00	1.00	1.00	1.00

Immediately mix by inversion. Place a marble over the top of the tubes and transfer the tubes to a boiling water bath. Incubate for 20 minutes. Remove from the boiling water bath and allow to cool to room temperature. Then add:

Reagent E (Ars-Mol Soln)	1.00	1.00	1.00	1.00	1.00	1.00	1.00	1.00

Shake or vortex the tubes until foaming stops and any precipitate present is dissolved. Then add:

Deionized Water	10.00	10.00	10.00	10.00	10.00	10.00	10.00	10.00

Mix and transfer to suitable cuvettes. Obtain the A540nm for Test, Blank and Standards, using a suitable spectrophotometer.

Calculations

Standard Curve: ΔA540nm Std = A540nm Std–A540nm Blank
Prepare a standard curve by plotting the ΔA540nm of the Standard versus the μmoles of glucose liberated.

Sample Determination: ΔA540nm Sample = A540nm Test–A540nm Test Blank
Determine the μmoles of glucose (reducing equivalent) liberated using the Standard curve.

$$\text{Units / ml enzyme} = \frac{(\mu\text{moles of glucose liberated})\ (2)}{(10)\ (0.1)}$$

2 = Volume (in milliliters) of assay
10 = Time of assay (in minutes) as per the Unit Definition
0.1 = Volume (in milliliter) of enzyme used

$$\text{Units / mg solid} = \frac{\text{units / ml enzyme}}{\text{mg solid / ml enzyme}}$$

$$\text{Units / mg protein} = \frac{\text{units / ml enzyme}}{\text{mg protein / ml enzyme}}$$

Unit Definition: *One unit will liberate 1.0 μmole of reducing sugar (measured as glucose) from nigeran per minute at pH 4.5 at 37°C.*

Final Assay Concentrations

In a 2.00 ml reaction mix, the final concentrations are 50 mM sodium acetate, 0.063% (w/v) nigeran, and 0.05–0.10 unit mycodextranase.

Notes

1. Sodium Sulfate, Sodium Carbonate, and Sodium Potassium Tartrate are dissolved in approximately 500 ml of deionized water. Cupric Sulfate is dissolved in approximately 100 ml of deionized water and slowly added to the above solution to avoid precipitation. Sodium Bicarbonate is dissolved first in deionized water and then added to the above solution. Dilute the solution to 1 liter. If a precipitate forms, it should be removed by filtration prior to use. Store in an amber bottle and avoid exposure to direct sunlight. Store at room temperature.
2. Molybdic Acid is dissolved in approximately 300 ml of deionized water. Add Sulfuric Acid slowly. Caution, this is an exothermic reaction! A solution of Arsenic acid is dissolved in approximately 300 ml of deionized water and is added to the above solution. The solution is diluted to a total volume of 1 liter and incubated at 37°C for 48–72 hours. If a precipitate forms, it should be removed by filtration prior to use. Store in an amber bottle and avoid exposure to direct sunlight. The solution expires six months after preparation. Store at room temperature in an exhaust hood.

Enzymatic Assay of MYELOPEROXIDASE (EC 1.11.1.7)

Principle

$$2\ \text{Guaiacol} + H_2O_2 \xrightarrow{\text{Myeloperoxidase}} 2\ \text{Oxidized Guaiacol} + 2H_2O$$

Conditions: T = 25°C, pH = 7.0, A470nm, Light path = 1 cm

Method: Continuous Spectrophotometric Rate Determination

Reagents

A. 50 mM Potassium Phosphate Buffer with 100 mM Guaiacol and 0.0017% (w/w) Hydrogen Peroxide, pH 7.0 at 25°C (Prepare 100 ml in deionized water using Potassium Phosphate, Monobasic, Anhydrous, Guaiacol, and Hydrogen Peroxide, 30% (w/w) Solution. Adjust to pH 7.0 at 25°C with 1 M KOH).

B. Myeloperoxidase Enzyme Solution (Immediately before use, prepare a solution containing 10 units/ml of Myeloperoxidase in cold deionized water).

Procedure

Pipette (in milliliters) the following reagents into suitable cuvettes:

	Test	Blank
Reagent A (Buffer)	3.00	3.00

Equilibrate to 25°C. Monitor the A470nm until constant, using a suitably thermostatted spectrophotometer. Then add:

Deionized water	—	0.035
Reagent B (Enzyme)	0.035	—

Immediately mix by inversion. At exactly one minute record the ΔA470nm for both the Test and Blank.

Calculations

$$\text{Units / ml enzyme} = \frac{(\Delta A470\text{nm Test at 1 min} - \Delta A470\text{ nm Blank at 1 min})\ (df)}{(1.0)\ (0.035)}$$

df = Dilution factor
1.0 = The increase in A470nm/minute per unit of enzyme
0.035 = Volume (in milliliter) of enzyme used

$$\text{Units / mg protein} = \frac{\text{units / ml enzyme}}{\text{mg protein / ml enzyme}}$$

Unit Definition: *One unit will produce an increase in absorbance (470nm) of 1.0 per minute at pH 7.0 and 25°C, calculated from the initial rate of reaction using guaiacol as substrate. Total reaction volume: 3.035 ml.*

Final Assay Concentrations

In a 3.035 ml reaction mix, the final concentrations are 49 mM potassium phosphate, 99 mM guaiacol, 0.0017% (w/w) hydrogen peroxide, and 0.35 unit myeloperoxidase.

Notes

1. Guaiacol will require approximately 30 minutes of stirring for dissolution. The (w/w) concentration of the Hydrogen Peroxide is calculated by diluting the Hydrogen Peroxide, 30% (w/w) Solution.
2. The calculations are based on a final reaction mix volume of 3.035 ml. If the final volume of the reaction mix is not 3.035 ml, a correction factor must be used to determine the correct activity.

Enzymatic Assay of MYOKINASE (EC 2.7.4.3)

Principle

$$2\ \text{ADP} + H_2O \xrightarrow{\text{Myokinase}} \text{AMP} + \text{ATP}$$

$$\beta\text{-D (+) Glucose} + \text{ATP} \xrightarrow{\text{Hexokinase}} \text{G-6-P} + \text{ADP}$$

$$\text{G-6-P} + \beta\text{-NADP} \xrightarrow{\text{G-6-PDH}} \text{6-Phosphoglucose-Lactone} + \beta\text{-NADPH}$$

Abbreviations Used

ADP = Adenosine 5'-Diphosphate
AMP = Adenosine 5'-Monophosphate
ATP = Adenosine 5'-Triphosphate
G-6-P = Glucose 6'-Phosphate
G-6-PDH = Glucose-6-Phosphate Dehydrogenase
β-NADP = β-Nicotinamide Adenine Dinucleotide Phosphate, Oxidized Form
β-NADPH = β-Nicotinamide Adenine Dinucleotide Phosphate, Reduced Form

Conditions: T= 37°C, pH 7.6, A340nm, Light path = 1 cm

Method: Continuous Spectrophotometric Rate Determination

Reagents

A. 250 mM Glycylglycine Buffer, pH 7.6 at 37°C (Prepare 100 ml in deionized water using Gly-Gly, Free Base. Adjust to pH 7.6 at 37°C with 1 M NaOH).

B. 40 mM Adenosine Diphosphate Solution (ADP) (Prepare 2 ml in deionized water using Adenosine 5'-Diphosphate, Sodium Salt).

C. 20 mM β-Nicotinamide Adenine Dinucleotide Phosphate, Oxidized Form Solution (β-NADP) (Prepare 5 ml in deionized water using β-Nicotinamide Adenine Dinucleotide Phosphate, Sodium Salt. *Prepare Fresh*).

D. 300 mM Magnesium Chloride Solution ($MgCl_2$) (Prepare 2.4 ml in deionized water using Magnesium Chloride, Hexahydrate).

E. 1000 mM β-D(+) Glucose Solution (Glucose) (Prepare 1 ml in deionized water using β-D(+)Glucose).

F. Hexokinase and Glucose-6-Phosphate Dehydrogenase Solution (G-6-PDH/Hex) (Immediately before use, prepare a solution containing 10 units/ml of Glucose-6-Phosphate Dehydrogenase activity and 20 units/ml of Hexokinase activity in cold deionized water using Hexokinase and Glucose-6-Phosphate Dehydrogenase (G-6-PDH)).

G. 0.1% (w/v) Bovine Serum Albumin Solution (BSA) (Prepare 100 ml in Reagent A using Albumin, Bovine or equivalent.

H. Myokinase Enzyme Solution (Myokinase) (Immediately before use, prepare a solution containing 0.2–0.6 unit/ml of Myokinase in cold Reagent G).

Procedure

Pipette (in milliliters) the following reagents into suitable cuvettes:

	Test	Blank
Deionized Water	1.57	1.57
Reagent A (Buffer)	0.60	0.60
Reagent B (ADP)	0.15	0.15
Reagent C (β-NADP)	0.35	0.35
Reagent D ($MgCl_2$)	0.10	0.10
Reagent E (Glucose)	0.03	0.03
Reagent F (G-6-PDH/Hex)	0.10	0.10

Mix by inversion and equilibrate to 37°C. Monitor the A340nm until constant, using a suitably thermostatted spectrophotometer. Then add:

Reagent H (Myokinase)	0.10	—
Reagent G (BSA)	—	0.10

Immediately mix by inversion and record the increase in A340nm for approximately 5 minutes. Obtain the ΔA340nm/minute using the maximum linear rate for both the Test and Blank.

Calculations

$$\text{Units / ml enzyme} = \frac{(\Delta A340\text{nm / min Test} - \Delta A340\text{nm Blank})\ (3)\ (df)}{(6.22)\ (0.1)}$$

3 = Total volume (in milliliters) of assay
df = Dilution factor
6.22 = Millimolar extinction coefficient of β-NADPH at 340nm
0.1 = Volume (in milliliter) of enzyme used

$$\text{Units / mg solid} = \frac{\text{units / ml enzyme}}{\text{mg solid / ml enzyme}}$$

$$\text{Units / mg protein} = \frac{\text{units / ml enzyme}}{\text{mg protein / ml enzyme}}$$

Unit Definition: *One unit will convert 2.0 μmoles of ADP to ATP + AMP per minute at pH 7.6 at 37°C.*

Final Assay Concentrations

In a 3.00 ml reaction mix, the final concentrations are 58 mM glycylglycine, 2.0 mM adenosine 5'-diphosphate, 2.3 mM β-nicotinamide adenine dinucleotide phosphate, 10 mM magnesium chloride, 10 mM glucose, 2 units hexokinase, 1 unit glucose-6-phosphate dehydrogenase, 0.003% (w/v) Bovine Serum Albumin, and 0.02–0.06 unit myokinase.

Notes

1. Unit Definition for Hexokinase: One unit will phosphorylate 1.0 μmole of D-glucose per minute at pH 7.6 at 25°C.
2. Unit Definition for Glucose-6-Phosphate Dehydrogenase: One unit will oxidize 1.0 μmole of D-glucose 6-phosphate to 6-phospho-D-gluconate per minute in the presence of NADP at pH 7.4 at 25°C.

Enzymatic Assay of MYOSIN ATPase Assay

Principle

$$ATP + H_2O \xrightarrow[Ca++]{ATPase} ADP + Pi$$

Abbreviations Used

ATP = Adenosine 5'-Triphosphate
ATPase = Adenosine 5'-Triphosphatase
ADP = Adenosine 5'-Diphosphate
Pi = Inorganic phosphate

Conditions: T = 25°C, pH = 9.0, A660nm, Light Path = 1 cm

Method: Spectrophotometric Stop Rate Determination

Reagents

A. 200 mM Glycine Buffer, pH 9.0 at 25°C (Prepare 100 ml in deionized water using Glycine, Free Base. Adjust to pH 9.0 at 25°C with 1 NaOH).
B. 100 mM Calcium Chloride Solution ($CaCl_2$) (Prepare 10 ml in deionized water using Calcium Chloride, Dihydrate).
C. 50 mM Adenosine 5'-Triphosphate Solution, pH 6.8 at 25°C (ATP) (Prepare 5 ml in deionized water using Adenosine 5'-Triphosphate, Disodium Salt. Adjust to pH 6.8 at 25°C with Sodium Carbonate, Anhydrous).
D. 10% (w/v) Ammonium Molybdate Solution (Prepare 15 ml in 10 N H_2SO_4 using Ammonium Molybdate Tetrahydrate).

E. Taussky-Shorr Reagent (Taussky-Shorr) (Prepare by adding 10 ml of Reagent D to 70 ml of deionized water, then add 5 g Ferrous Sulfate, Heptahydrate. Bring the volume to 100 ml with deionized water. Store in an amber container).

F. Phosphorus Standard (P Std) (Use Phosphorus Standard Solution. The concentration is 20 µg/ml, 0.645 µmoles/ml).

G. Myosin Solution (Myosin) (Immediately before use, prepare a solution containing 0.50–1.00 unit/ml of Myosin in cold deionized water).

Procedure

Step 1: Pipette (in milliliters) the following reagents into suitable containers:

	Test 1	Test 2	Test 3	Blank
Reagent A (Buffer)	1.00	1.00	1.00	1.00
Reagent B ($CaCl_2$)	0.20	0.20	0.20	0.20
Reagent C (ATP)	0.30	0.30	0.30	0.30
Deionized Water	0.42	0.41	0.40	0.40

Mix by swirling and equilibrate to 25°C. Then add:

Reagent G (Myosin)	0.08	0.09	0.10	—

Mix by swirling and incubate at 25°C for exactly 5 minutes. Then add:

Reagent E (Taussky-Shorr)	4.00	4.00	4.00	4.00
Deionized Water	2.00	2.00	2.00	2.00
Reagent G (Myosin)	—	—	—	0.10

Mix by swirling and incubate at room temperature for approximately 5 minutes. Transfer the solutions to suitable cuvettes and record the A660nm for both the Test and Blank.

Step 2: Prepare a standard curve by pipetting (in milliliters) the follow reagents into suitable containers:

	Std 1	Std 2	Std 3	Std 4	Std 5	Std Blank
Reagent E (Taussky-Shorr)	4.00	4.00	4.00	4.00	4.00	4.00
Reagent F (P Std)	0.25	0.05	0.75	1.00	1.50	—
Deionized Water	3.75	3.50	3.25	3.00	2.50	4.00

Mix by swirling and incubate at room temperature for approximately 5 minutes. Transfer the solutions to suitable cuvettes and record the A660nm for both the Standards and Standard Blank.

Calculations

Standard Curve: ΔA660nm Standard = A660nm Standard–A660nm Standard Blank

Prepare a standard curve by plotting the ΔA660nm Standard vs µmoles of phosphate.

Sample Determination: ΔA660nm Test = A660nm Test–A660nm Test Blank

Determine the micromoles of phosphate liberated using the Standard Curve.

$$\text{Units / ml enzyme} = \frac{(\mu\text{moles of phosphate released})\ (\text{df})}{(0.1)\ (5)}$$

df = Dilution factor
0.1 = Volume (in milliliter) of Myosin used
5 = Time (in minutes) of assay as per the Unit Definition

$$\text{Units / mg protein} = \frac{\text{units / ml enzyme}}{\text{mg protein / ml enzyme}}$$

Unit Definition: *One unit will liberate 1.0 μmole of inorganic phosphorus from adenosine 5'-triphosphate per minute at pH 9.0 at 25°C in the presence of calcium.*

Final Assay Concentrations

In a 2.00 ml reaction mixture, the final concentrations are 100 mM glycine, 10 mM calcium chloride, 7.5 mM adenosine 5'-triphosphate, and 0.1 unit of ATPase activity.

Enzymatic Assay of β-NAD SYNTHETASE (EC 6.3.5.1)

Principle

$$\text{ATP} + \text{demido-NAD}^+ + \text{NH}_3 \xrightarrow[\text{Mg}_2^+,\ \text{K}^+]{\beta\text{-NAD Synthetase}} \text{AMP} + \text{NAD}^+ + \text{ppi}$$

Abbreviations Used

ATP = Adenosine 5'-Triphosphate
Deamido-NAD^+ = Nicotinic Acid Adenine Dinucleotide
AMP = Adenosine 5'-Monophosphate
β-NAD = β-Nicotinamide Adenine Dinucleotide, Oxidized Form
ppi = Inorganic Pyrophosphate

Conditions: T = 37°C, pH = 8.5, A340nm, Light path = 1 cm

Method: Spectrophotometric Stop Rate Determination

Reagents

A. 120 mM Tris HCl Buffer, pH 8.5 at 37°C (Prepare 100 ml in deionized water using Trizma Base. Adjust to pH 8.5 at 37°C with 1 M HCl).
B. 30 mM Nicotinic Acid Adenine Dinucleotide Solution (deamido-NAD^+) (Prepare 1 ml in deionized water using Nicotinic Acid Adenine Dinucleotide).
C. 60 mM Adenosine 5'-Triphosphate Solution (ATP) (Prepare 1 ml in deionized water using Adenosine 5'-Triphosphate, Disodium Salt).
D. 30 mM Ammonium Sulfate Solution ($(NH_4)_2SO_4$) (Prepare 1 ml in deionized water using Ammonium Sulfate).
E. 1% (w/v) Bovine Serum Albumin Solution (BSA) (Prepare 1 ml in deionized water using Albumin, Bovine).
F. 240 mM Magnesium Chloride Solution ($MgCl_2$) (Prepare 10 ml in deionized water using Magnesium Chloride, Anhydrous).

G. 120 mM Potassium Chloride Solution (KCl) (Prepare 10 ml in deionized water using Potassium Chloride).

H. Alcohol Dehydrogenase Enzyme Solution (ADH) (Immediately before use, prepare a solution containing 200 units/ml of Alcohol Dehydrogenase in cold deionized water).

I. 95% (v/v) Ethanol (EtOH) (Prepare 1 ml in deionized water using 200 Proof Ethyl Alcohol).

J. β-NAD Synthetase Enzyme Solution (β-NAD Synth) (Immediately before use, prepare a solution containing 0.15–1.0 unit/ml of β-NAD Synthetase in cold deionized water).

Procedure

Pipette (in milliliters) the following reagents into suitable cuvettes:

	Test	Blank
Reagent A (Buffer)	1.55	1.55
Reagent B (deamido-NAD^+)	0.10	0.10
Reagent C (ATP)	0.10	0.10
Reagent D ($(NH_4)_2SO_4$)	0.10	0.10
Reagent E (BSA)	0.10	0.10
Reagent F ($MgCl_2$)	0.25	0.25
Reagent G (KCl)	0.50	0.50
Reagent H (ADH)	0.20	0.20
Reagent J (β-NAD Synth)	0.05	0.05

Mix by inversion and equilibrate to 37°C. Monitor the A340nm until constant, using a suitably thermostatted spectrophotometer. Then add:

Reagent I (EtOH)	0.05	—
Deionized Water	—	0.05

Immediately mix by swirling and incubate at 37°C for exactly 2 minutes. Immediately transfer to a boiling water bath for one minute. Remove the Test and Blank and place on ice until cooled. Filter to clarify using a 0.2 µm filter. Record the A340nm using a suitable spectrophotometer for both the Test and Blank.

Calculations

$$\text{Units / ml enzyme} = \frac{(\Delta A340\text{nm / min Test} - \Delta A340\text{nm / min Blank})\ (3)\ (df)}{(2)\ (6.22)\ (0.05)}$$

3 = Total volume (in milliliters) of assay
df = Dilution factor
2 = Time (in minutes) of assay as per the Unit Definition
6.22 = Millimolar extinction coefficient of β-NADH at 340nm
0.05 = Volume (in milliliter) of enzyme use

$$\text{Units / mg solid} = \frac{\text{units / ml enzyme}}{\text{mg solid / ml enzyme}}$$

$$\text{Units / mg protein} = \frac{\text{units / ml enzyme}}{\text{mg protein / ml enzyme}}$$

Unit Definition: *One unit will form 1.0 μmole of β-NAD from nicotinic acid adenine dinucleotide per minute at pH 8.5 at 37°C.*

Final Assay Concentrations

In a 3.00 ml reaction mix, the final concentrations are 62 mM Tris, 1 mM nicotinic acid adenine dinucleotide, 2 mM adenosine 5'-triphosphate, 1 mM ammonium sulfate, 0.03% (w/v) Bovine Serum Albumin, 20 mM magnesium chloride, 20 mM potassium chloride, 1.6% (v/v) ethanol, 40 units alcohol dehydrogenase, and 0.0075–0.05 unit β-NAD synthetase.

Notes

1. Glutamine can also act as a substrate.
2. The alcohol dehydrogenase used as a coupling enzyme must be suitable for the recycling assay of β-NAD and β-NADH. (It must contain less than 0.005 mole of β-NAD and β-NADH per mole of ADH).
3. Alcohol Dehydrogenase Unit Definition: One unit will convert 1.0 μmole of ethanol to acetaldehyde per minute at pH 8.8 at 25°C.

Enzymatic Assay of NAD⁺ KINASE (EC 2.7.1.23)

Principle

$$\beta\text{-NAD} + \text{ATP} \xrightarrow[\text{Mg}_2^{++}]{\text{NAD}^+\text{ Kinase}} \beta\text{-NADP} + \text{ADP}$$

$$\beta\text{-NADP} + \text{G-6-P} \xrightarrow{\text{G-6-PDH}} \beta\text{-NADPH} + \text{6-Phosphogluconate}$$

Abbreviations Used

ADP = Adenosine 5'-Diphosphate
ATP = Adenosine 5'-Triphosphate
G-6-PDH = Glucose 6-Phosphate Dehydrogenase
β-NAD = β-Nicotinamide Adenine Dinucleotide
β-NADP = β-Nicotinamide Adenine Dinucleotide Phosphate, Oxidized Form
β-NADPH = β-Nicotinamide Adenine Dinucleotide Phosphate, Reduced Form

Conditions: T = 37°C, pH = 7.5, A340nm, Light path = 1 cm

Method: Spectrophotometric Stop Rate Determination

Reagents

A. 1 M Tris Solution (Prepare 50 ml in deionized water using Trizma Base).
B. 28 mM β-Nicotinamide Adenine Dinucleotide and 33 mM ATP with 20 mM Magnesium Chloride, pH 7.5 at 37°C (Reaction Cocktail) (Prepare 25 ml in deionized water using β-Nicotinamide Adenine Dinucleotide, Adenosine 5'-Triphosphate, Disodium Salt and Magnesium Chloride Hexahydrate. Adjust the pH to 7.5 at 37°C using Reagent A).

C. 250 mM Glycylglycine Buffer, pH 7.5 at 25°C (Buffer) (Prepare 50 ml in deionized water using Glycylglycine. Adjust the pH to 7.5 at 25°C with 1 M NaOH).
D. 60 mM Glucose-6-Phosphate Solution (G-6-P) (Prepare 5.0 ml in deionized water using Glucose-6-Phosphate, Disodium Salt, Hydrate).
E. 300 mM Magnesium Chloride ($MgCl_2$) (Prepare 10 ml in deionized water using Magnesium Chloride Hexahydrate).
F. 10 mM Tris HCl Buffer, pH 7.5 at 37°C (Enzyme Diluent) (Prepare 50 ml in deionized water using Reagent A. Adjust to pH 7.5 at 37°C with 1 M HCl).
G. Glucose-6-Phosphate Dehydrogenase Solution (G-6-PDH) (Immediately before use prepare a solution containing 100 units per ml of Glucose-6-Phosphate Dehydrogenase in cold Reagent C).
H. NAD^+ Kinase Solution (NAD Kinase) (Immediately before use prepare a solution containing 10–30 units per ml in cold Reagent F).

Procedure

Step 1: Pipette (in milliliters) the following reagents into suitable tubes:

	Test	Blank
Reagent B (Reaction Cocktail)	1.00	1.00

Equilibrate to 37°C. Then add:

Reagent H (NAD+ Kinase)	1.00	—
Reagent F (Enzyme Diluent)	—	1.00

Immediately mix by swirling and incubate at 37°C for exactly 30 minutes. Immediately transfer to a boiling water bath for 2 minutes. Remove from the boiling water bath and let cool to room temperature. Centrifuge to clarify.

Step 2: Pipette (in milliliters) the following reagents into suitable cuvettes:

	Test	Blank
Reagent C (Buffer)	2.50	2.50
Reagent D (G-6-P)	0.10	0.10
Reagent E ($MgCl_2$)	0.10	0.10
Test Supernatant (from Step 1)	0.20	—
Blank Supernatant (from Step 1)	—	0.20

Mix by inversion and equilibrate to 37°C. Monitor the A340nm until constant, using a suitably thermostatted spectrophotometer. Record the initial A340nm for both the Test and Blank. Then add:

Reagent G (G-6-PDH Solution)	0.10	0.10

Immediately mix by inversion and record the increase in A340nm until complete (for approximately 5 minutes). Obtain the final A340nm for both the Test and Blank.

Calculations

ΔA340nm Test = (Final A340nm of Test)–(Initial A340nm of Test)

ΔA340nm Blank = (Final A340nm of Blank)–(Initial A340nm of Blank)

$$\text{Units / ml enzyme} = \frac{(\Delta\text{A340nm Test}) - (\Delta\text{A340nm Blank})\ (1000\ \text{nmoles}/\mu\text{mole})\ (3)\ (2)}{(6.22)\ (30)\ (1)\ (0.2)}$$

3 = Volume (in milliliters) of reaction in Step 2
2 = Volume (in milliliters) of reaction in Step 1
6.22 = Millimolar extinction coefficient of β-NADPH at 340nm
30 = Time (in minutes) of reaction for Step 1
1 = Volume (in milliliter) of enzyme used for Step 1
0.2 = Volume (in milliliter) of Test Supernatant (from Step 1) used in Step 2

$$\text{Units / mg solid} = \frac{\text{units / ml enzyme}}{\text{mg solid / ml enzyme}}$$

$$\text{Units / mg protein} = \frac{\text{units / ml enzyme}}{\text{mg protein / ml enzyme}}$$

Unit Definition: *One unit will phosphorylate 1.0 nanomole of β-NAD to β-NADP per minute at pH 7.5 at 37°C in the presence of ATP.*

Final Assay Concentrations

In a 2.00 ml reaction mix, the final concentrations are 14 mM β-nicotinamide adenine dinucleotide, 17 mM adenosine 5'-triphosphate, 10 mM magnesium chloride, 5 mM Tris, and 10–30 units NAD^+ Kinase.

Note

1. Unit definition for Glucose-6-Phosphate Dehydrogenase: One unit will oxidize 1.0 μmole of glucose-6-phosphate to 6-phosphogluconate per minute in the presence of β-NADP at pH 7.4 at 25°C.

Enzymatic Assay of NADase (EC 3.2.2.5)

Principle

$$\beta\text{-NAD} + H_2O \xrightarrow{\text{NADase}} \text{Nicotinamide} + \text{ADP-ribose ADH}$$

$$\beta\text{-NAD} + \text{EtOH} \rightarrow \beta\text{-NADH} + \text{Acetaldehyde}$$

Abbreviations Used

β-NAD = β-Nicotinamide Adenine Dinucleotide, Oxidized Form
ADP-ribose = Adenosine 5'-Diphosphate-Ribose
ADH = Alcohol Dehydrogenase
β-NADH = β-Nicotinamide Adenine Dinucleotide, Reduced Form

Conditions: T = 37°C, pH = 7.3, A340nm, Light path = 1 cm.

Method: Spectrophotometric Determination

Reagents

A. 100 mM Potassium Phosphate Buffer, pH 7.3 at 37°C (Prepare 100 ml in deionized water using Potassium Phosphate, Monobasic, Anhydrous. Adjust to pH 7.3 at 37°C with 1 M NaOH).

B. 7.5 mM β-Nicotinamide Adenine Dinucleotide, Oxidized Form Solution (β-NAD) (Prepare 5 ml in deionized water using β-Nicotinamide Adenine Dinucleotide).
C. 3 M Trichloroacetic Acid Solution (TCA) (Prepare 10 ml in deionized water using Trichloroacetic Acid, 6.1 N Solution, approximately 100% (w/v)).
D. 453 mM Glycine Buffer, pH 9.8 at 25°C (Prepare 100 ml in deionized water using Glycine, Free Base. Adjust to pH 9.8 at 25°C with 1 M NaOH).
E. 74 mM Pyrophosphate, 75 mM Semicarbazide and 1% (v/v) Ethanol (ADH React). (Prepare 100 ml in Reagent D using Pyrophosphate, Tetrasodium, Decahydrate, Semicarbazide Hydrochloride, and 200 Proof USP Ethyl Alcohol).
F. Alcohol Dehydrogenase Enzyme Solution (ADH) (Immediately before use, prepare a solution containing approximately 5000 units/ml of Alcohol Dehydrogenase, in cold Reagent A).
G. NADase Enzyme Solution (NADase) (Immediately before use, prepare a solution containing 0.1–0.2 unit/ml of NADase in cold Reagent A).

Procedure

Step 1: Pipette (in milliliters) the following reagents into suitable centrifuge tubes:

	Test	Blank
Reagent A (Buffer)	0.80	0.80
Reagent B (β-NAD)	0.20	0.20
Reagent C (TCA)	—	0.30

Mix by inversion and equilibrate to 37°C. Then add:

Reagent G (NADase)	0.20	0.20

Immediately mix by inversion and incubate for exactly 20 minutes at 37°C. Then add:

Reagent C (TCA)	0.30	—

Mix by inversion and centrifuge both the Test and Blank solutions.

Step 2: Pipette (in milliliters) the following reagents into suitable cuvettes:

	Test	Blank
Test Supernatant	0.40	—
Blank Supernatant	—	0.40
Reagent E (ADH React. Cocktail)	2.60	2.60

Mix by inversion and equilibrate to 25°C using a suitably thermostatted spectrophotometer. Record the initial A340nm for both the Test and Blank. Then add:

Test Blank Reagent F (ADH)	0.01	0.01

Immediately mix by inversion and monitor the increase in A340nm until constant. The maximum increase in A340nm should be obtained after 5–10 minutes at 25°C. Record the final A340nm for both the Test and Blank.

Calculations

$$A340nm = Final\ A340nm - Initial\ A340nm$$

$$Units/ml\ enzyme = \frac{(A340nm\ Blank - A340nm\ Test)(1.5)(3.01)(df)}{(20)(6.22)(0.2)(0.4)}$$

1.5 = Total volume (in milliliters) of Step 1
3.01 = Total volume (in milliliters) of Step 2
df = Dilution factor
20 = Time (in minutes) of Step 1 assay as per the Unit Definition
6.22 = Millimolar extinction coefficient of β-NADH at 340nm
0.2 = Volume (in milliliter) of enzyme used
0.4 = Volume (in milliliter) of Step 1 used in Step 2

$$Units/mg\ solid = \frac{units/ml\ enzyme}{mg\ solid/ml\ enzyme}$$

$$Units/mg\ protein = \frac{units/ml\ enzyme}{mg\ protein/ml\ enzyme}$$

Unit Definition: *One unit will hydrolyze 1.0 µmole of β-NAD to nicotinamide and ADP-ribose per minute at pH 7.3 at 37°C.*

Final Assay Concentrations

In a 1.20 ml reaction mix, the final concentrations are 83 mM potassium phosphate, 1.3 mM β-nicotinamide adenine dinucleotide and 0.02–0.04 unit NADase.

Note

1. The amount of β-NAD consumed in the first reaction catalyzed by NADase is determined by measuring the residual β-NAD using this reaction.

Enzymatic Assay of NADH OXIDASE

Principle

$$NADH + H^+ + O_2 \xrightarrow{NADH\ Oxidase} NAD + H_2O_2$$

Conditions: T = 30°C, pH = 7.0, A340nm, Light path = 1 cm.

Method: Continuous Spectrophotometric Rate Determination

Reagents

A. 250 mM Potassium Phosphate Buffer, pH 7.0 at 30°C (Prepare 100 ml in deionized water using Potassium Phosphate, Monobasic, Anhydrous. Adjust to pH 7.0 at 30°C with 1 M KOH).

B. 1 mM Flavin Adenine Dinucleotide Solution (FAD) (Prepare 10 ml in deionized water using Flavin Adenine Dinucleotide, Disodium Salt).

C. 2.0 mM β-Nicotinamide Adenine Dinucleotide, Reduced Form Solution NADH) (Dissolve the contents of one 10 mg vial of β-Nicotinamide Adenine Dinucleotide, Reduced Form, in the appropriate volume of deionized water. *Prepare Fresh*).

D. 30 mM Potassium Phosphate Buffer with 0.1% (w/v) Bovine Serum Albumin, pH 7.5 at 30°C (Enzyme Diluent) (Prepare 100 ml in deionized water using Potassium

Phosphate, Monobasic, Anhydrous, and Albumin, Bovine. Adjust to pH 7.5 at 30°C with 1 M KOH).

E. NADH Oxidase Enzyme Solution (Immediately before use, prepare a solution containing 0.2 unit/ml of NADH Oxidase in cold Reagent D).

Procedure

Pipette (in milliliters) the following reagents into suitable cuvettes:

	Test	Blank
Deionized Water	1.90	1.90
Reagent A (Buffer)	0.60	0.60
Reagent B (FAD)	0.30	0.30
Reagent C (NADH)	0.10	0.10

Mix by inversion and equilibrate to 30°C. Monitor the A340nm until constant, using a suitably thermostatted spectrophotometer. Then add:

	Test	Blank
Reagent E (Enzyme Solution)	0.10	—
Reagent D (Enzyme Diluent)	—	0.10

Immediately mix by inversion and record the decrease in A340nm for approximately 5 minutes. Obtain the A340nm/minute using the maximum linear rate for both the Test and Blank.

Calculations

$$\text{Units / ml enzyme} = \frac{(\text{A340nm / min Test} - \text{A340nm / min Blank})\ (3)\ (\text{df})}{(6.22)\ (0.10)}$$

3 = Total volume (in milliliters) of assay
df = Dilution factor
6.22 = Millimolar extinction coefficient of β-NADH at 340nm
0.10 = Volume (in milliliter) of enzyme used

$$\text{Units / mg solid} = \frac{\text{units / ml enzyme}}{\text{mg solid / ml enzyme}}$$

$$\text{Units / mg protein} = \frac{\text{units / ml enzyme}}{\text{mg protein / ml enzyme}}$$

Unit Definition: *One unit will oxidize 1.0 µmole of β-NADH per minute at pH 7.0 at 30°C.*

Final Assay Concentrations

In a 3.00 ml reaction mix, the final assay concentrations are 51 mM potassium phosphate, 0.07 mM β-nicotinamide adenine dinucleotide, reduced form, 0.1 mM flavin adenine dinucleotide, 0.003% (w/v) Bovine Serum Albumin, and 0.02 unit NADH oxidase.

Enzymatic Assay of NADH PEROXIDASE (EC 1.11.1.1)

Principle

$$\beta\text{-NADH} + H_2O_2 \xrightarrow{\text{NADH Peroxidase}} \beta\text{-NAD} + 2H_2O$$

Abbreviations Used

β-NADH = β-Nicotinamide Adenine Dinucleotide, Reduced Form
β-NAD = β-Nicotinamide Adenine Dinucleotide, Oxidized Form

Conditions: T = 25°C, pH 5.4, A340nm, Light path = 1 cm

Method: Continuous Spectrophotometric Rate Determination

Reagents

A. 100 mM Sodium Acetate Buffer with 0.30 mM Ethylenediaminetetraacetic Acid, pH 5.4 at 25°C (Prepare 100 ml in deionized water using Sodium Acetate, Trihydrate and Ethylenediaminetetraacetic Acid, Disodium Salt, Dihydrate. Adjust to pH 5.4 at 25°C with either 1 M HCl or 1 M NaOH).
B. 0.11% (v/v) Hydrogen Peroxide Solution (H_2O_2) (Prepare 100 ml in deionized water using Hydrogen Peroxide, 30% (w/w) Solution).
C 1000 mM Tris Solution (Tris) (Prepare 10 ml in deionized water using Trizma Base).
D. 23.4 mM β-Nicotinamide Adenine Dinucleotide, Reduced Form, Solution (β-NADH) (Dissolve the contents of one 10 mg vial of β-Nicotinamide Adenine Dinucleotide, Reduced Form, Sodium Salt in the appropriate volume of deionized water. Add one drop of Reagent C (Tris) per ml in order to neutralize the solution).
E. NADH Peroxidase Enzyme Solution (Immediately before use, prepare a solution containing 0.15–0.30 unit/ml of NADH Peroxidase in cold deionized water).

Procedure

Pipette (in milliliters) the following reagents into suitable cuvettes:

	Test	Blank
Reagent A (Buffer)	2.57	2.57
Reagent B (H_2O_2)	0.36	0.36
Deionized Water	—	0.10

Mix by inversion and equilibrate to 25°C. Then add:

Reagent D (β-NADH)	0.02	0.02
Reagent E (Enzyme Solution)	0.10	—

Immediately mix by inversion and monitor the decrease in A340nm for approximately 5 minutes. Obtain the ΔA340nm/minute using the maximum linear rate for both the Test and Blank.

Calculations

$$\text{Units / mg enzyme} = \frac{\Delta A340\text{nm / min Test} - \Delta A340\text{nm / min Blank}}{(6.22)\ (\text{mg enzyme / ml RM})}$$

6.22 = Millimolar extinction coefficient of β-NADH at 340nm.
RM = Reaction Mix

Unit Definition: *One unit will catalyze the oxidation of 1.0 μmole of β-NADH per min at pH 5.4 at 25°C.*

Final Assay Concentrations

In a 3.05 ml reaction mix, the final concentrations are 84.3 mM sodium acetate, 0.25 mM Ethylenediaminetetraacetic Acid, 0.013% (v/v) hydrogen peroxide, 0.2 mM NADH, and 0.015–0.030 unit NADH peroxidase.

Enzymatic Assay of NADH PEROXIDASE (EC 1.11.1.1)

Principle

$$\beta\text{-NADH} + H_2O_2 \xrightarrow{\text{NADH Peroxidase}} \beta\text{-NAD} + 2H_2O$$

Abbreviations Used

β-NADH = β-Nicotinamide Adenine Dinucleotide, Reduced Form
β-NAD = β-Nicotinamide Adenine Dinucleotide, Oxidized Form

Conditions: T = 25°C, pH 6.0, A365nm, Light path = 1 cm

Method: Continuous Spectrophotometric Rate Determination

Reagents

A. 200 mM Tris Acetate Buffer, pH 6.0 at 25°C (Prepare 100 ml in deionized water using Trizma Acetate. Adjust to pH 6.0 at 25°C with 1 M HCl).
B. 0.30% (w/w) Hydrogen Peroxide Solution (H_2O_2) (Prepare 10 ml in deionized water using Hydrogen Peroxide 30% (w/w) Solution).
C. 12 mM β-Nicotinamide Adenine Dinucleotide, Reduced Form, Solution (NADH) (Dissolve the contents of one 10 mg vial of β-Nicotinamide Adenine Dinucleotide, Reduced Form, Sodium Salt, in the appropriate volume of deionized water).
D. NADH Peroxidase Enzyme Solution (Immediately before use, prepare a solution containing 0.2–0.4 unit of NADH Peroxidase in cold deionized water).

Procedure

Pipette (in milliliters) the following reagents into suitable cuvettes:

	Test	Blank
Reagent A (Buffer)	3.00	3.00
Reagent C (NADH)	0.10	0.10
Reagent D (Enzyme Solution)	0.10	—
Deionized Water	—	0.10

Mix by inversion and equilibrate to 25°C. Monitor the baseline at A365nm for 5 minutes in order to determine any NADH oxidase activity which may be present. Then add:

	Test	Blank
Reagent B (H_2O_2)	0.10	0.10

Immediately mix by inversion and monitor the decrease in A365nm for approximately 5 minutes. Obtain the ΔA365nm/minute using the maximum linear rate for both the Test and Blank using a suitably thermostatted spectrophotometer.

Calculations

$$\text{Units / ml enzyme} = \frac{(\Delta A365nm / \min \text{Test} - \Delta A365nm / \min \text{Blank})(3.3)(df)}{(3.4)(0.1)}$$

3.3 = Total volume (in milliliters) of assay
df = Dilution factor
3.4 = Millimolar extinction coefficient2 of β-NADH at 365nm
0.1 = Volume (in milliliters) of enzyme used

$$\text{Units / mg solid} = \frac{\text{units / ml enzyme}}{\text{mg solid / ml enzyme}}$$

$$\text{Units / mg protein} = \frac{\text{units / ml enzyme}}{\text{mg protein / ml enzyme}}$$

Unit Definition: *One unit will decompose 1.0 μmole of H_2O_2 per minute at pH 6.0 at 25°C.*

Final Assay Concentrations

In a 3.30 ml reaction mix, the final concentrations are 182 mM tris acetate, 0.009% (w/w) hydrogen peroxide, 0.36 mM β-nicotinamide adenine dinucleotide, and 0.02–0.04 unit NADH peroxidase.

Note

1. The A365nm of both the Test and Blank must be corrected for the presence of any NADH oxidase activity.

Enzymatic Assay of NADPH-FMN OXIDOREDUCTASE (EC 1.6.99.1)

Principle

$$\beta\text{-NADPH} + \text{FMN} \xrightarrow{\text{NADPH-FMN Oxidoreductase}} \beta\text{-NADP} + \text{FMNH}_2$$

Abbreviations Used

β-NADPH = β-Nicotinamide Adenine Dinucleotide Phosphate, Reduced Form
FMN = Flavin Mononucleotide
β-NADP = β-Nicotinamide Adenine Dinucleotide Phosphate, Oxidized Form
$FMNH_2$ = Flavin Monomucleotide, (Reduced)

Conditions: T = 30°C, pH 7.0, A340nm, Light Path = 1 cm

Method: Continuous Spectrophotometric Rate Determination

Reagents

A. 50 mM Potassium Phosphate Buffer, pH 7.0 at 30°C (Prepare 100 ml in deionized water using Potassium Phosphate, Dibasic, Trihydrate. Adjust to pH 7.0 at 30°C with 50 mM Sodium Phosphate, Monobasic, Anhydrous).
B. 1.0 mM Flavin Mononucleotide Solution (FMN) (Prepare 10 ml in Reagent A using Flavin Mononucleotide, Sodium Salt. *Prepare Fresh*).

C. 2.0 mM β-Nicotinamide Adenine Dinucleotide Phosphate, Reduced Form Solution (β-NADPH) (Prepare 5 ml in Reagent A using β-Nicotinamide Adenine Dinucleotide, Phosphate, Reduced Form, Tetrasodium Salt or dissolve the contents of one 10 mg vial of β-Nicotinamide Adenine Dinucleotide Phosphate, Reduced Form, Tetrasodium Salt in the appropriate volume of Reagent A. *Prepare Fresh*).

D. NADPH-FMN Oxidoreductase Enzyme Solution (Immediately before use, prepare a solution containing 0.05–0.20 unit/ml of NADPH-FMN Oxidoreductase in cold Reagent A).

Procedure

Pipette (in milliliters) the following reagents into suitable cuvettes:

	Test	Blank
Reagent A (Buffer)	2.65	2.75
Reagent B (FMN)	0.05	0.05
Reagent C (β-NADPH)	0.20	0.20

Mix by inversion and equilibrate to 30°C. Monitor the A340nm until constant, using a suitably thermostatted spectrophotometer. Then add:

Reagent D (Enzyme Solution)	0.10	—

Immediately mix by inversion and record the decrease in A340 for approximately 5–10 minutes. Obtain the ΔA340nm/minute using the maximum linear rate for both the Test and Blank.

Calculations

$$\text{Units / ml enzyme} = \frac{(\Delta A340\text{nm Test} - \Delta A340\text{nm Blank})\ (3)\ (\text{df})}{(6.22)\ (0.1)}$$

3 = Total volume (in milliliters) of assay
df = Dilution factor
6.22 = Millimolar extinction coefficient of β-NADPH at 340nm
0.1 = Volume (in milliliter) of enyzme used at 340nm

$$\text{Units / mg solid} = \frac{\text{units / ml enzyme}}{\text{mg solid / ml enzyme}}$$

$$\text{Units / mg protein} = \frac{\text{units / ml enzyme}}{\text{mg protein / ml enzyme}}$$

Unit Definition: *One unit will reduce 1.0 μmole of FMN to $FMNH_2$ per minute at pH 7.0 at 30°C.*

Final Assay Concentrations

In a 3.00 ml reaction mix, the final concentrations are 50 mM potassium phosphate, 0.02 mM FMN, 0.13 mM β-NADPH and 0.005–0.02 unit NADPH-FMN oxidoreductase.

Note

1. Correct NADPH concentrations for purity and water content.

Enzymatic Assay of NAD-PYROPHOSPHORYLASE (EC 2.7.7.1)

Principle

$$\beta\text{-NMN} + \text{ATP} \xrightarrow{\text{NAD-Pyrophosphorylase}} \beta\text{-NAD} + \text{PP}$$

$$\beta\text{-NAD} + \text{Ethanol} \xrightarrow{\text{ADH}} \beta\text{-NADH} + \text{Acetaldehyde}$$

Abbreviations Used

ATP = Adenosine 5'-Triphosphate
ADH = Alcohol Dehydrogenase
β-NAD = β-Nicotinamide Adenine Dinucleotide, Oxidized Form
β-NADH = β-Nicotinamide Adenine Dinucleotide, Reduced Form
β-NMN = β-Nicotinamide Mononucleotide
PP = Pyrophosphate

Conditions: T = 37°C, pH = 7.4, A340nm, Light path = 1 cm

Method: Spectrophotometric Stop Rate Determination

Reagents

A. 250 mM Glycylglycine, Buffer pH 7.4 at 37°C. (Gly-Gly) (Prepare 10 ml in deionized water using Gly-Gly, Free Base. Adjust the pH to 7.4 at 37°C with 5 M NaOH).
B. 22 mM Adenosine Triphosphate Solution (ATP) (Prepare 1 ml in deionized water using Adenosine 5'-Triphosphate, Disodium Salt. Neutralize with Sodium Bicarbonate).
C. 60 mM Nicotinamide Mononucleotide Solution (NMN) (Prepare 1 ml in Reagent A using β-Nicotinamide Mononucleotide).
D. 150 mM Magnesium Chloride Solution ($MgCl_2$) (Prepare 10 ml in deionized water using Magnesium Chloride, Hexahydrate).
E. 75 mM Pyrophosphate Buffer with 75 mM Semicarbazide HCl, 23 mM Glycine, and 1% (v/v) Ethanol (Nondenatured), pH 8.7 at 25°C (Prepare 10 ml in deionized water using Pyrophosphate Tetrasodium, Anhydrous, Semicarbazide, Hydrochloride, Glycine, Free Base. Adjust the pH to 8.7 using 1 M HCl solution).
F. Alcohol Dehydrogenase Enzyme Solution (ADH) (Immediately before use prepare a solution containing 30 mg/ml of Alcohol Dehydrogenase in cold deionized water).
G. NAD-Pyrophosphorylase Enzyme Solution (NAD-PP) (Immediately before use prepare a solution containing 0.13–0.30 unit/ml of NAD-Pyrophosphorylase in cold deionized water).

Procedure

Step 1: Pipette (in milliliters) the following reagents into suitable containers:

	Test	Blank
Reagent A (Gly-Gly)	0.20	0.20
Reagent B (ATP)	0.10	0.10
Reagent C (NMN)	0.20	0.20
Reagent D ($MgCl_2$)	0.10	0.10
Deionized Water	0.30	0.30

Mix by swirling and equilibrate to 37°C. Then Add:

	Test	Blank
Reagent G (NAD-PP Enzyme solution)	0.10	—
Deionized Water	—	0.10

Immediately mix by swirling and incubate at 37°C for exactly 30 minutes. Stop the reaction by heating both the Test and Blank solutions for 1 minute at 100°C. Remove from the boiling water bath and let cool to room temperature. Filter the supernatant using 0.45 μm syringe filters.

Step 2: Pipette (in milliliters) the following reagents into suitable cuvettes:

	Test	Blank
Reagent E	2.49	2.49
Test Supernatant (Step 1)	0.50	—
Blank Supernatant (Step 1)	—	0.50

Mix by inversion and equilibrate to 25°C. Monitor the A340nm until constant using a suitable thermostatted spectrophotometer and record the initial A340nm for the Test and Blank solutions. Then add:

Reagent F (ADH)	0.01	0.01

Immediately mix by inversion and record the increase in A340nm for approximately 5–10 minutes until constant. Obtain the final A340nm for both the Test and Blank solutions.

Calculations

$$\Delta A340nm\ Test = A340nm\ Test\ Final - A340nm\ Test\ Initial$$

$$\Delta A340nm\ Blank = A340nm\ Blank\ Final - A340nm\ Blank\ Initial$$

$$\text{Units / ml enzyme} = \frac{(\Delta A340nm\ Test) - (\Delta A340nm\ Blank)\ (3.0)\ (1.0)\ (df)}{(6.22)\ (30)\ (0.1)\ (0.5)}$$

3.0 = Total volume (in milliliters) of reaction mixture in Step 2
1.0 = Total volume (in milliliters) of reaction mixure in Step 1
df = Dilution factor
6.22 = Millimolar extinction coefficient of β-NADPH at 340nm
30 = Reaction time (in minutes) of Step 1
0.5 = Volume (in milliliter) of Step 1 Test used in Step 2
0.1 = Volume (in milliliter) of enzyme used in Step 1

$$\text{Units / mg solid} = \frac{\text{units / ml enzyme}}{\text{mg solid / ml enzyme}}$$

$$\text{Units / mg protein} = \frac{\text{units / ml enzyme}}{\text{mg protein / ml enzyme}}$$

Unit Definition: *One unit will form 1.0 μmole of β-NAD from nicotinamide mononucleotide and ATP per minute at pH 7.4 at 37°C.*

Final Assay Concentrations

In a 1.00 ml reaction mix, the final concentrations are 100 mM glycylglycine, 2.2 mM adenosine 5'-triphosphate, 12 mM nicotinamide mononucleotide, 15 mM magnesium chloride, and 0.013–0.03 unit of NAD-Pyrophosphorylase.

Enzymatic Assay of NARINGINASE

Principle

$$\text{Naringin} + H_2O \xrightarrow{\text{Naringinase}} \text{Glucose}$$

Conditions: T = 40°C, pH 4.0, A540nm, Light path = 1 cm

Method: Colorimetric

Reagents

A. 50 mM Sodium Acetate Buffer, pH 4.0 at 40°C (Prepare 100 ml deionized water using Sodium Acetate, Trihydrate. Adjust to pH 4.0 at 40°C with 1 M HCl).

B. 0.125% Naringin Solution (Prepare 50 ml in Reagent A using Naringin. Adjust to pH 4.0 at 40°C with 1 M HCl or 1 N NaOH.)

C. Naringinase Enzyme Solution (Prepare a solution containing 0.07–0.10 units/ml Naringinase in cold deionized water).

D. 16 mM Copper Sulfate, 1.3 M Sodium Sulfate, 226 mM Sodium Carbonate, 190 mM Sodium Bicarbonate and 43 mM Sodium Potassium Tartrate (Copper Solution) (Prepare 1 liter in deionized water using Cupric Sulfate Pentahydrate, Sodium Bicarbonate, Sodium Sulfate, Anhydrous, Sodium Carbonate, Anhydrous and Sodium Tartrate Tetrahydrate).

E. 40 mM Molybdic Acid, 19 mM Arsenic Acid and 756 mM Sulfuric Acid (Arsenomolybdate Solution) (Prepare 1 liter in deionized water using Molybdic Acid, Ammonium Salt Tetrahydrate, Arsenic Acid, Sodium Salt and Sulfuric Acid).

F. 250 µg Glucose/ml Standard Solution (Stock) (Prepare 100 ml each in deionized water using D-(+)-Glucose with a Class A, 100 ml, volumetric flask).

G. Glucose Working Standard Solutions (Prepare 10 ml each in deionized water containing 25 µg/ml, 50 µg/ml, 100 µg/ml, 150 µg/ml and 250 µg/ml).

Procedure

Pipette (in milliliters) the following reagents into suitable containers:

	Test	Blank
Reagent B (Naringin Solution)	4.0	4.0
Deionized Water	—	1.0

Mix by swirling and equilibrate to 40°C. Then add:

Reagent C (Enzyme Solution)	1.0	—

Mix by swirling and incubate at 40°C for 30 minutes. Immediately transfer 1 ml of reaction mixture into a suitable, container containing 1 ml of Reagent D as indicated below and proceed with Somogyi's method of assay reducing sugars.

	Test	Blank	Std 1	Std 2	Std 3	Std 4	Std 5	Std Blank
Reagent D (Copper Soln)	1.0	1.0	1.0	1.0	1.0	1.0	1.0	1.0
Test Solution	1.0	—	—	—	—	—	—	—
Blank Solution	—	1.0	—	—	—	—	—	—
Deionized Water	—	—	—	—	—	—	—	1.0
Reagent G (Glucose Soln)	—	—	1.0	1.0	1.0	1.0	1.0	—

Mix and incubate in a boiling water bath for 10 minutes. Then add:

Reagent E (Arsenomolybdate Soln)	1.0	1.0	1.0	1.0	1.0	1.0	1.0	1.0

Invert and mix tubes until foaming stops or precipitate is dissolved. Then add:

Deionized Water	10.0	10.0	10.0	10.0	10.0	10.0	10.0	10.0

Mix by inversion and record the A540nm for Test, Blank and Standards in a suitable spectrophotometer.

Calculations

Standard Curve: A540nm Std = A540nm Std–A540 Std Blank

Prepare a standard curve by plotting the A540nm Standard versus glucose concentration. Use the slope to determine the concentration of the Test solution.

$$\text{Units / mg Protein} = \frac{(\text{A540nm Test} - \text{A540nm Blank})\,(5)}{(30)\,(\text{mg enzyme / RM})\,(1)}$$

30 = Time in minutes of reaction
RM = Reaction Mix
5 = Volume of Step 1 Reaction Mix
1 = Volume transferred from Step 1 to Step 2

Unit Definition: *One unit will liberate 1.0 μmole of reducing sugar (measured as glucose) from naringin per minute at pH 4.0 at 40°C.*

Final Assay Concentration

In a 5.0 ml reaction mix, the final concentrations are 40 mM sodium acetate, 0.1% naringin and 0.07–0.10 units naringinase.

Notes

1. Sodium Sulfate, Sodium Carbonate, and Sodium Potassium Tartrate are dissolved in approximately 500 ml of deionized water. Cupric Sulfate is dissolved in approximately 100 ml of deionized water and slowly added to the above solution to avoid precipitation. Sodium Bicarbonate is dissolved first in deionized water and then added to the above solution. Dilute the solution to 1 liter. If a precipitate forms, it should be removed by filtration prior to use. Store in an amber bottle and avoid exposure to direct sunlight. Store at room temperature.

2. Molybdic Acid is dissolved in approximately 300 ml of deionized water. Add Sulfuric Acid slowly. Caution, this is an exothermic reaction! Arsenic Acid is dissolved in approximately 300 ml of deionized water and is added to the above solution. The solution is diluted to a total volume of 1 liter and incubated at 37°C for 48–72 hours. If a precipitate forms, it should be removed by filtration prior to use. Store in an amber bottle and avoid exposure to direct sunlight. Solution expires six months after preparation. Store at room temperature in an exhaust hood.

Enzymatic Assay of NEURAMINIDASE (EC 3.2.1.18) Bovine Submaxillary Mucin as Substrate

Principle

$$\text{Mucin} + H_2O \xrightarrow{\text{Neuraminidase}} \text{NANA} + \text{Asialic Acid}$$

Abbreviation Used

NANA = N-Acetylneuraminic Acid

Conditions: T = 37°C, pH = 5.0, A550nm, Light path = 1 cm

Method: Colorimetric

Reagents

A. 100 mM Sodium Acetate Buffer with 2 mM Calcium Chloride, pH 5.0 at 37°C (Prepare 100 ml in deionized water using Sodium Acetate Anhydrous and Calcium Chloride, Dihydrate. Adjust to pH 5.0 at 37°C with 1 M HCl).

B. 0.4% (w/v) Mucin Solution (Mucin) (Prepare 5 ml in Reagent A using Mucin).

C. 0.2% (w/v) Bovine Serum Albumin Solution (BSA) (Prepare 100 ml in deionized water using Albumin, Bovine).

D. 5% (w/v) Phosphotungstic Acid Solution (PT) (Prepare 100 ml in 2.5 M HCl using Phosphotungstic Acid Free Acid).

E. Neuraminidase Enzyme Solution (Immediately before use, prepare a solution containing approximately 0.02–0.04 unit/ml of Neuraminidase in cold Reagent C).

F. 9 M Phosphoric Acid Solution (Prepare 20 ml in deionized water using Phosphoric Acid).

G. 200 mM Sodium Periodate Solution (Per) (Prepare 10 ml in Reagent F using Sodium m-Periodate).

H. 10% (w/v) Sodium Arsenite with 50 mM Sulfuric Acid and 500 mM Sodium Sulfate Solution (Ars) (Prepare 100 ml in deionized water using Sodium m-Arsenite Sulfuric Acid and Sodium Sulfate, Anhydrous).

I. 0.6% (w/v) Thiobarbituric Acid and 500 mM Sodium Sulfate Solution (TBA) (Prepare 100 ml in deionized water using 2-Thiobarbituric Acid and Sodium Sulfate, Anhydrous).

J. 5% (v/v) HCl in Butanol (But) (Prepare by combining 5 ml of concentrated Hydrochloric Acid and 95 ml of n-Butanol).

K. 0.16 mM N-Acetylneuraminic Acid Standard Solution (NANA) (Prepare 10 ml in deionized water using N-Acetylneuraminic Acid).

Procedure

Pipette (in milliliters) the following reagents into suitable containers:

	Test	Blank
Reagent A (Buffer)	0.20	0.20
Reagent B (Mucin)	0.20	0.20

Mix by inversion and equilibrate to 37°C. Then add:

Reagent E (Enzyme Solution)	0.10	—
Reagent C (BSA)	—	0.10

Mix by inversion immediately and incubate at 37°C for exactly 10 minutes. Then add:

Reagent D (PT)	0.50	0.50

Mix by inversion immediately and centrifuge for 3 minutes.

Colorimetric Assay

Standard Curve: Pipette (in milliliters) the following reagents into suitable containers:

	Std 1	Std 2	Std 3	Std 4	Std Blank
Reagent K (NANA)	0.05	0.10	0.20	0.30	0.00
Deionized water	0.45	0.40	0.30	0.20	0.50
Reagent D (PT)	0.50	0.50	0.50	0.50	0.50

Mix by inversion and centrifuge for 3 minutes. Pipette (in milliliters) the following reagents into suitable containers:

	Test	Test Blank	Std 1	Std 2	Std 3	Std 4	Std Blank
Test Supernatant	0.50	—	—	—	—	—	—
Test Blank Supernatant	—	0.50	—	—	—	—	—
Std 1 Supernatant	—	—	0.50	—	—	—	—
Std 2 Supernatant	—	—	—	0.50	—	—	—
Std 3 Supernatant	—	—	—	—	0.50	—	—
Std 4 Supernatant	—	—	—	—	—	0.50	—
Std Blank Supernatant	—	—	—	—	—	—	0.50
Reagent G (Per)	0.10	0.10	0.10	0.10	0.10	0.10	0.10

Mix by inversion and incubate for 20 minutes at 25°C. Then add:

Reagent H (Ars)	1.00	1.00	1.00	1.00	1.00	1.00	1.00

Vortex until the yellow-brown color disappears. Then add:

Reagent I (TBA)	3.00	3.00	3.00	3.00	3.00	3.00	3.00

Transfer to a boiling water bath and incubate for 15 minutes. Transfer to a cold water bath for 5 minutes. Then add:

	Test	Test Blank	Std 1	Std 2	Std 3	Std 4	Std Blank
Reagent J (But)	4.60	4.60	4.60	4.60	4.60	4.60	4.60

Shake or vortex until most of the pink color has been extracted from the lower layer. Centrifuge for 2–3 minutes. Transfer the upper layer to suitable cuvettes and record the A550nm for Samples, Tests, Blanks, and Standards.

Calculations

Standard Curve: ΔA550nm Standard = A550nm Standard–A550nm Standard Blank
Plot the ΔA550nm Standard vs µmoles n-Acetylneuraminic acid.
Sample Determination: ΔA550nm Test = A550nm Test–A550nm Test Blank
Determine the µmoles of n-Acetylneuraminic Acid (NANA) liberated using the Standard curve.

$$\text{Units / ml enzyme} = \frac{(\mu\text{moles of NANA liberated})\ (\text{df})}{(10)\ (0.1)}$$

df = Dilution factor
10 = Time (in minutes) of the assay as per the Unit Definition
0.1 = Volume (in milliliters) of enzyme used

$$\text{Units / mg solid} = \frac{\text{units / ml enzyme}}{\text{mg solid / ml enzyme}}$$

$$\text{Units / mg protein} = \frac{\text{units / ml enzyme}}{\text{mg protein / ml enzyme}}$$

Unit Definition: *One unit will liberate 1.0 µmole of N-acetylneuraminic acid per minute at pH 5.0 at 37°C using bovine submaxillary mucin as substrate.*

Final Assay Concentrations

In a 0.50 ml reaction mix, the final concentrations are 80 mM sodium acetate, 1.6 mM calcium chloride, 0.16% (w/v) mucin, 0.04% (w/v) Bovine Serum Albumin, and 0.002–0.004 unit neuraminidase.

Note

1. Only Neuraminidase from Vibrio Cholera requires calcium as a metal requirement, although calcium does not interfere with Neuraminidase from other sources.

Enzymatic Assay of NEURAMINIDASE (EC 3.2.1.18)

Principle

$$\text{3'-N - Acetylneuramin - Lactose} + H_2O \xrightarrow{\text{Neuraminidase}} \text{NANA} + \text{Lactose}$$

Abbreviation Used

NANA = N-Acetylneuraminic Acid

Conditions: T = 37°C, pH 6.5, A549nm, Light path = 1 cm

Method: Colorimetric

Reagents

A. 150 mM Sodium Acetate Buffer with 450 mM Sodium Chloride and 27 mM Calcium Chloride, pH 6.5 at 37°C (Prepare 100 ml in deionized water using Sodium Acetate, Trihydrate, Sodium Chloride and Calcium Chloride, Dihydrate. Adjust to pH 6.5 at 37°C with 1 M HCl).
B. 5.0 mM 3'-N-Acetylneuramin-Lactose Solution (NAN-Lactose) (Prepare 2 ml in Reagent A using 3'-N-Acetylneuramin-Lactose, Sodium Salt. *Prepare Fresh*).
C. 9 M Phosphoric Acid Solution (Prepare 20 ml in deionized water using Phosphoric Acid).
D. 250 mM Sodium m-Periodate Solution (Per) (Prepare 10 ml in Reagent C using Sodium m-Periodate).
E. 550 mM Sodium m-Arsenite and 500 mM Sodium Sulfate Solution (Ars) (Prepare 100 ml in deionized water using Sodium m-Arsenite and Sodium Sulfate, Anhydrous).
F. 56 mM Thiobarbituric Acid and 500 mM Sodium Sulfate Solution (TBA) (Prepare 100 ml in deionized water using 2-Thiobarbituric Acid and Sodium Sulfate, Anhydrous. In order for the solution to become clear, adjust the pH between 7.5 and 8.0 with either 1 M HCl or 1 M NaOH).
G. 0.13 mM N-Acetylneuraminic Acid Standard solution (NANA Std) (Prepare 10 ml in deionized water using N-Acetylneuraminic Acid).
H. 0.1% (w/v) Bovine Serum Albumin Solution (BSA) (Prepare 10 ml in Reagent A using Albumin, Bovine).
I. Cyclohexanone (Cyl Hex) (Use Cyclohexanone undiluted).
J. Neuraminidase Enzyme Solution (Immediately before use, prepare a solution containing approximately 0.01–0.02 unit/ml of Neuraminidase in Reagent H).

Procedure

Pipette (in milliliters) the following reagents into suitable containers:

	Test	Test Blank	Std 1	Std 2	Std 3	Std 4	Std 5	Std Blank
Reagent A (Buffer)	0.076	0.076	0.076	0.076	0.076	0.076	0.076	0.076
Reagent B (NAN-Lactose)	0.05	0.05	0.05	0.05	0.05	0.05	0.05	0.05
Reagent G (NANA Std)	—	—	0.025	0.050	0.075	0.100	0.125	—
Deionized Water	—	—	0.100	0.075	0.050	0.025	—	0.125

Mix by inversion and equilibrate to 37°C. Then add:

Reagent J (Enz Soln)	0.125	—	—	—	—	—	—	—
Reagent H (BSA)	—	0.125	—	—	—	—	—	—

Immediately mix by inversion and incubate at 37°C for exactly 10 minutes. Then add:

Reagent D (Per)	0.05	0.05	0.05	0.05	0.05	0.05	0.05	0.05

Mix by inversion and incubate at 25°C for 30 minutes. Then add:

	Test	Test Blank	Std 1	Std 2	Std 3	Std 4	Std 5	Std Blank
Reagent E (Ars)	0.50	0.50	0.50	0.50	0.50	0.50	0.50	0.50

Immediately mix by vortexing. Continue to agitate for 2–3 minutes. Then add:

Reagent F (TBA)	0.50	0.50	0.50	0.50	0.50	0.50	0.50	0.50

Mix by vortexing and incubate in a boiling water bath for 15 minutes. Cool on ice to room temperature. Then add:

Reagent I (Cyl Hex)	2.00	2.00	2.00	2.00	2.00	2.00	2.00	2.00

Mix by vortexing. Centrifuge for 2 minutes. Transfer the upper layer to suitable cuvettes and record the A549nm for Test, Test Blank, Standards, and Standard Blank.

Calculations

Standard Curve: ΔA549nm Standard = A549nm Standard–A549nm Standard Blank
Plot the ΔA549nm Standard vs μmoles of N-Acetylneuraminic Acid.

Sample Determination: ΔA549nm Test = A549nm Test–A549nm Test Blank
Determine the μmoles of n-Acetylneuraminic Acid (NANA) liberated using the Standard Curve.

$$\text{Units / ml enzyme} = \frac{(\mu\text{moles of NANA liberated})\,(\text{df})}{(0.125)\,(10)}$$

df = Dilution factor
0.125 = Volume (in milliliter) of enzyme used
10 = Time (in minutes) of assay as per the Unit Definition

$$\text{Units / mg solid} = \frac{\text{units / ml enzyme}}{\text{mg solid / ml enzyme}}$$

$$\text{Units / mg protein} = \frac{\text{units / ml enzyme}}{\text{mg protein / ml enzyme}}$$

Unit Definition: *One unit will liberate 1.0 μmole of N-acetylneuraminic acid per minute at pH 6.5 at 37°C using 3'-NAcetylneuramin-lactose as the substrate.*

Final Assay Concentrations

In a 0.251 ml reaction mix, the final concentrations are 150 mM sodium acetate, 450 mM sodium chloride, 27 mM calcium chloride, 1.0 mM 3'-N-acetylneuramin-lactose, 0.05% (w/v) Bovine Serum Albumin, and 0.0013–0.003 unit neuraminidase.

Enzymatic Assay of NEURAMINIDASE INSOLUBLE (EC 3.2.1.18) N-Acetylneuramin-Lactose as Substrate

Principle

$$\text{N-Acetylneuramin-Lactose} + H_2O \xrightarrow{\text{Neuraminidase}} \text{NANA} + \text{Lactose}$$

Abbreviation Used

NANA = N-Acetylneuraminic Acid

Conditions: T = 37°C, pH = 5.0, A550nm, Light path = 1 cm

Method: Colorimetric

Reagents

A. 100 mM Sodium Acetate Buffer with 2 mM Calcium Chloride, pH 5.0 at 37°C (Prepare 100 ml in deionized water using Sodium Acetate Anhydrous, and Calcium Chloride, Dihydrate. Adjust to pH 5.0 at 37°C with 1 M HCl).

B. 0.085% (w/v) N-Acetylneuramin-Lactose (2 >3) isomer Solution (NAN-Lactose) (Prepare 5 ml in Reagent A using N-Acetylneuramin-Lactose).

C. 0.2% (w/v) Bovine Serum Albumin Solution (BSA) (Prepare 100 ml in deionized water using Albumin, Bovine).

D. 5% (w/v) Phosphotungstic Acid Solution (PT) (Prepare 100 ml in 2.5 M HCl using Phosphotungstic Acid Free Acid).

E. Neuraminidase Insoluble Enzyme Solution (Immediately before use, mix by swirling and place 2–4 ml of suspension on a Whatman #50 filter paper and remove the suspension medium and soluble Neuraminidase by vacuum filtration. Stop the vacuum filtration before the gel cake becomes dry or begins to crack. It should be moist. Wash the gel cake with 10 volumes of deionized water, and again stop the vacuum filtration before the gel becomes dry or begins to crack. Resuspend 0.9 g of the washed gel cake with 1 ml of deionized water. Mix by swirling to form a homogenous gel slurry. Then dilute to 0.02–0.04 unit/ml of Neuraminidase in cold Reagent C).

F. 9 M Phosphoric Acid Solution (Prepare 20 ml in deionized water using Phosphoric Acid).

G. 200 mM Sodium m-Periodate Solution (Per) (Prepare 10 ml in Reagent F using Sodium m-Periodate).

H. 10% (w/v) Sodium m-Arsenite with 50 mM Sulfuric Acid and 500 mM Sodium Sulfate Solution (Ars) (Prepare 100 ml in deionized water using Sodium m-Arsenite, Sulfuric Acid, and Sodium Sulfate, Anhydrous).

I. 0.6% (w/v) Thiobarbituric Acid and 500 mM Sodium Sulfate Solution (TBA) (Prepare 100 ml in deionized water using 2-Thiobarbituric Acid and Sodium Sulfate, Anhydrous).

J. 5% (v/v) HCl in Butanol Solution (But) (Prepare by combining 5 ml of concentrated Hydrochloric Acid and 95 ml of n-Butanol).

K. 0.16 mM N-Acetylneuraminic Acid Standard Solution (NANA) (Prepare 10 ml in deionized water using N-Acetylneuraminic Acid).

Procedure

Pipette (in milliliters) the following reagents into suitable containers:

	Test	Blank
Reagent A (Buffer)	0.20	0.20
Reagent B (NAN-Lactose)	0.20	0.20

Mix by inversion and equilibrate to 37°C. Then add:

	Test	Blank
Reagent E (Enzyme Solution)	0.10	—
Reagent C (BSA)	—	0.10

Mix by inversion immediately and incubate at 37°C for exactly 10 minutes. Then add:

Reagent D (PT)	0.50	0.50

Mix by inversion immediately and centrifuge for 3 minutes.

Colorimetric Assay

Standard Curve: Pipette (in milliliters) the following reagents into suitable containers:

	Std 1	Std 2	Std 3	Std 4	Std Blank
Reagent K (NANA)	0.05	0.10	0.20	0.30	0.00
Deionized water	0.45	0.40	0.30	0.20	0.50
Rseagent D (PT)	0.50	0.50	0.50	0.50	0.50

Mix by inversion and centrifuge for 3 minutes. Pipette (in milliliters) the following reagents into suitable containers:

	Test	Test Blank	Std 1	Std 2	Std 3	Std 4	Std Blank
Test Supernatant	0.50	—	—	—	—	—	—
Test Blank Supernatant	—	0.50	—	—	—	—	—
Std 1 Supernatant	—	—	0.50	—	—	—	—
Std 2 Supernatant	—	—	—	0.50	—	—	—
Std 3 Supernatant	—	—	—	—	0.50	—	—
Std 4 Supernatant	—	—	—	—	—	0.50	—
Std Blank Supernatant	—	—	—	—	—	—	0.50
Reagent G (Per)	0.10	0.10	0.10	0.10	0.10	0.10	0.10

Mix by inversion and incubate for 20 minutes at 25°C. Then add:

Reagent H (Ars)	1.00	1.00	1.00	1.00	1.00	1.00	1.00

Vortex until the yellow-brown color disappears. Then add:

Reagent I (TBA)	3.00	3.00	3.00	3.00	3.00	3.00	3.00

Transfer to a boiling water bath and incubate for 15 minutes. Transfer to a cold water bath for 5 minutes. Then add:

Reagent J (But)	4.60	4.60	4.60	4.60	4.60	4.60	4.60

Shake or vortex the solution until most of the pink color has been extracted from the lower layer. Centrifuge for 2–3 minutes. Transfer the upper layer to suitable cuvettes and record the A550nm for Test, Test Blank, Standard, and Standard Blank.

Calculations

Standard Curve: ΔA550nm Standard = A550nm Standard–A550nm Standard blank
Plot the ΔA550nm Standard vs μmoles n-Acetylneuraminic acid.

Sample Determination: ΔA550nm Test = A550nm Test–A550nm Test Blank
Determine the μmoles of n-Acetylneuraminic Acid (NANA) liberated using the Standard curve.

$$\text{Units / ml enzyme} = \frac{(\mu\text{moles of NANA liberated})\ (\text{df})}{(0.1)\ (10)}$$

df = Dilution factor
0.1 = Volume (in milliliter) of enzyme used
10 = Time (in minutes) of the assay as per the Unit Definition
Units/ml of Packed Gel = Units/ml enzyme × 2

$$\text{Units / g agarose} = \frac{(\text{units / ml enzyme})\ (1000)}{15.7}$$

1000 = Conversion of mg to g
15.7 = mg agarose per ml of suspension

Unit Definition: *One unit will liberate 1.0 μmole of N-acetylneuraminic acid per minute at pH 5.0 at 37°C using NAN-lactose as substrate.*

Final Assay Concentrations

In a 0.50 ml reaction mix, the final concentrations are 80 mM sodium acetate, 1.6 mM calcium chloride, 0.034% (w/v) n-acetylneuramin-lactose (2 >3 isomer), 0.04% (w/v) Bovine Serum Albumin, and 0.002–0.004 unit neuraminidase.

Notes

1. This assay system is operating at less than the Km value for the substrate. It is important that the concentration of the 2>3 isomer is 0.085% (w/v).
2. Final volumes should be 0.50 ml for all test solutions performed.

Enzymatic Assay of NEURAMINIDASE (EC 3.2.1.18)
N-Acetylneuramin-Lactose as Substrate

Principle

$$\text{N - Acetylneuramin - Lactose} + H_2O \xrightarrow{\text{Neuraminidase}} \text{NANA} + \text{Lactose}$$

Abbreviation Used

NANA = N-Acetylneuraminic Acid

Conditions: T = 37°C, pH = 5.0, A550nm, Light path = 1 cm

Method: Colorimetric

Reagents

A. 100 mM Sodium Acetate Buffer with 2 mM Calcium Chloride, pH 5.0 at 37°C (Prepare 100 ml in deionized water using Sodium Acetate Anhydrous, and Calcium Chloride, Dihydrate. Adjust to pH 5.0 at 37°C with 1 M HCl).

B. 0.085% (w/v) N-Acetylneuramin-Lactose (2 >3) isomer Solution (NAN-Lactose) (Prepare 5 ml in Reagent A using N-Acetylneuramin- Lactose).

C. 0.2% (w/v) Bovine Serum Albumin Solution (BSA) (Prepare 100 ml in deionized water using Albumin, Bovine).

D. 5% (w/v) Phosphotungstic Acid Solution (PT) (Prepare 100 ml in 2.5 M HCl using Phosphotungstic Acid Free Acid).

E. Neuraminidase Enzyme Solution (Immediately before use, prepare a solution containing approximately 0.02–0.04 unit/ml of Neuraminidase in cold Reagent C).

F. 9 M Phosphoric Acid Solution (Prepare 20 ml in deionized water using Phosphoric Acid).

G. 200 mM Sodium m-Periodate Solution (Per) (Prepare 10 ml in Reagent F using Sodium m-Periodate).

H. 10% (w/v) Sodium m-Arsenite with 50 mM Sulfuric Acid and 500 mM Sodium Sulfate Solution (Ars) (Prepare 100 ml in deionized water using Sodium m-Arsenite, Sulfuric Acid and Sodium Sulfate, Anhydrous).

I. 0.6% (w/v) Thiobarbituric Acid and 500 mM Sodium Sulfate Solution (TBA) (Prepare 100 ml in deionized water using 2-Thiobarbituric Acid and Sodium Sulfate, Anhydrous).

J. 5% (v/v) HCl in Butanol Solution (But) (Prepare by combining 5 ml of concentrated Hydrochloric Acid and 95 ml of n-Butanol).

K. 0.16 mM N-Acetylneuraminic Acid Standard Solution (NANA) (Prepare 10 ml in deionized water using N-Acetylneuraminic Acid).

Procedure

Pipette (in milliliters) the following reagents into suitable containers:

	Test	Blank
Reagent A (Buffer)	0.20	0.20
Reagent B (NAN-Lactose)	0.20	0.20

Mix by inversion and equilibrate to 37°C. Then add:

Reagent E (Enzyme Solution)	0.10	—
Reagent C (BSA)	—	0.10

Mix by inversion immediately and incubate at 37°C for exactly 10 minutes. Then add:

Reagent D (PT)	0.50	0.50

Mix by inversion immediately and centrifuge for 3 minutes.

Colorimetric Assay

Standard Curve: Pipette (in milliliters) the following reagents into suitable containers:

	Std 1	Std 2	Std 3	Std 4	Std Blank
Reagent K (NANA)	0.05	0.10	0.20	0.30	0.00
Deionized water	0.45	0.40	0.30	0.20	0.50
Reagent D (PT)	0.50	0.50	0.50	0.50	0.50

Mix by inversion and centrifuge for 3 minutes. Pipette (in milliliters) the following reagents into suitable containers:

	Test	Test Blank	Std 1	Std 2	Std 3	Std 4	Std Blank
Test Supernatant	0.50	—	—	—	—	—	—
Test Blank Supernatant	—	0.50	—	—	—	—	—
Std 1 Supernatant	—	—	0.50	—	—	—	—
Std 2 Supernatant	—	—	—	0.50	—	—	—
Std 3 Supernatant	—	—	—	—	0.50	—	—
Std 4 Supernatant	—	—	—	—	—	0.50	—
Std Blank Supernatant	—	—	—	—	—	—	0.50
Reagent G (Per)	0.10	0.10	0.10	0.10	0.10	0.10	0.10

Mix by inversion and incubate for 20 minutes at 25°C. Then add:

Reagent H (Ars)	1.00	1.00	1.00	1.00	1.00	1.00	1.00

Vortex until the yellow-brown color disappears. Then add:

Reagent I (TBA)	3.00	3.00	3.00	3.00	3.00	3.00	3.00

Transfer to a boiling water bath and incubate for 15 minutes. Transfer to a cold water bath for 5 minutes. Then add:

Reagent J (But)	4.60	4.60	4.60	4.60	4.60	4.60	4.60

Shake or vortex the solution until most of the pink color has been extracted from the lower layer. Centrifuge for 2–3 minutes. Transfer the upper layer to suitable cuvettes and record the A550nm for Test, Test Blank, Standard, and Standard Blank.

Calculations

Standard Curve: ΔA550nm Standard = A550nm Standard–A550nm Standard blank
Plot the ΔA550nm Standard vs μmoles n-Acetylneuraminic acid.
Sample Determination: ΔA550nm Test = A550nm Test–A550nm Test Blank
Determine the μmoles of n-Acetylneuraminic Acid (NANA) liberated using the Standard curve.

$$\text{Units / ml enzyme} = \frac{(\mu\text{moles of NANA liberated})\ (\text{df})}{(0.1)\ (10)}$$

df = Dilution factor
0.1 = Volume (in milliliter) of enzyme used
10 = Time (in minutes) of the assay as per the Unit Definition

$$\text{Units / mg solid} = \frac{\text{units / ml enzyme}}{\text{mg solid / ml enzyme}}$$

$$\text{Units / mg protein} = \frac{\text{units / ml enzyme}}{\text{mg protein / ml enzyme}}$$

Unit Definition: *One unit will liberate 1.0 μmole of N-acetylneuraminic acid per minute at pH 5.0 at 37°C using NAN-lactose as substrate.*

Final Assay Concentrations

In a 0.50 ml reaction mix, the final concentrations are 80 mM sodium acetate, 1.6 mM calcium chloride, 0.034% (w/v) n-acetylneuramin-lactose (2 >3 isomer), 0.04% (w/v) Bovine Serum Albumin, and 0.002–0.004 unit neuraminidase.

Notes

1. Do not use this assay for Neuraminidase from Newcastle disease virus and Neuraminidase from *Streptococcus sp.*
2. Only Neuraminidase from Vibrio cholera requires calcium ions. However, the presence of calcium ions does not interfere with the activity of Neuraminidases from other sources.
3. This assay system is operating at less than the Km value for the substrate. It is important that the concentration of the 2>3 isomer is 0.085% (w/v).
4. Final volumes should be 0.50 ml for all test solutions performed.

Enzymatic Assay of NEURAMINIDASE (EC 3.2.1.18)

Principle

$$\text{MUN} + H_2O \xrightarrow{\text{Neuraminidase}} \text{NANA} + \text{MU}$$

Abbreviations Used

MUN = 2'-(4-Methylumbelliferyl)-a-D-N-Acetylneuraminic Acid
NANA = N-Acetylneuraminic Acid
MU = 4-Methylumbelliferone

Conditions: T = 37°C, pH 5.5, A364nm, Light path = 1 cm

Method: Stopped Spectrophotometric Rate Determination

Reagents

A. 1 M Sodium Acetate Buffer, pH 5.5 at 37°C (Prepare 100 ml in deionized water using Sodium Acetate, Trihydrate. Adjust to pH 5.5 at 37°C with 1 M HCl).
B. 100% (v/v) N,N-Dimethylformamide.
C. 250 mM Glycine Buffer, pH 10.4 at 37°C (Stop Buffer) (Prepare 25 ml in deionized water using Glycine, Free Base. Adjust to pH 10.4 at 37°C with 1 M NaOH).
D. 200 mM 2'-(4-Methylumbelliferyl)- α-D-NAcetylneuraminic Acid Stock Solution (Prepare 0.255 ml in Reagent B using 2'-(4-Methylumbelliferyl)-α-D-N-Acetylneuraminic Acid, Sodium Salt. Vortex to dissolve).

E. 1.2 mM 2'-(4-Methylumbelliferyl)- α-D-NAcetylneuraminic Acid Solution (MUN) (Prepare by adding 0.0125 ml of Reagent D to 0.125 ml of Reagent A. Bring to a volume of 2 ml with deionized water).

F. Neuraminidase Enzyme Solution (Immediately before use, prepare a solution containing approximately 0.08 unit/ml of Neuraminidase in cold Reagent A. Before use, warm the enzyme solution to 37°C).

Procedure

Step 1: Pipette (in milliliters) the following reagents into suitable containers:

	Test	Blank
Reagent E (MUN)	0.40	0.40

Incubate at 37°C for 10 minutes. Then add:

Reagent F (Enzyme Solution)	0.10	—
Reagent A (Buffer)	—	0.10

Mix by inversion and incubate at 37°C.

Step 2: At one minute, two minute and three minute time points, remove 0.10 ml from the Test and Blank and add to predispensed aliquots of 1.00 ml of Reagent C (Stop Buffer). Transfer the solutions to suitable cuvettes and record the A364nm for both the Tests and Blank.

Calculations

$$\text{Units / ml} = \frac{(\text{A364nm Test} - \text{A364nm Blank})\,(0.5)\,(1.1)\,(\text{df})}{(25.3)\,(T)\,(0.1)\,(0.1)}$$

0.5 = Volume (in milliliter) of assay in Step 1
1.1 = Volume (in milliliters) of stopped reaction in Step 2
25.3 = Millimolar extinction coefficient of 4-Methylumbelliferone at 364nm
df = Dilution factor
T = Time (in minutes) of assay as per the Unit Definition
0.1 = Volume (in milliliter) of enzyme used in Step 1
0.1 = Volume (in milliliter) of Step 1 used in Step 2

$$\text{Units / mg solid} = \frac{\text{units / ml enzyme}}{\text{mg solid / ml enzyme}}$$

$$\text{Units / mg protein} = \frac{\text{units / ml enzyme}}{\text{mg protein / ml enzyme}}$$

Unit Definition: *One unit will hydrolyze 1.0 μmole of 2'-(4-methylumbelliferyl) α-D-N-acetylneuraminic acid per minute at pH 5.5 at 37°C.*

Final Assay Concentrations

In a 0.50 ml reaction mix, the final concentrations are 250 mM sodium acetate, 0.5% (v/v) N,N-dimethylformamide, 1 mM 2'-(4-methylumbelliferyl)- α-D-N-acetylneuraminic acid, and 0.008 unit neuraminidase.

Enzymatic Assay of NITRATE REDUCTASE (EC 1.6.6.1)

Principle

$$\text{Nitrate} + \beta\text{-NADH} \xrightarrow{\text{Nitrate Reductase}} \text{Nitrite} + \beta\text{-NAD} + H_2O$$

$$\text{Nitrate} + \text{Sulfanilamide} + \text{NED} \rightarrow \text{Nitrite Color Complex}$$

Abbreviations Used

β-NADH = β-Nicotinamide Adenine Dinucleotide, Reduced Form
β-NAD = β-Nicotinamide Adenine Dinucleotide, Oxidized Form
NED = N-(1-Naphthyl) Ethylenediamine Dihydrochloride

Conditions: T = 30°C, pH = 7.3, A540nm, Light path = 1 cm

Method: Spectrophotometric Stop Rate Determination

Reagents

A. 25 mM Potassium Phosphate Buffer with 10 mM Potassium Nitrate and 0.05 mM Ethylenediaminetetraacetic Acid, pH 7.3 at 30°C (Prepare 100 ml in deionized water using Potassium Phosphate, Monobasic, Anhydrous, Potassium Nitrate and Ethylenediaminetetraacetic Acid, Disodium Salt, Dihydrate).
B. 2.0 mM β-Nicotinamide Adenine Dinucleotide, Reduced Form Solution (β-NADH) (Dissolve the contents of one 10 mg vial of β-Nicotinamide Adenine Dinucleotide, Reduced Form, Disodium Salt, in the appropriate volume of deionized water. *Prepare Fresh*).
C. 3 M Hydrochloric Acid Solution (HCl) (Prepare 100 ml in deionized water using Hydrochloric Acid).
D. 58 mM Sulfanilamide Solution (Sulf) (Prepare 100 ml in Reagent C using Sulfanilamide).
E. 0.77 mM N-(1-Naphthyl) Ethylenediamine Dihydrochloride Solution (NED) (Prepare 100 ml in deionized water using N-(1-Naphthyl)ethylenediamine Dihydrochloride).
F. 1.45 mM Nitrate Standard Solution (Std Soln) (Prepare 50 ml in Reagent A using Sodium Nitrite).
G. Nitrate Reductase Enzyme Solution (Immediately before use, prepare a soluiton containing 0.2–0.6 unit/ml of Nitrate Reductase in cold Reagent A).

Procedure

Pipette (in milliliters) the following reagents into suitable containers:

	Test	Blank	Std 1	Std 2	Std 3	Std 4	Std 5
Reagent A (Buffer)	1.80	1.90	1.89	1. 87	1.85	1.83	1.80
Reagent B (β-NADH)	0.10	0.10	0.10	0.10	0.10	0.10	0.10
Reagent F (Std Soln)	—	—	0.01	0.03	0.05	0.07	0.10

Mix by swirling and equilibrate to 30°C. Then add:

	Test	Blank	Std 1	Std 2	Std 3	Std 4	Std 5
Reagent G (Enz Soln)0.10	0.10	—	—	—	—	—	—

Immediately mix by swirling and incubate at 30°C for exactly 2 minutes. Stop the reaction by adding:

	Test	Blank	Std 1	Std 2	Std 3	Std 4	Std 5
Reagent D (Sulf)	1.00	1.00	1.00	1.00	1.00	1.00	1.00

Mix thoroughly by swirling, then add:

Reagent E (NED)	1.00	1.00	1.00	1.00	1.00	1.00	1.00

Mix by swirling and incubate for 10 minutes at 25°C. Transfer the solutions to suitable cuvettes and record the A540nm for the Test, Blank and Standards using a suitable spectrophotometer.

Calculations

Standard Curve: ΔA540nm Standard = A540nm Std–A540nm Std Blank
Prepare a standard curve by plotting the A540nm of the Standard versus μmoles of nitrite.

Sample Determination: ΔA540nm Sample = A540nm Test–A540nm Test Blank
Determine the μmoles of nitrite formed using the standard curve.

$$\text{Units / ml enzyme} = \frac{(\mu\text{mole Nitrite formed})\ (\text{df})}{(2)\ (0.1)}$$

df = Dilution factor
2 = Time of assay (in minutes)
0.1 = Volume of enzyme (in milliliter) used

$$\text{Units / mg solid} = \frac{\text{units / ml enzyme}}{\text{mg solid / ml enzyme}}$$

$$\text{Units / mg protein} = \frac{\text{units / ml enzyme}}{\text{mg protein / ml enzyme}}$$

Unit Definition: *One unit will reduce 1.0 μmole of nitrate to nitrite per minute in a β-NADH system at pH 7.3 at 30°C.*

Final Assay Concentrations

In a 2.00 ml reaction mix, the final concentrations are 24 mM potassium phosphate, 0.05 mM Ethylenediaminetetraacetic Acid, 9.5 mM potassium nitrate, 0.10 mM β-nicotinamide adenine dinucleotide, reduced form, and 0.02–0.06 unit nitrate reductase.

Note

1. Perform the color development of the Test and Standards at the same time in order to prevent deviations in color development from the standard curve.

Enzymatic Assay of NITRATE REDUCTASE (CYTOCHROME) (EC 1.9.6.1)

Principle

$$\text{Methyl Viologen (red)} + \text{Nitrate} \xrightarrow{\text{Nitrate Reductase}} \text{Methyl Viologen (ox)} + \text{Nitrite} + H_2O$$

Abbreviations Used

red = reduced
ox = oxidized

Conditions: T = 30°C, pH = 7.0, A540nm, Light path = 1 cm

Method: Colorimetric

Reagents

A. 150 mM Potassium Phosphate Buffer, pH 7.0 at 30°C (Prepare 100 ml in deionized water using Potassium Phosphate, Monobasic, Anhydrous. Adjust to pH 7.0 at 30°C with 1 M KOH. *Prepare Fresh*).
B. 100 mM Sodium Nitrate Solution ($NaNO_3$) (Prepare 10 ml in deionized water using Sodium Nitrate. *Prepare Fresh*).
C. 0.02% (w/v) Methyl Viologen Solution (Viologen) (Prepare 10 ml in deionized water using Methyl Viologen. *Prepare Fresh*).
D. 0.8% (w/v) Sodium Hydrosulfite Solution (HS) (Prepare 10 ml in deionized water using Sodium Hydrosulfite. *Prepare Fresh*).
E. 0.8% (w/v) Sodium Bicarbonate Solution (Bicarb) (Prepare 10 ml in deionized water using Sodium Bicarbonate).
F. Sodium Hydrosulfite/Sodium Bicarbonate Solution (Hydro/Bicarb) (Immediately before use, prepare by adding 5 ml of Reagent D and 5 ml of Reagent E to a suitable container. Do not shake).
G. 3 M Hydrochloric Acid Solution (HCl) (Prepare 10 ml in deionized water using Hydrochloric Acid. *Prepare Fresh*).
H. 1% (w/v) Sulfanilamide Solution (Sulfa) (Prepare 5 ml in Reagent G using Sulfanilamide. *Prepare Fresh*).
I. 0.01% (w/v) N-(1-Naphthyl) Ethylenediamine Solution (Naphthyl) (Prepare 100 ml in deionized water using N-(1-Naphthyl)ethylenediamine Dihydrochloride. *Prepare Fresh*).
J. Nitrate Reductase Enzyme Solution (Immediately before use, prepare a solution containing approximately 0.075 unit/ml of Nitrate Reductase in Cold deionized water).

Procedure

Pipette (in milliliters) the following reagents into suitable containers:

	Test	Blank
Deionized Water	—	0.10
Reagent A (Buffer)	0.10	0.10
Reagent B ($NaNO_3$)	0.10	0.10
Reagent C (Viologen)	0.10	0.10
Reagent J (Enzyme Solution)	0.10	—

Mix by inversion and equilibrate to 30°C. Then add:

	Test	Blank
Reagent F (Hydro/Bicarb)	0.10	0.10

Mix carefully (do not shake) until the blue color is uniform and incubate at 30°C for exactly 10 minutes. Shake vigorously until the blue color disappears. Then quickly add:

	Test	Blank
Reagent H (Sulfa)	0.50	0.50
Reagent I (Naphthyl)	0.50	0.50

Mix by inversion and incubate at 30°C for 10 minutes. Then add:

Deionized Water	1.50	1.50

Mix by inversion and record the A540nm for both the Test and Blank using a suitable spectrophotometer.

Calculations

$$\text{Units / ml enzyme} = \frac{(\text{A540nm Test} - \text{A540nm Blank})\,(3.0)\,(\text{df})}{(10)\,(17.2)\,(0.1)}$$

3.0 = Total volume of Colorimetric Assay
df = Dilution factor
10 = Time (in minutes) of the assay as per the Unit Definition
17.2 = Millimolar extinction coefficient of the Diazo-nitrite compound at 540nm
0.1 = Volume (in milliliter) of enzyme used in the assay

$$\text{Units / mg solid} = \frac{\text{units / ml enzyme}}{\text{mg solid / ml enzyme}}$$

Unit/g solid = units/mg solid × 1000

Unit Definition: *One unit will reduce 1.0 μmole of nitrate to nitrite per minute at pH 7.0 at 30°C in a methyl viologen system.*

Final Assay Concentrations

In a 0.50 ml reaction mix, the final concentrations are 30 mM potassium phosphate, 0.08% (w/v) sodium hydrosulfite, 0.08% (w/v) sodium bicarbonate, 0.004% (w/v) methyl viologen, 20 mM sodium nitrate, and 0.0075 unit nitrate reductase.

Notes

1. Methyl violgen is utilized as an electron donor in this reaction.
2. Dissolve all reagents in deionized water which have been boiled for 10 minutes and then cooled. After cooling, degas the water to further ensure the removal of oxygen. To dissolve the reagents mix gently to avoid aeration. Never shake any of the reagents. After preparing the different reagents, store in vials which are tightly capped. Store on ice at 0–5°C. It is best to have a minimum volume of air in these containers.
3. Allow the reagents to flow down the side of the vial. It is essential that all of the reagents be virtually free of oxygen in order to prevent the oxidation of the reduced methyl viologen, the substrate for the reaction. If the substrate is oxidized, no blue color will appear and no enzymatic reaction will occur.
4. Before the addition of Reagent F (Hydro/Bicarb) gently invert the container. Do not agitate or mix the solution since this may introduce some oxygen into the mixture.

Enzymatic Assay of NITRATE REDUCTASE (NAD(P)H) (EC 1.6.6.2)

Principle

$$\beta\text{-NADPH} + \text{Nitrate} \xrightarrow{\text{Nitrate Reductase}} \beta\text{-NADP} + \text{Nitrite} + H_2O$$

Abbreviations Used

β-NADPH = β-Nicotinamide Adenine Dinucleotide Phosphate, Reduced Form
β-NADP = β-Nicotinamide Adenine Dinucleotide Phosphate, Oxidized Form

Conditions: T = 25°C, pH = 7.5, A340nm, Light path = 1 cm

Method: Continuous Spectrophotometric Rate Determination

Reagents

A. 100 mM Potassium Phosphate Solution (Prepare 100 ml in deionized water using Potassium Phosphate, Monobasic, Anhydrous).
B. 100 mM Potassium Phosphate Buffer, pH 7.5 at 25°C (Prepare 100 ml in deionized water using Potassium Phosphate, Dibasic, Trihydrate. Adjust to pH 7.5 at 25°C with Reagent A).
C. 0.1 mM Flavin Adenine Dinucleotide Solution (FAD) (Prepare 100 ml in deionized water using Flavin Adenine Dinucleotide, Disodium Salt. *Prepare Fresh and Protect from Light*).
D. 12 mM β-Nicotinamide Adenine Dinucleotide Phosphate, Reduced Form Solution (β-NADPH) (Dissolve the contents of one 10 mg vial of β-Nicotinamide Adenine Dinucleotide Phosphate, Reduced Form, Tetrasodium Salt in the appropriate volume of deionized water. *Prepare Fresh*).
E. 100 mM Potassium Nitrate Solution (KNO_3) (Prepare 1 ml in deionized water using Potassium Nitrate).
F. Nitrate Reductase Enzyme Solution (Immediately before use, prepare a solution containing 0.13–0.25 unit/ml of Nitrate Reductase in cold Reagent B. *Store on Ice;* use within 5 minutes).

Procedure

Pipette (in milliliters) the following reagents into suitable cuvettes:

	Test	Blank
Deionized Water	0.80	1.10
Reagent B (Buffer)	1.60	1.60
Reagent C (FAD)	0.15	0.15
Reagent D (β-NADPH)	0.05	0.05
Reagent E (KNO_3)	0.30	—

Mix by inversion and equilibrate to 25°C. Monitor the A340nm until constant, using a suitably thermostatted spectrophotometer. Then add:

	Test	Blank
Reagent F (Enzyme Solution)	0.10	0.10

Immediately mix by inversion and record the decrease in A340nm for approximately 5 minutes. Obtain the ΔA340nm / minute using the maximum linear rate for both the Test and Blank.

Calculations

$$\text{Units / ml enzyme} = \frac{(\Delta A340nm / \text{min Test} - \Delta A340nm / \text{min Blank})\,(3)\,(df)}{(6.22)\,(0.1)}$$

3 = Total volume (in milliliters) of assay
df = Dilution factor
6.22 = Millimolar extinction coefficient of β-NADPH at 340nm
0.1 = Volume (in milliliter) of enzyme used

$$\text{Units / mg solid} = \frac{\text{units / ml enzyme}}{\text{mg solid / ml enzyme}}$$

$$\text{Units / mg protein} = \frac{\text{units / ml enzyme}}{\text{mg protein / ml enzyme}}$$

Unit Definition: *One unit will reduce 1.0 μmole of nitrate per minute in the presence of β-NADPH at pH 7.5 at 25°C.*

Final Assay Concentration

In a 3.00 ml reaction mix, the final concentrations are 57 mM potassium phosphate, 0.005 mM flavin adenine dinucleotide, 0.2 mM β-nicotinamide adenine dinucleotide phosphate, reduced form, 10 mM potassium nitrate and 0.013–0.025 unit nitrate reductase.

Enzymatic Assay of NITRIC OXIDE SYNTHETASE (EC 1.14.13.39)

Principle

$$\text{L-Arginine} + \beta\text{-NADPH} + O_2 \xrightarrow[\text{Ca}_2^+,\ \text{Tetrahydrobiopterin, Calmodulin}]{\text{Nitric Oxide Synthetase}} \text{Citrulline} + \text{NO} + \beta\text{-NADP}$$

Abbreviations Used

β-NADPH = β-Nicotinamide Adenine Dinucleotide Phosphate, Reduced Form
β-NADP = β-Nicotinamide Adenine Dinucleotide Phosphate, Oxidized Form

Conditions: T = 37°C, pH = 7.3

Method: HPLC Analysis of Products

Reagents

A. 80 mM HEPES Buffer with 1.0 mM Ethylenediaminetetraacetic Acid and 1.5 mM Calcium Chloride, pH 7.3 at 37°C (Prepare 100 ml in deionized water using HEPES, Free Acid Ethylenediaminetetraacetic Acid, Tetrasodium Salt, Hydrate and Calcium Chloride, Dihydrate. Adjust to pH 7.3 at 37°C with 1 M NaOH).

B. Calmodulin Solution (CaM) (Immediately before use, prepare a solution containing 5000 units/ml of Phosphodiesterase 3':5'-Cyclic Nucleotide Activator (Calmodulin) in cold Reagent A).

C. 20 mM DL-Dithiothreitol Solution (DTT) (Prepare 5 ml in Reagent A using DL-Dithiothreitol).
D. 1.0 mM Tetrahydrobiopterin Solution (THB) (Immediately before use, prepare 10 ml in Reagent A using (6R)-5,6,7,8-Tetrahydrobiopterin, Dihydrochloride. *Prepare Fresh, Keep on Ice*).
E. 10 μM Citrulline Standard Solution (Cit Std) (Prepare 10 ml in Reagent A using L-Citrulline).
F. 1.0 mM Flavin Adenine Dinucleotide Solution (FAD) (Prepare 10 ml in Reagent A using Flavin Adenine Dinucleotide, Disodium Salt).
G. 3.4 mM L-Arginine Solution (L-Arg) (Prepare 10 ml in Reagent A using L-Arginine, Hydrochloride).
H. 3.0 mM Nicotinamide Adenine Dinucleotide Phosphate, Reduced Form Solution (NADPH) (Prepare 10 ml in Reagent A using Nicotinamide Adenine Dinucleotide Phosphate, Reduced Form, Tetrasodium Salt).
I. 100% (v/v) Methanol (MeOH) (Use Absolute Methanol).
J. 10 mM Sodium Phosphate and 10 mM Triethylamine Buffer, pH 5.0 at 37°C (Buffer A) (Prepare 1 L in deionized water using Sodium Phosphate, Dibasic, Anhydrous and Triethylamine. Adjust to pH 5.0 at 37°C with 1 M HCl or 1 M NaOH. Before using, filter through a 0.22 μm filter).
K. Acetonitrile (Buffer B) (Acetonitrile).
L. o-Phthaldialdehyde Reagent Solution (OPA).
M. Nitric Oxide Synthetase Enzyme Solution (Immediately before use, prepare a solution containing 40–50 units/ml of Nitric Oxide Synthetase in cold Reagent A).

Procedure

Pipette (in milliliters) the following reagents into suitable containers:

	Test	Blank
Reagent A (Buffer)	1.95	1.75
Reagent B (CaM)	0.20	0.20
Reagent C (DTT)	0.20	0.20
Reagent D (THB)	0.20	0.20
Reagent F (FAD)	0.05	0.05

Mix by inversion and equilibrate to 37°C. Then add:

Reagent M (Enz Soln)	0.10	—
Reagent A (Buffer)	—	0.10

Incubate at 37°C for 30–60 seconds. Then add:

Reagent G (L-Arg)	0.10	0.10
Reagent H (NADPH)	0.20	0.20

Immediately mix by inversion and incubate for exactly 10 minutes at 37°C. After incubating for 10 minutes, pipette 0.5 ml of the Test and Blank into a 2 ml Eppendorf tube containing 0.5 ml of ice cold Reagent I (Methanol). *Store on ice.* Prepare Citrulline Standards by pipetting (in milliliters) the following reagents into 2 ml Eppendorf tubes:

	Std 1	Std 2	Std 3	Std 4	Std 5	Std 6	Std Blank
Deionized Water	0.49	0.45	0.40	0.30	0.20	—	0.50
Reagent E (Cit Std)	0.01	0.05	0.10	0.20	0.30	0.50	—
Reagent I (MeOH)	0.50	0.50	0.50	0.50	0.50	0.50	0.50

Store on ice. Centrifuge all samples to remove any precipitates.

Step 2: HPLC analysis of OPA derivatized samples:

1. **Column:** Supelcosil LC-18, 4.6 x 150 mm, C18, 5μm particle size.

Mobile Phase: Equilibrate the column with 5% Buffer B at 1.0 ml/min. Run a linear gradient of 5–15% Buffer B over 10 minutes, followed by 15–100% Buffer B over 3 minutes. Hold at 100% Buffer B for 5 minutes. Re-equilibrate the column for 15 minutes with 5% Buffer B. The gradient may need adjusting as the column ages.

Detection: Use fluorescence detection at an excitation wavelength of 340nm and an emission wavelength of 455 nm.

2. Warm a 10 ml aliquot of Reagent L (OPA) to room temperature (23°C).

3. Set up auto sampler to dispense 0.1 ml Reagent L (OPA) into 0.1 ml of sample in a HPLC vial. If auto sampler is not available then, accurately add 0.1 ml Reagent L (OPA) solution to 0.1 ml of the sample to be assayed. Mix for 1 minute, immediately inject 20 ul of derivatized sample onto the HPLC column. *This reaction is carried out at room temperature.* All of the reaction times need to be consistent with OPA + Sample before each injection. The OPA reagent will react with the primary amino acid(s) and the derivatized product detected by fluorescence.

Note: Ammonia, Proline and Cysteine are not detected by OPA.

4. Inject blank then standards of Citrulline and samples. A comparison can then be made between standard curve of Citrulline and Citrulline generated from the sample reaction.

Calculations

$$\text{Units / ml enzyme} = \frac{\text{(nmoles of citrulline) (df) (3)}}{(0.1)\,(10)}$$

df = Dilution factor
3 = Volume (in milliliters) of assay
0.1 = Volume (in milliliter) of enzyme used
10 = Time (in minutes) of assay as per the Unit Definition

$$\text{Units / mg solid} = \frac{\text{units / ml enzyme}}{\text{mg solid / ml enzyme}}$$

$$\text{Units / mg protein} = \frac{\text{units / ml enzyme}}{\text{mg protein / ml enzyme}}$$

Unit Definition: *One unit will convert 1.0 nmole of arginine to citrulline per minute at pH 7.3 at 37°C.*

Final Assay Concentration

In a 3.00 ml reaction mix, the final concentrations are 80 mM HEPES, 1 mM Ethylenediaminetetraacetic Acid, 1.5 mM calcium chloride, 1000 units calmodulin, 1.3 mM DL-dithiothreitol, 0.07 mM tetrahydrobiopterin, 0.02 mM flavin adenine dinucleotide, 0.1 mM arginine, 0.2 mM β-nicotinamide adenine dinucleotide phosphate, reduced form, and 4–5 units nitric oxide synthetase.

Note

1. Reagent G (L-Arg) and Reagent H (NADPH) should be prewarmed to 37°C before adding to the reaction.

Enzymatic Assay of NUCLEASE P1 (EC 3.1.30.1) RNA as Substrate

Principle

$$RNA + H_2O + \xrightarrow{\text{Nuclease P1}} \text{Acid Soluble Polynucleotides}$$

Abbreviation Used

RNA = Ribonucleic Acid

Conditions: T = 37°C, pH = 5.3, A260nm, Light path = 1 cm.

Method: Spectrophotometric Stop Rate Determination

Reagents

A. 0.59% (w/v) Barbital Acetate Buffer, pH 5.3 at 37°C (Prepare 25 ml in deionized water using Barbital Buffer. Adjust to pH 5.3 at 37°C with 1 M HOAc).
B. 0.2% (w/v) Ribonucleic Acid Solution (RNA) (Prepare 4 ml in deionized water using Ribonucleic Acid).
C. 2.5% (v/v) Perchloric Acid Solution ($HClO_4$) (Prepare 5 ml in deionized water using Perchloric Acid).
D. 0.25% (w/v) Uranyl Acetate Solution (Uran Acet) (Prepare 5 ml in Reagent C using Uranyl Acetate. Store on ice).
E. Nuclease P1 Enzyme Solution (Immediately before use, prepare a solution containing 0.1–0.2 unit/ml of Nuclease P1 in cold deionized water).

Procedure

Pipette (in milliliters) the following reagents into suitable containers:

	Test	Blank
Reagent A (Buffer)	0.20	0.20
Reagent B (RNA)	0.20	0.20

Mix by swirling and equilibrate to 37°C. Then add:

Reagent E (Enzyme Solution)	0.10	—
Deionized Water	—	0.10

Immediately mix by swirling and incubate for exactly 15 minutes at 37°C. Then add:

	Test	Blank
Reagent D (Uran Acet)	0.50	0.50

Mix by swirling and place in an ice bath for 20 minutes. Centrifuge for 5 minutes. Remove 0.80 ml of the supernatant and dilute to 2.80 ml with deionized water. Transfer the solutions to suitable cuvettes and obtain the A260nm for the Test and Blank.

Calculations

$$\text{Units / ml enzyme} = \frac{(\text{A260nm Test} - \text{A260nm Blank})\,(1)\,(2.8)\,(\text{df})}{(10.6)\,(15)\,(0.1)\,(0.8)}$$

1 = Volume (in milliliter) of stopped reaction.
2.8 = Final volume (in milliliters) of assay.
df = Dilution factor.
10.6 = Millimolar extinction coefficient of hydrolyzed ribonucleic acid at 260nm.
15 = Time (in minutes) of assay as per the Unit Definition.
0.1 = Volume (in milliliter) of enzyme used.
0.8 = Volume (in milliliter) of stopped reaction used in the spectrophotometric determination.

$$\text{Units / mg solid} = \frac{\text{units / ml enzyme}}{\text{mg solid / ml enzyme}}$$

$$\text{Units / mg protein} = \frac{\text{units / ml enzyme}}{\text{mg protein / ml enzyme}}$$

Unit Definition: *One unit will liberate 1.0 μmole of acid soluble nucleotides from RNA (R6625) per minute at pH 5.3 at 37°C.*

Final Assay Concentrations

In a 0.50 ml reaction mix, the final concentrations are 0.24% (w/v) barbital, 0.08% (w/v) ribonucleic acid, and 0.01–0.02 unit nuclease P1.

Enzymatic Assay of NUCLEASE P1 (EC 3.1.30.1) 3'-AMP as Substrate

Principle

$$\text{3'-AMP} + H_2O \xrightarrow{\text{Nuclease P1}} \text{Adenosine} + \text{Pi}$$

Abbreviations Used

3'-AMP = Adenosine 3'-Monophosphate
Pi = Orthophosphate

Conditions: T = 37°C, pH = 7.2, A660nm, Light path = 1 cm

Method: Spectrophotometric Stop Rate Determination

Reagents

A. 50 mM Tris HCl Buffer, pH 7.2 at 37°C (Prepare 100 ml in deionized water using Trizma Base. Adjust to pH 7.2 at 37°C with 1 M HCl).

B. 10 mM Adenosine 3'-Monophosphate Solution (3'-AMP) (Prepare 5 ml in Reagent A using Adenosine 3'-Monophosphate, Free Acid).
C. 10% (w/v) Ammonium Molybdate Solution (Prepare 25 ml in 10 N H_2SO_4, using Molybdate Acid, Ammonium Tetrahydrate Salt).
D. Taussky-Shorr Color Reagent (TSCR) (Prepare by adding 10 ml of Reagent C to 70 ml deionized water. Add 5 g Ferrous Sulfate, Heptahydrate. Bring the volume to 100 ml with deionized water. *Store in a Dark Container; Prepare Fresh Daily*).
E. Phosphorus Standard (P Std) (Use Phosphorus Standard Solution. The phosphorus concentration is 20 µg/ml, 0.645 µmoles).
F. 6% (w/v) Perchloric Acid Solution ($HClO_4$) (Prepare 50 ml in deionized water using Perchloric Acid).
G. Nuclease P1 Enzyme Solution (Immediately before use, prepare a solution containing 0.2–0.5 unit/ml in cold deionized water).

Procedure

Step 1: Pipette (in milliliters) the following reagent into suitable containers:

	Test	Blank
Reagent A (Buffer)	0.20	0.20
Reagent B (3'-AMP)	0.20	0.20

Equilibrate at 37°C for 5 minutes. Then add:

Reagent G (Enzyme Solution)	0.10	—

Incubate at 37°C for exactly 15 minutes. Then add:

Reagent F ($HClO_4$)	4.00	4.00
Reagent G (Enzyme Solution)	—	0.10
Deionized Water	0.50	0.50
Reagent D (TSCR)	5.00	5.00

Mix and incubate for 5 minutes at 25°C. Read the A660nm for both the Test and Blank using a suitable spectrophotometer.

Step 2: Standard Curve: Prepare a standard curve by pipetting (in milliliters) the following reagents into suitable container:

	Std 1	Std 2	Std 3	Std 4	Std 5	Std Blank
Reagent E (P Std)	0.10	0.30	0.50	0.70	1.00	—
Deionized Water	0.90	0.70	0.50	0.30	—	1.00
Reagent F ($HClO_4$)	4.00	4.00	4.00	4.00	4.00	4.00
Reagent D (TSCR)	5.00	5.00	5.00	5.00	5.00	5.00

Mix by incubate for 5 minutes at 25°C. Read the A660nm for the Standard and Standard Blank using a suitable spectrophotometer.

Calculations

Standard Curve: ΔA660nm Standard = A660nm Standard–A660nm Standard Blank

Prepare a standard curve by plotting ΔA660nm Standard vs µmoles of phosphate.

Sample Determination: ΔA660nm Test = A660nm Test–A660nm Blank

Determine the micromoles of Phosphate liberated using the Standard curve.

$$\text{Units / ml enzyme} = \frac{(\mu\text{moles of Phosphate released})\ (\text{df})}{(15)\ (0.1)}$$

df = Dilution factor
15 = Time (in minutes) of assay
0.1 = Volume (in milliliter) of enzyme used

$$\text{Units / mg solid} = \frac{\text{units / ml enzyme}}{\text{mg solid / ml enzyme}}$$

$$\text{Units / mg protein} = \frac{\text{units / ml enzyme}}{\text{mg protein / ml enzyme}}$$

Unit Definition: *One unit will hydrolyze 1.0 µmole of orthophosphate from 3'-AMP per min at pH 7.2 at 37°C.*

Final Assay Concentration

In a 0.5 ml reaction mix, the final concentrations are 40 mM Tris, 4 mM adenosine 3'-monophosphate and 0.1 unit nuclease P1.

Notes

1. Glassware must be clean and free of any soap residue, as soap contains phosphates and will contaminate, the assay.
2. TSCR should be added to the samples and standards at the same time and the time interval from TSCR addition to measuring the absorbance on the spectrophotometer must be the same for standards and samples.

Enzymatic Assay of NUCLEASE S1 as a Impurity Assay (EC 3.1.30.1)

Principle

$$\text{DNA} \xrightarrow{\text{Nuclease S1}} \text{Acid Soluble Polynucleotides}$$

Conditions: T = 37°C, pH = 4.6, A260nm, Light path = 1 cm

Method: Spectrophotometric Stop Rate Determination

Reagents

A. 30 mM Sodium Acetate Buffer with 50 mM Sodium Chloride, 0.10 mM Zinc Sulfate and 5.0% (v/v) Glycerol, pH 4.6 at 37°C (Prepare 100 ml in deionized water using Sodium Acetate, Trihydrate, Sodium Chloride, Zinc Sulfate, Heptahydrate, and Glycerol. Adjust to pH 4.6 at 37°C with 1 M HCl).

B. 10 mM Tris HCl Solution, pH 7.5 at 37°C (Tris) (Prepare 25 ml in deionized water using Trizma Base. Adjust to pH 7.5 at 37°C with 1 M HCl).

C. 0.2% (w/v) Deoxyribonucleic Acid Solution (DNA) (Prepare 25 ml in Reagent B using Deoxyribonucleic Acid, Sodium Salt. Solubilize by allowing to stand 12–16 hours at 2–8°C. Then stir gently at room temperature for 30–45 minutes).

D. Nuclease S1 Enzyme Solution (use neat, do not dilute).

E. 10% (v/v) Perchloric Acid Solution ($HClO_4$) (Prepare 50 ml in deionized water using Perchloric Acid).

Procedure

Pipette (in milliliters) the following reagents into suitable containers:

	Test	Blank
Reagent A (Buffer)	5.40	5.40
Reagent C (DNA)	0.60	0.60

Mix by swirling and equilibrate at 37°C. Then add:

Reagent D (Enzyme Solution)	0.10	—
Deionized Water	—	0.10

Mix by swirling and incubate for 15 minutes at 37°C. Remove 1.00 ml from each container at 0, 5, 10, and 15 minutes and add 1.0 ml each to separate containers containing 2.00 ml of Reagent E ($HClO_4$) each. Mix by swirling and incubate in an ice bath for 5 minutes. Centrifuge the samples for 10 minutes and transfer the supernatants to suitable cuvettes. Record the absorbance at 260nm for all the Test and Blank aliquots.

Calculations

$$\text{A260nm / min Test Tt} = \frac{\text{A260 nm Test Tt}}{\text{Tt}}$$

Tt = Time (in minutes) at which the reaction is stopped

$$\text{Units / ml enzyme} = \frac{(\text{A260nm / min Test Tt}) - (\text{A260nm / min Blank Tt})\ (6.1)\ (\text{df})}{(0.033)\ (1)\ (0.1)}$$

Tt = Time (in minutes) at which an aliquot is removed from the reaction mix
6.1 = Volume (in milliliters) of reaction mix
df = Dilution factor
0.033 = Absorbance for 1 microgram of nucleic acid in perchloric acid
1 = Volume (in milliliters) removed from the reaction mix
0.1 = Volume (in milliliter) of enzyme used

$$\text{Units / mg solid} = \frac{\text{units / ml enzyme}}{\text{mg solid / ml enzyme}}$$

$$\text{Units / mg protein} = \frac{\text{units / ml enzyme}}{\text{mg protein / ml enzyme}}$$

Unit Definition: *One unit will cause 1.0 microgram of single-stranded nucleic acid to become perchloric acid-soluble per minute at pH 4.6 at 37°C.*

Final Assay Concentration

In a 6.10 ml reaction mix, the final concentrations are 27 mM sodium acetate, 44 mM sodium chloride, 0.088 mM zinc sulfate, 4.4% (v/v) glycerol, 0.02% (w/v) deoxyribonucleic acid, 0.98 mM Tris and varying amounts of units of nuclease S1.

Enzymatic Assay of NUCLEASE, *MICROCOCCAL* (EC 3.1.31.1)

Principle

$$DNA + H_2O \xrightarrow{\text{Nuclease, } \textit{Micrococcal}} \text{Acid Soluble Polynucleotides}$$

Conditions: T = 37°C, pH = 8.8, A260nm, Light path = 1 cm

Method: Spectrophotometric Stop Rate Determination

Reagents

A. 100 mM Borate Buffer, pH 8.8 at 37°C (Prepare 25 ml in deionized water using Boric Acid. Adjust to pH 8.8 at 37°C with 1 M NaOH).
B. 10 mM Sodium Chloride Solution (NaCl) (Prepare 50 ml in deionized water using Sodium Chloride).
C. 0.25% (w/v) Deoxyribonucleic Acid Solution (DNA) (Prepare 5 ml in Reagent B using Deoxyribonucleic Acid, Sodium Salt. Gentle agitation may be required to obtain complete dissolution).
D. 10 mM Calcium Chloride Solution ($CaCl_2$) (Prepare 10 ml in deionized water using Calcium Chloride Dihydrate).
E. 0.1% (w/v) Bovine Serum Albumin Solution (BSA) (Prepare 100 ml in deionized water using Albumin, Bovine).
F. Nuclease, *Micrococcal* Enzyme Solution (Immediately before use, prepare a solution containing 0.04–0.08 unit/ml of *Micrococcal* Nuclease in cold Reagent E).
G. 7% (v/v) Perchloric Acid Solution ($HClO_4$) (Prepare 10 ml in deionized water using Perchloric Acid).

Procedure

Pipette (in milliliters) the following reagents into suitable tubes:

	Test	Blank
Reagent A (Buffer)	0.20	0.20
Reagent C (DNA)	0.20	0.20
Reagent D ($CaCl_2$)	0.10	0.10

Mix by swirling and equilibrate at 37°C. Then add:

Reagent E (BSA)	—	0.20
Reagent F (Enzyme)	0.20	—

Mix by swirling and incubate for exactly 30 minutes at 37°C. Then add:

	Test	Blank
Reagent G ($HClO_4$)	1.00	1.00

Place tubes in an ice bath for 10 minutes. Then add:

Deionized water	5.40	5.40

Mix by swirling and centrifuge for 10 minutes. Transfer the supernatants from both the Test and Blank to suitable cuvettes and record the A260nm for each using a suitable spectrophotometer.

Calculation

$$\text{Units / ml enzyme} = \frac{(\text{A260nm Test} - \text{A260nm Blank})\ (7.1)\ (\text{df})}{(10)\ (30)\ (0.2)}$$

7.1 = Total volume (in milliliters) of assay
df = Dilution factor
10 = Millimolar extinction coefficient of soluble polynucleotides at 260nm.
30 = Time (in minutes) of the assay as per the Unit Definition
0.2 = Volume (in milliliter) of enzyme used

$$\text{Units / mg solid} = \frac{\text{units / ml enzyme}}{\text{mg solid / ml enzyme}}$$

$$\text{Units / mg protein} = \frac{\text{units / ml enzyme}}{\text{mg protein / ml enzyme}}$$

Unit Definition: *One unit will produce 1 μmole of acid soluble polynucleotides from native DNA per minute at pH 8.8 at 37°C, based on EM 260 = 10,000 for the mixed nucleotides.*

Final Assay Concentrations

In a 0.70 ml reaction mix, the final concentrations are 29 mM borate, 2.9 mM sodium chloride, 0.07% (w/v) deoxyribonucleic acid, 1.4 mM calcium chloride, 0.03% (w/v) Bovine Serum Albumin and .008–.016 unit of nuclease, *micrococcal.*

Notes

1. Stir at 4°C until complete dissolution is obtained. This may take several hours. Do not store overnight to dissolve the DNA.
2. The enzyme concentration must be kept at approximately 1 mg protein per ml until just prior to dilution for use in the reaction mix.

Enzymatic Assay of NUCLEOSIDE 5'-DIPHOSPHATE KINASE (EC 2.7.4.6)

Principle

$$\text{ATP} + \text{TDP} \xrightarrow{\text{NDPK}} \text{ADP} + \text{TTP}$$

$$\text{ADP} + \text{PEP} \xrightarrow{\text{PK}} \text{ATP} + \text{Pyruvate}$$

$$\text{Pyruvate} + \beta\text{-NADH} \xrightarrow{\text{LDH}} \text{L-Lactate} + \beta\text{-NAD}$$

Abbreviations Used

ATP = Adenosine 5'-Triphosphate
ADP = Adenosine 5'-Diphosphate
TDP = Thymidine 5'-Diphosphate
NDPK = Nucleoside 5'-Diphosphate Kinase
β-NADH = β-Nicotinamide Adenine Dinucleotide, Reduced Form
β-NAD = β-Nicotinamide Adenine Dinucleotide, Oxidized Form
PEP = Phospho(enol) pyruvate
LDH = Lactic Dehydrogenase
PK = Pyruvate Kinase

Conditions: T = 25°C, pH = 7.6, A340nm, Light path = 1 cm

Method: Continuous Spectrophotometric Rate Determination

Reagents

A. 100 mM Triethanolamine Buffer, pH 7.6 at 25°C (Prepare 100 ml in deionized water using Triethanolamine Hydrochloride. Adjust to pH 7.6 at 25°C with 1 M NaOH).
B. 21 mM Thymidine 5'-Diphosphate Solution (TDP) (Prepare 2 ml in deionized water using Thymidine 5'-Diphosphate, Sodium Salt).
C. 33 mM Adenosine 5'-Triphosphate Solution (ATP) (Prepare 3 ml in deionized water using Adenosine 5'-Triphosphate, Disodium Salt).
D. 12 mM β-Nicotinamide Adenine Dinucleotide, Reduced Form Solution (β-NADH) (Prepare by dissolving the contents of one 10 mg vial of β-Nicotinamide Adenine Dinucleotide, Reduced Form, Disodium Salt in the appropriate volume of deionized water. *Prepare Fresh*).
E. 500 mM Magnesium Chloride and 2000 mM Potassium Chloride Solution ($MgCl_2$/KCl) (Prepare 10 ml in deionized water using Magnesium Chloride, 4.9 M Solution and Potassium Chloride).
F. 33 mM Phospho(enol) pyruvate Solution (PEP) (Prepare 2 ml in Reagent E using Phospho(enol) pyruvate, Tri(cyclohexylammonium Salt)).
G. PK/LDH Mixed Enzyme Solution (PK/LDH).
H. Nucleoside 5'-Diphosphate Kinase Enzyme Solution (NDPK) (Immediately before use, prepare a solution containing 0.5–1.0 unit/ml of Nucleoside 5'-Diphosphate Kinase in cold deionized water).

Procedure

Pipette (in milliliters) the following reagents into suitable cuvettes:

	Test	Blank
Reagent A (Buffer)	2.49	2.49
Reagent B (TDP)	0.10	0.10
Reagent C (ATP)	0.20	0.20
Reagent D (β-NADH)	0.05	0.05
Reagent F (PEP)	0.10	0.10
Reagent G (PK/LDH)	0.01	0.01

Mix by inversion and equilibrate to 25°C. Monitor the A340nm until constant, using a suitably thermostatted spectrophotometer. Then add:

	Test	Blank
Reagent H (NDPK)	0.05	—
Deionized Water	—	0.05

Immediately mix by inversion and record the decrease in A340nm for approximately 5 minutes. Obtain the ΔA340nm/minute using the maximum linear rate for both the Test and Blank.

Calculations

$$\text{Units / ml enzyme} = \frac{(\Delta A340\text{nm / min Test} - \Delta A340\text{nm / min Blank})\ (3)\ (\text{df})}{(6.22)\ (0.05)}$$

3 = Total volume (in milliliters) of assay
df = Dilution factor
6.22 = Millimolar extinction coefficient of β-NADH at 340nm
0.05 = Volume (in milliliters) of assay

$$\text{Units / mg solid} = \frac{\text{units / ml enzyme}}{\text{mg solid / ml enzyme}}$$

$$\text{Units / mg protein} = \frac{\text{units / ml enzyme}}{\text{mg protein / ml enzyme}}$$

Unit Definition: *One unit will convert 1.0 μmole each of TDP and ATP to TTP and ADP per minute at pH 7.6 at 25°C in a coupled system with PK/LDH.*

Final Assay Concentration

In a 3.00 ml reaction mix, the final concentrations are 83 mM triethanolamine, 0.70 mM thymidine 5'-diphosphate, 2.2 mM adenosine 5'-triphosphate, 0.2 mM β-NADH, 1.1 mM phospho(enol) pyruvate, 16.7 mM magnesium chloride, 66.7 mM potassium chloride, 10 units lactic dehydrogenase, 7 units pyruvate kinase and 0.025–0.050 unit nucleoside 5'-diphosphate kinase.

Notes

1. Unit Definition for L-Lactic Dehydrogenase: One unit will reduce 1.0 μmole of pyruvate to L-lactate per minute at pH 7.5 at 37°C.
2. Unit Definition for Pyruvate Kinase: One unit will convert 1.0 μmole of phospho(enol) pyruvate to pyruvate per minute at pH 7.6 at 37°C.

Enzymatic Assay of NUCLEOSIDE MONOPHOSPHATE KINASE (EC 2.7.4.4)

Principle

$$\text{ATP + UMP} \xrightarrow{\text{Nucleoside Monophosphate Kinase}} \text{ADP + UDP}$$

$$\text{ADP + UDP + 2PEP} \xrightarrow{\text{Pyruvate Kinase}} \text{2 Pyruvate + ATP + UTP}$$

$$\text{2 Pyruvate} + 2\beta\text{-NADH} \xrightarrow{\text{Lactic Dehydrogenase}} \text{2 Lactate} + 2\beta\text{-NAD}$$

Abbreviations Used

ATP = Adenosine 5'-Triphosphate
UMP = Uridine 5'-Monophosphate
ADP = Adenosine 5'-Diphosphate
UDP = Uridine 5'-Diphosphate
PEP = Phospho(enol) pyruvate
UTP = Uridine 5'-Triphosphate
β-NADH = β-Nicotinamide Adenine Dinucleotide, Reduced Form
β-NAD = β-Nicotinamide Adenine Dinucleotide, Oxidized Form

Conditions: T = 25°C, pH = 7.6, A340nm, Light path = 1 cm

Method: Continuous Spectrophotometric Rate Determination

Reagents

A. 100 mM Triethanolamine Buffer, pH 7.6 at 25°C (Prepare 50 ml in deionized water using Triethanolamine Hydrochloride. Adjust to pH 7.6 at 25°C with 1 N NaOH).
B. 27 mM Uridine 5'-Monophosphate Solution (UMP) (Prepare 10 ml in deionized water using Uridine 5'-Monophosphate, Disodium Salt).
C. 18 mM Adenosine 5'-Triphosphate Solution (ATP) (Prepare 10 ml in deionized water using Adenosine 5'-Triphosphate, Disodium Salt).
D. 19 mM Phospho(enol) pyruvate Solution (PEP) (Prepare 10 ml in deionized water using Phospho(enol) pyruvate, Mono(cyclohexylammonium) Salt. *Prepare Fresh*).
E. 500 mM Magnesium Sulfate with 2 M Potassium Chloride Solution ($MgSO_4$/KCl) (Prepare 10 ml in deionized water using Magnesium Sulfate, Heptahydrate and Potassium Chloride).
F. 14 mM β-Nicotinamide Adenine Dinucleotide, Reduced Form Solution (β-NADH) (Dissolve the contents of one 10 mg vial of β-Nicotinamide Adenine Dinucleotide, Reduced Form, Disodium Salt in the appropriate volume of deionized water. *Prepare Fresh*).
G. PK/LDH Enzymes Suspension.
H. Nucleoside Monophosphate Kinase Enzyme Solution. (Immediately before use, prepare a solution containing 0.1–0.2 unit/ml of Nucleoside Monophosphate Kinase in cold Reagent A).

Procedure

Prepare a reaction cocktail by pipetting (in milliliters) the following reagents into a suitable container:

Reagent A (Buffer)	21.7
Reagent B (UMP)	1.0
Reagent C (ATP)	3.0
Reagent D (PEP)	2.0
Reagent E ($MgSO_4$/KCl)	1.0
Reagent F (β-NADH)	0.8

Mix by swirling and adjust to pH 7.6 at 25°C with 1 M HCl or 1 M NaOH, if necessary. Pipette (in milliliters) the following reagents into suitable cuvettes:

	Test	Blank
Reaction Cocktail	2.95	2.95
Reagent G (PK/LDH)	0.05	0.05

Mix by inversion and equilibrate to 25°C. Monitor the A340nm until constant, using a suitably thermostatted spectrophotometer. Then add:

Reagent A (Buffer)	—	0.10
Reagent H (Enzyme Solution)	0.10	—

Immediately mix by inversion and record the decrease in A340nm for approximately 5 minutes. Obtain the ΔA340nm/minute using the maximum linear rate for both the Test and Blank.

Calculations

$$\text{Units / ml enzyme} = \frac{(\Delta A340\text{nm / min Test} - \Delta A340\text{nm / min Blank})\,(3.1)\,(\text{df})}{(2)\,(6.22)\,(0.1)}$$

3.1 = Total volume (in milliliters) of assay
df = Dilution factor
2 = 2 moles of β-NAD produced per mole of ATP used
6.22 = Millimolar extinction coefficient of β-NADH at 340nm
0.1 = Volume (in milliliters) of enzyme used

$$\text{Units / mg solid} = \frac{\text{units / ml enzyme}}{\text{mg solid / ml enzyme}}$$

$$\text{Units / mg protein} = \frac{\text{units / ml enzyme}}{\text{mg protein / ml enzyme}}$$

Unit Definition: *One unit will convert 1.0 μmole of UMP and ATP to UDP and ADP per minute at pH 7.6 at 25°C, using a coupled enzyme system with PK/LDH.*

Final Assay Concentrations

In a 3.10 ml reaction mix, the final concentrations are 73 mM triethanolamine, 0.87 mM uridine 5'-monophosphate, 1.7 mM adenosine 5'-triphosphate, 1.2 mM phospho(enol) pyruvate, 16 mM magnesium sulfate, 65 mM potassium chloride, 0.36 mM β-nicotinamide adenine dinucleotide, reduced form, 35 units pyruvate kinase, 50 units lactic dehydrogenase and 0.01–0.02 unit nucleoside monophosphate kinase.

Notes

1. Contains not less than 700 units/ml of Pyruvate Kinase and 1000 units/ml of Lactic Dehydrogenase.
2. Pyruvate Kinase Unit Definition: One unit will convert 1.0 μmole of phospho(enol) pyruvate to pyruvate per minute at pH 7.6 at 37°C.
3. Lactic Dehydrogenase Unit Definition: One unit will reduce 1.0 μmole of pyruvate to L-lactate per minute at pH 7.5 at 37°C.

Enzymatic Assay of NUCLEOSIDE PHOSPHORYLASE (EC 2.4.2.1)

Principle

$$\text{Inosine} + \text{Pi} \xrightarrow{\text{Nucleoside Phosphorylase}} \text{Hypoxanthine} + \text{Ribose-1-}PO_4$$

$$\text{Hypoxanthine} + 2H_2O + 2O_2 \xrightarrow{\text{Xanthine Oxidase}} \text{Uric Acid} + 2H_2O_2$$

Abbreviation Used

Pi = Inorganic Phosphate

Conditions: T = 25°C, pH = 7.4, A293nm, Light path = 1 cm

Method: Continuous Spectrophotometric Rate Determination

Reagents

A. 100 mM Potassium Phosphate Buffer, pH 7.4 at 25°C (Prepare 100 ml in deionized water using Potassium Phosphate, Monobasic. Adjust to pH 7.4 at 25°C with 1 M NaOH).

B. 7.5 mM Inosine Solution (Prepare 5 ml in deionized water using Inosine).

C. Xanthine Oxidase Enzyme Solution (Immediately before use, prepare a solution containing 10 units/ml of Xanthine Oxidase in ice cold deionized water).

D. Nucleoside Phosphorylase Enzyme Solution (Immediately before use, prepare a solution containing 0.125 units/ml of Nucleoside Phosphorylase in ice cold deionized water).

Procedure

Pipette (in milliliters) the following reagents into suitable cuvettes:

	Test	Blank
Reagent A (Buffer)	2.70	2.70
Reagent B (Inosine)	0.10	0.10
Reagent C (Xanthine Oxidase)	0.10	0.10

Mix by inversion and equilibrate to 25°C. Monitor the A293nm until constant, using a suitably thermostatted spectrophotometer. Then add:

	Test	Blank
Reagent D (Nucleoside Phosphorylase)	0.10	—
Deionized Water	—	0.10

Immediately mix by inversion and record the increase in A293nm for approximately 5 minutes. Obtain the ΔA293nm/minute using the maximum linear rate for both the Test and Blank.

Calculations

$$\text{Units / mg enzyme} = \frac{(\Delta A293\text{nm / min Test} - \Delta A293\text{nm / min Blank})}{(12.0)\ (\text{mg enzyme / ml RM})}$$

12.0 = Millimolar extinction coefficient of Uric Acid at 293nm

RM = Reaction Mix

Unit Definition: *One unit will cause the phosphorolysis of 1.0 µmole of inosine to hypoxanthine and ribose 1-phosphate per minute at pH 7.4 at 25°C.*

Final Assay Concentration

In a 3.0 ml reaction mix, the final concentrations are 90 mM potassium phosphate, 0.25 mM inosine, 1.0 units of xanthine oxidase and 0.0125 units of nucleoside phosphorylase.

Note

1. Xanthine Oxidase–One unit will convert 1.0 µmole of xanthine to uric acid per minute at pH 7.5 at 25°C.

Enzymatic Assay of 5'-NUCLEOTIDASE (EC 3.1.3.5)

Principle

$$\text{5'-AMP} + H_2O \xrightarrow{\text{5'-Nucleotidase}} \text{Adenosine} + \text{Pi}$$

Abbreviations Used

5'-AMP = Adenosine 5'-Monophosphate
Pi = Inorganic Phosphate

Conditions: T = 37°C, pH 7.5, A578nm, Light path = 1 cm

Method: Colorimetric

Reagents

A. 40 mM Veronal Buffer, pH 7.5 at 37°C (Prepare 100 ml in deionized water using Barbital, Free Acid. Adjust to pH 7.5 with 1 M NaOH).
B. 20 mM Manganese Sulfate Solution ($MnSO_4$) (Prepare 100 ml in deionized water using Manganese Sulfate).
C. 100 mM Nickel Chloride Solution ($NiCl_2$) (Prepare 100 ml in deionized water using Nickel Chloride, Hexahydrate).
D. 10 mM Adenosine 5'-Monophosphate Solution (5'-AMP) (Prepare 10 ml in deionized water using Adenosine 5'-Monophosphate, Sodium Salt).
E. 11% (w/v) Trichloroacetic Acid Solution (TCA) (Prepare 100 ml in deionized water using Trichloroacetic Acid).
F. 2.5 M Sodium Acetate Solution (NaOAc) (Prepare 250 ml in deionized water using Sodium Acetate, Trihydrate).
G. Acid Ammonium Molybdate Solution (Molyb).
H. Fiske and Subbarow Reducer Solution (F&S) (Use Fiske and Subbarow Reducer. Prepare by dissolving 158.7 mg of Fiske and Subbarow Reducer in 1 ml of deionized water).
I. Phosphorus Standard (P Std) (Use Phosphorus Standard Solution. The phosphorus concentration is 20 µg/ml, 0.645 µmoles/ml).
J. 0.01% (w/v) Bovine Serum Albumin (BSA) (Prepare 10 ml in deionized water using Albumin, Bovine).
K. 5'-Nucleotidase Enzyme Solution (Immediately before use, prepare a solution containing 0.03–0.12 unit/ml of 5'-Nucleotidase in cold Reagent J).

Procedure

Step 1: Pipette (in milliliters) the following reagents into suitable test tubes:

	Uninhibited Test	Inhibited Test
Reagent A (Buffer)	1.25	1.25
Reagent B ($MnSO_4$)	0.10	0.10
Reagent K (Enzyme Solution)	0.20	0.20
Reagent C ($NiCl_2$)	—	0.25
Deionized Water	0.25	—

Mix by swirling and equilibrate to 37°C. Then add:

Reagent D (5'-AMP)	0.20	0.20

Immediately mix by swirling and incubate for exactly 30 minutes at 37°C. Then add:

Reagent E (TCA)	2.00	2.00

Centrifuge for 5 minutes at 3000 rpm.

Step 2: Pipette (in milliliters) the following reagents into suitable containers:

	Uninhibited Test	Inhibited Test	Std 1	Std 2	Std Blank
Uninhibited Test Supernatant	0.80	—	—	—	—
Inhibited Test Supernatant	—	0.80	—	—	—
Reagent I (P Std)	—	—	0.10	0.40	—
Deionized Water	—	—	0.70	0.40	0.80
Reagent G (Molyb)	1.50	1.50	1.50	1.50	1.50
Reagent H (F&S)	0.20	0.20	0.20	0.20	0.20

Mix by swirling and allow to stand for 10 minutes at room temperature. Then add:

Reagent F (NaOAc)	0.50	0.50	0.50	0.50	0.50

Allow to stand at room temperature for 5 minutes. Transfer to suitable cuvettes and record the A578nm for the Uninhibited Test, Inhibited Test, Standards, and Standard Blank using a suitable spectrophotometer.

Calculations

Standard Curve: ΔA578nm Standard = A578nm Standard–A578nm Standard Blank

Prepare a standard curve by plotting ΔA578nm Standard versus µg of phosphate.

Sample Determination: ΔA578nm Test = A578nm Uninhibited Test–A578nm Inhibited Test

Determine the µg of phosphate liberated using the standard curve.

$$\text{Units / ml enzyme} = \frac{(\mu\text{g of Phosphate released})\ (4)\ (\text{df})}{(31)\ (30)\ (0.8)\ (0.2)}$$

4 = Volume (in milliliters) of stopped reaction in Step 1
df = Dilution factor
31 = Molecular weight of phosphorus
30 = Time (in minutes) of assay as per the Unit Definition
0.8 = Volume (in milliliter) of Step 1 used in Step 2
0.2 = Volume (in milliliter) of enzyme used in Step 1

$$\text{Units / mg solid} = \frac{\text{units / ml enzyme}}{\text{mg solid / ml enzyme}}$$

$$\text{Units / mg protein} = \frac{\text{units / ml enzyme}}{\text{mg protein / ml enzyme}}$$

Unit Definition: *One unit will hydrolyze 1.0 μmole of inorganic phosphorus from adenosine 5'-monophosphate per minute at pH 7.5 at 30°C.*

Final Assay Concentrations

In a 2.00 ml reaction mix, the final concentrations are 25 mM barbital, 1 mM manganese sulfate, 12.5 mM nickel chloride (when present), 1 mM adenosine 5'-monophosphate, 0.001% (w/v) Bovine Serum Albumin and 0.006–0.024 unit 5'-nucleotidase.

Note

1. The uninhibited test represents both non-specific phosphatase activity and 5'-nucleotidase activity. The 5'-nucleotidase is inhibited by nickel chloride. The difference between the uninhibited and inhibited enzyme activity represents the 5'-nucleotidase activity.

Enzymatic Assay of 5'-NUCLEOTIDASE (EC 3.1.3.5)

Principle

$$\text{5'-AMP} + H_2O \xrightarrow{\text{5'-Nucleotidase}} \text{Adenosine} + \text{Pi}$$

Abbreviations Used

5'-AMP = Adenosine 5'-Monophosphate
Pi = Inorganic Phosphate.

Conditions: T = 37°C, pH = 9.0, A660nm, Light path = 1 cm

Method: Colorimetric

Reagents

A. 200 mM Glycine Buffer, pH 9.0 at 37°C (Prepare 100 ml in deionized water using Glycine, Free Base. Adjust to pH 9.0 at 37°C with 1 M NaOH).
B. 66 mM Adenosine 5'-Monophosphate Solution (5'-AMP) (Psrepare 5 ml in deionized water using Adenosine 5'-Monophosphate, Sodium Salt).
C. 200 mM Magnesium Sulfate ($MgSO_4$) (Prepare 10 ml in deionized water using Magnesium Sulfate, Anhydrous).
D. Phosphorus Standard (P Std) (Use Phosphorus Standard Solution. The Phosphorus concentration is 20 μg/ml, 0.645 μmoles/ml).
E. 10% (w/v) Ammonium Molybdate Solution (Prepare 25 ml in 10 N H_2SO_4 using Molybdic Acid, Ammonium Tetrahydrate Salt).

F. Taussky-Shorr Color Reagent (TSCR) (Prepare 100 ml by adding 10 ml Reagent E to 70 ml of deionized water. Then add a 5 g vial of Ferrous Sulfate Heptahydrate, and mix until dissolved. Add enough deionized water for a final volume of 100 ml).

G. 5'-Nucleotidase Enzyme Solution (Immediately before use, prepare a solution containing 40–60 units/ml of 5'-Nucleotidase in cold deionized water).

Procedure

Pipette (in milliliters) the following reagents into suitable containers:

	Test 1	Test 2	Blank
Reagent A (Buffer)	1.50	1.50	1.50
Reagent B (5'-AMP)	0.15	0.15	0.15
Reagent C ($MgSO_4$)	0.10	0.10	0.10
Deionized Water	0.24	0.24	0.24

Mix and equilibrate to 37°C. Then add:

Reagent G (Enzyme Solution)	0.005	0.010	—

Immediately mix and incubate at 37°C for 1 minute (Test 1) and 2 minutes (Test 2). Then add:

Reagent F (TSCR)	4.00	4.00	4.00
Reagent G (Enzyme Solution)	0.005	—	0.010
Deionized Water	2.00	2.00	2.00

Mix by inversion and incubate for 5 minutes. Transfer to suitable cuvettes and record the A660nm for both Test 1, Test 2, and Blank using a suitable spectrophotometer.

Colorimetric Assay

Standard Curve: A standard curve is made by pipetting (in milliliters) the following reagents into suitable containers:

	Std 1	Std 2	Std 3	Std 4	Std 5	Std Blank
Deionized Water	3.80	3.60	3.40	3.20	3.00	4.00
Reagent D (P Std)	0.20	0.40	0.60	0.80	1.00	—
Reagent F (TSCR)	4.00	4.00	4.00	4.00	4.00	4.00

Mix by inversion and transfer the solutions to suitable cuvettes. Record the A660nm for both the Standards and Standard Blank.

Calculations

Standard Curve: A660nm Standard = A660nm Standard–A660nm Standard Blank

Prepare a standard curve by plotting A660nm Standard versus μmoles of Phosphate.

Sample Determination: A660nm Test = A660nm Test–A660nm Test Blank

Determine the μmoles of Phosphate liberated using the Standard curve.

$$\text{Units / ml enzyme} = \frac{(\mu\text{moles of Phosphate released})\ (\text{df})}{(0.005)\ (\text{T})}$$

df = Dilution factor.
0.005 = Volume (in milliliter) of enzyme used, or 0.01 ml.
T = Time (in minutes) of assay as per the Unit Definition

$$\text{Units / mg solid} = \frac{\text{units / ml enzyme}}{\text{mg solid / ml enzyme}}$$

$$\text{Units / mg protein} = \frac{\text{units / ml enzyme}}{\text{mg protein / ml enzyme}}$$

Unit Definition: *One unit will hydrolyze 1.0 µmole of inorganic phosphorus from adenosine 5'-monophosphate per min at pH 9.0 at 37°C.*

Final Assay Concentration

In a 2.00 ml reaction mix the final concentrations are 150 mM glycine, 5.0 mM adenosine 5'-monophosphate, 10 mM magnesium sulfate, and 0.2–0.6 unit 5'-nucleotidase.

Notes

1. This assay is not to be used to assay 5'-Nucleotidase, from bovine liver, or 5'-Nucleotidase, from Crotalus adamanteus, Insoluble.
2. The enzyme assay is to be run for both the 1.0 and 2.0 minutes incubation times.

Enzymatic Assay of 3'-NUCLEOTIDASE (EC 3.1.3.6)

Principle

$$\text{3'-AMP} + H_2O \xrightarrow{\text{3'-Nucleotidase}} \text{Adenosine} + \text{Pi}$$

Abbreviations Used

3'-AMP = Adenosine 3'-Monophosphate
Pi = Inorganic Phosphate

Conditions: T = 37°C, pH = 7.5, A660nm, Light path = 1 cm

Method: Colorimetric

Reagents

A. 100 mM Tris HCl Buffer, pH 7.5 at 37°C (Prepare 100 ml in deionized water using Trizma Base. Adjust to pH 7.5 at 37°C with 1 M HCl).
B. 40 mM Adenosine 3'-Monophosphate Solution (3'-AMP) (Prepare 5 ml in deionized water using Adenosine 3'-Monophosphate, Sodium Salt).
C. 20% (w/v) Trichloroacetic Acid Solution (TCA) (Prepare 25 ml in deionized water using Trichloroacetic Acid, 6.1 N Solution, approximately 100% (w/v)).
D. 10% (w/v) Ammonium Molybdate Solution (Prepare 25 ml in 10 N H_2SO_4 using Molybdic Acid, Ammonium Tetrahydrate Salt).
E. Taussky-Shorr Color Reagent (TSCR) (Prepare 100 ml by adding 10 ml of Reagent D to 70 ml of deionized water. Add 5 g Ferrous Sulfate Heptahydrate and mix until dissolved. Add enough deionized water for a final volume of 100 ml. Store in an amber container).

F. Phosphorus Standard Solution (P Std) (Use Phosphorus Standard Solution. The phosphorus concentration is 20 µg/ml, 0.645 µmoles/ml).

G. 3'-Nucleotidase Enzyme Solution (Immediately before use, prepare a solution containing 1.5–2.5 units/ml of 3'-Nucleotidase in cold Reagent A).

Procedure

Step 1: Pipette (in milliliters) the following reagents into suitable containers:

	Test	Blank
Reagent A (Buffer)	1.85	1.85
Reagent B (3'-AMP)	0.10	0.10

Mix by swirling and equilibrate to 37°C. Then add:

Reagent G (Enzyme Solution)	0.05	—

Immediately mix by swirling and incubate at 37°C for exactly 15 minutes. Then add:

Reagent C (TCA)	0.20	0.20
Reagent G (Enzyme Solution)	—	0.05

Mix by swirling and centrifuge to clarify.

Step 2: Pipette (in milliliters) the following reagents into suitable tubes:

	Test	Blank
Reagent E (TSCR)	1.50	1.50
Test Supernatant (from Step 1)	0.30	—
Blank Supernatant (from Step 1)	—	0.30
Deionized Water	1.20	1.20

Procedure

Mix by swirling and incubate at 25°C for 10 minutes. Transfer the solutions to suitable cuvettes and record the A660nm for both the Test and Blank using a suitable spectrophotometer.

Colorimetric Assay

Standard Curve: A standard curve is made by pipetting (in milliliters) the following reagents into suitable containers:

	Std 1	Std 2	Std 3	Std 4	Std 5	Std Blank
Reagent F (P Std)	0.10	0.25	0.35	0.60	1.00	—
Deionized Water	1.40	1.25	1.15	0.90	0.50	1.50
Reagent E (TSCR)	1.50	1.50	1.50	1.50	1.50	1.50

Mix by inversion and incubate at 25°C for 10 minutes. Transfer the solutions to suitable cuvettes and record the A660nm for the Standards and Standard Blank using a suitable spectrophotometer.

Calculations

Standard Curve: ΔA660nm Standard = A660nm Standard–A660nm Standard Blank
Prepare a Standard curve by plotting the ΔA660nm of the Standard us μmoles of Phosphate.

Sample Determination: ΔA660nm Test = A660nm Test–A660nm Test Blank
Determine the μmoles of Phosphate liberated using the Standard curve.

$$\text{Units / ml enzyme} = \frac{(\mu\text{moles of Phosphate released})\ (2.2)\ (\text{df})}{(0.05)\ (15)}$$

2.2 = Total volume (in milliliters) of Step 1
df = Dilution factor
0.05 = Volume (in milliliter) of enzyme used
15 = Time (in minutes) of assay as per the Unit Definition

Unit Definition: *One unit will hydrolyze 1.0 μmole of inorganic phosphorus from adenosine 3'-monophosphate per minute at pH 7.5 at 37°C.*

Final Assay Concentrations

In a 2.0 ml reaction mix the final concentrations are 95 mM Tris, 2 mM adenosine 3'-monophosphate, and 0.075–0.125 unit 3'-nucleotidase.

Enzymatic Assay of OCTOPINE DEHYDROGENASE (EC 1.5.1.11)

Principle

$$\text{Pyruvate} + \text{L-Arginine} + \beta\text{-NADH} \xrightarrow{\text{Octopine Dehydrogenase}} \text{Octopine} + H_2O + \beta\text{-NAD}$$

Abbreviations Used

β-NADH = β-Nicotinamide Adenine Dinucleotide, Reduced Form
β-NAD = β-Nicotinamide Adenine Dinucleotide, Oxidized Form

Conditions: T = 25°C, pH = 6.6, A340nm, Light path = 1 cm

Method: Continuous Spectrophotometric Rate Determination

Reagents

A. 150 mM Sodium Phosphate Buffer, pH 6.6 at 25°C (Prepare 200 ml in deionized water using Sodium Phosphate, Monobasic, Anhydrous. Adjust to pH 6.6 at 25°C with 1 M NaOH).
B. 60 mM Sodium Pyruvate Solution (Pyruv) (Prepare 10.0 ml in deionized water using Pyruvic Acid, Sodium Salt. *Prepare Fresh*).
C. 60 mM L-Arginine Solution (Arg) (Prepare 5 ml in deionized water using L-Arginine Free Base).
D. 4.2 mM β-Nicotinamide Adenine Dinucleotide, Reduced Form Solution (β-NADH) (Dissolve the contents of one 5 mg vial of β-Nicotinamide Adenine Dinucleotide, Reduced Form, Disodium Salt in the appropriate volume of cold Reagent A. *Prepare Fresh*).

E. Octopine Dehydrogenase Enzyme Solution (Immediately before use, prepare a solution containing 0.25–0.50 unit/ml of Octopine Dehydrogenase in cold Reagent A).

Procedure

Pipette (in milliliters) the following reagents into suitable cuvettes:

	Test	Blank
Reagent A (Buffer)	2.45	2.50
Reagent B (Pyruv)	0.20	0.20
Reagent C (Arg)	0.20	0.20
Reagent D (β-NADH)	0.10	0.10

Mix by inversion and equilibrate to 25°C. Monitor the A340nm until constant using a suitably thermostatted spectrophotometer. Then add:

Reagent E (Enzyme Solution)	0.05	—

Immediately mix by inversion and record the decrease in A340nm for approximately 5 minutes. Obtain the ΔA340nm/minute using the maximum linear rate for both the Test and Blank.

Calculations

$$\text{Units / mg enzyme} = \frac{(\Delta A340\text{nm / min Test} - \Delta A340\text{nm / min Blank})\ (3)\ (\text{df})}{(6.22)\ (0.05)}$$

3 = Total volume (in milliliters) of assay
df = Dilution factor
6.22 = Millimolar extinction coefficient of β-NADH at 340 nm
0.05 = Volume (in milliliter) of enzyme used

$$\text{Units / mg solid} = \frac{\text{units / ml enzyme}}{\text{mg solid / ml enzyme}}$$

$$\text{Units / mg protein} = \frac{\text{units / ml enzyme}}{\text{mg protein / ml enzyme}}$$

Unit Definition: *One unit will form 1.0 μmole of octopine from 1.0 μmole of pyruvate and 1.0 μmole of L-arginine per minute at pH 6.6 at 25°C.*

Final Assay Concentrations

In a 3.00 ml reaction mix, the final concentrations are 130 mM sodium phosphate, 4 mM sodium pyruvate, 4 mM L-arginine, 0.14 mM β-nicotinamide adenine dinucleotide, reduced form and 0.0125–0.025 unit octopine dehydrogenase.

Enzymatic Assay of ORNITHINE CARBAMYL TRANSFERASE (EC 2.1.3.3)

Principle

$$\text{Carbamyl Phosphate} + \text{L-Ornithine} \xrightarrow{\text{OCT}} \text{L-Citrulline} + \text{Pi}$$

Abbreviations Used

Pi = Inorganic Phosphate
OCT = Ornithine Carbamyl Transferase

Conditions: T = 37°C, pH = 8.5, A490nm, Light path = 1 cm

Method: Colorimetric

Reagents

A. 10 mM Tris HCl, pH 8.5 at 37°C (Prepare 100 ml in deionized water using Trizma Base. Adjust to pH 8.5 at 37°C with 1 M HCl).
B. 50 mM L-Ornithine Solution (L-Orn) (Prepare 10 ml in deionized water using L-Ornithine Hydrochloride. Adjust to pH 7.5 at 25°C with solid Sodium Bicarbonate. *Prepare Fresh*).
C. 50 mM Carbamyl Phosphate Solution (CP) (Prepare 5 ml in cold deionized water using Carbamyl Phosphate, Dilithium Salt. Keep in an ice bath while using. *Prepare Fresh*).
D. 6.25% (w/v) Trichloroacetic Acid Solution (TCA) (Use Trichloroacetic Acid, 0.38 N Solution, approximately 6.25% (w/v)).
E. Redox Buffer (Prepare by dissolving 0.9 g of Ferric Ammonium Sulfate, Dodecahydrate and 1.1 g Ferrous Ammonium Sulfate, Hexahydrate in 10 ml of Reagent F).
F. 1 M Sulfuric Acid Solution (H_2SO_4) (Prepare 50 ml in deionized water using Sulfuric Acid).
G. 7.4 M Phosphoric Acid with 3.0 M Sulfuric Acid Solution (Acid) (Prepare 60 ml in deionized water using Phosphoric Acid and Sulfuric Acid).
H. 75 mM 2,3-Butanedione Monoxime Solution (BuD) (Prepare 25 ml in deionized water using 2,3- Butanedione Monoxime).
I. 10 mM Citrulline Stock Solution (Citrul) (Prepare 100 ml in deionized water using L-Citrulline).
J. 1 mM Citrulline Standard Solution (STD) (Prepare by adding 1 ml of Reagent I to 9 ml deionized water. *Prepare Fresh*).
K. Ornithine Carbamyl Transferase Enzyme Solution (Immediately before use, prepare a solution containing 0.5–1 unit/ml in cold Reagent A).

Procedure

Pipette (in milliliters) the following reagents into suitable containers:

	Test	Blank
Reagent A (Buffer)	2.30	2.30
Reagent B (L-Orn)	0.30	0.30
Reagent C (CP)	0.30	0.30

Mix by inversion and equilibrate to 37°C. Then add:

Reagent K (Enzyme Solution)	0.10	—

Mix by inversion and incubate at 37°C for exactly 15 minutes. Then add:

	Test	Blank
Reagent D (TCA)	3.00	3.00
Reagent K (Enzyme Solution)	—	0.10

Mix by inversion.

Colorimetric Determination

Pipette (in milliliters) the following into suitable tubes (Mix by swirling after each addition):

	Test	Test Blank	Std 1	Std 2	Std 3	Std 4	Std 5	Std Blank
Deionized Water	—	—	0.90	0.80	0.60	0.40	0.20	1.00
Test	1.00	—	—	—	—	—	—	—
Test Blank	—	1.00	—	—	—	—	—	—
Reagent J (STD)	—	—	0.10	0.20	0.40	0.60	0.80	—
Reagent E (Redox)	0.25	0.25	0.25	0.25	0.25	0.25	0.25	0.25
Reagent G (Acid)	1.25	1.25	1.25	1.25	1.25	1.25	1.25	1.25
Reagent H (BuD)	0.50	0.50	0.50	0.50	0.50	0.50	0.50	0.50

Mix by swirling. Cover all tubes with a glass marble. Transfer tubes to a boiling water bath and incubate for 20 minutes. Transfer tubes to a cold water bath and cool to room temperature. *Protect from Light.* Then add:

Deionized Water	8.00	8.00	8.00	8.00	8.00	8.00	8.00	8.00

Mix by swirling, then transfer the solutions to suitable cuvettes and record the absorbance at A490nm for Test, Blank and Standards using a suitable spectrophotometer.

Calculations

Standard Curve: ΔA490nm Standard = A490nm Standard–A490nm Standard Blank

Plot the ΔA490nm Standard vs μmoles of citrulline.

Sample Determination: ΔA490nm Sample = A490nm Test–A490nm Test Blank

Determine the μmoles of citrulline liberated using the Standard curve.

$$\text{Units / ml enzyme} = \frac{(\mu\text{moles of citrulline liberated})\ (6)\ (\text{df})}{(15)\ (0.1)\ (1)}$$

6 = Volume (in milliliters) of assay

df = Dilution factor

15 = Time of assay (in minutes) as per the Unit Definition

0.1 = Volume (in milliliters) of enzyme used

1 = Volume (in milliliters) of assay used in Colorimetric Determination

$$\text{Units / mg solid} = \frac{\text{units / ml enzyme}}{\text{mg solid / ml enzyme}}$$

$$\text{Units / mg protein} = \frac{\text{units / ml enzyme}}{\text{mg protein / ml enzyme}}$$

Unit Definition: *One unit will form 1.0 μmole of citrulline from ornithine and carbamyl phosphate per minute at pH 8.5 at 37°C.*

Final Assay Concentrations

In a 3.00 ml reaction mix, the final concentrations are 8 mM Tris, 5 mM L-ornithine, 5 mM carbamyl phosphate and 0.05–0.1 unit ornithine carbamyl transferase.

Note

1. If the solutions are not clear, they may be clarified by filtration or by centrifugation.

Enzymatic Assay of L-ORNITHINE DECARBOXYLASE (EC 4.1.1.17)

Principle

$$\text{L-Orn} \xrightarrow[\text{PRPP}]{\text{Ornithine Decarboxylase}} \text{Putrescine} + CO_2$$

Abbreviations Used

L-Orn = L-Ornithine
PRP = Pyridoxal 5-Phosphate

Conditions: T = 37°C, pH 5.2

Method: Manometric Assay using Warburg Flasks

Reagents

A. 100 mM Citric Acid and 100 mM Potassium Phosphate Buffer, pH 5.2 at 37°C (Prepare 100 ml in deionized water using Citric Acid, Free Acid, Anhydrous and Potassium Phosphate, Tribasic. Adjust the pH to 5.2 at 37°C with 1 M KOH).
B. 100 mM L-Ornithine Hydrochloride Solution (Prepare 50 ml in Reagent A using L-Ornithine Hydrochloride. Adjust the pH to 5.2, with 1 M KOH, if necessary).
C. 10 mM Pyridoxal 5-Phosphate Solution (PRP) (Prepare 25 ml in Reagent A using Pyridoxal 5-Phosphate).
D. Ornithine Decarboxylase Enzyme Solution (Immediately before use, prepare a solution containing 2 units/ml of L-Ornithine Decarboxylase in cold deionized water).

Procedure

Pipette (in milliliters) the following reagents into Warburg flasks:

Main Chamber	Thermo-barometer Flask	Enzyme Blank	Test	Substrate Blank
Reagent A (Buffer)	2.80	2.80	0.30	0.30
Reagent B (Ornithine HCl)	—	—	2.50	2.50
Side Arm				
Reagent D (Enzyme Solution)	—	0.10	0.10	—
Reagent C (PRP)	—	0.10	0.10	0.10
Reagent A (Buffer)	0.20	—	—	0.10

Be sure to confirm the stability of the pressure with the flask sealed off and before proceeding with the assay. This is to ensure temperature equilibrium and the absence of leaks in the flask. The enzyme activity is determined by calculation of the rate of production of CO_2 at 37°C. The reaction rate should be linear for about 20 minutes.

Calculations

$$\frac{\text{Units}}{\text{ml Ornithine Decarboxylase}} = \frac{\text{(C) (K) (Dilution Factor)}}{\left(22.4\frac{1}{\text{mole}}1\right)\text{(ml Ornithine Decarboxylase)}}$$

C = mm of CO_2 gas evolved/minute
K = Warburg flask constant in μl/mm
22.4 lit = Volume gas occupies under STP conditions

Unit Definitions: *One unit will release 1.0 μmole of CO_2 from L-Ornithine per minute at pH 5.2 at 37°C.*

Final Assay Concentrations

In a 3.00 ml reaction mix, the final concentrations are 97 mM citric acid, 97 mM potassium phosphate, 83 mM L-ornithine, 0.33 mM pyridoxal 5-phosphate, and 0.2 unit L-ornithine decarboxylase.

Notes

1. The tests are done in triplicate, at least, since it is common for the flasks to have small leaks.
2. The mm of CO_2 gas evolved (C) is corrected for any temperature and barometric changes (Thermobarometer) during the experiment and is also corrected for the Substrate Blank and Enzyme Blanks:
   mm CO_2 corrected = mm CO_2 measured Test–mm CO_2 measured for [Thermobarometer + Substrate Blank + Enzyme Blank]
   Values of the corrected mm CO_2 produced are plotted versus time. The best straight line is drawn not necessarily through the origin. The slope, C = mm CO_2/time, is obtained.
3. The Flask constant, K, is calculated according to the equation:

$$K = \frac{\left[(V_g)\left(\frac{273}{T}2\right) + V_f\, a\right]}{P_o}$$

where
P_o = Standard pressure as mm of manometer fluid
V_g = Volume (in milliliters) of gas in flask and manometer
V_f = Volume (in milliliters) of liquid in flask
T = Absolute temperature
a = Solubility of gas (for CO_2 at 37°C, a = 0.57) The flask constant, K, must be calculated for each Warburg flask used.

Enzymatic Assay of OROTATE REDUCTASE (EC 1.3.1.14)

Principle

$$\text{Orotate} + \beta\text{-NADH} \xrightarrow{\text{Orotate Reductase}} \text{Dihydroorotate} + \beta\text{-NAD}$$

Abbreviations Used

β-NADH = β-Nicotinamide Adenine Dinucleotide, Reduced Form
β-NAD = β-Nicotinamide Adenine Dinucleotide, Oxidized Form

Conditions: T = 25°C, pH = 6.5, A340nm, Light Path = 1 cm

Method: Continuous Spectrophotometric Rate Determination

Reagents

A. 400 mM Sodium Phosphate Buffer, pH 6.5 at 25°C (Prepare 200 ml in deionized water using Sodium Phosphate, Monobasic, Anhydrous. Adjust to pH 6.5 at 25°C with 1 M NaOH).
B. 400 mM Cysteine Solution (Cys) (Prepare 10 ml in deionized water using L-Cysteine Hydrochloride, Monohydrate. Adjust to pH 6.5 at 25°C with 1 M NaOH).
C. 10 mM Orotic Acid Solution (Orotic) (Prepare 10 ml in 10 mM NaOH using Orotic Acid, Monohydrate. Adjust to pH 6.5 with 10 mM HCl).
D. 1.6 mM β-Nicotinamide Adenine Dinucleotide, Reduced Form Solution (β-NADH) (Prepare 2 ml in Reagent A using β-Nicotinamide Adenine Dinucleotide, Reduced Form, Disodium Salt or dissolve the contents of one 5 mg vial of β-Nicotinamide Adenine Dinucleotide, Reduced Form, Disodium Salt in the appropriate volume of Reagent A. *Prepare Fresh*).
E. 3 mM Flavin Adenine Dinucleotide Solution (FAD) (Prepare 1 ml in deionized water using Flavin Adenine Dinucleotide, Disodium Salt. *Prepare Fresh. Protect from Light*).
F. Orotate Reductase Enzyme Solution (Immediately before use, prepare a solution containing 1–2 units/ml of Orotate Reductase in cold Reagent A).

Procedure

Prepare a reaction cocktail by pipetting (in milliliters) the following reagents into a suitable container:

Deionized Water	6.40
Reagent A (Buffer)	10.00
Reagent B (Cys)	5.00
Reagent C (Orotic)	6.00
Reagent E (FAD)	0.10

Mix by stirring and adjust to pH 6.5 at 25°C with either 1 M HCl or 1 M NaOH. Pipette (in milliliters) the following reagents into suitable cuvette:

	Test	Blank
Reaction Cocktail	2.75	2.75
Reagent F (Enzyme Solution)	0.05	—
Reagent A (Buffer)	—	0.05

Mix by inversion and incubate for 5 minutes at 25°C to activate the enzyme. Monitor the A340nm until constant, using a suitably thermostatted spectrophotometer. Then add:

Reagent D (β-NADH)	0.20	0.20

Immediately mix by inversion and record the decrease in A340nm for approximately 5 minutes. Obtain the ΔA340nm/minute using the maximum linear rate for both the Test and Blank.

Calculations

$$\text{Units / ml enzyme} = \frac{(\Delta A340\text{nm / min Test} - \Delta A340\text{nm / min Blank})\ (3)\ (\text{df})}{(6.22)(0.05)}$$

3 = Total volume (in milliliters) of assay
df = Dilution factor
6.22 = Millimolar extinction coefficient of β-NADH at 340nm
0.05 = Volume (in milliliter) of enzyme used in the assay

$$\text{Units / mg solid} = \frac{\text{units / ml enzyme}}{\text{mg solid / ml enzyme}}$$

$$\text{Units / mg protein} = \frac{\text{units / ml enzyme}}{\text{mg protein / ml enzyme}}$$

Unit Definition: *One unit will convert 1.0 µmole of orotic acid to dihydroorotate per minute at pH 6.5 at 25°C.*

Final Assay Concentrations

In a 3.00 ml reaction mix, the final concentrations are 167 mM sodium phosphate, 67 mM cysteine hydrochloride, 2.0 mM orotic acid, 0.01 mM FAD, 0.11 mM β-NADH and 0.05–0.1 unit orotate reductase.

Enzymatic Assay of OROTIDINE 5'-MONOPHOSPHATE DECARBOXYLASE (EC 4.1.1.23)

Principle

$$\text{Orotidine 5'-Monophosphate} \xrightarrow{\text{OMP Decarboxylase}} \text{UMP} + CO_2$$

Abbreviations Used

OMP Decarboxylase = Orotidine 5'-Monophosphate Decarboxylase
UMP = Uridine 5'-Monophosphate

Conditions: T = 30°C, pH = 8.0, A295nm, Light path = 1 cm

Method: Continuous Spectrophotometric Rate Determination

Reagents

A. 30 mM Tris HCl Buffer, pH 8.0 at 30°C (Prepare 100 ml in deionized water using Trizma Base. Adjust to pH 8.0 at 30°C with 1 M HCl).
B. 75 mM Magnesium Chloride Solution ($MgCl_2$) (Prepare 10 ml in deionized water using Magnesium Chloride, Hexahydrate).
C. 18 mM Orotidine 5'-Monophosphate Solution (OMP) (Prepare 10 ml in deionized water using Orotidine 5'-Monophosphate, Sodium Salt. *Prepare Fresh*).

D. Orotidine-5'-Monophosphate Decarboxylase Enzyme Solution (Immediately before use, prepare a solution containing 30–60 units/ml of Orotidine 5'-Monophosphate Decarboxylase in cold deionized water).

Procedure

Pipette (in milliliters) the following reagents into suitable quartz cuvettes:

	Test	Blank
Reagent A (Buffer)	2.50	2.50
Reagent B ($MgCl_2$)	0.30	0.30
Reagent C (OMP)	0.10	0.10

Mix by inversion and equilibrate to 30°C. Monitor the A295nm until constant, using a suitably thermostatted spectrophotometer. Then add:

	Test	Blank
Deionized Water	—	0.10
Reagent D (Enzyme Solution)	0.10	—

Immediately mix by inversion and record the decrease in A295nm for approximately 5 minutes. Obtain the ΔA295nm/minute using the maximum linear rate for both the Test and Blank.

Calculations

$$\text{Units / ml enzyme} = \frac{(\Delta A295\text{nm / min Test} - \Delta A295\text{nm / min Blank})\ (60)\ (3)}{(0.50)\ (0.1)}$$

3 = Total volume (in milliliters) of assay
60 = Conversion from minutes to hours (Unit definition)
0.50 = The delta millimolar extinction coefficient between OMP and UMP at 295nm
0.1 = Volume (in milliliters) of enzyme used

Unit Definition: *One unit will convert 1.0 μmole of orotidine 5'-monophosphate to uridine 5'-monophosphate per hour at pH 8.0 at 30°C.*

Final Assay Concentrations

In a 3.00 ml reaction mix, the final concentrations are 25 mM Tris, 7.5 mM magnesium chloride, 0.6 mM orotidine 5'-monophosphate and 3–6 units orotidine 5'-monophosphate decarboxylase.

Enzymatic Assay of OROTIDINE-5'-MONOPHOSPHATE PYROPHOSPHORYLASE (EC 2.4.2.10)

Principle

$$\text{OMP} + \text{Pyrophosphate} \xrightarrow{\text{OMPP}} \text{Orotic Acid} + \text{PRPP}$$

Abbreviations Used

OMP = Orotidine 5'-Monophosphate
OMPP = Orotidine-5'-Monophosphate Pyrophosphorylase
PRPP = 5-Phosphorylribose-1-Pyrophosphate

Conditions: T = 30°C, pH = 8.0, A295nm, Light path = 1 cm

Method: Continuous Spectrophotometric Rate Determination

Reagents

A. 25 mM Tris HCl Buffer with 8 mM Magnesium Chloride, pH 8.0 at 30°C (Prepare 100 ml in deionized water using Trizma Base and Magnesium Chloride, Hexahydrate. Adjust to pH 8.0 at 30°C with 1 M HCl).
B. 35 mM Orotidine 5'-Monophosphate Solution (OMP) (Prepare 10 ml in deionized water using Orotidine 5'-Monophosphate, Sodium Salt. *Prepare Fresh*).
C. 50 mM Pyrophosphate Solution (PPi) (Prepare 10 ml in deionized water using Tetrasodium Pyrophosphate, Decahydrate. *Prepare Fresh*).
D. Orotidine-5'-Monophospháte Pyrophosphorylase Enzyme Solution (Immediately before use, prepare a solution containing 50–100 units/ml of Orotidine 5'-Monophosphate Pyrophosphorylase in cold Reagent A).

Procedure

Pipette (in milliliters) the following reagents into suitable quartz cuvettes:

	Test	Blank
Reagent (Buffer)	2.70	2.70
Reagent B (OMP)	0.10	0.10
Reagent D (Enzyme Solution)	0.10	0.10

Mix by inversion and equilibrate to 30°C. Monitor the A295nm until constant using a suitably thermostatted spectrophotometer. Then add:

Deionized Water	—	0.10
Reagent C (PPi)	0.10	—

Immediately mix by inversion and record the increase in A295nm for approximately 5 minutes. Obtain the ΔA295nm/minute using the maximum linear rate for both the Test and Blank.

Calculations

$$\text{Units / mg enzyme} = \frac{(\Delta A295\text{nm / min Test} - \Delta A295\text{nm / min Blank})\,(60)}{(3.5)\,(\text{mg enzyme / ml RM})}$$

60 = Conversion factor from minutes to hours as per the Unit Definition
3.5 = Millimolar extinction coefficient of orotic acid at 295nm
RM = Reaction Mix

Unit Definition: *One unit will convert 1.0 µmole of orotidine 5'-monophosphate to orotic acid in one hour at pH 8.0 at 30°C.*

Final Assay Concentrations

In a 3.00 ml reaction mix, the final concentrations are 23 mM Tris, 7 mM magnesium chloride, 1.2 mM orotidine 5'-monophosphate, 1.7 mM pyrophosphate and 5–10 units orotidine-5'-monophosphate pyrophosphorylase.

Enzymatic Assay of OROTIDINE-5'-PHOSPHATE PYROPHOSPHORYLASE and OROTIDINE-5'-PHOSPHATE DECARBOXYLASE

Principle

$$\text{Orotate} + \text{PRPP} \xrightarrow{\text{OMPP}} \text{OMP} + \text{PPi}$$

$$\text{OMP} \xrightarrow{\text{OMP Decarboxylase}} \text{Uridine 5'-Monophosphate} + CO_2$$

Abbreviations Used

PRPP = 5-Phosphorylribose-1-Pyrophosphate
OMPP = Orotidine-5'-Monophosphate Pyrophosphorylase
OMP = Orotidine 5'-Monophosphate
PPi = Inorganic Pyrophosphate
OMP Decarboxylase = Orotidine 5'-Monophosphate Decarboxylase

Conditions: T = 25°C, pH = 8.0, A295nm, Light path = 1 cm

Method: Continuous Spectrophotometric Rate Determination

Reagents

A. 25 mM Tris HCl Buffer with 8 mM Magnesium Chloride, pH 8.0 at 25°C (Prepare 100 ml in deionized water using Trizma Hydrochloride and Magnesium Chloride, Hexahydrate. Adjust to pH 8.0 at 25°C with 1 M NaOH).

B. 14 mM Orotic Acid Solution (Orotic Acid) (Prepare 2 ml in deionized water using Orotic Acid, Monosodium Salt. Heating may be required for solubilization).

C. 16 mM 5-Phosphorylribose-1-Pyrophosphate Solution (PRPP) (Prepare 2 ml in deionized water using 5-Phosphorylribose-1-Pyrophosphate, Sodium Salt).

D. Orotidine-5'-Phosphate Pyrophosphorylase and Orotidine-5'-Phosphate Decarboxylase Enzyme Solution (Immediately before use, prepare a solution containing approximately 26 units/ml of Orotidine-5'-Phosphate Pyrophosphorylase and Orotidine-5'-Phosphate Decarboxylase Mixed Enzymes in cold Reagent A).

Procedure

Pipette (in milliliters) the following reagents into suitable cuvettes:

	Test	Blank
Reagent A (Buffer)	2.75	2.75
Reagent B (Orotic Acid)	0.05	0.05
Reagent D (Enzyme Solution)	0.10	0.10

Mix by inversion and equilibrate to 25°C. Monitor the A295nm until constant using a suitably thermostatted spectrophotometer. Then add:

	Test	Blank
Reagent C (PRPP)	0.10	—
Deionized Water	—	0.10

Immediately mix by inversion and record the decrease in A295 for approximately 10 minutes. Obtain the A295nm/minute using the maximum linear rate for both the Test and Blank.

Calculations

$$\text{Units / ml enzyme} = \frac{(\text{A295nm / min Test} - \text{A295nm / min Blank})\,(3)\,(60)\,(\text{df})}{(3.95)\,(0.1)}$$

3 = Total volume (in milliliters) of assay
60 = Conversion factor from minutes to hours as per the Unit Definition
df = Dilution factor
3.95 = Millimolar extinction coefficient1 of orotic acid at 295nm
0.1 = Volume (in milliliter) of enzyme used

$$\text{Units / mg solid} = \frac{\text{units / ml enzyme}}{\text{mg solid / ml enzyme}}$$

$$\text{Units / mg protein} = \frac{\text{units / ml enzyme}}{\text{mg protein / ml enzyme}}$$

Unit Definition: *One unit will catalyze the phosphorylation of 1.0 µmole of orotic acid to 5'-OMP, which is then decarboxylated to 5'-UMP in one hour at pH 8.0 at 25°C, in a PRPP system.*

Final Assay Concentrations

In a 3.00 ml reaction mix, the final concentrations are 24 mM Tris, 8 mM magnesium chloride, 0.2 mM orotic acid, 0.5 mM 5'-phosphorylribose-1-pyrophosphate, 2.6 units orotidine-5'-phosphate pyrophosphorylase, and orotidine 5'-phosphate decarboxylase mixed enzymes.

Enzymatic Assay of OXALACETATE DECARBOXYLASE (EC 4.1.1.3)

Principle

$$\text{Oxalacetic Acid} \xrightarrow{\text{Oxalacetate Decarboxylase}} \text{Pyruvic Acid} + CO_2$$

$$\text{Pyruvic Acid} + \beta\text{-NADH} + H^+ \xrightarrow{\text{LDH}} \text{Lactic Acid} + \beta\text{-NAD}^+$$

Abbreviations Used

LDH = Lactate Dehydrogenase
β-NAD = β-Nicotinamide Adenine Dinucleotide
β-NADH = β-Nicotinamide Adenine Dinucleotide, Reduced Form

Conditions: T = 25°C, pH = 8.0, A340nm, Light path = 1 cm

Method: Continuous Spectrophotometric Rate Determination

Reagents

A. 100 mM Triethanolamine HCl pH 8.0 at 25°C (TEA Buffer) (Prepare in deionized water using Triethanolamine HCl. Adjust pH to 8.0 at 25°C with 1 M NaOH).

B. 10 mM Tris HCl (Diluent) (Prepare in deionized water using Trizma Base. Adjust pH to 8.0 at 25°C using 1 M HCl).
C. 200 mM Tris HCl pH 9.0 at 25°C (Prepare in deionized water using Trizma Base. Adjust pH to 9.0 using 5 M HCl).
D. 10 mM Manganese Chloride Solution ($MnCl_2$) (Prepare in deionized water using Manganese Chloride, Tetrahydrate).
E. 6.5 mM β-Nicotinamide Adenine Dinucleotide, Reduced Form (NADH) (Prepare in deionized water using β-Nicotinamide Adenine Dinucleotide, Reduced Form, Sodium Salt).
F. Lactic Dehydrogenase Solution (LDH Solution) (Immediately before use prepare a solution containing 550 units/ml Lactic Dehydrogenase, using cold Reagent B as the diluent).
G. 50 mM Oxalacetic Acid Solution (Substrate Solution) (Immediately before use prepare a solution of Oxalacetic Acid, in cold Reagent C. *Prepare Fresh*).
H. Oxalacetate Decarboxylase Solution (Immediately before use prepare a solution containing 0.5–1.0 unit per ml in cold Reagent B).

Procedure

Pipette (in milliliters) the following reagents into suitable cuvettes.

	Test	Blank
Reagent A (TEA Buffer)	0.90	0.90
Reagent D ($MnCl_2$)	0.10	0.10
Reagent E (β-NADH)	0.10	0.10
Deionized Water	0.90	0.90
Reagent F (LDH Solution)	0.02	0.02
Reagent H (Enzyme Solution)	0.05	—
Reagent B (Diluent)	—	0.05

Mix by inversion and equilibrate to 25°C. Monitor the A340nm until constant, using a suitably thermostatted spectrophotometer. Then add:

Reagent F (Substrate)	0.10	0.10

Immediately mix by inversion and record the decrease in A340nm from the linear portion of the curve. Obtain the ΔA340nm for both the Test and Blank.

Calculations

$$\text{Units / ml enzyme} = \frac{(\Delta A340\text{nm / min Test} - \Delta A340\text{nm / min Blank})\,(2.17)\,(\text{df})}{(6.22)\,(0.05)}$$

2.17 = Total volume (in milliliter) of enzyme assay
df = Dilution factor
6.22 = Millimolar extinction coefficient of β-NADH at 340nm.
0.05 = Volume (in milliliter) of enzyme used

$$\text{Units / mg solid} = \frac{\text{units / ml enzyme}}{\text{mg solid / ml enzyme}}$$

$$\text{Units / mg protein} = \frac{\text{units / ml enzyme}}{\text{mg protein / ml enzyme}}$$

Unit Definition: *One Unit will convert 1.0 μmole of oxalacetate to pyruvate and CO_2 per minute at pH 8.0 at 25°C.*

Final Assay Concentrations

In a 2.17 ml reaction mix, the final concentrations are 40 mM triethanolamine, 9.5 mM Trizma HCl, 0.5 mM manganese chloride, 0.3 mM β-nicotinamide adenine dinucleotide, reduced form, 9.2 μg/ml lactic dehydrogenase, 2.3 mM oxalacetic acid, and 0.05 units per ml oxalacetate decarboxylase.

Enzymatic Assay of OXALACETATE DECARBOXYLASE (EC 4.1.1.2)

Principle

$$\text{Oxalate} \xrightarrow{\text{Oxalate Decarboxylase}} \text{Formate} + CO_2$$

Conditions: T = 37°C, pH 3.0

Method: Manometric Assay using Warburg Flasks

Reagents:

A. 100 mM Potassium Citrate Buffer, pH 3.0 at 37°C (Prepare 100 ml in deionized water using Citric Acid, Tripotassium Salt. Adjust to pH 3.0 at 37°C with 1 M HCl).

B. 100 mM Oxalic Acid Solution, pH 3.0 at 37°C (Ox Acid) (Prepare 10 ml in Reagent A using Oxalic Acid, Dipotassium Salt, Monohydrate. Adjust the pH to 3.0 with 1 M HCl or 1 M KOH, if necessary).

C. Oxalate Decarboxylase Enzyme Solution (Immediately before use, prepare a solution containing 2 units/ml of Oxalate Decarboxylase in cold deionized water).

Procedure

Pipette (in milliliters) the following reagents into Warburg flasks:

Main Chamber	Thermo-barometer Flask	Enzyme Blank	Test	Substrate Blank
Reagent A (Buffer)	2.90	2.70	2.70	2.90
Reagent C (Enzyme Solution)	—	—	0.20	0.20
Reagent B (Ox Acid)	—	—	0.10	0.10
Deionized Water	0.10	0.10	—	—

Be sure to confirm the stability of the pressure with the flask sealed off before proceeding with the assay. This is to ensure temperature equilibrium and the absence of leaks in the flask. Equilibrate the closed system for 30 minutes at 37°C. The enzyme activity is determined by calculation of the rate of production of CO_2 at 37°C. The reaction rate should be linear for about 20 minutes.

Calculations

$$\frac{\text{Units}}{\text{ml Oxalate Decarboxylase}} = \frac{(C)(K)(df)}{\left(22.4\frac{1}{\text{mole}}1\right)(\text{ml Oxalate decarboxylase})}$$

C = mm of CO_2 gas evolved/minute
K = Warburg flask constant in µl/mm
22.4 l = Volume gas occupies under STP conditions

Unit Definitions: *One unit will convert 1.0 µmole of oxalate to formate and CO_2 per minute at pH 3.0 at 37°C.*

Final Assay Concentrations

In a 3.00 ml reaction mix, the final concentrations are 93 mM potassium citrate, 3.3 mM oxalic acid, and 0.4 unit oxalate decarboxylase.

Notes

1. The tests are done in triplicate, since it is common for the flasks to have small leaks.
2. The mm of CO_2 gas evolved (C) is corrected for any temperature and barometric changes (Thermobarometer) during the experiment and also for the Substrate Blank and Enzyme Blank:
   mm CO_2 corrected = mm CO_2 measured Test–mm CO_2 measured for
   [Thermobarometer + Substrate Blank + Enzyme Blank]
   Values of the corrected mm CO_2 produced are plotted versus time. The best straight line is drawn not necessarily through the origin. The slope, C = mm CO_2/time is obtained.
3. The Flask constant, K, is calculated according to the equation:

$$K = \frac{\left[(V_g)\left(\frac{273}{T}2\right) + V_f\, a\right]}{P_o}$$

where
P_o = Standard pressure as mm of manometer fluid
V_g = Volume (in milliliters) of gas in flask and manometer
V_f = Volume (in milliliters) of liquid in flask
T = Absolute temperature
a = Solubility of gas; (for CO_2 at 37°C, a = 0.57)

Enzymatic Assay of OXALACETATE DECARBOXYLASE (EC 4.1.1.2)

Principle

$$\text{Oxalate} \xrightarrow{\text{Oxalate Decarboxylase}} \text{Formate} + CO_2$$

$$\text{Formate} + \beta\text{-NAD Formate Dehydrogenase} \rightarrow CO_2 + \beta\text{-NADH}$$

Abbreviations Used

β-NAD = β-Nicotinamide Adenine Dinucleotide, Oxidized Form
β-NADH = β-Nicotinamide Adenine Dinucleotide, Reduced Form

Conditions: T = 37°C, pH = 5.0, A340nm, Light path = 1 cm

Method: Spectrophotometric Stop Rate Determination

Reagents

A. 100 mM Potassium Phosphate Solution (Prepare 100 ml in deionized water using Potassium Phosphate, Dibasic, Trihydrate).

B. 100 mM Potassium Phosphate Buffer, pH 5.0 at 37°C (Prepare 100 ml in deionized water using Potassium Phosphate, Monobasic, Anhydrous. Adjust to pH 5.0 at 37°C with Reagent A).

C. 200 mM Oxalic Acid Solution (Prepare 100 ml in deionized water using Oxalic Acid, Dihydrate).

D. 200 mM Oxalate Substrate Solution (Oxal) (Prepare 100 ml in deionized water using Oxalic Acid, Dipotassium Salt, Monohydrate. Adjust to pH 5.0 at 37°C with Reagent C).

E. 150 mM Potassium Phosphate Solution (Prepare 100 ml in deionized water using Potassium Phosphate, Monobasic, Anhydrous).

F. 150 mM Potassium Phosphate Buffer, pH 7.5 at 37°C (Stop Buffer) (Prepare 100 ml in deionized water using Potassium Phosphate, Dibasic, Trihydrate. Adjust to pH 7.5 at 37°C with Reagent E).

G. 57 mM β-Nicotinamide Adenine Dinucleotide, Oxidized Form Solution (β-NAD) (Prepare 5 ml in deionized water using β-Nicotinamide Adenine Dinucleotide, Lithium Salt).

H. Formate Dehydrogenase Enzyme Solution (FDH) (Immediately before use, prepare a solution containing 40 units/ml of Formate Dehydrogenase, in cold deionized water).

I. Oxalate Decarboxylase Enzyme Solution (Ox Decarb) (Immediately before use, prepare a solution containing 0.5–1.0 unit/ml of Oxalate Decarboxylase in cold Reagent B).

Procedure

Step 1: Pipette (in milliliters) the following reagents into suitable containers:

	Test	Blank
Reagent B (Buffer)	0.30	0.30
Reagent I (Ox Decarb)	0.10	—

Mix by swirling and equilibrate to 37°C. Then add:

Reagent D (Oxal)	0.20	0.20

Immediately mix by swirling and incubate at 37°C for exactly 2 minutes. Then add:

Reagent F (Stop Buffer)	2.00	2.00
Reagent I (Ox Decarb)	—	0.10

Mix by swirling.

Step 2: Transfer the Test and Blank solutions into suitable cuvettes. Then add:

	Test	Blank
Reagent G (β-NAD)	0.20	0.20

Mix by inversion and equilibrate to 37°C. Monitor the A340nm until constant, using a suitably thermostatted spectrophotometer. Record the initial A340nm for both the Test and Blank. Then add:

Reagent H (FDH)	0.20	0.20

Immediately mix by inversion and record the increase in A340nm until constant (for approximately 5 minutes). Obtain the final A340nm for both the Test and Blank.

Calculations

$$\Delta A340nm\ Test = A340nm\ Test\ Final - A340nm\ Test\ Initial$$

$$\Delta A340nm\ Blank = A340nm\ Blank\ Final - A340nm\ Blank\ Initial$$

$$Units/ml\ enzyme = \frac{(\Delta A340nm/min\ Test - \Delta A340nm/min\ Blank)(3)(df)}{(6.22)(0.1)(2)}$$

3 = Total volume (in milliliters) of assay
df = Dilution factor
6.22 = Millimolar extinction coefficient of β-NADH at 340nm
0.1 = Volume (in milliliter) of enzyme used
2 = Time (in minutes) as per the Unit Definition

$$Units/mg\ protein = \frac{units/ml\ enzyme}{mg\ protein/ml\ enzyme}$$

Unit Definition: *One unit will convert 1.0 μmole of oxalate to formate and* CO_2 *per minute at pH 5.0 at 37°C.*

Final Assay Concentrations

In a 0.60 ml reaction mix, the final concentrations are 67 mM potassium phosphate, 67 mM oxalate, and 0.05–0.10 unit oxalate decarboxylase.

Notes

1. Do not use the sodium salt of this product, as it will inhibit the reaction.
2. Formate Dehydrogenase Unit Definition: One unit will oxidize 1.0 μmole of formate to CO_2 per minute in the presence of β-NAD at pH 7.6 at 37°C.

Enzymatic Assay of OXALATE OXIDASE (EC 1.2.3.4)

Principle

$$Oxalate + O_2 \xrightarrow{Oxalate\ Oxidase} 2CO_2 + H_2O_2$$

$$MBTH + H_2O_2 + DMA \xrightarrow{POD} Indamine\ Dye\ (Purple\ Color) + 2H_2O$$

Abbreviations Used

MBTH = 3-Methyl-2-Benzothiazolinone Hydrazone
DMA = N, N-Dimethylaniline
POD = Peroxidase

Conditions: T = 37ºC, pH = 3.8, A600nm, Light path = 1 cm.

Method: Continuous Spectrophotometric Rate Determination

Reagents

A. 50 mM Succinate buffer, with 0.79 mM N, NDimethylaniline and 0.11 mM 3-Methyl-2-Benzothiazolinone Hydrazone, pH 3.8 at 37ºC (MBTH) (Prepare 100 ml in deionized water using Succinic Acid, 3-Methyl-2-Benzothiazolinone Hydrazone, Hydrochloride, and N,N-Dimethylaniline. Adjust to pH 3.8 at 37ºC with 1 M NaOH).

B. 100 mM Ethylenediaminetetraacetic Acid Solution (EDTA) (Prepare 10 ml in deionized water using Ethylenediaminetetraacetic Acid, Disodium Salt).

C. 200 mM Oxalic Acid Solution, pH 3.8 at 37ºC (Oxalic Acid) (Prepare 1 ml in deionized water using Oxalic Acid, Free Acid. Adjust to pH 3.8 at 37ºC with 1 M NaOH).

D. Peroxidase Enzyme Solution (POD) (Immediately before use, prepare a solution containing 1 mg/ml of Peroxidase, in cold deionized water).

E. Oxalate Oxidase Enzyme Solution (Ox Oxidase) (Immediately before use, prepare a solution containing 0.25–0.50 unit/ml of Oxalate Oxidase in cold deionized water).

Procedure

Pipette (in milliliters) the following reagents into suitable cuvettes:

	Test	Blank
Reagent A (MBTH)	2.70	2.70
Reagent B (EDTA)	0.10	0.10
Deionized Water	0.10	0.10
Reagent C (Oxalic Acid)	0.02	0.02
Reagent D (POD)	0.01	0.01

Mix by inversion and equilibrate to 37ºC. Monitor the A600nm until constant, using a suitably thermostatted spectrophotometer. Then add:

Reagent E (Ox Oxidase)	0.02	—
Deionized Water	—	0.02

Immediately mix by inversion and record the increase in A600nm for approximately 5 minutes. Obtain the A600nm/minute using the maximum linear rate for both the Test and Blank.

Calculations

$$\text{Units / ml enzyme} = \frac{(\text{A600nm / min Test} - \text{A600nm / min Blank})\ (2.95)\ (\text{df})}{(26.4)\ (0.02)}$$

2.95 = Total volume (in milliliters) of assay
df = Dilution factor
26.4 = Millimolar extinction coefficient of Indamine dye at 600nm.
0.02 = Volume (in milliliter) of Oxalate Oxidase used.

$$\text{Units / mg solid} = \frac{\text{units / ml enzyme}}{\text{mg solid / ml enzyme}}$$

Unit Definition: *One unit will form 1.0 μmole of H_2O_2 from oxalate per minute at pH 3.8 at 37°C.*

Final Assay Concentrations

In a 2.95 ml reaction mix, the final concentrations are 46 mM succinic acid, 0.1 mM 3-methyl-2-benzothiazolinone hydrazone, 0.72 mM N,N-dimethylaniline, 3.4 mM Ethylenediaminetetraacetic Acid, 1 mM oxalic acid, 0.01 mg peroxidase, and 0.005–0.01 unit oxalate oxidase.

Note

1. Peroxidase Unit Definition: One unit will form 1.0 mg purpurogallin from pyrogallol in 20 seconds at pH 6.0 at 20°C.

Enzymatic Assay of PAPAIN (EC 3.4.22.2)

Principle

$$\text{BAEE} + H_2O \xrightarrow{\text{Papain}} \text{Na - Benzoyl - L - Arginine} + \text{Ethanol}$$

Abbreviation Used

BAEE = Na-Benzoyl-L-Arginine Ethyl Ester.

Conditions: T = 25°C, pH = 6.2

Method: Titrimetric Rate Determination

Reagents

A. 80 mM Na-Benzoyl-L-Arginine Ethyl Ester (BAEE) (Prepare 50 ml in deionized water using Na-Benzoyl-L-Arginine Ethyl Ester Hydrochloride).
B. 20 mM Ethylenediaminetetraacetic Acid (EDTA) (Prepare 100 ml in deionized water using Ethylenediaminetetraacetic Acid, Disodium Salt, Dihydrate).
C. 50 mM L-Cysteine, pH 6.2 at 25°C (Prepare 25 ml in Reagent B using L-Cysteine Hydrochloride, Monohydrate. Adjust to pH 6.2 with 1 M NaOH.2 *Prepare Fresh*).
D. 3 M Sodium Chloride Solution (NaCl) (Prepare 25 ml in deionized water using Sodium Chloride .)
E. 20 mM Sodium Hydroxide Titrant, Standardized (NaOH) (Prepare 50 ml in cold deionized water using Sodium Hydroxide, Anhydrous. Standardize according to the ACS Reagent Procedure).
F. Papain Enzyme Solution (Immediately before use, prepare a solution containing approximately 1 unit/ml in cold deionized water).

Procedure

Using a suitable pH meter in conjunction with a magnetic stirrer, pipet the following reagents into a suitable titration vessel (in milliliters):

	Test
Reagent A (BAEE)	7.0
Reagent C (L-Cysteine)	1.0
Reagent D (NaCl)	1.0

Equilibrate at 25°C, and adjust to pH 6.3 by the addition of Reagent E (NaOH). Then add:

Reagent F (Enzyme Solution)	1.0

Monitor the pH of the reaction mix and record the time when the pH reaches 6.2. Maintain the pH of the reaction mix at pH 6.2 by the addition of small volumes of Reagent E and record the time (in minutes) when a total of 50 μl of Reagent E is consumed. Repeat this process for approximately 5–10 minutes. Plot the volume (ml) of 100 mM NaOH consumed vs the time (minutes) and determine the maximum linear rate from the plot. Use the time for the consumntion of 50 μl of 100 mM NaOH in the calculation.

Calculations

$$\text{Units / ml enzyme} = \frac{(0.05)\ (\text{Normality of NaOH})\ (1000)\ (\text{df})}{(\text{T})\ (1)}$$

0.05 = Volume (in milliliters) of Reagent E used to maintain the pH at 6.2
1000 = Conversion from millimoles to micromoles.
df = Dilution factor.
T = Time (in minutes) required to maintain the pH at 6.2 Per 50 μl aliquot
1 = Volume (in milliliter) of enzyme used.

$$\text{Units / mg solid} = \frac{\text{units / ml enzyme}}{\text{mg solid / ml enzyme}}$$

$$\text{Units / mg protein} = \frac{\text{units / ml enzyme}}{\text{mg protein / ml enzyme}}$$

Unit Definition: *One unit will hydrolyze 1.0 μmole of BAEE per minute at pH 6.2 at 25°C.*

Final Assay Concentrations

In a 10 ml reaction mix the final concentrations are 56 mM Na-benzoyl-L-arginine ethyl ester, 2.0 mM Ethylenediaminetetraacetic Acid, 5.0 mM L-cysteine, 300 mM sodium chloride and 1.0 unit papain.

Notes

1. Adjust the pH quickıy because L-cysteine readily oxidizes and forms a hazy solution.
2. Do not add the entire 50 μl aliquot in one delivery. This will raise the pH to greater than pH 6.3 which may result in an erroneous rate.
3. There must be at least 4 additions of 50 μl of NaOH during this time period. If there is not, increase the enzyme concentration and repeat.

Enzymatic Assay of PAPAIN/FICIN on Agarose (EC 3.4.22.2)

Principle

$$\text{BAEE} \xrightarrow{\text{Papain/Ficin}} \alpha\text{-N-Benzoyl-L-Arginine} + \text{Ethanol}$$

Abbreviation Used

BAEE = α-N-Benzoyl-L-Arginine Ethyl Ester

Conditions: T = 25°C, pH = 7.0

Method: Titrimetric

Reagents

A. 500 mM Potassium Chloride Solution (KCl) (Prepare 100 ml in deionized water using Potassium Phosphate).
B. 250 mM Benzoyl-L-Arginine Ethyl Ester Solution (BAEE) (Prepare 20 ml in Reagent A using α-N-Benzoyl-LArginine, Ethyl Ester).
C. 250 mM Ethylenediaminetetraacetic Acid (EDTA) (Prepare 100 ml in deionized water using Ethylenediaminetetraacetic Acid, Disodium, Dihydrate).
D. 250 mM L-Cysteine, pH 7.0 at 25°C (Prepare 25 ml in Reagent B using L-Cysteine Hydrochloride. Adjust to pH 7.0 with 1 M NaOH. *Prepare Fresh*).
E. 100 mM Sodium Hydroxide Titrant, Standardized (NaOH) (Prepare according to ACS Reagent Procedure)
F. Papain/Ficin Enzyme Solution (Immediately before use prepare a solution containing 2.5 mg/ml in cold deionized water. Allow to hydrate before use).

Procedure

Using a suitable pH meter in conjunction with a magnetic stirrer, pipette (in milliliters) the following reagents into a suitable titration vessel:

	Test
Reagent A (BAEE)	4.0
Reagent C (EDTA)	0.4
Reagent D (L-Cysteine)	4.0

Equilibrate at 25°C, and adjust to pH 7.0 by the addition of Reagent E. then add:

Reagent F (Enzyme Solution)	11.6

Monitor the pH of the reaction mix and record the time when the pH reaches 7.0. Add a known amount of Reagent E (NaOH) and record the time for the pH to return to 7.0. Repeat the process 3 times.

Calculation

$$\text{Units / g Solid} = \frac{(\text{Vol NaOH consumed / min})\ (\text{N NaOH})\ (1000\ \mu\text{Moles}/\mu\text{Mole})\ (1000)}{(\text{mg Sample / RM})}$$

N NaOH = Normality of Reagent E
1000 = Conversion of milligrams to grams (Unit definition)
RM = Reaction Mix (20 ml)

Unit Definition: *One unit will hydrolyze 1.0 μMole of BAEE per minute at pH 7.0 at 25°C.*

Final Assay Concentrations

In a 20 ml reaction mix the final concentrations are 100 mM potassium chloride, 50 mM BAEE, 5 mM EDTA, 50 mM L-cysteine and 25 mg papain/ficin agarose.

Note

1. Adjust the pH quickly because L-cysteine readily oxidizes and forms a hazy solution.

Enzymatic Assay of PAPAIN, INSOLUBLE (EC 3.4.22.2)

Principle

$$\text{BAEE} + H_2O \xrightarrow{\text{Papain}} \text{Na - Benzoyl - L - Arginine} + \text{Ethanol}$$

Abbreviation Used

BAEE = Na-Benzoyl-L-Arginine Ethyl Ester

Conditions: T = 30°C, pH = 7.0

Method: Titrimetric

Reagents

A. 500 mM Potassium Chloride and 250 mM Na-Benzoyl-L-Arginine Ethyl Ester Solution (BAEE) (Prepare 20 ml in deionized water using Potassium Chloride and Na-Benzoyl-L-Arginine Ethyl Ester, Hydrochloride).
B. 250 mM Ethylenediaminetetraacetic Acid Solution, pH 7.0 at 30°C (EDTA) (Prepare 100 ml in deionized water using Ethylenediaminetetraacetic Acid, Disodium Salt, Dihydrate. Adjust to pH 7.0 at 30°C with 1 M NaOH).
C. 250 mM L-Cysteine Solution, pH 7.0 at 30°C (L-Cys) (Prepare 20 ml in deionized water using L-Cysteine Hydrochloride, Monohydrate. Adjust to pH 7.0 at 30°C with 1 M NaOH. *Prepare Fresh*).
D. 100 mM Sodium Hydroxide Titrant, Standardized (NaOH) (Prepare 50 ml in deionized water using Sodium Hydroxide, Anhydrous. Standardize according to the ACS Reagent Procedure).
E. Papain Enzyme Suspension (Immediately before use, prepare a suspension containing 2–4 mg/ml of Insoluble Enzyme attached to Beaded Agarose or 25–50 mg/ml of Papain, Insoluble Enzyme attached to Carboxymethyl Cellulose in cold deionized water).

Procedure

Using a suitable pH meter in conjunction with a magnetic stirrer, pipette (in milliliters) the following reagents into a suitable titration vessel:

	Test
Reagent A (BAEE)	4.00
Reagent B (EDTA)	0.40
Reagent C (L-Cys)	4.00

Equilibrate to 30°C and adjust the pH to 7.0 by the addition of Reagent D (NaOH). Then add:

Reagent E (Enzyme Suspension)	11.60

Immediately mix by swirling. Readjust the pH to slightly above pH 7.0. When the pH reaches 7.0, begin the timer. Maintain the pH of the reaction mix at 7.0 by the addition of small volumes (0.05 ml) of Reagent D (NaOH). Record the volume of Reagent D (NaOH) used to maintain the pH at 7.0 and the time required for the consumption of Reagent D (NaOH). Run the reaction 1–5 minutes.

Calculation

$$\text{Units / g enzyme} = \frac{(\text{NaOH})\ (\text{Molarity of NaOH})\ (1000)}{(\text{T})\ (\text{mg of enzyme / RM})}$$

NaOH = Volume (in microliter) of NaOH used
1000 = Conversion factor from mg to g
T = Time (in minutes) required to consume the added Reagent D (NaOH) while maintaining the pH at 7.0
RM = Reaction Mix

Unit Definition: *One unit will hydrolyze 1.0 µmole of BAEE per minute at pH 7.0 at 30°C.*

Final Assay Concentrations

In a 20 ml reaction mix, the final concentrations are 100 mM potassium chloride, 50 mM Na-Benzoyl L-arginine ethyl ester, 5 mM Ethylenediaminetetraacetic Acid, 50 mM L-cysteine, 2–4 mg of insoluble papain attached to beaded agarose or 25–50 mg of insoluble papain attached to carboxymethyl cellulose.

Notes

1. Adjust the pH quickly because L-cysteine readily oxidizes and forms a hazy solution.
2. The enzyme suspension should be mixed by swirling before addition so that it is homogeneous.

Enzymatic Assay of PECTIN LYASE (EC 4.2.2.10)

Principle

$$\text{Pectin} + H_2O \xrightarrow{\text{Pectin Lyase}} \text{4-Deoxy-6-Methyl-4, 5-Galacturonic Acid Oligomers}$$

Conditions: T = 40°C, pH = 5.0, A235nm, Light path = 1 cm

Method: Continuous Spectrophotometric Rate Determination

Reagents

A. 100 mM Citrate and 100 mM Phosphate Buffer, pH 5.0 at 40°C (Prepare 100 ml in deionized water using Citric Acid, Free Acid, Anhydrous and Sodium Phosphate, Monobasic, Anhydrous. Adjust to pH 5.0 at 40°C with 1 N NaOH).
B. 0.5% (w/v) Pectin Solution (Pectin) (Prepare 50 ml in Reagent A using Pectin, from Citrus Fruits. Mild heating for 4–5 minutes is required to dissolve the pectin. Filter through a 0.22 μm filter to clarify if cloudy).
C. 100 mM Citrate and 100 mM Phosphate Buffer with 0.1% (w/v) Bovine Serum Albumin (Enzyme Diluent) (Prepare 25 ml in Reagent A using Albumin, Bovine).
D. Pectin Lyase Enzyme Solution (Immediately before use, prepare a solution containing 0.2–0.4 unit/ml in cold Reagent C).

Procedure

Pipette (in milliliters) the following reagents into suitable cuvettes:

	Test	Blank
Reagent B (Pectin)	1.80	1.80
Reagent A (Buffer)	0.20	0.20

Mix by inversion and equilibrate to 40°C. Monitor the A235nm until constant, using a suitably thermostatted spectrophotometer. Then add:

Reagent D (Enzyme Solution)	0.40	—
Reagent C (Enzyme Diluent)	—	0.40

Immediately mix by inversion and record the increase in A235nm for approximately 5 minutes. Obtain the ΔA235nm/minute using the maximum linear rate for both the Test and Blank.

Calculations

$$\text{Units / ml enzyme} = \frac{(\Delta A235nm / \min \text{ Test} - \Delta A235nm / \min \text{ Blank})\ (2.4)\ (df)}{(1.0)(0.4)}$$

2.4 = Total volume (in milliliters) of assay
df = Dilution factor
1.0 = Change in A235nm per minute at 40°C as per the Unit Definition
0.4 = Volume (in milliliter) of enzyme used

$$\text{Units / mg protein} = \frac{\text{units / ml enzyme}}{\text{mg protein / ml enzyme}}$$

Unit Definition: *One unit will cause a ΔA235nm of 1.0 per minute at 40°C due to the release of unsaturated products from Pectin.*

Final Assay Concentrations

In a 2.40 ml reaction mixture, the final concentrations are 100 mM citric acid, 100 mM sodium phosphate, 0.4% (w/v) pectin, 0.02% (w/v) Bovine Serum Albumin and 0.08–0.16 unit pectin lyase.

Enzymatic Assay of PECTIN LYASE (EC 4.2.2.10)

Principle

$$\text{Pectin} + H_2O \xrightarrow{\text{Pectin Lyase}} \text{4-Deoxy-6-Methyl-4, 5-Galacturonic Acid Oligomers}$$

Conditions: T = 40°C, pH = 6.0, A235nm, Light path = 1 cm

Method: Continuous Spectrophotometric Rate Determination

Reagents

A. 100 mM Citrate and 100 mM Phosphate Buffer, pH 6.0 at 40°C (Prepare 100 ml in deionized water using Citric Acid, Free Acid, Anhydrous and Sodium Phosphate, Monobasic, Anhydrous. Adjust to pH 6.0 at 40°C with 1 M NaOH).
B. 0.5% (w/v) Pectin Solution (Pectin) (Prepare 50 ml in Reagent A using Pectin, from Citrus Fruits. Mild heating for 4–5 minutes, with stirring, is required to dissolve the pectin. Filter through a 0.22 μm filter to clarify).
C. 100 mM Citrate and 100 mM Phosphate Buffer with 0.1% (w/v) Bovine Serum Albumin (Enzyme Diluent) (Prepare 25 ml in Reagent A using Albumin, Bovine).
D. Pectin Lyase Enzyme Solution (Immediately before use, prepare a solution containing 0.3–0.5 unit/ml in cold Reagent C).

Procedure

Pipette (in milliliters) the following reagents into suitable cuvettes:

	Test	Blank
Reagent B (Pectin)	1.80	1.80
Reagent A (Buffer)	0.20	0.20

Mix by inversion and equilibrate to 40°C. Monitor the A235nm until constant, using a suitably thermostatted spectrophotometer. Then add:

	Test	Blank
Reagent D (Enzyme Solution)	0.50	—
Reagent C (Enzyme Diluent)	—	0.50

Immediately mix by inversion and record the increase in A235nm for approximately 5 minutes. Obtain the ΔA235nm/minute using the maximum linear rate for both the Test and Blank.

Calculations

$$\text{Units / ml enzyme} = \frac{(\Delta\text{A235nm / min Test} - \Delta\text{A235nm / min Blank})\ (2.5)\ (\text{df})}{(1.0)\ (0.5)}$$

2.5 = Total volume (in milliliters) of assay
df = Dilution factor
1.0 = Change in A235nm per minute at 40°C as per the Unit Definition
0.5 = Volume (in milliliter) of enzyme used

$$\text{Units / mg protein} = \frac{\text{units / ml enzyme}}{\text{mg protein / ml enzyme}}$$

Unit Definition: *One unit will cause a ΔA235 of 1.0 per minute at 40°C due to the release of unsaturated products from Pectin.*

Final Assay Concentrations

In a 2.50 ml reaction mixture, the final concentrations are 100 mM citric acid, 100 mM sodium phosphate, 0.4% (w/v) pectin, 0.02% (w/v) Bovine Serum Albumin and 0.15–0.25 unit pectin lyase.

Enzymatic Assay of PECTINASE (EC 3.2.1.15)

Principle

$$\text{Polygalacturonic Acid} + H_2O \xrightarrow{\text{Pectinase}} \text{Galacturonic Acid}$$

$$I_2 + \text{Galacturonic Acid} \rightarrow \text{Oxidation products}$$

$$I_2 + 2Na_2S_2O_3 \rightarrow 2NaI + Na_2S_4O_6 \text{ (Titration of Excess Iodine)}$$

Conditions: T = 25°C, pH = 4.0

Method: Titrimetric

Reagents

A. 0.5% (w/v) Polygalacturonic Acid Solution (Polygalacturonic Acid) (Prepare 100 ml in deionized water using Polygalacturonic Acid. Adjust to pH 4.0 at 25°C with 1 M NaOH).

B. 50 mM Iodine with 200 mM Potassium Iodide (I_2/KI) (Prepare 500 ml in deionized water using Iodine and Potassium Iodide. Initially dissolve in 50 ml deionized water, allow to set at room temperature for 30 minutes and then QS to 500 ml with deionized water).

C. 1 M Sodium Carbonate (Na_2CO_3) (Prepare 100 ml in deionized water using Sodium Carbonate, Anhydrous).

D. 2.0 N Sulfuric Acid (H_2SO_4) (Prepare 200 ml in deionized water by diluting Sulfuric Acid, ACS Reagent).

E. 100 mM Sodium Thiosulfate, Standardized ($Na_2S_2O_3$) (Prepare 100 ml in deionized water using Sodium Thiosulfate, Pentahydrate. Standardize according to the ACS Reagent Procedure).

F. 1.0% (w/v) Starch Indicator (Prepare 100 ml in deionized water using Starch Potato, Soluble. *Prepare Fresh*).

G. Pectinase Enzyme Solution (Immediately before use, prepare a solution containing 100 units/ml of Pectinase in cold deionized water).

Procedure

Pipette (in milliliters) the following reagents into 50 ml Erlenmeyer flasks:

	Test	Blank
Reagent A (Polygalacturonic Acid)	4.90	5.00

Equilibrate to 25°C and then add the following reagent:

	Test	Blank
Reagent G (Pectinase)	0.10	—

Mix by swirling and incubate at 25°C for exactly 5.0 minutes. At the end of 5 minutes incubation, add:

Reagent B (I_2/KI)	5.0	5.0
Reagent C (Na_2CO_3)	1.0	1.0

Mix by swirling and store in the dark for 20 minutes. Then add:

Reagent D (H_2SO_4)	2.0	2.0

Mix by swirling and titrate the Test and Blank with Reagent E ($Na_2S_2O_3$) until it is light yellow. Then add 1 drop of Reagent F (Indicator) and continue to titrate with Reagent E ($Na_2S_2O_3$)until solutions are colorless. (Record the volume of Reagent E used).

Calculations

$$\text{Units / ml enzyme} = \frac{(1)\,(100)\,(\text{ml of Reagent E for Blank} - \text{ml of Reagent E for Test})\,(\text{df})}{(5)\,(0.1)\,(2)}$$

1 = One µmole galacturonic acid is oxidized by 1 microequivalent of I_2.
100 = microequivalents of S_2O_3/ml of Reagent E.
df = Dilution factor.
5 = Time of reaction in minutes.
0.1 = Volume (in milliliter) of enzyme used.
2 = microequivalents of S_2O_3 oxidized per microequivalent of I_2 reduced.

$$\text{Units / mg solid} = \frac{\text{units / ml enzyme}}{\text{mg solid / ml enzyme}}$$

$$\text{Units / mg protein} = \frac{\text{units / ml enzyme}}{\text{mg protein / ml enzyme}}$$

Unit Definition: *One unit will liberate 1.0 µmole of galacturonic acid from polygalacturonic acid per minute at pH 4.0 at 25°C.*

Final Assay Concentrations

In a 5.00 ml reaction mix, the final concentrations are 0.49% (w/v) polygalacturonic acid and 10 units pectinase.

Enzymatic Assay of PECTINESTERASE (EC 3.1.1.11)

Principle

$$\text{Pectin} + H_2O \xrightarrow{\text{Pectinesterase}} \text{Pectate} + CH_3OH$$

Conditions: T = 30°C, pH = 7.5

Method: Titrimetric

Reagents

A. 100 mM Sodium Chloride Solution (NaCl) (Prepare 300 ml in deionized water using Sodium Chloride).

B. 1.0% (w/v) Pectin Solution (Pectin) (Prepare 250 ml in Reagent A using Pectin. Dissolve the Pectin into approximately 150 ml of Reagent A which has been heated to boiling and stir with a magnetic stirrer. Cool to room temperature upon solubilizing and dilute to 250 ml with Reagent A).

C. 20 mM Sodium Hydroxide Solution–Standardized (NaOH) (Prepare 50 ml in deionized water using Sodium Hydroxide, Anhydrous. Standardize according to the ACS Reagent Procedure).

D. 1.7 M Sodium Chloride Solution (Enzyme Diluent) (Prepare 50 ml in deionized water using Sodium Chloride).

E. Pectinesterase Enzyme Solution (Immediately before use, prepare a solution containing 10–15 units/ml of Pectinesterase in cold Reagent D).

Procedure

Using a suitable pH meter in conjunction with a magnetic stirrer, pipette (in milliliters) the following reagents into a suitably thermostatted titration vessel:

	Test	Blank
Reagent B (Pectin)	20.00	20.00

Equilibrate to 30°C. Adjust to pH 7.5 with Reagent C (NaOH). Then add:

Reagent E (Enzyme Solution)	0.20	—
Reagent D (Enzyme Diluent)	—	0.20

Run the reaction for 1–5 minutes. Record the time when the pH reaches 7.5. Maintain the pH of the reaction mix at pH 7.5 by the addition of small volumes (0.05 ml) of Reagent C (NaOH). Record the volume of Reagent C (NaOH) used to maintain the pH at 7.5 and the time required.

Calculations

ml NaOH for Test – ml NaOH for Blank

$$\text{Units / ml enzyme} = \frac{(M)\,(1000)\,(df)}{(T)\,(0.20)}$$

M = Molarity of NaOH solution
1000 = Conversion from milliequivalent to microequivalents
df = Dilution factor
T = Time required to maintain the pH at 7.5
0.20 = Volume (in milliliter) of enzyme used

$$\text{Units / mg solid} = \frac{\text{units / ml enzyme}}{\text{mg solid / ml enzyme}}$$

$$\text{Units / mg protein} = \frac{\text{units / ml enzyme}}{\text{mg protein / ml enzyme}}$$

Unit Definition: *One unit will release 1.0 microequivalent of acid from pectin per minute at pH 7.5 at 30ºC.*

Initial Assay Concentrations

In a 21.00 ml reaction mix, the initial concentrations are 176 mM sodium chloride, 0.95% (w/v) pectin, and 2–3 units pectinesterase.

Note

1. The solution must be agitated during preparation to prevent pectin from adhering to the bottom of container and being burned.

Enzymatic Assay of PECTOLYASE (EC 3.2.1.15 and EC 4.2.2.10)

Principle

$$\text{Polygalacturonic Acid} + H_2O \xrightarrow{\text{Pectinase}} \text{Galacturonic Acid}$$

$$I_2 + 2Na_2S_2O_3 \rightarrow 2NaI + Na_2S_4O_6 \text{ (Titration of Acid)}$$

Conditions: T = 25°C, pH = 5.5

Method: Titrimetric

Reagents

A. 0.5% (w/v) Polygalacturonic Acid Solution (Prepare 100 ml in deionized water using Polygalacturonic Acid. Heat to solubilize. Adjust to pH 5.5 at 25°C with 1 M NaOH).

B. 100 mM Iodine with 200 mM Potassium Iodide (I_2/KI) (Prepare 500 ml in deionized water using Iodine and Potassium Iodide. Stir the solution, at room temperature on a magnetic stir plate overnight to dissolve the iodide completely).

C. 1 M Sodium Carbonate (Na_2CO_3) (Prepare 100 ml in deionized water using Sodium Carbonate, Anhydrous).

D. 2.0 N Sulfuric Acid (H_2SO_4) (Prepare 200 ml in deionized water by diluting Sulfuric Acid, ACS Reagent).

E. 50 mM Sodium Thiosulfate, Standardized ($Na_2S_2O_3$) (Prepare 100 ml in deionized water using Sodium Thiosulfate, Pentahydrate).

F. 1.0% (w/v) Starch Indicator (Indicator) (Prepare 100 ml in deionized water using Starch Potato, Soluble. Heat to solubilize. *Prepare Fresh*).

G. Pectinase Enzyme Solution (Immediately before use, prepare a solution containing 50–100 units/ml of Pectinase in cold deionized water).

Procedure

Pipette (in milliliters) the following reagents into 50 ml Erlenmeyer flasks:

	Test	Blank
Reagent A (Polygalacturonic Acid)	4.90	5.00

Equilibrate to 25°C. and then add the following reagent:

	Test	Blank
Reagent G (Enzyme Solution)	0.10	—

Mix by swirling and incubate at 25°C for exactly 5.0 minutes. Then add:

Reagent B (I_2/KI)	5.0	5.0
Reagent C (Na_2CO_3)	1.0	1.0

Mix by swirling and store in the dark for 20 minutes. Then add:

Reagent D (H_2SO_4)	2.0	2.0

Mix by swirling and titrate the Test and Blank with Reagent E ($Na_2S_2O_3$) until it is light yellow. Then add 1 drop of Reagent F (Indicator) and continue to titrate with Reagent E ($Na_2S_2O_3$) until the solutions are colorless. (Record the volume of Reagent E used).

Calculation

$$\mu\text{Moles Galacturonic Acid} = \frac{(\text{I·Eq I}_2)(50\ \mu\text{Eq / ml})\ (\text{ml of reagent E for blank - ml of reagent E for test})}{(2\ \text{Eq S}_2\text{O}_3)\ (5)\ (\text{mg enzyme RM})}$$

5 = Time of reaction in minutes
RM = Reaction Mix

Unit Definition: *One unit will liberate 1.0 μmole of galacturonic acid from polygalacturonic acid per minute at pH 5.5 at 25°C.*

Final Assay Concentrations

In a 5.00 ml reaction mix, the final concentrations are 0.5% (w/v) polygalacturonic acid and 5–10 units pectinase.

Enzymatic Assay of PENICILLIN AMIDASE Insolubilized Enzyme

Principle

$$\text{Benzylpenicillin} + H_2O \xrightarrow{\text{Penicillin Amidase}} \text{6-APA} + \text{Phenylacetic Acid}$$

Abbreviation Used

6-APA = 6-Aminopenicillanate

Conditions: T = 37°C, pH = 7.8

Method: Titrimetric

Reagents

A. 50 mM Sodium Phosphate Buffer, pH 7.8 at 37°C (Prepare 100 ml in deionized water using Sodium Phosphate, Monobasic, Anhydrous).

B. 4.0% (w/v) Penicillin-G Solution (PEN-G) (Prepare 10 ml in Reagent A using Penicillin-G, Procain Salt).

C. 500 mM Sodium Hydroxide Solution–Standardized (NaOH) (Prepare 100 ml in deionized water using Sodium Hydroxide, Anhydrous. Standardize according to the ACS Reagent Procedure).

D. Penicillin Amidase Insolubilized Enzyme Suspension (Prepare by suspending 200 mg of Penicillin Amidase, Insolubilized Enzyme, attached to macroporous oxirane acrylic beads (07628) in 20 ml of Reagent A. Filter through a 0.45 μm filter. Resuspend the filtered insoluble enzyme in 20 ml of Reagent A).

Procedure

Using a suitable pH meter in conjunction with a magnetic stirrer, pipette (in milliliters) the following reagents into a suitably thermostatted titration vessel:

	Test	Blank
Reagent D (Enzyme)	20.00	—
Reagent A (Buffer)	—	20.00

Mix by stirring at 37°C for 30 minutes. Then add:

Reagent B (Pen-G)	1.00	1.00

Immediately adjust to pH 7.8 with Reagent C (NaOH) using a burette. Run the reaction for 1–5 minutes. Record the time when the pH reaches 7.8. Maintain the pH of the reaction mix at pH 7.8 by the addition of small volumes (0.05 ml) of Reagent C. Record the volume of Reagent C used to maintain the pH at 7.8 and the time required.

Calculations

(NaOH) = Volume (in milliliters) of Reagent C required to maintain the pH at 7.8.

$$\text{Units / g enzyme} = \frac{[(\text{NaOH})\text{ Test} - (\text{NaOH})\text{ Blank}]\,(\text{M})\,(1000)}{(\text{T})\,(0.2)}$$

M = Molarity of Reagent C
1000 = Conversion factor from millimoles to micromoles
T = Time (in minutes) required to maintain the pH at 7.8 as per the Unit Definition
0.2 = Weight (in g) of solid penicillin amidase insolubilized enzyme used in the assay

Unit Definition: *One unit will hydrolyze 1.0 μmole of benzylpenicillin per minute at pH 7.8 at 37°C.*

Final Assay Concentration

In a 21.00 ml reaction mix, the initial concentrations are 50 mM potassium phosphate, 1.9% (w/v) penicillin-G, and 200 mg penicillin amidase insolubilized enzyme.

Enzymatic Assay of PENICILLIN AMIDASE1 (EC 3.5.1.11)

Principle

$$\text{Benzylpenicillin} + H_2O \xrightarrow{\text{Penicillin Amidase}} \text{6-APA} + \text{Phenylacetic Acid}$$

Abbreviation Used

6-APA = 6-Aminopenicillanate

Conditions: T = 37°C, pH = 7.8

Method: Titrimetric

Reagents

A. 50 mM Potassium Phosphate Buffer, pH 7.5 at 37°C (Prepare 50 ml in deionized water using Potassium Phosphate, Monobasic, Anhydrous. Adjust to pH 7.5 at 37°C with 1 M KOH).
B. 2.0% (w/v) Penicillin-G Solution (PEN-G) (Prepare 100 ml in Reagent A using Penicillin-G, Sodium Salt).
C. 500 mM Sodium Hydroxide Solution–Standardized (NaOH) (Prepare 100 ml in deionized water using Sodium Hydroxide, Anhydrous).
D. 100 mM Potassium Phosphate Buffer, pH 7.5 at 37°C (Enz Dil) (Prepare 25 ml in deionized water using Potassium Phosphate, Monobasic, Anhydrous. Adjust to pH 7.5 at 37°C with 1 M KOH).
E. Penicillin Amidase Enzyme Solution (Immediately before use, prepare a solution containing 30–40 units/ml of Penicillin Amidase in cold Reagent D).

Procedure

Using a suitable pH meter in conjunction with a magnetic stirrer, pipette (in milliliters) the following reagents into a suitably thermostatted titration vessel:

	Test
Reagent B (PEN-G)	20.00

Equilibrate to 37°C. Then add:

Reagent E (Enzyme Solution)	1.00

Immediately adjust to pH 7.8 with Reagent C (NaOH) using a burette. Monitor the pH of the reaction mix. Record the time when the pH reaches 7.8. Maintain the pH of the reaction mix at pH 7.8 by the addition of small volumes (50 µl) of Reagent C. Record the volume of Reagent C used to maintain the pH at 7.8 and the time required.

Calculations

(NaOH) = Volume (in milliliters) of Reagent C used in the Test required to maintain the pH at 7.8.

$$\text{Units / ml protein} = \frac{(\text{NaOH})\,(\text{M})\,(1000)\,(\text{df})}{(\text{T})\,(1)}$$

M = Molarity of Reagent C
1000 = Conversion factor from millimoles to micromoles
df = Dilution factor
T = Time (in minutes) required to maintain the pH at 7.8 as per the Unit Definition
1 = Volume (in milliliter) of enzyme used

$$\text{Units / mg protein} = \frac{\text{units / ml enzyme}}{\text{mg protein / ml enzyme}}$$

Unit Definition: *One unit will hydrolyze 1.0 µmole of benzylpenicillin per minute at pH 7.8 at 37°C.*

Initial Assay Concentration

In a 21.00 ml reaction mix, the initial concentrations are 52 mM potassium phosphate, 1.9% (w/v) penicillin-G, and 30–40 units penicillin amidase.

Note

1. This assay is not to be used to assay Penicillin Amidase, Insolubilized Enzyme, attached to macroporous oxirane acrylic beads.

Enzymatic Assay of PENICILLINASE (EC 3.5.2.6) Benzylpenicillin as Substrate

Principle

$$\text{Benzylpenicillin} + H_2O \xrightarrow{\text{Penicillinase}} \text{Benzylpenicilloic Acid}$$

Conditions: T = 25°C, pH = 7.0

Method: Titrimetric

Reagents

A. 100 mM Tris HCl Buffer with 0.1% (w/v) Bovine Serum Albumin, pH 7.0 at 25°C (Prepare 50 ml in deionized water using Trizma Base and Albumin, Bovine or equivalent. Adjust to pH 7.0 at 25°C with 1 M HCl).
B. Benzylpenicillin Solution (B-Pen) (Immediately before use, prepare 500 ml of a solution containing 200–300 units/ml in deionized water using Penicillin-G, Sodium Salt).
C. 10 mM Sodium Hydroxide Solution-Standardized (NaOH) (Prepare 400 ml in cold deionized water using Sodium Hydroxide, Anhydrous).
D. Penicillinase Enzyme Solution (Immediately before use, prepare a solution containing 3–7 units/ml of Penicillinase in cold Reagent A).

Procedure

Using a suitable pH meter in conjunction with a magnetic stirrer, pipette (in milliliters) the following reagents into a suitably thermostatted titration vessel:

	Test	Blank
Reagent B (B-Pen)	50.00	50.00

Adjust to pH 7.0 at 25°C with Reagent C (NaOH), using a burette. Then add:

	Test	Blank
Reagent A (Buffer)	—	0.10
Reagent D (Enzyme Solution)	0.10	—

Monitor the pH of the reaction mix. Record the time when the pH reaches 7.0. The time for the test procedure is 10 minutes. Maintain the pH of the reaction mix at pH 7.0 by the addition of small volumes (50 µl) of Reagent C. Record the volume of Reagent C used to maintain the pH at 7.0 and the time required.

Calculation

[NaOH] = Volume (in milliliters) of Reagent C used in the Test minus the Blank.

$$\text{Units / mg enzyme} = \frac{(\text{NaOH})\ (\text{M})\ (1000)}{(\text{T})\ (\text{mg enzyme / RM})}$$

M = Molarity of Reagent C
1000 = Conversion factor from millimoles to micromoles (Unit definition)
T = Time required to maintain the pH at 7.0 (in minutes) as per unit definition
RM = Reaction Mix

Unit Definition: *One unit will hydrolyze 1.0 µmole of benzylpenicillin to benzylpenicilloic acid per minute at pH 7.0 at 25°C.*

Final Assay Concentrations

In a 50.10 ml reaction mix, the Final concentrations are 0.2 mM Tris, 10,000–15,000 units benzylpenicillin, 0.0002% (w/v) BSA and 0.3–0.7 unit penicillinase.

Enzymatic Assay of PENICILLINASE (EC 3.5.2.6) Cephaloridine as Substrate

Principle

$$\text{Cephaloridine} + H_2O \xrightarrow{\text{Penicillinase}} \text{Hydrolyzed Cephaloridine}$$

Conditions: T = 25°C, pH = 7.0, A255nm, Light path = 1 cm

Method: Continuous Spectrophotometric Rate Determination

Reagents

A. 100 mM Potassium Phosphate Buffer, pH 7.0 at 25°C (Prepare 100 ml in deionized water using Potassium Phosphate, Monobasic, Anhydrous. Adjust to pH 7.0 at 25°C with 1 M KOH).

B. 1 mM Cephaloridine Solution (Ceph) (Prepare 10 ml in Reagent A using Cephaloridine).

C. Penicillinase Enzyme Solution (Immediately before use, prepare a solution containing 5–10 Cephaloridine units/ml of Penicillinase in cold Reagent A).

Procedure

Pipette (in milliliters) the following reagents into suitable quartz cuvettes:

	Test	Blank
Reagent A (Buffer)	2.70	2.71
Reagent B (Ceph)	0.30	0.30

Mix by inversion and equilibrate to 25°C. Monitor the A255nm until constant, using a suitably thermostatted spectrophotometer. Then add:

	Test	Blank
Reagent C (Enzyme Solution)	0.01	—

Immediately mix by inversion and record the decrease in A255nm for approximately 5 minutes. Obtain the ΔA255nm/minute using the maximum linear rate for both the Test and Blank.

Calculations

$$\text{Units / ml enzyme} = \frac{(\Delta A255nm / \min \text{ Test} - \Delta A255nm / \min \text{ Blank})\ (3.01)\ (df)}{(8)\ (0.01)}$$

3.01 = Volume (in milliliters) of assay
df = Dilution factor
8 = Change in the millimolar extinction coefficient between cephaloridine and the hydrolyzed β-lactam ring of cephaloridine
0.01 = Volume (in milliliter) of enzyme used

$$\text{Units / mg solid} = \frac{\text{units / ml enzyme}}{\text{mg solid / ml enzyme}}$$

$$\text{Units / mg protein} = \frac{\text{units / ml enzyme}}{\text{mg protein / ml enzyme}}$$

Unit Definition: *One unit will hydrolyze 1.0 µmole of cephaloridine per minute at pH 7.0 at 25°C.*

Final Assay Concentration

In a 3.01 ml reaction mix, the final concentrations are 100 mM potassium phosphate, 0.1 mM cephaloridine and 0.05–0.1 cephaloridine unit of penicillinase.

Enzymatic Assay of PEPSIN (EC 3.4.23.1)

Principle

$$\text{Hemoglobin} + H_2O \xrightarrow{\text{Pepsin}} \text{TCA - Soluble Peptides}$$

Conditions: T = 37°C, pH = 2.0, A280nm, Light path = 1 cm

Method: Spectrophotometric Stop Rate Determination

Reagents

A. 10 mM Hydrochloric Acid Solution (Enzyme Diluent) (Prepare 100 ml in deionized water using Hydrochloric Acid).
B. 300 mM Hydrochloric Acid Solution (Prepare 100 ml in deionized water using Hydrochloric Acid).
C. 2.5% (w/v) Hemoglobin Solution (Prepare 100 ml in deionized water using Hemoglobin, Bovine. Mix vigorously and filter through a glass wool filter).
D. 2.0% (w/v) Hemoglobin Substrate Solution (Hb) (Prepare by adding 20 ml of Reagent B to 80 ml of Reagent C).
E. 5% (w/v) Trichloroacetic Acid Solution (TCA) (Prepare 100 ml in deionized water using Trichloroacetic Acid Solution, 6.1 N, approximately 100% (w/v)).

F. Pepsin Enzyme Solution (Immediately before use, prepare a solution containing 0.01–0.05 mg/ml of Pepsin in cold Reagent A).

Procedure

Pipette (in milliliters) the following reagents into suitable containers:

	Test	Blank
Reagent D (Hb)	5.00	5.00

Equilibrate at 37°C. Then add:

Reagent F (Enzyme Solution)	1.00	—

Mix by swirling and incubate at 37°C for exactly 10 minutes. Then add:

Reagent E (TCA)	10.00	10.00
Reagent F (Enzyme Solution)	—	1.00

Mix by swirling and incubate at 37°C for 5 minutes. Filter the solutions through a Whatman #50 filter or a 0.8 μm syringe filter. Transfer the solutions to suitable quartz cuvettes and record the A280nm for both the Test and Blank.

Calculations

$$\text{Units / ml enzyme} = \frac{(\text{A280nm Test} - \text{A280nm Blank})\ (\text{df})}{(0.001)\ (10)\ (1)}$$

df = Dilution factor.
0.001 = Change in Absorbance at 280 nm per unit of Pepsin (Unit Definition).
10 = Time of assay (in minutes) as per the Unit Definition.
1 = Volume (in milliliter) of enzyme used.

$$\text{Units / mg solid} = \frac{\text{units / ml enzyme}}{\text{mg solid / ml enzyme}}$$

$$\text{Units / mg protein} = \frac{\text{units / ml enzyme}}{\text{mg protein / ml enzyme}}$$

Unit Definition: *One unit will produce a A 280nm of 0.001 per minute at pH 2.0 at 37°C measured as TCA –soluble products using hemoglobin as substrate. (Final volume = 16 ml. Light path = 1 cm).*

Final Assay Concentration

In a 6.00 ml reaction mix, the final concentrations are 52 mM hydrochloric acid, 1.7% (w/v) hemoglobin and 0.01–0.05 mg pepsin.

Enzymatic Assay of PEPSIN INSOLUBLE (EC 3.4.23.1)

Principle

$$\text{Hemoglobin} + H_2O \xrightarrow{\text{Pepsin}} \text{TCA - Soluble Peptides}$$

Conditions: T = 37°C, pH = 2.0, A280nm, Light path = 1 cm

Method: Spectrophotometric Stop Rate Determination

Reagents

A. 10 mM Hydrochloric Acid Solution (Enzyme Diluent) (Prepare 100 ml in deionized water using Hydrochloric Acid).
B. 300 mM Hydrochloric Acid Solution (Prepare 100 ml in deionized water using Hydrochloric Acid).
C. 2.5% (w/v) Hemoglobin Solution (Prepare 100 ml in deionized water using Hemoglobin, Bovine. Mix vigorously and filter through a glass wool filter).
D. 2.0% (w/v) Hemoglobin Substrate Solution (Hb) (Prepare by adding 20 ml of Reagent B to 80 ml of Reagent C).
E. 5% (w/v) Trichloroacetic Acid Solution (TCA) (Prepare 100 ml in deionized water using Trichloroacetic Acid Solution, 6.1 N, approximately 100% (w/v)).
F. Pepsin Insoluble Enzyme (Insol Enz) (Weigh between 0.20–0.50 g of insoluble enzyme into a graduated cylinder containing 25 ml of deionized water. Allow the solution to swell for 4–12 hours. Transfer 0.50 ml of gel suspension to a graduated micro-column. Wash the sample with deionized water. The amount of deionized water used should be 50X the ml of suspension used. Resuspend the washed suspension to 15–25 units/ml suspension using Reagent A).

Procedure

Pipette (in milliliters) the following reagents into suitable containers:

	Test	Blank
Reagent D (Hb)	5.00	5.00

Equilibrate at 37°C. Then add:

Reagent F (Enzyme Solution)	1.00	—

Incubate at 37°C for exactly 10 minutes in a shaker bath. Then add:

Reagent E (TCA)	10.00	10.00
Reagent F (Enzyme Solution)	—	1.00

Mix by swirling and incubate at 37°C for 5 minutes. Filter the solutions through a Whatman #50 filter or a 0.8 μm syringe filter. Transfer the solutions to suitable quartz cuvettes and record the A280nm for both the Test and Blank.

Calculations

$$\text{Units / ml suspension} = \frac{(\text{A280nm Test} - \text{A280nm Blank})\ (\text{df})}{(0.001)\ (10)\ (1)}$$

df = Dilution factor
0.001 = Change in Absorbance at 280 nm per unit of Pepsin (Unit Definition)
10 = Time of assay (in minutes) as per the Unit Definition
1 = Volume (in milliliter) of enzyme used

$$\text{Units / mg solid} = \frac{\text{units / ml suspension}}{\text{mg solid / ml suspension}}$$

$$\text{Units / mg Agarose} = \frac{\text{units / mg solid}}{\text{mg solid / mg Agarose}}$$

Unit Definition: *One unit will produce a ΔA280nm of 0.001 per minute at pH 2.0 at 37°C measured as TCA-soluble products using hemoglobin as substrate. (Final volume = 16 ml. Light path = 1 cm).*

Final Assay Concentration

In a 6.00 ml reaction mix, the final concentrations are 50 mM hydrochloric acid, 1.7% (w/v) hemoglobin and 5–15 units pepsin.

Enzymatic Assay of PEPSINOGEN

Principle

$$\text{Pepsinogen} \xrightarrow{\text{pH2}} \text{Pepsin}$$

$$\text{Hemoglobin} + H_2O \xrightarrow{\text{Pepsin}} \text{TCA - Soluble Peptides}$$

Conditions: T = 37°C, pH = 2.0, A280nm, Light path = 1 cm

Method: Spectrophotometric Stop Rate Determination

Reagents

A. 10 mM Hydrochloric Acid Solution (Enz Dil) (Prepare 100 ml in deionized water using Hydrochloric Acid. Adjust to pH 2.0 at 37°C with either 100 mM HCl or 100 mM NaOH).
B. 300 mM Hydrochloric Acid Solution (Prepare 100 ml in deionized water using Hydrochloric Acid).
C. 2.5% (w/v) Hemoglobin Solution (Prepare 100 ml in deionized water using Hemoglobin, Bovine. Mix vigorously and filter through a glass wool filter).
D. 2.0% (w/v) Hemoglobin Substrate Solution (Hb) (Prepare by adding 20 ml of Reagent B to 80 ml of Reagent C).
E. 5% (w/v) Trichloroacetic Acid Solution (TCA) (Prepare 100 ml in deionized water using Trichloroacetic Acid Solution, 6.1 N, approximately 100% (w/v)).
F. 1 μM Sodium Hydroxide Solution, pH 8.0 at 37°C (NaOH) (Prepare 10 ml in deionized water using 1 M NaOH. Adjust to pH 8.0 at 37°C with 1 mM HCl).
G. Pepsinogen (Use Pepsinogen, lyophilized powder).

Procedure

Step 1: Total Pepsinogen

A. Inactivation and Denaturation of Free Pepsin: Dissolve 1 mg of Pepsinogen in 1 ml of Reagent F (NaOH) which has been equilibrated at 37°C. Incubate at 37°C for exactly 30 minutes.

B. Activation of Pepsinogen to Pepsin After the 30 minute incubation in Step A, pipette (in milliliters) 0.20 ml of Step A solution into a 4 dram vial containing 10 ml of Reagent A (Enz Dil). Mix by swirling and let stand for 1 hour at 25°C.

Step 2: Total Available Pepsin

A. Activation of Pepsinogen to Pepsin (Dissolve 2 mg of Pepsinogen in 6 ml of Reagent A (Enz Dil) and let stand for 1 hour at 25°C. Then pipette 1 ml into a suitable container containing 24 ml of Reagent A (Enz Dil).

Step 3: Pepsin Assay

Pipette (in milliliters) the following reagents into suitable containers:

	Test 1	Test 2	Blank 1	Blank 2
ReagentD (Hb)	5.00	5.00	5.00	5.00

Equilibrate at 37°C. Then add:

Pepsin (from Step 1)	1.00	—	—	—
Pepsin (from Step 2)	—	1.00	—	—

Mix by swirling and incubate at 37°C for exactly 10 minutes. Then add:

	Test	Blank 1	Blank 2
Reagent E (TCA)	10.00	10.00	10.00
Pepsin (from Step 1)	—	1.00	—
Pepsin (from Step 2)	—	—	1.00

Mix by swirling and incubate at 37°C for 5 minutes. Filter the solutions through a Whatman #50 filter or a 0.8 μm syringe filter. Transfer the solutions to suitable quartz cuvettes and record the A280nm for both the Tests and Blanks.

Calculations

The calculation shown below can be used to determine both the Total Pepsinogen (Step 1) and also Total Pepsin (Step 2).

$$\text{Units / ml enzyme} = \frac{(\text{A280nm Test} - \text{A280nm Blank})\ (\text{df})}{(0.001)\ (10)\ (1)}$$

df = Dilution factor
0.001 = Change in Absorbance at 280nm per unit of Pepsin (Unit Definition)
10 = Time of assay (in minutes) as per the Unit Definition
1 = Volume (in milliliter) of enzyme used

$$\text{Units / mg solid} = \frac{\text{units / ml enzyme}}{\text{mg solid / ml enzyme}}$$

$$\text{Units / mg protein} = \frac{\text{units / ml enzyme}}{\text{mg protein / ml enzyme}}$$

Unit Definition: *One unit will produce a ΔA280nm of 0.001 per minute at pH 2.0 at 37°C measured as TCA-soluble products using hemoglobin as substrate. (Final volume = 16 ml. Light path = 1 cm).*

Final Assay Concentration

In a 6.00 ml reaction mix, the final concentrations are 52 mM hydrochloric acid, 1.7% (w/v) hemoglobin and 0.01 mg pepsinogen (total available pepsin assay).

Enzymatic Assay of PEPTIDASE

Principle

$$\text{L-Leucine-}\Delta\text{-Naphthylamide} + H_2O \xrightarrow{\text{Peptidase}} \text{L-Leucine} + \text{Naphthylamine}$$

$$\text{Naphthylamine} + NaNO_2HCl \rightarrow \text{Diazo Reagent}$$

$$\text{Diazo Reagent} + \text{N-(1-Naphthyl)ethylenediamine} \rightarrow \text{Blue Azo complex}$$

Conditions: T = 37°C, pH = 7.1, A580nm, Light path = 1 cm.

Method: Colorimetric

Reagents

A. 100 mM Potassium Phosphate Buffer, pH 7.1 at 37°C (Prepare 100 ml in deionized water using Potassium Phosphate, Monobasic, Anhydrous. Adjust to pH 7.1 at 37°C with 1 M KOH).
B. 0.02% (w/v) L-Leucine -Naphthylamide Solution (Leu-Nap) (Prepare 50 ml in Reagent A using L-Leucine-Naphthylamide Hydrochloride. L-0376, or use Leucine Aminopeptidase Substrate).
C. 2000 mM Hydrochloric Acid Solution (HCl) (Prepare 100 ml in deionized water using Hydrochloric Acid).
D. 0.5% (w/v) Ammonium Sulfamate Solution (AS) (Prepare 150 ml in deionized water using use Ammonium Sulfamate, 0.5% (w/v) Solution).
E. 0.2% (w/v) Sodium Nitrite Solution (Nit) (Prepare 10 ml in deionized water using Sodium Nitrite, or use Sodium Nitrite Tablets. *Prepare Fresh*).
F. 0.05% (w/v) N-(1-Naphthyl) ethylenediamine Solution (CLR) (Prepare 100 ml in Reagent H using N-(1-Naphthyl)ethylenediamine Dihydrochloride, or dissolve the contents of one bottle of N-(1-Naphthyl)ethylenediamine Dihydrochloride, in 110 ml of Ethanol (Reagent H)).
G. 0.1257 mM -Naphthylamine Solution (Std) (Prepare 10 ml in deionized water using Naphthylamine or use Leucine Aminopeptidase (LAP) Calibration Solution).
H. 95% Ethanol (v/v) (Use Ethanol, Nondenatured)
I. Peptidase Enzyme Solution (Immediately before use, prepare a solution containing 1–3 units/ml in cold Reagent A).

Procedure

Pipette (in milliliters) the following reagents into suitable containers:

	Test	Blank
Reagent A (Buffer)	—	0.50
Reagent B (Leu-Nap)	0.50	0.50

Mix by swirling and equilibrate to 37°C. Then add:

	Test	Blank
Reagent I (Enzyme Solution)	0.50	—

Incubate for exactly 60 minutes. Then add:

Reagent C (HCl)	0.50	0.50

Colorimetric Assay

Pipette (in milliliters) the following reagents into suitable containers:

	Test	Test Blank	Std 1	Std 2	Std 3	Std 4	Std 5	Std Blank
Test Soln	1.50	—	—	—	—	—	—	—
Blank Soln	—	1.50	—	—	—	—	—	—
Reagent C (HCl)	—	—	1.40	1.30	1.10	0.90	0.70	1.50
Reagent G (Std)	—	—	0.10	0.20	0.40	0.60	0.80	—

Mix by swirling. Then add:

Reagent E (Nit)	0.50	0.50	0.50	0.50	0.50	0.50	0.50	0.50

Mix by swirling and incubate at 25°C for 3 minutes. Then add:

Reagent D (AS)	1.00	1.00	1.00	1.00	1.00	1.00	1.00	1.00

Mix by swirling and incubate at 25C for 3 minutes. Then add:

Reagent F (CLR)	2.00	2.00	2.00	2.00	2.00	2.00	2.00	2.00

Mix by swirling and incubate at 25°C for 45 minutes. Transfer solutions to suitable cuvettes and record the A580nm for the Test, Blanks and Standards.

Calculations

Standard Curve: ΔA580nm Standard = A580nm Std–A580nm Std Blank

Prepare a standard curve by plotting ΔA580nm standard vs Δmoles of Δ-Naphthylamine.

Sample Determination: ΔA580nm Sample = A580nm Test–A580nm Blank

Determine the Δmoles of Δ-Naphthylamine liberated using the standard curve.

$$\text{Units / ml enzyme} = \frac{(\Delta\text{moles of - }\Delta\text{ - Naphthylamine liberated) (df)}}{(60)\ (0.5)}$$

df = Dilution factor

60 = Time of assay (in minutes) as per the unit definition.

0.5 = Volume (in milliliters) of enzyme used

$$\text{Units / mg solid} = \frac{\text{units / ml enzyme}}{\text{mg solid / ml enzyme}}$$

$$\text{Units / mg protein} = \frac{\text{units / ml enzyme}}{\text{mg protein / ml enzyme}}$$

Unit Definition: *One unit will liberate 1.0 μmole of Δ-naphthylamine from L-leucine naphthylamide per minute at pH 7.1 at 37°C.*

Final Assay Concentrations

In a 1.00 ml reaction mix, the final concentrations are 100 mM potassium phosphate, 0.01% (w/v) L-leucine -naphthylamide and 0.5–1.5 units peptidase.

Enzymatic Assay of PEPTIDYL ARGININE DEIMINASE (EC 3.5.3.15)

Principle

$$\text{BAEE} + H_2O \xrightarrow{\text{PAD}} \text{N}\alpha\text{ - Benzoyl - L - Citrulline} + \text{Ethanol}$$

Abbreviation Used

PAD = Peptidyl Arginine Deiminase

Conditions: T = 55°C, pH 7.2, A490nm, Light path = 1 cm

Method: Colorimetric

Reagents

A. 350 mM Tris HCl Buffer, pH 7.2 at 55°C (Prepare 100 ml in deionized water using Trizma Base. Adjust to pH 7.2 at 55°C using 1 M HCl).

B. 70 mM Calcium Chloride Solution ($CaCl_2$) (Prepare 10 ml in deionized water using Calcium Chloride, Dihydrate).

C. 70 mM Benzoyl Arginine Ethyl Ester Solution (BAEE) (Prepare 10 ml in deionized water using N-α-Benzoyl-L-Arginine Ethyl Ester, Hydrochloride).

D. 70 mM DL-Dithiothreitol Solution (DTT) (Prepare 10 ml in deionized water using DL-Dithiothreitol. *Prepare Fresh*).

E. 60% (w/v) Perchloric Acid Solution ($HClO_4$) (Prepare 100 ml in deionized water using Perchloric Acid).

F. Redox Reagent (Redox) (Prepare by dissolving 11 g of Ferrous Ammonium Sulfate Hexahydrate and 9 g of Ammonium Iron (III) Sulfate Dodecahydrate in 100 ml of 1 N H_2SO_4. Heat gently with stirring for 30 minutes to get a clear solution. Keep in a plastic bottle covered with aluminum foil to protect from light).

G. Acid Mixture (Acid) (Prepare 600 ml by adding 300 ml of Phosphoric Acid, to 200 ml of deionized water. Then add 100 ml of Sulfuric Acid. Do not add water to acid!)

H. 75 mM 2,3-Butanedione Monoxime Solution (BMO) (Prepare 100 ml in deionized water using 2,3-Butanedione Monoxime).

I. 1 mM Citrulline Standard Solution (Cit Std) (Prepare 100 ml in deionized water using L-Citrulline).

J. 0.1% (w/v) Bovine Serum Albumin Solution (Enz Dil) (Prepare 10 ml in deionized water using Albumin, Bovine).

K. Peptidyl Arginine Deiminase Enzyme Solution (Immediately before use, prepare a solution containing 0.5–2.5 units per ml in cold Reagent J).

Procedure

Pipette (in milliliters) the following reagents into an Eppendorf tube:

	Test	Blank
Reagent A (Buffer)	0.20	0.20
Reagent B ($CaCl_2$)	0.10	0.10
Reagent D (DTT)	0.10	0.10
Reagent K (Enz Sol)	0.20	—
Reagent J (Enz Dil)	—	0.20

Preincubate for 2 minutes at 55°C. Then add:

Reagent E ($HClO_4$)	—	0.10
Reagent C (BAEE)	0.10	0.10

Incubate for 30 minutes at 55°C. (Make sure the top of each tube is closed). Then add:

Reagent E ($HClO_4$)	0.10	—

Centrifuge to clarify. Remove 0.4 ml of the supernatant from both the Test and Blank and place each into a glass tube. Add 0.1 ml of Reagent F (REDOX) to each tube and mix by swirling. Boil for 10 minutes with the tubes covered with a glass onion. Remove and cool tubes. Add 0.5 ml of Reagent G (Acid) and 0.2 ml of Reagent H (BMO) to both the Test and Blank. Mix on a vortexer. Boil for 20 minutes and then cool. Transfer to suitable cuvettes and record the A490nm for both the Test and Blank using a suitable spectrophotometer.

Standard Curve: Prepare a standard curve by pipetting (in milliliters) the following reagents into suitable glass tubes.

	Std 1	Std 2	Std 3	Std Blank
Reagent I (Cit Std)	0.050	0.100	0.200	—
Deionized Water	0.35	0.30	0.200	0.40

Add 0.1 ml of Reagent F (REDOX) to each and mix on a vortexer. Boil for 10 minutes with the tubes covered with a glass marble. Remove and cool tubes. Add 0.5 ml of Reagent G (Acid) and 0.2 ml of Reagent H (BHO) to the Standard and Standard Blank. Mix on a vortexer. Boil for 20 minutes and then cool. Transfer to suitable cuvettes and record the A490nm for the Standards and Standard Blank, using a suitable spectrophotometer.

Calculations

Calculate the absorbance of 1 μmole of citrulline, using the Standard curve.

$$\text{Units / ml} = \frac{(\text{A490nm Test} - \text{A490nm Blank})\,(2)\,(\text{df})}{(\text{A490nm}/\mu\text{mole})\,(0.5)\,(0.1)}$$

2 = Conversion factor since only one-half of the reaction mixture is used in the colorimetric determination of citrulline
df = Dilution factor
A490nm/μmole = A490nm/μmole of citrulline
0.5 = Conversion factor from 30 minutes to 1 hour
0.1 = Volume (in milliliters) of enzyme used

$$\text{Units / mg solid} = \frac{\text{units / ml enzyme}}{\text{mg solid / ml enzyme}}$$

$$\text{Units / mg protein} = \frac{\text{units / ml enzyme}}{\text{mg protein / ml enzyme}}$$

Unit Definition: *One unit will produce 1 μmole of N-α-benzoylcitrulline ethyl ester from BAEE per hour at 55°C at pH 7.2.*

Final Assay Concentration

In a 0.7 ml reaction mix, the final concentrations are 100 mM Tris, 10 mM calcium chloride, 10 mM DL-dithiothreitol, and 0.1–0.5 unit peptidyl arginine deiminase.

Enzymatic Assay of PEROXIDASE (EC 1.11.1.7)

Principle

$$H_2O_2 + \text{Pyrogallol (donor)} \xrightarrow{\text{Peroxidase}} 2H_2O + \text{Purpurogallin (oxidized donor)}$$

Conditions: T = 20°C, pH = 6.0, A420nm, Light path = 1 cm

Method: Continuous Spectrophotometric Rate Determination

Reagents

A. 100 mM Potassium Phosphate Buffer, pH 6.0 at 20°C (Prepare 100 ml in deionized water using Potassium Phosphate, Monobasic, Anhydrous. Adjust to pH 6.0 at 20°C with 1.0 M KOH).

B. 0.50% (w/w) Hydrogen Peroxide Solution (H_2O_2) (Prepare 50 ml in deionized water using Hydrogen Peroxide, 30% (w/w) Solution. *Prepare Fresh*).

C. 5% (w/v) Pyrogallol Solution (Prepare 10 ml in deionized water using Pyrogallol. *Prepare Fresh and Keep away from Light*).

D. Peroxidase Enzyme Solution (Immediately before use, prepare a solution containing 0.4–0.7 unit/ml of Peroxidase in cold Reagent A).

Procedure

Pipette (in milliliters) the following reagents into suitable cuvettes:

	Test	Blank
Deionized Water	2.10	2.10
Reagent A (Buffer)	0.32	0.32
Reagent B (H_2O_2)	0.16	0.16
Reagent C (Pyrogallol)	0.32	0.32

Mix by inversion and equilibrate to 20°C. Monitor the A420nm until constant, using a suitably thermostatted spectrophotometer. Then add:

	Test	Blank
Reagent A (Buffer)	—	0.10
Reagent D (Enzyme Solution)	0.10	—

Immediately mix by inversion and record the increase in A420nm for approximately 5 minutes. Obtain the A420nm/20 seconds using the maximum linear rate for both the Test and Blank.

Calculations

$$\text{Units / ml enzyme} = \frac{(\text{A420nm / 20s Test} - \text{A420nm / 20s Blank})\,(3)\,(\text{df})}{(12)\,(0.1)}$$

s = seconds
3 = Volume (in milliliters) of assay
df = Dilution factor
12 = Extinction coefficient of 1 mg/ml of Purpurogallin at 420nm
0.1 = Volume (in milliliters) of enzyme used

$$\text{Units / mg solid} = \frac{\text{units / ml enzyme}}{\text{mg solid / ml enzyme}}$$

$$\text{Units / mg protein} = \frac{\text{units / ml enzyme}}{\text{mg protein / ml enzyme}}$$

Unit Definition: *One unit will form 1.0 milligram of purpurogallin from pyrogallol in 20 seconds at pH 6.0 at 20°C. This purpurogallin (20 seconds) unit is equivalent to approximately 18 μM units per minute at 25°C.*

Final Assay Concentrations

In a 3.00 ml reaction mix, the final concentrations are 14 mM potassium phosphate, 0.027% (w/w) hydrogen peroxide, 0.5% (w/v) pyrogallol and 0.04–0.07 unit peroxidase.

Notes

1. This assay procedure is not to be used to assay Peroxidase, Insoluble Enzyme attached to beaded agarose, or Peroxidase.
2. The enzyme concentration may have to be modified in order for the rate, A420nm/20 seconds, to be within the specified range of 0.16–0.28.

Enzymatic Assay of PEROXIDASE (EC 1.11.1.7) 2,2'-Azino-bis (3-Ethylbenzthiazoline-6-Sulfonic Acid) as a Substrate

Principle

$$H_2O_2 + \text{ABTS} \xrightarrow{\text{Peroxidase}} 2H_2O + \text{Oxidized ABTS}$$

Abbreviation Used

ABTS = 2, 2'-Azino-bis(3-Ethylbenzthiazoline-6-Sulfonic Acid)

Conditions: T = 25°C, pH = 5.0, A405nm, Light path = 1 cm

Method: Continuous Spectrophotometric Rate Determination

Reagents

A. 100 mM Potassium Phosphate Buffer, pH 5.0 at 25°C (Prepare 100 ml in deionized water using Potassium Phosphate, Monobasic. Adjust to pH 5.0 at 25°C using 1.0 M KOH).

B. 9.1 mM 2, 2'-Azino-bis(3-Ethylbenzthiazoline-6-Sulfonic Acid) Substrate Solution (ABTS1) (Prepare 30 ml in Reagent A using 2, 2'-Azino-bis (3-Ethylbenzthiazoline-6-Sulfonic Acid), Tablets Diammonium Salt. *Prepare Fresh*).

C. 0.3% (w/w) Hydrogen Peroxide Solution (H_2O_2) (Prepare 50 ml in deionized water using Hydrogen Peroxide, 30% (w/w) Solution. *Prepare Fresh*).

D. 40 mM Potassium Phosphate Buffer with 0.25% (w/v) Bovine Serum Albumin and 0.5% (v/v) Triton X-1002, pH 6.8 at 25°C (Enzyme Diluent) (Prepare 100 ml in deionized water using Potassium Phosphate, Monobasic, Albumin, Bovine, Triton X-100. Adjust to pH 6.8 at 25°C using 1 M KOH).

E. Peroxidase Enzyme Solution (Prepare an enzyme stock solution containing 10 mg/ml in cold Reagent D. Immediately before use, prepare a solution containing 0.20–0.80 unit/ml of Peroxidase in cold Reagent D).

Procedure

Pipette (in milliliters) the following reagents into suitable cuvettes:

	Test	Blank
Reagent B (ABTS)	2.90	2.90
Reagent D (Enzyme Diluent)	—	0.05
Reagent E (Enzyme Solution)	0.05	—

Mix by inversion and equilibrate to 25°C. Monitor the A405nm until constant, using a suitably thermostatted spectrophotometer. Then add:

Reagent C (H_2O_2)	0.10	0.10

Immediately mix by inversion and record the increase in A405nm for approximately 2 minutes. Obtain the ΔA405nm/minute using the maximum linear rate for both the Test and Blank.

Calculations

$$\text{Units / ml enzyme} = \frac{(\Delta A405\text{nm / min Test} - A405\text{nm / min Blank})\,(3.05)\,(\text{df})}{(36.8)\,(0.05)}$$

3.05 = Total volume (in milliliters) of assay
df = Dilution factor
36.8 = Millimolar extinction coefficient of oxidized ABTS at 405nm
0.05 = Volume (in milliliter) of enzyme used

$$\text{Units / mg solid} = \frac{\text{units / ml enzyme}}{\text{mg solid / ml enzyme}}$$

$$\text{Units / mg protein} = \frac{\text{units / ml enzyme}}{\text{mg protein / ml enzyme}}$$

Unit Definition: *One unit will oxidize 1.0 μmole of 2,2'-azino-bis (3-ethylbenzthiazoline-6-sulfonic acid) per minute at pH 5.0 at 25°C.*

Final Assay Concentrations

In a 3.05 ml reaction mix, the final concentrations are 96 mM potassium phosphate, 8.7 mM 2,2'-azino-bis(3-Ethylbenzthiazoline-6-sulfonic acid), 0.01% (w/w) hydrogen peroxide, 0.004% (w/v) Bovine Serum Albumin, 0.008% (v/v) Triton X-100 and 0.01–0.04 unit peroxidase.

Note

1. The maximum linear rate occurs within the first minute of the reaction.

Determination of the RZ Value for PEROXIDASE

Conditions: T = 25° C, A403nm and A275nm, Light path = 1 cm

Method: Spectrophotometric

Reagents

A. 100 mM Potassium Phosphate Buffer, pH 6.0 at 25°C (Prepare 50 ml in deionized water using Potassium Phosphate, Monobasic, Anhydrous. Adjust to pH 6.0 at 25°C with 1 M KOH).

B. Peroxidase Enzyme (POD) (Immediately before use, prepare 10 ml of a solution containing approximately 0.3 mg solid/ml of Peroxidase in Reagent A. Gently mix by inversion until the solids are completely dissolved).

Procedure

Pipette (in milliliters) the following reagents into suitable cuvettes:

	Test	Blank
Reagent B (POD)	3.00	—
Reagent A (Buffer)	—	3.00

Record the absorbance at 403nm and 275nm for both the Test and Blank using a suitable spectrophotometer.

Calculation

$$RZ = \frac{(\text{A403nm Test} - \text{A403nm Blank})}{(\text{A275nm Test} - \text{A275nm Blank})}$$

RZ = Reinheitszahl, A403/A275 value

Notes

1. The Test and Blank should be done in duplicate.
2. The difference in absorbance due to the cuvettes at 403nm and 275nm must be accounted for in the calculation of RZ.

Enzymatic Assay of PEROXIDASE (EC 1.11.1.7) from Soybean

Principle

$$H_2O_2 + \text{Pyrogallol (donor)} \xrightarrow{\text{Peroxidase}} 2H_2O + \text{Purpurogallin (oxidized donor)}$$

Conditions: T = 20°C, pH = 6.0, A420nm, Light path = 1 cm

Method: Continuous Spectrophotometric Rate Determination

Reagents

A. 100 mM Potassium Phosphate Buffer, pH 6.0 at 20°C (Prepare 100 ml in deionized water using Potassium Phosphate, Monobasic, Anhydrous. Adjust to pH 6.0 at 20°C with 1.0 M KOH).
B. 0.50% (w/w) Hydrogen Peroxide Solution (H_2O_2) (Prepare 50 ml in deionized water using Hydrogen Peroxide, 30% (w/w) Solution. *Prepare Fresh*).
C. 5% (w/v) Pyrogallol Solution (Prepare 10 ml in deionized water using Pyrogallol. *Prepare Fresh and Keep away from Light*).
D. 0.1% (w/v) Bovine Serum Albumin (Enzyme Diluent) (Prepare 50 ml in Reagent A using Albumin, Bovine).
E. Peroxidase Enzyme Solution (Immediately before use, prepare a solution containing 0.4–0.7 unit/ml1 of Peroxidase in cold Reagent D).

Procedure

Pipette (in milliliters) the following reagents into suitable cuvettes:

	Test	Blank
Deionized Water	2.10	2.10
Reagent A (Buffer)	0.32	0.32
Reagent B (H_2O_2)	0.16	0.16
Reagent C (Pyrogallol)	0.32	0.32

Mix by inversion and equilibrate to 20°C. Monitor the A420nm until constant, using a suitably thermostatted spectrophotometer. Then add:

	Test	Blank
Reagent D (Enzyme Diluent)	—	0.10
Reagent E (Enzyme Solution)	0.10	—

Immediately mix by inversion and record the increase in A420nm for approximately 5 minutes. Obtain the ΔA420nm/20 seconds using the maximum linear rate for both the Test and Blank.

Calculation

$$\text{Units / ml enzyme} = \frac{(\Delta A420\text{nm} / 20 \text{ sec Test} - \Delta A420 / 20 \text{ sec Blank})\ (3)\ (\text{df})}{(12)\ (0.1)}$$

sec = seconds
3 = Volume (in milliliters) of assay
df = Dilution factor
12 = Extinction coefficient of 1 mg/ml of Purpurogallin at 420 nm
0.1 = Volume (in milliliters) of enzyme used

$$\text{Units / mg solid} = \frac{\text{units / ml enzyme}}{\text{mg solid / ml enzyme}}$$

$$\text{Units / mg protein} = \frac{\text{units / ml enzyme}}{\text{mg protein / ml enzyme}}$$

Unit Definition: *One unit will form 1.0 milligram of purpurogallin from pyrogallol in 20 seconds at pH 6.0 at 20°C. This purpurogallin (20 seconds) unit is equivalent to approximately 18 μM units per minute at 25°C.*

Final Assay Concentrations

In a 3.00 ml reaction mix, the final concentrations are 14 mM potassium phosphate, 0.027% (w/w) hydrogen peroxide, 0.5% (w/v) pyrogallol, 0.003% (w/v) Bovine Serum Albumin, and 0.04–0.07 unit peroxidase.

Note

1. The enzyme concentration may have to be modified in order for the rate, A420nm/20 seconds, to be within the specified range of 0.7–0.9.

Enzymatic Assay of PEROXIDASE, INSOLUBLE

Principle

$$H_2O_2 + \text{Pyrogallol (donor)} \xrightarrow{\text{Peroxidase}} 2H_2O + \text{Purpurogallin (oxidized donor)}$$

Conditions: T = 30°C, pH 6.0, A420nm, Light path = 1 cm

Method: Continuous Spectrophotometric Rate Determination

Reagents

A. 100 mM Potassium Phosphate Buffer, pH 6.0 at 30°C (Prepare 100 ml in deionized water using Potassium Phosphate, Monobasic, Anhydrous. Adjust to pH 6.0 at 30°C with 1 M KOH).

B. 0.50% (w/w) Hydrogen Peroxide Solution (H_2O_2) (Prepare 50 ml in deionized water using Hydrogen Peroxide, 30% (w/w) Solution. *Prepare Fresh*).

C. 5.0% (w/v) Pyrogallol Solution (Pyrogallol) (Prepare 50 ml in deionized water using Pyrogallol. *Prepare Fresh and Keep away from Light*).

D. Peroxidase Insoluble Enzyme Suspension (Insol Enz) (Place sample on a small Buchner funnel. Using a suction flask and a light rate of suction, wash the agarose resin with about 10 times the sample volume of deionized water. Dry the moist gel with suction until the top of the packed gel cracks. Weigh the moist gel and place it into an appropriate container. Add one ml of deionized water per mg of sample gel. Immediately prior to assay dilute 0.1 ml of gel suspension to 2.1 ml with cold deionized water).

Procedure

Pipette (in milliliters) the following reagents into a suitable container while continuously mixing with a magnetic stirrer.

	Test
Reagent A (Buffer)	3.30
Reagent B (H_2O_2)	1.70
Reagent C (Pyrogallol)	3.30
Deionized Water	21.70

Equilibrate to 30°C. Then add:

Reagent D (Insol Enz)	0.10

Immediately filter enough of the reaction mix (3 ml) through a 0.45 µm syringe filter into a suitable cuvette and record the ΔA420nm using a suitably thermostatted spectrophotometer. This step should be repeated at least 4 times over seven minutes.

Calculation

$$\text{Units / g agarose} = \frac{(\Delta A420nm / min)(30.1)(21)(1000)}{(2.64)(0.1)(15.7)}$$

30.1 = Total volume (in milliliters) of the assay
21 = Dilution factor
1000 = Conversion factor from mg to g
2.64 = Millimolar extinction coefficient of purpurogallin at 420nm
0.1 = Volume (in milliliter) of enzyme used
15.7 = mg dry agarose/ml of suspension

Unit Definition: *One unit will form 1.0 µmole of purpurogallin from pyrogallol per minute at pH 6.0 at 30°C.*

Final Assay Concentrations

In a 30.1 ml reaction mix, the final concentrations are 11 mM potassium phosphate, 0.03% (w/w) hydrogen peroxide, 0.55% (w/v) pyrogallol, and 0.075 mg of dry agarose peroxidase, insoluble.

Enzymatic Assay of PHENOLASE (EC 1.10.3.1)

Principle

$$2\ \text{o-Diphenol} + \text{L-Asc Acid} + O_2 \xrightarrow{\text{Phenolase}} 2\ \text{o-Quinone} + 2H_2O + \text{D-Asc Acid}$$

Abbreviations Used

Asc Acid = L-Ascorbic Acid
D-Asc Acid = Dehydro-Ascorbic Acid

Conditions: T = 30°C, pH = 7.5, A265nm, Light path = 1 cm

Method: Continuous Spectrophotometric Rate Determination

Reagents

A. 50 mM Potassium Phosphate Buffer, pH 7.5 at 30°C (Prepare 100 ml in deionized water using Potassium Phosphate, Monobasic, Anhydrous. Adjust to pH 7.5 at 30°C with 1 M KOH).

B. 10 mM Chlorogenic Acid Solution (Chloro) (Prepare 10 ml in Reagent A using Chlorogenic Acid. *Prepare Fresh*).
C. 1 mM L-Ascorbic Acid Solution (Asc) (Prepare 10 ml in Reagent A using L-Ascorbic Acid, Free Acid. *Prepare Fresh*).
D. 100 mM Ethylenediaminetetraacetic Acid Solution (EDTA) (Prepare 10 ml in deionized water using Ethylenediaminetetraacetic Acid, Tetrasodium Salt, Hydrate).
E. Phenolase Enzyme Solution (Immediately before use, prepare a solution containing 200–300 units/ml of Phenolase in cold Reagent A).

Procedure

Pipette (in milliliters) the following reagents into suitable quartz cuvettes:

	Test	Blank
Deionized Water	0.67	0.67
Reagent A (Buffer)	2.09	2.10
Reagent B (Chloro)	0.10	0.10
Reagent C (Asc)	0.10	0.10
Reagent D (EDTA)	0.03	0.03

Mix by inversion and equilibrate to 30°C. Monitor the A265nm until constant, using a suitably thermostatted spectrophotometer. Then add:

Reagent E (Enzyme Solution)	0.01	—

Immediately mix by inversion and record the decrease in A265nm for approximately 5 minutes. Obtain the ΔA265nm/minute using the maximum linear rate for both the Test and Blank.

Calculations

$$\text{Units / mg enzyme} = \frac{(\Delta\text{A265nm / min Test} - \Delta\text{A265nm / min Blank})}{(0.01)\ (\text{mg enzyme / RM})}$$

0.01 = The change in A265nm/minute per unit Phenolase at pH 7.5 at 30°C in a 3 ml reaction mix
RM = Reaction Mix

Unit Definition: *One unit will produce a ΔA265nm of 0.01 per minute at pH 7.5 at 30°C, using chlorogenic acid as substrate in a 3 ml reaction volume.*

Final Assay Concentration

In a 3.00 ml reaction mix, the final concentrations are 38 mM potassium phosphate, 0.33 mM chlorøgenic acid, 0.03 mM L-ascorbic acid, 1 mM EDTA and 2.0–3.0 units of phenolase.

Note

1. The blank will have an absorbance of approximately 2.1.

Enzymatic Assay of PHENYLALANINE AMMONIA-LYASE (EC 4.3.1.5) L-Phenylalanine as a Substrate

Principle

$$\text{L-Phenylalanine} \xrightarrow{\text{PAL}} \text{trans-Cinnamate} + NH_3$$

Abbreviation Used

PAL = Phenylalanine Ammonia-Lyase

Conditions: T = 30°C, pH = 8.5, A270nm, Light path = 1 cm

Method: Continuous Spectrophotometric Rate Determination

Reagents

A. 150 mM Tris HCl Buffer, pH 8.5 at 30°C (Prepare 100 ml in deionized water using Trizma Base. Adjust to pH 8.5 at 30°C with 1 M HCl).
B. 3 mM L-Phenylalanine Solution (PHE) (Prepare 25 ml in Reagent A using L-Phenylalanine. Heat gently to dissolve).
C. Phenylalanine Ammonia-Lyase (PAL) (Immediately before use, prepare a solution containing 0.03–0.15 unit/ml of Phenylalanine-Lyase in Reagent A).

Procedure

Pipette (in milliliters) the following reagents into suitable cuvettes:

	Test	Blank
Reagent B (PHE)	2.00	2.00
Deionized Water	0.90	0.90

Mix by inversion and equilibrate to 30°C. Then add:

	Test	Blank
Reagent C (PAL)	0.10	—
Reagent A (Buffer)	—	0.10

Immediately mix by inversion and record the increase in A270nm for approximately 5 minutes. Obtain the ΔA270nm/minute using the maximum linear rate for both the Test and Blank.

Calculations

$$\text{Units/ml enzyme} = \frac{(\Delta A270\text{nm/min Test} - \Delta A270\text{nm/min Blank})\ (3)\ (\text{df})}{(19.73)}$$

3 = Total volume (in milliliters) of assay
df = Dilution factor
19.73 = Millimolar extinction coefficient of transcinnamate at 270nm

$$\text{Units/mg solid} = \frac{\text{units/ml enzyme}}{\text{mg solid/ml enzyme}}$$

$$\text{Units/mg protein} = \frac{\text{units/ml enzyme}}{\text{mg protein/ml enzyme}}$$

Unit Definition: *One unit will deaminate 1.0 μmole of L-phenylalanine to trans-cinnamate and NH_3 per minute at pH 8.5 at 30°C.*

Final Assay Concentrations

In a 3.00 ml reaction mix, the final concentrations are 150 mM Tris, 2 mM L-phenylalanine, and 0.003–0.015 unit phenylalanine ammonia-lyase.

Enzymatic Assay of PHENYLALANINE AMMONIA-LYASE (EC 4.3.1.5) L-Tyrosine as Substrate

Principle

$$\text{L-Tyrosine} \xrightarrow{\text{PAL}} \text{p-Coumarate} + NH_4^+$$

Abbreviation Used

PAL = Phenylalanine Ammonia-Lyase

Conditions: T = 30°C, pH = 8.5, A286nm, Light path = 1 cm

Method: Continuous Spectrophotometric Rate Determination

Reagents

A. 150 mM Tris HCl Buffer, pH 8.5 at 30°C (Prepare 100 ml in deionized water using Trizma Base. Adjust to pH 8.5 at 30°C with 1 M HCl).
B. 3 mM L-Tyrosine Solution (TYR) (Prepare 25 ml in Reagent A using L-Tyrosine, Free Base. Heat gently to dissolve).
C. Phenylalanine Ammonia-Lyase Enzyme Solution (PAL) (Immediately before use, prepare a solution containing 0.5–1.0 unit/ml of Phenylalanine Ammonia-Lyase in cold Reagent A).

Procedure

Pipette (in milliliters) the following reagents into suitable cuvettes:

	Test	Blank
Reagent B (TYR)	2.00	2.00
Deionized Water	0.95	0.95

Mix by inversion and equilibrate to 30°C. Then add:

Reagent C (PAL)	0.05	—
Reagent A (Buffer)	—	0.05

Immediately mix by inversion and record the increase in A286nm for approximately 5 minutes. Obtain the ΔA286nm/minute using the maximum linear rate for both the Test and Blank.

Calculations

$$\text{Units / ml enzyme} = \frac{(\Delta A286\text{nm / min Test} - \Delta A286\text{nm / min Blank})\ (3)\ (\text{df})}{(18.5)\ (0.05)}$$

3 = Total volume (in milliliters) of assay
df = Dilution factor
18.5 = Millimolar extinction coefficient of p-coumaric acid at 286nm
0.05 = Volume (in milliliter) of enzyme used

$$\text{Units / mg solid} = \frac{\text{units / ml enzyme}}{\text{mg solid / ml enzyme}}$$

$$\text{Units / mg protein} = \frac{\text{units / ml enzyme}}{\text{mg protein / ml enzyme}}$$

Unit Definition: *One unit will deaminate 1.0 μmole of L-tyrosine to p-coumarate and NH_3 per minute at pH 8.5 at 30°C.*

Final Assay Concentration

In a 3.00 ml reaction mix, the final concentrations are 103 mM Tris, 2 mM L-tyrosine, and 0.025–0.050 unit phenylalanine ammonia-lyase.

Enzymatic Assay of L-PHENYLALANINE DECARBOXYLASE (EC 4.1.1.53)

Principle

$$\text{L - Phenylalanine L - Phenylalanine} \xrightarrow[\text{PRP}]{\text{Decarboxylase}} \text{Phenylethylamine} + CO_2$$

Abbreviation Used

PRP = Pyridoxal 5-Phosphate

Conditions: T = 37°C, pH 5.5

Method: Radiolabelled Stop Reaction

Reagents

A. 200 mM Sodium Phosphate Solution (Prepare 50 ml in deionized water using Sodium Phosphate, Monobasic, Anhydrous).
B. 100 mM Citrate Solution (Prepare 100 ml in deionized water using Citric Acid, Free Acid, Anhydrous).
C. Mclvane's Buffer, pH 5.5 at 37°C (Buffer) (Prepare using 50 ml of Reagent A and adjust to pH 5.5 at 37°C with Reagent B).
D. 3.0 mM L-Phenylalanine Solution (Prepare 10 ml in Reagent B using L-Phenylalanine, Free Base. Heat (70°C) may be required in order to effect dissolution).
E. 0.3 mM Pyridoxal 5-Phosphate Solution (PRP) (Prepare 2 ml in Reagent B using Pyridoxal 5-Phosphate).
F. $_{14}$C[COOH] L-Phenylalanine (Use $_{14}$C[COOH] L-Phenylalanine, 50–60 μCi/μmol, 50 μCi/ml).
G. L-Phenylalanine Reaction Cocktail (L-Phe) (Prepare by adding 0.05 ml of Reagent E to 2 ml of Reagent C).
H. 1 M Methylbenzethonium Hydroxide Solution (MBH) (Use Benzethonium Hydroxide, approximately 1.0 M solution in Methanol).

I. Filter Paper (1 cm × 2 cm strips of filter paper are prepared using Whatman 3 MM paper. Add 0.050 ml of Reagent G to each strip and allow to air dry. This step is to be carried out in a fume hood).
J. Vacutainer Tubes (Use Vacutainer Blood Tubes, 13 × 75 mm, Becton Dickinson).
K. 25% (w/v) Trichloroacetic Acid Solution (TCA) (Prepare 10 ml in deionized water using Trichloroacetic Acid, 6.1 N Solution, approximately 100% (w/v)).
L. L-Phenylalanine Decarboxylase Enzyme Solution (Immediately before use, prepare a solution containing approximately 0.005–0.02 unit/ml of L-Phenylalanine Decarboxylase in cold Reagent C).
M. Scintillation Cocktail (Scint Fluid).

Procedure

Radiolabelled CO_2 is trapped by absorbing it onto Methylbenzethonium Hydroxide which is present on filter paper located above the reaction mixture in a vacutainer tube. Pipette (in milliliters) the following reagents into suitable vacutainer tubes.

	Test	Blank	PRP
Reagent G (L-Phe)	0.10	0.10	0.10
Reagent E (PRP)	0.10	0.10	—
Reagent C (Buffer)	—	—	0.10

Equilibrate to 37°C. Then add:

	Test	Blank	PRP
Reagent L (Enz Soln)	0.10	—	0.10
Reagent C (Buffer)	—	0.10	—

Immediately place Reagent H (Filter Paper) in the top of the vacutainer tube and seal the tubes by replacing the stopper. Incubate for exactly 10 or 20 minutes. Then add by injecting with a syringe and needle through the stopper (*do not remove stopper to add TCA*) (Reagent J):

	Test	Blank	PRP
Reagent K (TCA)	0.40	0.40	0.40

Mix by swirling. Allow to stand for 12–16 hours at room temperature to allow for all evolved CO_2 to be absorbed onto the filter paper. Remove filter papers from vacutainer tubes and place into scintillation vials containing 7 ml of Reagent L (Scint Fluid). This should be performed in a hood. Potential DPM (disintegrations per minute) are prepared by pipetting 0.05 ml of Reagent F (L-Phe) onto Reagent H (Filter Paper). Place into scintillation vials containing 7 ml of Reagent L (Scint Fluid).

Calculations

$$\text{Potential DPM}/\mu\text{mole} = \frac{\text{DPM of potential}}{\text{Total }\mu\text{moles of L-Phenylalanine}}$$

$$\text{Units/ml enzyme} = \frac{(\text{DPM Test} - \text{DPM Blank})(\text{df})}{(\text{T})(0.1)(\text{Potential DPM}/\mu\text{mole})}$$

DPM = Disintegrations per minute
df = Dilution factor
T = Time (in minutes) of assay as per the Unit Definition
0.1 = Volume (in milliliter) of enzyme used

$$\text{Units / mg solid} = \frac{\text{units / ml enzyme}}{\text{mg solid / ml enzyme}}$$

Unit Definition: *One unit will liberate 1.0 μmole of CO_2 from L-phenylalanine per minute at pH 5.5 at 37°C.*

Final Assay Concentration

In a 0.30 ml reaction mix, the final concentrations are 100 mM citric phosphate, 0.98 mM L-phenylalanine, 0.1 mM pyridoxal 5-phosphate, and 0.0005–0.002 unit L-phenylalanine decarboxylase.

Enzymatic Assay of L-PHENYLALANINE DEHYDROGENASE (EC 1.4.1.20)

Principle

$$\text{L-Phenylalanine} + H_2O + \beta\text{-NAD} \xrightarrow{\text{L-PD}} \text{Phenylpyruvate} + NH_3 + \beta\text{-NADH}$$

Abbreviations Used

L-PD= L- Phenylalanine Dehydrogenase
β-NAD = β-Nicotinamide Adenine Dinucleotide, Oxidized Form
β-NADH = β-Nicotinamide Adenine Dinucleotide, Reduced Form

Conditions: T = 30°C, pH = 10.5, A340nm, Light path = 1 cm

Method: Continuous Spectrophotometric Rate Determination

Reagents

A. 250 mM Glycine Buffer, pH 10.5 at 30°C (Gly Buff) (Prepare 100 ml in deionized water using Glycine, Free Base. Adjust to pH 10.5 at 30°C with 5 M NaOH).
B. 50 mM Potassium Phosphate, Monobasic Solution (Prepare 100 ml in deionized water using Potassium Phosphate, Monobasic, Anhydrous).
C. 50 mM Potassium Phosphate Buffer, pH 7.5 at 30°C (Enz Dil) (Prepare 50 ml in deionized water using Potassium Phosphate, Dibasic, Trihydrate. Adjust to pH 7.5 at 30°C with Reagent B).
D. 100 mM L-Phenylalanine Substrate Solution (Phe) (Prepare 10 ml in deionized water using L-Phenylalanine).
E. 30 mM β-Nicotinamide Adenine Dinucleotide, Oxidized Form Solution (β-NAD) (Dissolve the contents of one 20 mg vial of β-Nicotinamide Adenine Dinucleotide in the appropriate volume of deionized water or prepare 10 ml in deionized water using β- Nicotinamide Adenine Dinucleotide).
F. L-Phenylalanine Dehydrogenase Enzyme Solution (Prepare a solution containing 3–10 units of L-Phenylalanine Dehydrogenase in cold Reagent C. Store at 4°C. Immediately before use, dilute to 0.10–0.20 unit/ml in cold Reagent A).

Procedure

Pipette (in milliliters) the following reagents into suitable cuvettes:

	Test	Blank
Reagent A (Gly Buff)	2.30	2.30
Reagent D (Phe)	0.30	0.30
Reagent E (β-NAD)	0.30	0.30

Mix by inversion and equilibrate to 30°C. Monitor the A340nm until constant, using a suitably thermostatted spectrophotometer. Then add:

Reagent F (Enzyme Solution)	0.10	—
Reagent C (Enz Dil)	—	0.10

Immediately mix by inversion and record the increase in A340nm for approximately 5 minutes. Obtain the ΔA340nm/min using the maximum linear rate for both the Test and Blank.

Calculations

$$\text{Units / ml enzyme} = \frac{(\Delta A340\text{nm / min Test} - \Delta A340\text{nm / min Blank})\,(3)\,(\text{df})}{(6.22)\,(0.1)}$$

3 = Total volume (in milliliters) of assay
df = Dilution factor
6.22 = Millimolar extinction coefficient of β-NADH at 340nm
0.1 = Volume (in milliliter) of enzyme used

$$\text{Units / mg solid} = \frac{\text{units / ml enzyme}}{\text{mg solid / ml enzyme}}$$

$$\text{Units / mg protein} = \frac{\text{units / ml enzyme}}{\text{mg protein / ml enzyme}}$$

Unit Definition: *One unit will oxidize 1.0 μmole of L-phenylalanine per minute at pH 10.5 at 30°C in the presence of β-NAD.*

Final Assay Concentration

In a 3.00 ml reaction mix, the final concentrations are 192 mM glycine, 10 mM L-phenylalanine, 3 mM β-nicotinamide adenine dinucleotide, 1.7 mM potassium phosphate and 0.01–0.02 unit L-phenylalanine dehydrogenase.

Enzymatic Assay of L-PHENYLALANINE HYDROXYLASE (EC 1.14.16.1)

Principle

$$\text{6-MTHP} + \text{L-Phenylalanine} \xrightarrow{\text{PAH}} \text{6-MDHP} + \text{L-Tyrosine}$$

Abbreviations Used

6-MTHP = 6-Methyltetrahydropterine
6-MDHP = 6-Methyldihydropterine
PAH = L-Phenylalanine Hydroxylase

Conditions: T = 25°C, pH = 7.2, A450nm, Light Path = 1 cm

Method: Colorimetric

Reagents

A. 200 mM Tris HCl Buffer, pH 7.2 at 25°C (Prepare 100 ml in deionized water using Trizma Base. Adjust to pH 7.2 at 25°C with 1 M HCl).

B. 4.0 mM L-Phenylalanine Solution (PHE) (Prepare 30 ml in Reagent A using L-Phenylalanine).

C. Catalase Enzyme Solution (CAT) (Prepare 20 ml of a solution containing 4000 units/ml of Catalase in cold Reagent B (PHE). *Prepare Fresh*).

D. 16.65 mM DL-Dithiothreitol Solution (DTT) (Prepare 25 ml in deionized water using DL-Dithiothreitol).

E. 1.33 mM 6-Methyltetrahydropterine Solution (6-MTHP) (Prepare 10 ml in Reagent D (DTT) using DL-6-Methyl-5,6,7,8-Tetrahydropterine Dihydrochloride. *Prepare Fresh*).

F. 5% (v/v) Trichloroacetic Acid Solution (TCA) (Prepare 25 ml in deionized water using Trichloroacetic Acid, 6.1 N Solution).

G. 20% (v/v) Nitric Acid Solution with 0.05% (w/v) Sodium Nitrite (Nitric Acid) (Prepare 20 ml in deionized water using Nitric Acid and Sodium Nitrite. Dissolve the Sodium Nitrite in deionized water before adding to the Nitric Acid solution).

H. 0.1% (v/v) Nitrosonaphthol Solution (NNS) (Prepare 20 ml in Reagent I (NaOH) using 1-Nitroso-2-Naphthol).

I. 100 mM Sodium Hydroxide Solution (NaOH) (Prepare 50 ml in deionized water using Sodium Hydroxide Solution, 1.0 N).

J. 5.0 mM L-Tyrosine Standard Solution (TYR Std) (Prepare 10 ml in deionized water using L-Tyrosine Free Base).

K. Phenylalanine Hydroxylase Enzyme Solution (Immediately before use, prepare a solution containing 0.2–1.0 unit/ml of Phenylalanine Hydroxylase in cold Reagent A).

Procedure

Pipette (in milliliters) the following reagents into suitable containers:

	Test	Blank	Std 1	Std 2	Std 3	Std Blank
Reagent C (CAT)	0.50	0.50	0.50	0.50	0.50	0.50
Reagent K (Enzyme Soln)	0.02	0.02	—	—	—	—
Deionized Water	0.18	0.18	0.17	0.14	0.11	0.20
Reagent J (TYR Std)	—	—	0.03	0.06	0.09	—

Mix by swirling and equilibrate to 25°C. Then add:

Reagent E (6-MTHP)	0.30	0.30	0.30	0.30	0.30	0.30

Immediately mix by swirling, and incubate at 25°C for exactly 8 minutes. Then add:

	Test	Blank	Std 1	Std 2	Std 3	Std Blank
Reagent F (TCA)	1.00	1.00	1.00	1.00	1.00	1.00
Reagent E (6-MTHP)	—	0.30	—	—	—	—
Reagent G (Nitric Acid)	1.00	1.00	1.00	1.00	1.00	1.00
Reagent H (NNS)	1.00	1.00	1.00	1.00	1.00	1.00

Mix by swirling and heat at 55°C for 30 minutes. Cool to 25°C and centrifuge for 3 minutes. Transfer the solutions to suitable cuvettes and record the A450nm for the Test, Blank, Standards and Standard Blank with a suitable spectrophotometer.

Standard Curve: ΔA450nm Standard = A450nm Standard–A450nm Standard Blank

Prepare a standard curve by plotting the ΔA450nm for the Standards versus micromoles of Tyrosine.

Sample Determination: ΔA450nm Sample = A450nm Test–A450nm Blank

Determine the total micromoles of L-Tyrosine produced using the Standard curve.

$$\text{Units / ml enzyme} = \frac{(\mu\text{moles L-Tyrosine produced})\ (\text{df})}{(0.02)\ (8)}$$

df = Dilution factor

8 = Time correction factor (in minutes) as per Unit Definition

0.02 = Volume (in milliliters) of enzyme used in the assay

$$\text{Units / mg solid} = \frac{\text{units / ml enzyme}}{\text{mg solid / ml enzyme}}$$

$$\text{Units / mg protein} = \frac{\text{units / ml enzyme}}{\text{mg protein / ml enzyme}}$$

Unit Definition: *One unit will convert 1.0 μmole of L-phenylalanine to L-tyrosine per min at pH 7.2 and 25°C, using DL-methyl-5,6,7,8-tetrahydropterine as cofactor.*

Final Assay Concentration

In a 1.00 ml reaction mix, the final concentration are 104 mM Tris, 2 mM L-phenylalanine, 2000 units catalase, 5.0 mM DL-dithiothreitol, 0.40 mM DL-6-methyl-5,6,7,8-tetrahydropterine and 0.004–0.02 unit L-phenylalanine hydroxylase.

Note

1. Dissolve the 1-Nitroso-2-Naphthol in the 100 mM NaOH solution by heating to 100°C for 5 minutes and cooling to 25°C. The solution should be a dark green.

Enzymatic Assay of PHENYLETHANOLAMINE N-METHYLTRANSFERASE (EC 2.1.1.28)

Principle

$$\text{Normetanephrine} + {}_{14}\text{C}\,(\text{CH}_3)\text{-SAM} \xrightarrow{\text{PNMT}} {}_{14}\text{C-Metanephrine} + \text{SAM}$$

Abbreviations Used

${}_{14}C\,(CH_3)$-SAM = S-Adenosyl-L-[Methyl-${}_{14}$C]Methionine

PNMT = Phenylethanolamine N-Methyltransferase

SAM = S-Adenosyl-L-Methionine

Conditions: T = 37°C, pH = 8.5

Method: Radiolabelled Stop Reaction

Reagents

A. 400 mM Tris HCl Buffer, pH 8.5 at 37°C (Prepare 100 ml in deionized water using Trizma Base. Adjust to pH 8.5 at 37°C with 1 M HCl).
B. 300 mM DL-Normetanephrine Solution (Prepare 10 ml in deionized water using DL-Normetanephrine, Hydrochloride).
C. S-Adenosyl-L-[Methyl-$_{14}$C] Methionine ($_{14}$C (CH_3)-SAM) (Use 60 µCi/µmol, 25µCi/ml).
D. 0.2% (w/v) S-Adenosyl-L-Methionine (SAM) (Prepare 5 ml in deionized water using S-Adenosyl-L-Methionine, p-Toluenesulfonate Salt. *Prepare Fresh*).
E. 500 mM Borate Buffer, pH 10.0 at 37°C (Prepare 100 ml in deionized water using Boric Acid. Adjust to pH 10.0 at pH 37°C with 5 M NaOH).
F. Toluene:Isoamyl Alcohol (3:2) (Prepare 100 ml using Toluene and Isoamyl Alcohol).
G. Phenylethanolamine N-Methyltransferase Enzyme Solution (PNMT) (Immediately before use, prepare a solution containing approximately 100 units/ml in cold deionized water).
H. Scintillation Cocktail.

Procedure

Prepare a reaction cocktail by pipetting (in milliliters) the following reagents into a suitable container:

Reagent A (Tris HCl Buffer)	2.50
Reagent B (DL-Normetanephrine)	0.50
Reagent D (SAM)	0.148
Deionized Water	1.842
Reagent C ($_{14}$C (CH_3)-SAM)	0.010

Pipette (in milliliters) the following reagents into suitable plastic tubes:

	Test	Blank
Reaction Cocktail	0.25	0.25

Equilibrate for 5 minutes at 37°C. Then add:

Reagent G (PNMT)	0.015	—
Deionized Water	—	0.015

Immediately mix by swirling and incubate for 15 – 30 minutes at 37°C. Remove a 0.050 ml aliquot from the test mixture and add it to 0.50 ml of Reagent E to stop the reaction. Extract the $_{14}$C-Methylated product by the addition of 2.5 ml of Reagent F and mix (by vortexing) for 15 seconds. Centrifuge at 2500 RPM for 5 minutes. Remove 2 ml of the organic upper layer, and place in 7.0 ml glass scintillation vials with 5 ml of Reagent H. Count for 2 minutes with a suitable scintillation counter.

To determine the total potential decays per minute (dpm), pipette 0.02 ml of the reaction cocktail into 7.0 ml glass scintillation vials. Add 2.00 ml of Reagent F and 5.00 ml of Reagent H. Mix thoroughly and count for 2 minutes with a suitable scintillation counter.

Calculations

$$\text{nmoles of cold SAM} = \frac{(\text{mg / ml in 5.0 ml RM})\,(1000)\,(1000)}{\text{FW}}$$

1000 = Conversion factor from mmoles to µmoles
1000 = Conversion factor from µmoles to nmoles
FW = Formula Weight of SAM used
RM = Reaction Mixture

$$\text{nmoles of } {}_{14}\text{C-SAM} = \frac{(\mu\text{Ci / ml})\,(1000)\,(0.010)}{(\text{mCi / nmole})\,(5.0)}$$

1000 = Conversion factor from µmoles to nmoles
0.010 = Aliquot volume for 14C-SAM
5 = Volume of scintillation fluid
Total nmoles of SAM = nmole cold SAM plus nmole ${}_{14}$C-SAM

$$\text{Potential dpm / nmole} = \frac{\text{dpm of 0.02 ml RM}}{\text{Total nmoles of SAM}}$$

RM = Reaction Mixture

$$\text{units / mg enzyme} = \frac{(\text{dpm Test} - \text{dpm Blank})\,(60)\,(0.265)\,(2.5)}{(\text{potential dpm / nmole})\,(\text{T})\,(\text{mg enzyme / RM})\,(0.05)\,(2.0)}$$

T = Time of reaction
60 = Unit definition of 60 minutes
0.265 = Total volume of Reaction Mixture
2.5 = Total volume of organic layer
0.05 = Volume of Reaction Mixture removed and stopped
2.0 = Volume of organic layer counted

Unit Definition: *One unit will convert 1.0 nanomole of normetanephrine to metanephrine per hour at pH 8.5 at 37°C.*

Final Assay Concentrations

In a 0.265 ml reaction mix, the final concentrations are 189 mM Tris, 28 mM DL-normetanephrine, 0.006% (w/v) SAM, 0.047 µCi ${}_{14}$C (CH_3)-SAM and 1.5 units phenylethanolamine N-methyltransferase.

Enzymatic Assay of PHOSPHATASE, ALKALINE (EC 3.1.3.1) Glycine Assay

Principle

$$\text{p-Nitrophenyl Phosphate} + H_2O \xrightarrow{\text{Alkaline Phosphatase}} \text{p-Nitrophenol} + \text{Pi}$$

Abbreviation Used

Pi = Inorganic Phosphate

Conditions: T = 37°C, pH = 8.8, A410m, Light path = 1 cm

Method: Stopped Spectrophotometric Rate Determination

Reagents

A. 100 mM Glycine Buffer with 1.0 mM Magnesium Chloride, pH 8.8 at 37°C (Prepare 50 ml in deionized water using Glycine, Free Base and Magnesium Chloride, Hexahydrate. Adjust to pH 8.8 at 37°C with 1 M NaOH. *Prepare Fresh*).

B. 15.2 mM p-Nitrophenyl Phosphate Solution (PNPP) (Prepare 5 ml in deionized water Phosphatase Substrate. *Prepare Fresh*).

C. 20 mM Sodium Hydroxide Solution (NaOH) (Prepare 100 ml in deionized water using Sodium Hydroxide, Anhydrous).

D. Phosphatase, Alkaline Enzyme Solution (Immediately before use, prepare a solution containing 0.1–0.2 unit/ml of Alkaline Phosphatase in cold deionized water).

Procedure

Pipette (in milliliters) the following reagents into suitable containers:

	Test	Blank
Reagent A (Buffer)	0.50	0.50
Reagent B (PNPP)	0.50	0.50

Mix by inversion and equilibrate to 37°C. Then add:

Reagent D (Enzyme Solution)	0.10	—
Deionized Water	—	0.10

Immediately mix by inversion and incubate at 37°C for exactly 10 minutes. Then add:

Reagent C (NaOH)	10.00	10.00

Mix by swirling and transfer the solutions to suitable cuvettes and record the A410nm for both Test and Blank.

Calculations

$$\text{Units / ml enzyme} = \frac{(\text{A410nm Test} - \text{A410nm Blank})\ (11.1)\ (\text{df})}{(18.3)\ (0.1)\ (10)}$$

11.1 = Volume (in milliliters) of assay
df = Dilution factor
18.3 = Millimolar extinction coefficient of p-nitrophenol at 410nm
0.1 = Volume (in milliliters) of enzyme used
10 = Time of assay (in minutes) as per the Unit Definition

$$\text{Units / mg solid} = \frac{\text{units / ml enzyme}}{\text{mg solid / ml enzyme}}$$

$$\text{Units / mg protein} = \frac{\text{units / ml enzyme}}{\text{mg protein / ml enzyme}}$$

Unit Definition: *One unit will hydrolyze 1.0 μmole of p-nitrophenyl phosphate per minute at pH 8.8 at 37°C.*

Final Assay Concentrations

In a 1.10 ml reaction mix, the final concentrations are 45 mM glycine, 0.45 mM magnesium chloride, 6.9 mM p-nitrophenyl phosphate and 0.01–0.02 unit alkaline phosphatase.

Enzymatic Assay of PHOSPHATASE, ALKALINE (EC 3.1.3.1) Glycine with Zinc Assay

Principle

$$\text{p-Nitrophenyl Phosphate} + H_2O \xrightarrow{\text{Alkaline Phosphatase}} \text{p-Nitrophenol} + \text{Pi}$$

Abbreviation Used

Pi = Inorganic Phosphate

Conditions: T = 37°C, pH = 10.4, A405nm, Light path = 1 cm

Method: Continuous Spectrophotometric Rate Determination

Reagents

A. 100 mM Glycine Buffer with 1.0 mM Magnesium Chloride and 1.0 mM Zinc Chloride, pH 10.4 at 37°C (Prepare 50 ml in deionized water using Glycine, Free Base, Magnesium Chloride, Hexahydrate and Zinc Chloride. Adjust to pH 10.4 at 37°C with 1 M NaOH. *Prepare Fresh*).

B. 60 mM p-Nitrophenyl Phosphate Solution (PNPP) (Prepare 5 ml in deionized water Phosphatase Substrate. *Prepare Fresh*).

C. 1.0 mM Magnesium Chloride Solution ($MgCl_2$) (Prepare 50 ml in deionized water using Magnesium Chloride, Hexahydrate).

D. Phosphatase, Alkaline Enzyme Solution (Immediately before use, prepare a solution containing 0.1–0.2 unit/ml of Alkaline Phosphatase in cold Reagent C).

Procedure

Pipette (in milliliters) the following reagents into suitable cuvettes:

	Test	Blank
Reagent A (Buffer)	2.60	2.60
Reagent B (PNPP)	0.30	0.30

Mix by inversion and equilibrate to 37°C. Monitor the A405nm until constant, using a suitably thermostatted spectrophotometer. Then add:

Reagent C ($MgCl_2$)	—	0.10
Reagent D (Enzyme Solution)	0.10	—

Immediately mix by inversion and record the increase in A405nm for approximately 5 minutes. Obtain the ΔA405nm/minute using the maximum linear rate for both the Test and Blank.

Calculations

$$\text{Units / ml enzyme} = \frac{(\Delta A405nm / \text{min Test} - \Delta A405nm / \text{min Blank})\ (3)\ (df)}{(18.5)\ (0.1)}$$

3 = Volume (in milliliters) of assay
df = Dilution factor
18.5 = Millimolar extinction coefficient of p-nitrophenol at 405nm
0.1 = Volume (in milliliter) of enzyme used

$$\text{Units / mg solid} = \frac{\text{units / ml enzyme}}{\text{mg solid / ml enzyme}}$$

$$\text{Units / mg protein} = \frac{\text{units / ml enzyme}}{\text{mg protein / ml enzyme}}$$

Unit Definition: *One unit will hydrolyze 1.0 μmole of p-nitrophenyl phosphate per minute at pH 10.4 at 37°C.*

Final Assay Concentrations

In a 3.00 ml reaction mix, the final concentrations are 87 mM glycine, 0.90 mM magnesium chloride, 0.87 mM zinc chloride, 6.0 mM p-nitrophenyl phosphate and 0.01–0.02 unit alkaline phosphatase.

Note

1. This assay is not to be used for Phosphatase, Alkaline, Type XXIII from Trout Intestine, Phosphatase, Alkaline-Acrylic Beads, Phosphatase, Alkaline, Affinity Filtration Cartridge, Phosphatase, Alkaline-Agarose, Phosphatase, Alkaline-Biotinamidocaproyl, Phosphatase, Alkaline, from Shrimp or for Phosphatase, Alkaline, Bacterial (*Escherichia coli*).

Enzymatic Assay of PHOSPHATASE, ACID (EC 3.1.3.2)

Principle

$$\text{p-Nitrophenyl Phosphate} + H_2O \xrightarrow{\text{Phosphatase, Acid}} \text{p-Nitrophenol} + \text{Pi}$$

Abbreviation Used

Pi = Inorganic phosphate

Conditions: T = 37°C, pH = 4.8, A410nm, Light path = 1 cm

Method: Spectrophotometric Stop Rate Determination

Reagents

A. 90 mM Citrate Buffer, pH 4.8 at 37°C (Prepare 100 ml in deionized water using Citric Acid, Trisodium, Dihydrate, or Citrate Buffer Solution. Adjust to pH 4.8 at 37°C with 1 M NaOH or 1 M HCl).
B. 15.2 mM p-Nitrophenyl Phosphate (PNPP) (Prepare 5 ml in deionized water using Phosphatase Substrate).
C. 100 mM Sodium Hydroxide Solution (NaOH) (Prepare 50 ml in deionized water using Sodium Hydroxide, Anhydrous).

D. Acid Phosphatase Enzyme Solution (Immediately before use, prepare a solution containing 0.15–0.25 unit/ml of Phosphatase, Acid in cold deionized water).

Procedure

Pipette (in milliliters) the following reagents into suitable containers:

	Test	Blank
Reagent A (Buffer)	0.50	0.50
Reagent B (PNPP)	0.50	0.50

Mix by inversion and equilibrate to 37°C. Then add:

Reagent D (Enzyme Solution)	0.10	—

Immediately mix by inversion and incubate at 37°C for exactly 10 minutes. Then add:

Reagent C (NaOH)	4.00	4.00
Reagent D (Enzyme Solution)	—	0.10

Mix by inversion and record the A410nm for both the Test and Blank in a suitable spectrophotometer.

Calculations

$$\text{Units / ml enzyme} = \frac{(\text{A410 nm Test} - \text{A410nm Blank})\ (5.1)\ (\text{df})}{(10)\ (18.3)\ (0.1)}$$

5.1 = Total volume (in milliliters) of solution
df = Dilution factor
10 = Time of assay (in minutes) as per the Unit Definition
18.3 = Millimolar extinction coefficient of p-Nitrophenol at 410nm
0.1 = Volume (in milliliter) of enzyme used

$$\text{Units / mg solid} = \frac{\text{units / ml enzyme}}{\text{mg solid / ml enzyme}}$$

$$\text{Units / mg protein} = \frac{\text{units / ml enzyme}}{\text{mg protein / ml enzyme}}$$

Unit Definition: *One unit will hydrolyze 1.0 μmole of p-nitrophenyl phosphate per minute at pH 4.8 at 37°C.*

Final Assay Concentration

In a 1.10 ml reaction mix, the final concentrations are 41 mM citric acid, 6.9 mM p-nitrophenyl phosphate and 0.015–0.025 unit phosphatase, acid.

Enzymatic Assay of PHOSPHATASE, ACID, PROSTATIC (EC 3.1.3.2)

Principle

$$\text{p-Nitrophenyl Phosphate} + H_2O \xrightarrow{\text{Prostatic Acid Phosphatase}} \text{p-Nitrophenol} + \text{Pi}$$

Abbreviation Used

Pi = Inorganic phosphate

Conditions: T = 37°C, pH = 4.8, A410nm, Light path = 1 cm

Method: Spectrophotometric Stop Rate Determination

Reagents

A. 90 mM Citrate Buffer Solution, pH 4.8 at 37°C.
B. 40 mM Tartrate Buffer, with 90 mM Citrate, pH 4.8 at 37°C.
C. 11.3 mM p-Nitrophenyl Phosphate (PNPP) (Prepare 5 ml in deionized water using p-Nitrophenyl Phosphate Phosphatase Substrate).
D. 100 mM Sodium Hydroxide Solution (NaOH) (Prepare 50 ml in deionized water using Sodium Hydroxide, Anhydrous).
E. Prostatic Acid Phosphatase Enzyme Solution (Immediately before use, prepare a solution containing 0.015–0.02 unit/ml of Acid Phosphatase in cold deionized water).

Procedure

Pipette (in milliliters) the following reagents into suitable containers:

	Test 1	Test 2	Blank
Reagent A (Citrate Buffer)	0.50	—	0.50
Reagent B (Tartrate Buffer)	—	0.50	—
Reagent C (PNPP)	0.50	0.50	0.50

Mix and equilibrate to 37°C. Then add:

Reagent E (Enzyme Solution)	0.20	0.20	—

Immediately mix and incubate at 37°C for exactly 30 minutes. Then add:

Reagent D (NaOH)	3.80	3.80	3.80
Reagent E (Enzyme Solution)	—	—	0.20

Mix and record the A410nm for both the Tests and Blank in a suitable spectrophotometer.

Calculations

Total Acid Phosphatase Activity

$$\text{Units / ml Total enzyme} = \frac{(\Delta A410\text{nm / min Test 1} - \Delta A410\text{nm / min Blank})\ (5.0)\ (\text{df})}{(30)\ (18.3)\ (0.2)}$$

Nonprostatic Acid Phosphatase Activity

$$\text{Units / ml Nonprostatic enzyme} = \frac{(\Delta A410\text{nm / min Test 2} - \Delta A410\text{nm / min Blank})\ (5.0)\ (\text{df})}{(30)\ (18.3)\ (0.2)}$$

Prostatic Acid Phosphatase Activity

Units/ml Prostatic Enzyme = Units/ml Total Enzyme–Units/ml Nonprostatic Enzyme
5.0 = Total volume (in milliliters) of solution
30 = Time of assay (in minutes)
18.3 = Millimolar extinction coefficient of p-Nitrophenol at 410 nm at an Alkaline pH
0.2 = Volume (in milliliter) of enzyme used
df = Dilution factor

$$\text{Units / mg solid} = \frac{\text{units / ml Prostatic Enzyme}}{\text{mg solid / ml Total Enzyme}}$$

$$\text{Units / mg protein} = \frac{\text{units / ml Prostatic Enzyme}}{\text{mg protein / ml Total Enzyme}}$$

Unit Definition: *One unit will hydrolyze 1.0 μmole of p-nitrophenyl phosphate per minute at pH 4.8 at 37°C. Prostatic acid phosphatase activity is the difference between the total acid phosphatase activity and the acid phosphatase activity in the presence of 20 mM tartrate.*

Final Assay Concentration

In a 1.20 ml reaction mix, the final concentrations are 38 mM citric acid, 4.7 mM p-nitrophenyl phosphate and 0.004 unit prostatic acid phosphatase.

Enzymatic Assay of PHOSPHATASE, ALKALINE (EC 3.1.3.1) Diethanolamine Assay

Principle

$$\text{p-Nitrophenyl Phosphate} + H_2O \xrightarrow{\text{Alkaline Phosphatase}} \text{p-Nitrophenol} + \text{Pi}$$

Abbreviation Used

Pi = Inorganic Phosphate

Conditions: T = 37°C, pH = 9.8, A405nm, Light path = 1 cm

Method: Continuous Spectrophotometric Rate Determination

Reagents

A. 1.0 M Diethanolamine Buffer with 0.50 mM Magnesium Chloride, pH 9.8 at 37°C (Prepare 50 ml using Diethanolamine and Magnesium Chloride, Hexahydrate. Dissolve the Magnesium Chloride complete in deionized water before adding the Diethanolamine. Adjust to pH 9.8 at 37°C with 5 M HCl. *Prepare Fresh*).
B. 150 mM p-Nitrophenyl Phosphate Solution (PNPP) (Prepare 2 ml in deionized water using Sigma 104 Phosphatase Substrate. *Prepare Fresh*).
C. Phosphatase, Alkaline Enzyme Solution (Immediately before use, prepare a solution containing 0.1–0.2 unit/ml of Alkaline Phosphatase in cold Reagent A).

Procedure

Pipette (in milliliters) the following reagents into suitable cuvettes:

	Test	Blank
Reagent A (Buffer)	2.70	2.80
Reagent B (PNPP)	0.30	0.30

Mix by inversion and equilibrate to 37°C. Monitor the A405nm until constant, using a suitably thermostatted spectrophotometer. Then add:

Reagent C (Enzyme Solution)	0.10	—

Immediately mix by inversion and record the increase in A405nm for approximately 10 minutes. Obtain the ΔA405nm/minute using the maximum linear rate for both the Test and Blank.

Calculations

$$\text{Units / ml enzyme} = \frac{(\Delta A405nm / min\ Test - \Delta A405nm / min\ Blank)\ (3.1)\ (df)}{(18.5)\ (0.1)}$$

3.1 = Volume (in milliliters) of assay
df = Dilution factor
18.5 = Millimolar extinction coefficient of p-Nitrophenol at 405nm
0.1 = Volume (in milliliters) of enzyme used

$$\text{Units / mg solid} = \frac{\text{units / ml enzyme}}{\text{mg solid / ml enzyme}}$$

$$\text{Units / mg protein} = \frac{\text{units / ml enzyme}}{\text{mg protein / ml enzyme}}$$

Unit Definition: *One unit will hydrolyze 1.0 μmole of p-nitrophenyl phosphate per minute at pH 9.8 at 37°C.*

Final Assay Concentrations

In a 3.10 ml reaction mix, the final concentrations are 903 mM diethanolamine, 0.45 mM magnesium chloride, 14 mM p-nitrophenyl phosphate and 0.01–0.02 unit alkaline phosphatase.

Note

1. This enzyme assay is not to be used to assay Phosphatase, Alkaline, in which the specific activity is cited only in glycine units.

Enzymatic Assay of PHOSPHATASE, ALKALINE (EC 3.1.3.1) Glycine Assay

Principle

$$\text{p - Nitrophenyl Phosphate} \xrightarrow{\text{Alkaline Phosphatase}} \text{p - Nitrophenol} + \text{Pi}$$

Conditions: T = 37°C, pH = 10.4, A410nm, Light path = 1 cm

Method: Spectrophotometric Stop Rate Determination

Reagents

A. 100 mM Glycine Buffer with 1 mM Magnesium Chloride, pH 10.4 at 37°C (Prepare 50 ml in deionized water using Glycine and Magnesium Chloride Hexahydrate. Adjust to pH 10.4 at 37°C with 1 M NaOH. *Prepare Fresh*).
B. 15.2 mM p-Nitrophenyl Phosphate Solution (PNPP) (Prepare 2 ml in deionized water using Phosphatase Substrate. *Prepare Fresh*).
C. Phosphatase, Alkaline Enzyme Solution (Immediately before use prepare a solution containing 0.1–0.2 units/ml of Alkaline Phosphatase in cold deionized water).
D. 20 mM Sodium Hydroxide Solution (NaOH) (Prepare 100 ml in deionized water using Sodium Hydroxide).

Procedure

Pipette (in milliliters) the following reagents into suitable cuvettes:

	Test	Blank
Deionized Water	—	0.10
Reagent A (Buffer)	0.50	0.50
Reagent B (PNPP)	0.50	0.50

Mix by inversion and equilibrate to 37°C. Monitor the A410nm until constant, using a suitably thermostatted spectrophotometer. Then add:

Reagent C (Enzyme Solution)	0.10	—

Immediately mix by inversion and incubate for exactly 10 minutes. Then add:

Reagent D	10.00	10.00

Record the A410nm for both the test and blank.

Calculation

$$\text{Units / mg protein} = \frac{(\Delta\text{A410nm Test} - \Delta\text{A410nm Blank})\ (11.1)}{(10)\ (18.3)\ (\text{mg protein / RM})}$$

11.1 = total volume
10 = Time of Assay (Unit Definition)
18.3 = Millimolar extinction coefficient for p-nitrophenol
RM = Reaction Mix

Unit Definition: *One unit will hydrolyze 1.0 μmole of p-nitrophenyl phosphate per minute at pH 10.4 at 37°C.*

Final Assay Concentrations

In a 1.1 ml reaction mix, the final concentrations are 45 mM glycine, 0.45 mM magnesium chloride, 6.9 mM p-nitrophenyl phosphate and 0.01 to 0.02 units alkaline phosphatase.

Enzymatic Assay of PHOSPHO(ENOL)PYRUVATE CARBOXYLASE (EC 4.1.1.31)

Principle

$$PEP + CO_2 + H_2O\ PEP \xrightarrow{\text{Carboxylase}} \text{Oxalacetate} + Pi$$

$$\text{Oxalacetate} + \beta\text{-NADH} \xrightarrow{\text{Malic Dehydrogenase}} \text{Malate} + \beta\text{-NAD}$$

Abbreviations Used

PEP = Phospho(enol) pyruvate
PEP Carboxylase = Phospho(enol) pyruvate Carboxylase
Pi = Inorganic Phosphate
β-NADH = β-Nicotinamide Adenine Dinucleotide, Reduced Form
β-NAD = β-Nicotinamide Adenine Dinucleotide, Oxidized Form

Conditions: T = 25°C, pH = 8.5, A340nm, Light path = 1 cm

Method: Continuous Spectrophotometric Rate Determination

Reagents

A. 110 mM Tris Sulfate Buffer, pH 8.5 at 25°C (Prepare 100 ml by dissolving Trizma Base,in 110 mM Sulfuric Acid which was prepared in deionized water with Sulfuric Acid. Adjust to pH 8.5 at 25°C with 1 M NaOH).
B. 300 mM Magnesium Sulfate Solution ($MgSO_4$) (Prepare 10 ml in deionized water using Magnesium Sulfate, Heptahydrate).
C. 6 mM β-Nicotinamide Adenine Dinucleotide, Reduced Form Solution (β-NADH) (Dissolve the contents of one 10 mg vial of β-Nicotinamide Adenine Dinucleotide, Reduced Form, Disodium Salt, in the appropriate volume of Reagent A. *Prepare Fresh*).
D. Malic Dehydrogenase Enzyme Solution (MDH) (Immediately before use, prepare a solution containing 600 units/ml in Reagent A using Malic Dehydrogenase).
E. 100 mM Sodium Bicarbonate Solution (Bicarb) (Prepare 10 ml in deionized water using Sodium Bicarbonate).
F. 30 mM Phospho(enol) pyruvate Solution (PEP) (Prepare 1 ml in Reagent A using Phospho(enol) pyruvate, Monopotassium Salt).
G. Dioxane (Use low peroxide grade)
H. 300 mM Dithioerythritol Solution (DTE) (Prepare 1 ml in Reagent A using Dithioerythritol. *Prepare Fresh*).
I. 5.0 mM Magnesium Sulfate Solution (Enzyme Diluent) (Prepare 20 ml in Reagent A using Magnesium Sulfate, Heptahydrate).
J. Phospho(enol) pyruvate Carboxylase Enzyme Solution (Immediately before use, prepare a solution containing 0.5–1 mg/ml of Phospho(enol) pyruvate Carboxylase in Reagent I).

Procedure

Pipette (in milliliters) the following reagents into suitable cuvettes:

	Test	Blank
Reagent A (Buffer)	1.80	1.80
Reagent B ($MgSO_4$)	0.10	0.10
Reagent C (β-NADH)	0.10	0.10
Reagent E (Bicarb)	0.30	0.30
Reagent G (Dioxane)	0.30	0.30
Reagent H (DTE)	0.10	0.10

Mix by inversion and equilibrate to 25°C. Then add:

Reagent D (MDH)	0.01	0.01
Reagent I (Enzyme Diluent)	—	0.10
Reagent J (Enzyme Solution)	0.10	—

Mix by inversion and equilibrate to 25°C. Monitor the A340nm until constant, using a suitably thermostatted spectrophotometer. Then add:

Reagent F (PEP)	0.10	0.10

Immediately mix by inversion and record the decrease in A340nm for approximately 5 minutes. Obtain the ΔA340nm/minute using the maximum linear rate for both the Test and Blank.

Calculations

$$\text{Units / mg enzyme} = \frac{\Delta A340nm / min\ Test - \Delta A340nm\ Blank}{(6.22)\ (mg\ enzyme / ml\ RM)}$$

6.22 = Millimolar extinction coefficient of β-NADH at 340nm
RM = Reaction Mix

Unit Definition: *One unit will form 1.0 μmole of oxalacetate from phospho(enol) pyruvate and CO_2 per minute at pH 8.5 at 25°C.*

Final Assay Concentration

In a 2.91 ml reaction mix, the final concentrations are 80 mM Tris sulfate, 10.5 mM magnesium sulfate, 0.21 mM β-NADH, 10 mM sodium bicarbonate, 10% (v/v) dioxane, 10 mM DTE, 1.0 mM PEP, 6 units malic dehydrogenase and 0.05–0.1 mg of phospho(enol) pyruvate carboxylase.

Note

1. Malic Dehydrogenase Unit Definition: One unit will convert 1.0 μmole of oxalacetate and β-NADH to Lmalate and β-NAD per minute at pH 7.5 at 25°C.

Enzymatic Assay of PHOSPHODIESTERASE I (EC 3.1.4.1)

Principle

$$\text{bis (p-Nitrophenyl) Phosphate} + H_2O \xrightarrow{\text{Phosphodiesterase I}} \text{PNP} + \text{PNPP}$$

Abbreviations Used

PNP = p-Nitrophenol
PNPP = p-Nitrophenyl Phosphate

Conditions: T = 37°C, pH = 8.8, A410nm, Light path = 1 cm

Method: Continuous Spectrophotometric Rate Determination

Reagents

A. 100 mM Tris HCl Buffer, pH 8.8 at 37°C. (Prepare 100 ml in deionized water using Trizma Base. Adjust to pH 8.8 at 37°C with 1 M HCl).
B. 10 mM Ammonium Acetate Solution (NH_4OAc) (Prepare 20 ml in deionized water using Ammonium Acetate).
C. 1.0 mM bis(p-Nitrophenyl) Phosphate Solution (b-PNPP) (Prepare 10 ml in Reagent B using bis(p-Nitrophenyl) Phosphate, Hemicalcium Salt. Stir gently at a temperature of no greater than 50°C with sonication every 5 minutes).
D. 300 mM Magnesium Acetate Solution ($Mg(OAc)_2$) (Prepare 10 ml in deionized water using Magnesium Acetate, Tetrahydrate.
E. Phosphodiesterase I Enzyme Solution (Immediately before use, prepare a solution containing 0.1–0.2 unit/ml of Phosphodiesterase I in cold deionized water).

Procedure

Pipette (in milliliters) the following reagents into suitable cuvettes:

	Test	Blank
Deionized Water	0.60	0.70
Reagent A (Buffer)	1.00	1.00
Reagent C (b-PNPP)	1.00	1.00
Reagent D ($Mg(OAc)_2$)	0.30	0.30

Mix by inversion and equilibrate to 37°C. Monitor the A410nm until constant, using a suitably thermostatted spectrophotometer. Then add:

Reagent E (Enzyme Solution)	0.10	—

Immediately mix by inversion and record the increase at A410nm for approximately 10 minutes. Obtain the ΔA410nm/minute using the maximum linear rate for both the Test and Blank.

Calculations

$$\text{Units / ml enzyme} = \frac{(\Delta A410nm / \min \text{Test} - \Delta A410nm \text{ Blank})(3)(df)}{(18.3)(0.1)}$$

3 = Volume (in milliliters) of assay
df = Dilution factor
18.3 = Millimolar extinction coefficient of p-Nitrophenol at 410nm
0.1 = Volume (in milliliters) of enzyme used

$$\text{Units / mg solid} = \frac{\text{units / ml enzyme}}{\text{mg solid / ml enzyme}}$$

$$\text{Units / mg protein} = \frac{\text{units / ml enzyme}}{\text{mg protein / ml enzyme}}$$

Unit Definition: *One unit will hydrolyze 1.0 µmole of bis(p-nitrophenyl) phosphate per minute at pH 8.8 at 37°C.*

Final Assay Concentration

In a 3.00 ml reaction mix the final concentrations are 33 mM Tris, 3.3 mM ammonium acetate, 0.33 mM bis (p-nitrophenyl) phosphate, 30 mM magnesium acetate and 0.01–0.02 unit phosphodiesterase I.

Note

1. This assay is not to be used to assay Phosphodiesterase I from Crotalum durissus terrificus.

Enzymatic Assay of PHOSPHODIESTERASE I (EC 3.1.4.1)

Principle

$$\text{bis (p - Nitrophenyl) Phosphate} + H_2O \xrightarrow{\text{Phosphodiesterase I}} \text{PNP} + \text{PNPP}$$

Abbreviations Used

PNP = p-Nitrophenol
PNPP = p-Nitrophenyl Phosphate

Conditions: T = 25°C, pH = 8.9, A405nm, Light path = 1 cm.

Method: Continuous Spectrophotometric Rate Determination

Reagents

A. 200 mM Tris HCl Buffer, pH 8.9 at 25°C (Prepare 100 ml in deionized water using Trizma Base. Adjust to pH 8.9 at 25°C with 1 M HCl).
B. 6.0 mM bis(p-Nitrophenyl) Phosphate Solution (b-PNPP) (Prepare 10 ml in Reagent A using bis (p-Nitrophenyl) Phosphate, Hemicalcium Salt).
C. Phosphodiesterase I Enzyme Solution (Immediately before use, prepare a solution containing 0.15–0.30 unit/ml of Phosphodiesterase I in cold deionized water).

Procedure

Pipette (in milliliters) the following reagents into suitable cuvettes:

	Test	Blank
Reagent B (b-PNPP)	3.00	3.00

Equilibrate to 25°C. Monitor the A405nm until constant, using a suitably thermostatted spectrophotometer. Then add:

	Test	Blank
Reagent C (Enzyme Solution)	0.06	—
Deionized Water	—	0.06

Immediately mix by inversion and record the increase at A405nm for approximately 5 minutes. Obtain the A405nm/minute using the maximum linear rate for both the Test and Blank.

Calculations

$$\text{Units / ml enzyme} = \frac{(\text{A405nm / min Test} - \text{A405nm / min Blank})(3.06)(\text{df})}{(18.5)(0.06)}$$

3.06 = Volume (in milliliters) of assay
df = Dilution factor
18.5 = Millimolar extinction coefficient of p-Nitrophenol at 405nm
0.06 = volume (in milliliters) of enzyme used

$$\text{Units / mg solid} = \frac{\text{units / ml enzyme}}{\text{mg solid / ml enzyme}}$$

$$\text{Units / mg protein} = \frac{\text{units / ml enzyme}}{\text{mg protein / ml enzyme}}$$

Unit Definition: *One unit will hydrolyze 1.0 μmole of bis(p-nitrophenyl) phosphate per minute at pH 8.9 at 25°C.*

Final Assay Concentration

In a 3.06 ml reaction mix the final concentrations are 196 mM Tris, 5.9 mM bis (p-nitrophenyl) phosphate, and 0.009–0.02 unit phosphodiesterase I.

Enzymatic Assay of PHOSPHODIESTERASE II (EC 3.1.16.1)

Principle

$$\text{RNA - Core} \xrightarrow{\text{Phosphodiesterase II}} \text{Acid Soluble Nucleoside - 3'-Phosphates}$$

Abbreviation Used

RNA = Ribonucleic Acid

Conditions: T = 37°C, pH = 6.5, A260nm, Light path = 1 cm

Method: Spectrophotometric Stop Rate Determination

Reagents

A. 17 mg/ml Ribonucleic Acid-Core Solution (RNA Solution) (Prepare in deionized water using Ribonucleic Acid, Core).
B. 250 mM Succinic Acid Sodium Salt pH 6.5 at 37°C (Succinate Buffer) (Prepare in deionized water using Succinic Acid, Disodium Salt, Hexahydrate. Adjust the pH to 6.5 at 37°C using 1 M HCl).

C. 0.25% (w/v) Uranyl Acetate with 250 mM Perchloric Acid Solution (UA-PA Solution) (Prepare 50 ml in deionized water using Uranyl Acetate, Dihydrate, and Perchloric Acid).

D. Phosphodiesterase II Solution (Enzyme Solution) (Immediately before use prepare a solution containing 3–6 units/ml of Phosphodiesterase II in cold deionized water).

Procedure

Pipette (in milliliters) the following reagents into suitable tubes:

	Test	Blank
Reagent A (RNA Solution)	0.50	0.50
Reagent B (Succinate Buffer)	0.40	0.40
Deionized Water	0.90	1.10

Mix by swirling and equilibrate to 37°C. Then add:

Reagent D (Enzyme Solution)	0.20	—

Immediately mix by swirling and incubate both tubes at 37°C for exactly 30 minutes. Then add:

Reagent C (UA-PA Solution)	2.00	2.00

Immediately mix by inversion and chill in an ice bath for 5 minutes. Centrifuge both tubes. Dilute 0.2 ml of each supernatant to 8.0 ml using deionized water. Read the A260nm of the Test solution against the Blank solution using suitable cuvettes.

Calculations

$$\text{Units / ml enzyme} = \frac{(\text{A260nm Test} - \text{A260nm Blank})\ (8)\ (2)\ (\text{df})}{(16)\ (0.2)\ (0.2)}$$

8 = Volume (in milliliters) of diluted assay
2 = Dilution volume (in milliliters) after the addition of UA-PA
df = Dilution factor
16 = Increase in A260nm at 30 minutes as per the Unit Definition
0.2 = Volume (in milliliter) of enzyme used
0.2 = Volume (in milliliter) of reaction mix used in diluted assay

$$\text{Units / mg solid} = \frac{\text{units / ml enzyme}}{\text{mg solid / ml enzyme}}$$

$$\text{Units / mg protein} = \frac{\text{units / ml enzyme}}{\text{mg protein / ml enzyme}}$$

Unit Definition: *One unit will produce acid soluble nucleotides equivalent to a A260 of 16 in 30 minutes at pH 6.5 at 37°C, in a 2.0 ml reaction mixture.*

Final Assay Concentrations

In a 2.00 ml reaction mix, the final concentrations are 4.25 mg/ml ribonucleic acid-core, 50 mM succinic acid, and 0.6–1.2 units phosphodiesterase II.

Enzymatic Assay of 3'-PHOSPHODIESTERASE, 2' : 3'-CYCLIC NUCLEOTIDE (EC 3.1.4.37)

Principle

$$\text{Cyclic-}\beta\text{-NADP} + H_2O \xrightarrow{\text{3'-PDE}} \beta\text{-NADP}$$

$$\beta\text{-NADP} + \text{G 6-P} \xrightarrow{\text{G-6-PDH}} \text{6-Phosphogluconate} + \beta\text{-NADPH}$$

Abbreviations Used

Cyclic-β-NADP = β-Nicotinamide Adenine Dinucleotide 2' : 3'-Cyclic Monophosphate
3'-PDE = 3'-Phosphodiesterase, 2':3'-Cyclic Nucleotide
β-NADP = β-Nicotinamide Adenine Dinucleotide Phosphate, Oxidized Form
β-NADPH = β-Nicotinamide Adenine Dinucleotide Phosphate, Reduced Form
G 6-P = Glucose 6-Phosphate
G-6-PDH = Glucose-6-Phosphate Dehydrogenase

Conditions: T = 25°C, pH = 6.00, A340nm, Light path = 1 cm

Method: Continuous Spectrophotometric Rate Determination

Reagents

A. 200 mM 2-[N-Morpholino] Ethanesulfonic Acid (MES) Buffer, pH 6.00 at 25°C (Prepare 200 ml in deionized water using MES, Free Acid. Adjust to pH 6.00 at 25°C with 1 M KOH).

B. 10 mM β-Nicotinamide Adenine Dinucleotide 2' : 3'-Cyclic Monophosphate Solution (Cyclic-β-NADP) (Prepare 5 ml in Reagent A using β-Nicotinamide Adenine Dinucleotide 2':3'-Cyclic Monophosphate, Sodium Salt. *Prepare Fresh*).

C. 50 mM Glucose 6-Phosphate Solution (G 6-P) (Prepare 5 ml in Reagent A using Glucose 6-Phosphate, Monosodium).

D. 300 mM Magnesium Chloride Solution ($MgCl_2$) (Prepare 5 ml in Reagent A using Magnesium Chloride, Hexahydrate.

E. Glucose-6-Phosphate Dehydrogenase Enzyme Solution (G-6-PDH) (Immediately before use, prepare a solution containing 100–150 units/ml of Glucose-6-Phosphate Dehydrogenase in cold deionized water).

F. 3'-Phosphodiesterase, 2' : 3'-Cyclic Nucleotide Enzyme Solution (Immediately before use, prepare a solution containing 0.2 unit/ml of 3'-Phosphodiesterase, 2' : 3'-Cyclic Nucleotide in cold deionized water).

Procedure

Prepare a reaction cocktail by pipetting (in milliliters) the following reagents into a suitable container:

Reagent A (Buffer)	14.0
Reagent B (Cyclic-β-NADP)	2.0
Reagent C (G 6-P)	2.0
Reagent D ($MgCl_2$)	2.0

Mix by swirling and equilibrate to 25°C. Pipette (in milliliters) the following reagents into suitable cuvettes:

	Test	Blank
Reaction Cocktail	1.00	1.00
Reagent E (G-6-PDH)	0.05	0.05

Mix by inversion and equilibrate to 25°C. Monitor the A340nm until constant, using a suitably thermostatted spectrophotometer. Then add:

Deionized Water	—	0.05
Reagent F (Enzyme Solution)	0.05	—

Immediately mix by inversion and record the increase in A340nm for approximately 5 minutes. Obtain the ΔA340nm/minute using the maximum linear rate for both the Test and Blank.

Calculations

$$\text{Units / mg enzyme} = \frac{\Delta A340\text{nm / min Test} - \Delta A340\text{nm / min Blank}}{(6.22)\ (\text{mg enzyme / ml RM})}$$

6.22 = Millimolar extinction coefficient of β-NADPH at 340nm
RM = Reaction Mix

Unit Definition: *One unit will convert 1.0 µmole of 2':3'-cyclic NADP to NADP per minute at pH 6.0 at 25°C.*

Final Assay Concentrations

In a 1.10 ml reaction mix, the final concentrations are 182 mM MES, 0.91 mM β-nicotinamide adenine dinucleotide 2':3'-cyclic monophosphate, 4.5 mM glucose 6-phosphate, 27 mM magnesium chloride, 5.0–7.5 units glucose-6-phosphate dehydrogenase and 0.01 unit 3'-phosphodiesterase, 2':3'-cyclic nucleotide.

Note

1. Glucose 6-Phosphate Dehydrogenase Unit Definition: One unit will oxidize 1.0 µmole of D-glucose 6-phosphate to 6-phospho-D-gluconate per minute in the presence of NADP at pH 7.4 at 25°C.

Enzymatic Assay of PHOSPHODIESTERASE, 3':5'-CYCLIC NUCLEOTIDE Crude Complex

Principle

$$3':5'\text{-cAMP} + H_2O \xrightarrow{\text{PDE-3}':5'\text{-CN}} \text{AMP}$$

$$\text{AMP} + \text{ATP} \xrightarrow{\text{Myokinase}} 2\ \text{ADP}$$

$$2\ \text{ADP} + 2\ \text{PEP} \xrightarrow{\text{Pyruvate Kinase}} 2\ \text{ATP} + 2\ \text{Pyruvate}$$

$$2\ \text{Pyruvate} + 2\ \beta\text{-NADH} \xrightarrow{\text{Lactic Dehydrogenase}} 2\ \text{Lactate} + 2\ \beta\text{-NAD}$$

Abbreviations Used

3':5'-cAMP = Adenosine 3':5'-Cyclic Monophosphate
PDE-3':5'-CN = Phosphodiesterase, 3':5'-Cyclic Nucleotide
AMP = Adenosine 5'-Monophosphate
ATP = Adenosine 5'-Triphosphate
ADP = Adenosine 5'-Diphosphate
PEP = Phospho(enol) pyruvate
β-NADH = β-Nicotinamide Adenine Dinucleotide, Reduced Form
β-NAD = β-Nicotinamide Adenine Dinucleotide, Oxidized Form

Conditions: T = 30°C, pH = 7.5, A340nm, Light path = 1 cm

Method: Continuous Spectrophotometric Rate Determination

Reagents

A. 200 mM Tris HCl Buffer, pH 7.5 at 30°C (Prepare 50 ml in deionized water using Trizma Hydrochloride. Adjust to pH 7.5 at 30°C with 2 M NaOH).
B. 1 M Potassium Chloride Solution (KCl) (Prepare 1 ml in deionized water using Potassium Chloride).
C. 60 mM Magnesium Sulfate Solution ($MgSO_4$) (Prepare 10 ml in deionized water using Magnesium Sulfate, Anhydrous).
D. 0.665 mM Phospho(enol) pyruvate Solution (PEP) (Prepare by dissolving Phospho(enol) pyruvate, Monopotassium Salt in 3.7 ml of Reagent A. Then add 0.6 ml of Reagent B, 1.2 ml of Reagent C and 4.5 ml of deionized water. *Prepare Fresh*).
E. 30 mM Adenosine 5'-Triphosphate Solution (ATP) (Prepare 10 ml in deionized water using Adenosine 5'-Triphosphate, Disodium Salt. Adjust to pH 7.5 at 30°C with solid Sodium Bicarbonate. *Prepare Fresh*).
F. 60 mM Adenosine 3':5'-Cyclic Monophosphate Solution (3':5'-cAMP) (Prepare 1 ml in deionized water using Adenosine 3':5'-Cyclic Monophosphate, Sodium Salt. *Prepare Fresh*).
G. 0.70 mM Calcium Acetate Solution (Ca(OAc)2) (Prepare 10 ml in deionized water using Calcium Acetate).
H. β-Nicotinamide Adenine Dinucleotide, Reduced Form, Preweighed vial 1 mg (β-NADH) (Use β-Nicotinamide Adenine Dinucleotide, Reduced Form, Disodium Salt. *Prepare Fresh*).
I. PK/LDH Enzymes Suspension.
J. Myokinase Enzyme Solution (MK) (Immediately before use, prepare a solution containing 200 units/ml using Myokinase, in cold deionized water).
K. Phosphodiesterase, 3':5'-Cyclic Nucleotide Crude Complex Enzyme Solution (PDE-3':5'-CN) (Immediately before use, prepare a solution containing 0.05–0.1 unit/ml of Phosphodiesterase 3':5'-Cyclic Nucleotide, Crude Complex, in cold deionized water).

Procedure

Prepare a reaction cocktail by pipetting (in milliliters) the following reagents into Reagent H (β-NADH):

Reagent D (PEP)	9.00
Reagent E (ATP)	0.10
Reagent F (3':5'-cAMP)	0.10
Reagent G ($Ca(OAc)_2$)	0.15
Reagent I (PK/LDH)	0.05
Reagent J (MK)	0.05

Mix by inversion. Pipette (in milliliters) the following reagents into suitable cuvettes:

	Test	Blank
Reaction Cocktail	3.00	3.00

Equilibrate to 30°C. Monitor the A340nm until constant, using a suitably thermostatted spectrophotometer. Then add:

Reagent K (PDE-3':5'-CN)	0.10	—
Deionized Water	—	0.10

Mix by inversion and record the decrease in A340nm for approximately 15 minutes. Obtain the ΔA340nm/minute using the maximum linear rate for both the Test and Blank.

Calculations

$$\text{Units / ml enzyme} = \frac{(\Delta A340nm / \min \text{ Test} - \Delta A340nm / \min \text{ Blank})\,(3.1)\,(df)}{(2)\,(6.22)\,(0.1)}$$

3.1 = Total volume (in milliliters) of assay
df = Dilution factor
2 = 2 μmoles of β-NAD produced per μmole of 3':5'-cAMP hydrolyzed
6.22 = Millimolar extinction coefficient of β-NADH at 340nm
0.1 = Volume (in milliliter) of enzyme used in the assay

Unit Definition: *One unit will hydrolyze 1.0 μmole of 3':5'-cyclic-AMP to 5'-AMP per minute at pH 7.5 at 30°C.*

Final Assay Concentrations

In a 3.10 ml reaction mix, the final concentrations are 68 mM Tris, 55 mM potassium chloride, 7 mM magnesium sulfate, 0.61 mM phospho(enol) pyruvate, 0.31 mM adenosine 5'-triphosphate, 0.61 mM adenosine 3':5'-cyclic monophosphate, 11 units pyruvate kinase, 16 units lactic dehydrogenase, 3.1 units myokinase, 0.01 mM calcium acetate, 0.1 mM β-nicotinamide adenine dinucleotide, reduced form, and 0.005–0.01 unit phosphodiesterase, 3':5'-cyclic nucleotide, crude complex.

Notes

1. Contains not less than 700 Pyruvate Kinase units and 1000 L-Lactic Dehydrogenase units per ml.
2. L-Lactic Dehydrogenase Unit Definition: One unit will reduce 1.0 μmole of pyruvate to L-lactate per minute at pH 7.5 at 37°C.
3. Pyruvate Kinase Unit Definition: One unit will convert 1.0 μmole of phospho(enol) pyruvate to pyruvate per minute at pH 7.6 at 37°C.

4. Myokinase Unit Definition: One unit will convert 2.0 μmoles of ADP to ATP and AMP per minute at pH 7.6 at 37°C.
5. This reaction is very sensitive to chelators, which sequester the calcium ions at low Ca_2^+ concentrations. EDTA or EGTA concentrations of 0.005 mM in the cuvette will almost completely inhibit the reaction. The concentration of EDTA or EGTA should be less than 0.001 mM in the cuvette. Cuvettes should also be completely cleaned with 3% (w/v) NaOH solution before each use, as phosphodiesterase 3':5' cyclic activator can bind to quartz or glass cuvettes.

Enzymatic Assay of PHOSPHOGLUCOMUTASE (EC 5.4.2.2)

Principle

$$\alpha\text{-D-Glucose 1-Phosphate} \xrightarrow{\text{Phosphoglucomutase}} \alpha\text{-D-Glucose 6-Phosphate}$$

$$\alpha\text{-D-Glucose 6-Phosphate} + \beta\text{-NADP} \xrightarrow{\text{G-6-PDH}} \text{6-Phospho-D-gluconate} + \beta\text{-NADPH}$$

Abbreviations Used

β-NADP = β-Nicotinamide Adenine Dinucleotide Phosphate, Oxidized Form
β-NADPH = β-Nicotinamide Adenine Dinucleotide Phosphate, Reduced Form
G-6-PDH = Glucose-6-Phosphate Dehydrogenase

Conditions: T = 30°C, pH = 7.4, A340nm, Light path = 1 cm

Method: Continuous Spectrophotometric Rate Determination

Reagents

A. 250 mM Glycylglycine Buffer, pH 7.4 at 30°C (Prepare 50 ml in deionized water using Gly-Gly, Free Base. Adjust to pH 7.4 at 30°C with 1 M HCl).
B. 150 mM Glucose 1-Phosphate Solution (G 1-P) (Prepare 1 ml in deionized water using α-D-Glucose 1-Phosphate, Dipotassium Salt).
C. 20 mM β-Nicotinamide Adenine Dinucleotide Phosphate Solution (β-NADP) (Prepare 1 ml in deionized water using β-Nicotinamide Adenine Dinucleotide Phosphate, Sodium Salt or dissolve the contents of one 30 mg vial of β-Nicotinamide Adenine Dinucleotide Phosphate, Sodium Salt in the appropriate volume of deionized water. *Prepare Fresh*).
D. 1.2 mM Glucose 1,6-Diphosphate Solution (G 1,6-P) (Prepare 1 ml in deionized water using α-D-Glucose 1,6-Diphosphate, Cyclohexylammonium Salt, Hydrate).
E. 900 mM Magnesium Chloride Solution ($MgCl_2$) (Prepare 1 ml in deionized water using Magnesium Chloride, Hexahydrate).
F. 260 mM L-Cysteine HCl Solution (Cys) (Prepare 1 ml in deionized water using L-Cysteine Hydrochloride, Monohydrate. Adjust to pH 7.0 with solid Sodium Bicarbonate. *Prepare Fresh*).
G. Glucose-6-Phosphate Dehydrogenase Solution (G-6-PDH) (Immediately before use, prepare a solution containing 100 units/ml in cold Reagent A using Glucose-6-Phosphate Dehydrogenase).
H. Phosphoglucomutase Enzyme Solution (Immediately before use, prepare a solution containing 1–2 unit/ml of Phosphoglucomutase in cold Reagent A).

Procedure

Pipette (in milliliters) the following reagents into suitable cuvettes:

	Test	Blank
Reagent A (Buffer)	2.04	2.14
Reagent B (G 1-P)	0.10	0.10
Reagent C (β-NADP)	0.10	0.10
Reagent D (G 1,6-P)	0.05	0.05
Reagent E ($MgCl_2$)	0.10	0.10
Reagent F (Cys)	0.50	0.50
Reagent G (G-6-PDH)	0.01	0.01

Mix by inversion and equilibrate to 30°C. Monitor the A340nm until constant, using a suitably thermostatted spectrophotometer. Then add:

Reagent H (Enzyme Solution)	0.02	—

Immediately mix by inversion and record the increase in A340nm for approximately 5 minutes. Obtain the ΔA340nm/minute using the maximum linear rate for both the Test and Blank.

Calculations

$$\text{Units / ml enzyme} = \frac{(\Delta A340\text{nm / min Test} - \Delta A340\text{nm / min Blank})\ (3)\ (\text{df})}{(6.22)\ (0.02)}$$

3 = Total volume (in milliliters) of assay
df = Dilution factor
6.22 = Millimolar extinction coefficient of β-NADPH at 340 nm
0.02 = Volume (in milliliter) of phosphoglucomutase used

$$\text{Units / mg protein} = \frac{\text{units / ml enzyme}}{\text{mg protein / ml enzyme}}$$

Unit Definition: *One unit will convert 1.0 μmole of α-D-glucose 1-phosphate to α-D-glucose 6-phosphate per minute at pH 7.4 at 30°C.*

Final Assay Concentrations

In a 3.00 ml reaction mix, the final concentrations are 179 mM glycylglycine, 5.0 mM glucose 1-phosphate, 0.67 mM β-nicotinamide adenine dinucleotide phosphate, 0.02 mM glucose 1,6-diphosphate, 30 mM magnesium chloride, 43 mM L-cysteine, 1 unit glucose 6-phosphate dehydrogenase and 0.02–0.04 unit phosphoglucomutase.

Notes

1. Glucose 1,6-Diphosphate is required in order to obtain maximum activity.
2. Glucose-6-Phosphate Dehydrogenase unit definition: One unit will oxidize 1.0 μmole of D-glucose-6-phosphate to 6-phospho-D-gluconate per minute in the presence of β-NADP at pH 7.4 at 25°C.

Enzymatic Assay of 6-PHOSPHOGLUCONIC DEHYDROGENASE (EC 1.1.1.44)

Principle

$$6\text{-PGluc} + \beta\text{-NADP} \xrightarrow{6\text{-PGDH}} \text{D-Ribulose 5'-Phosphate} + CO_2 + \text{NADPH}$$

Abbreviations Used

6-PGluc = 6-Phosphogluconate
6-PGDH = 6-Phosphogluconic Dehydrogenase
β-NADP = β-Nicotinamide Adenine Dinucleotide Phosphate, Oxidized Form
β-NADPH = β-Nicotinamide Adenine Dinucleotide Phosphate, Reduced Form

Conditions: T = 37°C, pH 7.4, A340nm, Light path = 1 cm

Method: Continuous Spectrophotometric Rate Determination

Reagents

A. 250 mM Glycylglycine Buffer, pH 7.4 at 37°C. (Prepare 50 ml in deionized water using Glycylglycine, Hydrochloride. Adjust to pH 7.4 at 37°C with 1 M NaOH).
B. 60 mM 6-Phosphogluconate Solution (6-PGluc) (Prepare 1 ml in deionized water using 6-Phosphogluconic Acid, Trisodium Salt).
C. 20 mM β-Nicotinamide Adenine Dinucleotide Phosphate Solution (β-NADP) (Prepare 2 ml in deionized water using β-Nicotinamide Adenine Dinucleotide Phosphate, Sodium Salt. *Prepare Fresh*).
D. 300 mM Magnesium Chloride Solution ($MgCl_2$) (Prepare 1 ml in deionized water using Magnesium Chloride Hexahydrate).
E. 6-Phosphogluconic Dehydrogenase Enzyme Solution (Immediately before use, prepare a solution containing 0.3–0.5 unit/ml of 6-Phosphogluconic Dehydrogenase in cold deionized water).

Procedure

Pipette (in milliliter) the following reagents into suitable cuvettes:

	Test	Blank
Deionized Water	1.98	2.08
Reagent A (Buffer)	0.62	0.62
Reagent B (6-PGlu)	0.10	0.10
Reagent C (β-NADP)	0.10	0.10
Reagent D ($MgCl_2$)	0.10	0.10

Mix by inversion and equilibrate to 37°C. Monitor the A340nm until constant, using a suitably thermostatted spectrophotometer. Then add:

Reagent E (Enzyme Solution)	0.10	—

Immediately mix by inversion and record the increase in A340nm for approximately 5 minutes. Obtain the ΔA340nm/minute using the maximum linear rate for both the Test and Blank.

Calculations

$$\text{Units / mg enzyme} = \frac{\Delta A340\text{nm / min Test} - \Delta A340\text{nm / min Blank}}{(6.22)\ (\text{mg enzyme / ml RM})}$$

6.22 = Millimolar extinction coefficient of β-NADPH at 340nm
RM = Reaction Mix

Unit Definition: *One unit will oxidize 1.0 µmole of 6-phospho-D-gluconate to D-ribulose 5-phosphate and CO_2 per minute at pH 7.4 at 37°C in the presence of β-NADP.*

Final Assay Concentrations

In a 3.00 ml reaction mix, the final concentrations are 52 mM glycylglycine, 2.0 mM 6-phosphogluconate, 0.67 mM β-NADP, 10 mM magnesium chloride and 0.03–0.05 unit 6-phosphogluconic dehydrogenase.

Note

1. Not to be used with 6-Phosphogluconic Dehydrogenase, from Leuconostoc mesenteroides.

Enzymatic Assay of 6-PHOSPHOGLUCONIC DEHYDROGENASE (EC 1.1.1.44)

Principle

$$\text{6-PGluc} + \beta\text{-NADP} \xrightarrow{\text{6-PGDH}} \text{D-Ribulose 5'-Phosphate} + CO_2 + \beta\text{-NADH}$$

Abbreviations Used

6-PGluc = 6-Phosphogluconate
6-PGDH = 6-Phosphogluconic Dehydrogenase
β-NAD = β-Nicotinamide Adenine Dinucleotide, Oxidized Form
β-NADH = β-Nicotinamide Adenine Dinucleotide, Reduced Form

Conditions: T = 25°C, pH 7.5, A340nm, Light path = 1 cm

Method: Continuous Spectrophotometric Rate Determination

Reagents

A. 100 mM Glycylglycine Buffer, pH 7.5 at 25°C. (Prepare 50 ml in deionized water using Glycylglycine, Free Base. Adjust to pH 7.5 at 25°C with 1 M HCl).
B. 100 mM 6-Phosphogluconate Solution (6-PGluc) (Prepare 1 ml in deionized water using 6-Phosphogluconic Acid, Trisodium Salt).
C. 60 mM β-Nicotinamide Adenine Dinucleotide Solution (β-NAD) (Prepare 1 ml in deionized water using β-Nicotinamide Adenine Dinucleotide. *Prepare Fresh*).
D. 6-Phosphogluconic Dehydrogenase Enzyme Solution (Immediately before use, prepare a solution containing 1.5–3.0 units/ml of 6-Phosphogluconic Dehydrogenase in cold deionized water).

Procedure

Prepare (in milliliters) the following reagents into suitable cuvettes:

	Test	Blank
Reagent A (Buffer)	2.75	2.77
Reagent B (6-PGlu)	0.05	0.05
Reagent C (β-NAD)	0.10	0.10

Mix by inversion and equilibrate to 25°C. Monitor the A340nm until constant, using a suitably thermostatted spectrophotometer. Then add:

Reagent D (Enzyme Solution)	0.02	—

Immediately mix by inversion and record the increase in A340nm for approximately 5 minutes. Obtain the ΔA340nm/minute using the maximum linear rate for both the Test and Blank.

Calculations

$$\text{Units / mg enzyme} = \frac{\Delta A340nm / \text{min Test} - \Delta A340nm / \text{min Blank}}{(6.22)\ (\text{mg enzyme / ml RM})}$$

6.22 = Millimolar extinction coefficient of β-NADH at 340nm
RM = Reaction Mix

Unit Definition: *One unit will oxidize 1.0 μmole of 6-phospho-D-gluconate to D-ribulose 5-phosphate and CO_2 per minute to pH 7.5 at 25°C in the presence of β-NAD.*

Final Assay Concentrations

In a 2.92 ml reaction mix, the final concentrations are 94 mM glycylglycine, 1.7 mM 6-phosphogluconate, 2.0 mM β-NAD and 0.03–0.06 unit 6-phosphogluconic dehydrogenase.

Enzymatic Assay of PHOSPHOGLUCOSE ISOMERASE (EC 5.3.1.9)

Principle

$$\text{D-Fructose 6-Phosphate} \xrightarrow{\text{PGI}} \text{D-Glucose 6-Phosphate}$$

$$\text{D-Glucose 6-Phosphate} + \beta\text{-NADP} \xrightarrow{\text{G-6-PDH}} \text{6-Phosphogluconate} + \beta\text{-NADPH}$$

Abbreviations Used

PGI = Phosphoglucose Isomerase
β-NADPH = β-Nicotinamide Adenine Dinucleotide Phosphate, Reduced Form
β-NADP = β-Nicotinamide Adenine Dinucleotide Phosphate, Oxidized Form
G-6-PDH = Glucose-6-Phosphate Dehydrogenase

Conditions: T = 25°C, pH = 7.4, A340nm, Light path = 1 cm

Method: Continuous Spectrophotometric Rate Determination

Reagents

A. 250 mM Glycylglycine Buffer, pH 7.4 at 25°C (Prepare 100 ml in deionized water using Glycylglycine, Free Base. Adjust the pH to 7.4 with 1 M NaOH).

B. 100 mM D-Fructose 6-Phosphate Solution (F-6-P) (Prepare 1 ml in deionized water using D-Fructose 6-Phosphate, Disodium).

C. 20 mM β-Nicotinamide Adenine Dinucleotide Phosphate Solution (NADP) (Prepare 1 ml in deionized water using β-Nicotinamide Adenine Dinucleotide Phosphate, Sodium Salt. *Prepare Fresh*).

D. 100 mM Magnesium Chloride Solution ($MgCl_2$) (Prepare 10 ml in deionized water using Magnesium Chloride, Hexahydrate).

E. Glucose-6-Phosphate Dehydrogenase Enzyme Solution (G-6-PDH) (Immediately before use, prepare a solution containing 50 units/ml of Glucose-6-Phosphate Dehydrogenase in cold deionized water).

F. Phosphoglucose Isomerase Enzyme Solution (PGI) (Immediately before use, prepare a solution containing 0.3–0.7 unit/ml in cold deionized water).

Procedure

Pipette (in milliliters) the following reagents into suitable cuvettes:

	Test	Blank
Deionized Water	2.00	2.00
Reagent A (Buffer)	0.50	0.50
Reagent B (F-6-P)	0.10	0.10
Reagent C (NADP)	0.10	0.10
Reagent D ($MgCl_2$)	0.10	0.10
Reagent E (G-6-PDH)	0.10	0.10

Mix by inversion and equilibrate to 25°C. Monitor the A340nm until constant, using a suitably thermostatted spectrophotometer. Then add:

	Test	Blank
Reagent F (PGI)	0.10	—
Deionized Water	—	0.10

Immediately mix by inversion and record the increase in A340nm for approximately 5 minutes. Obtain the ΔA340nm/minute using the maximum linear rate for both the Test and Blank.

Calculations

$$\text{Units / ml enzyme} = \frac{(\Delta A340\text{nm / min Test} - \Delta A340\text{nm / min Blank})\,(3)\,(df)}{(6.22)\,(0.1)}$$

3 = Total volume (in milliliters) of assay
df = Dilution factor
6.22 = Millimolar extinction coefficient of β-NADPH at 340nm
0.1 = Volume (in milliliter) of enzyme used

$$\text{Units / mg solid} = \frac{\text{units / ml enzyme}}{\text{mg solid / ml enzyme}}$$

$$\text{Units / mg protein} = \frac{\text{units / ml enzyme}}{\text{mg protein / ml enzyme}}$$

Unit Definition: *One unit will convert 1.0 μmole of D-fructose 6-phosphate to D-glucose 6-phosphate per minute at pH 7.4 at 25°C.*

Final Assay Concentration

In a 3.00 ml reaction mix, the final concentrations are 42 mM glycylglycine, 3.3 mM D-fructose 6-phosphate, 0.67 mM β-nicotinamide adenine dinucleotide phosphate, 3.3 mM $MgCl_2$, 5.0 units Glucose-6-phosphate dehydrogenase and 0.03–0.07 unit phosphoglucose isomerase.

Notes

1. This assay procedure is not to be used to assay the activities of Phosphoglucose Isomerase from Bacillus stearothermophilus and Phosphoglucose Isomerase, Insoluble Enzyme attached to Polyacrylamide.
2. Glucose-6-Phosphate Dehydrogenase unit definition: One unit will oxidize 1.0 μmole of D-glucose 6-phosphate to 6-phospho-D-gluconate per minute in the presence of NADP at pH 7.4 at 25°C.

Enzymatic Assay of PHOSPHOGLUCOSE ISOMERASE (EC 5.3.1.9) from *Bacillus stearothermophilus*

Principle

$$\text{D-Fructose 6-Phosphate} \xrightarrow{\text{PGI}} \text{D-Glucose 6-Phosphate}$$

$$\text{D-Glucose 6-Phosphate} + \beta\text{-NADP} \xrightarrow{\text{G-6-PDH}} \text{6-Phosphogluconate} + \beta\text{-NADPH}$$

Abbreviations Used

PGI = Phosphoglucose Isomerase
β-NADPH = β-Nicotinamide Adenine Dinucleotide Phosphate, Reduced Form
β-NADP = β-Nicotinamide Adenine Dinucleotide Phosphate, Oxidized Form
G-6-PDH = Glucose-6-Phosphate Dehydrogenase

Conditions: T = 30°C, pH = 9.0, A340nm, Light path = 1 cm

Method: Continuous Spectrophotometric Rate Determination

Reagents

A. 100 mM Tris HCl Buffer, pH 9.0 at 30°C (Prepare 100 ml in deionized water using Trizma Base. Adjust the pH to 9.0 with 1 M HCl).
B. 100 mM D-Fructose-6-Phosphate Solution (F-6-P) (Prepare 1 ml in deionized water using D-Fructose 6-Phosphate, Disodium Salt).
C. 23 mM β-Nicotinamide Adenine Dinucleotide Phosphate Solution (NADP) (Prepare 1 ml in deionized water using β-Nicotinamide Adenine Dinucleotide Phosphate, Sodium Salt. *Prepare Fresh*).
D. 50 mM Tris HCl Buffer, pH 8.5 at 30°C (Enz Dil) (Prepare 50 ml in deionized water using Trizma Base. Adjust to pH 8.5 at 30°C with 1 M HCl).
E. Glucose-6-Phosphate Dehydrogenase Enzyme Solution (G-6-PDH) (Immediately before use, prepare a solution containing 1000 units/ml of Glucose-6-Phosphate Dehydrogenase in cold deionized water).

F. Phosphoglucose Isomerase Enzyme Solution (PGI) (Immediately before use, prepare a solution containing 0.2–0.4 unit/ml in Reagent D).

Procedure

Prepare a reaction cocktail by pipetting (in milliliters) the following reagents into a suitable container:

Reagent A (Buffer)	18.96
Reagent B (F-6-P)	0.60
Reagent C (NADP)	0.40

Pipette (in milliliters) the following reagents into suitable cuvettes:

	Test	Blank
Reaction Cocktail	2.994	2.994
Reagent E (G-6-PDH)	0.006	0.006

Mix by inversion and equilibrate to 30°C. Monitor the A340nm until constant, using a suitably thermostatted spectrophotometer. Then add:

Reagent F (PGI)	0.10	—
Reagent D (Enz Dil)	—	0.10

Immediately mix by inversion and record the increase in A340nm for approximately 5 minutes. Obtain the ΔA340nm/minute using the maximum linear rate for both the Test and Blank.

Calculations

$$\text{Units / ml enzyme} = \frac{(\Delta A340\text{nm / min Test} - \Delta A340\text{nm / min Blank})\ (3.1)\ (\text{df})}{(6.22)\ (0.1)}$$

3.1 = Total volume (in milliliters) of assay
df = Dilution factor
6.22 = Millimolar extinction coefficient of β-NADPH at 340nm
0.1 = Volume (in milliliter) of enzyme used

$$\text{Units / mg solid} = \frac{\text{units / ml enzyme}}{\text{mg solid / ml enzyme}}$$

$$\text{Units / mg protein} = \frac{\text{units / ml enzyme}}{\text{mg protein / ml enzyme}}$$

Unit Definition: *One unit will convert 1.0 μmole of D-fructose 6-phosphate to D-glucose 6-phosphate per minute at pH 9.0 at 30°C.*

Final Assay Concentrations

In a 3.10 ml reaction mix, the final concentrations are 93 mM Tris, 2.9 mM D-fructose 6-phosphate, 0.44 mM β-nicotinamide adenine dinucleotide phosphate, 6.0 units glucose-6-phosphate dehydrogenase and 0.02–0.04 unit phosphoglucose isomerase.

Note

1. Glucose-6-Phosphate Dehydrogenase Unit Definition: One unit will oxidize 1.0 μmole of D-glucose 6-phosphate to 6-phospho-D-gluconate per minute in the presence of NADP at pH 7.4 at 25°C.

Enzymatic Assay of PHOSPHOGLYCERATE MUTASE (EC 5.4.2.1)

Principle

$$3\text{-PGA} \xrightarrow[\text{DPGA1}]{\text{Phosphoglycerate Mutase}} 2\text{-PGA}$$

$$2\text{-PGA} \xrightarrow{\text{Enolase}} \text{PEP} + H_2O$$

$$\text{PEP} + \text{ADP} \xrightarrow{\text{PK}} \text{Pyruvate} + \text{ATP}$$

$$\text{Pyruvate} + \beta\text{-NADH} \xrightarrow{\text{LDH}} \text{L-Lactate} + \beta\text{-NAD}$$

Abbreviations Used

3-PGA = D(-)3-Phosphoglyceric Acid
DPGA = 2,3-Diphospho-D-Glyceric Acid
2-PGA = 2-Phosphoglyceric Acid
PEP = Phospho(enol) pyruvate
ADP = Adenosine 5'-Diphosphate
PK = Pyruvate Kinase
ATP = Adenosine 5'-Triphosphate
β-NADH = β-Nicotinamide Adenine Dinucleotide, Reduced Form
LDH = L-Lactic Dehydrogenase
β-NAD = β-Nicotinamide Adenine Dinucleotide, Oxidized Form

Conditions: T = 25°C, pH 7.6, A340nm, Light path = 1 cm

Method: Continuous Spectrophotometric Rate Determination

Reagents

A. 100 mM Triethanolamine Buffer, pH 7.6 at 25°C (Prepare 100 ml in deionized water using Triethanolamine Hydrochloride. Adjust to pH 7.6 at 25°C with 1 M NaOH).
B. 200 mM 3-Phosphoglyceric Acid Solution (3-PGA) (Prepare 1 ml in deionized water using D(-)3-Phosphoglyceric Acid, Trisodium Salt).
C. 21 mM Adenosine 5'-Diphosphate Solution (ADP) (Prepare 1 ml in deionized water using Adenosine 5'-Diphosphate, Sodium Salt. *Prepare Fresh*).
D. 40 mM 2,3-Diphosphoglyceric Acid Solution (DPGA) (Prepare 1 ml in deionized water using 2,3-Diphospho-D-Glyceric Acid, Pentacyclohexylammonium Salt. *Prepare Fresh*).
E. 6.4 mM β-Nicotinamide Adenine Dinucleotide, Reduced Form Solution (β-NADH) (Dissolve the contents of one 5 mg vial of β-Nicotinamide Adenine Dinucleotide, Reduced Form, Disodium Salt in the appropriate volume of Reagent A).
F. 50 mM Magnesium Sulfate Solution ($MgSO_4$) (Prepare 2 ml in deionized water using Magnesium Sulfate, Heptahydrate).

G. 2 M Potassium Chloride Solution (KCl) (Prepare 2 ml in deionized water using Potassium Chloride).
H. PK/LDH Enzymes (PK/LDH).
I. Enolase Enzyme Solution (Enolase) (Immediately before use, prepare a solution containing 100 units/ml of Enolase in cold deionized water).
J. Phosphoglycerate Mutase Enzyme Solution (Phosgly Mutase) (Immediately before use, prepare a solution containing 0.3–0.6 unit/ml of Phosphoglycerate Mutase in cold Reagent A).

Procedure

Prepare a reaction cocktail by pipetting (in milliliters) the following reagents into a suitable container:

Reagent A (Buffer)	22.00
Reagent B (3-PGA)	1.00
Reagent C (ADP)	1.00
Reagent D (DPGA)	1.00
Reagent E (β-NADH)	0.70
Reagent F ($MgSO_4$)	1.50
Reagent G (KCl)	1.50

Procedure

Mix by swirling and adjust to pH 7.6 at 25°C with either 1 M NaOH or 1 M HCl, if necessary. Pipette (in milliliters) the following reagents into suitable cuvettes:

	Test	Blank
Reaction Cocktail	2.87	2.87
Reagent H (PK/LDH)	0.02	0.02
Reagent I (Enolase)	0.03	0.03

Mix by inversion and equilibrate to 25°C. Monitor the A340nm until constant, using a suitably thermostatted spectrophotometer. Then add:

Reagent J (Phosgly Mutase)	0.10	—
Reagent A (Buffer)	—	0.10

Immediately mix by inversion and record the decrease in A340nm for approximately 5 minutes. Obtain the ΔA340nm/minute using the maximum linear rate for both the Test and Blank.

Calculations

$$\text{Units / ml enzyme} = \frac{(\Delta A340nm / \min \text{ Test} - \Delta A340nm / \min \text{ Blank})\ (3.20)\ (df)}{(6.22)\ (0.1)}$$

3.20 = Total volume (in milliliters) of assay
df = Dilution factor
6.22 = Millimolar extinction coefficient of β-NADH at 340nm
0.1 = Volume (in milliliter) of enzyme used

$$\text{Units / mg protein} = \frac{\text{units / ml enzyme}}{\text{mg protein / ml enzyme}}$$

Unit Definition: *One unit will convert 1.0 μmole of 3-phosphoglycerate to 2-phosphoglycerate per minute at pH 7.6 at 25°C in the presence of 1.3 mM 2,3-diphosphoglycerate.*

Final Assay Concentrations

In a 3.02 ml reaction mix, the final concentrations are 79 mM triethanolamine, 6.6 mM D(-)3-phosphoglyceric acid, 0.70 mM adenosine 5'-diphosphate, 1.3 mM 2,3-diphospho-D-glyceric acid, 0.15 mM β-nicotinamide adenine dinucleotide, reduced form, 2.5 mM magnesium sulfate, 99 mM potassium chloride, 14 units pyruvate kinase, 20 units L-lactic acid dehydrogenase, 3 units enolase, and 0.03–0.06 unit phosphoglycerate mutase.

Notes

1. 2,3-Diphospho-D-Glyceric Acid is a coenzyme in this reaction.
2. Contains not less than 700 units/ml of Pyruvate Kinase and 1000 units/ml of L-Lactic Dehydrogenase.
3. Pyruvate Kinase Unit Definition: One unit will convert 1.0 μmole of phospho(enol) pyruvate to pyruvate per minute at pH 7.6 at 25°C.
4. L-Lactic Dehydrogenase Unit Definition: One unit will reduce 1.0 μmole of pyruvate to L-lactate per minute at pH 7.5 at 37°C.

Enzymatic Assay of GLYCERALDEHYDE-3-PHOSPHATE DEHYDROGENASE/ 3-PHOSPHOGLYCERIC PHOSPHOKINASE (EC 2.7.2.3)

Principle

$$\text{3-PGA} + \text{ATP} \xrightarrow{\text{3-PGK}} \text{Glycerate-1, 3 Diphosphate} + \text{ADP}$$

$$\text{Glycerate-1, 3 Diphosphate} + \beta\text{-NADH} \xrightarrow{\text{GAPDH}} \text{G-3-P} + \beta\text{-NAD} + \text{Pi}$$

Abbreviations Used

3-PGA = 3-Phosphoglyceric Acid
ATP = Adenosine 5'-Triphosphate
3-PGK = 3-Phosphoglyceric Phosphokinase
ADP = Adenosine 5'-Diphosphate
β-NADH = β-Nicotinamide Adenine Dinucleotide, Reduced Form
GAPDH = Glyceraldehyde-3-Phosphate Dehydrogenase
G-3-P = Glyceraldehyde 3-Phosphate
β-NAD = β-Nicotinamide Adenine Dinucleotide, Oxidized Form
Pi = Inorganic Phosphate

Conditions: T = 25°C, pH = 7.6, A340nm, Light path = 1 cm

Method: Continuous Spectrophotometric Rate Determination

Reagents

A. 100 mM Triethanolamine Buffer with 4.0 mM L-Cysteine, and 0.5 mM Ethylendiaminetetra acetic acid, pH 7.6 at 25°C (Prepare 100 ml in deionized water using Triethanolamine Hydrochloride, L-Cysteine, Hydrochloride, Monohydrate and Ethylenediaminetetraacetic Acid, Disodium Salt, Dihydrate. Adjust to pH 7.6 at 25°C with 1 M NaOH).

B. 100 mM 3-Phosphoglyceric Acid Solution (3-PGA) (Prepare 2 ml in deionized water using D(-)3-Phosphoglyceric Acid, Disodium Salt).

C. 100 mM Magnesium Sulfate Solution ($MgSO_4$) (Prepare 10 ml in deionized water using Magnesium Sulfate, Heptahydrate).

D. 7.0 mM β-Nicotinamide Adenine Dinucleotide, Reduced Form Solution (β-NADH) (Prepare 1 ml in deionized water using β-Nicotinamide Adenine Dinucleotide, Reduced Form, Disodium Salt or dissolve the contents of one 5 mg vial of β-Nicotinamide Adenine Dinucleotide, Reduced Form, Disodium Salt in the appropriate volume of deionized water. *Prepare Fresh*).

E. 34 mM Adenosine 5'-Triphosphate Solution (ATP) (Prepare 1 ml in deionized water using Adenosine 5'-Triphosphate, Disodium Salt. *Prepare Fresh*).

F. 3-Phosphoglyceric Phosphokinase Enzyme Solution (3- PGK) (Immediately before use, prepare a solution containing 200 units/ml in cold deionized water using 3-Phosphoglyceric Phosphokinase).

G. Glyceraldehyde-3-Phosphate Dehydrogenase Enzyme Solution (GAPDH) (Immediately before use, prepare a solution containing 0.3–0.6 unit/ml of Glyceraldehyde-3-Phosphate Dehydrogenase in cold Reagent A).

Procedure

Pipette (in milliliters) the following reagents into suitable cuvettes:

	Test	Blank
Reagent A (Buffer)	2.45	2.55
Reagent B (3-PGA)	0.20	0.20
Reagent C ($MgSO_4$)	0.05	0.05
Reagent D (β-NADH)	0.05	0.05
Reagent E (ATP)	0.10	0.10
Reagent F (3-PGK)	0.05	0.05

Mix by inversion and equilibrate to 25°C. Monitor the A340nm until constant, using a suitably thermostatted spectrophotometer. Then add:

Reaction G (GAPDH)	0.10	—

Immediately mix by inversion and record the decrease in A340nm for approximately 5 minutes. Obtain the ΔA340nm/minute using the maximum linear rate for both the Test and Blank.

Calculations

$$\text{Units/ml enzyme} = \frac{(\Delta A340nm/\min \text{ Test} - \Delta A340nm/\min \text{ Blank})(3)(df)}{(6.22)(0.1)}$$

3 = Total Volume (in milliliters) of assay
df = Dilution factor
6.22 = Millimolar extinction coefficient of β-NADH at 340nm
0.1 = Volume (in milliliter) of enzyme used

$$\text{Units / mg solid} = \frac{\text{units / ml enzyme}}{\text{mg solid / ml enzyme}}$$

$$\text{Units / mg protein} = \frac{\text{units / ml enzyme}}{\text{mg protein / ml enzyme}}$$

Unit Definition: *One unit will reduce 1.0 μmole of 3-phosphoglycerate to D-glyceraldehyde 3-phosphate per minute in a coupled system with 3-phosphoglyceric phosphokinase at pH 7.6 at 25°C.*

Final Assay Concentrations

In a 3.00 ml reaction mix, the final concentrations are 85 mM triethanolamine, 6.7 mM 3-phosphoglyceric acid, 3.4 mM L-cysteine, 0.43 mM Ethylenediaminetetraacetic Acid, 1.7 mM magnesium sulfate, 0.1 mM β-nicotinamide adenine dinucleotide, reduced form, 1.1 mM adenosine 5'-triphosphate, 10 units 3-phosphoglyceric phosphokinase and 0.03–0.06 unit glyceraldehyde-3-phosphate dehydrogenase.

Note

1. 3-Phosphoglyceric Phosphokinase unit definition: One unit will convert 1.0 μmole of 1,3-diphosphoglycerate to 3-phosphoglycerate per minute at pH 6.9 at 25°C.

Enzymatic Assay of PHOSPHOLIPASE A2 (EC 3.1.1.4)

Principle

$$\text{Lecithin} \xrightarrow{\text{Phospholipase A2}} \text{Lysolecithin} + \text{Fatty Acid}$$

Conditions: T = 37°C, pH = 8.0

Method: Titrimetric

Reagents

A. 1000 mM Sodium Chloride Solution (Prepare 100 ml in deionized water using Sodium Chloride).
B. 100 mM Calcium Chloride Solution (Prepare 100 ml in deionized water using Calcium Chloride).
C. 10 mM Sodium Hydroxide Solution-Standardized (NaOH) (Prepare 50 ml in cold deionized water using Sodium Hydoxide. Standardized).
D. 2% Phosphatidylcholine Emulsion (Lecithin) (Prepare by adding 4 grams of Phosphatidylcholine to 30 ml of Reagent A and 10 ml of Reagent B. Stir 2–3 hours at 25°C to form an emulsion. Dilute to 200 ml with deionized water. Titrate with 1 M NaOH until rate of decrease in pH is about 0.01 pH units/minute).
E. Phospholipase A2 Enzyme Solution (Immediately before use, prepare a solution containing 2.5 units/ml of Phospholipase A2 in cold deionized water).

Procedure

Using a suitable pH meter in conjunction with a magnetic stirrer, pipette (in milliliters) the following reagents into a suitable titration vessel:

	Test	Blank
Reagent D (Lecithin)	10.00	10.00

Adjust to pH 8.0 at 37°C. then add:

Deionized Water	—	0.20
Reagent E (Enzyme Solution)	0.20	—

Run the reaction for 10 minutes. Maintain the pH of the reaction mix at pH 8.0 by the addition of small volumes of Reagent C. Record the volume of Reagent C used to maintain the pH.

Calculation

[NaOH] = ml NaOH used Test–ml NaOH used Blank

$$\text{Units / mg enzyme} = \frac{(\text{Molarity of NaOH})\,(\text{NaOH})\,(1000)}{(10)\,(\text{mg enzyme / RM})}$$

1000 = conversion from millimoles to micomoles (Unit definition)
10 = Time of assay (Unit Definition)
RM = Reaction Mix (initial volume = 10.2 ml)

Unit Definition: *One unit will hydrolyze 1.0 μmole of L-aphosphatidylcholine to L-α-lysophosphatidylcholine and a fatty acid per minute at pH 8.0 at 37°C.*

Initial Assay Concentrations

In a 10.2 ml reaction mix, the initial concentrations are 2% L-α-phosphatidylcholine, 147 mM sodium chloride, 4.9 mM calcium chloride and 0.5 units phospholipase A2.

Enzymatic Assay of PHOSPHOLIPASE A2 (I) (EC 3.1.1.4)

Principle

$$\text{L-}\alpha\text{-Lecithin} + H_2O \xrightarrow{\text{Phospholipase A2}} \text{L-}\alpha\text{-Lysolecithin} + \text{Fatty Acid}$$

Abbreviations Used

L-α-Lecithin = L-α-Phosphatidylcholine
L-α-Lysolecithin = L-α-Lysophosphatidylcholine

Conditions: T = 37°C, pH = 8.0

Method: Titrimetric

Reagents

A. 1000 mM Sodium Chloride Solution (Prepare 100 ml in deionized water using Sodium Chloride).

B. 100 mM Calcium Chloride Solution (Prepare 100 ml in deionized water using Calcium Chloride, Dihydrate).
C. 10 mM Sodium Hydroxide Solution-Standardized (NaOH) (Prepare 50 ml in cold deionized water using Sodium Hydroxide. Standardize it).
D. 2.0% (w/v) Phosphatidylcholine Emulsion (Lecithin) (Prepare by dissolving 4 grams of L-α-Phosphatidylcholine in a solution composed of 30 ml of Reagent A, 10 ml of Reagent B and 100 ml of deionized water. Stir 2–3 hours at 25°C to form an emulsion. Dilute to 200 ml with deionized water. Titrate with 1 M NaOH to pH 8.0 and continue to titrate to pH 8.0 (approximately 30 minutes) until the rate of decrease in pH is about 0.01 pH units/minute).
E. Phospholipase A2 Enzyme Solution (Immediately before use, prepare a solution containing 2.5 units/ml of Phospholipase A2 in cold deionized water).

Procedure

Using a suitable pH meter in conjunction with a magnetic stirrer, pipette (in milliliters) the following reagents into a suitable titration vessel:

	Test	Blank
Reagent D (Lecithin)	10.00	10.00

Adjust to pH 8.0 at 37°C if necessary. Then add:

Deionized Water	—	0.20
Reagent E (Enzyme Solution)	0.20	—

Run the reaction for approximately 10 minutes. Maintain the pH of the reaction mix at pH 8.0 by the addition of small volumes (0.05 ml) of Reagent C. Record the volume of Reagent C used to maintain the pH at 8.0 and the time required.

Calculation

[NaOH] = ml NaOH used for Test–ml NaOH used for Blank

$$\text{Units / mg enzyme} = \frac{\text{(Molarity of NaOH) (NaOH) (1000)}}{\text{(T) (mg enzyme / RM)}}$$

1000 = conversion from millimoles to micromoles (Unit definition)
T = Time required to maintain the pH at 8.0 (in minutes)
RM = Reaction Mix

Unit Definition: *One unit will hydrolyze 1.0 μmole of L-aphosphatidylcholine to L-α-lysophosphatidylcholine and a fatty acid per minute at pH 8.0 at 37°C.*

Initial Assay Concentrations

In a 10.2 ml reaction mix, the initial concentrations are 2.0% L-α-phosphatidylcholine, 147 mM sodium chloride, 4.9 mM calcium chloride and 0.5 units phospholipase A2.

Note

1. The overall volume of NaOH addition (enzyme minus blank) should not exceed 0.5 ml (0.05 × 10) in order to obtain linear results.

Enzymatic Assay of PHOSPHOLIPASE A2 (EC 3.1.1.4)

Principle

$$\text{L-}\alpha\text{-Lecithin} + H_2O \xrightarrow{\text{Phospholipase A2}} \text{L-}\alpha\text{-Lysolecithin} + \text{fatty acid}$$

Abbreviations Used

L-α-Lecithin = L-α-Phosphatidylcholine
L-α-Lysolecithin = L-α-Lysophosphatidylcholine

Conditions: T = 37°C, pH 8.5, A570nm, Light path = 1 cm

Method: Colorimetric

Reagents

A. 500 mM Tris HCl Buffer with 10 mM Calcium Chloride, pH 8.5 at 37°C (Prepare 100 ml in deionized water using Trizma Base and Calcium Chloride, Dihydrate. Adjust to pH 8.5 at 37°C with 1 M HCl).
B. 10 mM Calcium Chloride Solution ($CaCl_2$) (Prepare 50 ml in deionized water using Calcium Chloride, Dihydrate).
C. 2% (w/v) L-α-Phosphatidylcholine (L-α-Lecithin) (Prepare 10 ml in Reagent B using L-α-Phosphatidylcholine. This solution is used for both the Substrate and Standard).
D. 1.5% (w/v) Deoxycholate Solution (Prepare 10 ml in deionized water using Deoxycholic Acid, Sodium Salt).
E. 95% Ethanol (Nondenatured).
F. 25% (w/v) Ether (Prepare 20 ml in Reagent E using peroxidase-free Ether).
G. 2 M Hydroxylamine Solution (Prepare 10 ml in deionized water using Hydroxylamine, Hydrochloride).
H. 14% (w/v) Sodium Hydroxide Solution (NaOH) (Prepare 10 ml in deionized water using Sodium Hydroxide).
I. 3 N Hydrochloric Acid Solution (HCl) (Prepare 10 ml in deionized water using Hydrochloric Acid).
J. 10% (w/v) Ferric Chloride solution ($FeCl_3$) (Prepare 10 ml in deionized water using Ferric Chloride, Hexahydrate).
K. Phospholipase A2 Enzyme Solution (Immediately before use, prepare a solution containing 0.01 mg/ml of Phospholipase A2 in cold deionized water).

Procedure

Pipette (in milliliters) the following reagents into 4 dram vials:

	Test	Blank
Reagent A (Buffer)	0.70	0.70
Reagent C (L-α-Lecithin)	0.70	0.70
Reagent D (Deoxycholate Solution)	0.70	0.70

Equilibrate to 37°C. Then add:

	Test	Blank
Reagent K (Enzyme Solution)	0.05	—
Reagent B ($CaCl_2$)	—	0.05

Immediately mix by inversion and incubate at 37°C for 5 minutes at one minute intervals and remove 0.20 ml aliquots.

Test Mixture	0.20	—
Blank Mixture	—	0.20
Reagent F (Ether)	1.50	1.50
Reagent G (Hydroxylamine Solution)	0.20	0.20
Reagent H (NaOH)	0.20	0.20

Incubate at 25°C for 20 minutes. Then add:

	Test	Blank
Reagent I (HCl)	0.30	0.30
Reagent J ($FeCl_3$)	0.30	0.30

Mix by swirling and transfer to suitable cuvettes and record the A570nm for the Test and Blank.

Colorimetric Assay

Standard Curve: (Prepare a standard curve by pipetting (in milliliters) the following reagents into suitable containers.

	Std 1	Std 2	Std 3	Std 4	Std Blank
Reagent A (Buffer)	1.38	1.37	1.35	1.33	—
Reagent C (L-α-Lecithin)	0.02	0.03	0.05	0.07	—
Reagent D (Deoxycholate Soln)	0.70	0.70	0.70	0.70	0.70
Reagent B ($CaCl_2$)	0.05	0.05	0.05	0.05	0.05

Mix and incubate at 37°C for 5 minutes. Then pipette the following into suitable cuvettes.

Standard Mixture	0.20	0.20	0.20	0.20	0.20
Reagent F (Ether)	1.50	1.50	1.50	1.50	1.50
Reagent G (Hydroxylamine)	0.20	0.20	0.20	0.20	0.20
Reagent H (NaOH Soln)	0.20	0.20	0.20	0.20	0.20

Incubate at 25°C for 20 minutes. Then add:

Reagent I (HCl)	0.30	0.30	0.30	0.30	0.30
Reagent J ($FeCl_3$)	0.30	0.30	0.30	0.30	0.30

Mix by swirling and record the A570nm for both the Standards and Blank using a suitable spectrophotometer.

Calculations

$$\Delta A570nm\ Standard = A570nm\ Standard - A570nm\ Standard\ Blank$$

Prepare a standard curve by plotting the A570nm of the Standard vs the micromoles of L-α-Phosphatidylcholine. Use the slope (M) to determine the micromoles of L-α-Phosphatidylcholine of the Test mixture.

$$\Delta A570nm = \frac{A570nm\ Test\ -\ A570nm\ Blank}{Time\ in\ minutes}$$

$$Units\ /\ mg\ enzyme = \frac{A570nm\ /\ min}{(M)\ (0.5)\ (mg\ /\ enzyme\ /\ ml\ RM)\ (0.8)}$$

M = Slope of Std Curve
RM = Reaction Mix
0.5 = Conversion factor since the lysolecithin which is formed has one half the absorbance of lecithin at 570nm.
0.8 = Conversion factor since the MW is assumed to be 800 to obtain micromolar units from mg units.

Unit Definition: *One unit will hydrolyze 1.0 µmole of L-aphosphatidylcholine to L-α-lysophosphatidylcholine and a fatty acid per minute at pH 8.5 at 37°C.*

Final Assay Concentrations

In a 2.15 ml reaction mix, the final concentrations are 163 mM Tris, 0.07% (w/v) L-α-phosphatidylcholine, 0.49% (w/v) deoxycholate and 0.0005 mg phospholipase A2.

Enzymatic Assay of PHOSPHOLIPASE A2 (EC 3.1.1.4)

Principle

$$\text{L-}\alpha\text{-Lecithin} + H_2O \xrightarrow{\text{Phospholipase A2}} \text{L-}\alpha\text{-Lysolecithin} + \text{Fatty Acid}$$

Abbreviations Used

L-α-Lecithin = L-Phosphatidylcholine
L-α-Lysolecithin = L-α-Lysophosphatidylcholine

Conditions: T = 25°C, pH = 8.9

Method: Titrimetric

Reagents

A. 1000 mM Sodium Chloride Solution (Prepare 100 ml in deionized water using Sodium Chloride).
B. 100 mM Calcium Chloride Solution (Prepare 100 ml in deionized water using Calcium Chloride, Dihydrate).
C. 10 mM Sodium Hydroxide Solution-Standardized (NaOH) (Prepare 50 ml in cold deionized water using Sodium Hydroxide, Anhydrous. Standardize it).
D. 2.0% (w/v) Phosphatidylcholine Emulsion (Lecithin) (Prepare by dissolving 4 grams of L-α-Phosphatidylcholine or L-α-Phosphatidycholine in a solution consisting of 30 ml of Reagent A, 10 ml of Reagent B and 100 ml of deionized water. Stir 2–3 hours at 25°C to form an emulsion. Dilute to 200 ml with deionized water. Titrate with 1 M NaOH to pH 8.9 and continue to titrate to pH 8.9 until the rate of decrease in pH is about 0.01 pH units/minute).
E. Phospholipase A2 Enzyme Solution (Immediately before use, prepare a solution containing 2.5 units/ml of Phospholipase A2 in cold deionized water).

Procedure

Using a suitable pH meter in conjunction with a magnetic stirrer, pipette (in milliliters) the following reagents into a suitable titration vessel:

	Test	Blank
Reagent D (Lecithin)	10.00	10.00

Adjust to pH 8.9 at 25°C if necessary. Then add:

Deionized Water	—	0.20
Reagent E (Enzyme Solution)	0.20	—

Run the reaction for approximately 10 minutes. Maintain the pH of the reaction mix at pH 8.9 by the addition of small volumes (0.05 ml) of Reagent C. Record the volume of Reagent C used to maintain the pH at 8.9 and the time required.

Calculations

[NaOH] = ml NaOH used for Test–ml NaOH used for Blank

$$\text{Units / ml enzyme} = \frac{(\text{Molarity of NaOH})\,(\text{NaOH})\,(1000)\,(\text{df})\,(10.20)}{(\text{T})\,(0.20)}$$

df = Dilution factor
1000 = Conversion factor from millimoles to micromoles (Unit definition)
T = Time required to maintain the pH at 8.9 (in minutes)
0.20 = Volume (in milliliter) of enzyme used
10.20 = Total volume (in milliliter) of assay

$$\text{Units / mg solid} = \frac{\text{units / ml enzyme}}{\text{mg solid / ml enzyme}}$$

$$\text{Units / mg protein} = \frac{\text{units / ml enzyme}}{\text{mg protein / ml enzyme}}$$

Unit Definition: *One unit will hydrolyze 1.0 µmole of L-aphosphatidylcholine to L-α-lysophosphatidylcholine and a fatty acid per minute at pH 8.9 at 25°C.*

Initial Assay Concentrations

In a 10.2 ml reaction mix, the initial concentrations are 2.0% (w/v) L-α-phosphatidylcholine, 147 mM sodium chloride, 4.9 mM calcium chloride and 0.5 unit phospholipase A2.

Notes

1. This assay procedure is not to be used to assay Phospholipase A2 from Bee Venom,Insoluble, Phospholipase A2 from Porcine Pancreas, Phospholipase A2 from Bovine Pancreas and Phospholipase from Strepotmyces Violaceoruber.
2. The overall volume of NaOH addition (enzyme minus blank) should not exceed 0.5 ml (0.05 × 10) in order to obtain linear results.

Enzymatic Assay of PHOSPHOLIPASE B (EC 3.1.1.5)

Principle

$$\text{L-}\alpha\text{-Lysophosphatidylcholine} + H_2O \xrightarrow{\text{Phospholipase B}} \text{Glycerophosphorylcholine} + \text{Fatty Acid}$$

$$\text{Glycerophosphorylcholine} + H_2O \xrightarrow{\text{GPC-PDE}} \text{Glycerophosphate} + \text{Choline}$$

$$\text{Choline} + O_2 + H_2O \xrightarrow{\text{Choline Oxidase}} \text{Betaine} + 2H_2O_2$$

$$2H_2O_2 + \text{4-AAP} + \text{Phenol} \xrightarrow{\text{POD}} \text{Quinoneimine dye} + 2H_2O$$

Abbreviations Used

GPC-PDE = Glycerophosphorylcholine Phosphodiesterase
4-AAP = 4-Aminoantipyrine
POD = Peroxidase

Conditions: T = 37°C, pH = 8.0, A500nm, Light path = 1 cm

Method: Colorimetric

Reagents

A. 200 mM Tris HCl Buffer, pH 8.0 at 37°C (Buffer) (Prepare 100 ml in deionized water using Trizma Base. Adjust to pH 8.0 at 37°C with 1 M HCl).
B. 0.5% (w/v) L-α-Lysophosphatidylcholine Solution (LPC) (Prepare 1 ml in deionized water using L-α-Lysophosphatidylcholine).
C. 10 mM Calcium Chloride Solution ($CaCl_2$) (Prepare 10 ml in deionized water using Calcium Chloride, Dihydrate).
D. Glycerophosphorylcholine Phosphodiesterase Enzyme Solution (GPC-PDE) (Prepare a solution containing 1 unit/ml of Glycerophosphorylcholine Phosphodiesterase in cold deionized water).
E. 0.3% (w/v) 4-Aminoantipyrine Solution (4-AAP) (Prepare 5 ml in deionized water using 4- Aminoantipyrine).
F. 0.2% (w/v) Phenol Solution (Phenol) (Prepare 2 ml in deionized water using Phenol).
G. 100 mM Ethylenediaminetetraacetic Acid Solution (EDTA) (Prepare 5 ml in deionized water using Ethylenediaminetetraacetic Acid, Tetrasodium Salt, Hydrate).
H. Choline Oxidase Enzyme Solution (Chol Ox) (Prepare a solution containing 60 units/ml of Choline Oxidase in cold deionized water).
I. Peroxidase Enzyme Solution (POD) (Prepare a solution containing 50 units/ml of Peroxidase in cold deionized water).
J. 10 mM Tris HCl Buffer with 0.05% (w/v) Bovine Serum Albumin, pH 8.0 at 37°C (Enz Dil) (Prepare 10 ml in deionized water using Trizma Base and Albumin, Bovine. Adjust to pH 8.0 at 37°C with 1 M HCl).
K. Phospholipase B Enzyme Solution (Phospholipase B) (Immediately before use, prepare a solution containing 0.1–0.2 unit/ml of Phospholipase B in Reagent J).

Procedure

Prepare reaction cocktail I by pipetting (in milliliters) the following reagents into a suitable container:

Reagent A (Buffer)	1.00
Reagent B (LPC)	0.50
Reagent C ($CaCl_2$)	0.50
Reagent D (GPC-PDE)	1.00
Deionized Water	1.50

Mix by swirling. Adjust to pH 8.0 at 37°C using 1 M NaOH.Prepare reaction cocktail II by pipetting (in milliliters) the following reagents into a suitable container:

Reagent A (Buffer)	1.00
Reagent E (4-AAP)	1.00
Reagent F (Phenol)	1.00
Reagent G (EDTA)	2.00
Reagent H (Chol Ox)	1.00
Reagent I (POD)	1.00
Deionized Water	3.00

Mix by swirling. Adjust to pH 8.0 at 37°C using 1 M NaOH or 1 M HCl. Pipette (in milliliters) the following reagents into suitable containers:

	Test	Blank
Reaction Cocktail I	0.45	0.45

Equilibrate to 37°C. Then add:

Reagent K (Phospholipase B)	0.05	—
Reagent J (Enz Dil)	—	0.05

Immediately mix by inversion and incubate at 37°C for exactly 10 minutes. Then add:

Reaction Cocktail II	1.00	1.00

Mix by inversion and incubate at 37°C for 20 minutes. Then add:

Deionized Water	1.50	1.50

Mix by inversion and record the A500nm for both the Test and Blank using a suitable spectrophotometer.

Calculations

$$\text{Units / ml enzyme} = \frac{(\text{A500nm Test} - \text{A500nm Blank})\,(3.0)\,(\text{df})}{(0.5)\,(12)\,(2)\,(10)\,(0.05)}$$

3.0 = Total volume (in milliliters) of assay
df = Dilution factor
0.5 = Conversion factor derived from the fact that 2 μmoles of H_2O_2 produce 1 μmole of quinoneimine dye
12 = Millimolar extinction coefficient of quinoneimine dye at 500nm
2 = Conversion factor derived from the fact that 1 μmole of L-α-Lysophosphatidylcholine produces 2 μmoles of H_2O_2
10 = Time of assay (in minutes) as per the Unit Definition
0.05 = Volume (in milliliter) of enzyme used

$$\text{Units / mg solid} = \frac{\text{units / ml enzyme}}{\text{mg solid / ml enzyme}}$$

$$\text{Units / mg protein} = \frac{\text{units / ml enzyme}}{\text{mg protein / ml enzyme}}$$

Unit Definition: *One unit will produce 1.0 μmole of glycerophosphorylcholine from egg yolk L-α-lysophosphatidylcholine per minute at pH 8.0 at 37°C.*

Final Assay Concentration

In a 0.50 ml reaction mixture, the final concentrations are 41 mM Tris, 0.05% (w/v) L-α-lysophosphatidylcholine, 1 mM calcium chloride, 0.1 unit glycerophosphorylcholine phosphodiesterase, 0.005% (w/v) Bovine Serum Albumin and 0.005–0.01 unit phospholipase B.

Notes

1. Glycerophosphorylcholine Phosphodiesterase Unit Definition: One unit will produce 1.0 μmole of choline from L-α-glycerophosphorylcholine per minute at pH 8.0 at 37°C.
2. Choline Oxidase Unit Definition: One unit will form 1.0 μmole of H_2O_2 with oxidation of 1 μmole of choline to betaine aldehyde per minute at pH 8.0 at 37°C.
3. Peroxidase Unit Definition: One unit will form 1.0 mg purpurogallin from pyrogallol in 20 seconds at pH 6.0 at 20°C.

Enzymatic Assay of PHOSPHOLIPASE C (EC 3.1.4.3)

Principle

$$\text{Lecithin} + H_2O \xrightarrow{\text{Phospholipase C}} \text{Diglyceride} + \text{Choline Phosphate}$$

$$\text{Choline Phosphate} + H_2O \xrightarrow{\text{Alkaline Phosphatase}} \text{Choline} + \text{Pi}$$

$$\text{Choline} + O_2 \xrightarrow{\text{Choline Oxidase}} \text{Betaine Aldehyde} + H_2O_2$$

$$\text{Betaine Aldehyde} + O_2 \xrightarrow{\text{Choline Oxidase}} \text{Betaine} + H_2O_2$$

$$2H_2O_2 + \text{Phenol} + \text{4-Aminoantipyrine} \xrightarrow{\text{Peroxidase}} 4H_2O + \text{Red Dye}$$

Abbreviations Used

Lecithin = L-α-Phosphatidylcholine
Pi = Inorganic Phosphate

Conditions: T = 37°C, pH = 7.3, A500nm, Light path = 1 cm

Method: Colorimetric

Reagents

A. 100 mM Dimethylglutaric Acid Buffer, pH 7.3 at 37°C (Prepare 100 ml in deionized water using 3,3-Dimethylglutaric Acid. Adjust to pH 7.3 at 37°C with 2 M NaOH).
B. 5.0% (v/v) Triton X-100 Solution (Prepare 25 ml in deionized water using Triton X-100).
C. 26 mM Lecithin Substrate Solution (Lecithin) (Prepare 25 ml in Reagent B using L-α-Phosphatidylcholine).
D. 1 M Tris Buffer with 0.2% (v/v) SDS, pH 8.0 at 37°C (Tris-SDS) (Prepare 100 ml in deionized water using Trizma Base and Lauryl Sulfate, Sodium Salt. Adjust to pH 8.0 at 37°C using 5 M HCl).

E. 50 mM Tris Buffer, pH 8.0 at 37°C (Prepare 100 ml in deionized water using Trizma Base. Adjust to pH 8.0 at 37°C using 1 M HCl).
F. 150 mM 4-Aminoantipyrine Solution (4-AAP) (Prepare 1 ml in deionized water using 4-Aminoantipyrine).
G. 2.0% (w/v) Phenol Solution (Phenol) (Prepare 1 ml in deionized water using Phenol).
H. Choline Oxidase Enzyme Solution (Choline Oxidase) (Immediately before use, prepare a solution containing 50 units/ml in Reagent E using Choline Oxidase).
I. Peroxidase Enzyme Solution (Peroxidase) (Immediately before use, prepare a solution containing 2500 units/ml in deionized water using Peroxidase).
J. Alkaline Phosphatase Enzyme Solution (Alkaline Phosphatase) (Immediately before use, prepare a solution containing 1440 units/ml in deionized water using Phosphatase, Alkaline).
K. 11 mM Dimethylglutaric Acid Buffer with 0.1% (w/v) Bovine Serum Albumin, pH 7.3 at 37°C (Enzyme Diluent) (Prepare 100 ml in deionized water using 3,3-Dimethylglutaric Acid and Albumin, Bovine or equivalent. Adjust to pH 7.3 at 37°C with 2 M NaOH).
L. Phospholipase Enzyme Solution (Phospholipase) (Immediately before use, prepare a solution containing 0.2 unit/ml in cold Reagent K).

Procedure

Prepare the Chromogen Solution by pipetting the following reagents into a suitable container (in milliliters):

Reagent E (Tris Buffer)	4.50
Reagent F (4-AAP)	0.05
Reagent G (Phenol)	0.05
Reagent H (Choline Oxidase)	0.30
Reagent I (Peroxidase)	0.02
Deionized Water	0.08

Mix by swirling. Pipette the following reagents into suitable vials (in milliliter):

	Test	Blank
Deionized Water	0.20	0.20
Reagent A (Buffer)	0.40	0.40
Reagent C (Lecithin)	0.30	0.30

Mix by inversion and equilibrate to 37°C. Then add:

Reagent L (Phospholipase)	0.05	—

Mix by inversion and incubate at 37°C for exactly 10 minutes Then add:

Reagent D (Tris-SDS)	1.00	1.00
Reagent L (Phospholipase)	—	0.05

Mix by inversion.

Colorimetric Assay

Add the following reagents:

	Test	Blank
Chromogen Solution	1.00	1.00
Reagent J (Alkaline Phosphatase)	0.01	0.01

Mix by inversion and incubate at 37°C for 20 minutes. Cool to 25°C and transfer the solutions to suitable cuvettes. Read the A500nm for the Test and Blank.

Calculations

$$\text{Units / ml enzyme} = \frac{(\text{A500nm Test} - \text{A500nm Blank})\ (2.96)\ (\text{df})}{(12)\ (10)\ (0.05)}$$

2.96 = Total volume (in milliliters) of assay
df = Dilution factor
12 = Millimolar Extinction Coefficient of the Red dye
10 = Time of assay in (minutes) as per the Unit Definition
0.05 = Volume (in milliliter) of enzyme used

$$\text{Units / mg solid} = \frac{\text{units / ml enzyme}}{\text{mg solid / ml enzyme}}$$

$$\text{Units / mg protein} = \frac{\text{units / ml enzyme}}{\text{mg protein / ml enzyme}}$$

Unit Definition: *One unit will liberate 1.0 micromole of water soluble organic phosphorus from L-α-phosphatidylcholine per minute at pH 7.3 at 37°C.*

Final Assay Concentration

In a 0.95 ml reaction mix, the final concentrations are 43 mM dimethylglutaric acid, 0.005% (w/v) Bovine Serum Albumin, 8.2 mM lecthin, 1.6% (v/v) Triton X-100 and 0.01 unit phospholipase C.

Notes

1. Alkaline Phosphatase Unit Definition: One unit will hydrolyze 1.0 µmole of p-nitrophenyl phosphate per minute at 37°C.
2. Choline Oxidase Unit Definition: One unit will form 1.0 µmole of H_2O_2 from choline and H_2O per minute at pH 8.0 at 37°C.
3. Peroxidase Unit Definition: One unit will form 1.0 mg purpurogallin from pyrogallol in 20 seconds at pH 6.0 at 20°C.

Enzymatic Assay of PHOSPHOLIPASE C (EC 3.1.4.3) from *Bacillus cereus*

Principle

$$\text{Lecithin} + H_2O \xrightarrow{\text{Phospholipase C}} \text{Diglyceride} + \text{Choline Phosphate}$$

$$\text{Choline Phosphate} + H_2O \xrightarrow{\text{Alkaline Phosphatase}} \text{Choline} + \text{Pi}$$

$$\text{Choline} + O_2 \xrightarrow{\text{Choline Oxidase}} \text{Betaine} + H_2O_2$$

$$\text{Betaine Aldehyde} + O_2 \xrightarrow{\text{Choline Oxidase}} \text{Betaine} + H_2O_2$$

$$2H_2O_2 + \text{PHenol} + 4\text{-Aminoantipyrine} \xrightarrow{\text{Peroxidase}} 4H_2O + \text{Quinoneimine Dye}$$

Abbreviations Used

Lecithin = L-α-Phosphatidylcholine
Pi = Inorganic Phosphate

Conditions: T = 37°C, pH = 7.3, A500nm, Light path = 1 cm

Method: Colorimetric

Reagents

A. 100 mM Dimethylglutaric Acid Buffer, pH 7.3 at 37°C (Prepare 100 ml in deionized water using 3,3-Dimethylglutaric Acid. Adjust to pH 7.3 at 37°C with 2 M NaOH).
B. 5.0% (v/v) Triton2 X-100 Solution (Prepare 25 ml in deionized water using Triton X-100).
C. 4.0% (w/v) Lecithin Substrate Solution (Lecithin) (Prepare 25 ml in Reagent B using L-α-Phosphatidylcholine. *Prepare Fresh*).
D. 1 M Tris Buffer with 0.2% (w/v) Sodium Dodecyl Sulfate, pH 8.0 at 37°C (Tris-SDS) (Prepare 100 ml in deionized water using Trizma Base and Lauryl Sulfate, Sodium Salt. Adjust to pH 8.0 at 37°C using 5 M HCl).
E. 50 mM Tris Buffer, pH 8.0 at 37EC (Tris Buffer) (Prepare 100 ml in deionized water using Trizma Base. Adjust to pH 8.0 at 37°C using 1 M HCl).
F. 148 mM 4-Aminoantipyrine Solution (4-AAP) (Prepare 1 ml in deionized water using 4-Aminoantipyrine).
G. 2.0% (w/v) Phenol Solution (Phenol) (Prepare 1 ml in deionized water using Phenol).
H. Choline Oxidase Enzyme Solution (Choline Oxidase) (Immediately before use, prepare a solution containing 50 units/ml in Reagent E using Choline Oxidase).
I. Peroxidase Enzyme Solution (Peroxidase) (Immediately before use, prepare a solution containing 2500 units/ml in deionized water using Peroxidase).
J. Alkaline Phosphatase Enzyme Solution (Alkaline Phosphatase) (Immediately before use, prepare a solution containing 1440 units/ml in deionized water using Phosphatase, Alkaline).
K. 10 mM Dimethylglutaric Acid Buffer with 0.1% (w/v) Bovine Serum Albumin, pH 7.3 at 37°C (Enzyme Diluent) (Prepare 100 ml in deionized water using 3,3-Dimethylglutaric Acid and Albumin, Bovine. Adjust to pH 7.3 at 37°C with 2 M NaOH).
L. Phospholipase Enzyme Solution (Phospholipase) (Immediately before use, prepare a solution containing 0.08–0.2 unit/ml in cold Reagent K).

Procedure

Prepare the Chromogen Solution by pipetting (in milliliters) the following reagents into a suitable container:

Reagent E (Tris Buffer)	4.50
Reagent F (4-AAP)	0.05
Reagent G (Phenol)	0.05
Reagent H (Choline Oxidase)	0.30
Reagent I (Peroxidase)	0.02
Deionized Water	0.08

Mix by swirling. Pipette (in milliliters) the following reagents into suitable containers:

Reaction Cocktail	
Deionized water	2.00
dReagent A (Buffer)	4.00
Reagent C (Lecithin)	3.00

Mix by inversion and adjust pH to 7.3 at 37°C with either 1 N HCl or 1 N NaOH. Pipette (in milliliters) the following reagents into a suitable vial:

	Test	Blank
Reaction Cocktail	0.90	0.90

Mix by inversion and equilibrate to 37°C. Then add:

Reagent L (Phospholipase C)	0.05	—

Mix by inversion and incubate at 37°C for exactly 10 minutes. Then add:

Reagent D (Tris-SDS)	1.00	1.00
Reagent L (Phospholipase C)	—	0.05

Mix by inversion. Then add:

Chromogen Solution	1.00	1.00
Reagent J (Alkaline Phosphatase)	0.01	0.01

Mix by inversion and incubate at 37°C for 20 minutes. Cool to 25°C and filter through a AP-20 borosilicate glass fiber, pore sizes: 0.45 μm membrane and a Cellulose Acetate, pore size: 0.20 μm membrane combination, respectively. Transfer solutions to suitable cuvettes and read the A500nm for the Test and Blank.

Calculations

$$\text{Units / ml enzyme} = \frac{(\text{A500nm Test} - \text{A500nm Blank})\ (2.96)\ (\text{df})}{(12)\ (10)\ (0.05)}$$

2.96 = Total volume (in milliliters) of assay
df = Dilution factor
12 = Millimolar extinction coefficient4 of quinoneimine dye at 500nm
10 = Time (in minutes) of the assay as per the Unit Definition
0.05 = Volume (in milliliter) of enzyme used

$$\text{Units / mg solid} = \frac{\text{units / ml enzyme}}{\text{mg solid / ml enzyme}}$$

$$\text{Units / mg protein} = \frac{\text{units / ml enzyme}}{\text{mg protein / ml enzyme}}$$

Unit Definition: *One unit will liberate 1.0 micromole of water soluble organic phosphorus from L-α-phosphatidylcholine per minute at pH 7.3 at 37°C.*

Final Assay Concentration

In a 0.95 ml reaction mix, the final concentrations are 42 mM dimethylglutaric acid, 1.6% (v/v) Triton X-100, 0.63% (w/v) lecithin, and 0.004–0.01 unit phospholipase C.

Notes

1. This product is also assayed using L-α-Phosphatidylcholine Type IX-E, from Fresh Frozen Egg Yolk.
2. Alkaline Phosphatase Unit Definition: One unit will hydrolyze 1.0 µmole of p-nitrophenyl phosphate per minute at 37°C.
3. Choline Oxidase Unit Definition: One unit will form 1.0 µmole of H_2O_2 from choline and H_2O per minute at pH 8.0 at 37°C.
4. Peroxidase Unit Definition: One unit will form 1.0 mg purpurogallin from pyrogallol in 20 seconds at pH 6.0 at 20°C.

Enzymatic Assay of PHOSPHOLIPASE C (EC 3.1.4.3) from *C. perfringens*

Principle

$$\text{L-}\alpha\text{-Lecithin} + H_2O \xrightarrow{\text{Phospholipase C}} \text{1,2-Diglyceride} + \text{Choline Phosphate}$$

$$\text{Choline phosphate} + H_2O \xrightarrow{\text{Alkaline Phosphatase}} \text{Choline} + \text{Pi}$$

Abbreviations Used

L-α-Lecithin = L-α-Phosphatidylcholine
Pi = Inorganic Phosphate

Conditions: T = 37°C, pH 7.3, A660nm, Light path = 1 cm

Method: Colorimetric

Reagents

A. 50 mM Tris Maleate Buffer, pH 7.3 at 37°C (Prepare 100 ml in deionized water using Trizma Maleate. Adjust to pH 7.3 at 37°C with 10 M NaOH).
B. 50 mM Calcium Chloride Solution ($CaCl_2$) (Prepare 25 ml in Reagent A using Calcium Chloride, Dihydrate).
C. 2.0% (w/v) L-α-Phosphatidylcholine (Lecithin) (Prepare 25 ml in deionized water using L-α-Phosphatidylcholine, from Fresh Frozen Egg Yolk).
D. 50 mM Tris Maleate Buffer with 1.0% (w/v) Bovine Serum Albumin, pH 7.3 at 37°C (Enzyme Dil) (Prepare 25 ml in Reagent A using Albumin, Bovine).

E. 270 mM Ethylenediaminetetraacetic Acid Solution, pH 7.3 at 37°C (EDTA) (Prepare 25 ml in deionized water using Ethylenediaminetetraacetic Acid, Tetrasodium Salt, Hydrate. Adjust to pH 7.3 at 37°C with 5 M HCl).

F. Alkaline Phosphatase Enzyme Solution (Alk Phos) (Immediately before use, prepare a solution containing 40 units/ml in deionized water using Phosphatase, Alkaline).

G. 20% (w/v) Lauryl Sulfate Solution (SDS) (Prepare 25 ml in deionized water using Lauryl Sulfate, Sodium Salt).

H. Phosphorus Std (P Std) (Use Phosphorus Standard Solution. The Phosphorus concentration is 20 μg/ml, 0.645 μmole).

I. 10% (w/v) Ascorbic Acid Solution (Ascorbic Acid) (Prepare 25 ml in deionized water using L-Ascorbic Acid, Sodium Salt).

J. 4.2% (w/v) Molybdic Acid Solution (Molyb Acid) (Prepare 25 ml in 10 N H_2SO_4 using Molybdic Acid, Ammonium Salt, Tetrahydrate and Sulfuric Acid).

K. Ames Color Reagent (Clr Rgt) (Prepare by combining 10 ml of Reagent I (Ascorbic Acid), 6 ml Reagent J (Molyb Acid) and 54 ml of deionized water. Mix by swirling and store in the dark at room temperature. This reagent should be prepared 30 minutes before use).

L. Phospholipase C Enzyme Solution (Phospholipase C) (Immediately before use, prepare a solution containing 0.1–1.0 unit/ml in cold Reagent D).

Procedure

Pipette (in milliliters) the following reagents into 4 dram vials:

	Test	Test Blank	Std 1	Std 2	Std 3	Std 4	Std Blank
Reagent A (Buffer)	2.00	2.00	2.00	2.00	2.00	2.00	2.00
Reagent B ($CaCl_2$)	0.50	0.50	0.50	0.50	0.50	0.50	0.50
Reagent C (Lacithin)	1.50	1.50	1.50	1.50	1.50	1.50	1.50
Reagent D (Enzyme Dil)	0.90	0.90	—	—	—	—	—
Reagent H (P Std)	—	—	0.25	0.50	0.75	1.00	—
Deionized Water	—	—	0.75	0.50	0.25	—	1.00

Mix by swirling and equilibrate at 37°C. Then Add:

Reagent L (Phospholipase C)	0.10	—	—	—	—	—	—

Mix by swirling and incubate at 37°C for exactly 15 minutes. Then add:

Reagent E (EDTA)	0.90	0.90	0.90	0.90	0.90	0.90	0.90

Mix by swirling. Then add:

Reagent L (Phospholipase C)	—	0.10	—	—	—	—	—
Reagent F (Alk Phos)	0.10	0.10	0.10	0.10	0.10	0.10	0.10

Mix by swirling and incubate at 37°C for 120 minutes. Pipette (in milliliters) the following reagents into 4 dram vials:

Test Solution	1.00	—	—	—	—	—	—
Test Blank Solution	—	1.00	—	—	—	—	—
Standard 1	—	—	1.00	—	—	—	—
Standard 2	—	—	—	1.00	—	—	—
Standard 3	—	—	—	—	1.00	—	—
Standard 4	—	—	—	—	—	1.00	—
Standard Blank	—	—	—	—	—	—	1.00

Mix by swirling. Then add:

Reagent G (SDS)	0.50	0.50	0.50	0.50	0.50	0.50	0.50
Reagent K (Clkr Rgt)	3.00	3.00	3.00	3.00	3.00	3.00	3.00

Mix by swirling and incubate at 37°C for 60 minutes. Centrifuge and transfer the Test, Test Blank, Standards, and Standard Blank to suitable cuvettes. Read the A660nm for each of the samples and blanks using a suitable spectrophotometer.

Calculations

Standard Curve: ΔA660nm Standard = A660nm Standard–A660nm Standard Blank

Prepare a Standard curve by plotting the ΔA660nm Standard versus micromoles of phosphorus.

Sample Determination: ΔA660nm Test = A660nm Test–A660nm Test Blank

Determine the micromoles of Phosphorus liberated using the Standard curve.

$$\text{Units / ml enzyme} = \frac{(\mu\text{moles of phosphate released})\ (\text{df})}{(15)\ (0.1)}$$

df = Dilution factor

15 = Time of reaction (in minutes) as per the Unit Definition

0.1 = Volume (in milliliter) of enzyme used

$$\text{Units / mg solid} = \frac{\text{units / ml enzyme}}{\text{mg solid / ml enzyme}}$$

$$\text{Units / mg protein} = \frac{\bullet\text{units / ml enzyme}}{\text{mg protein / ml enzyme}}$$

Unit Definition: *One unit will liberate 1.0 μmole of water soluble organic phosphorus from L-α-phosphatidylcholine per min at pH 7.3 at 37°C.*

Final Assay Concentrations

In a 5.00 ml reaction mix, the final concentrations are 35 mM Tris maleate, 5 mM calcium chloride, 0.6% (w/v) L-α-phosphatidylcholine, 0.2% (w/v) Bovine Serum Albumin, and 0.01–0.10 unit phospholipase c.

Notes

1. Break up any large lumps of L-α-phosphatidylcholine and dissolve in deionized water. Sonicate at 30 second intervals for approximately 5 minutes or until the product is completely dissolved. The resulting suspension should be homogeneous.
2. Alkaline Phosphatase Unit Definition: One unit will hydrolyze 1.0 μmole of p-nitrophenyl phosphate per minute at pH 9.8 at 37°C.

Enzymatic Assay of PHOSPHOLIPASE C, PHOSPHATIDYLINOSITOL-SPECIFIC (EC 3.1.4.10)

Principle

Step 1

$$\underset{\text{(membrane stroma bound)}}{\text{Acetylcholinesterase}} \xrightarrow{\text{PLP C}} \underset{\text{(unbound)}}{\text{Acetylcholinesterase}}$$

Step 2

$$\text{ATI} + \text{DTNB} \xrightarrow{\text{Acetylcholinesterase}} \text{ATI derivative} + \text{TNB}$$

Abbreviations Used

PLP C = Phospholipase C, Phosphatidylinositol-Specific
ATI = Acetylthiocholine Iodide
DTNB = 5,5'-Dithio-bis(2-Nitrobenzoic Acid)
TNB = 5-Thio-2-Nitrobenzoic Acid

Conditions: T = 30°C, pH = 7.4, A412nm, Light path = 1 cm

Method: Stopped Spectrophotometric Rate Determination

Reagents

A. 10 mM Tris-HCl with 144 mM Sodium Chloride Buffer, pH 7.4 at 30°C (Prepare 100 ml in deionized water using Trizma Hydrochloride and Sodium Chloride. Adjust to pH 7.4 at 30°C with 1 M NaOH).

B. 200 mM Sodium Acetate Solution, pH 4.5 at 30°C (NaOAc) (Prepare 50 ml in deionized water using Sodium Acetate, Trihydrate. Adjust to pH 4.5 at 30°C with 1 M HCl).

C. 100 mM Sodium Phosphate with 0.1% (v/v) Triton X-100 Solution, pH 7.4 at 30°C (Phosphate/X-100) (Prepare 50 ml in deionized water using Sodium Phosphate, Monobasic, Anhydrous and Triton X-100. Adjust to pH 7.4 at 30°C with 1 M NaOH).

D. 1 mM Acetylthiocholine Iodide with 0.2 mM 5,5'-Dithio-bis(2-Nitrobenzoic Acid) Solution (ATI/DTNB) (Prepare 50 ml in Reagent C using Acetylthiocholine Iodide and 5,5'- Dithio-bis(2-Nitrobenzoic Acid). Prepare this solution fresh every 4 hours since there is a spontaneous yellowing of the solution due to degeneration).

E. 0.001% (v/v) Phenylmethylsulfonyl Fluoride Solution (PMSF) (Prepare 100 ml in Isopropanol, Anhydrous using Phenylmethylsulfonyl Fluoride).

F. Acetylcholinesterase Membrane-Bound Enzyme Solution (AChE) (Immediately before use, prepare a solution containing 6.25 mg/ml in Reagent A using Cholinesterase, Acetyl. Equilibrate to 37°C prior to use).

G. 0.05% (w/v) Bovine Serum Albumin Solution (Enz Dil) (Prepare 20 ml in Reagent A using Albumin, Bovine).

H. Phospholipase C, Phosphatidylinositol-Specific Enzyme Solution (PLPC) (Immediately before use, prepare a solution containing 0.05–0.10 unit/ml of Phospholipase C, Phosphatidylinositol-Specific in Reagent G).

Procedure

Step 1: Pipette (in milliliters) the following reagents into suitable microcentrifuge tubes:

	Test	Blank
Reagent G (Enz Dil)	—	0.10
Reagent H (PLPC)	0.10	—

Equilibrate to 30°C. Then add:

Reagent F (AChE)	0.10	0.10

Mix by swirling and incubate at 30°C for exactly 10 minutes. Then add:

Reagent B (NaOAc)	0.20	0.20

Mix by swirling and incubate at 4°C for 15 minutes. Centrifuge for 3 minutes. Place the supernatant from the Test and Blank into separate microcentrifuge tubes.

Step 2: Pipette (in milliliters) the following reagents into suitable containers:

Reagent D (ATI/DTNB)	1.45	1.45

Equilibrate to 30°C. Then add:

Test Supernatant	0.05	—
Blank Supernatant	—	0.05

Mix by swirling and incubate for exactly 10 minutes. Then add:

Reagent E (PMSF)	0.10	0.10

Mix by swirling and transfer the Test and Blank to suitable cuvettes. Record the A412nm for both the Test and Blank using a suitable spectrophotometer.

Calculations

$$\text{Units / ml enzyme} = \frac{(\text{A412nm Test} - \text{A412nm Blank})\,(0.4)\,(1.6)\,(\text{df})}{(0.05)\,(10)\,(13.6)\,(0.1)}$$

1.6 = Total volume (in milliliters) of assay in Step 2
0.4 = Total Volume in (in milliliters) of assay in Step 1
df = Dilution factor
0.05 = Volume (in milliliter) of Step 1 used in Step 2
10 = Time of assay (in minutes) as per the Unit Definition
13.6 = Millimolar extinction coefficient of TNB at 412nm
0.1 = Volume (in milliliter) of enzyme used

$$\text{Units / mg solid} = \frac{\text{units / ml enzyme}}{\text{mg solid / ml enzyme}}$$

$$\text{Units / mg protein} = \frac{\text{units / ml enzyme}}{\text{mg protein / ml enzyme}}$$

Unit Definition: *One unit will liberate 1.0 unit of acetylcholinesterase per minute from a membrane-bound crude preparation at pH 7.4 at 30°C (10 minute incubation).*

Final Assay Concentrations

In a 0.20 ml reaction mix, the final concentrations are 10 mM Tris, 144 mM sodium chloride, 0.625 mg acetylcholinesterase, membrane-bound, 0.02% (w/v) Bovine Serum Albumin, and 0.005–0.010 unit phospholipase C, phosphatidylinositol-specific.

Note

1. Reagent F (AChE) should be equilibrated to 30°C before adding to the reaction mixture.

Enzymatic Assay of PHOSPHOLIPASE D (EC 3.1.4.4)

Principle

$$\text{L-}\alpha\text{-Phosphatidylcholine} + 2H_2O \xrightarrow{\text{Phospholipase D}} \text{Choline} + \text{Phosphatidic Acid}$$

$$2\ \text{Choline} + O_2 \xrightarrow{\text{Choline Oxidase}} \text{Betaine Aldehyde} + H_2O_2$$

$$2H_2O_2 + \text{4-AAP} + \text{Phenol} \xrightarrow{\text{Peroxidase}} 4H_2O + \text{Quinoneimine Dye}$$

Abbreviation Used

4-AAP = 4-Aminoantipyrine

Conditions: T = 30°C, pH = 5.6, A500nm, Light path = 1 cm

Method: Colorimetric

Reagents

A. 50 mM Sodium Lauryl Sulfate Solution (SDS) (Prepare 10 ml in deionized water using Lauryl Sulfate, Sodium Salt).
B. 1 M Sodium Acetate Buffer, pH 5.6 at 30°C (NaOAc) (Prepare 100 ml in deionized water using Sodium Acetate, Trihydrate. Adjust to pH 5.6 at 30°C with 1 M HCl).
C. 17.9% (v/v) Ethanol Solution (EtOH) (Prepare 1 ml in deionized water using 200 Proof USP Ethyl Alcohol).
D. 0.46% (w/v) L-α-Phosphatidylcholine Substrate Solution (Prepare by transferring 2.2 ml (220 mg) of L-α-Phosphatidylcholine to a 50 ml Erlenmeyer flask. Evaporate off the hexane by bubbling nitrogen gas through the liquid. Place the Erlenmeyer flask containing the substrate into a desiccator connected to a vacuum line for 4 hours. Add in order: 3 ml of Reagent A (SDS), 6 ml of Reagent B (Buffer), and 39 ml of deionized water. Mix, using a magnetic stirrer, until a uniform suspension is obtained. Add 0.272 ml of Reagent C (EtOH) to obtain a 0.1% (v/v) ethanol concentration in the substrate solution. *Prepare Fresh*).
E. 500 mM Calcium Chloride Solution ($CaCl_2$) (Prepare 25 ml in deionized water using Calcium Chloride, Dihydrate).
F. 100 mM Tris HCl Buffer, pH 8.0 at 30°C (Prepare 100 ml in deionized water using Trizma Base. Adjust to pH 8.0 at 30°C with 1 M HCl).

G. 10 mM Tris HCl Buffer with 2 mM Ethylenediaminetetraacetic Acid and 1.0% (w/v) Potassium Chloride (Enzyme Diluent) (Prepare 10 ml in Reagent F using Ethylenediaminetetraacetic Acid, Disodium Salt, Dihydrate and Potassium Chloride).
H. Choline Oxidase Enzyme Solution (COD) (Prepare a solution containing 10 units/ml of Choline Oxidase in cold Reagent G).
I. 1 mM Choline Chloride Standard (Chol Std Soln) (Prepare 50 ml in deionized water using Choline Chloride Salt. *Prepare Fresh*).
J. Choline Color Reagent Mixture (CCRM) (Prepare by dissolving 39 mg of 4-Aminoantipyrine, Free Base 80 mg of Phenol and 8 mg of Peroxidase in 5.5 ml of Reagent F (Tris Buffer, pH 8.0). Store in an amber bottle to protect from light).
K. 2 M Tris HCl Buffer, pH 9.0 at 25°C (Prepare 25 ml in deionized water using Trizma Base. Adjust to pH 9.0 at 25°C with 1 M HCl).
L. Phospholipase D Enzyme Solution (PLD) (Immediately before use, prepare a solution containing 10–20 units/ml in cold deionized water).

Procedure

Pipette (in milliliters) the following reagents into suitable containers:

	Test	Blank
Reagent D (Substrate Soln)	2.40	2.40
Reagent E ($CaCl_2$)	0.30	0.30
Deionized Water	0.20	0.30

Mix by swirling and equilibrate to 30°C using a thermostatted water bath. Then add:

Reagent L (PLD)	0.10	—

Immediately mix by swirling and incubate the containers for exactly 10 minutes at 30°C. The containers should be swirled several times during the reaction. At the end of 10 minutes, transfer the Test and Blank to a boiling water bath. Remove tubes from the water bath after 5 minutes and let cool to room temperature. Add 0.05 ml of Reagent K (Tris HCl Buffer). Mix, centrifuge and filter both Test and Blank through a 0.45 μm filter. Pipette (in milliliters) the following reagents into suitable containers.

Test Filtrate	2.00	—
Blank filtrate	—	2.00
Reagent J (CCRM)	0.10	0.10
Reagent H (COD)	0.10	0.10

Mix by inversion and let stand 2–3 hours at room temperature. Then add:

Deionized Water	2.00	2.00

Centrifuge to clarify and then transfer the solutions to suitable cuvettes. Record the A500nm for both Test and Blank using a suitable spectrophotometer.

Colorimetric Assay

Standard Curve: Pipette (in milliliters) the following reagents into suitable containers:

	Std 1	Std 2	Std 3	Std 4	Std Blank
Reagent D (Substrate Soln)	2.40	2.40	2.40	2.40	2.40
Reagent E ($CaCl_2$)	0.30	0.30	0.30	0.30	0.30
Reagent I (Chol Std Soln)	0.05	0.10	0.20	0.30	—
Deionized Water	0.25	0.20	0.10	—	0.30

Mix vigorously by vortexing and then place Standard and Standard Blank in a boiling water bath. Remove tubes after 5 minutes from the water bath and let cool to room temperature. Add 0.05 ml of Reagent K (Tris HCl Buffer), centrifuge and filter the Standards and Standard Blank through a 0.45 μm filter. Pipette (in milliliters) the following reagents into suitable containers:

Std 1 Filtrate	2.00	—	—	—	—
Std 2 Filtrate	—	2.00	—	—	—
Std 3 Filtrate	—	—	2.00	—	—
Std 4 Filtrate	—	—	—	2.00	—
Blank Filtrate	—	—	—	—	2.00
Reagent J (CCRM)	0.10	0.10	0.10	0.10	0.10
Reagent H (COD)	0.10	0.10	0.10	0.10	0.10

Mix by inversion and let stand 2–3 hours at room temperature. Then add:

Deionized Water	2.00	2.00	2.00	2.00	2.00

Clarify the solutions by centrifugation. Transfer the solutions to cuvettes and record the A500nm for both Standards and Standard Blank using a suitable spectrophotometer.

Calculations

Standard Curve: ΔA500nm Standard = A500nm Standard–A500nm Standard Blank

Prepare a standard curve by plotting ΔA500nm Standard versus the micromoles of Choline.

Sample Determination: ΔA500nm Sample = A500nm Test–A500nm Blank

Determine the total micromoles of Choline liberated using the Standard curve.

$$\text{Units / ml enzyme} = \frac{\text{(micromoles choline liberated) (6) (df)}}{(0.1)}$$

6 = Time conversion factor for one hour (as per the Unit Definition)
0.1 = Volume (in milliliter) of enzyme used
df = Dilution factor

$$\text{Units / mg solid} = \frac{\text{units / ml enzyme}}{\text{mg solid / ml enzyme}}$$

$$\text{Units / mg protein} = \frac{\text{units / ml enzyme}}{\text{mg protein / ml enzyme}}$$

Unit Definition: *One unit will liberate 1.0 µmole of choline from L-α-phosphatidylcholine (egg yolk) per hour at pH 5.6 at 30°C.*

Final Assay Concentration

In a 3.00 ml reaction mixture, the final concentrations are 0.37% (w/v) L-α-phosphatidylcholine, 0.08% (v/v) ethanol, 99 mM sodium acetate, 2 mM sodium lauryl sulfate, 50 mM calcium chloride, and 1–2 units phospholipase D.

Notes

1. This assay should not be used to assay Phospholipase D.
2. Choline Oxidase Unit Definition: One unit will form 1.0 µmole of H_2O_2 per minute from choline and H_2O at pH 8.0 at 37°C.
3. Peroxidase Unit Definition: One unit will form 1.0 mg purpurogallin from pyrogallol in 20 sec at pH 6.0 at 20°C.

Enzymatic Assay of PHOSPHOLIPASE D (EC 3.1.4.4)

Principle

$$\text{L-}\alpha\text{-Phosphatidylcholine} + H_2O \xrightarrow{\text{Phospholipase D}} \text{Choline} + \text{Phosphatidic Acid}$$

$$\text{Choline} + O_2 + H_2O \xrightarrow{\text{Choline Oxidase}} \text{Betaine Aldehyde} + H_2O_2$$

$$2H_2O_2 + 4\text{AAP} + \text{Phenol} \xrightarrow{\text{Peroxidase}} 4H_2O + \text{Quinoneimine Dye}$$

Abbreviation Used

4-AAP = 4-Aminoantipyrine

Conditions: T = 30°C, pH = 8.0, A500nm, Light path = 1 cm

Method: Colorimetric

Reagents

A. 100 mM Tris HCl Buffer, pH 8.0 at 30°C (0.1 M Tris Buffer) (Prepare 100 ml in deionized water using Trizma Base. Adjust to pH 8.0 at 30°C with 1 M HCl).

B. 50 mM Sodium Lauryl Sulfate Solution (SDS) (Prepare 10 ml in deionized water using Lauryl Sulfate, Sodium Salt).

C. 17.9% (v/v) Ethanol Solution (EtOH) (Prepare 1 ml in deionized water using Ethyl Alcohol Denatured).

D. 0.46% (w/v) L-α-Phosphatidylcholine Substrate Solution (Prepare by transferring 2.2 ml (220 mg) of L-α-Phosphatidylcholine to a 50 ml Erlenmeyer flask. Evaporate off the hexane by bubbling nitrogen gas through the liquid. Place the Erlenmeyer flask containing the substrate into a desiccator connected to a vacuum line for 4 hours. Add 6 ml of Reagent A (Buffer), 3 ml of Reagent B (SDS), and 39 ml of deionized water. Mix, using a magnetic stirrer, until a uniform suspension is obtained. Add 0.272 ml of Reagent C (EtOH) to obtain a 0.1% (v/v) ethanol concentration in the substrate solution. *Prepare Fresh*).

E. 500 mM Calcium Chloride Solution ($CaCl_2$) (Prepare 25 ml in deionized water using Calcium Chloride, Dihydrate).

F. 10 mM Tris HCl Buffer pH 8.0, with 2 mM Ethylenediaminetetraacetic Acid and 1.0% (w/v) Potassium Chloride (Enzyme Diluent) (Prepare 10 ml in Reagent A using Ethylenediaminetetraacetic Acid. Disodium Salt, Dihydrate and Potassium Chloride).
G. Choline Oxidase Enzyme Solution (COD) (Prepare a solution containing 10 units/ml of Choline Oxidase in cold Reagent F).
H. 1.0 mM Choline Chloride Standard (Chol Std Soln) (Prepare 50 ml in deionized water using Choline Chloride. *Prepare Fresh*).
I. Choline Color Reagent Mixture (CCRM) (Prepare by dissolving 39 mg of 4-Aminoantipyrine, Free Base, 80 mg of Phenol and 4 mg of Peroxidase in 5.5 ml of Reagent A. Store in an amber bottle to protect from light).
J. 2 M Tris HCl Buffer, pH 9.0 at 25°C (Prepare 25 ml in deionized water using Trizma Base. Adjust to pH 9.0 at 25°C with 1 M HCl).
K. Phospholipase D Enzyme Solution (PLD) (Immediately before use, prepare a solution containing 10–20 units/ml in cold deionized water).

Procedure

Pipette (in milliliters) the following reagents into suitable containers:

	Test	Blank
Reagent D (Substrate Soln)	2.40	2.40
Reagent E ($CaCl_2$)	0.30	0.30
Deionized Water	0.20	0.30

Mix vigorously by vortexing and equilibrate to 30°C using a thermostatted metabolic shaker. Then add:

Reagent K (PLD)	0.10	—

Immediately mix by swirling and incubate the containers, for exactly 10 minutes at 30°C. The vials should be swirled several times during the reaction. Transfer the Test and Blank to a boiling water bath. Remove tubes from the water bath after 5 minutes and let cool to room temperature. Add 0.05 ml of Reagent J (Tris HCl Buffer), centrifuge and filter both Test and Blank through a 0.45 μm filter. Pipette (in milliliters) the following reagents into suitable containers.

Test Filtrate	2.00	—
Blank Filtrate	—	2.00
Reagent I (CCRM)	0.10	0.10
Reagent G (COD)	0.10	0.10

Mix by inversion and let stand between 2–3 hours at room temperature. Then add:

Deionized Water	2.00	2.00

Centrifuge to clarify and then transfer the solutions to suitable cuvettes. Record the A500nm for both Test and Blank using a suitable spectrophotometer.

Colorimetric Assay

Standard Curve: Pipette (in milliliters) the following reagents into suitable containers:

	Std 1	Std 2	Std 3	Std 4	Std Blank
Reagent D (Substrate Soln)	2.40	2.40	2.40	2.40	2.40
Reagent E ($CaCl_2$)	0.30	0.30	0.30	0.30	0.30
Reagent H (Chol Std Soln)	0.05	0.10	0.20	0.30	—
Deionized Water	0.25	0.20	0.10	—	0.30

Mix vigorously by vortexing and then place Standard and Standard Blank in a boiling water bath. Remove tubes after 5 minutes from the water bath and let cool to room temperature. Add 0.05 ml of Reagent J (Tris HCl Buffer), centrifuge and filter the Standards and Standard Blank through a 0.45 μm filter. Pipette (in milliliters) the following reagents into suitable containers:

Std 1 Filtrate	2.00	—	—	—	—
Std 2 Filtrate	—	2.00	—	—	—
Std 3 Filtrate	—	—	2.00	—	—
Std 4 Filtrate	—	—	—	2.00	—
Blank Filtrate	—	—	—	—	2.00
Reagent I (CCRM)	0.10	0.10	0.10	0.10	0.10
Reagent G (COD)	0.10	0.10	0.10	0.10	0.10

Mix by inversion and let stand between 2–3 hours at room temperature. Then add:

Deionized Water	2.00	2.00	2.00	2.00	2.00

Clarify the solutions by centrifugation. Transfer the solutions to cuvettes and record the A500nm for both Standards and Standard Blank using a suitable spectrophotometer.

Calculations

Standard Curve: ΔA500nm Standard = A500nm Standard–A500nm Standard Blank

Prepare a standard curve by plotting ΔA500nm Standard versus the micromoles of Choline.

Sample Determination: ΔA500nm Sample = A500nm Test–A500nm Blank

Determine the total micromoles of Choline liberated using the Standard curve.

$$\text{Units / ml enzyme} = \frac{(\text{micromoles choline liberated})\,(6)\,(\text{df})}{(0.1)}$$

6 = Time conversion factor for one hour (as per Unit Definition)
0.1 = Volume (in milliliter) of enzyme used
df = Dilution factor

$$\text{Units / mg solid} = \frac{\text{units / ml enzyme}}{\text{mg solid / ml enzyme}}$$

$$\text{Units / mg protein} = \frac{\text{units / ml enzyme}}{\text{mg protein / ml enzyme}}$$

Unit Definition: *One unit will liberate 1.0 μmole of choline from L-α-phosphatidylcholine (egg yolk) per hour at pH 8.0 at 30°C.*

Final Assay Concentrations

In a 3.00 ml reaction mixture, the final concentrations are 0.37% (w/v) L-α-phosphatidylcholine, 0.08% (v/v) ethanol, 9.9 mM Tris, 2 mM sodium lauryl sulfate, 50 mM calcium chloride, and 1–2 units phospholipase D.

Notes

1. Choline Oxidase Unit Definition: One unit will form 1.0 μmole of H_2O_2 per minute from choline and H_2O at pH 8.0 at 37°C.
2. Peroxidase Unit Definition: One unit will form 1.0 mg purpurogallin from pyrogallol in 20 sec at pH 6.0 at 20°C.

Enzymatic Assay of PHOSPHOMANNOSE ISOMERASE

Principle

$$\text{D-Mannose 6-Phosphate} \xrightarrow{\text{PMI}} \text{D-Fructose 6-Phosphate}$$

$$\text{D-Fructose 6-Phosphate} \xrightarrow{\text{PGI}} \text{D-Glucose 6-Phosphate}$$

$$\text{D-Glucose 6-Phosphate} + \beta\text{-NADP} \xrightarrow{\text{G6PDH}} \text{Gluconate-6-P} + \text{NADPH}$$

Abbreviations Used

PMI = Phosphomannose Isomerase
PGI = Phosphoglucose Isomerase
G6PDH = Glucose-6-Phosphate Dehydrogenase
β-NADP = β-Nicotinamide Adenine Dinucleotide Phosphate, Oxidized Form
β-NADPH = β-Nicotinamide Adenine Dinucleotide Phosphate, Reduced Form

Conditions: T = 25°C, pH 7.6, A340nm, Light path = 1 cm

Method: Continuous Spectrophotometric Rate Determination

Reagents

A. 100 mM Triethanolamine HCl Buffer, pH 7.6 at 25°C (Prepare 100 ml in deionized water using Triethanolamine Hydrochloride. Adjust to pH 7.6 at 25°C with 1 M NaOH).
B. 66 mM D-Mannose 6-Phosphate Solution (Mann-6-P) (Prepare 10 ml in deionized water using D-Mannose 6-Phosphate Disodium Salt).
C. 13.5 mM β-Nicotinamide Adenine Dinucleotide Phosphate, Oxidized Form, Solution (β-NADP) (Prepare 5 ml in deionized water using β-Nicotinamide Adenine Dinucleotide Phosphate, Sodium salt. *Prepare Fresh*).
D. Phosphoglucose Isomerase Enzyme Solution (PGI) (Immediately before use, prepare a solution containing 1000 units/ml of Phosphoglucose Isomerase in cold deionized water).
E. Glucose-6-Phosphate Dehydrogenase Enzyme Solution (G6PDH) (Immediately before use, prepare a solution containing 100 units/ml of Glucose-6-Phosphate Dehydrogenase in cold deionized water).

F. Phosphomannose Isomerase Enzyme Solution (PMI) (Immediately before use, prepare a solution containing 0.075–0.150 units/ml of Phosphomannose Isomerase in cold Reagent A).

Procedure

Pipette (in milliliters) the following reagents into suitable cuvettes:

	Test	Blank
Reagent A (Buffer)	2.52	2.52
Reagent B (Mann-6-P)	0.25	0.25
Reagent C (β-NADP)	0.10	0.10
Reagent D (PGI)	0.02	0.02
Reagent E (G6PDH)	0.01	0.01

Mix by inversion and equilibrate to 25°C. Monitor the A340nm until constant, using a suitably thermostatted spectrophotometer. Then add:

Reagent F (PMI)	0.10	—
Reagent A (Buffer)	—	0.10

Immediately mix by inversion and record the increase in A340nm for approximately 5 minutes. Obtain the ΔA340nm/minute using the maximum linear rate for both the Test and Blank.

Calculations

$$\text{Units / ml enzyme} = \frac{(\Delta\text{A340nm / min Test} - \Delta\text{A340nm / min Blank})\,(3)\,(\text{df})}{(6.22)\,(0.1)}$$

3 = Total volume (in milliliters) of assay
df = Dilution factor
6.22 = Millimolar extinction coefficient of β-NADPH at 340nm
0.1 = Volume (in milliliter) of enzyme used

$$\text{Units / mg solid} = \frac{\text{units / ml enzyme}}{\text{mg solid / ml enzyme}}$$

$$\text{Units / mg protein} = \frac{\text{units / ml enzyme}}{\text{mg protein / ml enzyme}}$$

Unit Definition: *One unit will convert 1.0 μmole of D-mannose 6-phosphate to D-fructose 6-phosphate per minute at pH 7.6 at 25°C, using a coupled enzyme system with phosphoglucose isomerase and glucose-6-phosphate dehydrogenase.*

Final Assay Concentrations

In a 3.00 ml reaction mix, the final concentrations are 87 mM triethanolamine, 5.5 mM D-mannose 6-phosphate, 0.45 mM nicotinamide adenine dinucleotide phosphate, 20 units phosphoglucose isomerase, 1 unit glucose-6-phosphate dehydrogenase and 0.0075–0.015 unit phosphomannose isomerase.

Notes

1. Glucose-6-Phosphate Dehydrogenase unit definition: One unit will oxidize 1.0 μmole of D-glucose 6-phosphate to 6-phospho-D-gluconate per minute in the presence of β-NADP at pH 7.4 at 25°C.
2. Phosphoglucose Isomerase unit definition: One unit will convert 1.0 μmole of fructose 6-phosphate to glucose 6-phosphate per minute at pH 7.4 at 25°C.

Enzymatic Assay of PHOSPHORIBOISOMERASE (EC 5.3.1.6)

Principle

$$\text{D-Ribose 5-P} \xrightarrow{\text{PRI}} \text{D-Ribulose 5-P}$$

$$\text{D-Ribulose 5-P} \xrightarrow{\text{Ru-5-P-3-Epim.}} \text{D-Xylulose 5-P}$$

$$\text{D-Xylulose 5-P} + \text{Ribose} \xrightarrow[\text{Mg}^{++}\text{,Cocarboxylase}]{\text{5-P TK}} \text{GAP} + \text{Sedoheptulose 7-P}$$

$$\text{GAP} \xrightarrow{\text{TPI}} \text{DHAP}$$

$$\text{DHAP} + \beta\text{-NADH} + \text{H}^{+} \xrightarrow{\alpha\text{-GDH}} \beta\text{-NAD} + \alpha\text{-Glycerophosphate}$$

Abbreviations Used

PRI = Phosphoriboisomerase
Ru-5-P-3-Epim. = Ribulose 5-Phosphate-3-Epimerase
GAP = Glyceraldehyde 3-Phosphate
TK = Transketolase
TPI = Triosephosphate Isomerase
β-NADH = β-Nicotinamide Adenine Dinucleotide, reduced form
α-GDH = α-Glycerophosphate Dehydrogenase
DHAP = Dihydroxyacetone Phosphate
β-NAD = β-Nucleotide Adenine Dinucleotide, oxidized form

Conditions: T = 30°C, pH = 7.7, A340nm, Light path = 1 cm

Method: Continuous Spectrophotometric Rate Determination

Reagents

A. 250 mM Glycylglycine Buffer, pH 7.7 at 30°C (Prepare 100 ml in deionized water using Glycylglycine Buffer. Adjust to pH 7.7 at 30°C using 5 N NaOH).
B. 100 mM D-Ribose 5-Phosphate Solution (R-5-P) (Prepare 1.0 ml in deionized water using D-Ribose 5'-Phosphate Disodium).
C. 2.0 mM Cocarboxylase Solution (Prepare 1.0 ml in deionized water using Cocarboxylase. *Prepare Fresh*).
D. 2.5 mM β-NADH Solution (Dissolve the contents of 1 vial of β-NADH Disodium in 2.5 ml in Reagent A).
E. 300 mM Magnesium Chloride Solution ($MgCl_2$) (Prepare 1 ml in deionized water using Magnesium Chloride, Hexahydrate).

F. Ribulose 5-Phosphate-3-Epimerase Enzyme Solution(Ru 5-P-3-Epim). (Immediately before use, prepare a solution containing 10 units/ml of D-Ribulose 5-Phosphate-3-Epimerase in deionized water).

G. Transketolase Enzyme Solution (TK) (Immediately before use, prepare a solution containing 10 units/ml of Transketolase Enzyme Solution in deionized water).

H. α-Glycerophosphate Dehydrogenase-Triosephosphate Isomerase Solution (α-GDH-TPI) (Immediately before use, prepare a solution containing 100 units/ml of α-Glycerophosphate Dehydrogenase-Triosephosphate Isomerase, in deionized water. The 100 units/ml is based on α-GDH units).

I. Phosphoriboisomerase Enzyme Solution (PRI) (Immediately before use, prepare a solution containing 0.25 to 1.0 of Phosphoriboisomerase using cold Reagent A).

Procedure

Pipette (in milliliters) the following reagents into suitable cuvettes:

	Test	Blank
Reagent A (Buffer)	2.35	2.35
Reagent B (R-5-P)	0.10	0.10
Reagent C (Cocarboxylase)	0.05	0.05
Reagent D (β-NADH)	0.15	0.15
Reagent E ($MgCl_2$)	0.10	0.10
Reagent F (Ru-5-P-3-Epim)	0.05	0.05
Reagent G (TK)	0.05	0.05
Reagent H (α-GDH-TPI)	0.05	0.05

Mix by inversion and equilibrate to 30°C. Monitor the A340nm until constant, using a suitably thermostatted spectrophotometer. Then add:

Reagent I	0.10	—
Reagent A	—	0.10

Immediately mix by inversion and record the decrease in A340nm for approximately 15 minutes. The maximum linear rate usually occurs between 10–15 minutes. Obtain the ΔA340nm/minute using the maximum linear rate for both the Test and Blank.

Calculations

$$\text{Units / mg enzyme} = \frac{\Delta A340nm / \text{min Test} - \Delta A340nm / \text{min Blank}}{(6.22)\ (\text{mg enzyme / ml RM})}$$

6.22 = Millimolar extinction coefficient of β-NAD at 340nm
RM = Reaction Mix

Unit Definition: *One unit will convert 1.0 μmole of D-ribose 5-phosphate to D-ribulose 5-phosphate per minute at pH 7.7 at 30°C.*

Final Assay Concentrations

In a 3.0 ml reaction mix, the final concentrations are 217 mM glycylglycine, 3.33 mM D-ribose 5-phosphate, 0.033 mM cocarboxylase, 0.125 mM β-NADH, 10.0 mM magnesium chloride, 0.5 units ribulose 5-phosphate-3-epimerase, 0.5 units transketolase, 5.0 units α-glycerophosphate dehydrogenase, approximately 50.0 units triosephosphate isomerase, and 0.025 to 0.1 units phosphoriboisomerase.

Notes

1. Ribulose-5-Phosphate-3-Epimerase unit definition: One unit will convert 1 μmole of D-ribulose-5-phosphate to D-xylulose-5-phosphate per minute at pH 7.7 at 25°C in a coupled system with ribose-5-phosphate, β-NADH, transketolase, α-glycerophosphate dehydrogenase/triosephosphate isomerase, and cocarboxylase.
2. Transketolase unit definition: One unit will produce 1 μmole of glyceraldehyde-3-phosphate from xylulose-5-phosphate per minute of pH 7.7 at 25°C in the presence of ribose-5-phosphate, cocarboxylase, and magnesium in a coupled system with triosephosphate isomerase and α-glycerophosphate dehydrogenase.
3. α-Glycerophosphate Dehydrogenase unit definition: One unit will convert 1.0 μmole of dihydroxyacetone phosphate to α-glycerophosphate per minute at pH 7.4 and 25°C.
4. Triosephosphate Isomerase unit definition: One unit will convert 1.0 μmole of D-glyceraldehyde-3-phosphate to dihydroxyacetone phosphate per minute at pH 7.6 and 25°C.

Enzymatic Assay of PHOSPHORIBOSYL-PYROPHOSPHATE SYNTHETASE (EC 2.7.6.1)

Principle

$$\text{ATP} + \text{Ribose 5-Phosphate} \xrightarrow{\text{PRPP Synthetase}} \text{PRPP} + \text{AMP}$$

$$\text{AMP} + \text{ATP} \xrightarrow{\text{MK}} 2\ \text{ADP}$$

$$2\ \text{PEP} + 2\text{ADP} \xrightarrow{\text{PK}} 2\ \text{Pyruvate} + 2\text{ATP}$$

$$\text{Pyruvate} + 2\ \beta\text{-NADH} \xrightarrow{\text{LDH}} 2\ \text{Lactate} + 2\ \beta\text{-NAD}$$

Abbreviations Used

ATP = Adenosine 5'-Triphosphate
PRPP = Phosphoribosyl-Pyrophosphate
AMP = Adenosine 5'-Monophosphate
MK = Myokinase
ADP = Adenosine 5'-Diphosphate
PEP = Phospho(enol) pyruvate
PK = Pyruvate Kinase
β-NADH = β-Nicotinamide Adenine Dinucleotide, Reduced Form
LDH = Lactic Dehydrogenase
β-NAD = β-Nicotinamide Adenine Dinucleotide, Oxidized Form

Conditions: T = 37°C, pH = 7.6, A340nm, Light path = 1 cm

Method: Continuous Spectrophotometric Rate Determination

Reagents

A. 125 mM Sodium Phosphate buffer with 7 mM Magnesium Chloride, pH 7.6 at 37°C (Prepare 100 ml in deionized water using Sodium Phosphate, Monobasic, Anhydrous and Magnesium Chloride, 4.9 M Solution. Adjust to pH 7.6 at 37°C with 1 M NaOH).

B. 60 mM Ribose 5-Phosphate Solution (R-5-P) (Prepare 5 ml in Reagent A using D-Ribose 5-Phosphate, Disodium Salt).
C. 1.19 M Sodium Bicarbonate Solution (Prepare 5 ml in deionized water using Sodium Bicarbonate).
D. 120 mM Adenosine 5'-Triphosphate Solution (ATP) (Prepare 1 ml in Reagent C using Adenosine 5'-Triphosphate, Disodium Salt).
E. 70 mM Phospho(enol) pyruvate Solution (PEP) (Prepare 1 ml in deionized water using Phospho(enol) pyruvate, Mono(cyclohexylammonium) Salt).
F. β-Nicotinamide Adenine Dinucleotide, Reduced Form, (β-NADH) (Use one 10 mg vial of β-Nicotinamide Adenine Dinucleotide, Reduced Form, Disodium Salt).
G. PK/LDH Enzyme Suspension (PK/LDH).
H. Myokinase Enzyme Suspension (MK).
I. Phosphoribosyl-Pyrophosphate Synthetase Enzyme Solution (PRPP Syn) (Immediately before use, prepare a solution containing 0.1–1.0 unit/ml of Phosphoribosyl-Pyrophosphate Synthetase in cold Reagent A. *Store on ice*).

Procedure

Prepare a reaction cocktail by pipetting (in milliliters) the following reagents into Reagent F (β-NADH):

Reagent D (ATP)	1.00
Reagent E (PEP)	1.00
Reagent A (Buffer)	1.00

Pipette (in milliliters) the following reagents into suitable cuvettes:

	Test	Blank
Reaction Cocktail	0.10	0.10
Reagent A (Buffer)	1.00	1.10
Reagent B (R-5-P)	0.10	—
Reagent G (PK/LDH)	0.01	0.01
Reagent H (MK)	0.005	0.005

Mix by inversion and equilibrate to 37°C. Then add:

Reagent I (PRPP Syn)	0.05	0.05

Immediately mix by inversion and record the decrease in A340nm for approximately 5 minutes. Obtain the ΔA340nm/minute using the maximum linear rate for both the Test and Blank.

Calculations

$$\text{Units / ml enzyme} = \frac{(\Delta\text{A340nm / min Test} - \Delta\text{A340nm / min Blank})\ (1.265)\ (\text{df})}{(2)\ (6.22)\ (0.05)}$$

1.265 = Total volume (in milliliters) of assay
df = Dilution factor

2 = Conversion factor required since 2 moles of β-NADH are oxidized per mole of phosphoribosylpyrophosphate produced
6.22 = Millimolar extinction coefficient of β-NADH at 340nm
0.05 = Volume (in milliliter) of PRPP synthetase used

Unit Definition: *One unit will catalyze the formation of 1.0 µmole of AMP from ATP and ribose-5-phosphate per minute at pH 7.6 at 37°C.*

Final Assay Concentrations

In a 1.265 ml reaction mix, the final concentrations are 117 mM sodium phosphate, 4.7 mM ribose 5-phosphate, 3.2 mM adenosine 5'-triphosphate, 1.8 mM phospho(enol) pyruvate, 0.34 mM β-nicotinamide adenine dinucleotide, reduced form, 6.5 mM magnesium chloride, 31 mM sodium bicarbonate, 7 units pyruvate kinase, 10 units lactic dehydrogenase, 10 units myokinase, and 0.005–0.05 unit phosphoribosylpyrophosphate synthetase.

Notes

1. Contains not less than 700 Pyruvate Kinase units and 1000 L-Lactic Dehydrogenase units per ml.
2. Contains not less than 2000 Myokinase units per ml.
3. Pyruvate Kinase Unit Definition: One unit will convert 1.0 µmole of phos(enol) pyruvate to pyruvate per minute at pH 7.6 at 37°C.
4. L-Lactic Dehydrogenase Unit Definition: One unit will reduce 1.0 µmole of pyruvate to L-lactate per minute at pH 7.5 at 37°C.
5. Myokinase Unit Definition: One unit will convert 2.0 µmoles of ADP to ATP and AMP per minute at pH 7.6 at 37°C.

Enzymatic Assay of PHOSPHORIBULOKINASE (EC 2.7.1.19)

Principle

$$\text{D-Ribulose 5-Phosphate} + \text{ATP} \xrightarrow{\text{PRK}} \text{D-Ribulose 1, 5-Bisphosphate} + \text{ADP}$$

$$\text{ADP} + \text{PEP} \xrightarrow{\text{PK}} \text{Pyruvate} + \text{ATP}$$

$$\text{Pyruvate} + \beta\text{-NADH} \xrightarrow{\text{LDH}} \text{Lactate} + \beta\text{-NAD}$$

Abbreviations Used

ATP = Adenosine 5'-Triphosphate
PRK = Phosphoribulokinase
ADP = Adenosine 5'-Diphosphate
PEP = Phospho(enol) pyruvate
PK = Pyruvate Kinase
β-NADH = β-Nicotinamide Adenine Dinucleotide, Reduced Form
LDH = Lactic Dehydrogenase
β-NAD = β-Nicotinamide Adenine Dinucleotide, Oxidized Form

Conditions: T = 37°C, pH = 7.9, A340nm, Light path = 1 cm

Method: Continuous Spectrophotometric Rate Determination

Reagents

A. 67 mM Tris HCl Buffer, pH 7.9 at 37°C (Prepare 100 ml in deionized water using Trizma Base. Adjust to pH 7.9 at 37°C with 1 M HCl).
B. 50 mM Potassium Chloride Solution (KCl) (Prepare 10 ml in Reagent A using Potassium Chloride).
C. 6 mM Magnesium Sulfate Solution ($MgSO_4$) (Prepare 10 ml in Reagent A using Magnesium Sulfate, Heptahydrate).
D. 0.54 mM Phospho(enol) pyruvate Solution (PEP) (Prepare 10 ml in Reagent A using Phospho(enol) pyruvate, Monopotassium Salt).
E. 0.36 mM Ethylenediaminetetraacetic Acid Solution (EDTA) (Prepare 10 ml in Reagent A using Ethylenediaminetetraacetic Acid, Disodium Salt, Dihydrate).
F. β-Nicotinamide Adenine Dinucleotide, Reduced Form Solution (β-NADH) (Use β-Nicotinamide Adenine Dinucleotide, Disodium, 1 mg vial).
G. 50 mM Tris HCl Buffer, pH 7.1 at 37°C (Enz Dil) (Prepare 25 ml in deionized water using Trizma Base. Adjust to pH 7.1 at 37°C with 1 M HCl).
H. 20 mM D-Ribulose 5-Phosphate Solution (RU-5-P) (Prepare 1 ml in deionized water using D-Ribulose 5-Phosphate, Sodium Salt).
I. 9.1 mM Adenosine 5'-Triphosphate Solution (ATP) (Prepare 1 ml in deionized water using Adenosine 5'-Triphosphate Disodium Salt).
J. 82 mM Glutathione, Reduced Form Solution (GSH) (Prepare 1 ml in deionized water using Glutathione, Reduced Form, Free Acid).
K. PK/LDH Enzymes Suspension.
L. Phosphoribulokinase Enzyme Solution (PRK) (Immediately before use, prepare a solution containing 0.3–0.6 unit/ml of Phosphoribulokinase in cold Reagent G).

Procedure

Prepare an initial reaction cocktail by pipetting (in milliliters) the following reagents into a suitable container:

Reagent A (Buffer)	60.00
Reagent B (KCl)	10.00
Reagent C ($MgSO_4$)	10.00
Reagent D (PEP)	10.00
Reagent E (EDTA)	10.00

Mix by swirling and adjust to pH 7.9 at 37°C with either 1 M HCl or 1 M NaOH, if necessary. Prepare a final reaction cocktail by pipetting 10 ml of the initial reaction cocktail into Reagent F (β-NADH). Pipette (in milliliters) the following reagents into suitable cuvettes:

	Test	Blank
Final Reaction Cocktail	2.50	2.50
Reagent I (ATP)	0.10	0.10
Reagent J (GSH)	0.10	0.10
Reagent K (PK/LDH)	0.03	0.03
Reagent L (PRK)	0.10	0.10

Mix by inversion and equilibrate to 37°C. Monitor the ΔA340nm until constant, using a suitably thermostatted spectrophotometer. Then add:

	Test	Blank
Reagent H (RU-5-P)	0.10	—
Deionized Water	—	0.10

Immediately mix by inversion and record the decrease in A340nm for approximately 5 minutes. Obtain the ΔA340nm/minute using the maximum linear rate for both the Test and Blank.

Calculations

$$\text{Units / ml enzyme} = \frac{(\Delta A340nm/\min \text{ Test} - \Delta A340nm/\min \text{ Blank})(2.93)(df)}{(6.22)(0.1)}$$

2.93 = Total volume (in milliliters) of assay
df = Dilution factor
6.22 = Millimolar extinction coefficient of β-NADH at 340nm
0.1 = Volume (in milliliter) of enzyme used

$$\text{Units / mg solid} = \frac{\text{units / ml enzyme}}{\text{mg solid / ml enzyme}}$$

$$\text{Units / mg protein} = \frac{\text{units / ml enzyme}}{\text{mg protein / ml enzyme}}$$

Unit Definition: *One unit will transfer 1.0 μmole of phosphate from ATP to D-ribulose 5-phosphate per minute at pH 7.9 at 37°C.*

Final Assay Concentrations

In a 2.93 ml reaction mix, the final concentrations are 57 mM Tris, 4.3 mM potassium chloride, 0.5 mM magnesium sulfate, 0.046 mM phospho(enol) pyruvate, 0.031 mM Ethylenediaminetetraacetic Acid, 0.31 mM adenosine 5'-triphosphate, 2.8 mM glutathione, reduced form, 0.12 mM β-nicotinamide adenine dinucleotide, reduced form, 0.68 mM D-ribulose 5-phosphate, 21 units pyruvate kinase, 30 units lactic dehydrogenase, and 0.03–0.06 unit phosphoribulokinase.

Notes

1. Contains not less than 700 Pyruvate Kinase units and 100 Lactic Dehydrogenase units per ml.
2. Pyruvate Kinase Unit Definition: One unit will convert 1.0 μmole of phospho(enol) pyruvate to pyruvate per minute at pH 7.6 at 37°C.
3. L-Lactic Dehydrogenase Unit Definition: One unit will reduce 1.0 μmole of pyruvate to L-lactic per minute at pH 7.5 at 37°C.

Enzymatic Assay of PHOSPHORYLASE a (EC 2.4.1.1)

Principle

$$\text{Glycogen}_n + \text{Pi} \xrightarrow{\text{Phosphorylase a}} \text{Glycogen}_{n-1} + \alpha\text{-D-Glucose 1-Phosphate}$$

$$\alpha\text{-D-Glucose 1-Phosphate} \xrightarrow{\text{PGLUM}} \alpha\text{-D-Glucose 6-Phosphate}$$

$$\alpha\text{-D-Glucose 6-Phosphate} + \beta\text{-NADP} \xrightarrow{\text{G-6-PDH}} \text{6-Phosphogluconate} + \beta\text{-NADPH}$$

Abbreviations Used

Pi = Inorganic Phosphate
β-NADP = β-Nicotinamide Adenine Dinucleotide Phosphate, Oxidized Form
β-NADH = β-Nicotinamide Adenine Dinucleotide Phosphate, Reduced Form
G-6-PDH = Glucose-6-Phosphate Dehydrogenase
PGLUM = Phosphoglucomutase

Conditions: T = 30°C, pH = 6.8, A340nm, Light path = 1 cm

Method: Continuous Spectrophotometric Rate Determination

Reagents

A. 500 mM Potassium Phosphate Buffer, pH 6.8 at 30°C (Prepare 100 ml in deionized water using Potassium Phosphate, Monobasic, Anhydrous. Adjust to pH 6.8 at 30°C with 1 M KOH).
B. 4% (w/v) Glycogen Solution (Glycogen) (Prepare 10 ml in deionized water using Glycogen).
C. 300 mM Magnesium Chloride Solution ($MgCl_2$) (Prepare 5 ml in deionized water using Magnesium Chloride, Hexahydrate).
D. 100 mM Ethylenediaminetetraacetic Acid Solution (EDTA) (Prepare 2 ml in deionized water using Ethylenediaminetetraacetic Acid, Tetrasodium, Hydrate).
E. 6.5 mM β-Nicotinamide Adenine Dinucleotide Phosphate Solution (β-NADP) (Prepare 15 ml in deionized water using β-Nicotinamide Adenine Dinucleotide, Phosphate, Sodium. *Prepare Fresh*).
F. 0.1% (w/v) α-D-Glucose 1,6-Diphosphate Solution (G 1,6-DiP) (Prepare 1 ml in deionized water using α-D-Glucose 1,6-Diphosphate, Cyclohexylammonium Salt, Hydrate).
G. Glucose-6-Phosphate Dehydrogenase Solution (G-6-PDH) (Immediately before use, prepare a solution containing 10 units/ml in cold deionized water using Glucose-6-Phosphate Dehydrogenase).
H. Phosphoglucomutase Solution (PGLUM) (Immediately before use, prepare a solution containing 10 units/ml in cold deionized water using Phosphoglucomutase).
I. 40 mM β-Glycerophosphate with 80 mM Cysteine Solution, pH 6.8 at 30°C (Diluent) (Prepare 20 ml in deionized water using β-Glycero-phosphate, Disodium Salt, Hydrate and L-Cysteine, Hydrochloride, Monohydrate. Adjust to pH 6.8 with 1 M NaOH).
J. Phosphorylase a Enzyme Solution (Immediately before use, prepare a solution containing 0.25 unit/ml of Phosphorylase a in cold Reagent I).

Procedure

Prepare a reaction cocktail by pipetting (in milliliters) the following reagents into a suitable container:

Deionized water	99.50
Reagent A (Buffer)	15.00
Reagent B (Glycogen)	7.50
Reagent C ($MgCl_2$)	0.67
Reagent D (EDTA)	0.15
Reagent E (β-NADP)	10.00
Reagent F (G 1,6-DiP)	0.50

Mix and adjust to pH 6.8 at 30°C with 100 mM HCl or 100 mM NaOH, if necessary. Pipette (in milliliters) the following reagents into suitable cuvettes:

	Test	Blank
Reaction Cocktail	2.70	2.70
Reagent G (G-6-PDH)	0.10	0.10
Reagent H (PGLUM)	0.10	0.10

Mix by inversion and equilibrate to 30°C. Monitor the A340nm until constant, using a suitably thermostatted spectrophotometer. Then add:

Reagent I (Diluent)	—	0.10
Reagent J (Enzyme Solution)	0.10	—

Immediately mix by inversion and record the increase in A340nm for approximately 10 minutes. Obtain the ΔA340nm/minute using the maximum linear rate for both the Test and Blank.

Calculations

$$\text{Units / ml enzyme} = \frac{(\Delta\text{A340nm / min Test} - \Delta\text{A340nm / min Blank})\ (3)\ (\text{df})}{(6.22)\ (0.1)}$$

3 = Total volume (in milliliters) of the assay
df = Dilution factor
6.22 = Millimolar extinction coefficient of β-NADH at 340nm
0.1 = Volume (in milliliter) of enzyme used

$$\text{Units / mg solid} = \frac{\text{units / ml enzyme}}{\text{mg solid / ml enzyme}}$$

$$\text{Units / mg protein} = \frac{\text{units / ml enzyme}}{\text{mg protein / ml enzyme}}$$

Unit Definition: *One unit will form 1.0 µmole of a-D-glucose 1-phosphate from glycogen and orthophosphate per minute at pH 6.8 at 30°C, measured in a system containing phosphoglucomutase, β-NADP, and glucose-6-phosphate dehydrogenase.*

Final Assay Concentrations

In a 3.00 ml reaction mix, the final concentrations are 51 mM potassium phosphate, 0.20% (w/v) glycogen, 1.4 mM magnesium chloride, 0.10 mM Ethylenediaminetetraacetic Acid, 0.44 mM β-Nicotinamide Adenine Dinucleotide Phosphate, Oxidized form, 0.0003%

(w/v) α-D-glucose 1,6-diphosphate, 1.3 mM β-glycerophosphate, 2.7 mM cysteine, 1 unit glucose-6-phosphate dehydrogenase, 1 unit phosphoglucomutase, and 0.025 unit phosphorylase a.

Notes

1. The maximum linear rate should not exceed A340nm/minute of 0.1.
2. Glucose-6-Phosphate Dehydrogenase Unit Definition: One unit will oxidize 1.0 µmole of glucose 6-phosphate to 6-phospho-D-gluconate per minute in the presence of β-NADP at pH 7.4 at 25°C.
3. Phosphoglucomutase Unit Definition: One unit will convert 1.0 µmole of α-D-glucose 1-phosphate to α-D-glucose 6-phosphate per minute at pH 7.4 at 30°C.

Enzymatic Assay of PHOSPHORYLASE b (EC 2.4.1.1)

Principle

$$\text{(Glycogen)n} + \text{Pi} \xrightarrow{\text{Phosphorylase a}} \text{(Glycogen)n-1} + \alpha\text{-D-Glucose 1-Phosphate}$$

$$\text{(Glycogen)n} + \text{Pi} \xrightarrow[\text{5'AMP}]{\text{Phosphorylase b}} \text{(Glycogen)n-1} + \alpha\text{-D-Glucose 1-Phosphate}$$

$$\alpha\text{-D-Glucose 1-Phosphate} \xrightarrow{\text{Phosphoglucomutase}} \alpha\text{-D-Glucose 6-Phosphate}$$

$$\alpha\text{-D-Glucose 6-Phosphate} + \beta\text{-NADP} \xrightarrow{\text{G-6-PDH}} \text{6-Phosphogluconate} + \beta\text{-NADPH}$$

Abbreviations Used

5'-AMP = Adenosine 5'-Monophosphate
Pi = Inorganic Phosphate
PGLUM = Phosphoglucomutase
β-NADP = β-Nicotinamide Adenine Dinucleotide Phosphate, Oxidized Form
G-6-PDH = Glucose-6-Phosphate Dehydrogenase
β-NADPH = β-Nicotinamide Adenine Dinucleotide Phosphate, Reduced Form

Conditions: T = 30°C, pH = 6.8, A340nm, Light path = 1 cm

Method: Continuous Spectrophotometric Rate Determination

Reagents

A. 500 mM Potassium Phosphate Buffer, pH 6.8 at 30°C (Prepare 100 ml in deionized water using Potassium Phosphate, Monobasic, Anhydrous. Adjust to pH 6.8 at 30°C with 1 M KOH).
B. 4% (w/v) Glycogen Solution (Glycogen) (Prepare 10 ml in deionized water using Glycogen Type III. This may require stirring and heat to solubilize).
C. 300 mM Magnesium Chloride Solution ($MgCl_2$) (Prepare 5 ml in deionized water using Magnesium Chloride, Hexahydrate).
D. 100 mM Ethylenediaminetetraacetic Acid Solution (EDTA) (Prepare 2 ml in deionized water using Ethylenediaminetetraacetic Acid, Tetrasodium Salt).
E. 6.5 mM β-Nicotinamide Adenine Dinucleotide Phosphate Solution (β-NADP) (Prepare 15 ml in deionized water using β-Nicotinamide Adenine Dinucleotide Phosphate, Sodium Salt. *Prepare Fresh*).

F. 0.1% (w/v) α-D-Glucose 1,6-Diphosphate Solution (G 1,6-DiP) (Prepare 1 ml in deionized water using α-D-Glucose 1,6-Diphosphate, Cyclohexylammonium Salt, Hydrate).

G. Glucose-6-Phosphate Dehydrogenase Solution (G-6-PDH) (Immediately before use, prepare a solution containing 10 units/ml in cold deionized water using Glucose-6-Phosphate Dehydrogenase).

H. Phosphoglucomutase Solution (PGLUM) (Immediately before use, prepare a solution containing 10 units/ml in cold deionized water using Phosphoglucomutase).

I. 40 mM β-Glycerophosphate with 80 mM Cysteine, pH 6.8 at 30°C (Enzyme Diluent) (Prepare 50 ml in deionized water using β-Glycerophosphate Disodium Salt, Hydrate and L-Cysteine, Hydrochloride, Monohydrate. Adjust to pH 6.8 with 1 M NaOH).

J. Phosphorylase b Enzyme Solution (Phosphor b) (Immediately before Step 1: Prepare a solution containing 5–10 mg solid/ml of Phosphorylase b in cold Reagent I).

K. Diluted Phosphorylase b Enzyme Solution (Dil Phosphor b) (Immediately before Step 2, prepare a solution containing 0.20–0.40 unit/ml of Phosphorylase B in cold Reagent I).

L. 100 mM Adenosine 5'-Monophosphate Solution (5'-AMP) (Prepare 1 ml in deionized water using Adenosine 5'-Monophosphate, Sodium Salt. *Prepare Fresh*).

Procedure

Prepare a reaction cocktail by pipetting (in milliliters) the following reagents into a suitable container:

Deionized water	99.50
Reagent A (Buffer)	15.00
Reagent B (Glycogen)	7.50
Reagent C ($MgCl_2$)	0.67
Reagent D (EDTA)	0.15
Reagent E (β-NADP)	10.00
Reagent F (G 1,6-DiP)	0.50

Mix by stirring and adjust to pH 6.8 at 30°C with 100 mM HCl or 100 mM NaOH, if necessary.

Step 1: Pipette (in milliliters) the following reagents into suitable cuvettes:

	Test	Blank
Reaction Cocktail	2.70	2.70
Reagent G (G-6-PDH)	0.10	0.10
Reagent H (PGLUM)	0.10	0.10

Mix by inversion and equilibrate to 30°C. Monitor the A340nm until constant, using a suitably thermostatted spectrophotometer. Then add:

Reagent J (Phosphor b)	0.10	—
Reagent I (Enzyme Diluent)	—	0.10

Mix by inversion and record the increase in A340nm for approximately 10 minutes. Obtain the ΔA340nm/minute using the maximum linear rate for both the Test and Blank.

Step 2: Pipette (in milliliters) the following reagents into suitable cuvettes:

	Test	Blank
Reaction Cocktail	2.70	2.70
Reagent G (G-6-PDH)	0.10	0.10
Reagent H (PGLUM)	0.10	0.10
Reagent K (5'-AMP)	0.050	0.050

Mix by inversion and equlibrate to 30°C. Monitor the A340nm until constant, using a suitably thermostatted spectrophotometer. Then add:

	Test	Blank
Reagent K (Dil Phosphor b)	0.10	—
Reagent I (Enzyme Diluent)	—	0.10

Immediately mix by inversion and record the increase in A340nm for approximately 5 minutes. Obtain the ΔA340nm/minute using the maximum linear rate for both the Test and Blank.

Calculations

$$\text{Phosphorylase a units / ml enzyme} = \frac{(\Delta A340\text{nm / min Test without 5' AMP} - \Delta A340\text{nm / min Blank without 5' AMP})\,(3.00)\,(df)}{(6.22)\,(0.1)}$$

$$\text{Phosphorylase a units / mg solid} = \frac{\text{Phosphorylase a units / ml enzyme}}{\text{mg solid / ml enzyme}}$$

$$\text{Phosphorylase a and b units / ml enzyme} = \frac{(\Delta A340\text{nm / min Test with 5' AMP} - \Delta A340\text{nm / min Blank with 5' AMP})\,(3.00)\,(df)}{(6.22)\,(0.1)}$$

$$\text{Phosphorylase a and b units / mg solid} = \frac{\text{Phosphorylase a and b units / ml enzyme}}{\text{mg solid / ml enzyme}}$$

Phosphorylase b units/mg solid = Phosphorylase a and b units/mg solid–Phosphorylase a units/mg solid

3.00 = Total volume (in milliliters) of Phosphorylase A assay

df = Dilution factor

6.22 = Millimolar extinction coefficient of β-NADPH at 340 nm

0.1 = Volume (in milliliter) of enzyme used

$$\text{Units / mg protein} = \frac{\text{units / ml enzyme}}{\text{mg protein / ml enzyme}}$$

Unit Definition: *One unit will form 1.0 μmole of α-D-glucose 1-phosphate from glycogen and orthophosphate in the presence of adenosine 5'-monophosphate per minute at pH 6.8 at 30°C, measured in a system containing phosphoglucomutase, β-nicotinamide adenine dinucleotide phosphate, and glucose-6-phosphate dehydrogenase.*

Final Assay Concentrations

In a 3.05 ml reaction mix, the final concentrations are 50 mM potassium phosphate, 0.2% (w/v) glycogen, 1.3 mM magnesium chloride, 0.10 mM Ethylenediaminetetraacetic Acid, 0.43 mM β-nicotinamide adenine dinucleotide phosphate, 0.0003% (w/v) α-D-glucose 1,6-diphosphate, 1 unit glucose-6-phosphate dehydrogenase, 1 unit phosphoglucomutase, 1.6 mM adenosine 5'-monophosphate, and 0.02–0.04 unit phosphorylase b.

Notes

1. Phosphorylase a activity is present in the phosphorylase b preparation and must be accounted for in the enzyme assay.
2. α-D-Glucose 1,6-Diphosphate is an activator of phosphoglucomutase.
3. Phosphorylase b is enzymatically inactive in the absence of 5'-AMP.
4. The maximal rate is reached within approximately 3 minutes. This maximal rate is due to any phosphorylase a activity which may be present.
5. The maximum linear rate should not exceed a ΔA340nm of 0.1 absorbance unit/minute.
6. Glucose-6-Phosphate Dehydrogenase Unit Definition: One unit will oxidize 1.0 μmole of D-glucose 6-phosphate to 6-phospho-D-gluconate per minute in the presence of β-NADP at pH 7.4 at 25°C.
7. Phosphoglucomutase Unit Definition: One unit will convert 1.0 μmole of α-D-glucose 1-phosphate to α-D-glucose 6-phosphate per minute at pH 7.4 at 30°C.

Enzymatic Assay of PHOSPHORYLASE KINASE (EC 2.7.1.38)

Principle

$$\text{2 Phosphorylase b} + \text{4 ATP} \xrightarrow{\text{Phosphorylase Kinase}} \text{Phosphorylase a} + \text{4 ADP}$$

$$\text{Glycogenn} + \text{Pi} \xrightarrow{\text{Phosphorylase a}} \text{Glycogenn - 1} + \alpha\text{-D-Glucose 1-Phosphate}$$

$$\alpha\text{-D-Glucose 1-Phosphate} \xrightarrow{\text{PGLUM}} \alpha\text{-D-Glucose 6-Phosphate}$$

$$\alpha\text{-D-Glucose 6-Phosphate} + \beta\text{-NADP} \xrightarrow{\text{G-6-PDH}} \text{6-Phosphogluconate} + \beta\text{-NADPH}$$

Abbreviations Used

ATP = Adenosine 5'-Triphosphate
ADP = Adenosine 5'-Diphosphate
Pi = Inorganic Phosphate
PGLUM = Phosphoglucomutase
β-NADP = β-Nicotinamide Adenine Dinucleotide Phosphate, Oxidized Form
G-6-PDH = Glucose-6-Phosphate Dehydrogenase
β-NADPH = β-Nicotinamide Adenine Dinucleotide Phosphate, Reduced Form

Conditions: T = 30°C, pH = 7.7, A340nm, Light path = 1 cm

Method: Continuous Spectrophotometric Rate Determination

Reagents

A. 30 mM L-Cysteine Solution, pH 7.0 at 30°C (L-Cys, pH 7.0) (Prepare 50 ml in deionized water using L-Cysteine, Hydrochloride, Monohydrate. Adjust to pH 7.0 at 30°C with solid Sodium Bicarbonate. *Prepare Fresh*).

B. 80 mM L-Cysteine Solution, pH 6.8 at 30°C (L-Cys, pH 6.8) (Prepare 50 ml in deionized water using L-Cysteine, Hydrochloride, Monohydrate. Adjust to pH 6.8 at 30°C with solid Sodium Bicarbonate. *Prepare Fresh*).

C. 250 mM Tris HCl Buffer, pH 7.7 at 30°C (Prepare 100 ml in deionized water using Trizma Hydrochloride. Adjust to pH 7.7 at 30°C with 1 M NaOH).

D. 250 mM β-Glycerophosphate Solution (β-Glycerophos) (Prepare 10 ml in Reagent C using β-Glycerophosphate, Disodium Salt, Hydrate).

E. 40 mM β-Glycerophosphate Solution, pH 6.8 at 30°C (β-Glycerophos, pH 6.8) (Prepare 50 ml in Reagent B using β-Glycerophosphate, Disodium Salt, Hydrate. Adjust to pH 6.8 at 30°C with 1 M HCl or 1 M NaOH, if necessary).

F. 100 mM Magnesium Acetate Solution, pH 7.7 at 30°C (Mg Acet) (Prepare 10 ml in deionized water using Magnesium Acetate, Tetrahydrate. Adjust to pH 7.7 at 30°C with 1 M NaOH).

G. Phosphorylase b Enzyme Solution (Phos b) (Immediately before use, prepare a solution containing approximately 300 units/ml of Phosphorylase b in cold Reagent A).

H. 30 mM Adenosine 5'-Triphosphate Solution, pH 7.7 at 30°C (ATP) (Prepare 10 ml in deionized water using Adenosine 5'-Triphosphate, Disodium Salt. Adjust to pH 7.7 at 30°C with 1 M NaOH).

I. Phosphorylase Kinase Enzyme Solution (Phos Kinase) (Immediately before use, prepare a solution containing approximately 75 units/ml of Phosphorylase Kinase in cold Reagent A).

Procedure

Pipette (in milliliters) the following reagents into suitable cuvettes:

	Test	Blank
Reagent A (L-Cys, pH 7.0)	0.29	0.30
Reagent D (β-Glycerophos)	0.25	0.25
Reagent F (Mg Acet)	0.10	0.10
Reagent G (Phos b)	0.25	0.25
Reagent I (Phos Kinase)	0.01	—

Mix by inversion and equilibrate to 30°C. Then add:

Reagent H (ATP)	0.10	0.10

Immediately mix by inversion and incubate at 30°C. After exactly 5 and 10 minutes, remove 0.1 ml from both the Test and Blank solutions and place into suitable tubes containing 1.90 ml of Reagent E (β-Glycerophos, pH 6.8) to stop the reaction. Assay for Phosphorylase a activity according to the attached procedure (Enzymatic Assay of Phosphorylase a), substituting 0.10 ml from each of the aforementioned stopped reactions in place of Reagent J (Phosphorylase a Enzyme Solution) in the attached assay. Proceed with the calculation.

Calculations

Phosphorylase a Activity

$$\text{Units / ml enzyme} = \frac{(\Delta A340nm / \min \text{ Test} - \Delta A340nm / \min \text{ Blank})(3)(df)}{(6.22)(0.1)}$$

3 = Total volume (in milliliters) of assay
df = Dilution factor
6.22 = Millimolar extinction coefficient of β-NADH at 340nm
0.1 = Volume (in milliliter) of enzyme used

$$\text{Units / mg solid} = \frac{\text{units / ml enzyme}}{\text{mg solid / ml enzyme}}$$

Phosphorylase Kinase Activity

$$\text{Units / ml enzyme} = \frac{(\text{units of Phos a Test / ml}) - (\text{units of Phos a Blank / ml})(2)}{(T)(0.01)(0.1)}$$

2 = Volume (in milliliters) of the stopped reaction (Phosphorylase Kinase Assay)
T = Time (in minutes) of the assay (Phosphorylase Kinase Assay) as per the Unit Definition
0.1 = Volume (in milliliter) of reaction mix used in the stopped reaction
0.01 = Volume (in milliliter) of enzyme used in Phosphorylase Kinase assay

$$\text{Units / mg solid} = \frac{\text{units / ml enzyme}}{\text{mg solid / ml enzyme}}$$

$$\text{Units / mg protein} = \frac{\text{units / ml enzyme}}{\text{mg protein / ml enzyme}}$$

Unit Definition: *One unit will form 1.0 μmolar unit of phosphorylase a from phosphorylase b per minute at pH 7.7 at 30°C in the presence of ATP.*

Final Assay Concentrations

In a 1.00 ml reaction mix, the final concentrations are 17 mM L-cysteine, 63 mM β-glycerophosphate, 10 mM magnesium acetate, 75 units phosphorylase b, 63 mM Tris, and 0.75 unit phosphorylase kinase.

Notes

1. β-Glycerophosphate is a stabilizer for phosphorylase kinase.
2. Phosphorylase b Unit Definition: One unit will form 1.0 μmole of α-D-glucose 1-phosphate from glycogen and orthophosphate in the presence of 5'-AMP, per minute at pH 6.8 at 30°C measured in a system containing phosphoglucomutase, β-NADP, and glucose 6-phosphate dehydrogenase.

Enzymatic Assay of PHOSPHOTRANSACETYLASE (EC 2.3.1.8)

Principle

$$\text{CoA} + \text{Acetyl Phosphate} \xrightarrow{\text{Phosphotransacetylase}} \text{Acetyl-CoA} + \text{Pi}$$

Abbreviations Used

CoA = Coenzyme A
Acetyl-CoA = Acetyl Coenzyme A
Pi = Inorganic phosphate

Conditions: T = 25°C, pH = 7.4, A233nm, Light path = 1 cm

Method: Continuous Spectrophotometric Rate Determination

Reagents

A. 100 mM Tris HCl Buffer, pH 7.4 at 25°C (Prepare 100 ml in deionized water using Trizma Base. Adjust to pH 7.4 at 25°C with 5 M HCl).

B. 100 mM Glutathione Solution (Gluth) (Prepare 10 ml in Reagent A using Glutathione, Free Acid, Reduced Form. *Prepare Fresh*).

C. 6.5 mM Coenzyme A Solution (CoA) (Prepare 2 ml in Reagent A using Coenzyme A, Sodium Salt. *Prepare Fresh*).

D. 220 mM Acetyl Phosphate Solution (Acet Phos) (Prepare 1 ml in Reagent A using Acetyl Phosphate, Lithium Potassium Salt. *Prepare Fresh*).

E. 1 M Ammonium Sulfate Solution ($(NH_4)2SO_4$) (Prepare 50 ml in deionized water using Ammonium Sulfate).

F. 25 mM Tris HCl Buffer with 500 mM Ammonium Sulfate, pH 8.0 at 25°C (Enzyme Diluent) (Prepare 25 ml in deionized water using Trizma Base and Ammonium Sulfate. Adjust to pH 8.0 at 25°C with 1 M HCl).

G. Phosphotransacetylase Enzyme Solution (Immediately before use, prepare a solution containing 1–2 units of phosphotransacetylase in cold Reagent F).

Procedure

Pipette (in milliliters) the following reagents into suitable cuvettes.

	Test	Blank
Reagent A (Buffer)	2.60	2.60
Reagent B (Gluth)	0.05	0.05
Reagent C (CoA)	0.20	0.20
Reagent D (Acet Phos)	0.10	0.10
Reagent E ($(NH_4)2SO_4$)	0.03	0.03

Mix by inversion and equilibrate to 25°C. Moniter the A233nm until constant, using a suitably thermostatted spectrophotometer. Then add:

Reagent G (Enzyme Solution)	0.02	—
Reagent F (Enzyme diluent)	—	0.02

Immediately mix by inversion and record the increase in A233nm for approximately 5 minutes. Obtain the ΔA233nm/minute using the maximum linear rate for both the Test and Blank.

Calculations

$$\text{Units / mg enzyme} = \frac{(\Delta A233\text{nm / min Test} - \Delta A233\text{nm / min Blank})\ (3.0)\ (\text{df})}{(4.44)\ (0.02)}$$

3.0 = Total volume (in milliliters) of assay

df = Dilution factor

4.44 = Millimolar extinction coefficient of Acetyl-CoA at 233nm

0.02 = Volume (in milliliter) of enzyme used

$$\text{Units / mg solid} = \frac{\text{units / ml enzyme}}{\text{mg solid / ml enzyme}}$$

$$\text{Units / mg protein} = \frac{\text{units / ml enzyme}}{\text{mg protein / ml enzyme}}$$

Unit Definition: *One unit will convert 1.0 µmole of CoA to acetyl-CoA per minute at pH 7.4 at 25°C using acetyl phosphate as substrate.*

Final Assay Concentrations

In a 3.00 ml reaction mix, the final concentrations are 98.5 mM Tris, 1.6 mM glutathione, 0.43 mM coenzyme A, 7.23 mM acetyl phosphate, 13.3 mM ammonium sulfate and 0.02–0.04 unit phosphotransacetylase.

Enzymatic Assay of PHYTASE (EC 3.1.3.8)

Principle

$$\text{Phytic Acid} + H_2O \xrightarrow{\text{Phytase}} \text{D-myo-inositol 1, 2, 4, 5, 6-PKP} + \text{Pi}$$

Abbreviations Used

D-myo-inositol 1,2,4,5,6-PKP = D-myo-inositol
1,2,3,4,5-Pentakisphosphate
Phytic Acid = Myo-Inositol Hexakis[dihydrogen phosphate]
Pi = Inorganic Phosphate

Conditions: T = 37°C, pH 2.5, A400nm, Light path = 1 cm

Method: Colorimetric

Reagents

A. 200 mM Glycine Buffer, pH 2.8 at 37°C (Prepare 100 ml in deionized water using Glycine, Hydrochloride. Adjust to pH 2.8 at 37°C with either 1 M HCl or 1 M NaOH).
B. 44.1 mM Phytic Acid Solution, pH 2.5 at 37°C (Phytic Acid) (Prepare 10 ml in Reagent A, using Phytic Acid, Magnesium-Potassium Salt. Adjust to pH 2.5 at 37°C with 1 M HCl).1 C. 5% (w/v) Ammonium Molybdate Solution (Amm Moly) (Prepare 100 ml in deionized water using Molybdic Acid, Ammonium Tetrahydrate Salt).
D. 5 N Sulfuric Acid Solution (H_2SO_4) (Prepare 100 ml in deionized water using Sulfuric Acid).
E. Acetone (Use Acetone, ACS Reagent Grade).
F. Color Reagent Solution (CRS) (Prepare 100 ml by adding 25 ml of Reagent C to 25 ml of Reagent D and 50 ml of Reagent E. Mix immediately prior to use. *Prepare Fresh*).
G. 50 mM Potassium Phosphate Solution (P Std) (Prepare 10 ml in deionized water using Potassium Phosphate, Monobasic, Anhydrous).

H. Phytase Enzyme Solution (Immediately before use, prepare a solution containing 0.5–2.0 units/ml of Phytase in cold Reagent A).

Procedure

Pipette (in milliliters) the following reagents into a suitable container:

	Test	Blank
Reagent B (Phytic Acid)	0.50	0.50

Equilibrate to 37°C. Then add:

Reagent H (Enzyme Solution)	0.025	—
Reagent A (Buffer)	—	0.025

Immediately mix by inversion and incubate at 37°C for exactly 30 minutes. Then add:

Reagent F (CRS)	4.00	4.00
Deionized Water	0.025	0.025

Mix by swirling. Transfer to suitable cuvettes and record the A400nm for both the Test and Blank using a suitable spectrophotometer.

Colorimetric Assay

Standard Curve: A standard curve is made by pipetting (in milliliters) the following reagents into suitable containers:

	Std 1	Std 2	Std 3	Std 4	Std 5	Std Blank
Deionized Water	0.04	0.03	0.02	0.01	—	0.05
Reagent G (P Std)	0.01	0.02	0.03	0.04	0.05	—
Reagent B (Phytic Acid)	0.05	0.05	0.05	0.05	0.05	0.05

Mix by swirling and incubate at 37°C for 30 minutes. Then add:

Reagent F (CRS)	4.00	4.00	4.00	4.00	4.00	4.00

Mix by swirling. Transfer to suitable cuvettes and record the A400nm for both the Standards and the Standard Blank using a suitable spectrophotometer.

Calculations

Standard Curve: A400nm Standard = A400nm Standard–A400nm Standard Blank

Prepare a standard curve by plotting A400nm Standard versus µmoles of phosphate.

Sample Determination: A400nm Test = A400nm Test–A400nm Test Blank

Determine the µmoles of phosphate liberated using the Standard curve.

$$\text{Units / ml enzyme} = \frac{(\mu\text{moles of Phosphate released})\ (\text{df})}{(30)\ (0.025)}$$

df = Dilution factor

30 = Time (in minutes) of assay per the Unit Definition

0.025 = volume (in milliliter) of enzyme used

$$\text{Units / mg solid} = \frac{\text{units / ml enzyme}}{\text{mg solid / ml enzyme}}$$

Unit Definition: *One unit will liberate 1.0 μmole of inorganic P from 4.2 × 10–2 M Mg·K phytate (P-7660) per minute at pH 2.5 at 37°C.*

Final Assay Concentration

In a 0.525 ml reaction mix, the final concentrations are 200 mM glycine, 42 mM phytic acid and 0.013–0.050 unit phytase.

Enzymatic Assay of PHYTASE (EC 3.1.3.26)

Principle

$$\text{Phytic Acid} + H_2O \xrightarrow{\text{Phytase}} \text{1L - myo - inositol 1, 2, 3, 4, 5 - PKP} + \text{Pi}$$

Abbreviations Used

L-Myo-inositol 1,2,3,4,5-PKP = L-Myo-inosital 1,2,3,4,5-Pentakisphosphate
Phytic Acid = Myo-Inositol Hexakisphosphate
Pi = Inorganic Phosphate

Conditions: T = 55°C, pH 5.15, A660nm, Light path = 1 cm

Method: Colorimetric

Reagents

A. 200 mM Sodium Acetate Buffer, pH 5.15 at 55°C (Prepare 100 ml in deionized water using Sodium Acetate, Trihydrate. Adjust to pH 5.15 at 55°C with 1 M Acetic Acid).
B. 100 mM Magnesium Sulfate Solution ($MgSO_4$) (Prepare 100 ml in deionized water, using Magnesium Sulfate, Heptahydrate).
C. 6.82 mM Phytic Acid Solution, pH 5.15 at 55°C (Prepare 25 ml in Reagent A using Phytic Acid, Dodecasodium Salt. Adjust the pH to 5.15 at 55°C with 0.1 N Acetic Acid).
D. 10% (v/v) Trichloroacetic Acid Solution (TCA) (Prepare 50 ml in deionized water using Trichloroacetic Acid, 6.1 N Solution).
E. Phosphorus Standard Solution (P Std) (Use Phosphorus Standard Solution. The phosphorus concentration is 20 μg/ml, 0.645 μmoles/ml).
F. 10% (w/v) Ammonium Molybdate Solution (Amm Moly) (Prepare 25 ml in 10 N Sulfuric Acid using Molybdic Acid, Ammonium Tetrahydrate Salt and Sulfuric Acid).
G. Tausky-Shorr Color Reagent Solution (TSCR) (Prepare by adding 10 ml Reagent F to 70 ml of deionized water. Then add 5 g Ferrous Sulfate, Heptahydrate and mix until dissolved. Add enough deionized water to a final volume of 100 ml. Store in the dark).
H. Phytase Enzyme Solution (Immediately before use, prepare a solution containing 20 mg solid/ml in cold deionized water).

Procedure

Prepare the following Standards (in milliliters):

	Std Blank	Std 1	Std 2	Std 3	Std 4	Std 5
Reagent E (P Std)	0.00	0.20	0.40	0.60	0.80	1.00
Deionized Water	2.00	1.80	1.60	1.40	1.20	1.00
Reagent D (TCA)	1.00	1.00	1.00	1.00	1.00	1.00

Pipette (in milliliters) the following reagents into a 50 ml Erlenmeyer flask:

	Test
Deionized Water	5.00
Reagent A (Buffer)	10.00
Reagent B ($MgSO_4$)	0.40
Reagent C (Phytic Acid)	4.40

Mix by swirling and equilibrate at 55°C for several minutes. At time zero, add:

Reagent H (Enzyme)	0.20

Immediately mix gently by swirling, stopper the flasks, and incubate at 55°C. At each of the times 0, 10, 20, 30, 40, 50 and 60 minutes, remove a 2.0 ml aliquot and transfer to an appropriately labeled test tube that contains 1.0 ml of Reagent D (TCA). After collecting all the aliquots, add to each of the aliquots and the phosphorus standards (labeled Std Blank to Std6) the following reagents (in milliliters):

Deionized Water	2.0
Reagent G (TSCR)	5.0

Mix by swirling. After approximately 5 minutes, read the A660nm of each of the test tubes.

Calculations

Standard Curve: ΔA660nm Standard = A660nm Standard–A660nm Standard Blank

Prepare a standard curve by plotting ΔA660nm Standard versus μmoles of phosphate.

Sample Determination: ΔA660nm Test = A660nm Test–A660nm Test Blank

Determine the μmoles of phosphate liberated using the Standard curve.

$$\text{Units / ml enzyme} = \frac{(\mu\text{moles of Phosphate released})\ (\text{df})\ (20)}{(\text{T})\ (0.20)\ (2.0)}$$

df = Dilution factor

T = 10, 20, 30, 40, 50, or 60 minutes

0.20 = Volume (in milliliters) of enzyme used

20 = Volume of (milliliters) of initial Reaction Mixture volume

2.0 = Volume of Reaction Mixture taken for Phosphorus quantitation

$$\text{Units / mg solid} = \frac{\text{units / ml enzyme}}{\text{mg solid / ml enzyme}}$$

Unit Definition: *One unit will liberate 1.0 μmole of inorganic phosphorus from 1.5 × 10–3 M phytate per minute at pH 5.15 at 55°C.*

Final Assay Concentration

In a 20.00 ml reaction mix, the final concentrations are 144 mM acetate, 2.0 mM magnesium sulfate, 1.5 mM phytate and 4 mg phytase.

Notes

1. It is important that the phytic acid concentrations be corrected for purity and water content since the substrate concentrations are critical.
2. The color is not stable, so each test tube should be read at approximately the same time after the addition of Reagent G (TSCR).

Enzymatic Assay of PLASMA AMINE OXIDASE (EC 1.4.3.6)

Principle

$$\text{Benzylamine} + O_2 + H_2O \xrightarrow{\text{Monoamine Oxidase}} + \text{Benzaldehyde} + NH_3 + H_2O_2$$

Conditions: T = 25°C, pH = 7.4 , A250nm, Light path = 1 cm

Method: Continuous Spectrophotometric Rate Determination

Reagents

A. 200 mM Potassium Phosphate Buffer, pH 7.4 at 25°C (Prepare 100 ml in deionized water using Potassium Phosphate Monobasic. Adjust to pH 7.4 at 25°C with 1 M NaOH).

B. 1 M Sulfuric Acid Solution (H_2SO_4) (Prepare 10 ml in dieonized water using Sulfuric Acid).

C. 100 mM Benzylamine Sulfate Solution (Prepare 100 ml in deionized water. Facilitate solubilization by first dissolving into 5 ml of Reagent B using Benzylamine, Hydrochloride. Adjust to pH 7.4 with 1 M NaOH).

D. Plasma Amine Oxidase Enzyme Solution (Immediately before use, prepare a solution containing approximately 0.15–0.20 units/ml of Plasma Amine Oxidase in cold Reagent A).

Procedure

Pipette (in milliliters) the following reagents into suitable quartz cuvettes:

	Test	Blank
Reagent A (Buffer)	2.80	2.80
Reagent C (Benzylamine Sulfate)	0.10	0.10

Mix by inversion and equilibrate to 25°C. Monitor the A250nm until constant, using a suitably thermostatted spectrophotometer. Then add:

Reagent D (Enzyme Solution)	0.10	—

Immediately mix by inversion and record the increase in A250nm for approximately 5 minutes. Obtain the ΔA250nm/minute by using the maximum linear rate for both the Test and Blank.

Calculations

$$\text{Units / ml enzyme} = \frac{(\Delta A250nm / \text{min Test} - \Delta A250nm / \text{min Blank})\,(3)\,(df)}{(11.3)\,(0.1)}$$

3 = Total volume (in milliliters) of assay
df = Dilution factor
11.3 = Difference between the millimolar extinction coefficients of benzylamine and benzaldehyde at 250nm
0.1 = Volume (in milliliter) of enzyme used

$$\text{Units / mg solid} = \frac{\text{units / ml enzyme}}{\text{mg solid / ml enzyme}}$$

$$\text{Units / mg protein} = \frac{\text{units / ml enzyme}}{\text{mg protein / ml enzyme}}$$

Unit Definition: *One unit will oxidize 1.0 micromole of benzylamine to benzaldehyde per minute at pH 7.4 at 25°C.*

Final Assay Concentrations

In a 3.00 ml reaction mix, the final concentrations are 187 mM potassium phosphate buffer, 3.33 mM benzylamine sulfate and 0.015–0.02 unit plasma amine oxidase.

Enzymatic Assay of PLASMIN (EC 3.4.21.7)

Principal

$$\alpha\text{-Casein (Perchloric Acid Insoluble)} \xrightarrow{\text{Plasmin}} \text{Perchloric Acid Soluble Casein Fragments}$$

Conditions: T = 37°C, pH = 7.5, A275nm, Light Path = 1 cm

Method: Spectrophotometric Stop Rate Determination

Reagents

A. 10 mM Potassium Phosphate with 70 mM Sodium Phosphate and 100 mM Lysine, pH 7.5 at 37°C (Phosphate Lysine Buffer) (Prepare 100 ml in deionized water using Potassium Phosphate, Monobasic, Anhydrous, Sodium Phosphate, Dibasic, Anhydrous and L-Lysine, Monohydrochloride. Adjust to pH 7.5 at 37°C with 1 M NaOH).

B. 400 mM Sodium Phosphate Buffer, pH 7.5 at 37°C (Phosphate Buffer) (Prepare 350 ml in deionized water using Sodium Phosphate, Monobasic, Anhydrous. Adjust to pH 7.5 at 37°C using 1 M NaOH).

C. 1% (w/v) Perchloric Acid Solution (Prepare 1000 ml by diluting 14.3 ml of Perchloric Acid to 1000 ml with deionized water).

D. 3% (w/v) α-Casein Solution (α-Casein) (Prepare 330 ml by dispersing 10 g of α-Casein in 85 ml of deionized water. Add 7 ml of 1 M NaOH, and heat on a hot plate until the Casein is dissolved. Then cool the solution and adjust the volume to 700 ml using deionized water. Add 3 ml of 1 M HCl to the solution with stirring and then add 13 ml more. Slowly transfer the Casein solution to 1000 ml of 1% Perchloric Acid (Reagent C) with stirring. Allow the solution to stand overnight at room temperature. Centrifuge the resulting suspension at low speed the next day). Wash the precipitate three times with 150 ml each of deionized water and suspend the washed precipitate in 120 ml of Phosphate Buffer (Reagent B). Adjust the pH of the suspension to 7.5 using 1 M NaOH and stir until the Casein is completely dissolved. Readjust the pH to 7.5 and dilute to 330 ml using Phosphate Buffer (Reagent B). Freeze in aliquots of 5–10 ml. This substrate is *very unstable*; keep frozen until ready to use. Thaw out in a 37°C water bath and use as soon as the last ice crystals melt).

E. 1.7 M Perchloric Acid Solution (Perchloric Acid) (Prepare 100 ml in deionized water using Perchloric Acid).

F. Plasmin Solution (Plasmin) (Immediately before use prepare a solution containing 1 unit/ml Plasmin in cold deionized water).

Procedure

Pipette (in milliliters) the following reagents into suitable tubes.

	Test	Blank
Reagent A (Phosphate Lysine Buffer)	0.50	0.50
Reagent F (Plasmin)	0.30	—
Deionized Water	0.20	0.20

Mix by swirling and then add:

Reagent D (α-Casein)	1.00	1.00

Immediately mix by swirling and incubate at 37°C for exactly 20 minutes. Then add:

Reagent E (Perchloric Acid)	3.00	3.00

Mix by swirling gently. Then add:

Reagent F (Plasmin)	—	0.30

Mix by swirling and allow to stand at 25°C for 20 minutes. Centrifuge and then filter the solutions through #54 Whatman filter paper. If the solutions are still hazy, filter through 0.45 μm syringe filters. Transfer the solutions to suitable cuvettes and record the A275nm for both the Test and Blank. The absorbance for the blank should be between 0.075 and 0.10. The corrected absorbance for both the samples and controls should be below 0.50.

Calculations

$$\text{Units / ml enzyme} = \frac{(\text{A275nm Test} - \text{A275nm Blank})\ (\text{df})}{(0.30)\ (1.0)}$$

df = Dilution factor
0.30 = Volume (in milliliter) of enzyme used
1.0 = Change in absorbance at A275nm as per the Unit Definition

$$\text{Units / mg solid} = \frac{\text{units / ml enzyme}}{\text{mg solid / ml enzyme}}$$

$$\text{Units / mg protein} = \frac{\text{units / ml enzyme}}{\text{mg protein / ml enzyme}}$$

Unit Definition: *One unit will produce a ΔA275nm of 1.0 from α-casein in 20 minutes at pH 7.5 at 37°C, when measuring perchloric acid soluble products in a volume of 5.0 ml.*

Final Assay Concentrations

In a 2.00 ml reaction mix, the final concentrations are 2.5 mM potassium phosphate, 1.5% (w/v) α-casein, 218 mM sodium phosphate, 25 mM lysine, and 0.3 unit plasmin.

Note

1. Varying volumes of plasmin (not exceeding 0.50 ml) may be added to the test reaction mix. Bring the volume of plasmin up to 0.50 ml with deionized water. For each test a blank must be run containing the same amount of plasmin.

Enzymatic Assay of PLASMINOGEN

Procedure

$$\text{Plasminogen} \xrightarrow{\text{Urokinase}} \text{Plasmin}$$

$$\alpha\text{-Casein (Perchloric Acid Insoluble)} \xrightarrow{\text{Plasmin}} \text{Perchloric Acid Soluble Casein Fragments}$$

Conditions: T = 37°C, pH = 7.5, A275nm, Light path = 1 cm

Method: Spectrophotometric Stop Rate Determination

Reagents

A. 400 mM Potassium Phosphate Buffer, pH 7.5 at 37°C (Phosphate Buffer) (Prepare 350 ml in deionized water using Potassium Phosphate, Monobasic, Anhydrous. Adjust to pH 7.5 at 37°C with 1 M KOH).

B. 10 mM Potassium Phosphate with 70 mM Sodium Phosphate and 100 mM Lysine, pH 7.5 at 37°C (Phosphate/Lysine Buffer) (Prepare 100 ml in deionized water using Potassium Phosphate, Monobasic, Anhydrous, Sodium Phosphate, Dibasic, Anhydrous and L-Lysine, Monohydrochloride. Adjust to pH 7.5 at 37°C with either 1 M HCl or 1 M NaOH).

C. 1% (w/v) Perchloric Acid Solution (Perchloric Acid) (Prepare 1000 ml by diluting 14.3 ml of Perchloric Acid to 1000 ml with deionized water).

D. 3% (w/v) α-Casein Solution (α-Casein) (Prepare 330 ml by dispersing 10 g of α-Casein in 85 ml of deionized water. Add 7 ml of 1 M NaOH and heat on a hot plate until the α- Casein is dissolved. Then cool the solution and adjust the volume to 700 ml using deionized water. Add 3 ml of 1 M HCl to the solution with stirring and then add an additional 13 ml of 1 M HCl. Slowly transfer the α-Casein solution to

1000 ml of Reagent C (Perchloric Acid) with stirring. Allow the solution to stand overnight at room temperature. Centrifuge the resulting suspension at room temperature. Wash the precipitate three times with 150 ml of deionized water and suspend the washed precipitate in 120 ml of Phosphate Buffer (Reagent A). Adjust the pH of the suspension to 7.5 using 1 M NaOH and stir until the α-Casein is completely dissolved. Readjust to pH 7.5 and dilute to 330 ml using Phosphate Buffer (Reagent A). Freeze in aliquots of 5 to 10 ml. The substrate is *very unstable*; keep frozen until ready to use. Thaw out in a 37°C water bath and use as soon as the last ice crystals melt).

E. 1.7 M Perchloric Acid Solution (PA) (Prepare 100 ml in deionized water using Perchloric Acid).

F. Urokinase Enzyme Solution (Urokinase) (Use Urokinase, undiluted, approximately 25 units/ml).

G. Plasminogen Solution (Immediately before use, prepare a solution containing 1.0 unit/ml of Plasminogen in cold deionized water).

Procedure

Pipette (in milliliters) the following reagents into suitable containers:

	Test	Blank
Reagent B (Phosphate/Lysine Buffer)	0.50	0.50
Reagent G (Plasminogen)	0.30	0.30
Reagent F (Urokinase)	0.01	—
Deionized Water	0.20	0.20

Mix by swirling and incubate at 37°C for 15 minutes. Then add:

Reagent D (α-Casein)	1.00	1.00

Mix by swirling and incubate at 37°C for 20 minutes. Then add:

Reagent E (PA)	3.00	3.00

Mix by swirling. Then add:

Reagent F (Urokinase)	—	0.01

Mix by swirling and allow to stand at 25°C for 20 minutes. Centrifuge both the Test and Blank solutions. Then filter through Whatman #54 filter paper. If the solutions are still hazy, filter through a 0.45 μm syringe filter. Transfer the filtered solutions to suitable cuvettes and record the A275nm for both the Test and Blank. The corrected absorbance for the Test should be below 0.5.

Calculation

$$\text{Units/ml enzyme} = \frac{(\text{A275nm Test} - \text{A275nm Blank})\,(\text{df})}{(0.3)\,(1)}$$

df = Dilution factor
0.3 = Volume (in milliliter) of plasminogen
1 = Change in absorbance at A275nm as per the Unit Definition

$$\text{Units / mg solid} = \frac{\text{units / ml enzyme}}{\text{mg solid / ml enzyme}}$$

$$\text{Units / mg protein} = \frac{\text{units / ml enzyme}}{\text{mg protein / ml enzyme}}$$

Unit Definition: *One unit will produce a ΔA275 of 1.0 from α-casein in 20 minutes at pH 7.5 at 37°C, when measuring perchloric acid soluble products in a volume of 5.0 ml. Activity is determined after activation to plasmin with urokinase.*

Final Assay Concentrations

In a 2.01 ml reaction mix, the final concentrations are 17 mM sodium phosphate, 202 mM potassium phosphate, 25 mM L-lysine, 1.5% (w/v) α-casein, 0.25 unit urokinase, and 0.3 unit plasminogen.

Note

1. Varying volumes of Plasminogen (not exceeding 0.50 ml) may be added to the test reaction mix. Bring the volume of plasminogen up to 0.50 ml with deionized water. For each test a blank must be run containing the same amount of plasminogen.

Enzymatic Assay of POLYGALACTURONASE (EC 3.2.1.15)

Principle

$$\text{Polygalacturonic Acid} + H_2O \xrightarrow{PG} \text{Reducing Sugars}$$

Abbreviation Used

PG = Polygalacturonase

Conditions: T = 30°C, pH 5.0, A540nm, Light path = 1 cm

Method: Colorimetric

Reagents

A. 50 mM Sodium Acetate Buffer, pH 5.0 at 30°C (Prepare 100 ml in deionized water using Sodium Acetate, Trihydrate. Adjust to pH 5.0 at 37°C with 1 M HCl).

B. 0.50% (w/v) Polygalacturonic Acid Solution (Poly Gal) (Prepare 25 ml in Reagent A (Buffer) using Polygalacturonic Acid, Sodium Salt).

C. 50 mM Sodium Acetate with 0.05% (w/v) Bovine Serum Albumin, pH 5.0 at 30°C (Enzyme Diluent) (Prepare 25 ml in Reagent A (Buffer) using Albumin, Bovine).

D. Polygalacturonase Enzyme Solution (Immediately before use, prepare a solution containing 0.2–1.0 unit/ml Polygalacturonase in cold Reagent C).

E. 16 mM Copper Sulfate, 1300 mM Sodium Sulfate, 226 mM Sodium Carbonate, 190 mM Sodium Bicarbonate and 43 mM Sodium Potassium Tartrate Solution (Copper Solution) (Prepare 1 liter in deionized water using Cupric Sulfate Pentahydrate, Sodium Bicarbonate, Sodium Sulfate, Anhydrous, Sodium Carbonate, Anhydrous and Sodium Potassium Tartrate Tetrahydrate).

F. 40 mM Molybdic Acid, 19 mM Arsenic Acid and 756 mM Sulfuric Acid Solution (Arsenomolybdate Soln) (Prepare 1 liter in deionized water using Molybdic Acid, Ammonium Salt, Tetrahydrate, Arsenic Acid, Sodium Salt and Sulfuric Acid).

G. 0.1% (w/v) D-Galacturonic Acid Standard Solution (Gal Std) (Prepare 10 ml in deionized water using D-Galacturonic Acid, Monohydrate).

Procedure

Pipette (in milliliters) the following reagents into suitable tubes:

	Test	Blank
Reagent B (Poly Gal)	1.90	1.90

Equilibrate to 30°C. Then add:

Reagent D (Enzyme Soln)	0.10	—
Reagent C (Enzyme Diluent)	—	0.10

Immediately mix by inversion and incubate at 30°C for 10 minutes. Then add:

Reagent E (Copper Soln)	2.00	2.00

Cover the tubes with glass marbles and place in a boiling water bath for 15 minutes. Cool to room temperature. Then add:

Reagent F (Arsenomolybdate Soln)	2.00	2.00

Mix by inversion until the foaming stops or the precipitate is dissolved. Centrifuge the tubes to clarify the solution. Transfer the supernatant to suitable cuvettes and record the A540nm for both the Test and Blank using a suitable spectrophotometer.

Standard Curve: A standard curve is made by pipetting (in milliliters) the following reagents into suitable containers.

	Std 1	Std 2	Std 3	Std Blank
Reagent G (Gal Std)	0.00	0.03	0.07	0.10
Reagent A (Buffer)	2.00	1.97	1.93	1.90

Mix by inversion and incubate at 30°C for 10 minutes.Then add:

Reagent E (Copper Soln)	2.00	2.00	2.00	2.00

Cover the tubes with glass marbles and place in a boiling water bath for 15 minutes. Cool to room temperature. Then add:

Reagent F (Arsenomolybdate Soln)	2.00	2.00	2.00	2.00

Mix by inversion until the foaming stops or the precipitate is dissolved. Centrifuge the tubes to clarify. Transfer to suitable cuvettes and record the A540nm for the Standards and Standard Blank using a suitable spectrophotometer.

Calculations

Standard Curve: ΔA540nm Standard = A540nm Standard–A540nm Standard Blank

Prepare a standard curve by plotting ΔA540nm Standard vs µmoles of D-Galacturonic Acid.

Sample Determination: ΔA540nm Sample = A540nm Test–A540nm Blank

Determine the µmoles of D-Galacturonic Acid liberated using the standard curve.

Calculations

$$\text{Units / ml enzyme} = \frac{(\mu\text{moles of D - Galacturonic Acid released})\,(\text{df})}{(10)\,(0.1)}$$

df = Dilution factor
10 = Time of assay (in minutes) as per the Unit Definition
0.1 = Volume (in milliliters) of enzyme used

$$\text{Units / mg solid} = \frac{\text{units / ml enzyme}}{\text{mg solid / ml enzyme}}$$

$$\text{Units / mg protein} = \frac{\text{units / ml enzyme}}{\text{mg protein / ml enzyme}}$$

Unit Definition: *One unit will release 1.0 µmole of reducing sugar measured as D-galacturonic acid from polygalacturonic acid per minute at pH 5.0 at 30°C.*

Final Assay Concentration

In a 2.00 ml reaction mix, the final concentrations are 0.5% (w/v) polygalacturonic acid, 50 mM sodium acetate, 0.002% (w/v) Bovine Serum Albumin and 0.02–0.10 unit polygalacturonase.

Notes

1. Sodium Sulfate, Sodium Carbonate, and Sodium Potassium Tartrate are dissolved in approximately 500 ml of deionized water. Cupric Sulfate is dissolved in approximately 100 ml of deionized water and slowly added to the above solution to avoid precipitation. Sodium Bicarbonate is dissolved first in deionized water and then added to the above solution. Dilute the solution to 1 liter. If a precipitate forms, it should be removed by filtration prior to use. Store in an amber bottle and avoid exposure to direct sunlight. Store at room temperature.
2. Molybdic Acid is dissolved in approximately 300 ml of deionized water. Add Sulfuric Acid slowly. Caution, this is an exothermic reaction! Arsenic Acid is dissolved in approximately 300 ml of deionized water and is added to the above solution. The solution is diluted to a total volume of 1 liter and incubated at 37°C for 48–72 hours. If a precipitate forms, it should be removed by filtration prior to use. Store in an amber bottle and avoid exposure to direct sunlight. The solution expires six months after its preparation. Store at room temperature in an exhaust hood.

Enzymatic Assay of POLYNUCLEOTIDE PHOSPHORYLASE (EC 2.7.7.8)

Principle

$$\text{ADP} + \text{ADP} \xrightarrow{\text{Polynucleotide Phosphorylase}} \text{ppApA} + \text{Pi}$$

Abbreviations Used

ADP = Adenosine 5'-Diphosphate
ppApA = Adenylyl(3'-5')Adenosine 5'-Diphosphate
Pi = Inorganic Phosphate

Conditions: T = 37°C, pH = 9.0, A660nm, Light path = 1 cm

Method: Colorimetric

Reagents

A. 1 M Tris HCl Buffer, pH 9.0 at 37°C (Prepare 100 ml in deionized water using Trizma Base. Adjust to pH 9.0 at 37°C with 1 M HCl).
B. 125 mM Adenosine 5'-Diphosphate Solution (ADP) (Prepare 20 ml in deionized water using Adenosine 5'-Diphosphate, Sodium Salt. *Prepare Fresh*).
C. 62.5 mM Magnesium Chloride Solution ($MgCl_2$) (Prepare 100 ml in deionized water using Magnesium Chloride, Hexahydrate).
D. 4 mM Ethylenediaminetetraacetic Acid Solution (EDTA) (Prepare 100 ml in deionized water using Ethylenediaminetetraacetic Acid, Tetrasodium Salt, Hydrate. Adjust to pH 7.0 at 25°C with 100 mM HCl).
E. 0.2% (w/v) Bovine Serum Albumin Solution (BSA) (Prepare 10 ml in deionized water using Albumin, Bovine).
F. 2.5% (v/v) Perchloric Acid (PCA) (Prepare 500 ml in deionized water using Perchloric Acid).
G. Acid Molybdate Solution (Moly).
H. Fiske & Subbarow Reducer (FS) (Use Fiske & Subbarow Reducer. Prepare as per instructions on bottle).
I. Phosphorus Standard Solution (Std) (Use Phosphorus Standard Solution. The concentration is 20 µg/ml).
J. 0.1% (w/v) Bovine Serum Albumin Solution (BSA) (Prepare 10 ml in deionized water using Albumin, Bovine).
K. Polynucleotide Phosphorylase Enzyme Solution (Immediately before use, prepare a solution containing 5–15 units/ml in cold Reagent J).

Procedure

Prepare a stock reaction cocktail by pipetting (in milliliters) the following reagents into a suitable container:

Deionized Water	1.50
Reagent A (Buffer)	2.50
Reagent B (ADP)	4.00
Reagent C ($MgCl_2$)	2.00
Reagent D (EDTA)	2.50

Mix by swirling and adjust to pH 9.0 at 37°C with either 100 mM HCl or 100 mM NaOH, if necessary. Prepare a reaction cocktail by pipetting (in milliliters) the following reagents into a suitable container:

Deionized Water	0.80
Stock Reaction Cocktail	2.00
Reagent E (BSA)	0.40

Mix by swirling and equilibrate to 37°C. Pipette (in milliliters) the following reagents into suitable containers:

	Test	Blank
Reaction Cocktail	0.08	0.08
Reagent K (Enzyme Solution)	0.02	—

Mix by swirling and incubate at 37°C for exactly 15 minutes. Then add:

	Test	Blank
Reagent F (PCA)	0.90	0.90
Reagent K (Enzyme Solution)	—	0.02

Mix by swirling. Incubate in an ice bath for 10 minutes. Centrifuge both Test and Blank.

Colorimetric Determination

Pipette (in milliliters) the following into suitable tubes (Mix by swirling after each addition):

	Test	Test Blank	Std 1	Std 2	Std 3	Std 4	Std Blank
Deionized Water	—	—	0.70	0.60	0.40	0.20	0.80
Test Supernatant	0.80	—	—	—	—	—	—
Test Blank	—	0.80	—	—	—	—	—
Reagent I (Std)	—	—	0.10	0.20	0.40	0.60	—
Reagent G (Moly)	1.50	1.50	1.50	1.50	1.50	1.50	1.50

Mix by swirling and incubate at 25°C for 5 minutes. Then add:

Reagent H (FS)	0.20	0.20	0.20	0.20	0.20	0.20	0.20

Mix by swirling. Incubate at 25°C for 10 minutes. Record the absorbance at 660 nm for Test, Blank and Standards using a suitable spectrophotometer.

Calculations

Standard Curve: ΔA660nm Standard = A660nm Standard–A660nm Standard blank
Plot the ΔA660nm Standard vs μmoles of phosphorus.

Sample Determination: ΔA660nm Sample = A660nm Test–A660nm Test Blank
Determine the μmoles of phosphorus liberated using the Standard curve.

$$\text{Units / ml enzyme} = \frac{(\mu\text{moles of phosphorus liberated}) (\text{df})}{(0.8)(0.02)}$$

df = Dilution factor
0.8 = Volume of assay (in milliliters) used in Colorimetric Determination
0.02 = Volume (in milliliters) of enzyme assay

Unit Definition: *One unit will polymerize 1.0 μmole of ADP releasing 1.0 μmole of inorganic phosphate in 15 minutes at pH 9.0 at 37°C.*

Final Assay Concentrations

In a 0.10 ml reaction mix, the final concentrations are 100 mM Tris, 20 mM adenosine 5'-diphosphate, 5 mM magnesium chloride, 0.4 mM Ethylenediaminetetraacetic Acid, 0.04% (w/v) Bovine Serum Albumin, and 0.1–0.3 unit polynucleotide phosphorylase.

Enzymatic Assay of POLYOL DEHYDROGENASE (EC 1.1.1.14)

Principle

$$\text{Xylitol} + \beta\text{-NAD} \xrightarrow{\text{Polyol Dehydrogenase}} \text{D-Xylulose} + \beta\text{-NADH}$$

Abbreviations Used

β-NAD = β-Nicotinamide Adenine Dinucleotide, Oxidized Form
β-NADH = β-Nicotinamide Adenine Dinucleotide, Reduced Form

Conditions: T = 25°C, pH = 8.6, A340nm, Light path = 1 cm

Method: Continuous Spectrophotometric Rate Determination

Reagents

A. 100 mM Glycine Buffer, pH 8.6 at 25°C (Prepare 100 ml in deionized water using Glycine, Free Base. Adjust to pH 8.6 at 25°C with 1 M NaOH).
B. 2.4 M Xylitol Solution (Xylitol) (Prepare 3 ml in deionized water using Xylitol. *Prepare Fresh*).
C. 186 mM β-Nicotinamide Adenine Dinucleotide, Oxidized Form, Solution (β-NAD) (Prepare 2 ml in deionized water using β-Nicotinamide Adenine Dinucleotide, Sodium Salt).
D. 10 mM 2-Mercaptoethanol Solution (2-ME) (Prepare 4 ml in deionized water using 2-Mercaptoethanol).
E. Polyol Dehydrogenase Enzyme Solution (Immediately before use, prepare a solution containing 1–2 units/ml of Polyol Dehydrogenase in cold deionized water).

Procedure

Prepare a reaction cocktail by pipetting (in milliliters) the following reagents into a suitable container:

Reagent A (Buffer)	23.00
Reagent B (Xylitol)	2.50
Reagent C (β-NAD)	1.00
Reagent D (2-ME)	3.00

Mix by swirling and adjust to pH 8.6 at 25°C with 1 M NaOH. Pipette (in milliliters) the following reagents into suitable cuvettes:

	Test	Blank
Reaction Cocktail	2.90	2.90

Equilibrate to 25°C. Monitor the A340nm until constant, using a suitably thermostatted spectrophotometer. Then add:

	Test	Blank
Reagent E (Enzyme Solution)	0.02	—
Deionized Water	—	0.02

Immediately mix by inversion and record the increase in A340nm for approximately 5 minutes. Obtain the Δ340nm/minute using the maximum linear rate for both the Test and Blank.

Calculations

$$\text{Units / ml enzyme} = \frac{(\Delta A340\text{nm / min Test} - \Delta A340\text{nm / min Blank})\ (2.92)\ (\text{df})}{(6.22)\ (0.02)}$$

2.92 = Total volume (in milliliters) of assay
df = Dilution factor
6.22 = Millimolar extinction coefficient of β-NADH at 340nm
0.02 = Volume (in milliliter) of enzyme used

$$\text{Units / mg solid} = \frac{\text{units / ml enzyme}}{\text{mg solid / ml enzyme}}$$

$$\text{Units / mg protein} = \frac{\text{units / ml enzyme}}{\text{mg protein / ml enzyme}}$$

Unit Definition: *One unit will convert 1.0 μmole of xylitol to D-xylulose per minute at pH 8.6 at 25°C.*

Final Assay Concentrations

In a 2.92 ml reaction mix, the final concentrations are 77 mM glycine, 202 mM xylitol, 6.3 mM β-nicotinamide adenine dinucleotide, 1 mM 2-mercaptoethanol, and 0.02–0.04 unit polyol dehydrogenase.

Enzymatic Assay of POLYPHENOL OXIDASE (EC 1.14.18.1)

Principle

$$\text{L-DOPA} + O_2 \xrightarrow{\text{PPO}} \text{Benzoquinone derivative} + H_2O$$

$$\text{Benzoquinone derivative} + \text{L-Asc Acid} \rightarrow \text{L-DOPA} + \text{Dehydro-Asc Acid}$$

Abbreviations Used

L-DOPA = L-3,4-Dihydroxyphenylalanine
PPO = Polyphenol Oxidase
L-Asc = L-Ascorbic Acid
Dehydro-Asc Acid =Dehydro-Ascorbic Acid

Conditions: T = 25°C, pH = 6.5, A265nm, Light path = 1 cm

Method: Continuous Spectrophotometric Rate Determination

Reagents

A. 50 mM Potassium Phosphate Buffer, pH 6.5 at 25°C (Prepare 100 ml in deionized water using Potassium Phosphate, Monobasic. Adjust to pH 6.5 at 25°C with 1 M NaOH).

B. 5 mM L-3,4-Dihydroxyphenylalanine Solution (L-DOPA) (Prepare 10 ml in Reagent A using L-3,4-Dihydroxyphenylalanine. *Prepare Fresh*).

C. 2.1 mM L-Ascorbic Acid Solution (Prepare 10 ml in Reagent A using L-Ascorbic Acid, Sodium Salt. *Prepare Fresh*).

D. 0.065 mM Ethylenediaminetetraacetic Acid Solution (EDTA) (Prepare 10 ml in Reagent A using Ethylenediaminetetraacetic Acid, Disodium Salt).

E. Polyphenol Oxidase Enzyme Solution (Immediately before use, prepare a solution containing 500–1000 units/ml of Polyphenol Oxidase in cold Reagent A).

Procedure

Pipette (in milliliters) the following reagents into suitable quartz cuvettes:

	Test	Blank
Reagent A (Buffer)	2.60	2.80
Reagent B (L-DOPA)	0.10	0.10
Reagent C (Ascorbic Acid)	0.10	—
Reagent D (EDTA)	0.10	0.10

Mix by inversion and equilibrate to 25°C. Monitor the A265nm until constant, using a suitably thermostatted spectrophotometer. Then add:

Reagent E (Enzyme Solution)	0.10	—

Immediately mix by inversion and record the decrease in A265nm for approximately 5 minutes. Obtain the ΔA265nm/minute using the maximum linear rate for both the Test and Blank.

Calculations

$$\text{Units / mg enzyme} = \frac{(\Delta A265nm / \min \text{ Test} - \Delta A265nm / \min \text{ Blank})}{(0.001)\ (\text{mg enzyme / RM})}$$

0.001 = the change in A265nm/minute per unit Polyphenol Oxidase at pH 6.5 at 25°C in a 3 ml reaction mix

RM = Reaction Mix (final volume = 3 ml)

Unit Definition: *One unit will cause the change in A265nm of 0.001 per minute at pH 6.5 at 25°C in a 3 ml reaction mix containing L-3,4-dihydroxyphenylalanine and L-ascorbic acid.*

Final Assay Concentrations

In a 3 ml reaction mix, the final concentrations are 50 mM potassium phosphate, 0.17 mM L-3,4-dihydroxyphenylalanine, 0.07 mM L-ascorbic acid and 50–100 units of polyphenol oxidase.

Note

1. Ascorbic acid is not included in the blank to produce a true blank. Possible interaction with L-Dopa and ascorbic acid might give false readings.

Enzymatic Assay of PROLIDASE (EC 3.4.13.9)

Principle

$$\text{Glycyl-L-Proline} + H_2O \xrightarrow{\text{Prolidase}} \text{Glycine} + \text{L-Proline}$$

Conditions: T = 40°C, pH = 8.0, A242nm, Light path = 1 cm

Method: Continuous Spectrophotometric Rate Determination

Reagents

A. 50 mM Tris HCl Buffer, pH 8.0 at 40°C (Prepare 100 ml in deionized water using Trizma Base. Adjust to pH 8.0 at 40°C with 1 M HCl).
B. 30 mM Glutathione Solution (GSH) (Prepare 5 ml in deionized water using Glutathione, Fresh Acid, Reduced Form. *Prepare Fresh*).
C. 200 mM Manganese Chloride Solution ($MnCl_2$) (Prepare 10 ml in deionized water using Manganese Chloride, Tetrahydrate).
D. 25 mM Glycyl-L-Proline Solution (Gly-Pro) (Prepare 30 ml in Reagent A using Gly-Pro. Adjust to pH 8.0 at 40°C with 1 M HCl or 1 M NaOH, if necessary).
E. Prolidase Enzyme Solution (Immediately before use prepare a solution containing approximately 5 mg/ml of Prolidase in cold Reagent A).

Procedure

Prepare an activated enzyme (activation mix with enzyme) by pipetting (in milliliters) the following reagents into a suitable container in the order specified:

Reagent A (Buffer)	2.40
Reagent C ($MnCl_2$)	0.40
Reagent B (GSH)	0.10
Reagent E (Enzyme Solution)	0.20

Mix and incubate at 40°C for 20, 25, and 30 minutes. Immediately after completing the activation step, pipette (in milliliters) the following reagents into suitable quartz cuvettes:

	Test	Blank
Reagent D (Gly-Pro)	2.70	2.70
Reagent C ($MnCl_2$)	0.20	0.20

Mix by inversion and equilibrate to 40°C. Monitor the A242nm until constant, using a suitably thermostatted spectrophotometer. Then add:

Activation Mix with Enzyme	0.50	—
Activation Mix without Enzyme	—	0.50

Immediately mix by inversion and record the decrease in A242nm for approximately 5 minutes. Obtain the ΔA242nm using the maximum linear rate for both the Test and Blank.

Calculations

$$\text{Units / ml enzyme} = \frac{(\Delta A242nm / min\ Test - \Delta A242nm / min\ Blank)\ (3.1)\ (3.4)}{(0.0254)\ (0.5)\ (0.2)}$$

3.1 = Volume (in milliliters) of activation mix
3.4 = Volume (in milliliters) of reaction mix
0.0254 = Millimolar extinction coefficient for Glycyl-L-proline at 242nm
0.5 = Volume (in milliliter) of activated enzyme used in assay
0.2 = Volume (in milliliter) of enzyme used in activation mix

$$\text{Units / mg solid} = \frac{\text{units / ml enzyme}}{\text{mg solid / ml enzyme}}$$

$$\text{Units / mg protein} = \frac{\text{units / ml enzyme}}{\text{mg protein / ml enzyme}}$$

Unit Definition: *One unit will hydrolyze 1.0 μmole of glycyl-L-proline per minute at pH 8.0 at 40°C.*

Final Assay Concentrations

In a 3.40 ml reaction mix, the final concentrations are 5.7 mM Tris, 16 mM manganese chloride, 0.14 mM glutathione, 20 mM glycyl-L-proline and 0.16 mg prolidase.

Notes

1. Do not adjust the pH of the Activation Mixture.
2. Use the activated enzyme from all three time points. The time point which provides the largest A242nm should be used in the calculation.
3. The A242nm/min should be between 0.03–0.05.

Enzymatic Assay of PROLINE IMINOPEPTIDASE (EC 3.4.11.5)

Principle

$$\text{L-Proline p-Nitroanilide} \xrightarrow{\text{PIP}} \text{L-Proline + p-Nitroaniline}$$

Abbreviation Used

PIP = Proline Iminopeptidase

Conditions: T = 30°C, pH = 8.0, A410nm, Light path = 1 cm

Method: Spectrophotometric Stop Rate Determination

Reagents

A. 100 mM Tris HCl Buffer, pH 8.0 at 30°C (Prepare 100 ml in deionized water using Trizma Base. Adjust to pH 7.2 at 30°C with 1 M HCl).
B. 1.0 mM L-Proline p-Nitroanilide Solution (PPNA) (Prepare 10 ml in deionized water using L-Proline p-Nitroanilide, Trifluoroacetic Acid, Salt).

C. 1000 mM Sodium Acetate Buffer, pH 4.0 at 30°C (Acet) (Prepare 100 ml in deionized water using Sodium Acetate, Trihydrate. Adjust to pH 4.0 at 30°C with 5 N HCl).
D. Proline Iminopeptidase Enzyme Solution (Immediately before use, prepare a solution containing 0.075–0.150 unit/ml of Proline Iminopeptidase in cold Reagent A).

Procedure

Pipette (in milliliters) the following reagents into suitable cuvettes:

	Test	Blank
Reagent A (Buffer)	1.00	1.25
Reagent B (PPNA)	0.25	0.25

Equilibrate to 30°C. Then add:

Reagent D (Enzyme Solution)	0.25	—

Immediately mix by swirling and incubate for exactly 10 minutes. Then add:

Reagent C (Acet)	0.50	0.50

Mix by swirling and transfer to suitable cuvettes. Record the A410nm for both the Test and Blank using a suitable spectrophotometer.

Calculations

$$\text{Units / ml enzyme} = \frac{(\text{A410nm Test} - \text{A410nm Blank})\,(2.0)\,(\text{df})}{(10)\,(5.57)\,(0.25)}$$

2.0 = Total volume (in milliliters) of assay
df = Dilution factor
10 = Time of assay (in minutes)
5.57 = Millimolar extinction coefficient of p-Nitroaniline under the conditions of this assay
0.25 = Volume (in milliliter) of enzyme used

$$\text{Units / mg solid} = \frac{\text{units / ml enzyme}}{\text{mg solid / ml enzyme}}$$

$$\text{Units / mg protein} = \frac{\text{units / ml enzyme}}{\text{mg protein / ml enzyme}}$$

Unit Definition: *One unit will hydrolyze 1.0 μmole of proline pnitroanilide per minute at pH 8.0 at 30°C.*

Final Assay Concentration

In a 1.50 ml reaction mix, the final concentrations are 83 mM Tris, 0.17 mM L-proline p-nitroanilide and 0.019–0.038 unit proline iminopeptidase.

Enzymatic Assay of PROTEASE (EC 3.4.21.19)

Principle

$$\text{N-t-BOC-GAPE} + H_2O \xrightarrow{\text{Protease}} \text{N-t-BOC-L-Glutamic Acid} + \text{Phenol}$$

Abbreviations Used

BOC = N-tert-Butoxy-Carbonyl
N-t-BOC-GAPE = N-t-BOC-L-Glutamic Acid α-Phenyl Ester

Conditions: T = 37°C, pH = 7.8, A270nm, Light path = 1 cm.

Method: Continuous Spectrophotometric Rate Determination

Reagents

A. 200 mM Tris-Phosphate Buffer, pH 7.8 at 37°C (Prepare 100 ml in deionizsed water using Trizma Base. Adjust to pH 7.8 at 37°C with Phosphoric Acid).
B. 60 mM N-t-BOC-L-Glutamic Acid α-Phenyl Ester Substrate Solution (N-t-BOC-GAPE) (Prepare by dissolving 97 mg of N-t-BOC-L-Glutamic Acid α-Phenyl Ester, in 5 ml of 1,4-Dioxane. Make sure that the solid is at room temperature before opening. Any moisture in this substrate will cause immediate hydrolysis. *Prepare Fresh*).
C. Protease Enzyme Solution (Immediately before use, prepare a solution containing 0.1–2.0 units/ml of Protease in deionized water).

Procedure

Pipette (in milliliters) the following reagents into suitable cuvettes:

	Test	Blank
Reagent A (Buffer)	2.85	2.85
Reagent B (N-t-BOC-GAPE)	0.05	0.05

Mix by inversion and equilibrate to 37°C. Monitor the A270nm until constant, using a suitably thermostatted spectrophotometer. Then add:

Reagent C (Enzyme Solution)	0.10	—
Deionized Water	—	0.10

Immediately mix by inversion and record the increase in A270nm for approximately 5 minutes. Obtain the A270nm/minute using the maximum linear rate for both the Test and Blank.

Calculations

$$\text{Units / ml enzyme} = \frac{(\text{A270nm / min Test} - \text{A270nm / min Blank})\ (3)\ (\text{df})}{(1.5)\ (0.1)}$$

3 = Volume (in milliliters) of assay
df = Dilution factor
1.5 = Millimolar extinction coefficient of phenol at 270nm
0.1 = Volume (in milliliter) of enzyme used

$$\text{Units / mg solid} = \frac{\text{units / ml enzyme}}{\text{mg solid / ml enzyme}}$$

$$\text{Units / mg protein} = \frac{\text{units / ml enzyme}}{\text{mg protein / ml enzyme}}$$

Unit Definition: *One unit will hydrolyze 1 μmole of N-t-BOC-L-glutamic acid α-phenyl ester per minute at pH 7.8 at 37°C. One unit is equivalent to approximately 0.004 casein digestion unit.*

Final Assay Concentrations

In a 3.00 ml reaction mix, the final concentrations are 190 mM Tris, 1 mM N-t-BOC-L-glutamic acid α-phenyl ester, 1.7% 1,4-dioxane and 0.01–0.2 unit protease.

Enzymatic Assay of PROTEASE (EC 3.4.21.40)

Principle

$$\text{TAME} + H_2O \xrightarrow{\text{Protease}} \text{Na - p - Tosyl - L - Arginine} + \text{Methanol}$$

Abbreviation Used

TAME = Na-p-Tosyl-L-Arginine Methyl Ester

Conditions: T = 25°C, pH 8.0, A247nm, Light path = 1 cm

Method: Continuous Spectrophotometric Rate Determination

Reagents

A. 100 mM Potassium Phosphate Buffer, pH 8.0 at 25°C (Prepare 100 ml in deionized water using Potassium Phosphate, Dibasic Trihydrate. Adjust to pH 8.0 at 25°C with Reagent B).

B. 100 mM Potassium Phosphate Solution (Prepare 100 ml in deionized water using Potassium Phosphate, Monobasic, Anhydrous).

C. 0.80 mM Na-p-Tosyl-L-Arginine Methyl Ester Solution (TAME) (Prepare 50 ml in Reagent A using Na-p-Tosyl-LArginine Methyl Ester).

D. Protease Enzyme Solution (Immediately before use, prepare a solution containing 1.7–2.1 units/ml of Protease in cold deionized water).

Procedure

Pipette (in milliliters) the following reagents into suitable cuvettes:

	Test	Blank
Reagent C (TAME)	2.90	2.90

Equilibrate to 25°C. Then add:

Reagent D (Enzyme Solution)	0.10	—
Deionized Water	—	0.10

Immediately mix by inversion and record the increase in A247nm for approximately 10 minutes. Obtain the A247nm/minute using the maximum linear rate for both the Test and Blank.

Calculations

$$\text{Units / ml enzyme} = \frac{(\text{A247nm / min Test} - \text{A247nm / min Blank})\ (3)\ (\text{df})}{(0.54)\ (0.1)}$$

3 = Volume (in milliliters) of enzyme used
df = Dilution factor
0.54 = Millimolar extinction coefficient of Na-p-Tosyl-L-Arginine at 247nm
0.1 = Volume (in milliliters) of enzyme used

$$\text{Units / mg solid} = \frac{\text{units / ml enzyme}}{\text{mg solid / ml enzyme}}$$

$$\text{Units / mg protein} = \frac{\text{units / ml enzyme}}{\text{mg protein / ml enzyme}}$$

Unit Definition: *One unit will hydrolyze 1.0 μmole of Na-p-tosyl-L-arginine methyl ester per minute at pH 8.0 at 25°C.*

Final Assay Concentrations

In a 3.00 reaction mix, the final concentrations are 97 mM potassium phosphate, 0.77 mM Na-p-tosyl-L-arginine methyl ester and 0.17–0.21 unit protease.

Enzymatic Assay of PROTEASE (Casein as a Substrate)

Principle

$$\text{Casein} + H_2O \xrightarrow{\text{Protease}} \text{Amino Acids}$$

Conditions: T = 37°C, pH = 7.5, A660nm, Light path = 1 cm.

Method: Colorimetric

Reagents

A. 50 mM Potassium Phosphate buffer, pH 7.5 at 37°C. (Prepare 200 ml in deionized water using Potassium Phosphate, Dibasic, Trihydrate. Adjust to pH 7.5 at 37°C with 1 M HCl).

B. 0.65% (w/v) Casein Solution (Casein) (Prepare 125 ml in Reagent A using Casein. Heat gently (do not boil) to 80–90°C for 10 minutes with stirring. Adjust the pH to 7.5 at 37°C, if necessary, with either 1 M NaOH or 1 M HCl).

C. 110 mM Trichloroacetic Acid Reagent (TCA) (Dilute 9 ml of Trichloroacetic Acid, 6.1 N, approximately 100% (w/v), to 500 ml with deionized water).

D. Folin & Ciocalteu's Phenol Reagent (F-C) (Dilute 10 ml of Folin & Ciocalteu's Phenol Reagent, to 40 ml with deionized water).

E. 500 mM Sodium Carbonate Solution (Na_2CO_3) (Prepare 500 ml in deionized water using Sodium Carbonate Anhydrous).

F. 10 mM Sodium Acetate Buffer with 5 mM Calcium Acetate, pH 7.5 at 37°C (Enzyme Diluent) (Prepare 100 ml in deionized water using Sodium Acetate, Trihydrate, and Calcium Acetate. Adjust the pH to 7.5 at 37°C with 0.1 M Acetic acid or 0.1 M NaOH).

G. 1.1 mM L-Tyrosine Standard (Std Soln) (Prepare 100 ml in deionized water using L-Tyrosine, Free Base. Heat gently (do not boil) until tyrosine dissolves and cool to room temperature).

H. Protease Enzyme Solution (Immediately before use, prepare a solution containing 0.1–0.2 unit/ml of Protease in cold Reagent F).

Procedure

Pipette the following reagents into suitable vials (in milliliters):

	Test	Blank
Reagent B (Casein)	5.00	5.00

Equilibrate to 37°C. Then add:

Reagent H (Enzyme Solution)	1.00	—

Mix by swirling and incubate at 37°C for exactly 10 minutes. Then add:

Reagent C (TCA)	5.00	5.00
Reagent H (Enzyme Solution)	—	1.00

Mix by swirling and incubate at 37°C for about 30 minutes. Filter through Whatman #50 filter paper or a 0.45 m filter and use the filtrate in color development.

Color Development

Standard Curve: Prepare a standard curve by pipetting the following reagents into suitable vials (in milliliters):

	Std 1	Std 2	Std 3	Std 4	Std Blank
Reagent G (Std Soln)	0.05	0.10	0.20	0.40	0.00
Deionized Water	1.95	1.90	1.80	1.60	2.00
Reagent E (Na_2CO_3)	5.00	5.00	5.00	5.00	5.00
Reagent D (F-C)	1.00	1.00	1.00	1.00	1.00

Sample: Pipette the following reagents into 4 dram vials (in milliliters):

	Test	Blank
Test Filtrate	2.00	—
Blank Filtrate	—	2.00
Reagent E (Na_2CO_3)	5.00	5.00
Reagent D (F-C)	1.00	1.00

Mix by swirling and incubate at 37°C for 30 minutes. Remove the vials and allow them to cool to room temperature. Filter through a 0.45 m filter immediately prior to reading. Read the absorbance at 660nm for each of the vials in suitable cuvettes.

Calculations

Standard Curve: A660nm Standard = A660nm Standard–A660nm Standard Blank
Plot the A660nm Standard vs moles of Tyrosine.
Sample Determination: A660nm Sample = A660nm Test–A660nm Sample Blank
Determine the moles of Tyrosine equivalents liberated using the Standard curve.

$$\text{Units / ml enzyme} = \frac{\text{(mole Tyrosine equivalents released) (11)}}{(1)\ (10)\ (2)}$$

11 = Total volume (in milliliters) of assay
10 = Time of assay (in minutes) as per the Unit Definition
1 = Volume of enzyme (in milliliter) of enzyme used
2 = Volume (in milliliters) used in Colorimetric Determination

$$\text{Units / mg solid} = \frac{\text{units / ml enzyme}}{\text{mg solid / ml enzyme}}$$

$$\text{Units / mg protein} = \frac{\text{units / ml enzyme}}{\text{mg protein / ml enzyme}}$$

Unit Definition: *One unit will hydrolyze casein to produce color equivalent to 1.0 µmole (181 g) of tyrosine per minute at pH 7.5 at 37ºC (color by Folin & Ciocalteu's reagent).*

Final Assay Concentrations

In a 6.00 ml reaction mix, the final concentrations are 42 mM potassium phosphate, 0.54% (w/v) casein, 1.7 mM sodium acetate, 0.8 mM calcium acetate, and 0.1–0.2 unit protease.

Enzymatic Assay of PROTEASE (Endoproteinase Lys-C) from *Lysobacter enzymogenes*

Principle

$$\text{N-p-Tosyl-Gly-Pro-Lys-pNA} + H_2O \xrightarrow{\text{Protease}} \text{N-p-Tosyl-Gly-Pro-Lys} + \text{p-Nitroaniline}$$

Abbreviation Used

pNA = p-Nitroanilide

Conditions: T = 25°C, pH 7.7, A405nm, Light path = 1 cm

Method: Continuous Spectrophotometric Rate Determination

Reagents

A. 25 mM Tris HCl Buffer with 1 mM Ethylenediaminetetraacetic Acid, pH 7.7 at 25°C (Prepare 100 ml in deionized water using Trizma Base and Ethylenediaminetetraacetic Acid, Disodium Salt, Dihydrate. Adjust to pH 7.7 at 25°C with 1 M HCl).
B. 14 mM N-p-Tosyl-Gly-Pro-Lys p-Nitroanilide Substrate Solution (Prepare 1 ml in Reagent A using N-p-Tosyl-Gly-Pro-Lys p-Nitroanilide).
C. Protease Enzyme Solution (Immediately before use, prepare a solution containing 0.075–0.15 unit/ml of Protease in cold Reagent A).

Procedure

Pipette (in milliliters) the following reagents into suitable cuvettes:

	Test	Blank
Reagent A (Buffer)	2.70	2.70
Reagent B (Substrate Solution)	0.15	0.15

Mix by inversion and equilibrate to 25°C. Monitor the A405nm until constant, using a suitably thermostatted spectrophotometer. Then add:

Reagent C (Enzyme Solution)	0.10	—
Reagent A (Buffer)	—	0.10

Immediately mix by inversion and record the increase in A405nm for approximately 5 minutes. Obtain the ΔA405nm/minute using the maximum linear rate for both the Test and Blank.

Calculations

$$\text{Units / ml enzyme} = \frac{(\Delta A405\text{nm / min Test} - \Delta A405\text{nm / min Blank})\ (2.95)\ (\text{df})}{(10.4)\ (0.1)}$$

2.95 = Volume (in milliliters) of assay
df = Dilution factor
10.4 = Millimolar extinction coefficient of p-nitroaniline at 405nm
0.1 = Volume (in milliliter) of enzyme used

Unit Definition: *One unit will hydrolyze 1.0 µmole of N-p-Tosyl-Gly-Pro-Lys p-Nitroanilide per minute at 7.7 at 25°C.*

Final Assay Concentrations

In a 2.95 ml reaction mix, the final concentrations are 25 mM Tris, 1 mM Ethylenediaminetetraacetic Acid, 0.71 mM N-p-tosyl-gly-pro-lys p-nitroanilide, and 0.0075–0.015 unit protease.

Enzymatic Assay of PROTEASE (EC 3.4.23.6)

Principle

$$\text{Hemoglobin} + H_2O \xrightarrow{\text{Protease}} \text{Amino Acids}$$

Conditions: T = 37°C, pH = 2.8, A660nm, Light path = 1 cm

Method: Colorimetric

Reagents

A. 50 mM Potassium Phthalate Buffer, pH 2.8 at 37°C. (Prepare by dissolving 4.03 g of Phthalic Acid, Monopotassium Salt in 100 ml of deionized water. Then add 53 ml of 200 mM HCl. Adjust to pH 2.8 at 37°C with 1 M HCl. Dilute to 400 ml with deionized H_2O).

B. 2.5% (w/v) Hemoglobin Solution (Prepare by dissolving 2.5 g of Hemoglobin, Bovine in 70 ml of deionized water. To this solution add 20 ml of 0.5 M HCl. Adjust the pH to 2.8 at 37°C with 1 M NaOH. Dilute to 100 ml with deionized water. *Note*: Hemoglobin slowly dissolves and clumps if it is all added at one time).

C. 110 mM Trichloroacetic Acid Reagent (TCA) (Dilute 9 ml of Trichloroacetic Acid, 6.1 N Solution, approximately 100% (w/v) to 500 ml with deionized water).

D. Folin & Ciocalteu's Phenol Reagent (F-C) (Dilute 10 ml of Folin & Ciocalteu's Phenol Reagent to 40 ml with deionized water).

E. 500 mM Sodium Carbonate Solution (Na_2CO_3 Soln) (Prepare 500 ml in deionized water using Sodium Carbonate Anhydrous).

F. 1.1 mM L-Tyrosine Standard Solution (Std Soln) (Prepare 100 ml in deionized water using L-Tyrosine, Free Base. Heat gently (do not boil) until tyrosine dissolves and cool to room temperature).

G. Protease Enzyme Solution (Immediately before use, prepare a solution containing 0.5–1.0 unit/ml of Protease in cold Reagent A).

Procedure

Pipette (in milliliters) the following reagents into suitable vials:

	Test	Blank
Reagent B (Hemoglobin Solution)	5.00	5.00

Incubate the vials at 37°C until equilibrated. Then add:

	Test	Blank
Reagent G (Enzyme Solution)	1.00	—

Mix by swirling and incubate at 37°C for exactly 10 minutes. Then add:

	Test	Blank
Reagent C (TCA)	10.00	10.00
Reagent G (Enzyme Solution)	—	1.00

Mix by swirling and incubate at 37°C for approximately 30 minutes. Filter through a Whatman #50 filter paper and use the filtrate in the color development.

Colour Development

Standard Curve: Prepare a standard curve by pipetting (in milliliters) the following reagents into suitable vials:

	Std 1	Std 2	Std 3	Std 4	Std Blank
Reagent F (Std Soln)	0.05	0.100	0.20	0.40	0.00
Deionized Water	1.95	1.90	1.80	1.60	2.00
Reagent E ($Na2CO_3$ Soln)	5.00	5.00	5.00	5.00	5.00
Reagent D (F-C)	1.00	1.00	1.00	1.00	1.00

Sample: Pipette (in milliliters) the following reagents into a suitable vial:

	Test or Blank
Filtrate	2.00
Reagent E ($Na2CO_3$ Soln)	5.00
Reagent D (F-C)	1.00

Mix by swirling and incubate at 37°C for 30 minutes. Remove the vials and allow them to cool to room temperature. If the solutions are hazy, filter through a 0.45 μm filter prior to reading. Read the absorbance at 660nm for each of the vials in suitable cuvettes.

Calculations

Standard Curve: ΔA660nm Standard = A660nm Standard–A660nm Standard Blank
Plot the ΔA660nm Standard vs μmoles Tyrosine.
Sample Determination: ΔA660nm Sample = A660nm Test–A660nm Test Blank
Determine the μmoles of Tyrosine equivalents liberated using the Standard curve.

$$\text{Units / ml enzyme} = \frac{(\mu\text{mole Tyrosine equivalents released})\ (16)}{(1)\ (10)\ (2)}$$

16 = Total volume (in milliliters) of stopped reaction
2 = Volume (in milliliters) used in Colorimetric Determination
1 = Volume (in milliliters) of enzyme used
10 = Time of reaction (minutes) as per the Unit Definition

$$\text{Units / mg solid} = \frac{\text{units / ml enzyme}}{\text{mg solid / ml enzyme}}$$

Unit Definition: *One unit will hydrolyze hemoglobin to produce color equivalent to 1.0 μmole (181 μg) of tyrosine per minute at pH 2.8 at 37°C (color by Folin & Ciocalteu's reagent).*

Final Assay Concentrations

In a 6.00 ml reaction mix, the final concentrations are 8.3 mM potassium phthalate, 2.1% (w/v) hemoglobin, and 0.5–1.0 unit protease.

Enzymatic Assay of PROTEIN DISULFIDE-ISOMERASE (EC 5.3.4.1)

Principle

$$\text{Ribonuclease A (Scrambled Disulfide Bonds)} \xrightarrow{\text{PDI}} \text{Ribonuclease A (Unscrambled Disulfide Bonds)}$$

Abbreviation Used

PDI = Protein Disulfide-Isomerase

Conditions: T = 30°C, pH 7.5, A260nm, Light path = 1 cm

Method: Continuous Spectrophotometric Rate Determination

Reagents

A. 50 mM Sodium Phosphate Buffer with 5.0 mM Ethylenediaminetetraacetic Acid, pH 7.5 at 30°C (Prepare 100 ml in deionized water using Sodium Phosphate, Monobasic, Anhydrous and Ethylenediaminetetraacetic Acid Disodium Salt, Dihydrate. Adjust to pH 7.5 at 30°C with 1 M NaOH).

B. 50 mM Tris HCl Buffer with 25 mM Potassium Chloride and 5 mM Magnesium Chloride, pH 7.5 at 30°C (TKM) (Prepare 100 ml in deionized water using Trizma Hydrochloride, Potassium Chloride and Magnesium Chloride, 4.9 M Solution. Adjust to pH 7.5 at 30°C with 1 M NaOH).
C. 10 mM Acetic Acid Solution (HOAc) (Prepare 25 ml in deionized water using Acetic Acid, Glacial).
D. 100 mM DL-Dithiothreitol Solution (DTT) (Prepare 10 ml in deionized water using DL-Dithiothreitol).
E. 0.0083% (w/v) Ribonucleic Acid Solution (RNA) (Prepare 5 ml in Reagent B using Ribonucleic Acid).
F. Scrambled Ribonuclease A Enzyme Solution (SRNase A) (Immediately before use, prepare a solution containing 0.05% (w/v) of Ribonuclease A, with Scrambled Disulfide Bonds, in cold Reagent C. Keep at 4°C. This solution is stable for several weeks at 4°C).
G. Protein Disulfide-Isomerase Enzyme Solution (PDI) (Immediately before use, prepare a solution containing 0.02 mg/ml of Protein Disulfide-Isomerase in cold Reagent A. Keep at 4°C).

Procedure

Step 1: PDI Assay Prepare a reaction cocktail by pipetting (in milliliters) the following reagents into a suitable Eppendorf tube:

	Test	Blank
Reagent A (Buffer)	0.94	0.95
Reagent D (DTT)	0.01	0.01
Reagent G (PDI)	0.01	—

Mix by inversion and incubate at 30°C for 5 minutes. Then add:

Reagent F (SRNase A)	0.05	0.05

Immediately start to record time. Mix by inversion and remove 0.010 ml aliquots at 1, 5, and 9 minutes time points (Reaction Mix T1, T5, and T9). Proceed with the RNase assay (Step 2).

Step 2: RNase Assay Pipette (in milliliters) the following reagents into suitable cuvettes:

	Test 1	Test 2	Test 3	Blank
Reagent E (RNA)	3.00	3.00	3.00	3.00

Incubate at 25°C for 5 minutes. Then add:

PDI Reaction Mix T1	0.01	—	—	—
PDI Reaction Mix T5	—	0.01	—	—
PDI Reaction Mix T9	—	—	0.01	—
Blank Mixture	—	—	—	0.01

Mix by inversion and record the increase in A260nm for approximately 5 minutes. Obtain the ΔA260nm/minute using the maximum linear rate for both the Tests and Blank.

Calculations

$$\text{Units / ml enzyme} = \frac{(\Delta A260\text{nm / min Test} - \Delta A260\text{nm / min Blank})\,(\text{df})\,(1.01)\,(100)}{(T)\,(1)}$$

100 = Dilution of enzyme resulting from using 0.01 ml of the PDI reaction mixture
df = Dilution factor
1.01 = Volume (in milliliters) of assay
T = Time (in minutes) of assay
1 = Change in A260nm which is equal to 1 unit

$$\text{Units / mg solid} = \frac{\text{units / ml enzyme}}{\text{mg solid / ml enzyme}}$$

$$\text{Units / mg protein} = \frac{\text{units / ml enzyme}}{\text{mg protein / ml enzyme}}$$

Unit Definition: *One unit is the amount of enzyme which reactivates one unit of ribonuclease from scrambled ribonuclease per minute at pH 7.5 at 30°C.*

Final Assay Concentrations

In a 1.01 ml reaction mix, the final concentrations are 47 mM sodium phosphate, 4.7 mM Ethylenediaminetetraacetic Acid, 1 mM DL-dithiothreitol, 0.0025% (w/v) scrambled ribonuclease A, 0.5 mM acetic acid, and 0.2 µg protein disulfide-isomerase.

Notes

1. Average the values for the time points at 5 and 9 minutes. The 1 minute time point is not reliable enough to be included in the average.
2. One unit of ribonuclease is defined as the amount of enzyme that will release 1.0 A260nm absorbance unit per minute at pH 7.5 at 25°C.

Enzymatic Assay of PROTEIN KINASE 3':5'-CYCLIC AMP DEPENDENT Phosphorylating Activity

Principle

$$\text{Casein} + \gamma\text{-}_{32}\text{P-ATP} \xrightarrow{\text{Protein Kinase}} [_{32}\text{P}]\text{-Phosphorylated Casein} + \text{ADP Cyclic AMP}$$

Conditions: T = 30°C, pH = 6.5

Method: Radioactive

Reagents

A. 1000 mM Potassium Phosphate Buffer, pH 6.5 at 30°C (Prepare 100 ml in deionized water using Potassium Phosphate, Monobasic, Anhydrous. Adjust to pH 6.5 at 30°C with 2 M KOH).

B. 5.0% (w/v) Casein Solution (Casein) (Use Casein from Bovine Milk, 5% (w/v) Solution).

C. 500 mM Magnesium Acetate Solution ($Mg(OAc)_2$) (Prepare 10 ml in deionized water using Magnesium Acetate, Tetrahydrate).

D. 250 mM Aminophylline Solution (AP) (Prepare 10 ml in deionized water using Aminophylline, Hydrate).

E. 330 mM Dithiothreitol Solution (DTT) (Prepare 10 ml in deionized water using DL-Dithiothreitol. *Prepare Fresh*).

F. 1 mM Adenosine 3':5'-Cyclic Monophosphate Solution (cAMP) (Prepare 1 ml in deionized water using Adenosine 3':5'-Cyclic Monophosphate, Sodium Salt. *Prepare Fresh*).

G. 10.0 mM Adenosine 5'-Triphosphate Solution (ATP) (Prepare 1 ml in deionized water using Adenosine 5'-Triphosphate, Disodium Salt. *Prepare Fresh*).

H. γ-$_{32}$P-Adenosine 5'-Triphosphate Solution (γ-32P-ATP) (Use product with minimum radioactive concentration of 30 curies/mmole and 2 millicuries/ml).

I. 13.5% (v/v) Trichloroacetic Acid Solution (TCA) (Prepare 20 ml in deionized water using Trichloroacetic Acid, 6.1 N Solution, approximately 100% (w/v)).

J. Protein Kinase 3':5'-Cyclic AMP Dependent Enzyme Solution (Immediately before use, prepare a solution containing 200–400 units/ml of Protein Kinase 3':5'-Cyclic AMP Dependent in cold deionized water).

K. 6.75% (v/v) Trichloroacetic Acid Solution (Wash) (Prepare 20 ml in deionized water using Trichloroacetic Acid, 6.1 N Solution, approximately 100% (w/v)).

L. Methylethyl Cellosolve (Prepare 500 ml by adding 250 ml of Ethylene Glycol Monoethyl Ether to 250 ml of Ethylene Glycol Monomethyl Ether).

M. Scintillation Cocktail.

Procedure

Prepare a reaction cocktail by pipetting (in milliliters) the following reagents into a suitable container:

Deionized Water	1.95
Reagent A (Buffer)	0.50
Reagent C ($Mg(OAc)_2$)	0.25
Reagent D (AP)	0.10
Reagent E (DTT)	0.10

Pipette (in milliliters) the following reagents into 2.90 ml of the reaction cocktail.

Reagent F (cAMP)	0.05
Reagent G (ATP)	0.05

Mix by swirling. Transfer 1 ml to a suitable container and add enough Reagent H (γ-$_{32}$P-ATP) to yield approximately 150,000–200,000 counts/minute (cpm) in 0.05 ml of the solution. Then add 0.50 ml of Reagent B (Casein). This is the Reaction Cocktail. Pipette 0.05 ml aliquots of Reagent J (Enzyme Solution) into a multiwell disposable titerplate. Place in a 4°C ice bath. Add 0.05 ml of Reaction Cocktail to each well and mix by air injection. Immediately transfer the titerplate to a 30° C water bath. Incubate at 30°C for 10 minutes. Then add 0.10 ml of Reagent I (TCA) to each well. Filter the material in the wells through 0.45 µm Millipore HA Type filters. Wash 3 times with Reagent K (Wash). Transfer the filters to suitable dram scintillation vials containing 2.00 ml of Reagent L (Methylethyl Cellosolve). To each scintillation vial, add 5 ml of Reagent M (Scintillation Cocktail). Count the radioactivity in a suitable scintillation counter.

Calculations

The total number of picomoles (pMoles) of ATP in the reaction mixture is calculated as follows:

$$\frac{(0.05)\,(0.01)\,(109)}{(3.00)\,(1.5)} = 1.1111 \times 105 \text{ pmole / ml Reaction Cocktail}$$

Find **cpm/pmole** by counting 0.05 ml (5555 pmoles) of the Reaction Cocktail.
0.05 = Volume (in milliliters) of ATP used in the Reaction Cocktail
0.01 = Millimolar concentration of ATP (Reagent G)
109 = Conversion of millimoles to picomoles
3.00 = Intermediate volume (in milliliters) of Reaction Cocktail
1.5 = Total volume (in milliliters) of Reaction Cocktail

$$\text{Units / ml} = \frac{\text{CPM Counted}}{(\text{cpm / pmole})\,(10)\,(0.05)}$$

CPM counted = Actual count–background on filters
10 = Time of Assay (in minutes) as per the Unit Definition
0.05 = Volume (in milliliters) of enzyme used

Unit Definition for Protein Kinase

One unit will transfer 1.0 picomole (10–12 mole) of phosphate from γ-$_{32}$*P-ATP to hydrolyzed and partially dephosphorylated casein (C-4765) per minute at pH 6.5 at 30°C, in the presence of 0.006 mM cyclic-AMP (A6885).*

Final Assay Concentrations

In a 0.10 ml solution, the final assay concentrations are 55 mM potassium phosphate, 14 mM magnesium acetate, 3 mM aminophylline, 4 mM dithiothreitol, 0.006 mM adenosine 3':5'-cyclic monophosphate, 0.055 mM adenosine 5'-triphosphate, 0.83% (w/v) casein, and 10–20 units protein kinase 3':5'-cyclic AMP dependent.

Enzymatic Assay of PROTEIN KINASE C

Principle

$$\text{Histone} + {}_{32}\text{P - ATP} \xrightarrow{\text{Protein Kinase}} [{}_{32}\text{P}]\text{ - Phosphorylated Histone} + \text{ADP}$$

Abbreviations Used

$_{32}$P-ATP = Adenosine 5'-Triphosphate $_{32}$P-labelled
ADP = Adenosine 5'-Diphosphate

Conditions: T = 30°C, pH = 7.4

Method: Radiolabelled Stop Reaction

Reagents

A. 1 M Tris HCl Buffer, pH 7.4 at 25°C (Prepare 100 ml in deionized water using Trizma Base. Adjust to pH 7.4 at 25°C with 1 M HCl. Store at 4°C).

B. 4.9 M Magnesium Chloride Solution ($MgCl_2$) (Use Magnesium Chloride, 4.9 M Solution).

C. 100 mM Ethylene Glycol-bis(β-Aminoethyl Ether)N,N,N',N'-Tetraacetic Acid Solution, pH 7.4 at 25°C (EGTA) (Prepare 50 ml in Reagent A using Ethylene Glycol-bis (β-Aminoethyl Ether) N,N,N',N'-Tetraacetic Acid. Adjust to pH 7.4 at 25°C with 1 M NaOH. Store at 4°C).

D. 500 mM Calcium Chloride Solution ($CaCl_2$) (Prepare 50 ml in deionized water using Calcium Chloride, Dihydrate. Store at 4°C).

E. 0.5 mM Adenosine 5'-Triphosphate Solution, pH 7.0 at 30°C (Prepare 10 ml in deionized water using Adenosine 5'-Triphosphate, Disodium Salt. Adjust to pH 7.0 at 30°C with either 1 M HCl or 1 M NaOH. Confirm the concentration by using a molar extinction coefficient1 of 15,400 measured at 259 mM. Store at–20°C).

F. 0.1% (w/v) Triton X-100 Solution (X-100) (Prepare 10 ml in deionized water using Triton X-100).

G. 1 mg/ml Histone Type III-S Solution (Histone) (Prepare 1 ml in deionized water using Histone, Type III-S. The histone used should have a lysine and arginine content greater than or equal to 25.8% and 2.2%, respectively. *Prepare Fresh*).

H. 250 mM Tris HCl Buffer with 100 mM Magnesium Chloride and 10 mM Ethylene Glycol-bis (β-Aminoethyl Ether) N,N,N',N'-Tetraacetic Acid, pH 7.4 at 30°C (Buffer I) (Prepare 100 ml using Reagents A, B, and C).

I. 250 mM Tris HCl Buffer with 100 mM Magnesium Chloride and 5 mM Calcium Chloride, pH 7.4 at 30°C (Buffer II) (Prepare 100 ml using Reagents A, B, and D).

J. 50 mM Tris HCl Buffer with 5 mM DL-Dithiothreitol, pH 7.4 at 30°C (Enz Dil) (Prepare 25 ml in deionized water using Reagent A and DL-Dithiothreitol).

K. 2 mg/ml Phosphatidylserine Solution (PS) (Use L-α-Phosphatidyl-L-Serine, Chloroform: Methanol (95:5) Solution).

L. Chloroform Solution.

M. 0.5 mg/ml 1,2-Dioleoyl-rac-Glycerol Solution (DAG) (Prepare 4 ml in Reagent L using 1,2-Dioleoyl-rac-Glycerol (C18:1,[cis]-9). Store under nitrogen in a brown bottle).

N. Lipid Activator Solution (Lipids) (Immediately before use, prepare by pipetting 0.25 ml of Reagent K and 40 μl of Reagent M into a suitable container. Close the container immediately while sealing it under nitrogen. Dry the lipid solution under nitrogen while rotating the container gently. Then add 0.5 ml of Reagent F to the dried lipids. Sonicate in a cold bath sonicator for 1 minute. The solution will become cloudy).

O. $_{32}$P-Adenosine 5'-Triphosphate Solution ($_{32}$P-ATP) (Use $_{32}$P-Adenosine 5'-Triphosphate, 3000 Ci/mmole, 10 mCi/ml).

P. ATP Reaction Mix (ATP) (Immediately before use, prepare by adding 1 μl of Reagent O to 100 μl of Reagent E for each of the 9 wells).

Q. Protein Kinase C Enzyme Solution (Immediately before use, prepare a solution containing 3.3–5.0 units/ml of Protein Kinase C in cold Reagent J).

R. Scintillation Fluid (Scin Fluid).

S. 10% (v/v) Trichloroacetic Acid Solution (TCA) (Prepare 1 liter in deionized water using Trichloroacetic Acid, 6.1 N Solution, approximately 100% (w/v)).

T. Nitrocellulose Paper (Use Cellulose Nitrate).

Procedure

This assay is performed in triplicate in a final volume of 0.05 ml. Each sample is assayed with and without activators. The assay components are added as 0.01 ml aliquots and the last component added is the enzyme solution. The reaction is conducted in a 96 well microtiter plate. A second microtiter plate is used to contain the enzyme solution prior to pipetting with a multichannel pipette. Plan the assay on a 8 × 12 chart (as indicated) since a 12 row microtiter plate will be used. The first microtiter plate is set up for assaying samples, I-IV, 3 replications each, with (+) and without (-) activators. The second microtiter plate is for the Protein Kinase C samples.

Plate No. 1

	1	2	3	4	5	6	7	8	9	10	11	12
A –	I	I	I	II	II	II	III	III	III	IV	IV	IV
B +	I	I	I	II	II	II	III	III	III	IV	IV	IV

Plate No. 2

	1	2	3	4	5	6	7	8	9	10	11	12
A	I	I	I	II	II	II	III	III	III	IV	IV	IV
	30μl	30μl	30μl	30μl	30μl	30μl	30μl	30μl	30μl	30μl	30μl	30μl

Before conducting the enzyme assay, set up a dot blot blotting apparatus in order that the wells from the microtiter plate can be blotted onto Reagent T (nitrocellulose paper) on top of Whatman 3 mm paper. Place the microtiter plate on ice. Pipette (in milliliters) the following reagents into 24 of the 96 wells of microtiter Plate No. 1 as previously indicated:

	Test (-)Activator	Test (+)Activator
Reagent G (Histone)	0.01	0.01
Reagent H (Buffer I)	0.01	—
Reagent I (Buffer II)	—	0.01
Reagent F (X-100)	0.01	—
Reagent P (ATP)	0.01	0.01

In a second 96-well microtiter plate, pipette (in milliliter) 0.03 ml, per well, of Reagent Q (Enzyme Solution) as indicated in Plate No. 2. For controls, pipette 0.03 ml of Reagent J (Enz Dil) into several wells. In the event there are more than 4 samples (including the control) leave one row of space between each row of samples (Plate No. 2). Volumes are to be added with a multichannel pipette from plate No. 2. Prepare Reagent N (Lipids). Then add:

Reagent N (Lipids)	—	0.01
Reagent Q (Enzyme Soln)	0.01	0.01

Incubate at 30°C for exactly 3 minutes; then place on ice. Connect the blot apparatus to a vacuum and transfer 40 μl of the 50 μl contents of the microtiter plate (Plate No. 1) to the parallel place on the blotter. Leave for 5 minutes with vacuum applied. Disassemble the blot apparatus. Soak the nitrocellulose paper in 200–300 ml of Reagent S (TCA). Shake gently for 10–15 minutes. Change to Fresh Reagent S (TCA). Shake for another 10–15 minutes. Remove the nitrocellulose paper and place on filter paper to soak up the excess Reagent S (TCA). Cut out the circles (transferred samples) from the nitrocellulose paper and place into scintillation vials containing 1 ml of Reagent R (Scin Fluid). Count the radioactivity in a suitable scintillation counter.

Calculations

$$\text{Units / ml enzyme} = \frac{(\text{cpm})\ (\text{df})}{(\text{SR})\ (3)\ (0.01)} \times \frac{5}{4}$$

cpm = actual count-background on filter
df = Dilution factor
5/4 = Correction factor for removing 40 μl from the 50 μl reaction mixture
SR = Specific radioactivity (cpm/nmole)
3 = Time (in minutes) of assay as per the Unit Definition
0.01 = Volume (in milliliter) of enzyme used

Unit Definition: *One unit will transfer 1 nanomole of phosphate from ATP to Histone I (Histone Type III-S) per minute at pH 7.4 at 30°C.*

Final Assay Concentrations

In a 0.05 ml reaction mix, the final concentrations are 200 μg/ml histone, 60 mM Tris, 20 mM magnesium chloride, 1 mM calcium chloride, 200 μg/ml phosphatidylserine, 8 μg/ml 1,2-dioleoyl-rac-glycerol (C18:1, [cis]-9), 0.02% (w/v) Triton X-100, 0.1 mM adenosine 5'-triphosphate, 1 mM DL-dithiothreitol, and 0.03–0.05 unit protein kinase C.

Note

1. For Protein Kinase C, initially reconstitute the vial in 100 ml of deionized water or 10% (w/v) glycerol if the solution is to be stored frozen.

Enzymatic Assay of PROTEIN KINASE CATALYTIC SUBUNIT Phosphorylating Activity

Principle

$$\text{Casein} + \gamma\text{-}_{32}\text{P - ATP} \xrightarrow[\text{Catalytic Subunit}]{\text{Protein Kinase}} [_{32}\text{P}]\text{ - Phosphorylated Casein} + \text{ADP}$$

Abbreviations Used

γ-$_{32}$P-ATP = Adenosine 5'-Triphosphate γ-$_{32}$P label
ADP = Adenosine 5'-Diphosphate

Conditions: T = 30°C, pH = 6.5

Method: Radioactive

Reagents

A. 1000 mM Potassium Phosphate Buffer, pH 6.5 at 30°C (Prepare 100 ml in deionized water using Potassium Phosphate, Monobasic, Anhydrous. Adjust to pH 6.5 at 30°C with 2 M KOH).

B. 5.0% (w/v) Casein Solution (Casein) (Use Casein, from Bovine Milk, 5% (w/v) Solution).

C. 500 mM Magnesium Acetate Solution ($Mg(OAc)_2$) (Prepare 10 ml in deionized water using Magnesium Acetate, Tetrahydrate).

D. 250 mM Aminophylline Solution (AP) (Prepare 10 ml in deionized water using Aminophylline, Hydrate).

E. 330 mM Dithiothreitol Solution (DTT) (Prepare 10 ml in deionized water using DL-Dithiothreitol. *Prepare Fresh*).

F. 10.0 mM Adenosine 5'-Triphosphate Solution (ATP) (Prepare 1 ml in deionized water using Adenosine 5'-Triphosphate, Disodium Salt. *Prepare Fresh*).

G. γ-$_{32}$P-Adenosine 5'-Triphosphate Solution (γ-$_{32}$P-ATP) (Use product with minimum radioactive concentration of 30 curies/mmole and 2 millicuries/ml).

H. 13.5% (w/v) Trichloroacetic Acid Solution (TCA) (Prepare 20 ml in deionized water using Trichloroacetic Acid, 6.1 N Solution, approximately 100% (w/v)).

I. Protein Kinase Catalytic Subunit Enzyme Solution (Immediately before use, prepare a solution containing 200–400 units/ml of Protein Kinase Catalytic Subunit in cold Reagent J).

J. 39 mM DL-Dithiothreitol Solution (Enzyme Diluent) (Prepare 10 ml in deionized water using DL-Dithiothreitol. *Prepare Fresh*).

K. 6.75% (w/v) Trichloroacetic Acid Solution (Wash Solution) (Prepare 20 ml in deionized water using Trichloroacetic Acid, 6.1 N Solution, approximately 100% (w/v)).

L. Methylethyl Cellosolve (Prepare by adding equal volumes of Ethylene Glycol Monoethyl Ether to Ethylene Glycol Monomethyl Ether).

M. Scintillation Cocktail .

Procedure

Prepare a reaction cocktail by pipetting (in milliliters) the following reagents into a suitable container:

Deionized Water	2.00
Reagent A (Buffer)	0.50
Reagent C ($Mg(OAc)_2$)	0.25
Reagent D (AP)	0.10
Reagent E (DTT)	0.10
Reagent F (ATP)	0.05

Mix by swirling. Transfer 1 ml to a suitable container and add enough Reagent G (γ-$_{32}$P-ATP) to yield approximately 150,000–200,000 counts/minute (cpm) in 0.05 ml of the solution. Then add 0.50 ml of Reagent B (Casein). This is the Reaction Cocktail. Pipette 0.05 ml aliquots of Reagent I (Enzyme Solution) into a multiwell disposable titerplate. Place in a 4°C ice bath. Add 0.05 ml of the Reaction Cocktail to each well and mix by air injection. Immediately transfer the titerplate to a 30°C water bath. Incubate at 30°C for 10 minutes. Then add 0.10 ml of Reagent H (TCA) to each well. Filter the material in the wells through 0.45 μm Millipore HA Type filters. Wash 3 times with Reagent K (Wash Solution). Transfer the filters to suitable 2 dram scintillation vials containing 2.00 ml of Reagent L (Methylethyl Cellosolve). To each scintillation vial, add 5 ml of Reagent M. Count the radioactivity in a suitable scintillation counter.

Calculations

The total number of picomoles (pMoles) of ATP in the reaction mixture is calculated as follows:

$$= 1.1 \ \frac{(0.05)\,(0.01)\,(109)}{(3.00)\,(1.5)} \times 105 \text{ pmole / ml Reaction Cocktail}$$

Find **cpm/pmole** by counting 0.05 ml (5555 pmoles) of the Reaction Cocktail.
0.05 = Volume (in milliliters) of ATP used in the Reaction Cocktail
0.01 = Millimolar concentration of ATP (Reagent F)
109 = Conversion of millimoles to picomoles
3.00 = Intermediate volume (in milliliters) of Reaction Cocktail
1.5 = Total volume (in milliliters) of Reaction Cocktail CPM Counted

$$\text{Units/ml} = (\text{cpm/pmole})\ (10)\ (0.05)$$

CPM counted = Actual count–background on filters
10 = Time of Assay (in minutes) as per the Unit Definition
0.05 = Volume (in milliliters) of Protein Kinase Catalytic Subunit used

$$\text{Units / μg protein} = \frac{\text{units / ml enzyme}}{\text{μg protein / ml enzyme}}$$

Unit Definition: *One unit will transfer 1.0 picomole (10–12 mole) of phosphate from* γ*-*$_{32}$*P-ATP to hydrolyzed, partially dephosphorylated casein per minute at pH 6.5 at 30°C.*

Final Assay Concentrations

In a 0.10 ml solution, the final assay concentrations are 55 mM potassium phosphate, 14 mM magnesium acetate, 3 mM aminophylline, 4 mM dithiothreitol, 0.055 mM adenosine 5'-triphosphate, 0.83% (w/v) casein, and 10 – 20 units protein kinase catalytic subunit.

Note

1. The concentration of dithiothreitol does not include that contributed by the protein kinase catalytic subunit diluent.

Enzymatic Assay of PROTEINASE A (EC 3.4.23.6)

Principle

$$\text{Oxidized Chain B Insulin} \xrightarrow{\text{Proteinase A}} \text{TCA Soluble Peptide Fragments}$$

Abbreviations Used

TCA = Trichloroacetic Acid

Conditions: T = 25°C, pH 6.0, A500nm, Light path = 1 cm

Method: Colorimetric

Reagents

A. 50 mM MES Buffer, pH 6.0 at 25°C (Prepare 100 ml in deionized water using MES, Free Acid. Adjust to pH 6.0 at 25°C With 1 M NaOH).
B. 652 mM Sodium Nitrite Solution ($NaNO_2$) (Prepare 100 ml in deionized water using Sodium Nitrite. *Store on ice*).
C. 943 mM Sodium Carbonate Solution (Na_2CO_3) (Prepare 100 ml in deionized water using Sodium Carbonate, Anhydrous).
D. 13.4% (v/v) Trichloroacetic Acid Solution (TCA) (Prepare 100 ml in deionized water using Trichloroacetic Acid, 6.1 N Solution, approximately 100% (w/v)).
E. 120 mM Hydrochloric Acid Solution (HCl) (Prepare 100 ml in deionized water using Hydrochloric Acid, 1.0 N).
F. 0.9% (w/v) Sulfanilic Acid Solution (Sol Acid) (Prepare 100 ml in Reagent E using Sulfanilic Acid, Anhydrous. *Store on ice*).
G. Diazonium Color Reagent (DCR) (Prepare 5.5 ml by adding 5 ml of ice cold Reagent F to 0.5 ml of ice cold Reagent B. Let react in an ice bath for 15 minutes.
H. 0.20% (w/v) Insulin, Chain B, Oxidized Suspension (Insulin) (Prepare 1 ml in Reagent A using Insulin, Chain B, Oxidized).
I. Proteinase A Enzyme Solution (Immediately before use, prepare a solution containing 0.02 mg/ml of Proteinase A in Reagent A).

Procedure

Pipette (in milliliters) the following reagents into suitable containers:

	Test	Blank
Reagent H (Insulin)	0.25	0.25
Reagent A (Buffer)	—	0.025

Equilibrate in a 25°C shaking water bath, rpm=200, for 5 minutes. Then add:

Reagent I (Enzyme Solution)	0.025	—

Immediately mix by swirling and incubate at 25°C in a shaking water bath, rpm = 200, for exactly 10 minutes. Then add:

Reagent D (TCA)	0.25	0.25

Mix by swirling and let stand at 5 minutes at room temperature. Centrifuge for 5 minutes at 20,000 rpm in a microcentrifuge. Prepare Reagent G (DCR). Pipette (in milliliters) the following reagents into suitable containers.

Test Supernatant	0.15	—
Blank Supernatant	—	0.15
Reagent C (Na_2CO_3)	0.75	0.75

Mix by swirling, then add at 20 second intervals:

	Test	Blank
Reagent G (DCR)	0.10	0.10

Mix by swirling and let react for exactly three minutes at room temperature. Record the A500nm at 20 second intervals for both the Test and Blank using a suitable spectrophotometer.

Standard Curve: Prepare a Standard Curve by pipetting (in milliliters) the following reagents into suitable cuvettes:

	Std 1	Std 2	Std 3	Std Blank
Reagent H (Insulin)	0.025	0.05	0.075	—
Reagent A (Buffer)	0.05	0.025	—	0.075
Reagent D (TCA)	0.075	0.075	0.075	0.075
Reagent C (Na_2CO_3)	0.075	0.075	0.075	0.075

Mix by swirling, then add at 20 second intervals:

Reagent G (DCR)	0.10	0.10	0.10	0.10

Mix by swirling and let react for exactly 3 minutes at room temperature. Read the A500nm at 20 second time intervals for the Standards and Standard Blank using a suitable spectrophotometer.

Calculations

$$\text{Insulin Color Factor} = \frac{\text{A500nm Standard Test} - \text{A500nm Standard Blank}}{\mu\text{g insulin in reaction}}$$

$$\text{Units / ml enzyme} = \frac{(\text{A500nm Test} - \text{A500nm Blank})\ (0.525)\ (\text{df})}{(\text{Insulin Color Factor})\ (10)\ (1000)\ (0.15)\ (0.025)}$$

0.525 = Volume (in milliliter) of stopped reaction
df = Dilution factor
10 = Time (in minutes) of assay as per the Unit Definition
1000 = Conversion factor from micrograms to milligrams as per the Unit Definition
0.15 = Volume (in milliliter) of stopped reaction used in the colorimetric determination
0.025 = Volume (in milliliter) of enzyme used in the stopped reaction

$$\text{Units / mg solid} = \frac{\text{units / ml enzyme}}{\text{mg solid / ml enzyme}}$$

$$\text{Units / mg protein} = \frac{\text{units / ml enzyme}}{\text{mg protein / ml enzyme}}$$

Unit Definition: *One unit will hydrolyze one mg of Insulin Chain B (Oxidized) per minute at pH 6.0 at 25°C.*

Final Assay Concentrations

In a 0.275 ml reaction mix, the final concentrations are 0.18% (w/v) insulin, chain B, oxidized, 50 mM MES, and 0.0005 mg proteinase A.

Enzymatic Assay of PROTEINASE K (EC 3.4.21.64) from Tritirachium album

Principle

$$\text{Casein} + H_2O \xrightarrow{\text{Proteinase K}} \text{Amino Acids}$$

Conditions: T = 37°C, pH = 7.5, A660nm, Light path = 1 cm

Method: Colorimetric

Reagents

A. 50 mM Potassium Phosphate buffer, pH 7.5 at 37°C. (Prepare 200 ml in deionized water using Potassium Phosphate, Dibasic, Trihydrate. Adjust to pH 7.5 at 37°C with 1 M HCl).

B. 0.65% (w/v) Casein Solution (Casein) (Prepare 125 ml in Reagent A using Casein. Heat gently (do not boil) until a homogenous dispersion is obtained. Adjust the pH to 7.5 at 37°C if necessary with either 1 M NaOH or 1 M HCl).

C. 110 mM Trichloroacetic Acid Reagent (TCA) (Dilute 9 ml of Trichloroacetic Acid, 6.1 N, approximately 100% (w/v), to 500 ml with deionized water).

D. Folin & Ciocalteu's Phenol Reagent (F-C) (Dilute 10 ml of Folin & Ciocalteu's Phenol Reagent, to 40 ml with deionized water).

E. 500 mM Sodium Carbonate Solution (Na_2CO_3) (Prepare 500 ml in deionized water using Sodium Carbonate Anhydrous).

F. 10 mM Sodium Acetate Buffer with 5 mM Calcium Acetate, pH 7.5 at 37°C (Enzyme Diluent) (Prepare 100 ml in deionized water using Sodium Acetate, Trihydrate and Calcium Acetate. Adjust the pH to 7.5 at 37°C with 0.1 M Acetic acid or 0.1 M NaOH).

G. 1.1 mM L-Tyrosine Standard (Std Soln) (Prepare 100 ml in deionized water using L-Tyrosine, Free Base. Heat gently (do not boil) until the Tyrosine dissolves and cool to room temperature).

H. Proteinase K Enzyme Solution (Immediately before use, prepare a solution containing 0.1–0.2 unit/ml of Proteinase K in cold Reagent F).

Procedure

Pipette (in milliliters) the following reagents into suitable vials:

	Test	Blank
Reagent B (Casein)	5.00	5.00

Equilibrate to 37EC. Then add:

Reagent H (Enzyme Solution)	1.00	—

Mix by swirling and incubate at 37°C for exactly 10 minutes. Then add:

Reagent C (TCA)	5.00	5.00
Reagent H (Enzyme Solution)	—	1.00

Mix by swirling and incubate at 37°C for about 30 minutes. Filter through a Whatman #50 filter paper and use the filtrate in the color development.

Color Development

Standard Curve: Prepare a standard curve by pipetting (in milliliters) the following reagents into suitable vials:

	Std 1	Std 2	Std 3	Std 4	Std Blank
Reagent G(Std Soln)	0.05	0.10	0.20	0.40	0.00
Deionized Water	1.95	1.90	1.80	1.60	2.00
Reagent E(Na_2CO_3)	5.00	5.00	5.00	5.00	5.00
Reagent D (F-C)	1.00	1.00	1.00	1.00	1.00

Sample: Pipette (in milliliters) the following reagents into 4 dram vials:

	Test	Blank
Test Filtrate	2.00	—
Blank Filtrate	—	2.00
Reagent E (Na_2CO_3)	5.00	5.00
Reagent D (F-C)	1.00	1.00

Mix by swirling and incubate at 37°C for 30 minutes. Remove the vials and allow them to cool to room temperature. If the solutions are hazy, filter through a 0.45 μm filter prior to measuring the absorbances. Transfer the solutions to suitable cuvettes and obtain the A660nm for the Standard, Standard Blank, Test, and Blank using a suitable spectrophotometer.

Calculations

Standard Curve: ΔA660nm Standard = A660nm Standard–A660nm Standard Blank
Plot the ΔA660nm Standard vs μmoles of Tyrosine.

Sample Determination: ΔA660nm Sample = A660nm Test–A660nm Sample Blank
Determine the μmoles of Tyrosine equivalents liberated using the Standard curve.

$$\text{Units / ml enzyme} = \frac{(\mu\text{mole Tyrosine equivalents released})\ (11)}{(1)\ (10)\ (2)}$$

11 = Total volume (in milliliters) of stopped reaction
10 = Time of assay (in minutes) as per the Unit Definition
1 = Volume of enzyme (in milliliter) of enzyme used
2 = Volume (in milliliters) used in the Colorimetric Determination

$$\text{Units / mg solid} = \frac{\text{units / ml enzyme}}{\text{mg solid / ml enzyme}}$$

$$\text{Units / mg protein} = \frac{\text{units / ml enzyme}}{\text{mg protein / ml enzyme}}$$

Unit Definition: *One unit will hydrolyze casein to produce color equivalent to 1.0 μmole (181 μg) of tyrosine per minute at pH 7.5 at 37°C (color by Folin & Ciocalteu's reagent).*

Final Assay Concentrations

In a 6.00 ml reaction mix, the final concentrations are 42 mM potassium phosphate, 0.54% (w/v) casein, 1.7 mM sodium acetate, 0.8 mM calcium acetate, and 0.1–0.2 unit protease.

Enzymatic Assay of PROTOCATECHUATE 3,4-DIOXYGENASE (EC 1.13.11.3)

Principle

$$\text{Protocatechuate} + O_2 \xrightarrow{\text{PCD}} \text{3-Carboxy-cis, cis-Muconate}$$

Abbreviation Used

PCD = Protocatechuate 3,4-Dioxygenase

Conditions: T = 37°C, pH = 7.5, A290nm, Light path = 1 cm

Method: Continuous Spectrophotometric Rate Determination

Reagents

A. 200 mM Acetic Acid Solution (Acet Acid) (Prepare 100 ml in deionized water using Acetic Acid, Glacial).
B. 50 mM Tris Acetate Buffer, pH 7.5 at 25°C (Prepare 100 ml in deionized water using Trizma Base. Adjust to pH 7.5 at 25°C with Reagent A).
C. 0.40 mM Protocatechuic Acid Solution (PCA) (Prepare 50 ml in Reagent B using Protocatechuic Acid).
D. Protocatechuate 3,4-Dioxygenase Enzyme Solution (Immediately before use, prepare a solution containing 0.15 mg/ml of Protocatechuate 3,4-Dioxygenase in ice cold Reagent B).

Procedure

Pipette (in milliliters) the following reagents into suitable cuvettes:

	Test	Blank
Reagent C (PCA)	3.00	3.00

Equilibrate to 37°C and monitor the A290nm until constant using a suitably thermostatted spectrophotometer. Then add:

	Test	Blank
Reagent D (Enzyme Soln)	0.10	—
Reagent B (Buffer)	—	0.10

Immediately mix by inversion and record the decrease in A290nm for approximately 5 minutes. Obtain the ΔA290nm/minute using the maximum linear rate for both the Test and Blank.

Calculations

$$\text{Units / ml enzyme} = \frac{(\Delta A290nm / \text{min Test} - \Delta A290nm / \text{min Blank})\ (3.1)\ (df)}{(3.8)\ (0.1)}$$

3.1 = Total volume (in milliliters) of assay
df = Dilution factor
3.8 = Millimolar extinction coefficient of protocatechuic acid at 290nm
0.1 = Volume (in milliliter) of enzyme used

$$\text{Units / mg solid} = \frac{\text{units / ml enzyme}}{\text{mg solid / ml enzyme}}$$

$$\text{Units / mg protein} = \frac{\text{units / ml enzyme}}{\text{mg protein / ml enzyme}}$$

Unit Definition: *One unit will oxidize 1.0 μmole of protocatechuate to 3-carboxy-cis, cis-muconate per minute at pH 7.5 at 37°C.*

Final Assay Concentrations

In a 3.10 ml reaction mix, the final concentrations are 50 mM Tris, 0.39 mM protocatechuic acid, and 0.015 mg protocatechuate 3,4-dioxygenase.

Enzymatic Assay of PULLULANASE (EC 3.2.1.41)

Principle

$$\text{Pullulan} + H_2O \xrightarrow{\text{Pullulanase}} \text{Maltotriose}$$

Conditions: T = 25°C, pH 5.0, A546nm, Light path = 1 cm

Method: Colorimetric

Reagents

A. 20 mM Sodium Acetate Buffer, pH 5.0 at 25°C (Prepare 100 ml in deionized water using Sodium Acetate, Trihydrate. Adjust to pH 5.0 at 25°C with 1 M HCl).
B. 2.0% (w/v) Pullulan Solution (Prepare 10 ml in Reagent A using Pullulan).
C. 2.0 mM Glucose Standard Solution (Prepare 10 ml in deionized water using D-(+)Glucose, Anhydrous).
D. 16 mM Copper Sulfate, 1.3 mM Sodium Sulfate, 226 mM Sodium Carbonate, 190 mM Sodium Bicarbonate and 43 mM Sodium Potassium Tartrate (Copper Soln) (Prepare 1 liter in deionized water using Cupric Sulfate Pentahydrate; Sodium Bicarbonate; Sodium Sulfate, Anhydrous; Sodium Carbonate, Anhydrous; and Sodium Potassium Tartrate Tetrahydrate).
E. 40 mM Molybdic Acid, 19 mM Arsenic Acid and 756 mM Sulfuric Acid (Ars-Mol) (Prepare 1 liter in deionized water using Molybdic Acid, Ammonium Salt Tetrahydrate; Arsenic Acid, Sodium Salt and Sulfuric Acid).
F. Pullulanase Enzyme Solution (Prepare a solution containing 0.40–0.80 unit/ml Pullulanase in Reagent A).

Procedure

Pipette (in milliliters) the following reagents into 4- dram vials:

	Test	Blank	Std 1	Std 2	Std 3	Std 4	Std 5	Std Blank
Reagent A (Buffer)	0.35	0.35	0.40	0.30	0.20	0.10	—	0.50
Reagent B (Pullulan Soln)	0.05	0.05	—	—	—	—	—	—

Mix by swirling and equilibrate to 25°C. Then add:

	Test	Blank	Std 1	Std 2	Std 3	Std 4	Std 5	Std Blank
Reagent F (Enzyme Soln)	0.10	—	—	—	—	—	—	—
Reagent C (Glucose Std)	—	—	0.10	0.20	0.30	0.40	0.50	—

Mix by swirling and incubate for exactly 10 minutes. Then add:

	Test	Blank	Std 1	Std 2	Std 3	Std 4	Std 5	Std Blank
Reagent D (Copper Soln)	0.50	0.50	0.50	0.50	0.50	0.50	0.50	0.50
Reagent F (Enzyme Soln)	—	0.10	—	—	—	—	—	—

Mix by swirling and incubate in a boiling water bath for 60 minutes. Remove the vials and let cool to 20°C. Then add:

	Test	Blank	Std 1	Std 2	Std 3	Std 4	Std 5	Std Blank
Reagent E (Arsenomolybdate soln)	0.50	0.50	0.50	0.50	0.50	0.50	0.50	0.50

Shake and mix the vials until the foaming stops and any precipitate present is dissolved. Then add:

	Test	Blank	Std 1	Std 2	Std 3	Std 4	Std 5	Std Blank
Deionized Water	11.00	11.00	11.00	11.00	11.00	11.00	11.00	11.00

Mix by inversion and record the A546nm for the Test, Blank, Standards and Standard Blank using a suitable spectrophotometer.

Calculations

ΔA546nm Std = A546nm Std–A546nm Std Blank

Prepare a standard curve by plotting the ΔA546nm of the Glucose Standards versus micromoles of glucose.

Sample Determination: ΔA546nm Test = A546nm Test–A546nm Blank

Determine the μmoles of maltotriose (measured as glucose reducing equivalent) using the Standard curve.

$$\text{Units / mg protein} = \frac{(\mu\text{moles of maltotriose released})\ (\text{df})}{(10)\ (0.1)}$$

df = Dilution factor

10 = Time (in minutes) of assay as per the Unit Definition

0.1 = Volume (in milliliter) of enzyme used

$$\text{Units / mg solid} = \frac{\text{units / ml enzyme}}{\text{mg solid / ml enzyme}}$$

$$\text{Units / mg protein} = \frac{\text{units / ml enzyme}}{\text{mg protein / ml enzyme}}$$

Unit Definition: *One unit will liberate 1.0 μmole of maltotriose (measured as glucose) from pullulan per minute at pH 5.0 at 25°C.*

Final Assay Concentrations

In a 0.50 ml reaction mix, the final concentrations are 20 mM sodium acetate, 0.2% (w/v) pullulan and 0.04–0.08 unit pullulanase.

Notes

1. Sodium Sulfate, Sodium Carbonate, and Sodium Potassium Tartrate are dissolved in approximately 500 ml of deionized water. Cupric Sulfate is dissolved in approximately 100 ml of deionized water and slowly added to the above solution to avoid precipitation. Sodium Bicarbonate is dissolved first in deionized water and then added to the above solution. Dilute the solution to 1 liter. If a precipitate forms, it should be removed by filtration prior to use. Store in an amber bottle and avoid exposure to direct sunlight. Store at room temperature.
2. Molybdic Acid is dissolved in approximately 300 ml of deionized water. Add Sulfuric Acid slowly. Caution, this is an exothermic reaction! Arsenic Acid is dissolved in approximately 300 ml of deionized water and is added to the above solution. The solution is diluted to a total volume of 1 liter and incubated at 37°C for 48–72 hours. If a precipitate forms, it should be removed by filtration prior to use. Store in an amber bottle and avoid exposure to direct sunlight. The solution expires six months after preparation. Store at room temperature in an exhaust hood.

Enzymatic Assay of L-PYROGLUTAMATE AMINOPEPTIDASE (EC 3.4.19.3)

Principle

$$\text{PGA-2-N Nap} + \text{Water} \xrightarrow{\text{PGAP}} \text{PGA} + \text{2-N Nap}$$

Abbreviations Used

PGA-2-N Nap = L-Pyroglutamate-β-Naphthyamide
PGA = L-Pyroglutamate
2-N Nap = β-Naphthylamine

Conditions: T = 37°C, pH 8.00, A580nm, Light path = 1 cm

Method: Colorimetric

Reagents

A. 100 mM Potassium Phosphate with 10 mM EDTA, 5% Glycerol and 5 mM DTT, pH 8.00 at 37°C. (Prepare 100 ml in deionized water using Potassium Phosphate, Dibasic, EDTA, Disodium, Glycerol and DTT. Adjust to pH 8.00 at 37°C with 2 N NaOH).
B. 20 mM L-Pyrrolidonyl-β-Naphthylamide. (PNP) (Prepare 10 ml in Methanol, using L-Pyrrolidonyl-β-Naphthylamide).
C. 25% Trichloroacetic Acid. (Prepare 10 ml in deionized water using Trichloroacetic Acid).
D. 0.2% Sodium Nitrite Solution ($NaNO_2$) (Prepare 10 ml in deionized water using Sodium Nitrite).
E. 0.5% Ammonium Sulfamate Solution (Sulfamate) (Use Ammonium Sulfamate).
F. N-1-Naphthylethylenediamine Solution (NED) (Prepare by adding 110 ml of 95% Ethyl Alcohol to contents of bottle of N-1-Naphthylethylenediamine).

G. LAP Calibration Solution.

H. L-Pyroglutamate Aminopeptidase Enzyme Solution (Immediately before use, prepare a solution containing 10 mg solid/ml of L-Pyroglutamate Aminopeptidase in cold Reagent A).

Procedure

Pipette (in milliliters) the following reagents into suitable containers:

	Test	Blank
Reagent A (Buffer)	1.00	1.00
Reagent B (PNP)	0.10	0.10

Mix by swirling and equilibrate to 37°C. Then add:

Reagent H (Enzyme Solution)	0.10	—

Immediately mix and incubate at 37°C for exactly 15 minutes. Then add:

Reagent C (TCA)	1.00	1.00
Reagent H (Enzyme Solution)	—	0.10

Prepare standards and samples by pipetting (in milliliters) the following reagents into suitable containers:

	Std 1	Std 2	Std 3	Std 4	Std Blank	Test	Test Blank
Test	—	—	—	—	—	1.00	—
Blank	—	—	—	—	—	—	1.00
Reagent G (Std)	0.20	0.40	0.60	0.80	—	—	—
Deionized Water	0.80	0.60	0.40	0.20	1.00	—	—
Reagent D ($NaNO_2$)	1.00	1.00	1.00	1.00	1.00	1.00	1.00

Mix quickly and incubate at 25°C for 3 minutes. Then add:

Reagent E (NH_4Sulf)	1.00	1.00	1.00	1.00	1.00	1.00	1.00

Mix quickly and incubate at 25°C for 3 minutes. Then add:

Reagent F (NED)	2.00	2.00	2.00	2.00	2.00	2.00	2.00

Mix quickly and incubate for 45 minutes at 25°C. Transfer the material into suitable cuvettes and record the A580 for each using a suitable spectrophotometer.

Calculations

Find the slope plotting the A580nm Standards vs LAP concentration. Use the slope (M) to determine the concentration of the test mixture.

$$\text{Units / mg protein} = \frac{(\text{A580nm Test} - \text{A580nm Blank})\ (2.2)}{(15)\ (\text{mg protein / RM})}$$

2.2 = Volume of test Mixture
15 = Time of assay
RM = Reaction Mix (Total Volume = 1.2 ml)

Unit Definition: *One unit will hydrolyze 1.0 nanomole of L-pyroglutamic acid β-naphthylamide to L-pyroglutamic acid and β-naphthylamine per min at pH 8.0 at 37°C.*

Final Assay Concentrations

In a 1.2 ml reaction mix, the final concentrations are 92 mM potassium phosphate, 9.2 mM EDTA, 4.6% glycerol, 4.6 mM DTT, 1.7 mM L-pyrrolidonyl-β-naphthylamide, 8.3% methanol And 1 mg solid L-pyroglutamate aminopeptidase.

Enzymatic Assay of PYROGLUTAMATE AMINOPEPTIDASE (EC 3.4.19.3)

Principle

$$\text{p Glu - }\beta\text{ - Naphthylamide} + H_2O \xrightarrow{\text{PGLUA}} \text{p Glu} + \beta\text{ - Naphthylamine}$$

$$\beta\text{ - Naphthylamine} + NaNO_2 \xrightarrow{H+} \text{Diazo Reagent}$$

$$\text{Diazo Reagent} + \text{N - (1 - Naphthyl) - Ethylenediamine} \rightarrow \text{Blue Azo Dye}$$

Abbreviations Used

p Glu = L-Pyroglutamic acid
PGLUA = Pyroglutamate Aminopeptidase
p Glu-β-Naphthylamide = L-Pyroglutamic acid β-Naphthylamide

Conditions: T = 37°C, pH 8.0, A580nm, Light path = 1 cm

Method: Colorimetric

Reagents

A. 100 mM Potassium Phosphate Buffer with 10 mM Ethylenediaminetetraacetic Acid, 5.0% (v/v) Glycerol, and 5.0 mM Dithiothreitol, pH 8.0 at 37°C (Prepare 100 ml in deionized water using Potassium Phosphate, Dibasic, Trihydrate, Ethylenediaminetetraacetic Acid, Disodium Salt, Dihydrate, Glycerol and DL-Dithothreitol. Adjust to pH 8.0 at 37°C with 1 M HCl).
B. Methanol (Use Absolute Methanol).
C. 22 mM L-Pyroglutamic Acid β-Naphthylamide Solution (p Glu-β-Nap) (Prepare 5 ml in Reagent B using L-Pyroglutamic Acid β-Naphthylamide).
D. 25% (w/v) Trichloroacetic Acid Solution (TCA) (Prepare 25 ml in deionized water using Trichloroacetic Acid, 6.1 N Solution, approximately 100% (w/v)).
E. 0.1% (w/v) Sodium Nitrite Solution ($NaNO_2$) (Prepare 10 ml in deionized water using Sodium Nitrite).
F. 0.5% (w/v) Ammonium Sulfamate Solution (NH_4 Sulf) (Use Ammonium Sulfamate, 0.5% (w/v)).
G. 95% (v/v) Ethanol (Prepare 125 ml in deionized water using Ethyl Alcohol, Denatured).

H. 0.05% (w/v) N-1-Naphthylethylenediamine Solution (NED) (Prepare by adding 110 ml of Reagent G to the contents of 1 bottle of N-1-Naphthylethylenediamine).

I. 0.0018% (w/v) β-Naphthylamine Standard Solution.

J. Pyroglutamate Aminopeptidase Enzyme Solution (Immediately before use, prepare a solution containing 35–70 units/ml of Pyroglutamate Aminopeptidase in cold Reagent A).

Procedure

Pipette (in milliliters) the following reagents into suitable containers:

	Test	Blank
Reagent A (Buffer)	1.00	1.00
Reagent C (p Glu-β-Nap)	0.10	0.10

Mix by swirling and equilibrate to 37°C. Then add:

	Test	Blank
Reagent J (Enzyme Solution)	0.10	—

Immediately mix by swirling and incubate at 37°C for exactly 15 minutes. Then add:

	Test	Blank
Reagent D (TCA)	1.00	1.00
Reagent J (Enzyme Solution)	—	0.10

Mix by swirling.

Colorimetric Assay

Pipette (in milliliters) the following reagents into suitable containers:

	Test	Test Blank	Std 1	Std 2	Std 3	Std 4	Std 5	Std Blank
Test Solution	1.00	—	—	—	—	—	—	—
Blank Solution	—	1.00	—	—	—	—	—	—
Reagent I (Std)	—	—	0.10	0.20	0.30	0.50	0.80	—
Deionized Water	—	—	0.90	0.80	0.70	0.50	0.20	1.00
Reagent E ($NaNO_2$)	1.00	1.00	1.00	1.00	1.00	1.00	1.00	1.00

Mix quickly by swirling and incubate at room temperature for 3 minutes. Then add:

	Test	Test Blank	Std 1	Std 2	Std 3	Std 4	Std 5	Std Blank
Reagent F (NH_4 Sulf)	1.00	1.00	1.00	1.00	1.00	1.00	1.00	1.00

Mix quickly by swirling and incubate at room temperature for 3 minutes. Then add:

	Test	Test Blank	Std 1	Std 2	Std 3	Std 4	Std 5	Std Blank
Reagent H (NED)	2.00	2.00	2.00	2.00	2.00	2.00	2.00	2.00

Mix quickly by swirling and incubate at room temperature for 45 minutes. Transfer the solutions to suitable cuvettes and record the A580nm for the Test, Test Blank, Standards, and Standard Blank using a suitable spectrophotometer.

Calculations

Standard Curve: ΔA580nm Standard = A580nm Standard–A580nm Standard Blank

Prepare a standard curve by plotting the ΔA580nm Standard vs nanomoles of β-Naphthylamine.

Sample Determination: ΔA580nm Sample = A580nm Test–A580nm Blank

Determine the nanomoles of β-Naphthylamine liberated using the standard curve.

$$\text{Units / ml enzyme} = \frac{(\text{nanomoles of } \beta\text{ - Naphthylamine liberated})\ (2.2)\ (\text{df})}{(15)\ (0.1)}$$

2.2 = Total volume (in milliliters) of stopped reaction
df = Dilution factor
15 = Time (in minutes) of assay as per the Unit Definition
0.1 = Volume (in milliliter) of enzyme used

$$\text{Units / mg solid} = \frac{\text{units / ml enzyme}}{\text{mg solid / ml enzyme}}$$

$$\text{Units / mg protein} = \frac{\text{units / ml enzyme}}{\text{mg protein / ml enzyme}}$$

Unit Definition: *One unit will hydrolyze 1.0 nanomole of L-pyroglutamic acid β-naphthylamide to L-pyroglutamic acid and β-naphthylamine per minute at pH 8.0 at 37°C.*

Final Assay Concentrations

In a 1.20 ml reaction mix, the final concentrations are 92 mM potassium phosphate, 9.2 mM Ethylenediaminetetraacetic Acid, 4.6% (v/v) glycerol, 4.6 mM DL-dithiothreitol, 1.8 mM L-pyroglutamic acid β-naphthylamide, and 3.5–7.0 unit pyroglutamate aminopeptidase.

Enzymatic Assay of PYROPHOSPHATASE, ACID

Principle

$$\text{ATP} + H_2O \xrightarrow{\text{Acid Pyrophosphatase}} \text{ADP} + \text{Pi}$$

Abbreviations Used

ATP = Adenosine 5'-Triphosphate
ADP = Adenosine 5'-Diphosphate
Pi = Inorganic Phosphate

Conditions: T = 37°C, pH 5.0, A660nm, Light path = 1 cm.

Method: Colorimetric

Reagents

A. 50 mM Sodium Acetate Buffer with 10 mM 2-Mercaptoethanol, 1 mM Ethylenediaminetetraacetic Acid, pH 5.0 at 37°C (Prepare 25 ml in deionized water using Sodium Acetate, Trihydrate, 2-Mercaptoethanol, and Ethylenediaminetetraacetic Acid, Disodium Salt, Dihydrate. Adjust to pH 5.0 at 37°C with 1 M HCl).

B. 2 mM Adenosine 5'-Triphosphate Solution, pH 5.0 (Substrate Solution) (Prepare 10 ml in cold Reagent A using Adenosine 5'-Triphosphate, Disodium Salt, Dihydrate).

C. 10 N Sulfuric Acid Solution (H_2SO_4) (Prepare 50 ml in deionized water using Sulfuric Acid).

D. Taussky-Shorr Reagent (TSCR) (Prepare by adding 10 ml of 10% (w/v) Ammonium Molybdate, Tetrahydrate, in 10 N H_2SO_4, to 70 ml deionized water. Add 5 g of Ferrous Sulfate, Heptahydrate. Bring the volume to 100 ml with deionized water).

E. 0.2 mM Phosphorus Standard (Std) (Prepare 5 ml in deionized water using Phosphorus Standard Solution. The Phosphorus concentration is 20 μg/ml, 0.645 μmoles/ml).

F. Acid Pyrophosphatase Enzyme Solution (The enzyme solution should be used undiluted and should be approximately 2000–4000 units/ml).

Procedure

Pipette (in milliliters) the following reagents into suitable microcentrifuge tubes:

	Test	Blank
Reagent B (Substrate Solution)	0.045	0.050

Equilibrate to 37°C. Then add:

Reagent F (Enzyme Solution)	0.005	—

Immediately mix by swirling and incubate at 37°C for exactly 30 minutes. Then add:

Reagent D (TSCR)	0.900	0.900
Deionized Water	0.050	0.050

Mix by swirling and incubate at 25°C for 10 minutes. Transfer to suitable cuvettes and record the A660nm for both Test and Blank in a suitable spectrophotometer.

Standard Preparation: Prepare standards by pipetting (in milliliters) the following reagents into suitable microcentrifuge tubes:

	Std 1	Std 2	Std 3	Std 4	Std 5	Std Blank
Deionized Water	0.095	0.090	0.080	0.050	—	0.100
Reagent E (Std)	0.005	0.010	0.020	0.050	0.100	—
Reagent D (TSCR)	0.900	0.900	0.900	0.900	0.900	0.900

Mix and incubate at 25°C for 10 minutes. Transfer to suitable cuvettes and record the A660nm for Standards and Standard Blank.

Standard Curve: A660nm Standard = A660nm Standard–A660nm Standard Blank

Prepare a standard curve by plotting the A660nm Standard vs nanomoles of Phosphorus.

Sample Determination: A660nm Sample = A660nm Test–A660nm Test Blank

Determine the nanomoles of Phosphate liberated using the Standard Curve.

$$\text{Units / ml enzyme} = \frac{\text{(nanomoles of Phosphate released)}}{(0.005)\ (30)}$$

0.005 = Volume (in milliliters) of enzyme used
30 = Time (in minutes) as per the Unit Definition

Unit Definition: *One unit will release 1.0 nanomole of inorganic phosphorus from ATP in 30 minutes at pH 5.0 at 37°C.*

Final Assay Concentrations

In a 0.050 milliliter reaction mix, the final concentrations are 45 mM sodium acetate, 9 mM 2-mercaptoethanol, 1.8 mM adenosine 5'-triphosphate, 0.9 mM Ethylenediaminetetraacetic Acid, and 10–20 units acid pyrophosphatase.

Notes

1. Since very small volumes are utilized, accurate pipetting of solutions is critical.
2. The time period between the addition of Taussky-Shorr reagent and measuring the absorbance in a spectrophotometer must be consistent for all standards and samples.

Enzymatic Assay of PYROPHOSPHATASE, NUCLEOTIDE (EC 3.6.1.9)

Principle

$$\beta\text{-NAD} + H_2O \xrightarrow{\text{Pyrophosphatase, Nucleotide}} \text{NMN} + \text{AMP}$$

$$\text{Ethanol} + \beta\text{-NAD} \xrightarrow{\text{Alcohol Dehydrogenase 1}} \text{Acetaldehyde} + \beta\text{-NADH}$$

Abbreviations Used

β-NAD = β-Nicotinamide Adenine Dinucleotide, Oxidized Form
NMN = Nicotinamide Mononucleotide
AMP =Adenosine 5'-Monophosphate
β-NADH = β-Nicotinamide Adenine Dinucleotide, Reduced Form.

Conditions: T = 37°C, pH = 7.4, A340nm, Light path = 1 cm.

Method: Spectrophotometric Stop Rate Determination

Reagents

A. 4.9 mM β-Nicotinamide Adenine Dinucleotide, Oxidized Form, Solution (β-NAD) (Dissolve the contents of one 20 mg vial of β-Nicotinamide Adenine Dinucleotide, in 5.8 ml of reaction cocktail. Step 1).
B. 300 mM Magnesium Chloride Solution ($MgCl_2$) (Prepare 10 ml in deionized water using Magnesium Chloride Hexahydrate).
C. 1000 mM Sodium Bicarbonate Solution ($NaHCO_3$) (Prepare 50 ml in deionized water using Sodium Bicarbonate).
D. 50 mM Sodium Pyrophosphate Buffer, pH 9.8 at 25°C (Prepare 50 ml in deionized water using Sodium Pyrophosphate, Decahydrate. Adjust to pH 9.8 at 25°C with 1 M NaOH).
E. 95% (v/v) Ethanol (Nondenatured) (Prepare 1 ml in deionized water using Nondenatured Ethanol).

F. Alcohol Dehydrogenase Solution (ADH) (Immediately before use, prepare a solution containing 10 mg/ml in cold deionized water using Alcohol Dehydrogenase).

G. Pyrophosphatase, Nucleotide Enzyme Solution (Immediately before use, prepare a solution containing 0.05–0.10 mg/ml in cold deionized water).

Procedure

Step 1: Prepare a reaction cocktail by pipetting (in milliliters) the following reagents into Reagent A (β-NAD):

Deionized Water	3.80
Reagent C ($NaHCO_3$)	1.50
Reagent B ($MgCl_2$)	0.50

Mix by inversion. Pipette (in milliliters) the following reagents into suitable cuvettes:

	Test	Blank
Reaction Cocktail	2.90	2.90

Equilibrate to 37°C. Then add:

Reagent G (Enzyme Solution)	0.10	—
Deionized Water	—	0.10

Immediately mix by inversion and incubate at 37°C for exactly 30 minutes.

Step 2: Pipette (in milliliters) the following reagents into suitable cuvettes:

	Test	Blank
Deionized Water	1.35	1.35
Reagent D (Buffer)	1.50	1.50
Reagent E (EtOH)	0.05	0.05
Reagent F (ADH)	0.01	0.01

Mix by inversion and equilibrate to 25°C. Monitor the A340nm until constant, using a suitably thermostatted spectrophotometer. Then add:

Test Solution (Step 1)	0.10	—
Blank Solution (Step 1)	—	0.10

Mix by inversion and incubate at 25°C for approximately 15 minutes (until there is no increase in A340nm). Obtain the final A340nm for both the Test and the Blank.

Calculations

$$\text{Units / mg enzyme} = \frac{\text{A340nm Blank} - \text{A340nm Test (3.01)}}{(6.22)\,(30)\,(\text{mg enzyme / ml RM})\,(0.1)}$$

6.22 = Millimolar extinction coefficient of β-NADH at 340nm

3.01 = Final volume of Step 2

30 = Time of assay (in minutes) as per Unit Definition
0.1 = volume from Step 1 used in Step 2

Unit Definition: *One unit will hydrolyze 1.0 μmole of β-NAD to NMN and AMP per minute at pH 7.4 at 37°C in the presence of Mg ions.*

Final Assay Concentrations

In a 3.00 ml reaction mix, the final concentrations are 250 mM sodium bicarbonate, 25 mM magnesium chloride, 4.7 mM β-NAD, and 0.005–0.010 unit pyrophosphatase, nucleotide.

Notes

1. β-NAD which is not hydrolyzed by pyrophosphatase, nucleotide, is measured in the reaction which alcohol dehydrogenase.
2. This should be done prior to the completion of Step 1.

Enzymatic Assay of PYRUVATE CARBOXYLASE (EC 6.4.1.1)

Principle

$$\text{Pyruvate} + \text{ATP} + \text{HCO-3} \xrightarrow{\text{PC}} \text{Oxalacetate} + \text{ADP} + \text{Pi AcCoA}$$

$$\text{Oxalacetate} + \text{NADH} \xrightarrow{\text{MDH}} \text{Malate} + \text{NAD}$$

Abbreviations Used

ATP = Adenosine 5'-Triphosphate.
PC = Pyruvate Carboxylase.
AcCOA = Acetyl Coenzyme A.
ADP = Adenosine 5'-Diphosphate.
Pi = Inorganic Phosphate.
NADH = Nicotinamide Adenine Dinucleotide, Reduced Form.
MDH = Malic Dehydrogenase.
NAD = Nicotinamide Adenine Dinucleotide, Oxidized Form

Conditions: T = 30°C, pH = 7.8, A340nm, Light path = 1 cm

Method: Continuous Spectrophotometric Rate Determination

Reagents

A. 135 mM Triethanolamine Buffer with 7 mM Magnesium Sulfate, 9 mM Pyruvic Acid, and 0.15% (w/v) Bovine Serum Albumin, pH 8.0 at 30°C (Substrate) (Prepare 100 ml in deionized water using Triethanolamine Hydrochloride, Magnesium Sulfate, Anhydrous, Pyruvic Acid, Sodium Salt and Albumin, Bovine. Adjust to pH 8.0 at 30°C with 1 M HCl).
B. 0.3 mM Acetyl Coenzyme A Solution (Prepare 10 ml in deionized water using Acetyl Coenzyme A, Sodium Salt. *Prepare Fresh*).
C. Malic Dehydrogenase Enzyme Solution (AcCoA/MDH) (Immediately before use, add 150 units of Malic Dehydrogenase 4 to 5 ml of Reagent B. Bring the solution to a total volume of 10.0 ml with deionized water).

D. 100 mM Triethanolamine Buffer with 30 mM Adenosine 5'-Triphosphate and 450 mM Potassium Bicarbonate, pH 8.0 at 30°C (ATP/$KHCO_3$) (Prepare 10 ml in deionized water using Triethanolamine Hydrochloride, Adenosine 5'-Triphosphate, Disodium Salt, and Potassium Bicarbonate. Adjust to pH 8.0 at 30°C with 1 M KOH).

E. 2.6 mM -Nicotinamide Adenine Dinucleotide, Reduced Form (NADH) (Prepare 10 ml in deionized water using -Nicotinamide Adenine Dinucleotide, Reduced Form, Disodium Salt).

F. 50 mM Tris HCl Buffer with 50% (v/v) Glycerol, 2 mM Magnesium Acetate, and 1 mM Ethylenediaminetetraacetic Acid, pH 7.4 at 30°C (Enz Dil) (Prepare 50 ml in deionized water using Trizma Base, Glycerol, Magnesium Acetate, Tetrahydrate, and Ethylenediaminetetraacetic Acid, Tetrasodium Salt, Hydrate. Adjust to pH 7.4 at 30°C with 1 M HCl).

G. Pyruvate Carboxylase Enzyme Solution (PC) (Immediately before use, prepare a solution containing approximately 30–90 units/ml of Pyruvate Carboxylase in cold Reagent F).

Procedure

Prepare a reaction cocktail by pipetting (in milliliters) the following reagents into a suitable container:

Reagent A (Substrate)	20.00
Reagent C (AcCoA/MDH)	5.00
Reagent E (-NADH)	2.50

Mix by swirling. Adjust to pH 7.8 at 30°C if necessary, with either 1 M HCl or 1 M KOH. Pipette (in milliliters) the following reagents into a suitable container:

	Test	Blank
Reaction Cocktail	2.90	2.90
Reagent G (PC)	0.005	—
Reagent F (Enz Dil)	—	0.005

Mix by inversion and equilibrate to 30°C. Monitor the A340nm until constant using a suitably thermostatted spectrophotometer. Record the A340nm/minute of the Test. Then add:

Reagent D (ATP/$KHCO_3$)	0.10	0.10

Immediately mix by inversion and record the decrease in A340nm for approximately 5 minutes. Obtain the A340nm/minute using the maximum linear rate for both the Test and Blank.

Calculations

$$\text{Units / ml enzyme} = \frac{(\text{A340nm / min Test} - \text{A340nm / min Blank})\ (3.005)\ (\text{df})}{(6.22)\ (0.05)}$$

3.005 = Total volume (in milliliters) of the assay
df = Dilution factor

6.22 = Millimolar extinction coefficient of NADH at 340nm
0.005 = Volume (in milliliter) of pyruvate carboxylase used in the assay

$$\text{Units / mg protein} = \frac{\text{units / ml enzyme}}{\text{mg protein / ml enzyme}}$$

Unit Definition: *One unit will convert 1.0 mmole of pyruvate and CO_2 to oxalacetate per minute at pH 7.8 at 30ºC.*

Final Assay Concentrations

In a 3.005 ml reaction mix, the final concentrations are 134 mM triethanolamine, 5 mM magnesium sulfate, 7 mM pyruvic acid, 0.12% (w/v) Bovine Serum Albumin, 0.23 mM nicotinamide adenine dinucleotide, reduced form, 0.05 mM acetyl coenzyme A, 2.63 units malic dehydrogenase, 1 mM adenosine 5'-triphosphate, 15 mM potassium bicarbonate, 0.05% (v/v) glycerol, 0.002 mM magnesium acetate, 0.001 mM Ethylenediaminetetraacetic Acid, 0.05 mM Tris, 0.15–0.45 unit pyruvate carboxylase.

Notes

1. Lactic Dehydrogenase is the principle contaminant that may interfere with the assay for pyruvate carboxylase. If the A340nm/min is not zero, it must be subtracted from the A340nm/min for the Test after the addition of Reagent D (ATP/$KHCO_3$).
2. Malic Dehydrogenase Unit Definition: One unit will convert 1.0 mole of oxalacetate and NADH to L-malate and NAD per minute at pH 7.5 at 25ºC.

Enzymatic Assay of PYRUVATE DECARBOXYLASE (EC 4.1.1.1)

Principle

$$\text{Pyruvate} \xrightarrow{\text{PDC}} \text{Acetaldehyde} + CO_2$$

$$\text{Acetaldehyde} + \beta\text{-NADH} \xrightarrow{\text{ADH}} \text{Ethanol} + \beta\text{-NAD}$$

Abbreviations Used

PDC = Pyruvate Decarboxylase
β-NADH = β-Nicotinamide Adenine Dinucleotide, Reduced Form
ADH = Alcohol Dehydrogenase
β-NAD = β-Nicotinamide Adenine Dinucleotide, Oxidized Form

Conditions: T = 25ºC, pH 6.0, A340nm, Light path = 1 cm

Method: Continuous Spectrophotometric Rate Determination

Reagents

A. 200 mM Citrate Buffer, pH 6.0 at 25ºC (Prepare 100 ml in deionized water, using Citric Acid, Free Acid, Anhydrous. Adjust to pH 6.0 at 25ºC with 1 M NaOH).
B. 1 M Sodium Pyruvate Solution (Pyr) (Prepare 5 ml in deionized water using Pyruvic Acid, Sodium Salt).
C. 6.4 mM β-Nicotinamide Adenine Dinucleotide, Reduced Form, Solution (β-NADH) (Dissolve the contents of one 5 mg vial of β-Nicotinamide Adenine Dinucleotide, Reduced Form, Disodium, in the appropriate volume of deionized water).

D. Alcohol Dehydrogenase Enzyme Solution (ADH) (Immediately before use, prepare a solution containing 200 units/ml of Alcohol Dehydrogenase, in cold deionized water).

E. Pyruvate Decarboxylase Enzyme Solution (PDC). Immediately before use, prepare a solution containing 0.3–0.6 unit/ml of Pyruvate Decarboxylase in cold Reagent A).

Procedure

Pipette (in milliliters) the following reagents into suitable cuvettes:

	Test	Blank
Reagent A (Buffer)	2.70	2.90
Reagent B (Pyr)	0.10	0.10
Reagent C (β-NADH)	0.05	—
Reagent D (ADH)	0.05	—

Mix by inversion and equilibrate to 25°C. Monitor the A340nm until constant using a suitably thermostatted spectrophotometer. Then add:

Reagent E (PDC)	0.10	—

Immediately mix by inversion and record the decrease in A340nm for approximately 5 minutes. Obtain the A340nm/minute using the maximum linear rate for both the Test and Blank.

Calculations

$$\text{Units / ml enzyme} = \frac{(\text{A340nm / min Test} - \text{A340nm / min Blank})\ (3)\ (\text{df})}{(6.22)\ (0.1)}$$

3 = Total volume (in milliliters) of assay
df = Dilution factor
6.22 = Millimolar extinction coefficient of β-NADH at 340nm
0.1 = Volume (in milliliter) of enzyme used

$$\text{Units / mg protein} = \frac{\text{units / ml enzyme}}{\text{mg protein / ml enzyme}}$$

Unit Definition: *One unit will convert 1.0 μmole of pyruvate to acetaldehyde per minute at pH 6.0 at 25°C.*

Final Assay Concentrations

In a 3.00 ml reaction mix, the final concentrations are 187 mM citric acid, 33 mM sodium pyruvate, 0.11 mM β-nicotinamide adenine dinucleotide, reduced form, 10 units alcohol dehydrogenase, and 0.03–0.06 unit pyruvate decarboxylase.

Note

1. Alcohol Dehydrogenase Unit Definition: One unit will convert 1.0 μmole of ethanol to acetaldehyde per minute at pH 8.8 at 25°C.

Enzymatic Assay of PYRUVATE DEHYDROGENASE

Principle

$$\text{Pyruvate} + \text{CoA} + \text{NAD} + \text{TPP} \xrightarrow{\text{Pyruvate Dehydrogenase}} \text{Acetyl CoA} + \text{NADH} + CO_2$$

Abbreviations Used

CoA = Coenzyme A
NAD = Nicotinamide Adenine Dinucleotide, Oxidized Form
TPP = Thiamine Pyrophosphate
NADH = -Nicotinamide Adenine Dinucleotide, Reduced Form
Acetyl CoA = Acetyl Coenzyme A

Conditions: T = 30°C, pH = 7.4, A340nm, Light path = 1 cm

Method: Continuous Spectrophotometric Rate Determination

Reagents

A. 150 mM MOPS HCl Buffer, pH 7.4 at 30°C (Prepare 100 ml in deionized water using MOPS, Sodium Salt. Adjust to pH 7.4 at 30°C with 1 M HCl).
B. 12 mM Magnesium Chloride Solution ($MgCl_2$) (Prepare 5 ml in deionized water using Magnesium Chloride, Hexahydrate).
C. 0.6 mM Calcium Chloride Solution ($CaCl_2$) (Prepare 5 ml in deionized water using Calcium Chloride, Dihydrate).
D. 18 mM Thiamine Pyrophosphate (Cocarboxylase) Solution (TPP) (Prepare 5 ml in deionized water using Cocarboxylase. *Prepare Fresh*).
E. 0.72 mM Coenzyme A Solution (CoA) (Prepare 5 ml in deionized water using Coenzyme A, Sodium Salt. *Prepare Fresh*).
F. 20 mM Nicotinamide Adenine Dinucleotide Solution (NAD) (Prepare 4 ml in deionized water using Nicotinamide Adenine Dinucleotide. *Prepare Fresh*).
G. 15.6 mM L-Cysteine Solution, pH 7.4 at 30°C (L-Cys) (Prepare 20 ml in deionized water using L-Cysteine, Hydrochloride, Monohydrate. Adjust to pH 7.4 at 30°C with 1 M NaOH. *Prepare Fresh*).
H. 75 mM Pyruvic Acid Solution (PYR) (Prepare 10 ml in deionized water using Pyruvic Acid, Sodium Salt. *Prepare Fresh*).
I. 50 mM MOPS HCl Buffer, pH 7.4 at 30°C (Enzyme Diluent) (Prepare 100 ml in deionized water using MOPS, Sodium Salt. Adjust to pH 7.4 at 30C with 1 M HCl).
J. Pyruvate Dehydrogenase Enzyme Solution (Immediately before use, prepare a solution containing 0.4–0.6 unit/ml of Pyruvate Dehydrogenase in cold Reagent I).

Procedure

Pipette (in milliliters) the following reagents into suitable containers:

	Test	Blank
Reagent A (Buffer)	1.00	1.00
Reagent B ($MgCl_2$)	0.05	0.05
Reagent C ($CaCl_2$)	0.05	0.05
Reagent D (TPP)	0.05	0.05
Reagent E (CoA)	0.50	0.50
Reagent F (NAD)	0.30	0.30
Reagent G (L-Cys)	0.50	0.50
Deionized Water	0.30	0.30
Reagent J (Enzyme Solution)	0.05	—
Reagent I (Enzyme Diluent)	—	0.05

Mix by inversion and equilibrate to 30°C. Monitor the A340nm until constant, using a suitably thermostatted spectrophotometer. Then add:

Reagent H (PYR)	0.20	0.20

Immediately mix by inversion and record the increase in A340nm for approximately 5 minutes. Obtain the A340nm/minute using the maximum linear rate for both the Test and Blank.

Calculations

$$\text{Units / ml enzyme} = \frac{(\text{A340nm / min Test} - \text{A340nm / min Blank})\,(\text{df})\,(3.00)}{(6.22)\,(0.05)}$$

df = Dilution factor
3.00 = Total volume (in milliliters) of assay
6.22 = Millimolar extinction coefficient of NADH at 340nm
0.05 = Volume (in milliliters) of enzyme used

$$\text{Units / mg protein} = \frac{\text{units / ml enzyme}}{\text{mg protein / ml enzyme}}$$

Unit Definition: *One unit will convert 1.0 mole of NAD to NADH per minute at pH 7.4 at 30°C in the presence of saturating levels of Coenzyme A.*

Final Assay Concentrations

In a 2.958 ml reaction mix, the final concentrations are 51 mM MOPS, 0.20 mM magnesium chloride, 0.01 mM calcium chloride, 0.30 mM cocarboxylase, 0.12 mM coenzyme A, 2.0 mM -NAD, 2.64 mM L-cysteine hydrochloride, 5.1 mM pyruvate and 0.003–0.005 unit pyruvate dehydrogenase.

Enzymatic Assay of PYRUVATE KINASE (EC 2.7.1.40)

Principle

$$\text{ADP} + \text{PEP} \xrightarrow{\text{Pyruvate Kinase}} \text{ATP} + \text{Pyruvate}$$

$$\text{Pyruvate} + \beta\text{-NADH} \xrightarrow{\text{Lactic Dehydrogenase}} \text{Lactate} + \beta\text{-NAD}$$

Abbreviations Used

ADP = Adenosine 5'-Diphosphate
PEP = Phospho(enol) Pyruvate

ATP = Adenosine 5'-Triphosphate
β-NADH = β-Nicotinamide Adenine Dinucleotide, Reduced Form
β-NAD = β-Nicotinamide Adenine Dinucleotide, Oxidized Form

Conditions: T = 30°C, pH = 7.2, A340nm, Light path = 1 cm

Method: Continuous Spectrophotometric Rate Determination

Reagents

A. 100 mM Imidazole HCl Buffer, pH 7.2 at 30°C. (Prepare 50 ml in deionized water using Imidazole. Adjust to pH 7.2 with 1 M HCl).
B. 100 mM Adenosine 5'-Diphosphate Solution (ADP) (Prepare 10 ml in deionized water using Adenosine 5'-Diphosphate, Sodium Salt. *Prepare Fresh*).
C. 1000 mM Magnesium Chloride Solution ($MgCl_2$) (Prepare 10 ml in deionized water using Magnesium Chloride, Hexahydrate).
D. 2500 mM Potassium Chloride Solution (KCl) (Prepare 10 ml in deionized water using Potassium Chloride).
E. 155 mM Phospho(enol) pyruvate Solution (PEP) (Prepare 10 ml in deionized water using Phospho(enol) pyruvate, Mono(Cyclohexylammonium) Salt. *Prepare Fresh*).
F. 13.1 mM β-Nicotinamide Adenine Dinucleotide, Reduced Form Solution (β-NADH) (Dissolve the contents of one 10 mg vial of β-Nicotinamide Adenine Dinucleotide, Reduced Form, Disodium Salt in the appropriate volume of deionized water *or* prepare 1 ml in deionized water using β-Nicotinamide Adenine Dinucleotide, Reduced Form, Disodium Salt. *Prepare Fresh*).
G. L-Lactic Dehydrogenase Enzyme Solution (LDH) (Immediately before use, prepare a solution containing 400 units/ml in deionized water using Llactic Dehydrogenase).
H. Pyruvate Kinase Enzyme Solution (PK) (Immediately before use, prepare a solution containing 0.15–0.30 unit/ml of Pyruvate Kinase in cold deionized water).

Procedure

Prepare a reaction cocktail by pipetting (in milliliters) the following reagents into a suitable container:

Reagent A (Buffer)	19.00
Reagent B (ADP)	2.00
Reagent C ($MgCl_2$)	0.40
Reagent D (KCl)	0.75
Deionized water	2.50
Reagent F (β-NADH)	0.38

Mix and adjust to pH 7.2 at 30°C with 100 mM HCl or 100 mM KOH, if necessary. Pipette (in milliliters) the following reagents into suitable cuvettes:

	Test	Blank
Reaction Cocktail	2.75	2.85
Reagent G (LDH)	0.05	0.05
Reagent E (PEP)	0.10	—

Mix by inversion and equilibrate to 30°C. Monitor the A340nm until constant, using a suitably thermostatted spectrophotometer. Then add:

	Test	Blank
Deionized Water	—	0.10
Reagent H (PK)	0.10	—

Immediately mix by inversion and record the decrease in A340nm for approximately 10 minutes. Obtain the ΔA340nm/minute using the maximum linear rate for both the Test and Blank.

Calculations

$$\text{Units / ml enzyme} = \frac{(\Delta A340nm / \text{min Test} - \Delta A340nm / \text{min Blank})\ (3)\ (df)}{(6.22)\ (0.1)}$$

3 = Total volume (in milliliters) of assay
df = Dilution factor
6.22 = Millimolar extinction coefficient of β-NADH at 340nm
0.1 = Volume (in milliliters) of enzyme used

$$\text{Units / mg solid} = \frac{\text{units / ml enzyme}}{\text{mg solid / ml enzyme}}$$

$$\text{Units / mg protein} = \frac{\text{units / ml enzyme}}{\text{mg protein / ml enzyme}}$$

Unit Definition: *One unit will convert 1.0 μmole of phospho(enol) pyruvate to pyruvate per minute at pH 7.2 at 30°C.*

Final Assay Concentrations

In a 3.00 ml reaction mix, the final concentrations are 72 mM imidazole, 7.6 mM adenosine 5'- diphosphate, 15.2 mM magnesium chloride, 71.2 mM potassium chloride, 5.2 mM phospho(enol) pyruvate, 0.19 mM β-nicotinamide adenine dinucleotide, reduced form, 20 units L-lactic dehydrogenase and 0.015–0.030 unit pyruvate kinase.

Note

1. L-Lactic Dehydrogenase Unit Definition: One unit will reduce 1.0 μmole of pyruvate to L-lactate per minute at pH 7.5 at 37°C.

Enzymatic Assay of PYRUVATE KINASE (EC 2.7.1.40)

Principle

$$\text{Phospho(enol) pyruvate} + \text{ADP} \xrightarrow[\text{Mg}_2^+,\ \text{K}^+]{\text{Pyruvate Kinase}} \xrightarrow{\text{Pyruvate Kinase}} \text{Pyruvate} + \text{ATP}$$

$$\text{Pyruvate} + \beta\text{-NADH} \xrightarrow{\text{L-Lactic Dehydrogenase}} \text{L-Lactate} + \beta\text{-NAD}$$

Abbreviations Used

ADP = Adenosine 5'-Diphosphate
ATP = Adenosine 5'-Triphosphate

β-NADH = β-Nicotinamide Adenine Dinucleotide, Reduced Form
β-NAD = β-Nicotinamide Adenine Dinucleotide, Oxidized Form

Conditions: T = 37°C, pH = 7.6, A340nm, Light path = 1 cm

Method: Continuous Spectrophotometric Rate Determination

Reagents

A. 100 mM Potassium Phosphate Buffer, pH 7.6 at 37°C. (Prepare 100 ml in deionized water using Potassium Phosphate, Monobasic, Anhydrous. Adjust to pH 7.6 at 37°C with 1 M KOH).
B. 17 mM Phospho(enol) pyruvate Solution (PEP) (Prepare 1 ml in deionized water using Phospho(enol) pyruvate, Trisodium Salt, Hydrate. *Prepare Fresh*).
C. 1.3 mM β-Nicotinamide Adenine Dinucleotide, Reduced Form Solution (β-NADH) (Dissolve the contents of a 5 mg vial of β-Nicotinamide Adenine Dinucleotide, Reduced Form, Disodium Salt in the appropriate volume of Reagent A. *Prepare Fresh*).
D. 100 mM Magnesium Sulfate Solution ($MgSO_4$) (Prepare 1 ml in deionized water using Magnesium Sulfate, Heptahydrate).
E. 44 mM Adenosine Diphosphate Solution (ADP) (Prepare 1 ml in deionized water using Adenosine 5'-Diphosphate, Sodium Salt).
F. L-Lactic Dehydrogenase Solution (LDH) (Immediately before use, prepare a solution containing 5000 units/ml in cold Reagent A using L-Lactic Dehydrogenase).
G. Pyruvate Kinase (Immediately before use, prepare a solution containing 0.15–0.30 unit/ml of Pyruvate Kinase in cold Reagent A).

Procedure

Pipette (in milliliters) the following reagents into suitable cuvettes:

	Test	Blank
Deionized Water	1.40	1.40
Reagent A (Buffer)	0.80	0.90
Reagent B (PEP)	0.10	0.10
Reagent C (β-NADH)	0.25	0.25
Reagent D ($MgSO_4$)	0.20	0.20
Reagent E (ADP)	0.10	0.10
dReagent F (LDH)	0.002	0.002

Mix by inversion and equilibrate to 37°C. Monitor the A340nm until constant, using a suitably thermostatted spectrophotometer. Then add:

	Test	Blank
Reagent G (Enzyme Solution)	0.10	—

Immediately mix by inversion and record the decrease in A340nm for approximately 5 minutes. Obtain the ΔA340nm/minute using the maximum linear rate for both the Test and Blank.

Calculations

$$\text{Units / ml enzyme} = \frac{(\Delta A340\text{nm / min Test} - \Delta A340\text{nm / min Blank})\ (2.952)\ (\text{df})}{(6.22)\ (0.1)}$$

2.952 = Total volume (in milliliters) of assay
df = Dilution factor
6.22 = Millimolar extinction coefficient of β-NADH at 340nm
0.1 = Volume (in milliliters) of enzyme used

$$\text{Units / mg solid} = \frac{\text{units / ml enzyme}}{\text{mg solid / ml enzyme}}$$

$$\text{Units / mg protein} = \frac{\text{units / ml enzyme}}{\text{mg protein / ml enzyme}}$$

Unit Definition: *One unit will convert 1.0 μmole of phospho(enol) pyruvate to pyruvate per minute at pH 7.6 at 37°C.*

Final Assay Concentrations

In a 2.952 ml reaction mix, the final concentrations are 39 mM potassium phosphate, 0.58 mM phospho(enol) pyruvate, 0.11 mM β-nicotinamide adenine dinucleotide, reduced form, 6.8 mM magnesium sulfate, 1.5 mM adenosine 5'-diphosphate, 10 units lactic dehydrogenase and 0.015–0.030 unit pyruvate kinase.

Note

1. Lactic Dehydrogenase Unit Definition: One unit will reduce 1.0 μmole of pyruvate to L-lactate per minute at pH 7.5 at 37°C.

Enzymatic Assay of PYRUVATE KINASE (EC 2.7.1.40) From Rabbit Liver

Principle

$$\text{Phospho(enol) pyruvate} + \text{ADP} \xrightarrow[\text{Mg}_2^+]{\text{Pyruvate Kinase}} \text{Pyruvate} + \text{ATP}$$

$$\text{Pyruvate} + \beta\text{-NADH} \xrightarrow{\text{Lactic Dehydrogenase}} \text{Lactate} + \beta\text{-NAD}$$

Abbreviations Used

ADP = Adenosine 5'-Diphosphate
ATP = Adenosine 5'-Triphosphate
β-NADH = β-Nicotinamide Adenine Dinucleotide, Reduced Form
β-NAD = β-Nicotinamide Adenine Dinucleotide, Oxidized Form

Conditions: T = 37°C, pH = 7.6, A340nm, Light path = 1 cm

Method: Continuous Spectrophotometric Rate Determination

Reagents

A. 100 mM Potassium Phosphate Buffer, pH 7.6 at 37°C. (Prepare 100 ml in deionized water using Potassium Phosphate, Monobasic, Anhydrous. Adjust to pH 7.6 at 37°C with 1 M KOH).
B. 8.0 mM Phospho(enol) pyruvate Solution (PEP) (Prepare 1 ml in deionized water using Phospho(enol) Pyruvate, Monopotassium Salt).
C. 3 mM β-Nicotinamide Adenine Dinucleotide, Reduced Form Solution (β-NADH) (Dissolve the contents of a 10 mg vial of β-Nicotinamide Adenine Dinucleotide,

Reduced Form, Disodium Salt in the appropriate volume of Reagent A. *Prepare Fresh*).

D. 100 mM Magnesium Sulfate Solution (Prepare 1 ml in deionized water using Magnesium Sulfate, Heptahydrate).

E. 40 mM Adenosine Diphosphate Solution (ADP) (Prepare 1 ml in deionized water using Adenosine 5'-Diphosphate, Di(Monocyclohexylammonium) Salt).

F. L-Lactic Dehydrogenase Solution (LDH) (Immediately before use, prepare a solution containing 500 units/ml of L-Lactic Dehydrogenase in cold Reagent A).

G. 30 mM Fructose 1,6-Diphosphate Solution (F 1,6-P) (Prepare 2 ml in deionized water using Fructose 1,6-Diphosphate, Sodium Salt).

H. Pyruvate Kinase (Immediately before use, prepare a solution containing 0.3–0.6 unit/ml of Pyruvate Kinase in cold Reagent A).

Procedure

Pipette (in milliliters) the following reagents into suitable cuvettes:

	Test	Blank
Deionized Water	1.30	1.30
Reagent A (Buffer)	0.80	0.90
Reagent B (PEP)	0.16	0.16
Reagent C (β-NADH)	0.20	0.20
Reagent D (Magnesium Sulfate)	0.20	0.20
Reagent E (ADP)	0.10	0.10
Reagent F (LDH)	0.04	0.04
Reagent G (F 1,6-P)	0.10	0.10

Mix by inversion and equilibrate to 37°C. Monitor the A340nm until constant, using a suitably thermostatted spectrophotometer. Then add:

Reagent H (Enzyme Solution)	0.10	—

Immediately mix by inversion and record the decrease in A340nm for approximately 5 minutes. Obtain the ΔA340nm/minute using the maximum linear rate for both the Test and Blank.

Calculations

$$\text{Units / ml enzyme} = \frac{(\Delta A340\text{nm / min Test} - \Delta A340\text{nm / min Blank})\ (3)\ (\text{df})}{(6.22)\ (0.1)}$$

3 = Total volume (in milliliters) of assay
df = Dilution factor
6.22 = Millimolar extinction coefficient of β-NADH at 340nm
0.1 = Volume (in milliliters) of enzyme used

$$\text{Units / mg solid} = \frac{\text{units / ml enzyme}}{\text{mg solid / ml enzyme}}$$

$$\text{Units / mg protein} = \frac{\text{units / ml enzyme}}{\text{mg protein / ml enzyme}}$$

Unit Definition: *One unit will convert 1.0 µmole of phospho(enol) pyruvate to pyruvate per minute at pH 7.6 at 37°C in the presence of 1.0 mM of fructose 1,6-diphosphate.*

Final Assay Concentrations

In a 3.00 ml reaction mix, the final concentrations are 38 mM potassium phosphate, 0.43 mM phospho(enol) pyruvate, 0.2 mM β-nicotinamide adenine dinucleotide, 6.7 mM magnesium sulfate, 1.3 mM adenosine 5'-diphosphate, 20 units lactic dehydrogenase, 1 mM fructose 1,6-diphosphate and 0.03 to 0.06 unit pyruvate kinase.

Note

1. Lactic Dehydrogenase Unit Definition: One unit will reduce 1.0 micromole of pyruvate to L-lactate per minute at pH 7.5 at 37°C.

Enzymatic Assay of PYRUVATE KINASE from Rabbit Muscle, Dog Muscle and Chicken Muscle (EC 2.7.1.40)

Principle

$$\text{Phospho(enol) pyruvate} + \text{ADP} \xrightarrow[Mg_2^+,\ K^+]{\text{Pyruvate Kinase}} \text{Pyruvate} + \text{ATP}$$

$$\text{Pyruvate} + \beta\text{-NADH} \xrightarrow{\text{Lactic Dehydrogenase}} \text{Lactate} + \beta\text{-NAD}$$

Abbreviations Used

ADP = Adenosine 5'-Diphosphate
ATP = Adenosine 5'-Triphosphate
β-NADH = β-Nicotinamide Adenine Dinucleotide, Reduced Form
β-NAD = β-Nicotinamide Adenine Dinucleotide, Oxidized Form

Conditions: T = 37°C, pH = 7.6, A340nm, Light path = 1 cm

Method: Continuous Spectrophotometric Rate Determination

Reagents

A. 100 mM Potassium Phosphate Buffer, pH 7.6 at 37°C. (Prepare 100 ml in deionized water using Potassium Phosphate, Monobasic, Anhydrous. Adjust to pH 7.6 at 37°C with 1 M KOH).

B. 8.0 mM Phospho(enol) pyruvate Solution (PEP) (Prepare 1 ml in deionized water using Phospho(enol) pyruvate, Monopotassium Salt. *Prepare Fresh*).

C. 3.0 mM β-Nicotinamide Adenine Dinucleotide, Reduced Form Solution (β-NADH) (Prepare 2 ml in deionized water using β-Nicotinamide Adenine Dinucleotide, Reduced Form, Disodium Salt. *Prepare Fresh*).

D. 100 mM Magnesium Sulfate Solution ($MgSO_4$) (Prepare 1 ml in deionized water using Magnesium Sulfate, Heptahydrate).

E. 40 mM Adenosine Diphosphate Solution (ADP) (Prepare 1 ml in deionized water using Adenosine 5'-Diphosphate, Sodium Salt).

F. L-Lactic Dehydrogenase Solution (LDH) (Immediately before use, prepare a solution containing 500 units/ml in cold Reagent A using L-Lactic Dehydrogenase).

G. Pyruvate Kinase Enzyme Solution (Immediately before use, prepare a solution containing 0.3–0.6 unit/ml of Pyruvate Kinase in cold Reagent A).

Procedure

Pipette (in milliliters) the following reagents into suitable cuvettes:

	Test	Blank
Deionized Water	1.40	1.40
Reagent A (Buffer)	0.80	0.90
Reagent B (PEP)	0.16	0.16
Reagent C (β-NADH)	0.20	0.20
Reagent D ($MgSO_4$)	0.20	0.20
Reagent E (ADP)	0.10	0.10
Reagent F (LDH)	0.04	0.04

Mix by inversion and equilibrate to 37°C. Monitor the A340nm until constant, using a suitably thermostatted spectrophotometer. Then add:

Reagent G (Enzyme Solution)	0.10	—

Immediately mix by inversion and record the decrease in A340nm for approximately 5 minutes. Obtain the ΔA340nm/minute using the maximum linear rate for both the Test and Blank.

Calculations

$$\text{Units / ml enzyme} = \frac{(\Delta A340\text{nm / min Test} - \Delta A340\text{nm / min Blank})\ (3)\ (\text{df})}{(6.22)\ (0.1)}$$

3 = Total volume (in milliliters) of assay
df = Dilution factor
6.22 = Millimolar extinction coefficient of β-NADH at 340nm
0.1 = Volume (in milliliter) of enzyme used

$$\text{Units / mg solid} = \frac{\text{units / ml enzyme}}{\text{mg solid / ml enzyme}}$$

$$\text{Units / mg protein} = \frac{\text{units / ml enzyme}}{\text{mg protein / ml enzyme}}$$

Unit Definition: *One unit will convert 1.0 μmole of phospho(enol) pyruvate to pyruvate per minute at pH 7.6 at 37°C.*

Final Assay Concentrations

In a 3.00 ml reaction mix, the final concentrations are 31 mM potassium phosphate, 0.43 mM phospho(enol) pyruvate, 0.20 mM β-nicotinamide adenine dinucleotide, reduced form, 6.7 mM magnesium sulfate, 1.3 mM adenosine 5'-diphosphate, 20 units lactic dehydrogenase and 0.03–0.06 unit pyruvate kinase.

Notes

1. This assay is not to be used to assay Pyruvate Kinase from Rabbit Liver, Pyruvate Kinase, from Bacillus stearothermophilus, Pyruvate Kinase, from Porcine Heart, and Pyruvate Kinase, Insoluble Enzyme.

2. Unit Definition for L-Lactic Dehydrogenase: One unit will reduce 1.0 μmole of pyruvate to L-lactate per minute at pH 7.5 at 37°C.

Enzymatic Assay of PYRUVATE OXIDASE (EC 1.2.3.3) (pH 5.7)

Principle

$$\text{Pyruvate} + O_2 + \text{Pi} \xrightarrow[\text{FAD, TPP, } MG_2^+]{\text{Pyruvate Oxidase}} \text{Acetylphosphate} + CO_2 + H_2O_2$$

$$2H_2O_2 + 4\text{-Aminoantipyrine} + \text{EHSPT} \xrightarrow{\text{POD}} \text{Quinoneimine dye} + 4H_2O$$

Abbreviations Used

Pi = Inorganic Phosphate
FAD = Flavin Adenine Dinucleotide
TPP = Thiamine Pyrophosphate
EHSPT = N-Ethyl-N-(2-Hydroxy-3-Sulfopropyl)-m-Toluidine
POD = Peroxidase

Conditions: T = 37°C, pH 5.7, A550nm, Light path = 1 cm

Method: Spectrophotometric Rate Determination

Reagents

A. 150 mM Potassium Phosphate Buffer, pH 5.9 at 37°C (Prepare 100 ml in deionized water using Potassium Phosphate, Monobasic, Anhydrous. Adjust to pH 5.9 at 37°C with 1 M KOH).

B. 0.15 mM Flavin Adenine Dinucleotide Solution (FAD) (Prepare 10 ml in deionized water using Flavin Adenine Dinucleotide, Disodium Salt. *Prepare Fresh*).

C. 3 mM Cocarboxylase (Thiamine Pyrophosphate) Solution (TPP) (Prepare 10 ml in deionized water using Cocarboxylase. *Prepare Fresh*).

D. 7.4 mM 4-Aminoantipyrine Solution (4-AAP) (Prepare 25 ml in deionized water using 4-Aminoantipyrine, Free Base).

E. Peroxidase Enzyme Solution (POD) (Immediately before use, prepare a solution containing 50 purpurogallin units/ml in deionized water using Peroxidase).

F. 150 mM Magnesium Sulfate Solution ($MgSO_4$) (Prepare 5 ml in deionized water using Magnesium Sulfate, Heptahydrate).

G. 300 mM Sodium Pyruvate Solution (Pyr) (Prepare 1 ml in deionized water using Pyruvic Acid, Sodium Salt).

H. 0.3% (w/v) N-Ethyl-N-(2-Hydroxy-3-Sulfopropyl)-m- Toluidine Solution (EHSPT) (Prepare 25 ml in deionized water using N-Ethyl-N-(2-Hydroxy-3-Sulfopropyl)-m-Toluidine, Sodium Salt).

I. 50 mM Potassium Phosphate Buffer, pH 5.7 at 37°C (Enzyme Diluent) (Prepare 50 ml in deionized water using Potassium Phosphate, Monobasic, Anhydrous. Adjust to pH 5.7 at 37°C with 1 M KOH).

J. 15 mM Ethylenediaminetetraacetic Acid Solution (EDTA) (Prepare 25 ml in deionized water using Ethylenediaminetetraacetic Acid, Disodium Salt, Dihydrate).

K. Pyruvate Oxidase Enzyme Solution (Immediately before use, prepare a solution containing 0.2–0.4 unit/ml of Pyruvate Oxidase in cold Reagent I).

Procedure

Prepare a reaction cocktail by pipetting (in milliliters) the following reagents into a suitable container:

Reagent A (Buffer)	10.00
Reagent D (4-AAP)´	2.00
Reagent H (EHSPT)	2.00
Reagent C (TPP)	2.00
Reagent B (FAD)	2.00
Reagent J (EDTA)	2.00
Reagent F ($MgSO_4$)	2.00
Reagent E (POD)	3.00

Mix by swirling. Pipette (in milliliters) the following reagents into suitable containers:

	Test	Blank
Reaction Cocktail	2.50	2.50
Reagent G (Pyr)	0.50	0.50

Mix by inversion and equilibrate to 37°C. Monitor the A550nm until constant, using a suitably thermostatted spectrophotometer. Then add:

Reagent I (Enzyme Diluent)	—	0.10
Reagent K (Enzyme Solution)	0.10	—

Immediately mix by inversion and record the increase in A550nm for approximately 5 minutes. Obtain the ΔA550nm/min using the maximum linear rate for both the Test and Blank.

Calculations

$$\text{Units / ml enzyme} = \frac{(\Delta\text{A550nm / min Test} - \Delta\text{A550nm / min Blank})\ (3.1)\ (\text{df})}{(36.88)\ (0.1)\ (0.5)}$$

3.1 = Volume (in milliliters) of assay
df = Dilution factor
36.88 = Millimolar extinction coefficient of quinoneimine dye under the assay conditions
0.1 = Volume (in milliliter) of enzyme used
0.5 = Factor based on the equation that one mole of H_2O_2 produces half a mole of quinoneimine dye

$$\text{Units / mg solid} = \frac{\text{units / ml enzyme}}{\text{mg solid / ml enzyme}}$$

$$\text{Units / mg protein} = \frac{\text{units / ml enzyme}}{\text{mg protein / ml enzyme}}$$

Unit Definition: *One unit will produce 1.0 µmole of H_2O_2 per minute at pH 5.7 at 37°C during the conversion of pyruvate and phosphate to acetylphosphate and CO_2.*

Final Assay Concentrations

In a 3.10 ml reaction mix, the final concentrations are 50 mM potassium phosphate, 0.48 mM 4-aminoantipyrene, 0.02% (w/v) N-ethyl-N-(2-hydroxy-3-sulfopropyl)-mtoluidine, 0.2 mM cocarboxylase, 0.0097 mM flavin adenine dinucleotide, 0.97 mM Ethylenediaminetetraacetic Acid, 9.7 mM magnesium sulfate, 48 mM sodium pyruvate, 15 units peroxidase, and 0.02–0.04 unit pyruvate oxidase.

Enzymatic Assay of PYRUVATE OXIDASE (EC 1.2.3.3) From *Pediococcus species*

Principle

$$\text{Pyruvate} + O_2 + \text{Pi} \xrightarrow[\text{FAD, TPP, } Mg_2^+]{\text{Pyruvate Oxidase}} \text{Acetylphosphate} + CO_2 + H_2O_2$$

$$2H_2O_2 + \text{4-Aminoantipyrine} + \text{N, N-Dimethylaniline} \xrightarrow{\text{POD}} \text{Quinonedimine dye} + 4H_2O$$

Abbreviations Used

Pi = Inorganic Phosphate
FAD = Flavin Adenine Dinucleotide
TPP = Thiamine Pyrophosphate
POD = Peroxidase

Conditions: T = 37°C, pH = 6.7, A565nm, Light path = 1 cm.

Method: Spectrophotometric Stop Rate Determination

Reagents

A. 1 M Potassium Phosphate Buffer, pH 6.7 at 37°C (Prepare 100 ml in deionized water using Potassium Phosphate, Monobasic, Anhydrous. Adjust to pH 6.7 at 37°C with 1 M NaOH).
B. 1 mM Flavin Adenine Dinucleotide Solution (FAD) (Prepare 1 ml in deionized water using Flavin Adenine Dinucleotide, Disodium Salt, *Prepare Fresh*).
C. 10 mM Cocarboxylase (Thiamine Pyrophosphate) Solution (TPP) (Prepare 1 ml in deionized water using Cocarboxylase. *Prepare Fresh*).
D. 15 mM 4-Aminoantipyrine Solution (4-AAP) (Prepare 100 ml in deionized water using 4-Aminoantipyrine Free Base).
E. Peroxidase Enzyme Solution (POD) (Immediately before use, prepare a solution containing 50 purpurogallin units/ml in deionized water using Peroxidase).
F. 100 mM Magnesium Chloride Solution ($MgCl_2$) (Prepare 5 ml in deionized water using Magnesium Chloride, Hexahydrate).
G. 0.2% (v/v) N,N-Dimethylaniline Suspension (DMA) (Prepare 100 ml in deionized water using N,N-Dimethylaniline, N,N-Dimethylaniline is insoluble in deionized water. Prepare by adding 0.2 ml of N,N-Dimethylaniline to 80 ml of deionized water. Mix by swirling and bring to a total volume of 100 ml with deionized water).
H. Reaction Mixture II (Rxn II) (Prepare by combining 1 ml of Reagent F and 2 ml of Reagent G. *Prepare Fresh*).

I. 1 M Sodium Pyruvate Solution (Pyr) (Prepare 1 ml in deionized water using Pyruvic Acid, Sodium Salt).

J. 200 mM Sodium Phosphate Solution (Prepare 100 ml in deionized water using Sodium Phosphate, Monobasic, Anhydrous).

K. 100 mM Citric Acid Solution (Prepare 100 ml in deionized water using Citric Acid, Monohydrate).

L. McIlvain Buffer (Prepare 100 ml by adding 63.15 ml of Reagent J and 36.85 ml of Reagent K).

M. 100 mM Ethylenediaminetetraacetic Acid Solution (EDTA) (Prepare 100 ml in Reagent L using Ethylenediaminetetraacetic Acid, Disodium Salt, Dihydrate. Adjust to pH 5.5 with either 1 M NaOH or 1 M HCl).

N. 10 mM Potassium Phosphate with 10 μM Flavin Adenine Dinucleotide Solution, pH 7.0 at 37°C (Enzyme Diluent) (Prepare 100 ml in deionized water using Potassium Phosphate, Monobasic, Anhydrous, and Flavin Adenine Dinucleotide, Disodium Salt. Adjust to pH 7.0 at 37°C with 1 M KOH).

O. Pyruvate Oxidase Enzyme Solution (Immediately before use, prepare a solution containing 0.1–0.2 unit/ml of Pyruvate Oxidase in Reagent N).

Procedure

Prepare a reaction cocktail by pipetting (in milliliters) the following reagents into a suitable container:

Deionized Water	1.70
Reagent A (Buffer)	2.00
Reagent B (FAD)	0.10
Reagent C (TPP)	0.20
Reagent D (4-AAP)	1.00
Reagent E (POD)	1.00

Mix by swirling and equilibrate to 37°C. Pipette (in milliliters) the following reagents into suitable containers:

	Test	Blank
Reaction Cocktail	0.60	0.60
Reagent H (Rxn II)	0.30	0.30
Reagent I (Pyr)	0.10	0.10

Mix by inversion and equilibrate to 37°C. Monitor the A565nm until constant, using a suitably thermostatted spectrophotometer. Then add:

Reagent N (Enzyme Diluent)	—	0.02
Reagent O (Enzyme Solution)	0.02	—

Immediately mix by inversion and incubate at 37°C for exactly 10 minutes. Then add:

Reagent M (EDTA)	2.00	2.00

Mix by inversion and incubate at 25°C for exactly 5 minutes. Transfer to suitable cuvettes and record the A565nm for both the Test and Blank using a suitable spectrophotometer.

Calculations

$$\text{Units / ml enzyme} = \frac{(\text{A565nm Test} - \text{A565nm Blank})\ (2)\ (3.02)\ (\text{df})}{(10)\ (23.56)\ (0.02)}$$

3.02 = Total volume (in milliliters) of assay
2 = 2 moles H_2O_2 used per mole of dye
df = Dilution factor
10 = Time (in minutes) of assay as per the Unit Definition
23.56 = Millimolar extinction coefficient of quinonediimine dye at 565nm
0.02 = Volume (in milliliters) of enzyme used

$$\text{Units / mg solid} = \frac{\text{units / ml enzyme}}{\text{mg solid / ml enzyme}}$$

$$\text{Units / mg protein} = \frac{\text{units / ml enzyme}}{\text{mg protein / ml enzyme}}$$

Unit Definition: *One unit will produce 1.0 μmole of H_2O_2 per minute during the conversion of pyruvate and phosphate to acetylphosphate and CO_2.*

Final Assay Concentrations

In a 1.02 ml reaction mix, the final concentrations are 196 mM potassium phosphate, 0.01 mM flavin adenine dinucleotide, 0.20 mM cocarboxylase, 1.5 mM 4-aminoantipyrine, 0.04% (v/v) N,N-dimethylaniline, 9.8 mM magnesium chloride, 5 purpurogallin units peroxidase and 0.05–0.1 unit pyruvate oxidase.

Note

1. Peroxidase Unit Definition: One unit will form 1.0 mg purpurogallin from pyrogallol in 20 seconds at pH 6.0 at 20°C.

Enzymatic Assay of RENIN (EC 3.4.99.19)

Principle

$$\text{Angiotensinogen} \xrightarrow{\text{Renin}} \text{Angiotensin} + \text{Peptide Fragment}$$

Conditions: T = 37°C, pH = 6.0

Method: Radioimmunoassay

Reagents

A. 0.84% (w/v) Dimercaprol with 1.87% (w/v) Benzyl Benzoate (Dimercaprol Soln).
B. 3.3% (w/v) 8-Hydroxyquinoline Solution (8-Hydroxyquin).
C. 2.0% (w/v) Maleate Buffer, pH 6.0 at 25°C (Use Maleate Buffer pH 6.0).
D. 5% (w/v) Bovine Serum Albumin (Enzyme Diluent) (Prepare 10 ml in deionized water using Albumin).

E. Angiotensinogen Solution (Angiotensinogen Soln) (Immediately before use, prepare a solution containing 0.2 unit/ml of Angiotensinogen in cold deionized water).
F. "0" Standard.
G. Renin Solution (Immediately before use prepare a solution containing 0.0001 unit/ml in Reagent D by adding 3 ml of Reagent D to a 1 unit vial of Renin. Continue diluting (3000X) with Reagent D).
H. Angiotensin Standards (Angiotensin Std) (Use Angiotensin Standards).
I. Angiotensin II labeled.
J. Blank Antiserum.
K. Angiotensin I Antiserum.
L. Angiotensin I Second Antibody.

Procedure

Step 1: Pipette (in milliliters) the following reagents in the following order into a 12 × 75 mm polystyrene tube:

Reagent G (Enzyme Solution)	0.500
Reagent A (Dimercaprol Soln)	0.005
Reagent B (8-Hydroxyquin)	0.005
Reagent E (Angiotensinogen Soln)	0.100
Reagent C (Maleate Buffer)	1.000

Vortex the tube for 3 seconds, then immediately pipette 0.1 ml of this solution into a suitable tube labeled "Cold Mix" and store in an ice bath. Incubate the remainder of the reaction mix at 37°C in a suitable water bath. Pipette 0.1 ml of this reaction mix at the end of 10 minutes, 20 minutes, and 30 minutes, into suitable tubes labeled "10 minutes", "20 minutes", "30 minutes", respectively and store in a ice bath.

Step 2: Pipette (in milliliters) the following reagents in the order listed on page 832 into suitable 12 × 75 mm polystyrene tubes using reagents supplied with the kit.
Once all the reagents have been added to the appropriate tubes, vortex each tube for 5 seconds and incubate for two hours at room temperature. At the end of the two hour incubation, mix the Angiotensin I Second Antibody (Reagent L) well and pipette 0.5 ml of Reagent L into each tube except in the tubes labeled "Total Counts." Vortex each tube and incubate for 20–30 minutes at room temperature. Centrifuge all tubes at 1000 × g for 15 minutes. Remove the supernatant from each tube (except the tubes labeled "Total Counts") by aspiration into a I liquid waste container. Cap and count each tube in a gamma counter for one minute per tube.

Tube number	O Std reagent F	Angiotensin Std reagent H	Samples	Angiotensin (I)	Blank Anti-serum reagent J	Angiotensin I Anti-serum reagent K
Total counts 1				0.1 ml		
Total counts 2				0.1 ml		
Blanks 3	0.1 ml			0.1 ml	0.1 ml	
Blanks 4	0.1 ml			0.1 ml	0.1 ml	
"O" std 5	0.1 ml			0.1 ml		0.1 ml
"O" Std 6	0.1 ml			0.1 ml		0.1 ml
0.01 ng 7		0.1 ml		0.1 ml		0.1 ml
0.01 ng 8		0.1 ml		0.1 ml		0.1 ml
0.25 ng 9		0.1 ml		0.1 ml		0.1 ml
0.25 ng 10		0.1 ml		0.1 ml		0.1 ml
0.05 ng 11		0.1 ml		0.1 ml		0.1 ml
0.05 ng 12		0.1 ml		0.1 ml		0.1 ml
0.10 ng 13		0.1 ml		0.1 ml		0.1 ml
0.10 ng 14		0.1 ml		0.1 ml		0.1 ml
0.25 ng 15		0.1 ml		0.1 ml		0.1 ml
0.25 ng 16		0.1 ml		0.1 ml		0.1 ml
0.50 ng 17		0.1 ml		0.1 ml		0.1 ml
0.50 ng 18		0.1 ml		0.1 ml		0.1 ml
1.00 ng 19		0.1 ml		0.1 ml		0.1 ml
1.00 ng 20		0.1 ml		0.1 ml		0.1 ml
Samples			0.1 ml	0.1 ml		0.1 ml
10 min 21			0.1 ml	0.1 ml		0.1 ml
20 min 22			0.1 ml	0.1 ml		
30 min 23			0.1 ml	0.1 ml		0.1 ml
Cold Mix Sample 24			0.1 ml	0.1 ml		0.1 ml

Calculations

Determine the ng Angiotensin I of each reacted sample tube and cold mix sample tube of each set from the standard curve. The standard curve is generated from a plot of percentage bound of each standard relative to that bound for the "0" standard (%B/Bo) versus ng angiotensin I of the standard.

%B/Bo = (Net counts for standard/Net counts for "0" standard) × 100

If the gamma counter is not equipped with a data reproduction unit which will generate a standard curve and final sample results, it will be necessary to plot the standard curve on semi-log graph paper from which the sample ng amount of angiotensin I levels can be obtained.

$$\text{Units/ml} = \frac{(\text{ng Sample} - \text{ng Cold Mix})\,(1.61)\,(3000)}{(0.5)\,(0.1)\,(\text{Time})\,(100{,}000)}$$

1.61 = Final reaction volume in ml
3000 = Dilution factor for the Renin
0.5 = Volume of Renin sample added to reaction mix in ml

0.1 = Volume of reaction mix removed in ml
Time = Hours expressed in fractions
100,000 = Conversion of DuPont to Sigma Units

Unit Definition: *One unit will liberate 100 μg of angiotensin I from angiotensinogen (A2283) per hour at pH 6.0 at 37°C.*

Final Assay Concentration

In a 1.61 ml reaction mixture the final concentrations are 0.003% (w/v) dimercaprol, 0.006% (w/v) benzyl benzoate, 0.01% (w/v) 8-hydroxyquinoline, 1.2% (w/v) maleate, 0.02 unit angiotensinogen, 2% (w/v) Bovine Serum Albumin, and 3 × 10–5 unit renin.

Note

1. The amount of generated angiotensin I is measured by radioimmunoassay. The basic principle of this procedure is the competition between radioactive and non-radioactive antigen for a fixed number of antibody binding sites. If increasing amounts of nonradioactive antigen (i.e. standards or unknowns) and a fixed amount of tracer are allowed to react with a constant and limiting amount of antibody, a decreasing quantity of radioactive antigen is bound to the antibody. The separation of bound from free antigen is achieved by the addition of a second antibody which precipitates the primary antigen-antibody complex. Polyethylene glycol is used to accelerate this reaction. The second antibody is prepared against gamma globulin of the species used for raising the primary antibody. After incubation and centrifugation, the supernatant is discarded and the antigen-antibody complex is counted to quantitate the bound tracer. The data are used to construct a standard curve from which the values of the unknowns may be obtained by interpolation. Plasma renin is expressed as ng/ml/hr of angiotensin I generated.

Enzymatic Assay of RENNIN (EC 3.4.23.4)

Principle

$$\text{Milk} \xrightarrow{\text{Rennin}} \text{Clotted Milk}$$

Conditions: T = 30°C

Method: Clotting time

Reagents

A. 1 M Calcium Chloride Solution (Prepare 5 ml in deionized water using Calcium Chloride, Dihydrate).
B. 10.4% (w/v) Milk Solution with 10 mM Calcium Chloride (Milk) (Prepare by dissolving 20.9 g of Carnation Instant Nonfat Dry Milk Powder in 200 ml of deionized water or use skimmed cow's milk. Then add 2 ml of Reagent A).
C. 0.1% (w/v) Rennet Standard Solution (Rennet Std) (Prepare 10 ml in deionized water using Rennet).
D. Rennin Enzyme Solution (Rennin) (Immediately before use, prepare a solution containing 0.01–0.05 mg/ml of Rennin in cold deionized water. Dilute accordingly so that the clotting time is 0.75–1.5 times that of Reagent C).

Procedure

Step 1: Pipette (in milliliters) the following reagents into a 50 ml Erlenmeyer flask:

	Control
Reagent B (Milk)	10.00

Incubate at 30°C in a water bath for 45 minutes. At t_0 add:

Reagent C (Rennet Std)	1.00

Swirl gently (Erlenmeyer flask) at 30°C in a water bath. Stop timing and swirling when a white-translucent semiliquified film appears on the side of the flask above the miik. This is t_1. After t_1, the milk will continue to congeal.

Step 2: Pipette (in milliliters) the following reagents into a suitable test tube.

	Test
Reagent B (Milk)	10.00

Incubate at 30°C in a water bath for 45 minutes. Then add:

Reagent D (Rennin)	1.00

Swirl gently (Erlenmeyer flask) at 30°C in a water bath. Stop timing and swirling when a white-translucent semi-liquified film appears on the side of the flask above the milk. This is t_2.

Calculations

$$\text{Units / ml enzyme} = \frac{(t_1)(df)}{(t_2)(1)}$$

t_1 = Clotting time of Rennet
df = Dilution factor
t_2 = Clotting time of Rennin
1 = Volume (in milliliter) of enzyme

$$\text{Units / mg solid} = \frac{\text{units / ml enzyme}}{\text{mg solid / ml enzyme}}$$

$$\text{Units / mg protein} = \frac{\text{units / ml enzyme}}{\text{mg protein / ml enzyme}}$$

Unit Definition: *One unit will coagulate 10 ml of milk per minute at 30°C.*

Final Assay Concentrations

In a 11.00 ml reaction mix, the final concentrations are 9.5% (w/v) milk, 9 mM calcium chloride, and 0.01–0.05 mg rennin.

Notes

1. Rennin clots milk by cleavage of a single Ser- Phe105+Met-Ala bond in the α-chain of casein.
2. It is important that the enzyme solution is assayed within 15 minutes of the standard rennet solution using the same substrate preparation incubated for approximately the same time.

Enzymatic Assay of RHODANESE (EC 2.8.1.1)

Principle

$$\text{Thiosulfate} + \text{Cyanide} \xrightarrow{\text{Rhodanese}} \text{Sulfite} + \text{Thiocyanate}$$

Conditions: T = 25°C, pH = 8.6, A460nm, Light path = 1 cm

Method: Colorimetric

Reagents

A. 200 mM Potassium Phosphate Buffer, pH 8.6 at 25°C (Prepare 100 ml in deionized water using Potassium Phosphate, Monobasic, Anhydrous. Adjust to pH 8.6 at 25°C with 1 M KOH).

B. 125 mM Sodium Thiosulfate Solution (Prepare 10 ml in deionized water using Sodium Thiosulfate, Pentahydrate).

C. 250 mM Potassium Cyanide Solution (Sodium Cyanide) (Prepare 10 ml in Reagent A using Potassium Cyanide ACS Reagent Grade).

D. 14% (w/v) Nitric Acid Solution (Prepare 50 ml in deionized water using concentrated Nitric Acid).

E. 410 mM Ferric Nitrate Solution (Ferric Nitrate) (Prepare 25 ml in Reagent D using Ferric Nitrate, Nonahydrate).

F. 37% (w/w) Formaldehyde Solution (Formaldhyde) (Use Formaldehyde, 37% Solution Formalin).

G. 12.5 mM Sodium Thiosulfate Solution with 0.025% (w/v) Bovine Serum Albumin (Enzyme Diluent) (Prepare 25 ml in deionized water using Sodium Thiosulfate, Pentahydrate and Albumin, Bovine).

H. Rhodanese Enzyme Solution (Immediately before use, prepare a solution containing 2.0–3.0 units/ml Rhodanese in Reagent G).

Procedure

Pipette (in milliliters) the following reagents into 4 dram vials:

	Test	Blank
Reagent A (Buffer)	0.30	0.30
Reagent B (Sodium Thiosulfate)	0.50	0.50
Reagent C (Sodium Cyanide)	0.25	0.25

Mix by swirling and equilibrate to 25°C. Then add:

Reagent H (Enzyme Solution)	0.20	—

Mix by swirling and incubate at 25°C for exactly 5 minutes. Then add:

Reagent F (Formaldehyde)	0.25	0.25
Reagent H (Enzyme Solution)	—	0.20

Mix by swirling and then add:

	Test	Blank
Reagent E (Ferric Nitrate)	1.25	1.25
Deionized Water	12.50	12.50

Mix by swirling and transfer the solutions to suitable cuvettes and record the A460nm for both the Test and Blank using a suitably thermostatted spectrophotometer.

Calculations

$$\text{Units / ml enzyme} = \frac{(\text{A460nm Test} - \text{A460nm Blank})\,(15.25)\,(\text{df})}{(5)\,(3.17)\,(0.2)}$$

5 = Conversion factor for 5 minutes to 1 minute
3.17 = Millimolar extinction coefficient of Thiocyanate at 460nm
df = Dilution factor
15.25 = Total volume of assay
0.2 = Volume (in milliliters) of enzyme used

$$\text{Units / mg solid} = \frac{\text{units / ml enzyme}}{\text{mg solid / ml enzyme}}$$

$$\text{Units / mg protein} = \frac{\text{units / ml enzyme}}{\text{mg protein / ml enzyme}}$$

Unit Definition: *One unit will convert 1.0 μmole of cyanide to thiocyanate per minute at pH 8.6 at 25°C.*

Final Assay Concentrations

In a 1.25 ml reaction mix, the final concentrations are 88 mM potassium phosphate, 52 mM sodium thiosulfate, 50 mM sodium cyanide, 0.004% (w/v) BSA, and 0.4–0.6 unit rhodanese.

Note

1. This assay should be run in a fume hood with a mechanical exhaust system since small quantities of HCN may be liberated.

Fluorometric Assay of RIBOFLAVIN BINDING PROTEIN, Apo Form

Principle

$$\text{RBP} + \text{Riboflavin (Fluorescent)} \rightarrow \text{RBP - Riboflavin Complex (Quenched)}$$

Abbreviation Used

RBP = Riboflavin Binding Protein

Conditions: Excitation = 450 nm, Emission = 530 nm, Light path = 1 cm, T = 25°C, pH = 5.5

Method: Fluorimetric Determination

Reagents

A. 40 ug/ml Riboflavin Stock Solution (Prepare 500 ml in deionized water containing 1 ml of concentrated HCl using Riboflavin).

B. 1.0 ug/ml Riboflavin Solution (Prepare 40 ml in deionized water using Reagent A. *Prepare Fresh*).

C. 50 mM Sodium Acetate Buffer pH 5.5 at 25°C (Prepare 50 ml in deionized water using Acetic Acid, Glacial and adjust pH to 5.5 with NaOH).

D. Riboflavin-Binding Protein Solution (Prepare a solution containing 1 mg/ml in cold deionized water).

E. 1% (w/v) Bovine Serum Albumin (BSA Solution) (Prepare 100 ml in deionized water using Albumin, Bovine).

Procedure

Pipette (in milliliters) the following reagents into suitable tubes.

	Blank	Control	Sample
Reagent E (BSA Solution)	1.00	1.00	1.00
Reagent D (RBP)	—	—	0.050
Reagent B (Riboflavin Solution)	—	1.00	1.00
Deionized Water	1.00	—	—
Reagent C (Sodium Acetate Buffer)	10.00	10.00	10.00

Transfer solutions into suitable cuvettes. Zero the fluorimeter on the Blank cuvette. Read the excitation of the Control and Sample.

Calculations

F1 = Fluorescence reading of Sample

F2 = Fluorescence reading of Control

If F1/F2 <0.1, sample must be diluted.

$$\mu g \text{ Riboflavin/mg Protein} = [1-(F1/F2)] / (\text{mg Protein/RM})$$

RM = Reaction mix

Binding Capacity: Expressed as μg riboflavin bound per mg protein.

Final Assay Concentrations

In a 12.0 ml reaction mix the final concentrations are 0.083% Bovine Serum Albumin, 4.2 μg/ml Riboflavin Binding Protein, 0.083 μg/ml Riboflavin, and 42 mM Sodium Acetate.

Enzymatic Assay of RIBOKINASE (EC 2.7.1.15)

Principle

$$\text{ATP} + \text{D-Ribose} \xrightarrow{\text{Ribokinase}} \text{ADP} + \text{D-Ribose 5-P}$$

$$\text{D-Ribose 5-P} \xrightarrow{\text{PRI}} \text{D-Ribulose 5-P}$$

$$\text{D-Ribulose 5-P} \xrightarrow{\text{Ru-5-P-3-Epim}} \text{D-Xylulose 5-P}$$

$$\text{D-Xylulose 5-P} + \text{D-Ribose 5-P} \xrightarrow[\text{Mg}^{++}\text{, Cocarboxylase}]{\text{TK}} \text{GAP} + \text{Sedoheptulose 7-P}$$

$$\text{GAP} \xrightarrow{\text{TPI}} \text{DHAP}$$

$$\text{DHAP} + \beta\text{-NADH} \xrightarrow{\alpha\text{-GDH}} \beta\text{-NAD} + \alpha\text{-Glycerophosphate}$$

Abbreviations Used

ADP = Adenosine 5'-Phosphate
PRI= Phosphoriboisomerase
Ru-5-P-3-Epim = Ribulose-5-Phosphate-3-Epimerase
TK = Transketolase
GAP = Glyceraldehyde 3-Phosphate
TPI = Triosephosphate Isomerase
DHAP = Dihydroxyacetone Phosphate
β-NADH = β-Nicotinamide Adenine Dinucleotide, Reduced Form
α-GDH = α-Glycerophosphate Dehydrogenase
β-NAD = β-Nicotinamide Adenine Dinucleotide, Oxidized Form

Conditions: T = 37°C, pH = 7.7, A340nm, Light path = 1 cm

Method: Continuous Spectrophotometric Rate Determination

Reagent

A. 250 mM Glycylglycine Buffer, pH 7.7 at 37°C (Prepare 100 ml in deionized water using Glycylglycine. Adjust to pH 7.7 at 37°C with 1 M NaOH).
B. 100 mM Ribose Solution (Ribose) (Prepare 1 ml in deionized water using D(-)Ribose).
C. 2 mM Cocarboxylase Solution (Cocarboxylase) (Prepare 1 ml in deionized water using Cocarboxylase).
D. 2.5 mM β-Nicotinamide Adenine Dinucleotide Solution, Reduced Form (β-NADH) (Dissolve the contents of one 5 mg vial of β-Nicotinamide Adenine Dinucleotide, Reduced Form, Disodium Salt in the appropriate volume of deionized water).
E. 300 mM Magnesium Chloride Solution ($MgCl_2$) (Prepare 1 ml in deionized water using Magnesium Chloride, Hexahydrate).
F. 180 mM Adenosine 5'-Triphosphate Solution (ATP) (Prepare 1 ml in deionized water using Adenosine 5'-Triphosphate, Disodium Salt).
G. Ribulose-5-Phosphate-3-Epimerase Enzyme Solution (Ru-5-P-3-Epim) (Immediately before use, prepare a solution containing 10 units/ml of D-Ribulose 5-Phosphate-3-Epimerase in cold deionized water).
H. Transketolase Enzyme Solution (TK) (Immediately before use, prepare a solution containing 10 units/ml of Transketolase in cold deionized water).
I. α-Glycerophosphate Dehydrogenase/Triosephosphate Isomerase Enzyme Solution (α-GDH/TPI) (Immediately before use, prepare a solution containing 100 units/ml of α-Glycerophosphate Dehydrogenase-Triosephosphate Isomerase in cold deionized water. The 100 units/ml is based on α-GDH units).

J. Phosphoriboisomerase Enzyme Solution (PRI) (Immediately before use, prepare a solution containing 50 units/ml of Phosphoriboisomerase in cold deionized water).
K. Ribokinase Enzyme Solution (Ribokinase) (Immediately before use, prepare a solution containing 0.3–0.5 unit/ml of Ribokinase in cold Reagent A).

Procedure

Pipette (in milliliters) the following reagents into suitable cuvettes:

	Test	Blank
Reagent A (Buffer)	2.25	2.35
Reagent B (Ribose)	0.10	0.10
Reagent C (Cocarboxylase)	0.05	0.05
Reagent D (β-NADH)	0.15	0.15
Reagent E ($MgCl_2$)	0.10	0.10
Reagent F (ATP)	0.05	0.05
Reagent G (Ru-5-P-3-Epim)	0.05	0.05
Reagent H (TK)	0.05	0.05
Reagent I (α-GDH/TPI)	0.05	0.05
Reagent J (PRI)	0.05	0.05

Mix by inversion and equilibrate to 37°C. Monitor the A340nm until constant, using a suitably thermostatted spectrophotometer. Then add:

Reagent K (Ribokinase)	0.10	—

Immediately mix by inversion and record the decrease in A340nm for approximately 15 minutes. The maximum linear rate usually occurs between 10–15 minutes. Obtain the ΔA340nm/min using the maximum linear rate for both the Test and Blank.

Calculations

$$\text{Units / ml enzyme} = \frac{(\Delta A340\text{nm / min Test} - \Delta A340\text{nm / min Blank})\ (3)\ (\text{df})}{(6.22)\ (0.1)}$$

3 = Total volume (in milliliters) of assay
df = Dilution factor
6.22 = Millimolar extinction coefficient of β-NADH at 340nm
0.1 = Volume (in milliliter) of enzyme used

$$\text{Units / mg protein} = \frac{\text{units / ml enzyme}}{\text{mg protein / ml enzyme}}$$

Unit Definition: *One unit will convert 1.0 μmole of D-ribose to D-ribose 5-phosphate per minute at pH 7.7 at 37°C in the presence of ATP and magnesium.*

Final Assay Concentration

In a 3.00 ml reaction mix, the final concentrations are 196 mM glycylglycine, 3.3 mM D-ribose, 0.03 mM cocarboxylase, 0.13 mM β-nicotinamide adenine dinucleotide, reduced form, 10 mM magnesium chloride, 3 mM adenosine 5'-triphosphate, 0.5 unit ribulose-5-phosphate-3-epimerase, 0.5 unit transketolase, 5 units α-glycerophosphate dehydrogenase, approximately 50 units triosephosphate isomerase, 2.5 units phosphoriboisomerase, and 0.03–0.05 unit ribokinase.

Notes

1. Phosphoriboisomerase Unit Definition: One unit will convert 1.0 μmole of D-ribose 5-phosphate to D-ribulose 5-phosphate per minute at pH 7.7 at 30°C.
2. D-Ribulose-5-Phosphate-3-Epimerase Unit Definition: One unit will convert 1 μmole of D-ribulose 5-phosphate to D-xylulose 5-phosphate per minute at pH 7.7 at 25°C in a coupled system with ribose-5-phosphate, β-NADH, transketolase, α-glycerophosphate dehydrogenase/ triosephosphate isomerase and cocarboxylase.
3. Transketolase Unit Definition: One unit will produce 1 μmole of glyceraldehyde 3-phosphate from xylulose 5-phosphate per minute at pH 7.7 at 25°C in the presence of ribose 5-phosphate, cocarboxylase, and magnesium in a coupled system with triosephosphate isomerase and α-glycerophosphate dehydrogenase.
4. α-Glycerophosphate Dehydrogenase Unit Definition: One unit will convert 1.0 μmole of dihydroxyacetone phosphate to α-glycerophosphate per minute at pH 7.4 and 25°C.
5. Triosephosphate Isomerase Unit Definition: One unit will convert 1.0 μmole of D-glyceraldehyde 3-phosphate to dihydroxyacetone phosphate per minute at pH 7.6 and 25°C.

Enzymatic Assay of RIBONUCLEASE A (EC 3.1.27.5)

Principle

$$RNA + H_2O \xrightarrow{\text{RNase A}} \text{Oligonucleotides}$$

Abbreviations Used

RNA = Ribonucleic Acid

RNase A = Ribonuclease A

Conditions: T = 25°C, pH = 5.0, A300nm, Light path = 1 cm

Method: Spectrophotometric Stop Rate Determination

Reagents

A. 100 mM Sodium Acetate Buffer, pH 5.0 at 25°C (Prepare approximately 100 ml. Prepare 50 ml of 100 mM Sodium Acetate Buffer using Sodium Acetate, Trihydrate, in deionized water. Adjust to pH 5.0 at 25°C with 100 mM Acetic Acid. This will require approximately 50 ml of Acetic Acid).

B. 0.1% (w/v) Ribonucleic Acid Solution (RNA) (Prepare 25 ml in Reagent A using Ribonucleic Acid, Type VI from Torula Yeast).

C. Ribonuclease A Enzyme Stock Solution (Prepare a solution containing 50–75 Kunitz units/ml in cold deionized water).

D. Ribonuclease A Enzyme Solution (RNase A, Step 1) (Immediately before use, prepare a solution containing 0.2–0.3 Kunitz unit/ml of Ribonuclease A in cold deionized water).

E. Ribonuclease A Enzyme Solution (RNase A, Step 2) (Immediately before use, prepare a solution containing 0.50–0.75 Kunitz unit/ml of Ribonuclease A in cold deionized water).

Procedure

Step 1: Rate Determination (Et): Pipette (in milliliters) the following reagents into suitable cuvettes:

	Test
Reagent B (RNA)	1.50
Deionized Water	1.30

Mix by inversion and equilibrate to 25°C using a suitably thermostatted spectrophotometer. Monitor the A300nm until constant. Then add:

Reagent D (RNase A, Step 1)	0.2

Immediately mix by inversion and record the decrease in A300nm for 10 minutes at 1 minute intervals. These values are used in the calculation as Et.

Step 2: Pipette (in milliliters) the following reagents into suitable cuvettes:

	Test
Reagent B (RNA)	1.50
Deionized Water	—

Mix by inversion and equilibrate to 25°C. Monitor the A300nm until constant, using a suitably thermostatted spectrophotometer. Then add:

Reagent E (RNase A, Step 2)	1.50

Mix by inversion and incubate at 25°C for 120 minutes. Record the A300nm until constant. The final A300nm is the total hydrolysis determination (Ef).

Calculations

Plot ln(Et–Ef) versus time (minutes) and determine the slope of the line.
ln(Et–Ef)
Slope = t (slope) (3) (df)
Kunitz units/ml enzyme = 0.2
3 = Volume (in milliliters) of assay
df = Dilution factor
0.2 = Volume (in milliliters) of enzyme

$$\text{Units / mg solid} = \frac{\text{units / ml enzyme}}{\text{mg solid / ml enzyme}}$$

$$\text{Units / mg protein} = \frac{\text{units / ml enzyme}}{\text{mg protein / ml enzyme}}$$

Unit Definition: *One unit will cause a decrease of 100% per minute in the value of Et–Ef at pH 5.0 at 25°C.*

Final Assay Concentrations

In a 3.00 ml reaction mix, the final concentrations are 50 mM sodium acetate, 0.05% (w/v) ribonucleic acid, 0.04–0.06 unit of RNase A (Step 1) and 0.75–1.13 units/ml RNase A (Step 2).

Notes

1. The substrate concentration must be verified prior to running the assay. Add 1.5 ml of the RNA solution to 1.5 ml deionized water in a quartz or methacrylate cuvette. Mix by inversion and read the absorbance at A300nm. This absorbance must be prior to beginning the assay.
2. Perform the assay in triplicate. Average the results from the three tests in order to obtain the final value.
3. When performing RNase impurity assays, the RNase control must be used for Step 2 (total hydrolysis).
4. This assay is not to be used to assay Ribonuclease A, Insoluble enzyme attached to beaded Agarose, Ribonuclease A, Insoluble enzyme attached to Polyacrylamide, and Ribonuclease A, Insoluble enzyme attached to Macroporous Acrylic Beads.

Enzymatic Assay of RIBONUCLEASE B (EC 3.1.27.5)

Principle

$$\text{RNA} + H_2O \xrightarrow{\text{RNase B}} \text{Oligonucleotides}$$

Abbreviations Used

RNA = Ribonucleic Acid
RNase B = Ribonuclease B

Conditions: T = 25°C, pH = 5.0, A300nm, Light path = 1 cm

Method: Spectrophotometric Stop Rate Determination

Reagents

A. 100 mM Sodium Acetate Buffer, pH 5.0 at 25°C (Prepare 100 ml. Initially prepare 100 mM Sodium Acetate Buffer using Sodium Acetate, Trihydrate, in 50 ml of deionized water. Adjust to pH 5.0 at 25°C with 0.1 M Acetic Acid. This will require approximately 50 ml of Acetic Acid).

B. 0.1% (w/v) Ribonucleic Acid Solution (RNA) Prepare 25 ml in Reagent A using Ribonucleic Acid, Type VI from Torula Yeast).

C. Ribonuclease B Enzyme Stock Solution (Prepare a solution containing 50–75 Kunitz units/ml in cold deionized water).

D. Ribonuclease B Enzyme Solution (RNase B, Step 1) (Immediately before use, prepare a solution containing 0.2–0.3 Kunitz units/ml in cold deionized water by diluting 0.05 ml of Reagent C to 12.05 ml with cold deionized water).

E. Ribonuclease B Enzyme Solution (RNase B, Step 2) (Immediately before use, prepare a solution containing 0.50–0.75 Kunitz units/ml in cold deionized water by diluting 0.1 ml of Reagent C to 10.0 ml with cold deionized water).

Procedure

Step 1: Rate Determination (Et): Pipette (in milliliters) the following reagents into suitable tubes:

	Test	Blank
Reagent B (RNA)	1.50	—
Deionized Water	1.30	3.0

Mix by inversion and equilibrate to 25°C using a suitably thermostatted spectrophotometer. Monitor the A300nm until constant. Then add:

Reagent D (RNase B)	0.2	—

Immediately mix by inversion and record the decrease in A300nm for 10 minutes at 1 minute intervals. These values are used in the calculation as Et.

Step 2: Pipette (in milliliters) the following reagents into suitable cuvettes:

	Test	Blank
Reagent B (RNA)	1.50	—
Deionized Water	—	3.0

Mix by inversion and equilibrate to 25°C. Monitor the A300nm until constant, using a suitably thermostatted spectrophotometer. Then add:

Reagent E (RNase B)	1.50	—

Mix by inversion and incubate at 25°C for 120 minutes. Record the A300nm until constant. The final A300nm is for the total hydrolysis determination (Ef).

Calculations

Plot ln (Et–Ef) versus t (min) and determine the slope of the line.
ln (Et–Ef)
Slope = t
Kunitz unit/mg enzyme = mg enzyme/ml RM
RM = Reaction Mixture

Unit Definition: *One unit will cause a decrease of 100% per minute in the value of Et–Ef at pH 5.0 at 25°C.*

Final Assay Concentrations

In a 3 ml reaction mix, the final concentrations are 50 mM sodium acetate, 0.05% (w/v) ribonucleic acid, 0.04–0.06 unit of RNase B (Step 1) and 0.75–1.13 units RNase B (Step 2).

Enzymatic Assay of RIBONUCLEASE C

Principle

$$\text{Polycytidylic Acid (5') } + H_2O \xrightarrow{\text{Ribonuclease C}} \text{Acid Soluble Oligonucleotides}$$

Conditions: T = 25°C, pH 7.5, A260nm, Light path = 1 cm.

Method: Spectrophotometric Stop Rate Determination

Reagents

A. 100 mM Tris HCl Buffer with 0.05% (w/v) Bovine Serum Albumin, pH 7.5 at 25°C (Prepare 100 ml in deionized water using Trizma Hydrochloride, and Albumin, Bovine. Adjust to pH 7.5 at 25°C with 1 M KOH).

B. 0.075% (w/v) Polycytidylic Acid (5') Solution (Poly C) (Prepare 10 ml in Reagent A using Polycytidylic Acid (5'), Potassium Salt).

C. 50 mM Potassium Phosphate Buffer with 2 M Potassium Chloride, pH 8.0 at 25°C (Enz Dil) (Prepare 50 ml in deionized water using Potassium Phosphate, Dibasic, Trihydrate, and Potassium Chloride. Adjust to pH 8.0 at 25°C with 1 M HCl).

D. 2 M Perchloric Acid Solution ($HClO_4$) (Prepare 15 ml in deionized water using Perchloric Acid).

E. Ribonuclease C Enzyme Solution (Immediately before use, prepare a solution containing 100 units/ml of Ribonuclease C in Reagent C).

Procedure

Step 1: Pipette (in milliliters) the following reagents into suitable tubes:

	Test	Blank
Reagent B (Poly C)	0.996	0.996

Equilibrate to 25°C. Then add:

Reagent E (Enzyme Solution)	0.004	—
Reagent C (Enz Dil)	—	0.004

Immediately mix by swirling and incubate at 25°C for exactly 15 minutes. Then add:

Reagent D ($HClO_4$)	1.00	1.00

Immediately mix by swirling and transfer the solutions to an ice bath and let chill for 2 minutes. Centrifuge for 2 minutes.

Step 2: Pipette (in milliliters) the following reagents into suitable tubes:

	Test	Blank
Test Supernatant	0.10	—
Blank Supernatant	—	0.10
Deionized Water	0.90	0.90

Mix by swirling and transfer the solutions to suitable cuvettes. Record the A260nm for both the Test and Blank using a suitable spectrophotometer.

Calculations

$$\text{Units / ml enzyme} = \frac{(\text{A260nm Test} - \text{A260nm Blank})\,(2)\,(1)\,(\text{df})}{(2)\,(0.004)\,(0.1)}$$

2 = Total volume (in milliliters) of stopped reaction (Step 1)
1 = Volume (in milliliters) of Step 2
df = Dilution factor
2 = The increase in A260nm (Arbitrary value) per unit of enzyme as per the Unit Definition
0.004 = Volume (in milliliter) of enzyme used in Step 1
0.1 = Volume (in milliliter) of stopped reaction used in Step 2

$$\text{Units / mg protein} = \frac{\text{units / ml enzyme}}{\text{mg protein / ml enzyme}}$$

Unit Definition: *One unit will produce acid soluble oligonucleotides equivalent to a A260 of 2.0 in 15 minutes at pH 7.5 at 25°C, in a reaction volume of 1.0 ml.*

Final Assay Concentrations

In a 1.00 ml reaction mix, the final concentrations are 100 mM Tris, 0.075% (w/v) mM polycytidylic acid (5'), 0.05% (w/v) Bovine Serum Albumin, 0.2 mM potassium phosphate, 8 mM potassium chloride, and 0.4 unit ribonuclease C.

Enzymatic Assay of RIBONUCLEASE from *Aspergillus clavatus*

Principle

$$\text{RNA} + H_2O \xrightarrow{\text{Ribonuclease}} \text{Acid Soluble Oligonucleotides}$$

Abbreviation Used

RNA = Ribonucleic Acid

Conditions: T = 37°C, pH = 7.5, A260nm, Light path = 1 cm

Method: Spectrophotometric Stop Rate Determination

Reagents

A. 50 mM Potassium Phosphate Buffer, pH 7.5 at 37°C (Prepare 100 ml in deionized water using Potassium Phosphate, Monobasic, Anhydrous. Adjust to pH 7.5 at 37°C with 1 M KOH).
B. 0.2% (w/v) Ribonucleic Acid Solution (RNA) (Prepare 5 ml in cold Reagent A using Ribonucleic Acid).
C. 25% (v/v) Perchloric Acid Solution ($HClO_4$) (Prepare 25 ml in deionized water using Perchloric Acid).
D. 17.7 mM Uranyl Acetate Solution (Uran Acet) (Prepare 25 ml in Reagent C using Uranyl Acetate, Dihydrate).
E. Ribonuclease Enzyme Solution (Immediately before use, prepare a solution containing 75–100 units/ml of Ribonuclease in cold Reagent A).

Procedure

Pipette (in milliliters) the following reagents into suitable tubes:

	Test	Blank
Reagent B (RNA)	1.40	1.40

Equilibrate to 37°C. Then add:

Reagent E (Enz Soln)	0.08	—
Reagent A (Buffer)	0.02	0.10

Immediately mix by swirling and incubate at 37°C for exactly 30 minutes. Then add:

Reagent D (Uran Acet)	0.25	0.25

Mix by swirling, place in an ice bath for 30 minutes. Centrifuge for 10 minutes. Pipette (in milliliters) the following reagents into suitable tubes:

Test Supernatant	0.20	—
Blank Supernatant	—	0.20
Deionized Water	4.80	4.80

Mix by swirling and transfer the solutions to suitable cuvettes. Record the A260nm for both the Test and Blank using a suitable spectrophotometer.

Calculations

$$\text{Units / ml enzyme} = \frac{(\text{A260nm Test} - \text{A260nm Blank})\,(5)\,(1.75)\,(\text{df})}{(0.08)\,(1)\,(0.2)}$$

5 = Total volume (in milliliters) of assay
1.75 = Volume (in milliliters) of stopped reaction
df = Dilution factor
0.08 = Volume (in milliliter) of enzyme used
1 = Extinction coefficient (Arbitrary value) as per the Unit Definition
0.2 = Volume (in milliliter) of stopped reaction used in the assay

$$\text{Units / mg solid} = \frac{\text{units / ml enzyme}}{\text{mg solid / ml enzyme}}$$

$$\text{Units / mg protein} = \frac{\text{units / ml enzyme}}{\text{mg protein / ml enzyme}}$$

Unit Definition: *One unit will produce acid soluble oligonucleotides equivalent to a A260 of 1.0 in 30 minutes at pH 7.5 at 37°C in a 1.5 ml reaction volume.*

Final Assay Concentrations

In a 1.50 ml reaction mix, the final concentrations are 50 mM potassium phosphate, 0.2% (w/v) ribonucleic acid, and 6–8 units ribonuclease.

Note

1. This reaction is linear between 0 and 0.5 absorbance unit.

Enzymatic Assay of RIBONUCLEASE from Chicken Liver Polycytidylic Acid (5') as Substrate

Principle

$$\text{Polycytidylic Acid (5')} + H_2O \xrightarrow{\text{Ribonuclease}} \text{Acid Soluble Oligonucleotides}$$

Conditions: T = 37°C, pH 6.5, A260nm, Light path = 1 cm

Method: Spectrophotometric Stop Rate Determination

Reagents

A. 1 M Potassium Phosphate Monobasic Solution (Prepare 100 ml in deionized water using Potassium Phosphate, Monobasic, Anhydrous).

B. 1 M Potassium Phosphate, Dibasic, Solution (Prepare 100 ml in deionized water using Potassium Phosphate, Dibasic, Trihydrate).

C. 10 mM Potassium Phosphate Buffer, pH 6.5 at 37°C (Prepare 100 ml by adjusting the pH of Reagent A with Reagent B to 6.5 at 37°C. Then dilute 1:100 with deionized water).

D. 0.02% (w/v) Polycytidylic Acid (5') Solution (Poly C) (Prepare 5 ml in Reagent C using Polycytidylic Acid (5'), Potassium Salt).

E. 200 mM Hydrochloric Acid Solution (HCl) (Prepare 50 ml in deionized water using Hydrochloric Acid).

F. 20 mM Lanthanum Nitrate Solution ($La(NO_3)_3$) (Prepare 50 ml in Reagent E using Lanthanum Nitrate, Hexahydrate. Store on ice).

G. Ribonuclease Enzyme Solution (Immediately before use, prepare a solution containing 200 units/ml of Ribonuclease in cold Reagent C).

Procedure

Pipette (in milliliters) the following reagents into suitable tubes:

	Test	Blank
Reagent D (Poly C)	0.50	0.50

Equilibrate to 37°C. Then add:

	Test	Blank
Reagent G (Enzyme Solution)	0.01	—
Reagent C (Buffer)	—	0.01

Immediately mix by swirling and incubate at 37°C for exactly 15 minutes. Then add:

	Test	Blank
Reagent F ($La(NO_3)_3$)	0.50	0.50

Store on ice for 15 minutes. Centrifuge for about 3–5 minutes at room temperature. Transfer 0.8 ml each of the Test and Blank supernatants to suitable cuvettes and record the A260nm for both the Test and Blank using a suitable spectrophotometer.

Calculations

$$\text{Units / ml enzyme} = \frac{(\text{A260nm Test} - \text{A260nm Blank})\,(\text{df})}{(1)\,(0.01)}$$

df = Dilution factor
1 = Extinction coefficient (Arbitrary value) as per the Unit Definition
0.01 = Volume (in milliliter) of enzyme used

$$\text{Units / mg protein} = \frac{\text{units / ml enzyme}}{\text{mg protein / ml enzyme}}$$

Unit Definition: *One unit will produce acid soluble oligonucleotides equivalent to a change in A260nm of 1.0 in 15 minutes at pH 6.5 at 37°C, in a reaction volume of 1.1 ml.*

Final Assay Concentrations

In a 0.51 ml reaction mix, the final concentrations are 10 mM potassium phosphate, 0.02% (w/v) polycytidylic acid (5'), and 2 units ribonuclease.

Enzymatic Assay of RIBONUCLEASE from Chicken Liver Polyuridylic Acid (5') as Substrate

Principle

$$\text{Polyuridylic Acid (5')} + H_2O \xrightarrow{\text{Ribonuclease}} \text{Acid Soluble Oligonucleotides}$$

Conditions: T = 37°C, pH 6.5, A260nm, Light path = 1 cm

Method: Spectrophotometric Stop Rate Determination

Reagents

A. 1 M Potassium Phosphate Monobasic Solution (Prepare 100 ml in deionized water using Potassium Phosphate, Monobasic, Anhydrous).
B. 1 M Potassium Phosphate, Dibasic, Solution (Prepare 100 ml in deionized water using Potassium Phosphate, Dibasic, Trihydrate).
C. 10 mM Potassium Phosphate Buffer, pH 6.5 at 37°C (Prepare 100 ml by adjusting the pH of Reagent A with Reagent B to 6.5 at 37°C. Then dilute 1:100 with deionized water).
D. 0.02% (w/v) Polyuridylic Acid (5') Solution (Poly U) (Prepare 5 ml in Reagent C using Polyuridylic Acid (5'), Potassium Salt).
E. 200 mM Hydrochloric Acid Solution (HCl) (Prepare 50 ml in deionized water using Hydrochloric Acid).
F. 20 mM Lanthanum Nitrate Solution ($La(NO_3)_3$) (Prepare 50 ml in Reagent E using Lanthanum Nitrate, Hexahydrate. Store on ice).
G. Ribonuclease Enzyme Solution (Immediately before use, prepare a solution containing 13 units/ml of Ribonuclease in cold Reagent C).

Procedure

Pipette (in milliliters) the following reagents into suitable tubes:

	Test	Blank
Reagent D (Poly U)	0.50	0.50

Equilibrate to 37°C. Then add:

Reagent G (Enzyme Solution)	0.01	—
Reagent C (Buffer)	—	0.01

Immediately mix by swirling and incubate at 37°C for exactly 15 minutes. Then add:

Reagent F ($La(NO_3)_3$)	0.50	0.50

Store on ice for 15 minutes. Centrifuge for about 3–5 minutes at room temperature. Transfer 0.8 ml each of the Test and Blank supernatants to suitable cuvettes and record the A260nm for both the Test and Blank using a suitable spectrophotometer.

Calculations

$$\text{Units / ml enzyme} = \frac{(\text{A260nm Test} - \text{A260nm Blank})\,(\text{df})}{(1)\,(0.01)}$$

df = Dilution factor
1 = Extinction coefficient (arbitrary value) as per the Unit Definition
0.01 = Volume (in milliliter) of enzyme used

$$\text{Units / mg protein} = \frac{\text{units / ml enzyme}}{\text{mg protein / ml enzyme}}$$

Unit Definition: *One unit will produce acid soluble oligonucleotides equivalent to a change in A260nm of 1.0 in 15 minutes at pH 6.5 at 37°C, in a reaction volume of 1.1 ml.*

Final Assay Concentrations

In a 0.51 ml reaction mix, the final concentrations are 10 mM potassium phosphate, 0.02% (w/v) polyuridylic acid (5'), and 0.13 unit ribonuclease.

Enzymatic Assay of RIBONUCLEASE T1 (EC 3.1.27.3)

Principle

$$\text{RNA} + H_2O \xrightarrow{\text{Ribonuclease T1}} \text{Acid Soluble Oligonucleotides} + \text{3'-GMP}$$

Abbreviations Used

RNA = Ribonucleic Acid
3'-GMP = 3'-Guanosine Monophosphate

Conditions: T = 37°C, pH = 7.5, A260nm, Light path = 1 cm

Method: Spectrophotometric Stop Rate Determination

Reagents

A. 200 mM Tris HCl Buffer, pH 7.5 at 37°C (Prepare 100 ml in deionized water using Trizma Base. Adjust to pH 7.5 at 37°C with 1 M HCl).

B. 1.2% (w/v) Ribonucleic Acid Solution (RNA) (Prepare 5 ml in Reagent A using Ribonucleic Acid).

C. 20 mM Ethylenediaminetetraacetic Acid Solution (EDTA) (Prepare 25 ml in deionized water using Ethylenediaminetetraacetic Acid, Disodium Salt, Dihydrate).

D. 25% (v/v) Perchloric Acid Solution ($HClO_4$) (Prepare 25 ml in deionized water using Perchloric Acid).

E. 17.6 mM Uranyl Acetate Solution (Uran. Acet). (Prepare 5 ml in Reagent D using Uranyl Acetate, Dihydrate).

F. Ribonuclease T1 Enzyme Solution (Immediately before use, prepare a solution containing 70–80 units/ml of Ribonuclease T1 in cold Reagent A).

Procedure

Step 1: Pipette (in milliliters) the following reagents into suitable containers:

	Test	Blank
Reagent A (Buffer)	0.25	0.25
Deionized Water	0.30	0.40
Reagent B (RNA)	0.25	0.25
Reagent C (EDTA)	0.10	0.10

Mix by swirling and equilibrate to 37°C. Then add:

Reagent F (Enzyme Solution)	0.10	—

Immediately mix by swirling and incubate at 37°C for exactly 15 minutes. Then add:

Reagent E (Uran. Acet).	0.25	0.25
Reagent F (Enzyme Solution)	—	0.10
Deionized Water	0.10	—

Mix by swirling and allow to stand at room temperature for 15 minutes. Centrifuge for 10 minutes.

Step 2: Pipette (in milliliters) the following reagents into suitable tubes:

Test Supernatant	0.20	—
Blank Supernatant	—	0.20
Deionized Water	4.80	4.80

Mix by swirling and transfer the solutions to suitable cuvettes. Record the A260nm for both the Test and Blank using a suitable spectrophotometer.

Calculations

$$\text{Units / ml enzyme} = \frac{(\text{A260nm Test} - \text{A260nm Blank})\,(1.35)\,(5)\,(\text{df})}{(1)\,(0.1)\,(0.2)}$$

1.35 = Volume (in milliliters) of stopped reaction in Step 1
5 = Total volume (in milliliters) of assay in Step 2
df = Dilution factor
1 = Extinction coefficient (arbitrary value) as per the Unit Definition
0.1 = Volume (in milliliter) of enzyme used in Step 1
0.2 = Volume (in milliliter) of reaction mix used in Step 2

$$\text{Units / mg solid} = \frac{\text{units / ml enzyme}}{\text{mg solid / ml enzyme}}$$

$$\text{Units / mg protein} = \frac{\text{units / ml enzyme}}{\text{mg protein / ml enzyme}}$$

Unit Definition: *One unit will produce acid soluble oligonucleotides equivalent to a A260 of 1.0 in 15 minutes at pH 7.5 at 37°C, in a reaction volume of 1 ml. Substrate: Yeast RNA.*

Final Assay Concentrations

In a 1.00 ml reaction mix, the final concentrations are 70 mM Tris, 0.30% (w/w) ribonucleic acid, 2 mM Ethylenediaminetetraacetic Acid, and 7–8 units ribonuclease T1.

Enzymatic Assay of RIBONUCLEASE T2 (EC 3.1.27.1)

Principle

$$\text{Polyadenylic Acid (5')} + H_2O \xrightarrow{\text{Ribonuclease T2}} \text{Acid Soluble Oligonucleotides}$$

Conditions: T = 37°C, pH 4.5, A260nm, Light path = 1 cm

Method: Spectrophotometric Stop Rate Determination

Reagents

A. 200 mM Sodium Acetate Buffer, pH 4.5 at 37°C (Prepare 100 ml in deionized water using Sodium Acetate, Trihydrate. Adjust to pH 4.5 at 37°C with 1 M HCl).
B. 1.2% (w/v) Polyadenylic (5') Solution (Poly A) (Prepare 5 ml in deionized water using Polyadenylic Acid (5'), Potassium Salt).
C. 20 mM Ethylenediaminetetraacetic Acid Solution (EDTA) (Prepare 25 ml in deionized water using Ethylenediaminetetraacetic Acid, Disodium Salt, Dihydrate).
D. 25% (v/v) Perchloric Acid Solution ($HClO_4$) (Prepare 25 ml in deionized water using Perchloric Acid).
E. 17.7 mM Uranyl Acetate Solution (Uran Acet) (Prepare 5 ml in Reagent D using Uranyl Acetate, Dihydrate).
F. Ribonuclease T2 Enzyme Solution (Immediately before use, prepare a solution containing 40–80 units/ml in cold Reagent A).

Procedure

Pipette (in milliliters) the following reagents into suitable tubes:

	Test	Blank
Deionized Water	0.30	0.30
Reagent A (Buffer)	0.25	0.35
Reagent B (Poly A)	0.25	0.25
Reagent C (EDTA)	0.10	0.10

Mix by swirling and equilibrate to 37°C. Then add:

Reagent F (Enzyme Solution)	0.10	—

Immediately mix by swirling and incubate at 37°C for exactly 15 minutes. Then add:

Reagent E (Uran Acet)	0.25	0.25

Mix by swirling and allow to stand at room temperature for 15 minutes. Centrifuge for 10 minutes. Pipette (in milliliters) the following reagents into suitable tubes:

Test Supernatant	0.20	—
Blank Supernatant	—	0.20
Deionized Water	4.80	4.80

Mix by swirling and transfer the solutions to suitable cuvettes. Record the A260nm for both the Test and Blank using a suitable spectrophotometer.

Calculations

$$\text{Units / ml enzyme} = \frac{(\text{A260nm Test} - \text{A260nm Blank})\,(5)\,(1.25)\,(\text{df})}{(0.1)\,(1)\,(0.2)}$$

5 = Total volume (in milliliters) of assay
1.25 = Volume (in milliliters) of stopped reaction
df = Dilution factor
0.1 = Volume (in milliliter) of enzyme used
1 = Extinction coefficient (arbitrary value) as per the Unit Definition
0.2 = Volume (in milliliter) of reaction mix used in the assay

$$\text{Units / mg solid} = \frac{\text{units / ml enzyme}}{\text{mg solid / ml enzyme}}$$

$$\text{Units / mg protein} = \frac{\text{units / ml enzyme}}{\text{mg protein / ml enzyme}}$$

Unit Definition: *One unit will produce acid soluble oligonucleotides equivalent to a A260nm of 1.0 in 15 minutes at pH 4.5 at 37°C in a 1.0 ml reaction volume. Substrate: Polyadenylic Acid (5').*

Final Assay Concentrations

In a 1.00 ml reaction mix, the final concentrations are 70 mM sodium acetate, 0.3% (w/v) polyadenylic acid (5'), 2 mM Ethylenediaminetetraacetic Acid, and 4–8 units ribonuclease T2.

Enzymatic Assay of RIBONUCLEASE U1 (EC 3.1.27.3)

Principle

$$RNA + H_2O \xrightarrow{\text{Ribonuclease U1}} \text{Acid Soluble Oligonucleotides}$$

Abbreviation Used

RNA = Ribonucleic Acid

Conditions: T = 37°C, pH = 7.5, A260nm, Light path = 1 cm

Method: Spectrophotometric Stop Rate Determination

Reagents

A. 200 mM Tris HCl Buffer, pH 7.5 at 30°C (Prepare 100 ml in deionized water using Trizma Hydrochloride. Adjust to pH 7.5 at 37°C with 1 M NaOH).

B. 1.2% (w/v) Ribonucleic Acid Solution (RNA) (Prepare 5 ml in deionized water using Ribonucleic Acid. Adjust to pH 7.5 at 37°C with 1 M NaOH).

C. 20 mM Ethylenediaminetetraacetic Acid Solution (EDTA) (Prepare 5 ml in deionized water using Ethylenediaminetetraacetic Acid, Tetrasodium Salt, Hydrate).

D. 25% (v/v) Perchloric Acid Solution ($HClO_4$) (Prepare 10 ml in deionized water using Perchloric Acid).

E. 17.7 mM Uranyl Acetate Solution (Uran Acet) (Prepare 5 ml in Reagent D using Uranyl Acetate, Dihydrate).

F. Ribonuclease U1 Enzyme Solution (Immediately before use, prepare a solution containing 67–100 units/ml of Ribonuclease U1 in cold Reagent A).

Procedure

Step 1: Pipette (in milliliters) the following reagents into suitable containers:

	Test	Blank
Deionized Water	0.25	0.25
Reagent A (Buffer)	0.25	0.25
Reagent B (RNA)	0.25	0.25
Reagent C (EDTA)	0.10	0.10

Mix by swirling and equilibrate to 37°C. Then add:

Reagent F (Enzyme Soln)	0.15	—

Immediately mix by swirling and incubate at 37°C for exactly 15 minutes. Then add:

Reagent E (Uran Acet)	0.25	0.25
Reagent F (Enzyme Soln)	—	0.15

Mix by swirling and incubate to 37°C for 15 minutes. Mix by swirling and centrifuge for 10 minutes.

Step 2: Pipette (in milliliters) the following reagents into suitable containers:

	Test	Blank
Test Supernatant	0.20	—
Blank Supernatant	—	0.20
Deionized Water	4.80	4.80

Mix by swirling and transfer the solutions to suitable cuvettes. Record the A260nm for both the Test and Blank using a suitable spectrophotometer.

Calculations

$$\text{Units / ml enzyme} = \frac{(\text{A260nm Test} - \text{A260nm Blank})\,(1.25)\,(5)\,(\text{df})}{(1)\,(0.15)\,(0.2)}$$

1.25 = Volume (in milliliters) of stopped reaction (Step 1)
5 = Total volume (in milliliters) of assay (Step 2)
df = Dilution factor
1 = Extinction coefficient (arbitrary value) as per the Unit Definition
0.15 = Volume (in milliliter) of enzyme used in Step 1
0.2 = Volume (in milliliter) of stopped reaction used in Step 2

$$\text{Units / mg protein} = \frac{\text{units / ml enzyme}}{\text{mg protein / ml enzyme}}$$

Unit Definition: *One unit will produce acid soluble oligonucleotides equivalent to a A260 of 1.0 in 15 minutes at pH 7.5 at 37°C in a 1.0 ml reaction volume. Substrate: Yeast RNA.*

Final Assay Concentrations

In a 1.00 ml reaction mix, the final concentrations are 80 mM Tris, 0.30% (w/v) ribonucleic acid, 2 mM Ethylenediaminetetraacetic Acid, and 10–15 units Ribonuclease U1.

Enzymatic Assay of RIBONUCLEASE U2 (EC 3.1.27.4)

Principle

$$\text{RNA} + H_2O \xrightarrow{\text{Ribonuclease U2}} \text{Acid Soluble Oligonucleotides}$$

Abbreviation Used

RNA = Ribonucleic acid

Conditions: T = 37°C, pH 4.5, A260nm, Light path = 1 cm

Method: Spectrophotometric Stop Rate Determination

Reagents

A. 200 mM Sodium Acetate Buffer, pH 4.5 at 37°C (Prepare 100 ml in deionized water using Sodium Acetate, Trihydrate. Adjust to pH 4.5 at 37°C with 5 M HCl).
B. 1.2% (w/v) Ribonucleic Acid Solution (RNA) (Prepare 3 ml in Reagent A using Ribonucleic Acid).

C. 20 mM Ethylenediaminetetraacetic Acid Solution (EDTA) (Prepare 25 ml in deionized water using Ethylenediaminetetraacetic Acid, Disodium Salt, Dihydrate).
D. 25% (v/v) Perchloric Acid Solution ($HClO_4$) (Prepare 25 ml in deionized water using Perchloric Acid).
E. 0.75% (w/v) Uranyl Acetate Solution (Uran Acet) (Prepare 5 ml in Reagent D using Uranyl Acetate, Dihydrate).
F. Ribonuclease U2 Enzyme Solution (Immediately before use, prepare a solution containing approximately 100 units/ml of Ribonuclease U2 in cold Reagent A).

Procedure

Step 1: Pipette (in milliliters) the following reagents into suitable tubes:

	Test	Blank
Deionized Water	0.25	0.25
Reagent A (Buffer)	0.25	0.40
Reagent B (RNA)	0.25	0.25
Reagent C (EDTA)	0.10	0.10

Mix by swirling and equilibrate to 37°C. Then add:

Reagent F (Enzyme Solution)	0.15	—

Immediately mix by swirling and incubate at 37°C for exactly 15 minutes. Then add:

Reagent E (Uran Acet)	0.25	0.25

Mix by swirling and allow to stand at 37°C for 15 minutes. Centrifuge for 10 minutes.

Step 2: Pipette (in milliliters) the following reagents into suitable tubes:

Test Supernatant	0.20	—
Blank Supernatant	—	0.20
Deionized Water	4.80	4.80

Mix by swirling and transfer the solutions to suitable cuvettes. Record the A260nm for both the Test and Blank using a suitable spectrophotometer.

Calculations

$$\text{Units / ml enzyme} = \frac{(\text{A260nm Test} - \text{A260nm Blank})\,(5)\,(1.25)\,(\text{df})}{(0.15)\,(1)\,(0.2)}$$

5 = Total volume (in milliliters) of assay (Step 2)
1.25 = Volume (in milliliters) of stopped reaction (Step 1)
df = Dilution factor
0.15 = Volume (in milliliter) of enzyme used in Step 1
1 = The increase in A260nm/minute (arbitrary value) per unit of enzyme as per the Unit Definition
0.2 = Volume (in milliliter) of stopped reaction used in Step 2

$$\text{Units / mg solid} = \frac{\text{units / ml enzyme}}{\text{mg solid / ml enzyme}}$$

$$\text{Units / mg protein} = \frac{\text{units / ml enzyme}}{\text{mg protein / ml enzyme}}$$

Unit Definition: *One unit will produce acid soluble oligonucleotides equivalent to a A260nm of 1.0 in 15 minutes at pH 4.5 at 37°C in a 1.0 ml reaction volume. Substrate: Yeast RNA.*

Final Assay Concentrations

In a 1.00 ml reaction mix, the final concentrations are 130 mM sodium acetate, 0.3% (w/v) ribonucleic acid, 2 mM Ethylenediaminetetraacetic Acid, and 15 units ribonuclease U2.

Enzymatic Assay of RIBONUCLEIC ACID POLYMERASE (EC 2.7.7.6)

Principle

$$\text{DNA} + \text{NTP} \xrightarrow{\text{RNA Polymerase}} \text{DNA} + \text{RNA} + \text{Ppi}$$

$$\text{PPi} + \text{UDPG} \xrightarrow{\text{Pyrophosphorylase}} \text{UTP} + \text{Glucose 1-Phosphate}$$

$$\text{Glucose 1-Phosphate} \xrightarrow{\text{Phosphoglucomutase}} \text{Glucose 6-Phosphate}$$

$$\text{Glucose 6-Phosphate} + \beta\text{-NADP} \xrightarrow{\text{G-6-PDH}} \text{6-PG} + \beta\text{-NADPH}$$

Abbreviations Used

DNA = Deoxyribonucleic Acid
NTP = Nucleotide Triphosphate
RNA = Ribonucleic Acid
PPi = Inorganic Pyrophosphate
UDPG = Uridine 5'-Diphosphoglucose
UTP = Uridine 5'-Triphosphate
β-NADP = β-Nicotinamide Adenine Dinucleotide Phosphate, Oxidized Form
G-6-PDH = Glucose-6-Phosphate Dehydrogenase
6-PG = 6-Phospho-D-Gluconate
β-NADPH = β-Nicotinamide Adenine Dinucleotide Phosphate, Reduced Form

Conditions: T = 37°C, pH = 7.9, A340nm, Light path = 1 cm

Method: Continuous Spectrophotometric Rate Determination

Reagents

A. 190 mM Tris Buffer with 75 mM Magnesium Chloride, 450 mM Potassium Chloride and 4 mM Dithiothreitol, pH 7.9 at 37°C (Prepare 100 ml in deionized water using Trizma Base, Magnesium Chloride, Hexahydrate, Potassium Chloride and DL-Dithiothreitol. Adjust to pH 7.9 at 37°C with 1 M HCl).
B. 1 M Tris Base Solution (Prepare 50 ml in deionized water using Trizma Base).
C. 10 mM Nucleotide 5'-Triphosphate Solution (NTP) (Prepare the Nucleotide 5'-Triphosphate Solution by combining equal volumes of the following components:

1. 40 mM Cytidine 5'-Triphosphate Solution (CTP) (Prepare 10 ml in deionized water using Cytidine 5'-Triphosphate, Sodium Salt).
2. 40 mM Guanosine 5'-Triphosphate Solution (GTP) (Prepare 10 ml in deionized water using Guanosine 5'-Triphosphate, Lithium Salt).
3. 40 mM Adenosine 5'-Triphosphate Solution (ATP) (Prepare 10 ml in deionized water using Adenosine 5'-Triphosphate, Disodium Salt).
4. 40 mM Uridine 5'-Triphosphate Solution (UTP) (Prepare 10 ml in deionized water using Uridine 5'-Triphosphate, Sodium Salt).

Adjust to pH 7.9 with a small volume of Reagent B. *Prepare Fresh*).

D. 10 mM Tris Buffer with 0.1 mM Ethylenediaminetetraacetic Acid, pH 7.9 at 25°C (DNA Buffer) (Prepare 100 ml in deionized water using Trizma Base and Ethylenediaminetetraacetic Acid, Tetrasodium Salt, Hydrate. Adjust to pH 7.9 at 25°C with 1 M HCl).

E. 0.05% (w/v) Deoxyribonucleic Acid with 0.5% (w/v) Bovine Serum Albumin Solution (DNA/BSA) (Prepare 50 ml in Reagent D using Deoxyribonucleic Acid (DNA) and Albumin, Bovine, Essentially Fatty Acid Free).

F. 90 mM Uridine 5'-Diphosphoglucose with 2 mM Glucose 1,6-Diphosphate Solution (UDPG/G1,6DP) (Prepare 3 ml in deionized water using Uridine 5'-Diphosphoglucose, Disodium Salt and α-D-Glucose 1,6-Diphosphate, Cyclohexylammonium Salt, Hydrate).

G. 10 mM β-Nicotinamide Adenine Dinucleotide Phosphate Solution (β-NADP) (Dissolve the contents of one 10 mg vial of β-Nicotinamide Adenine Dinucleotide Phosphate, Sodium Salt in the appropriate volume of deionized water *or* prepare 2 ml in deionized water using β-Nicotinamide Adenine Dinucleotide Phosphate, Sodium Salt. *Prepare Fresh*).

H. 20 mM Sodium Phosphate Solution, pH 7.5 at 25°C. (Prepare 100 ml in deionized water using Sodium Phosphate, Monobasic, Anhydrous. Adjust to pH 7.5 at 25°C with 1 M NaOH).

I. 50% (v/v) Glycerol Solution (Prepare 200 ml in Reagent H using Glycerol).

J. Phosphoglucomutase/Glucose-6-Phosphate Dehydrogenase4 Solution (PGM/G-6-PDH) (Prepare a solution containing 500 units/ml of Phosphoglucomutase and 500 units/ml of Glucose-6-Phosphate Dehydrogenase in Reagent I).

K. Uridine-5'-Diphosphoglucose Pyrophosphorylase Solution (UDPG-PP) (Prepare a solution containing 100 units/ml of Uridine-5'-Diphosphoglucose Pyrophosphorylase in cold Reagent I).

L. 100 mM Manganese Chloride Solution ($MnCl_2$) (Prepare 10 ml in deionized water using Manganese Chloride, Tetrahydrate).

M. Ribonucleic Acid Polymerase Enzyme Solution (Immediately before use, prepare a solution containing 900 units/ml of Ribonucleic Acid Polymerase in cold Reagent A).

Procedure

Pipette (in milliliters) the following reagents into suitable 1 ml cuvettes:

	Test	Blank
Reagent A (Buffer)	0.30	0.30
Reagent C (NTP)	0.05	0.05
Reagent D (DNA Buffer)	—	0.40
Reagent E (DNA/BSA)	0.40	—
Reagent F (UDPG/G1,6DP)	0.05	0.05
Reagent G (β-NADP)	0.05	0.05
Reagent J (PGM/G-6-PDH)	0.01	0.01
Reagent K (UDPG-PP)	0.01	0.01
Reagent L ($MnCl_2$)	0.01	0.01

Mix by inversion and equilibrate to 37°C. Monitor the A340nm until constant, using a suitably thermostatted spectrophotometer. Then add:

Reagent M (Enzyme Solution)	0.10	0.10

Immediately mix by inversion and record the increase in A340nm for approximately 15 minutes. Obtain the ΔA340nm/minute using the maximum linear rate for both the Test and Blank.

Calculations

$$\text{Units / ml enzyme} = \frac{(\Delta A340nm / min\ Test - \Delta A340nm / min\ Blank)\ (10)\ (0.98)\ (df)}{(0.00622)\ (0.1)}$$

10 = Unit definition specifies 10 minute reaction
0.98 = Total volume (in milliliters) of assay
df = Dilution factor
0.00622 = Micromolar extinction coefficient of β-NADPH at 340nm
0.1 = Volume (in milliliters) of enzyme used

$$\text{Units / mg solid} = \frac{\text{units / ml enzyme}}{\text{mg solid / ml enzyme}}$$

$$\text{Units / mg protein} = \frac{\text{units / ml enzyme}}{\text{mg protein / ml enzyme}}$$

Unit Definition: *One unit will release 1.0 nanomole (10–9 mole) of PPi in 10 minutes at pH 7.9 at 37°C using calf thymus DNA.*

Final Assay Concentrations

In a 0.98 ml reaction mix, the final concentrations are 82 mM Tris6, 31 mM magnesium chloride, 184 mM potassium chloride, 2 mM DL-dithiothreitol, 0.5 mM adenosine 5'-triphosphate, 0.5 mM cytidine 5'-triphosphate, 0.5 mM guanosine 5'-triphosphate, 0.5 mM uridine 5'-triphosphate, 0.04 mM Ethylenediaminetetraacetic Acid, 0.2% (w/v) Bovine Serum Albumin, 0.02% (w/v) deoxyribonucleic acid, 5 mM uridine 5'-diphosphoglucose, 0.1 mM glucose 1,6-diphosphate, 0.5 mM β-nicotinamide adenine dinucleotide phosphate, 5 units phosphoglucomutase, 5 units glucose-6-phosphate dehydrogenase, 1 unit uridine-5'-diphosphoglucose pyrophosphorylase, 1 mM manganese chloride, 90 units ribonucleic acid polymerase, 1% (v/v) glycerol, and 0.4 mM sodium phosphate.

Notes

1. This procedure is not to be used to assay the activity of Ribonucleic Acid Polymerase.
2. For the assay of RNA Polymerase from wheat germ, place the DNA solution in a boiling water bath for 10 minutes to denature the DNA. Remove from the water bath and quickly cool in an ice/water bath to below 20°C to obtain single stranded DNA. Do not heat the DNA solution used for assaying the E. coli enzyme. Add the albumin to the solution after the DNA has been heat denatured.
3. Glucose-1,6-Diphosphate is added to ensure that the glucose 1-phosphate continues to glucose 6-phosphate. It is required in order for the phosphoglucomutase to be optimally active.
4. Glucose-6-phosphate dehydrogenase is inhibited by ammonium sulfate.
5. The two enzymes listed are ammonium sulfate suspensions. Add 1000 units of phosphoglucomutase and glucose-6-phosphate dehydrogenase to a centrifuge tube. Centrifuge and remove most of the supernatant. Dissolve the pellet in 2 ml of Reagent H (Glycerol Solution) and store at 0–5°C. The solution is stable for 1 week. This will give a solution containing 500 units/ml. Sulfate free enzymes of equal specifications can be used without centrifugation; dissolve them in 10 mM sodium phosphate, pH 7.5 at 25°C, containing 50% glycerol.
6. This concentration does not include the concentration of Trizma Base used to adjust the pH of the NTP solution (Reagent C).
7. Phosphoglucomutase Unit Definition: One unit will convert 1.0 μmole of α-D-glucose 6-phosphate per minute at pH 7.4 at 30°C.
8. Glucose-6-Phosphate Dehydrogenase Unit Definition: One unit will oxidize 1.0 μmole of D-glucose 6- phosphate to 6-phospho-D-gluconate per minute in the presence of NADP at pH 7.4 at 25°C.
9. Uridine-5'-Diphosphoglucose Pyrophosphorylase Unit Definition: One unit will cause the formation of 1.0 μmole of glucose 1-phosphate from uridine-5'-diphosphoglucose and inorganic pyrophosphate per minute at pH 7.6 at 25°C.

Enzymatic Assay of D-RIBULOSE 1,5-DIPHOSPHATE CARBOXYLASE (EC 4.1.1.39)

Principle

$$\text{RuDP} + CO_2 \xrightarrow[Mg_2^+]{\text{RUBISCO}} 2 \text{ moles of 3-phosphoglycerate}$$

$$\text{3-Phosphoglycerate} + \text{ATP} \xrightarrow{\text{PGK}} \text{Glycerate 1,3-Diphosphate} + \text{ADP}$$

$$\text{Glycerate 1,3-Diphosphate} + \beta\text{-NADH} \xrightarrow{\text{GAPDH}} \text{Glyceraldehyde 3-Phosphate} + \text{NAD} + \text{Pi}$$

$$\text{Glyceraldehyde 3-Phosphate} \xrightarrow{\text{TPI}} \text{DHAP}$$

$$\text{DHAP} + \beta\text{-NADH} \xrightarrow{\alpha\text{-GDH}} \alpha\text{-Glycerophosphate} + \beta\text{-NAD}$$

Abbreviations Used

RUDP = D-Ribulose 1,5-Diphosphate
RUBISCO = D-Ribulose 1,5-Diphosphate Carboxylase
ATP = Adenosine 5'-Triphosphate
PGK = 3-Phosphoglyceric Phosphokinase
ADP = Adenosine 5'-Diphosphate
β-NADH = β-Nicotinamide Adenine Dinucleotide, Reduced Form

GAPDH = Glyceraldehyde 3-Phosphate Dehydrogenase
Pi = Inorganic Phosphate
TPI = Triosephosphate Isomerase
DHAP = Dihydroxyacetone Phosphate
β-NAD = β-Nicotinamide Adenine Dinucleotide, Oxidized Form
α-GDH = α-Glycerophosphate Dehydrogenase

Conditions: T = 25°C, pH = 7.8, A340nm, Light path = 1 cm

Method: Continuous Spectrophotometric Rate Determination

Reagents

A. 1 M Tris HCl Buffer, pH 7.8 at 25°C (Prepare 25 ml in deionized water using Trizma Base. Adjust to pH 7.8 at 25°C with 10 M HCl).
B. 100 mM Magnesium Chloride Solution ($MgCl_2$) (Prepare 5 ml in deionized water using Magnesium Chloride, Hexahydrate).
C. 2 M Potassium Bicarbonate Solution ($KHCO_3$) (Prepare 5 ml in deionized water using Potassium Bicarbonate).
D. 2.5 mM β-Nicotinamide Adenine Dinucleotide, Reduced Form Solution (β-NADH) (Prepare 5 ml in Reagent A using β-Nicotinamide Adenine Dinucleotide, Reduced Form, Disodium Salt or dissolve the contents of one 1 mg vial of β-Nicotinamide Dinucleotide, Reduced Form, Disodium Salt in the appropriate volume of Reagent A).
E. 100 mM Adenosine 5'-Triphosphate Solution (ATP) (Prepare 5 ml in deionized water using Adenosine 5'-Triphosphate, Disodium Salt).
F. 100 mM Glutathione, Reduced Form, Solution (GSH) (Prepare 5 ml in deionized water using Glutathione, Reduced Form, Free Acid).
G. 25 mM D-Ribulose 1,5-Diphosphate Solution (RuDP) (Prepare 2 ml in deionized water using D-Ribulose 1,5-Diphosphate, Sodium Salt, Hydrate).
H. α-Glycerophosphate Dehydrogenase-Triosephosphate Isomerase Enzyme Suspension (α-GDH/TPI) (Immediately before use prepare a solution containing 50 units/ml of α-Glycerophosphate Dehydrogenase-Triosephosphate Isomerase in cold deionized water).
I. Glyceraldehyde-3-Phosphate Dehydrogenase/3- Phosphoglyceric hosphokinase Enzyme Solution (GAPDH/3-PGK) (Immediately before use prepare a solution containing 50 units/ml of Glyceraldehyde-3-Phosphate Dehydrogenase/3-Phosphoglyceric Phosphokinase, in cold deionized water).
J. D-Ribulose 1,5-Diphosphate Carboxylase (RUBISCO) (Immediately before use, prepare a solution containing 0.07–0.15 unit/ml of D-Ribulose 1,5-Diphosphate Carboxylase in cold Reagent A).

Procedure

Prepare a reaction cocktail by pipetting (in milliliters) the following reagents into a suitable container:

Reagent A (Buffer)	4.40
Reagent B ($MgCl_2$)	1.50
Reagent C ($KHCO_3$)	1.00
Reagent D (β-NADH)	2.40
Reagent E (ATP)	1.50
Reagent F (GSH)	1.50
Reagent G (RuDP)	0.60

Mix by swirling. Adjust to pH 7.8 at 25°C, if necessary, with either 0.1 M NaOH or 0.1 M HCl. Pipette (in milliliters) the following reagents into suitable cuvettes:

	Test	Blank
Reaction Cocktail	1.29	1.29
Reagent H (α-GDH/TPI)	0.10	0.10
Reagent I (GAPDH/3-PGK)	0.10	0.10
Deionized Water	1.41	1.41

Mix by inversion and equilibrate to 25°C. Monitor the A340nm until constant, using a suitably thermostatted spectrophotometer. Then add:

Reagent J (RUBISCO)	0.10	—
Reagent A (Buffer)	—	0.10

Immediately mix by inversion and record the decrease in A340nm for approximately 5 minutes. Obtain the ΔA340nm/minute using the maximum linear rate for both the Test and Blank.

Calculations

$$\text{Units / ml enzyme} = \frac{(\Delta A340\text{nm / min Test} - \Delta A340\text{nm / min Blank})\,(3)\,(\text{df})}{(4)\,(6.22)\,(0.1)}$$

3 = Total volume (in milliliters) of assay
df = Dilution factor
4 = 4 μmoles of β-NADH are oxidized for each μmole of D-Ribulose 1,5-diphosphate utilized
6.22 = Millimolar extinction coefficient for β-NADH at 340nm
0.1 = Volume (in milliliter) of enzyme used

$$\text{Units / mg solid} = \frac{\text{units / ml enzyme}}{\text{mg solid / ml enzyme}}$$

$$\text{Units / mg protein} = \frac{\text{units / ml enzyme}}{\text{mg protein / ml enzyme}}$$

Unit Definition: *One unit will convert 1.0 μmole of D-RuDP and CO_2 to 2.0 μmoles of D-3-phosphoglycerate per minute at pH 7.8 at 25°C.*

Final Assay Concentrations

In a 3.00 ml reaction mix, the final concentrations are 259 mM Tris, 5 mM magnesium chloride, 67 mM potassium bicarbonate, 0.2 mM β-nicotinamide adenine dinucleotide, reduced form, 5 mM adenosine 5'-triphosphate, 5 mM glutathione, reduced form, 0.5 mM D-ribulose 1,5-diphosphate, 5 units α-glycerophosphate dehydrogenasetriose-phosphate isomerase, 5 units glyceraldehyde-3-phosphate dehydrogenase/3-phosphoglyceric phosphokinase, and 0.007–0.015 unit D-ribulose 1,5-diphosphate carboxylase.

Notes

1. 3-Phosphoglyceric Phosphokinase Unit Definition: One unit will convert 1.0 μmole of 3-phosphoglycerate to 1,3-diphosphoglycerate per minute at pH 7.5 at 25°C.
2. Glyceraldehyde-3-Phosphate Dehydrogenase Unit Definition: One unit will reduce 1.0 μmole of 3-phosphoglycerate to D-glyceraldehyde 3-phosphate per minute in a coupled system with 3-phosphoglyceric phosphokinase at pH 7.6 at 25°C.
3. Triosephosphate Isomerase Unit Definition: One unit will convert 1.0 μmole of D-glyceraldehyde-3-phosphate to dihydroxyacetone phosphate per minute at pH 7.6 and 25°C.
4. α-Glycerophosphate Dehydrogenase Unit Definition: One unit will convert 1.0 μmole of dihydroxyacetone phosphate to α-glycerophosphate per minute at pH 7.4 at 25°C.

Enzymatic Assay of D-RIBULOSE 5-PHOSPHATE 3-EPIMERASE (EC 5.1.3.1)

Principle

$$\text{D-Ribulose 5-Phosphate} \xrightarrow{\text{Ru-5-P-3-Epi}} \text{D-Xylulose 5-Phosphate}$$

$$\text{D-Xylulose 5-Phosphate} + \text{R5-P} \xrightarrow[\text{TPP}]{\text{TK}} \text{G 3-P} + \text{Sedoheptulose 7-Phosphate}$$

$$\text{G 3-P} \xrightarrow{\text{TPI}} \text{Dihydroxyacetone Phosphate}$$

$$\text{Dehydroxyacetone Phosphate} + \beta\text{-NADH} \xrightarrow{\alpha\text{-GDH}} \alpha\text{-Glycerol Phosphate} + \beta\text{-NAD}$$

Abbreviations Used

RU-5-P-3-Epi = D-Ribulose 5-Phosphate 3-Epimerase
TK = Transketolase
TPP = Thiamine Pyrophosphate
R5-P = D-Ribose 5-phosphate
G 3-P = Glyceraldehyde 3-Phosphate
TPI = Triosephosphate Isomerase
α-GDH = α-Glycerophosphate Dehydrogenase
β-NADH = β-Nicotinamide Adenine Dinucleotide, Reduced Form
β-NAD = β-Nicotinamide Adenine Dinucleotide, Oxidized Form

Conditions: T = 25°C, pH = 7.7, A340nm, Light path = 1 cm

Methods: Continuous Spectrophotometric Rate Determination

Reagents

A. 250 mM Glycylglycine Buffer, pH 7.7 at 25°C (Prepare 100 ml in deionized water using Glycylglycine, Free Base. Adjust to pH 7.7 at 25°C with 1 M NaOH).
B. 100 mM D-Ribulose 5-Phosphate Solution (RU 5-P) (Prepare 1 ml in deionized water using D-Ribulose 5-Phosphate, Sodium Salt in deionized water).

C. 100 mM D Ribose 5-Phosphate Solution (R 5-P) (Prepare 1 ml in deionized water using D-Ribose 5-Phosphate, Disodium Salt).

D. 0.10% (w/v) Cocarboxylase/Thiamine Pyrophosphate Solution (TPP) (Prepare 1 ml in cold deionized water using Cocarboxylase. *Prepare Fresh*).

E. 300 mM Magnesium Chloride Solution ($MgCl_2$) (Prepare 5 ml in deionized water using Magnesium Chloride, Hexahydrate).

F. 2.6 mM β-Nicotinamide Adenine Dinucleotide, Reduced Form Solution (β-NADH) (Dissolve the contents of one 5 mg vial of β-Nicotinamide Adenine Dinucleotide, Reduced Form, Disodium Salt, in the appropriate volume of Reagent A. *Prepare Fresh*).

G. α-Glycerophosphate Dehydrogenase/Triosephosphate Isomerase Enzyme Solution (α-GDH/TPI) (Immediately before use, prepare a solution containing 10 α-GDH units/ml of α-Glycerophosphate Dehydrogenase/ Triosephosphate Isomerase in cold deionized water).

H. Transketolase Enzyme Solution (TK) (Immediately before use, prepare a solution containing 10 units/ml of Transketolase, in cold deionized water).

I. D-Ribulose 5-Phosphate 3-Epimerase Enzyme Solution (RU-5-P-3-EPI) (Immediately before use, prepare a solution containing 0.25–0.50 unit/ml of D-Ribulose 5-Phosphate 3-Epimerase in cold deionized water).

Procedure

Pipette (in milliliters) the following reagents into suitable cuvettes:

	Test	Blank
Deionized Water	1.80	1.90
Reagent A (Buffer)	0.55	0.55
Reagent B (Ru-5-P)	0.05	0.05
Reagent C (R 5-P)	0.05	0.05
Reagent D (TPP)	0.05	0.05
Reagent E ($MgCl_2$)	0.15	0.15
Reagent F (β-NADH)	0.15	0.15
Reagent G (α-GDH/TPI)	0.05	0.05
Reagent H (TK)	0.05	0.05

Mix by inversion and equilibrate to 25°C. Monitor the A340nm until constant, using a suitably thermostatted spectrophotometer. Then add:

Reagent I (Ru-5-P-3-Epi)	0.10	—

Immediately mix by inversion and record the decrease in A340nm for approximately 5 minutes. Obtain the ΔA340nm/minute using the maximum linear rate for both the Test and Blank.

Calculations

$$\text{Units / ml enzyme} = \frac{(\Delta\text{A340nm / min Test} - \Delta\text{A340nm / min Blank})\ (3)\ (\text{df})}{(6.22)\ (0.1)}$$

3 = Total volume (in milliliters) of assay
df = Dilution factor
6.22 = Millimolar extinction coefficient of β-NADH at 340nm
0.1 = Volume (in milliliter) of enzyme used

$$\text{Units / mg solid} = \frac{\text{units / ml enzyme}}{\text{mg solid / ml enzyme}}$$

$$\text{Units / mg protein} = \frac{\text{units / ml enzyme}}{\text{mg protein / ml enzyme}}$$

Unit Definition: *One unit will convert 1 μmole of D-ribulose 5-phosphate to D-xylulose 5-phosphate per minute at pH 7.7 at 25°C, in a coupled system with ribose 5-phosphate, β-NADH, transketolase, α-glycerophosphate dehydrogenase, triosephosphate isomerase and cocarboxylase.*

Final Assay Concentrations

In a 3.00 ml reaction mix, the final concentrations are 58 mM glycylglycine, 1.7 mM ribulose 5-phosphate, 1.7 mM ribose 5-phosphate, 0.002% (w/v) cocarboxylase, 15 mM magnesium chloride, 0.13 mM β-nicotinamide adenine dinucleotide, reduced form, 0.5 unit aglycerophosphate dehydrogenase, 5 units triosephosphate isomerase, 0.5 unit transketolase, and 0.025–0.05 unit D-ribulose 5-phosphate 3-epimerase.

Notes

1. There may be a sudden decrease in β-NADH within the first two minutes. This decrease may be approximately 0.1–0.2 per minute. Use the linear portion of the curve, after the sudden decrease, to obtain the maximum linear rate.
2. Triosephosphate Isomerase Unit Definition: One unit will convert 1.0 μmole of D-glyceraldehyde 3-phosphate to dihydroxyacetone phosphate per minute at pH 7.6 at 25°C.
3. α-Glycerophosphate Dehydrogenase Unit Definition: One unit will convert 1.0 μmole of dihydroxyacetone phosphate to α-glycerophosphate per minute at pH 7.4 at 25°C.
4. Transketolase Unit Definition: One unit will produce 1.0 μmole of glyceraldehydes 3-phosphate from xylulose 5-phosphate per minute at pH 7.7 at 25°C, in the presence of ribose 5-phosphate, thiamine pyrophosphate and Mg^{++} using a coupled system with α-GDH/TPI.

Enzymatic Assay of SACCHAROPINE DEHYDROGENASE, NAD⁺, Lysine Forming (EC 1.5.1.7)

Principle

$$\text{L-Lysine} + \alpha\text{-Ketoglutaric} + \beta\text{-NADH} \xrightarrow{\text{Saccharopine Dehydrogenase}} \text{Saccharopine} + \beta\text{-NAD}$$

Abbreviations Used

β-NADH = β-Nicotinamide Adenine Dinucleotide, Reduced Form
β-NAD = β-Nicotinamide Adenine Dinucleotide, Oxidized Form

Conditions: T = 25°C, pH = 6.8, A340nm, Light path = 1 cm

Method: Continuous Spectrophotometric Method

Reagents

A. 100 mM Potassium Phosphate Buffer with 1 mM Ethylenediaminetetraacetic Acid (EDTA), pH 6.8 at 25°C (Prepare 100 ml in deionized water using Potassium Phosphate, Monobasic and Ethylenediaminetetraacetic Acid, Tetrasodium Hydrate. Adjust to pH 6.8 at 25°C with 1 M NaOH).

B. 0.23 mM β-Nicotinamide Adenine Dinucleotide, Reduced Form (β-NADH) (Prepare 50 ml in Reagent A using β-Nicotinamide Adenine Dinucleotide, Reduced Form, Disodium Salt).

C. 79.8 mM α-Ketoglutarate Solution (Prepare 1.0 ml in Reagent A using α-Ketoglutaric Acid, Monopotassium Salt. *Prepare Fresh*).

D. 300 mM L-Lysine Solution (L-Lysine) (Prepare 10 ml in Reagent A using L-Lysine Monohydrochloride).

E. Saccharopine Dehydrogenase Enzyme Solution (Immediately before use, prepare a solution containing 0.1–0.5 units/ml of Saccharopine Dehydrogenase in cold Reagent A).

Procedure

Pipette (in milliliters) the following reagents into suitable cuvettes:

	Test	Blank
Reagent B (β-NADH)	2.75	2.75
Reagent C (α-Ketoglutarate)	0.10	0.10
Reagent D (L-Lysine)	0.10	0.10

Mix by inversion and equilibrate to 25°C. Monitor the A340nm until constant, using a suitably thermostatted spectrophotometer. Then add:

Reagent E (Enzyme Solution)	0.10	—
Reagent A (Buffer)	—	0.10

Immediately mix by inversion and record the decrease in the A340nm for approximately 5 minutes. Obtain the ΔA340nm/minute using the maximum linear rate for both the Test and Blank.

Calculations

$$\text{Units / mg enzyme} = \frac{(\Delta A340\text{nm / min Test} - \Delta A340\text{nm / min Blank})\,(3.05)\,(\text{df})}{(6.22)\,(0.1)}$$

3.05 = Volume (in milliliters) of assay
df = Dilution factor
6.22 = Millimolar extinction coefficient of β-NADH at 340nm
0.1 = Volume (in milliliters) of enzyme used

$$\text{Units / mg solid} = \frac{\text{units / ml enzyme}}{\text{mg solid / ml enzyme}}$$

$$\text{Units / mg protein} = \frac{\text{units / ml enzyme}}{\text{mg protein / ml enzyme}}$$

Unit Definition: *One unit will catalyze the conversion of 1.0 μmole of L-lysine and α-ketoglutaric acid to saccharopine per minute at pH 6.8 at 25°C.*

Final Assay Concentrations

In a 3.05 ml reaction mix, the final concentrations are 100 mM potassium phosphate, 1 mM Ethylenediaminetetraacetic Acid, 0.21 mM β-icotinamide adenine dinucleotide, reduced form, 2.6 mM α-ketoglutaric acid, 9.8 mM L-lysine, and 0.01–0.05 units saccharopine dehydrogenase.

Note

1. Preweigh and refrigerate vials containing 14.7 mg of α-ketoglutaric acid. Dissolve each vial with 1.0 ml of Reagent A. (Prepare Fresh, Store in Ice).

Enzymatic Assay of SALICYLATE HYDROXYLASE (EC 1.14.13.1)

Principle

$$\text{Salicylate} + \beta\text{-NADH} + O_2 \xrightarrow{\text{Salicylate Hydroxylase}} \text{Catechol} + \beta\text{-NAD} + H_2O + CO_2$$

Abbreviations Used

β-NADH = β-Nicotinamide Adenine Dinucleotide, Reduced Form
β-NAD = β-Nicotinamide Adenine Dinucleotide, Oxidized Form

Conditions: T = 30°C, pH = 7.6, A340nm, Light path = 1 cm

Method: Continuous Spectrophotometric Rate Determination

Reagents

A. 20 mM Potassium Phosphate Buffer with 1.0 mM Ethylenediaminetetraacetic Acid, pH 7.6 at 30°C (Prepare 100 ml in deionized water using Potassium Phosphate, Dibasic, Trihydrate and Ethylenediaminetetraacetic Acid, Disodium Salt, Dihydrate. Adjust to pH 7.6 at 30°C with 1 M HCl).
B. 3 mM Sodium Salicylate Solution (Salicylate) (Prepare 10 ml in Reagent A using Salicylic Acid, Sodium Salt).
C. 5.4 mM β-Nicotinamide Adenine Dinucleotide, Reduced Form Solution (β-NADH) (Dissolve the contents of one 5 mg vial of β-Nicotinamide Adenine Dinucleotide, Reduced Form, Disodium Salt in the appropriate volume of cold Reagent A. *Prepare Fresh*).
D. Salicylate Hydroxylase Enzyme Solution (Immediately before use, prepare a solution containing 0.25–0.5 unit/ml of Salicylate Hydroxylase in cold deionized water).

Procedure

Pipette (in milliliters) the following reagents into suitable cuvettes:

	Test	Blank
Reagent A (Buffer)	2.70	2.70
Reagent B (Salicylate)	0.10	0.10
Reagent C (β-NADH)	0.10	0.10

Mix by inversion and equilibrate to 30°C. Monitor the A340nm until constant, using a suitably thermostatted spectrophotometer. Then add:

	Test	Blank
Deionized Water	—	0.10
Reagent D (Enzyme Solution)	0.10	—

Immediately mix by inversion and record the decrease in A340nm for approximately 5 minutes. Obtain the ΔA340nm/minute using the maximum linear rate for both the Test and Blank.

Calculations

$$\text{Units/ml} = \frac{(\Delta A340\text{nm/min Test} - \Delta A340\text{nm/min Blank})(3.0)(\text{df})}{(6.22)(0.10)}$$

3.0 = Total volume (in milliliters) of assay
df = Dilution factor
6.22 = Millimolar extinction coefficient of β-NADH at 340nm
0.10 = Volume (in milliliters) of enzyme used

$$\text{Units/mg solid} = \frac{\text{units/ml enzyme}}{\text{mg solid/ml enzyme}}$$

$$\text{Units/mg protein} = \frac{\text{units/ml enzyme}}{\text{mg protein/ml enzyme}}$$

Unit Definition: *One unit will convert 1.0 μmole of salicylate and β-NADH to catechol and β-NAD per minute at pH 7.6 at 30°C.*

Final Assay Concentrations

In a 3.00 ml reaction mix, the final concentrations are 19 mM potassium phosphate, 0.97 mM Ethylenediaminetetraacetic Acid, 0.1 mM sodium salicylate, 0.18 mM β-NADH, and 0.025–0.05 unit salicylate hydroxylase.

Enzymatic Assay of SARCOSINE DEHYDROGENASE (EC 1.5.99.1)

Principle

$$\text{Sarcosine} + H_2O + \text{PMS} \xrightarrow{\text{Sarcosine Dehydrogenase}} \text{Glycine} + \text{HCHO} + \text{PMSH}_2$$

$$2\text{PMSH}_2 + \text{NTB} \rightarrow 2\text{PMS} + \text{Diformazan}$$

Abbreviations Used

PMS = Phenazine Methosulfate
NTB = Nitro Blue Tetrazolium
$PM\,SH_2$ = Phenazine Methosulfate (Reduced Form)

Conditions: T = 37°C, pH = 7.5, A570nm, Light path = 1 cm

Method: Spectrophotometric Stop Rate Determination

Reagents

A. 50 mM Potassium Phosphate Buffer with 500 mM Sarcosine and 0.5% (v/v) Triton, X-100, pH 7.5 at 37°C (Substrate Solution) (Prepare 25 ml in deionized water using Potassium Phosphate, Monobasic, Anhydrous, Sarcosine, Free Base and Triton X-100. Adjust to pH 7.5 at 37°C with 1 M KOH).

B. 0.01% (w/v) Phenazine Methosulfate and 0.1% (w/v) Nitro Blue Tetrazolium Color Reagent (PMS-NBT) (Prepare 5 ml in deionized water using Phenazine Methosulfate and Nitro Blue Tetrazolium).

C. 300 mM Hydrochloric Acid Solution (HCl) (Prepare 25 ml in deionized water using Hydrochloric Acid).

D. 50 mM Potassium Phosphate Buffer, pH 7.5 at 37°C (Enzyme Diluent) (Prepare 25 ml in deionized water using Potassium Phosphate, Monobasic, Anhydrous. Adjust to pH 7.5 at 37°C with 1 M KOH).

E. Sarcosine Dehydrogenase Enzyme Solution (Immediately before use, prepare a solution containing 0.013–0.025 units/ml of Sarcosine Dehydrogenase in Reagent D. Immediately prior to starting the reaction, equilibrate to 37°C).

Procedure

Pipette (in milliliters) the following reagents into suitable containers:

	Test	Blank
Reagent A (Substrate Solution)	0.50	0.50
Reagent B (PMS-NBT)	0.10	0.10

Mix by swirling and equilibrate to 37°C. Then add:

Reagent E (Enzyme Solution)	0.50	—

Immediately mix by swirling and incubate for exactly 15 minutes at 37°C. Then add:

Reagent C (HCl)	3.00	3.00
Reagent E (Enzyme Solution)	—	0.50

Mix by swirling and transfer to suitable cuvettes. Record the A570nm for both the Test and Blank in a suitable spectrophotometer.

Calculations

$$\text{Units/ml enzyme} = \frac{(\text{A570nm Test} - \text{A570nm Blank})\,(4.1)\,(\text{df})}{(15)\,(20.1)\,(0.5)}$$

4.1 = Total volume (in milliliters) of assay
df = Dilution factor
15 = Time of assay (in minutes) as per the Unit Definition
20.1 = Millimolar extinction coefficient of diformazan at 570nm
0.5 = Volume (in milliliter) of enzyme used

$$\text{Units/mg solid} = \frac{\text{units/ml enzyme}}{\text{mg solid/ml enzyme}}$$

$$\text{Units / mg protein} = \frac{\text{units / ml enzyme}}{\text{mg protein / ml enzyme}}$$

Unit Definition: *One unit will convert 1.0 μmole of sarcosine to glycine and formaldehyde per minute at pH 7.5 at 37°C.*

Final Assay Concentrations

In a 1.10 ml reaction mix, the final concentrations are 45 mM potassium phosphate, 227 mM sarcosine, 0.2% (v/v) Triton X-100, 0.0009% (w/v) phenazine methosulfate, 0.009% (w/v) nitro blue tetrazolium, and 0.0065–0.0125 unit sarcosine dehydrogenase.

Note

1. Triton X-100 is a registered trademark of Union Carbide.

Enzymatic Assay of SARCOSINE OXIDASE (EC 1.5.3.1)

Principle

$$\text{Sarcosine} + O_2 + H_2O \xrightarrow{\text{Sarcosine Oxidase}} \text{Glycine} + \text{Formaldehyde} + H_2O_2$$

$$2H_2O_2 + \text{Phenol} + \text{4-AAP} \xrightarrow{\text{Peroxidase}} 4H_2O + \text{Quinoneimine Dye}$$

Abbreviation Used

4-AAP = 4-Aminoantipyrine

Conditions: T = 37°C, pH = 8.3, A480nm, Light path = 1 cm

Method: Colorimetric

Reagents

A. 200 mM Tris HCl Buffer, pH 8.3 at 37°C. (Prepare 100 ml in deionized water using Trizma Hydrochloride. Adjust to pH 8.3 at 37°C with 1 M NaOH).
B. 15 mM 4-Aminoantipyrine Solution (4-AAP) (Prepare 1 ml in deionized water using 4- Aminoantipyrine Free Base).
C. 0.2% Phenol Solution (Phenol) (Prepare 1 ml in deionized water using Phenol).
D. Peroxidase Enzyme Solution (POD) (Immediately before use, prepare a solution containing 50 purpurogalin units/ml in deionized water using).
E. 1000 mM Sarcosine Solution (Sarc) (Prepare 1 ml in deionized water using Sarcosine, Free Base).
F. Ethanol (EtOH) (Use Reagent Grade Ethanol).
G. 10 mM Potassium Phosphate Solution, pH 7.5 at 37°C (Prepare 100 ml in deionized water using Potassium Phosphate, Monobasic. Adjust to pH 7.5 at 37°C with 1 M NaOH).
H. Sarcosine Oxidase Enzyme Solution (Immediately before use, prepare a solution containing 0.25–1.00 units/ml in cold Reagent G).

Procedure

Pipette (in milliliters) the following reagents into suitable cuvettes:

	Test	Blank
Deionized Water	0.20	0.20
Reagent A (Buffer)	0.05	0.05
Reagent B (4-AAP)	0.05	0.05
Reagent C (Phenol)	0.05	0.05
Reagent D (POD)	0.05	0.05
Reagent E (Sarc)	0.10	0.10

Mix and equilibrate to 37°C. Then add:

Reagent G (Diluent)	—	0.01
Reagent H (Enzyme Solution)	0.01	—

Mix and incubate at 37°C for exactly 5 min. Then add:

Reagent F (EtOH)	2.50	2.50

Mix. Record the A480nm for both Test and Blank using a suitable spectrophometer.

Calculations

$$\text{Units / mg Protein} = \frac{(\text{A480nm Test} - \text{A480nm Blank})\,(3.01)\,(2)}{(17.14)\,(5)\,(\text{mg Protein / RM})}$$

3.01 = Total Volume of Assay
2 = Two Hydrogen Peroxide molecules used
17.14 = Millimolar Extinction Coefficient of the Quinoneimine dye at 480nm
5 = Time of assay in minutes
RM = Reaction Mix

Unit Definition: *One unit will form 1.0 μmole of formaldehyde from sarcosine per minute at pH 8.3 at 37 C.*

Final Assay Concentrations

In a 0.51 ml reaction mix, the final concentrations are 49 mM Tris HCl, 1.5 mM 4-aminoantipyrine, 0.02% phenol, 2.5 purpurogallin units peroxidase, 196 mM sarcosine, 0.2 mM potassium phosphate and 0.0023–0.01 units sarcosine oxidase.

Enzymatic Assay of SARCOSINE OXIDASE (EC 1.5.3.1)

Principle

$$\text{Sarcosine} + H_2O + O_2 \xrightarrow{\text{Sarcosine Oxidase}} \text{Glycine} + \text{Formaldehyde} + H_2O_2$$

Conditions: T = 37°C, pH = 8.3, A410nm, Light path = 1 cm

Method: Spectrophotometric Stop Rate Determination

Reagents

A. 60 mM Glycylglycine Buffer, pH 8.3 at 37°C (Prepare 100 ml in deionized water using Gly-Gly. Adjust to pH 8.3 at 37°C with 1 M NaOH).

B. 300 mM Sarcosine Solution (Sarcosine) (Prepare 10 ml in Reagent A using Sarcosine).
C. 500 mM Acetic Acid (CH_3COOH) (Prepare 25 ml in deionized water using Acetic Acid).
D. 2 M Ammonium Acetate Solution (Amm Acet) (Prepare 40 ml in deionized water using Ammonium Acetate).
E. Isopropanol (Use Isopropanol, Anhydrous).
F. Acetylacetone.
G. Color Reagent (Prepare by adding 20 ml of Reagent D to 40 ml of Reagent E. Then add 0.15 ml of Reagent F. Mix by swirling and store overnight at 0–5°C).
H. Sarcosine Oxidase Enzyme Solution (Immediately before use, prepare a solution containing 0.5–1.0 unit/ml of Sarcosine Oxidase in cold Reagent A).

Procedure

Pipette (in milliliters) the following reagents into suitable containers:

	Test	Blank
Reagent A (Buffer)	0.20	0.30
Reagent B (Sarcosine)	0.20	0.20

Mix by inversion and equilibrate to 37°C. Then add:

Reagent H (Enzyme Solution)	0.10	—

Immediately mix by swirling and incubate for exactly 10 minutes at 37°C. Then add:

Reagent C (CH_3COOH)	0.50	0.50

Mix by swirling. Then add:

Reagent G (Color Reagent)	3.00	3.00

Mix by swirling and incubate at 60°C for 30 minutes. Transfer the solution to suitable cuvettes and record the A410nm for both the Test and Blank in a suitable spectrophotometer.

Calculations

$$\text{Units / ml enzyme} = \frac{(\text{A410nm Test} - \text{A410nm Blank})\ (4)\ (\text{df})}{(10)\ (8)\ (0.1)}$$

4 = Total volume (in milliliters) of assay
df = Dilution factor
10 = Time of assay (in minutes) as per the Unit Definition
8 = Millimolar extinction coefficient of colored product
0.1 = Volume (in milliliters) of enzyme used

$$\text{Units / mg solid} = \frac{\text{units / ml enzyme}}{\text{mg solid / ml enzyme}}$$

$$\text{Units / mg protein} = \frac{\text{units / ml enzyme}}{\text{mg protein / ml enzyme}}$$

Unit Definition: *One unit will form 1.0 μmole of formaldehyde from sarcosine per minute at pH 8.3 at 37°C.*

Final Assay Concentration

In a 0.50 ml reaction mix, the final concentrations are 60 mM glycylglycine, 120 mM sarcosine, and 0.05–0.10 unit sarcosine oxidase.

Enzymatic Assay of SIALYLTRANSFERASE (EC 2.4.99.1)

Principle

$$\text{CMP-Sialic Acid} + \text{Asialomucin} \xrightarrow{\text{Sialyltransferase}} [_{14}C]\text{ Mucin} + \text{CMP}$$

Abbreviations Used

CMP-Sialic Acid = Cytidine 5'-Monophosphosialic Acid
CMP = Cytidine 5'-Monophosphate

Conditions: T = 37°C, pH = 6.5

Method: Radiometric Stopped Reaction

Reagents

A. 500 mM Sodium Cacodylate, 5% (w/v) Triton CF-54, 0.5% (w/v) Bovine Serum Albumin, pH 6.5 at 37°C (Buffer) (Prepare 50 ml in deionized water using Cacodylic Acid, Sodium Salt, Triton CF-54, and Albumin, Bovine. Adjust the pH to 6.5 at 37°C with either 1 N HCl or 1 N NaOH).
B. 31.6 mM Bovine Submaxillary Asialomucin (Asialomucin) (Prepare 1 ml in deionized water using Asialomucin).
C. Cytidine 5'-Monophospho[$_{14}$C] sialic acid (CMP-Sia) (Use Cytidine 5'-Monophospho [$_{14}$C]sialic Acid, Ammonium Salt, 25 μCi/ml, 150–310 mCi/mmol).
D. 7.4 mM Cytidine 5'-Triphosphate (CTP) (Prepare 1 ml in deionized water using Cytidine 5'-Triphosphate, Sodium Salt).
E. 200 mM Sodium Chloride (NaCl) (Prepare 1 liter in deionized water using Sodium Chloride).
F. 4 ml Sephadex G-50 (Fine) column (Column) (Prepare 40 ml of swelled resin in Reagent E using Sephadex G-50 Fine. Pack 4 ml of swelled resin into a Column, Liquid Chromatography, 1.0 × 10 cm Luer-Lock).
G. Sialyltransferase Enzyme Solution (Enzyme) (Immediately before use, prepare a solution containing approximately 0.18 unit/ml of Sialyltransferase in cold deionized water).
H. Liquid Scintillation Cocktail (LSC).

Procedures

Prepare a reaction cocktail by pipetting (in milliliters) the following reagents into a suitable container:

Reagent A (Buffer)	0.150
Reagent B (Asialomucin)	0.075
Reagent C (CMP-Sia)	0.075
Deionized water	0.300

Mix thoroughly. Pipette (in milliliters) the following into suitable tubes:

	Test	Blank
Reaction cocktail	0.040	0.040

Equilibrate to 37°C, then add:

Deionized water	—	0.010
Reagent G (Enzyme)	0.010	—

At exactly 15 minutes, stop the reaction by adding:

Reagent D (CTP)	0.005	0.005

Immediately place the reaction tubes on ice. Carefully transfer the contents of a reaction tube to the Sephadex G-50 column and begin elution with Reagent E (NaCl). Collect all column fractions in 7 ml scintillation vials. Collect the first 1.5 ml and set aside. The initial fraction should contain only background counts. Collect the next 1.5 ml. This second fraction should be the one that contains the sialated asialomucin. Collect the next 1.5 ml fraction. The third fraction should contain approximately half of the remaining radiolabel as unused CMP[$_{14}$C]sialic acid.

Determine the potential by pipetting 0.02 ml of the reaction cocktail in 1.5 ml of reagent E into a scintillation vial. Add 5.5 ml of Reagent H (LSC) to each scintillation vial and immediately mix by vortexing. Count the vials for 2 to 5 minutes on a suitable scintillation counter and correct to Decays Per Minute (DPM).

Calculations

$$\text{DPM / nmole of Potential} = \frac{(\text{DPM of potential})\,(0.60)}{(0.02)\,(\text{nmoles of CMP-Sia})\,(0.075)}$$

DPM = Decays Per Minute
0.60 = Volume (in milliliters) of reaction cocktail
0.02 = ml of Reaction cocktail used to determine total potential nmoles of CMP-Sia = Total nmoles/ml of Cytidine
5'-Monophospho[$_{14}$C]sialic acid in the Reaction cocktail
0.075 = Volume (in milliliters) of CMP-Sia in the reaction cocktail

$$\text{Units / mg Enzyme} = \frac{(\text{DPM Test} - \text{DPM Blank})\,(0.05)\,(5.5)}{(\text{DPM / nmole})\,(15)\,(\text{mg enzyme / RM})}$$

0.05 = Reaction volume
5.5 = Factor to correct for inhibition due to Triton CF-54
15 = Reaction time
RM = Reaction Mix

Unit Definition: *One unit will transfer 1.0 μmole of N-acetylneuraminic acid from CMP-N-acetylneuraminic acid to asialomucin per minute at pH 6.5 at 37°C.*

Final Assay Concentrations

In a 0.05 ml reaction mix, the final concentrations are 100 mM sodium cacodylate, 1% (v/v) Triton CF-54, 0.1% (w/v) Bovine Serum Albumin, 3.2 mM asialomucin, 0.008–0.015 mM cytidine 5'-monophospho[$_{14}$C]sialic acid, and 0.002 unit sialytransferase.

Notes

1. The Cacodylate buffer initially appears very cloudy. This is due to the solubility of the Triton. After approximately 45 minutes at room temperature with constant stirring the solution becomes clear. This solution can be stored at 0–5°C for several weeks.
2. The concentration is based on N-acetylgalactosamine residues, which is approximately 17 mg/ml.
3. Degas the swelled resin by aspiration prior to packing the column. The void volume of the column can be checked by using a Hemoglobin: Vitamin B12 solution at 17 mg/ml Hemoglobin, Bovine and 5 mg/ml Vitamin B12. The Hemoglobin will elute where the Mucin should elute and the Vitamin B12 will elute where any remaining CMP[$_{14}$C]sialic acid should. Fraction volumes may vary and must be verified before use.
4. The result will be an opaque slurry. Remove any large bubbles from the vial by gentle shaking.
5. This factor was determined empirically by kinetic analysis of the pure enzyme.

Enzymatic Assay of SORBITOL DEHYDROGENASE (EC 1.1.1.14)

Principle

$$\text{D-Fructose} + \beta\text{-NADH} \xrightarrow{\text{Sorbitol Dehydrogenase}} \text{D-Sorbitol} + \beta\text{-NAD}$$

Abbreviations Used

β-NADH = β-Nicotinamide Adenine Dinucleotide, Reduced Form
β-NAD = β-Nicotinamide Adenine Dinucleotide, Oxidized Form

Conditions: T = 25°C, pH 7.6, A340nm, Light path = 1 cm

Method: Continuous Spectrophotometric Rate Determination

Reagents

A. 100 mM Triethanolamine Buffer, pH 7.6 at 25°C (Prepare 100 ml in deionized water using Triethanolamine, Hydrochloride. Adjust to pH 7.6 at 25°C with 5 M NaOH).
B. 1.1 M D-Fructose Solution (Fructose) (Prepare 5 ml in deionized water using D(-)Fructose).
C. 12.8 mM β-Nicotinamide Adenine Dinucleotide, Reduced Form, Solution (β-NADH) (Dissolve the contents of one 10 mg vial of β-Nicotinamide Adenine Dinucleotide, Reduced Form, Preweighed Vial, Disodium Salt in the appropriate volume of deionized water. *Prepare Fresh*).
D. 1.0% (w/v) Bovine Serum Albumin (BSA) (Prepare 10 ml in deionized water using Albumin, Bovine).
E. Sorbitol Dehydrogenase Enzyme Solution (Prepare a solution containing 70–150 units/ml of Sorbitol Dehydrogenase in cold deionized water. Store at 4°C for 1 hour. Immediately before use, dilute to a final concentration of 0.55–0.75 unit/ml with cold Reagent D).

Procedure

Pipette (in milliliters) the following reagents into suitable cuvettes:

	Test	Blank
Reagent A (Buffer)	2.35	2.35
Reagent B (Fructose)	0.50	0.50
Reagent C (β-NADH)	0.05	0.05

Mix by inversion and equilibrate to 25°C. Monitor the A340nm until constant, using a suitably thermostatted spectrophotometer. Then add:

Reagent E (Enzyme Solution)	0.10	—
Reagent D (BSA)	—	0.10

Immediately mix by inversion and record the decrease in A340nm for approximately 5 minutes. Obtain the ΔA340nm/minute using the maximum linear rate for both the Test and Blank.

Calculations

$$\text{Units / ml enzyme} = \frac{(\Delta A340nm / \min \text{ Test} - \Delta A340nm / \min \text{ Blank})\ (3)\ (df)}{(6.22)\ (0.1)}$$

3 = Total volume (in milliliters) of assay
df = Dilution factor
6.22 = Millimolar extinction coefficient of β-NADH at 340nm
0.1 = Volume (in milliliter) of enzyme

$$\text{Units / mg solid} = \frac{\text{units / ml enzyme}}{\text{mg solid / ml enzyme}}$$

$$\text{Units / mg protein} = \frac{\text{units / ml enzyme}}{\text{mg protein / ml enzyme}}$$

Unit Definition: *One unit will convert 1.0 µmole of D-fructose to D-sorbitol per minute at pH 7.6 at 25°C.*

Final Assay Concentrations

In a 3.00 ml reaction mix, the final concentrations are 78 mM triethanolamine, 183 mM D-fructose, 0.2 mM β-nicotinamide adenine dinucleotide, reduced form, 0.033% (w/v) Bovine Serum Albumin and 0.055–0.075 unit sorbitol dehydrogenase.

Enzymatic Assay of SPHINGOMYELINASE from Bacterial Sources (EC 3.1.4.12)

Principle

$$\text{TNPAL - Sphingomyelin} + H_2O \xrightarrow{\text{Sphingomyelinase}} \text{Choline Phosphate} + \text{TNPAL - N - Achlsphingosine}$$

Abbreviations Used

TNPAL-Sphingomyelin = Trinitrophenylaminolauroyl-Sphingomyelin
TNPAL-N-Acylsphingosine = Trinitrophenylaminolauroyl-N-Acylsphingosine

Conditions: T = 37°C, pH = 7.4, A330nm, Light path = 1 cm

Method: Colorimetric

Reagents

A. 500 mM Tris HCl Buffer with 20 mM Magnesium Chloride, pH 7.4 at 37°C (Prepare 100 ml in deionized water using Trizma Base and Magnesium Chloride, Hexahydrate. Adjust to pH 7.4 at 37°C with 1 M HCl).
B. Trinitrophenylaminolauroyl-Sphingomyelin Substrate Solution (TNPAL-Sphingomyelin) (Use Trinitrophenylaminolauroyl-Sphingomyelin. Immediately before use, pipette 0.045 ml (0.05 mg) of T-1014 into a one-dram glass vial on ice. Evaporate the solvent from the material with nitrogen gas. Turn the nitrogen on low, and equip the hose with a 200 µl pipet tip. Insert the pipet tip into the vial about half-way. Hold the vial at a 45° angle and rotate slowly until the substrate is a yellow film on the glass vial. Immediately return the vial to the ice and cap the vial).
C. 0.2% (v/v) Triton X-100 Solution (Triton X-100) (Prepare 10 ml in deionized water using Triton X-100).
D. Isopropanol: Heptane: H_2SO_4 Solution (40:10:1) (ISOPRO) (Prepare 25 ml using Isopropanol, Anhydrous, n-Heptane and Sulfuric Acid).
E. Heptane (Heptane).
F. Sphingomyelinase Enzyme Solution (Immediately before use, prepare a solution containing 1.0 unit/ml in cold Reagent A).

Procedure

Pipette (in milliliters) the following reagents into a one-dram vial of Reagent B (TNPAL Sphingomyelin):

	Test	Blank
Reagent A (Buffer)	0.18	0.18
Reagent C (Triton X-100)	0.20	0.20

Mix by swirling to remove any yellow substrate affixed to the glass. The solution should be clear yellow. Equilibrate at 37°C for exactly 2 minutes. Then add:

Reagent F (Enzyme Solution)	0.01	—
Reagent A (Buffer)	—	0.01

Mix by swirling and incubate for exactly 2 minutes. Then add:

Reagent D (ISOPRO)	1.50	1.50

Put caps on vials and place the vials in ice. Uncap vials right before addition. Then add:

	Test	Blank
Deionized Water	0.80	0.80
Reagent E (Heptane)	1.80	1.80

Replace the caps and shake the vials vigorously for 5 minutes by hand. Allow the vials to sit at 25°C for 5 minutes. Remove the upper layer and transfer to suitable cuvettes. Record the A330nm for both the Test and Blank using a suitable spectrophotometer.

Calculations

$$\text{Units / ml enzyme} = \frac{(\text{A330nm Test} - \text{A330nm Blank})\,(2.4)}{(14.0)\,(2)\,(0.01)}$$

14.0 = Millimolar extinction coefficient of TNPAL-N-Acylsphingosine at 330nm
2 = Time of assay in minutes (Unit Definition)
0.01 = Volume (in milliliters) of enzyme used
2.4 = Volume (in milliliters) of organic phase used in assay

$$\text{Units / mg solid} = \frac{\text{units / ml enzyme}}{\text{mg solid / ml enzyme}}$$

$$\text{Units / mg protein} = \frac{\text{units / ml enzyme}}{\text{mg protein / ml enzyme}}$$

Unit Definition: *One unit will hydrolyze 1.0 μmole of TNPAL-sphingomyelin per minute at pH 7.4 at 37°C.*

Final Assay Concentrations

In a 0.39 ml reaction mix, the final concentrations are 250 mM Tris, 10 mM $MgCl_2$, 12% (w/v) TNPAL-sphingomyelin, 1.0% (v/v) Triton X-100, and 0.01 unit sphingomyelinase.

Notes

1. This assay is not to be used for Sphingomyelinase, from Human Placenta.
2. The substrate is unstable and the solution should be stored capped and on ice until just prior to running the assay.

Enzymatic Assay of SPHINGOMYELINASE (EC 3.1.4.12)

Principle

$$[\text{N-Methyl-}_{14}\text{C}]\ \text{Sphingomyelin} \xrightarrow{\text{SM}} {}_{14}\text{C-Choline Phosphate} + \text{N-Acylsphingosine}$$

Abbreviation Used

SM = Sphingomyelinase

Conditions: T = 37°C, pH = 5.0

Method: Radiolabelled Stop Reaction

Reagents

A. 100 mM Potassium Acetate Buffer with 0.5% (w/v) Triton X-100 and 0.5% (w/v) Human Serum Albumin, pH 5.0 at 37°C (THS) (Prepare 100 ml in deionized water using Potassium Acetate, Triton X-100 and Albumin, Human. Adjust to pH 5.0 at 37°C with 1 M HCl).

B. 6.1 N Trichloroacetic Acid Solution (TCA) (Use Trichloroacetic Acid, 6.1 N Solution, approximately 100% w/v).

C. 1.0% (w/v) Human Serum Albumin Solution (HSA) (Prepare 100 ml in deionized water using Albumin, Human).

D. 0.08 mM Sphingomyelin Solution (Sphingomyelin) (Prepare by adding 1 mg of Sphingomyelin and 30 µl of $_{14}$C-Sphingomyelin (50–62 Ci/mmol, 25mCi/ml) to 15.6 ml of Reagent A. Sonicate to ensure complete dissolution).

E. 0.25% (w/v) Triton X-100 with 0.1% (w/v) Human Serum Albumin (Enz Dil) (Prepare 25 ml in deionized water using Triton X-100 and Reagent C).

F. Sphingomyelinase Enzyme Solution (Immediately before use, prepare a solution containing 5–10 units/ml of Sphingomyelinase in Reagent E).

G. Scintillation Cocktail.

H. 1:1 Methylethyl Cellosolve (Prepare by adding equal volumes of Ethylene Glycol Monoethyl Ether and Ethylene Glycol Monomethyl Ether).

Procedure

Pipette (in milliliters) the following reagents into suitable containers:

	Test	Blank
Reagent D (Sphingomyelin)	0.150	0.150

Equilibrate to 37°C. Then add:

Reagent F (Enzyme Solution)	0.05	—
Reagent E (Enz Dil)	—	0.05

Mix by swirling and incubate at 37°C for exactly 20 minutes. Then add:

Reagent C (HSA)	1.00	1.00
Reagent B (TCA)	0.10	0.10

Mix by swirling and centrifuge for 5 minutes. Remove 0.10 ml of the supernatant from both the Test and Blank and place each into a scintillation vial containing 2 ml of Reagent H (Methylethyl Cellosolve) followed by the addition of 5 ml of Reagent G (Scintillation Cocktail). Count the radioactivity in a suitable scintillation counter. Prepare potential counts by pipetting 0.10 ml of Reagent D (Sphingomyelin) into 2.0 ml of Reagent H (Methylethyl Cellosolve) followed by the addition of 5 ml of Reagent G (Scintillation Cocktail). Count the radioactivity in a suitable scintillation counter. Calculate the DPM/nmole sphingomyelin.

Calculations

$$\text{DPM / nmole} = \frac{\text{DPM of Potential}}{\text{nmoles of Sphingomyelin in potential}}$$

$$\text{Units / ml enzyme} = \frac{(\text{DPM Test} - \text{DPM Blank})\,(3)\,(1.3)\,(\text{df})}{(\text{DPM / nmole of Sphingomyelin})\,(0.1)\,(0.05)}$$

DPM = Disintegrations per minute
3 = Conversion factor from 20 minutes to 1 hour as per the Unit Definition
1.3 = Total volume (in milliliters) of stopped reaction
df = Dilution factor
0.1 = Volume (in milliliter) of supernatant which is added to the scintillation fluid
0.05 = Volume (in milliliter) of enzyme used

$$\text{Units / mg protein} = \frac{\text{units / ml enzyme}}{\text{mg protein / ml enzyme}}$$

Unit Definition: *One unit will hydrolyze 1.0 nanomole of sphingomyelin to N-acylsphingosine and choline phosphate per hour at pH 5.0 at 37°C.*

Final Assay Concentrations

In a 0.20 ml reaction mix, the final concentrations are 75 mM potassium acetate, 0.4% (w/v) Triton X-100, 0.4% (w/v) human serum albumin, 0.06 mM sphingomyelin, and 0.25–0.50 unit sphingomyelinase.

Note

1. Not more than 20% of the available counts should be converted during the reaction to ensure linearity.

Enzymatic Assay of STREPTOKINASE

Principle

$$\text{Fibrinogen} \xrightarrow{\text{Thrombin}} \text{Fibrin}$$

$$\text{Fibrin (insoluble) Plasmin Complex} \xrightarrow{\text{Streptokinase}} \text{Fibrin Fragments (soluble)}$$

Conditions: T = 37°C, pH 7.5

Method: Visual Determination of Clot Liquification

Reagents

A. 197 mM Borate/Borax Buffer with 0.9% (w/v) Sodium Chloride, pH 7.5 at 37°C (Borate Buffer) (Prepare 100 ml in deionized water using 1.11 gram of Boric Acid, 0.7 gram of Borax, and 0.9 g of Sodium Chloride. Adjust to pH 7.5 at 37°C with either 1 M HCl or 1 M NaOH).
B. 9.0% (w/v) Fibrinogen Solution (Fibrinogen) (Prepare 10 ml in Reagent A using Fibrinogen. To solubilize place in a 37°C water bath for 1–2 hours).
C. Plasminogen Solution (Plasminogen) (Prepare approximately 10.0 ml of a solution containing 1.7 units/ml in Reagent A using Plasminogen. Store on ice).
D. Thrombin Solution (Prepare 2.5 ml of a solution containing 100 units/ml in Reagent A using Thrombin, Store on ice).
E. 0.50% (w/v) Gelatin Solution with 100 mM Potassium Phosphate, 0.1% (w/v) Sodium Chloride, and 0.01% (w/v) Thimerosal (Gelatin Diluent) (Prepare 100 ml in deionized water using Gelatin, Potassium Phosphate, Monobasic, Anhydrous, Sodium Chloride, and Thimerosal).

F. Streptokinase Standard (Std) (Immediately before use, prepare a solution containing approximately 175 units/ml Streptokinase in Reagent E. World Health Organization (WHO) Standards for Streptokinase or a WHO traceable lot should be used as the standard).

G. Streptokinase Solution (Immediately before use, prepare a solution containing approximately 175 units/ml Streptokinase in Reagent E).

H. 4 mm Glass Beads (Glass Bead) (Use 4 mm Glass Bead).

Procedure

A. Clot Formation Check: Pipette (in milliliters) the following reagents into a suitable container:

	Test
Reagent A (Borate Buffer)	0.250
Reagent E (Gelatin Diluent)	0.150
Reagent B (Fibrinogen)	0.400
Reagent C (Plasminogen)	0.250

Mix by swirling, then equilibrate at 37°C for 3 minutes. Then add:

Reagent D (Thrombin)	0.100

Mix by swirling and incubate at 37°C for approximately 2–3 minutes to allow for clot formation. Gently add a glass bead to the top of the test reaction mixture. If the glass bead remains on top of the reaction test mixture, the system is working properly. Proceed to the Clot Lysing Check. If the bead falls through the clot increase the concentration of Fibrinogen in Reagent B and repeat Clot Formation Check.

B. Clot Lysing Check: Pipette (in milliliters) the following reagents into a suitable container:

	Test
Reagent A (Borate Buffer)	0.250
Reagent E (Gelatin Diluent)	0.100
Reagent B (Fibrinogen)	0.400
Reagent C (Plasminogen)	0.250

Mix by swirling, equilibrate at 37°C for 3 minutes. Then add:

Reagent F (STD)	0.050

Mix by swirling, equilibrate at 37°C for 1 minute. Then add:

Reagent D (Thrombin)	0.100

Record initial time (T_I) in minutes, and mix by swirling and incubate at 37°C for approximately 2–3 minutes to allow for clot formation. Then gently add a glass bead to the top of the test reaction mixture. Record time as final time (T_F) when the glass bead touches the bottom of the tube. If the change in time is between 5 to 10 minutes, then proceed with the standard curve. The change in time is the difference between the recorded time upon addition of thrombin (T_I: Initial Time) and the time it takes for the glass bead to touch the bottom of the container (T_F: Final Time). If the change in time is not between 5 to 10 minutes, then prepare an appropriate dilution of the standard and repeat Clot Lysing Check.

C: Standard Curve: Pipette (in milliliters) the following reagents into a suitable container:

	Std 1	Std 2	Std 3	Std 4	Std 5	Std 6
Reagent A (Borate Buffer)	0.250	0.250	0.250	0.250	0.250	0.250
Reagent E (Gelatin Diluent)	0.145	0.140	0.138	0.135	0.130	0.120
Reagent B (Fibrinogen)	0.400	0.400	0.400	0.400	0.400	0.400
Reagent C (Plasminogen)	0.250	0.250	0.250	0.250	0.250	0.250

Mix by swirling, equilibrate at 37°C for 3 minutes. Then add:

Reagent F (STD)	0.005	0.010	0.012	0.015	0.020	0.030

Mix by swirling, equilibrate at 37°C for 1 minute. Then add:

Reagent D (Thrombin)	0.100	0.100	0.100	0.100	0.100	0.100

Mix by swirling and incubate at 37°C for 2–3 minutes and record the initial time (T_i) in minute after addition of Reagent D. Gently add the glass bead to each container starting with Std 1. Record time as the final time (T_F) when the glass bead touches the bottom of the container. Pipette (in milliliters) the following reagents into a suitable container:

	Std 1	Std 2	Std 3	Std 4
Reagent A (Borate Buffer)	0.250	0.250	0.250	0.250
Reagent E (Geltain Diluent)	0.140	0.138	0.135	0.130
Reagent B (Fibrinogen)	0.400	0.400	0.400	0.400
Reagent C (Plasminogen)	0.250	0.250	0.250	0.250

Equilibrate at 37°C for 3 minutes. Then add:

Reagent G (Streptokinase Soln)	0.010	0.012	0.015	0.020

Equilibrate at 37°C for 1 minute. Then add:

Reagent D (Thrombin)	0.100	0.100	0.100	0.100

Mix by swirling and incubate at 37°C for 2–3 minutes and record the initial time (T_i) in minutes after addition of Reagent F. Gently add the glass bead to each container starting with Sample 1. Record time as the final time (T_F) when the glass bead touches the bottom of the container.

Calculations

Standard Curve: T (STD) = T_f(STD)–T_i(STD)
Plot the Log (T) of the Standards vs the Log of the units of Streptokinase per reaction mixture, [Log(Units/ml)].
Sample Determination: T (Sample) = T_f(Sample)–T_i(Sample)
Calculate the Log (T) of the sample and use the standard curve to determine the Log of the units/ml of Streptokinase in the reaction mixture [Log(units/ml)].
Units/ml = {Antilog[Log(units/ml)Streptokinase from Curve]}(DF) Enzyme
df = Dilution factor
units units/ml enzyme = mg solid mg solid/ml enzyme

Unit Definition: *One unit will liquify a standard clot of fibrinogen, plasminogen and thrombin at pH 7.5 at 37°C in 10 minutes.*

Final Assay Concentrations

In a 1.15 ml reaction mix, the final concentrations are 171 mM borate, 0.065% (w/v) gelatin, 3.1% (w/v) fibrinogen, 0.43 unit plasminogen, 0.88–5.3 units streptokinase, 10 units thrombin, 0.79% (w/v) sodium chloride, 0.001% (w/v) thimerosal, and 13 mM potassium phosphate.

Notes

1. The concentration of Fibrinogen required will vary from lot to lot. This solution should be yellow in color. It may require filtering before use.
2. Due to the variances in the concentration of fibrinogen used in the assay, it may be required to add additional higher or lower standards to the standard curve in order to obtain reasonable clot lysing times. Clot lysing times for the standard curve are generally between 7 and 45 minutes.
3. Due to variances in the standard curve, altering the amount of sample aliquot used in the assay may be required.

Enzymatic Assay of STREPTOLYSIN O

Principle

$$\text{Intact Red Blood Cells} \xrightarrow{\text{Streptolysin O}} \text{Lysed Red Blood Cells}$$

Conditions: T = 37°C, pH = 7.4

Method: Visual determination of hemolytic activity

Reagents

A. 10 mM Sodium Phosphate Buffer with 154 mM Sodium Chloride, pH 7.4 at 37°C (PBS-A) (Prepare 50 mL in deionized water using Sodium Phosphate, Monobasic, Monohydrate, Sodium Phosphate, Dibasic, Anhydrous and Sodium Chloride).
B. 10 mM Sodium Phosphate Buffer with 103 mM Sodium Chloride, pH 7.4 at 37°C (PBS-B) (Prepare 50 mL in deionized water using Sodium Phosphate, Monobasic, Monohydrate, Sodium Phosphate, Dibasic and Sodium Chloride).
C. 10 mM Sodium Phosphate with 103 mM Sodium Chloride, 1 mg/mL Bovine Serum Albumin and 100 mM Dithioerythritol (Buffer I) (Prepare 50 mL in Reagent B using Bovine Serum Albumin and Dithioerythritol).

D. 2% (v/v) Suspension of human red blood cells. (2% RBCs) (Prepare 5.0 mL by centrifuging human whole blood at low speed for 5 minutes. Decant the supernatant and gently resuspend the pellet in Reagent A in a ratio of 1:10. Centrifuge at low speed for 5 minutes. Repeat resuspension and centrifugation two more times. Dilute red blood cells to 2% (v/v) in Reagent B).

E. Streptolysin O Solution (Immediately before use prepare a solution containing approximately 100,000 units per mL Streptolysin O in cold deionized water).

Procedure

Pipette (in milliliters) the following reagents into approximately 40 wells of a suitable microtiter plate.

	Test	Blank
Reagent C (Buffer I)	0.10	0.10

Add 0.1 ml Reagent E to the first well. Incubate at 25°C for 15 minutes. Make serial dilutions of test solutions by consecutive transfers of 0.1 mL. Leave one well containing Reagent C only as a blank. Then add:

Reagent D (2 % RBCs)	0.05	0.05

Incubate at 37°C for 30 minutes, then refrigerate for 30 minutes at 4°C. Record hemolysis for both test (serial dilutions) and blank.

Calculation

$$\text{Units / mg enzyme} = \frac{1}{\text{Mg Streptolysin O in well giving 50\% lysis}}$$

Unit Definition: *One unit will cause 50% lysis of 2% red blood cell suspensions in phosphate buffered saline, pH 7.4, after incubation at 37°C for 30 minutes.*

Final Assay Concentrations

In a 0.25 mL reaction mix, the final concentrations are 10 mM Sodium Phosphate, 103 mM Sodium Chloride, 0.4% (v/v) Red Blood Cells, 0.4 mg/mL Bovine Serum Albumin, and 40 mM Dithioerythritol.

Enzymatic Assay of SUCCINIC SEMIALDEHYDE DEHYDROGENASE (EC 1.2.1.16)

Principle

$$\text{Succinic Semialdehyde} + \text{NADP} + H_2O \xrightarrow{\text{SSADH}} \text{Succinate} + \text{NADPH}$$

Abbreviations Used

NADP = Nicotinamide Adenine Dinucleotide, Oxidized Form
SSADH = Succinic Semialdehyde Dehydrogenase
NADPH = Nicotinamide Adenine Dinucleotide Phosphate, Reduced Form

Conditions: T = 25°C, pH 8.6, A340nm, Light path = 1 cm

Method: Continuous Spectrophotometric Rate Determination

Reagents

A. 100 mM Potassium Pyrophosphate Buffer, pH 8.6 at 25°C (Prepare 100 ml in deionized water using Tetrapotassium Pyrophosphate, Anhydrous. Adjust to pH 8.6 at 25°C with 1 M H_3PO_4).
B. 100 mM 2-Mercaptoethanol Solution (2-ME) (Prepare 10 ml in deionized water using 2-Mercaptoethanol).
C. 25 mM β-Nicotinamide Adenine Dinucleotide Phosphate Solution (NADP) (Prepare 10 ml in deionized water using β-Nicotinamide Adenine Dinucleotide Phosphate, Sodium Salt or dissolve the contents of one 10 mg vial of β-Nicotinamide Adenine Dinucleotide Phosphate, Sodium Salt in the appropriate volume of deionized water).
D. 50 mM Succinic Semialdehyde Substrate Solution (Substrate) (Prepare 1 ml in Reagent A using Succinic Semialdehyde).
E. 75 mM Potassium Phosphate Buffer, with 25% (v/v) Glycerol, pH 7.2 at 25°C (Enzyme Diluent) (Prepare 25 ml in deionized water using Potassium Phosphate, Monobasic, Anhydrous and Glycerol. Adjust to pH 7.2 at 25°C using 1 M KOH).
F. Succinic Semialdehyde Dehydrogenase Enzyme Solution (Immediately before use, prepare a solution containing 0.25–0.5 unit/ml of Succinic Semialdehyde Dehydrogenase in cold Reagent E).

Procedure

Prepare a reaction cocktail by pipetting (in milliliters) the following reagents into a suitable container:

Reagent A (Buffer)	23.10
Reagent B (2-ME)	0.90

Mix by swirling and equilibrate to 25°C. Pipette (in milliliters) the following reagents into suitable cuvettes:

	Test	Blank
Reaction Cocktail	2.40	2.40
Reagent C (NADP)	0.15	0.15
Reagent D (Substrate)	0.30	0.30

Mix by inversion and equilibrate to 25°C. Monitor the A340nm until constant, using a suitably thermostatted spectrophotometer. Then add:

	Test	Blank
Reagent F (Enzyme Solution)	0.10	—
Reagent E (Enzyme Diluent)	—	0.10

Immediately mix by inversion and record the increase in A340nm for approximately 5 minutes. Obtain the ΔA340nm/minute using the maximum linear rate for both the Test and Blank.

Calculations

$$\text{Units / ml enzyme} = \frac{(\Delta A340nm / \min \text{ Test} - \Delta A340nm / \min \text{ Blank})(3)(df)}{(6.22)(0.1)}$$

3 = Total volume (in milliliters) of assay
df = Dilution factor
6.22 = Millimolar extinction coefficient of β-NADP at 340nm
0.1 = Volume (in milliliters) of enzyme used

$$\text{Units / mg solid} = \frac{\text{units / ml enzyme}}{\text{mg solid / ml enzyme}}$$

$$\text{Units / mg protein} = \frac{\text{units / ml enzyme}}{\text{mg protein / ml enzyme}}$$

Unit Definition: *One unit will convert 1.0 µmole of succinic semialdehyde to succinate per minute with a stoichiometric reduction of 1.0 µmole of NADP$^+$ at pH 8.6 at 25°C.*

Final Assay Concentrations

In a 3.00 ml reaction mix, the final concentrations are 87 mM potassium pyrophosphate, 3 mM 2-mercaptoethanol, 1.3 mM β-nicotinamide adenine dinucleotide phosphate, 5.0 mM succinic semialdehyde, 0.83% (v/v) glycerol, 2.5 mM potassium phosphate, and 0.025–0.05 unit succinic semialdehyde dehydrogenase.

Enzymatic Assay of SUCCINIC THIOKINASE (EC 6.2.1.4)

Principle

$$\text{GTP} + \text{Succinate} + \text{CoA} \rightarrow \text{GDP} + \text{Pi} + \text{Succinyl-CoA}$$

Abbreviations Used

GTP = Guanosine 5'-Triphosphate
GDP = Guanosine 5'-Diphosphate
CoA = Coenzyme A
Pi = Orthophosphate

Conditions: T = 30°C, pH = 7.4, A235nm, Light path = 1 cm

Method: Continuous Spectrophotometric Rate Determination

Reagents

A. 50 mM Tris-Succinate Buffer, pH 7.4 at 30°C (Prepare 50 ml in deionized water using Trizma Succinate. Adjust to pH 7.4 at 30°C with 2 M NaOH).
B. 10 mM Guanosine 5'-Triphosphate Solution (GTP) (Prepare 1 ml in deionized water using Guanosine 5'-Triphosphate, Sodium Salt. *Prepare Fresh*).
C. 3 mM Coenzyme A Solution (CoA) (Prepare 1 ml in deionized water using Coenzyme A, Sodium Salt. *Prepare Fresh*).
D. 300 mM Magnesium Chloride Solution ($MgCl_2$) (Prepare 1 ml in deionized water using Magnesium Chloride, Hexahydrate).

E. Succinic Thiokinase Solution (Immediately before use, prepare a solution containing 1.25–2.5 units/ml of Succinic Thiokinase in cold deionized water).

Procedure

Pipette (in milliliters) the following reagents into suitable cuvettes:

	Test	Blank
Reagent A (Buffer)	2.825	2.825
Reagent B (GTP)	0.025	0.025
Reagent C (CoA)	0.05	0.05
Reagent D ($MgCl_2$)	0.10	0.10

Mix by inversion and equilibrate to 30°C. Monitor the A235nm until constant, using a suitably thermostatted spectrophotometer. Then add:

	Test	Blank
Deionized Water	—	0.010
Reagent E (Enzyme Solution)	0.010	—

Immediately mix by inversion and record the increase in A235nm for approximately 5 minutes. Obtain the ΔA235nm/minute using the maximum linear rate for both the Test and Blank.

Calculations

$$\text{Units / mg enzyme} = \frac{(\Delta A235nm / min\ Test - \Delta A235nm / min\ Blank)\ (df)}{(4.0)\ (mg\ enzyme / ml\ RM)}$$

df = Dilution factor
4.0 = Change in the millimolar extinction coefficient of Succinyl-CoA to CoA at 235nm
RM = Reaction Mix

Unit Definition: *One unit will convert 1.0 µmole of succinate to succinyl-CoA per minute at pH 7.4 at 30°C in the presence of coenzyme A and GTP.*

Final Assay Concentrations

In a 3.01 ml reaction mix, the final concentrations are 47 mM tris succinate, 0.08 mM GTP, 0.05 mM CoA, 10 mM $MgCl_2$ and 0.0125–0.025 units succinic thiokinase.

Enzymatic Assay of SUCCINYL COENZYME A TRANSFERASE (EC 2.8.3.5)

Principle

$$\text{Acetoacetyl CoA} + \text{Succinate} \xrightarrow{\text{SCT}} \text{Succinyl CoA} + \text{Acetoacetate}$$

Abbreviations Used

SCT = Succinyl Coenzyme A Transferase
Acetoacetyl CoA = Acetoacetyl Coenzyme A
Succinyl CoA = Succinyl Coenzyme A

Conditions: T = 25°C, pH = 8.1, A310nm, Light path = 1 cm

Method: Continuous Spectrophotometric Rate Determination

Reagents

A. 67 mM Tris H_2SO_4 Buffer with 10 mM Succinate and 5.0 mM Magnesium Sulfate, pH 8.1 at 25°C (Succ) (Prepare 10 ml in deionized water using Trizma Base, Succinic Acid, Disodium Salt, Hexahydrate and Magnesium Sulfate, Heptahydrate. Adjust to pH 8.1 at 25°C with 1 M H_2SO_4).

B. 3.0 mM Acetoacetyl CoA Solution (Acet CoA) (Prepare 0.1 ml in deionized water using Acetoacetyl Coenzyme A, Sodium Salt).

C. 67 mM Tris H_2SO_4 Buffer, pH 8.1 at 25°C (Enz Dil) (Prepare 25 ml in deionized water using Trizma Base. Adjust to pH 8.1 at 25°C with 1 M H_2SO_4).

D. Succinyl Coenzyme A Transferase Enzyme Solution (Immediately before use, prepare a solution containing 0.04–0.1 unit/ml of Succinyl Coenzyme A Transferase in cold Reagent C).

Procedure

Pipette (in milliliters) the following reagents into suitable cuvettes:

	Test	Blank
Reagent A (Succ)	0.94	0.94
Reagent B (Acet CoA)	0.01	0.01

Mix by inversion and equilibrate to 25°C. Monitor the A310nm until constant, using a suitably thermostatted spectrophotometer. Then add:

	Test	Blank
Reagent D (Enzyme Solution)	0.05	—
Reagent C (Enz Dil)	—	0.05

Immediately mix by inversion and record the decrease in A310nm for approximately 5 minutes. Obtain the ΔA310nm/minute using the maximum linear rate for both the Test and Blank.

Calculations

$$\text{Units / ml enzyme} = \frac{(\Delta A310nm / \min \text{ Test} - \Delta A310nm / \min \text{ Blank})\,(1)\,(df)}{(9.25)\,(0.05)}$$

1 = Total volume (in milliliter) of assay
df = Dilution factor
9.25 = Millimolar extinction coefficient of acetoacety coenzyme A at 310nm
0.05 = Volume (in milliliter) of enzyme used

$$\text{Units / mg protein} = \frac{\text{units / ml enzyme}}{\text{mg protein / ml enzyme}}$$

Unit Definition: *One unit will form 1.0 μmole of succinyl CoA from acetoacetyl CoA and succinic acid per minute at pH 8.1 at 25°C.*

Final Assay Concentrations

In a 1.00 ml reaction mix, the final concentrations are 66 mM Tris, 9.4 mM succinic acid, 4.7 mM magnesium sulfate, 0.03 mM acetoacetyl coenzyme A, and 0.002–0.005 unit succinyl coenzyme A transferase.

Enzymatic Assay of SUCROSE PHOSPHATE SYNTHETASE (EC 2.4.1.14)

Principle

$$\text{UDPG} + \text{Fructose-6-PO}_4 \xrightarrow{\text{SPS}} \text{UDP} + \text{Sucrose-6-PO}_4$$

$$\text{UDP} + \text{PEP} \xrightarrow{\text{PK}} \text{Pyruvate} + \text{UTP}$$

$$\text{Pyruvate} + \beta\text{-NADH} \xrightarrow{\text{LDH}} \text{Lactate} + \beta\text{-NAD}$$

Abbreviations Used

UDPG = Uridine 5'-Diphosphoglucose
UDP = Uridine 5'-Diphosphate
SPS = Sucrose Phosphate Synthetase
PEP = Phospho(enol) pyruvate
PK = Pyruvate Kinase
UTP = Uridine 5'-Triphosphate
LDH = Lactic Dehydrogenase
β-NADH = β-Nicotinamide Adenine Dinucleotide, Reduced Form
β-NAD = β-Nicotinamide Adenine Dinucleotide, Oxidized Form

Conditions: T = 37°C, pH = 7.5, A340nm, Light path = 1 cm

Method: Spectrophotometric Stop Rate Determination

Reagents

A. 10 mM HEPES NaOH Buffer with 10 mM Magnesium Chloride and 0.004% (w/v) Albumin, pH 7.5 at 37°C (Prepare 100 ml in deionized water using HEPES, Free Acid, Magnesium Chloride, Hexahydrate and Albumin Bovine Serum. Adjust to pH 7.5 at 37°C with 1 M NaOH).
B. 139 mM Uridine 5'-Diphosphoglucose Solution (UDPG) (Prepare 1 ml in deionized water using Uridine 5'-Diphosphoglucose, Disodium Salt).
C. 132 mM D-Fructose 6-Phosphate Solution (Fructose-6-PO_4) (Prepare 1 ml in deionized water using D-Fructose 6-Phosphate, Disodium Salt).
D. PK/LDH Enzyme Suspension .
E. 200 mM Tris HCl Buffer, pH 7.5 at 37°C (Prepare 100 ml in deionized water using Trizma Base. Adjust to pH 7.5 at 37°C with 1 M HCl).
F. 1000 mM Potassium Chloride Solution (KCl) (Prepare 10 ml in deionized water using Potassium Chloride).
G. 60 mM Magnesium Sulfate Solution ($MgSO_4$) (Prepare 10 ml in deionized water using Magnesium Sulfate, Heptahydrate).
H. 40 mM Phospho(enol) pyruvate Solution (PEP) (Prepare 5 ml in deionized water using Phospho(enol) pyruvate, Monopotassium Salt).
I. 0.14 mM β-Nicotinamide Adenine Dinucleotide, Reduced Form Solution (β-NADH) (Prepare the reaction cocktail in a 1 mg vial of β-Nicotinamide Adenine Dinucleotide, Reduced Form. The total volume will be 10 ml).
J. 100 mM Ethylenediaminetetraacetic Acid Solution (EDTA) (Prepare 10 ml in deionized water using Ethylenediaminetetraacetic Acid, Tetrasodium Salt, Hydrate).

K. Sucrose Phosphate Synthetase Enzyme Solution (SPS) (Immediately before use, prepare a solution containing 1–2 units/ml of Sucrose Phosphate Synthetase in cold Reagent A).

Procedure

Step 1: Prepare a reaction cocktail by pipetting (in milliliters) the following reagents into Reagent I (β-NADH):

Reagent E (Tris HCl Buffer)	3.40
Reagent F (KCl)	0.50
Reagent G ($MgSO_4$)	1.00
Reagent H (PEP)	0.15
Reagent J (EDTA)	0.05
Reagent D (PK/LDH)	0.15

Mix by inversion. Pipette (in milliliters) the following reagents into suitable cuvettes:

	Test	Blank
Reagent A (HEPES Buffer)	0.50	0.50
Reagent B (UDPG)	0.10	—
Reagent C (Fructose-6-PO_4)	0.40	0.40

Mix by inversion and equilibrate to 37°C. Then add:

Reagent K (SPS)	1.00	1.00

Immediately mix by inversion and incubate for exactly 30 minutes at 37°C. Stop the reaction by heating both the Test and Blank solutions for 10 minutes at 100°C. Cool with running tap water. Then add:

Reagent B (UDPG)	—	0.10

Step 2: Pipette (in milliliters) the following reagents into suitable cuvettes:

Reaction Cocktail	2.90	2.90

Equilibrate to 37°C. Monitor the A340nm until constant using a suitably thermostatted spectrophotometer and record the initial A340nm for both the Test and Blank. Then add:

Test Solution	0.10	—
Blank Solution	—	0.10

Immediately mix by inversion and record the decrease in A340nm until complete for approximately 5 minutes. Obtain the final A340nm for both the Test and Blank.

Calculations

ΔA340nm Test = A340nm Test Final–A340nm Test Initial

ΔA340nm Blank = A340nm Blank Final–A340nm Blank Initial

$$\text{Units / mg enzyme} = \frac{\text{A340nm Test} - \text{A340nm Blank (3)}}{(6.22)\ (\text{mg enzyme / ml RM})\ (0.1)}$$

6.22 = Millimolar extinction coefficient of β-NADH at 340nm
3 = final volume of Step 2
0.1 = Volume from Step 1 used in Step 2
RM = Reaction Mixture

Unit Definition: *One unit will convert 1.0 µmole each of UDPGlucose and D-fructose 6-phosphate to UDP and sucrose 6-phosphate in 30 minutes at pH 7.5 at 37°C, assayed in a coupled assay system with PK/LDH.*

Final Assay Concentrations

In a 2 ml reaction mix, the final concentrations are 7.5 mM HEPES, 7.5 mM magnesium chloride, 0.003% Bovine Serum Albumin, 7.0 mM UDPG, 26 mM D-fructose 6-phosphate, and 1–2 units sucrose phosphate synthetase.

Notes

1. Contains not less than 700 pyruvate kinase units and 1000 lactic dehydrogenase units per ml.
2. The initial absorbance should be between 0.6–0.7.

Enzymatic Assay of SUCROSE PHOSPHORYLASE (EC 2.4.1.7)

Principle

$$\text{Sucrose} + \text{Pi} \xrightarrow{\text{Sucrose Phosphorylase}} \text{Glucose 1-Phosphate} + \text{Fructose}$$

$$\text{Glucose 1-Phosphate} \xrightarrow{\text{Phosphoglucomutase}} \text{Glucose 6-Phosphate}$$

$$\text{Glucose 6-Phosphate} + \beta\text{-NAD} \xrightarrow{\text{G-6-PDH}} \text{6-PG} + \beta\text{-NADH}$$

Abbreviations Used

Pi = Inorganic Phosphate
β-NAD = β-Nicotinamide Adenine Dinucleotide, Oxidized Form
G-6-PDH = Glucose-6-Phosphate Dehydrogenase
6-PG = 6-Phospho-D-Gluconate
β-NADH = β-Nicotinamide Adenine Dinucleotide, Reduced Form

Conditions: T = 30°C, pH = 7.0, A340nm, Light path = 1 cm

Method: Continuous Spectrophotometric Rate Determination

Reagents

A. 60 mM Imidazole Buffer with 12 mM Magnesium Chloride, pH 7.0 at 30°C (Prepare 200 ml in deionized water using Imidazole, and Magnesium Chloride, Hexahydrate. Adjust to pH 7.0 at 30°C with 1 M HCl).
B. 120 mM Sucrose Solution (Sucrose) (Prepare 100 ml in Reagent A using Sucrose).
C. 5 mM β-Nicotinamide Adenine Dinucleotide, Oxidized Form Solution (β-NAD) Dissolve the contents of one 10 mg vial of Nicotinamide Adenine Dinucleotide, in the appropriate volume of eagent B. *Prepare Fresh*).

D. Phosphoglucomutase/Glucose-6-Phosphate Dehydrogenase1 olution (PGM/G-6-PDH) (Prepare a solution containing 20–25 units/ml of hosphoglucomutase, and 0–25 units/ml of Glucose-6-Phosphate Dehydrogenase, in Reagent B).
E. 2 mM α-D-Glucose 1,6-Diphosphate Solution (G1,6DP) (Prepare 2 ml in Reagent B using α-D-Glucose 1,6-Diphosphate, Potassium Salt, Hydrate).
F. 1.5 M Potassium Phosphate Solution, pH 7.0 at 30°C (Phosphate) (Prepare 10 ml in deionized water using Potassium Phosphate, Monobasic, Anhydrous. Adjust to pH 7.0 at 30°C with 1 M KOH).
G. Sucrose Phosphorylase Enzyme Solution (Immediately before use, prepare a solution containing 1.0–2.0 units/ml of Sucrose Phosphorylase in cold Reagent B).

Procedure

Pipette (in milliliters) the following reagents into suitable cuvettes:

	Test	Blank
Reagent B (Sucrose)	2.60	2.70
Reagent C (β-NAD)	0.10	0.10
Reagent D (PGM/G-6-PDH)	0.05	0.05
Reagent E (G1,6DP)	0.05	0.05
Reagent G (Enzyme Solution)	0.10	—

Mix by inversion and equilibrate to 30°C. Monitor the A340nm until constant, using a suitably thermostatted spectrophotometer. Then add:

Reagent F (Phosphate)	0.10	0.10

Immediately mix by inversion and record the increase in A340nm for approximately 5 minutes. Obtain the A340nm/minute using the maximum linear rate for both the Test and Blank.

Calculations

$$\text{Units / ml enzyme} = \frac{(\text{A340nm / min Test} - \text{A340nm / min Blank})\ (3)\ (\text{df})}{(6.22)\ (0.1)}$$

3 = Total volume (in milliliters) of assay
df = Dilution factor
6.22 = Millimolar extinction coefficient of β-NADH at 340nm
0.1 = Volume (in milliliters) of enzyme used

$$\text{Units / mg solid} = \frac{\text{units / ml enzyme}}{\text{mg solid / ml enzyme}}$$

$$\text{Units / mg protein} = \frac{\text{units / ml enzyme}}{\text{mg protein / ml enzyme}}$$

Unit Definition: *One unit will convert 1.0 μmole each of sucrose and phosphate to glucose 1-phosphate and fructose per minute at pH 7.0 at 30°C.*

Final Assay Concentrations

In a 3.00 ml reaction mix, the final concentrations are 58 mM Imidazole, 12 mM magnesium chloride, 116 mM sucrose, 0.2 mM β-nicotinamide adenine dinucleotide, 2.5 units phosphoglucomutase, 2.5 units glucose-6-phosphate dehydrogenase, 0.10–0.20 unit sucrose phosphorylase.

Notes

1. G-6-PDH is inhibited by ammonium sulfate.
2. The two enzymes listed come as ammonium sulfate suspensions. Add 1000 units each of phosphoglucomutase and glucose-6-phosphate dehydrogenase to a centrifuge tube. Centrifuge and remove most of the supernatant. Dissolve the pellet in 20 ml of Reagent.
3. Glucose 1,6-Diphosphate is added to ensure that the glucose 1-phosphate continues to glucose 6-phosphate. It is required in order for the phosphoglucomutase to be optimally active.
4. Phosphoglucomutase Unit Definition: One unit will convert 1.0 μmole of α-D-glucose 1-phosphate to α-D-glucose 6-phosphate per minute at pH 7.4 at 30°C.
5. Glucose-6-Phosphate Dehydrogenase Unit Definition: One unit will oxidize 1.0 μmole of glucose 6-phosphate To 6-phospho-D-gluconate per minute in the presence of NAD at pH 7.8 at 30°C. Either NAD or NADP may be used as the coenzyme. Under optimal conditions, the activity found with NAD is approximately 1.8 times that found with NADP).

Enzymatic Assay of SUCROSE SYNTHETASE (EC 2.4.1.13)

Principle

$$\text{UDPG} + \text{D-Fructose} \xrightarrow{\text{SS}} \text{UDP} + \text{Sucrose}$$

$$\text{UDP} + \text{PEP} \xrightarrow{\text{PK}} \text{Pyruvate} + \text{UTP}$$

$$\text{Pyruvate} + \beta\text{-NADH} \xrightarrow{\text{LDH}} \text{Lactate} + \beta\text{-NAD}$$

Abbreviations Used

UDPG = Uridine 5'-Diphosphoglucose
SS = Sucrose Synthetase
UDP = Uridine 5'-Diphosphate
PEP = Phospho(enol) pyruvate
PK = Pyruvate Kinase
UTP = Uridine 5'-Triphosphate
β-NADH = β-Nicotinamide Adenine Dinucleotide, Reduced Form
LDH = Lactic Dehydrogenase
β-NAD = β-Nicotinamide Adenine Dinucleotide, Oxidized Form

Conditions: T = 37°C, pH = 7.5, A340nm, Light path = 1 cm

Method: Spectrophotometric Stop Rate Determination

Reagents

A. 10 mM HEPES Buffer with 10 mM Magnesium Chloride and 0.004% (w/v) Bovine Serum Albumin, pH 7.5 at 37°C (Prepare 100 ml in deionized water using HEPES,

Free Acid, Magnesium Chloride, 4.9 M Solution and Albumin Bovine Serum. Adjust to pH 7.5 at 37°C with 1 M NaOH).

B. 139 mM Uridine 5'-Diphosphoglucose Solution (UDPG) (Prepare 1 ml in deionized water using Uridine 5'-Diphosphoglucose, Disodium Salt).

C. 222 mM D-Fructose Solution (Fructose) (Prepare 1 ml in deionized water using D(-)Fructose).

D. 200 mM Tris HCl Buffer, pH 7.5 at 37°C (Tris HCl) (Prepare 25 ml in deionized water using Trizma Hydrochloride. Adjust to pH 7.5 at 37°C with 1 M NaOH).

E. 1 mM Potassium Chloride Solution (KCl) (Prepare 5 ml in deionized water using Potassium Chloride).

F. 60 mM Magnesium Sulfate Solution ($MgSO_4$) (Prepare 10 ml in deionized water using Magnesium Sulfate, Heptahydrate).

G. 40 mM Phospho(enol) pyruvate Solution (PEP) (Prepare 1 ml in deionized water using Phospho(enol) pyruvate, Monopotassium Salt).

H. 100 mM Ethylenediaminetetraacetic Acid Solution (EDTA) (Prepare 1 ml in deionized water using Ethylenediaminetetraacetic Acid, Tetrasodium Salt, Hydrate).

I. β-Nicotinamide Adenine Dinucleotide, Reduced Form (β-NADH) (Use β-Nicotinamide Adenine Dinucleotide, Reduced Form, Disodium Salt).

J. PK/LDH Enzymes Suspension (PK/LDH).

K. Sucrose Synthetase Enzyme Solution (SS) (Immediately before use, prepare a solution containing 1–2 units/ml of Sucrose Synthetase in cold Reagent A).

Procedure

Step 1: Pipette (in milliliters) the following reagents into suitable containers:

	Test	Blank
Reagent A (Buffer)	0.50	0.50
Reagent B (UDPG)	0.10	—
Reagent C (Fructose)	0.40	0.40

Mix by swirling and equilibrate to 37°C. Then add:

Reagent K (SS)	1.00	1.00

Immediately mix by inversion and incubate for exactly 30 minutes at 37°C. Stop the reaction by heating both the Test and Blank Solutions for 10 minutes at 100°C. Cool with running tap water. Then add:

Reagent B (UDPG)	—	0.10

Mix by swirling.

Step 2: Prepare a reaction cocktail by pipetting (in milliliters) the following reagents into a suitable container:

Reagent D (Tris HCl)	20.40
Reagent E (KCl)	3.00
Reagent F ($MgSO_4$)	6.00
Reagent G (PEP)	0.90
Reagent H (EDTA)	0.30
Reagent J (PK/LDH)	0.90
Deionized Water	28.50
Reagent I (β-NADH)	6 mg

Mix by swirling. Pipette (in milliliters) the following reagents into suitable cuvettes:

	Test	Blank
Reaction Cocktail	2.90	2.90

Equilibrate to 37°C. Monitor the A340nm until constant (should be between 0.6 to 0.7) using a suitably thermostatted spectrophotometer and record the initial A340nm for both the Test and Blank. Then add:

Test Solution (from Step 1)	0.10	—
Blank Solution (from Step 1)	—	0.10

Immediately mix by inversion and record the decrease in A340nm until complete (5–10 minutes). Obtain the final A340nm for both the Test and Blank.

Calculations

$$\Delta A340nm\ Test = A340nm\ Test\ Initial - A340nm\ Test\ Final$$
$$\Delta A340nm\ Blank = A340nm\ Blank\ Initial - A340nm\ Blank\ Final$$

$$\text{Units / ml enzyme} = \frac{(\text{A340nm Test} - \text{A340nm Blank})\,(2)\,(3)\,(df)}{(6.22)\,(0.1)}$$

2 = Volume (in milliliters) of assay (Step 1)
3 = Volume (in milliliters) of assay (Step 2)
df = Dilution factor
6.22 = Millimolar extinction coefficient of β-NADH at 340nm
0.1 = Volume (in milliliter) of Test Solution of Step 1 used in Step 2

$$\text{Units / mg protein} = \frac{\text{units / ml enzyme}}{\text{mg protein / ml enzyme}}$$

Unit Definition: *One unit will convert 1.0 µmole each of UDP glucose and D-fructose to UDP and sucrose in 30 minutes at pH 7.5 at 37°C, assayed in a coupled assay system with PK/LDH.*

Final Assay Concentrations

In a 2.00 ml reaction mix, the final concentrations are 7 mM uridine 5'-diphosphoglucose, 44 mM D-fructose, 7.5 mM HEPES, 7.5 mM magnesium chloride, 0.003% (w/v) Bovine Serum Albumin, and 1–2 units sucrose synthetase.

Notes

1. Pyruvate Kinase Unit Definition: One unit will convert 1.0 µmole of phospho(enol) pyruvate to pyruvate per minute at pH 7.6 at 37°C.

2. L-Lactic Dehydrogenase Unit Definition: One unit will reduce 1.0 µmole of pyruvate to L-lactate per minute at pH 7.5 at 37°C.

Enzymatic Assay of SULFATASE (EC 3.1.6.1)

Principle

$$\text{p-Nitrocatechol Sulfate} + H_2O \xrightarrow{\text{Sulfatase}} \text{p-Nitrocatechol} + \text{Sulfate}$$

Conditions: T = 37°C, pH = 5.0, A515nm, Light path = 1 cm

Method: Colorimetric

Reagents

A. 200 mM Sodium Acetate Buffer, pH 5.0 at 37°C (Prepare 50 ml in deionized water using Sodium Acetate, Trihydrate. Adjust to pH 5.0 at 37°C with 5 M HCl).

B. 6.25 mM p-Nitrocatechol Sulfate Solution (PNCS) (Prepare 10 ml in deionized water using p-Nitrocatechol Sulfate, Dipotassium Salt).

C. 1 M Sodium Hydroxide (NaOH) (Prepare 50 ml in deionized water using Sodium Hydroxide, Anhydrous).

D. 0.2% (w/v) Sodium Chloride Solution (Prepare 50 ml in deionized water using Sodium Chloride).

E. Sulfatase Enzyme Solution (Immediately before use, prepare a solution containing 2.5–5.0 units/ml of Sulfatase in cold Reagent D).

Procedure

Pipette (in milliliters) the following reagents into suitable containers:

	Test	Blank
Reagent A (Buffer)	0.50	0.50
Reagent B (PNCS)	0.40	0.40

Mix by swirling and equilibrate to 37°C. Then add:

Reagent E (Enzyme Solution)	0.10	—

Mix by swirling and incubate at 37°C for exactly 30 minutes. Then add:

Reagent C (NaOH)	5.00	5.00
Reagent E (Enzyme Solution)	—	0.10

Immediately mix by swirling. Transfer the solutions to suitable cuvettes and record the A515nm for both the Test and Blank using a suitable spectrophotometer.

Calculations

$$\text{Units / ml enzyme} = \frac{(\text{A515nm Test} - \text{A515nm Blank})\,(2)\,(\text{df})\,(6)}{(12.6)\,(0.1)}$$

2 = Time factor correction (Unit Definition Time = 1 hour)
df = Dilution factor

6 = Total volume (in milliliters) of the assay
12.6 = Millimolar extinction coefficient of 4-nitrocatechol at 515nm.
0.1 = Volume (in milliliter) of enzyme used

$$\text{Units / mg solid} = \frac{\text{units / ml enzyme}}{\text{mg solid / ml enzyme}}$$

Unit Definition: *One unit will hydrolyze 1.0 μmole of p-nitrocatechol sulfate per hour at pH 5.0 at 37°C.*

Final Assay Concentrations

In a 1.00 ml reaction mix, the final concentrations are 100 mM sodium acetate, 2.5 mM p-nitrocatechol, 0.02% (w/v) sodium chloride, and 0.25–0.50 unit of sulfatase.

Note

1. This enzyme assay is not to be used to assay Sulfatase from *Aerobacter aerogenes.*

Enzymatic Assay of SULFATASE (EC 3.1.6.1) from *Aerobacter aerogenes*

Principle

$$\text{p-Nitrophenyl Sulfate} + H_2O \xrightarrow{\text{Sulfatase}} \text{p-Nitrophenol} + \text{Sulfate}$$

Conditions: T = 37°C, pH = 7.1, A420nm, Light path = 1 cm

Method: Continuous Spectrophotometric Rate Determination

Reagents

A. 200 mM Tris Buffer Solution, pH 7.1 at 37°C .
B. 120 mM p-Nitrophenyl Sulfate Solution (Prepare 10 ml in deionized water using p-Nitrophenyl Sulfate, Potassium Salt).
C. Sulfatase Enzyme Solution (Immediately before use, prepare a solution containing 0.5 unit/ml of Sulfatase in cold Reagent A).

Procedure

Pipette (in milliliters) the following reagents into suitable cuvettes:

	Test	Blank
Reagent A (Buffer)	2.70	2.80
Reagent B (p-Nitrophenyl Sulfate)	0.20	0.20

Mix by inversion and equilibrate to 37°C. Monitor the A420nm until constant, using a suitably thermostatted spectrophotometer. Then add:

	Test	Blank
Reagent C (Enzyme Solution)	0.10	—

Immediately mix by inversion and record the increase in the A420nm for approximately 5 minutes. Obtain the ΔA420nm/minute using the maximum linear rate for both the Test and Blank.

Calculations

$$\text{Units / ml enzyme} = \frac{(\Delta A420\text{nm / min Test} - \Delta A420\text{nm / min Blank})\ (3.0)\ (\text{df})}{(9.5)\ (0.10)}$$

3.0 = Total volume (in milliliters) of assay
df = Dilution factor
9.5 = Millimolar extinction coefficient of p-nitrophenol at 420nm
0.10 = Volume (in milliliters) of enzyme used

$$\text{Units / mg solid} = \frac{\text{units / ml enzyme}}{\text{mg solid / ml enzyme}}$$

$$\text{Units / mg protein} = \frac{\text{units / ml enzyme}}{\text{mg protein / ml enzyme}}$$

Unit Definition: *One unit will hydrolyze 1.0 μmole of p-nitrophenyl sulfate per minute at pH 7.1 at 37°C.*

Final Assay Concentrations

In a 3.00 ml reaction mix, the final concentrations are 187 mM Tris, 8.0 mM p-nitrophenyl sulfate and 0.05 unit of sulfatase.

Enzymatic Assay of SULFITE OXIDASE (EC 1.8.3.1)

Principle

$$\text{Slfite} + O_2 + H_2O + \text{Cyt-c (Oxidized)} \xrightarrow{\text{Sulfite Oxidase}} \text{Sulfate} + \text{Cyt-c} + H_2O_2\ \text{(Reduced)}$$

Abbreviation Used

Cyt-c = Cytochrome-c

Conditions: T = 25°C, pH = 8.5, A550nm, Light path = 1 cm

Method: Continuous Spectrophotometric Rate Determination

Reagents

A. 100 mM Tris HCl Buffer, pH 8.5 at 25°C (Prepare 500 ml in deionized water using Trizma Base. Adjust to pH 8.5 at 25°C with 1 M HCl).
B. 33 mM Sodium Sulfite Solution (Prepare 10 ml in Reagent A using Sodium Sulfite, Anhydrous).
C. 2 mM Cytochrome-c (Prepare 10 ml in Reagent A using Cytochrome-c, from Chicken Heart).
D. Sulfite Oxidase Enzyme Solution (Immediately before use, prepare a solution containing 0.05–0.07 units/ml of Sulfite Oxidase in cold Reagent A).

Procedure

Pipette (in milliliters) the following reagents into suitable cuvettes:

	Test	Blank
Reagent A (Buffer)	2.77	2.77
Reagent B (Sodium Sulfite)	0.03	—
Reagent C (Cytochrome-c)	0.10	0.10
Deionized Water	—	0.03

Equilibrate to 25°C. Monitor the A550nm until constant, using a suitably thermostatted spectrophotometer. Then add:

Reagent D (Enzyme Solution)	0.10	0.10

Immediately mix by inversion and record the increase in A550nm for approximately 5 minutes. Obtain the ΔA550nm/minute using the maximum linear rate for both the Test and Blank.

Calculations

$$\text{Units / mg enzyme} = \frac{(\Delta A550\text{nm / min Test} - \Delta A550\text{nm / min Blank})}{(21)\ (\text{mg enzyme / ml RM})}$$

21 = Millimolar extinction coefficient between oxidized and reduced Cytochrome-c at pH 8.5
RM = Reaction Mix

Unit Definition: *One unit will oxidize 1.0 μmole of sulfite to sulfate during the reduction of cytochrome c per minute at pH 8.5 at 25°C.*

Final Assay Concentrations

In a 3.0 ml reaction mix, the final concentrations are 100 mM Tris, 0.33 mM sodium sulfite, 0.07 mM cytochrome-c, and 0.005–0.007 unit sulfite oxidase.

Enzymatic Assay of SUPEROXIDE DISMUTASE (EC 1.15.1.1)

Principle

Superoxide Dismutase (SOD) catalyzes the reaction:

$$2O_2^- + 2H^+ \xrightarrow{\text{SOD}} O_2 + H_2O_2$$

Superoxide Dismutase activity is measured as the inhibition of the rate of reduction of Cytochrome c by the superoxide radical, observed at 550 nm:

$$\text{Cytochrome - c (oxidized)} + O_2^- \rightarrow \text{Cytochrome c (reduced)} + O_2$$

The superoxide radical is produced enzymatically by the reaction:

$$\text{Xanthine} + O_2 + H_2O \xrightarrow{\text{XOD}} \text{Uric acid} + O_2^- + H^+$$

Abbreviations Used

SOD = Superoxide Dismutase
XOD = Xanthine Oxidase

Condition: T = 25EC, pH = 7.8, A550nm, Light path = 1 cm

Method: Continuous Spectrophotometric Rate Determination

Reagents

A. 216 mM Potassium Phosphate Buffer, pH 7.8 at 25°C (Prepare 50 ml in deionized water using Potassium Phosphate, Monobasic, Anhydrous. Adjust to pH 7.8 at 25°C with 1 M KOH).

B. 10.7 mM Ethylenediaminetetraacetic Acid Solution (EDTA) (Prepare 50 ml in deionized water using Ethylenediaminetetraacetic Acid, Disodium Salt, Dihydrate).

C. 1.1 mM Cytochrome c Solution (Cyt c) (Prepare 2 ml in deionized water using Cytochrome c, from Horse Heart).

D. 0.108 mM Xanthine Solution (Xanthine) (Prepare 500 ml in deionized water using Xanthine. Titrate into solution with 1 M KOH).

E. Xanthine Oxidase Enzyme Solution (XOD) (Immediately before use, prepare a solution containing 0.05 unit/ml of Xanthine Oxidase in cold deionized water).

F. Superoxide Dismutase Enzyme Solution (Immediately before use, prepare a solution containing 10 units/ml of Superoxide Dismutase in cold deionized water).

Procedure

Prepare a reaction cocktail by pipetting (in milliliters) the following reagents into a suitable container:

Deionized Water	23.0
Reagent A (Buffer)	25.0
Reagent B (EDTA)	1.0
Reagent C (Cyt c)	1.0
Reagent D (Xanthine)	50.0

Mix and adjust to pH 7.8 at 25°C, if necessary, with 1 M HCl or 1 M KOH. Pipette (in milliliters) the following reagents into suitable cuvettes:

	Test 1 Uninhibited	Test 2 Inhibited	Blank
Reaction Cocktail	2.80	2.80	2.80

Equilibrate to 25°C. Monitor the absorbance, A550nm, until constant, using a suitably thermostatted spectrophotometer. Then add:

Deionized Water	0.10	—	0.20
Reagent E (XOD)	0.10	0.10	—
Reagent F (Enzyme Solution)	—	0.10	—

Immediately mix by inversion and record the increase in A550nm for approximately 5 minutes. Obtain the ΔA550nm/minute by using the maximum linear rate for both the Uninhibited (Test 1), Inhibited (Test 2) and Blank.

Calculations

$$\%\ \text{Inhibition} = \frac{\Delta A550\text{nm}/\text{min Uninhibited} - \Delta A550\text{nm}/\text{min Inhibited}}{\Delta A550\text{nm}/\text{min Uninhibited} - \Delta A550\text{nm}/\text{min Blank}} \times 100$$

$$\text{Units / ml enzyme} = \frac{\text{\% Inhibition (df)}}{(50\%)\ (0.1)}$$

df = Dilution factor
0.1 = Volume (in milliliter) of enzyme used
50% = Inhibition of the rate of cytochrome-c reduction as per the Unit Definition

$$\text{Units / mg solid} = \frac{\text{units / ml enzyme}}{\text{mg solid / ml enzyme}}$$

$$\text{Units / mg protein} = \frac{\text{units / ml enzyme}}{\text{mg protein / ml enzyme}}$$

Unit Definitions: *One unit will inhibit the rate of reduction of cytochrome c by 50% in a coupled system, using xanthine and xanthine oxidase at pH 7.8 at 25°C in a 3.0 ml reaction volume. The xanthine oxidase concentration should produce an initial (uninhibited) ΔA550nm of 0.025–0.005 per minute.*

Final Assay Concentrations

In a 3.00 ml reaction mix, the final concentrations are 50 mM potassium phosphate, 0.1 mM Ethylenediaminetetraacetic Acid, 0.01 mM cytochrome C, 0.05 mM xanthine, 0.005 unit xanthine oxidase and 1 unit superoxide dismutase.

Notes

1. The ΔA550nm/minute of Uninhibited (Test 1) should be 0.025–0.005. If not, repeat the above procedure and adjust the xanthine oxidase concentration so that the ΔA550nm/minute of Uninhibited (Test 1) is 0.025–0.005. (The quantity of xanthine oxidase added to the Inhibited (Test 2) must be the same as that added to the Uninhibited (Test 1)).
2. The decrease in the rate of the Inhibited (Test 2) must be between 40–60% of that of the Uninhibited (Test 1). A value as close to 50% as possible is desired.
3. Xanthine Oxidase Unit Definition: One unit will convert 1.0 μmole of xanthine to uric acid per minute at pH 7.5 at 25°C.

Enzymatic Assay of THIOGLUCOSIDASE (EC 3.2.3.1)

Principle

$$\text{Thioglucoside} + H_2O \xrightarrow{\text{Thioglucosidase}} \text{Thiol} + \text{Glucose}$$

$$\text{Glucose} + \text{ATP} \xrightarrow{\text{Hexokinase}} \text{Glucose 6-Phosphate} + \text{ADP}$$

$$\text{Glucose 6-Phosphate} + \beta\text{-NAD} \xrightarrow{\text{G-6-P-DH}} \text{6-Phosphogluconate} + \beta\text{-NADH}$$

Abbreviations Used

ATP = Adenosine 5'-Triphosphate
ADP = Adenosine 5'-Diphosphate
β-NAD = β-Nicotinamide Adenine Dinucleotide, Oxidized Form
G-6-P-DH = Glucose 6-Phosphate Dehydrogenase
β-NADH = β-Nicotinamide Adenine Dinucleotide, Reduced Form

Conditions: T = 25°C, pH = 6.0, A340nm, Light path = 1 cm

Method: Spectrophotometric Stop Rate Determination

Reagents

A. 126 mM Sodium Phosphate Buffer with 37 mM Citric Acid and 1.0 mM Ascorbic Acid, pH 6.0 at 25°C (Prepare 1000 ml in deionized water using Sodium Phosphate, Dibasic, Anhydrous, Citric Acid, Free Acid, Monohydrate and Ascorbic Acid, Sodium Salt. Adjust to pH 6.0 at 25°C with Either 1 M NaOH or 1 M HCl).
B. 4.8 mM Sinigrin Solution (Sinigrin) (Prepare 100 ml in Reagent A using Sinigrin, Monohydrate).
C. Glucose Determination Kit (16–10).
D. Thioglucosidase Enzyme Solution (Immediately before use, prepare a solution containing 0.5–1.0 unit/ml of Thioglucosidase in cold deionized water).

Procedure

Step 1: Pipette (in milliliters) the following reagents into suitable containers:

	Test	Blank
Reagent B (Sinigrin)	9.00	9.00

Equilibrate to 25°C. Then add:

Reagent D (Enzyme Solution)	1.00	—
Deionized Water	—	1.00

Immediately mix by inversion and incubate for exactly 10 minutes.

Step 2: Pipette (in milliliters) the following reagents into suitable cuvettes:

Reagent C (16–10)	3.00	3.00

Equilibrate to 25°C. Monitor the A340nm until constant using a suitably thermostatted spectrophotometer. Then add:

Test Solution (From Step 1)	0.10	—
Blank Solution (From Step 1)	—	0.10

Immediately mix by inversion and record the increase in A340nm for approximately 5–10 minutes until constant. Obtain the A340nm for both the Test and Blank.

Calculations

$$\text{Units / ml enzyme} = \frac{(\Delta\text{A340nm Test} - \Delta\text{A340nm Blank})\,(10)\,(3.1)\,(\text{df})}{(6.22)\,(0.1)\,(10)\,(1)}$$

10 = Total volume (in milliliters) of assay in Step 1
3.1 = Total volume (in milliliters) of assay in Step 2
df = Dilution factor
6.22 = Millimolar extinction coefficient of β-NADH at 340nm
1 = Volume (in milliliters) of enzyme used in Step 1

0.1 = Volume (in milliliters) from Step 1 and Step 2
10 = Time of assay (in minutes) of Step 1 as per the Unit Definition

$$\text{Units / mg solid} = \frac{\text{units / ml enzyme}}{\text{mg solid / ml enzyme}}$$

$$\text{Units / mg protein} = \frac{\text{units / ml enzyme}}{\text{mg protein / ml enzyme}}$$

Unit Definition: *One unit will produce 1.0 µmole glucose per minute from sinigrin at pH 6.0 at 25°C.*

Final Assay Concentrations

In a 10.00 ml reaction mix, the final concentrations are 113 mM sodium phosphate, 33 mM citric acid, 0.8 mM ascorbic acid, 4.3 mM sinigrin and 0.5–1.0 unit thioglucosidase.

Notes

1. Step 2 should be done while the enzyme solution is incubating at 25°C in Step 1.
2. Hexokinase Unit Definition: One unit will phosphorylate 1.0 µmole of D-glucose per minute at pH 7.6 at 25°C.
3. Glucose-6-Phosphate Dehydrogenase Unit Definition: One unit will oxidize 1.0 µmole of D-glucose 6-phosphate to 6-phospho-D-gluconate per minute in the presence of NADP or NAD at pH 7.8 at 30°C.

Enzymatic Assay of THIOREDOXIN REDUCTASE (EC 1.6.4.5)

Principle

$$\text{Thioredoxin} - S_2 + \text{DTNB} \xrightarrow[\text{NADPH}]{\text{Thioredoxin Reductase}} \text{TNB} + \text{Thioredoxin} - (SH_2)$$

Abbreviations Used

DTNB = 5,5'-Dithio-bis(2-Nitrobenzoic Acid)
TNB = 5-Thio-2-Nitrobenzoic Acid
NADPH = β-Nicotinamide Adenine Dinucleotide Phosphate, Reduced Form

Conditions: T = 25°C, pH = 7.0, A412nm, Light path = 1 cm

Method: Continuous Spectrophotometric Rate Determination

Reagents

A. 500 mM Potassium Phosphate Buffer, pH 7.5 at 25°C (Prepare 100 ml in deionized water using Potassium Phosphate, Monobasic, Anhydrous).
B. 100 mM Ethylenediaminetetraacetic Acid Solution (EDTA) (Prepare 10 ml in deionized water using Ethylenediaminetetraacetic Acid, Disodium Salt, Dihydrate).
C. 7 mM β-Nicotinamide Adenine Dinucleotide Phosphate, Reduced Form, Solution (NADPH) (Prepare 1 ml in deionized water using β-Nicotinamide Adenine Dinucleotide Phosphate, Reduced Form, Tetrasodium Salt).
D. 0.5% (w/v) Bovine Serum Albumin Solution (BSA) (Prepare 5 ml in deionized water using Albumin, Bovine).
E. 0.1% (w/v) Thioredoxin Solution (Thioredoxin).
F. 100% Ethanol (Use 200 Proof USP Ethyl Alcohol).

G. 100 mM 5'5-Dithio-bis(2-Nitrobenzoic Acid Solution) (DTNB) (Prepare 1 ml in Reagent F using 5,5'-Dithio-bis(2-Nitrobenzoic Acid).

H. Thioredoxin Reductase Enzyme Solution (Immediately before use, prepare a solution containing 3 units/ml of Thioredoxin Reductase in Reagent D).

Procedure

Pipette (in milliliters) the following reagents into suitable cuvettes.

	Test	Blank
Reagent A (Buffer)	0.14	0.14
Reagent B (EDTA)	0.07	0.07
Reagent C (NADPH)	0.02	0.02
Reagent D (BSA)	0.02	0.07
Reagent E (Thioredoxin)	0.10	0.10
Deionized water	0.265	0.265
Enzyme Solution	0.05	—

Mix by inversion and equilibrate to 25°C. Monitor the A412nm until constant, using a suitably thermostatted spectrophotometer. Then add:

Reagent G (DTNB)	0.035	0.035

Immediately mix by inversion and record the increase in A412nm for approximately 2 minutes. Obtain the ΔA412nm/minute using the maximum linear rate for both the Test and Blank.

Calculation

$$\text{Units / ml enzyme} = \frac{(\Delta A412\text{nm / min Test} - \Delta A412\text{nm / min Blank})\ (0.7)\ (\text{df})}{(1)\ (0.05)}$$

0.7 = Volume (in milliliter) of assay
df = Dilution factor
1 = Change in absorbance at 412nm that is equivalent to one unit
0.05 = Volume (in milliliter) of enzyme used

$$\text{Units / mg solid} = \frac{\text{units / ml enzyme}}{\text{mg solid / ml enzyme}}$$

$$\text{Units / mg protein} = \frac{\text{units / ml enzyme}}{\text{mg protein / ml enzyme}}$$

Unit Definition: *One unit will cause an increase in absorbance of 1.0 at 412nm (when measured in a coupled assay with E. coli thioredoxin and DTNB) per minute per ml at pH 7.0 at 25°C.*

Final Assay Concentrations

In a 0.70 ml reaction mix, the final concentrations are 100 mM potassium phosphate, 10 mM Ethylenediaminetetraacetic Acid, 0.2 mM β-nicotinamide adenine dinucleotide phosphate, reduced form, 0.05% (w/v) Bovine Serum Albumin, 0.014% (w/v) thioredoxin, 5 mM 5, 5'-dithio-bis(2-nitrobenzoic acid), 0.15 unit thioredoxin reductase.

Suitability Assay of THIOREDOXIN

Principle

$$DTT\text{-}(SH_2) + Insulin\text{-}S_2 \xrightarrow{Thioredoxin} Insulin\text{-}(SH)_2 + DTT\text{-}S_2$$

Abbreviations Used

$DTT\text{-}(SH_2)$ = DL-Dithiothreitol
$Insulin\text{-}S_2$ = Insulin (native disulfide bridge)
$Insulin\text{-}(SH)_2$ = Insulin (reduced disulfides)
$DTT\text{-}S_2$ = DL-Dithiothreitol (disulfide bridge)

Conditions: T = 25°C, pH 7.0, A650nm, Light path = 1 cm

Method: Turbidimetric Rate Determination

Reagents

A. 100 mM Sodium Phosphate Buffer, pH 7.0 at 25°C (Prepare 100 ml in deionized water using Sodium Phosphate, Monobasic, Anhydrous. Adjust to pH 7.0 at 25°C with 1 M NaOH).
B. 100 mM Ethylenediaminetetraacetic Acid Solution (EDTA) (Prepare 10 ml in deionized water using Ethylenediaminetetraacetic Acid, Disodium Salt, Dihydrate).
C. 100 mM DL-Dithiothreitol Solution (DTT) (Prepare 1 ml in deionized water using DL-Dithiothreitol).
D. 50 mM Tris HCl Buffer, pH 7.5 at 25°C (Prepare 10 ml in deionized water using Trizma Base. Adjust to pH 7.5 at 25°C with 1 M HCl).
E. 1% (w/v) Insulin Solution (Insulin) (Prepare by dissolving 10 mg of Insulin in 1 ml of Reagent D. This will produce a milky solution. Adjust the pH to below 3 at 25°C with 5 N HCl. The solution should begin to clear. Titrate again with 5 N NaOH. The solution should become turbid at pH 7 and then clear at pH 7.5. The final pH should be 7.5 at 25°C).
F. Thioredoxin Solution (Thioredoxin) (Immediately before use prepare a solution containing 0.025–0.1 mg/ml of Thioredoxin (protein) in cold deionized water).

Procedure

Prepare a reaction cocktail by pipetting (in milliliters) the following reagents into a suitable container:

Reagent A (Buffer)	7.56
Reagent B (EDTA)	0.24
Reagent E (Insulin)	1.20

Mix by swirling. Pipette (in milliliters) the following reagents into suitable containers:

	Test	Blank
Reaction Cocktail	0.75	0.75
Reagent F (Thioredoxin)	0.08	—
Deionized Water	0.16	0.24

Mix by swirling and then add:

	Test	Blank
Reagent C (DTT)	0.01	0.01

Immediately mix by swirling and incubate at 25°C for 20 minutes. Then begin to record the increase in A650nm every 5 minutes for a total of 15 minutes. Obtain the ΔA650nm/minute using the maximum linear rate for both the Test and Blank.

Calculations

$$\text{S.A.} = \frac{(\Delta\text{A650nm / min Test} - \Delta\text{A650nm / min Blank})\,(1)\,(\text{df})}{(\text{mg})}$$

S.A. = Specific Activity
df = Dilution factor
1 = Total volume (in milliliter) of assay
mg = mg Thioredoxin

Specification

Specific Activity (using insulin): ΔA650nm = 1–3 per mg protein per minute in the presence of 1 mM EDTA.

Final Assay Concentrations

In a 1.00 ml reaction mix, the final concentrations are 5 mM Tris, 63 mM sodium phosphate, 0.1% (w/v) insulin, 1 mM DL-Dithiothreitol, 2 mM Ethylenediaminetetraacetic Acid, and 0.002–0.008 mg thioredoxin.

Notes

1. The turbidity of the reaction mixture begins to increase after approximately 20 minutes.
2. If the third A650nm reading (at 10 minutes) exceeds 0.8, dilute the thioredoxin solution (Reagent F) and run the assay again because rates above this value are not linear.

Enzymatic Assay of THROMBIN (EC 3.4.21.5)

Principle

$$\text{Fibrinogen} \xrightarrow{\text{Thrombin}} \text{Fibrin}$$

Conditions: T = 37°C, pH 7.35.

Method: Fibrometer

Reagents

A. 143 mM Diethyl Barbiturate and 143 mM Sodium Acetate Solution (Soln A) (Prepare 275 ml in deionized water using Sodium Acetate, Trihydrate, and Barbital, Sodium Salt).

B. 4.25% (w/v) Sodium Chloride Solution (Prepare 300 ml in deionized water using Sodium Chloride).

C. 36 mM Diethyl Barbiturate, 36 mM Sodium Acetate, 0.85% (w/v) Sodium Chloride Solution, pH 7.35 at 25°C (Soln B) (Prepare 1 liter by combining 250 ml of Reagent

A, 200 ml of Reagent B, 217 ml of 0.1 M HCl, and 333 ml of deionized water. Adjust the pH to 7.35 at 37°C with either 1 M HCl or 1 M NaOH).

D. 25.7 mM Sodium Citrate Solution (Prepare 100 ml in deionized water using Citric Acid Trisodium Salt, Dihydrate).

E. 0.85% (w/v) Sodium Chloride Solution (NaCl) (Prepare 300 ml in deionized water using Reagent B).

F. 7.2 mM Diethyl Barbiturate, 7.2 mM Sodium Acetate, 5.1 mM Sodium Citrate, 0.68% (w/v) Sodium Chloride, 1.0% (w/v) Bovine Serum Albumin, and 0.5% (w/v) Polyethylene Glycol, pH 7.35 at 37°C (Enz Dil) (Prepare 500 ml by combining 100 ml of Reagent C, 100 ml of Reagent D, 300 ml of Reagent E, Albumin Bovine, and Polyethylene Glycol. Adjust the pH to 7.35 at 37°C with either 1 M HCl or 1 M NaOH).

G. Plasma Solution (Plasma) (Immediately before use, reconstitute 2.0 ml of Human Plasma, lyophilized with 2 ml of deionized water. Then add 2 ml of Reagent E. Keep at room temperature).

H. Thrombin Enzyme Solution (Thrombin) (Immediately before use, reconstitute Thrombin vial with 1 ml of deionized water. Further dilute to 2.7–5.1 units/ml of Thrombin in Reagent F, in glass tubes. Serial dilutions should be made in Reagent F so that clotting times are between 15–25 seconds. If the clotting time recorded is too long, increase the Thrombin concentration. If the time recorded is too short, decrease the Thrombin concentrations. Subsequent dilutions are also to be made in glass tubes).

I. NIH Thrombin Standard Solution (Std) (Use Thrombin Reference Standard (Lot J) which has been diluted in Reagent F.
Note: Current Standard Curve has been made with N.I.H. Thrombin standard diluted in glass test tubes).

Procedure

Step 1: Pipette (in milliliters) the following reagents into fibrometer cups containers, which have been prewarmed to 37°C. Test Reagent G (Plasma) 0.20. Allow to stand at 37°C for thirty seconds. Then add:

Reagent H (Thrombin)	0.10

Record the clotting time using a suitably thermostatted fibrometer. A 0.3 ml Fibrometer probe should be used. Compare the clotting times of the Tests against a NIH Standard curve for Thrombin clotting times versus units of Thrombin/ml. Determine the units of Thrombin of the Test from the Standard Curve.

Calculations

$$\text{Units/ml enzyme} = (\text{Units of Thrombin from Std Curve})\ (\text{df})$$

df = Dilution factor

$$\text{Units / mg solid} = \frac{\text{units / ml enzyme}}{\text{mg solid / ml enzyme}}$$

$$\text{Units / mg protein} = \frac{\text{units / ml enzyme}}{\text{mg protein / ml enzyme}}$$

Unit Definition: *Activity is expressed in NIH units obtained by direct comparison to a NIH Thrombin Reference Standard, Lot J.*

Final Assay Concentrations

In a 0.30 ml reaction mix, the final concentrations are 33% (v/v) plasma, and 0.27–0.50 unit thrombin (components of the enzyme diluent are not included).

Note

1. Standards should be used immediately and kept on ice after they are dissolved.

Enzymatic Assay of THROMBOCYTIN

Principle

ne Nitroanili - p + $\xrightarrow{\text{Arg-Pro-Gly-Tosyl-p-N}}$ in Thrombocyt de Nitroanili - p Arg - Pro - Gly - Tosyl - p - N

Conditions: T = 37°C, pH 8.4, A405nm, Light path = 1 cm.

Method: Continuous Spectrophotometric Rate Determination

Reagents

A. 100 mM Tris and 100 mM Imidazole Buffer with 150 mM Sodium Chloride, pH 8.4 at 37°C (Prepare 100 ml in deionized water using Trizma Base, Imidazole, and Sodium Chloride. Adjust to pH 8.4 at 37°C with 1 M HCl).
B. 1.5 mM N-p-Tosyl-Gly-Pro-Arg p-Nitroanilide Solution (Substrate) (Prepare 5 ml in deionized water using N-p-Tosyl-Gly-Pro-Arg p-Nitroanilide, Acetate).
C. 0.1% (w/v) Bovine Serum Albumin (BSA) (Prepare 10 ml in Reagent A).
D. Thrombocytin Enzyme Solution (Immediately before use, prepare a solution containing 50 units/ml of Thrombocytin in Reagent C).

Procedure

Pipette (in milliliters) the following reagents into suitable quartz cuvettes:

	Test	Blank
Reagent A (Buffer)	2.55	2.55
Reagent B (Substrate Solution)	0.30	0.30

Equilibrate to 37°C. Monitor the A405nm until constant, using a suitably thermostatted spectrophotometer. Then add:

Reagent D (Enzyme Solution)	0.15	—
Reagent C (BSA)	—	0.15

Immediately mix by inversion and record the increase in A405nm for approximately 5 minutes. Obtain the A405nm/minute using the maximum linear rate for both the Test and Blank.

Calculations

$$\text{Units / mg enzyme} = \frac{(\text{A405nm / min Test} - \text{A405nm / min Blank})}{(0.0106)\ (\text{mg enzyme / ml RM})}$$

0.0106 = Micromolar extinction coefficient of p-nitroaniline at 405nm
RM = Reaction Mix

Unit Definition: *One unit will release 1.0 nmole of p-nitroaniline from N-p-tosyl-Gly-Pro-Arg p-nitroanilide per minute at pH 8.4 at 37°C.*

Final Assay Concentrations

In a 3.0 ml reaction mix, the final concentrations are 90 mM Tris, 90 mM imidazole, 135 mM sodium chloride, 0.005% BSA, 0.15 mM N-p-tosyl-Gly-Pro-Arg p-nitroanilide.

Enzymatic Assay of THYMIDINE PHOSPHORYLASE (EC 2.4.2.4)

Principle

$$\text{Thymidine} + \text{Orthophosphate} \xrightarrow{\text{TP}} \text{Thymine} + \text{2-Deoxy-R-P}$$

Abbreviations Used

TP = Thymidine Phosphorylase
2-Deoxy-R-P = 2-Deoxy-D-Ribose 1-Phosphate

Conditions: T = 25°C, pH = 7.4, A290nm, Light path = 1 cm

Method: Continuous Spectrophotometric Rate Determination

Reagents

A. 200 mM Potassium Phosphate Buffer, pH 7.4 at 25°C. (Prepare 500 ml in deionized water using Potassium Phosphate, Monobasic, Anhydrous. Adjust to pH 7.4 at 25°C with 1 M NaOH).
B. 1 mM Thymidine Solution (Prepare 100 ml in Reagent A using Thymidine).
C. 10 mM Potassium Phosphate Buffer, pH 7.0 at 25°C (Enzyme Diluent) (Prepare 100 ml in deionized water using Potassium Phosphate, Monobasic, Anhydrous. Adjust to pH 7.0 at 25°C with 1 M NaOH).
D. Thymidine Phosphorylase Enzyme Solution (Immediately before use, prepare a solution containing 1–2 units/ml of Thymidine Phosphorylase in cold Reagent C).

Procedure

Pipette (in milliliters) the following reagents into suitable quartz cuvettes:

	Test	Blank
Reagent B (Thymidine)	3.00	3.00

Equilibrate to 25°C. Monitor the A290nm until constant, using a suitably thermostatted spectrophotometer. Then add:

Reagent C (Enzyme Diluent)	—	0.03
Reagent D (Enzyme Solution)	0.03	—

Immediately mix by inversion and record the decrease in A290nm for approximately 5 minutes. Obtain the ΔA290nm/minute using the maximum linear rate for both the Test and Blank.

Calculations

$$\text{Units / mg enzyme} = \frac{(\Delta A290\text{nm / min Test} - \Delta A290\text{nm / min Blank})\,(3.03)\,(\text{df})}{(1.0)\,(0.03)}$$

3.03 = Total volume (in milliliters) of assay
df = Dilution factor
1.0 = Difference in the millimolar extinction coefficient between thymidine and thymine under the assay conditions
0.03 = Volume (in milliliters) of enzyme used

Unit Definition: *One unit will convert 1.0 µmole each of thymidine and phosphate to thymine and 2-deoxyribose 1-phosphate per minute at pH 7.4 at 25°C.*

Final Assay Concentration

In a 3.03 ml reaction mix, the final concentrations are 198 mM potassium phosphate, 1 mM thymidine and 0.03–0.06 unit thymidine phosphorylase.

Note

1. The initial A290nm should be between 1.8 and 1.95. If the absorbance is not within the specific range, then fresh reagents should be prepared.

Enzymatic Assay of TISSUE PLASMINOGEN ACTIVATOR SINGLE CHAIN T-PA Amidolytic Activity

Principle

$$\text{D-Ile-Pro-Arg-pNA} + H_2O \xrightarrow{\text{T-PA}} \text{D-Ile-Pro-Arg} + \text{p-Nitroaniline}$$

Abbreviation Used

pNA = p-Nitroanilide

Conditions: T = 25°C, pH 6.0 and 9.0, A405nm, Light path = 1 cm

Method: Continuous Spectrophotometric Rate Determination

Reagents

A. 500 mM Potassium Phosphate Buffer, pH 6.0 at 25°C (Phos Buffer) (Prepare 100 ml in deionized water using Potassium Phosphate, Monobasic, Anhydrous. Adjust to pH 6.0 at 25°C with 1 M NaOH).
B. 500 mM Tris HCl Buffer, pH 9.0 at 25°C (Tris Buffer) (Prepare 100 ml in deionized water using Trizma Base. Adjust to pH 9.0 at 25°C with 1 M HCl).
C. 0.40 mM D-Ile-Pro-Arg p-Nitroanilide Solution (Substrate pH 6) (Prepare 20 ml in Reagent A using D-Ile-Pro-Arg p-Nitroanilide Dihydrochloride).
D. 0.40 mM D-Ile-Pro-Arg p-Nitroanilide Solution (Substrate pH 9) (Prepare 20 ml in Reagent B using D-Ile-Pro-Arg p-Nitroanilide Dihydrochloride).
E. 1 M Potassium Bicarbonate Solution (Enzyme Diluent) (Prepare 25 ml in deionized water using Potassium Bicarbonate).
F. Tissue Plasminogen Activator Single Chain T-PA Enzyme Solution (T-PA) (Immediately before use, prepare a solution containing 0.5–1.0 unit/ml of Tissue Plasminogen Activator Single Chain T-PA in cold Reagent E).

Procedure

Pipette (in milliliters) the following reagents into suitable cuvettes:

	Test 1	Blank 1	Test 2	Blank 2
Reagent C (Substrate pH 6)	0.90	0.90	—	—
Reagent D (Substrate pH 9)	—	—	0.90	0.90
Reagent F (T-PA)	0.10	—	0.10	—
Reagent E (Enzyme Diluent)	—	0.10	—	0.10

Immediately mix by inversion and record the increase in A405nm for approximately 10 minutes using a suitably thermostatted spectrophotometer. Obtain the ΔA405nm/minute using the maximum linear rate for both the Tests and Blanks. Compare the activity of Test 2 to that of a control sample. The activity should be similar.

Calculations

$$\text{Units/ml} = \frac{(\Delta A405)\,(df)\,(1)}{(0.1)\,(10)}$$

1 = Volume (in milliliter) of assay
df = Dilution factor
0.1 = Volume (in milliliter) of enzyme used
10 = Millimolar extinction coefficient of pnitroaniline at 405nm under the conditions of the assay

Final Assay Concentrations

In a 1.00 ml reaction mix the final concentrations are 0.36 mM D-Ile-Pro-Arg p-nitroanilide, 450 mM Tris or 450 mM potassium phosphate, 100 mM potassium bicarbonate, and 0.05–0.10 unit tissue plasminogen activator single chain T-PA.

Note

1. The assay is run at pH 6.0 and pH 9.0. There should be very little activity observed at pH 6.0 with the maximum activity observed at pH 9.0.

Enzymatic Assay of Tissue Plasminogen Activator Fibrinolytic Assay

Principle

Fibrinogen	Plasminogen
↓	↓
↓Thrombin	↓ Tissue Plasminogen Activator
↓	↓
↓	↓ Plasmin

Fibrin (Clot) →→→→→→→→→ Soluble Proteins

Conditions: T = 37°C, pH = 7.3

Method: Visual Fibrolytic Endpoint Determination

Reagents

A. 0.1 M Sodium Phosphate/0.01% Tween 80 (BUF) (Prepare 100 ml in deionized water using Sodium Phosphate, Monobasic and Polyoxyethylenesorbitan, Monooleate (Tween 80). Adjust to pH 7.3 at 37°C with 2 NaOH).

B. 9% (w/v) Fibrinogen Solution (FIB) (Prepare 10 ml in Reagent A using Fibrinogen, Type I-S from Bovine Plasma. Allow solution to stir for at least one hour at 37°C).

C. 0.15 M Sodium Chloride Solution (NaCl) (Prepare 50 ml in deionized water using Sodium Chloride).

D. 0.5 M Sodium Chloride Solution (Prepare 50 ml in deionized water using Sodium Chloride).

E. 1.0 M Potassium Bicarbonate Solution (Prepare 50 ml in deionized water using Potassium Bicarbonate).

F. Thrombin Solution (THR) (Prepare 4 ml at 250 units/ml in cold Reagent C using Thrombin from Human Plasma. Keep on ice).

G. Plasminogen Solution (PLS) (Prepare by dissolving a five unit vial in 1.0 ml of cold Reagent D using Plasminogen from Human Plasma. After dissolution, dilute to 0.05 units/ml with cold Reagent A. Keep on ice).

H. Tissue Plasminogen Activator Standard (t-PA Std) (Prepare by dissolving a 10 µg vial, approximately 5000 units of t-PA per vial, in 1.0 ml of cold Reagent E using Chromogenix t-PA standard. Place on ice for 15 minutes and then dilute to approximately 2500 units/ml in cold Reagent A0

I. Tissue Plasminogen Activator Test Solutions (Prepare solution containing approximately 500 units/ml in cold Reagent E. Place on ice for 15 minutes. Immediately before use, prepare a solution containing approximately 2500 units/ml with cold Reagent A).

Procedure

Clot Formation Check: Pipette (in milliliters) the following reagents into a suitable container:

	Test
Reagent B (FIB)	0.200
Reagent G (PLS)	0.125

Equilibrate at 37°C for 3 minutes. Then add:

Reagent A (BUF)	0.125

Equilibrate at 37°C for 1 minute. Then add:

Reagent F (Thrombin)	0.050

Mix by swirling and incubate at 37°C for approximately 2–3 minutes to allow for clot formation. Then gently add a glass bead to the top of the test reaction mixture. If the glass bead remains on top of the reaction test mixture, then the system is working properly. Proceed to the Clot Lysing Check. If the bead falls through the clot, increase the concentration of Fibrinogen in Reagent B and repeat Clot Formation Check.

Clot Lysing Check: Pipette (in milliliters) the following reagents into a suitable containers:

	Test
Reagent B (FIB)	0.200
Reagent G (PLS)	0.125

Equilibrate at 37°C for 3 minutes. Then add:

Reagent H (t-PA Std)	0.125

Equilibrate at 37°C for 1 minutes. Then add:

Reagent F (Thrombin)	0.050

Record initial time (T_I) in minutes, and mix by swirling and incubate at 37°C for approximately 2–3 minutes to allow for clot lysis. Then gently add a glass bead to the top of the test reaction mixture. Record time as final time (T_F) when the glass bead touches the bottom of the tube. If the change in time is between 5 to 10 minutes, then proceed with standard curve. The change in time is the difference between the recorded time upon addition of thrombin (T_I: Initial Time) and the time it takes for the glass bead to touch the bottom of the container (T_F: Final Time). If the change in time is not between 5 to 10 minutes, then prepare an appropriate dilution of the standard and repeat Clot Lysing Check.

Standard Curve: Pipette (in milliliters) the following reagents into a suitable container:

	Std 1	Std 2	Std 3	Std 4	Std 5	Std 6	Std 7
Reagent B (FIB)	0.200	0.200	0.200	0.200	0.200	0.200	0.200
Reagent G (PLS)	0.125	0.125	0.125	0.125	0.125	0.125	0.125

Mix by swirling and equilibrate at 37°C for 3 minutes. Then add:

Reagent A (BUF)	—	0.025	0.045	0.065	0.085	0.105	0.115
Reagent H (TPA)	0.125	0.100	0.080	0.060	0.040	0.020	0.010

Mix by swirling and equilibrate at 37°C for 1 minute. Then add:

Reagent F (THR)	0.050	0.050	0.050	0.050	0.050	0.050	0.050

Mix by swirling and incubate at 37°C for 2–3 minutes and record the initial time (T_I) in minutes and record the initial time (T_I) in minutes after addition of Reagent F. Gently add the glass bead to each container starting with Std 1. Record time as the final time (T_F) when the glass bead touches the bottom of the container.

Test: Pipette (in milliliters) the following in a suitable container:

	Test 1	Test 2	Test 3
Reagent B (FIB)	0.200	0.200	0.200
Reagent G (PLS)	0.125	0.125	0.125

Mix by swirling and equilibrate at 37°C for 3 minutes. Then add:

Reagent A (BUF)	0.075	0.065	0.055
Reagent I (Test)	0.050	0.060	0.070

Mix by swirling and equilibrate at 37°C for 1 minute. Then add:

Reagent F (THR)	0.050	0.050	0.050

Mix by swirling and incubate at 37°C for 2–3 minutes and record the initial time (T_I) in minutes after addition of Reagent F. Gently add the glass bead to each container starting with Test-1. Record time as the final time (T_F) when the glass bead touches the bottom of the container.

Calculations

Analysis of Standard-1: ΔT (Std 1) = T_F (Std 1)–T_I (Std 1)
Log ΔT (Std 1)

$$\text{Log} = \frac{\text{Units}}{\text{Std Reaction Mixture}} = \text{Log}\ \frac{\text{Units / vial} \times \text{DF}}{0.125}$$

units/vial = units/vial claimed by Chromogenix
DF = Dilution factor
Plot the log ΔT (T_F–T_I) versus log units of Tissue Plasminogen Activator per reaction mixture as a linear standard curve. Determine the linear equation for the standard curve.
Analysis of Test–1

$$\frac{\text{Units}}{\text{Test Reaction Mixture}} = \frac{\text{(units liberated from Std Curve)} \times \text{DF}}{0.050}$$

Units liberated = y
DF = Dilution factor
0.050 = Milliliters of Test used in reaction mixture

Enzymatic Assay of TRANSALDOLASE (EC 2.2.1.2)

Principle

$$\text{D-Fructose 6-Phosphate} + \text{D-Erythrose 4-Phosphate} \xrightarrow{\text{TA}} \text{S-7-P} + \text{GAP}$$

$$\text{GAP} \xrightarrow{\text{TPI}} \text{Dihydroxyacetone Phosphate}$$

$$\text{DHAP} + \beta\text{-NADH}\ \alpha\text{-GDH} \rightarrow \alpha\text{-Glycerophosphate} + \beta\text{-NAD}$$

Abbreviations Used

TA = Transaldolase
S-7-P = Sedoheptulose 7-Phosphate
GAP = Glyceraldehyde 3-Phosphate
TPI = Triosephosphate Isomerase

DHAP = Dihydroxyacetone Phosphate
β-NADH = β-Nicotinamide Adenine Dinucleotide, Reduced Form
α-GDH = α-Glycerophosphate Dehydrogenase
β-NAD = β-Nicotinamide Adenine Dinucleotide, Oxidized Form

Conditions: T = 25°C, pH 7.7, A340nm, Light path = 1 cm

Method: Continuous Spectrophotometric Rate Determination

Reagents

A. 250 mM Glycylglycine Buffer, pH 7.7 at 25°C (Prepare 100 ml in deionized water using Glycylglycine, Free Base. Adjust to pH 7.7 at 25°C with 1 M NaOH).
B. 100 mM D-Erythrose 4-Phosphate Solution (E-4-P) (Prepare 1 ml in deionized water using D-Erythrose 4-Phosphate, Sodium Salt).
C. 200 mM D-Fructose 6-Phosphate Solution (F-6-P) (Prepare 2 ml in deionized water using D-Fructose 6-Phosphate, Disodium Salt).
D. 300 mM Magnesium Chloride Solution ($MgCl_2$) (Prepare 10 ml in deionized water using Magnesium Chloride, Hexahydrate).
E. 2.6 mM β-Nicotinamide Adenine Dinucleotide, Reduced Form, Solution (β-NADH) (Dissolve the contents of one 5 mg vial of β- Nicotinamide Adenine Dinucleotide, Reduced Form, Disodium Salt in the appropriate volume of Reagent A).
F. α-Glycerophosphate Dehydrogenase/Triosephosphate Isomerase Enzyme Solution (α-GDH/TPI) (Immediately before use, prepare a solution containing 0.1 mg/ml of α-Glycerophosphate Dehydrogenase/ Triosephosphate Isomerase in cold Reagent A).
G. Transaldolase Enzyme Solution (TA) (Immediately before use, prepare a solution containing 0.25–0.50 unit/ml of Transaldolase in cold deionized water).

Procedure

Pipette (in milliliters) the following reagents into suitable cuvettes:

	Test	Blank
Deionized Water	1.80	1.80
Reagent A (Buffer)	0.55	0.55
Reagent B (E-4-P)	0.05	0.05
Reagent C (F-6-P)	0.10	0.10
Reagent E (β-NADH)	0.15	0.15
Reagent D ($MgCl_2$)	0.15	0.15
Reagent F (α-GDH/TPI)	0.10	0.10

Mix by inversion and equilibrate to 25°C. Monitor the A340nm until constant, using a suitably thermostatted spectrophotometer. Then add:

	Test	Blank
Reagent G (TA)	0.10	—
Deionized Water	—	0.10

Immediately mix by inversion and record the decrease in A340nm for approximately 5 minutes. Obtain the ΔA340nm/minute using the maximum linear rate for both the Test and Blank.

Calculations

$$\text{units / ml enzyme} = \frac{(\Delta A340\text{nm / min Test} - \Delta A340\text{ nm / min Blank})(3)(\text{df})}{(6.22)(0.1)}$$

3 = Total volume (in milliliters) of assay
df = Dilution factor
6.22 = Millimolar extinction coefficient of β-NADH at 340nm
0.1 = Volume (in milliliter) of transaldolase used

$$\text{Units / mg solid} = \frac{\text{units / ml enzyme}}{\text{mg solid / ml enzyme}}$$

$$\text{Units / mg protein} = \frac{\text{units / ml enzyme}}{\text{mg protein / ml enzyme}}$$

Unit Definition: *One unit will produce 1.0 µmole of D-glyceraldehyde 3-phosphate from D-fructose 6-phosphate per minute in the presence of D-erythrose 4-phosphate, at pH 7.7 at 25°C in a coupled system with α-GDH/TPI and β-NADH.*

Final Assay Concentrations

In a 3.00 ml reaction mix, the final concentrations are 67 mM glycylglycine, 2 mM D-erythrose 4-phosphate, 6.7 mM D-fructose 6-phosphate, 15 mM magnesium chloride, 0.13 mM α-nicotinamide adenine dinucleotide, 0.01 mg of α-glycerophosphate ehydrogenase/triosephosphate isomerase, 0.025–0.050 units transaldolase.

Enzymatic Assay of TRANSAMINASE, Broad-Range

Principle

$$\text{L-Aspartate} + \alpha\text{-Ketoglutarate} + \beta\text{-NADH} \xrightarrow[\text{PRP}]{\text{TG}} \text{Oxaloacetate} + \text{L-Glutamate} + \beta\text{-NADH}$$

Conditions: T=25°C, pH=7.0, A340nm, Light path=1 cm

Method: Continuous Spectrophotometric Rate Determination

Reagents

A. 50 mM potassium Phosphate Buffer, pH 7.0 at 25°C (Prepare 100 ml in deionized water using Potassium Phosphate, Monobasic, Anhydrous. Adjust to pH 7.0 at 25°C with 1 M KOH).
B. 200 mM L-Aspartate Solution, pH 7.0 at 25°C (L-Asp) (Prepare 10 ml in Reagent A using L-Aspartic Acid, Monosodium Salt, Hydrate).
C. 200 mM α-Ketoglutaric Acid Solution (α-KG) (Prepare 10 ml in Reagent A using α-Ketoglutaric Acid, Monopotassium Salt).
D. 10 mM Pyridoxal 5-Phosphate Solution (PRP) (Prepare 10 ml in deionized water using Pyridoxal 5-Phosphate).
E. 6.4 mM β-Nicotinamide Adenine Dinucleotide, Reduced Form, Solution (β-NADH) (Prepare 5 ml in deionized water using β-Nicotinamide Adenine Dinucleotide, Reduced Form, Disodium Salt).

F. Malic Dehydrogenase Enzyme Solution (MDH).

G. Transaminase Broad-Range Enzyme Solution (TBR) (Immediately before use, prepared solution containing approximately 1.5 units/ml in cold Reagent A).

Procedure

Pipette (in milliliters) the following reagents into a suitable cuvette:

	Test	Blank
Reagent A (Buffer)	2.30	2.30
Reagent C (α-KG)	0.30	0.30
Reagent B (L-ASP)	0.30	0.30
Reagent D (PRP)	0.06	0.06
Reagent E (β-NADH)	0.045	0.045
Reagent F (MDH)	0.01	0.01

Mix by inversion and equilibrate to 25°C. Monitor the A340nm until constant using a suitably thermostatted spectrophotometer. Then add:

Reagent G (TBR)	0.025	—
Reagent A (Buffer)	—	0.025

Immediately mix by inversion and record decrease in A340nm for approximately 5 minutes. Obtain the ΔA340/min using the maximum linear rate for both the Test and the Blank.

Calculations

$$\text{units / ml enzyme} = \frac{(\Delta A340\text{nm / min Test} - \Delta A340\text{ nm / min Blank})\ (3.04)\ (\text{df})}{(6.22)\ (0.025)}$$

3.04 = Total volume (in milliliters) of assay
df = Dilution factor
6.22 = Millimolar extinction coefficient of β-NADH at 340nm
0.025 = Volume (in milliliter) of enzyme used

$$\text{Units / mg solid} = \frac{\text{units / ml enzyme}}{\text{mg solid / ml enzyme}}$$

$$\text{Units / mg protein} = \frac{\text{units / ml enzyme}}{\text{mg protein / ml enzyme}}$$

Unit Definition: *One unit will produce 1.0 μmole of oxaloacetate from L-aspartate during the transamination of 2-ketoglutarate per minute at pH 7.0 at 25°C in the presence of pyridoxal phosphate.*

Final Assay Concentrations

In a 3.04 ml reaction mix, the final concentrations are 38 mM potassium phosphate, 19.7 mM L-aspartate, 19.7 mM α-ketoglutaric acid, 0.2 mM pyridoxal 5-phosphate, 0.1 mM β-nicotinamide adenine dinucleotide, reduced form, 20 units malic dehydrogenase, 0.2 and 0.04 units transaminase.

Enzymatic Assay of TRANSGLUTAMINASE (EC 2.3.2.13)

Principle

$$\text{CBZ - Gln - Gly + Hydroxylamine} \xrightarrow{\text{Transglutaminase}} \text{CBZ - Gln - Gly - Hydroxamate}$$

Abbreviation Used

CBZ = N-Carbobenzoxy

Conditions: T = 37°C, pH = 6.0, A525nm, Light path = 1 cm

Method: Colorimetric

Reagents

A. 1000 mM Tris Buffer, pH 6.0 at 37°C (Prepare 50 ml in deionized water using Trizma Base. Adjust to pH 6.0 at 37°C with Glacial Acetic Acid).
B. CBZ-Glutaminylglycine (CBZ-Gln-Gly).
C. 200 mM Hydroxylamine with 20 mM Glutathione, Reduced Form Solution (HA/Glut) (Prepare 10 ml in deionized water using Hydroxylamine Hydrochloride and Glutathione, Reduced Form. *Prepare Fresh*).
D. 1000 mM Calcium Chloride Solution ($CaCl_2$) (Prepare 1 ml in deionized water using Calcium Chloride Dihydrate).
E. 10 mM L-Glutamic Acid γ-Monohydroxamate Solution (Std) (Prepare 10 ml in deionized water using L-Glutamic Acid γ-Monohydroxamate).
F. 12% (v/v) Trichloroacetic Acid Solution (TCA) (Prepare 100 ml in deionized water using Trichloroacetic Acid, 6.1 N Solution).
G. 5% (w/v) Ferric Chloride Solution ($FeCl_3$) (Prepare 100 ml in Reagent H using Ferric Chloride, Hexahydrate).
H. 100 mM Hydrochloric Acid (Prepare 100 ml in deionized water using Hydrochloric Acid).
I. Transglutaminase Enzyme Solution (Immediately before use, prepare a solution containing 2 units/ml of Transglutaminase in cold deionized water).

Procedure

Prepare a reaction cocktail by combining the following reagents into a suitable container:

Reagent B (CBZ-Gln-Gly)	120mg

Then add (in milliliters):

Reagent A (Buffer)	2.00
Reagent C (HA/Glut)	5.00

Mix by inversion. Then add:

Reagent D ($CaCl_2$)	0.05

Mix by inversion. Adjust to pH 6.0 at 37°C with 100 mM NaOH. Then add enough deionized water to make a final volume of 10.0 ml. Pipette (in milliliters) the following reagents into suitable containers:

	Test	Test Blank	Std	Std Blank
Reaction Cocktail	0.20	—	—	—

Equilibrate to 37°C. Then add:

	Test	Test Blank	Std	Std Blank
Reagent I (Enzyme Solution)	0.03	—	—	—

Mix by inversion and incubate at 37°C for exactly 10 minutes. Then add:

	Test	Test Blank	Std	Std Blank
Deionized Water	—	—	—	0.10
Reaction Cocktail	—	0.20	—	—
Reagent E (Std)	—	—	0.10	—
Reagent F (TCA)	0.50	0.50	0.50	0.50
Reagent I (Enzyme Solution)	—	0.03	—	—

Mix by inversion. Then add:

	Test	Test Blank	Std	Std Blank
Reagent G ($FeCl_3$)	0.50	0.50	0.50	0.50

Mix by inversion. Centrifuge for 5 minutes. Transfer the solutions to suitable cuvettes. Record the A525nm for the Standard, Test and Blanks.

Calculations

$$\text{EmM} = (\text{A525nm Std} - \text{A525nm Std. Blank})\,(1.1)$$

$$\text{units / mg enzyme} = \frac{(\text{A525nm Test} - \text{A525nm Blank})\,(1.23)}{(\text{emM})\,(\text{mg enzyme / RM})\,(10)}$$

1.1 = Volume of Standard (in milliliters)
1.23 = Volume of Color Mix
RM = Reaction Mix (volume = 0.23 ml)
10 = Time of reaction in minutes

Unit Definition: *One unit of enzyme will catalyze the formation of 1.0 μmole of hydroxamate per minute from Na-CBZGlutaminylglycine and hydroxylamine at pH 6.0 at 37°C. (L-Glutamic acid γ-monohydroxamate is the standard).*

Final Assay Concentrations

In a 0.23 ml reaction mix, the final concentrations are 174 mM Tris, 31 mM CBZ-glutaminylglycine, 87 mM hydroxylamine, 8.7 mM glutathione, reduced form, 4 mM calcium chloride and 0.06 unit transglutaminase.

Note

1. There may be lot to lot variation in the extinction coefficient of L-glutamic acid-monohydroxamate; therefore, an extinction coefficient must be calculated for each lot. This calculation is based on reading the absorbance of a 1.1 ml standard solution which contains 0.1 ml of Reagent E (Std).

Enzymatic Assay of TRANSKETOLASE (EC 2.2.1.1)

Principle

$$X\,5\text{-}P + + R\,5\text{-}P \xrightarrow[\text{TPP, } Mg_2^+]{\text{Transketolase}} G\,3\text{-}P + \text{Sedoheptulose 7-Phosphate}$$

$$G\,3\text{-}P \xrightarrow{\text{TPI}} \text{Dihydroxyacetone Phosphate}$$

$$\text{Dihydroxyacetone Phosphate} + \beta\text{-NADH} \xrightarrow{\alpha\text{-GDH}} \alpha\text{-Glycerol Phosphate} + \beta\text{-NAD}$$

Abbreviations Used

X 5-P = Xylulose 5-Phosphate
R 5-P = Ribose 5-Phosphate
TPP = Thiamine Pyrophosphate
G 3-P = Glyceraldehyde 3-Phosphate
TPI = Triosephosphate Isomerase
α-GDH = α-Glycerophosphate Dehydrogenase
β-NADH = β-Nicotinamide Adenine Dinucleotide, Reduced Form
β-NAD = β-Nicotinamide Adenine Dinucleotide, Oxidized Form

Conditions: T = 25°C, pH = 7.7, A340nm, Light path = 1 cm

Methods: Continuous Spectrophotometric Rate Determination.

Reagents

A. 250 mM Glycylglycine Buffer, pH 7.7 at 25°C (Prepare 50 ml in deionized water using Glycylglycine, Free Base. Adjust to pH 7.7 at 25°C with 1 M NaOH).
B. 100 mM Xylulose 5-Phosphate Solution (X 5-P) (Prepare 1 ml in deionized water using D-Xylulose 5-Phosphate, Sodium Salt).
C. 50 mM Ribose 5-Phosphate Solution (R 5-P) (Prepare 1 ml in deionized water using D-Ribose 5-Phosphate, Disodium Salt).
D. 0.10% (w/v) Cocarboxylase (Thiamine Pyrophosphate) Solution (Prepare 1 ml in cold deionized water using Cocarboxylase. *Prepare Fresh*).
E. 4.3 mM β-Nicotinamide Adenine Dinucleotide, Reduced Form Solution (β-NADH) (Dissolve the contents of one 5 mg vial of β-Nicotinamide Adenine Dinucleotide, Reduced Form, Disodium Salt. *Prepare Fresh*).
F. 300 mM Magnesium Chloride Solution ($MgCl_2$) (Prepare 1 ml in deionized water using Magnesium Chloride Sterile Filtered Solution).
G. α-Glycerophosphate Dehydrogenase/Triosephosphate Isomerase Enzyme Solution. (α-GDH/TPI) (Immediately before use, prepare a solution containing 2000 TPI units/ml of α-Glycerophosphate Dehydrogenase/ Triosephosphate Isomerase in cold deionized water).
H. Transketolase Enzyme Solution (Immediately before use, prepare a solution containing 5.0 units/ml of Transketolase in cold Reagent A).

Procedure

Pipette (in milliliters) the following reagents into suitable cuvettes:

	Test	Blank
Reagent A (Buffer)	2.48	2.49
Reagent B (X 5-P)	0.10	0.10
Reagent C (R 5-P)	0.10	0.10
Reagent D (Cocarboxylase)	0.05	0.05
Reagent E (α-NADH)	0.10	0.10
Reagent F ($MgCl_2$)	0.15	0.15
Reagent G (α-GDH/TPI)	0.01	0.01

Mix by inversion and equilibrate to 25°C. Monitor the A340nm until constant, using a suitably thermostatted spectrophotometer. Then add:

Reagent H (Enzyme Solution)	0.01	—

Immediately mix by inversion and record the decrease in A340nm for approximately 10 minutes. Obtain the ΔA340nm/minute using the maximum linear rate for both the Test and Blank.

Calculations

$$\text{Units / ml enzyme} = \frac{(\Delta A340nm / min\ Test - \Delta A340nm / min\ Blank)\ (3)\ (df)}{(6.22)\ (0.01)}$$

3 = Total volume (in milliliters) of assay
df = Dilution factor
6.22 = Millimolar extinction coefficient of β-NADH at 340nm
0.01 = Volume (in milliliters) of enzyme used

$$\text{Units / mg solid} = \frac{\text{units / ml enzyme}}{\text{mg solid / ml enzyme}}$$

$$\text{Units / mg protein} = \frac{\text{units / ml enzyme}}{\text{mg protein / ml enzyme}}$$

Unit Definition: *One unit will produce 1.0 μmole of glyceraldehyde 3-phosphate from xylulose 5-phosphate per minute at pH 7.7 at 25°C, in the presence of ribose 5-phosphate, thiamine pyrophosphate and Mg^{++} using a coupled system with α-GDH/TPI.*

Final Assay Concentrations

In a 3.00 ml reaction mix, the final concentrations are 216 mM glycylglycine, 3.3 mM xylulose 5-phosphate, 1.7 mM ribose 5-phosphate, 0.002% (w/v) cocarboxylase, 0.14 mM β-nicotinamide adenine dinucleotide, reduced form, 15 mM magnesium chloride, 20 units α-glycerophosphate dehydrogenase/triosephosphate isomerase (based on triosephosphate isomerase units) and 0.05 unit transketolase.

Notes

1. Triosephosphate Isomerase Unit Definition: One unit will convert 1.0 μmole of glyceraldehyde 3-phosphate to dihydroxyacetone phosphate per minute at pH 7.6 at 25°C.
2. α-Glycerophosphate Dehydrogenase Unit Definition: One unit will convert 1.0 μmole of dihydroxyacetone phosphate to α-glycerophosphate per minute at pH 7.4 at 25°C.

Enzymatic Assay of TREHALASE (EC 3.2.1.28)

Principle

$$\text{Trehalose} + H_2O \xrightarrow{\text{Trehalase}} 2\ \text{Glucose}$$

Conditions: T = 37°C, pH = 5.7, A340nm, Light path = 1 cm

Method: Spectrophotometric Stop Rate Determination

Reagents

A. 135 mM Citric Acid Buffer, pH 5.7 at 37°C (Prepare 100 ml in deionized water using Citric Acid, Free Acid, Monohydrate. Adjust to pH 5.7 at 37°C with 1 M NaOH).
B. 140 mM D-Trehalose Solution (Prepare 10 ml in Reagent A using D(+)Trehalose, Dihydrate).
C. 500 mM Tris Buffer, pH 7.5 at 37°C (Prepare 100 ml in deionized water using Trizma Base. Adjust to pH 7.5 at 37°C with 1 M HCl).
D. Trehalase Enzyme Solution (Immediately before use, prepare a solution containing 0.1–0.3 unit/ml of Trehalase in cold Reagent A).
E. Glucose Determination Vial (Use Glucose (HK) 10 Reagent. Dissolve the contents in 10 ml of deionized water).

Procedure

Step 1: Pipette (in milliliters) the following reagents into suitable cuvettes:

	Test	Blank
Reagent A (Citrate Buffer)	0.3	0.3
Reagent D (Enzyme Solution)	0.1	0.1

Mix by inversion and equilibrate to 37°C using a suitably thermostatted spectrophotometer. Then add:

Reagent B (D-Trehalose)	0.1	—

Immediately mix by inversion and incubate at 37°C for exactly 15 minutes. Then add:

Reagent C (Tris Buffer)	0.5	0.5
Reagent B (D-Trehalose)	—	0.1

Step 2: Pipette (in milliliters) the following reagents into suitable cuvettes:

	Test	Blank
Reagent E (16–10)	3.0	3.0

Equilibrate to 37°C. Monitor the A340nm until constant, using a suitably thermostatted spectrophotometer. Record the initial A340nm for both Test and Blank. Then add:

Test Solution	0.1	—
Blank Solution	—	0.1

Immediately mix by inversion and record the increase in A340nm until complete (approximately 5 minutes). Obtain the final A340nm for both the Test and Blank.

Calculations

$$\Delta A340nm\ Test = A340nm\ Test\ Final - A340nm\ Test\ Initial$$
$$\Delta A340nm\ Blank = A340nm\ Blank\ Final - A340nm\ Blank\ Initial$$

$$\text{Units/ml enzyme} = \frac{(\Delta A340nm\ Test - \Delta A340nm\ Blank)(1.0)(3.1)}{(6.22)(2)(15)(0.1)(0.1)}$$

6.22 = Millimolar extinction coefficient of β-NADH at 340nm
2 = Number of Glucose molecules per molecule of Trehalose
15 = Reaction time (in minutes) of Step 1
1.0 = Final volume (in milliliters) of Step 1
3.1 = Final volume (in milliliters) of Step 2
0.1 = Volume From Step 1 used in Step 2
0.1 = Volume (in milliliters) of enzyme used

$$\text{Units/mg protein} = \frac{\text{units/ml enzyme}}{\text{mg protein/ml enzyme}}$$

Unit Definition: *One unit will convert 1.0 μmole of trehalose to 2.0 μmoles of glucose per minute at pH 5.7 at 37°C (liberated glucose determined at pH 7.5).*

Final Assay Concentrations

In a 0.50 ml reaction mix, the final concentrations are 135 mM citric acid, 28 mM D-trehalose, and 0.01–0.03 unit of trehalase.

Enzymatic Assay of TRYPSIN (EC 3.4.21.4)

Principle

$$\text{BAEE} + H_2O \xrightarrow{\text{Trypsin}} \text{Na-Benzoyl-L-Arginine} + \text{Ethanol}$$

Abbreviation Used

BAEE = Na-Benzoyl-L-Arginine Ethyl Ester

Conditions: T = 25°C, pH = 7.6, A253nm, Light path = 1 cm.

Method: Continuous Spectrophotometric Rate Determination

Reagents

A. 67 mM Sodium Phosphate Buffer, pH 7.6 at 25°C (Prepare 100 ml in deionized water using Sodium Phosphate, Monobasic, Anhydrous. Adjust to pH 7.6 at 25°C with 1 M NaOH).
B. 0.25 mM Na-Benzoyl-L-Arginine Ethyl Ester Solution (BAEE) (Prepare 50 ml in Reagent A using Na-Benzoyl-L-Arginine Ethyl Ester, Hydrochloride).
C. Trypsin Enzyme Solution (Immediately before use, prepare a solution containing 350–700 BAEE units/ml of Trypsin in cold Reagent A).

Procedure

Pipette (in milliliters) the following reagents into suitable quartz cuvettes:

	Test	Blank
Reagent B (BAEE)	3.00	3.00

Equilibrate to 25°C. Monitor the A253nm until constant, using a suitably thermostatted spectrophotometer. Then add:

Reagent A (Enzyme Diluent)	—	0.20
Reagent C (Enzyme Solution)	0.20	—

Immediately mix by inversion and record the increase in A253nm for approximately 5 minutes. Obtain the A253nm/minute using the maximum linear rate for both the Test and Blank.

Calculations

$$\text{BAEE Units / ml enzyme} = \frac{(\text{A253nm / min Test} - \text{A253nm Blank})\,(\text{df})}{(0.001)\,(0.2)}$$

df = Dilution factor

0.001 = The change in A253nm/minute per unit of Trypsin at pH 7.6 at 25°C in a 3.2 ml reaction mix

0.2 = Volume (in milliliters) of enzyme used

$$\text{Units / mg solid} = \frac{\text{units / ml enzyme}}{\text{mg solid / ml enzyme}}$$

$$\text{Units / mg protein} = \frac{\text{units / ml enzyme}}{\text{mg protein / ml enzyme}}$$

Unit Definition: *One BAEE unit will produce a A253nm of 0.001 per minute with BAEE as substrate at pH 7.6 at 25°C in a reaction volume of 3.2 ml.*

Final Assay Concentration

In a 3.20 ml reaction mix, the final concentrations are 67 mM sodium phosphate, 0.23 mM BAEE, 70–140 units trypsin.

Enzymatic Assay of TRYPSIN INSOLUBLE (EC 3.4.21.4)

Principle

$$\text{BAEE} + H_2O \xrightarrow{\text{Trypsin}} \text{Na - Benzoyl - L - Arginine} + \text{Ethanol}$$

Abbreviation Used

BAEE = Na-Benzoyl-L-Arginine Ethyl Ester

Conditions: T = 30°C, pH = 8.0

Method: Titrimetric

Reagents

A. 50 mM Na-Benzoyl-L-Arginine Ethyl Ester and 100 mM Calcium Chloride Substrate Solution, pH 8.0 at 30°C (Substrate) (Prepare 125 ml in deionized water using

Na-Benzoyl-L-Arginine Ethyl Ester, Hydrochloride and Calcium Chloride, Dihydrate. Adjust to pH 8.0 at 30°C with 0.1 M NaOH).

B. 20 mM Standardized Sodium Hydroxide Solution (NaOH) (Prepare 100 ml in deionized water using Sodium Hydroxide, Anhydrous).

C. Trypsin Insoluble Enzyme Solution prepare the trypsin insoluble enzyme solution as follows:

1 Add appropriate amount (ml) of suspension to a graduated (0.1 ml increment) microcolumn or Buchner Funnel.
2. Wash sample with Milli-8 water under pressure. The amount of milli-q water used should be 50 X the volume of suspension used.
3. Resuspend the gel in an appropriate amount of water. If using a Buchner Funnel, transfer the suspension to a graduated micro-column. Once the gel is suspended in the micro-column, allow the excess water to drain from the column.
4. Determine the amount of packed gel present from the graduations on the column.
5. Resuspend the packed gel (in micro-column) with the appropriate enzyme diluent to the units required to perform the assay. If the volume required does not fit into the microcolumn, perform subsequent dilutions by making the first dilution in the micro column, then transferring the suspension to a plastic container. Continue dilutions in plastic container until the appropriate units are reached. Results shall be reported as unit/ml packaged gel.

Procedure

Using a suitable pH meter in conjunction with a magnetic stirrer, pipette (in milliliters) the following reagents into a suitable titration vessel:

	Test
Reagent A (Substrate)	20.00

Equilibrate to 30°C. Adjust to pH 8.0 at 30°C with 20 mM NaOH (do not allow pH to exceed 8.35). Readjust pH to 8.35. When pH reaches 8.0 begin timing. Run blank by adding 0.5 X equivalents NaOH (X = estimated units of activity in test aliquot, Ideally 1). Time the minutes required to return to pH 8.0. Repeat until time is relatively constant between injections (<10% deviation)

Reagent C (Insol Enz)	0.100

Aliquot should be adjusted to produce 1–2 units of activity). Readjust pH to approximately 8.35. When the pH reaches 8.0 begin timing. Maintain the pH of the reaction mix at 8.0 by the addition of small volumes of NaOH (0.5 equivalents NaOH per unit of activity in aliquot). Record the volumes of NaOH used to maintain the pH at 8.0 and the time required for the consumption of NaOH. Ten injections of NaOH should not exceed a total time of 5 minutes, and pH should not exceed 8.35 at anytime during the titration.

Calculation

$$\text{Units / ml enzyme solution} = \frac{(\text{NaOH})\ (\text{molarity of NaOH})\ (1000)}{(\text{T})\ (0.1)}$$

NaOH = Volume (in milliliters) of NaOH used
1000 = Conversion factor from millimoles to μmoles as per the Unit Definition
T = Time (in minutes) required for the pH to reach 8.0
0.1 = Volume (in milliliter) of insoluble enzyme added

Units/ml of Packed Gel = Units/ml enzyme solution × 2

Unit Definition: *One unit will hydrolyze 1.0 μmole of BAEE per minute at pH 8.0 at 30°C.*

Initial Assay Concentrations

In a 20.10 ml reaction mix, the initial concentrations are 50 mM Na-benzoyl-L-arginine ethyl ester, 100 mM calcium chloride and 2–4 units trypsin insoluble enzyme.

Note

1. Aliquots of Trypsin and normality of NaOH solution need to be adjusted to insure a pH of 8.35 at all times during titration. Total time of titration should be less than 5 minutes. If these conditions are not met pH induced hydrolysis will effect the rate determination. For Trypsin, agarose (T1763) having an activity of approximately 20 units/ml packed gel a 100 μl aliquot (of a 1 to 2 dilution of packed gel) requires 25 μl injections of 20 mM NaOH to maintain the above conditions. This translates into injections equaling 0.5 equivalents of NaOH for every 1 unit of activity in the aliquot. The blank determination should be run using the same size injection and same normality of NaOH as used in test. The blank will not null out pH induced hydrolysis correctly unless this condition is met.

Enzymatic Assay of TRYPSIN TPCK/DITC GLASS (EC 3.4.21.4)

Principle

$$\text{BAEE} + H_2O \xrightarrow{\text{Trypsin}} \text{Na-Benzoyl-L-Arginine} + \text{Ethanol}$$

Abbreviation Used

BAEE = Na-Benzoyl-L-Arginine Ethyl Ester

Conditions: T = 25°C, pH = 7.6, A253nm, Light path = 1 cm

Method: Spectrophotometric Rate Determination

Reagents

A. 400 mM Sodium Phosphate Buffer, pH 7.6 at 25°C (Prepare 100 ml in deionized water using Sodium Phosphate, Dibasic, Anhydrous. Adjust to pH 7.6 with 1 M HCl).
B. 0.26 mM Na-Benzoyl-L-Arginine Ethyl Ester Solution (BAEE) (Prepare 50 ml in Reagent A using Na-Benzoyl-Larginine Ethyl Ester, Hydrochloride).
C. Trypsin Enzyme Standard Solution (STD) (Immediately before use, prepare a solution containing 0.01 mg/ml of Trypsin, in cold Reagent A).
D. Trypsin Insoluble Enzyme (Trypsin/DITC) (Use Trypsin, TPCK Bound to DITC glass).

Procedure

Step 1: Standard Determination Pipette (in milliliters) the following reagents into suitable containers:

	Standard	Blank
Reagent B (BAEE)	3.10	3.10

Equilibrate to 25°C. Then add:

Reagent C (Std)	0.10	—
Reagent A (Buffer)	—	0.10

Immediately mix by inversion and incubate at 25°C for exactly 1.5 minutes. Record the A253nm for the Standard and Blank using a suitably thermostatted spectrophotometer.

Step 2: Sample Determination: Weigh (in milligrams) the following reagent into a suitable container.

	Test	Blank
Reagent D (Trypsin/DITC)	5–10	—

Pipette (in milliliters) the following reagents into suitable containers:

Reagent A (Buffer)	0.10	0.10
Reagent B (BAEE)	3.10	3.10

Immediately vortex at a medium speed for exactly 1.5 minutes. Then filter the Test and Blank through a 0.45 μm syringe filter. Record the A253nm for both the Test and Blank filtrate using a suitable spectrophotometer.

Calculations

Standard

$$\text{Units / ml enzyme} = \frac{(\text{A253nm Test} - \text{A253nm Blank})\ (\text{df})}{(0.001)\ (1.5)}$$

df = Dilution factor
0.001 = The change in A253nm/minute per unit of Trypsin at pH 7.6 at 25°C in a 3.2 ml reaction mixture
1.5 = Time (in minutes) of assay

$$\text{Units / mg solid} = \frac{\text{units / ml enzyme}}{\text{mg solid / ml enzyme}}$$

Compare the activity of the standard to its release value (activity). This is to ensure that the assay is functioning correctly.

Test

$$\text{Units / g enzyme} = \frac{(\text{A253nm Test} - \text{A253nm Blank})\ (1000)}{(0.001)\ (1.5)\ (\text{mg Trypsin bound to DITC glass})}$$

1000 = Conversion from mg to g
0.001 = The change in A253nm/minute per unit of Trypsin at pH 7.6 at 25°C in a 3.2 ml reaction mix
1.5 = Time (in minutes) of assay

Unit Definition: *One BAEE unit will produce A253nm of 0.001 per minute with BAEE as substrate at pH 7.6 at 25°C in a reaction volume of 3.2 ml.*

Final Assay Concentration

In a 3.2 ml reaction mix, the final concentrations are 400 mM sodium phosphate, 0.25 mM Na-benzoyl-L-arginine ethyl ester, 5–10 mg of trypsin, TPCK bound to DITC glass.

Enzymatic Assay of TRYPTOPHANASE (EC 4.1.99.1)

Principle

$$\text{PRP} + \text{L-Tryptophan} + H_2O \xrightarrow{\text{Tryptophanase}} \text{Indole} + \text{Pyruvate} + NH_3$$

Abbreviation Used

PRP = Pyridoxal 5-Phosphate

Conditions: T = 37°C, pH 8.3, A540nm, Light path = 1 cm

Method: Colorimetric

Reagents

A. 1000 mM Potassium Phosphate, Dibasic, Solution (Prepare 100 ml in deionized water using Potassium Phosphate, Dibasic, Trihydrate).
B. 1000 mM Potassium Phosphate, Monobasic, Solution (Prepare 100 ml in deionized water using Potassium Phosphate, Monobasic, Anhydrous).
C. 1000 mM Potassium Phosphate Buffer, pH 8.3 at 37°C (Prepare by equilibrating 100 ml of Reagent A to 37°C and adjusting the pH to 8.3 at 37°C with Reagent B).
D. 0.81 mM Pyridoxal 5-Phosphate (PRP) (Prepare 5 ml in deionized water using Pyridoxal 5-Phosphate. *Prepare Fresh*).
E. 50 mM L-Tryptophan Solution, pH 10.8 at 37°C (Prepare 25 ml in deionized water using L-Tryptophan. Adjust to pH 10.8 at 37°C with 1 M NaOH).
F. 100% (w/v) Trichloroacetic Acid (TCA) (Use Trichloroacetic Acid, 6.1 N Solution).
G. Toluene.
H. 95% Ethanol (Nondenatured) I. 5% (w/v) p-Dimethylaminobenzaldehyde Solution (DMAB) (Prepare 20 ml in Reagent H using p-Dimethylaminobenzaldehyde. *Prepare Fresh* and *protect from light*).
J. 859 mM Hydrochloric Acid-Alcohol Reagent (Acid-Alcohol) (Prepare by adding 8 ml of Reagent M to 100 ml of Reagent H).
K. 0.43 mM Indole Standard Solution (Indole Std) (Prepare 100 ml in deionized water using Indole).
L. Tryptophanase Enzyme Solution (Immediately before use, prepare a solution containing 2–4 mg/ml in Reagent C).
M. Hydrochloric Acid (Use Hydrochloric Acid).

Procedure

Pipette (in milliliters) the following reagents into suitable containers:

	Test	Substrate	Blank
Reagent C (Buffer)	0.20	0.20	0.40
Reagent D (PRP)	0.10	0.10	0.10
Reagent L (Enzyme Solution)	0.20	0.20	—
Deionized Water	1.30	1.50	1.30

Mix by inversion and equilibrate to 37°C. Then add:

Reagent E (L-Tryptophan Soln)	0.20	—	0.20

Immediately mix by inversion and incubate at 37°C for exactly 10 minutes. Then add:

Reagent F (TCA)	0.20	0.20	0.20

Mix by inversion. Add 2.0 ml of Reagent G (Toluene) to the Test, Test Blank, and Substrate Blank. Mix vigorously to phase extract the liberated indole. Allow the phases to separate and then pipette (in milliliters) the following reagents into suitable containers:

Aliquot from Toluene Layer	0.20	0.20	0.20
Reagent I (DMAB)	1.00	1.00	1.00
Reagent J (Acid-Alcohol)	8.80	8.80	8.80

Mix by inversion and allow to equilibrate at 25°C for 10 minutes. Transfer the solutions to suitable cuvettes and record the A540nm for the Test, Test Blank, and Substrate and Blank using a suitably thermostatted spectrophotometer.

Colorimetric Assay

Prepare a standard curve by pipetting (in milliliters) the following reagents into suitable containers:

	Std 1	Std 2	Std 3	Std Blank
Reagent C (Buffer)	0.20	0.20	0.20	0.20
Reagent K (Indole Std)	0.10	0.20	0.30	—
Deionized Water	1.70	1.60	1.50	1.80

Mix by inversion and then add:

Reagent F (TCA)	0.20	0.20	0.20	0.20

Mix by inversion and add 2.0 ml of Reagent G (Toluene) to each container. Mix vigorously to phase extract the indole. Allow the phases to separate and then pipette (in milliliters) the following reagents into suitable containers.

Aliquot from				
Toluene Layer	0.20	0.20	0.20	0.20
Reagent I (DMAB)	1.00	1.00	1.00	1.00
Reagent J (Acid-Alcohol)	8.80	8.80	8.80	8.80

Mix by inversion and allow to equilibrate at 25°C for 10 minutes. Transfer the solutions to suitable cuvettes and record the A540nm for the Standards and Standard Blank, using a suitably thermostatted spectrophotometer.

Calculation

Standard Curve: A540nm Standard = A540nm Standard–A540nm Standard Blank

Prepare a standard curve by plotting the A540nm of the Standard vs micrograms of indole.

Sample Determination: A540nm = A540nm Test–(A540nm Test Blank + A540nm Substrate Blank)

Determine the total micrograms of indole liberated using the Standard curve.

$$\text{Units / mg enzyme} = \frac{(\mu\text{g indole released})}{(\text{mg enzyme / RM})}$$

RM = Reaction Mix

Unit Definition: *One mg will release 15–40 µg of indole from L-tryptophan in 10 minutes at pH 8.3 at 37°C.*

Final Assay Concentration

In a 2.00 ml reaction mix, the final concentrations are 200 mM potassium phosphate, 0.041 mM pyridoxal 5-phosphate, 5 mM tryptophan, and 0.4–0.8 mg tryptophanase.

Enzymatic Assay of TYRAMINE OXIDASE (EC 1.4.3.4)

Principle

$$\text{Tyramine} + H_2O + O_2 \xrightarrow{\text{Tyramine Oxidase}} \text{p-HPA} + NH_3 + H_2O_2$$

$$2H_2O_2 + \text{4-AAP} + \text{Phenol} \xrightarrow{\text{POD}} \text{Quinoneimine Dye} + 2H_2O$$

Abbreviations Used

p-HPA = p-Hydroxyphenylacetaldehyde

4-AAP = 4-Aminoantipyrine

POD = Peroxidase

Conditions: T = 37°C, pH 7.5, A480nm, Light path = 1 cm

Method: Colorimetric

Reagents

A. 20 mM Potassium Phosphate Buffer with 2.12 mM Phenol, 1.48 mM 4-Aminoantipyrine, and 0.1 mM Tyramine, pH 7.5 at 37°C (Reaction Mix) (Prepare

20 ml in deionized water using Potassium Phosphate, Monobasic, Anhydrous Phenol, 4-Aminoantipyrine and Tyramine, Free Base. Adjust to pH 7.5 at 37°C with 1 M KOH).

B. 100% (v/v) Ethanol (EtOH) (Use 200 Proof USP Ethyl Alcohol).

C. Peroxidase (POD).

D. 20 mM Potassium Phosphate Buffer, pH 7.5 at 37°C (Enz Dil) (Prepare 25 ml in deionized water using Potassium Phosphate, Monobasic, Anhydrous. Adjust to pH 7.5 at 37°C with 1 M KOH).

E. Tyramine Oxidase Enzyme Solution (Immediately before use, prepare a solution containing 0.09–0.18 unit/ml of Tyramine Oxidase in cold Reagent D).

Procedure

Prepare a reaction cocktail by adding Reagent C (POD) to Reagent A (Reaction Mix) for a final concentration of 5–15 units/ml of Peroxidase. Adjust to pH 7.5 at 37°C with 1 M KOH or 1 M HCl, if necessary. Pipette (in milliliters) the following reagents into suitable cuvettes:

	Test	Blank
Reaction Cocktail	1.00	1.00

Equilibrate to 37°C. Then add:

Reagent E (Enzyme Solution)	0.03	—
Reagent D (Enz Dil)	—	0.03

Immediately mix by inversion and incubate at 37°C for exactly 10 minutes. Then add:

Reagent B (EtOH)	2.00	2.00

Immediately mix by inversion and record the A480nm for both the Test and Blank using a suitable spectrophotometer.

Calculations

$$\text{Units / ml enzyme} = \frac{(\text{A480nm Test} - \text{A480nm Blank})\,(3.03)\,(\text{df})}{(17.17)\,(0.5)\,(10)\,(0.03)}$$

3.03 = Total volume (in milliliters) of stopped reaction
df = Dilution factor
17.17 = Millimolar extinction coefficient of quinoneimine dye under the conditions of this assay
0.5 = Conversion factor derived from the fact that 2 μmoles of H_2O_2 produce 1 μmole of quinoneimine dye at 480nm
10 = Time (in minutes) of the assay as per the Unit Definition
0.03 = Volume (in milliliter) of enzyme used in assay

$$\text{Units / mg solid} = \frac{\text{units / ml enzyme}}{\text{mg solid / ml enzyme}}$$

$$\text{Units / mg protein} = \frac{\text{units / ml enzyme}}{\text{mg protein / ml enzyme}}$$

Unit Definition: *One unit will oxidize 1.0 μmole of tyramine to p-hydroxyphenylacetaldehyde per minute at pH 7.5 at 37°C.*

Final Assay Concentrations

In a 1.03 ml reaction mix, the final concentrations are 20 mM potassium phosphate, 2.06 mM phenol, 1.44 mM 4-aminoantipyrine, 0.1 mM tyramine, 5–15 units peroxidase, and 0.0027–0.0054 unit tyramine oxidase.

Note

1. Peroxidase Unit Definition: One unit will form 1.0 mg purpurogallin from pyrogallol in 20 seconds at pH 6.0 at 20°C.

Enzymatic Assay of TYROSINASE (EC 1.14.18.1)

Principle

$$\text{L-Tyrosine} + O_2 \xrightarrow{\text{Tyrosinase}} \text{L-DOPA}$$

$$\text{L-DOPA} \xrightarrow{\text{Tyrosinase}} \text{L-DOPA-quinone} + H_2O$$

Abbreviation Used

L-DOPA = L-3,4-Dihydroxyphenylalanine

Conditions: T = 25°C, pH = 6.5, A280nm, Light path = 1 cm

Method: Continuous Spectrophotometric Rate Determination

Reagents

A. 50 mM Potassium Phosphate Buffer, pH 6.5 at 25°C (Prepare 50 ml in deionized water using Potassium Phosphate, Monobasic, Anhydrous. Adjust to pH 6.5 at 25°C with 1 M KOH).

B. 1 mM L-Tyrosine Solution (Prepare 100 ml in deionized water using L-Tyrosine, Free Base).

C. Tyrosinase Enzyme Solution (Immediately before use, prepare a solution containing 500–1,000 units/ml of Tyrosinase in cold Reagent A).

Procedure

Prepare a reaction cocktail by pipetting (in milliliters) the following reagents into a suitable container:

Deionized Water	9.00
Reagent A (Buffer)	10.00
Reagent B (Tyrosine)	10.00

Mix and adjust to pH 6.5 at 25°C with 1 M HCl or 1 M NaOH, if necessary. Immediately before use, oxygenate by bubbling 99.9% pure O_2 through the reaction cocktail for 3 to 5 minutes. Pipette (in milliliters) into suitable quartz cuvettes:

	Test	Blank
Reaction Cocktail	2.90	2.90

Equilibrate to 25°C. Monitor the A280nm until constant, using a suitably thermostatted spectrophotometer. Then add:

Reagent A (Buffer)	—	0.10
Reagent C (Enzyme Solution)	0.10	—

Immediately mix by inversion and record the increase in A280nm for approximately 10 minutes. Obtain the A280nm/minute using the maximum linear rate for both the Test and Blank.

Calculations

$$\text{Units / ml enzyme} = \frac{(\text{A280nm Test} - \text{A280nm Blank})\ (\text{df})}{(0.001)\ (0.1)}$$

df = Dilution factor

0.001 = The change in A280nm/minute per unit of Tyrosinase at pH 6.5 at 25°C in a 3 ml reaction mix

0.1 = Volume (in milliliters) of enzyme used

$$\text{Units / mg solid} = \frac{\text{units / ml enzyme}}{\text{mg solid / ml enzyme}}$$

$$\text{Units / mg protein} = \frac{\text{units / ml enzyme}}{\text{mg protein / ml enzyme}}$$

Unit Definition: *One unit will cause an increase in A280nm of 0.001 per minute at pH 6.5 at 25°C in a 3 ml reaction mix containing L-tyrosine.*

Final Assay Concentrations

In a 3.00 ml reaction mix, the final concentrations are 18 mM potassium phosphate, 0.3 mM L-tyrosine and 50–100 units tyrosinase.

Note

1. Final volume of all cuvettes must equal 3 ml as stated in the Unit Definition.

Enzymatic Assay of TYROSINASE Catechol Oxidase Activity (EC 1.14.18.1)

Principle

$$\text{Catechol} + \text{Ascorbic Acid} \xrightarrow{\text{Catechol Oxidase}} \text{Dehydro - Ascorbic Acid} + \text{o - enzoquinone}$$

Conditions: T = 25°C, pH = 6.5, A265nm, Light path = 1 cm

Method: Continuous Spectrophotometric Rate Determination

Reagents

A. 50 mM Potassium Phosphate Buffer, pH 6.5 at 25°C (Prepare 100 ml in deionized water using Potassium Phosphate, Monobasic, Anhydrous. Adjust to pH 6.5 at 25°C with 2 M NaOH).

B. 5.0 mM Catechol Solution (Catechol) (Prepare 10 ml in Reagent A using Catechol).
C. 2.1 mM Ascorbic Acid Solution (Asc Acid) (Prepare 10 ml in Reagent A using L-Ascorbic Acid, Sodium Salt. *Prepare Fresh*).
D. 0.065 mM Ethylenediaminetetraacetic Acid Solution (EDTA) (Prepare 10 ml in Reagent A using Ethylenediaminetetraacetic Acid, Disodium Salt, Dihydrate).
F. Tyrosinase Enzyme Solution (Immediately before use, prepare a solution containing 500–1000 units/ml of Tyrosinase in Reagent A).

Procedure

Pipette (in milliliters) the following reagents into suitable quartz cuvettes:

	Test	Blank
Reagent A (Buffer)	2.60	2.80
Reagent B (Catechol)	0.10	0.10
Reagent C (Asc Acid)	0.10	—
Reagent D (EDTA)	0.10	0.10

Mix by inversion and equilibrate to 25°C. Monitor the A265nm until constant, using a suitably thermostatted spectrophotometer. Then add:

Reagent F (Enzyme Solution)	0.10	—

Immediately mix by inversion and record the decrease in A265nm for approximately 5 minutes. Obtain the ΔA265nm using the maximum linear rate for both the Test and Blank.

Calculations

$$\text{Units / mg enzyme} = \frac{\Delta A265\text{nm / min Test} - \Delta A265\text{nm / min Blank}}{(0.001)\ (\text{mg enzyme / RM})}$$

0.001 = The change in A265nm per unit of Tyrosinase in a 3.00 ml reaction mixture at pH 6.5 at 25°C.
RM = Reaction Mix

Unit Definition: *One unit is equal to a ΔA265nm of 0.001 per minute at pH 6.5 at 25°C in a 3 ml reaction mix containing catechol and L-ascorbic acid.*

Final Assay Concentrations

In a 3.00 ml reaction mix, the final concentrations are 50 mM potassium phosphate, 0.17 mM catechol, 0.070 mM L-ascorbic acid, 0.0022 mM EDTA, and 50–100 units of catechol oxidase.

Enzymatic Assay of TYROSINASE Polyphenol Oxidase Activity (EC 1.14.18.1)

Principle

$$\text{L-DOPA} + O_2 + \text{Ascorbic Acid} \xrightarrow{\text{TYR}} \text{o-Benzoquinone} + H_2O + \text{Dehydro-Ascorbic Acid}$$

Abbreviations Used

TYR = Tyrosinase
L-DOPA = L-3,4-Dihydroxyphenylalanine

Conditions: T = 25°C, pH 6.5, A265nm, Light path = 1 cm

Method: Continuous Spectrophotometric Rate Determination

Reagents

A. 50 mM Potassium Phosphate Buffer, pH 6.5 at 25°C (Prepare 100 ml in deionized water using Potassium Phosphate, Monobasic, Anhydrous. Adjust to pH 6.5 at 25°C with 1 M NaOH).
B. 5.0 mM L-3,4-Dihydroxyphenylalanine (L-DOPA) (Prepare 10 ml in Reagent A using L-3,4-Dihydroxyphenylalanine).
C. 2.1 mM Ascorbic Acid Solution (Prepare 10 ml in Reagent A using L-Ascorbic Acid, Sodium Salt. *Prepare Fresh*).
D. 0.065 mM Ethylenediaminetetraacetic Acid (EDTA) (Prepare 10 ml in Reagent A using Ethylenediaminetetraacetic Acid, Disodium, Dihydrate Salt).
E. Tyrosinase Enzyme Solution (PPO) (Immediately before use, prepare a solution containing 500–1000 u/ml Tyrosinase in Reagent A).

Procedure

Pipette (in milliliters) the following reagents into suitable quartz cuvettes:

	Test	Blank
Reagent A (Buffer)	2.60	2.80
Reagent B (L-DOPA)	0.10	0.10
Reagent C (Asçorbic Acid)	0.10	—
Reagent D (EDTA)	0.10	0.10

Mix by inversion and equilibrate to 25°C. Monitor the A265nm until constant, using a suitably thermostatted spectrophotometer. Then add:

Reagent E (TYR)	0.10	—

Immediately mix by inversion and record the decrease in A265nm for approximately 5 minutes. Obtain the ΔA265nm/minute using the maximum linear rate for both the Test and Blank.

Calculations

$$\text{Units / mg enzyme} = \frac{(\Delta\text{A265nm / min Test} - \Delta\text{A265nm / min Blank})\ (\text{df})}{(0.001)\ (0.1)}$$

df = Dilution factor
0.001 = The change in A265nm/min per unit of polyphenol oxidase in a 3.0 ml reaction mixture at pH 6.5 at 25°C.
0.1 = Volume (in milliliter) of enzyme used

$$\text{Units / mg solid} = \frac{\text{units / ml enzyme}}{\text{mg solid / ml enzyme}}$$

$$\text{Units / mg protein} = \frac{\text{units / ml enzyme}}{\text{mg protein / ml enzyme}}$$

Unit Definition: *One unit is equal to a ΔA265nm of 0.001 per min at pH 6.5 at 25°C in 3 ml reaction mix containing L-DOPA and L-ascorbic acid.*

Final Assay Concentrations

In a 2.92 ml reaction mix, the final concentrations are 50 mM potassium phosphate, 0.17 mM L-3,4-dihydroxyphenylalanine, 0.072 mM ascorbic acid, .0022 mM Ethylenediaminetetraacetic Acid, and 50–100 units of tyrosinase.

Enzymatic Assay of L-TYROSINE DECARBOXYLASE (EC 4.1.1.25)

Principle

$$\text{L-Tyrosine L-Tyrosine} \xrightarrow[\text{PRP}]{\text{Decarboxylase}} \text{Tyramine} + CO_2$$

Abbreviation Used

PRP = Pyridoxal 5-Phosphate

Conditions: T = 37°C, pH 5.5

Method: Radiolabelled Stop Reaction

Reagents

A. 200 mM Sodium Phosphate Solution (Prepare 50 ml in deionized water using Sodium Phosphate, Dibasic, Anhydrous).

B. 100 mM Citrate Phosphate Buffer, pH 5.5 at 37°C (Buffer) (Prepare 100 ml in deionized water using Citric Acid, Free Acid, Anhydrous. Adjust to pH 5.5 at 37°C with Reagent A).

C. McIlvaine's Buffer, pH 5.5 at 37°C (Buffer) (Prepare using 50 ml of Reagent A and adjust to pH 5.5 at 37°C with Reagent B).

D. 2.65 mM L-Tyrosine Solution (Prepare 10 ml in Reagent C using L-Tyrosine, Free Base. Heat (70°C) may be required in order to solubilize).

E. 0.3 mM Pyridoxal 5-Phosphate Solution (PRP) (Prepare 2 ml in Reagent C using Pyridoxal 5-Phosphate).

F. $_{14}$C[COOH] L-Tyrosine (Use $_{14}$C[COOH] L-Tyrosine, 50–60 mCi/mmol, 50 µCi/ml).

G. L-Tyrosine Reaction Cocktail (L-Tyr) (Prepare by adding 0.05 ml of Reagent E to 2 ml of Reagent C).

H. 1 M Methylbenzethonium Hydroxide Solution (MBH) (Use Benzethonium Hydroxide, approximately 1.0 M solution in Methanol).

I. Filter Paper (1 cm × 2 cm strips of filter paper are prepared using Whatman 3 MM paper. Add 0.050 ml of Reagent G to each strip and allow to air dry. This step is to be carried out in a fume hood).

J. Vacutainer Tubes (Use Vacutainer Blood Tubes, 13 × 75 mm, Becton Dickinson).

K. 25% (w/v) Trichloroacetic Acid Solution (TCA) (Prepare 10 ml in deionized water using Trichloroacetic Acid, 6.1 N Solution, approximately 100% (w/v).

L. L-Tyrosine Decarboxylase Enzyme Solution (Immediately before use, prepare a solution containing approximately 0.005–0.02 units/ml of L-Tyrosine Decarboxylase in cold Reagent C).

M. Scintillation Cocktail (Scint Fluid).

Procedure

Radiolabelled CO_2 is trapped by adsorbing it on an organic solvent which is present on filter paper located above the reaction mixture in a vacutainer tube. Pipette (in milliliters) the following reagents into suitable vacutainer tubes.

	Test	Blank	PRP
Reagent G (L-Tyr)	0.10	0.10	0.10
Reagent E (PRP)	0.10	0.10	—
Reagent C (Buffer)	—	—	0.10

Equilibrate to 37°C. Then add:

Reagent L (Enz Soln)	0.10	—	0.10
Reagent C (Buffer)	—	0.10	—

Immediately place Reagent H (Filter Paper) in the top of the vacutainer tube and seal the tubes by replacing the stopper. Incubate for exactly 10 or 20 minutes. Then add by injecting with a syringe and needle through the stopper. (*Do not remove stopper to add TCA*).

Reagent K (TCA)	0.40	0.40	0.40

Mix by swirling. Allow to stand for 12–16 hours at room temperature to allow for all evolved CO_2 to be absorbed onto the filter paper. Remove filter papers from vacutainer tubes and place into scintillation vials containing 7 ml of Reagent L (Scint). This should be performed in a hood. Potential DPM (disintegration per minute) are prepared by pipetting 0.05 ml of Reagent F (L-Tyr) onto Reagent H (Filter Paper). Place into scintillation vi al containing 7 ml of Reagent L (Scint Fluid).

Calculations

$$\text{Potential DPM / } \mu\text{mole} = \frac{\text{DPM of potential}}{\text{Total } \mu\text{moles of L - Tyrosine}}$$

$$\text{Units / ml enzyme} = \frac{(\text{DPM Test} - \text{DPM Blank})\,(\text{df})}{(\text{T})\,(0.1)\,(\text{potential DPM / } \mu\text{mole})}$$

DPM = Disintegration per minute
df = Dilution factor
T = Time (in minutes) of assay as per the Unit Definition
0.1 = Volume (in milliliter) of enzyme used

$$\text{Units / mg solid} = \frac{\text{units / ml enzyme}}{\text{mg solid / ml enzyme}}$$

Unit Definition: *One unit will liberate 1.0 µmole of CO_2 for L-tyrosine per minute at pH 5.5 at 37°C.*

Final Assay Concentrations

In a 0.30 ml reaction mix, the final concentrations are 100 mM citric acid, 0.86 mM L-tyrosine, 0.1 mM pyridoxal 5-phosphate, and 0.0005–0.002 units L-tyrosine decarboxylase.

Notes

1. The enzyme activity is also measured without any Reagent D(PRP) being present.
2. This concentration is only an estimate since it is not known what exact volume of sodium phosphate is required to adjust the pH to 5.5 at 37°C.

Enzymatic Assay of TYROSINE HYDROXYLASE (EC 1.14.16.2)

Principle

$$\text{L-Tyrosine} + \text{DMTHP} + O_2 \xrightarrow{\text{TH}} \text{L-DOPA} + \text{Quinonoid Dihydropterin} + H_2O$$

$$\text{Quinonoid Dihydropterin} + \beta\text{-NADH} \xrightarrow{\text{DHT-RDase}} \beta\text{-NAD} + \text{DMTHP}$$

Abbreviations Used

DMTHP = 6,7-Dimethyl-5,6,7,8-Tetrahydropterine
TH = Tyrosine Hydroxylase
L-DOPA = L-3,4-Dihydroxyphenylalanine
β-NADH = β-Nicotinamide Adenine Dinucleotide, Reduced Form
DHT-RDase = Dihydropteridine Reductase
β-NAD = β-Nicotinamide Adenine Dinucleotide, Oxidized Form

Conditions: T = 37°C, pH = 7.0, A340nm, Light path = 1 cm.

Method: Continuous Spectrophotometric Rate Determination

Reagents

A. 100 mM Tris HCl Buffer, pH 7.0 at 37°C (Buffer) (Prepare 100 ml in deionized water using Trizma Base. Adjust to pH 7.0 at 37°C with 1 M HCl).
B. 30 mM L-Tyrosine Solution, pH 7.0 at 37°C (L-Tyr) (Prepare 10 ml in deionized water using L-Tyrosine. Adjust to pH 7.0 at 37°C with 1 M NaOH).
C. 10 mM 6,7-Dimethyl-5,6,7,8-Tetrahydropterine Solution (DMTHP) (Immediately before use prepare 2 ml in deionized water using 6,7-Dimethyl-5,6,7,8-etrahydropterine, Monohydrochloride).
D. 0.43 mM β-Nicotinamide Adenine Dinucleotide, Reduced Form, Solution (β-NADH) (Dissolve the contents of one 10 mg vial of β-Nicotinamide Adenine Dinucleotide, Reduced Form, Disodium Salt, in the appropriate volume of deionized water. *Prepare Fresh*).
E. Catalase Enzyme Solution (Catalase) (Immediately before use, prepare a solution containing 10,000 units/ml of Catalase, in cold Reagent A).
F. Dihydropteridine Reductase Enzyme Solution (DHT-RDase) (Immediately before use, prepare a solution containing 5 units/ml of Dihyropteridine Reductase, in cold Reagent A).

G. Tyrosine Hydroxylase Enzyme Solution (TH) (Immediately before use, prepare a solution containing 150–450 units/ml of Tyrosine Hydroxylase in cold Reagent A).

Procedure

Pipette (in milliliters) the following reagents into suitable cuvettes:

	Test	Blank
Reagent A (Buffer)	1.50	1.50
Reagent B (L-Tyr)	0.03	0.03
Reagent C (DMTHP)	0.05	0.05
Reagent D (β-NADH)	1.00	1.00
Reagent E (Catalase)	0.05	0.05
Reagent F (DHT-RDase)	0.10	0.10
Deionized Water	0.17	0.20

Mix by inversion and equilibrate to 37°C. Monitor the A340nm until constant, using a suitably thermostatted spectrophotometer. Then add:

Reagent G (TH)	0.03	—

Immediately mix by inversion and record the decrease in A340nm for approximately 5 minutes. Obtain the A340nm/minute using the maximum linear rate for both the Test and Blank.

Calculations

$$\text{Units / ml enzyme} = \frac{(\text{A340nm / min Test} - \text{A340nm / min Blank})\ (2.93)\ (\text{df})\ (1000)}{(6.22)\ (0.03)}$$

2.93 = Total volume (in milliliters) of assay
df = Dilution factor
1000 = Conversion factor from μmoles to nanomoles as per the Unit Definition
6.22 = Millimolar extinction coefficient of β-NADH at 340nm
0.03 = Volume (in milliliter) of enzyme used

$$\text{Units / mg protein} = \frac{\text{units / ml enzyme}}{\text{mg protein / ml enzyme}}$$

Unit Definition: *One unit will form 1.0 nanomole of L-DOPA from tyrosine per minute at pH 7.0 at 37°C.*

Final Assay Concentrations

In a 2.93 ml reaction mix, the final concentrations are 57 mM Tris, 0.3 mM L-tyrosine, 0.2 mM 6,7-dimethyl- 5,6,7,8-tetrahyropterine, 0.15 mM β-nicotinamide adenine dinucleotide, reduced form, 500 units catalase, 0.5 unit dihydropteridine reductase, and 4.5–14 units tyrosine hydroxylase.

Notes

1. Catalase is added to protect the assay system against peroxide-mediated inactivation.
2. A significant rate for the blank (A340nm/min = 0.03–0.04) will occur.

3. Catalase Unit Definition: One unit will decompose 1.0 µmole of H_2O_2 per minute at pH 7.0 at 25°C, while the H_2O_2 concentration falls from 10.3 to 9.2 mM. The rate of is appearance of H_2O_2 is followed by observing the rate of decrease in absorbance at 240 nm.
4. Dihydropteridine Reductase Unit Definition: One unit will oxidize 1.0 µmole of β-NADH to β-NAD with 6,7- dimethyldihydropterine (quinonoid isomer) as the nonnucleotide substrate per minute at pH 7.2 at 25°C.

Enzymatic Assay of UREASE (EC 3.5.1.5)

Principle

$$\text{Urea} + H_2O \xrightarrow{\text{Urease}} CO_2 + 2NH_3$$

Conditions: T = 30°C, pH = 8.2, A480nm, Light Path = 1 cm

Method: Spectrophotometric Stop Rate Determination.

Reagents

A. 10 mM Potassium Phosphate Buffer, pH 8.2 at 30°C with 10 mM Lithium Chloride and 1 mM Ethylenediaminetetraacetic Acid. (Prepare 100 ml in deionized water using Potassium Phosphate, Dibasic, Trihydrate, Lithium Chloride, Anhydrous, and Ethylenediaminetetraacetic Acid, Tetrasodium Salt, Hydrate. Adjust to pH 8.2 at 30°C with 1 M HCl).
B. 66 mM Urea Solution. (Prepare 25 ml in Reagent A using Urea).
C. 0.4% (w/v) Ficoll Solution (Prepare 20 ml in deionized water using Ficoll, Type 400-DL).
D. Nessler's Color Reagent (NCR) (Prepare 75 ml using 15 ml Aldrich Nessler's Reagent, 15 ml Reagent C, and 45 ml deionized water).
E. 2.50 mM Ammonia Standard Solution (Amm Std) (Prepare 50 ml in deionized water using Ammonium Sulfate).
F. Urease Enzyme Solution (Immediately before use, prepare a solution containing 100–150 units/ml of Urease in cold Reagent A).

Procedure

Pipette (in milliliters) the following reagents into a suitable container:

	Test
Reagent B (Urea Solution)	10.00

Incubate at 30°C. Then at time zero add:

Reagent F (Enzyme Solution)	0.025

Immediately mix by inversion and remove 1.0 ml aliquot from the Test solution. Transfer to a suitable container containing 5.0 ml of Reagent D (NCR) and mix by inversion. This is to be used as the Reagent Blank. Continue to incubate at 30°C. Then at 2, 4 and 6 minutes (time increments), transfer 1.0 ml Test solution into suitable containers containing 5.0 ml of Reagent D (NCR) and mix by inversion. Transfer the solutions to suitable cuvets and record the A480nm for the Reagent Blank and Tests using a suitable spectrophotometer.

Colorimetric

Standards: Pipette (in milliliters) the following reagents into suitable containers.

	Std Blank	Std 1	Std 2	Std 3	Std 4	Std 5	Std 6
Deionized water	1.00	0.90	0.80	0.60	0.40	0.20	—
Reagent E (Amm Std)	—	0.10	0.20	0.40	0.60	0.80	1.00
Reagent D (NCR)	5.00	5.00	5.00	5.00	5.00	5.00	5.00

Mix by inversion. Transfer to suitable cuvets, measure and record the A480nm for the Standard Blank and Standards.

Calculations

A480nm Standard = A480nm Standard–A480nm Standard Blank
Plot µmoles of NH_3 vs A480nm to obtain A480nm/µmole NH_3
A480nm Test = A480nm Test–A480nm Reagent Blank
A480nm Test/min = A480nm Test/Time increment (in minutes)·

$$\text{Units / g enzyme} = \frac{\text{A480nm Test / minute (1000)}}{(\text{A480nm / }\mu\text{mole NH}_3)\text{ (mg enzyme / ml RM)}}$$

RM = Reaction mixture
1000 = Conversion factor for converting milligrams to grams

Unit Definition: *One unit will liberate 1.0 µmole of NH_3 from urea per minute at pH 8.2 at 30°C.*

Final Assay Conditions

In a 10.025 ml reaction mixture the final concentrations are 10 mM potassium phosphate, 10 mM lithium chloride, 1 mM Ethylenediaminetetraacetic Acid, 66 mM urea and 2.5–3.75 units urease.

Enzymatic Assay of UREASE (ATP-Hydrolyzing) (EC 3.5.1.45)

Principle

$$\text{Urea} + \text{ATP} \xrightarrow{\text{UA}} 2\text{NH}_3 + \text{CO}_2 + \text{ADP} + \text{Pi}$$

$$\text{PEP} + \text{ADP} \xrightarrow{\text{PK}} \text{Pyruvate} + \text{ATP}$$

$$\text{Pyruvate} + \text{NADH} \xrightarrow{\text{LDH}} \text{Lactate} + \text{NAD}$$

Abbreviations Used

ATP = Adenosine 5'-Triphosphate
UA = Urease, ATP-Hydrolyzing
ADP = Adenosine 5'-Diphosphate

PEP = Phospho(enol) pyruvate
PK = Pyruvate Kinase
β-NADH = β-Nicotinamide Adenine Dinucleotide, Reduced Form
LDH = Lactic Dehydrogenase
β-NAD = β-Nicotinamide Adenine Dinucleotide, Oxidized Form
Pi = Inorganic Phosphate

Conditions: T = 30°C, pH = 8.0, A340nm, Light path = 1.0 cm

Method: Continuous Spectrophotometric Rate Determination

Reagents

A. 100 mM Tris HCl Buffer, pH 8.0 at 30°C (Prepare 100 ml in deionized water using Trizma Base. Adjust to pH 8.0 at 30°C with 1 M HCl).
B. 11 mM Adenosine 5'-Triphosphate Solution (ATP) (Prepare 15 ml in Reagent A using Adenosine 5'-Triphosphate, Disodium Salt).
C. 17 mM Phospho(enol) pyruvate Solution (PEP) (Prepare 15 ml in Reagent A using Phospho(enol) pyruvate, Tri(cyclohexylammonium) Salt).
D. 555 mM Potassium Chloride Solution (KCl) (Prepare 15 ml in Reagent A using Potassium Chloride).
E. 89 mM Magnesium Sulfate Solution ($MgSO_4$) (Prepare 15 ml in Reagent A using Magnesium Sulfate, Heptahydrate).
F. 85 mM Potassium Bicarbonate Solution ($KHCO_3$) (Prepare 15 ml in Reagent A using Potassium Bicarbonate).
G. 4.5 mM β-Nicotinamide Adenine Dinucleotide, Reduced Form Solution (β-NADH) (Prepare 1 ml in deionized water using β-Nicotinamide Adenine Dinucleotide, Reduced Form, Disodium Salt).
H. 300 mM Urea Solution (Urea) (Prepare 10 ml in deionized water using Urea).
I. PK/LD Enzymes Suspension (PK/LDH).
J. 50 mM Tris HCl, pH 8.0 at 30°C (Enzyme Diluent) (Prepare 50 ml in deionized water using Trizma Base. Adjust to pH 8.0 at 30°C with 1 M HCl).
K. Urease (ATP-Hydrolyzing) Enzyme Solution (UA) (Immediately before use, prepare a solution containing 0.2–0.4 unit/ml of Urease (ATP-Hydrolyzing) in cold Reagent J).

Procedure

Prepare a reaction cocktail by pipetting (in milliliters) the following reagents into a suitable containers:

Reagent A (Buffer)	49.50
Reagent B (ATP)	10.00
Reagent C (PEP)	10.00
Reagent D (KCl)	10.00
Reagent E ($MgSO_4$)	10.00
Reagent F ($KHCO_3$)	10.00
Reagent I (PK/LDH)	0.50

Mix by swirling. Pipette (in milliliters) the following reagents into suitable cuvettes:

	Test	Blank
Reaction Cocktail	2.70	2.70
Reagent G (β-NADH)	0.10	0.10
Reagent H (Urea)	0.10	0.10

Mix by inversion and equilibrate to 30°C. Monitor the A340nm until constant, using a suitably thermostatted spectrophotometer. Then add:

	Test	Blank
Reagent K (UA)	0.10	—
Reagent J (Enzyme Diluent)	—	0.10

Immediately mix by inversion and record the decrease in A340nm for approximately 5 minutes. Obtain the ΔA340nm/minute using the maximum linear rate for both the Test and Blank.

Calculations

$$\text{Units / mg enzyme} = \frac{(\Delta A340\text{nm / min Test} - \Delta A340\text{nm / min Blank})}{(6.22)\ (\text{mg enzyme / ml RM})}$$

6.22 = Millimolar extinction coefficient of β-NADH at 340nm
RM = Reaction Mix

Unit Definition: *One unit will liberate 2 μmoles of NH_3 from 1 μmole of urea per minute at pH 8.0 at 30°C in a coupled system with PK/LDH.*

Final Assay Concentrations

In a 3.00 ml reaction mix, the final concentrations are 91 mM Tris, 1.0 mM ATP, 1.5 mM phospho(enol) pyruvate, 50 mM potassium chloride, 8.0 mM magnesium sulfate, 7.7 mM potassium bicarbonate, 0.15 mM β-NADH, 10 mM urea, 3.2 units pyruvate kinase, 4.5 units lactic dehydrogenase, and0.02–0.04 unit urease (ATP-hydrolyzing).

Notes

1. Contains not less than 700 Pyruvate Kinase units and 1000 Lactic Dehydrogenase units per ml.
2. Unit Definition for L-Lactic Dehydrogenase: One unit will reduce 1.0 μmole of pyruvate to L-lactate per minute at pH 7.5 at 37°C.
3. Unit Definition for Pyruvate Kinase: One unit will convert 1.0 μmole of phospho(enol) pyruvate to pyruvate per minute at pH 7.6 at 37°C.

Enzymatic Assay of UREASE from Jack Beans (EC 3.5.1.5)

Principle

$$\text{Urea} + H_2O \xrightarrow{\text{Urease}} CO_2 + 2NH_3$$

Conditions: T = 25°C, pH = 7.0

Method: Titrimetric

Reagents

A. 750 mM Sodium Phosphate Buffer, pH 7.0 at 25°C (Prepare 200 ml in deionized water using Sodium Phosphate, Monobasic, Anhydrous. Adjust to pH 7.0 at 25°C with 1 M NaOH. Store at room temperature).

B. 500 mM Urea with 0.05% (w/v) Bovine Serum Albumin Solution (Urea) (Prepare 50 ml in Reagent A using Urea and Albumin, Bovine. Adjust to pH 7.0 at 25°C with 100 mM HCl or 100 mM NaOH, if necessary).

C. 0.10% (w/v) 3-(4-Dimethylamino-1-Naphthylazo)-4-Methoxybenzenesulfonic Acid Solution (Indicator) (Prepare 100 ml in deionized water using 3-(4-Dimethylamino-1-Naphthylazo)-4-Methoxybenzenesulfonic Acid Solution. Facilitate dissolving by first adding 2.6 ml of 100 mM NaOH then adjusting the final volume to 100 ml with deionized water).

D. 100 mM Standardized Hydrochloric Acid Solution (HCl) (Prepare 100 ml in cold deionized water using Concentrated Hydrochloric Acid. Standardize according to the ACS Reagent Procedure).

E. 20 mM Sodium Phosphate Buffer, pH 7.0 at 25°C (Enzyme Diluent) (Prepare 100 ml in deionized water using Sodium Phosphate, Monobasic, Anhydrous. Adjust to pH 7.0 at 25°C with 1 M NaOH).

F. Urease Enzyme Solution (Immediately before use, prepare a solution containing 200–400 units/ml of Urease in cold Reagent E).

Procedure

Pipette (in milliliters) the following reagents into suitable containers:

	Test	Blank
Reagent B (Urea)	1.00	1.00

Equilibrate to 25°C. Then add:

	Test	Blank
Reagent F (Enzyme Solution)	0.10	—

Mix by stirring and incubate at 25°C for exactly 5 minutes. Then add:

	Test	Blank
Reagent D (HCl)	3.00	3.00
Reagent F (Enzyme Solution)	—	0.10
Reagent C (Indicator)	2 drops	2 drops

Using a suitable magnetic stirrer, titrate immediately with Reagent D (HCl) by adding small amounts until the color of the indicator turns from orange to pink. This is the end point. Record the volume (Vol.) of Reagent D (HCl) used for both the Test and Blank solutions.

Calculations

$$\text{Vol. HCl} = \text{Vol. HCl Test} - \text{Vol. HCl Blank}$$

$$\text{Units / ml enzyme} = \frac{(\text{HCl Molarity})\ (\text{Vol. HCl})\ (1000)\ (\text{df})}{(5)\ (0.1)}$$

Vol. HCl = Volume (in milliliters) of Reagent D used in the titration
1000 = Conversion factor from millimoles to micromoles as per the Unit definition
df = Dilution factor
5 = Time of assay (in minutes) as per the Unit Definition
0.1 = Volume (in milliliter) of enzyme used

$$\text{Units / mg solid} = \frac{\text{units / ml enzyme}}{\text{mg solid / ml enzyme}}$$

Units/g solid = Units/mg solid × 1000

Unit Definition: *One unit will liberate 1.0 µmole of ammonia from urea per minute at pH 7.0 at 25°C, under our assay conditions. It is equivalent to 1.0 I.U. or 0.054 Sumner unit (1.0 mg ammonia nitrogen in 5 minutes at pH 7.0 at 20°C).*

Final Assay Concentrations

In a 1.10 ml reaction mix, the final concentrations are 684 mM sodium phosphate, 455 mM urea, 0.05% (w/v) Bovine Serum Albumin and 25–50 units urease.

Note

1. This assay is not to be used to determine the activity of Urease, from Bacillus pasteurii.

Enzymatic Assay of URICASE (EC 1.7.3.3)

Principle

$$\text{Urate} + O_2 + H_2O \xrightarrow{\text{Uricase}} \text{Allantoin} + H_2O_2 + CO_2$$

Conditions: T = 25°C, pH = 8.5, A293nm, Light path = 1 cm

Method: Continuous Spectrophotometric Rate Determination

Reagents

A. 20 mM Borax Solution (Prepare 100 ml in deionized water using Borax, Sodium Tetraborate).
B. 20 mM Boric Acid Buffer, pH 8.5 at 25°C (Prepare 200 ml in deionized water using Boric Acid. Adjust to pH 8.5 at 25°C with Reagent A).
C. 48 µM Uric Acid Solution (Uric Acid) (Prepare 125 ml in Reagent B using Uric Acid Sodium Salt. The solution may require heat to effect complete dissolution. The A293nm of this solution should be from 0.6–0.7 in a 1 cm light path. Dilute accordingly with Reagent B. *Prepare Fresh*).
D. Uricase Enzyme Solution (Immediately before use, prepare a solution containing approximately 0.25 unit/ml of Uricase in Reagent B).

Procedure

Pipette (in milliliters) the following reagents into suitable cuvettes:

	Test	Blank
Reagent C (Uric Acid)	2.90	2.90

Equilibrate to 25°C. Monitor the A293nm until constant using a suitably thermostatted spectrophotometer. Then add:

	Test	Blank
Reagent D (Enzyme Solution)	0.10	—
Reagent B (Buffer)	—	0.10

Immediately mix by inversion and record the decrease in A293nm for approximately 5 minutes. Obtain the ΔA293nm/minute using the maximum linear rate for both the Test and Blank.

Calculations

$$\text{Units / ml enzyme} = \frac{(\Delta A293nm / min\ Test - \Delta A293nm / min\ Blank)\ (3)\ (df)}{(12.3)\ (0.1)}$$

3 = Total volume (in milliliters) of assay
df = Dilution factor
12.3 = Millimolar extinction coefficient of Uric Acid at 293nm
0.1 = Volume (in milliliter) of enzyme used

$$\text{Units / mg solid} = \frac{\text{units / ml enzyme}}{\text{mg solid / ml enzyme}}$$

$$\text{Units / mg protein} = \frac{\text{units / ml enzyme}}{\text{mg protein / ml enzyme}}$$

Unit Definition: *One unit will convert 1.0 μmole of uric acid to allantoin per minute at pH 8.5 at 25°C.*

Final Assay Concentration

In a 3.00 ml reaction mix, the final concentrations are 20 mM boric acid, 46 μM uric acid, and 0.025 unit uricase.

Note

1. This assay is not to be used to assay the activity of Uricase, Microbial source, from Arthrobacter globiformis.

Enzymatic Assay of URICASE (EC 1.7.3.3)

Principle

$$\text{Urate} + H_2O + O_2 \xrightarrow{\text{Uricase}} \text{Allantoin} + H_2O_2 + CO_2$$

Conditions: T = 25°C, pH = 9.0, A293nm, Light path = 1 cm

Method: Continuous Spectrophotometric Rate Determination

Reagents

A. 20 mM Boric Acid Buffer, pH 9.0 at 25°C (Prepare 100 ml in deionized water using Boric Acid. Adjust to pH 9.0 at 25°C with 1 M HCl or 1 M NaOH).
B. 3.57 mM Uric Acid Solution (Urate) (Prepare 10 ml in Reagent A using Uric Acid, Sodium Salt. The solution may require heat and vortexing in order to effect complete dissolution).
C. Uricase Enzyme Solution (Immediately before use, prepare a solution containing 0.5–1.0 unit/ml of Uricase in cold Reagent A).

Procedure

Pipette (in milliliters) the following reagents into suitable cuvettes:

	Test	Blank
Reagent A (Buffer)	3.00	3.00
Reagent B (Urate)	0.10	0.10

Mix by swirling and equilibrate to 25°C. Monitor the A293nm until constant using a suitably thermostatted spectrophotometer. Then add:

Reagent C (Enzyme Solution)	0.02	—
Reagent A (Buffer)	—	0.02

Immediately mix by inversion and record the decrease in A293nm for approximately 5 minutes. Obtain the ΔA293nm/minute using the maximum linear rate for both the Test and Blank.

Calculations

$$\text{Units / ml enzyme} = \frac{(\Delta A293\text{nm / min Test} - \Delta A293\text{nm / min Blank})\ (3.12)\ (\text{df})}{(12.6)\ (0.02)}$$

3.12 = Total volume (in milliliters) of assay
df = Dilution factor
12.6 = Millimolar extinction coefficient of Uric Acid at 293nm
0.02 = Volume (in milliliter) of enzyme used

$$\text{Units / mg solid} = \frac{\text{units / ml enzyme}}{\text{mg solid / ml enzyme}}$$

$$\text{Units / mg protein} = \frac{\text{units / ml enzyme}}{\text{mg protein / ml enzyme}}$$

Unit Definition: *One unit will convert 1.0 μmole of uric acid to allantoin per minute at pH 9.0 at 25°C.*

Final Assay Concentrations

In a 3.12 ml reaction mix, the final concentrations are 20 mM boric acid, 0.11 mM uric acid, and 0.01–0.02 unit uricase.

Note

1. If the ΔA293 nm/min is 70.06, prepare a fresh dilutions of the enzyme with a higher dilution factor so that the ΔA293nm/min is <0.06.

Enzymatic Assay of URIDINE 5'-DIPHOSPHOGALACTOSE 4-EPIMERASE (EC 5.1.3.2)

Principle

$$\text{UDP - Gal UDP - Gal} \xrightarrow{\text{4-Epimerase}} \text{UDPG}$$

$$\text{UDPG} + 2\beta\text{-NAD} + \text{H}_2\text{O UDPG} \xrightarrow{\text{Dehydrogenase}} \text{UDP-Glucuronate} + 2\beta\text{-NADH}$$

Abbreviations Used

UDP-Gal = Uridine 5'-Diphosphogalactose
UDPG = Uridine 5'-Diphosphoglucose
UDP-Gal 4-Empimerase = Uridine 5'-Diphosphogalactose 4-Epimerase
β-NAD = β-Nicotinamide Adenine Dinucleotide, Oxidized Form
β-NADH = β-Nicotinamide Adenine Dinucleotide, Reduced Form
UDP-Glucuronate = Uridine 5'-Diphosphoglucuronate

Conditions: T = 25°C, pH = 8.8, A340nm, Light path = 1 cm

Method: Continuous Spectrophotometric Rate Determination

Reagents

A. 1000 mM Glycine Buffer, pH 8.8 at 25°C (Prepare 100 ml in deionized water using Glycine, Free Base. Adjust to pH 8.8 at 25°C with 1 M NaOH).
B. 5 mM Uridine 5'-Diphosphogalactose Solution (UDP-Gal) (Prepare 5 ml in deionized water using Uridine 5'-Diphosphogalactose, Sodium Salt).
C. 50 mM β-Nicotinamide Adenine Dinucleotide Solution (β-NAD) (Prepare 2 ml in deionized water using β-Nicotinamide Adenine Dinucleotide or dissolve the contents of one 50 mg vial of β-Nicotinamide Adenine Dinucleotide in the appropriate volume of deionized water. *Prepare Fresh*).
D. Uridine 5'-Diphosphoglucose Dehydrogenase Enzyme Solution (UDPGDH) (Immediately before use, prepare a solution containing 2 units/ml in cold deionized water using Uridine 5'- Diphosphoglucose Dehydrogenase).
E. 100 mM Citric Acid Solution, pH 7.0 at 25°C (Enzyme Diluent) (Prepare 100 ml in deionized water using Citric Acid, Free Acid, Monohydrate. Adjust to pH 7.0 at 25°C with 1 M NaOH).
F. Uridine 5'-Diphosphogalactose 4-Epimerase Enzyme Solution (Immediately before use, prepare a solution containing 0.06–0.13 unit/ml of Uridine 5'-Diphosphogalactose 4-Epimerase in cold Reagent E).

Procedure

Pipette (in milliliters) the following reagents into suitable cuvettes:

	Test	Blank
Deionized Water	2.38	2.38
Reagent A (Buffer)	0.30	0.30
Reagent B (UDP-Gal)	0.06	0.06
Reagent C (β-NAD)	0.06	0.06
Reagent D (UDPGDH)	0.10	0.10

Mix by inversion and equilibrate to 25°C. Monitor the A340nm until constant, using a suitably thermostatted spectrophotometer. Then add:

	Test	Blank
Reagent E (Enzyme Diluent)	—	0.10
Reagent F (Enzyme Solution)	0.10	—

Immediately mix by inversion and record the increase in A340nm for approximately 5 minutes. Obtain the ΔA340nm/minute using the maximum linear rate for both the Test and Blank.

Calculations

$$\text{Units / ml enzyme} = \frac{(\Delta A340nm\ /\ min\ Test\ -\ \Delta A340nm\ /\ min\ Blank)\ (3.0)\ (df)}{(2)\ (6.22)\ (0.10)}$$

3.0 = Total volume (in milliliters) of assay
df = Dilution factor
2 = 2 Moles of β-NADH produced per mole UDPG oxidized
6.22 = Millimolar extinction coefficient of β-NADH at 340nm
0.10 = Volume (in milliliters) of enzyme used

$$\text{Units / mg solid} = \frac{\text{units / ml enzyme}}{\text{mg solid / ml enzyme}}$$

$$\text{Units / mg protein} = \frac{\text{units / ml enzyme}}{\text{mg protein / ml enzyme}}$$

Unit Definition: *One unit will convert 1.0 μmole of UDP-galactose to UDP-glucose per minute at pH 8.8 at 25°C.*

Final Assay Concentrations

In a 3.00 ml reaction mix, the final concentrations are 100 mM glycine, 0.1 mM uridine 5'-diphosphogalactose, 1 mM β-NAD, 0.2 unit uridine-5'-diphosphoglucose dehydrogenase, 3.3 mM citric acid and 0.006–0.013 unit uridine 5'-diphosphogalactose 4- epimerase.

Note

1. Uridine 5'-Diphosphoglucose Dehydrogenase Unit Definition: One unit will oxidize 1.0 μmole of UDP-glucose to UDP-glucuronic acid per min at pH 8.7 at 25°C.

Enzymatic Assay of URIDINE-5'-DIPHOSPHOGLUCOSE PYROPHOSPHORYLASE (EC 2.7.7.9)

Principle

$$\text{UDPG + PPi + UDPG} \xrightarrow{\text{Pyrophosphorylase}} \text{UTP + Glucose 1 - Phosphate}$$

$$\text{Glucose 1 - Phosphate} \xrightarrow{\text{Phosphoglucomutase}} \text{Glucose 6 - Phosphate}$$

$$\text{Glucose 6 - Phosphate} + \beta\text{ - NADP} \xrightarrow{\text{G-6-PDH}} \text{6 - Phosphogluconate} + \beta\text{ - NADPH}$$

Abbreviations Used

UDPG = Uridine 5'-Diphosphoglucose
PPi = Inorganic Pyrophosphate

UTP = Uridine 5'-Triphosphate
β-NADP = β-Nicotinamide Adenine Dinucleotide Phosphate, Oxidized Form
β-NADPH = β-Nicotinamide Adenine Dinucleotide Phosphate, Reduced Form
G-6-PDH = Glucose-6-Phosphate Dehydrogenase

Conditions: T = 25°C, pH = 7.6, A340nm, Light path = 1 cm

Method: Continuous Spectrophotometric Rate Determination

Reagents

A. 100 mM Tris HCl Buffer, pH 7.6 at 25°C (Prepare 50 ml in deionized water using Trizma Base. Adjust to pH 7.6 at 25°C with 1 M HCl).
B. 4.0 mM Uridine 5'-Diphosphoglucose Solution (UDPG) (Prepare 5 ml in deionized water using Uridine 5'-Diphosphoglucose, Disodium Salt. *Prepare Fresh*).
C. 300 mM Magnesium Chloride Solution ($MgCl_2$) (Prepare 5 ml in deionized water using Magnesium Chloride, Hexahydrate).
D. 250 mM L-Cysteine Solution (Cys) (Prepare 5 ml in deionized water using L-Cysteine Hydrochloride, Monohydrate. Adjust to pH 7.0 with solid Sodium Bicarbonate).
E. 20 mM β-Nicotinamide Adenine Dinucleotide Phosphate Solution (β-NADP) (Dissolve the contents of one 30 mg vial of β-Nicotinamide Adenine Dinucleotide Phosphate in the appropriate volume of deionized water or prepare 1 ml in deionized water using β-Nicotinamide Adenine Dinucleotide Phosphate, Sodium Salt. *Prepare Fresh*).
F. 0.6 mM Glucose 1,6-Diphosphate (G 1,6-P) (Prepare 1 ml in deionized water using α-D-Glucose 1,6-Diphosphate, Cyclohexylammonium Salt, Hydrate).
G. Phosphoglucomutase Enzyme Solution (PGLUM) (Immediately before use, prepare a solution containing 15 units/ml in cold deionized water using Phosphoglucomutase).
H. Glucose-6-Phosphate Dehydrogenase (G-6-PDH) (Immediately before use, prepare a solution containing 15 units/ml in cold deionized water using Glucose-6- Phosphate Dehydrogenase).
I. 10 mM Tris HCl with 10 mM Magnesium Chloride, pH 7.6 at 25°C (Enzyme Diluent) (Prepare 25 ml in deionized water using Trizma Base and Magnesium Chloride, Hexahydrate).
J. Uridine-5'-Diphosphoglucose Pyrophosphorylase Enzyme Solution (Immediately before use, prepare a solution containing 0.25–0.50 unit/ml of Uridine-5'-Diphosphoglucose Pyrophosphorylase in cold Reagent I).
K. 50 mM Sodium Pyrophosphate Solution, pH 7.6 at 25°C (PPi) (Prepare 25 ml in deionized water using Tetrasodium Pyrophosphate, Decahydrate. Adjust to pH 7.6 at 25°C with 1 M HCl. *Prepare Fresh*).

Procedure

Pipette (in milliliters) the following reagents into suitable cuvettes:

	Test	Blank
Deionized Water	0.27	0.27
Reagent A (Buffer)	1.50	1.50
Reagent B (UDPG)	0.50	0.50
Reagent C ($MgCl_2$)	0.16	0.16
Reagent D (Cys)	0.12	0.12
Reagent E (β-NADP)	0.10	0.10
Reagent F (G 1,6-P)	0.05	0.05
Reagent G (PGLUM)	0.05	0.05
Reagent H (G-6-PDH)	0.05	0.05
Reagent J (Enzyme Solution)	0.10	—
Reagent I (Enzyme Diluent)	—	0.10

Mix by inversion and equilibrate to 25°C. Monitor the A340nm until constant, using a suitably thermostatted spectrophotometer. Then add:

Reagent K (PPi)	0.10	0.10

Immediately mix by inversion and record the increase in A340nm for approximately 5 minutes. Obtain the ΔA340nm/minute using the maximum linear rate for both the Test and Blank.

Calculations

$$\text{Units / ml enzyme} = \frac{(\Delta A340\text{nm / min Test} - \Delta A340\text{nm / min Blank})\ (3)\ (\text{df})}{(6.22)\ (0.1)}$$

3 = Volume (in milliliters) of assay
df = Dilution factor
6.22 = Millimolar extinction coefficient of β-NADPH at 340nm
0.1 = Volume (in milliliter) of enzyme used

$$\text{Units / mg solid} = \frac{\text{units / ml enzyme}}{\text{mg solid / ml enzyme}}$$

$$\text{Units / mg protein} = \frac{\text{units / ml enzyme}}{\text{mg protein / ml enzyme}}$$

Unit Definition: *One unit will cause the formation of 1.0 μmole of glucose 1-phosphate from uridine 5'-diphosphoglucose and inorganic pyrophosphate per minute at pH 7.6 at 25°C.*

Final Assay Concentrations

In a 3.00 ml reaction mix, the final concentrations are 50 mM Tris, 0.67 mM uridine 5'-diphosphoglucose, 16 mM magnesium chloride, 10 mM L-cysteine, 0.67 mM β-nicotinamide adenine dinucleotide phosphate, 0.01 mM Glucose 1,6-diphosphate, 0.75 unit phosphoglucomutase, 0.75 unit glucose 6-phosphate dehydrogenase, 1.7 mM sodium pyrophosphate and 0.025–0.05 unit uridine-5'-diphosphoglucose pyrophosphorylase.

Notes

1. Glucose-6-Phosphate Dehydrogenase Unit Definition: One unit will oxidize 1.0 μmole of D-glucose 6-phosphate to 6-phospho-D-gluconate per minute in the presence of β-NADP at pH 7.4 at 25°C.
2. Phosphoglucomutase Unit Definition: One unit will convert 1.0 μmole of α-D-glucose 1-phosphate to α-D-glucose 6-phosphate per minute at pH 7.4 at 30°C.

Enzymatic Assay of URIDINE-5'-DIPHOSPHOGLUCURONYL TRANSFERASE (EC 2.4.1.17)

Principle

$$\text{UDPGA} + \text{Phenolphthalein} \xrightarrow{\text{UDPGT}} \text{UDP} + \text{Phenolphthalein Glucuronide}$$

Abbreviations Used

UDPGA = Uridine 5'-Diphosphoglucuronic Acid
UDPGT = Uridine-5'-Diphosphoglucuronyl Transferase
UDP = Uridine 5'-Diphosphate

Conditions: T = 37°C, pH = 8.0, A552nm, Light path = 1 cm

Method: Colorimetric

Reagents

A. 1000 mM Tris HCl Buffer, pH 8.0 at 37°C (Prepare 100 ml in deionized water using Trizma Hydrochloride. Adjust to pH 8.0 at 37°C with 1 M NaOH).
B. 30 mM Uridine 5'-Diphosphoglucuronic Acid Solution (UDPGA) (Prepare 5 ml in deionized water using Uridine 5'-Diphosphoglucuronic Acid, Trisodium Salt).
C. 1.0% (w/v) Phenolphthalein Solution (PPT) (Prepare 5 ml in Reagent I using Phenolphthalein).
D. 1000 mM Magnesium Chloride Solution ($MgCl_2$) (Prepare 5 ml in deionized water using Magnesium Chloride, Hexahydrate).
E. 10% (w/v) Bovine Serum Albumin Solution (BSA) (Prepare 50 ml in deionized water using Albumin, Bovine or equivalent).
F. 200 mM 2-Mercaptoethanol Solution (2-ME) (Prepare 5 ml in deionized water using 2-Mercaptoethanol).
G. 200 mM Glycine Solution, pH 10.4 at 25°C (Gly) (Prepare 200 ml in deionized water using Glycine, Free Base. Adjust to pH 10.4 at 25°C with 1 M NaOH).
H. Uridine-5'-Diphosphoglucuronyl Transferase Enzyme Solution (Immediately before use, prepare a solution containing 100–200 mg/ml in cold deionized water).
I. 95% Ethanol (Nondenatured) (Prepare 50 ml in deionized water using 200 Proof USP Ethyl Alcohol).

Procedure

Prepare a reaction cocktail by pipetting (in milliliters) the following reagents into a suitable container:

Deionized Water	17.00
Reagent A (Buffer)	20.00
Reagent C (PPT)	2.00
Reagent D ($MgCl_2$)	3.00
Reagent E (BSA)	20.00
Reagent F (2-ME)	1.00

Mix by swirling and equilibrate to 37°C. Adjust to pH 8.0 with 10 mM HCl or 10 mM NaOH, if necessary. Pipette (in milliliters) the following reagents into a suitable container:

	Test
Reaction Cocktail	6.30
Reagent H (Enzyme Solution)	0.20

Mix by swirling and equilibrate to 37°C. Then add:

Reagent B (UDPGA)	0.50

Immediately mix by swirling, start timing, and place a 0.5 ml aliquot into a suitable container containing 2.5 ml of Reagent I (95% ethanol). This stopped sample is T_0. Incubate the remaining 6.5 ml of reaction mixture at 37°C. At approximately 2, 4, 6, 8, and 10 minutes place 0.5 ml aliquots into 2.5 ml of 95% ethanol. Note the exact time (T) the aliquot is placed into the Reagent I (95% ethanol) (the stop reaction). Centrifuge all stop reactions for 5 minutes.

Colorimetric Determination

Sample: Pipette (in milliliters) the following into suitable containers:

	Time 0 Test	Time 2 Test	Time 4 Test	Time 6 Test	Time 8 Test	Time 10 Test
Test T0	0.50	—	—	—	—	—
Test T2	—	0.50	—	—	—	—
Test T4	—	—	0.50	—	—	—
Test T6	—	—	—	0.50	—	—
Test T8	—	—	—	—	0.50	—
Test T10	—	—	—	—	—	0.50
Reagent G (Gly)	4.00	4.00	4.00	4.00	4.00	4.00

Mix by swirling and record the absorbance at A552nm for each of the Tests at the different time points using a suitable spectrophotometer.

Calculations

Plot A552nm versus the time (T) for each of the centrifuged stop reactions. Analyze the data by Linear Regression Analysis.

$$\text{Units / ml enzyme} = \frac{(\text{Slope})\,(7)\,(3)\,(4.5)}{(26)\,(0.2)\,(0.5)\,(0.5)}$$

7 = Volume (in milliliters) of Reaction Mix
3 = Volume (in milliliters) of Stop Reaction
4.5 = Volume (in milliliters) of Colorimetric Development Mix
26 = Millimolar extinction coefficient of Phenolphthalein at 552nm
0.2 = Volume (in milliliters) of enzyme used
0.5 = Volume (in milliliters) of reaction mix used in stop reaction
0.5 = Volume of (in milliliters) stop reaction used in Colorimetric Determination

$$\text{Units / mg solid} = \frac{\text{units / ml enzyme}}{\text{mg solid / ml enzyme}}$$

$$\text{Units / mg protein} = \frac{\text{units / ml enzyme}}{\text{mg protein / ml enzyme}}$$

Unit Definition: *One unit will transfer 1.0 μmole of glucuronic acid from uridine 5'-diphosphoglucuronic acid to phenolphthalein per minute at pH 8.0 at 37°C.*

Final Assay Concentrations

In a 7.00 ml reaction mix, the final concentrations are 286 mM Tris, 0.029% (w/v) phenolphthalein, 2.7% (v/v) ethanol, 42.8 mM magnesium chloride, 2.9% (w/v) BSA, 2.85 mM 2-mercaptoethanol, 2.1 mM uridine 5'-diphosphoglucuronic acid, and 20–40 mg uridine-5'-diphosphoglucuronyl transferase.

Enzymatic Assay of UROKINASE (EC 3.4.21.73)

Principle

$$\text{Plasminogen} + H_2O \xrightarrow{\text{Urokinase}} \text{Plasmin}$$

$$\text{Casein} \xrightarrow{\text{Plasmin}} \text{Perchloric Acid Soluble Amino Acids}$$

Conditions: T = 37°C, pH = 7.5, A275nm, Light path = 1 cm

Method: Spectrophotometric Stop Rate Determination.

Reagents

A. 60 mM Tris HCl Buffer with 90 mM Sodium Chloride, pH 7.5 at 37°C. (Prepare 100 ml in deionized water using Trizma Base, and Sodium Chloride. Adjust to pH 7.5 at 37°C with 1 M HCl).
B. Porcine Plasminogen Solution (Plas) (Immediately before use, prepare a solution containing 2 units/ml Plasminogen, in deionized water).
C. 1.4% (w/v) α-Casein Suspension (Casein) (Prepare 100 ml in Reagent A using α-Casein. Do not heat. Stir to make a homogenous suspension).
D. 500 mM Perchloric Acid Reagent (PCA) (Prepare 100 ml in deionized water using Perchloric Acid).
E. Urokinase Enzyme Solution (Immediately before use, prepare a solution containing 1–2 units/ml of Urokinase in cold Reagent A).

Procedure

Pipette (in milliliters) the following reagents into suitable containers:

	Test	Blank
Reagent A (Buffer)	0.90	0.90
Reagent B (Plas)	1.00	1.00

Mix by swirling and equilibrate to 37°C. Then add:

Reagent E (Enzyme Solution)	0.10	—

Immediately following enzyme addition, add:

Reagent C (Casein)	2.00	2.00

Mix by swirling and incubate at 37°C for exactly 15 minutes. Then add:

Reagent D (PCA)	6.00	6.00
Reagent E (Enzyme Solution)	—	0.10

Mix by swirling and incubate at 25°C for 60 minutes. Filter the solution through a Whatman #50 filter paper and transfer the filtrate to suitable quartz cuvettes. Record the A275nm for both the Test and Blank using a suitable spectrophotometer.

Calculations

$$\text{Units / ml enzyme} = \frac{(\text{A275nm Test} - \text{A275nm Blank})\,(10)}{(1)\,(15)\,(0.1)}$$

10 = Volume (in milliliters) of assay
1 = Change in Absorbance as per the Unit Definition
15 = Time of reaction (in minutes) as per the Unit Definition
0.1 = Volume (in milliliters) of enzyme used

$$\text{Units / mg solid} = \frac{\text{units / ml enzyme}}{\text{mg solid / ml enzyme}}$$

$$\text{Units / mg protein} = \frac{\text{units / ml enzyme}}{\text{mg protein / ml enzyme}}$$

Unit Definition: *One unit will activate that amount of porcine plasminogen which will produce a A275nm of 1.0 per ml per minute at pH 7.5 at 37°C, when measuring perchloric acid soluble products from α-casein (1 cm light path).*

Final Assay Concentrations

In a 4 ml reaction mix, the final concentrations are 45 mM Tris, 68 mM sodium chloride, 2 units plasminogen, 0.70% (w/v) α-casein and 0.1–0.2 unit urokinase.

Notes

1. Filtering the solution by vacuum or with 0.45 μm syringe filters will result in lower activity.
2. Higher activities are obtained when human plasminogen is used as a substrate.

Enzymatic Assay of XANTHINE OXIDASE (EC 1.1.3.22)

Principle

$$\text{Xanthine} + H_2O + O_2 \xrightarrow{\text{Xanthine Oxidase}} \text{Uric Acid} + H_2O_2$$

Conditions: T = 25°C, pH = 7.5, A290nm, Light Path = 1 cm

Method: Continuous Spectrophotometric Rate Determination

Reagents

A. 50 mM Potassium Phosphate Buffer, pH 7.5 at 25°C (Prepare 200 ml in deionized water using Potassium Phosphate, Monobasic, Anhydrous. Adjust to pH 7.5 at 25°C with 1 M KOH).

B. 0.15 mM Xanthine Solution (Prepare 100 ml by initially dissolving Xanthine, in a minimal volume of NaOH. Add approximately 90 ml of deionized water. Adjust to pH 7.5 at 25°C with either 1 M NaOH or 1 M HCl. Dilute to a final volume of 100 ml. *Prepare Fresh*).

C. Xanthine Oxidase Enzyme Solution (Immediately before use, prepare a solution containing 0.1–0.2 unit/ml of Xanthine Oxidase in cold Reagent A).

Procedure

Pipette (in milliliters) the following reagents into suitable quartz cuvettes:

	Test	Blank
Reagent A (Buffer)	1.90	1.90
Reagent B (Xanthine)	1.00	1.00
Deionized Water	—	0.10

Mix by inversion and equilibrate to 25°C. Monitor the A290nm until constant, using a suitably thermostatted spectrophotometer. Then add:

Reagent C (Enzyme Solution)	0.10	—

Immediately mix by inversion and record the increase in A290nm for approximately 5 minutes. Obtain the A290nm/minute using the maximum linear rate for both the Test and Blank.

Calculations

$$\text{Units / ml enzyme} = \frac{(\text{A290nm / min Test} - \text{A290nm / min Blank})\,(3)\,(\text{df})}{(12.2)\,(0.1)}$$

3 = Total volume (in milliliters) of assay

df = Dilution factor

12.2 = Millimolar extinction coefficient of Uric Acid at 290nm

0.1 = Volume (in milliliter) of enzyme used

$$\text{Units / mg solid} = \frac{\text{units / ml enzyme}}{\text{mg solid / ml enzyme}}$$

$$\text{Units / mg protein} = \frac{\text{units / ml enzyme}}{\text{mg protein / ml enzyme}}$$

Unit Definition: *One unit will convert 1.0 μmole of xanthine to uric acid per minute at pH 7.5 at 25°C.*

Final Assay Concentrations

In a 3.00 ml reaction mix, the final concentrations are 33 mM potassium phosphate, 0.050 mM xanthine and 0.01–0.02 unit xanthine oxidase.

Enzymatic Assay of XYLANASE (EC 3.2.1.8)

Principle

$$\text{Xylan} + H_2O \xrightarrow{\text{Xylanase}} \text{Reducing Sugar (measured as Xylose)}$$

Conditions: T = 30°C, pH = 4.5, A540nm, Light path = 1 cm

Method: Colorimetric

Reagents

A. 50 mM Sodium Acetate Buffer, pH 4.5 at 30°C (Prepare 50 ml in deionized water using Sodium Acetate, Trihydrate. Adjust to pH 4.5 at 30°C with 1 M HCl).
B. 1.0% (w/v) Xylan Substrate Solution (Xylan) (Prepare 5 ml in Reagent A using Xylan).
C. 0.05% (w/v) Bovine Serum Albumin (Enz Diluent) (Prepare 25 ml in Reagent A using Albumin, Bovine).
D. Xylanase Enzyme Solution (Immediately before use, prepare a solution containing 5–10 units/ml of Xylanase in cold Reagent C).
E. 16 mM Copper Sulfate, 1.3 M Sodium Sulfate, 226 mM Sodium Carbonate, 190 mM Sodium Bicarbonate and 43 mM Sodium Potassium Tartrate Solution (Copper Soln) (Prepare 1 liter in deionized water using Cupric Sulfate Pentahydrate, Sodium Bicarbonate, Sodium Sulfate, Anhydrous, Sodium Carbonate, Anhydrous, and Sodium Potassium Tartrate Tetrahydrate).
F. 40 mM Molybdic Acid, 19 mM Arsenic Acid and 756 mM Sulfuric Acid Solution (Ars-Mol) (Prepare 1 liter in deionized water using Molybdic Acid, Ammonium Salt Tetrahydrate, Arsenic Acid, Sodium Salt and Sulfuric Acid).
G. 1 mg/ml Xylose Standard Solution (Xylose Std) (Prepare 10 ml in deionized water using D(+)Xylose).

Procedure

Pipette (in milliliters) the following reagents into suitable containers:

	Test	Blank	Std 1	Std 2	Std 3	Std 4	Std Blank
Reagent B (Xylan)	1.90	1.90	—	—	—	—	—
Reagent G (Xylose Std)	—	—	0.02	0.05	0.07	0.10	—
Reagent C (Enz Diluent)	—	—	1.98	1.95	1.93	1.90	2.00

Mix by swirling and equilibrate to 30°C. Then add:

	Test	Blank	Std 1	Std 2	Std 3	Std 4	Std Blank
Reagent D (Enzyme Soln)	0.10	—	—	—	—	—	—
Reagent C (Enz Diluent)	—	0.10	—	—	—	—	—

Mix by swirling and incubate at 30°C for exactly 10 minutes. Then add:

Reagent E (Copper Soln)	2.00	2.00	2.00	2.00	2.00	2.00	2.00

Mix by swirling. Place a marble over the top of the tubes and transfer the tubes to a boiling water bath. Incubate for 10 minutes. Remove the tubes from the boiling water bath and allow to cool to room temperature. Then add:

Reagent F (Ars-Mol)	2.00	2.00	2.00	2.00	2.00	2.00	2.00

Shake or vortex the tubes until foaming stops and any precipitate present is dissolved. Centrifuge to clarify. Transfer the solutions to suitable cuvettes. Obtain the A540nm for Test, Blank and Standards, using a suitable spectrophotometer.

Calculations

Standard Curve: A540nm Std = A540nm Std–A540nm Std Blank

Prepare a standard curve by plotting the A540nm Standard vs the μmoles of Xylose.

Sample Determination: A540nm Sample = A540nm Test–A540nm Blank

Determine the μmoles of xylose using the Standard Curve.

$$\text{Units / ml enzyme} = \frac{(\mu\text{moles of xylose liberated})\ (\text{df})}{(10)\ (0.1)}$$

df = Dilution factor
10 = Time of assay (in minutes) as per Unit Definition
0.1 = Volume (in milliliters) of enzyme used

$$\text{Units / mg solid} = \frac{\text{units / ml enzyme}}{\text{mg solid / ml enzyme}}$$

$$\text{Units / mg protein} = \frac{\text{units / ml enzyme}}{\text{mg protein / ml enzyme}}$$

Unit Definition: *One unit will liberate 1.0 μmole of reducing sugar measured as xylose equivalents from xylan (X-0627) per minute at pH 4.5 at 30°C.*

Final Assay Concentrations

In a 2.00 ml reaction mix, the final concentrations are 50 mM sodium acetate, 0.95% (w/v) xylan, 0.003% (w/v) Bovine Serum Albumin, and 0.5–1.0 unit xylanase.

Notes

1. Sodium Sulfate, Sodium Carbonate, and Sodium Potassium Tartrate are dissolved in approximately 500 ml of deionized water. Cupric Sulfate is dissolved in approximately 100 ml of deionized water and is slowly added to the above solution to avoid precipitation. Sodium Bicarbonate is

dissolved first in deionized water and then added to the above solution. Dilute the solution to 1 liter. If a precipitate forms, it should be removed by filtration prior to use. Store in an amber bottle and avoid exposure to direct sunlight. Store at room temperature.

2. Molybdic Acid is dissolved in approximately 300 ml of deionized water. Add Sulfuric Acid slowly. Caution, this is an exothermic reaction! Arsenic Acid is dissolved in approximately 300 ml of deionized water and is added to the above solution. The solution is diluted to a total volume of 1 liter and incubated at 37°C for 48–72 hours. If a precipitate forms, it should be removed by filtration prior to use. Store in an amber bottle and avoid exposure to direct sunlight. The solution expires six months after preparation. Store at room temperature in an exhaust hood.

Enzymatic Assay of XYLANASE Activity in DRISELASE (EC 3.2.1.8)

Principle

$$\text{Xylan} + H_2O \xrightarrow{\text{Xylanase}} \text{Reducing Sugar (measured as glucose)}$$

Conditions: T = 37°C, pH = 4.5, A410nm, Light path = 1 cm

Method: Colorimetric

Reagents

A. 100 mM Sodium Acetate Buffer, pH 4.5 at 37°C (Prepare 50 ml in deionized water using Sodium Acetate, Trihydrate. Adjust to pH 4.5 at 37°C with 5 M HCl).

B. 2.5% (w/v) Xylan Substrate solution (Xylan) (Prepare 15 ml in Reagent A using Xylan).

C. Driselase (Xylanase) Enzyme Solution (Immediately before use, prepare a solution containing 0.075 – 0.15 units/ml of Xylanase in cold deionized water).

D. 500 mM Sodium Hydroxide Solution (Prepare 200 ml using deionized water and Sodium Hydroxide (1.0 N)).

E. 0.5% (w/v) p-Hydroxybenzoic Acid Hydrazide Solution (PAHBAH) (Prepare 200 ml in Reagent D using p-Hydroxybenzoic Acid Hydrazide. *Prepare Fresh before Use*).

F. 0.2 mg/ml Glucose Standard Solution (Glucose Std) (Prepare by diluting Glucose Standard Solution with deionized water).

Procedure

Pipette (in milliliters) the following reagents into suitable containers:

	Test	Blank
Deionized Water	—	1.00
Reagent A (Buffer)	3.00	3.00
Reagent B (Xylan)	1.00	1.00

Mix by swirling and equilibrate to 37°C. Then add:

Reagent C (Enzyme Solution)	1.00	—

Mix by swirling and incubate at 37°C for exactly 60 minutes. Pipette (in milliliters) the following reagents into suitable test tubes:

	Test	Blank	Std 1	Std 2	Std 3	Std 4	Std 5	Std Blank
Deionized water	0.97	0.97	0.95	0.90	0.80	0.60	0.40	1.00
Test Solution	0.03	—	—	—	—	—	—	—
Blank Solution	—	0.03	—	-	—	-	—	—
Reagent F (Std)	—	—	0.05	0.10	0.20	0.40	0.60	—
Reagent E (PAHBAH)	9.00	9.00	9.00	9.00	9.00	9.00	9.00	9.00

Immediately mix by swirling and transfer the tubes to a boiling water bath. Incubate for 5 minutes. Remove the tubes from the boiling water bath and allow to cool to room temperature. Mix by inversion and transfer the solutions to suitable cuvettes. Obtain the A410nm for the Test, Blank, and Standards using a suitable spectrophotometer.

Calculations

Standard Curve: ΔA410nm Std = A410nm Std–A410nm Std Blank

Prepare a standard curve by plotting the DA410nm Standard vs the milligrams of Glucose.

Sample Determination: ΔA410nm Sample = A410nm Test–A410nm Blank

Determine the milligrams of glucose using the standard curve.

$$\text{Units / ml enzyme} = \frac{\text{(milligrams of glucose liberated) (5) (df)}}{(60)\ (1)\ (0.03)}$$

5 = Volume (in milliliters) of assay
df = Dilution factor
60 = Time of assay (in minutes) as per Unit Definition
1 = Volume (in milliliters) of enzyme used
0.03 = Volume (in milliliter) used for reducing sugar determination

$$\text{Units / mg solid} = \frac{\text{units / ml enzyme}}{\text{mg solid / ml enzyme}}$$

$$\text{Units / mg protein} = \frac{\text{units / ml enzyme}}{\text{mg protein / ml enzyme}}$$

Unit Definition: *One unit will liberate 1.0 milligram of reducing sugar from xylan (measured as glucose) per minute at pH 4.5 at 37°C.*

Final Assay Concentrations

In a 5.00 ml reaction mix, the final concentrations are 60 mM sodium acetate, 0.50% (w/v) xylan and 0.075–0.15 unit xylanase.

Enzymatic Assay of β-XYLOSIDASE (EC 3.2.1.37)

Principle

$$\text{ONP-}\beta\text{-D-Xylopyranoside} \xrightarrow{\beta\text{-Xylosidase}} \text{ONP} + \beta\text{-D-Xylopyranoside}$$

Abbreviation Used

ONP = o-Nitrophenol

Conditions: T = 25°C, pH = 5.0, A410nm, Light path = 1 cm

Method: Spectrophotometric Stop Rate Determination

Reagents

A. 100 mM Citrate Buffer, pH 5.0 at 25°C (Prepare 100 ml in deionized water using Citric Acid, Free Acid. Adjust to pH 5.0 at 25°C with 1 M NaOH).
B. 10 mM o-Nitrophenyl β-D-Xylopyranoside Solution (ONP-β-D-Xylopyranoside) (Prepare 10 ml in deionized water using o-Nitrophenyl β-D-Xylopyranoside).
C. 200 mM Borate Buffer, pH 9.8 at 25°C (Prepare 100 ml in deionized water using Boric Acid. Adjust to pH 9.8 at 25°C with 1 M NaOH).
D. β-Xylosidase Enzyme Solution (Prepare a solution containing 1–2 units/ml of β-Xylosidase in cold deionized water).

Procedure

Pipette (in milliliters) the following reagents into suitable cuvettes:

	Test	Blank
Reagent A (Citrate Buffer)	0.45	0.45
Reagent B (ONP-β-D-Xylopyranoside)	0.50	0.50

Mix by inversion and equilibrate to 25°C. Then add:

Reagent D (Enzyme Solution)	0.05	—
Deionized Water	—	0.05

Immediately mix by inversion and incubate at 25°C for 10 minutes. Then add:

Reagent C (Borate Buffer)	3.00	3.00

Mix by inversion and record the A410nm for both the Test and Blank, using a suitably thermostatted spectrophotometer.

Calculations

$$\text{Units / mg enzyme} = \frac{(\text{A410nm Test} - \text{A410nm Blank})(4.0)}{(10)(4.6)(\text{mg enzyme / RM})}$$

4.0 = Total volume (in milliliters) of solution.
10 = Time of assay in minutes (unit definition).
4.6 = Millimolar extinction coefficient of o-nitrophenol at 410nm and pH 9.8
RM = Reaction Mix

Unit Definition: *One unit will hydrolyze 1.0 μmole of o-nitrophenyl β-D-xyloside to o-nitrophenol and D-xylose per minute at pH 5.0 at 25°C.*

Final Assay Concentration

In a 1.00 ml reaction mix, the final concentrations are 45 mM citric acid, 5.0 mM ONP-β-D-xylopyranoside, and 0.05–0.10 unit β-xylosidase.

Enzymatic Assay of L-XYLULOSE REDUCTASE (EC 1.1.1.10)

Principle

$$\text{Lylitol} + \beta\text{-NADP} \xrightarrow{\text{Xylulose Reductase}} \text{L-Xylulose} + \beta\text{-NADPH}$$

Abbreviations Used

β-NADPH = β-Nicotinamide Adenine Dinucleotide Phosphate, Reduced Form
β-NADP = β-Nicotinamide Adenine Dinucleotide Phosphate, Oxidized Form

Conditions: T = 25°C, pH = 10.0, A340nm, Light path = 1 cm

Method: Continuous Spectrophotometric Rate Determination

Reagents

A. 100 mM Glycine Buffer, pH 10.0 at 25°C (Prepare 100 ml in deionized water using Glycine, Free Base. Adjust to pH 10.0 at 25°C with 1 M NaOH).
B. 100 mM Magnesium Chloride Solution ($MgCl_2$) (Prepare 5 ml in deionized water using Magnesium Chloride, 4.9 M Solution).
C. 657 mM Xylitol Solution (Xylitol) (Prepare 5 ml in deionized water using Xylitol).
D. 12.5 mM β-Nicotinamide Adenine Dinucleotide Phosphate Solution (β-NADP) (Prepare 2 ml in deionized water using β-Nicotinamide Adenine Dinucleotide Phosphate, Sodium Salt).
E. L-Xylulose Reductase Enzyme Solution (Immediately before use, prepare a solution containing 0.1–0.2 unit/ml of L-Xylulose Reductase in cold Reagent A).

Procedure

Pipette (in milliliters) the following reagents into suitable cuvettes:

	Test	Blank
Reagent A (Buffer)	2.50	2.50
Reagent B ($MgCl_2$)	0.10	0.10
Reagent C (Xylitol)	0.20	0.20
Reagent D (β-NADP)	0.10	0.10

Mix by inversion and equilibrate to 25°C. Monitor the A340nm until constant, using a suitably thermostatted spectrophotometer. Then add:

Reagent E (Enzyme Solution)	0.10	—
Reagent A (Buffer)	—	0.10

Immediately mix by inversion and record the increase in A340nm for approximately 5 minutes. Obtain the A340nm/minute using the maximum linear rate for both the Test and Blank.

Calculations

$$\text{Units / mg enzyme} = \frac{(\text{A340nm / min Test} - \text{A340nm / min Blank})\ (3)\ (\text{df})}{(6.22)\ (0.1)}$$

3 = Total volume (in milliliters) of assay
df = Dilution factor
6.22 = Millimolar extinction coefficient of β-NADPH at 340nm
0.1 = Volume (in milliliter) of enzyme used.

$$\text{Units / mg protein} = \frac{\text{units / ml enzyme}}{\text{mg protein / ml enzyme}}$$

Unit Definition: *One unit will oxidize 1.0 μmole of xylitol to L-xylulose per minute at pH 10.0 at 25°C.*

Final Assay Concentrations

In a 3.00 ml reaction mix, the final concentrations are 87 mM glycine, 3.3 mM magnesium chloride, 44 mM xylitol, 0.42 mM β-nicotinamide adenine dinucleotide phosphate, and 0.01–0.02 unit L-xylulose reductase.

Enzyme Activator Assays

Suitability Assay for COLIPASE

Principle

$$\text{Triglyceride} + H_2O \xrightarrow[\text{Colipase}]{\text{Lipase}} \text{Diglyceride} + \text{Fatty Acid}$$

Conditions: T = 25°C, pH = 8.8, A365nm, Light path = 1 cm

Method: Turbidimetric Determination

Reagents

A. 25 mM Tris HCl Buffer with 25 mM Sodium Deoxycholate, pH 8.8 at 25°C (Deoxycholate) (Prepare 50 ml in deionized water using Trizma Base, and Deoxycholic Acid, Sodium Salt. Adjust to pH 8.8 at 25°C with 1 M HCl).

B. 95% (v/v) Ethanol (Prepare 20 ml in deionized water).

C. 2% (v/v) Triolein Substrate Solution (Triolein) (Prepare by dissolving 0.5 ml of Triolein in 25 ml of Reagent B).

D. 0.04% (v/v) Substrate Emulsion (Prepare by adding dropwise (0.05 ml), 1 ml of Reagent C to 50 ml of Reagent A. The resulting emulsion should have approximately 100 absorbance units when measured at 365 nm).

E. Lipase Enzyme Solution (Lipase) (Immediately before use, prepare a solution containing 1 mg/ml of Lipase in cold deionized water. Then dilute with cold deionized water to approximately 2000 units/ml).

F. Colipase Solution (Colipase) (Immediately before use, prepare a solution containing 0.5–1.0 mg/ml in cold deionized water).

Procedure

Pipette (in milliliters) the following reagents into suitable cuvettes:

	Test	Blank
Reagent D (Substrate Emulsion)	1.00	1.00

Equilibrate to 25°C. Then add:

	Test	Blank
Reagent E (Lipase)	0.05	0.05

Mix by inversion and monitor the A365nm for both the Test and Blank using a suitably thermostatted spectrophotometer. There will be a decrease in A365nm of approximately 0.03 absorbance unit/minute due to some Colipase present in the lipase. After 3–5 minutes, add:

	Test	Blank
Reagent F (Colipase)	0.01	—
Deionized Water	—	0.01

Immediately mix by inversion and record the decrease in A365nm for approximately 5 minutes. Obtain the A365nm/minute using the maximum linear rate for both the Test and Blank.

Calculation

A365nm Sample = A365nm Test–A365nm Blank

Compare the A365nm of the Test Sample to that of a Control Sample. The rates should be similar.

Final Assay Concentration

In a 1.06 ml reaction mix, the final concentrations are 23 mM Tris, 23 mM sodium deoxycholate, 0.038% (v/v) triolein, 1.8% (v/v) ethanol, 100 units lipase, 5–10 µg colipase.

Note

1. 5–10 µg of Colipase will yield a maximum rate of lipase activity under these conditions.

Enzymatic Assay of PHOSPHODIESTERASE, 3':5'-CYCLIC NUCLEOTIDE ACTIVATOR DEFICIENT

Principle

$$\text{3':5-cAMP} + H_2O \xrightarrow{\text{PDE}} \text{AMP}$$

$$\text{AMP} + \text{ATP} \xrightarrow{\text{Myokinase}} \text{2 ADP}$$

$$\text{2 ADP} + \text{2 PEP} \xrightarrow{\text{Pyruvate Kinase}} \text{2 ATP} + \text{2Pyruvate}$$

$$\text{2 Pyruvate} + 2\beta\text{-NADH} \xrightarrow{\text{Lactic Dehydrogenase}} \text{2 Lactate} + 2\beta\text{-NAD}$$

Abbreviations Used

3':5'-cAMP = Adenosine 3':5'-Cyclic Monophosphate
PDE = Phosphodiesterase, 3':5'-Cyclic Nucleotide
AMP = Adenosine 5'-Monophosphate
ATP = Adenosine 5'-Triphosphate
ADP = Adenosine 5'-Diphosphate
PEP = Phospho(enol) pyruvate
β-NADH = β-Nicotinamide Adenine Dinucleotide, Reduced Form
β-NAD = β-Nicotinamide Adenine Dinucleotide, Oxidized Form

Conditions: T = 30°C, pH = 7.5, A340nm, Light path = 1 cm

Method: Continuous Spectrophotometric Rate Determination

Reagents

A. 100 mM Tris HCl Buffer with 80 mM Potassium Chloride and 9.7 mM Magnesium Sulfate, pH 7.5 at 30°C (Prepare 50 ml in deionized water using Trizma Base, Potassium Chloride and Magnesium Sulfate, Heptahydrate. Adjust to pH 7.5 at 30°C with 1 M NaOH).

B. 0.9 mM Phospho(enol) pyruvate Solution (PEP) (Prepare 50 ml in Reagent A using Phospho(enol) pyruvate, Monopotassium Salt. *Prepare Fresh*).

C. 30 mM Adenosine 5'-Triphosphate Solution (ATP) (Prepare 2 ml in deionized water using Adenosine 5'-Triphosphate, Disodium Salt. Adjust to pH 7.5 at 30°C with Sodium Bicarbonate. *Prepare Fresh*).

D. 120 mM Adenosine 3':5'-Cyclic Monophosphate Solution (3':5'-cAMP) (Prepare 2 ml in deionized water using Adenosine 3':5'-Cyclic Monophosphate, Sodium Salt. *Prepare Fresh*).

E. PK/LDH Enzymes Suspension.

F. Myokinase Enzyme Solution (MK) (Immediately before use, prepare a solution containing 200 units/ml in cold deionized water using Myokinase).

G. 1.75 mM Calcium Acetate Solution ($Ca(OAc)_2$) (Prepare 10 ml in deionized water using Calcium Acetate).

H. 6.4 mM β-Nicotinamide Adenine Dinucleotide, Reduced Form Solution (β-NADH) (Dissolve the contents of one 5 mg vial of β-Nicotinamide Adenosine Dinucleotide, Reduced Form, Disodium Salt in the appropriate volume of deionized water. *Prepare Fresh*).

I. Phosphodiesterase, 3':5'-Cyclic Nucleotide Activator Solution (Calmodulin) (Immediately before use, prepare a solution containing 300 units/ml in deionized water using Phosphodiesterase, 3':5'-Cyclic Nucleotide Activator).

J. Phosphodiesterase, 3':5'-Cyclic Nucleotide Enzyme, Activator Deficient Solution (Immediately before use, prepare a solution containing 0.2 unit/ml of Phosphodiesterase, 3':5'- Cyclic Nucleotide, Activator Deficient, in cold deionized water).

Procedure

Prepare a reaction cocktail by pipetting (in milliliters) the following reagents into a suitable container:

Deionized Water	6.45
Reagent B (PEP)	19.95
Reagent C (ATP)	0.30
Reagent D (3':5'-cAMP)	0.30
Reagent E (PK/LDH)	0.15
Reagent F (MK)	0.30
Reagent G ($Ca(OAc)_2$)	0.45
Reagent H (β-NADH)	0.60

Mix by swirling and adjust to pH 7.5 at 30°C with 1 M HCl or 1 M NaOH, if necessary. Pipette (in milliliters) the following reagents into suitable cuvettes:

	Non-Activated (NA)		Activated (A)	
	Test-NA	Blank-NA	Test-A	Blank-A
Deionized Water	0.10	0.20	—	0.10
Reaction Cocktail	3.00	3.00	3.00	3.00
Reagent I (Calmodulin)	—	—	0.10	0.10

Mix by inversion and equilibrate to 30°C. Monitor the A340nm until constant, using a suitably thermostatted spectrophotometer. Then add:

	Non-Activated (NA)		Activated (A)	
	Test-NA	Blank-NA	Test-A	Blank-A
Reagent J (Enzyme Solution)	0.10	—	0.10	—

Immediately mix by inversion and record the decrease in A340nm for approximately 15 minutes. Obtain the A340nm/minute using the maximum linear rate for both the Test and Blank of the activated and non-activated systems.

Calculations

Non-Activated Enzyme

$$\text{Units / mg enzyme} = \frac{(\Delta A340nm / \text{min Test - NA}) - (\Delta A340nm / \text{min Blank - NA})\ (3.2)\ (df)}{(2)\ (6.22)\ (0.1)}$$

Activated Enzyme

$$\text{Units / mg enzyme} = \frac{(\Delta A340nm / \text{min Test - A}) - (\Delta A340nm / \text{min Blank - A})\ (3.2)\ (df)}{(2)\ (6.22)\ (0.1)}$$

3.2 = Total volume (in milliliters) of assay
df = Dilution factor
2 = 2 μmoles of β-NAD produced per μmole of 3':5'-cAMP hydrolyzed
6.22 = Millimolar extinction coefficient of β-NADH at 340nm
0.1 = Volume (in milliliters) of enzyme ued in the assay

$$\text{Units / mg solid} = \frac{\text{units / ml enzyme}}{\text{mg solid / ml enzyme}}$$

$$\text{Units / mg protein} = \frac{\text{units / ml enzyme}}{\text{mg protein / ml enzyme}}$$

Unit Definition: *One unit will hydrolyze 1.0 μmole of 3':5'-cyclic-AMP to 5'-AMP per minute at pH 7.5 at 30°C.*

Final Assay Concentrations

In a 3.20 ml reaction mix, the final concentrations are 66 mM Tris, 53 mM potassium chloride, 6.4 mM magnesium sulfate, 0.59 mM phospho(enol) pyruvate, 0.30 mM adenosine 5'-triphosphate, 1.2 mM adenosine 3':5'-cyclic monophosphate, 11 units pyruvate kinase, 16 units lactic dehydrogenase, 6.3 units myokinase, 0.026 mM calcium acetate, 0.13 mM β-NADH and 0.02 unit phosphodiesterase, 3':5'-cyclic nucleotide, activator deficient. (In the activated test 30 units of calmodulin are present).

Notes

1. L-Lactic Dehydrogenase Unit Definition: One unit will reduce 1.0 μmole of pyruvate to L-lactate per minute at pH 7.5 at 37°C.
2. Pyruvate Kinase Unit Definition: One unit will convert 1.0 μmole of phospho(enol) pyruvate to pyruvate per minute at pH 7.6 at 37°C.
3. Myokinase Unit Definition: One unit will convert 2.0 μmoles of ADP to ATP and AMP per minute at pH 7.6 at 37°C.
4. Phosphodiesterase 3':5'-Cyclic Nucleotide Activator Unit Definition: One unit will stimulate 0.016 activated units of phosphodiesterase 3':5'-cyclic nucleotide, P-0520, to 50% of the maximum activity of the enzyme when saturated with activator, in the presence of 0.01 mM Ca^{++}, pH 7.5 at 30°C.

Enzymatic Assay of PHOSPHODIESTERASE, 3':5'-CYCLIC NUCLEOTIDEACTIVATOR (DEFICIENT)

Principle

$$3':5'\text{-cAMP} + H_2O \xrightarrow{\text{PDE-3':5'-CN}} \text{AMP} + \text{PDE - 3':5'-CNA}$$

$$\text{AMP} + \text{ATP} \xrightarrow{\text{Myokinase}} 2\ \text{ADP}$$

$$2\ \text{ADP} + 2\ \text{PEP} \xrightarrow{\text{Pyruvate Kinase}} 2\ \text{ATP} + 2\ \text{Pyruvate}$$

$$2\ \text{Pyruvate} + 2\beta\text{-NADH} \xrightarrow{\text{Lactic Dehydrogenase}} 2\ \text{Lactate} + 2\beta\text{-NAD}$$

Abbreviations Used

3':5'-cAMP = Adenosine 3':5'-Cyclic Monophosphate
PDE-3':5'-CNA = Phosphodiesterase, 3':5'-Cyclic Nucleotide Activator (Calmodulin)
PDE-3':5'-CN = Phosphodiesterase, 3':5'-Cyclic Nucleotide Activator Deficient
AMP = Adenosine 5'-Monophosphate
ATP = Adenosine 5'-Triphosphate
ADP = Adenosine 5'-Diphosphate
PEP = Phospho(enol) pyruvate
β-NADH = β-Nicotinamide Adenine Dinucleotide, Reduced Form
β-NAD = β-Nicotinamide Adenine Dinucleotide, Oxidized Form

Conditions: T = 30°C, pH = 7.5, A340nm, Light path = 1 cm

Method: Continuous Spectrophotometric Rate Determination

Reagents

A. 200 mM Tris HCl Buffer, pH 7.5 at 30°C (Prepare 50 ml in deionized water using Trizma Base. Adjust to pH 7.5 at 30°C with 1 M HCl).
B. 1.0 M Potassium Chloride Solution (KCl) (Prepare 1 ml in deionized water using Potassium Chloride).
C. 60 mM Magnesium Sulfate Solution ($MgSO_4$) (Prepare 10 ml in deionized water using Magnesium Sulfate, Heptahydrate).
D. 0.665 mM Phospho(enol) pyruvate Solution (PEP) (Prepare by dissolving Phospho(enol) pyruvate, Monopotassium Salt in 3.7 ml of Reagent A. Then add: 0.6 ml of Reagent B, 1.2 ml of Reagent C, and 4.5 ml of deionized water. *Prepare Fresh*).
E. 30 mM Adenosine 5'-Triphosphate Solution (ATP) (Prepare 10 ml in deionized water using Adenosine 5'-Triphosphate, Disodium Salt. Adjust to pH 7.5 with solid Sodium Bicarbonate. *Prepare Fresh*).
F. 120 mM Adenosine 3':5'-Cyclic Monophosphate Solution (3':5'-cAMP) (Prepare 1 ml in deionized water using Adenosine 3':5'-Cyclic Monophosphate, Sodium Salt. *Prepare Fresh*).
G. PK/LDH Enzymes Suspension.
H. Myokinase Enzyme Solution (MK) (Immediately before use, prepare a solution containing 200 units/ml of Myokinase in cold deionized water).

I. 1.75 mM Calcium Acetate Solution ($Ca(OAc)_2$) (Prepare 10 ml in deionized water using Calcium Acetate).

J. β-Nicotinamide Adenine Dinucleotide, Reduced Form Solution (β-NADH) (Use β-Nicotinamide Adenine Dinucleotide, Reduced Form, Disodium Salt. *Prepare Fresh*).

K. Phosphodiesterase, 3':5'-Cyclic Nucleotide-Activator Solution (PDE-3':5'-CNA) (Immediately before use, prepare a solution containing 300 units/ml of Phosphodiesterase, 3':5'-Cyclic Nucleotide-Activator in cold deionized water).

L. Phosphodiesterase, 3':5'-Cyclic Nucleotide Enzyme, Solution (PDE-3':5'-CN) (Immediately before use prepare a solution containing 0.2 unit/ml of Phosphodiesterase 3':5'cyclic Nucleotide, Activator Deficient, in deionized water).

Procedure

Prepare a reaction cocktail by pipetting (in milliliters) the following reagents into a 1 mg vial of:

Reagent J	1.00
Reagent D (PEP)	9.00
Reagent E (ATP)	0.10
Reagent F (3':5'-cAMP)	0.10
Reagent G (PK/LDH)	0.05
Reagent H (MK)	0.10
Reagent I ($Ca(OAc)_2$)	0.15

Mix by inversion. Pipette (in milliliters) the following reagents into suitable cuvettes:

	Non-Activated	Activated	Blank
Reaction Cocktail	3.00	3.00	3.00
Deionized Water	0.10	—	0.10
Reagent K (PDE-3':5'-CNA)	—	0.10	0.10

Mix by inversion and equilibrate to 30°C. Monitor the A340nm until constant, using a suitably thermostatted spectrophotometer.

Reagent L (PDE-3':5'-CN)	0.10	0.10	—

Immediately mix by inversion and record the decrease in A340nm for approximately 15 minutes. Obtain the maximum linear rate for both the Non-Activated and Activated.

Calculations

I. PDE-3':5'-CN

$$\text{PDE-3':5'-CN Non-Activated Units/ml} = \frac{(\Delta A340nm/\min \text{ Non-Activated} - \Delta A340nm/\min \text{ Blank})(3.2)(df)}{(2)(6.22)(0.1)}$$

df = Dilution factor

3.2 = Total volume (in milliliters) of assay

2 = 2 μmoles of β-NAD produced per mole of 3':5'-cAMP hydrolyzed

6.22 = Millimolar extinction coefficient of β-NADH at 340nm
0.1 = Volume (in milliliter) of PDE-3':5'-CN Non-Activated enzyme used

$$\text{Units / mg solid} = \frac{\text{units / ml enzyme}}{\text{mg solid / ml enzyme}}$$

$$\text{Units / mg protein} = \frac{\text{units / ml enzyme}}{\text{mg protein / ml enzyme}}$$

$$\text{PDE - 3':5'-CN Activated Units / ml} = \frac{(\Delta A340\text{nm / min Activated} - \Delta A340\text{nm / min Blank})\,(3.2)\,(\text{df})}{(2)\,(6.22)\,(0.1)}$$

3.2 = Total volume (in milliliters) of assay
df = Dilution factor
2 = 2 μmoles of β-NAD produced per mole of 3':5'-cAMP hydrolyzed
6.22 = Millimolar extinction coefficient of β-NADH at 340nm
0.1 = Volume (in milliliter) of PDE-3':5'-CN Activated enzyme used

$$\text{Units / mg solid} = \frac{\text{units / ml enzyme}}{\text{mg solid / ml enzyme}}$$

$$\text{Units / mg protein} = \frac{\text{units / ml enzyme}}{\text{mg protein / ml enzyme}}$$

Unit Definition: *One unit will hydrolyze 1.0 μmole of 3':5' cyclic-AMP to 5'-AMP per minute at pH 7.5 at 30°C.*

Final Assay Concentrations

In a 3.20 ml reaction mix (Test), the final concentrations are 66 mM Tris, 54 mM potassium chloride, 6.4 mM magnesium sulfate, 0.6 mM phospho(enol) pyruvate, 0.30 mM adenosine 5'-triphosphate, 1.2 mM adenosine 3':5'-cyclic monophosphate, 0.03 mM calcium acetate, 11 units pyruvate kinase, 16 units L-lactic dehydrogenase, 6 units myokinase, 0.1 mM β-nicotinamide adenine dinucleotide, reduced form, 0.02 unit phosphodiesterase 3':5'-cyclic nucleotide, and 30 units phosphodiesterase 3':5'-cyclic nucleotide activator (when present).

Notes

1. Contains not less than 700 Pyruvate Kinase units and 1000 L-Lactic Dehydrogenase units each per ml.
2. This reaction is very sensitive to chelators which sequester Ca^{++} ions at low Ca^{++} concentrations. EDTA or EGTA concentrations of 0.005 mM in the cuvette will almost completely inhibit the reaction. The concentration of EDTA or EGTA should be less then 0.001 mM in the cuvette. Cuvettes should be completely cleaned with 3% (w/v) NaOH before each use, since PDE-3':5'-CNA can bind to glass or quartz cuvettes.
3. L-Lactic Dehydrogenase Unit Definition: One unit will reduce 1.0 μmole of pyruvate to L-lactate per minute at pH 7.5 at 37°C.
4. Pyruvate Kinase Unit Definition: One unit will convert 1.0 μmole of phospho(enol) pyruvate to pyruvate per minute at pH 7.6 at 37°C.
5. Myokinase Unit Definition: One unit will convert 2.0 μmoles of ADP to ATP and AMP per minute at pH 7.6 at 37°C.
6. Phosphodiesterase 3':5'-Cyclic Nucleotide Activator Unit Definition: One unit will stimulate 0.008 activated units of phosphodiesterase 3':5'-cyclic nucleotide, in a 3 ml reaction volume at pH 7.5 at 30°C to 50% of the maximum activity of the enzyme when saturated with activator, in the presence of 0.01 mM Ca_2^+.

Protein Inhibitor Assays

Protease Inhibitor Assays Suitability Assay of 4-(2-AMINOETHYL) BENZENESULFONYL FLUORIDE HYDROCHLORIDE as a Serine Protease Inhibitor

Principle

The reaction shown below is inhibited by AEBSF

$$\text{ATEE} + H_2O \xrightarrow{\text{Chymotrypsin}} \text{N-Acetyl-L-Tyrosine} + \text{Ethanol}$$

Abbreviations Used

AEBSF = 4-(2-Aminoethyl)Benzenesulfonyl Fluoride
ATEE = N-Acetyl-L-Tyrosine Ethyl Ester

Conditions: T = 25°C, pH = 7.0, A237nm, Light path = 1 cm

Method: Continuous Spectrophotometric Rate Determination

Reagents

A. 50 mM Potassium Phosphate Solution (Prepare 100 ml in deionized water using Potassium Phosphate, Monobasic, Anhydrous).

B. 50 mM Potassium Phosphate Buffer, pH 7.0 at 25°C (Prepare 100 ml in deionized water using Potassium Phosphate, Dibasic, Trihydrate. Adjust to pH 7.0 at 25°C with Reagent A).

C. 1.0 mM N-Acetyl-L-Tyrosine Ethyl Ester Solution (ATEE) (Prepare 20 ml in Reagent B using N-Acetyl-L-Tyrosine Ethyl Ester).

D. 4.2 mM 4-(2-Aminoethyl)Benzenesulfonyl Fluoride Solution (AEBSF) (Prepare 10 ml in deionized water using 4-(2-Aminoethyl)Benzenesulfonyl Fluoride Solution).

E. α-Chymotrypsin Enzyme Solution (Immediately before use, prepare a solution containing approximately 8 units/ml of α-Chymotrypsin, in cold deionized water).

Procedure

Step 1: Pipette (in milliliters) the following reagents into suitable containers:

	Test 1 Inhibited	Test 2 Uninhibited	Blank
Reagent E (Enz Soln)	2.00	2.00	—
Reagent D (AEBSF)	0.20	—	—
Deionized Water	1.80	2.00	—

Mix by swirling and incubate at 25°C for exactly 1 minute. Then proceed with Step 2.

Step 2: Pipette (in milliliters) the following reagents into suitable cuvettes:

Reagent C (ATEE)	3.00	3.00	3.00

Equilibrate to 25°C. Then add:

Test 1 (from Step 1)	—	—	—
Test 2 (from Step 1)	0.10	—	—
Reference (from Step 1)	—	0.10	—
Deionized Water	—	—	0.10

Immediately mix by inversion and record the decrease in A237nm for approximately 5 minutes. Obtain the ΔA237nm/minute using the maximum linear rate for Test 1, Test 2, and the Reference Solution.

Calculations

$$\% \text{ Inhibition} = \frac{(\Delta A237nm / \min \text{ Test} - \Delta A237nm / \min \text{ Blank})}{(\Delta A237nm / \min \text{ Test uninhibited} - \Delta A237nm / \min \text{ Blank})} \times 100\%$$

Specification

Compare the % inhibition of serine protease activity to that of a control. The values should be similar.

Final Assay Concentrations

In a 3.10 ml reaction mix, the final concentrations are 48 mM potassium phosphate, 0.007–0.07 mM 4-(2-aminoethyl)benzenesulfonyl fluoride, 0.97 mM N-acetyl-L-tyrosine ethyl ester, and 0.4 unit α-chymotrypsin.

Enzymatic Assay of-α-1-ANTITRYPSIN Inhibition of Trypsin Activity

Principle

$$\text{BAEE} + H_2O \xrightarrow{\text{Trypsin}} \text{Na - Benzoyl - L - Arginine} + \text{Ethanol}$$

Abbreviation Used

BAEE = Na-Benzoyl-L-Arginine Ethyl Ester α-1-Antitrypsin inhibits this reaction.

Conditions: T = 25°C, pH = 7.6, A253nm, Light path = 1 cm

Method: Continuous Spectrophotometric Rate Determination

Reagents

A. 67 mM Sodium Phosphate Buffer, pH 7.6 at 25°C (Prepare 100 ml in deionized water using Sodium Phosphate, Dibasic, Anhydrous. Adjust to pH 7.6 at 25°C with 1 M HCl).
B. 0.25 mM Na-Benzoyl-L-Arginine Ethyl Ester Solution (BAEE) (Prepare 50 ml in Reagent A using Na-Benzoyl-L-Arginine Ethyl Ester, Hydrochloride).
C. 1 mM Hydrochloric Acid Solution (HCl) (Prepare 50 ml in deionized water using concentrated Hydrochloric Acid).
D. Trypsin Enzyme Solution (Trypsin) (Immediately before use, prepare a solution containing 0.5 mg/ml of Trypsin in cold Reagent C).
E. α-1-Antitrypsin Solution (α-1-Antitryp) (Immediately before use, prepare a solution containing 2 mg solid/ml of α-1-Antitrypsin in cold Reagent A.

Procedure

Step 1: Pipette (in milliliters) the following reagents into suitable containers:

	Uninh	Test 1	Test 2	Test 3	Test 4	Test 5
Reagent E (α-1-Antitryp)	—	0.10	0.20	0.30	0.40	0.50
Reagent D (Trypsin)	0.50	0.50	0.50	0.50	0.50	0.50
Reagent C (HCl)	9.50	9.40	9.30	9.20	9.10	9.00

Mix by inversion.

Step 2: Pipette (in milliliters) the following reagents into suitable containers:

	Uninh	Test 1	Test 2	Test 3	Test 4	Test 5	Blank
Reagent B (BAEE)	3.00	3.00	3.00	3.00	3.00	3.00	3.00
Reagent C (HCl)	—	—	—	—	—	—	0.20

Mix by inversion and equilibrate to 25°C. Monitor the A253nm until constant, using a suitably thermostatted spectrophotometer. Then add:

	Uninh	Test 1	Test 2	Test 3	Test 4	Test 5	Blank
Deionized Water	—	—	0.90	0.80	0.70	0.60	1.00
Uninh (Step 1)	0.20	—	—	—	—	—	—
Test 1 (Step 1)	—	0.20	—	—	—	—	—
Test 2 (Step 1)	—	—	0.20	—	—	—	—
Test 3 (Step 1)	—	—	—	0.20	—	—	—
Test 4 (Step 1)	—	—	—	—	0.20	—	—
Test 5 (Step 1)	—	—	—	—	—	0.20	—

Immediately mix by inversion and record the increase in A253nm for approximately 5 minutes. Obtain the A253nm/minute using the maximum linear rate for the Tests, Blank, and Uninhibited Solution.

Calculations

Trypsin Activity

$$\text{BAEE units / mg enzyme} = \frac{(\text{A253nm / min Test} - \text{A253nm / min Blank})\,(\text{df})}{(0.001)\,(\text{mg Trypsin / RM})}$$

df = Dilution factor

0.001 = The change in A253nm/minute per unit of Trypsin as per the Unit Definition (One BAEE unit = ΔA253 of 0.001 per minute with BAEE as substrate at pH 7.6 at 25°C.

Reaction volume = 3.2 ml (1 cm light path))

RM = Reaction Mixture

Plot the Trypsin activity versus ml of α-1-Antitrypsin. Determine from the plot the amount of α-1-Antitrypsin needed for 100% inhibition.

The amount (in mg) of α-1-Antitrypsin needed for 100% inhibition = (ml of α-1-Antitrypsin for 100% inhibition) (conc. of α-1-Antitrypsin (mg/ml)) mg of α-1-Antitrypsin needed for 100% Inhibition mg α-1-Antitrypsin to inhibit 1 mg Trypsin = mg Trypsin/Reaction Mix.

Final Assay Concentrations

In a 3.20 reaction mix, the final concentrations are 0.23 mM Na-benzoyl-L-arginine ethyl ester, 63 mM sodium phosphate, 0.06 mM hydrochloric acid, 0.005 mg trypsin, 0.004–0.02 mg α-1-antitrypsin.

Enzymatic Assay of APROTININ

Principle

Aprotinin will inhibit the following reaction:

$$\text{BAPNA} \xrightarrow{\text{Trypsin}} \text{N}-\alpha-\text{Benzoyl}-\text{DL}-\text{Arginine} + \text{p}-\text{Nitroaniline}$$

Abbreviation Used

BAPNA = N-α-Benzoyl-DL-Arginine-p-Nitroanilide

Conditions: T = 25°C, pH = 7.8, A405nm, Light path = 1 cm

Method: Continuous Spectrophotometric Rate Determination

Reagents

A. 200 mM Triethanolamine Buffer with 20 mM Calcium Chloride, pH 7.8 at 25°C (Prepare 100 ml in deionized water using Triethanolamine Hydrochloride and Calcium Chloride, Dihydrate. Adjust to pH 7.8 at 25°C with 1 M NaOH).

B. 0.1% (w/v) Nα-Benzoyl-DL-Arginine-p-Nitroanilide Solution (BAPNA) (Prepare 25 ml in deionized water using Nα-Benzoyl-DL-Arginine-p-Nitroanilide, Hydrochloride. Heat solution (to not greater than 65°C to facilitate solubilization. *Prepare Fresh. The Complete Solubilization of Substrate is Critical*).

C. 1 mM Hydrochloric Acid Solution (HCl) (Prepare 50 ml in deionized water using Hydrochloric Acid).

D. Trypsin Enzyme Solution (Prepare a solution containing 1040–1560 units/ml of Trypsin, Type III in cold Reagent C. *Prepare Fresh*).

E. 0.9% (w/v) Sodium Chloride Solution (NaCl) (Prepare 100 ml in deionized water using Sodium Chloride).

F. Aprotinin Inhibitor Solution (Prepare three separate Aprotinin solutions in Reagent E, each containing 0.047–0.06 Trypsin Inhibitor Units per ml. Use a separate solution for each replication of assay).

Procedure

Pipette (in milliliters) the following reagents into suitable cuvettes:

	Uninhibited Test	Inhibited Test	Blank
Reagent A (Buffer)	1.60	1.60	1.60
Reagent C (Hcl)	—	—	0.20
Reagent D (Trypsin)	0.20	0.20	—
Reagent E (NaCl)	0.20	—	0.20
Reagent F (Inhibitor)	—	0.20	—

Mix by inversion and equilibrate to 25°C. Monitor the A405nm until constant, using a suitably thermostatted spectrophotometer. Then add:

Reagent B (BAPNA)	1.00	1.00	1.00

Immediately mix by inversion and record the increase in A405nm for approximately 5 minutes. Obtain the ΔA405nm/minute using the maximum linear rate for the Uninhibited, Inhibited, and Blank Solutions.

Calculations

$$\text{TIU / ml} = \frac{(\Delta A405\text{nm / minute Uninhibited} - \Delta A405\text{nm / minute Inhibited})\ (\text{df})}{(9.96)\ (\text{ml Aprotinin / ml RM})}$$

TIU = Trypsin Inhibitor Units
df = Dilution factor
9.96 = The millimolar extinction coefficient of p-Nitroaniline at 405nm
RM = Reaction Mix

$$\text{TIU / mg solid} = \frac{\text{TIU / ml}}{\text{mg solid / ml}}$$

$$\%\ \text{Inhibition} = \frac{\Delta A405\text{nm / minute Uninhibited} - \Delta A405\text{nm / minute Inhibited} \times 100}{\Delta A405\text{nm / minute Uninhibited} - \Delta A405\text{nm / minute Blank}}$$

Unit Definition: *One trypsin inhibitor unit (TIU) will decrease the activity of two trypsin units by 50% where one trypsin unit will hydrolyze 1.0 μmole of Na-benzoyl-DL-Arginine-p-Nitroanilide (BAPNA) per minute at pH 7.8 at 25°C.*

Final Assay Concentrations

In a 3.00 ml reaction mix, the final concentrations are 107 mM triethanolamine, 11 mM calcium chloride, 0.03% (w/v) Nα-benzoyl-DL-arginine p-nitroanilide, 0.07 mM hydrochloric acid, 208–312 units trypsin, 0.12% (w/v) sodium chloride, 0.0003% (w/v) thimerosal and 0.0094–0.012 trypsin inhibitor unit of aprotinin.

Notes

1. If the solution is hazy, continue to stir over gentle heat (to not greater that 65°C) until the solution becomes clear. Do not use the solution if it turns yellow (this indicates possible chemical decomposition of the substrate due to overheating).
2. The % inhibition must be between 40 and 60 percent, for the assay to be valid. Adjust the concentration of the inhibitor solution so that the results are obtained in this range.
3. In cases where there is variability in the results, ensure that the uninhibited rate has a ΔA of 0.08–0.12. This may be required to reduce the variance caused by the range in specific activity of the Trypsin used. Adjust the concentration of the trypsin solution so that the results are in the range.

Enzymatic Assay of BROMELAIN INHIBITOR

Principle

Bromelain Inhibitor will inhibit the following reaction:

$$\text{LNPE} \xrightarrow{\text{Bromelain}} \text{Na - CBZ - L - Lysine} + \text{p - Nitrophenol}$$

Abbreviations Used

LNPE = Na-CBZ-L-Lysine p-Nitrophenyl Ester
CBZ = Carbobenzoxy

Conditions: T = 25°C, pH = 4.6, A317nm, Light path = 1 cm

Method: Continuous Spectrophotometric Rate Determination

Reagents

A. 20 mM Sodium Acetate Buffer with 100 mM Potassium Chloride and 1.0 mM L-Cysteine, pH 4.6 at 25°C (Prepare 100 ml in deionized water using Sodium Acetate, Trihydrate, Potassium Chloride and L-Cysteine, Hydrochloride, Monohydrate. Adjust to pH 4.6 at 25°C with 1 M HCl. *Prepare Fresh*).
B. 90% (v/v) Acetonitrile (Prepare 1.0 ml in deionized water using Acetonitrile).
C. 1.25 mM Na-CBZ-L-Lysine p-Nitrophenyl Ester Solution (LNPE) (Prepare 1.0 ml in Reagent B using Na-CBZ-L-Lysine p-Nitrophenyl Ester, Hydrochloride).
D. 10 mM Sodium Acetate with 100 mM Potassium Chloride, pH 4.6 at 25°C (Enz Dil) (Prepare 100 ml in deionized water using Sodium Acetate, Trihydrate and Potassium Chloride. Adjust to pH 4.6 at 25°C with 1 M HCl).
E. Bromelain Enzyme Solution (Bromelain) (Immediately before use, prepare a solution containing 0.5–0.7 unit/ml of Bromelain, in Reagent D).
F. Bromelain Inhibitor Solution (Inhib) (Immediately before use, prepare a solution containing 0.14 mg/ml of Bromelain Inhibitor in Reagent D).

Procedure

Pipet (in milliliters) the following reagents into suitable cuvettes:

	Uninh	Test 1	Test 2	Test 3	Test 4	Blank
Reagent A (Buffer)	3.00	3.00	3.00	3.00	3.00	3.00
Reagent D (Enz Dil)	0.05	0.03	0.02	0.01	—	0.10
Reagent E (Bromelain)	0.05	0.05	0.05	0.05	0.05	—
Reagent F (Inhib)	—	0.02	0.03	0.04	0.05	—

Mix by inversion and equilibrate to 25°C. Allow the solution to incubate for a minimum of 2 minutes at 25°C. This is required for full activation of the bromelain. Monitor the A317nm until constant, using a suitably thermostatted spectrophotometer. Then add:

Reagent C (LNPE)	0.05	0.05	0.05	0.05	0.05	0.05

Immediately mix by inversion and record the increase in A317nm for approximately 5 minutes. Obtain the ΔA317nm/minute using the maximum linear rate for both the Tests and Blank.

Calculations

$$\%\ \text{Inhibition} = \frac{(\Delta A317\text{nm Uninh} - \Delta 317\text{nm Test})\ (100)}{\Delta A317\text{nm Uninh} - \Delta A317\text{nm Blank}}$$

ΔA317nm Uninh = the uninhibited rate (Bromelain Inhibitor is not present)

Plot the Percent Inhibition versus the milligrams of Inhibitor in the reaction mix. Obtain the milligrams of Inhibitor at 50% inhibition from the graph.

$$\text{mg Inhibitor / mg Bromelain} = \frac{(\text{mg Inhibitor})}{(\text{mg Bromelain / RM})}$$

mg Inhibitor = The milligrams of inhibitor at 50% inhibition (obtained from the graph)
RM = Reaction mix (Total volume = 3.15 ml)

Specifications

Approximately 0.6 mg will inhibit the activity of 1 mg purified bromelain by 50%.

Final Assay Concentrations

In a 3.15 ml reaction mix, the final concentrations are 19 mM sodium acetate, 98 mM potassium chloride, 0.95 mM L-cysteine, 1.4% (v/v) acetonitrile, 0.02 mM Na-CBZ-L-lysine p-nitrophenyl ester, 0.025–0.035 unit bromelain, and 3–7 µg bromelain inhibitor.

Notes

1. The concentration of the substrate solution *Must* be precise. This solution is stable for several days when stored at 0–5°C.
2. Bromelain is used because other bromelain enzymes will give different results.
3. The concentration of the Bromelain Inhibitor solution *Must* be precise.
4. Bromelain Unit Definition: One unit will release 1.0 µmole of p-nitrophenol from Na-CBZ-L-lysine p-nitrophenyl ester per minute at pH 4.6 at 25°C.

Enzymatic Assay of CARBOXYPEPTIDASE INHIBITOR

Principle

Carboxypeptidase Inhibitor will inhibit the reaction shown below:

$$\text{Hippuryl-L-Phe} + H_2O \xrightarrow{\text{Carboxypeptidase A}} \text{Hippuric Acid} + \text{L-Phenylalanine}$$

Conditions: T = 25°C, pH = 7.5, A254nm, Light path = 1 cm

Method: Continuous Spectrophotometric Rate Determination

Reagents

A. 20 mM Tris HCl Buffer with 500 mM Sodium Chloride, pH 7.5 at 25°C (Prepare 100 ml in deionized water using Trizma Base and Sodium Chloride. Adjust to pH 7.5 at 25°C with 1 M HCl).

B. 1.0 mM Hippuryl-L-Phenylalanine Solution (Hippuryl-L-Phe) (Prepare 30 ml in Reagent A using Hippuryl-Phe. Dissolution may require 30 minutes).

C. Carboxypeptidase A Enzyme Solution (CPA) (Immediately before use, prepare a solution containing approximately 0.1 mg/ml " 50% of Carboxypeptidase A, in cold Reagent A. The protein concentration should be determined using the E280nm 0.1% = 1.88).

D. Carboxypeptidase Inhibitor (CPI) (Immediately before use, prepare a solution in Reagent A containing Carboxypeptidase Inhibitor so that a 0.05 ml aliquot produces 20–70% inhibition of Carboxypeptidase A).

Procedure

Preincubate 0.05 ml of Reagent D (CPI) and 0.05 ml of Reagent C (CPA) at 25°C for 30 seconds. Pipette in (milliliters) the following reagents into suitable quartz cuvettes.

	CPI/CPA Test	CPA Control	Blank
Reagent B (Hippuryl-L-Phe)	3.00	3.00	3.00
CPI/CPA Mixture	0.10	—	—
Reagent C (CPA)	—	0.05	—
Reagent A (Buffer)	—	0.05	0.10

Immediately mix by inversion and record the increase in A254nm for approximately 3 minutes using air as the reference. Obtain the ΔA254nm/minute using the maximum linear rate for the Test, Control, and Blank.

Calculations

A.

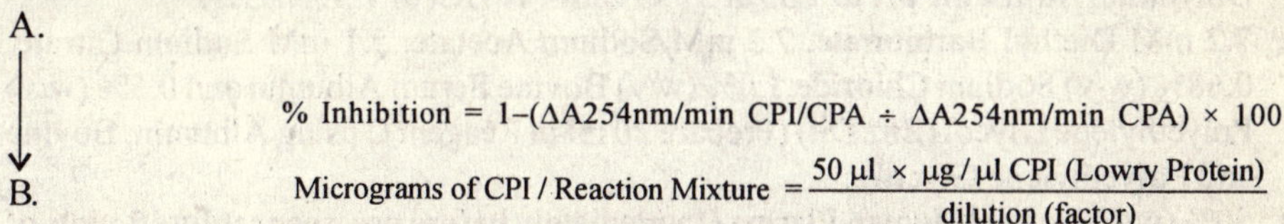

$$\% \text{ Inhibition} = 1 - (\Delta A254\text{nm/min CPI/CPA} \div \Delta A254\text{nm/min CPA}) \times 100$$

B.

$$\text{Micrograms of CPI / Reaction Mixture} = \frac{50\ \mu l \times \mu g/\mu l \text{ CPI (Lowry Protein)}}{\text{dilution (factor)}}$$

C. Since the assay is linear at 20–70% inhibition, use the following equation to determine how many micrograms of CPI are needed to produce 50% inhibition.

EX:

$$\frac{\mu g \text{ CPI}}{\%\text{Inhibition}} = \frac{X}{50}$$

D. Micrograms of CPA/Reaction Mixture = 50 µl × µg/µl CPA (Use E280nm 0.1% = 1.88)

E.

$$\text{Units} / \mu g \text{ CPI protein} = \frac{\mu g \text{ CPA / RM}}{\mu g \text{ CPI needed to produce 50\% inhibition}}$$

RM = Reaction Mixture

Specification

One µg of inhibitor protein will inhibit the activity of 2–8 µg of carboxypeptidase A by 50% using hippuryl-L-phenylalanine as substrate at pH 7.5 at 25°C.

Final Assay Concentrations

In a 3.10 ml reaction mix, the final concentrations are 20 mM Tris, 500 mM sodium chloride, 0.97 mM hippuryl-L-phenylalanine, approximately 0.005 mg carboxypeptidase A, and varying amounts of carboxypeptidase inhibitor.

Note

1. Determine the protein concentration of the Carboxypeptidase Inhibitor by the Lowry Method. Dilutions must be made so that the sample does not exceed 0.025 mg of Bovine Serum Albumin equivalents.

Enzymatic Assay of HIRUDIN

Principle

Hirudin + Thrombin → Hirudin / Thrombin complex + Thrombin (excess)

$$\text{Plasma (Fibrinogen)} \xrightarrow{\text{Thrombin (excess)}} \text{Clot (Fibrin)}$$

Conditions: T = 37°C, pH = 7.35

Method: Fibrometer

A. 36 mM Diethyl Barbiturate and 36 mM Sodium Acetate, 0.85% (w/v) Sodium Chloride Solution, pH 7.35 at 37°C (Prepare 100 ml using Sodium Acetate, Trihydrate, Barbital, Sodium Salt and Sodium Chloride. Adjust the pH to 7.35 at 37°C with 1 M HCl).

B. 0.85% (w/v) Sodium Chloride Solution (Prepare 100 ml in deionized water using Sodium Chloride).

C. 7.2 mM Diethyl Barbiturate, 7.2 mM Sodium Acetate, 5.1 mM Sodium Citrate, and 0.68% (w/v) Sodium Chloride (Prepare 100 ml by combining 20 ml of Reagent A, 60 ml of Reagent B, and the appropriate amount of Citric Acid, Trisodium Salt, Dihydrate. Adjust the pH to 7.35 at 37°C with 1 N HCl or 1 M NaOH).

D. 7.2 mM Diethyl Barbiturate, 7.2 mM Sodium Acetate, 5.1 mM Sodium Citrate, 0.68% (w/v) Sodium Chloride, 1.0% (w/v) Bovine Serum Albumin and 0.5% (w/v) Polyethylene Glycol (Enz Dil) (Prepare 20 ml in Reagent C using Albumin, Bovine and Polyethylene Glycol).

E. 50% (w/v) Normal Human Plasma (Immediately before use, reconstitute 2 vials of Human Plasma, lyophilized with 2 ml each of deionized water. Let stand 5 minutes. Then add 2 ml of Reagent B (NaCl) to each. Combine vials and keep at room temperature).

F. Thrombin Enzyme Solution (Thrombin) (Immediately before use, reconstitute a vial of Thrombin, with 1 ml of deionized water. Assay for units/ml of Thrombin using the Thrombin assay. After determining units/ml of Thrombin of this solution, further dilute to 9.8–10.8 units/ml of Thrombin using Reagent D. Assay this solution again for units/ml of Thrombin and use this volume as the working Thrombin Solution).

G. NIH Thrombin Standard Solution (Std) (Use Thrombin Reference Standard (Lot J) which has been diluted in Reagent D.
Note: Current Standard Curve has been made with N.I.H. Thrombin standard diluted in glass test tubes. Next standard curve will use Plastic test tubes).

H. Hirudin Solution (Hirudin) (Immediately before use, prepare a solution containing approximately 10 units/ml using Reagent D at room temperature).

Procedure

Pipette (in milliliters) the following reagents into a test tube:

	Test
Reagent F (Thrombin)	0.20
Reagent H (Hirudin)	0.10

Incubate at 37°C in a water bath for minimum of 1 minute. After 1 minute, but before 3 minutes, remove 0.1 ml of the assay mix and assay it for excess thrombin activity using the thrombin assay. Repeat the assay with another 0.1 ml aliquot. The excess thrombin is stable over the 1–3 minute time period used for the assay. Using the time it takes to stop the fibrometer (the time should be in the 15–25 second range and give a % inhibition of thrombin of between 45 and 55%) and the thrombin standard curve, determine the thrombin concentration in units/ml.

$$\text{Units / ml H} = \frac{[(\text{T units / assay}) - (\text{Excess T units})]}{(0.1)\,(\text{ml Hirudin / ml assay mix})} \times \text{df}$$

$$\text{Units / vial H} = \frac{\text{Units / ml H}}{\text{vial H / ml}}$$

$$\%\ \text{Inhibition} = \frac{[(\text{T units / assay}) - (\text{Excess T units})] \times 100}{(\text{T units / assay})}$$

T Units/assay = (Units/ml working Thrombin) (0.2)
Excess T units = (Units/ml T) (0.3 Incubation mix)
0.1 = Volume (in milliliter) of Hirudin used
df = Dilution factor
T units/assay = Thrombin units used in Incubation Mix
Units/ml Working Thrombin = Final thrombin (units/ml) concentration of Thrombin Enzyme solution
0.2 = Volume (in milliliter) of working Thrombin incubated with Hirudin in Incubation Mix
Excess T Units = Thrombin units in Incubation Mix not inhibited by Hirudin
Units/ml T = Measured thrombin concentration of Test mix from the Thrombin standard curve
0.3 Incubation Mix = Volume (in milliliter) of Test Mix Volume
Units/ml = Thrombin Inhibitor Units per ml of Hirudin
vial H/ml = Hirudin dilution in vials/ml
Units/vial H = Thrombin Inhibitor Units per vial of Hirudin
Units/ml H = Units/mL Hirudin

Unit Definition: *One unit will neutralize 1 NIH unit of thrombin at 37°C, based on direct comparison to an NIH thrombin reference standard (Lot J).*

Final Assay Concentrations

In a 0.30 ml reaction mix, the final concentrations are 33% (v/v) plasma (see thrombin assay), 0.67 unit thrombin and 0.33 unit hirudin (components of the enzyme diluent are not included).

Notes

1. The enzymatic assay for thrombin must accompany this assay when it is sent to a customer.
2. Reagent A may be stored in frozen aliquots of 20 ml for up to three years.
3. Reagent C may be stored up to six months at 0–5°C.
4. Prepare a minimum of two vials of NHP and combine in a four dram glass vial.
5. Each 10 unit vial contains enough Thrombin for approximately 5 assays. Prepare multiple vials of Thrombin and combine in a four dram glass vial.
6. Concentration of Hirudin should be approximately 10 units/ml. If clotting time of excess thrombin assay is not in the 15–25 second range, the Hirudin concentration must be adjusted. Dilute Hirudin if the time is greater than 25 seconds. Increase the Hirudin concentration if the time is less than 15 seconds. The Hirudin concentration should be ½ of the thrombin concentration in reaction mixture.
7. The most acceptable linear range to measure Hirudin activity is 45–55% inhibition of thrombin.

Enzymatic Assay of PROTEASE INHIBITOR, of Calcium Activated Neutral Protease

Principle

$$\text{N, N - Dimethylated Casein} + H_2O \xrightarrow{\text{Protease, Calcium Activated Neutral}} \text{Amino Acids}$$

This reaction is inhibited by the protease inhibitor.

Conditions: T = 30°C, pH = 7.5, A293nm, Light path = 1 cm

Method: Stopped Spectrophotometric Rate Determination

Reagents

A. 200 mM Tris HCl Buffer with 1.5 mM β-Mercaptoethanol and 2.0 mM Calcium Chloride, pH 7.5 at 30°C (Substrate Buffer) (Prepare 100 ml in deionized water using Trizma Base, 2-Mercaptoethanol and Calcium Chloride, Dihydrate. Adjust to pH 7.5 t 30°C with 1 M HCl).

B. 0.5% (w/v) Casein, N,N-Dimethylated Solution (Casein) (Prepare 15 ml in Reagent A using Casein, N,N-Dimethylated).

C. 200 mM Tris HCl Buffer, pH 7.5 at 30°C (Protease Buffer) (Prepare 100 ml in deionized water using Trizma Base. Adjust to pH 7.5 at 30°C with 1 M HCl).

D. 100 mM Ethylenediaminetetraacetic Acid Solution (EDTA) (Prepare 2 ml in deionized water using Ethylenediaminetetraacetic Acid, Tetrasodium Salt, Hydrate).

E. 6.1 N Trichloroacetic Acid Solution (TCA) (Use Trichloroacetic Acid, 6.1 N Solution, approximately 100% (w/v)).

F. Protease Solution (Protease) (Immediately before use, prepare a solution containing 2.5 units/ml of Protease, Calcium Activated Neutral in cold Reagent C).

G. Protease Inhibitor Solution (Inhibitor) (Immediately before use, prepare a solution containing 20 mg/ml of Protease Inhibitor in cold Reagent B).

Procedure

Pipette (in milliliters) the following reagents into suitable containers, with the exception of Reagent F (Protease):

Tube Number	Reagent B (Casein)	Reagent G (Inhibitor)	Reagent D (EDTA)	Deionized Water	Reagent F (Protease)
1 (Blank)	1.00	—	0.10	—	0.50
2 (Control)	1.00	—	—	0.10	0.50
3 (Blank)	0.80	0.20	0.10	—	0.50
4 (Test)	0.80	0.20	—	0.10	0.50
5 (Blank)	0.60	0.40	0.10	—	0.50
6 (Test)	0.60	0.40	—	0.10	0.50
7 (Blank)	0.40	0.60	0.10	—	0.50
8 (Test)	0.40	0.60	—	0.10	0.50
9 (Blank)	0.20	0.80	0.10	—	0.50
10 (Test)	0.20	0.80	—	0.10	0.50
11 (Blank)	—	1.00	0.10	—	0.50
12 (Test)	—	1.00	—	0.10	0.50

Mix by inversion and then add 0.50 ml of Reagent F (Protease) to each tube. Mix by inversion and incubate at 30°C for exactly 30 minutes. Then add 0.20 ml of Reagent E (TCA) to each tube. Mix by inversion and allow the tubes to stand at room temperature for 10 minutes. Centrifuge in a clinical centrifuge at maximum speed in order to clarify the samples. Remove the clear supernatant from tubes 1 and 2 and record the absorbance at 280nm using water as the blank. Save these supernatants.
Pipette 1.0 ml of each of the 12 clear, acidic supernatant into 0.4 ml of 5 N NaOH. Allow to stand for 5 minutes and clarify by centrifugation. Record the absorbance at 293 nm for each of the supernatants, using water as the Blank.

Calculations

$$\text{Units of protease per assay} = \frac{(\text{A280nm Control} - \text{A280nm Blank No. 1})}{(0.5)\,(0.5)}$$

0.5 = Change in absorbance at A280nm per unit as per the Unit Definition
0.5 = Volume (in milliliter) of protease used

$$\%\ \text{Inhibition} = \left[1 - \frac{(\Delta\text{A293nm Test}}{\Delta\text{A293nm Control}}\right] \times 100$$

ΔA293nm Test = A293nm Test–A293nm Blank
ΔA293nm Control = A293nm Control–A293nm Blank

Plot the % inhibition versus mg of protease inhibitor per assay. Calculate from the graph the mg of protease inhibitor required for 50% inhibition. Divide the mg of inhibitor required for 50% inhibition by the units of protease per assay. This is equivalent to the mg of inhibitor required to inhibit 1 unit of protease.

$$\text{Units / mg solid} = \frac{1}{(\text{mg inhibitor required to reduce the activity of 1 units of protease by 50\%})}$$

Unit Definition: *One unit will reduce the activity of one unit of calciumactivated neutral protease (P4533) by 50% at pH 7.5 at 30°C. (Final volume = 1.8 ml, Light path = 1 cm)*

Final Assay Concentrations

In a 1.60 ml reaction mix, the final concentrations are 188 mM Tris, 0.9 mM 2-mercaptoethanol, 1.3 mM calcium chloride, 0.3% (w/v) casein, N,N-dimethylated, 1.25 units protease and 4–20 mg of protease inhibitor.

Enzymatic Assay of PROTEASE INHIBITOR COCKTAIL Casein as a Substrate

Principle

$$\text{Casein} + H_2O \xrightarrow{\text{Protease}} \text{Amino Acids}$$

This reaction is inhibited by the protease inhibitor cocktail.

Conditions: T = 37°C, pH = 7.5, A660nm, Light path = 1 cm

Method: Colorimetric

Reagents

A. 50 mM Potassium Phosphate buffer, pH 7.5 at 37°C. (Prepare 200 ml in deionized water using Potassium Phosphate, Dibasic, Trihydrate. Adjust to pH 7.5 at 37°C with 1 M HCl).

B. 0.65% (w/v) Casein Solution (Casein) (Prepare 125 ml in Reagent A using Casein. Heat gently (do not boil) until a homogenous dispersion is obtained. Adjust the pH to 7.5 at 37°C, if necessary, with either 1 M NaOH or 1 M HCl).

C. 110 mM Trichloroacetic Acid Reagent (TCA) (Dilute 9 ml of Trichloroacetic Acid, 6.1 N, approximately 100% (w/v) to 500 ml with deionized water).

D. Folin & Ciocalteu's Phenol Reagent (F-C) (Dilute 10 ml of Folin & Ciocalteu's Phenol Reagent to 40 ml with deionized water).

E. 500 mM Sodium Carbonate Solution (Na_2CO_3) (Prepare 500 ml in deionized water using Sodium Carbonate Anhydrous).

F. 10 mM Sodium Acetate Buffer with 5 mM Calcium Acetate, pH 7.5 at 37°C (Enzyme Diluent) (Prepare 100 ml in deionized water using Sodium Acetate, Trihydrate and Calcium Acetate. Adjust the pH to 7.5 at 37°C with 0.1 M Acetic acid or 0.1 M NaOH).

G. 1.1 mM L-Tyrosine Standard (Std Soln) (Prepare 100 ml in deionized water using L-Tyrosine, Free Base. Heat gently (do not boil) until tyrosine dissolves and cool to room temperature).

H. Protease Enzyme Solution (Immediately before use, prepare a solution containing 10 mg/ml of Pancreatin in cold Reagent F).

I. Protease Inhibitor Cocktail (Inhibitor) (Prepare a 10 X stock, solution of the Protease Inhibitor Cocktail by adding 10 ml of deionized water to a 100 ml bottle of lyophilized cocktail).

Procedure

Protease Inhibitor Complex: Pipette (in milliliters) the following reagents into suitable containers:

	Unhib. Test	Inhib. Test	Blank
Reagent A (Buffer)	2.00	1.80	1.80
Reagent I (Inhibitor)	—	0.20	0.20
Reagent H (Protease)	0.10	0.10	—
Reagent F (Enzyme Diluent)	—	—	0.10

Mix by inversion and incubate at room temperature for one hour. Pipette the following reagents into suitable vials (in milliliters):

Reagent B (Casein)	5.00	5.00	5.00

Equilibrate to 37°C. Then add:

Protease Inhibitor Complex	1.00	1.00	—

Mix by swirling and incubate at 37°C for exactly 10 minutes. Then add:

	Unhib. Test	Inhib. Test	Blank
Reagent C (TCA)	5.00	5.00	5.00
Protease Inhibitor Complex	—	—	1.00

Mix by swirling and incubate at 37°C for exacatly 30 minutes. Filter through Whatman #50 filter paper and use the filtrate in color development.

Color Development

Standard Curve: Prepare a standard curve by pipetting the following reagents into suitable vials (in milliliters).

	Std 1	Std 2	Std 3	Std 4	Blank
Reagent G (Std Soln)	0.05	0.10	0.20	0.40	0.00
Deionized Water	1.95	1.90	1.80	1.60	2.00
Reagent E (Na_2CO_3)	5.00	5.00	5.00	5.00	5.00
Reagent D (F-C)	1.00	1.00	1.00	1.00	1.00

Sample: Pipette the following reagents into 4 dram vials (in milliliters):

	Inhib. Test	Unhib. Test	Blank
Inhib. Test Filtrate	2.00	—	—
Unhib. Test Filtrate	—	2.00	2.00
Blank Filtrate	—	—	2.00
Reagent E (Na_2CO_3)	5.00	5.00	5.00
Reagent D (F-C)	1.00	1.00	1.00

Mix by swirling and incubate at 37°C for 30 minutes. Remove the vials and allow them to cool to room temperature. If the solutions are hazy, filter through a 0.45 μm filter prior to reading. Read the absorbance at A660nm for each of the vials in suitable cuvettes.

Calculations

Standard Curve: ΔA660nm Standard = A660nm Standard–A660nm Standard Blank

Plot the ΔA660nm Standard vs μmoles of Tyrosine.

Sample Determination: ΔA660nm Sample = A660nm Uninhib. Test–A660nm Sample Blank

ΔA660nm Sample = A660nm Inhib. Test–A660nm Sample Blank

Determine the μmoles of Tyrosine equivalents liberated using the Standard curve.

$$\text{Units / ml pancreatin} = \frac{(\mu\text{mole Tyrosine equivalents released})}{(1)(10)(2)}$$

10 = Time of assay (in milliliters) as per the Unit Definition

1 = Volume of enzyme (in milliliter) of enzyme used

2 = Volume (in milliliters) used in Colorimetric Determination

$$\text{Units / mg pancreatin} = \frac{\text{units / ml pancreatin}}{\text{mg solid / ml pancreatin}}$$

Inhibition of Pancreatin = Uninhib.Units/mg S of pancreatin — Inhib. units/mg S of Pancreatin
Uninhibited units/mg S of Pancreatin
S = Solid

Unit Definition: *One unit will hydrolyze casein to produce color equivalent to 1.0 μmole (181 μg) of tyrosine per minute at pH 7.5 at 37°C (color by Folin and Ciocalteu's reagent).*

Final Assay Concentration

In a 6.00 ml reaction mix, the final concentrations are 42 mM potassium phosphate, 0.54% (w/v) casein, 1.7 mM sodium acetate, 0.8 mM calcium acetate, and 0.48 mg pancreatin.

Enzymatic Assay of TRYPSIN INHIBITOR

Principle

$$\text{BAEE} + H_2O \xrightarrow{\text{Trypsin}} \text{N-}\alpha\text{-Benzoyl-L-Arginine} + \text{Ethanol}$$

Abbreviation Used

BAEE = N-α-Benzoyl-L-Arginine Ethyl Ester

Conditions: T = 25°C, pH = 7.6, A253nm, Light path = 1 cm

Method: Continuous Spectrophotometric Rate Determination

Reagents

A. 67 mM Sodium Phosphate buffer, pH 7.6 at 25°C (Prepare 100 ml in deionized water using Sodium Phosphate, Monobasic, Anhydrous. Adjust to pH 7.6 at 25°C with 1 M NaOH).
B. 0.25 mM N-α-Benzoyl-L-Arginine Ethyl Ester Solution (BAEE) (Prepare 50 ml in Reagent A using N-α-Benzoyl-L-Arginine Ethyl Ester, Hydrochloride).
C. 1 mM Hydrochloric Acid Solution (HCl) (Prepare 50 ml in deionized water using concentrated Hydrochloric Acid).
D. Trypsin Enzyme Solution (Trypsin) (Immediately before use, prepare a solution containing 1 mg protein/ml of Trypsin in cold Reagent C).
E. Trypsin Inhibitor Solution (Inhib) (Immediately before use, prepare a solution containing 1.0 mg/ml of Trypsin Inhibitor in cold Reagent A).

Procedure

Pipette (in milliliters) the following reagents into suitable containers: Part A:

	Uninh	Test 1	Test 2	Test 3	Test 4	Test 5
Reagent E (Inhib)	—	0.10	0.15	0.20	0.25	0.30
Reagent D (Trypsin)	0.50	0.50	0.50	0.50	0.50	0.50
Reagent C (HCl)	9.50	9.40	9. 35	9.30	9.25	9.20

Mix by inversion and pipette (in milliliters) the following reagents into suitable cuvettes: Part B:

	Uninh	Test 1	Test 2	Test 3	Test 4	Test 5
Reagent B (BAEE)	3.00	3.00	3.00	3.00	3.00	3.00
Reagent C (HCl)	0.10	0.10	0.10	0.10	0.10	0.10

Mix by inversion and equilibrate to 25°C. Monitor the A253nm until constant, using a suitably thermostatted spectrophotometer. Immediately mix by inversion and record the increase in A253nm for approximately 5 minutes. Obtain the ΔA253nm/minute using the maximum linear rate for the Tests, Blank, and Uninhibited Solution.

Calculations

$$\text{Trypsin Activity in BAEE units / mg enzyme} = \frac{(\Delta A253\text{nm / min Test} - \Delta A253\text{nm / min Blank})\,(\text{df})\,(10.0)}{(0.001)\,(0.10)\,(0.5)}$$

df = Dilution factor
0.001 = The change in A253nm/minute per unit of Trypsin at pH 7.6 at 25°C in a 3.2 ml reaction mix
0.10 = Volume (in milliliters) enzyme used (Part B)
10.0 = Total volume in milliliters of assay (Part A)
0.5 = Volume (in milliliters) of enzyme used (Part A)

$$\text{Units / mg solid} = \frac{\text{units / ml enzyme}}{\text{mg solid / ml enzyme}}$$

Plot the Trypsin activity (BAEE units/mg protein) vs ml of Trypsin Inhibitor/RM
Mg Trypsin Inhibitor = (ml of Trypsin Inhibitor) (Conc. of Trypsin Inhibitor, mg/ml)

$$\text{Mg Trypsin Inhibited by 1 mg Trypsin Inhibitor} = \frac{\text{mg Trypsin / RM}}{\text{mg Trypsin Inhabitor (from plot)}}$$

Unit Definition: *One trypsin unit will produce a ΔA253nm of 0.001 per minute with BAEE as substrate at pH 7.6 at 25°C in a reaction volume of 3.2 ml.*

Final Assay Concentrations

In a 3.20 ml reaction mix, the final concentrations are 63 mM sodium phosphate, 0.23 mM BAEE, 0.002 mM HCl, 0.005 mg trypsin, and 0.0003–0.001 mg trypsin inhibitor.

Enzymatic Assay of TRYPSIN-CHYMOTRYPSIN INHIBITOR CHYMOTRYPSIN INHIBITION

Principle

$$\text{BTEE} + H_2O \xrightarrow{\text{Chymotrypsin}} \text{N-Benzoyl-L-Tyrosine} + \text{Ethanol}$$

Abbreviation Used

BTEE = N-Benzoyl-L-Tyrosine Ethyl Ester
This reaction is inhibited by Trypsin-Chymotrypsin Inhibitor

Conditions: T = 25°C, pH = 7.8, A256nm, Light path = 1 cm

Method: Continuous Spectrophotometric Rate Determination

Reagents

A. 80 mM Tris HCl Buffer, pH 7.8 at 25°C (Buffer I) (Prepare 100 ml in deionized water using Trizma Base. Adjust to Ph 7.8 at 25°C with 1 M HCl).

B. 50% (v/v) Methanol (Prepare 50 ml in deionized water using Methanol, Absolute).

C. 1.18 mM N-Benzoyl-L-Tyrosine Ethyl Ester Solution (BTEE) (Prepare 50 ml in Reagent B using N-Benzoyl-L-Tyrosine Ethyl Ester).

D. 2 M Calcium Chloride Solution ($CaCl_2$) (Prepare 5 ml in deionized water using Calcium Chloride, Dihydrate).

E. 1 mM Hydrochloric Acid Solution (HCl) (Prepare 50 ml in deionized water using Hydrochloric Acid).

F. 10 mM Tris HCl Buffer, pH 7.8 at 25°C (Buffer II) (Prepare 25 ml in deionized water using Reagent A. Adjust to pH 7.8 at 25°C with either 0.1 M NaOH or 0.1 M HCl).

G. Chymotrypsin Enzyme Solution (Chymo) (Immediately before use, prepare a solution containing 2 mg/ml of α-Chymotrypsin in cold Reagent E).

H. Trypsin-Chymotrypsin Inhibitor Solution (TCI) (Immediately before use, prepare a solution containing 1 mg/ml of Trypsin-Chymotrypsin Inhibitor in cold Reagent F).

Procedure

Pipette (in milliliters) the following reagents into suitable containers: Part A:

	Uninh	Test 1	Test 2	Test 3	Test 4	Test 5
Reagent H (TCI)	—	0.05	0.075	0.10	0.15	0.20
Reagent G (Chymo)	0.50	0.50	0.50	0.50	0.50	0.50
Reagent F (BufferII)	9.50	9.45	9.425	9.40	9.35	9.30

Mix by inversion and pipette (in milliliters) the following reagents into suitable cuvettes: Part B:

Reagent C (BTEE)	1.40	1.40	1.40	1.40	1.40	1.40	1.40
Reagent A (BufferI)	1.42	1.42	1.42	1.42	1.42	1.42	1.42
Reagent D ($CaCl_2$)	0.08	0.08	0.08	0.08	0.08	0.08	0.08
Reagent F (BufferII)	—	—	—	—	—	—	0.10

Mix by inversion and equilibrate to 25°C. Monitor the A256nm until constant, using a suitably thermostatted spectrophotometer. Then add:

Uninh (Part A)	0.10	—	—	—	—	—	—
Test 1 (Part A)	—	0.10	—	—	—	—	—
Test 2 (Part A)	—	—	0.10	—	—	—	—
Test 3 (Part A)	—	—	—	0.10	—	—	—
Test 4 (Part A)	—	—	—	—	0.10	—	—
Test 5 (Part A)	—	—	—	—	—	0.10	—

Immediately mix by inversion and record the increase in A256nm for approximately 5 minutes. Obtain the ΔA256nm/minute using the maximum linear rate for the Tests, Blank, and Uninhibited Solution.

Calculations

$$\text{BTEE units / ml} = \frac{(\Delta A256nm / \text{min Test} - \Delta A256nm / \text{min Blank})\,(3)\,(0.5)\,(df)}{(0.964)\,(0.1)\,(10)}$$

3 = Volume (in milliliters) of Part B of the Assay
0.5 = Volume (in milliliter) of enzyme used in Part A of the Assay
df = Dilution factor
0.964 = Millimolar extinction coefficient of N-Benzoyl-L-Tyrosine at 256 nm
0.1 = Volume (in milliliter) of enzyme from Part A used in Part B of the Assay
10 = Volume (in milliliters) of Part A of the Assay

Plot the Chymotrypsin activity (BTEE units/ml) versus volume (in milliliter) of Trypsin-Chymotrypsin Inhibitor/Reaction Mixture in Step A. Determine the point at which the line intercepts the abscissa. This point is the X-intercept.

Y0 = Mg of Trypsin-Chymotrypsin Inhibitor/ml, in Part B of the Assay that results in complete inhibition of Chymotrypsin: = (X-Intercept) (mg TCI/ml in Reagent H) CT = Mg of Chymotrypsin/ml in Part B of the Assay = (0.5) (mg Chymotrypsin/ml in Reagent G) CT
Mg of Chymotrypsin inhibited by 1 mg Trypsin-chymotrypsin Inhibitor = Y0

Unit Definition: *One Chymotrypsin unit will hydrolyze 1.0 μmole of BTEE per minute at pH 7.8 at 25°C.*

Specifications

One mg of protein will inhibit 2–5 mg of chymotrypsin with activity of approximately 40 BTEE units per mg protein.

Final Assay Concentration

In a 3.00 ml reaction mix, the final concentrations are 38 mM Tris, 0.55 mM N-benzoyl-L-tyrosine ethyl ester, 23% (v/v) methanol, 53 mM calcium chloride, 1.7 μM hydrochloric acid, 3.3 μg chymotrypsin, and 0.17–0.67 μg trypsin-chymotrypsin inhibitor.

Enzymatic Assay of TRYPSIN-CHYMOTRYPSIN INHIBITOR Trypsin Inhibitory Activity

Principle

$$\text{BAEE} + H_2O \xrightarrow{\text{Trypsin}} \text{N-}\alpha\text{-Benzoyl-L-Arginine} + \text{Ethanol}$$

Abbreviation Used

BAEE = N-α-Benzoyl-L-Arginine Ethyl Ester

This reaction is inhibited by Trypsin-Chymotrypsin Inhibitor

Conditions: T = 25°C, pH = 7.6, A253nm, Light path = 1 cm

Method: Continuous Spectrophotometric Rate Determination

Reagents

A. 400 mM Sodium Phosphate Buffer, pH 7.6 at 25°C (Buffer I) (Prepare 100 ml in deionized water using Sodium Phosphate, Monobasic, Anhydrous. Adjust to pH 7.6 at 25°C with 1 M NaOH).

B. 0.25 mM N-α-Benzoyl-L-Arginine Ethyl Ester Solution (BAEE) (Prepare 50 ml in Reagent A using N-α-Benzoyl-L-Arginine Ethyl Ester, Hydrochloride).

C. 1 mM Hydrochloric Acid Solution (HCl) (Prepare 50 ml in deionized water using concentrated Hydrochloric Acid).

D. 5 mM Sodium Phosphate Buffer, pH 7.6 at 25°C (Buffer II) (Prepare 100 ml in deionized water using Sodium Phosphate, Monobasic, Anhydrous. Adjust to pH 7.6 at 25°C with 1 M NaOH).

E. Trypsin Enzyme Solution (Trypsin) (Immediately before use, prepare a solution containing 2.0 mg/ml of Trypsin, in cold Reagent C).

F. Trypsin-Chymotrypsin Inhibitor Solution (Inhib) (Immediately before use, prepare a solution containing 1.0 mg/ml of Trypsin-Chymotrypsin Inhibitor in cold Reagent D).

Procedure

Pipette (in milliliters) the following reagents into suitable containers: Part A:

	Uninh	Test 1	Test 2	Test 3	Test 4	Test 5
Reagent F (Inhib)	—	0.05	0.075	0.10	0.15	0.20
Reagent E (Trypsin)	0.50	0.50	0.50	0.50	0.50	0.50
Reagent D (Buffer II)	9.50	9.45	9.425	9.40	9.35	9.30

Mix by inversion and pipette (in milliliters) the following reagents into suitable cuvettes: Part B:

Reagent B (BAEE)	3.00	3.00	3.00	3.00	3.00	3.00	3.00
Reagent D (Buffer II)	0.10	0.10	0.10	0.10	0.10	0.10	0.20

Mix by inversion and equilibrate to 25°C. Monitor the A253nm until constant, using a suitably thermostatted spectrophotometer. Then add:

Uninh (Part A)	0.10	—	—	—	—	—	—
Test 1 (Part A)	—	0.10	—	—	—	—	—
Test 2 (Part A)	—	—	0.10	—	—	—	—
Test 3 (Part A)	—	—	—	0.10	—	—	—
Test 4 (Part A)	—	—	—	—	0.10	—	—
Test 5 (Part A)	—	—	—	—	—	0.10	—

Immediately mix by inversion and record the increase in A253nm for approximately 5 minutes. Obtain the ΔA253nm/minute using the maximum linear rate for the Tests, Blank, and Uninhibited Solutions.

Calculations

$$\text{BTEE units / ml} = \frac{(\Delta A253\text{nm / min Test} - \Delta A253\text{nm / min Blank})\ (0.5)\ (\text{df})}{(0.001)\ (0.1)\ (10)}$$

0.5 = Volume (in milliliter) of enzyme used in Part A of the Assay
df = Dilution factor

0.001 = The change in A253nm/minute per unit of Trypsin at pH 7.6 at 25°C in a 3.2 ml reaction mix (as per the Unit Definition)
0.1 = Volume (in milliliter) of enzyme from Part A used in Part B of the Assay
10 = Volume (in milliliters) of Part A of the Assay

Plot the Trypsin activity (BAEE units/ml) versus volume (in milliliter) of Trypsin-Chymotrypsin Inhibitor/Reaction Mixture in Step A. Determine the point at which the line intercepts the abscissa. This point is the X-intercept.

Y0 = Mg of Trypsin-Chymotrypsin Inhibitor/ml in Part B of the Assay that results in complete inhibition of Trypsin: = (X-Intercept) (mg TCI/ml in Reagent F)
CT = Mg of Trypsin/ml Part B of the Assay = (0.5) (mg Trypsin/ml in Reagent E) CT
Mg of Trypsin inhibited by 1 mg Trypsin-chymotrypsin Inhibitor = YO

Unit Definition: *One trypsin unit will produce a ΔA253nm of 0.001 per minute with BAEE as substrate at pH 7.6 at 25°C in a reaction volume of 3.2 ml.*

Specification

One mg protein will inhibit 3–5 mg of Trypsin with an activity of approximately 10,000 BAEE units per mg protein.

Final Assay Concentrations

In a 3.20 ml reaction mix, the final concentrations are 375 mM sodium phosphate, 0.23 mM N-α-benzoyl-L-arginine ethyl ester, 1.6 μM HCl, 10 μg trypsin, and 0.5–2.0 μg trypsinchymotrypsin inhibitor.

Enzymatic Assay of TRYPSINOGEN

Principle

$$\text{Trypsinogen} \xrightarrow{\text{Trypsin}} \text{Trypsin} + \text{Hexapeptide}$$

$$\text{BAEE} + H_2O \xrightarrow{\text{Trypsin}} \text{Na-Benzoyl-A-Arginine} + \text{Ethanol}$$

Abbreviation Used

BAEE = N-α-Benzoyl-L-Arginine Ethyl Ester

Conditions: T = 25°C, pH = 7.6, A253nm, Light path = 1 cm

Method: Continuous Spectrophotometric Rate Determination

Reagents

A. 67 mM Sodium Phosphate Buffer, pH 7.6 at 25°C (Prepare 100 ml in deionized water using Sodium Phosphate, Monobasic, Anhydrous. Adjust to pH 7.6 at 25°C with 1 M NaOH).
B. 0.25 mM Nα-Benzoyl-L-Arginine Ethyl Ester Solution (BAEE) (Prepare 50 ml in Reagent A using N-α-Benzoyl-Larginine Ethyl Ester, Hydrochloride).
C. 5 mM Hydrochloric Acid Solution (HCl) (Prepare 50 ml in deionized water using Hydrochloric Acid).
D. 1 M Calcium Chloride Solution ($CaCl_2$) (Prepare 10 ml in deionized water using Calcium Chloride, Dihydrate).
E. 1 mM Hydrochloric Acid Solution (HCl) (Prepare 50 ml in deionized water using Reagent C).

F. 400 mM Tris HCl Buffer, pH 8.4 at 25°C (Buffer) (Prepare 50 ml in deionized water using Trizma Hydrochloride. Adjust to pH 8.4 at 25°C using 1 M NaOH).

G. 0.02% (w/v) Trypsin Enzyme Solution (Trypsin) (Immediately before use, prepare 3 ml in Reagent C using Trypsin, Type III).

H. Trypsinogen Enzyme Solution (Trypsinogen) (Immediately before use, prepare a solution containing 5 mg/ml of Trypsinogen in Reagent C).

Procedure

Prepare an Activating Mixture by pipetting (in milliliters) the following reagents into a suitable container:

Reagent D ($CaCl_2$)	2.00
Reagent F (Buffer)	38.00
Reagent G (Trypsin)	2.00

Mix by swirling.

Step 1: Total Trypsin Activity: At zero time add 0.1 ml of Reagent H (Trypsinogen) to 1 ml of the Activating Mixture and incubate at 5°C for 24 – 28 hours. Then dilute 0.2 ml to 10.2 ml with Reagent E (HCl). Proceed with the trypsin assay.

Step 2: Free Trypsin Activity: Allow the Activating Mixture and Reagent H (Trypsinogen) to stand separately for 24–28 hours at 0–5°C. After 24–28 hours, add 0.10 ml of Reagent H (Trypsinogen) to 1.00 ml of the Activating Mixture. *Immediately* measure the trypsin activity.

Step 3: Trypsin Assay: Pipette (in milliliters) the following reagents into suitable cuvettes:

	Test	Blank
Reagent B (BAEE)	3.00	3.00

Equilibrate to 25°C. Monitor the A253nm until constant, using a suitably thermostatted spectrophotometer. Then add:

	Test	Blank
Activating Mixture	—	0.20
Enzyme Solution (from Step 1 or Step 2)	0.20	—

Immediately mix by inversion and record the increase in A253nm for approximately 5 minutes. Obtain the A253nm/minute using the maximum linear rate for both the Test and Blank.

Calculations

The calculation shown below can be used to determine both the total Trypsin activity of the Trypsinogen (Step 1) and the free Trypsin activity of the Trypsinogen (Step 2).

$$\text{BAEE units / ml enzyme} = \frac{(\Delta A253\text{nm / minute Test} - \Delta A253\text{nm / minute Blank})\ (\text{df})}{(0.001)\ (0.20)}$$

df = Dilution factor
0.001 = The change in A253nm/minute per unit of Trypsin at pH 7.6 at 25°C in a 3.2 ml reaction mix
0.2 = Volume (in milliliter) of enzyme used

$$\text{Units / mg solid} = \frac{\text{units / ml enzyme}}{\text{mg solid / ml enzyme}}$$

$$\text{Units / mg protein} = \frac{\text{units / ml enzyme}}{\text{mg protein / ml enzyme}}$$

Unit Definition: *One BAEE unit will produce a ΔA253nm of 0.001 per minute with BAEE as substrate at pH 7.6 at 25°C in a reaction volume of 3.2 ml (1 cm light path).*

Final Assay Concentrations

In a 3.20 ml reaction mix, the final concentrations are 63 mM sodium phosphate, 0.23 mM N-α-benzoyl-L-arginine ethyl ester, 0.4 mM Tris, 0.002 mM hydrochloric acid, 0.06 mM calcium chloride, 0.034 μg trypsin (Step 1), and 1.7 μg trypsin (Step 2), 0.6 μg trypsinogen (Step 1), and 0.03 mg trypsinogen (Step 2).

Note

1. Measure the trypsin activity for enzyme solutions from both Step 1 (Total Trypsin Activity) and Step 2 (Free Trypsin Activity).

Non Specifis Protease Assays

Suitability Assay for AZO DYE IMPREGNATED COLLAGEN

Principle

$$\text{Azocoll} + H_2O \xrightarrow{\text{Protease}} \text{Colored Reaction Products}$$

Abbreviation Used

Azocoll = Azo Dye Impregnated Collagen

Conditions: T = 37°C, pH = 7.0, A520nm, Light path = 1 cm

Method: Colorimetric

Reagents

A. 100 mM Potassium Phosphate Buffer, pH 7.0 at 37°C (Prepare 100 ml in deionized water using Potassium Phosphate, Monobasic. Adjust to pH 7.8 at 37°C with 1 M KOH).

B. Azo Dye Impregnated Collagen (Azocoll).

C. Protease Enzyme Solution (Protease) Step 1 (Immediately before use, prepare a solution containing 0.09–0.13 unit/ml of Protease, in cold Reagent A. Store at 0–4°C).

D. Protease Enzyme Solution (Protease) Step 2 (Immediately before use, prepare a solution containing 10.0–12.00 units/ml of Protease, in cold Reagent A. Store at 0–4°C.

Procedure

Weigh (in milligrams) the following reagents into 4 dram vials:

	Test 1	Test 2	Test 3	Test 4	Test 5	Test 6	Blank
Reagent B (Azocoll)	25	25	25	25	25	25	25

Then add (in milliliters):

	Test 1	Test 2	Test 3	Test 4	Test 5	Test 6	Blank
Reagent A (Buffer)	5.08	5.06	5.04	5.02	5.00	5.00	5.00

Equilibrate to 37°C. Then add at 1 minute time intervals:

	Test 1	Test 2	Test 3	Test 4	Test 5	Test 6	Blank
Reagent C (Protease)	0.02	0.04	0.06	0.08	0.10	—	—
Reagent D (Protease)	—	—	—	—	—	0.10	—

Incubate at 37°C for exactly 15 minutes in a shaking water bath (60 cycles/minute). Remove the vials and immediately filter the suspension through 0.45 micron syringe filters. Transfer the filtered solutions to suitable cuvettes and record the A520nm using a suitable spectrophotometer.

Calculations

Sample Curve: ΔA520nm Sample = A520nm Test–A520nm Test Blank
Plot the ΔA520nm Sample vs units of protease per assay.

Specifications

The sample is suitable if it has similar results compared to that of a control. The plot of absorbance versus units of protease should be nonlinear. The ΔA520nm of Test 6 >1.0.

Final Assay Concentrations

In a 5.00 ml reaction mix, the final concentrations are 100 mM phosphate, 0.5% (w/v) azo dye impregnated collagen, and 0.009–0.013 unit protease.

Suitability Assay for AZOALBUMIN as a Substrate for Trypsin

Principle

$$\text{Azoalbumin} + H_2O \xrightarrow{\text{Trypsin}} \text{Colored Reaction Products}$$

Conditions: T = 37°C, pH 8.3, A440nm, Light path = 1 cm

Method: Colorimetric

Reagents

A. 0.50% (w/v) Sodium Bicarbonate Buffer ($NaHCO_3$), pH 8.3 at 37°C (Prepare 50 ml in deionized water using Sodium Bicarbonate. Adjust to pH 8.3 at 37°C with 1 M HCl).
B. 2.5% (w/v) Azoalbumin Solution (Azoalbumin) (Prepare 25 ml in Reagent A using Azoalbumin. Gentle heating and stirring may be needed to form a solution. Adjust to pH 8.3 at 37°C, if necessary, with 1 M HCl).
C. 5.0% (w/v) Trichloroacetic Acid Solution (TCA) (Prepare 50 ml in deionized water using Trichloroacetic Acid, 6.1 N Solution, approximately 100% (w/v)).
D. 500 mM Sodium Hydroxide Solution (NaOH) (Prepare 20 ml in deionized water using Sodium Hydroxide Solution, 1.0 N).
E. Trypsin Enzyme Solution (Trypsin) (Immediately before use, prepare a solution containing containing 5 mg solid/ml of Trypsin in cold Reagent A).

Procedure

Pipette (in milliliters) the following reagents into suitable containers:

	Test	Blank
Reagent A (Buffer)	1.50	1.50
Reagent B (Azoalbumin)	2.50	2.50

Mix by swirling and equilibrate to 37°C. Then add:

Reagent E (Trypsin)	1.00	—

Mix by swirling and incubate at 37°C for exactly 10 minutes. Then add:

Reagent E (Trypsin)	—	1.00

Immediately remove a 1 ml aliquot from both the Test and Blank solutions and place in a suitable containers. Then add:

	Test	Blank
Reagent C (TCA)	4.0	4.0

Mix by swirling and centrifuge solutions for five minutes. Remove a 1 ml aliquot from the supernatants of both the Test and Blank and place in suitable containers. Then add:

Reagent D (NaOH)	3.0	3.0

Mix by swirling and transfer the Test and Blank solutions to suitable cuvettes. Record the A440nm for both the Test and Blank in a suitable spectrophotometer.

Calculation

$$\Delta A440nm = A440nm\ Test - A440nm\ Blank$$

Compare the A440nm of the Test to that of a control sample. The absorbance should be similar.

Final Assay Concentrations

In a 5.00 ml reaction mix, the final concentrations are 0.50% (w/v) sodium bicarbonate, 1.25% (w/v) azoalbumin, and 5 mg trypsin.

Suitability Assay for AZOCASEIN as a Substrate for Trypsin

Principle

$$\text{Azocasein} + H_2O \xrightarrow{\text{Trypsin}} \text{Colored Reaction Products}$$

Conditions: T = 37°C, pH 8.3, A440nm, Light path = 1 cm

Method: Colorimetric

Reagents

A. 0.50% (w/v) Sodium Bicarbonate Buffer ($NaHCO_3$), pH 8.3 at 37°C (Prepare 100 ml in deionized water using Sodium Bicarbonate. Adjust to pH 8.3 at 37°C with 1 M HCl).
B. 2.50% (w/v) Azocasein Solution (Azocasein) (Prepare 15 ml in Reagent A using Azocasein. Gentle heating and stirring may be needed to form a solution).
C. 5.0% (v/v) Trichloroacetic Acid Solution (TCA) (Prepare 40 ml in deionized water using Trichloroacetic Acid, 6.1 N Solution, 100% w/v).
D. 500 mM Sodium Hydroxide (NaOH) Solution (Prepare 50 ml in deionized water using Sodium Hydroxide Solution, 1.0 Normal).
E. Trypsin Enzyme Solution (Trypsin) (Prepare a solution containing 10 mg solid/ml of Trypsin in Reagent A).

Procedure

Pipette (in milliliters) the following reagents into suitable containers:

	Test	Blank
Reagent B (Azocasein)	2.50	2.50
Reagent A (Buffer)	1.50	2.50

Mix by swirling and equilibrate to 37°C. Then add:

Reagent E (Trypsin)	1.00	—

Mix by swirling and incubate at 37°C for exactly 30 minutes. Remove 1 ml aliquot from both the Test and Blank solutions and place in a suitable container. Then add:

Reagent C (TCA)	4.0	4.0

Mix by swirling and filter using Whatman #54 paper or 0.45 µm syringe filters. Remove 1 ml aliquot from both the Test and Blank solutions and place in a suitable container. Then add:

Reagent D (NaOH)	3.0	3.0

Mix by swirling and transfer the Test and Blank solutions to suitable cuvettes. Record the A440nm for both the Test and Blank using a suitable spectrophotometer.

Calculations

$$\Delta A440nm = A440nm\ Test - A440nm\ Blank$$

Compare the A440nm of the Test to that of a control sample. The absorbance should be similar.

Final Assay Concentrations

In a 5.00 ml reaction mix, the final concentrations are 0.50% (w/v) sodium bicarbonate, 1.25% (w/v) azocasein, and 10 mg trypsin.

Enzymatic Assay of CASEINASE (Collagenase Products)

Principle

$$\text{Azocasein} + H_2O \xrightarrow{\text{Protease}} \text{Amino Acids}$$

Conditions: T = 37°C, pH = 7.5, A660nm, Light path = 1 cm

Method: Colorimetric

Reagents

A. 50 mM Sodium Phosphate Buffer, pH 7.5 at 37°C. (Prepare 200 ml in deionized water using Sodium Phosphate, Dibasic, Anhydrous. Adjust to pH 7.5 at 37°C with 1 M HCl).

B. 0.65% (w/v) Casein Solution (Casein) (Prepare 125 ml in Reagent A using Casein. Heat gently to 80–85°C (do not boil) until a homogenous dispersion is obtained. Allow the solution to cool to 37°C. Adjust the pH to 7.5 at 37°C with 0.1 M HCl or 0.1 M NaOH, if necessary).

C. 6.1 N Trichloroacetic Acid Reagent (TCA) (Use Trichloroacetic Acid, 6.1 N Solution, approximately 100% (w/v)).
D. Folin & Ciocalteu's Phenol Reagent (F-C) (Dilute 10 ml of Folin & Ciocalteu's Phenol Reagent, 2.0 N, to 40 ml with deionized water).
E. 500 mM Sodium Carbonate Solution (Na_2CO_3) (Prepare 500 ml in deionized water using Sodium Carbonate, Anhydrous).
F. 50 mM TES Buffer with 0.36 mM Calcium Chloride, pH 7.4 at 37°C Enzyme Diluent) (Prepare 100 ml in deionized water using TES Free Acid and Calcium Chloride, Dihydrate. Adjust the pH to 7.4 at 37°C with 1 M NaOH).
G. 1.1 mM L-Tyrosine Standard (Std Soln) (Prepare 100 ml in deionized water using L-Tyrosine, Free Base. Heat gently until tyrosine dissolves and cool to room temperature).
H. Collagenase Enzyme Solution (Immediately before use, prepare a solution containing 0.05–0.10 mg/ml of Collagenase in cold Reagent F).

Procedure

Pipette (in milliliters) the following reagents into suitable containers:

	Test	Blank
Reagent B (Casein)	5.00	5.00

Equilibrate to 37°C. Then add:

Reagent H (Enzyme Solution)	1.00	—

Mix by inversion and incubate at 37°C for exactly 30 minutes. Then add:

Reagent C (TCA)	0.50	0.50
Reagent H (Enzyme Solution)	—	1.00

Filter through Whatman #50 filter paper or 0.8 μm syringe filters and use the filtrate in the color development.

Color Development

Standard Curve: Prepare a standard curve by pipetting (in milliliters) the following reagents into suitable vials:

	Std 1	Std 2	Std 3	Std 4	Std 5	Std Blank
Reagent G (Std Soln)	0.05	0.10	0.20	0.40	0.50	—
Deionized Water	1.95	1.90	1.80	1.60	1.50	2.00
Reagent E (Na_2CO_3)	5.00	5.00	5.00	5.00	5.00	5.00
Reagent D (F-C)	1.00	1.00	1.00	1.00	1.00	1.00

Mix vigorously by inversion.

Sample: Pipette (in milliliters) the following reagents into suitable vials:

	Test	Blank
Test Filtrate	2.00	—
Blank Filtrate	—	2.00
Reagent E (Na_2CO_3)	5.00	5.00
Reagent D (F-C)	1.00	1.00

Mix vigorously by inversion. Incubate sample and standard vials at 37°C for 30 minutes. Remove and allow the vials to cool to room temperature. Transfer to suitable cuvettes. (If the solutions are hazy either centrifuge or filter through a 0.45 µm filter prior to determining the A660nm). Determine the A660nm for the Test, Test Blank, Standards, and Standard Blank.

Calculations

Standard Curve: ΔA660nm Standard = A660nm Standard–A660nm Standard Blank
Plot the ΔA660nm Standard vs µmoles Tyrosine.
Sample Determination: ΔA660nm Sample = A660nm Test–A660nm Test Blank
Determine the µmoles of Tyrosine equivalents liberated using the Standard curve.

$$\text{Units / ml enzyme} = \frac{(\mu\text{mole Tyrosine equivalents released})\,(10)\,(6.5)\,(\text{df})}{(1)\,(2)}$$

10 = Time conversion from 30 minutes to 5 hours (Unit Definition)
6.5 = Total volume (in milliliters) of stopped reaction
df = Dilution factor
2 = Volume (in milliliters) of sample used in Colorimetric Assay
1 = Volume (in milliliters) of enzyme used

$$\text{Units / mg solid} = \frac{\text{units / ml enzyme}}{\text{mg solid / ml enzyme}}$$

Unit Definition: *One unit will hydrolyze casein to produce color equivalent to 1.0 µmole (181 µg) of tyrosine per 5 hours at pH 7.5 at 37°C (color by Folin & Ciocalteu reagent).*

Final Assay Concentrations

In a 6.00 ml reaction mix, the final concentrations are 42 mM potassium phosphate, 0.54% (w/v) casein, 8.3 mM TES, 0.06 mM calcium chloride and 0.05 mg–0.10 mg collagenase.

Suitability Assay for Cellulose Azure as a Substrate for CELLULASE

Principle

$$\text{Cellulose Azure} + H_2O \xrightarrow{\text{Cellulose}} \text{Colored Reaction Products}$$

Conditions: T = 37°C, pH = 5.0, A575nm, Light path = 1 cm

Method: Colorimetric

Reagents

A. 10 mM Sodium Citrate Buffer, pH 5.0 at 37°C (Prepare 100 ml in deionized water using Citric Acid, Free acid, Anhydrous. Adjust to pH 5.0 at 37°C with 1 M NaOH).

B. Cellulose Azure (Weigh 20 mg of Cellulose Azure, each into three 4 dram vials).
C. Methanol Saturated with Calcium Chloride (Prep... 0.5 ml by saturating Methanol, Absolute with Calcium Chloride, Anhydrous).
D. Ethanol (Use Ethanol, Denatured).
E. Alcohol Precipitation Reagent (APR) (Prepare by combining 100 ml of Reagent D and 0.5 ml of Reagent C).
F. Cellulase Enzyme Solution (Immediately before use, prepare a solution containing 50 units/ml of Cellulase in cold Reagent A).

Procedure

Pipette (in milliliters) the following reagents into 4 dram vials containing Reagent B (Cellulose Azure):

	Test	Std 1	Std 2	Blank
Reagent A (Buffer)	2.00	2.00	2.00	2.00

Cap and equilibrate to 37°C in a shaker bath at zero time. Then add:

Reagent E (APR)	—	—	—	4.00
(Enzyme Solution)	0.10	0.10	0.10	0.10

Remove Blank from shaker bath and set aside. Incubate at 37°C. Exactly 5 minutes after the enzyme addition, add:

Reagent E (APR)	4.00	—	—	—

Remove Test 1 from shaker bath and set aside. Exactly 10 minutes after the enzyme addition, add:

Reagent E (APR)	—	4.00	—	—

Remove Test 2 from shaker bath and set aside. Exactly 10 minutes after the enzyme addition, add:

Reagent E (APR)	—	—	4.00	—

Remove Test 3 from Shaker bath. Mix by swirling and filter the Blank, Test 1, Test 2, and Test 3 through Whatman #54 filter paper. Transfer the filtered solutions to suitable cuvettes. Record the A575nm for both the Tests and Blank using a suitable spectrophotometer.

Calculations

$$\Delta A575nm\ \text{Test 1, 2, 3} = A575nm\ \text{Test 1, 2, 3} - A575nm\ \text{Blank}$$

$$\Delta A575nm / min / mg\ \text{Cellulase} = \frac{\Delta A575nm}{(T)\ (\text{mg solid cellulase / RM})}$$

T = Time (in minutes) of assay. Time is 5, 10, or 20 minutes
RM = Reaction Mix

The activity is expressed as the change in absorbance at 575nm per minute per mg solid of cellulase. Compare the activity of the Test1 to that of a control sample. The activity should be similar.

Final Assay Concentrations

In a 2.10 ml reaction mix, the final concentrations are 10 mM citrate, 20 mg cellulose azure and 5 units cellulase.

Note

1. Different times of incubation are utilized in order to evaluate the linearity of the reaction and the quantity of unbound dye present.

Enzymatic Assay of ELASTIN-CONGO RED

Principle

$$\text{Insoluble Elastin - Congo Red} + H_2O \xrightarrow{\text{Elastase}} \text{Soluble Hydrolysis Products}$$

Conditions: T = 37°C, pH = 8.8, A495nm, Light path = 1 cm

Method: Spectrophotometric Stop Rate Determination

Reagents

A. 200 mM Tris HCl Buffer, pH 8.8 at 37°C (Prepare 100 ml in deionized water using Trizma Base. Adjust to pH 8.8 at 37°C with 1 M HCl).
B. Elastin-Congo Red Substrate (El-CR).
C. Elastase Enzyme Solution (Immediately before use, prepare a solution containing approximately 1500–1800 units/ml of Elastase, in cold Reagent A).

Procedure

Step 1: Standard Curve: Weigh (in milligrams) the following reagent:

	Std 1	Blank 1	Std 2	Blank 2	Std 3	Blank 3	Std 4	Blank 4
Reagent B (EL-CR)	5.00	5.00	10.00	10.00	15.00	15.00	20.00	20:00

Then add (in milliliters):

Reagent A (Buffer)	6.00	6.01	6.00	6.01	6.00	6.01	6.00	6.01

Mix and equilibrate to 37°C. Then add:

Reagent C (Enzyme)	0.01	—	0.01	—	0.01	—	0.01	—

Mix by swirling and place all the containers in a suitably thermostatted metabolic shaker. Incubate the Standards and Standard Blanks at 37°C for 12–16 hours. Filter the solutions through 0.8 μm syringe filter. Record the A495nm for the Standards and Standard Blanks using a suitable spectrophotometer.

Calculations

Standard Curve: ΔA495nm Standard = A495nm Std–A495nm Std Blank

Prepare a standard curve by plotting the A495nm of the Standards vs milligrams of solubilized Elastin-Congo Red.

$$\Delta A495 / \text{mg Elastin - Congo Red} = \frac{\Delta A495\text{nm}}{\text{mg Elastin - Congo Red}}$$

Compare the result of the Test to that of a control sample. The activity should be similar.

Final Assay Concentrations

In a 6.01 ml reaction mix, the final concentrations are 200 mM Tris, 5–20 mg elastin-congo red and 15–18 units elastase.

Notes

1. The Metabolic shaker should be adjusted to approximately 60 shake cycles/minute.
2. If the filtrate is hazy, it can be centrifuged to remove the haziness.
3. Elastase Unit Definition: One unit will solubilize 1 mg of elastin in 20 minutes at pH 8.8 at 37°C.

Suitability Assay for KERATIN AZURE as a Substrate for Proteinase K

Principle

$$\text{Keratin Azure} + H_2O \xrightarrow{\text{Proteinase K}} \text{Colored Reaction Product}$$

Conditions: T = 37°C, pH = 7.5, A595nm, Light path = 1 cm

Method: Colorimetric

Reagents

A. 50 mM Sodium Phosphate Buffer, pH 7.5 at 37°C (Prepare 100 ml in deionized water using Sodium Phosphate, Dibasic, Anhydrous. Adjust to pH 7.5 at 37°C with 1 M HCl).

B. Keratin Azure .

C. Proteinase K Enzyme Solution (Immediately before use, prepare a solution containing 100 units/ml of Proteinase K, in cold deionized water).

Procedure

Weigh (in milligrams) the following reagents into suitable containers:

	Test 1	Test 2	Test 3	Test 4	Blank
Reagent B (Keratin Azure)	20	20	20	20	20

Pipette (in milliliters) the following reagents into suitable containers:

	Test 1	Test 2	Test 3	Test 4	Blank
Reagent A (Buffer)	4.00	4.00	4.00	4.00	4.00
Deionized Water	0.40	0.30	0.20	0.10	0.50

Mix by swirling and equilibrate at 37°C for 10 minutes in a water bath. Then add:

	Test 1	Test 2	Test 3	Test 4	Blank
Reagent B (Keratin Azure)	0.10	0.20	0.30	0.40	—

Mix by swirling and incubate at 37°C for exactly 1 hour on a metabolic shaker where constant mixing can be maintained. Centrifuge the solutions for 10 minutes and transfer the supernatants to suitable cuvettes. Record the A595nm for both the Tests and Blanks with a suitable spectrophotometer.

Calculations

$$\Delta A595nm\ Test = A595nm\ Test - A595nm\ Blank$$

$$\Delta A595nm / hour / mg\ enzyme = \frac{\Delta A595nm\ Test}{mg\ enzyme / RM}$$

RM = Reaction Mixture

Specification

Compare the ΔA595nm/hour/mg enzyme of the test to that of a control sample. These values should be similar.

Final Assay Concentrations

In a 4.50 ml reaction mix, the final concentrations are 44 mM sodium phosphate, 0.44% (w/v) keratin azure, and 10–40 units proteinase K.

Enzymatic Assay of PROTEASE Casein as a Substrate

Principle

$$Casein + H_2O \xrightarrow{Protease} Amino\ Acids$$

Conditions: T = 37°C, pH = 7.5, A660nm, Light path = 1 cm

Method: Colorimetric

Reagents

A. 50 mM Potassium Phosphate buffer, pH 7.5 at 37ºC. (Prepare 200 ml in deionized water using Potassium Phosphate, Dibasic, Trihydrate. Adjust to pH 7.5 at 37ºC with 1 M HCl).
B. 0.65% (w/v) Casein Solution (Casein) (Prepare 125 ml in Reagent A using Casein. Heat gently (do not boil) to 80–90°C for 10 minutes with stirring. Adjust the pH to 7.5 at 37°C, if necessary, with either 1 M NaOH or 1 M HCl).
C. 110 mM Trichloroacetic Acid Reagent (TCA) (Dilute 9 ml of Trichloroacetic Acid, 6.1 N, approximately 100% (w/v), to 500 ml with deionized water).
D. Folin & Ciocalteu's Phenol Reagent (F-C) (Dilute 10 ml of Folin & Ciocalteu's Phenol Reagent, to 40 ml with deionized water).
E. 500 mM Sodium Carbonate Solution (Na_2CO_3) (Prepare 500 ml in deionized water using Sodium Carbonate Anhydrous).

F. 10 mM Sodium Acetate Buffer with 5 mM Calcium Acetate, pH 7.5 at 37°C (Enzyme Diluent) (Prepare 100 ml in deionized water using Sodium Acetate, Trihydrate and Calcium Acetate. Adjust the pH to 7.5 at 37°C with 0.1 M Acetic acid or 0.1 M NaOH).
G. 1.1 mM L-Tyrosine Standard (Std Soln) (Prepare 100 ml in deionized water using L-Tyrosine, Free Base. Heat gently (do not boil) until tyrosine dissolves and cool to room temperature).
H. Protease Enzyme Solution (Immediately before use, prepare a solution containing 0.1–0.2 unit/ml of Protease in cold Reagent F).

Procedure

Pipette the following reagents into suitable vials (in milliliters):

	Test	Blank
Reagent B (Casein)	5.00	5.00

Equilibrate to 37°C. Then add:

Reagent H (Enzyme Solution)	1.00	—

Mix by swirling and incubate at 37°C for exactly 10 minutes. Then add:

Reagent C (TCA)	5.00	5.00
Reagent H (Enzyme Solution)	—	1.00

Mix by swirling and incubate at 37°C for about 30 minutes. Filter through Whatman #50 filter paper or a 0.45 μm filter and use the filtrate in color development.

Color Development

Standard Curve: Prepare a standard curve by pipetting the following reagents into suitable vials (in milliliters):

	Std 1	Std 2	Std 3	Std 4	Std Blank
Reagent G (Std Soln)	0.05	0.10	0.20	0.40	0.00
Deionized Water	1.95	1.90	1.80	1.60	2.00
Reagent E (Na_2CO_3)	5.00	5.00	5.00	5.00	5.00
Reagent D (F-C)	1.00	1.00	1.00	1.00	1.00

Sample: Pipette the following reagents into 4 dram vials (in milliliters):

	Test	Blank
Test Filtrate	2.00	—
Blank Filtrate	—	2.00
Reagent E (Na_2CO_3)	5.00	5.00
Reagent D (F-C)	1.00	1.00

Mix by swirling and incubate at 37°C for 30 minutes. Remove the vials and allow them to cool to room temperature. Filter through a 0.45 µm filter immediately prior to reading. Read the absorbance at 660nm for each of the vials in suitable cuvettes.

Calculations

Standard Curve: ΔA660nm Standard = A660nm Standard–A660nm Standard Blank
Plot the ΔA660nm Standard vs µmoles of Tyrosine.
Sample Determination: ΔA660nm Sample = A660nm Test–A660nm Sample Blank
Determine the µmoles of Tyrosine equivalents liberated using the Standard curve.

$$\text{Units / ml enzyme} = \frac{(\mu\text{mole Tyrosine equivalents released})\ (11)}{(1)\ (10)\ (2)}$$

11 = Total volume (in milliliters) of assay
10 = Time of assay (in minutes) as per the Unit Definition
1 = Volume of enzyme (in milliliter) of enzyme used
2 = Volume (in milliliters) used in Colorimetric Determination

$$\text{Units / mg solid} = \frac{\text{units / ml enzyme}}{\text{mg solid / ml enzyme}}$$

$$\text{Units / mg protein} = \frac{\text{units / ml enzyme}}{\text{mg protein / ml enzyme}}$$

Unit Definition: *One unit will hydrolyze casein to produce color equivalent to 1.0 µmole (181 µg) of tyrosine per minute at pH 7.5 at 37°C (color by Folin & Ciocalteu's reagent).*

Final Assay Concentrations

In a 6.00 ml reaction mix, the final concentrations are 42 mM potassium phosphate, 0.54% (w/v) casein, 1.7 mM sodium acetate, 0.8 mM calcium acetate, and 0.1–0.2 unit protease.

Enzymatic Assay of PROTEASE

Principle

$$\text{Casein} + H_2O \xrightarrow{\text{Protease}} \text{Amino Acids}$$

Conditions: T = 37°C, pH = 3.0, A660nm, Light path = 1 cm

Methods: Colorimetric

Reagents

A. 100 mM Lactic Acid Buffer, pH 3.0 at 37°C (Lactic Acid) (Prepare 100 ml in deionized water using L(+)-Lactic Acid, Free Acid. Adjust the pH to 3.0 at 37°C with 1 M NaOH).

B. 1.5% (w/v) Casein Solution, pH 3.0 at 37°C (Casein) (Prepare 100 ml by dissolving 1.5 g of Casein, into 60 ml of Reagent A. Heat until completely dissolved. Cool to

37°C and adjust the pH to 3.0 with 0.1 M NaOH. Then add 20 ml of Reagent A and deionized water to a total volume of 100 ml).

C. 400 mM Trichloroacetic Acid Reagent (TCA) (Dilute 6.6 ml of Trichloracetic Acid, 6.1 N, approximately 100% (w/v), to 100 ml with deionized water).

D. Folin and Ciocalteu's Phenol Reagent (F-C) (Dilute 10 ml of Folin and Ciocalteu's Phenol Reagent, to 100 ml with deionized water).

E. 400 mM Sodium Carbonate Solution (Na_2CO_3) (Prepare 500 ml in deionized water using Sodium Carbonate, Anhydrous).

F. 2 mM Hydrochloric Acid Solution (Enzyme Diluent) (Prepare 25 ml in deionized water using Hydrochloric Acid).

G. 1.1 mM L-Tyrosine Standard (Std Soln) (Prepare 100 ml in deionized water using L-Tyrosine, Free Base. Heat gently (*Do Not Boil*), until the Tyrosine dissolves and then cool to room temperature).

H. Protease Enzyme Solution (Immediately before use, prepare a solution containing 0.0045–0.009 unit/ml of Protease in cold Reagent F).

Procedure

Pipette (in milliliters) the following reagents into suitable containers:

	Test	Blank
Reagent B (Casein)	1.00	1.00

Equilibrate to 37°C, then add:

Reagent H (Enzyme Solution)	1.00	—

Mix by swirling and incubate at 37°C for exactly 60 minutes. Then add:

Reagent C (TCA)	2.00	2.00
Reagent H (Enzyme Solution)	—	1.00

Mix by swirling and incubate at 37°C for 25 minutes. Filter through Whatman #50 filter paper and use the filtrate in the color development step.

Color Development

Standard Curve: Prepare a standard curve by pipetting (in milliliters) the following reagents into suitable containers:

	Std 1	Std 2	Std 3	Std 4	Std Blank
Reagent G (Std Soln)	0.05	0.10	0.20	0.40	0.00
Deionized Water	0.95	0.90	0.80	0.60	1.00
Reagent E (Na_2CO_3)	5.00	5.00	5.00	5.00	5.00
Reagent D (F-C)	1.00	1.00	1.00	1.00	1.00

Sample: Pipette (in milliliters) the following reagents into suitable containers:

	Test	Blank
Test Filtrate	1.00	—
Blank Filtrate	—	1.00
Reagent E (Na_2CO_3)	5.00	5.00
Reagent D (F-C)	1.00	1.00

Mix by swirling and incubate at 37°C for 20 minutes. Remove the containers and allow them to cool to room temperature. If the solutions are hazy, filter through a 0.45 µm filter prior to reading. Transfer the solutions to suitable cuvettes and record the absorbance at 660nm for the Test, Blank, and Standards using a suitable spectrophotometer.

Calculations

Standard Curve: ΔA660nm Sample = A660nm Test–A660nm Blank

Determine the µmoles of Tyrosine equivalents liberated using the Standard curve.

$$\text{Units / ml enzyme} = \frac{(\mu\text{mole Tyrosine equivalents released})\ (4)\ (\text{df})}{(1)\ (60)\ (1)}$$

4 = Total volume (in milliliters) of assay
df = Dilution factor
1 = Volume (in milliliters) of enzyme used
60 = Time of assay (in minutes) as per the Unit Definition
1 = Volume (in milliliters) used in the Colorimetric Determination

$$\text{Units / mg solid} = \frac{\text{units / ml enzyme}}{\text{mg solid / ml enzyme}}$$

$$\text{Units / mg protein} = \frac{\text{units / ml enzyme}}{\text{mg protein / ml enzyme}}$$

Unit Definition: *One unit will hydrolyze casein to produce color equivalent to 1.0 µmole (181 µg) of tyrosine per minute at pH 3.0 at 37°C (color by Folin and Ciocalteu's reagent).*

Final Assay Concentrations

In a 2.0 ml reaction mix, the final concentrations are 40 mM lactic acid, 0.75% (w/v) casein, 1 mM hydrochloric acid and 0.0045–0.009 unit protease.

Enzymatic Assay of PROTEASE IMPURITY using Fluorescein Isothiocyanate-Casein

Principle

$$\text{FITC-Casein} + H_2O \xrightarrow{\text{Protease}} \text{FITC} + \text{Casein}$$

Abbreviations Used

FITC-Casein = Fluorescein Isothiocyanate-Casein
FITC = Fluorescein Isothiocyanate

Conditions : Part A: T = 37°C, pH 7.6, Part B: T = 25°C, pH 8.5, Excitation = 490nm, Emission = 525 nm, Light path = 1 cm

Method: Fluorometric Stopped Reaction

Reagents

A. 20 mM Sodium Phosphate Buffer with 150 mM Sodium Chloride, pH 7.6 at 37°C (Prepare 200 ml in deionized water using Sodium Phosphate, Dibasic, Anhydrous and Sodium Chloride. Adjust to pH 7.6 at 37°C with 1 M HCl).

B. 0.5% (w/v) Fluorescein Isothiocyanate-Casein Solution (FITC-C) (Prepare 1.0 ml in Reagent A using Casein-Fluorescein Isothiocyanate. *Keep away from Light*).

C. Protease Enzyme Solution (Immediately before use, prepare a solution containing 10 μg/ml in cold Reagent A using Protease Type VIII).

D. 10% (v/v) Trichloroacetic Acid Solution (TCA) (Prepare 10 ml in deionized water using Trichloroacetic Acid, 6.1 N Solution).

E. 500 mM Tris HCl Solution, pH 8.5 at 25°C (Prepare 500 ml in deionized water using Trizma Base. Adjust to pH 8.5 at 25°C with 5 M HCl).

F. 0.001% (w/v) Fluorescein Isothiocyanate Standard Solution (FITC) (Prepare 1 ml in Reagent E using Fluorescein Isothiocyanate. *Keep away from Light*).

G. Sample Solution (Sample) (Prepare solution in cold deionized water as per the requested assignment).

Procedure

Part A: To prepare Protease Standards (PS), pipette (in milliliters) the following reagents into suitable containers:

	PS 1	PS 2	PS 3	PS 4
Reagent A (Buffer)	1.90	0.90	0.70	0.50
Reagent C (Protease)	0.10	0.10	0.30	0.50

Mix by swirling. Pipette (in milliliters) the following reagents into suitable microcentrifuge tubes:

	Test	Test Blank	PS 1A	PS 2A	PS 3A	PS 4A	PS Blank
Reagent A (Buffer)	0.02	0.02	0.02	0.02	0.02	0.02	0.02
Reagent B (FITC-C)	0.02	0.02	0.02	0.02	0.02	0.02	0.02

Equilibrate to 37°C. Then add:

	Test	Test Blank	PS 1A	PS 2A	PS 3A	PS 4A	PS Blank
Reagent G (Sample)	0.01	—	—	—	—	—	—
PS 1	—	—	0.01	—	—	—	—
PS 2	—	—	—	0.01	—	—	—
PS 3	—	—	—	—	0.01	—	—
PS 4	—	—	—	—	—	0.01	—

Immediately mix by swirling and incubate at 37°C for exactly 60 minutes. Then add:

	Test	Test Blank	PS 1A	PS 2A	PS 3A	PS 4A	PS Blank
Reagent D (TCA)	0.15	0.15	0.15	0.15	0.15	0.15	0.15
Reagent G (Sample)	—	0.01	—	—	—	—	—
PS 4 (Protease Standard)	—	—	—	—	—	—	0.01

Mix by swirling and incubate at 37°C for exactly 60 minutes. Centrifuge all tubes for 15 minutes.

Part B: Pipette (in milliliters) the following reagents into suitable containers:

	Test	Test Blank	PS 1A	PS 2A	PS 3A	PS 4A	PS Blank
Reagent E (Tris)	10.00	10.00	10.00	10.00	10.00	10.00	10.00
Supernatant-Test	0.10	—	—	—	—	—	—
Supernatant-Test Blank	—	0.10	—	—	—	—	—
Supernatant-PS-1A	—	—	0.10	—	—	—	—
Supernatant-PS-2A	—	—	—	0.10	—	—	—
Supernatant-PS-3A	—	—	—	—	0.10	—	—
Supernatant-PS-4A	—	—	—	—	—	0.10	—
Supernatant-PS Blank	—	—	—	—	—	—	0.10

Mix by swirling and transfer the solutions to suitable cuvettes. Record the fluorescence intensity at the excitation wavelength of 490 nm and the emission wavelength of 525nm for all the containers in a suitable fluorometer at 25°C. Prepare the FITC (FS) standards by pipetting (in milliliters) the following reagents into suitable containers:

	FS 1	FS 2	FS 3	FS 4	PS Blank
Reagent E (Tris)	10.00	10.00	10.00	10.00	10.00
Reagent F (FITC)	0.01	0.05	0.07	0.10	—

Mix by swirling and transfer to suitable cuvettes. Record the fluorescence intensity at the excitation wavelength of 490nm and the emission wavelength of 525 nm for all the containers in a suitable fluorometer at 25°C.

Calculations

F = Fluorescence at the excitation wavelength of 490nm and the emission wavelength of 525nm

$$\Delta\text{FFS Standard} = \text{FFS Standard} - \text{FFS Blank}$$

$$\Delta\text{FPS Standard} = \text{FPS Standard} - \text{FPS Blank}$$

Specifications

The change in fluorescence for the sample must be less than the 10 mg fluorescein isothiocyanate (FITC) standard and the 5 ng protease standard. The detection limit for this assay is 2.56 × 10–5 µmoles FITC released per minute.

Final Assay Concentrations

In a 0.05 ml reaction mix, the final concentrations are 16 mM sodium phosphate, 120 mM sodium chloride and 0.2% (w/v) casein-fluorescein isothiocyanate. The units of protease will vary.

Note

1. At this point, 0.10 ml of the supernatant from all samples and standards (FITC standard must be included) can be pipetted into 4 dram vials, wrapped in foil and placed in the cooler for up to a week before diluting with 10 ml of Reagent E (Tris) and read on the fluorometer.

Enzymatic Assay of PROTEASE INSOLUBLE

Principle

$$\text{N-t-BOC-GAPE} + H_2O \xrightarrow{\text{Protease}} \text{N-t-BOC-L-Glutamic Acid} + \text{Phenol}$$

Abbreviations Used

BOC = N-tert-Butoxy-Carbonyl
N-t-BOC-GAPE = N-t-BOC-L-Glutamic Acid a-Phenyl Ester

Conditions: T = 37°C, pH = 7.8, A270nm, Light path = 1 cm

Method: Spectrophotometric Determination

Reagent

A. 200 mM Tris-Phosphate Buffer, pH 7.8 at 37°C (Prepare 100 ml in deionized water using Trizma Base. Adjust to pH 7.8 at 37°C with Phosphoric Acid).
B. 60 mM N-t-BOC-L-Glutamic Acid α-Phenyl Ester Substrate Solution (N-t-BOC-GAPE) (Prepare by dissolving 97 mg of N-t-BOC-L-Glutamic Acid α-Phenyl Ester, in 5 ml of 1,4-Dioxane. Make sure that the solid is at room temperature before opening. Any moisture in this substrate will cause immediate hydrolysis of the substrate. *Prepare Fresh*).
C. Protease Insoluble Enzyme Suspension (Insol Enz). Accurately weigh approximately 0.5 g of insoluble protease and suspend in 10 ml of deionized water in a conical centrifuge tube. Mix by swirling gently and allow to swell for 1 hour. Then centrifuge at 500–1000 g for 5 minutes. Measure the packed gel volume. Resuspend the gel and place 2–4 ml of suspension on a Whatman #50 filter paper and remove the suspension medium and soluble protease by vacuum filtration. Stop the vacuum filtration before the gel becomes dry or begins to crack. It should be moist. Wash the gel cake with approximately 10 volumes of deionized water, and again stop the vacuum filtration before the gel becomes dry or begins to crack. Resuspend 0.9 g of the washed gel cake with 1 ml of deionized water. Mix by swirling to form a homogenous slurry. Then dilute to 3–5 units/ml of protease in cold deionized water.

Procedure

Pipette (in milliliters) the following reagents into suitable 4 dram vials:

	Test	Blank
Reagent A (Buffer)	4.87	4.87
Reagent C (Insoluble Enzyme)	0.05	—
Deionized Water	—	0.05

Mix by swirling and equilibrate to 37°C. Then add:

Reagent B (N-t-BOC-GAPE)	0.083	0.083

Immediately mix by stirring and incubate at 37°C for exactly 5 minutes. Stir continuously with a magnetic stirrer. Remove 1 ml from the reaction mix, and filter through a 0.45 μm syringe filter. Immediately transfer the supernatant into a suitable cuvette and record the absorbance at A270nm versus Reagent A using a suitable spectrophotometer.

Calculations

$$\text{Units / ml suspension} = \frac{(\text{A270nm Test} - \text{A270nm Blank})\ (5.003)\ (\text{df})}{(5)\ (1)\ (1.5)\ (0.05)}$$

5.003 = Total volume (in milliliters) of assay
df = Dilution factor
5 = Time (in minutes) of assay
1 = Volume (in milliliters) removed from the reaction mixture for absorbance measurement
1.5 = Millimolar extinction coefficient of phenol at 270nm
0.05 = Volume (in milliliter) of enzyme used

$$\text{Units / gram solid} = \frac{(\text{units / ml suspension})\ (1000)}{(15.7)}$$

1000 = Conversion factor from mg to g
15.7 = mg dry agarose/ml suspension
Units/ml packed gel = Units/ml suspension × 2

Unit Definition: *One unit will hydrolyze 1.0 μmole of N-t-BOC-L-glutamic acid α-phenyl ester per minute at pH 7.8 at 37°C.*

Final Assay Concentrations

In a 5.003 ml reaction mix, the final concentrations are 195 mM Tris, 1 mM N-t-BOC-L-glutamic acid α-phenyl ester, and 0.15–0.25 units of protease insoluble.

Note

1. Mix the enzyme suspension gently. Elevating the suspension surface along the sides of the vessel may result in the attachment of agarose beads to the walls of the vessel effectively reducing the concentration of the enzyme in the suspension. Whenever an insoluble enzyme suspension is sampled, it must be done while the enzyme suspension is stirring.

Enzymatic Assay of PROTEASE INSOLUBLE

Principle

$$\text{BAEE} + H_2O \xrightarrow{\text{Protease}} \text{Na - Benzoyl - L - Arginine} + \text{Ethanol}$$

Abbreviation Used

BAEE = Na-Benzoyl-L-Arginine Ethyl Ester

Conditions: T = 30°C, pH 7.0

Method: Titrimetric

Reagents

A. 1 M Potassium Chloride Solution (KCl) (Prepare 15 ml in deionized water using Potassium Chloride).

B. 500 mM Na-Benzoyl-Arginine Ethyl Ester (BAEE) (Prepare 15 ml in deionized water using Na-Benzoyl-L-Arginine Ethyl Ester, Hydrochloride. Adjust to pH 7.0 at 30°C with 1 M NaOH).
C. 100 mM Sodium Hydroxide Titrant, Standardized (NaOH) (Prepare 50 ml in deionized water using Sodium Hydroxide, Anhydrous).
D. Protease Insoluble Enzyme Solution (Insol Enz) (Immediately before use, prepare a suspension containing approximately 20 mg of Protease, Insoluble Enzyme, 8 ml of deionized water and 1 ml of Reagent A. Mix by swirling. Filter on Whatman #50 filter paper and remove any soluble protease by vacuum filtration. Stop the vacuum filtration before the gel becomes dry or begins to crack. It should be moist).

Procedure

Using a suitable pH meter in conjunction with a magnetic stirrer, pipette (in milliliters) the following reagents into a suitable titration vessel:

	Test
Deionized Water	8.00
Reagent A (KCl)	1.00
Reagent B (BAEE)	1.00

Mix by swirling and equilibrate to 30°C. Adjust to pH 7.0 by the addition of Reagent C (NaOH). Then add:

Reagent D (Insol Enz)	20 mg

Start a timer when the pH reaches 7.0. Run the reaction for 2–10 minutes. Maintain the pH of the reaction mix at pH 7.0 by the addition of small volumes (0.05 ml) of Reagent C (NaOH). Record the volume of Reagent C (NaOH used to maintain the pH at 7.0) and the time (in minutes) required to consume the added Reagent C (NaOH).

Calculations

$$\text{Units / g solid} = \frac{(\text{Molarity of NaOH})\ (\text{NaOH})\ (1000)}{(\text{T})\ (\text{mg of insoluble enzyme})}$$

NaOH = Volume (in microliters) of Reagent C (NaOH)
1000 = Conversion factor from mg to g
T = Time required (in minutes) to consume the added Reagent C (NaOH) while maintaining the pH at 7.0

Unit Definition: *One unit will hydrolyze 1.0 μmole of BAEE per minute at pH 7.0 at 30°C.*

Final Assay Concentrations

In a 10.00 ml reaction mix, the final concentrations are 100 mM potassium chloride, 50 mM Na-benzoyl-L-arginine ethyl ester, and 20 mg of protease insoluble enzyme.

Enzymatic Assay of PROTEASE

Principle

$$\text{Casein} + H_2O \xrightarrow{\text{Protease}} \text{Amino Acids}$$

Conditions: T = 30°C, pH = 11.0, A275nm, Light path = 1 cm

Method: Spectrophotometric Stop Rate Determination

Reagents

A. 100 mM Sodium Tetraborate Solution (Prepare 10 ml in deionized water using Borax).

B. 0.6% (w/v) Casein with 10.0 mM Sodium Tetraborate, pH 11.0 at 30°C (Casein). (Prepare by dissolving 600 mg of Casein in 4 ml of 0.1 M NaOH in a hot water bath (60°C). Cool to room temperature and then add 10 ml of Reagent A and 80 ml of deionized water. Adjust to pH 11.0 at 30°C with 1 M NaOH and bring to a volume of 100 ml with deionized water).

C. 110 mM Trichloroacetic Acid with 220 mM Sodium Acetate and 330 mM Acetic Acid (TCA) (Prepare 40 ml in deionized water using Trichloroacetic Acid, 6.1 N, approximately 100% (w/v), Sodium Acetate Trihydrate and Acetic Acid, Glacial).

D. 10.0 mM Sodium Tetraborate Buffer, pH 11.0 at 30°C (Enz Dil I) (Prepare 10 ml in deionized water using Borax).

E. 2.0 mM Calcium Acetate Solution (Enz Dil II) (Prepare 10 ml in deionized water using Calcium Acetate).

F. Protease Enzyme Solution (Prepare by dissolving 1 mg of Protease in 2 ml of ice-cold Reagent D. Immediately before use, dilute to 0.20–0.30 unit/ml of Protease with Reagent E).

Procedure

Pipette (in milliliters) the following reagents into suitable containers:

	Test	Blank
Reagent B (Casein)	3.00	3.00

Equilibrate to 30°C. Then add:

	Test	Blank
Reagent F (Enzyme Solution)	0.50	—

Immediately mix by swirling and incubate at 30°C for exactly 10 minutes. Then add:

	Test	Blank
Reagent C (TCA)	3.20	3.20
Reagent F (Enzyme Solution)	—	0.50

Mix by swirling and incubate at 37°C for 20 minutes. Filter the solutions through Whatman #50 filter paper and transfer the solutions to suitable cuvettes and record the A275nm for both the Test and Blank using a suitable spectrophotometer.

Calculations

$$\text{Units / ml enzyme} = \frac{(\text{A275nm Test} - \text{A275nm Blank})\,(6.7)\,(\text{df})}{(10)\,(1.34)\,(0.5)}$$

6.7 = Total volume (in milliliters) of assay
df = Dilution factor
10 = Time (in minutes) of the assay per the Unit Definition
1.34 = Millimolar extinction coefficient of tyrosine under the conditions of this assay
0.5 = Volume (in milliliter) of enzyme used

$$\text{Units / mg solid} = \frac{\text{units / ml enzyme}}{\text{mg solid / ml enzyme}}$$

$$\text{Units / mg protein} = \frac{\text{units / ml enzyme}}{\text{mg protein / ml enzyme}}$$

Unit Definition: *One unit will hydrolyze casein to produce peptide equivalent to 1.0 µmole (181 µg) of Tyrosine per minute at pH 11.0 at 30°C. This protease is an alkaline protease and is approximately twice as active at pH 11.0, as described above, than at the usual assay conditions for protease, pH 7.5 and 37°C.*

Final Assay Concentrations

In a 3.50 ml reaction mix, the final concentrations are 0.5% (w/v) casein, 9 mM sodium tetraborate, 0.3 mM calcium acetate, and 0.10–0.15 unit protease.

Protein Quantitation Assays

Preparation of Working Protein Standard for Use in Protein Determination

Conditions: T = 25°C, A280nm, Light path = 1 cm

Method: Spectrophotometric Concentration Determination

Reagents

A. 0.85% (w/v) Sodium Chloride Solution (NaCl) (prepare 100 ml in deionized water using Sodium Chloride).

B. 0.5% (w/v) Working Protein Standard (WPS) (Prepare 50 ml in Reagent A using Albumin Bovine Serum or Protein Standards, Bovine Serum Albumin).

Procedure

Pipette (in milliliter) the following reagents into a suitable cuvette:

	Test
Reagent A (NaCl)	0.90

Equilibrate to 25°C using a suitably thermostatted spectrophotometer. Monitor the A280nm until constant. Record the Initial A280nm for the Test. Then add:

Reagent B (WPS)	0.10

Mix by inversion. Monitor the increase in A280nm until constant. Record the Final A280nm for the Test.

Calculations

$$\text{mg protein / ml of WPS} = \frac{(\text{Final A280nm Test} - \text{Initial A280nm Test})}{(0.667)\ (0.1)}$$

WPS = Working Protein Standard
0.667 = Extinction coefficient of a 0.1% (w/v) BSA Solution at 280nm
0.1 = Volume of Working Protein Standard used

PROTEIN DETERMINATION Bicinchoninic Acid Method

Principle

$$\text{Copper} + \text{Protein} \xrightarrow{\text{Alkaline pH}} \text{Copper - Protein Complex}$$

Conditions: T = 25°C, A562nm, Light path = 1 cm

Method: Colorimetric

Reagents

A. 0.85% (w/v) Sodium Chloride Solution (NaCl) (Use Sodium Chloride Solution, 0.85% or prepare 100 ml in deionized water using Sodium Chloride).
B. 0.1% (w/v) Working Protein Standard (WPS).
C. Bicinchoninic Acid Solution (BCA).
D. 4% (w/v) Copper (II) Sulfate Pentahydrate Solution ($CuSO_4$) (Use Copper (II) Sulfate Pentahydrate).
E. BCA-Copper Solution (BCA/Cu) (Prepare 51 ml by adding 1 ml of Reagent D to 50 ml of Reagent C. Mix by stirring; the solution should be light green in color).
F. Protein Sample Solution (Protein) (Prepare a solution containing 0.3–1 mg/ml of protein in Reagent A).

Procedure

Pipette (in milliliters) the following reagents into suitable containers:

	Test	Std 1	Std 2	Std 3	Std 4	Blank
Reagent A (NaCl)	—	0.075	0.050	0.025	—	0.100
Reagent B (WPS)	—	0.025	0.050	0.075	0.100	—
Reagent F (Protein)	0.10	—	—	—	—	—

Mix by inversion. Then add:

	Test	Std 1	Std 2	Std 3	Std 4	Blank
Reagent E (BCA/Cu)	2.00	2.00	2.00	2.00	2.00	2.00

Mix by inversion and incubate at 37°C for 30 minutes. Cool to room temperature in an ice bath. Transfer the solutions to suitable cuvettes and record the A562nm for the Test, Standards, and Blank with the Blank as the reference.

Calculations

Standard Curve: ΔA562nm Standard = A562nm Std–A562nm Blank

Prepare a Standard curve by plotting the ΔA562nm of the Standard vs µg of protein.

Sample Determination: ΔA562nm Sample = A562nm Test–A562nm Blank

Determine the mg of protein using the Standard Curve.

mg Protein = (mg of protein from standard curve) (df)
df = Dilution factor

$$\% \text{ Protein} = \frac{(\text{mg of protein})\,(100)}{\text{mg solid / ml Reagent A}}$$

For products that are liquid:

$$\text{mg protein / ml} = \frac{(\text{mg protein})}{\text{ml Reagent A}}$$

100 = Conversion to percentage

PROTEIN DETERMINATION Biuret Method

Principle

$$\text{Copper} + \text{Protein} \xrightarrow{\text{Alkaline pH}} \text{Copper - Protein Complexes}$$

Conditions: T = 25°C, A540nm, Light path = 1 cm

Method: Colorimetric

Reagents

A. 0.85% (w/v) Sodium Chloride Solution (NaCl) (Use Sodium Chloride Solution, 0.85% or prepare 100 ml in deionized water using Sodium Chloride).
B. 0.5% (w/v) Working Protein Standard (WPS).
C. Biuret Reagent (Biuret) .
D. Protein Sample Solution (Protein) (Prepare a solution containing 0.5–4 mg protein/ ml of Sample in Reagent A).

Procedure

Pipette (in milliliters) the following reagents into suitable containers:

	Test	Std 1	Std 2	Std 3	Std 4	Std 5	Blank
Reagent A (NaCl)	—	0.96	0.90	0.80	0.40	—	1.00
Reagent B (WPS)	—	0.04	0.10	0.20	0.60	1.00	—
Reagent D (Protein)	1.00	-	—	—	—	—	—

Mix by swirling. Then add:

Reagent C (Biuret)	4.00	4.00	4.00	4.00	4.00	4.00	4.00

Mix thoroughly by vortexing and incubate for 30 minutes at 25°C. Transfer to suitable cuvettes and record the absorbance at 540nm for Test, Standards, and Blank.

Calculations

$$\Delta A540nm\ Standard = A540nm\ Std - A540nm\ Std\ Blank$$

Prepare a standard curve by plotting the A540nm of the Standards vs mg of protein.

Sample Determination: ΔA540nm Sample = ΔA540nm Test–A540nm Test Blank

Determine the mg of protein using the Standard curve.

mg Protein = (mg of protein from the Standard curve) (df)

$$\%\ Protein = \frac{(mg\ protein)\ (100)}{(mg\ solid\ /\ ml\ Reagent\ D)}$$

For products that are liquid:

$$mg\ protein\ /\ ml = \frac{(mg\ protein)}{(ml\ Reagent\ D)}$$

100 = Conversion to percentage

PROTEIN DETERMINATION Biuret Method using TCA Precipitation

Conditions: T = 25°C, A540nm, Light path = 1 cm

Method: Colorimetric

Reagents

A. 0.85% Sodium Chloride Solution (NaCl) (prepare 100 ml in deionized water using Sodium Chloride).
B. 1% Working Protein Standard (WPS).
C. Biuret Reagent (Biu).
D. Trichloroacetic Acid (TCA).
E. Protein Sample Solution (Pro) (Prepare a solution containing 1–3 mg protein/ml of Sample in Reagent A).

Procedure

Pipette (in milliliters) the following reagents into suitable containers:

	Test	Std 1	Std 2	Std 3	Std 4	Std 5	Blank
Reagent A (NaCl)	—	0.98	0.95	0.90	0.70	0.50	1.00
Reagent B (WPS)	—	0.02	0.05	0.10	0.30	0.50	—
Reagent E (Pro)	1.00	—	—	—	—	—	—

Mix. Then add:

Reagent D (TCA)	0.10	0.10	0.10	0.10	0.10	0.10	0.10

Mix thoroughly and place the containers in an ice bath for 10 minutes. Centrifuge for 20 minutes in a clinical centrifuge. Remove and discard the supernatant. Then add:

Reagent A (NaCl)	1.00	1.00	1.00	1.00	1.00	1.00	1.00

Mix thoroughly until pellet dissolves. Then add:

	Test	Std 1	Std 2	Std 3	Std 4	Std 5	Blank
Reagent C (Biu)	4.00	4.00	4.00	4.00	4.00	4.00	4.00

Mix thoroughly and seal the containers. Incubate for 30 minutes at 25°C. Transfer to suitable cuvettes and record the absorbance at 540nm for Test, Standards, and Blank.

Calculations

A540nm Standard = A540nm Std–A540nm Blank

Find the slope (M) by plotting the A540nm Standards vs Protein concentration.

Sample Determination: Determine the concentration of protein using the following equation:

$$\text{mg Protein} = \frac{\text{A540nm Test} - \text{A540nm Blank}}{(\text{M})}$$

$$\%\ \text{Protein} = \frac{(\text{mg Protein})\ (100)}{(\text{mg solid / ml Reagent D})}$$

For Products that are liquid:

$$\text{mg Protein / ml} = \frac{(\text{mg Protein})\ (\text{Dilution})}{(\text{ml Reagent D})}$$

100 = Conversion to percentage

PROTEIN DETERMINATION Lowry/TCA Method

Conditions: T = 25°C, A750nm, Light path = 1 cm

Method: Colorimetric

Reagents

A. 0.85% Sodium Chloride Solution (NaCl) (prepare 100 ml in deionized water using Sodium Chloride).
B. 0.01% Lowry Protein Standard (LoPS).
C. 100 mM Sodium Hydroxide Solution (NaOH) (Prepare 1 liter in deionized water using Sodium Hydroxide).
D. 2% (w/v) Sodium Carbonate Solution ($NaCO_3$) (Prepare 1 liter in Reagent C using Sodium Carbonate).
E. 2% (w/v) Sodium Potassium Tartrate Solution (SPT) (Prepare 10 ml in deionized water using Sodium Potassium Tartrate, Tetrahydrate).
F. 1% (w/v) Cupric Sulfate Solution ($CuSO_4$) (Prepare 10 ml in deionized water using Cupric Sulfate, Pentahydrate).
G. 50% Folin-Ciocalteu's Reagent (FC) (Prepare 100 ml in deionized water using Folin-Ciocalteu Reagent. *Store in an amber bottle—Sensitive to light. Prepare Fresh*).
H. Trichloroacetic Acid (TCA)
I. Protein Sample Solution (Pro) (Prepare a solution containing 0.03–0.080 mg protein/ml of Sample in Reagent A).

Procedure

Prepare the Lowry Reagent by pipetting (in milliliters) the following reagents into a suitable container:

Reagent D ($NaCO_3$)	100.0
Reagent E (SPT)	1.0
Reagent F ($CuSO_4$)	1.0

Mix. Pipette (in milliliters) the following reagents into suitable containers:

	Test	Std 1	Std 2	Std 3	Std 4	Std 5	Blank
Reagent A (NaCl)	—	0.90	0.80	0.60	0.40	0.20	1.00
Reagent B (LoPS)	—	0.10	0.20	0.40	0.60	0.80	—
Reagent I (Pro)	1.00	—	—	—	—	—	—

Mix. Then add:

Reagent H (TCA)	0.10	0.10	0.10	0.10	0.10	0.10	0.10

Mix thoroughly and place in an ice bath for 10 minutes. Centrifuge at 6000 rpm for 20 minutes. Discard the supernatant. Then add:

Reagent A (NaCl)	1.00	1.00	1.00	1.00	1.00	1.00	1.00

Mix thoroughly. Then add:

Lowry Reagent	5.00	5.00	5.00	5.00	5.00	5.00	5.00

Mix thoroughly by vortexing and incubate for 10 minutes at 25°C. Then add:

Reagent F (FC)	0.50	0.50	0.50	0.50	0.50	0.50	0.50

Mix thoroughly by vortexing and incubate for 30 minutes at 25°C. Transfer to suitable cuvettes and record the absorbance at 750nm for Test, Standards and Blank.

Calculations

$$\Delta A750nm\ Standard = A750nm\ Std - A750nm\ Blank$$

Plotting the ΔA750nm Standards vs Protein concentration.

Sample Determination: ΔA750nm Test = A750nm Test–A750nm Blank

Determine the mg protein from the Standard curve.

$$\%\ Protein = \frac{(mg\ Protein)\ (100)}{(mg\ solid\ /\ ml\ Reagent\ D)}$$

For Products that are liquid:

$$mg\ Protein\ /\ ml = \frac{(mg\ Protein)\ (Dilution)}{(ml\ Reagent\ D)}$$

100 = Conversion to percentage

PROTEIN DETERMINATION Lowry Method

Conditions: T = 25°C, A750nm, Light path = 1 cm

Method: Colorimetric

Reagents

A. 0.85% Sodium Chloride Solution (NaCl) (prepare 100 ml in deionized water using Sodium Chloride).
B. 0.01% Lowry Protein Standard (LoPS) (Prepare 10 ml in Reagent A using WPS prepared as per Working Protein Standard Procedure).
C. 100 mM Sodium Hydroxide Solution (NaOH) (Prepare 1 liter in deionized water using Sodium Hydroxide).
D. 2% (w/v) Sodium Carbonate Solution ($NaCO_3$) (Prepare 1 liter in Reagent C using Sodium Carbonate).
E. 2% (w/v) Sodium Potassium Tartrate Solution (SPT) (Prepare 10 ml in deionized water using Sodium Potassium Tartrate, Tetrahydrate).
F. 1% (w/v) Cupric Sulfate Solution ($CuSO_4$) (Prepare 10 ml in deionized water using Cupric Sulfate, Pentahydrate).
G. 50% Folin-Ciocalteu's Reagent (FC) (Prepare 100 ml in deionized water using Folin-Ciocalteu Reagent. *Store in an amber bottle-Sensitive to light. Prepare Fresh*).
H. Protein Sample Solution (Pro) (Prepare a solution containing 0.03–0.080 mg protein/ml of Sample in Reagent A).

Procedure

Prepare the Lowry Reagent by pipetting (in milliliters) the following reagents into a suitable container:

Reagent D ($NaCO_3$)	100.0
Reagent E (SPT)	1.0
Reagent F ($CuSO_4$)	1.0

Mix. Pipette (in milliliters) the following reagents into suitable containers:

	Test	Std 1	Std 2	Std 3	Std 4	Std 5	Blank
Reagent A (NaCl)	—	0.90	0.80	0.60	0.40	0.20	1.00
Reagent B (LoPS)	—	0.10	0.20	0.40	0.60	0.80	—
Reagent H (Pro)	1.00	—	—	—	—	—	—

Mix. Then add:

Lowry Reagent	5.00	5.00	5.00	5.00	5.00	5.00	5.00

Mix thoroughly by vortexing and incubate for 10 minutes at 25°C. Then add:

Reagent G (FC)	0.50	0.50	0.50	0.50	0.50	0.50	0.50

Mix thoroughly by vortexing and incubate for 30 minutes at 25°C. Transfer to suitable cuvettes and record the absorbance at 750 nm for Test, Standards and Blank.

Calculations

$$\Delta A750nm\ Standard = A750nm\ Std - A750nm\ Blank$$

Plotting the ΔA750nm Standards vs Protein concentration.

Sample Determination: ΔA750nm Test = A750nm Test–A750nm Blank

Determine the mg protein from the Standard curve.

$$\%\ Protein = \frac{(mg\ Protein)\ (100)}{(mg\ solid\ /\ ml\ Reagent\ D)}$$

For Products that are liquid:

$$mg\ Protein\ /\ ml = \frac{(mg\ Protein)\ (Dilution)}{(ml\ Reagent\ D)}$$

100 = Conversion to percentage

PROTEIN DETERMINATION Modified Lowry Method

Principle

$$Copper + Protein \xrightarrow{Alkaline\ pH} Copper - Protein\ Complex\ (Blue\ color)\ phenol\ reagent$$

Conditions: T = 25°C, A750nm, Light path = 1 cm

Method: Colorimetric

Reagents

A. 0.85% Sodium Chloride Solution (NaCl) (Use Sodium Chloride Solution, 0.85 or prepare 100 ml in deionized water using Sodium Chloride).
B. 0.01% Lowry Protein Standard (WPS)
C. Modified Lowry Reagent (LOW) (Prepare by reconstituting in deionized water per label instructions a vial of Lowry Reagent, Modified).
D. Protein Sample Solution (Sample) (Prepare a solution containing 10–100 µg protein/ml of sample in Reagent A).
E. 330 mM Folin and Ciocalteu's Phenol Reagent (F & C) (Prepare by diluting an appropriate volume of Folin and Ciocalteu's Phenol Reagent in deionized water and mix thoroughly. Prepare fresh immediately prior to use).

Procedure

Pipette (in milliliters) the following reagents into suitable containers:

	Test	Std 1	Std 2	Std 3	Std 4	Std 5	Blank
Reagent A (NaCl)	—	0.90	0.70	0.50	0.30	—	1.00
Reagent B (WPS)	—	0.10	0.30	0.50	0.70	1.00	—
Reagent D (Sample)	1.00	—	—	—	—	—	—
Reagent C (Low)	1.00	1.00	1.00	1.00	1.00	1.00	1.00

Mix thoroughly by swirling and incubate for 10 minutes at 25°C. Then add:

	Test	Std 1	Std 2	Std 3	Std 4	Std 5	Blank
Reagent E (F & C)	0.50	0.50	0.50	0.50	0.50	0.50	0.50

Mix thoroughly by vortexing and incubate for 30 minutes at 25°C. Transfer the solutions to suitable cuvettes and record the A750nm for Test, Standards and Blank using a suitable spectrophotometer.

Calculations

$$\Delta A750nm\ Standard = A750nm\ Std - A750nm\ Blank$$

Prepare a standard curve by plotting the ΔA750nm of the Standards vs μg of protein.

Sample Determination: ΔA750nm Test = A750nm Test–A750nm Blank

Determine the μg protein from the Standard curve.

μg Protein = (μg of protein from the Standard curve) (df)

df = Dilution factor

$$\%\ Protein = \frac{(\mu g\ Protein)\ (100)}{(\mu g\ solid\ /\ ml\ Reagent\ D)}$$

100 = Conversion to percentage

For Products that are liquid:

$$\mu g\ Protein\ /\ ml = \frac{(\mu g\ Protein)}{(mg\ sample\ /\ ml\ Reagent\ D)}$$

Notes

1. This solution can be stored for approximately one month or until the solution becomes cloudy.
2. Incubation time must be at least 10 minutes to allow for the copper solution to react with peptide bonds. This time can be as long as 24 hours if needed.
3. Incubation time must be at least 30 minutes but nolonger than 60 minutes, because color development will continue after 30 minutes. All standards, samples, and blank absorbances must be read on spectrophotometer within a short time of each other.

Quantitative Metabolite Assays

Determination of the Concentration and Molecular Weight of ADENOSINE 5'-DIPHOSPHATE

Principle

$$PEP + ADP \xrightarrow{PK} Pyruvate + ATP$$

$$Pyruvate + \beta\text{-}NADH \xrightarrow{LDH} Lactate + \beta\text{-}NAD$$

Abbreviations Used

PEP = Phospho(enol) pyruvate

ADP = Adenosine 5'-Diphosphate

PK = Pyruvate Kinase
ATP = Adenosine 5'-Triphosphate
β-NADH = β-Nicotinamide Adenine Dinucleotide, Reduced Form
LDH = Lactate Dehydrogenase
β-NAD = β-Nicotinamide Adenine Dinucleotide, Oxidized Form

Conditions: T = 25°C, pH = 7.6, A340nm, Light path = 1 cm

Method: Spectrophotometric Determination

Reagents

A. 150 mM Triethanolamine HCl Buffer, pH 7.6 at 25°C (Prepare 100 ml in deionized water using Triethanolamine, Hydrochloride. Adjust to pH 7.6 at 25°C with 1 M NaOH).
B. 32 mM Phospho(enol) pyruvate, 83 mM Magnesium Sulfate, and 135 mM Potassium Chloride Solution (PEP) (Prepare 2 ml in deionized water using Phospho(enol) pyruvate, Tri(cyclohexylammonium)Salt, Magnesium Sulfate and Potassium Chloride).
C. 3.8 mM β-Nicotinamide Adenine Dinucleotide, Reduced Form (β-NADH) (Prepare 2 ml in Reagent A using β-Nicotinamide Adenine Dinucleotide, Reduced Form, Disodium Salt).
D. Adenosine 5'-Diphosphate Solution (ADP) (Weigh approximately 2.5 mg of Adenosine 5'- Diphosphate and dissolve in 25 ml of deionized water).
E. L-Lactic Dehydrogenase Enzyme Suspension (LDH) (Use L-Lactic Dehydrogenase).
F. Pyruvate Kinase Enzyme Suspension (PK).

Procedure

Pipette (in milliliters) the following reagents into suitable cuvettes:

	Test	Blank
Reagent A (Buffer)	1.70	1.80
Reagent B (PEP)	0.15	0.15
Reagent C (β-NADH)	0.10	—
Reagent D (ADP)	1.00	1.00
Reagent E (LDH)	0.01	0.01

Mix by inversion and equilibrate to 25°C using a suitably thermostatted spectrophotometer. Record the initial A340nm for both the Test and Blank. Then add:

Reagent F (PK)	0.02	0.02

Immediately mix by inversion and allow the reaction to proceed to completion (approximately 5 minutes). Record the final A340nm for both the Test and Blank.

Calculation

$\Delta A = A_i - A_f$ Ai = Initial Absorbance
Af = Final Absorbance

$$\text{Micromoles ADP / weighed sample} = \frac{(\Delta A \text{ Test} - \Delta A \text{ Blank})(2.98)(25)}{6.22}$$

2.98 = Total volume (in milliliters) of assay
25 = Dilution factor
6.22 = Millimolar extinction coefficient of β-NADH at 340nm

$$\text{Apparent Molecular Weight} = \frac{\text{mg sample weighed} \times 1000}{\mu\text{moles ADP / weighed sample}}$$

Final Assay Concentrations

In a 2.98 ml reaction mix, the final concentrations are 91 mM triethanolamine, 1.6 mM phospho(enol) pyruvate, 4.2 mM magnesium sulfate, 6.8 mM potassium chloride, 0.13 mM β-nicotinamide adenine dinucleotide, reduced form, 40 units pyruvate kinase, 100 units lactic dehydrogenase, and varying amounts of adenosine 5'-diphosphate.

Notes

1. Lactic Dehydrogenase Unit Definition: One unit will reduce 1.0 μmole of pyruvate to L-lactate per minute at pH 7.5 at 37°C.
2. Pyruvate Kinase Unit Definition: One unit will convert 1.0 μmole of phospho(enol) pyruvate to pyruvate per minute at pH 7.6 at 37°C.

Determination of the Concentration and Molecular Weight of ARGININOSUCCINIC ACID

Principle

$$\text{L-Argininosuccinate} \xrightarrow{\text{AL}} \text{L-Arginine} + \text{Fumarate}$$

$$\text{L-Arginine Arginase} \rightarrow \text{Ornithine} + \text{Urea}$$

Abbreviation Used

AL = Argininosuccinate Lyase

Conditions: T = 37°C, pH = 7.5, A240nm, Light path = 1 cm

Method: Spectrophotometric Rate Determination

Reagents

A. 100 mM Potassium Phosphate Buffer, pH 7.5 at 37°C (Prepare 100 ml in deionized water using Potassium Phosphate, Monobasic, Anhydrous. Adjust to pH 7.5 at 37°C with 1 M KOH).
B. 40 mM Potassium Sulfate Solution (K_2SO_4) (Prepare 15 ml in deionized water using Potassium Sulfate).
C. Argininosuccinate Acid Substrate Solution (ASA) (Weigh approximately 2 mg of Argininosuccinic Acid and dissolve in 4.5 ml of deionized water. Then add 0.5 ml of Reagent B. Centrifuge and save the supernatant).
D. Arginase Enzyme Solution (Arginase) (Immediately before use, prepare a solution containing 0.8 mg/ml of Arginase in cold deionized water).

E. Argininosuccinate Lyase Enzyme Solution (AL) (Immediately before use, prepare a solution containing 20 mg/ml of Argininosuccinate Lyase in cold deionized water).

Procedure

Pipette (in milliliters) the following reagents into suitable cuvettes:

	Test	Blank
Deionized Water	—	0.50
Reagent A (Buffer)	2.20	2.20
Reagent C (ASA)	0.50	—
Reagent D (Arginase)	0.20	0.20

Mix by inversion and equilibrate to 37°C using a suitably thermostatted spectrophotometer. Record the initial A240nm for both the Test and Blank. Then add:

	Test	Blank
Reagent E (AL)	0.10	0.10

Immediately mix by inversion and allow the reaction to proceed to completion (approximately 5 minutes). Record the final A240nm for both the Test and Blank.

Calculations

$$\Delta A = Af - Ai$$

Af = Final Absorbance
Ai = Initial Absorbance

$$\text{Micromoles ASA / weighed sample} = \frac{(\Delta A \text{ Test} - \Delta A \text{ Blank})(3)(5)}{(0.5)(2.44)}$$

ASA = Argininosuccinic Acid
3 = Total volume (in milliliters) of assay
5 = Dilution factor
0.5 = Volume (in milliliter) of argininosuccinic acid used in assay
2.44 = Millimolar extinction coefficient of fumarate at 240nm

$$\text{Apparent Moledular Weight} = \frac{\text{mg sample weighed} \times 1000}{\mu\text{moles of ASA / weighed sample}}$$

1000 = Conversion factor from mg to μg
ASA = Argininosuccinic Acid

Final Assay Concentrations

In a 3.00 ml reaction mix, the final concentrations are 73 mM potassium phosphate, 0.7 mM potassium sulfate, 0.16 mg arginase, 2 mg argininosuccinate lyase, and varying amounts of argininosuccinic acid.

Determination of the Concentration and Molecular Weight of COENZYME A using PHOSPHOTRANSACETYLASE

Principle

$$\text{Acetyl Phosphate} + \text{CoA} \rightarrow \text{Acetyl-CoA} + \text{Pi}$$

Abbreviations Used

CoA = Coenzyme A
Acetyl-CoA = Acetyl Coenzyme A
Pi = Inorganic Phosphate

Conditions: T = 25°C, pH = 7.6, A233nm, Light path = 1 cm

Methods: Spectrophotometric

Reagents

A. 100 mM Tris Buffer, pH 7.6 at 25°C (Prepare 100 ml in deionized water using Trizma Base. Adjust to pH 7.6 at 25°C with 1 M HCl).
B. 100 mM Acetyl Phosphate Solution (Prepare 10 ml in deionized water using Acetyl Phosphate, Lithium Potassium Salt. *Prepare Fresh*).
C. Coenzyme A Solution (CoA) (Weigh two samples accurately, approximately 3.0 mg and dissolve each in 10 ml of deionized water which has been purged with nitrogen).
D. Phosphotransacetylase Enzyme Solution (Prepare a solution containing 100 units/ml of Phosphotransacetylase, from *Clostridium kluyveri* in Reagent A).

Procedure

Pipette (in milliliters) the following reagents into suitable cuvettes:

	Test	Blank
Reagent A (Buffer)	1.90	1.90
Reagent B (Acetyl Phosphate)	0.10	—
Reagent C (CoA)	1.00	1.00
Deionized Water	—	0.10

Mix by inversion and equilibrate to 25°C. Monitor the A233nm until constant, using a suitably thermostatted spectrophotometer. Then add:

Reagent D (Enzyme Solution)	0.02	0.02

Mix by inversion and allow the reaction to proceed for 10 minutes. Upon completion of the reaction record the A233nm and calculate the A233nm and the apparent molecular weight.

Calculations

$\Delta A = Ai - Af$ Ai = Initial absorbance
Af = Final absorbance

$$\text{Micromoles CoA / weighed sample} = \frac{\Delta A\ (3.02)\ (10)}{4.44}$$

3.02 = Volume of assay
10 = Sample dilution factor
4.44 = Extinction coefficient of acetyl coenzyme A at 233nm

$$\text{apparent mw} = \frac{\text{mg sample weighed} \times 1000}{\mu\text{moles of CoA / weighed sample}}$$

Final Assay Concentrations

In a 3.02 ml reaction mix, the final concentrations are 63.6 mM tris HCl, 3.3 mM acetyl phosphate, and 2.0 unit phosphotransacetylase.

Determination of the Concentration and Molecular Weight of DIHYDROXYACETONE PHOSPHATE DIMETHYLKETAL

Principle

$$\text{DHAP - DMA - CHA} \xrightarrow{\text{pH = 1}} \text{DHAP}$$

$$\text{DHAP} + \beta\text{-NADH} \xrightarrow[\alpha\text{-GDH}]{37°\text{C}} \alpha\text{-Glycerophosphate} + \beta\text{-NAD}$$

Abbreviations Used

DHAP-DMK-CHA = Dihydroxyacetone Phosphate Dimethylketal, Di(Monocyclohexylammonium)
DHAP = Dihydroxyacetone Phosphate
β-NADH = β-Nicotinamide Adenine Dinucleotide, Reduced Form
α-GDH = α-Glycerophosphate Dehydrogenase
β-NAD = β-Nicotinamide Adenine Dinucleotide, Oxidized Form

Conditions: T = 25°C, pH = 7.6, A340nm, Light path = 1 cm

Method: Spectrophotometric

Reagents

A. 200 mM Triethanolamine HCl Buffer, pH 7.6 at 25°C (Prepare 100 ml in deionized water using Triethanolamine Hydrochloride. Adjust to pH 7.6 at 25°C with 1 M NaOH).
B. 3.8 mM β-Nicotinamide Adenine Dinucleotide, Reduced Form (β-NADH) (Prepare 1 in Reagent A using β-Nicotinamide Adenine Dinucleotide, Reduced Form, Disodium Salt).
C. Dihydroxyacetone Phosphate Solution (DHAP) (Weigh approximately 1.25 mg of Dihydroxyacetone Phosphate Dimethylketal, Di (monocyclohexylammonium Salt:Monohydrate) and dissolve in 5 ml of 0.1 N HCl. Incubate at 37°C for 1.5–2.0 hours. Then dilute a 1.0 ml aliquot to 5 ml with deionized water).
D. α-Glycerophosphate Dehydrogenase Enzyme Solution (α-GDH) (Immediately before use, prepare a solution containing approximately 50 units/ml of α-Glycerophosphate Dehydrogenase in Reagent A).

Procedure

Pipette (in milliliters) the following reagents into suitable cuvettes:

	Test	Blank
Reagent A (Buffer)	1.90	2.00
Reagent B (β-NADH)	0.10	—
Reagent C (DHAP)	1.00	1.00

Mix by inversion and equilibrate at 25°C for 5 minutes. Record the initial A340nm using a suitably thermostatted spectrophotometer. Then add:

	Test	Blank
Reagent D (α-GDH)	0.02	0.02

Mix by inversion and allow the reaction to proceed for 5–10 minutes. Record the final A340nm for both the Test and Blank.

Calculations

ΔA = Ai X–Af Ai = Initial absorbance
3.02 Af = Final Absorbance
3/3.02 = Correction for volume change

$$\text{Micromoles DHAP / weighed sample} = \frac{(\Delta A \text{ Test} - \Delta A \text{ Blank})(3.02)(25)}{6.22}$$

3.02 = Total volume (in milliliters) of assay
25 = Dilution factor
6.22 = Millimolar extinction coefficient of β-NADH at 340nm

$$\text{Apparent molecular weight} = \frac{\text{mg sample weighed} \times 1000}{\mu\text{moles DHAP / weighed sample}}$$

1000 = Conversion factor from mg to μg
DHAP = Dihydroxyacetone phosphate

Final Assay Concentrations

In a 3.02 ml reaction mix, the final concentrations are 134 mM triethanolamine, 0.13 mM β-nicotinamide adenine dinucleotide, reduced form, and 1 unit α-glycerophosphate dehydrogenase.

Note

1. α-Glycerophosphate Dehydrogenase Unit Definition: One unit will convert 1.0 μmole of dihydroxyacetone phosphate to α-glycerophosphate per minute at pH 7.4 at 25°C.

Determination of the Concentration and Molecular Weight of FRUCTOSE-1,6-DIPHOSPHATE

Procedure

$$\text{F-1,6-P2} \xrightarrow{\text{Aldolase}} \text{GAP} + \text{DAP}$$

$$\text{GAP} \xrightarrow{\text{TPI}} \text{DAP}$$

$$\text{DAP} + \beta\text{-NADH } \alpha\text{-Glycerophosphate} \xrightarrow{\text{Dehydrogenase}} \alpha\text{-GOP}^{\cdot} + \beta\text{-NAD}$$

Abbreviations Used

F-1,6-P2 = Fructose-1,6-Diphosphate
GAP = DL-Glyceraldehyde 3-Phosphate
DAP = Dihydroxyacetone Phosphate Dihydroxyacetone
TPI = Triosephosphate Isomerase
β-NADH = β-Nicotinamide Adenine Dinucleotide, Reduced Form

β-NAD = β-Nicotinamide Adenine Dinucleotide, Oxidized Form
α-GOP = L-α-Glycerophosphate

Conditions: T = 25°C, pH 7.6, A340nm, Light path = 1 cm

Method: Spectrophotometric

Reagents

A. 400 mM Triethanolamine Buffer, pH 7.6 at 25°C (TEA) (Prepare 100 ml in deionized water using Triethanolamine, Hydrochloride. Adjust to pH 7.6 at 25°C with 1 M NaOH).
B. Fructose-1,6-Diphosphate Solution (F-1,6-P2) (Weigh two samples accurately, approximately 1.5 mg, and dissolve each in 25.0 ml of deionized water).
C. 4.2 mM β-Nicotinamide Adenine Dinucleotide, Reduced Form Solution (β-NADH) (Prepare 10 ml in Reagent A using β-Nicotinamide Adenine Dinucleotide, Reduced Form, Disodium Salt).
D. α-Glycerophosphate Dehydrogenase-Triosephosphate Isomerase Enzyme Solution (α-GDH/TPI) (Prepare a solution containing approximately 250–500 units/ml of TPI activity with α-Glycerophosphate Dehydrogenase-Triosephosphate Isomerase in cold Reagent A).
E. Aldolase Enzyme Solution (Prepare a solution containing approximately 20 units/ml of Aldolase, Type IV from Rabbit Muscle in cold Reagent A).

Procedure

Pipette (in milliliters) the following reagents into suitable cuvettes:

	Test	Blank
Reagent A (TEA)	1.50	1.60
Reagent B (F-1,6-P2)	1.00	—
Reagent C (β-NADH)	0.10	—
Reagent D (α-GDH/TPI)	0.01	0.01
Deionized Water	0.40	1.40

Mix by inversion and obtain the A340nm using a suitable thermostatted spectrophotometer at 25°C. Then add:

Reagent E (Aldolase)	0.01	0.01

Mix by inversion and allow the reaction to proceed for 5 minutes. Upon completion of the reaction record the A340nm and calculate the A340nm and apparent molecular weight.

$$\Delta A = A_t \frac{3.01}{3.02} A_f$$

A_i = Initial absorbance
A_f = Final absorbance

$$\text{micromoles F-1, 6-P}_2 \text{ / weighed sample} = \frac{\Delta A \times 3.02 \times 25}{(6.22)(2)}$$

3.02 = Total volume of Reaction Mixture
25 = Dilution factor
6.22 = Millimolar extinction coefficient of β-NADH at 340nm
2 = Number of moles of a-glycerophosphate per mole of fructose-1,6-P/weighed sample

$$\text{Apparent molecular weight} = \frac{\text{mg sample weighed} \times 1000}{\mu\text{moles F-1, 6-P}_2 \text{ / weighed sample}}$$

Final Assay Concentrations

In a 3.02 ml reaction mix, the final concentrations are 215 mM TEA, 0.14 mM β-NADH, 0.2 units aldolase and 2.5–5.0 units α-GDH/TPI.

Determination of the Concentration and Molecular Weight of D-FRUCTOSE 6-PHOSPHATE

Principle

$$\text{D-Fructose 6-Phosphate} \xrightarrow{\text{PGI}} \text{D-Glucose 6-Phosphate}$$

$$\text{D-Glucose 6-Phosphate} + \beta\text{-NADP} \xrightarrow{\text{G-6-PDH}} \text{6-Phosphogluconate} + \beta\text{-NADPH}$$

Abbreviations Used

PGI = Phosphoglucose Isomerase
β-NADP = β-Nicotinamide Adenine Dinucleotide Phosphate, Oxidized Form
G-6-PDH = Glucose-6-Phosphate Dehydrogenase
β-NADPH = β-Nicotinamide Adenine Dinucleotide Phosphate, Reduced Form

Conditions: T = 25°C, pH = 7.6, A340nm, Light path = 1 cm

Method: Spectrophotometric Endpoint Determination

Reagents

A. 300 mM Triethanolamine HCl Buffer, pH 7.6 at 25°C (Prepare 100 ml in deionized water using Triethanolamine, Hydrochloride. Adjust to pH 7.6 at 25°C with 1 M NaOH).

B. 300 mM Magnesium Chloride Solution ($MgCl_2$) (Prepare 5 ml in deionized water using Magnesium Chloride, Sterile Filtered Solution).

C. 20.0 mM β-Nicotinamide Adenine Dinucleotide Phosphate Solution (β-NADP) (Prepare 1 ml in deionized water using β-Nicotinamide Adenine Dinucleotide Phosphate, Sodium Salt. *Prepare Fresh*).

D. 0.25 mM D-Fructose 6-Phosphate Solution (F-6-P) (Prepare 25 ml in deionized water, using D-Fructose 6-Phosphate, in a volumetric flask).

E. Glucose-6-Phosphate Dehydrogenase Enzyme Solution (G-6-PDH) (Immediately before use, prepare a solution containing 100 units/ml of Glucose-6-Phosphate Dehydrogenase in cold Reagent A).

F. Phosphoglucose Isomerase Enzyme Solution (PGI) (Immediately before use, prepare a solution containing 100 units/ml of Phosphoglucose Isomerase in cold Reagent A).

Procedure

Pipette (in milliliters) the following reagents into suitable cuvettes:

	Test	Blank
Reagent A (Buffer)	1.80	1.80
Reagent B ($MgCl_2$)	0.10	0.10
Reagent C (β-NADP)	0.05	0.05
Reagent D (F-6-P)	1.00	—
Deionized Water	—	1.00
Reagent E (G-6-PDH)	0.01	0.01

Mix by inversion and equilibrate to 25°C using a suitably thermostatted spectrophotometer. Record the initial A340nm for both the Test and Blank. Then add:

Reagent F (PGI)	0.01	0.01

Immediately mix by inversion and allow the reaction to proceed to completion (approximately 10 minutes). Record the final A340nm for both the Test and Blank.

Calculations

$$\Delta A = Af - Ai$$

Ai = Initial Absorbance
Af = Final Absorbance

$$\text{Micromoles F-6-P / Weighed sample} = \frac{(\Delta A\ \text{Test} - \Delta A\ \text{Blank})(2.97)(\text{df})}{6.22}$$

F-6-P = D-Fructose 6-Phosphate
2.97 = Total volume (in milliliters) of assay
df = Dilusion factor
6.22 = Millimolar extinction coefficient of β-NADPH at 340nm

$$\text{Apparent Molecular Weight} = \frac{\text{mg sample weighed} \times 1000}{\mu\text{moles of F-6-P / weighed sample}}$$

1000 = Conversion from mg to μg
F-6-P = D-Fructose 6-Phosphate

Final Assay Concentrations

In a 2.97 ml reaction mix, the final concentrations are 184 mM triethanolamine, 10 mM magnesium chloride, 0.34 mM β-nicotinamide adenine dinucleotide phosphate, 1 unit glucose-6-phosphate dehydrogenase, 1 unit phosphoglucose isomerase and varying amounts of fructose-6-phosphate.

Note

1. When determining the molecular weight for D-Fructose 6-Phosphate, dissolve in 1.0 mM Ethylenediaminetetraacetic Acid, using Ethylenediaminetetraacetic Acid, Tetrasodium Salt. Also, correct all weigh-up's for purity water content and salt.
2. Glucose-6-Phosphate Dehydrogenase Unit Definition: One unit will oxidize 1.0 μmole of D-glucose 6-phosphate to 6-phospho-D-gluconate per minute in the presence of β-NADP at pH 7.4 at 25°C.
3. Phosphoglucose Isomerase Unit Definition: One unit will convert 1.0 μmole of D-fructose 6-phosphate to D-glucose 6-phosphate per minute at pH 7.4 at 25°C.

Determination of the Concentration and Molecular Weight of α-D-GALACTOSE 1-PHOSPHATE

Principle

$$\text{Galactose - 1 - Phosphate} + \text{UDP - Glucose} \xrightarrow{\text{UT}} \text{UDP - Galactose} + \text{Glucose 1 - Phosphate}$$

$$\text{Glucose - 1 - Phosphate} \xrightarrow{\text{PGLUM}} \text{Glucose - 6 - Phosphate}$$

$$\text{Glucose - 6 - Phosphate} + \beta\text{ - NADP} \xrightarrow{\text{G-6-PDH}} \text{6 - Phosphoglucono lactone} + \beta\text{ - NADPH}$$

Abbreviations Used

UDP-Glucose = Uridine Diphosphoglucose
UT = Galactose-1-Phosphate Uridyl Transferase
UDP-Galactose = Uridine Diphosphogalactose
PGLUM = Phosphoglucomutase
β-NADP = β-Nicotinamide Adenine Dinucleotide Phosphate, Oxidized Form
G-6-PDH = Glucose-6-Phosphate Dehydrogenase
β-NADPH = β-Nicotinamide Adenine Dinucleotide Phosphate, Reduced Form

Conditions: T = 25°C, pH 8.7, A340nm, Light path = 1 cm

Method: Spectrophotometric Determination

Reagents

A. 200 mM Glycine buffer, pH 8.7 at 25°C (Prepare 100 ml in deionized water using Glycine, Free Base. Adjust to pH 8.7 at 25°C with 1 M NaOH).
B. 3.5 mM Uridine Diphosphoglucose Solution (UDPG) (Prepare 5 ml in deionized water using Uridine 5'-Diphosphoglucose, Disodium Salt).
C. 0.59 mM Glucose 1,6-Diphosphate Solution (G 1,6P) (Prepare 1 ml in deionized water using Glucose 1,6-diphosphate, Cyclohexylammonium Salt, Hydrate).
D. 13 mM β-Nicotinamide Adenine Dinucleotide Phosphate (β-NADP) (Prepare 2 ml in cold Reagent A using β-Nicotinamide Adenine Dinucleotide, Phosphate, Sodium Salt. *Prepare Fresh*).
E. 0.3 mM Galactose-1-Phosphate solution (G-1-P) (Prepare in deionized water, using α-D-Galacatose-1-Phosphate, Dipotassium Salt, in a volumetric, class A flask).
F. Glucose-6-Phosphate Dehydrogase Enzyme Solution (G-6-PDH) (Immediately before use, prepare a solution containing 50 units/ml of Glucose-6-Phosphate Dehydrogenase, in cold deionized water).
G. Phosphoglucomutase Enzyme Solution (PGLUM) (Immediately before use, prepare a solution containing 50 units/ml of Phosphoglucomutase in cold deionized water).
H. Galactose-1-Phosphate Uridyl Transferase Enzyme Solution (UT) (Immediately before use, prepare a solution containing 10 units/ml of Galactose-1-Phosphate Uridyl Transferase in cold Buffer A.
I. 100 mM Magnesium Chloride Solution (Prepare 1 ml in deionized water using Magnesium Chloride, Hexahydrate).

Procedure

Pipette (in milliliters) the following reagents into suitable cuvettes:

	Test	Blank
Reagent A (Buffer)	1.60	1.60
Reagent B (UDPG)	0.10	0.10
Reagent C (6–1,6-P)	0.05	0.05
Reagent D (β-NADP)	0.10	0.10
Deionized Water	—	1.00
Reagent I	0.05	0.05
Reagent E (G-1-P)	1.00	—
Reagent F (G-6-PDH)	0.05	0.05
Reagent G (PGLUM)	0.05	0.05

Mix by inversion and equilibrate to 25°C, using a suitably thermostatted spectrophotometer. Record the initial A340nm for both the Test and Blank versus air. Then add.

	Test	Blank
Reagent H (UT)	0.10	0.10

Immediately mix by inversion and allow the reaction to proceed until the ΔA340nm/minute is 0.0020, and this rate is maintained for a minimum of five minutes. Record the final A340nm for both the Test and Blank.

Calculations

$$\Delta A = Af - Ai$$

Af = Final Absorbance
Ai = Initial Absorbance

$$\text{Micromoles G-6-1P sample} = \frac{(\Delta A\ \text{Test} - \Delta A\ \text{Blank})\ (3.10)\ (df)}{6.22}$$

G-1-P = α-D-Galactose-1-Phosphate
3.10 = Total volume (in milliliters) of assay
df = Dilution factor
6.22 = Millimolar extinction coefficient of β-NADPH at 340nm

$$\text{Apparent Molecular Weight} = \frac{\text{mg sample weighed} \times 1000}{\mu\text{moles of G-6-P / weighed sample}}$$

1000 = Conversion factor form mg to μg
G-1-P = α-D-Galacatose-1-Phosphate

Final Assay Concentrations

In a 3.10 ml reaction mix, the final concentrations are 116 mM glycine, 0.11 mM uridine diphosphoglucose, 9.5 μM glucose 1,6-diphosphate, 0.42 mM β-nicotinamide adenine dinucleotide phosphate, oxidized form, 2.5 units glucose-6-phosphate dehydrogenase, 2.5 units phosphoglucomutase, 1 unit galactose-1-phosphate uridyl transferase and varying amounts of α-D-galactose 1-phosphate.

Notes

1. If the test material is α-Galactose-1-Phosphate, Barium, prepare in 30 mM Ethylenediamine-tetraacetic Acid, Tetrasodium, Tetrahydrate.

2. Galactose-1-Phosphate Uridyl Transferase Unit Definition: One unit will form 1.0 μmole of glucose 1-phosphate from UDP-glucose, galactose 1-phosphate and NADP per minute at pH 8.7 at 25°C as detected by a coupled system using phosphoglucomutase.
3. Phosphoglucomutase Unit Definition: One unit will convert 1.0 μmole of α-D-glucose 1-phosphate to α-D-glucose 6-phosphate per minute at pH 7.4 at 30°C.
4. Glucose 6-Phosphate Dehydrogenase Unit Definition: One unit will oxidize 1.0 μmole of Dglucose 6-phosphate to G-phospho-D-gluconate per minute in the presence of NADP at pH 7.4 at 25°C.

Impurity Assay of α-D-GLUCOSE 1-PHOSPHATE present in α-D-GALACTOSE 1-PHOSPHATE

Principle

$$\alpha\text{-D-Glucose 1-Phosphate} \xrightarrow{\text{Phosphoglucomutase}} \alpha\text{-D-Glucose 6-Phosphate}$$

$$\alpha\text{-D-Glucose 6-Phosphate} + \beta\text{-NADP} \xrightarrow{\text{G-6-PDH}} \text{6-Phospho-D-gluconate} + \beta\text{-NADPH}$$

Abbreviations Used

β-NADP = β-Nicotinamide Adenine Dinucleotide Phosphate, Oxidized Form
β-NADPH = β-Nicotinamide Adenine Dinucleotide Phosphate, Reduced Form
G-6-PDH = Glucose-6-Phosphate Dehydrogenase

Conditions: T = 25°C, pH = 7.4, A340nm, Light path = 1 cm

Method: Continuous Spectrophotometric Rate Determination

Reagents

A. 250 mM Glycylglycine Buffer, pH 7.4 at 30°C (Prepare 50 ml in deionized water using Gly-Gly, Free Base. Adjust to pH 7.4 at 25°C with 1 M HCl).
B. 10 mg/ml Galactose 1-Phosphate Solution, Potassium (Gal-1-P) (Prepare 2 ml in deionized water using α-D-Galactose 1-Phosphate, Dipotassium Salt).
C. 20 mM β-Nicotinamide Adenine Dinucleotide Phosphate Solution (β-NADP) (Prepare 1 ml in deionized water using β-Nicotinamide Adenine Dinucleotide Phosphate, Sodium Salt or dissolve the contents of one 30 mg vial of β-Nicotinamide Adenine Dinucleotide Phosphate, Sodium Salt, in the appropriate volume of deionized water. *Prepare Fresh*).
D. 1.2 mM Glucose 1,6-Diphosphate Solution (G 1,6-P) (Prepare 1 ml in deionized water using α-D-Glucose 1,6-Diphosphate, Cyclohexylammonium Salt, Hydrate).
E. 300 mM Magnesium Chloride Solution ($MgCl_2$) (Prepare 1 ml in deionized water using Magnesium Chloride, Hexahydrate).
F. Glucose-6-Phosphate Dehydrogenase Solution (G-6-PDH) (Immediately before use, prepare a solution containing 25 units/ml in cold Reagent A using Glucose-6-Phosphate Dehydrogenase).
G. Phosphoglucomutase Enzyme Solution (PGLUM) (Immediately before use, prepare a solution containing 25 units/ml of Phosphoglucomutase, in cold Reagent A).

Procedure

Pipette (in milliliters) the following reagents into suitable cuvettes:

	Test	Blank
Reagent A (Buffer)	1.40	1.40
Deionized Water	—	1.00
Reagent B (Gla 1-P)	1.00	—
Reagent D (G 1, 6-P)	0.05	0.05
Reagent C (β-NADP)	0.10	0.10
Reagent E (MgCl$_2$)	0.25	0.25
Reagent F (G-6-PDH)	0.10	0.10

Mix by inversion and equilibrate to 25°C, using a suitably thermostatted spectrophotometer. Record the initial A340nm for both the Test and Blank vs air. Then add:

Reagent G (PGLUM)	0.10	0.10

Immediately mix by inversion and allow the reaction to proceed until ΔA340/minute is 0.0020, (and this rate is maintained for a minimum of five minutes). Record the final A340nm for both the Test and Blank vs air.

Calculations

$$\Delta A = Af - Ai$$

Af = Final Absorbance
Ai = Initial Absorbance

$$\text{Micromoles of Glu-1-P / 1 ml of Gal-1-P} = \frac{(\Delta A\ \text{Test} - \Delta A\ \text{Blank})\ (3)}{6.22}$$

Glu-1-P = α-D-Glucose-1-Phosphate
Gal-1-P = α-D-Galactose-1-Phosphate
3 = Total volume (in milliliters) of assay
6.22 = Millimolar extinction coefficient of β-NADPH at 340nm

$$\text{Molar \% Glu-1-P} = \frac{(\mu\text{moles of Glu-1-P / 1 ml of Gal-1-P})\ (100)}{\mu\text{moles of Gal-1-P / 1 ml of Gal-1-P}}$$

$$\frac{\mu\text{moles Glu-1P}}{1.0\ \text{ml of Gal-1-P}} = \frac{\text{mgs} \times 1000}{\text{Enzymatic MW}}$$

Final Assay Concentrations

In a 3.00 ml reaction mix, the final concentrations are 117 mM glycylglycine, 0.67 mM β-nicotinamide adenine dinucleotide phosphate, 0.02 mM glucose 1,6-diphosphate, 25 mM magnesium chloride, 2.5 units glucose 6-phosphate dehydrogenase and 2.5 units phosphoglucomutase and varing concentrations of α-D-galactose 1-phosphate.

Notes

1. If the test material is α-Galactose-1-Phosphate, Barium Salt, prepare in 30 mM EDTA, using Ethylenediaminetetraacetic Acid, Tetrasodium, Hydrate.
2. Glucose 1,6-Diphosphate is required in order to obtain maximum activity.
3. Glucose-6-Phosphate Dehydrogenase unit definition: One unit will oxidize 1.0 µmole of D-glucose 6-phosphate to 6-phospho-D-gluconate per minute in the presence of β-NADP at pH 7.4 at 25°C.
4. Phosphoglucomutase Unit Definition: One unit will convert 1.0 µmole of α-D-glucose 1-phosphate to α-D-glucose 6-phosphate per minute at pH 7.4 at 30°C.

Determination of the Concentration and Molecular Weight of L-α-GLYCEROPHOSPHATE

Principle

$$\text{L-}\alpha\text{-Glycerophosphate} + \beta\text{-NAD} + \text{Hydrazine} \xrightarrow{\text{GDH}} \text{DAP Hydrazone} + \text{NADH}$$

Abbreviations Used

β-NAD = β-Nicotinamide Adenine Dinucleotide, Oxidized Form
GDH = α-Glycerophosphate Dehydrogenase
DAP = Dihydroxyacetone Phosphate
β-NADH = β-Nicotinamide Adenine Dinucleotide, Reduced Form

Conditions: T = 25°C, pH = 9.5, A340nm, Light path = 1 cm

Method: Spectrophotometric Determination

Reagents

A. 1 M Glycine Buffer with 609 mM Hydrazine Sulfate and 5.3 mM Ethylenediaminetetraacetic Acid, pH 9.5 at 25°C (Prepare 50 ml by dissolving Glycine, Free Base, Hydrazine Sulfate and Ethylenediaminetetraacetic Acid, Tetrasodium Salt, Hydrate, in 2 M NaOH until dissolved. Adjust to pH 9.5 with either 1 M NaOH or 1 M HCl. Then dilute to 50 ml with deionized water. *Prepare Fresh*).
B. 20 mM β-Nicotinamide Adenine Dinucleotide, Oxidized Form, Solution (β-NAD) (Prepare by dissolving the contents of one 50 mg vial β-Nicotinamide Adenine Dinucleotide, in the appropriate volume of deionized water. *Prepare Fresh*).
C. Glycerophosphate Dehydrogenase Enzyme Solution (GDH) (Immediately before use, prepare a solution containing 1200 units/ml of α-Glycerophosphate Dehydrogenase, in cold deionized water).
D. L-α-Glycerophosphate Solution (L-α-GP) (Immediately before use, prepare a solution containing approximately 0.3 µmoles/ml of L-α-Glycerophosphate in deionized water).

Procedure

Pipette (in milliliters) the following reagents into suitable cuvettes:

	Test	Blank
Deionized Water	0.22	1.22
Reagent A (Buffer)	1.35	1.35
Reagent D (L-α-GP)	1.00	—
Reagent B (β-NAD)	0.38	0.38

Mix by inversion and equilibrate to 25°C using a suitably thermostatted spectrophotometer. Record the initial A340nm for the Test and Blank. Then add:

Reagent C (Enzyme Solution)	0.05	0.05

Immediately mix by inversion and allow the reaction to proceed to completion (approximately 5 minutes). Record the final A340nm for both the Test and Blank.

Calculations

$$\Delta A = Af - Ai$$

Af = Final Absorbance
Ai = Initial Absorbance

$$\text{Micromoles L-}\alpha\text{-GP / ml RM} = \frac{(\Delta A \text{ Test} - \Delta A \text{ Blank})(3)}{(6.22)}$$

L-α-GP = L-α-Glycerophosphate
3 = Total volume (in milliliters) of assay
6.22 = Millimolar extinction coefficient of β-NADH at 340nm
RM = Reaction Mixture

$$\text{Apparent Molecular Weight} = \frac{\text{mg / ml L-}\alpha\text{-GP} \times 1000}{\mu\text{moles of L-}\alpha\text{-GP / ml RM}}$$

1000 = Conversion factor from mg to μg
L-α-GP = L-α-Glycerophosphate

Final Assay Concentration

In a 3.00 ml reaction mix, the final concentrations are 450 mM glycine, 274 mM hydrazine sulfate, 2.4 mM Ethylenediaminetetraacetic Acid, 2.5 mM β-nicotinamide adenine dinucleotide, 60 units aglycerophosphate dehydrogenase, and varying amounts of L-α-glycerophosphate.

Note

1. Once the pH of this solution has been adjusted to 9.5 at 25°C, it must be used within two hours. After that period of time, a baseline drift will occur when the solution is measured spectrophotometrically.

Determination of the Concentration and Molecular Weight of (D)-(-)-β-HYDROXYBUTYRIC ACID

Principle

$$\text{D(-)-}\beta\text{-Hydroxybutyrate} + \beta\text{-NAD} \xrightarrow[\text{Hydrazine}]{\beta\text{-HBDH}} \text{Acetoacetate} + \beta\text{-NADH}$$

Abbreviations Used

β-NAD = β-Nicotinamide Adenine Dinucleotide, Oxidized Form
β-HBDB = β-Hydroxybutyrate Dehydrogenase
β-NADH = β-Nicotinamide Adenine Dinucleotide, Reduced Form

Conditions: T = 25°C, pH = 8.5, A340nm, Light path = 1 cm

Method: Spectrophotometric Determination

Reagents

A. 100 mM Tris HCl Buffer, pH 8.5 at 25°C (Prepare 100 ml in deionized water using Trizma Base. Adjust to pH 8.5 at 25°C with 1 M HCL).

B. 1.0 M Hydrazine Hydrate Solution (Hydrazine) (Prepare 20 ml by adding 1 ml of Hydrazine Hydrate to 5 ml of 1 M HCl. Mix and dilute to 20 ml with deionized water. The pH of this solution should be approximately 8.5 at 25°C).

C. 15.1 mM β-Nicotinamide Adenine Dinucleotide Solution (β-NAD) (Dissolve the contents of one 20 mg vial of β-Nicotinamide Adenine Dinucleotide in the appropriate volume of deionized water).

D. 0.4 mM D-(-)-β-Hydroxybutyric Acid Solution (HBA) (Immediately before use, dissolve approximately 50 mg of D-β-Hydroxybutyric Acid, Sodium Salt in 100 ml of deionized water. Further dilute 10:1 with deionized water).

E. β-Hydroxybutyrate Dehydrogenase Enzyme Solution (Immediately before use, prepare a solution containing 25–50 units/ml of β-Hyroxybutyrate Dehydrogenase in cold deionized water).

Procedure

Pipette (in milliliters) the following reagents into suitable containers:

	Test	Blank
Reagent A (Buffer)	1.00	1.00
Reagent B (Hydrazine)	1.00	1.00
Reagent C (β-NAD)	0.10	0.10
Reagent D (HBA)	1.00	—
Deionized Water	—	1.00

Mix by inversion and equilibrate to 25°C using a suitably thermostatted spectrophotometer. Record the initial A340nm for both the Test and Blank. Then add:

Reagent E (Enzyme Solution)	0.04	0.04

Immediately mix by inversion and monitor the increase in A340nm until the reaction is complete (approx. 40–60 minutes). Record the final A340nm for both the Test and Blank.

Calculations

ΔA = Af–Ai Ai = Initial Absorbance

Af = Final Absorbance

$$\mu\text{moles D(-) }\beta\text{-HBA / ml Reaction Mixture} = \frac{(\Delta A\text{ Test} - \Delta A\text{ Blank})\,(3.14)\,(\text{df})}{(6.22)\,(1)}$$

D(-)HBA = D-(-)-β-Hydroxybutryic Acid

df = Dilution factor

3.14 = Total volume (in milliliters) of assay

6.22 = Millimolar extinction coefficient of β-NADH at 340nm

1 = Volume (in milliliters) of D(-)β-HBA

$$\text{Apparent Molecular Weight} = \frac{\mu\text{g of D(-)HBA / ml Reaction Mixture}}{\mu\text{moles of D(-)HAB / ml Reaction Mixture}}$$

$$\%\text{ Purity} = \frac{\text{Theoretical Molecular Weight}}{\text{Apparent Molecular Weight}}$$

Theoretical Molecular Weight = 126.1

Final Assay Concentrations

In a 3.14 ml reaction mix, the final concentrations are 32 mM Tris, 319 mM hydrazine hydrate, 0.48 mM β-nicotinamide adenine dinucleotide, oxidized form, 1–2 units of β-hydroxybutyrate dehydrogenase, and varying amounts of (D) (-)-β-hydroxybutyric acid.

Note

1. β-Hydroxybutyrate Dehydrogenase Unit Definition: One unit will oxidize 1.0 µmole of D(-)-β-hydroxybutyrate to acetoacetate per minute at pH 7.8 at 37°C.

Determination of the Concentration and Molecular Weight of D(-)LACTIC ACID

Principle

$$\text{D(-)Lactate} + \beta\text{-NAD} + \text{Hydrazine} \xrightarrow{\text{LDH}} \text{Pyruvate Hydrazone} + \beta\text{-NADH}$$

Abbreviations Used

β-NAD = β-Nicotinamide Adenine Dinucleotide, Oxidized Form
β-NADH = β-Nicotinamide Adenine Dinucleotide, Reduced Form
LDH = D-Lactic Dehydrogenase

Conditions: T = 25°C, pH = 9.5, A340nm, Light path = 1 cm

Method: Spectrophotometric Rate Determination

Reagents

A. 600 mM Hydrazine Buffer with 1 M Glycine and 5.6 mM Ethylenediaminetetraacetic Acid (EDTA), pH 9.5 at 25°C (Prepare 50 ml in deionized water using Hydrazine Sulfate and Glycine, Free Base and Ethylenediaminetetraacetic Acid, Trisodium. Adjust to pH 9.5 at 25°C with 10 M NaOH. *Prepare Fresh. Use within 2 Hours*).
B. 0.3 mM D(-)Lactic Acid Solution (Immediately before use, prepare 100 ml in deionized water).
C. 50 mM β-Nicotinamide Adenine Dinucleotide (β-NAD) (Dissolve the contents of one 50 mg vial of β-Nicotinamide Adenine Dinucleotide, in the appropriate volume of deionized water or prepare 2 ml in deionized water using β-Nicotinamide Adenine Dinucleotide. *Prepare Fresh*).
D. D-Lactic Dehydrogenase (Immediately before use, prepare a solution containing 5000 units/ml of D(-)Lactic Dehydrogenase in cold deionized water).

Procedure

Pipette (in milliliters) the following reagents into suitable cuvettes:

	Test	Blank
Deionized Water	0.35	1.35
Reagent A (Buffer)	1.40	1.40
Reagent B (Lactate)	1.00	—
Reagent C (β-NAD)	0.15	0.15

Mix by inversion and equilibrate to 25°C. Monitor the A340nm until constant, using a suitably thermostatted spectrophotometer. Record the initial absorbance at A340nm (Ai) of both the Test and Blank. Then add:

	Test	Blank
Reagent D (D-Lactic Dehydrogenase)	0.10	0.10

Immediately mix by inversion and record the increase in absorbance at A340nm until the reaction is complete. Record the final absorbance at A340nm (Af) of both the Test and Blank.

Calculations

$$\Delta A = Af - Ai$$

Ai = Initial Absorbance
Af = Final Absorbance

$$\text{Micromoles Lactic Acid / weighed sample} = \frac{(\Delta A340\text{nm Test} - \Delta A340\text{ Blank})\,(3)\,(\text{df})}{6.22}$$

3 = Volume (in milliliters) of assay
df = Dilution factor of D(-)Lactic Acid
6.22 = Millimolar extinction coefficient of β-NADH at 340nm

$$\text{Apparent Molecular Weight} = \frac{(\text{mg sample weighed})\,(1000)}{\mu\text{moles Lactic Acid / weighed sample}}$$

1000 = Conversion factor from mg to μg

Final Assay Concentrations

In a 3.00 ml reaction mix, the final concentrations are 280 mM hydrazine, 467 mM glycine, 4.2 mM Ethylenediaminetetraacetic Acid, 2.5 mM β-nicotinamide adenine dinucleotide, 500 units D(-)lactic dehydrogenase, and 0.1 mM D(-)lactic acid.

Notes

1. The reaction is considered complete when a A340nm/min of 0.002 is achieved over a 5 minute time period. If bubbles are present in the reaction mixture, remove them before taking final absorbance readings (Af).
2. D-Lactic Dehydrogenase Unit Definition: One unit will reduce 1.0 μmole of pyruvate to D-lactate per minute at pH 7.0 at 25°C.

Determination of the Concentration and Molecular Weight of L(+)LACTIC ACID

Principle

$$\text{L(+) Lactate} + \beta\text{-NAD} + \text{Hydrazine} \xrightarrow{\text{LDH}} \text{Pyruvate Hydrazone} + \beta\text{-NADH}$$

Abbreviations Used

β-NAD = β-Nicotinamide Adenine Dinucleotide, Oxidized Form
β-NADH = β-Nicotinamide Adenine Dinucleotide, Reduced Form
LDH = L-Lactic Dehydrogenase

Conditions: T = 25°C, pH = 9.5, A340nm, Light path = 1 cm

Method: Spectrophotometric Rate Determination

Reagents

A. 600 mM Hydrazine Buffer with 1 M Glycine and 5.6 mM Ethylenediaminetetraacetic Acid (EDTA), pH 9.5 at 25°C (Prepare 50 ml in deionized water using Hydrazine Sulfate and Glycine, Free Base and Ethylenediaminetetraacetic Acid. Adjust to pH 9.5 at 25°C with 10 M NaOH).
B. 0.3 mM L(+)Lactic Acid Solution (Immediately before use, prepare 100 ml dionized water).
C. 50 mM β-Nicotinamide Adenine Dinucleotide (β-NAD) (Dissolve the contents of one 50 mg vial of β-Nicotinamide Adenine Dinucleotide, in the appropriate volume of deionized water or prepare 2 ml in deionized water using β-Nicotinamide Adenine Dinucleotide. *Prepare Fresh*).
D. L-Lactic Dehydrogenase (Immediately before use, prepare a solution containing approximately 5000 units/ml in cold deionized water using L-Lactic Dehydrogenase).

Procedure

Pipette (in milliliters) the following reagents into suitable cuvettes:

	Test	Blank
Deionized Water	0.35	1.35
Reagent A (Buffer)	1.40	1.40
Reagent B (Lactate)	1.00	—
Reagent C (β-NAD)	0.15	0.15

Mix by inversion and equilibrate to 25°C. Monitor the A340nm until constant, using a suitably thermostatted spectrophotometer. Record the initial absorbance at A340nm (Ai) of both the Test and Blank. Then add:

	Test	Blank
Reagent D (L-Lactic Dehydrogenase)	0.10	0.10

Immediately mix by inversion and record the increase in absorbance at A340nm until the reaction is complete. Record the final absorbance at A340nm (Af) of both the Test and Blank.

Calculations

$$\Delta A = Af - Ai$$

Ai = Initial Absorbance
Af = Final Absorbance

$$\text{Micromoles Lactic Acid / weighed sample} = \frac{(\Delta A340\text{nm Test} - \Delta A340\text{ Blank})\,(3)\,(df)}{6.22}$$

df = Dilution factor of L(+)Lactic Acid
3 = Volume (in milliliters) of assay
6.22 = Millimolar extinction coefficient of β-NADH at 340nm

$$\text{Apparent Molecular Weight} = \frac{(\text{mg sample weighed})\ (1000)}{\mu\text{moles Lactic Acid / weighed sample}}$$

1000 = Conversion factor from mg to µg

Final Assay Concentrations

In a 3.00 ml reaction mix, the final concentrations are 280 mM hydrazine, 467 mM glycine, 2.6 mM Ethylenediaminetetraacetic Acid, 2.5 mM β-nicotinamide adenine dinucleotide, 500 units L-lactic dehydrogenase, and varying amounts of L(+)lactic acid.

Notes

1. Reaction time is approximately 30 minutes.
2. L-Lactic Dehydrogenase Unit Definition: One unit will reduce 1.0 µmole of pyruvate to L-lactate per minute at pH 7.5 at 37°C.

Determination of Concentration and Molecular Weight of β-NICOTINAMIDE ADENINE DINUCLEOTIDE

Principle

$$\beta\text{-NAD} + \text{Ethanol} \xrightarrow{\text{ADH}} \text{Acetaldehyde} + \beta\text{-NADH}$$

Abbreviations Used

β-NAD = β-Nicotinamide Adenine Dinucleotide, Oxidized Form
ADH = Alcohol Dehydrogenase
β-NADH = β-Nicotinamide Adenine Dinucleotide, Reduced Form

Conditions: T = 25°C, pH 8.8, A340nm, Light path = 1 cm

Method: Spectrophotometric

Reagents

A. 100 mM Pyrophosphate Buffer with 45 mM Semicarbazide HCl pH 8.8 at 25°C (Prepare 100 ml in deionized water using Tetrasodium Pyrophosphate, Decahydrate and Semicarbazide, Hydrochloride.Adjust to pH 8.8 at 25°C with 1 M HCl or 1 M NaOH).
B. 100% (v/v) Ethanol (EtOH) (Use 200 Proof USP Ethyl Alcohol).
C. 0.25 mM β-Nicotinamide Adenine Dinucleotide, Oxidized Form (NAD) (Immediately before use, prepare 10 ml in cold deionized water with β-Nicotinamide Adenine Dinucleotide using a Class A, volumetric flasks).
D. Alcohol Dehydrogenase Enzyme Solution (ADH) (Immediately before use, prepare a solution containing 300–400 units/ml of Alcohol Dehydrogenase, in cold Reagent A).

Procedure

Pipette (in milliliters) the following reagents into suitable cuvettes:

	Test	Blank
Reagent A (Buffer)	1.50	1.50
Reagent B (EtOH)	0.05	0.05
Reagent C (NAD)	1.00	—
Deionized Water	0.40	1.40

Mix by inversion and equilibrate for 5 minutes at 25°C. Record the initial absorbance, Ai, 340nm, using a suitably thermostatted spectrophotometer. Then add:

Reagent D (ADH)	0.10	0.10

Mix by inversion and allow the reaction to proceed for 5–10 minutes until the reaction is complete. Record the final absorbance, Af, 340nm for both the Test and Blank.

Calculations

$$\Delta A = (\text{Af Test} - \text{Ai Test}) - (\text{Af Blank} - \text{Ai Blank})$$

Af = Final Absorbance
Ai = Initial Absorbance

$$\text{Micromoles NAD} = \frac{(\Delta A)\,(3.05)\,(\text{df})}{(6.22)\,(1)}$$

3.05 = Total volume (in milliliters) of assay
df = Dilution factor
6.22 = Millimolar extinction coefficient of NADH
1 = Volume (in milliliters) of NAD sample

$$\text{Apparent MW} = \frac{\text{mg sample weighed} \times 1000}{\mu\text{moles NAD / sample}}$$

Final Assay Concentrations

In a 3.05 ml reaction mix, the final concentrations are 53 mM pyrophosphate, 24 mM semicarbazide, 2% (v/v) ethanol, and 30–40 units alcohol dehydrogenase and 0.08 mM of β-nicotinamide adenine dinucleotide.

Notes

1. Correct for purity, water content, and salts when determining concentration.
2. The reaction is complete when the ΔA340nm of the Test-Blank is less than 0.002 A340nm/min.

Determination of the Concentration and Molecular Weight of PHOSPHO(ENOL) PYRUVATE

Principle

$$\text{PEP} + \text{ADP} \xrightarrow{\text{PK}} \text{Pyruvate} + \text{ATP}$$

$$\text{Pyruvate} + \beta\text{-NADH} \xrightarrow{\text{LDH}} \text{Lactate} + \beta\text{-NAD}$$

Abbreviations Used

PEP = Phospho(enol) pyruvate
ADP = Adenosine-5'-Diphosphate

ATP = Adenosine-5'-Triphosphate
PK = Pyruvate Kinase
LDH = Lactate Dehydrogenase
β-NADH = β-Nicotinamide Adenine Dinucleotide, Reduced Form
β-NAD = β-Nicotinamide Adenine Dinucleotide, Oxidized Form

Conditions: T = 37°C, pH = 7.6, A340nm, Light path = 1 cm

Method: Spectrophotometric

Reagents

A. 100 mM Potassium Phosphate Buffer, pH 7.6 at 37°C (Prepare 100 ml in deionized water using Potassium Phosphate, Monobasic, Anhydrous. Adjust to pH 7.6 at 37°C with 1 M KOH).
B. Phospho(enol) pyruvate Solution (PEP) (Weigh approximately 1.5 mg of Phospho(enol) pyruvate and dissolve in 10 ml of deionized water. Dilute to 0.06 mg/ml in deionized water).
C. 2.5 mM β-Nicotinamide Adenine Dinucleotide, Reduced Form Solution (β-NADH) (Dissolve the contents of one vial of β-Nicotinamide Adenine Dinucleotide, Reduced Form Disodium in 2.5 ml of cold Reagent A).
D. 1000 mM Potassium Chloride Solution (KCl) (Prepare 5 ml in deionized water using Potassium Chloride).
E. 50 mM Magnesium Sulfate Solution ($MgSO_4$) (Prepare 5 ml in deionized water using Magnesium Sulfate, Heptahydrate).
F. 40 mM Adenosine-5'-Diphosphate Solution (ADP) (Prepared 5 ml in deionized water using Adenosine 5'-Diphosphate, Di (Monocyclohexylammonium) Salt).
G. PK/LDH Mixed Enzymes (Use PK/LDH Enzymes Suspension, undiluted).

Procedure

Pipette (in milliliters) the following reagents into suitable cuvettes:

	Test	Blank
Reagent A (Buffer)	1.48	2.48
Reagent B (PEP)	1.00	—
Reagent C (β-NADH)	0.13	0.13
Reagent D (KCl)	0.08	0.08
Reagent E ($MgSO_4$)	0.16	0.16
Reagent F (ADP)	0.10	0.10

Mix by inversion and equilibrate for 5 minutes at 37°C and record the initial A340nm using a suitably thermostatted spectrophotometer. Then add:

Reagent G (PK/LDH)	0.05	0.05

Mix by inversion and allow the reaction to proceed for 15 minutes. Record the final A340nm for both the Test and Blank.

Calculations

$\Delta A = Ai - Af$ Ai = Initial absorbance

Af = Final absorbance

$$\text{Micromoles PEP / weighed sample} = \frac{(\Delta A)(3.00)(25)}{(6.22)}$$

3.00 = Total volume of Reaction Mix

25 = Dilution factor

6.22 = Millimolar extinction coefficient of β-NADH at 340nm

$$\text{Apparent molecular weight} = \frac{\text{mg sample weighed} \times 1000}{\mu\text{moles PEP / weighed sample}}$$

Final Assay Concentration

In a 3.0 ml reaction mix, the final concentrations are 54 mM potassium phosphate, 0.11 mM β-NADH, 27 mM KCl, 2.7 mM $MgSO_4$, 1.3 mM ADP, 35 units PK and 50 units LDH.

Determination of the Concentration and Molecular Weight of PHOSPHOCREATINE

Principle

$$\text{Phosphocreatine} + \text{ADP} \xrightarrow{\text{Creatine Phosphokinase}} \text{Creatine} + \text{ATP}$$

$$\text{D-Glucose} + \text{ATP} \xrightarrow{\text{Hexokinase}} \text{Glucose 6-Phosphate} + \text{ADP}$$

$$\text{Glucose 6-Phosphate} + \beta\text{-NADP} \xrightarrow{\text{G-6-PDH}} \text{6-PG} + \beta\text{-NADPH}$$

Abbreviations Used

ADP = Adenosine 5'-Diphosphate

ATP = Adenosine 5'-Triphosphate

β-NADP = β-Nicotinamide Adenine Dinucleotide Phosphate, Oxidized Form

G-6-PDH = Glucose-6-Phosphate Dehydrogenase

6-PG = 6-Phospho-D-Gluconate

β-NADPH = β-Nicotinamide Adenine Dinucleotide Phosphate, Reduced Form

Conditions: T = 30°C, pH = 7.4, A340nm, Light path = 1 cm

Method: Spectrophotometric Determination

Reagents

A. 250 mM Glycylglycine Buffer with 0.1% (w/v) Bovine Serum Albumin, pH 7.4 at 30°C (Prepare 100 ml in deionized water using Glycylglycine, Free Base and Albumin Bovine. Adjust to pH 7.4 at 30°C with 1 M NaOH).

B. 20 mM Adenosine 5'-Diphosphate Solution (ADP) (Prepare 1 ml in deionized water using Adenosine 5'-Diphosphate, Sodium Salt).

C. 1 M D-Glucose Solution (Glucose) (Prepare 1 ml in deionized water using β-D(+)Glucose).

D. 20 mM β-Nicotinamide Adenine Dinucleotide Phosphate Solution (β-NADP) Dissolve the contents of one 30 mg vial of-Nicotinamide Adenine Dinucleotide

Phosphate, Sodium Salt, in the appropriate volume of deionized water. *Prepare Fresh*).

E. 300 mM Magnesium Chloride Solution ($MgCl_2$) (Prepare 1 ml in deionized water using Magnesium Chloride, Hexahydrate).

F. Glucose-6-Phosphate Dehydrogenase and Hexokinase Enzyme Solution (G-6-PDH/HK) (Immediately before use, prepare a solution containing 50 units/ml of Hexokinase and Glucose-6-Phosphate Dehydrogenase, in cold Reagent A. The 50 units are based on the Glucose-6-Phosphate Dehydrogenase Activity).

G. Creatine Phosphokinase Enzyme Solution (CPK) (Immediately before use, prepare a solution containing 50 units/ml of Creatine Phosphokinase, in cold Reagent A).

H. Phosphocreatine Solution (Phosphocreatine) Immediately before using, weigh approximately 2 mg of Phosphocreatine and dissolve in 25 ml of deionized water).

Procedure

Pipette (in milliliters) the following reagents into suitable cuvettes:

	Test	Blank
Reagent A (Buffer)	0.60	0.60
Reagent B (ADP)	0.10	0.10
Reagent C (Glucose)	0.10	0.10
Reagent D (β-NADP)	0.06	0.06
Reagent E ($MgCl_2$)	0.04	0.04
Deionized Water	1.00	2.00
Reagent F (G-6-PDH/HK)	0.10	0.10
Reagent H (Phosphocreatine)	1.00	—

Mix by inversion and equilibrate to 30°C using a suitably thermostatted spectrophotometer. Record the initial A340nm for the Test and Blank. Then add:

	Test	Blank
Reagent G (CPK)	0.10	0.10

Immediately mix by inversion and allow the reaction to proceed to completion (approximately 5 minutes). Record the final A340nm for both the Test and Blank.

Calculations

$$\Delta A = Af - Ai$$

Af = Final Absorbance
Ai = Initial Absorbance

$$\text{Micromoles Phosphocreatine / weighed sample} = \frac{(\Delta A\ \text{Test} - \Delta A\ \text{Blank})\ (3.1)\ (25)}{(6.22)}$$

3.1 = Total volume (in milliliters) of assay
25 = Dilution factor
6.22 = Millimolar extinction coefficient of β-NADPH at 340nm

$$\text{Apparent Molecular Weight} = \frac{\text{mg sample weighed} \times 1000}{\mu\text{moles PC / weighed sample}}$$

1000 = Conversion factor from mg to μg
PC = Phosphocreatine

Final Assay Concentration

In a 3.10 ml reaction mix, the final concentrations are 65 mM glycylglycine, 0.03% (w/v) Bovine Serum Albumin, 0.65 mM adenosine 5'-diphosphate, 32 mM D-glucose, 0.4 mM β-nicotinamide adenine dinucleotide phosphate, 4 mM magnesium chloride, 5 units creatine phosphokinase, 10 units hexokinase, 5 units glucose 6-phosphate dehydrogenase, and varying amounts of phosphocreatine.

Notes

1. Hexokinase and Glucose 6-Phosphate Dehydrogenase, has twice as many units of Hexokinase as it does Glucose 6-Phosphate Dehydrogenase.
2. Hexokinase Unit Definition: One unit will phosphorylate 1.0 μmole of D-glucose per minute at pH 7.6 at 25°C.
3. Glucose-6-Phosphate Dehydrogenase Unit Definition: One unit will oxidize 1.0 μmole of D-glucose 6-phosphate to 6-phospho-D-gluconate per minute in the presence of β-NADP at pH 7.4 at 25°C.
4. Creatine Phosphokinase Unit Definition: One unit will transfer 1.0 μmole of phosphate from phosphocreatine to ADP per minute at pH 7.4 at 30°C.

Determination of the Concentration and Molecular Weight of PHOSPHORYLRIBOSE-1-PYROPHOSPHATE using OROTATE, OROTIDINE-5'-PHOSPHATE PYROPHOSPHORYLASE and OROTIDINE-5'-PHOSPHATE DECARBOXYLASE

Principle

$$\text{PRPP} + \text{Orotate} \xrightarrow[\text{Orotidine-5'-Phosphate Pyrophosphorylase}]{} \text{Orotidine-5'-Monophosphate} + \text{PPi}$$

$$\text{Orotidine-5'-Monophosphate} \xrightarrow[\text{Orotidine-5'-Phosphate Pyrophosphorylase}]{} \text{Uridine-5'-Monophosphate} + CO_2$$

Abbreviations Used

PRPP = Phosphorylribose-1-Pyrophosphate
PPi = Inorganic Pyrophosphate

Conditions: T = 25°C, pH = 8.0, A295nm, Light path = 1 cm

Methods: Spectrophotometric

Reagents

A. 100 mM Tris Buffer, pH 8.0 at 25°C (Prepare 100 ml in deionized water using Trizma Base. Adjust to pH 8.0 with 1 M HCl).
B. 10 mM Orotic Acid Sodium (Substrate) (Prepare 50 ml in Reagent A using Orotic Acid Sodium. Heat may be necessary to form the solution).
C. 100 mM Magnesium Chloride Hexahydrate ($MgCl_2 \cdot 6H_2O$) (Prepare 50 ml in deionized water using Magnesium Chloride Hexahydrate).

D. Phosphorylribose-1-Pyrophosphate Solution (PRPP) (Weigh two samples accurately, approximately 2.5 mg, and dissolve in 10 ml deionized water).
E. 25 mM Glycylglycine Buffer, pH 7.0 at 25°C (Prepare 100 ml in deionized water using Glycylglycine. Adjust to pH 7.0 with 1 M HCl).
F. Orotidine-5'-Phosphate Pyrophosphorylase and Orotidine-5'-Phosphate Decarboxylase Enzyme Solution (Prepare approximately 5 units/ml in Reagent E. Centrifuge the solution and decant the supernatant.

Procedure

Pipette (in milliliters) the following reagent into suitable vials:

	Test	Blank
Deionized Water	—	1.10
Reagent A (Buffer)	1.65	1.65
Reagent B (Substrate)	0.10	—
Reagent C ($MgCl_2 \cdot 6H_2O$)	0.05	0.05
Reagent D (PRPP Solution)	1.00	—

Mix by inversion and obtain the A295nm using a suitable thermostatted spectrophotometer at 25°C. Then add:

Reagent F (Enzyme Solution)	0.20	0.20

Mix by inversion and allow the reaction to proceed for 2–3 hours. Upon completion of the reaction record the A295nm and calculate the A295nm and apparent molecular weight.

Calculations

$$\Delta A = A_j \times \frac{2.80}{3.00} A_f;$$

$$\frac{A_j = \text{initial absorbance}}{A_f = \text{final absorbance}}$$

$$\text{micromoles PRPP / weighed sample} = \frac{\Delta A \times 3.00 \times 10}{3.95}$$

3.00 = Total volume of Reaction Mixture
10 = Dilution factor
3.95 = Extinction coefficient of orotate at 295nm

$$\text{Apparent molecular weight} = \frac{\text{mg sample weighed} \times 1000}{\text{micromoles PRPP / weighed sample}}$$

Final Assay Concentrations

In a 3.0 ml reaction mix, the final concentrations are 58.3 mM tris, 0.33 mM orotic acid, 1.67 mM $MgCl_2 \cdot 6H_2O$, 1.67 mM glycylglycine. and 1.0 unit orotidine-5'-phosphate pyrophosphorylase.

Determination of the Concentration and Molecular Weight of PYRUVIC ACID

Principle

$$\beta\text{-NADH} + \text{Pyruvate} \xrightarrow{\text{LDH}} \text{Lactate} + \beta\text{-NAD}$$

Abbreviations Used

β-NADH = β-Nicotinamide Adenine Dinucleotide, Reduced Form
LDH = Lactic Dehydrogenase
β-NAD = β-Nicotinamide Adenine Dinucleotide, Oxidized Form

Conditions: T = 25°C, pH 7.5, A340nm, Light path = 1 cm

Method: Spectrophotometric Determination

Reagents

A. 100 mM Potassium Phosphate Buffer, pH 7.5 at 25°C (Prepare 100 ml in deionized water using Potassium Phosphate, Monobasic, Anhydrous. Adjust to pH 7.5 with 1 M KOH).
B. 0.25 mM Pyruvic Acid Solution (Pyruvate) (Immediately before use, prepare 25 ml in cold deionized water).
C. 0.30 mM β-Nicotinamide Adenine Dinucleotide, Reduced Form (β-NADH) (Prepare 50 ml by dissolving the contents of a 10 mg vial of β-Nicotinamide Adenine Dinucleotide, Reduced Form, Disodium Salt in the appropriate volume of Reagent A. *Prepare Fresh.* Alternatively, prepare 20 ml in Reagent A using β-Nicotinamide Adenine Dinucleotide, Reduced Form, Disodium Salt).
D. Lactic Dehydrogenase Enzyme Solution (LDH) (Immediately before use, prepare a solution containing 500–1000 units of L-Lactic Dehydrogenase, in cold Reagent A).

Procedure

Pipette (in milliliters) the following reagents into suitable cuvettes:

	Test	Blank
Reagent B (Pyruvate)	1.00	—
Deionized Water	—	1.00
Reagent C (β-NADH)	1.90	1.90

Mix by inversion and equilibrate to 25°C using a suitably thermostatted spectrophotometer. Record the initial A340nm for both the Test and Blank. Then add:

	Test	Blank
Reagent D (LDH)	0.10	0.10

Immediately mix by inversion and monitor the decrease in A340nm until the reaction is complete. Record the final A340nm for both the Test and Blank.

Calculations

$\Delta A = Ai - Af$ Ai = Initial Absorbance
Af = Final Absorbance

$$\text{Micromoles Pyruvic acid / ml sample} = \frac{(\Delta A\ \text{Test} - \Delta A\ \text{Blank})\ (3)\ (df)}{(6.22)}$$

3 = Total volume of Reaction Mix
df = Dilution factor
6.22 = Millimolar extinction coefficient of β-NADH at 340nm

$$\text{Apparent molecular weight} = \frac{\mu g\ \text{sample / ml sample}}{\mu \text{moles of pyruvic acid / ml sample}}$$

1000 = Conversion from mg to µg

$$\%\ \text{Purity} = \frac{\text{theoretical molecular weight}}{\text{apparent molecular weight}} \times 100\%$$

Final Assay Concentrations

In a 3.00 reaction mix, the final concentrations are 66.7 mM potassium phosphate, 0.19 mM β-nicotinamide adenine dinucleotide, reduced form, 50–100 unit L-lactic dehydrogenase, and 0.083 mM pyruvic acid.

Notes

1. L-Lactic Dehydrogenase Unit Definition: One unit will reduce 1.0 µmole of pyruvate to L-lactate per minute at pH 7.5 at 37°C.
2. This procedure is not to be used for the pyruvate impurity in α-ketoglutaric acid.

Determination of the Molecular Weight of TETRAHYDROFOLIC ACID

Principle

$$\text{THF} + \text{FIGLU} \xrightarrow{\text{FIGLU Transferase}} \text{L-Glutamate} + \text{5-Formimino-THF}$$

Abbreviations Used

THF = Tetrahydrofolic Acid
FIGLU = Formimino-L-Glutamic Acid
5-Formimino-THF = 5-Formimino-Tetrahydrofolic Acid

Conditions: T = 25°C, pH = 7.2, A350nm, Light path = 1 cm

Method: Spectrophotometric

Reagents

A. 200 mM Potassium Phosphate Buffer, pH 7.2 at 25°C (Prepare 100 ml in deionized water using Potassium Phosphate, Monobasic, Anhydrous. Adjust to pH 7.2 at 25°C with 1 M KOH).
B. 54 mM Formimino-L-Glutamic Acid Solution (FIGLU) (Prepare by dissolving 65 mg of Formimino-L-Glutamic Acid, Hemibarium Salt, in 4.5 ml of deionized water. Add 0.50 ml of 0.5 M Sodium Sulfate, Anhydrous, to precipitate Barium Sulfate. Centrifuge and save the supernatant).
C. Tetrahydrofolic Acid Solution (THF) (Immediately before use, weigh approximately 4.0 mg of Tetrahydrofolic Acid and dissolve in 10 ml of Reagent D).

D. 20 mM Potassium Hydroxide Solution with 1000 mM 2-Mercaptoethanol (KOH/ 2-ME) (Prepare 25 ml in deionized water using Potassium Hydroxide and 2-Mercaptoethanol).

E. Formimino-L-Glutamic Acid Transferase Enzyme Solution (FIGLU Transferase) (Immediately before use, prepare a solution containing 0.3 unit/ml of Formimino-L-Glutamic Acid Transferase in cold deionized water).

F. 10% (v/v) Perchloric Acid Solution (Per-Acid) (Prepare 25 ml in deionized water using Perchloric Acid).

Procedure

Pipette (in milliliters) the following reagents into suitable cuvettes:

Substrate Blank	Test	Enzyme	Blank
Reagent A (Buffer)	2.50	2.50	2.50
Reagent B (FIGLU)	0.50	0.50	0.50
Reagent C (THF)	0.50	0.50	—
Deionized Water	1.30	1.50	1.80

Mix by inversion and equilibrate for 5 minutes at 25°C. Record the initial (Ai, 350nm) for the Test, Enzyme Blank, and Substrate Blank using a suitably thermostatted spectrophotometer. Then add:

Reagent E (FIGLU Transferase)	0.20	—	0.20

Mix by inversion and allow the reaction to proceed for 1–2 hours at 25°C. Then add:

Reagent F (Per Acid)	1.50	1.50	1.50

Mix by swirling and place in a boiling water bath for 1 minute. Then cool in an ice bath and centrifuge to clarify. Transfer the solutions to suitable cuvettes and record the final (Af, 350nm) for the Test, Enzyme Blank and Substrate Blank using a suitable Spectrophotometer.

Calculations

$$\Delta A350nm = Af\ 350nm - Ai\ 350nm$$

$$\text{Corrected } \Delta A350nm = \Delta A350nm\ \text{Test} - (\Delta A350nm\ \text{Enzyme Blank} + \Delta A350nm\ \text{Substrate Blank})$$

$$\text{Micromoles THF / weighed sample} = \frac{(\text{Corrected } \Delta A350nm)\ (6.5)\ (10)}{(25)\ (0.5)}$$

6.5 = Total volume of colorimetric assay
0.5 = Volume of THF used in assay
10 = Dilution factor of weighed sample
25 = Millimolar extinction coefficient of 5,10-methenyltetrahydrofolic Acid at 350nm (5-Formimino-THF is converted to 5,10-Methenyltetrahydrofolic Acid under acidic conditions).

$$\text{Apparent molecular weight} = \frac{\text{mg sample weighed} \times 1000}{\mu\text{moles of THF / weighed sample}}$$

Final Assay Concentrations

In a 5.00 ml reaction mix, the final concentrations are 100 mM potassium phosphate, 5.4 mM FIGLU, 100 mM 2-mercaptoethanol, 2.0 mM potassium hydroxide and 0.06 unit formimino-L-glutamic acid transferase.

Suitability Assay of THIOREDOXIN

Principle

$$\text{DTT-(SH}_2\text{)} + \text{Insulin-S2} \xrightarrow{\text{Thioredoxin}} \text{Insulin-(SH)}_2 + \text{DTT-S2}$$

Abbreviations Used

DTT-(SH_2) = DL-Dithiothreitol
Insulin-S2 = Insulin (native disulfide bridge)
Insulin-$(SH)_2$ = Insulin (reduced disulfides)
DTT-S2 = DL-Dithiothreitol (disulfide bridge)

Conditions: T = 25°C, pH 7.0, A650nm, Light path = 1 cm

Method: Turbidimetric Rate Determination

Reagents

A. 100 mM Sodium Phosphate Buffer, pH 7.0 at 25°C (Prepare 100 ml in deionized water using Sodium Phosphate, Monobasic, Anhydrous. Adjust to pH 7.0 at 25°C with 1 M NaOH).
B. 100 mM Ethylenediaminetetraacetic Acid Solution (EDTA) (Prepare 10 ml in deionized water using Ethylenediaminetetraacetic Acid, Disodium Salt, Dihydrate).
C. 100 mM DL-Dithiothreitol Solution (DTT) (Prepare 1 ml in deionized water using DLDithiothreitol).
D. 50 mM Tris HCl Buffer, pH 7.5 at 25°C (Prepare 10 ml in deionized water using Trizma Base. Adjust to pH 7.5 at 25°C with 1 M HCl).
E. 1% (w/v) Insulin Solution (Insulin) (Prepare by dissolving 10 mg of Insulin, in 1 ml of Reagent D. This will produce a milky solution. Adjust the pH to below 3 at 25°C with 5 N HCl. The solution should begin to clear. Titrate again with 5 N NaOH. The solution should become turbid at pH 7 and then clear at pH 7.5. The final pH should be 7.5 at 25°C).
F. Thioredoxin Solution (Thioredoxin) (Immediately before use prepare a solution containing 0.025–0.1 mg/ml of Thioredoxin (protein) in cold deionized water).

Procedure

Prepare a reaction cocktail by pipetting (in milliliters) the following reagents into a suitable container:

Reagent A (Buffer)	7.56
Reagent B (EDTA)	0.24
Reagent E (Insulin)	1.20

Mix by swirling. Pipette (in milliliters) the following reagents into suitable containers:

	Test	Blank
Reaction Cocktail	0.75	0.75
Reagent F (Thioredoxin)	0.08	—
Deionized Water	0.16	0.24

Mix by swirling and then add:

Reagent C (DTT)	0.01	0.01

Immediately mix by swirling and incubate at 25°C for 20 minutes. Then begin to record the increase in A650nm every 5 minutes for a total of 15 minutes. Obtain the ΔA650nm/minute using the maximum linear rate for both the Test and Blank.

Calculations

$$\text{S.A.} = \frac{(\Delta A650\text{nm / min Test} - \Delta A650\text{nm / min Blank})\ (1)\ (\text{df})}{(\text{mg})}$$

S.A. = Specific Activity
df = Dilution factor
1 = Total volume (in milliliter) of assay
mg = mg Thioredoxin

Specification

Specific Activity (using insulin): ΔA650nm = 1–3 per mg protein per minute in the presence of 1 mM EDTA.

Final Assay Concentrations

In a 1.00 ml reaction mix, the final concentrations are 5 mM Tris, 63 mM sodium phosphate, 0.1% (w/v) insulin, 1 mM DL-dithiothreitol, 2 mM Ethylenediaminetetraacetic Acid, and 0.002–0.008 mg thioredoxin.

Notes

1. The turbidity of the reaction mixture begins to increase after approximately 20 minutes.
2. If the third A650nm reading (at 10 minutes) exceeds 0.8, dilute the thioredoxin solution (Reagent F) and run the assay again because rates above this value are not linear.

References

Adler, A.J. and Kistiakowsky, G.B.; (1962) *Journal of the American Chemicals Society* **84:**695–703.

Agranoff, B.W. and Brady, R.O.; (1956) *Journal of Biological Chemistry* **219:**221–229.

Alexander, M., Heppel L.A. and Hurwitz J.; (1961) *Journal of Biological Chemistry* **236:**3014–3019.

Allain, C.C., Poon L.S., Chan, C.S.G., Richmond, W. and Fu, P.C.; (1974) Clinical Chemistry, **20:** 470–475.

Ameyama, M.; (1982) *Methods in Enzymology* **89:**Part D, 20–29.

Ando, M., Yoshimoto T., Ogushi S., Rikitake, K., Shibata, S., and Tsuru, D.; (1979) *Journal of Biochemistry* **82:**1165–1172 .

Appleby, C.A., and Morton, R.K.; (1959) *Biochemical Journal* **71:**492–499.

Arima, T., Uchida, T., and Egami, F.; (1968) *Biochemical Journal* **106:**601–607.

Arnon, R.; (1970) *Methods in Enzymology*, **XIX,** 226–228.

Avigad, G. and Englard, S.; (1975) *Methods in Enzymology*, XLI, 142–147.

Bacon, J.S.D.; (1955) *Methods in Enzymology*, Volume I, 258–262.

Bahl, O.P. and Agrawal, K.M.L (1969) *Journal of Biological Chemistry* **244:**2970–2978.

Bajkowski, A.S. and Frankfater, A.; (1975) *Analytical Biochemistry* **68:**119–127.

Baker, A.L. and Tolbert, N.E.; (1966) *Methods in Enzymology* **IX**, 339–340.

Baker, A.L. and Tolbert, N.E.; (1966) *Methods in Enzymology*, **IX,** 338–342.

Baratti, J., Maroux, S. Louvard, D., and Desnuelle, P.; (1973) *Biochimica et Biophysica Acta* **315:**147–161.

Barrett, A. J. and Kirschke, H.; (1981) *Methods in Enzymology*, Volume 80, 535–538.

Barrett, A.J.; (1970) *Biochemical Journal* 117, 601–607.

Barrett, A.J.; (1981) *Methods in Enzymology* **80:**Part C, 561–565.

Bauer S., Lamed, R. and Lapidot, Y.; (1972) *Biotechnology and Bioengineering* **XIV,** 861–870.

Beisenherz, G., Bucher, T., and Garbade, K.H.; (1955) *Methods in Enzymology* **I**, 391–397.

Benveniste, R. and Davies, J.; (1971) *Biochemistry* **10:**1787–1796.

Berg, P.; (1956) *Journal of Biological Chemistry*, **222:**991–1013.

Bergmeyer, H. U., Gawehn K., and Grassl, M.; (1974) *Methods of Enzymatic Analysis* (Bergmeyer, H.U. ed.). Second Edition, Volume I, 434–435.

Bergmeyer, H. U.; (1974) *Methods of Enzymatic Analysis*, 2nd ed., Volume II, 486.

Bergmeyer, H. U.; (1974) *Methods of Enzymatic Analysis*, Volume I, 2nd ed., 451.

Bergmeyer, H.U. and Bernt, E.; (1974) in *Methods of Enzymatic Analysis*, (Bergmeyer, H.U. ed.). Volume 2, 574–578, Academic Press, New York, NY.

Bergmeyer, H.U. and Bernt, E.; (1974) in *Methods of Enzymatic Analysis* (Bergmeyer, H.U. ed.). Volume III, 2nd ed., 1205–1212, Academic Press, Inc., New York, NY.

Bergmeyer, H.U. and Bernt, E.; (1974) in *Methods of Enzymatic Analysis* (Bergmeyer, H.U. ed.) Volume III, 2nd ed., 1205–1212.

Bergmeyer, H.U. and Gawehn, K. and Grassl, M.; (1974) in *Methods of Enzymatic Analysis* (Bergmeyer, H.U. ed.) Volume I, 2nd ed., 456.

Bergmeyer, H.U., Gawehn, K. and Grassl, M.; (1974) in *Methods of Enzymatic Analysis* (Bergmeyer, H.U. ed.). Volume I, Second Edition, 442–443, Academic Press, Inc., New York, NY.

Bergmeyer, H.U., Gawehn, K., and Grassl, M.; (1974) in *Methods of Enzymatic Analysis* (Bergmeyer, H.U. ed.). Second Edition, Volume I, 509–510, Academic Press, Inc., New York, NY.

Bergmeyer, H.U., Gawehn, K., and Grassl, M.; (1974) in *Methods of Enzymatic Analysis* (Bergmeyer, H.U. ed.). Second Edition, Volume I, 521–522, Academic Press Inc., New York.

Bergmeyer, H.U., Gawehn, K., and Grassl, M.; (1974) in *Methods of Enzymatic Analysis* (Bergmeyer, H.U. ed.). 2nd ed., Volume I, 473–474.

Bergmeyer, H.U., Gawehn, K., and Grassl, M.; (1974) in *Methods of Enzymatic Analysis* (Bergmeyer, H.U). Vol. I, 2nd ed., 495–496, Academic Press, Inc., New York, NY.

Bergmeyer, H.U., Gawehn, K., and Grassl, M.; (1974) in *Methods of Enzymatic Analysis* (Bergmeyer, H.U. ed.). 2nd edition, Volume I, 466–467, Academic Press, Inc., New York, NY.

Bergmeyer, H.U., Gawehn, K., and Grassl, M.; (1974) in *Methods of Enzymatic Analysis* (Bergmeyer, H.U. ed.). Volume I, 438, Academic Press, Inc, New York, NY.

Bergmeyer, H.U., Gawehn, K., and Grassl, M.; (1974) in Methods of Enzymatic Analysis (Bergmeyer, H.U. ed.). 2nd ed., Volume I, 431–432, Academic Press, New York, NY.

Bergmeyer, H.U., Gawehn, K., and Grassl, M.; (1974) in Methods of Enzymatic Analysis (Bergmeyer, H.U., ed.). 2nd ed., Volume I, 407–413, Academic Press, New York, NY.

Bergmeyer, H.U., Gawehn, K., and Grassl, M.; (1974) in *Methods of Enzymatic Analysis* (Bergmeyer, H.U. ed.). Volume I, 2nd ed., 456, Academic, Press, Inc., New York, NY.

Bergmeyer, H.U., Gawehn, K., Grassl, M.; (1974) in *Methods of Enzymatic Analysis* (Bergmeyer, H.U. ed.). 2nd ed., Volume I, 508, Academic Press, New York, NY.

Bergmeyer, H.U., Grassl, M. and Walter, H.E.; (1983) in *Methods of Enzymatic Analysis* (Bergmeyer, H.U. ed.). 3rd ed., Volume II, 157–158.

Bergmeyer, H.U., Grassl, M., and Walter, H.E.; (1983) in *Methods of Enzymatic Analysis* (Bergmeyer, H.U. ed.) 3rd ed., Volume II, 222–223, Verlag Chemie, Deerfield Beach, FL.

Bergmeyer, H.U., Grassl, M., and Walter, H.E.; (1983) in *Methods of Enzymatic Analysis* (Bergmeyer, H.U. ed.). 3rd ed., Volume II, 261–262, Verlag Chemie, Deerfield Beach, FL (1993) *Reagents Chemicals ACS Specification*, 8th ed., 95, American Chemical Society, Washington DC.

Bergmeyer, H.U.; (1974) *Methods of Enzymatic Analysis*, 2nd ed., Volume 1, 453–454.

Bergmeyer, H.U.; (1974) *Methods in Enzymatic Analysis*, 2nd ed., Vol. II, 449.

Bergmeyer, H.U.; (1974) *Methods of Enzymatic Analysis* **I**, 468.

Bergmeyer, H.U.; (1974) *Methods of Enzymatic Analysis*, 2nd edition, Volume II, 488–489.

Bergmeyer, H.U.; (1974) *Methods of Enzymatic Analysis*, Vol. 1, 2nd edition, 431.

Bergmeyer, H.U.; (1974) *Methods of Enzymatic Analysis*, Vol. I, 2nd ed., 459–460.

Bergmeyer, H.U.; (1983) *Methods of Enzymatic Analysis*, 2nd edition, Volume I, 427.

Berman T. and Magasanik B.; (1966) *Journal of Biological Chemistry* **241**:800–806.

Bessey, O.A. et al., (1946) *Journal of Biological Chemistry* **164**:35–41.

Bessey, O.A., Lowry, O.H., and Brock, M.J., (1946) *Journal of Biological Chemistry* **164**:321–329.

Bessey, O.A., Lowry, O.H., and Brock, M.J.; (1946) *Journal of Biological Chemistry* **164**:301–308.

Blethen, S.; (1970) *Methods in Enzymology*, XVIIA, 330–335 .

Bonting, S.L., Simon, K.A., and Hawkins, N.M.; (1961) *Archives of Biochemistry and Biophysics* **95**:416–423.

Borooah J., Leaback, D.H. and Walker, P.G.; (1961) *Biochem J.* **78**:106–110.

Borooah, J., Leaback, D.H. and Walker, P.G.; (1961) *Biochem. J.* **78**:93–99.

Borooah, J., Leaback, D.H. and Walker, P.G.; (1961) *Biochemical Journal* **78**:56–67.

Borooah, J., Leaback, D.H. and Walker, P.G.; (1961) *Biochemical Journal* **78**:100–103.

Borooah, J., Leakback, D.H., and Walker, P.G.; (1961) *Biochemical Journal* **78**:106–110.

Bowers, Jr., G.N., McComb, R.B., Christensen, R.G., Jr., and Schaffer, R.; (1980) *Clin. Chem.* **26**:724–729.

Bredt, D.S. and Snyder, S.H.; (1990) *Proceedings National Academy of Sciences*, USA, **87**:682–685.

Bridger, W.A. and Cohen, L.H.; (1968) *Journal of Biological Chemistry* **243**:644–650.

Brodbeck, U. and Ebner, K.E.; (1966) *Journal of BiologicalChemistry* **241**:762–764.

Brooks, S.P.J. and Storey, K.B., (1993) *Analytical Biochemistry* **212**:452–456.

Bryant, F. and Ljungdahl, L.G.; (1981) *Biochemical and Biophysical Research Communications* **100**:793–799.

Buchanan, B.B. and Arnon, D.I.; (1971) *Methods in Enzymology*, XXIII, Part A, 417–418.

Burton, R. M.; (1955) *Methods in Enzymology*, Volume I, 397.

Calandra, G.B. and Cole, R.M.; (1980) *Infection and Immunity* **28**:1033–1037.

Cardenas, J.M. and Dyson, R.D.; (1973) *Journal of Biological Chemistry* **248**:6938–6944.

Cardini, C.E., LeLoir, L.F., and Chiriboga, J.; (1955) *Journal of Biological Chemistry* **214**:149–155.

Carlberg I. and Mannervik, B.; (1977) *Biochimica et Biophysica Acta* **484**:268–274.

Carlsen, R.B. and Pierce, J.G.; (1972) *Journal of Biological Chemistry* **247**:23–32.

Carter C.E. and Cohen L.H.; (1956) *Journal of Biological Chemistry* **222**:17–30.

Cassidy, J.T., Jourdian, G.W. and Roseman, S.; (1965) *J. Biol. Chem.* **240**:3501–3506.

Chance, B. and Maehly, A.C.; (1955) *Methods in Enzymology*, II, 773–775.

Chell, R.M., Sundaram, T.K., and Wilkinson, A.E.; (1978) *Biochemical Journal* **173**:165–177.

Chell, R.M., Sundaram, T.K., and Wilkinson, A.E.; (1978), *Biochemical Journal* **173**:165–177.

Chen, W.P., Matsuo, M., and Yasui, T.; (1986) *Agric. Biol. Chem.* **50**:1183–1194.

Cho, H., Ramer, S.E., Itoh, M., Winkler, D.G., Kitas, E., Bannwarth, W., Burn, P., Saito, H., and Walsh, C.T.; (1991) *Biochemistry*, 30, 6210–6216.

Comb, D.G. and Roseman, S.; (1960) *Journal of Biological Chemistry* **235**:2480.

Combie, J., Blake, J.W., Nugent, T.E., and Tobin, T.; (1982) *Clin. Chem.* **28**:83–86.

Craine, J.E. et al.; (1972) *Journal of Biological Chemistry* **247**:6082–6091.

Craven, G.R., Steers, E., Jr. and Anfinsen, C.B.; (1965) *J. Biol. Chem.* **240**:2468–2477.

Craven, G.R., Steers, E., Jr., and Anfinsen, C.B.; (1965) *Journal of Biological Chemistry* **240**:2468–2477.

Curthoys, N.P. and Weiss, R.F.; (1974) *Journal of Biological Chemistry* **249**:3261–3266.

Curzon, G. and Vallet, L.; (1960) *Biochem J.* **74**:279–287.

Cushman, D.W., and Cheung, H.S.; (1971) *Biochem. Pharm.* **20**:1637–1648.

Danforth, W.H.; (1965) *Journal of Biological Chemistry* **240**:588.

Darrow, R.A., and Colowick, S.P.; (1962) *Methods in Enzymology* **V**, 226–227.

Davis, K.A. and Williams, G.R.; (1969) *Canadian Journal of Biochemistry* **47**:553.

Dawson, C.R., and Magee, R.J.; (1955) *Methods in Enzymology* **II**, 817–821.

Dawson, R.M.C., Elliott, D.C., Elliott, W.H. and Jones, K.M.; (1990) in *Data for Biochemical Research*, 3rd ed., 4–5, Oxford University Press, New York, NY.

Dawson, R.M.C., Elliott, D.C., Elliott, W.H., Jones, K.M.; (1987) *Data for Biochemical Research*, pp. 460, Oxford University Press, Oxford.

DePaoli-Roach, A.A.; (1984) *Journal of Biological Chemistry*, 259, 12144–12152.

Desser, R.K., Himmelhoch, S.R., Evans, W.H., Januska, M., Mage, M., and Shelton, E.; (1972) *Arch. Biochem. Biophys.*, **148**:452–465.

Distler, J. J., and Jourdian, G. W.; (1973) *J. Biol. Chem.* **248**:6772–6780.

Dodgson, K.S.; (1961) *Biochemical Journal* **78**:312–319 .

Dolin, M.I.; (1957) *Journal of Biological Chemistry* **225**:557–573.

Dorfman, A.; (1955) *Methods in Enzymology*, Volume I, 166–173.

Dupont.; (1991) *Angiotensin I [125I] Radioimmunoassay Kit*, NEA-104, 105 Instruction Manual.

Easterby, J.S., O'Brien, M.J.; (1973) *Eur. J. Biochem.* **38**:201–211.

Esders, T.W. and Lynn, S.Y.; (1985) *Journal of Biological Chemistry* **260**:3915–3922.

Esders, T.W. and Michrina, C.A.; (1979) *Journal of Biological Chemistry* **254**:2710–2715.

Fishman, W.H. and Bernfeld, P.; (1955) *Methods in Enzymology*, Volume I, 262–269.

Fishman, W.H.; (1974) in *Methods of Enzymatic Analysis* (Bergmeyer, H.U. ed.) 2nd ed., Volume II, 929–932, Academic Press, New York.

Fitzgerald, D.K., Brodbeck, U., Kiyosawa, I., Mawal, R., Colvin, B., and Ebner, K.E.; (1970) *Journal of Biological Chemistry* **245**:2103–2108.

Flaks, J. G.; (1963) *Methods in Enzymology* **VI**, 148–152.

Folk, J.E., Piez, K.A., Carroll, W.R. and Gladner, J.A.; (1960) *J. Biol. Chem.* **235**:2272–2277.

Fonda, M.L.; (1985) *Methods in Enzymology* **113**:11–16 .

Forster, G., Bernt, E. and Bergmeyer, H.U.; (1974) in *Methods of Enzymatic Analysis* (Bergmeyer, H.U. ed.) Volume II, 2nd ed., 789–793:Academic Press, Inc., New York, NY.

Fraser, M.J.; (1980) *Methods in Enzymology* **65**:Part 1, 255-263.

Friedmann, H.C. and Vennesland, B.; (1958) *Journal of Biological Chemistry* **233:**1398–1406.

Fujisawa, H. and Hayaishi, O.; (1968) *Journal of Biological Chemistry* **243:**2673–2681.

Gatt, S., Dinur, T., and Barenholz, Y.; (1978) *Biochimica et Biophysica Acta* **530:**503–507.

Geer, B.W., Krochko, D., Oliver, M.J., Walker, V.K. and Williamson, J.H.; (1980) *Comp. Biochem. Physiol.* 65B, 25–34.

Giacomello, A. and Salerno, C.; (1977) *Analytica Biochemistry* **79:**263–267.

Gibson, T.D., Parker, S.M., and Woodward, J.R.; (1991) *Enzyme Microb. Technol.* **13:**171–178.

Gilliam, M.B., Sherman, M.P., Griscavage, J.M., and Ignarro, L.J.; (1993) *Analytical Biochemistry* **212**, 359–365.

Grant, D.A.W. and Hermon-Taylor, J.; (1975) *Biochem. J.* **147:**363–366.

Gunsalus, I.C., Galeener, C.C., and Stamer, J.R.; (1955) *Methods in Enzymology*, Volume II, 238–242.

Habig, W.H., Pabst, M.J., and Jakoby, W.B.; (1974) *J. Biol. Chem.* **249:**7130–7139.

Hall, S.W. and Kuhn, H.; (1986) *Eur. J. Biochem.* **161:**551–556.

Hamberg, M. and Samuelsson, B.; (1967) *J. Biol. Chem.* **242:**5329.

Hash, J.H.; (1975) *Methods of Enzymology*, XLIII, 737–755.

Hassall, H.; (1971) *Methods in Enzymology*, XVII B, 895–897.

Hathaway G.M., Lubben, T.H., and Traugh, J.A.; (1980) *Journal of Biological Chemistry* **255:**8038–8041.

Hayaishi, O.; (1955) *Methods in Enzymology*, Volume I 660–672.

Hiwada, K., Ito, T., Yokoyama, M., and Kokubu, T.; (1980) *European Journal of Biochemistry* **104:**155–165.

Holmgren, A.; (1979) *Journal of Biological Chemistry* **254:**9627–9632.

Hong, Y.-N. and Schopfer, P.; (1981) *Planta* **152:**325–335.

Hooper, N.M., and Turner, A.J.; (1987) Biochemical Journal **241:**625–633.

Hopner, T. and Knappe, J.; (1974) in *Methods of Enzymatic Analysis*, (Bergmeyer, H.U. ed.). Volume III, 1551–1555, Academic Press Inc., New York, NY.

Hopner, T. and Knappe, J.; (1974) in *Methods of Enzymatic Analysis* (Bergmeyer, H.U. ed.). 2nd ed., Volume III, 1551–1555.

Horiuchi, T., Suzuki, T., Hiruma, M. and Saito, N.; (1989) *Agricultural and Biological Chemistry*, 53, 1493–1501.

Hosokawa, K. and Stanier, R.Y.; (1966) *Journal of Biological Chemistry* **241:**2453–2460.

Hovingh, P. and Linker, A.; (1974) *Carbohydrate Research* **37:**181–192.

Hovingh, P. and Linker, A.; (1974) *Carbohydrate Research* **37:**181–192.

Hovingh, P., and Linker, A.; (1974) *Carbohydrate Research* **37:**181–192.

Human Blood Coagulation, Haemostasis, and Thrombosis (1976) p. 721–722, 2nd ed., R. Biggs, ed., Blackwell Scientific Publications.

Iversen, O.J. and Grov, A.; (1973) *Eur. J. Biochem.* **38:**293–300.

Jablonski, E., and DeLuca, M.; (1977) *Biochemistry* **16:**2932–2936.

Jakoby, W.B.; (1962) *Methods in Enzymology*, V, 771–774.

Janson, J., and Porath, J.; (1966) *Methods in Enzymology* **VIII**, 615.

Johnson, J.C., Shanoff, M., Bass, S.T., Boezi, J.A. and Hansen, R.G.; (1968) *Anal. Biochem.* **26:**137–145.

Jordan, P.M. and Seehra, J.S. *Methods in Enzymology*, Volume 123, 427–434.

Jorns, M.S.; (1979) *Journal of Biological Chemistry* **254:**12145–12152.

Joshi, J.G.; (1982) *Methods in Enzymology* **89:**Part D, 599–605.

Kalckar, H.M.; (1947) *Journal of Biological Chemistry* **167:**461–475.

Kaneda, M. and Tominaga, N.; (1975) *Journal of Biochemistry* **78:**1287–1296.

Keesey, J.; (1987) *Biochemica Information*, 1st ed., pp 19–20, Boehringer Mannheim Biochemicals, IN.

Keesey, J.; (1987) in *Biochemica Information*, pp. 49, First Edition, Boehringer Mannheim Biochemicals, Indianapolis, IN.

Kingdon, H.S., Hubbard, J.S., and Stadtman, E.R.; (1968) *Biochemistry* **7:**2136–2142.

Knox, W. E. and Pitt, B. M.; (1957) *Journal of Biological Chemistry* **225:**675–688.

Kobayashi, M. and Matsuda, K.; (1980) *Biochimica et Biophysica Acta* **614:**46–62.

Kofron, J.L., Kuzmi, P., Kishore, V., Colón-Bonilla, E., and Rich, D.H.; (1991) *Biochemistry* **30:**6127–6134.

Kohn, L.D., Warren, W.A., and Carroll, W.R.; (1970). *J.Biol. Chem.* **245:**3821–3830.

Kuby, S.A. and Lardy, H.S.; (1953) *Journal of the American Chemical Society* **75:**890–896.

Kunitz, M.; (1950) *Journal of General Physiology* **33:**349–362.

Kunitz, M.; (1950) *Journal of General Physiology* **33:**363–377.

Kusakabe, H., Kodama, K., Yoshino, H. Kuninaka, A., H., Misono, H., and Soda, K.; (1980) *Journal of Biological Chemistry* **255:**976–981.

Kusakabe, H., Midorikawa, Y., Kuninaka, A., and Yoshino, H.; (1983) *Agricultural Biological Chemistry* **47:**179–182.

Laskowski, M.; (1955) *Methods in Enzymology*, Volume II, 2–26.; (1993) *Reagent Chemicals ACS Specification*, 8th ed., p. 95, American Chemical Society, Washington, D.C.

Lawson, A. and Quinn, A.G.; (1967) *Biochemical Journal* **105:**483–490.

Leach, F.R. and Webster, J.J.; (1986) *Methods in Enzymology*, 133, Part B, 51–70.

Leder, I.G.; (1957) *Journal of Biological Chemistry*, 225, 125–136.

Lee, H.C. and Aarhus, R.; (1991) *Cell Regulation* **2:**203–209.

Lee, H.C., Walseth, T.F., Bratt, G.T., Hayes, R.N., and Clapper, D.L.; (1989) *J. Biol. Chem.* **264:**1608–1615.

Lenman, M., et al.; (1990) *European Journal of Biochemistry* **194:**747–753.

Lever, M.; (1972) *Analytical Biochemistry*, Vol. 47, 273–279.

Levvy, G.A., Conchie, J.; (1966) *Methods in Enzymology* **VIII:**571–584.

Levy, C.C. and Goldman, P.; (1967) *Journal of Biological Chemistry* **242:**2933–2938.

Levy, C.C. and Karpetsky, T.P.; (1980) *Journal of Biological Chemistry* **255:**2153–2159.

Li, S.-C. and Li, Y.-T.; (1970) *Journal of Biological Chemistry* **245:**5153–5160.

Lieberman, I., Kornberg, A., and Simms, E.S.; (1955) *J. Biol. Chem.* **215:**403–415.

Liener, I.E. and Friedenson, B.; (1970) *Methods in Enzymology*, XIX, 261–273.

Lin, S. and Cohen, H.P.; (1968) *Analytical Biochemistry* **24:**531–540.

Lin, S.H. and Van Wart, H.E.; (1982) *Biochemistry* **21:**5528–5533.

Lindberg, U.; (1964) *Biochimica et Biophysica Acta* **82:**237–248.

Linker, A. and Hovingh, P.; (1972) *Methods in Enzymology* **28**:902–911.

Littlewood, G.M., Hooper, N.M., and Turner, A.J.; (1989) *Biochemical Journal* **257**:361–367.

Lowe, R.H. and Evans, H.J.; (1964) *Biochim. Biophys. Acta.* **85**:377–389.

Lynen, F. and Wieland, O.; (1955) *Methods in Enzymology*, Volume I, 566–573.

MacDonald, D.L., Stanier, R.Y. and Ingraham, J.L.; (1954) *Journal of Biological Chemistry* **210**:809–820.

MacLeod, R.M., Farkas, W., Fridovich, I. and Handler, P.; (1961) J. Biol. Chem. **236**:1841.

Mandl, I., MacLennan, J.D., Howes, E.L., DeBellis, R.H., and Sohler, A.; (1953) *Journal of Clinical Investigation* **32**:1323–1329.

Marshall, M. and Cohen, P.P.; (1972) *Journal of Biological Chemistry* **247**:1641–1653.

Martinez, G., Barker, H.A. and Horecker, B.L.; (1963) *Journal of Biological Chemistry*, 238, 1598–1603.

Marumo, K., and Waite, J.H.; (1986) *Biochim. Biophys. Acta* **872**:98–103.

Massey, T.H. and Deal, Jr., W.C.; (1973) *Journal of Biological Chemistry* **248**:56–62.

Massey, V. and Curti, B.; (1966) *Journal of Biological Chemistry* **241**:3417–3423.

Masters, B.S.S., Williams, C.H., Kamin, H.; (1967) *Methods in Enzymology*, X, 565–573.

Masurekar, P.S. and Goodhue, C.T.; (1978) United States Patent No. 4, 093, 517, Eastman Kodak Company, Rochester, NY.

Matsushita, K., Shinagawa, E., Adachi, O., and Ameyama, M.; (1979) *Journal of Biochemistry* **85**:1173–1181.

Mavis, R.D. and Stellwagen, E.; (1968) *Journal of Biological Chemistry* **243**:809–814.

Mayes, J.S. and Hanson R.G.; (1966) *Methods in Enzymology* **IX**, 708–713.

McCord, J. M. and Fridovich, I.; (1969) *J. Biol. Chem.* **244**:6049–6055.

McCullough, J.C., Chabner, B.A. and Bertino, J.R.; (1971) *Journal of Biological Chemistry* **246**:7207–7213.

Metrione, R.M. Neves, A.G., and Fruton, J.S.; (1966) *Biochemistry* **5**:1597.

Mitchell, W.M., and Harrington, W.F.; (1968) *Journal of Biological Chemistry* **243**:4683–4692.

Mitz, M.A. and Schlueter, R.J.; (1958) *Biochimica Et Biophysica Acta* **27**:134–143.

Moellering H. and Gruber, W.; (1966) *Analytical Biochemistry* **17**:369–376.

Moore, S. and Stein, W.H.; (1948) *J. Biol. Chem.* **176**:367–388.

Morin, A.; (1993), *Enzyme Microbial Technology* **15**:208–214.

Murao, S. and Tanaka, N.; (1981) *Agric. Biol. Chem.* **45**:2383–2384.

Murphy, J., Baker, D.C., Behling, C., and Turner, R.A. (1982) *Analytical Biochemistry* **122**:328–337.

Nair, P.P., Gordon M., and Reback, J.; (1967) *Journal of Biological Chemistry* **242**:7–11.

Nakanishi, T., Matsumura, Y., Minamiura, N., and Yamamoto, T.; (1974) *Agricultural Biological Chemistry* **38**:37–44.

Nelson N.; (1944) *J. Biol. Chem.* **153**:375–380.

Nelson, N.; (1944) *Journal of Biological Chemistry* **153**:375–380.

Netheler, H.G.; (1974) in *Methods of Enzymatic Analysis* (Bergmeyer, H.U. ed.). 2nd ed., Volume I, 184–187, Academic Press, Inc., New York, NY.

Nikodejevic, B., Senoh, S., Daly, J.W., and Creveling, C.R.; (1970) *The Journal of Pharmacology and Experimental Therapeutics* **174**:83–93.

Nishimura, S., Harada, F., Narushima, U., and Seno, T.; (1967) *Biochimica Et Biophysica Acta* **142:**133–148.

Noda, L., Nihei, T. and Morales, M.F.; (1960) *J. Biol. Chem.* **235:**2830–2834.

Noltmann, E.A., Gubler, C.J., and Kuby, S.A.; (1961) *Journal of Biological Chemistry* **236:**1225–1230.

Nordlie, R.C. and Arion, W. J.; (1966) *Methods in Enzymology* **9:**619–625.

Ohshima, H., Oguchi, S., Adachi, H., Iida, S., Suzuki, H., Sugimura, T., and Esumi, H.; (1992) *Biochemical and Biophysical Research Communications* **183:**238–244.

Ohshima, T., Misono, H. and Soda, K.; (1978) *Journal of Biological Chemistry* **253:**5719.

Ohtakara, A.; (1988) *Methods in Enzymology*, 161 Part B, 505–510.

Okabe, H., Sagesaka, K., Nakajima, N., and Noma, A.; (1977) *Clinica Chimica Acta* **80:**87–94.

Osswald, W.F., McDonald, R.E., Niedz, R.P., Shapiro, J.P., and Mayer, R.T.; (1992) *Analytical Biochemistry* **204:**40–46.

Ototani, N. and Yosizawa, Z.; (1979) Carbohydrate Research, 70, 295 (1993) *Reagent Chemicals ACS Specification*, Volume II, 2–26, Americal Chemical Society, Washington, DC.

P.F. Hallenberg and L.P. Hager.; (1978) *Methods in Enzymology*, Volume LII, Part C, 521–529.

Palmieri, S., Leoni, O., and Iori, R.; (1982) *Analytical Biochemistry* **123:**320–324.

Paul, K.-G. and Avi-Dor, Y.; (1954) *Acta Chemica Scandinavica* **8:**649–657.

Pentchev, P.G., Brady, R.O., Gal, A.E., and Hibbert, S.R.; (1977) *Biochimica et Biophysica Acta* **488:**312–321.

Perry, S.V.; (1955) *Methods in Enzymology*, II, 582–588.

Pfleiderer, G.; (1970) *Methods in Enzymology*, XIX, 514–521.

Phillips, J.P., Feuer, H. and Thyagarajan, B.S.; (1967) *Organic Spectral Data*, Volume IX, 64.

Pontremoli, S., Traniello, S., Luppis, B., and Wood, W.A., (1965) *Journal of Biological Chemistry* **240:**3459–3463.

Potier, M., Mameli, L., Belisle, M., Dallaire, L., and Melancon, S.B.; (1979) *Analytical Biochemistry* **94:**287–296.

Prescott, L.M. and Jones, M.E.; (1969) *Analytical Biochemistry* **32:**408–419.

Price, C.P., Hammond, P.M., and Scawen, M.D.; (1983) *Clinical Chemistry* **29:**358–361.

Pütter, J. and Becker, R.; (1983) in *Methods of Enzymatic Analysis* (Bergmeyer, H.U., ed.). 3rd ed., Vol III, pp. 286–293, Verlag Chemie, Deerfield Beach, FL.

Quinn, D.M., Shirai, K., Jackson, R.L., and Harmony, J.K.; (1982) *Biochemistry* **21:**6872–6879.

Racker, E.; (1951) *Journal of Biological Chemistry* **190:**685.

Racker, E.; (1957) *Archives of Biochemistry and Biophysics* **69:**300–310.

Racker, E.; (1962) *Methods in Enzymology* V, 266–270.

Racker, E.; (1974) in *Methods of Enzymatic Analysis* (Bergmeyer, H.U. ed.) 2nd ed., Volume III, 1342–1345, Academic Press, New York, NY.

Ramasastri, B.V. and Blakley, R.L.; (1964) *Journal of Biological Chemistry* **239:**112–114.

Ratner, S.; (1970) *Methods in Enzymology*, Vol. XVIIA, 304–309.

Redinbaugh, M.G. and Campbell, W.H.; (1985) *Journal of Biological Chemistry* **260:**3380–3385.

Reese, E.T. and Mandels, M.; (1964) *Canadian Journal of Microbiology* **10:**103–114.

Richmond, W.; (1973) *Clinical Chemistry* **19:**1350–1356.

Rick, W.; (1974) in *Methods of Enzymatic Analysis*, 2nd ed., Volume II, pp 1009–1012, Academic Press Inc., New York, NY.

Rinderknecht, H., Wilding, P., and Haverback, B.J.; (1967) *Experientia* **23:**805.

Rose, Z. B. and Racker, E.; (1966) *Methods in Enzymology*, Volume IX, 357–360.

Rush, C., Willetts, A., Davies, G. Dauter, Z, Watson, H., and Littlechild, J.; (1995) Febs Letters, 359, 244–246.

Sabo, D.L., Boeker, E.A., Byers, B., Waron, H., and Fischer, E.H.; (1974) *Biochemistry* **13:**662–670.

Sabularse, D.C. and Anderson, R.L.; (1981) *Biochemical and Biophysical Research Communications* **103:**848–855.

Sadler, J.E., Beyer, T.A., Oppenheimer, C.L., Paulson, J.C., Prieels, J.P., Rearick, J.I., and Hill, R.; (1982) in *Methods in Enzymology* (Ginsberg, V, ed.). Vol. 83, pp. 480–481.

Saito, H., Yamagata, T. and Suzuki, S.; (1968) *J. Biol. Chem.* **243:**1536–1542.

Saito, H., Yamagata, T. and Suzuki, S.; (1968) *Journal of Biological Chemistry* **243:**1536–1542.

Saito, H., Yamagata, T., and Suzuki, S.; (1968) *J. Biol. Chem.* **243:**1536.

Sanchez, M.A., Romero, C., Manjon, A., and Iborra, J.L.; (1987) *Biotechnology Letters* **9:**871–874.

Schachter, H. et al.; (1969) *Journal of Biological Chemistry*, *244*, 4785–4792.

Schimke, R.T., Berlin, C.M., Sweeney, E.W., and Carrol, W.R.; (1966) *Journal of Biological Chemistry* **241:**2228–2236.

Schneir, M.L. and Rafelson, M.E. Jr.; (1966) *Biochim. Biophys. Acta.* **130:**1–11.

Scott, J.H. and Schekman, R.; (1980) *Journal of Bacteriology* **142:**414–423.

Sedewitz, B., Schleifer, K.H. and Gotz, F.; (1984) *Journal of Bacteriology* **160;**273–278.

Shea, P.A. and Aprison, M.H.; (1973) *Analytical Biochemistry* **56:**165.

Shimazono, H. and Hayaishi, O., (1957) *Journal of Biological Chemistry* **227:**151–159.

Shimizu, H. et al.; (1979) *J. Ferment. Technol.* **57:**428.

Shimizu, H., Kuratsu, T., and Hirata, F.; (1979) *J.Ferment. Technol.* **57:**428–433.

Shirai, K. and Jackson, R. L.; (1982) *Journal of Biological Chemistry* **257:**1253–1258.

Shugar, D.; (1952) *Biochimica et Biophysica Acta* **8:**302–309.

Silverstein, R.M.; (1975) *Analytical Biochemistry* **63:**281–282.

Silverstein, R.M.; (1975). *Anal. Biochem.* **63:**281–282.

Simons, P.C., and Vander Jagt, D.L.; (1977) *Anal. Biochem.* **82:**334–341.

Smith, L.D., Budgen, N., Bungard, S.J., Danson, M.J., and Hough, D.W.; (1989) *Biochemical Journal* **261:**973–977.

Soda K., Misono, H., Mori, K. and Sakato, H.; (1971) *Biochemical and Biophysical Research Communication* **44:**931.

Soda, K.; (1968) *Analytical Biochemistry* **25:**228–235.

Somogyi M.; (1945) *J. Biol. Chem.* **160:**61–68.

Somogyi M.; (1952) *J. Biol. Chem.* **195:**19–23.

Somogyi, M.; (1945) *Journal of Biological Chemistry* **160:**61–68.

Somogyi, M.; (1952) *Journal of Biological Chemistry* **195:**19–23.

Srere, P.A.; (1969), *Methods in Enzymology*, XIII, 1–11.

Stoops, J.K., Horgan, D.J., Runnegar, T.C., Jersey, J., Webb, E. and Zerner, B.; (1969) *Biochemistry* **8:**2026–2033 (1993) *Reagent Chemicals ACS Specifications*, 8th ed., p 95, American Chemical Society, Washington, DC.

Strecker, H.J.; (1955) *Methods in Enzymology* **1:**335.

Strehler, B.L.; (1974) in *Methods of Enzymatic Analysis* (Bergmeyer, H.U. ed.). 2nd ed., Vol. 4, 2112–2121.

Strominger, J.L., Heppel, L.A., and Maxwell, E.S.; (1959) *Biochim. Biophys. Acta* **32:**412–421.

Suato, F.K.; (1993) *Reagent Chemicals ACS Specifications*, 8th edition, 95.

Sullivan, B. and Howe, M.A.; (1933) *Journal of the American Chemical Society* **55:**320–324.

Szewczuk, A. and Mulczyk, M.; (1969) *European J.Biochemistry* **8:**63–67.

Tabor, H. and Wyngarden, L.; (1958) *Journal of Clinical Investigation* **37:**824–828.

Tanaka, H.I., Imahara, H., Esaki, N., and Soda, K.; (1980) *Journal of Applied Biochemistry* **2:**439–444.

Taussky, H. H. and Shorr, E.; (1953) *J. Biol. Chem.* **202:**675–685.

Taussky, H.H. and Shorr, E.; (1953) *Journal of Biological Chemistry* **202:**644–651.

Taussky, H.H. and Shorr, E.; (1953) *Journal of Biological Chemistry* **202:**675–685.

Taussky, H.H. and Shorr, E.; (1953) *Journal of Biological Chemistry* **202:**588–595.

Trautschold, I., Werle, E. and Schweitzer, G.; (1974) in *Methods of Enzymatic Analysis* (Bergmeyer, H.U. ed.). Vol 2, 2nd ed., 1031–1034, Academic Press, Inc., New York, NY.

Traverso-Cori, A., Chaimovich, H., and Cori, O.; (1965) *Archives of Biochemistry and Biophysics* **109:**173–184.

Uchida, T.; (1966) *Journal of Biochemistry* **60:**115–132.

Uda, Y., Li, S.C., and Li, Y.-.; (1977) *Journal of Biological Chemistry* **252:**5194–5200.

Umbreit W.W., Burris R.H., and Stauffer, J.F.; (1951) in *Manometric Techniques and Tissue Metabolism*, Burgess Publishing Co. Minneapolis, MN.

Umetsu, H., Abe, M., Sugawara, Y. and Nakai, T.; (1981) *Food Chemistry* **7:**125–138.

Van Schaftingen, E., Lederer, B., Bartrons, R., and Hers, H.-G.; (1982) *European Journal of Biochemistry*, **129:**191–195.

Van Wart, H.E., and Steinbrink, D. R.; (1981) *Analytical Biochemistry* **113:**356–365.

Verlag Chemie, Deerfield Beach, FL Shirfrin, S., Parrott, C.L. and Luborsky, S.W.; (1974) *Journal of Biological Chemistry* **249:**1335–1340.

Vogt, V.M.; (1973) *European Journal of Biochemistry* **33:**192–200.

Wang, Q.M., Fiol, C.J., DePaoli-Roach, A.A., and Roach, P.J.; (1994) *Journal of Biological Chemistry*, 269, 14566–14574.

Warren L.; (1959) *J. Biol. Chem.* **234:**1971–1975.

Webber, S. et al.; (1978) *Analytical Biochemistry* **84:**491–503.

Weinstein, J., de Souza-e-Silva, U, and Paulson, J. C.; (1982) *Journal of Biological Chemistry* **257:**13835–13844.

Weinstein, J., de Souza-e-Silva, U, and Paulson, J. C.; (1982) *Journal of Biological Chemistry* **257:**pp 13845–13853.

Wellner, D. and Lichtenber, L. A.; (1971) *Methods in Enzymology*, XVII B, 593–596.

Wendel, A.; (1980) *Enzymatic Basis of Detoxication*, Volume 1, p. 333, Academic Press NY.

Wharton, D. C. and Tzagoloff, A.; (1967) *Methods in Enzymology* **10:**245–250.

Whatley, F.R. and Arnon, D.I.; (1963) *Methods in Enzymology*, VI, 308–311.

White, H. and Jencks, W.P.; (1976) *Journal of Biological Chemistry* **251:**1688–1699.

White, H., Solomon, F., and Jencks, W.P.; (1976) *Journal of Biological Chemistry* **251:**1700–1707.

White-Stevens, R.H. and Kamin, H.; (1972) *Journal of Biological Chemistry* **247:**2358–2370.

Wilbur, K.M. and Anderson, N.G.; (1948) *Journal of Biological Chemistry* **176:**147–154.

Wilkinson, A.P., Rhodes, M.J., and Fenwick, G.R.; (1984) *Analytical Biochemistry* **139:**284–291.

Williams, V.R. and Lartigue, D.J.; (1969) *Methods in Enzymology*, XIII, 354–361.

Wirnt, R.; (1974) in *Methods of Enzymatic Analysis*, 2nd ed., Volume II, pp 1009–1012, Academic Press Inc., New York, NY.

Wittenberg, J. and Kornberg, A.; (1953) *Journal of Biological Chemistry* **202:**431–444.

Wood, T.; (1970) *Analytical Biochemistry* **33:**297–306.

Worthington Biochemical Corporation, Freehold, NJ. Bergmeyer, H.U., Gawehn, K., and Grassl, M.; (1974) in *Methods of Enzymatic Analysis* (Bergmeyer, H.U., ed.) Volume I, 2nd ed., 436–437, Academic Press, Inc. New York, NY.

Worthington, C.C.; (1988) in *Worthington Enzyme Manual* (Worthington, C.C. ed.). 57–59.

Worthington, C.E.; (1988) in *Worthington Enzyme Manual* **240–241:**Worthington Biochemical Corporation, Freehold, NJ.

Worthington, V.; (1993), *Worthington Enzyme Manual* **241:**Freehold, NJ.

Wünsch, E. and Heidrich, H.-G.; (1963) *Zeitschrift für Physiologische Chemie* **333:**149–151.

Yamagata, T. et al., (1968) *Journal of Biological Chemistry* **243:**1523–1535.

Yamagata, T., Saito, H., Habuchi, O. and Suzuki, S.; (1968) *Journal of Biological Chemistry* **243:**1523–1535.

Yamagata, T., Saito, H., Habuchi, O., and Suzuki, S.; (1968) *J. Biol. Chem.* **243:**1523–1535.

Yamamoto, K., et al.; (1986) *Agricultural Biological Chemistry*, **50:**1689.

Yamamoto, K., Kondo, Y., Kumagai, H., and Tochikura, T.; (1985) *Agric. Biol. Chem.* **49:**2463–2464.

Yamanaka, K.; (1975) *Methods in Enzymology 41*, Part B, 138–142.

Yang, J.K. and Epstein, W.; (1983) *Journal of Biological Chemistry* **258:**3750–3758.

Yano, T., et al.; (1985) *Agricultural Biological Chemistry* **49:**3179.

Yokobayashi, K., Misaki A., and Harada, T.; (1970) *Biochimica et Biophysica Acta* **212:**458–469.

Yu, C.K. and Dietrich, L.S.; (1972) *Journal of Biological Chemistry* **247:**4794–4802.

Zekri, M., Harb, J., Bernard, S., and Meflah, K.; (1988) *European Journal of Biochemistry* **172:**93–99.